Mineral Deposits of Finland

Mineral Deposits of Finland

Edited by

Wolfgang D. Maier

Raimo Lahtinen

Hugh O'Brien

ELSEVIER

AMSTERDAM • BOSTON • HEIDELBERG • LONDON • NEW YORK • OXFORD
PARIS • SAN DIEGO • SAN FRANCISCO • SINGAPORE • SYDNEY • TOKYO

Elsevier
Radarweg 29, PO Box 211, 1000 AE Amsterdam, Netherlands
The Boulevard, Langford Lane, Kidlington, Oxford OX5 1GB, UK
225 Wyman Street, Waltham, MA 02451, USA

ISBN: 978-0-12-410438-9

British Library Cataloguing in Publication Data
A catalogue record for this book is available from the British Library

Library of Congress Cataloging-in-Publication Data
A catalog record for this book is available from the Library of Congress

Contents

PART 4 DEPOSITS RELATED TO CARBONATITIES AND KIMBERLITES

PART 5 GOLD DEPOSITS

PART 7

PART 8

Contributors

W. Brownscombe
The Natural History Museum, Cromwell Road, London, UK

N.D.J. Cook
Mawson Resources, Vancouver, BC, Canada

J. Coppard
Consultant Geologist, Letty Green, Hertford, UK

L. Dahlenborg
Department of Geology, GeoBiosphere Science Centre, Lund University, Lund, Sweden

P. Eilu
Geological Survey of Finland, Espoo, Finland

I. Haapala
University of Helsinki, Helsinki, Finland

T. Halkoaho
Geological Survey of Finland, Kuopio, Finland

E. Hanski
Oulu Mining School, Oulu, Finland

A. Hartikainen
Geological Survey of Finland, Kuopio, Finland

C. Hartshorne
AngloAmerican Exploration, London, UK

T. Havela
Mawson Resources, Finland

E. Heilimo
Geological Survey of Finland, Kuopio, Finland

P. Heino
Yara Suomi Oy, Siilinjärvi Mine, Siilinjärvi, Finland

R.J. Herrington
The Natural History Museum, Cromwell Road, London, UK

M.R. Hudson
United States Geological Survey, US

T. Huhtelin
QP-FRB Outokumpu Ferrochrome Kemi Mine, Kemi, Finland

E. Hyvönen
Geological Survey of Finland, Rovaniemi, Finland

C. Ihlenfeld
AngloAmerican Exploration, London, UK

M. Iljina
Markku Iljina GeoConsulting Oy, Rovaniemi, Finland

M. Imaña
Lithogeochemical and Exploration Consulting, Avalidlandet, Sweden

S. Jones
First Quantum Minerals, Ltd, Sodankylä, Finland

T. Karinen
Geological Survey of Finland, Rovaniemi, Finland, Mustavaara Mine Ltd, Oulu, Finland

J. Kinnunen
Mawson Resources, Ltd., Finland

S. Klatt
Magnus Minerals Oy, Pelkosenniemi, Finland

J. Konnunaho
Geological Survey of Finland, Rovaniemi, Finland

A. Kontinen
Geological Survey of Finland, Kuopio, Finland

K. Koppström
Agnico Eagle Finland, Kiistala, Finland

V.J. Kortelainen
Agnico Eagle Finland, Kiistala, Finland

J. Kousa
Geological Survey of Finland, Kuopio, Finland

I. Lahti
Geological Survey of Finland, Kuopio, Finland

M. Lappalainen
First Quantum Minerals, Ltd, Sodankylä, Finland

M.J. Lehtinen
Senior Geologist, Lappeenranta, Finland

J. Liikanen
Endomines Oy, Hattu, Finland

T. Lindborg
Sotkamo Silver Oy, Vuokatti, Finland

K. Luolavirta
Oulu Mining School, Oulu, Finland

J. Luukas
Geological Survey of Finland, Kuopio, Finland

W.D. Maier
Cardiff University, Cardiff, UK;
University of Oulu, Oulu, Finland

T. Mäki
Pyhäsalmi Mine Oy, Pyhäsalmi, Finland

H.V. Makkonen
Geological Survey of Finland, Kuopio, Finland

M. Moilanen
Oulu Mining School, Oulu, Finland

F. Molnár
Geological Survey of Finland, Espoo, Finland

N. Kärkkäinen
Geological Survey of Finland, Espoo, Finland

T. Niiranen
Geological Survey of Finland, Rovaniemi, Finland

P.A. Nurmi
Geological Survey of Finland, Espoo, Finland

V. Nykänen
Geological Survey of Finland, Rovaniemi, Finland

H. O'Brien
Geological Survey of Finland, Espoo, Finland

N.H.S. Oliver
Holcombe Coughlin Oliver Trading as HCO Associates Pty Ltd,
Queensland, Australia

O. Sarapää
Geological Survey of Finland, Rovaniemi, Finland

H. Papunen
University of Turku, Turku, Finland

J. Parkkinen
Parkkinen Geoconsulting, Espoo, Finland

N.L. Wyche
Geological Survey of Western Australia, East Perth, Australia

P. Peltonen
Northland Resources S.A., Kolari, Finland;
First Quantum Minerals, Sodankylä, Finland

E. Pohjolainen
Geological Survey of Finland, Espoo, Finland

A.R. Prave
School of Geography and Geosciences, University of St. Andrews, UK

J.-P. Ranta
Oulu Mining School, Oulu, Finland

K. Rasilainen
Geological Survey of Finland, Espoo, Finland

F. Santaguida
First Quantum Minerals, Ltd, Sodankylä, Finland

P. Sarala
Geological Survey of Finland, Rovaniemi, Finland

P. Schaubs
CSIRO Earth Resources and Engineering Science, Australian Resources Research Centre, Bentley, Australia

J.K. Siikaluoma
AngloAmerican Exploration, AA Sakatti Mining Oy, Sodankylä, Finland

P.B. Sorjonen-Ward
Geological Survey of Finland, Kuopio, Finland

A. Taipale
Palsatech Oy, Rovaniemi, Finland Mustavaara Mine Ltd, Oulu, Finland, Finland

O.T. Rämö
University of Helsinki, Helsinki, Finland

T. Al-Ani
Geological Survey of Finland, Espoo, Finland

T. Ahtola
Geological Survey of Finland, Espoo, Finland

T. Törmänen
Geological Survey of Finland, Rovaniemi, Finland

I. Tuokko
Sotkamo Silver Oy, Vuokatti, Finland

J. Välimaa
Agnico Eagle Finland, Kiistala, Finland

E. Vanhanen
Mawson Resources, LTD, Finland

T. Voipio
First Quantum Minerals, Ltd, Sodankylä, Finland

N.L. Wyche
Geological Survey of Western Australia, East Perth, Australia

J. Ylinen
First Quantum Minerals, Ltd, Sodankylä, Finland

Y. Zhang
CSIRO Earth Resources and Engineering Science, Australian Resources Research Centre, Bentley, Australia

Preface

W.D. Maier, R. Lahtinen, H. O'Brien

The world faces heightened competition for key natural resources in a global environment of increasing demand and decreasing search space. These factors are the main drivers of the ongoing transformation from a linear to a circular economy. However, the limitations to recycling caused, for example, by product complexity and logistics suggest that demand for primary raw materials will remain high for decades. The European Union (EU), in particular, is highly dependent on metals, notably PGE for the automotive sector, but also REE, Nb, Ta and other metals for a range of high tech applications, especially new renewable energy solutions. Increasing amounts of phosphate are required to meet rising food demand globally, at decreasing availability of arable land due to climate change and increasing populations. Countries such as China, Russia, South Africa, the Democratic Republic of Congo and Zimbabwe are key supplier countries of many "critical" metals, i.e., those that are essential for modern high technology but whose supply is not assured. The capacity of some of these primary producer countries to provide secure supplies throughout the coming decades is uncertain. For example, underground mining of narrow PGE reefs in South Africa is currently uneconomic and may cease within the next decades, which could lead to sharp price increases for these metals. On the other hand, many European deposits that are amenable to relatively low cost surface mining may become economic as a result, particularly since they have the advantage of proximity to mature infrastructure (roads, railways, power grids, etc.) and a well-trained workforce.

Superimposed on potential short-term supply problems caused by socioeconomic factors or political decisions are long-term trends that threaten the availability of critical metals. The ambitious "20-20-20" energy and climate targets of the EU will have a major impact on metal supplies. By 2020, greenhouse gas emissions must be reduced to 20% below 1990 levels, with 20% of energy coming from renewable resources. The fuel cells, photovoltaic cells, and wind turbines required for a transition to a low carbon society will trigger a >100% increase in the demand for many key metals over the next decades; One fossil, non-renewable resource (metals) will replace another (hydrocarbons)

In view of the considerable threats to future supplies of critical metals, the efficiency of exploration, mining, and beneficiation, as well as the search for alternative sources (e.g., PGE in road dust, a range of metals in deep-sea polymetallic nodules, Ni-Co-PGE in laterites, REE and Nb in fenites, etc.) needs to improve and accelerate. The available resources of many of the metals mentioned are too small to satisfy increasing demand solely by recycling.

The EU, in particular, needs to make better use of its internal mineral resources, as outlined within EU strategic actions such as the "Raw Materials Initiative – Meeting our Critical Needs for Jobs and Growth in Europe" and the "European Technology Platform on Sustainable Mineral Resources." The mineral endowment of a country can create sustainable national wealth. Examples are the silver mines of medieval Sweden, and the modern day iron, coal, gold, and nickel mines of Australia. The Fennoscandian Shield, comprising Finland, Sweden, Norway, and Russian Karelia and Kola, are among Europe's most mineralized regions, and the similarity in geology with mineral giants like South Africa, Western Australia, and Canada suggests an enormous untapped mineral potential for, e.g., Ni, Cu, PGE, Cr, V, Zn, REE, and gold.

Recent years have seen several important new mineral discoveries in Finland (Suurikuusikko Au, Sakatti Ni-Cu-PGE) as well as developments of previously discovered ore deposits (Talvivaara polymetallic black shales, Kevitsa Ni-Cu-PGE). The present book provides an overview of these and many other deposits, including magmatic Ni-Cu-Cr-PGE-V deposits (Part 3), deposits associated with carbonatites and kimberlites (Part 4), gold deposits (Part 5), Fe deposits (Part 6), volcanogenic massive sulphide deposits hosting Zn and Cu (Part 7), as well as examples of a range of other deposits (Part 8 and 9). In addition, there is an overview on the history of mineral exploration in Finland (Part 1), a synthesis on the geological evolution and metallogeny of Finland (Part 2), and sections on exploration methods (Part 10) and mineral resources (Part 11). The book is targeted at geologists working for industry, Geological Survey organisations and universities as well as students aiming for a career in the minerals sector.

We would like to thank the authors who contributed their invaluable expertise on Finnish ore geology. Many of these geologists are employed by the Geological Survey of Finland (GTK) and this book would not have been possible without its assistance and collaboration. In addition, we would like to thank the reviewers for their essential contribution: Rodney Allen, Steve Barnes, Roberto Dall´Agnol, Pasi Eilu, Marco Fiorentini, Rich Goldfarb, David Groves, Eero Hanski, Jock Harmer, Murray Hitzman, Michel Houlé, Hannu Huhma, Asko Käpyaho, Bernd Lehmann, Ferenc Molnar, Nick Oliver, Petri Peltonen, Alain Plouffe, Kauko Puustinen, and Olli Äikäs. Finally, we would like to thank Marisa LaFleur for her editorial support and Mohanapriyan Rajendran for the production of several sets of proofs.

WDM dedicates this book to his son Randolph, whose nascent fascination with mineral exploration has been a great inspiration.

PART

1

A HISTORY OF EXPLORATION FOR AND DISCOVERY OF FINLAND'S ORE DEPOSITS

1

I. Haapala, H. Papunen

ABSTRACT

Historically documented mining in Finland started in the 1530s when the area formed part of Sweden. The post of commissioner of mines was founded in 1638. The Swedish government activated mineral exploration in Finland in the seventeenth century and, during the Age of Utility, in the eighteenth century. Numerous small iron, copper, and lead occurrences were found, as well as the larger Orijärvi copper deposit. During 1809-1917 when Finland was a Grand Duchy within the Russian Empire, exploration was reorganized and strengthened. New, generally small iron deposits were discovered to supply iron works. Deposits found in the Pitkäranta area were mined for copper, tin, zinc, and iron. The Geological Survey of Finland was established in 1885. The large and rich Outokumpu copper deposit was discovered in 1910 as a result of scientific exploration by Otto Trüstedt of the Geological Survey. Since then, boulder tracing (later also till geochemistry), geophysical measurements, and diamond drilling have been the standard exploration methods. After Finland declared independence in 1917, mineral exploration was considered important for the development of domestic industry. In 1921 the Geological Survey discovered the nickel–copper ore field of Petsamo, and by the end of the 1930s, several other deposits, including the Otanmäki iron–titanium–vanadium deposit, were discovered. After the Second World War, the Geological Survey was reorganized and strengthened, and in the following decades it discovered numerous important ore deposits, including the Vihanti zinc–copper deposit (1951) and the Kemi chromium deposit (1960). The state-owned mining companies Outokumpu Oy and Rautaruukki Oy established their own exploration departments in the 1950s. Outokumpu Oy discovered the Kotalahti (1956), Vammala, and Kylmäkoski nickel–copper deposits, the Pyhäsalmi copper-zinc deposit (1958), the Vuonos (1965) and Kylylahti (1984) Outokumpu-type deposits, and several gold deposits. Otanmäki Oy/Rautaruukki Oy discovered several iron deposits in Lapland, as well as the large apatite-rich Sokli carbonatite complex. Rautaruukki Oy terminated their exploration activities in 1985 and

Mineral Deposits of Finland. http://dx.doi.org/10.1016/B978-0-12-410438-9.00001-7

Outokumpu Oy in 2003. Finnish private companies also carried out successful mineral exploration. After becoming a member of the European Union in 1995, the role of international mining and exploration companies has strongly increased in Finland, the main interest being in precious metal deposits.

Keywords: exploration; Finland; history; ore deposits; Geological Survey of Finland; Outokumpu Oy; Rautaruukki Oy.

INTRODUCTION

The activity and success of exploration for mineral deposits in Finland has, over the centuries, varied markedly depending on the political, cultural, and economic situation, as well as on the more or less fortuitous availability of persons capable to plan, lead, and carry out prospecting and exploitation of deposits. A major historical background is provided by the changes in the form of government. The eastern part of Norden (or geographic Fennoscandia) now known as Finland was gradually incorporated into the kingdom of Sweden during the twelfth and thirteenth centuries. Over the subsequent five centuries, the eastern border of Sweden shifted several times as a result of wars between Sweden and Novgorod/Russia, until 1809 when Finland was ceded to Russia. Since December 1917 Finland has been an independent state.

Several extensive volumes have been published on the history of the mining industry and exploration in Finland. The first that should be mentioned is the work of Tekla Hultin, which deals with the evolution of the mining industry and exploration during the time of Swedish rule (Hultin, 1896, 1897). The exploration and mining industry during the period of the Russian rule has been documented in detail by Dr. Eevert Laine (e.g., Laine, 1950, 1952). A comprehensive account of the mining industry in Finland from 1530–2001 was recently published by Kauko Puustinen (1997, 2003). Hyvärinen and Eskola (1986) published a chapter on mineral exploration in Finland, including selected case histories of discoveries. A shorter article, "One hundred years of ore exploration in Finland," by one of the current authors (Papunen, 1986), has been an important source, for the present review.

TIME OF SWEDISH RULE

The skill to collect limonitic bog and lake iron ore nodules and produce iron from it in simple furnaces was known in Finnish villages in ancient times, and such iron occurrences were still harvested in small quantities at the end of the nineteenth century. Some lake ore deposits south of Iisalmi in eastern Finland were still studied as possible manganese ores in 1940s and 1950s.

During the reign of King Gustaf I Vasa (1496–1560), the crown supported exploration and mining in the eastern part of Sweden as well. The first clearly documented mine in Finland was the Ojamo skarn-type iron deposit in Karjaa, southern Finland. The deposit was mined in the 1530s by Erik Fleming, the councilor of southern Finland, although it was not until 1542 that King Gustaf I awarded him the formal privilege to exploit the deposits (Puustinen, 1997, 2010). The ore was first worked up in Siuntio, and later in the nearby Mustio ironworks. After Fleming's time mining slowed down. Under the reign of Gustaf I some other minor deposits (e.g., copper ore of Remojärvi in Juva in eastern Finland and iron ore in Siuntio) were exploited. How the deposits were discovered is not known.

In the seventeenth century, under the reign of Gustaf II Adolf Vasa (1594–1632) the mining industry rose to a new level, driven by mercantilism and wars in central Europe, especially the Thirty Years' War in 1618–1648. One reason for the success of the Swedish army in this war was the large and rich Dannemora iron and the Falun copper mines. Gustaf II Adolf had a keen interest in ores and mining. As the crown prince, he had stayed in Finland in 1611 and collected ore specimens that he brought to Stockholm for closer studies (Salokorpi, 1999). As king he traveled in Finland during 1614–1616 and gave personal orders to open or reopen mines and to establish new ironworks (Hultin, 1896, 1897). He had an optimistic view of Finland's ore potential. After his death in the battle of Lützen in 1632 mining policies were continued by his successors. The Board of Mines was established in Stockholm in 1637, and Finland received a permanent post of commissioner of mines in 1638. Finland's first university, Academia Aboensis (Academy of Turku), was founded in 1640 in Turku.

The mining and manufacturing of the large ore deposits in central Sweden seriously depleted Swedish forests. As Finland contained huge tracts of virgin forest that could be exploited for the production of charcoal, as well as rivers for transport and rapids to run furnaces and hammers, experts were sent to Finland to explore for ore deposits and set up metal works in suitable places. Clergymen and other officials were urged to assist in these efforts. The Ojamo mine was reopened and several new ironworks were established: Mustio or Svartå in Karjaa (1550–1901); Antskog (1630–1880), Billnäs (1641–1905), and Fiskars (1649–1904) in Pohja; Fagervik in Inkoo (1646–1904); Orisberg in Isokyrö (1676–1900); Skogby in Tammisaari (1682–1908); and Tykö in Perniö (1686–1908). As raw material, these ironworks mainly used ore or raw iron from the Utö deposits in the Stockholm archipelago. If possible, hard-rock iron deposits from Finland were used, including Ojamo (exploited periodically in 1533–1863), Vittinki in Ylistaro (1563–1920), Sådö (1610–1863) and Långvik (1662–1863) in Inkoo, Juvakaisenmaa in Kolari (1662–1917), Morbacka in Lohja (1668–1873), Malmberg in Kisko (1670–1866), Vihiniemi in Perniö (1690–1865), and Kelkkala in Tammisaari (1690–1900), as well as many other relatively small deposits. In some cases, Finnish lake and bog iron ores were used. The copper and lead occurrences found in the seventeenth century were generally relatively small (Puustinen, 2003).

The beginning of the eighteenth century was overshadowed by the Great Nordic War (1702–1721) as a result of which Sweden lost parts of eastern Finland to Russia. In order to strengthen the Swedish and Finnish mining industry after the war, a royal statute of 1723 promised significant rewards for discovery of ore deposits and benefits for their exploitation.

The Board of Mines sent geological experts to work in Finland, including Daniel Tilas (1712–1772) and Magnus Linder (1709–1799). In the report of his travel to Finland in 1737–1738, Tilas described the bedrock and mineral occurrences of southern Finland. His visits to larger villages were announced in churches, and people were urged to report their findings. Tilas himself carried out exploration for ore deposits at several sites, including the newly discovered Ansomäki iron deposit near **Haveri** in Viljakkala (Puustinen, 2006). In Tammela he visited a site where in 1733 a farmer had found a large boulder containing high-grade copper ore. Tilas had recognized that erratic boulders are generally situated southeast of their sources, which enabled him to find chalcopyrite-bearing veins in outcrops at Hopiavuori (later known as Tilasinvuori). The deposit was mined in 1740–1749 (Puustinen, 2014). This is the first time that "boulder tracing" was successfully applied to the exploration of ore deposits, long before the theory of the great Nordic continental glaciation (Ice Age) was presented.

Tilas' colleague Magnus Linder was appointed to the post of deputy commissioner of mines in Finland in 1741, but because of renewed war with Russia he could not travel to Finland until 1744. He held the permanent post of commissioner of mines in 1747–1787. When Linder started his work, the few existing mines and ironworks were stagnant and the mining administration was thoroughly decayed. Linder set about reorganizing the mining industry. In 1744 he discovered the skarn-type Sillböle iron deposit in Helsinki, which was in production in 1744–1770 and 1823–1866. In the same limestone-skarn horizon, a few kilometers west of Sillböle, other skarn iron deposits were found in Hämeenkylä and Jupperi, and they were mined periodically from 1786 to 1860 (Saltikoff et al., 1994). In 1757 the owner of the **Orijärvi** estate found metallic ore minerals in a rock outcrop in his back forest, and the rock turned out to be good copper ore. The deposit was mined with some interruptions from 1758 to 1954, and during the last 30 years zinc and lead were also produced (Nikander, 1929; Turunen, 1957; Poutanen, 1996). Orijärvi was the first significant copper deposit found in Finland (Figs. 1.1 and 1.2).

PERIOD OF RUSSIAN RULE

ORGANIZATION OF GEOLOGICAL RESEARCH AND MAIN RESULTS OF MINERAL EXPLORATION

During the Napoleonic wars the Russian army occupied Finland in 1808–1809. Russia annexed Finland and conveyed it the status of a Grand Duchy. Because of the war, mining and exploration had largely ceased. The only significant discovery was that of the Kulonsuonmäki titaniferous iron deposit near Karkkila. It was mined in 1817–1888, and the Högfors ironworks was built to process the ore. The office of the commissioner of mines was weak, but the situation improved when General Count Fabian Steinheil was appointed Governor General in 1810. Steinheil, who himself was a skillful mineralogist, wanted to promote geological research, exploration, and the mining industry in Finland. On his recommendation, the young lawyer and mineralogist Nils Gustaf Nordenskiöld (1792–1866) was appointed at commissioner of mines in 1818. Nordenskiöld had studied law and chemistry under Johan Gadolin in the Academy of Turku as well as science of mining and metallurgy in Uppsala, and he was familiar with the ore deposits of central Sweden. Steinheil arranged financing for a three-year expedition to central and western Europe, during which time Nordenskiold got acquainted with mineral exploration and its organization as well as other branches of the mining industry. The office of commissioner of mines was rearranged in 1821. The leader of the office became superintendent, and the staff was increased. Nils Gustaf Nordenskiöld was appointed the first superintendent in 1823, a post he retained until 1855. In 1858 the office was reorganized and became the Board of Mines.

Nordenskiöld reorganized the mineral exploration and the training of his staff. He gave prospectors detailed directions on how to compile geological maps and collect samples. All known ore fields were restudied and some of them were mined. Numerous new ore deposits, mainly iron deposits, were found, and the mining compass came into use. The remarkable **Jussarö** iron ore field in the archipelago near Tammisaari was discovered in 1834, and the deposit was mined in 1834–1861. Most of the other occurrences were quite small and were mined only briefly. A few deposits, such as Haukia (exploited 1839–1864) and Pahalahti (1826–1854) in Kisko, and Väsby in Korppoo (1839–1864) produced thousands of tons of iron ore. The old deposits of Ojamo, Sillböle, Hämevaara, Malmberg,

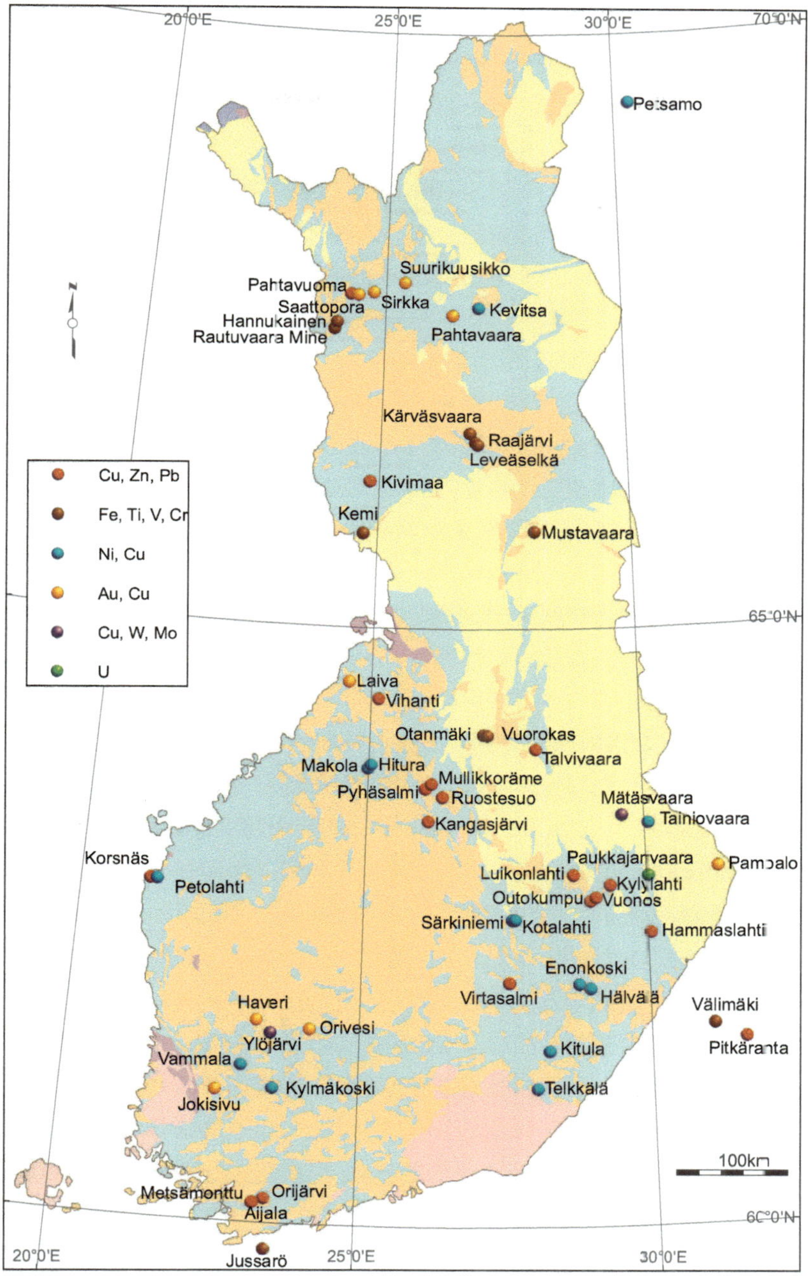

FIGURE 1.1 Location of the metallic ore deposits mined in Finland in 1900–2013.

Source: *Data are from Puustinen (2002), from later ore data files of the Geological Survey of Finland, and from the publications cited in the text.*

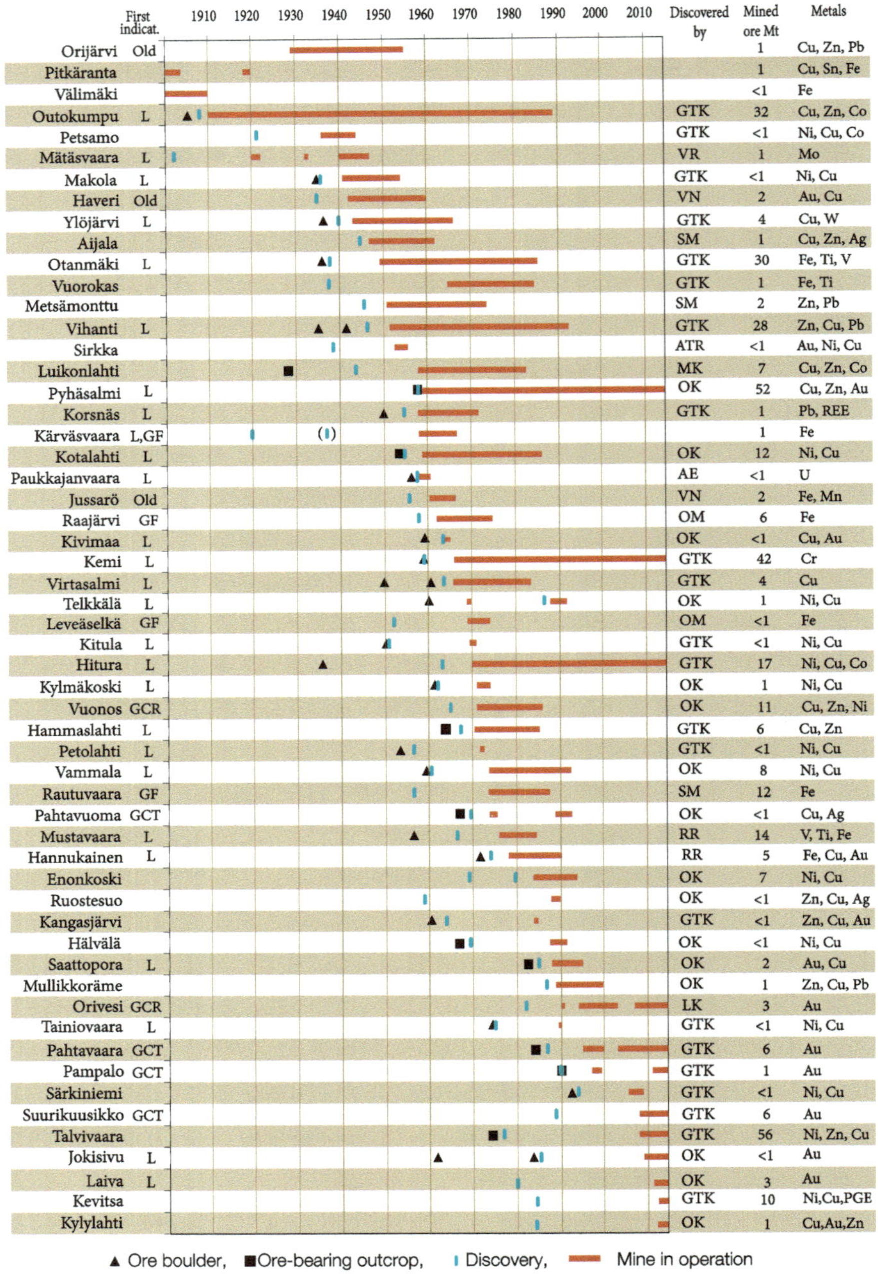

FIGURE 1.2 The life span of the metallic ore mines in Finland, 1900–2013, from exploration to exploitation.

Explanations: First indication of the deposit: GF, geophysical anomaly (airborne or ground); GCR, rock geochemistry; GCT, till or stream sediment geochemistry; L, layman-sent sample; Old, old mine. Discovered by: AE, Atomienergia Oy; ATR, Atri Oy; GTK, Geological Survey of Finland; LK, Oy Lohja Ab; MK, Malmikaivos Oy; OK, Outokumpu Oy; OM, Otanmäki Oy; SM, Suomen Malmi Oy; VN, Oy Vuoksenniska Ab; VR, Oy Värtsilä Ab.

Source: *Data from same references as in Fig. 1.1.*

and Vihiniemi were again mined, and special attention was paid to the deposits of Haveri and Juvakaisenmaa. The aim was to make Finland self-sufficient with respect to iron, but this goal was not attained. Iron ore and raw iron were imported from the Utö mines in Sweden to the ironworks of southern and western Finland. When large limonitic lakes and bog ores were found in central and eastern Finland, a number of ironworks were established in those areas to work the lake ores. The Juantehdas (Strömsdal) ironworks in Juankoski, founded in 1746, was in operation until 1911. Of the ironworks established in eastern Finland during the nineteenth century, those of particular importance included Möhkö in Ilomantsi (in operation 1837–1907), Värtsilä in Tohmajärvi (1851–1920), Varkaus (1815–1908), and Annantehdas in Suojärvi (1809–1905). In the nineteenth century, exploitation of the lake ores was less expensive than mining of hard rock ores of similar iron contents, but their utilization was hampered by high phosphorus contents. Many ironworks in eastern Finland developed into large enterprises, which sold their products mainly to St Petersburg (Laine, 1952). The smelting of lake and bog ores reached a peak in the 1860s and 70s. Subsequent international development of iron metallurgy decreased the profitability of the production of iron from lake and bog ores, and mining ceased in the 1920s.

In 1810, copper showings and remnants of earlier mining attempts were found in the Pitkäranta area by Russian mining officer Anton Furman on the northeastern side of Lake Ladoga, and later cassiterite was also identified (Laine, 1952). The deposits represent skarn-type occurrences at the margin of the Salmi rapakivi granite batholith. The deposits were mined, mainly by Russian enterprises, for copper, tin, and iron in 1842–1904. The geology and ore deposits of the Pitkäranta area were studied in detail by Trüstedt (1907) (see Chapter 8).

In 1837, gold-bearing boulders were found near the mouth of the Kemijoki River, and Nordenskiöld enthusiastically began systematic exploration. Prospecting was continued for several years, first in the Kemi, Tornio, and Rovaniemi areas, and then, in 1847–1848, in Kuusamo under the leadership of Henrik Holmberg. Minor amounts of gold were obtained by panning in several places in Kuusamo, but not enough to continue the extensive prospecting. When rumors of the California gold discoveries reached Finland in 1849, Holmberg left in autumn 1849 for America to continue gold prospecting in California and Alaska. The disappointing results of gold prospecting in Lapland inspired Finland's national poet Johan Ludvig Runeberg to his well-known words in the Finnish national anthem (presented for the first time in 1848): *Our land is poor, and so shall be to him who gold will crave*. It took more than 100 years to realize that Runeberg's views were overly pessimistic.

From the modest success of exploration, Nordenskiöld concluded that to improve the efficiency of prospecting the whole country should be mapped geologically. On his suggestion, geological mapping was started in southern Finland in 1860s, but it was interrupted because of new placer-type gold findings in Lapland.

In 1867 and 1868, mining engineer Tellef Dahll from the Geological Survey of Norway had found gold flakes by panning in the Tana (Teno) and Enare (Inari) rivers. These rivers mark the international boundary between Norway and Finland, and gold was found on both sides of the boundary. In 1868, a Finnish expedition led by mining engineer Konrad Lihr was sent to study the gold potential on the Finnish side the Tana River. The occurrences proved to be uneconomic. On their return to Rovaniemi the group performed test panning at various sites, and a promising gold occurrence was found on the banks of the Ivalojoki River.

During the following summer, two sailors, Jakob Ervast and Nils Lepistö, panned within one month 2 kg gold at Saariporttikoski in the Ivalojoki River. This discovery led to a gold rush to Ivalojoki. A "crown station" was constructed at Ivalojoki River in 1870, and it became the center for the gold diggers—in 1871 up to 500 men—who worked there. The staff of the Board of Mines superintended the mining. Geologist A.M. Jernström prepared a geological map of the area at the scale of 1:800 000. The amount of gold produced was highest in 1871 and 1872, about 56 and 53 kg, respectively, but then the results declined and the crown station was closed in 1890. Prospecting and washing for gold moved to the tributaries of Palsinoja and Sotajoki, then at the beginning of the twentieth century to the Laanila area, and to Lemmenjoki River in 1945. Gold washing has continued to a minor degree. Attempts to find economic primary gold deposits in the Laanila area in early twentieth century failed (Stigzelius, 1986).

In 1877, a geological office was founded under the Board of Mines with the aim of producing geological maps of the whole country, and financing was arranged for 10 years. The mapping that had been started in 1865 in southern Finland could now be continued. In 1885, the responsibilities of the geological office were transformed to a new government agency within the Ministry of Industry, the Geological Commission, later known as the Geological Survey.

Of the ore deposits found in the nineteenth century, the **Välimäki** titaniferous iron ore on the northeastern side of Lake Ladoga (Blankett, 1896) was of note. Indications of the ore, ore boulders, and a magnetic anomaly were detected already in 1855 during H.J. Holmberg's research trip to the area, but the actual ore body was not found until 1889, by prospectors of a Russian company. The deposit was mined in 1889–1910 and contained about 0.35 Mt of ore with 32% Fe and 5.5% Ti (Figs. 1.1 and 1.2). The iron ore was transported to the Vitele ironworks in Aunus.

In connection with a railway construction, a low-grade molybdenite deposit was discovered in 1902 at **Mätäsvaara** in Lieksa, eastern Finland, at the contact between Archean granitoid gneiss and Svecofennidic potassium granite. After several attempts to exploit the deposit, it was effectively mined by Oy Vuoksenniska Ab in 1940–1947, producing 1.15 million tons of ore with 0.14% MoS_2 (Kranck, 1945; Zeidler, 1949).

DISCOVERY OF THE OUTOKUMPU (KERETTI) COPPER DEPOSIT

The discovery of the Outokumpu copper ore in North Karelia was an epoch-making achievement in the history of the Finnish mining industry. The story of the discovery has been described in numerous articles in Finland and abroad. The short review presented here is based mainly on Saksela (1948).

In March 1908, a large (approximately 5 m^3) boulder containing ore minerals was found at three meters depth in soil when dredging the Kivisalmi canal in Rääkkylä, eastern Finland. The discoverers, building engineer Montin and machine operators Eskelinen and Asplund, thought that the boulder could be a meteorite; at the turn of the century, three meteorites had fallen down in Finland, including the famous Bjurböle stone meteorite that fell in 1899 between Porvoo and Helsinki. They sent a sample of the boulder to the Geological Survey, where it was identified as a high-grade copper ore that contained the minerals pyrite, pyrrhotite, and chalcopyrite. Otto Trüstedt, a mining engineer and geologist who then worked for the Survey, visited the site of discovery. He observed that the quartzitic ore boulder, which was blasted into several pieces, had been rounded at the edges and situated in a till bed. No other ore boulders were discovered at the locality.

At that time, the approximate movements of the continental ice sheet were already known from the directions of glacial striations on outcrops. In North Karelia, there existed two directions: 105–113°

and 165–177°. Trüstedt decided not to continue prospecting solely on the basis of glacial striations. At the beginning of the twentieth century, the fieldwork for the geological mapping at a scale of 1:400,000 had largely been completed in southeastern Finland, and some results had been published. The geologists who had conducted the mapping, B. Frosterus and W.W. Wilkman, were of the opinion that the quartzite of the boulder could only be the so-called "Kalevian quartzite," that was found in several places in North Karelia. Trüstedt studied in detail the Kalevian quartzite formations in Liperi and Rääkkylä, north and northwest of Kivisalmi. In some cases the rock contained a little pyrite and pyrrhotite, but no chalcopyrite was found.

Trüstedt now returned to the site of the Kivisalmi ore boulder and carefully studied the composition of the gravel associated with the ore boulder during dredging of the canal. His aim was to find some clue to characterize the source locality. He found several boulders containing tremolite fels (skarn) and black, sulfide- and graphite-rich schist. Trüstedt returned to the Geological Office and showed the samples to Frosterus and Wilkman. Frosterus remembered that in 1899 they had seen and studied similar rocks near Outokumpu Hill in Kuusjärvi, notably a hundreds-of-meters-wide mica-bearing quartzite schist that contained intercalations of talc and tremolite schists, carbonate rock, and skarn, as well as lenses of serpentinized dunite. Wilkman had written in his diary that the northern slope of Outokumpu Hill contains "banded, ore-bearing quartzite."

In the autumn of 1908, exploration moved to Kuusjärvi. Numerous chalcopyrite-bearing ore boulders were found in the Outokumpu area. Some ore boulders also contained uvarovite and other chrome silicates. The area was poorly exposed, but systematic magnetic measurements revealed an anomalous, northeast-trending zone near the assumed contact between the mica schist and quartzite schist. In March 1909, the Senate of Finland approved 8000 marks to continue and intensify exploration.

In the summer of 1909 several test pits were dug, first at Sänkivaara and the Outokumpu hills and then to the eastern side of Outokumpu. No bedrock hosted ore was found, but rocks of the quartzite-carbonate rock association, and some copper ore boulders. Trüstedt concluded that the ore must be situated in quartzite between the northernmost ore boulders and the assumed northern contact of the quartzite against the mica schist. Because of the thick overburden, he decided to use the remaining funds for diamond drilling. The first two drill holes did not hit ore, so Trüstedt paid for the third hole from his own pocket. On March 16, 1910, the drill penetrated 9 m of high-grade copper ore containing approximately 6% Cu. The Outokumpu deposit was discovered, about 50 km north-northwest of the original Kivisalmi boulder. Nearly simultaneously, on March 18, the Senate appropriated 2000 marks to continue drilling.

After the discovery of the ore it was necessary to continue exploration to determine the size, shape, and grade of the deposits utilizing geological and geophysical (magnetic and electrical equipotential) methods and diamond drilling. The work was financed by the State (the finder) and Hackman & Co. (the land owner). It was soon ascertained that the ore body was at least 1400 m long. The State and Hackman & Co. formed a general partnership named "Outokumpu Kopparverk" to exploit the deposit. Pilot mining was carried out in 1910–1912, and more effective mining was started in late 1913. In 1917 the mine was lent to a Norwegian-Finnish mining company Ab Outokumpu Oy, but in 1921 the State of Finland bought up the shares of the Norwegian partner and in 1924 also the shares of Hackman & Co. Under the leadership of Eero Mäkinen, mining became productive. Later on, new mines were opened paving the way for Outokumpu Oy to become a significant internationally operating mining company.

EARLY YEARS OF INDEPENDENCE: 1917–1944

The discovery of the Outokumpu copper deposit in 1910 had to some extent changed the generally pessimistic view about the possibilities of finding economic mineral deposits in Finland. The discovery had also shown the usefulness of new exploration methods, boulder tracing, geophysical measurements, and diamond drilling. Further optimism resulted from the discovery of some minor sulfide ore deposits in eastern Finland, including the pyrite deposits of Otravaara in Eno and Tipasjärvi in Sotkamo. The Otravaara pyrite ore body was discovered in 1918 on the basis of an ore boulder sent to Suomen Malmitutkimus Oy (Finnish Ore Exploration Co.) in Helsinki. Aarne Laitakari, the director of the company, found more ore boulders, had small trenches dug through the till cover near the boulders and across the strike of foliation, and found the ore body hosted by sericite quartzite (Saxen, 1923). The deposit was exploited in 1919–1924 by Suomen Mineraali Oy. The important nickel–copper deposits of Petsamo were discovered a few years later (1921).

To promote exploration for mineral deposits, Pentti Eskola, Eero Mäkinen, and some other geologists from the Geological Survey wrote in Finnish and Swedish articles and booklets on Finnish mineral deposits and their exploration, as well as possibilities for evolution of the mining industry.

DISCOVERY OF THE PETSAMO (PECHENGA) NICKEL–COPPER ORE DEPOSITS

In 1920, in the Peace Treaty of Tartu, Petsamo (Pechenga), a strip of the Arctic coast between the Soviet Union and Norway, was incorporated into the new Republic of Finland. In spring 1921 the Geological Survey of Finland sent two students, Alppi Talvia and Hugo Törnqvist, to study if the banded iron formations of Sydvaranger in Norway continue across the border to Petsamo. No extension of the iron ores could be found, and the group moved from Salmijärvi east to the fells of Petsamo. In Kotseljoki, on the western end of Kammikivitunturi, Törnqvist found a rusty sulfide-bearing rock that could be followed on the ground as a magnetic anomaly. Later the rock was identified by the Geological Survey as nickel–copper ore. In 1922 the studies were led by Dr. Hans Hausen, an experienced geologist, and Törnqvist was his assistant. During the short Arctic summer they succeeded in establishing the main geological frameworks of the nickel ore. Nickel occurrences were found in an 80-km-long bent schist belt (greenstone belt) that extended from the Norwegian border across Petsamo to the border with the Soviet Union. The belt included mafic–ultramafic metavolcanic and intrusive rocks and metasediments. The Kotseljoki Ni-Cu occurrence was interpreted to be hosted by a dike, whereas the other occurrences were disseminations in ultramafic rocks (Hausen, 1926).

The next expedition took place in 1924 under the leadership of Dr. Väinö Tanner; other members were Hans Hausen, H. Törnqvist, and J.N. Soikero. Tanner also led the annual field campaigns during the following years until 1931, followed by Dr. Heikki Väyrynen until 1934. During these years, the exploration comprised detailed research in the long but narrow ore-bearing zone characterized by phyllites or black schists and mafic–ultramafic (ferropicritic) volcanic and plutonic rocks (Väyrynen, 1938). Magnetic, electric equipotential, and, since 1927, electromagnetic geophysical techniques were used along with geological mapping. Diamond drilling was carried out at selected localities. Due to the lack of local roads, the research equipment had to be carried to the fells by men or reindeers. For several years the researchers also had to prepare basic maps. The work became much easier in the early 1930s when airborne photographs became available.

When the Geological Survey finished exploration in Petsamo in 1934, more than 10 ore deposits had been found, including the Kaulatunturi, Kammikivitunturi, Onkitunturi, Ortoaivitunturi, and Pilgujärvi deposits. The ore resources of the Kaulatunturi deposit were estimated to be at least 5 million tons (Haapala, 1945).

By 1933 it had become clear that mining would be profitable at the Kaulatunturi deposit. At the time, Outokumpu Oy, the country's only company mining metallic ore, did not have the financial resources to found a new mine in the wilds of Lapland. Under these circumstances, it was logical to seek a foreign partner. After long and thorough negotiations, in the summer of 1934 the Government of Finland concluded an agreement with the Mond Nickel Company (London), a subsidiary of the Canadian International Nickel Company (INCO), for exploitation of the ore. For operations in Finland, Mond Nickel Company established the company Petsamon Nikkeli Oy. Exploration continued under the leadership of Dr. Paavo Haapala, and construction for mining of the Kaulatunturi deposit began. From autumn 1940 to autumn 1944 Petsamon Nikkeli Oy was in partnership with the German company I.G. Farbenindustrie AG, and after the Second World War the district of Petsamo was incorporated into the Soviet Union (Autere and Liede, 1989; Vuorisjärvi, 1989).

OTHER EXPLORATION FOR ORE DEPOSITS

In the mid-1930s, when Finland's economy slowly started to recover from the global economic depression, the government aimed to support mineral exploration in order to find new ore deposits and thereby strengthen economic development. The Geological Survey received new statutes in 1936, which, without neglecting basic research, emphasized the role of practical geology. Three research fields were defined: ore geology, bedrock geology (petrology, Precambrian geology), and geology of surficial deposits (Quaternary geology). The number of staff was increased. The new director, Aarne Laitakari, supported public enlightenment on geology by giving lectures and writing newspaper articles and books directed to a broad audience. University professors Pentti Eskola and Matti Sauramo wrote popular textbooks and in their presentations highlighted geology as an important future field of science. The number of geology students increased, and many of them found their places in ore exploration and the mineral industry.

Outokumpu Oy carried out exploration for ore deposits in North Karelia in 1935–1939 under the project named Suuri Malminuotta (the Great Ore Seine Operation). A number of students from the University of Helsinki participated in the program under the leadership of Paavo Haapala. The students received both theoretical and practical training in exploration, and many of them were subsequently employed by Outokumpu Oy or the Geological Survey.

In 1935, a new company, Suomen Malmi Oy, was established to carry out ore exploration, including geological and geophysical methods and deep drillings. In the 1930s, the company participated in exploration in the Pitkäranta ore field and several other areas (e.g., the iron ore fields of Porkonen-Pahtavaara, Misi, and **Kärväsvaara** in Lapland). Local villagers had already discovered the Kärväsvaara occurrence in 1921 on the basis of magnetic anomalies and digging. In 1937–1938, Suomen Malmi carried out geological mapping, magnetometric measurements, and diamond drilling at the locality, and sank a 20-m-deep shaft into the ore. Although the iron content was reasonably high (≥50% Fe), the locally high sulfide content and scattered distribution of iron decreased the quality of the magnetite ore. When negotiations of the mining rights failed, Suomen Malmi Oy stopped exploration at Kärväsvaara (Olson, 1937).

In the second half of the 1930s, the Geological Survey received two important ore boulder specimens from local farmers. In 1936, Kusti Ainasoja sent rusty, sulfide-bearing samples from Hosionperä in Järvikylä, Nivala, which caused a short visit to the locality to obtain unaltered samples. An analysis of an unaltered sample from the original serpentinite boulder showed 0.63% Ni and 1.7% Cu. The area was poor in outcrops and no serpentinite rocks were known to exist in Nivala. The succeeding field studies in 1936 and 1937 revealed a narrow, several kilometers long ore boulder fan in the area. Magnetic and electromagnetic measurements showed anomalies near the northwest end of the boulder fan, and soon thereafter the **Makola** nickel deposit was found by diamond drilling. In 1939 biogeochemical studies— based on the ash of birch leaves—showed anomalous Ni contents in the same area. The ore was exploited by Outokumpu Oy from 1941–1954.

In 1937, Kalle Leppänen sent a pyrite-rich boulder specimen from a Järvenpää farm in **Ylöjärvi**, about 20 km northwest of Tampere. State geologist Martti Saksela became interested in the sample: it contained abundant pyrite in a sericite schist and resembled the ores of Otravaara and Karhunsaari that he had studied earlier. He visited the locality and recognized in the schist a zone of sericite-quartzite (metasomatic "ore quartzite") that contained small amounts of pyrite. A prospecting team was moved to the area. Numerous boulders containing sulfides in sericite schist were found, and a 2-km-long sericite schist zone was outlined. About 5 km south of the Järvenpää farm, the team found several breccia boulders, which contained fragments of mafic metavolcanic rocks in a matrix composed of tourmaline, quartz, chalcopyrite, and arsenopyrite.

The boulder fan guided prospectors to the shore of the small Paronen Lake where tourmaline breccia was found in outcrop, but without chalcopyrite. Field studies, including magnetic surveys and diggings, were continued in summer 1938 under the leadership of Dr. Erkki Mikkola, and the Ylöjärvi or Paronen copper deposit was found near the southern end of the lake. Diamond drilling was started at the end of 1938, and in spring 1940 the research material was assigned to Outokumpu Oy. Mining of the copper ore started in 1942. In 1945, a mineral resembling quartz was found in the ore and identified as scheelite, and after this discovery tungsten became an important by-product. In 1942–1966, about 4 million tons of ore with 0.76% Cu were mined, producing 28,000 tons Cu, 427 tons W, 49 tons Ag, and 270 kg Au (Himmi et al., 1979).

In 1937, two summer assistants of the Geological Survey found two high-grade Fe-Ti ore boulders near the Sukeva prison in the municipality of Sonkajarvi, between Iisalmi and Kajaani. The following summer, detailed studies began to find the source of the boulders. Neither new ore boulders nor ore-bearing outcrops were found. At the same time, Quaternary geologists were mapping surficial deposits in Vuolijoki, some 30 km northwest of Sukeva, and they observed strong compass disturbances in a place called **Otanmäki**. When the explorer group, led by Veikko Pääkkönen, heard of this, they moved from Sukeva to Otanmäki, and soon found strong magnetic anomalies and iron ore outcrops (Pääkkönen, 1952). The ore consisted of similar titaniferous magnetite ore as the Sukeva boulders. In the next few years, the area was surveyed in detail. Several Fe-Ti-V ore bodies were located in a gabbro-anorthosite complex, and diamond drilling verified the economic significance of the deposit. Because of metamorphic recrystallization, magnetite and ilmenite occurred as separate grains, which was an advantage in processing the ore (Pääkkönen, 1956). The Otanmäki and adjacent **Vuorokas** deposits were mined in 1953–1985, first by Otanmäki Oy and, since 1968, by Rautaruukki Oy. Altogether, approximately 31 Mt of Fe-Ti-V ore was mined, and significant resources still remained.

In addition to the activities described, two private companies, Atri Oy and Oy Vuoksenniska Ab, prospected for ore deposits in the 1930s and 40s. Atri Oy was especially active in Lapland, where the most important occurrence was the **Sirkka** gold prospect ("Sirkka mine") in Kittilä. The Sirkka prospect was composed of hydrothermal gold and sulfide-bearing veinlets in the greenschist facies rocks of the greenstone belt of central Lapland. After Atri Oy, the prospect was studied intensely by Oy Vuoksenniska Ab, Outokumpu Oy, and the Geological Survey (Eilu et al., 2007). Vuoksenniska studied the Mo deposit of Mätäsvaara and the old **Haveri** iron deposit. During a visit to the old Haveri iron mine in 1935, Professor Hans Hausen from Åbo Akademi University and mining counselor Bengt Grönblom from Oy Vuoksenniska Ab took specimens from the mining waste. Analyses showed that the specimens contained considerable contents of copper and a few grams of gold per ton. This led to intense new geological and geophysical studies with diggings, diamond drilling, and to sinking of an exploratory shaft in the iron ore at Kruuvanmäki, under the leadership of Kurt Lupander. Vuoksenniska started mining in 1942, and when the mine was closed in 1960 it had produced about 1.5 Mt of ore at 2.8 ppm Au and 0.37% Cu.

The prospecting company Malmikaivos Oy was founded in 1941. It was first a subsidiary of Yhtyneet Paperitehtaat Oy (the United Paper Mills Company) and later of Myllykoski Oy. Dr. Martti Saksela was appointed the director and Dr. Erkki Aurola the chief geologist of the new company. They placed extensive advertisements in newspapers requesting ore samples from laypersons, and the company soon received abundant specimens. The **Luikonlahti** copper occurrence in Kaavi became the main exploration target. The prospect had been studied since 1910 by Hackman & Co., the Geological Survey, Outokumpu Oy, and Ruskealan Marmori Oy, before Malmikaivos Oy started long-lasting detailed research in early 1940s.

THE YEARS 1945–2013

After the two lost wars, the Winter War in 1939–1944 and the Continuation War in 1941–1944, Finland was smaller and poorer than before, and heavy war reparations had to be paid to the Soviet Union. Without copper production from the Outokumpu mines, the metal industry of Finland would not have been able to fulfill the reparation contracts (Kuisma, 1985, p. 235). There was a global shortage of raw materials, and Finland needed metals for industry and foreign currency. In this difficult situation the government wanted to support and strengthen the mineral industry and ore exploration, which turned out to be a far-sighted decision.

EXPLORATION BY THE GEOLOGICAL SURVEY

Evolution of the exploration organization

The Geological Survey got new statutes in 1946. The Survey was divided into four departments: Exploration Geology (headed by Dr. Aarno Kahma), Petrology (headed by Dr. Ahti Simonen), Quaternary Geology (headed by Dr. Esa Hyyppä), and Chemistry (headed by Lauri Lokka). The number of staff was increased from 14 in 1935 to 143 at the end of 1959, reaching the maximum 944 persons in 1988 (Kauranne, 2010). The Exploration Department, had a central position in the Survey, but all departments had good collaboration. For example, the Petrologic Department selected the areas for the geological mapping on a scale of 1:100,000 partly on the basis of ore-geological aspects. Kahma and geophysicist Mauno Puranen developed and applied airborne geophysical and ground geophysical methods for exploration.

Airborne magnetic mapping from the flight altitude of 150 m was started in 1951, electromagnetic mapping in 1954, and radiometric mapping in 1956. Since 1972, airborne geophysical mapping has also been conducted from the altitude of 30 m. The airborne maps have been useful in mineral exploration and bedrock mapping. The airborne geophysical surveys were started by the Exploration Department, but in 1972 a new Geophysical Department was established for these and related surveys. Maunu Puranen was appointed the head of the new department. The first geochemical studies for local ore exploration were conducted in the late 1930s, and in the 1950s till geochemical studies became a part of the exploration program. A separate unit, the Geochemical Department, was established in 1973 to carry out systematic geochemical mapping, and Professor Kalevi Kauranne was appointed as the first head. In the mid-1970s, in keeping with the national decentralization program, the Geological Survey established offices for fieldwork in different parts of the country. In the late 1970s, they were grouped into three regional offices, one in Rovaniemi for Northern Finland, one in Kuopio for Mid-Finland, and one (with headquarters) in Espoo for Southern Finland. In 2004 the fourth regional office was established in Kokkola for Western Finland. Aarno Kahma was succeeded as the head of the Exploration Department by Dr. Lauri Hyvärinen in 1979–1981, Dr. Jouko Talvitie in 1982–1995, and Dr. Pekka Nurmi in 1995–1998.

Ore Exploration

The reorganized Geological Survey had considerable exploration success. A number of deposits were discovered during the time Aarno Kahma acted as the department head in 1946–1977: the Vihanti Zn-Cu deposit, the Korsnäs Pb deposit, the Petolahti Ni-Cu deposit, the Kemi Cr deposit, the Hitura Ni-Cu deposit, the Virtasalmi Cu-deposit, the Pyhäselkä (Hammaslahti) Cu deposit, and the Talvivaara Ni-Cu-Zn deposit.

In 1936, the Geological Survey received several sulfide ore samples from the Alpua village in **Vihanti**, sent by Juho and Atte Lumiaho and Jaakko Salo. Dr. Sampo Kilpi's team carried out boulder tracing and geological mapping in the area. The new Zn-rich boulder sample sent in 1939 by Edvard Kesäläinen from Törmäperä and the pyrite ore samples sent in 1941 by Atte Lumiaho from silty basal till in Rantala reactivated the research. Because of the glacial overburden, the exploration was started with detailed Quaternary geological studies led by Simo Kaitaro and Esa Hyyppä. The studies indicated that the glacial transport had been from west-northwest to east-southeast and that the boulders had not traveled long distance from their sources (Hyyppä, 1948). The prospecting continued in 1945–1949 in the possible source area (ca. 18 km^2) with magnetic, electromagnetic, and gravimetric surveys led by Drs. Paavo Haapala, Aarno Kahma, and Aimo Mikkola. Diamond drilling in 1947–1950 into geophysically promising sites led to the finding of an important Zn-Cu deposit at Lampinsaari. Outokumpu Oy exploited the Vihanti deposit in 1954–1992, and the total production of zinc ore was 27.9 Mt.

Exploration in **Korsnäs** was started in 1951 on the basis of the galena-rich carbonate rock specimen that Gottfrid Pistol had found in 1950 from the bank of Poickelbäck creek. The exploration was led first by Max Kulonpalo and later by Dr. Oke Vaasjoki. Again, the studies were started with boulder tracing and other geological observations. This resulted in a collection of different types of carbonate rock boulders in a wide area where the prevailing rocks were mica gneiss and mica schist. Lauri Hyvärinen was able to construct from mutually similar carbonate rock boulders four different, narrow boulder fans, which had the axial direction of approximately 165°, corresponding to the youngest of the three glacial striae directions known from the Vaasa region. Systematic geochemical studies of till were started in 1953. They showed four lead anomaly fans that coincided well with the boulder fans (Hyvärinen, 1958).

Diamond drilling was started in 1953 after magnetic, electromagnetic, and gravimetric measurements. The so-called Häppelträsk lead ore body was discovered in 1955 at the tip of the strongest till anomaly, and the Poickelbäck prospect was founded in 1956 at the tip of the most well-defined boulder and lead anomaly fan. The Häppelträsk deposit was a galena-bearing calcite and pegmatite vein complex in a shear zone. It was estimated to contain at least 0.7 Mt of ore with 3.5–5.5% Pb. The deposit was mined in 1957–1972 by Outokumpu Oy. A by-product, apatite concentrate, was obtained for separation of lanthanides (Himmi, 1975). The small **Petolahti** nickel–copper deposit in Maalahti was discovered in 1958 on the basis of a boulder sample found by G. Pistol. It was a small deposit in a differentiated Subjotnian diabase dike (Ervamaa, 1962). The deposit was exploited by Outokumpu Oy in 1972–1973.

In 1953, Jopi Malinen sent to the Geological Survey a boulder sample containing native antimony as large, cm-sized crystals and disseminations in silicified rock, discovered in Törnävä, Seinäjoki. Various geological, geophysical, and geochemical methods, as well as excavations and diamond drilling, were applied in search for the source, under the direction of Veikko Pääkkönen. As a result of the studies in 1953–1958, two ore zones, about 50 m apart, were found in schists, totaling about 0.5 Mt of ore with 0.51–1.37% Sb and variable Au contents (Pääkkönen, 1966). During renewed exploration led by Veijo Yletyinen in the 1970s, another deposit with about 0.4 Mt of ore averaging 0.7% Sb was discovered, as well as a cassiterite-bearing pegmatite lens with 0.14 Mt of rock containing 0.3% Sn. The prospects of Törnävä have not been mined.

An important chain of events in the mineral history of Finland took place in spring 1959 when a diver and amateur prospector Martti Matikainen sent a pyrite- and magnetite-bearing specimen from **Kemi** to the Geological Survey. When geologist Pentti Ervamaa visited Kemi, Matikainen showed him other specimens he had collected from a new freshwater canal in Elijärvi, near the city of Kemi. A heavy ultramafic rock sample turned out to represent chromium ore. Exploration was started under the leadership of Valto Veltheim, and soon a chromite-rich layer was discovered in an altered ultramafic rock. In the winter of 1959–1960, all available working resources of the Geological Survey were concentrated on this research area; by the end of May 1960, 29 holes had been drilled into the prospect. It was calculated that the deposit contained at least 7.6 Mt of ore with more than 30% Cr_2O_3, situated in a layered mafic–ultramafic intrusion (Kahma et al., 1962). The results were passed to Outokumpu Oy in May 1963 (Alapieti et al., 1989). After follow-up studies and process development, the decision was made to open a mine in 1964, and the production started in 1969. By 2010, approximately 37 Mt of ore had been mined, and the mining operations continue.

In connection with the exploration around the Makola deposit in Nivala in late 1930s, a low-grade Ni-Cu dissemination was found in serpentinite about 5 km northeast of the Makola deposit. In the 1960s, the Geological Survey returned to the **Hitura** locality. Airborne geophysical surveys indicated that the Hitura ultramafic intrusion is more extensive than previously supposed. After detailed ground magnetic and electromagnetic surveys, diamond drilling was started at the northern margin of the serpentinite intrusion, and a new semicircular Ni-Cu deposit following the margins of the serpentinite body was found in 1963. The deposit was estimated to contain 5 Mt of ore with 0.75% Ni and 0.28% Cu. The exploration at Hitura was led by Veijo Yletyinen. The prospect was handed over to Outokumpu Oy in 1964. After continued drilling, underground test mining at the 205 m level with related concentration tests, and feasibility studies, mining was started in 1970 (Papunen et al., 1997).

In the 1960s, the Geological Survey carried out extensive exploration, led by Lauri Hyvärinen, in the Juva-Virtasalmi area in eastern Finland to find the source of the numerous copper ore boulders found in this area. The chalcopyrite-bearing skarn boulders contained reddish-brown garnet of similar composition, suggesting that they all were of the same source. The prospecting was directed to the northeastern end of the long boulder fan. A chalcopyrite-bearing skarn outcrop was discovered in 1964 at Hällinmäki, a hamlet of Narila in the municipality of Virtasalmi, and in 1965–1966 87 holes were drilled in the occurrence. The deposit was estimated to contain at least 2 Mt of ore with 0.5% Cu (Hyvärinen, 1969). The **Hällinmäki (Virtasalmi)** deposit was handed over to Outokumpu Oy in 1966. The deposit was mined from the end of 1966 until 1984 producing 4.18 Mt ore averaging 0.78% Cu.

A chalcopyrite-bearing ore boulder discovered in 1960 at Petäjävesi in the municipality of Pihtipudas, led to systematic exploration. Magnetic and electromagnetic anomalies in the Murtoselkä area in the Säviä schist sequence, in the municipality of Pielavesi, led to drilling in 1966–1967 and the discovery of the Säviä Zn-Cu deposit. The estimated ore reserves are about 4 Mt of copper ore with 1.1% Cu and zinc-bearing pyrite ore with 2% Zn and 33% S. Because the deposit is located beneath a lake, it has not been exploited, but it is a future reserve. Aatto Laitakari was in charge of the exploration.

In connection with geological bedrock mapping, in 1966 Osmo Nykänen from the Petrology Department of the Geological Survey detected chalcopyrite in graywacke schist at **Hammaslahti** in the municipality of Pyhäselkä, and he informed the Exploration Department of the matter. Detailed exploration, including geological and geophysical studies, was undertaken under the leadership of Lauri Hyvärinen in 1967–1968 in the locality. It was observed that the Cu-mineralized rock was situated in an electromagnetic anomaly zone. The first drill holes immediately below and on the sides of the first chalcopyrite showing yielded hardly any copper or contained only iron sulfides. Before finishing the exploration, it was decided to drill one more hole, which should meet the possible continuation of the chalcopyrite-bearing zone at a depth of 50–60 m. This "last" hole penetrated better quality chalcopyrite dissemination than the surface samples, and the deep drilling was continued. The deposit was investigated by 88 drill holes, and the estimated resources amounted to 3 Mt of ore containing on average 1.2% Cu. Besides chalcopyrite, the deposit contained pyrrhotite, pyrite, and sphalerite. The deposit was exploited by Outokumpu Oy in 1973–1986.

In 1968 the Geological Survey received an albite-rich sample with 3.8% Cu, sent by Matti Vänni and taken from an outcrop at Riikonkoski in the municipality of Kittilä, within the greenstone belt of central Lapland. Geological mapping and ground geophysical survey were carried out the following year, and on the basis of received information, 33 holes were drilled in 1970. It was estimated that the so-called eastern ore contained about 2.5 Mt of rock with 0.68% Cu, and the western ore about 6.5 Mt of rock with 0.41% Cu. The prospect has not been exploited.

In 1969, the Geological Survey started systematic exploration in the areas of differentiated mafic intrusions in Lapland. The first targets were the Koitelainen and Akanvaara intrusions that, according to recent age determinations, are 2.45 Ga old, and the 2.05-Ga-old Kevitsa (Keivitsa) ultramafic intrusion, all in the Neoarchean–Palaeoproterozoic volcanic-sedimentary greenstone belt of central Lapland, in the municipality of Sodankylä. A number of researchers participated in the studies, but the main merits of the geological achievements go to Tapani Mutanen.

The exploration in the Koitelainen area was initiated by the discovery of a Ni-anomalous weathered ultramafic rock in spring 1969 in Rookkijärvi, in the western part of the Koitelainen intrusive complex. The exploration work continued with varying intensity and different targets for more than 20 years. In 1969, the main interest was in the Pd-anomalous rocks in the lower parts of the intrusion. In 1973–1977, the roughly 20 × 27 km intrusion was mapped geologically, and showings of different ore types were discovered, including magnetite- and ilmenite-rich ultramafic pegmatoid pipes and V-rich magnetite gabbros enriched in platinum-group elements (PGE). The magnetite gabbro also contained a 2-m-thick layer with 2–2.5 ppm Au. A low-altitude airborne magnetic survey and ground geophysical measurements were used in the exploration. The most interesting discovery was in 1977 when the last two of the 24 drill holes planned for the Koitelainen intrusion intersected a chromitite layer (UC) in the northeastern part of the intrusion. In 1978–1982, the UC chromitite layer was drilled in different parts of the intrusion. Chromitite layers (LC) were found also in pyroxenite near the base of the Koitelainen intrusion. In 1983–1985, Mutanen continued the geological mapping. In 1985–1989, the main interest was in the exploration and study of the PGE showings in different parts of the intrusion.

Mutanen (1997) identified three main zones in the layered mafic–ultramafic series of the Koitelainen intrusive complex: the ultramafic Lower Zone (LZ; dunites, peridotites, pyroxenites), the mafic Main Zone (MZ; gabbros, norites), and the Upper Zone (UZ; gabbros, anorthosites, magnetite gabbros). Above the layered series is granophyre. The total thickness of the intrusive complex is 3.2 km. Of the different ore types of the Koitelainen intrusion, the upper chromitite layer (UC) appears to be the most promising. It extends along a strike 56 km, is 1.2–1.3 m wide, and its average Cr_2O_3 content is about 20%, with 1.1 ppm PGE. It may be regarded as a future reserve.

The Akanvaara layered mafic intrusion is located about 80 km southeast of Koitelainen. It is 15 km × 7 km in size and resembles the Koitelainen intrusion in occurrence and layered structure. Lower and upper chromitite layers and local PGE enrichments were found there.

The exploration of the **Kevitsa** (Keivitsa) ultramafic–mafic complex was prompted by the glacial erratics of ultramafic rocks with disseminated Cu-Ni—sulfides that Mutanen discovered in 1973. The first diamond drill holes in 1984 intersected several meters of sulfide-rich rock at the basal contact of the complex, but it proved to be “false ore,” pyrrhotite-rich sulfides with low base and precious metal values. Outokumpu Oy’s exploration department had previously studied this occurrence with similar negative results. Systematic ground magnetic and electromagnetic surveys and geochemical sampling of the basal till in 1984–1987 gave anomalies that were drilled in 1987. The two last of the planned 24 drill holes intersected 22–24 m of disseminated ore with 0.36% Ni, 0.40% Cu, and 1–1.4 ppm PGE + Au. Subsequent drilling programs at Kevitsa in 1990 and 1992–1995 delineated a large low-grade Cu-Ni-PGE-Au deposit. The deposit was composed of three main types: the irregular ore that makes up the main ore and typically contains 0.4–0.6% Cu, 0.2–0.4% Ni, and 0.015% Co, as well as 0.6–1.0 ppm Pt + Pd + Au; the false ore that contains >5% S but generally <0.1% Ni; and the Ni-PGE ore that occurs as pipe-like bodies and contains >0.5% Ni and 1–27 ppm PGE (Mutanen, 1997; Törmänen and Iljina, 2007).

After the Geological Survey left the final research report to the Ministry of Trade and Industry in 1994, the rights to the Kevitsa deposit were sold to Outokumpu Oy. Outokumpu made follow-up and feasibility studies in 1995–1998, but resigned the rights to the large but low-grade deposit. Scandinavian Gold Prospecting AB (later Scandinavian Minerals Ltd.) claimed the deposit in 2000 and planned to open the Kevitsa multimetal mine. First Quantum Minerals Ltd. (FQM) bought Scandinavian Minerals Ltd. in 2008, and started production in 2012. In December 2012, the FQM Kevitsa Mine published the proven and probable mineral resources as 157 Mt, with Ni 0.31% and Cu 0.41%, and precious metals Pd 0.18, Pt 0.24, and Au 0.12 ppm.

The black schists or graphite-bearing phyllites of the **Talvivaara** area in the municipality of Sotkamo have been known for their high sulfide contents since regional geological mapping began in 1902–1926. They were explored briefly by Oy Prospector Ab and the Geological Survey in 1930s, and in more detail by Suomen Malmi Oy in 1961–1962, paying special attention to their high contents of Cu, Ni, and Zn. More serious exploration was carried out by the Geological Survey in 1977–1984 to decipher the ore potential of the metal-bearing black schist. The leader of the exploration team was Dr. Pentti Ervamaa. Two test holes drilled in 1977 showed promising and uniformly distributed Cu, Ni, and Zn contents in the black schist, and in the following years an extensive exploration program was carried out. The exploration comprised, besides geological field and laboratory studies, geophysical (magnetic, electromagnetic, gravimetric) measurements and lithogeochemical studies. Altogether, 95 holes were drilled in the anomalous black schist, mainly with 170–200 m profile distances. Of the geophysical methods used, only the multifrequency slingram measurements could be applied to outline the mineralized zones from barren black schist.

The main ore minerals were pyrite, pyrrhotite, sphalerite, chalcopyrite, pentlandite, galena, alabandite, ullmannite, stannite, and molybdenite. Scintillometric measurements of the drill cores indicated relatively low uranium contents. For the main mineralized zones of the Talvivaara area, Kolmisoppi and Kuusilampi, the following estimates were reported in 1980 and 1986: Kolmisoppi, 81 Mt of rock averaging Cu 0.14%, Ni 0.27%, Zn 0.54%, Co 0.02%, Mn 0.32%, S 10.37%, C 7.1%, V 657 ppm, and Mo 103 ppm; and Kuusilampi, 221.4 Mt rock with Cu 0.14%, Ni 0.26%, Zn 0.54%, Co 0.02%, Mn 0.36%, S 8.54%, C 7.8%, V 614 ppm, and Mo 101 ppm. The geochemistry of the black schist has been studied by Loukola-Ruskeeniemi (1991).

In 1986, Outokumpu Oy received the mining rights to the Talvivaara deposits. Diamond drilling was continued and extensive feasibility studies were carried out. Although the deposit was large, it was considered too low grade for conventional technologies. Outokumpu tested and developed bioheap leaching technology to extract nickel and other metals. Outokumpu sold the mining rights and research results at a formal price to Pekka Perä, a former Outokumpu employee, and the Talvivaara Mining Company was established in 2004. After the permits to commence mining were obtained and necessary constructions were made, metal production was started in 2008 utilizing the bioheap leaching technology. Since 2012 the company has had serious problems, mainly because of environmental accidents and unfavorable metal price evolution.

In the late 1970s, the Geochemical Department of the Geological Survey carried out regional geochemical till sampling (1 composited sample / 16 km^2) in the area of the Archean Hattu schist (greenstone) belt in eastern Finland. Emphasis was on base metals, but no significant occurrences were found. Reanalysis of the finest faction (grain size < 0.06 mm) of the till samples in 1982 revealed distinct W and Mo anomalies in the southern part of the Hattu belt, and scheelite grains were observed in the heavy mineral fractions of the corresponding samples. A more detailed geochemical sampling focused on a promising anomaly in the Kuittila area. Subsequent excavations and deep drillings resulted in the discovery of tonalite-hosted quartz veins carrying scheelite and molybdenite, and soon also gold disseminations were found in the quartz and sericite-rich shear zones (Nurmi et al., 1993). There was additional interest in the area from 1985 to 1987 when Outokumpu discovered a promising gold occurrence in a shear zone at Rämepuro, approximately 15 km north of Kuittila, on the basis of a copper- and gold-bearing sample taken by an amateur prospector from a small creek bank outcrop. The geochemical till anomalies and gold discoveries indicated gold potential of the Kuittila sub-belt and the whole Hattu schist belt.

In 1986–1992, the Geological Survey carried out the extensive and successful Ilomantsi Gold Project, which covered large parts of the Hattu schist belt and comprised detailed geological-petrological studies (geological mapping, geochemical and isotopic studies) and ore exploration (till and rock geochemical studies, geophysical exploration, mineralogical, isotopic and fluid inclusion studies). Till geochemistry turned out to be a very effective method in prospecting for gold. Regional-scale sampling (1 composite sample / 16 km^2) was sufficient for finding gold provinces; sampling density of 16 bottom till samples / km^2 was suitable for delineating gold-bearing zones; and for prospect-scale exploration, sampling of bottom till and underlying rock specimens at 10 m intervals along traverses 100–300 m apart was needed (Hartikainen and Nurmi, 1993). A number of significant gold deposits, prospects, and showings were found along the >40-km-long Au-anomalous zone, the Karelian Gold Line, in the Hattu schist (greenstone) belt. Of the nine gold prospects (Kuittila, Kelokorpi, Korvilansuo, Kivisuo, Elinsuo, Muurinsuo, Rämepuro, Ward, and Korpilampi) described by Nurmi et al. (1993), all but the Rämepuro prospect were found from sites indicated by till geochemistry. The Ilomantsi Gold Project was coordinated by Pekka Nurmi, while practical exploration was managed by Martti Damsten and Aimo Hartikainen. The research results of the project were published in a volume of 15 articles, edited by Nurmi and Sorjonen-Ward (1993).

After the end of the Ilomantsi Gold Project, the so-called Ward prospect was exploited under the name **Pampalo** Mine by Outokumpu Mining in 1996–1999, producing 0.114 Mt of ore at 15.3 ppm Au. When Outokumpu withdrew from mining activities, the rights to the deposit were sold in 1994 to Polar Mining Oy, a Finnish subsidiary of Australian mining company Dragon Mining, and two years later the rights were transformed to Endomines Oy. Endomines has constructed the necessary ore processing plants and continued inventory of the deposits. Plans exist to continue mining at Pampalo and start exploitation of some other deposits in the Hattu schist belt.

In the 1970s, the Geochemical Department carried out regional geochemical mapping along lines ("line geochemistry") in the Central Lapland Greenstone Belt. The concentration of Si, Al, Fe, Mg, Ca, Na, K, Ti, V, Cr, Mn, Co, Ni, Cu, Zn, Pb, and Ag were analyzed. The area of the Sattasvaara komatiite complex was characterized by elevated contents of Mg, Cr, Ni, and Co, and several local Cu anomalies appeared in the monotonous komatiitic environment indicating sulfide mineralization. Additional geochemical till sampling was carried out with a grid of 50 × 100 m in the winter of 1984–1985 to check the Cu anomalies, and also Au was analyzed. A distinct Au anomaly was found in **Pahtavaara**, and follow-up studies, including sampling of the bedrock surface by percussion drilling and excavated trenches, unearthed an altered zone containing visible gold in 1985 between komatiitic lavas and tuffites (Pulkkinen et al., 1986; Korkiakoski, 1992). The exploration was continued with diamond drilling and geophysical measurements to delineate the occurrence. The deposit was mined by Terra Mining in 1996–2000 and by Scan Mining in 2003–2014. The ore resourced were estimated by Scan Mining in 2005 as 3 Mt at 3.2 ppm Au.

The Geological Survey launched a new project in 1986, led by Ilkka Härkönen, to work up the gold potential of the Palaeoproterozoic greenstone belt of central Lapland. On the basis of low-altitude airborne surveys and geochemical sampling, four targets were selected for detailed studies in the Kittilä area: Soretiavuoma, Suurikuusikko, Kuotko, and Petäjäselkä. Soon a small pocket containing visible gold in a quartz-carbonate vein was discovered in a road cut about 4 km south-southwest of Suurikuusikko, which gave an additional impetus to the project. Ground geophysical (magnetic, electromagnetic, gravimetric) and geochemical sampling carried out in the selected areas indicated that the target areas of Soretiavuoma, Suurikuusikko, and Kuotko were all located in the same, north-trending Kiistala shear zone. Diamond drilling across this shear zone was started in 1987 at **Suurikuusikko**, and the first drill hole penetrated the silicified and albitized shear and breccia zone, which contained pyrite and arsenopyrite as well as 6.8 ppm Au / 8 m (Härkönen and Keinänen, 1989). Exploration and drilling continued in 1998, but were interrupted after 21 drill holes. The deposit appeared to be very scattered, and most of the gold ore was refractory. Gold occurred mainly in the lattices of arsenopyrite and pyrite (Kojonen and Johanson, 1999). After finding a suitable bioleaching processing method for the refractory gold ore, the drilling program was continued in 1995. By the end of 1996, a total of 77 holes had been drilled outlining a resource of 1.5 Mt with an average grade of 5.9 ppm Au and cutoff value of 1 ppm (Patison et al., 2007 and references therein). After an international tendering process, the deposit was sold in 1998 to Riddarhyttan Resources AB, which continued exploration and feasibility studies. In 2005 Agnico-Eagle acquired

Riddarhyttan, and the mine was opened in 2006 as the Kittilä Gold Mine. At that time, the probable reserves were estimated at 2.4 Mt averaging 5.16 ppm Au, but in the course of the mining the reserves have increased so that at the end of 2011, when 3.05 Mt of ore had been mined, the combined reserves and resources were estimated to be 55 Mt at 4.13 ppm Au (Geological Survey of Finland, Kittilä Mine database).

The Geological Survey started exploration for sulfide ores in Kuusamo in 1983, when an Au-bearing Co-Cu occurrence was found in Kouvervaara in connection with uranium prospecting. Because the occurrence was visible in new high- and low-altitude airborne magnetic and electromagnetic maps, a systematic survey of electromagnetic and magnetic anomalies in the sericite quartzite formation was started. During the first stage, 30 anomalies corresponding to that of Kouvervaara were selected for closer exploration. After ground geophysical studies, 15 anomalies were selected for follow-up studies. The Juomasuo Au deposit was discovered by diamond drilling in 1985 of a geophysical anomaly about 25 km NE from Kouvervaara (Pankka et al., 1991). The deposit is situated in a shear zone hosted by an altered sericite quartzite formation. Altogether, 44 holes were drilled into the Juomasuo deposit in 1985–1989. The main gold deposit was estimated to contain about 0.7 Mt of ore grading 5–6 ppm Au, or, if Co is emphasized, 1.8 Mt of rock with 0.2% Co and 3 ppm Au. In the immediate surroundings of the main deposit were six smaller satellite ore bodies. When elevated contents of uranium were observed to be present in the Juomasuo deposit, radiometric measurements were also applied in the prospecting. The Geological Survey discovered more than 20 Au-bearing sulfidic ore occurrences in the Kuusamo area, the most promising being Juomasuo, Sivakkaharju, Hangaslampi, Pohjasvaara, and Meurastuksenaho (Vanhanen, 2001).

In 1990, Outokumpu Oy acquired the exploration rights over the area encompassing those five gold deposits and continued exploration and feasibility studies, including test mining, until 2004. Exploration activities have been continued since 2010 by Dragon Mining Oy. A continued drilling program has increased the ore resources (at the end of 2012 the inferred resource estimate for the five Kuusamo gold deposits was 3.4 Mt grading 4.2 ppm Au), and plans have been made for opening of mines.

Systematic up-to-date exploration carried out in the Leppävirta-Juva area since the 1990s brought to light several Ni occurrences in association with the 1.9 Ga mafic–ultramafic intrusions, including the **Särkiniemi** (discovered 1994) and Rytky (2000) deposits in Leppävirta, near the Kotalahti mine (Mäkinen and Makkonen, 2004). The Särkiniemi deposit was estimated to contain 0.29 Mt of ore with 0.91% Ni and 0.53% Cu (Geological Survey of Finland, Särkiniemi–nickel database), and the Rytky deposit 0.90 Mt ore with 0.75% Ni and 0.47% Cu (Geological Survey of Finland, Rytky nickel database). Suomen Nikkeli Oy (Finn Nickel Oy) mined the Särkiniemi deposit in 2007–2008.

In 1979, low-altitude airborne geophysical mapping showed a strong radiometric anomaly at Lake Palmottu in the municipality of Nummi-Pusula. Field studies and diamond drillings from 1980–1984 unearthed 1–30-m-wide dikes of uraninite-bearing pegmatite and sheared granite that cut the migmatites of the area and contain 0.1–0.2% U. It was estimated that the prospect contains 1 Mt of rock with 0.11% U to the depth of about 250 m (Geological Survey of Finland, Palmottu–uranium database).

Since the 1980s, the search for gold has been the focus of the Survey's exploration programs. A large number of gold occurrences have been discovered in the Central Lapland Greenstone Belt (Korkalo, 2006; Eilu and Nykänen, 2011; Nykänen et al., 2011), but also in the Palaeoproterozoic Svecofennian schist belts in central and southern Finland (Grönholm and Kärkkäinen, 2012; Eilu, 2012).

ACTIVITIES OF SUOMEN MALMI OY

The exploration company Suomen Malmi Oy undertook exploration in 1945 in the leptite zone of southwestern Finland, which, in view of the previously known prospects and old, shut-down mines, was thought to contain still exploitable ore deposits. First, three areas were selected for investigation—the surroundings of the Orijärvi and old Aijala mines and Lohjansaari—but by 1950, the region under exploration had expanded from the eastern side of Lohjansaari all the way to the village of Perniö, an area of no less than 500 km^2. The project was the first systematic regional exploration program conducted in Finland on such a wide scale. The operations included geological bedrock mapping on a scale of 1:4000 on the basis of airborne photographs and a network of staked-out lines, magnetic and electromagnetic surveying, as well as diamond drilling of observed geophysical ore indicators. The procedure of investigation was believed to be suitable for exploration of every type of ore.

In the very first year of operations, interesting geophysical anomalies were detected on the area of the old Aijala silver mine and closer investigation of the eastern extension of the geophysical anomaly led to the discovery of the **Aijala** copper ore (Turunen, 1953). The deposit was estimated to contain 0.815 Mt of ore at 2.13% Cu, and in 1948 it was sold to Outokumpu Oy, which exploited the deposit in 1949–1958 with total production of 0.835 Mt of ore averaging Cu 1.59%, Zn 0.7%, Au 0.7 ppm, and Ag 14 ppm.

In the spring of 1946, a hole was drilled into the western extension of the electromagnetic anomaly of Aijala, and a sulfide-bearing zone was observed to exist also there. The **Metsämonttu** deposit was estimated to contain 0.600 Mt of ore at 4.6% Zn extending down to the 150 m level (Turunen, 1953). The deposit was likewise passed on to Outokumpu Oy. The mine was worked in two stages, 1952–1958 and 1964–1974, and it yielded a total of 1.5 Mt of ore with Zn 3.5%, Cu 0.3%, Pb 0.8%, Au 1.43 ppm, and Ag 25 ppm.

Several other ore prospects and showings were discovered in the leptite belt, but they proved to be of no economic significance. It became obvious that the leptite belt in Finland had only a few significant ore deposits compared to the Bergslagen Province in Sweden, although the belts displayed many geological similarities.

In the 1950s, Suomen Malmi Oy's main prospecting targets were the iron ores of the Misi and Kolari districts in northern Finland. In the 1960s, Suomen Malmi turned its attention to the Kuusamo district, where, among other prospects, the Au-Cu occurrence of Apajalahti was explored in the shore zone of Yli-Kitkajärvi. In the 1970s, Outokumpu Oy, and in the 1980s, the Geological Survey studied the same ore province. In 1960–1962 the company analyzed and mapped the Talvivaara sulfide occurrence in black schist and recognized the high nickel concentrations in the sulfides. The deep extension of the deposit was confirmed with two diamond drill holes. In 1962–1963 Suomen Malmi conducted regional exploration in the Archean Kuhmo greenstone belt, and in 1963 they investigated with geological mapping, geophysical surveys, and diamond drilling the Arola nickel occurrence discovered earlier by Outokumpu Oy. The deposit was transferred in 1972 to Malmikaivos Oy, which extended the studies and finally estimated the deposit to contain 1.52 Mt of mineralized rock at 0.56% Ni. Hekki Tuominen was the chief geologist of Suomen Malmi Oy in 1945-1957, and Toivo Mikkola in 1957-1966.

In 1966, the Ministry of Trade and Industry, which was in charge of the state-owned company, decided that Suomen Malmi Oy had to terminate the geological investigations. After that the company continued as a contractor for ore exploration and mining. One of its most important tasks was to carry out airborne geophysical surveys for the Outokumpu Oy and Rautaruukki Oy. In 1990, the Ministry sold the shares of the company to a private entrepreneur. The exploration data file with reports, maps, and analyses were transferred to the Geological Survey where they are available for review.

EXPLORATION BY OUTOKUMPU OY

Evolution of the organization

The shortage of metals and raw materials was a great challenge for the Finnish heavy industry after the Second World War. The opening of the Otanmäki mine in 1950 came at the right time to improve the iron supply of the domestic steel industry. At the end of the 1940s, the governmental authorities encouraged Outokumpu Oy to take a major economic responsibility of ore exploration and development of new metal mining. After protracted discussions between the Ministry of Trade and Industry and the Director General Dr. Eero Mäkinen, the Outokumpu Company finally decided to establish its own exploration department in 1951.

At its founding stage Mäkinen invited Veikko Vähätalo, an experienced economic geologist, to take charge of the exploration operations. In 1951, the team in the newly established exploration department gained its first members when Olavi Helovuori and Olavi Kouvo were hired as company geologists. The personnel were enlarged in succeeding years, when Matti Laurila joined as geophysicist and Erkki Viluksela and Aarto and Maija Huhma came on board as geologists. Dr. Paavo Haapala was appointed the chief geologist and member of the board of directors of the company in 1954.

Dr. Veikko Vähätalo was the head of the exploration department until 1965, followed by Pauli Isokangas until 1975, Dr. Pentti Rouhunkoski until 1984, and then Dr. Matti Ketola. Since 1990, the responsibility of exploration in Finland and Nordic countries was transferred to Outokumpu Finnmines Oy, an affiliated company of Outokumpu Oyj, with Matti Ketola as the head. Dr. Juhani Nuutilainen was the head of exploration in 1990–1992, followed by Tuomo Korkalo until 2002.

In the 1970s the permanent staff numbered 210–250, with an additional 30 summer assistants in the field. A total of 55 geologists and geophysicists were engaged in exploration. In the early years of the 1980s, international mining and exploration operations took the main role in Outokumpu Oy, but at the same time, the domestic exploration activity and personnel gradually decreased. In 2002, the Outokumpu Company left the mining business, with the exception of the Kemi chromite mine, and concentrated on developing stainless steel production. The ore exploration data files with reports, maps and analyses, were transferred to the Archives of the Geological Survey, where they are available for review.

Exploration operations

The first deposit discovered was at **Kotalahti** in the municipality of Leppävirta. Sulfide-bearing black schist samples sent in for inspection in 1954 caused Olavi Helovuori to visit the site. The samples proved to have no significance, but he observed an interesting mafic rock type on the shore of the small Lake Huuhtijärvi. Field assistant Jouko Talvitie, then a student of geology, was sent to study the area in more detail. In a newly constructed road cut of highway NR 5, he found a sulfide-bearing ultramafic rock, which was found to contain considerable concentrations of nickel and copper. A follow-up geophysical survey revealed an electromagnetic anomaly caused by a sulfide-bearing mafic–ultramafic intrusion. After a period of intensive diamond drilling, the ore body was delineated and mining operations were started in 1959.

The **Pyhäsalmi** deposit was discovered in 1958. That year, August was very dry, and a farmer named Erkki Ruotanen attempted to deepen his well through the glacial till. However, underneath the till he intersected bedrock composed of massive pyrite ore. The samples sent to Outokumpu Oy contained considerable concentrations of copper and zinc. The operations to open a mine started the following summer, and production from an open pit got under way in 1962. The deposit has now been exploited for more than 50 years.

Following the discovery of Pyhäsalmi, Outokumpu Oy conducted airborne geophysical surveys over an extensive area around the deposit, and a number of interesting anomalies were registered. Many of them contained zinc and copper-bearing sulfides, and a few proved to be mineable deposits, such as **Ruostesuo** and **Kangasjärvi**.

In the postwar period, the role of ore samples sent by laymen to exploration organizations was of great significance in bringing to light new prospective areas. A good example is the location of a zone with potential for nickel ores in southwestern Finland, from where Outokumpu Oy received, since 1960, a number of nickel sulfide samples in the area extending from Ahlainen in the west to Vammala and Kylmäkoski in the southeast. With the accumulation of findings, the zone as a whole appeared to hold potential for nickel ores in mafic and ultramafic intrusive rocks, and it became a target of continuous attention with field studies and geophysical and geochemical surveys (Papunen, 1976). Intensive exploration initiated by boulder tracing led to the discovery of the Kovero-oja deposit in **Vammala**, Taipale in **Kylmäkoski**, and finally the "deep orebody" of the Stormi intrusion in Vammala, which all led to mining activity. In all, about a dozen deposits were unearthed, but most of them were small in tonnage or of too low grade to be mined.

The **Vuonos** ore body in the Outokumpu area has a different story of discovery. While analyzing trace element contents in the rocks around the Keretti ore body, Maija and Aarto Huhma observed that the ore proper and its extensions along the ore horizon were characterized by Co/Ni > 1, whereas in all other rock types of the Outokumpu association the concentrations of Co were much less than those of Ni. This knowledge was confirmed by checking the analyzed drill cores of the Outokumpu association (serpentinite, carbonate rocks, tremolite/diopside skarn, sulfide ore), and in the area of Vuonos, roughly 6 km northeast of the Keretti deposit, anomalously high Co/Ni values were registered in a drilled section. A new test hole drilled in this section penetrated a massive Outokumpu-type Cu-Co-Zn ore, mined from 1972 to 1985.

The belt was a target of intense exploration from 1979 to 1985. Dr. Markku Mäkelä initiated and led the program in the first years and Jyry Saastamoinen was the leader in the last years. A notable result of this project was the discovery of the **Kylylahti** deposit in Polvijärvi in 1984. Outokumpu continued exploration, but did not decide to start mining. The Australian company Vulcan Resources Ltd. bought the deposit in 2004. In 2010 the rights were transferred to Altona Mining Ltd., which immediately started sinking an inclined shaft, inventory drilling, and feasibility studies, and opened the Kylylahti mine in 2012. The probable reserves were at that time estimated as 4.34 Mt of ore with 1.56% Cu, 0.29% Co, 0.17% Ni, 0.58% Zn, and 0.65 ppm Au (Geological Survey of Finland, Kylylahti–Copper Database). In 2014 the mine was sold to Boliden Ab.

Besides the discovery of Vuonos, basic geochemical research led in the 1960s also to some other practical applications in the ore prospecting of Outokumpu Oy. An analytical and research laboratory was founded by the exploration department in 1960, and three years later the leader of the laboratory, Dr. T.A. Häkli, published a basic study on the distribution of nickel between coexisting silicates and sulfides (Häkli, 1963). The work relied on the geochemical and mineralogical data collected from a number of Finnish ore-bearing and barren mafic and ultramafic intrusions, and it was concluded that the nickel concentrations of coexisting silicate minerals can be applied in the classification of ore potential of the intrusion even if sulfide-bearing samples were not available.

Based on this conclusion, a program was organized to collect representative samples of all the known mafic and ultramafic intrusions in Finland. The samples were analyzed for sulfur and sulfide-bound Fe, Ni, Cu, Co, and Zn, as well as for the Fe and Ni contents in mafic silicates. Since nickel tenors of sulfides and silicates were proven to be in equilibrium, it was possible to predict, on the basis of the nickel content of the silicates, the composition of any undiscovered sulfide. The studies indicated that it is possible to delineate areas likely to contain nickel ore (Häkli, 1970, 1971). The method was applied at Outokumpu exploration for a long time, and it was instrumental in the location of numerous minor nickel sulfide occurrences and the mineable deposit of Laukunkangas (Grundstrom, 1980). Lamberg (2005) published the summary of the collected comprehensive data and presented the scientific basis for the computerized programs for the calculations of nickel ore potential.

In 1962, Nickel-bearing boulder samples were obtained from Laukunkangas, in the municipality of **Enonkoski**, but no ore outcrops were found at that time. In connection with the regional nickel program in 1969, the Laukunkangas mafic intrusion was found to be sulfide-bearing. The intrusion was exposed and mapped in detail as well as drilled to ascertain the nature of deeper portions. On the basis of the findings, the intrusion was estimated to contain 4.5 Mt of rock with 0.33% Ni and 0.10% Cu. At that time these contents were too low for mining operations.

At the end of the 1970s, Leo Grundstrom studied the Laukunkangas occurrence for his academic thesis. In the course of the work, it became clear that the small ultramafic rock unit, penetrated in the drilling of the eastern part of the intrusion, contained a noticeably higher grade of nickeliferous sulfides than the gabbroic rocks that predominated in the occurrence. In follow-up work, the holes drilled into the eastern part of the occurrence led to the discovery of the ultramafic portion of the intrusion and a high-grade Ni-Cu ore body associated with it. By 1983, the ore deposit had been proven by drillings to be promising enough to warrant sinking an inclined shaft for further investigation underground. The decision to start mining was made in 1984, and it was based on the amount of 4.3 Mt of ore at 1.1% Ni and 0.29% Cu. In the mining period 1984–1994 the total amount of ore including the low-grade disseminated ore was calculated to be 7.9 Mt at 0.72% Ni and 0.20% Cu of which 6.7 Mt at 0.87% Ni and 0.22% Cu was exploited.

The bedrock mapping in 1962 revealed nickel-bearing sulfides in a mafic intrusion in Makkola, some 10 km southeast of Laukunkangas. The drillings in 1970–1972 showed that the host noritic intrusion contains disseminated sulfides. Massive sulfides were discovered in the drilling of the Hälvälä intrusion nearby. The **Hälvälä** occurrence was estimated to contain 0.448 Mt of ore at 1.5% Ni and 0.36% Cu, and it was exploited in 1988–1992 as a satellite mine of Laukunkangas. In 1989 Outokumpu mined selected parts of the small **Tainiovaara** nickel deposit, hosted by serpentinized komatiite and found in 1975 by the Geological Survey.

Outokumpu Oy started exploration in central Lapland in 1960. The greenstone belt of Kittilä was one of the early targets of field studies and geophysical surveys, and also soil and stream-sediment geochemistry were applied in 1966. Tracking down geochemical copper anomalies, a prospecting team led by Tuomo Korkalo discovered in the fall of 1970 an occurrence of chalcopyrite in bedrock at the edge of a mire called Pahtavuoma in the municipality of Kittilä. A detailed study of the occurrence was undertaken using geological mapping, geophysics, and drilling. In addition to the Cu and Cu-Zn mineralizations, the drillings also brought to light a uranium mineralization within the same rock assemblage. The Pahtavuoma ores were investigated underground in 1974–1976, when an inclined shaft and drifts were made for a total length of nearly 1.5 km. The deposit was estimated to contain more than 4 Mt of ore at 1% Cu and ~20 g/ton Ag.

In Outokumpu's stream-sediment study the **Saattopora** area also proved anomalous. The presence of sulfides was established in the bedrock in follow-up investigations in 1965–1967, but the prospect was not then regarded worth drilling. After the discovery of the **Pahtavuoma** deposit, the area attracted renewed attention and in drillings in 1972 a schist zone containing chalcopyrite was penetrated. The amount of ore was estimated to be some 7.4 Mt at 0.69% Cu and 0.10% Ni. Later in the mid-1980s, Tuomo Korkalo observed high gold values in the ore, and after drilling a resource estimate in 1988–1989 indicated 0.68 Mt of ore at 3.6 g/t Au and 0.3% Cu. The ore was mined in 1989 - 1995 and a total of 6277 kg gold and 5177 t copper was produced.

Saattopora was the first gold discovery in the Kittilä greenstone belt that led to mining. Erkki Ilvonen from Outokumpu exploration discovered in the 1980s several other gold indications in the area, which led to regional studies by the Geological Survey of the Sirkka gold line that indicated the great gold ore potential of central Lapland. In 1965, Outokumpu discovered the small **Kivimaa** Cu-Au deposit in a quartz-carbonate vein system cutting metavolcanics in Tervola, and mined it in 1969 (Rouhunkoski and Isokoski, 1974).

Exploration for gold ore at **Jokisivu** in the municipality of Huittinen started in 1985 on the basis of two gold-bearing boulder samples received in 1964 and 1984. The exploration included detailed geological mapping, geochemical till investigations, rock sampling along profiles by drilling and blasting, geophysical measurements, and diamond drilling. The Jokisivu deposit was located with drilling in 1985. It comprised gold-bearing sulfide mineralization in sheared and quartz-veined zones hosted by metamorphosed diorite (Luukkonen, 1994). Outokumpu continued studies in several phases until 2003, after which Dragon Mining Oy (earlier Polar Mining Oy) continued with extensive drilling, feasibility studies, and exploitation. The resource estimate was in 2005 1.47 Mt grading 6.8 ppm Au (Grönholm, 2006). The mining was commenced in 2009.

Since the late 1950s, Outokumpu Oy explored for uranium, focusing on the areas of Palaeoproterozoic quartzites in eastern and northern Finland. Several occurrences were discovered, including the Nuottijärvi prospect in the municipality of Paltamo (discovered 1959) and the Kesänkitunturi prospect in Kolari (1965). The Nuottijärvi prospect is a mineralized breccia in a carbonate-apatite zone between the Jatulian quartzite and overlying Kalevian mica schists, and contains 1–2.5 Mt rock with 0.04% U (Sarikkola, 1979; Geological Survey of Finland, Nuottijärvi–Uranium Database), the Kesänki prospect consists of uraninite-bearing zones (with 0.06% U) in the quartzite (Sarikkola, 1979).

In the 1970s, Outokumpu also carried out follow-up studies of the porphyry-type Mo-Cu and Au occurrences of Pohjanmaa, including the known Mo-Cu prospects of Rautio (Kalajoki) and the Au-deposit of Kopsa (Haapavesi) (Gaál and Isohanni, 1979). The **Laivakangas** (or **Laiva**) Au-deposit near Raahe was discovered as a result of detailed exploration based on a boulder sample that was sent to Outokumpu in 1980. Nordic Mines Ab started to mine the deposit in 2011.

In addition to its own ore prospecting in the 1960s, the geological staff of Outokumpu Oy carried out feasibility studies and mining plans for numerous deposits handed over to the company by the Geological Survey. The most laborious and time-consuming of these tasks was the investigation of the technical properties involved in the mining of the Hitura deposit, which finally led to the opening of the nickel mine in 1970. A positive result was achieved also with the Hammaslahti copper occurrence discovered by the survey in 1966. Outokumpu started the feasibility study in 1971 and the mine was active in 1973–1986, and 5.59 Mt of ore at 1.11% Cu and 1.26% Zn was then mined.

The discovery of the Laukunkangas deposit was the only one of economic value made in Finland for a long time, and it maintained faith in the productivity of exploration. In the mining industry, however, the costs of ore exploration were included in expenditures of research and development (RD), and the exploration was considered to take a disproportionately large portion of the funds of RD in the whole production chain from ores to metals. As a consequence, it was decided in the mid-1980s to cut down substantially the costs spent in ore exploration. Exploration continued, but the responsibility shifted to the Geological Survey to a greater extent than ever.

In order to cut exploration costs the state-owned companies Outokumpu and Rautaruukki joined forces in Lapland at the beginning of 1982 by establishing a joint venture named Lapin Malmi. Up to the mid-1980s, the most important of its projects was the investigation of the Pd-Pt occurrences discovered in the layered intrusions of Kemi-Penikat-Suhanko. Jarmo Lahtinen of Outokumpu exploration and Tuomo Alapieti from Oulu University identified numerous Pd-Pt occurrences of different types (e.g., Alapieti and Lahtineni, 1989; Alapieti and Kärki, 2005). During the International Platinum Symposia organized in Finland in 1989 and 2005, the occurrences became known to specialists around the world, and after Finland joined the EU in 1995 and the mining rights became open for international exploration companies, the Pd-Pt occurrences have been targets for extensive exploration studies.

The Pd-Pt-Ni-Cu-Au occurences in the Suhanko intrusion, at Konttijarvi and Ahmavaara, had already been discovered by Outokumpu Oy in 1964. The exploration continued periodically until the 1980s with Konttijärvi as the main target (161 drill holes) and showed that pyrrhotite-chalcopyrite-pentlandite-PGE mineralization existed in the basal part (marginal series) of the layered intrusion, some also as offshoots in the basement rocks near the basal contact of the intrusion (Alapieti et al., 1989). The exploration and feasibility studies were continued in 2000–2003 by Arctic Platinum Partnership, a joint venture of Gold Fields Finland Oy and Outokumpu Mining Oy. When Outokumpu withdrew from the partnership, Gold Fields Arctic Platinum Oy continued as a subsidiary of the South African company Gold Fields Ltd. Exploitation of the Suhanko deposits was not considered profitable in the 2000s due to low metal prices and problems in processing (flotation), but in the 2010s a massive drilling program increased the ore resources and the development of a suitable hydrometallurgical Platsol method for processing the ore have led to a more positive outlook.

International activities were the main focus of Outokumpu's exploration from the mid-1980s (Kuisma 1985), and domestic exploration was targeted to massive sulfide deposits and nickel to guarantee the raw materials for the steel works. Besides extensive studies of layered mafic intrusions and their PGE potential, also basic geochemical and volcanological studies were undertaken on ultramafic rocks with komatiitic affinities in the Archean provinces of eastern and northern Finland. The nickel discoveries of Arola in Kuhmo and Vaara in the Suomussalmi greenstone belts as well as Ruossakero and Sarvisoaivi in the Enontekiö Archean greenstone belts, and several occurrences of nickel sulfides at the Proterozoic Pulju greenstone belt in Kittilä were surveyed in detail, but the resources were not regarded economic for mining.

The most notable event in the exploration of the 1990s was the discovery of the deep extension of the Pyhäsalmi sulfide deposit. In 1996 the known ore reserves of the Pyhäsalmi mine were so small that the mine was facing termination of its operations in the next few years. However, in December 1996, deep drilling instigated by mine geologist Timo Mäki intersected a considerable thickness of massive sulfides, and the inventory work following this discovery increased the life of the mine by more than 20 years. The Pyhäsalmi deep extension, together with the discoveries of the Vuonos, Laukunkangas, and Saattopora deposits, were all good examples of successful innovative actions in the history of the Outokumpu exploration department.

EXPLORATION ACTIVITIES OF OTANMÄKI OY/RAUTARUUKKI OY

When mining operations in Otanmäki began in 1952, the ore proved to be more heterogeneous than the experts had predicted. Instead of a large, uniform ore body that was indicated in early reports, the deposit more resembled a school of herrings, with hundreds of small, separate ore pockets. In the first estimate of the ore reserves made on the basis of mining technology, the actual amount of ore fell well below original expectations. The ore grade confronted the geologists with challenging problems, and many novel geophysical measuring methods and apparatus were developed to determine the economic limits of the Otanmäki ore deposit.

In the 1950s, the establishment of ironworks was planned using the titanium-bearing magnetite concentrate as a raw material. The Otanmäki ore grade was not suitable as such for the blast furnace, and the company turned to the Geological Survey and Suomen Malmi Oy with an appeal for the intensification of iron ore prospecting in northern Finland. However, the level of planned activities did not satisfy Otanmäki Oy, and it decided in 1957 to establish an exploration department of its own with the aim of increasing the ore reserves. Heikki Paarma was appointed the head, as he had been in charge of geological investigations from the very inception of mining operations at Otanmäki and he also felt deeply concerned about the procurement of ore for the planned ironworks.

Otanmäki Oy's request to the Geological Survey led to airborne magnetic surveys in the Misi area, in the eastern part of the Peräpohja schist belt, in the municipality of Kemijärvi, where the **Kärväsvaara** iron occurrence was known since 1921. On the basis of geophysical surveys, Suomen Malmi Oy investigated the area and discovered in 1955 the small Sääski magnetite deposit. In 1957, Otanmäki Oy obtained the mining rights to the Kärväsvaara deposit, and with the opening of the mine, the district became an exploration target of the company's own exploration department. The **Raajärvi** magnetite deposit was discovered the following year. The deposit was covered with a thick sand layer, and magnetic gradient measurements with correct interpretation played a significant role in the discovery of the ore. In 1958, the company carried out a low-altitude aeromagnetic survey that led to the discovery of the **Leveäselkä** magnetic anomaly, and a deposit containing 1 Mt of magnetite ore was inventoried in 1960. The discovered ores of the Misi district (Nuutilainen, 1968; Niiranen et al., 2003) contained no harmful additional components, but, on the contrary, the concentrate contained magnesium, which was advantageous for mixing with the titanium-bearing iron concentrates. Hence the new discoveries facilitated the decision to establish a blast furnace in northern Finland dependent on local ore reserves.

The mining rights to the **Rautuvaara** ore discovered by Suomen Malmi Oy in the Kolari district were transferred in 1960, along with the title to surrounding tracts, to Otanmäki Oy. As a result of intensive exploration, including low-altitude airborne magnetic surveys, the Laurinoja open cut mine was established in 1982 in the **Hannukainen** ore field (Hiltunen, 1982).

The Otanmäki Company merged into Rautaruukki Oy in 1969. Already in 1959 an exploration office was opened in Rovaniemi, and the exploration operations were concentrated in northern Finland. From there the staff in charge of ore exploration moved in 1966 to Oulu, where the head office of the Rautaruukki Oy was later transferred.

Earlier, in 1957, Otanmäki Oy received laymen samples of magnetite gabbro associated with the layered mafic intrusions of Koillismaa. The region was studied with low-altitude airborne magnetic surveys in 1961, and as a result, magnetite gabbro horizons were located in the intrusions. The Mustavaara deposit was drilled and inventoried in 1970–1971, and in 1976 the **Mustavaara** mine was opened. However, at Mustavaara, magnetite was so intensely intergrown with ilmenite that the production of a magnetite concentrate was not possible, but the high vanadium content of the ore was extracted in the vanadium works at the site (Juopperi, 1977). In the 1980s, based on the Otanmäki and Mustavaara vanadium works, the Rautaruukki Company produced about 10% of global vanadium consumption.

The first indication of a nickel occurrence at Oravainen was obtained in the summer of 1972 when a local prospector sent a sample of a glacial boulder with a nickel content of 2.24% for inspection. By means of boulder tracing, geophysical surveys, and geological mapping, the provenance of the boulders was found in February 1973. On the basis of diamond drilling, 1.3 Mt of ore with 0.95% Ni and 0.16% Cu were inventoried by the end of 1974. The ore deposit was in 1976 transferred to the ownership of Outokumpu Oy, and in 1976–1991 Outokumpu Finnmines evaluated the possibilities of mining the ore, but the resource was not indicated suitable for mining.

Rautaruukki Oy's ore exploration project of the longest duration was in the Sokli area of northeastern Finnish Lapland known as Itäkaira (Eastern Wilds). Many factors contributed to the discovery of this carbonatite, which ranks among the world's largest in area. In the years preceeding the discovery, Heikki Paarma had become acquainted with the geology and alkaline rock intrusions of the Kola Peninsula. He also knew that on the Russian side, close to the Finnish border, was the Kovdor magnetite mine in a carbonatitic alkaline rock intrusion. Itäkaira was selected for prospection since it is located adjacent to the Kovdor magnetite-rich carbonatite deposit. A mosaic of good aerial photographs were obtained from the area, and from it Jouko Talvitie located a ring structure, which, however, was proven to be a granite intrusion. Low-altitude airborne geophysical measurements were started in May 1967, and the first two profiles showed a strong magnetic anomaly in the Sokli area.

The first samples of carbonatite were found in a boulder field of the Sokli area. At the same time, the heavy mineral investigations carried out in the area yielded pyrochlore, a Nb-mineral characteristically present in carbonatites. Subsequently, a low-altitude airborne geophysical survey was flown over the whole area, and the results were useful in determining the extent and structure of the carbonatite sub-outcrop. The Sokli carbonatite complex was investigated by Rautaruukki Oy until 1980, and Dr. Heikki Vartiainen was in charge of the studies. The mineral of main economic interest was apatite. Besides directing the exploration and mining tests, Vartiainen extensively studied the petrology and mineralogy of the occurrence, collaborating with domestic and international experts, including Dr. Allan Woolley from the British Museum and Soviet geologists, geophysicists, and dressing/process engineers. Heikki Vartiainen published the scientific results in his extensive doctoral thesis (Vartiainen, 1980) and several shorter articles. He also compiled the exploration history of Sokli in Finnish (Vartiainen, 2012).

Heikki Paarma was in charge of research and ore exploration until 1977, and since then the operations were directed by Dr. Juhani Nuutilainen. The ore exploration by the Otanmäki and Rautaruukki companies was traditionally concentrated on the search for iron ore and the compound metals in steel making. Abundant and adventurous experimentation and application of geophysical methods characterized the work, and many new methods came into general use in Finland through their development and testing in prospecting. These include, for example, the investigation of large-scale structures by means of satellite pictures and aerial photographs, the use of false color photography, and the use of heavy minerals in mineral exploration. In order to cut exploration costs, from 1982–1985, the Rautaruukki and Outokumpu companies had a joint venture called Lapin Malmi. In 1979–1984, Rautaruukki Oy, together with Soviet experts, carried out extensive research on iron ore potential of the territory of Finland called the RAETSU project. As a result of the research the potential to find new exploitable iron ore reserves was considered to be low, and so Rautaruukki Oy decided to terminate exploration in 1985. The personnel were joined with Outokumpu exploration, and the ore exploration data file moved to Outokumpu's data file.

EXPLORATION BY PRIVATE COMPANIES

The Oy Prospektor Ab was founded in the late 1930s. Its predecessor was Avtalsgruppen, which comprised a number of enterprises interested in ore prospecting. Among them were about a dozen firms, including the leading cement manufacturers Paraisten Kalkki Oy and Lohjan Kalkkitehdas Oy. The prominent mining and industrial tycoon Petter Forström was the chief promoter of the Prospektor Company, and Professor Hans Hausen acted in the beginning as the geological expert. In the 1930s Oy Prospektor Ab investigated the Talvivaara black schist occurrence in Sotkamo, but after the war the main attention was focused on the chalcocite occurrence at Kiiminki. In 1954, the company started regular exploratory work and hired Rolf Bosröm and Matti Tavela as geologists. In 1956, the name of the firm was changed to Oy Malminetsijä Ab; this Finnish version was thought then to be more suitable for the popular ore prospecting competitions. In 1964, Oy Malminetsijä Ab was taken over by Paraisten Kalkkivuori Oy, and its operations were terminated in 1971, when it was absorbed by the parent organization. Oy Malminetsijä Ab operated all over the country, but major prospects in addition to the aforementioned Kiiminki and Sotkamo districts were the nickel occurrences at Parikkala in southeastern Finland and at Sääksjärvi, east of Pori, as well as the Huhus iron ore deposit in the frontier municipality of Ilomantsi.

Hackman & Co. had engaged in intermittant ore prospecting mainly in North Karelia, ever since the discovery of the Outokumpu deposit. At the beginning of the century it had claims at Luikonlahti, for example. In the 1950s, the company had its own organization for ore exploration, and the targets were at Kotalahti, which later became an active mining area by Outokumpu Oy, and at Varpaisjärvi, where Hackman & Co. discovered the molybdenum occurrence of Kylmämäki. The geologists working for the company were Juhani Nuutilainen and, at some stage, Ossi Näykki.

Preliminary mineral exploration was undertaken in the provinces of Oulu and Lapland by Pohjois-Suomen tutkimussäätiö (Research Foundation of Northern Finland) between 1950 and 1957. The executive director of the foundation was Veikko Loppi, and the board of directors included prominent citizens interested in the research of Lapland, including Governor Kalle Määttä, Director General Kaarlo Hillilä, and Dr. Paavo Haapala. The foundation organized an ore prospecting competition in the region, and within its framework geologists Erkki Aurola and Väinö Makkonen, as well as geophysicist Holger Jalander, carried out exploration programs. The foundation withdrew from the sponsorship of prospecting competitions at the beginning of 1957, but after that the Lapin Maakuntaliitto (Provincial Federation of Lapland) and the Lapin Tutkimusseura (Research Society of Lapland) continued the operations between the years 1958 and 1962.

In the summer of 1954, to secure the domestic reserves of iron ore, the steel-making company Oy Vuoksenniska Ab undertook an extensive exploration program in the **Jussarö** region (Mikkola et al., 1966). Airborne magnetic surveys were made over an area of approximately 70 km^2, and the details of observed anomalies were examined either from a boat or from the ice in winter. Divers collected required samples from anomalous areas under the sea, and diamond drilling was done using a tower structure erected on the sea bottom. In the beginning of 1957, a test shaft was driven on the island and the underground investigation of the banded iron ores continued until the mining operations began in 1961. Vuoksenniska closed the mine in 1967.

A similar investigation was conducted also close to the island of Nyhamn to the south of the Åland Islands simultaneously with the exploration of Jussarö. In addition to the geophysical surveys carried out by boat and on the ice, a diver took samples from the exposure at the sea bottom. The work continued with sinking a test shaft on the island of Nyhamn, and from there a tunnel was dug into the ore body. A test sample was mined from the ore, but as the Jussarö deposit was found to be of better grade, the exploration at Nyhamn was discontinued in 1959.

Malmikaivos Oy started exploration in the **Luikonlahti** area in 1940, and the promising Asuntotalo ore body was discovered by boulder tracing and excavations in 1944. Eleven staked-out claims were secured since 1947. To retain the claims, the company continued the prospecting in 1957. The Kunttisuo deposit was intersected by diamond drilling, and in the following year the deep extensions of the largest ore body, the Asuntotalo deposit, were intersected by drilling. In 1958, the landowner, Ruskealan Marmori Oy, and Malmikaivos Oy decided jointly to sink a test shaft into the Kuparivuori (also known as Copper Mountain) to investigate the Asuntotalo ore underground. In 1965, the decision to open a mine was made, and production commenced in 1968. When the operations came to an end in 1983, a total of approximately 7.7 Mt of ore had been mined with a content of 1.2% Cu, 0.12% Co, 0.09% Ni, and 0.65% Zn. When the mine was closed, the company started to exploit talc ores, and the exploration department continued to investigate talc deposits in North Karelia.

The discovery of kimberlite boulders in the Kaavi area in the early 1980s started an interesting period of exploration, and, as a result, the company had a test mine at Lahtojoki. In 1984, Malmikaivos Oy established a joint venture with the Australian company Ashton Mine Ltd., which had special experience in diamond prospecting. The operations to find kimberlite pipes were kept secret, and it was a big surprise in 1996 when Malmikaivos Oy finally revealed that the joint venture had discovered 24 kimberlite pipes in eastern Finland and that diamonds were discovered in 14 of them. Despite interesting results, Malmikaivos Oy sold its rights in the joint venture to Ashton Mine, which continued the exploration for some time. Later on, the claims were transferred to international enterprises, which have continued the diamond exploration but without major new discoveries.

In 1955, private companies that were dependent on electric power in their operations founded Atomienergia Oy to investigate the feasibility of utilizing nuclear power. Erkki Aalto headed the new company at its initial stage. The search for raw materials became the central task. Professor Kalervo Rankama, a specialist in nuclear geochemistry, was consulting for the company, and on his recommendation, Atomienergia Oy undertook exploration for uranium deposits in Finland. In summer 1956, Tauno Piirainen was hired to conduct ground investigations, including studies of prospective geological formations and inspection of ore specimens collected by laymen. In the next year, the brothers Eino and Matti Justander, armed with a Geiger counter, ran across radioactive boulders at Hutunvaara, in the municipality of Eno. The samples initiated intensive investigations between Koli and Kaltimo. Accumulations of uranium-bearing quartzite boulders were found at Herajärvi and Riutta.

At Hutunvaara, the ore was found in the bedrock, and diamond drillings indicated the extent of the occurrence. The company decided to extract uranium ore concentrate on a small scale from the Paukkajanvaara mine. Altogether 30,700 tons of ore was extracted in 1960, but the **Paukkajanvaara** deposit turned out to have insignificant resources. Since no other deposits were available, mining operations were discontinued in the fall of 1960 (Piirainen, 1968). Other investigation targets of the Atomienergia organization were in the municipality of Kisko, where minor pitchblende dikes were discovered in the old Malmberg iron mine. From 1956, the company employed five geologists, headed by chief geologist Heikki Wennervirta. When Atomienergia Oy finished the active exploration, Outokumpu Oy continued the uranium exploration in eastern and northern Finland, and the Geological Survey in southern Finland, until the 1980s.

The paper and pulp company Kajaani Oy was reopening the old Tipasjärvi pyrite mine in northeastern Finland during a sulfur shortage in 1952. It founded its own ore exploration organization in 1971 and started systematic prospecting in the Kainuu region. Timo Kopperoinen was engaged as chief geologist. The Tipasjärvi area in Sotkamo became the most important target. A Zn-rich boulder sample from the Hiidenkirkko area led the company to start geochemical till exploration in 1980. At the same time, a bedrock research project of the University of Oulu discovered zinc ore boulders and also a mineralized outcrop in the Taivaljärvi area northeast of Hiidenkirkko.

Kajaani Oy got the exploration claims to the Taivaljärvi area and soon after the discovery of the Oulu University team, Kajaani Oy localized with diamond drilling the Taivaljärvi Ag-Zn-Pb deposit near the mineralized outcrop. The deposit was drilled from the surface to the depth of 150 m, and the result was so encouraging that an inclined shaft was constructed to take a major ore sample from the deep portion of the ore for process testing. In the final stage of Tipasjärvi exploration in 1989, Kajaani Oy had a joint venture with Outokumpu Oy called Taivalhopea (Papunen et al., 1989). It made additional inventory drilling underground, but the feasibility study indicated that opening a mine there would not be economic at that time.

With that decision, Kajaani Oy gave up exploration. However, the exploration of the Taivaljärvi deposit was reactivated after a hiatus of 20 years, when Silver Resources Oy, now Sotkamo Silver Oy, acquired the mining rights in 2005 and completed the mineral resource estimate with additional studies. A mining license was granted in 2011 and the production was expected to start in 2015. With a cutoff grade of 50 ppm Ag (eq) the deposit contains 3.52 Mt measured and indicated resources with 0.72% Zn, 0.27% Pb, 91 ppm Ag, and 0.27 ppm Au, and additional inferred resources of 1.5 Mt with equal composition (Lindborg et al., 2015).

Partek (Paraisten Kalkkivuori Oy, Paraisten Kalkki Oy) and Lohja Oy (Oy Lohjan Kalkki Ab), two conglomerates that had their main interest in industrial minerals and rocks, also carried out some exploration for certain ore deposits. In the 1960–1980s, initiated by a discovery of spodumene-bearing pegmatite boulders, Paraisten Kalkki Oy/Partek carried out exploration for lithium (spodumene) in the Kaustinen–Ullava area, and discovered promising spodumene pegmatite dikes. Since 1999, Keliber Oy has continued exploration and made preparations for exploiting the deposits. In the 2000s, the Geological Survey contributed by exploring the Kaustinen-Ullava area and discovering new spodumene pegmatites.

The occurrence of sericite schist as an alteration product of felsic volcanic rocks in the Orivesi area, Tampere schist belt, has been known by geologists since late 1940s. Between 1946 and the 1980s, several companies studied the occurrence located northwest of Lake **Kutemajärvi** as a sericite deposit, including Viento Oy, Renlundin Tiili Oy, Suomen Mineraali Oy, Kemira Oy, and Oy Lohja Ab. The investigations of Kemira Oy included detailed geological mapping and diamond drilling. It was estimated that the deposit contained 14 Mt of sericite schist containing 37% sericite, 6% topaz, 8% kaolinite, 1% andalusite, and 48% quartz. Kemira also considered the possibility that this F-rich alteration assemblage was caused by fluorine-rich hydrothermal fluids, and analyzed the schist for metals known to follow fluorine, such as Li, Sn, Rb, Au, and Ag. The concentrations were not encouraging, and the company abandoned the claim in 1974. Lohja Oy had the claim to the deposit after Kemira Oy from 1975 to 1990, and studied it as a possible source of industrial minerals sericite and topaz. After detailed inventory studies and process tests, it decided that exploitation would not be economic.

The Department of Geology at the University of Helsinki had from 1980–1983 the so-called Porphyry Project to explore for porphyry-type Mo, Cu and Au deposits in association with the

Svecofennian granitoids (Nurmi, 1985). The project was led by Professor Gabor Gaál and Pekka Nurmi. The project had several targets in the Tampere schist belt, and the Kutemajärvi sericite schist in Orivesi was one of them, selected because the alteration assemblage resembled those of porphyry-type deposits. The exploration was conducted in collaboration with Lohja Oy. Together, 72 rock samples were taken along three N-S profiles across the sericitized zone. The profiles showed a zoned anomaly pattern with barren sericite schist in the north, gold in the middle, and base metals in the south. Two diamond drill holes made by Lohja Oy in 1982 on the anomalies intersected an ore-grade rock that became known as Pipe 2. Later, diamond drilling unearthed a deposit consisting of three subvertical pipes of gold ore. With a cutoff grade of 2 ppm Au, the deposit was estimated to contain 0.63 Mt of ore with 9.4 ppm Au (Ollila et al., 1990; Kinnunen, 2008). Outokumpu Finnmines Oy purchased the mining rights to the deposit in 1990 and continued technical and feasibility studies, including pilot mining in 1990–1993. The Orivesi Gold Mine was officially opened in 1994 and was operated by Outokumpu until 2003, when the mine was sold to Polar Mining Oy/Dragon Mining. Mining was resumed in 2007, with Dragon Mining focusing efforts on the Sarvisuo lode system, which was discovered in 2002. Sarvisuo is located 300 m from the Kutema deposit.

When the ore reserves of Finland's mines declined in the early 1970s and several mines had to be closed, the Ministry of Trade and Industry established in 1971 a special fund to support ore geological research in universities targeted especially at northern Finland. Originally, the funds were available from 1971–1977, but because the results were encouraging, the program was extended to the 1990s. The geological departments of the Universities of Oulu, Helsinki, Turku, and Åbo Akademi, as well as the Helsinki University of Technology, had research projects financed by this fund. The projects produced results that have been useful in planning and carrying out successful exploration programs in different organizations, and in two cases (Orijärvi and Taivaljärvi) the projects have directly contributed to the discovery of ore deposits. The projects have also contributed to the training of exploration geologists, who have subsequently taken responsibility for exploration in the Geological Survey and mining enterprises.

Exploration by foreign companies

Finland's entry into the European Union in 1995 and changes in the mining law made the country open for international mining and exploration enterprises. This led to a rush of international mining companies to Finland, and large areas in northern Finland were blanketed with applications for ore prospecting permits. Much of their activity comprised follow-up and feasibility studies of known ore prospects, discovered previously by Finnish exploration enterprises and the Geological Survey, as described earlier in this chapter, but in addition, important new discoveries have also been made.

The Sakatti Cu-Ni-PGE prospect near Sodankylä is one of the most promising new discoveries. It was found in 2009 by an exploration team of Anglo American Ltd., led by geologist Jim Coppard. The Sakatti area was one of the targets selected for closer study on the basis of geological models of the Central Lapland Greenstone Belt and related mafic–ultramafic igneous intrusions and extrusions, and existing basic geological, airborne geophysical, and till geochemical maps from the Geological Survey. Several targets were tested before the diamond drilling hit the Sakatti occurrence. The exact size and grade of the deposit is not yet known, but the existing data indicate that it is a large and rich magmatic Cu-Ni-PGE deposit. Unfortunately, the deposit is located at the margin of the Viiankiaapa Mire Reserve, a Natura 2000 area, which complicates the exploration and exploitation of the deposit.

The Rompas Au-U ore field in Ylitornio was discovered by France's Areva NC in 2008 on the basis of an airborne radiometric survey. In the follow-up studies, a large number of hydrothermal high-grade gold and uraninite-bearing pockets with mineralized veins, breccias, and disseminations were found in the rocks of the Peräpohja schist belt, within an area of approximately 10 km × 10 km. The Canadian company Mawson Resources Ltd. purchased the property in 2010 and has continued the exploration.

SUMMARY

The preceding description of mineral exploration in Finland covers a long timespan from the seventeenth century to the present time. Initially, traditions and activity of mineral exploration in the territory of Finland relied mainly on the skills of Swedish mining experts, who knew the characteristics of the Swedish mineral deposits and applied their knowledge in looking for analogous formations in the bedrock of Finland. A simple device for magnetic surveys, the mine compass, was the only instrument to help in the search in the areas covered by overburden. Also the geology of Finland was poorly known until the end of the nineteenth century when the Geological Survey was established and regional bedrock mapping was started.

The first attempt to take samples from the deep part of the bedrock with diamond drilling was made at the end of the 1890s in Pitkäranta, where the initiator of multidiscipline mineral exploration in Finland, Otto Trüstedt, studied the feasibility and extension of the Fe-Cu-Sn-Zn deposits. Later on, Otto Trüstedt was employed by the Geological Survey of Finland, and in this capacity he successfully located the provenance of the high-grade copper ore boulder of Kivisalmi, eastern Finland, and discovered the Outokumpu copper deposit in 1910. The exploration process and discovery was a turning point in the mining history of Finland, and the discovery became the backbone of the domestic mining industry, metal works, and new methodology in mineral exploration. In the follow-up exploration operations, glacial boulder tracing, geophysical surveys, diamond drilling, and application of geological bedrock mapping were in common use, although the development of suitable equipment for drilling and geophysical surveys took some time. Another significant discovery of the Geological Survey took place in 1924 in Petsamo (Pechenga), and the exploration in difficult remote conditions took all the available resources of the Survey for more than 10 years.

In the mid-1930s, the government of Finland increased its interest in geological studies, and Suomen Malmi Oy was founded to improve the capacity of exploration. The Geological Survey discovered several new deposits before the outbreak of the Second World War. The knowledge of regional geology was still imperfect and greenfields exploration was largely based on ore-bearing samples sent by laypeople. After the war, several extensive campaigns were organized to intensify the layperson's prospecting, resulting in a great number of samples, which initiated the study of new discoveries.

After the war, the efficiency of the Geological Survey was increased with funding, new organization, and addition of personnel from 14 to 143 in 1959, which gradually increased to 944 in 1988. By 1951, the Geological Survey had started regional airborne geophysical surveys, and in the timespan of about 50 years the whole of Finland has been covered with accurate geophysical information. Systematic geochemical mapping since the 1970s has produced important data for mineral exploration and environmental welfare. Mineral exploration was a key area of the Survey until the 1990s, but the role of environmental questions and other services of the Survey have grown since then.

State-owned mining companies, Outokumpu Oy and Rautaruukki Oy, established their own exploration departments in the 1950s, and besides geological mapping, geophysical, and geochemical surveys, they participated in the development of methodology and instrumentation of exploration. These companies also had access to brownfields exploration around the existing mines, since they had experience in the geology of mineralized areas. Over time there was discussion about the overlapping of the exploration duties of the Geological Survey and the state-owned companies, and also the competition between different state organizations to claim mineral deposits begun to be questioned. Finland, however, was one of few countries where grassroot mineral exploration was preserved in the national Geological Survey, whereas during the 1980s the surveys in the neighboring Nordic countries, for example, left the exploration activities or transferred them to a specific company.

Until 1995, when Finland joined the European Union, private exploration companies played a relatively minor role, but a few important metallic ore deposits, such as Haveri (gold), Jussarö, Luikonlahti, and Taivaljärvi, were discovered and developed until that time. From 1995, foreign exploration companies can stake claims in Finland, which led to the discovery of the Sakatti and Rompas deposits. In contrast, the exploration and development of industrial mineral deposits has largely been conducted by private companies.

In 1985, Rautaruukki Oy abandoned mineral exploration and its personnel and data were transferred to Outokumpu Oy. At the same time, Outokumpu strengthened its exploration abroad and its domestic activity decreased. At the turn of the millennium, Outokumpu Company had become an international mining giant and had mineral rights in many parts of the world, its exploration division had discovered several new mineral deposits, and it had active mining in Ireland, Australia, Chile, and Canada. The company, however, decided in 2002 to exit the mining business and with the means acquired from its exploration and mining assets, it developed the stainless steel works at Tornio.

The Geological Survey continued its role as an exploration organization. At present, however, grassroots exploration plays a minor role, and the Survey has changed to a databank of geological information. In this way it services all the exploration enterprises. In the role of geological databank, the Geological Survey of Finland is a world leader. The Canadian Fraser Institute annually ranks countries on the basis of their attractiveness in exploration and mining via surveys that are sent to approximately 4100 exploration, development, and other mining-related companies around the world. In the 2013 annual survey, Finland was ranked second for attractive mineral exploration, and Finland, Alberta, and Nevada have been consistently ranked in the top 10 over the last five surveys.

ACKNOWLEDGMENTS

The authors are indebted to Dr. Kauko Puustinen, official reviewer of the manuscript, for his constructive suggestions that have without doubt improved the text, and to Markku Tiainen, Harri Kutvonen, and Kirsti Keskisaari for preparing Figs. 1.1 and 1.2.

REFERENCES

Alapieti, T.T., Kujanpää, J., Lahtinen, J.J., Papunen, H., 1989. The Kemi chromitite deposit, Northern Finland. Economic Geology 84, 1057–1077.

Alapieti, T.T., Lahtinen, J.J., 1989. Early Proterozoic layered intrusions in the north-eastern part of the Fennoscandian shield. In: Alapieti, T.T. (Ed.), Proceedings of the Fifth International Platinum Symposium, Guide to the Post-Symposium Field Trip, pp. 3–41. Geological Survey of Finland, Guide 29.

Alapieti, T.T., Kärki, A.J. (Eds.), 2005. Early Palaeoproterozoic (2.5-2.4) Tornio–Näränkävaara layered intrusion belt and related chrome and platinum-group element mineralization, Northern Finland, p. 110. Field Trip Guidebook. Geological Survey of Finland, Guide 51a.

Alapieti, T., Lahtinen, J., Huhma, H., et al., 1989. Platinum-group element-bearing Cu-Ni sulphide mineralisaton in the marginal series of the early Proterozoic Suhanko-Konttijärvi layered intrusion, Northern Finland. In: Prendergast, M.D., Jones, M.J. (Eds.), Magmatic Sulphides–Zimbabwe Volume. Institute of Mining and Metallurgy, London, pp. 177–187.

Autere, E., Liede, J.E. (Eds.), 1989. Petsamon nikkeli, taistelu metallista. Vuorimiesyhdistys Bergsmannaföreningen r.y, Helsinki, p. 305.

Blankett, H., 1896. Om Välimäki malmfält, jämte några andra geologiska data från Sordavala socken i Östra Finland. Geologiska Föreningens i Stockholm Förhandlingar 18, 201–227.

Eilu, P., 2012. Gold mineralisation in southwestern Finland. In: Grönholm, S., Kärkkäinen, N. (Eds.), Gold in southern Finland—Results of GTK studies in 1998–2011, pp. 11–22. Geological Survey of Finland, Special Paper 52.

Eilu, P., Nykänen, V., 2011. Active and ongoing gold exploration and mining in northern Finland. In: Excursion guide in the Proceedings of the 25th International Applied Geochemistry Symposium 2011. Vuorimiesyhdistys–Finnish Association of Mining and Metallurgical Engineers r.y., Series B92-7, Rovaniemi, Finland, p. 48.

Eilu, P., Pankka, H., Keinänen, V., et al., 2007. Characteristics of gold mineralisation in the greenstone belts of northern Finland. In: Ojala, V.J. (Ed.), Gold in the Central Lapland Greenstone Belt, Finland, pp. 57–106 Geological Survey of Finland, Special Paper 44.

Ervamaa, P., 1962. The Petolahti diabase, and associated nickel-copper-pyrrhotite ore, Finland. Geological Survey of Finland Bulletin 199, 42 p.

Gaal, G., Isohanni, M., 1979. Characteristics of igneous intrusions and various wall rocks in some Precambrian porphyry copper-molybdenum deposits in Pohjanmaa, Finland. Economic Geology 74, 1198–1210.

Grönholm, P., 2006. The Jokisivu gold deposit, southwest Finland. Bulletin Geological Society of Finland, Special Issue 1, 42.

Grönholm, S., Kärkkäinen, N. (Eds.), 2012. Gold in southern Finland: Results of GTK studies 1998–2011, p. 276. Geological Survey of Finland, Special Paper 52.

Grundstöm, L., 1980. The Laukunkangas nickel-copper occurrence in southeastern Finland. Bulletin Geological Society of Finland 52, 23–53.

Haapala, P., 1945. Petsamon nikkelimalmialueen löytöhistoria, tutkimukset ja geologia. Vuoriteollisuus–Berghanteringen 4, 16–21.

Häkli, A., 1963. Distribution of nickel between the silicate and sulphide phases in some basic intrusions in Finland. Bulletin Commission Géologique Finlande 2009, 54.

Häkli, A., 1970. Factor analysis of the sulphide phase in mafic–utramafic rocks in Finland. Bulletin of the Geological Society of Finland 42, 109–118.

Häkli, A., 1971. Silicate nickel and its application to the exploration of nickel ores. Bulletin of the Geological Society of Finland 43, 247–263.

Härkönen, I., Keinänen, V., 1989. Exploration of structrally controlled gold deposits in the Central Lapland Greenstone Belt. Geological Survey of Finland, Special Paper 10, 79–82.

Hartikainen, A., Nurmi, P.A., 1993. Till geochemistry in gold exploration in late Archaean Hattu schist belt, Ilomantsi, Eastern Finland. In: Nurmi, P.A., Sorjonen-Ward (Eds.), Geological Development, Gold Mineralization and Exploration Methods in the Late Archean Hattu schist belt, Ilomantsi, Eastern Finland, pp. 323–352. Geological Survey of Finland, Special Paper 17.

Hausen, H., 1926. Über die prequartäre des Petsamo-Gebites am Eismeere. Bulletin Commission Géologique Finlande 76, 101.

Hiltunen, A., 1982. The Precambrian geology and skarn iron ores of the Rautuvaara area, Northern Finland. Geological Survey of Finland, Bulletin 318, 133.

Himmi, R., 1975. Outokumpu Oy:n Korsnäsin ja Petolahden kaivosten vaiheita. Vuoriteollisuus - Bergshanteringen 33, 35–38.

Himmi, R., Huhma, M., Häkli, T.A., 1979. Mineralogy and metal distribution in the copper-tungsten deposit of Ylöjärvi, southwest Finland. Economic Geology 74, 1183–1197.

Hultin, T., 1896. Historiska upplysningar om berghanteringen i Finland under svenska tiden. 1, Jernbruken. University of Helsinki dissertation, 247 p.

Hultin, T., 1897. Historiska upplysningar om bergshanteringen i Finland under svenska tiden 2, 92.

Hyvärinen, L., 1958. Lyijymalmin geokemiallisesta prospektoinnista Korsnäsissä. Geologinen tutkimuslaitos. Geoteknillisiä Julkaisuja 61, 7–23.

Hyvärinen, L., 1969. On the geology of the copper ore field in the Virtasalmi area, eastern Finland. Bulletin of the Geological Survey of Finland 240, 82.

Juopperi, A., 1977. The magnetite gabbro and related Mustavaara vanadium ore deposit in the Porttivaara layered intrusion, northeastern Finland. Geological Survey of Finland, Bulletin 288, 68.

Hyvärinen, L., Eskola, L., 1986. Malminetsintä. In: Papunen, H., Haapala, I., Rouhunkoski, P. (Eds.), Suomen Malmigeologia, metalliset malmiesiintymät. Suomen Geologinen Seura ry, pp. 215–289.

Hyyppä, E., 1948. Tracing the source of the pyrite-stones from Vihanti on the basis of glacial geology. Geological Survey of Finland Bulletin 142, 97–122.

Kahma, A., Siikala, T., Veltheim, V., et al., 1962. On the prospecting and geology of the Kemi chromite deposit, Finland, a preliminary report. Bulletin Commission Géologique Finlande 194, 91.

Kauranne, L.K., 2010. Ikuisesti nuori, Geologisen tutkimuskeskuksen 125-vuotishistoriikki. Geologiska forskningcentralens historia i sammandrag 1886–2011. Geological Survey of Finland—A Brief History 1886–2011. Geologian Tutkimuskeskus, Espoo, 185 p.

Kinnunen, A., 2008. A Palaeoproterozoic high-sulphidation epithermal gold deposit at Orivesi, southern Finland. Acta Universitatis Ouluensis A 507, 183.

Kojonen, K., Johanson, B., 1999. Determination of refractory gold distribution by microanalysis, diagnostic leaching and image analysis. Mineralogy and Petrology 67, 1–19.

Korkalo, T., 2006. Gold and copper deposits in central Lapland, northern Finland, with special reference to their exploration and exploitation. Acta Universitatis Ouluensis A 461, 122.

Korkiakoski, E.A., 1992. Geology and geochemistry of the metakomatiite-hosted Pahtavaara gold deposit in Sodankylä, northern Finland, with emphasis on hydrothermal alteration. Geological Survey of Finland, Bulletin 360, 96.

Kranck, E.H., 1945. The molybdenum deposits at Mätäsvaara in Carelia (E. Finland). Geologiska Föreningen i Stockholm Förhandlingar 67, 325–350.

Kuisma, M., 1985. Kuparikaivoksesta suuryritykseksi, Outokumpu 1910–1985. Outokumpu Oy, Espoo. 463 p.

Laine, E., 1950. Malminetsintä Suomessa 1809–1884. Geologinen tutkimuslaitos. Geoteknillisiä julkaisuja 49, 103.

Laine, E., 1952. Suomen vuoritoimi 1809–1884, III: Harkkohytit, kaivokset ja konepajat. Historiallisia tutkimuksia XXXI, 3. Suomen Historiallinen Seura, Helsinki. 569 p.

Lamberg, P., 2005. Nickel and cobalt numbers—Novel methodology for tracing critical processes of Ni-Cu-PGE ore formation. In: Tormänen, T.O., Alapieti, T.T. (Eds.), Platinum-Group Elements—From Genesis to Beneficiation and Environmental Impact. August 8–11, Oulu, Finland. Extended Abstracts, pp. 449–452.

Lindborg, T., Papunen, H., Parkkinen, J., Tuokko, I., 2015. The Taivaljärvi (Silver Mine) Ag-Au-Zn deposit in the Archaean Tipasjärvi greenstone belt, eastern Finland. pp. 633–655.

Loukola-Ruskeeniemi, K., 1991. Geochemical evidence for the hydrothermal origin of sulphur, base metals and gold in Proterozoic metamorphose black shales Kainuu and Outokumpu areas, Finland. Mineralium Deposita 26, 152–164.

Luukkonen, A., 1994. Main geological features, metallogeny and hydrothermal alteration phenomena of certain gold and gold-tin-tungsten prospects in southern Finland. Geological Survey of Finland, Bulletin 377, 135 Appendices 1–8.

Mäkinen, J., Makkonen, H.V., 2004. Petrology and structure of the Palaeoprterozoic (1.9 Ga) Rytky nickel sulphide deposit, Central Finland: a comparison with the Kotalahti nickel deposit. Mineralium Deposita 39, 405–421.

Mikkola, A., Strandström, G., Johanson, H., Raja-Halli, H., 1966. Jussarön malmikenttä - Jussarö malmfält. Vuoriteollisuus–Bergshanteringen 24, 55–72.

Mutanen, T., 1997. Geology and ore petrology of the Akanvaara and Koitelainen mafic layered intrusions and the Kevitsa-Satovaara layered complex, northern Finland. Geological Survey of Finland, Bulletin 395, 233 Appendices 1–5.

Niiranen, T., Hanski, E., Eilu, P., 2003. General geology, alteration, and iron deposits in the Palaeoproterozoic Misi region, northern Finland. Bulletin Geological Survey of Finland 75 (1–2), 69–92.

Nikander, G., 1929. Fiskars bruks historia 1630-1924. Åbo: Minneskrift Utgiven av Fiskars Aktiebolag 200.

Nurmi, P.A., 1985. Lithogeochemistry in exploration for porphyry-type molybdenum and copper deposits, southern Finland. Journal of Geochemical Exploration 23 163–191.

Nurmi, P.A., Sorjonen-Ward, P. (Eds.), 1993. Geological development, gold mineralization and exploration methods in the late Archean Hattu schist belt, Ilomantsi, eastern Finland, pp. 193–231. Geological Survey of Finland, Special Paper 17.

Nurmi, P.A., Sorjonen-Ward, P., Damsten, M., 1993. Geological setting, characteristics and exploration history of mesothermal gold occurrences in the late Archean Hattu schist belt, Ilomantsi, eastern Finland. In: Nurmi, P.A., Sorjonen-Ward, P. (Eds.), Geological Development, Gold Mineralization and Exploration Methods in the Late Archean Hattu Schist Belt, Ilomantsi, Eastern Finland, pp. 193–231. Geological Survey of Finland, Special Paper 17.

Nuutilainen, J., 1968. On the geology of the Misi iron ore province, northern Finland. Annales Academiae Scientiarum Fennicae, Series A(III) 96, 98.

Nykänen, V., Karinen, T., Niiranen, T., Lahti, I., 2011. Modelling the gold potential of central Lapland, Northern Finland. In: Nenonen, K., Nurmi, P. (Eds.), Geoscience for Society, pp. 71–82. Geological Survey of Finland, Special Paper 49.

Ollila, H., Saikkonen, R., Moisio, J., Kojonen, K., 1990. Oriveden Kutemajärven kultaesiintymä. Vuoriteollisuus-Berghanteringen 1/1990, 26–30.

Olson, E.O., 1937. Kärväsvaaran rautamalmiesiintymän tutkiminen 1937. *Suomen Malmi Osakeyhtiö*. Unpublished report in the Archives of the Geological Survey of Finland, 55 p.

Pääkkönen, V., 1952. Otanmäen titaani-rautamalmialueen löytöhistoria ja tutkimusten alkuvaiheet. Vuoriteollisuus–Berghanterinen 29–30.

Pääkkönen, V., 1956. Otanmäki, the ilmenite-magnetite ore field of Finland. Bulletin Commission géologique Finlade 171, 71.

Pääkkönen, V., 1966. On the geology and minereralogy of the occurrence of native antimony at Seinäjoki, Finland. Bulletin Commission Géologique Finlande 225, 70.

Papunen, H., 1976. Outokumpu Oy:n Kylmäkosken kaivoksen geologiasta. Summary: On the geology of the Kylmäkoski mine. Vuoriteollisuus–Bergshanteringen 34, 119–123.

Papunen, H., 1986. One hundred years of ore exploration in Finland. In: Tanskanen, H. (Ed.), The Development of Geological Sciences in Finland. Geological Survey of Finland, Bulletin, 336, pp. 165–203.

Papunen, H., Isomäki, O.-P., Penttilä, V.-J., 1997. Geology and mineral deposits of the Central Ostrobothnia, metallogeny, The Hitura deposit. Geological Survey of Finland, Guide 41, 61–66.

Papunen, H., Kopperoinen, T., Tuokko, I., 1989. The Taivaljärvi Ag Zn deposit in the Archean greenstone belt, eastern Finland. Economic Geology 84, 1262–1276.

Pankka, H., Puustinen, K., Vanhanen, E., 1991. Kuusamon liuskealueen kulta-koboltti-uraaniesiintymät. In: Summary: Au-Co-U deposits in the Kuusamo volcano-sedimentary belt, Finland, p. 53. Geological Survey of Finland, Report of Investigation 101.

Patison, N.L., Salamis, G., Kortelainen, V.J., 2007. The Suurikuusikko gold deposit: project development summary of Northern Europe's largest gold deposit. In: Ojala, V.J. (Ed.), Gold in Central Lapland Greenstone Belt, Finland, pp. 125–134. Geological Survey of Finland, Special Paper 44.

Piirainen, T., 1968. Die Petrologie und die Uranlagerstätten des Koli-Kaltimogebietes im finnischen Nordkarelien. Bulletin Commission géologique Finlande 237, 99.

Poutanen, P., 1996. Suomalaisen kuparin ja sinkin juurilla, Orijärven kaivos 1757–1957. Outokumpu. Gummerus Kirjapaino Oy, Jyväskylä. 147 p.

Pulkkinen, E., Ollila, J., Manner, R., Koljonen, T., 1986. Geochemical exploration for gold in the Sattasvaara comatiite complex, Finnish Lapland. In: Prospecting in Areas of Glaciated Terrain. Institution of Mining and Metallurgy, London, pp. 129–137.
Puustinen, K., 1997. Mining in Finland during the period 1530–1995. Geological Survey of Finland, Special Paper 23. 43–54.
Puustinen, K., 2003. Suomen kaivosteollisuus ja mineraalisten luonnonvarojen tuotanto vuosina 1530–2001, historiallinen katsaus erityisesti tuotantolukujen valossa. In: Geologian Tutkimuskeskus, Arkistoraportti M10.1/2003/3, p. 578.
Puustinen, K., 2006. Suomen vuoritoimi 1700-luvulla. Geologi 58, 103–108.
Puustinen, K., 2010. Erik Fleming sekä Ojamon kaivoskirja vuodelta 1542 ja raudantuotanto Siuntiossa. Geologi 62, 70–76.
Puustinen, K., 2014. Tilasinvuoren kuparikaivos Tammelassa 1740–1749. Summary: Tilasinvuori copper mine in Tammela during 1740–1749. Geologi 66, 80–85.
Rouhunkoski, P., Isokoski, P., 1974. The copper-gold vein deposit of Kivimaa at Tervola, N-Finland. Bulletin of the Geological Society of Finland 46, 29–35.
Saksela, M., 1948. *Outokummun kuparimalmin löytö*. English summary: The discovery of Outokumpu ore field. Geologinen Tutkimuslaitos. Geoteknillisiä Julkaisuja 47, 36.
Salokorpi, A., 1999. Suomen rautaruukit. Otavan Kirjapaino Oy, Keuruu. 143 p.
Saltikoff, B., Laitakari, I., Kinnunen, K., Oivanen, P., 1994. Helsingin seudun vanhat kaivokset ja louhokset. Geologian tutkimuskeskus, Opas 35, 64.
Sarikkola, R., 1979. Paltamon Nuottijärven ja Kolarin Kesänkitunturin uraaniesiintymät. In: Parkkinen, M. (Ed.), Uraaniraaka–Ainesymposiumi, B 27. Vuorimiesyhdistys–Bergsmannaföreningen r.y., Sarja, pp. 61–64.
Saxén, M., 1923. Über die Petrologie des Otravaaragebietes im östliscen Finland. Bulletin Commission géologique Finlande 65, 63 Jyväskylä, 251 p.
Stigzelius, H., 1986. Kultakuume, Lapin kullan historia. Suomen Matkailuliitto ry, Helsinki. Gummerus Kirjapaino Oy, Jyväskylä. 251 p.
Törmänen, T., Iljina, M., 2007. Stop 2: The Kevitsa intrusion and associater Ni-Cu-PGE deposit. In: Ojala, V.J., Patison, N., Eilu, P. (Eds.), Geological Survey of Finland, Opas-Guide 54, pp. 48–54.
Trüstedt, O., 1907. Die Erzlagerstätten von Pitkäranta am Ladoga-See. Bulletin Commision géologique Finlande 19, 333.
Turunen, E., 1953. Aijalan ja Metsämontun kaivokset. Vuoriteollisuus–Bergshanteringen 11, 16–28.
Turunen, E., 1957. Orijärven kaivos 1757–1957. Vuoriteollisuus–Bergshanteringen 15, 13–24.
Vanhanen, E., 2001. Geology, mineralogy and geochemistry of the Fe-Co-Au(-U) deposits of the Paleoproterozoic Kuusamo schist belt, northeastern Finland. Geological Survey of Finland, Bulletin 399, 229.
Vartiainen, H., 1980. The petrography, mineralogy and petrochemistry of the Sokli carbonatite massif, Northern Finland. Geological Survey of Finland, Bulletin 313, 126.
Vartiainen, H., 2012. Sokli malminetsintätyömaana ja kaivoshankkeena, Osa 1, Otanmäki Oy/Rautaruukki Oy:n aika 1067–85, kertomus-ja kuvadokumentti. Espoo, 187 p.
Väyrynen, H., 1938. Petrologie des Nickelerzfeld Kaulatunturi-Kammikivitunturi in Petsamo. Geological Survey of Finland, Bulletin 116, 198.
Vuorijärvi, E., 1989. Petsamon nikkeli kansainvälisessä politiikassa 1939–1944. Kustannusosakeyhtiö Otava, Helsinki. 267.
Zeidler, W., 1949. Om gruvdriften i Mätäsvaara åren 1940–47. Vuoriteollisuus–Berghanteringen 1949 (1), 20–33.

PART

2

SYNTHESIS OF THE GEOLOGICAL EVOLUTION AND METALLOGENY OF FINLAND

2

E. Hanski

ABSTRACT

The bulk of the Finnish bedrock evolved during four time periods, encompassing (1) the 3200–2700 Ma Archean cratonic basement, (2) its Paleoproterozoic, 2500–1920 Ma volcano-sedimentary cover and related intrusive magmatism, (3) the 1930–1770 Ma Svecofennian composite orogen, and (4) the 1640–1460 Ma rapakivi magmatism. Each of these periods displays its own metallogenic characteristics. In the Archean, orogenic gold and epithermal Ag-Au, komatiite-related Ni-Cu, and porphyry Mo have exceeded or approached the economic threshold, and a single carbonatite body is exploited for apatite. The most important deposit type in the Archean part of the Fennoscandian Shield is banded iron formation, but the major deposits occur outside Finland. Incipient rifting of the Archean supercontinent was accompanied by emplacement of large 2440 Ma mafic–ultramafic intrusions hosting stratiform Cr and Fe-Ti-V deposits, PGE-enriched reefs, and Ni-Cu-PGE contact and offset mineralization. At a later stage, when deep-water, pelitic continental margin sediments were deposited, intruding mafic magma produced Ni-Cu deposits (2050 Ma) in relatively small magma chambers. In addition, sedimentary black shale-hosted Ni-Zn-Cu deposits were formed. Continental breakup resulted in the formation of new oceanic crust, and Outokumpu-type volcanic massive sulfide (VMS) Cu-Co-Ni deposits were generated by hydrothermal processes affecting exposed ultramafic mantle rocks on the seafloor. The most prolific ore-forming stage ensued during the assembly of the Svecofennian orogen, composed of a collage of island-arc terranes and microcontinents. Many ore types typical of convergent plate boundaries were formed, including prominent VMS Pb-Zn, ultramafic-hosted Ni-Cu, and orogenic gold deposits. In addition, porphyry Cu and Mo, epithermal Au-Ag, iron oxide-copper-gold

Mineral Deposits of Finland. http://dx.doi.org/10.1016/B978-0-12-410438-9.00002-9

(IOCG) and pegmatite-hosted Li styles of mineralization are locally significant. So far, the anorogenic rapakivi magmatism has not provided exploitable metal occurrences, but shows potential for several high-tech metal deposits. Diamondiferous kimberlite pipes, approximately 600 Ma and 1200 Ma in age, have been found in eastern Finland, and the phosphorous and high-tech metal resources of a large Devonian carbonatite are currently assessed.

Keywords: metallogeny; ore deposits; geological evolution; Precambrian; Fennoscandian Shield; Finland.

INTRODUCTION

Over the last 100 years, approximately 50 metallic mines have been in operation in Finland. Since the 1970s, systematic collection of data on mined deposits and other explored metal occurrences has been carried out by the Geological Survey of Finland. This has resulted in the generation of extensive metal-specific databases containing information on, for example, Ni, Au, and Zn occurrences (e.g., Eilu and Pankka, 2009). Recently, the shield-wide metallic deposit map was updated (Eilu et al., 2013) and a new industrial mineral deposit map compiled (Gautneb et al., 2013). The explanatory notes for the metallogenic map were published in a recent Geological Survey of Finland Special Paper volume edited by Eilu (2012a), describing the main metallogenic areas in Fennoscandia, among them 47 examples being located entirely or mostly in Finland. This compilation does not contain in-depth geological data on major deposits, but rather a collection of spatially grouped information from currently and previously mined deposits and all other known significant metal occurrences that have been promising enough for drilling or other detailed studies, with their total number amounting to about 350 in Finland.

The explanatory volume and related map are particularly helpful in accessing the country's mineral potential or planning exploration strategy in different parts of the country. Altogether, 26 metals plus the Rare Earth Element (REE) group are included, reflecting the fact that the number of sought-after mineral commodities has been widening along with the technical development of the society. In addition to traditional base metal and gold exploration in northern Europe, much effort has recently been devoted to finding new potential sources of high-tech metals (Sarapää et al., 2015), as the European high-technology industry is currently heavily dependent on imports of most of these metals.

With increasing knowledge of the distribution of ore deposits in Finland and in other parts of the Fennoscandian Shield, a metallogenic provinciality has become apparent, with styles of mineralization varying in time and space at different scales. On the other hand, advances in geochronology have been important in resolving many of the uncertainties in the timing of formation of metal deposits and their host rocks (Huhma et al., 2011). These, together with the better understanding of the crustal evolution of Finland (e.g., Lehtinen et al., 2005), have enabled the relationship between different classes of ore deposits and their geotectonic settings, though there still remain many uncertainties. In general, the metallogenic history of Finland corresponds well with the global pattern seen in Precambrian shields, but the degree of metal endowment is not necessarily similar to that of other shields, and some peculiarities, such as the existence of Outokumpu-type volcanic massive sulfide (VMS) deposits, are unique to the Fennoscandian Shield.

Although many articles have been published that synthesize the geological evolution and related metallogeny of the Finnish part of the Fennoscandian Shield (e.g., Weihed et al., 2005; Lahtinen et al., 2010), it was regarded as justified to provide the reader of this volume with an introductory account to the geology of Finland and its most prevalent mineralization with an emphasis on the deposits that have been mined or are otherwise significant in characterizing the metallogenic epochs or provinces that they represent. The main focus of this chapter is on metallic ore deposits. Finnish industrial mineral occurrences are dealt with in Chapter 9.5. First, a summary of the general geological evolution of Finland is presented, followed by a description of different ore types occurring in major geotectonic units of the Finnish bedrock. The most important ore deposits are shown on two maps representing the southern and northern parts of Finland, respectively. A larger selection of deposits is presented in the supplementary online material, including information on the main metals, total tonnages, etc.

GEOLOGICAL EVOLUTION OF THE BEDROCK OF FINLAND

The bedrock of Finland is composed almost entirely of Precambrian rocks that are part of the Fennoscandian Shield. Exposed portions of the shield are found in Finland, Sweden, Norway, and northwest Russia, which, together with the Ukrainian Shield, represent the oldest exposed rocks in Europe. In the west, the Precambrian bedrock is overthrust by the Caledonian orogenic belt and in the south and east, it is overlain by Phanerozoic and Neoproterozoic platform sediments (Fig. 2.1A).

The crustal architecture of the Precambrian bedrock in the central and eastern part of the shield is defined by three large-scale crustal units: (1) the 3500–2500 Ma Archean basement, (2) its Paleoproterozoic (2500–1900 Ma) sedimentary-volcanic cover, and (3) the Svecofennian orogenic belt, 1930–1800 Ma in age. The Archean basement has usually been divided into two main crustal segments, the Karelian and Kola cratons (also called blocks, provinces, or domains), which are separated by the Belomorian mobile belt and Lapland granulite belt (Fig. 2.1A) and are thought to have collided with each other during the Paleoproterozoic Lapland-Kola orogeny (Lahtinen et al., 2005). In Finland, the Kola craton rocks only occur in northernmost Finland, north of the Lapland granulite belt, whereas the rest of the Archean bedrock belongs to the Karelian craton. The hidden southwestern boundary of the Karelian craton, running in the northwest–southeast direction across central Finland (Fig. 2.1), has been precisely delineated by measuring Nd isotopic compositions of granitic rocks (Huhma, 1986; Huhma et al., 2011). For more detailed information of the distribution of Archean crust in Finland and its division into distinct terranes or complexes, the reader is referred to Sorjonen-Ward and Luukkonen (2005) and Hölttä et al. (2012a).

As is typical of Archean granitoid-greenstone terranes around the world, in the Karelian craton relatively narrow, linear supracrustal belts are surrounded by voluminous, broadly tonalitic to granodioritic granitoids and migmatites (Fig. 2.1B). However, as recent studies have shown, the "granite-gneiss basement" of the Archean part of the shield has a complicated lithology and tectonomagmatic history. For example, in the Karelian craton, several magmatic suites have been distinguished including a tonalite-trondhjemite-granodiorite (TTG) suite (around 2830–2720 Ma, rarely 2950 Ma), sanukitoids (around 2720–2740 Ma), a quartz diorite-quartz monzodiorite (QQ) suite (around 2700 Ma), and a granodiorite-granite-monzogranite (GGM) suite (around 2730–2660 Ma) (Mikkola et al., 2011; Hölttä et al., 2012b). Migmatization of older rocks seems to be coeval with the emplacement of leucogranitoids of the CGM suite at around 2700–2690 Ma. Still other major components of the Archean basement consist of paragneiss complexes, derived from immature graywackes with depositional ages of around 2690–2710 Ma (Kontinen et al., 2007), and blocks of mafic to felsic high-grade granulites containing the oldest, Mesoarchean rocks of Finland, such as the ~3500 Ma Siurua trondhjemitic gneisses (Mutanen and Huhma, 2003; Lauri et al., 2011) and ~3200 Ma paleosomes of the Varpaisjärvi granulites (Mänttäri and Hölttä, 2002). Neodymium isotope compositions of the TTG suite indicate that most of the rocks represent relatively juvenile crust (Huhma et al., 2012b) and were generated from

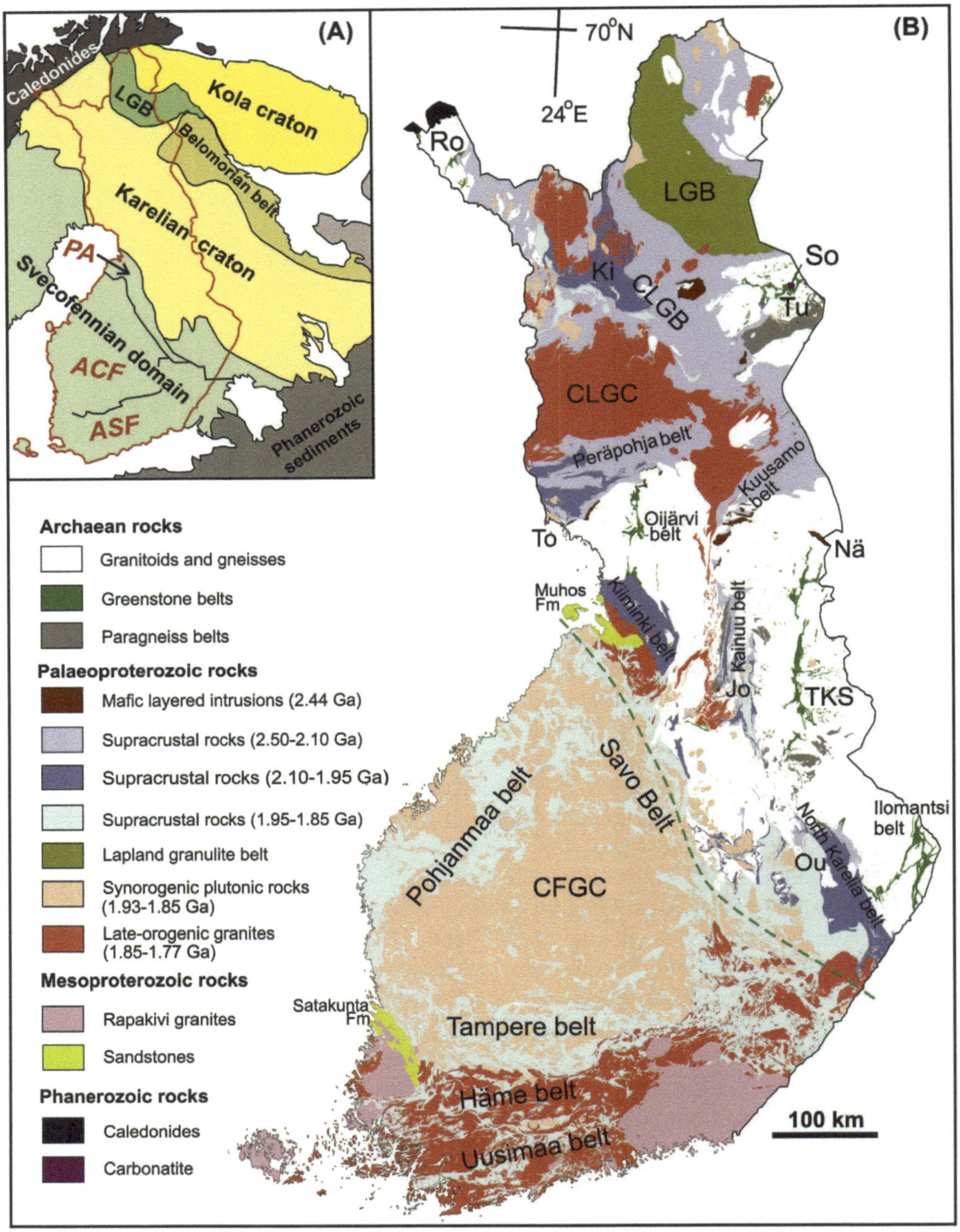

FIGURE 2.1 (A) Main bedrock units of the eastern part of the Fennoscandian Shield. (B) Geological map of Finland.

A abbreviations: PA = Primitive arc complex of central Finland, ACF = Accretionary arc complex of central and western Finland, ASF =Accretionary arc complex of southern Finland.

B abbreviations: CFGC = Central Finland Granitoid Complex, CLGC = Central Lapland Granitoid Complex, Jo = Jormua ophiolite, Ki = Kittilä greenstone complex, Nä = Näränkävaara, Ou = Outokumpu ophiolite, Ro = Rommaeno complex, So = Sokli carbonatite, TKS = Tipasjärvi-Kuhmo-Suomussalmi belt, To = Tornio, Tu = Tulppio.

Sources: *(A) Modified after Vaasjoki et al. (2005) and Lahtinen et al. (2005). (B) Based on DigiKP, the digital map database of the Geological Survey of Finland, Version 1.0 (available at www.geo.fi/en/bedrock.html). The green dashed line shows the southwestern boundary of the Karelian craton, based on Nd isotopic compositions of intrusive rocks (modified after Huhma et al., 2011).*

mafic protoliths (oceanic crust) by dehydration melting (Hölttä et al., 2012b). Sanukitoids are interpreted as melting products of metasomatized subcontinental lithospheric mantle (Halla, 2005) and magmas of the GGM suite as partial melts of preexisting TTG crust (Hölttä et al., 2012b).

Among the Neoarchean greenstone belts, the Tipasjärvi-Kuhmo-Suomussalmi (TKS) belt (Papunen et al., 2009) and Ilomantsi belt (Nurmi and Sorjonen-Ward, 1993), both in eastern Finland, are the most extensively investigated (Fig. 2.1B). Other, less studied and/or less exposed greenstone belts include Oijärvi in southern Finnish Lapland and Ropi in the northwest corner of Finland, and Tulppio in northeast Finland (Sorjonen-Ward and Luukkonen, 2005). In general, the greenstone belts are characterized by abundant tholeiitic basaltic submarine volcanism accompanied by komatiitic volcanic and cumulate rocks. In addition, felsic volcanic rocks and volcanoclastics are common, with intermediate varieties being less abundant. Among sedimentary rocks, turbiditic graywackes predominate.

The oldest U-Pb zircon ages of volcanic rocks have been measured in around 2940 Ma intermediate volcanic rocks in the lower part of the Suomussalmi greenstone belt, but the bulk of the volcanic and sedimentary rocks in the Finnish greenstone belts seem to be younger, with their ages falling in the range of 2800–2880 Ma (Huhma et al., 2012a). In general, direct evidence for an older sialic basement on which the supracrustal rocks of the greenstone belts were deposited is absent as the exposed contacts with the surrounding granitoids are intrusive or tectonic. Thus, the tectonic setting of the greenstone belts has been controversial. It is possible that different greenstone belts were formed in different geotectonic settings, or they form composite terranes consisting of allochthonous segments generated in different environments (e.g., Puchtel et al., 1998). For example, the geochemistry and isotopic characteristics of the mafic to ultramafic volcanics of the Tipasjärvi-Kuhmo-Suomussalmi belt suggest that they represent intraoceanic plateau magmatism, whereas an arc signature in the chemistry of the Ilomantsi belt volcanic rocks is consistent with a subduction zone environment (Hölttä et al., 2012b). Hölttä et al. (2012b) discussed the tectonic evolution of the Karelian craton and concluded that the basic structure of the granite–greenstone terranes was produced by Neoarchean accretion of exotic terranes at around 2830–2750 Ma and subsequent major collisional orogeny at around 2730–2670 Ma. These collisional events were linked to the assembly of Kenorland, one of the earliest inferred supercontinents, consisting of the Hearne Domain, Superior Province, and Fennoscandian Shield (Barley et al., 2005; Bleeker and Ernst, 2006).

After stabilization by around 2600 Ma, the Archean craton of the Fennoscandian Shield started to experience rifting events approximately during the Archean–Proterozoic transition. Subsequently, it served as a substrate for long-lasting intracratonic to continental margin sedimentation and volcanism that produced the supracrustal cover sequence collectively assigned to the *Karelian formations* (shown in Fig. 2.1B as two groups of supracrustal rocks ranging in age between 2.50 and 1.95 Ga). The record of these formations spans a geological evolution of around 600 Ma and has been traditionally subdivided by Finnish and Russian researchers into several major informal stratigraphic units. Hanski and Melezhik (2012) redefined these units as systems and proposed time limits for them based on the correlation with globally recognized and radiometrically constrained Paleoproterozoic events. The time periods of formation are the following: Sumian (2505–2430 Ma), Sariolian (2430–2300 Ma), Jatulian (2300–2060 Ma), Ludicovian (2060–1960 Ma), and Kalevian (1960–1900 Ma). Episodes of continental mafic magmatism that took place between around 2500 and 1970 Ma are well constrained by several generations of mafic dike swarms in the Archean basement (Vuollo and Huhma, 2005). For detailed summaries of the sedimentological and magmatic evolution during this time period; see Hanski and Melezhik (2012) and Hanski (2012).

The incipient rifting of the supercontinent was accompanied by emplacement of ~2500–2440 Ma layered gabbro-norite intrusions and mafic dike swarms, which are assigned to the Sumian magmatism in the Fennoscandian Shield. In Finland, the most prominent examples include the Tornio-Näränkävaara belt of intrusions (Fig. 2.1B) at the southern edge of the Peräpohja and Kuusamo schist belts in central Finland (Iljina and Hanski, 2005; Iljina et al., 2015) and the Koitelainen and Akanvaara intrusions in central Finnish Lapland (Mutanen, 1997; Hanski et al., 2001b). This magmatism and associated continental extension is commonly attributed to the impact of a mantle plume head at the base of the craton (Amelin et al., 1995; Hanski et al., 2001b), though this model in its simplest form meets some difficulties if the prolonged magmatic history of more than 50 Ma is real. Currently, it seems that there were two magmatic episodes, at around 2500 Ma and around 2435–2445 Ma, but more precise dating is needed (see Hanski, 2012, for a discussion on geochronology). Mantle-derived magmas experienced strong interaction with continental crust, which appears to have played a key role in ore formation, but also generated felsic magmas occurring now as small A-type granitic intrusions (Lauri et al., 2006). Supracrustal rocks of the Sumian system are locally abundant in eastern Finnish Lapland, particularly in the Salla belt from where they can be followed to central Finnish Lapland. They are dominated by subaerially erupted, heavily crustally contaminated lava flows, varying from basaltic andesites to dacites and rhyolites.

By definition, the Sariolian supracrustal rocks are younger than the ~2440 Ma layered intrusions. This is consistent with conglomerates and mafic volcanic rocks lying on partly eroded blocks of mafic intrusions, for example, in the Peräpohja schist belt. However, it is not always easy to distinguish between Sumian and Sariolian rocks as both can be deposited unconformably on Archean basement rocks. In Finnish Lapland, Sariolian volcanic rocks comprise komatiites, tholeiitic basalts, and andesites carrying a strong crustal signature in their chemistry. Sedimentary rocks include polymict, immature conglomerates and associated arkosic sandstones linked to an ancient, Huronian glaciation event, with the type locality occurring at Urkkavaara, North Karelia in eastern Finland (Marmo and Ojakangas, 1984).

The subsequent Jatulian system is characterized by deposition of quartzitic sandstones including red-bed deposits, eruption of up to 0.5-km-thick sequences of ~2100 Ma continental flood basalts varying from depleted MORB-types to enriched Fe-tholeiites, and accumulation of sedimentary carbonates that show the largest positive carbon isotopic excursion (Lomagundi-Jatuli event) in Earth's history (Karhu, 2005; Melezhik et al., 2013a). This time was also favorable for evaporitic sedimentation as evidenced by a thick anhydrite–halite sequence recovered by drilling in the Onega Lake region, Russian Karelia (Morozov et al., 2010). The Jatulian rock record indicates strong chemical weathering in a warm climate, oxygenation of the atmosphere as the result of the Great Oxygenation Event (GOE; see Holland, 2006) and peneplanation of the earlier rifted continental crust. An intrusive magmatic phase produced differentiated mafic–ultramafic sills at around 2220 Ma widely in eastern and northern Finland.

The Ludicovian system starting at around 2060 Ma bears witness for the first significant accumulations of organic matter as carbonaceous black shale deposits, which together with other pelitic sediments indicate deepening of the sedimentary basins. The earlier major change in atmospheric and ocean chemistry related to the GOE was accompanied by the buildup of sulfate in seawater (e.g., Strauss et al., 2013), creating prerequisites for bacterial sulfate reduction and the establishment of H_2S-rich anoxic water columns. Though not located in Finland, it is worth mentioning that the most conspicuous rocks of this period (Shunga event) are the extremely carbon-rich rocks in Russian Karelia, called maksovites and shungites, which are interpreted to represent petrified oil-rich deposits (Melezhik et al., 2013b). Another interesting feature is the reappearance in the geological record of highly magnesian volcanic rocks after a break of around 350 Ma. These komatiites and picrites are found in central Finnish Lapland, differing clearly in the mode of occurrence and chemistry from the older Sariolian komatiites in the same region (Hanski et al., 2001a). It should also be pointed out that there are differentiated mafic–ultramafic intrusions belonging to this stage, of which some contain economic mineral deposits, as described in the following. Although it is not clear whether the Archean craton was already totally disrupted at this stage, most mafic magmas do not show isotopic signatures indicative of interaction with continental crust (Hanski, 2012), with the remarkable exceptions being some Ni-Cu sulfide-bearing intrusions (e.g., Kevitsa; Hanski et al., 1997).

In the Central Lapland Greenstone Belt, central Finnish Lapland, there is a large amount of mafic volcanic and associated marine sedimentary rocks, such as Banded Iron Formations (BIFs), that are assigned to the Kittilä Group (Hanski and Huhma, 2005) and called the Kittilä greenstone complex in this chapter (Fig. 2.1B). These rocks form a special case because, although being around 2000 Ma in age and hence conforming geochronologically with the definition of Ludicovian rocks, they cannot be correlated with any other unit in the stratigraphy of the Karelian formations. The presence of ophiolitic ultramafic rocks and depleted mantle-like isotopic characteristics of most volcanic rocks led Hanski and Huhma (2005) to conclude that the Kittilä Group is an allochthonous rock unit, which at least partly represents a piece of an ancient oceanic crust. In that sense, they differ in their geotectonic setting from coeval Karelian sedimentary and volcanic rocks, which were deposited in a continental margin setting.

The Kalevian supracrustal system (1970–1900 Ma) contains abundant pelitic metasedimentary rocks and turbiditic graywackes developed mostly at the southwestern margin of the Karelian craton in the Peräpohja, Kiiminki, Kainuu, and North Karelia belts (Fig. 2.1B), representing Karelian sediments deposited in deepest water. Kontinen (1987) divided the Kalevian rocks into the autochthonous-parautochthonous Lower Kaleva and dominantly allochthonous Upper Kaleva. Turbiditic sequences and black schists are typical for the Lower Kalevian rocks but, in some areas, mafic submarine volcanic rocks or banded iron formations occur as well. In situ dating has revealed a dominantly Archean provenance for detrital zircons for the sedimentary units of the Lower Kaleva (Lahtinen et al., 2010). The Upper Kalevian rocks are lithologically more homogeneous consisting of thick-bedded graywackes with minor intercalations of black schists and without volcanic interbeds. The source of detrital zircon grains included a major Paleoproterozoic component falling in age between 2160 and 1920 Ma (Lahtinen et al., 2010).

The exact timing of the final breakup of the Kenorland supercontinent and separation of Superior and Hearne from Fennoscandia is not well defined but current estimates place it at around 2060–2050 Ma (Lahtinen et al., 2005, 2009), which also represents a time of substantial shield-scale and global events such as the end of the Lomagundi-Jatuli anomaly, the first appearance of phosphorites, and the reappearance of komatiitic magmatism (Melezhik et al., 2012). The breakup resulted in the formation of new oceanic crust, pieces of which are locally preserved in more or less complete ophiolitic complexes. The presence of Paleoproterozoic ophiolitic complexes is one of the most unique features of the Finnish bedrock. Of these, Jormua in the Kainuu schist belt, central Finland, presents the lithologically most complete ophiolitic sequence, whereas Outokumpu in the North Karelia belt, eastern Finland, contains mainly metamorphosed mantle peridotites and minor 1960 Ma gabbroic rocks, and Nuttio in central Lapland represents the most dismembered type constituting highly serpentinized refractory peridotites (Kontinen, 1987; Hanski, 1997; Peltonen, 2005a; Peltonen et al., 2008). The Jormua ophiolite comprises mantle tectonites, gabbros, plagiogranites, sheeted dikes, pillow lavas, and hyaloclastites. This ophiolite is peculiar in being a composite in the sense that the mantle rocks originated from Neoarchean subcontinental lithospheric mantle, whereas the overlying segment of oceanic crust is Paleoproterozoic, having been formed at around 1950 Ma. The tectonically emplaced slivers of ophiolitic rocks are confined to the Upper Kalevian sedimentary rocks both in Kainuu and North Karelia (Kontinen, 1987; Peltonen et al., 2008), suggesting that at least part of the Upper Kalevian turbidites were deposited on a newly formed oceanic crust (Lahtinen et al., 2010).

The youngest supracrustal rocks of the Paleoproterozoic cover sequence in the Karelian craton are coarse clastic sandstones and polymictic conglomerates (Kumpu Group; see Fig. 3) lying unconformably on deformed Karelian supracrustal rocks in central Lapland. They represent molasse-type sediments deposited soon after the ~1880–1890 Ma synorogenic plutonism (Hanski and Huhma, 2005).

While a more or less similar, prolonged geological evolution can be pictured for the Karelian formations over an area covering almost half of the shield, such a common history cannot be envisaged for the Svecofennian domain in southern and western Finland. The reason is that the Svecofennian rocks were formed within a shorter time interval in a different, more dynamic convergent margin setting where multiple collisional events brought together arc complexes and microcontinents with independent earlier evolutionary histories (Lahtinen et al., 2005). Soon after the advent of the plate tectonic theory it was realized that the origin of the Svecofennian domain is related to subduction processes (Hietanen, 1975), but as opposed to earlier models with a single subduction zone, the geological history of the domain has turned out to be much more complicated. In the Finnish part of the Svecofennian domain, Vaasjoki et al. (2005) distinguished three separate arc complexes (Fig. 2.1A):

(1) The primitive arc complex of central Finland (PA) adjacent to the Archean craton margin, containing volcanic and related plutonic rocks with ages in the range of around 1930–1920 Ma.
(2) The accretionary arc complex of central and western Finland (ACF) with volcanism dated at around 1910–1890 Ma (Kähkönen, 2005).
(3) The approximately coeval accretionary arc complex of southern Finland (ASF).

These three complexes are also known as the *Savo belt*, *central Svecofennia* and *southern Svecofennia*, respectively (Kähkönen, 2005). Furthermore, the PA (Savo belt) overlaps with the *Raahe-Ladoga zone*, a name that has often been used in the literature in connection with ore deposit studies as this zone has been regarded as a suture zone marking one of the most important metallogenic provinces in Finland (e.g., Gaál, 1990; Ekdahl, 1993).

Though being juxtaposed to the Archean craton, the rocks of the Savo belt are juvenile showing no Nd isotopic or inherited zircon evidence for an Archean component in them (Lahtinen and Huhma, 1997; Vaasjoki et al., 2003). The oldest dated rocks are ~1930–1920 Ma gneissic tonalites and associated volcanic rocks in the PA, but based on isotopic studies, it has been proposed that there are still older, 2100–2000 Ma, buried microcontinents (e.g., Keitele) in the Svecofennian domain (Lahtinen et al., 2005). The age relationships have become more complex after publication of the dating results obtained by Rutland et al. (2004) and Williams et al. (2008). Based on in situ zircon analyses, the former concluded that there is a major unconformity in the Vammala area, where a migmatite complex metamorphosed at around 1920 Ma is overlain by rocks of the Tampere belt (refer to Fig. 2.1B). The supracrustal rocks of the migmatite complex contain partly Archean detritus and were likely deposited as far back as around 1970 Ma, being thus older than any of the PA rocks. A similar unconformity was later established by Williams et al. (2008) in the Himanka area, Pohjanmaa belt (Fig. 2.1B), close to the western coast of Finland.

According to Kähkönen (2005), the Svecofennian supracrustal rocks typically comprise calc-alkaline volcanic rocks of volcanic arc magmatism, ranging in composition from low-K to shoshonitic basalts to rhyolites, and associated accretionary prism sediments, such as turbiditic graywackes and mudrocks. Besides island arc basalts, MORB-, EMORB-, and WPB-type pillow lavas are present locally. Sedimentary carbonate rocks are mostly confined to the southernmost arc complex.

Due to the complicated tectonic history, granitic magmatism has taken place at several stages. Nironen (2005) gives the following classification of orogenic granitoid rocks:

- Preorogenic (1950–1910 Ma)
- Synorogenic (1890–1860 Ma) divided into two subclasses:
  - Synkinematic (1890–1870 Ma)
  - Postkinematic (1880–1860 Ma)
- Late-orogenic (1840–1800 Ma)
- Postorogenic (1810–1770 Ma)

In Fig. 2.1B, for simplicity, Svecofennian granitoids are divided into two groups: synorogenic (1.93–1.85 Ga) and late-orogenic (1.85–1.77 Ga). In addition, mafic–ultramafic plutonism occurred as multiple pulses. Peltonen (2005b) distinguishes three groups of intrusions of which Group I are found around the Central Finland Granitoid Complex (CFGC). They were emplaced close to the peak of the Svecofennian orogeny at around 1890 Ma and can host Ni-Cu sulfide deposits. The Group II intrusions are large synvolcanic layered gabbro complexes restricted to the arc complex of southern Finland (ASF), and the less abundant Group III intrusions include Ti-Fe-P-rich gabbros within the CFGC.

A detailed accretionary history of the Svecofennian composite orogenic belt has been described by Lahtinen et al. (2005, 2009) as follows: (1) microcontinent accretion stage (1920–1870 Ma), (2) continental extension stage (1860–1840 Ma), (3) continent–continent collision stage (1840–1790 Ma), and (4) orogenic collapse and stabilization stage (1790–1770 Ma) followed by orogenic collapse and stabilization (1.79–1.77 Ga). During the prolonged accretionary history, the Karelian craton acted as the foreland upon which Svecofennian nappes were emplaced. Widespread orogenic to postorogenic plutonism is found within the Archean craton several hundreds of kilometers northeast of the collisional zone. Thermal effects of the Svecofennian orogenies are also observed as isotopic resetting of minerals and disturbance of Paleomagnetic signatures in the Archean basement and its Paleoproterozoic cover. It needs to be remembered that during the broadly coeval Kola–Lapland orogeny, the craton was also subjected to thrusting from the northeast, leading to the development of the Lapland–Kola orogen including the Lapland granulite belt. Syn- to late-orogenic reactivation of the foreland area between the two orogens has an important bearing on the formation of orogenic gold deposits.

After the intrusion of the latest postorogenic granites at around 1770 Ma, a tectonic and magmatic quiescence (crustal stabilization) of more than 100 Ma ensued, being interrupted by the emplacement of anorogenic A-type rapakivi granites in southern Finland (Fig. 2.1B) at around 1650–1540 Ma (Rämö and Haapala, 1995, 2005). The magmatism produced several large rapakivi batholiths and a number of smaller plutons as well as leucocratic gabbroic bodies and tholeiitic diabase dike swarms. The bimodal nature of the magmatism has been explained by a mantle-plume model in which felsic magmas were produced by thermal perturbation of the lower crust due to underplating of mafic magmas from a deep mantle plume.

The subsequent Mesoproterozoic to middle Paleozoic evolution of the Finnish bedrock has been described by Kohonen and Rämö (2005). After the emplacement of rapakivi granites, intracratonic rift basins (around 1600–1300 Ma) started to develop. These are preserved in restricted areas in southwestern and western Finland as two tectonic depressions filled by the Satakunta and Muhos formation sandstones, respectively (Fig. 2.1B). Mafic dikes and sills recording an extensional stage at around 1265 Ma are abundant in the Satakunta sandstones and adjacent rapakivi granites. A few mafic dikes were emplaced in northern Finland at around 1100–1000 Ma. Neoproterozoic magmatism is represented by ~600 Ma kimberlites in eastern Finland, and the still younger (ca. 365 Ma) Sokli carbonatite and Iivaara ijolite massif in northern Finland represent the westernmost igneous bodies of the Kola alkaline province (O'Brien et al., 2005). Caledonidic Paleozoic sedimentary rocks are found in Finland only as an overthrust nappe in the far northwestern part of the country.

METALLOGENY OF ARCHEAN GRANITE-GREENSTONE TERRANES

In general, Archean cratons are well endowed with certain metal ore deposits, including Fe in banded iron formations, Ni-Cu in komatiites, Au in orogenic gold deposits, and Cu-Zn in volcanogenic massive sulfides (VMS). Of these, banded iron formations, whose deposition took place dominantly during the Neoarchean and Paleoproterozoic before the appearance of the oxygenated atmosphere (Trendall and Blockley, 2004), are not well developed in the Archean greenstone belts in Finland, only forming thin Algoma-type banded quartz-magnetite-amphibole/pyroxene layers between mafic or felsic volcanic rocks or mica schists, reaching commonly a few meters in thickness (Lehto and Niiniskorpi, 1977). The largest deposits are found in the Ilomantsi greenstone belt (Fig. 2.2), with resources of up to 36 Mt (Sorjonen-Ward, 2012), thus being rather small. However, the large Kostomuksha BIF deposit, located on the eastern side of the Finnish-Russian border in Russian Karelia (Gor'kovets et al., 1981), has been in production for tens of years, and the currently active Sydvaranger BIF mine in northern Norway is also located close to the Finnish-Norwegian border.

The metals that have turned out to be the most promising exploration targets in the Archean bedrock of Finland are Ni, Au, Ag, and Mo. Unlike in some other Archean greenstone belts, such as the Kalgoorlie terrane in Western Australia, only a few occurrences of komatiite-hosted Ni-Cu sulfides have been discovered in the Finnish greenstone belts (Konnunaho et al., 2015). They are found mostly in the Kuhmo and Suomussalmi greenstone belts in eastern Finland and in the Rommaeno complex (Ruossakero deposit, Fig. 2.3) in northwestern Finland. They contain sulfide disseminations and less abundant massive sulfides in olivine-rich cumulate rocks. None of the mineralized komatiites has so far proved to be economic, with the exception of the small Tainiovaara deposit that was exploited in the 1980s. In some deposits such as Vaara (Fig. 2.3), postmagmatic alteration has upgraded the metal content of the sulfide fraction (Konnunaho et al., 2013), which enhances the potential of their exploitation.

Many reasons have been suggested for the scarcity of significant komatiitic Ni-Cu deposits in Finland (Konnunaho et al., 2013; Maier et al., 2013): the parental magma of the Finnish komatiites was more differentiated and cooler than the most ore-producing komatiites elsewhere, and dynamic lava channel environments where significant amounts of sulfide-bearing substrate could be assimilated were uncommon. Also, the total volume of the mafic–ultramafic magmatism was small compared to Western Australia or Canada, and the small size or absence of VMS and other sulfidic deposits restricts the amount of potential sources of external sulfur (cf. Bekker et al., 2009), although clear evidence for

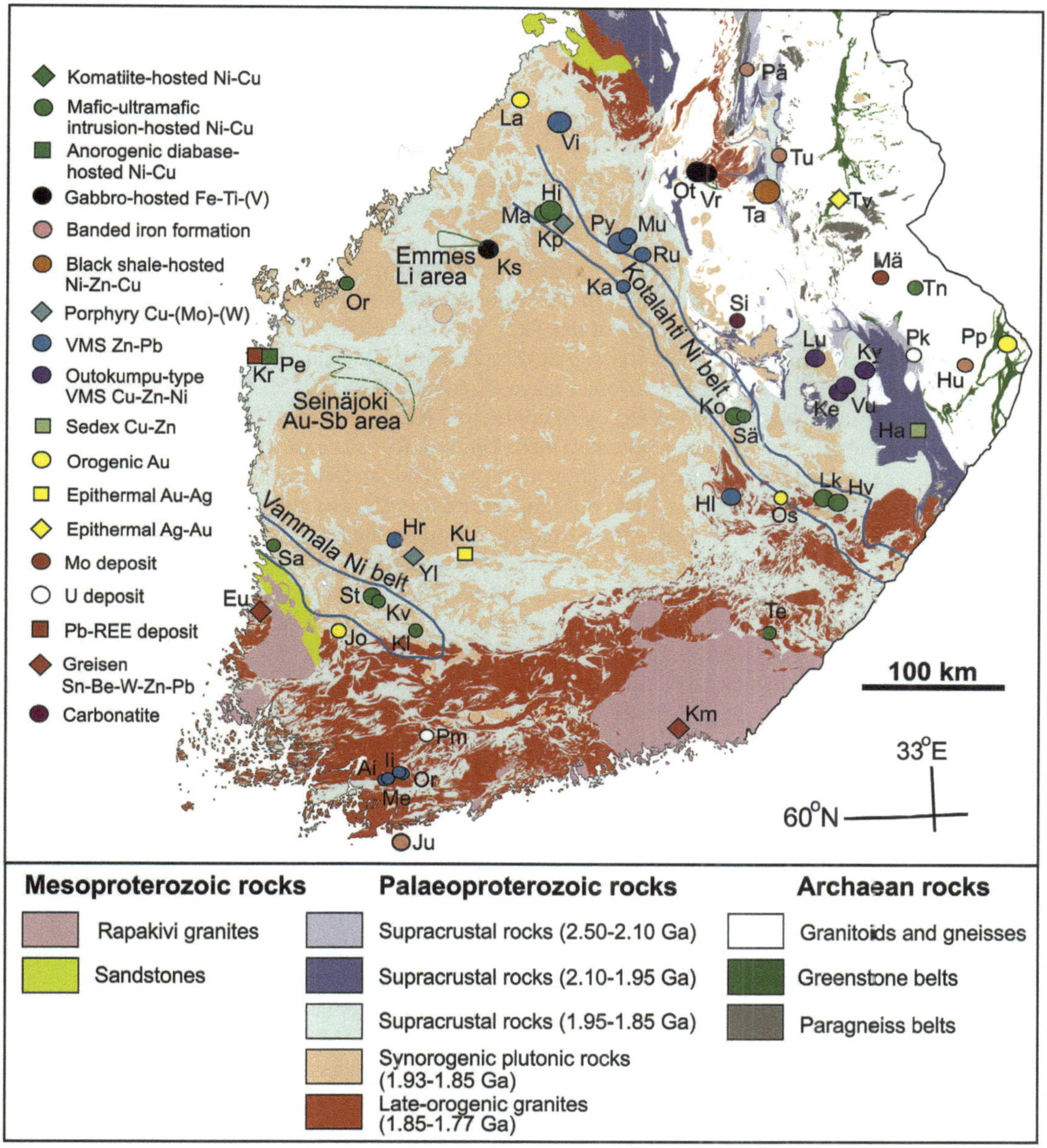

FIGURE 2.2 Main ore deposits in southern part of Finland.

The boundaries of the Kotalahti and Vammala Ni belts after Makkonen (2015) and those of the Seinäjoki Au-Sb area and Emmes Li area are from Eilu and Kärkkäinen (2012) and Ahtola (2012), respectively. Deposit names: Ai = Aijala, Eu = Eurajoki, Ha = Hammaslahti, Hi = Hitura, Hl = Hällinmäki, Hr = Haveri, Hv = Hälvälä, Hu = Huhus, Il = Iilijärvi, Jo = Jokisivu, Ju = Jussarö, Ka = Kangasjärvi, Ke = Keretti, Kl = Kymäkoski, Km = Kymi, Ko = Kotalahti, Kp = Kopsa, Ks = Koivusaarenneva, Kr = Korsnäs, Ku = Kutemajärvi, Kv = Kovero-oja, Ky = Kylylahti, La = Laivakangas, Lk = Laukunkangas, Lu = Luikonlahti, Ma = Makola, Mä = Mätäsvaara, Me = Metsämonttu, Mu = Mullikkoräme, Or = Oravainen, Ot = Otanmäki, Or = Orijärvi, Os = Osikonmäki, Pä = Pääkkö, Pe = Petolahti, Pk = Paukkajanvaara, Pm = Palmottu, Pp = Pampalo, Py = Pyhäsalmi, Ru = Ruostesuo, Sa = Sahalahti, Sä = Särkinniemi, Si = Siilinjärvi, St = Stormi, Ta = Talvivaara, Te = Telkkälä, Tn = Tainiovaara, Tu = Tuomivaara, Tv = Taivaljärvi, Vi = Vihanti, Vr = Vuorokas, Vu = Vuonos, Yl = Ylöjärvi.

Source: *The geological map is based on DigiKP, the digital map database of the Geological Survey of Finland, Version 1.0 (available at www.geo.fi/en/bedrock.html).*

assimilation of crustal sulfur can be found, for example, in the Vaara deposit in the Suomussalmi belt (Konnunaho et al., 2013). In addition, the Finnish komatiites seem to be around 100 Ma older than the ~2700 Ma mineralized komatiites elsewhere, though whether this has a direct bearing on the Ni prospectivity remains unknown.

Volcanic massive sulfide (VMS) deposits, which are commonly found in Archean terranes, especially in the Superior province in Canada, are rare in the Finnish greenstone belts. Subeconomic Zn-Pb mineralization is found in the northernmost part of the TKS belt (Kopperoinen and Tuokko, 1988) and in the Ilomantsi belt (Sorjonen-Ward, 2012). The Ag-Au-Zn deposit at Taivaljärvi in the southern part of the TKS belt is currently at an advanced feasibility stage (Lindborg et al., 2015). The Taivaljärvi deposit is regarded as an ancient metamorphosed analog of epithermal Au-Ag deposits (Papunen et al., 1989) and, as such, is a rare ore type in the Archean because these kinds of deposits have a low preservation potential and are almost exclusively limited to Mesozoic or younger island-arc terranes (e.g., Kerrich et al., 2005). The ore occurs within a sequence of felsic volcanic breccias, tuffs, and tuffites dated at 2798 ± 2 Ma (Huhma et al., 2012a) and is characterized by strong premetamorphic leaching and K-Mg alteration.

Orogenic gold mineralization has been discovered in most Archean greenstone belts in Finland, including TKS, Oijärvi, and Ilomantsi, but as in the case of Ni, the number of economically viable gold deposits is still low compared to some other Archean cratons. Currently, there is one gold mine in production (mining started in 2011), exploiting the Pampalo gold deposit (Fig. 2.2) in the Hattu schist belt (Nurmi et al., 1993; Sorjonen-Ward et al., 2015), which is the eastern branch of the Ilomantsi greenstone belt in eastern Finland. Gold is principally hosted by mica schists, feldspathic sediments, and tonalitic intrusions in zones that have undergone strong hydrothermal alteration including hydration, silicification, potassic alteration, tourmalinization, sulfidation, and carbonation (Nurmi et al., 1993). The main time period of gold mineralization in the Archean greenstone belts in Finland falls in the range of 2720–2640 Ma (Eilu, 2015), coinciding with one of the major periods of global orogenic gold ore formation (Goldfarb et al., 2001) related to the amalgamation of Kenorland (e.g., Bierlein et al., 2009). In spite of the general Svecofennian metamorphic overprint (cf. Kontinen et al., 1992), no evidence has been found for Paleoproterozoic gold mineralization in Archean greenstone belts in Finland (Eilu, 2015).

The rare Mesoarchean metallic deposits that have been exploited in Finland include the Mätäsvaara molybdenite deposit in eastern Finland (Fig. 2.2), which was mined during World War II (Haapala and Rämö, 2015). The ore is associated with red aplitic and white pegmatitic granite dikes cutting gray granitoid gneisses and could be classified as an Archean porphyry Mo deposit. Another notable Mo occurrence is the unexploited Aittojärvi deposit (Fig. 2.3) at the northern end of the TKS belt (Haapala and Rämö, 2015). An Archean carbonatite at Siilinjärvi (Fig. 2.2) in central Finland represents an unusual magmatic event, which took place in an anorogenic setting at around 2600 Ma, producing a major apatite deposit (O'Brien and Heilimo, 2015).

METALLOGENY OF THE PALEOPROTEROZOIC CRATONIC SEDIMENTATION AND MAGMATISM

Intracratonic or epicontinental environments in fault-controlled basins and troughs are conducive for formation of stratiform sediment-hosted Cu (SSC) deposits (red-bed Cu deposits) and SEDEX-type Zn and Pb deposits, and a subsequent marine transgression may create favorable conditions for deposition of Kupferschiefer-type shale-hosted ores enriched in Cu and other metals (Ag, Pb, Zn, Au, PGE) (e.g.,

Hitzman et al., 2010). Furthermore, evaporitic sedimentary beds can act as sources of hypersaline basinal brines that are good agents for metal transportation. After breakup of a continent, the shelf of the passive continental margin can act as a site of Fe or Mn ore deposition (e.g., Müller et al., 2005). During their prolonged development lasting hundreds of millions of years, continental and epicontinental sedimentary basins may experience repeated episodes of mafic volcanism and intrusive magmatism as a result of the movement of the continental plate over ascending mantle plumes derived from the deep mantle, resulting in formation of continental flood basalt-related magmatic Ni-Cu deposits. Thus, many metallogenic events may take place in a cratonic or epicratonic environment. The following section deals with syngenetic metallic mineralization associated with ~2500–1960 Ma Karelian sedimentary and igneous rocks constituting the Paleoproterozoic cover of the Archean cratonic basement. The same areas may also comprise epigenetic mineralization, particularly orogenic gold deposits, and these are discussed in the next section dealing with the Svecofennian period.

In general, mafic intrusive magmas are integral to formation of Cr, Ni-Cu-PGE, and Fe-Ti-V deposits in differentiated mafic layered intrusions. This is also the case in the Paleoproterozoic of northern Finland. Fracturing of the Archean craton caused by its initial rifting produced channels through which a large amount of mafic magma ascended and ponded near the unconformity between the Archean basement and the earliest Paleoproterozoic supracrustal rocks. Large differentiated layered intrusions were formed, including the ~2440 Ma Tornio-Näränkävaara belt of intrusions, where ore-forming processes generated various types of oxide and sulfide mineralization (Iljina and Hanski, 2005; Iljina et al., 2015; Karinen et al., 2015). The largest chrome deposit in Europe is hosted by the 2440 Ma Kemi mafic–ultramafic intrusion (Fig. 2.3) (Alapieti et al., 1989; Alapieti and Huhtelin, 2005; Huhtelin, 2015), which has been mined since 1966. At Kemi, chromium is enriched in the lower part of the ultramafic lower zone with the thickness of the chromite-rich cumulates reaching 60 m.

The ore has a relatively low Cr/Fe but the Cr resources are immense, securing the extent of mining operations far into the future. Several styles of PGE-bearing sulfidic mineralization are encountered in the Tornio-Näränkävaara intrusion belt (Halkoaho et al., 1990; Iljina, 2005; Andersen et al., 2006; Iljina et al., 2015):

- Up to five PGE-bearing low-sulfide reefs have been encountered at different stratigraphic levels in the Penikat intrusion (Fig. 2.3) and the dismembered Portimo complex, but have proved uneconomic so far.
- The contact-type disseminated mineralization of the Ahmavaara and Konttijärvi deposits (Fig. 2.3) in the Portimo complex and the Haukiaho and Kaukua deposits (Fig. 2.3) in the Koillismaa intrusion are under active research and inventory.
- Offset-type set of base metal- and PGE-rich, massive sulfide veins occur in the footwall rocks of the Kilvenjärvi block of the Portimo complex (Fig. 2.3).

The Mustavaara Fe-Ti-V deposit (Fig. 2.3), which is hosted by one of the blocks of the dismembered Koillismaa intrusion in eastern Finland, is described by Karinen et al. (2015). It was mined between 1976 and 1985, and the mine is currently in the process of being reopened. The mineralized magnetite-rich cumulate layer in the upper part of the igneous stratigraphy is up to 200 m thick and 19 km long, along strike. So far, the Mustavaara deposit is the only iron ore body of this age group that has been mined, but magnetite-rich cumulates also occur in other coeval intrusions such as Koitelainen and Akanvaara in Lapland (Fig. 2.3). These intrusions also host stratiform chromite deposits, but these are currently uneconomic.

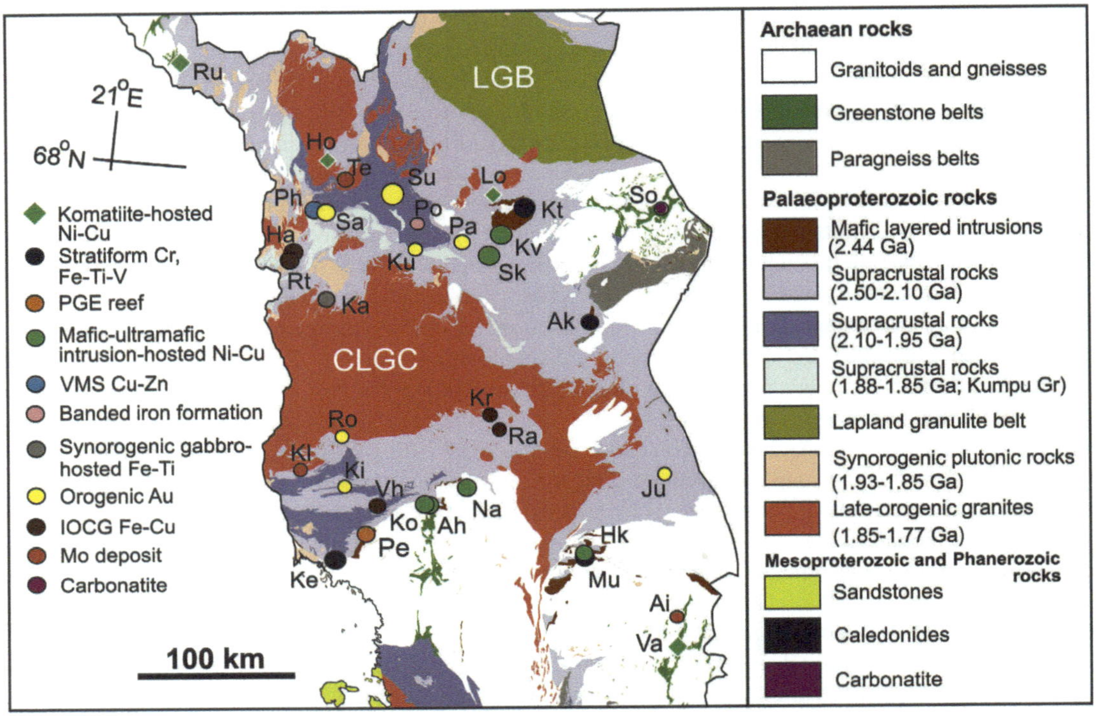

FIGURE 2.3 Major ore deposits in the northern part of Finland.

Deposit names: Ah = Ahmavaara, Ai = Aittojärvi, Ak = Akanvaara, Ha = Hannukainen, Hk = Haukiaho, Ho = Hotinvaara, Ju = Juomasuo, Ka = Karhujupukka, Ke = Kemi, Ki = Kivimaa, Kl = Kallijärvi, Ko = Konttijärvi, Kr = Kärväsvaara, Ku = Kutuvuoma, Kt = Koitelainen, Kv = Kevitsa, Lo = Lomalampi, Mu = Mustavaara, Na = Narkaus, Pa = Pahtavaara, Pe = Penikat, Ph = Pahtavuoma, Po = Porkonen, Ra = Raajärvi, Ro = Rompas, Rt = Rautuvaara, Ru = Ruossakero, Sa = Saattopora, Sk = Sakatti, So = Sokli, Su = Suurikuusikko, Te = Tepasto, Va = Vaara, Vh = Vähäjoki.

Source: *The map is based on DigiKP, the digital map database of the Geological Survey of Finland, Version 1.0 (available at: www.geo.fi/en/bedrock.html).*

Jatulian sedimentary and volcanic rocks do not host notable syngenetic metal enrichment in Finland. There are apparently small uranium occurrences in the Jatulian quartzites and conglomerates in southeast Finland (Piirainen, 1968). The only period of uranium mining in Finland took place between 1958 and 1961 when a small deposit in the Jatulian sequence at Paukkajanvaara (refer to Fig. 2.2) in the Koli area was mined. The Paukkajanvaara deposit consists of stratiform precipitates of colloform pitchblende and secondary uranophane in the matrix of the metasediments and epigenetic veins and breccia infillings at the contact of a diabase (Äikäs and Sarikkola, 1987; Pohjolainen, 2015).

Ludicovian (ca. 2050 Ma) komatiitic and picritic volcanic rocks and related subvolcanic intrusions in Finnish Lapland are associated with sulfide-bearing black shales and therefore were emplaced in a favorable environment for assimilation of external sulfur. Despite this setting, no economic komatiite-related Ni-Cu deposits have so far been discovered in Lapland. Nevertheless, there are two subeconomic occurrences, Hotinvaara and Lomalampi (Fig. 2.3), showing that this magmatic episode has some Ni potential (Konnunaho et al., 2015). The Lomalampi disseminated mineralization, which is hosted by an olivine cumulate lens, is interesting in showing an exceptionally high ratio of Pt to base metals (Konnunaho et al., 2015). In contrast to komatiites, the ~2050 Ma intrusive mafic magmatism has produced several prominent Ni-Cu deposits in central Lapland. The Kevitsa differentiated mafic–ultramafic intrusion contains a large Ni-Cu-PGE deposit (Fig. 2.3) that is located within the ultramafic lower part of the intrusion (Mutanen, 1997; Santaguida et al., 2015). The Kevitsa mine started its metal production in 2012. Mineralization consists of disseminated sulfides and is divided into several types due to a very large variation in metal tenors (Mutanen, 1997; Yang et al., 2013). The newly discovered, very promising Sakatti Ni-Cu deposit (Fig. 2.3), which is located not far from Kevitsa, is another example of mineralized, possibly coeval mafic–ultramafic intrusions (Coppard et al., 2013, Brownscombe et al., 2015).

Close to Kevitsa in age is the ~2060 Ma Otanmäki Fe-Ti-V deposit (Fig. 2.2), which is hosted by layered gabbroic rocks in central Finland (Lindholm and Anttonen, 1980; Kuivasaari et al., 2012). The Otanmäki mine was an important V producer on a global scale from 1953–1985. The gabbroic intrusions are spatially associated with gneissic peralkaline-alkaline A-type granites, which locally show Nb, Ta, Zr, Sc, and REE mineralization (Sarapää et al., 2015).

The giant, low-grade black shale-hosted Ni-Zn-Cu-Co deposit at Talvivaara (Fig. 2.2), extending along strike for around 12 km, occurs in Lower Kalevian sedimentary rocks in the Kainuu schist belt, central Finland (Loukola-Ruskeeniemi and Lahtinen, 2012; Kontinen and Hanski, 2015). Other metals that are found in sufficiently high concentrations for potential commercial exploitation at Talvivaara include uranium and manganese. The low metal grade of the ore has been circumvented by applying the bioleaching heap technique for metal extraction, which commenced in 2008. The sedimentary rocks hosting the stratabound deposit are very rich in graphite (5–15 wt% C) and sulfides (5–30 wt% S), and the black shale-bearing sedimentary sequence is devoid of any volcanic intercalations. It is generally accepted that the black shales represent organic carbon-rich muds accumulated under anoxic and euxinic conditions, and that nickel was derived from abnormally Ni-rich seawater, but the ultimate source of Ni is not yet well understood.

Another base metal sulfide mineralization occurring within Kalevian sedimentary rocks is the Hammaslahti Cu-Zn deposit in eastern Finland (Fig. 2.2). The mine was in operation between 1973 and 1986. The ore body is located in a turbidite-black schist environment and hosted by hydrothermally altered quartz graywacke and arkosite. The origin of the deposit is still under discussion; it seems to share features of both Besshi-type volcanogenic massive sulfide deposits and sediment-hosted stratiform Cu deposits (SEDEX) (Loukola-Ruskeeniemi et al., 1991). The relationship between the deposit and mafic volcanic rocks of the Tohmajärvi igneous complex that occur approximately 20 km south of Hammaslahti also remains debated.

Sedimentary ore deposits in Kalevian sedimentary rocks are not confined to sulfidic mineralization though only sulfide ores have reached the economic threshold. Laajoki and Saikkonen (1977) described Lake Superior-type oxide, carbonate, sulfide, and silicate facies banded iron formations, up to 50 m in thickness, in association with black schists and mica schists at Pääkkö (Fig. 2.2) in the Kainuu schist belt, central Finland. Laajoki (2005) assigned them to Lower Kalevian formations, being thus broadly correlative with the host rocks of the Talvivaara Ni-Zn-Cu-Co deposit. Another similar BIF occurrence is known at Tuomivaara (Fig. 2.2), about 65 km south-southeast of Pääkkö (Gehör and Havola, 1988).

In addition to the BIF occurrences in the Kainuu schist belt, Paleoproterozoic sedimentary iron ores are found in the Kittilä greenstone complex, which, in spite of being roughly coeval with the Kainuu

BIF, are not correlative with them due to the likely allochthonous nature of the Kittilä volcanic complex. This is why ore deposits in the complex can also be discussed in the following section dealing with ophiolitic complexes. The most notable of the iron formations is the Porkonen deposit (Fig. 2.3) which, as the Kainuu BIFs, contains many mineralogical varieties, including magnetite-quartz banded sediments in the oxide facies and manganosiderites in the carbonate facies (Paakkola, 1971). According to Korkalo (2006), there are more than 100 noneconomic iron occurrences in BIFs within the Central Lapland Greenstone Belt (CLGB), of which most are found in the Kittilä greenstone complex. The other main area of occurrence of BIF is the Jauratsi area in the eastern part of the CLGB. In terms of timing, the ~2000 Ma CLGB and Kainuu belt banded iron formations do not fit with any major peak of BIF occurrence in Earth's history (cf. Holland, 2006).

Other syngenetic ore deposits of the Kittilä greenstone complex are a number of minor VMS-style base metal occurrences in the southern part of the complex. The Pahtavuoma Cu-Zn deposit (Fig. 2.3) (intermittent production between 1974 and 1993) is the only one that has led to commercial mining (Korkalo, 2006). The Pahtavuoma deposit is hosted by a sedimentary formation, containing graphite-bearing phyllites and mica schists, that is correlative with the metasediments hosting the Kevitsa Ni-Cu bearing-intrusion.

METALLOGENY OF OPHIOLITIC COMPLEXES

Globally, ophiolitic complexes provide important resources of podiform chromitites and Cyprus-type volcanogenic massive sulfides (e.g., Robb, 2005) and these metals can also be expected to be found in Finnish ophiolites. Indeed, the ore deposits that have had the most substantial impact on developing the metallurgical industry in Finland are the Cu-Zn-Co deposits associated with the 1960 Ma Outokumpu ophiolitic rocks in eastern Finland (Fig. 2.1B), the first one of which was discovered in 1910 (Peltola, 1978). These deposits have often been classified as VMS deposits and, more specifically, Cyprus-type VMS deposits (e.g., Robb, 2005). However, in contrast to most VMS deposits elsewhere, the Outokumpu sulfides are closely associated with ultramafic (mantle rock) massifs and their highly silicified fringes, as opposed to volcanic rocks. In addition, the sulfides at Outokumpu are locally abnormally rich in Ni. These features led Papunen (1987) to propose a separate VMS subclass, the Outokumpu type, for these ores. The genesis of the Outokumpu-type deposits was recently reinterpreted by Peltonen et al. (2008) using a polygenetic model in which a Cu-rich proto-ore was generated by hydrothermal processes on serpentinized ultramafic seafloor, whereas nickel enrichment in sulfides took place during subsequent obduction, due to alteration and interaction of ultramafic massifs and adjacent sulfide-bearing black schists.

Mining operations continued at the two mines of the Outokumpu area, Keretti and Vuonos (Fig. 2.2), until 1989 and 1986, respectively, and at Luikonlahti, where an analogous deposit is located, until 1983. A new Outokumpu-type Cu-Co-Ni-Zn deposit was discovered at Kylylahti in 1984, where mining started in 2011. In addition to base metals, gold has been produced as a by-product at Outokumpu. In addition, talc and soapstone production has raised the economic importance of altered ophiolitic ultramafic rocks in eastern Finland (Lehtinen, 2015). Small podiform chromitite occurrences are known to be associated with ultramafic mantle rocks in the Outokumpu, Jormua, and Nuttio ophiolitic complexes (Vuollo et al., 1995; Hanski, 1997; Tsuru et al., 2000); however, no economic deposit has been discovered.

METALLOGENY OF THE SVECOFENNIAN ARC COMPLEXES AND OROGENIC MAGMATISM

Accretionary orogenic belts of all ages host diverse mineral deposit types developed at different stages of orogeny (Kerrich et al., 2005; Bierlein et al., 2009), including deposits formed during the arc magmatism, such as porphyry Cu-Au-Mo and epithermal Au-Ag deposits, and volcanogenic massive sulfide (VMS) deposits that form in extensional back-arc basins. Orogenic gold deposits are also typical and are generated later in the history of the convergent margin during and after terrane accretion. There also are other ore types that can be related to subduction processes but not detected in Finland, such as Carlin-type gold deposits (Muntean et al., 2011). Thus, accretionary plate margins throughout Earth's history represent productive settings for the formation of different mineral deposits, and this has also occurred in Finland where the Svecofennian orogeny at around 1900–1800 Ma produced a large number of deposits of diverse genetic types.

In terms of the total amount of metals, the Pyhäsalmi Zn-Cu volcanogenic massive sulfide (VMS) deposit (Fig. 2.2) is the most productive among the Svecofennian mineral deposits (Mäki and Puustjärvi, 2003; Roberts et al., 2004; Mäki et al., 2015). Including the three satellite mines (Mullikkoräme, Ruostesuo, and Kangasjärvi; Fig. 2.2), the ore in the Pyhäsalmi area has been mined continuously since 1962. The deposits occur in the oldest (ca. 1920–1930 Ma) of the three arc complexes (Savo belt) that collided with the Archean craton margin. The deposits are associated with a bimodal suite of volcanic rocks and display typical characteristics of Kuroko-style VMS mineralization, including intense hydrothermal alteration of the felsic host rocks to cordierite-sericite schists recording enrichment in Ba, K, and Mg. Located in the same metallogenic area, approximately 100 km northwest of Pyhäsalmi, the Vihanti VMS deposit (Fig. 2.2) (mined from 1954–1992) shares many features with Pyhäsalmi, but its average Zn/Cu ratio is twice that of the Pyhäsalmi deposit (Västi, 2012; Mäki et al., 2015). The associated volcanic rocks have a lesser mafic component than at Pyhäsalmi, being mainly intermediate to felsic. Other occurrences of VMS mineralization in the Savo belt were described by Ekdahl (1993). Of these, the Hällinmäki Cu deposit (Fig. 2.2), hosted by mafic volcanic rocks, was exploited from 1966–1984. Given the composite nature of the Svecofennian orogen, it is not surprising that the correlation of the different VMS-bearing zones, the Pyhäsalmi-Vihanti area in Finland and Skellefte and Bergslagen districts in Sweden, has not been straightforward; all seem to have different ages and geodynamic characters (Weihed et al., 2005).

Further south, on the southern side of the Central Finland Granitoid Complex, potential VMS mineralization is found in the western part of the Tampere belt where the Haveri Cu-Au deposit is located (Fig. 2.2). Since the 1730s, it was intermittently mined in a very small scale for iron, as it is locally enriched in magnetite, but after discovering gold in the deposit in the 1930s, it became a Cu-Au mine until its closure in 1962. The host rocks are submarine tholeiitic mafic volcanic rocks and hence different from those of the Pyhäsalmi and Vihanti deposits. The origin of the Haveri deposit has been uncertain due to strong alteration, deformation, and metamorphism. Most recently, Eilu (2012c, 2015) has classified it as a VMS deposit formed in a back-arc setting at around 1900 Ma, with no signs of postdepositional (orogenic) gold overprint.

The VMS deposits just discussed belong to the Savo belt and central Svecofennia. Several occurrences of VMS-style base and precious metal mineralization have also been discovered in the Uusimaa belt belonging in the southernmost of the three arc complexes, the southern Svecofennia. They form Cu-Zn-Pb ores in the Aijala-Orijärvi area, as the area was called by Latvalahti (1979), or in the Orijärvi Zn-Cu metallogenic area as defined by Eilu and Tontti (2012). Three mines have been in production in this region: Aijala, Metsämonttu, and Orijärvi (Fig. 2.2). The first two were in operation from 1948–1961 and 1951–1974, respectively, while Orijärvi started to produce Cu in 1758 and was intermittently mined until 1958. They are associated with strongly altered felsic to mafic volcanic rocks and chemical metasediments such as iron formation and chert, with the alteration being characterized by enrichment in K, Mg, and Fe. It is also worth mentioning the small Iilijärvi Ag-Au-Cu-Pb-Zn deposit (Fig. 2.2), which is peculiar in showing characteristics of a gold-rich VMS deposit or a Cu-Pb-Zn-bearing epithermal Ag-Au deposit (Mäkelä, 1989; Eilu, 2012b). Whereas the Pyhäsalmi-Vihanti belt cannot be extended to Sweden across the Gulf of Bothnia, the Orijärvi belt has been correlated to the Bergslagen VMS district in central Sweden (Lahtinen et al., 2005).

Porphyry Cu and Mo systems are among the most widely distributed mineralization types along Andean-type continental margins. Most deposits range in age from Paleozoic to early Pleistocene, but in addition, several Neoarchean and some Paleoproterozoic deposits are known. In Fennoscandia, the most notable example is the huge Aitik Cu-Au-Ag deposit in northern Sweden hosted by ~1900 Ma volcaniclastics, which are closely associated with the subduction-related quartz monzodiorites of the Haparanda suite (Wanhainen et al., 2012). In Finland, this category of mineralization is not significant in terms of past metal production, but according to Nurmi et al. (1984) there are tens of small occurrences throughout the Svecofennian domain (see also Haapala and Rämö, 2015). Nurmi et al. (1984) distinguished five main types based on their ore element association: Mo-Cu, Mo, Cu, Cu-Au, and Cu-W. Two notable examples are the Kopsa Cu-Au deposit (Fig. 2.2) in the northwest part of the Raahe-Ladoga zone, comprising a quartz-vein stockwork and sulfide dissemination in a calc-alkaline tonalite stock (Gaál and Isohanni, 1979), and the Ylöjärvi Cu-W deposit (Fig. 2.2) in the Tampere belt in southern Finland, which is confined to tourmaline breccias in country rocks of an ~1885 Ma batholith (Himmi et al., 1979; Haapala and Rämö, 2015). The Ylöjärvi deposit was mined from 1943–1966.

Epithermal Au-Ag deposits are regarded as near-surface or volcanic manifestations of the same processes that are responsible for the formation of porphyry deposits, representing a characteristic deposit type of the Andean or oceanic subduction zones in the circum-Pacific belt in the recent Earth (Simmons et al., 2005). Most known deposits are younger than Paleozoic in age, with the scarcity of deposits in ancient rocks being due to poor preservation potential rather than secular changes in ore-forming processes (Wilkinson and Kesler, 2007). In Finland, epithermal Au-Ag deposits are relatively rare. One of them is the well-documented Kutemajärvi deposit (Fig. 2.2) in the Tampere belt (Luukkonen, 1994; Poutiainen and Grönholm, 1996; Kinnunen, 2008), where mining started in 1994 and is still ongoing. The ore is hosted by metasomatic quartz rock in an environment of intense sericitization, which is characterized by a strong depletion of nearly all major and trace elements, and enrichment in Ag, Au, As, Bi, F, S, Si, and Te.

Although orogenic Ni-Cu deposits have been distinguished as a separate subclass of magmatic Ni-Cu sulfide deposits, by Naldrett (2004), for example, they have not always been listed among the typical ore deposits associated with convergent plate margins or collision boundaries (e.g., Laznicka, 2010), likely due to their high depth of origin and therefore nonexposure in young orogenic belts. Nevertheless, Svecofennian orogenic Ni-Cu sulfide deposits have been the most important source of Ni during Finland's mining history, though the more recent discovery of Kevitsa and Sakatti, formed in a continental setting, has changed this situation.

Svecofennian Ni-Cu deposits hosted by small mafic–ultramafic intrusions have been identified in two major zones (Peltonen, 2005b; Makkonen, 2015), as indicated in Fig. 2.2. These have been called the *Kotalahti nickel belt*, which follows the PA (Savo belt) adjacent to the Archean craton margin, and the

Vammala nickel belt, which occurs in the southern to southwestern part of the arc complex of central and western Finland (ACF), on the southern side of the Central Finland Granitoid Complex. The latter is thus located far from the craton margin (Papunen and Gorbunov, 1985; Peltonen, 2005b). It is noteworthy that some intrusions also occur in the Archean crust close to its southwest margin (Makkonen, 2015). There is also a Ni-Cu-bearing intrusion, Telkkälä (Fig. 2.2), on the northern side of the large rapakivi granite massif in southeast Finland, being located off the two major Ni belts. The fertility of the orogenic mafic magmatism with respect to Ni-Cu sulfides is confirmed by the fact that, since 1941, 10 Ni-Cu mines have been in operation and many subeconomic deposits have been investigated. The most prolific deposits in the southeastern part of the Kotalahti nickel belt are Kotalahti and Laukunkangas (Fig. 2.2) and in the northwestern part, Hitura (Fig. 2.2). In the Vammala nickel belt, the Kylmäkoski, Kovero-oja, and Stormi mines (Fig. 2.2) have been in operation. Most of the mines were closed by the middle of the 1990s, but the Hitura mine is still intermittently in production. Aside from the two mentioned nickel belts, there is a minor Ni-Cu deposit at Oravainen, close to the western Finnish coast (Isohanni, 1985).

The Ni deposits are mostly located within the Group I orogenic mafic–ultramafic intrusion type as defined by Peltonen (2005b). Further subclassification assigns the intrusions in the Kotalahti belt (i.e., those located close to the craton) to Group Ia, and those in the Vammala belt, far from the craton boundary, to Group Ib. The latter type is mostly dominated by weakly differentiated bodies rich in olivine and clinopyroxene as the main pyroxene, whereas the former type more commonly consists of differentiated orthopyroxene-bearing ultramafic–mafic bodies. Intensely deformed and migmatized, amphibolite-facies metaturbidites are the most common country rocks, some of them containing interbeds of black schist. Most of the Group I intrusions fall in the age range of around 1875–1885 Ma, being broadly coeval with the peak of the regional metamorphism and deformation of the orogeny (Peltonen, 2005b).

The parental magma that generated the ore-bearing intrusions was basalt or picritic basalt with the estimated MgO content varying in the range of 7–15 wt% (Makkonen, 1996; Lamberg, 2005; Peltonen, 2005b; Makkonen and Huhma, 2007), but being most commonly between 10 and 12 wt.% (Makkonen, 2015). High-MgO volcanic rocks found in both Ni belts are thought to be comagmatic with the mineralized intrusions, but since their separation from their mantle sources, they did not necessarily have the same history as the magma that produced the intrusions (Barnes et al., 2009). The intrusions are regarded as remnants of cumulate-rich magma conduits of tholeiitic arc magmatism. They were emplaced at high confining pressures (up to 6 kbar) at mid-crustal levels of the Svecofennian arc terrane and boudinaged in concomitant deformation into small lenses and fragments (Peltonen, 2005b).

Since the pioneering work by Häkli (1963, 1971), many studies have been devoted to comparison of Svecofennian economic, subeconomic, and barren intrusions using mineral chemistry, whole-rock major and trace element chemistry, and Nd isotopes (e.g., Mäkinen, 1987; Makkonen, 1996; Peltonen, 1995; Lamberg, 2005; Makkonen and Huhma, 2007; Makkonen et al., 2008). Various degrees of crustal contamination seem to be a ubiquitous feature in all mineralized intrusions as expressed by chemical and isotopic parameters. Makkonen and Huhma (2007) did not find a correlation between the degree of nickel mineralization and initial εNd values but, instead, between the latter and the country rock type.

The Group III of Peltonen (2005b) produced small gabbroic bodies within the Central Finland Granitoid Complex. Their age is similar to the Group I intrusions, as indicated by the age of 1881 ± 6 Ma measured for the Koivusaarenneva intrusion (Fig. 2.2) (Kärkkäinen, 1999; Sarapää et al., 2015). However, they are normally regarded as postkinematic intrusions, emplaced together with K-rich granites as a bimodal suite. Koivusaarenneva is the most important of the Fe-Ti-V-mineralized intrusions of this magmatic event. It forms a sill-like, layered gabbro-leucogabbro body, a few kilometers long, intruded into tonalitic rocks. It hosts significant resources of ilmenite and vanadium-rich magnetite (Kärkkäinen and Bornhorst, 2003). The Karhujupukka Fe-Ti-V deposit (Fig. 2.3) in western Finnish Lapland, hosted by deformed anorthositic gabbro, is likely a roughly coeval deposit associated with the synorogenic magmatism (Karvinen et al., 1988; Sarapää et al., 2015).

Other types of iron ore are found in southernmost Finland, where a number of small- and medium-sized deposits of skarn iron ore and banded iron formation have been known since the sixteenth century. Of these, the most significant is Jussarö (Kahma, 1979), which occurs mostly under the sea and was large enough to allow commercial utilization in two periods lasting from 1834–1861 and from 1961–1967.

In the Kolari-Pajala region in western Lapland, on both sides of the Finnish-Swedish border, there are several epigenetic iron ore deposits, of which some also contain Cu and Au and were therefore classified as iron oxide-copper-gold (IOCG) deposits by Niiranen et al. (2007) (cf. Williams et al., 2005). Iron ore was mined at Rautuvaara and Hannukainen from 1974–1992, and a new mine is planned to be opened at Hannukainen (Fig. 2.3). This last deposit is described by Moilanen and Peltonen (2015) . The ores are located at contact zones between ~1860 Ma synorogenic monzonite intrusions and supracrustal rocks of the Central Lapland Greenstone Belt and were previously regarded as skarn ores genetically related to the emplacement of monzonitic plutons (Hiltunen, 1982).

The current understanding is that the mineralization event took place later, at around 1800 Ma, and was controlled by fluids transported through shear zones related to postpeak metamorphic D_3 thrusting in northern Finland (Niiranen et al., 2007). Moilanen and Peltonen (2015) attributed the origin of the ore-forming saline fluids to dissolution of evaporates deposited earlier in the intracontinental rifting stage. The genesis thus differs from that of the magmatic-hydrothermal Olympic Dam Fe-Cu-Au-U deposit in Australia, a type example of IOCG deposits, which many researchers regard as genetically linked to A-type anorogenic felsic magmatism (Haapala, 1995; Williams et al., 2005; Pirajno, 2009). Other Fe deposits in northern Finland, though without elevated Cu and Au, which Niiranen et al. (2003, 2005) classified as IOCG deposits, are the small iron ores deposits of the Raajärvi area (mined between 1964 and 1975) in the eastern part of the Peräpohja belt (Fig. 2.3).

These rather pure magnetite ores occur in association with Jatulian-age dolomites and are thought to be hydrothermal replacement deposits that formed somewhere between 2020 and 2060 Ma under the influence of fluids derived from a deep-seated magmatic source (Niiranen et al., 2005). If the estimated timing is correct, the mineralization processes in the Raajärvi area took place during the deposition of the Karelian sedimentary rocks and is not related to the Svecofennian orogeny but to the final breakup of the Archean craton during the early Ludicovian. A further example of possible IOCG deposits in the southern part of the Peräpohja belt is the Vähäjoki deposit (Fig. 2.3) (Liipo and Laajoki, 1991). Bearing in mind the ongoing debate on the genesis of the Kiruna-type Fe ores in Sweden, which sometimes are included in the IOCG deposit type, the previous discussion highlights the disputed nature of the ore deposits included into the IOCG class (e.g., Groves et al., 2010).

The majority of the Paleoproterozoic gold deposits and prospects in Finland can be classified as orogenic Au deposits (load gold), with their genesis related to the Svecofennian orogenic processes (Eilu et al., 2003; Sundblad, 2003; Eilu, 2015). In Finland, the resources of about 30 Paleoproterozoic orogenic gold occurrences have been assessed. In addition to the proper Svecofennian domain, a large number of deposits are found in northern Finland, as a result of the effects of Svecofennian tectonothermal processes on the cratonic foreland (Eilu et al., 2007).

Eilu (2015) presents a summary of the most salient features of the Finnish orogenic gold deposits. They are structurally controlled following second- or third-order faults or shear zones branching from the major structures. Greenschist-facies to amphibolite-facies host rocks show a wide variation in lithology, from ultramafic and mafic volcanic rocks to tonalitic intrusions, and are commonly highly altered due to albitization, carbonatization, etc. In many cases in northern Finland, part of the alteration is considered to have taken place prior to gold mineralization, rendering the host rocks more competent and conducive to fluid flow. Iron sulfides and sulfarsenides are the most abundant ore minerals with pyrite ± arsenopyrite being dominant in greenschist-facies rocks and pyrrhotite ± löllingite and arsenopyrite in amphibolite-facies rocks. Most of the gold is in its native form but can also occur as refractory gold in pyrite or arsenopyrite.

In contrast to most of the orogenic gold deposits around the world, in which gold is the only commodity, many Finnish occurrences display atypical mineralogical and chemical features, including anomalous concentrations of Ag, Cu, Co, Ni, or Sb hosted by phases such as chalcopyrite and sulfosalts. Some deposits are enriched in U or Ba in the form of uraninite and barite, respectively. Because of these peculiarities, Eilu (2015) classifies these gold deposits, including those in the Kuusamo or Peräpohja belts, as "Orogenic gold with anomalous metal association." For more details, see Eilu (2015).

Dozens of gold occurrences have been identified in the supracrustal rocks in the Kittilä and Sodankylä areas in northern Finland (Korkalo, 2006). Many of them are closely linked to a major tectonic shear zone, the so-called Sirkka Line, trending northeast–southwest immediately to the south of the Kittilä greenstone complex. Another major gold-controlling shear zone runs roughly north–south in the middle part of the Kittilä greenstone complex. This is where the Suurikuusikko gold deposit (Fig. 2.3) is located (Patison, 2007; Patison et al., 2015). Gold production from this deposit, which is the largest known gold resource in northern Europe, commenced in 2008. The deposit is characterized by arsenopyrite-pyrite dissemination, carrying refractory gold, in strongly hydrothermally altered host rocks that were originally mafic volcanic rocks and intervening marine sediments.

Other notable examples of orogenic gold occurrences (Fig. 2.3) include the Saattopora deposit (Korkalo, 2006) (mined from 1988–1995) and the komatiite-hosted Pahtavaara deposit (Korkiakoski, 1992), which started to produce gold in 1996, and the Kutuvuoma deposit (production from 1998–2000) (Korkalo, 2006). Elsewhere in northern Finland (see Fig. 2.3), orogenic gold deposits occur in the Peräpohja belt including the small, mined-out (1969–1970) Cu-Au deposit at Kivimaa (Fig. 2.3) (Rouhunkoski and Isokangas, 1974) and the promising Rompas-area occurrences (Vanhanen et al., 2015), and in the Kuusamo belt where the Juomasuo Au-Co deposit (Fig. 2.3) has been test-mined (Vanhanen, 2001; Eilu, 2015).

The timing of the orogenic gold systems in northern Finland is not precisely constrained, though most of the mineralization is thought to have taken place during a continental collision event at 1840–1790 Ma (Eilu et al., 2007). If this is true, then orogenic gold deposits were not yet exposed to erosion when the molasse-like sediments (the Kumpu Group) were deposited. Minor paleoplacer gold mineralization has been identified in these sediments, including the Outapää and Kaarestunturi prospects (Härkönen, 1984), which must have derived their gold from other sources such as VMS exemplified by the Pahtavuoma Cu-Zn deposit.

An overview of orogenic and other gold mineralization in the southwestern part of the Svecofennian domain was published recently by Eilu (2012b), and in other papers of the volume edited by Grönholm and Kärkkäinen (2012), new results of exploration for gold in southern Finland are reported. Another gold-bearing zone in southern Finland is the Vammala migmatite belt, where the Jokisivu deposit (Fig. 2.2) is located (Eilu, 2012b). It started producing gold in 2009. The Jokisivu deposit shows typical characteristics of orogenic gold deposits, with gold occurring in quartz veins hosted by synorogenic quartz diorites and gabbros (Saalmann et al., 2010).

Farther north, the Seinäjoki region in the Pohjanmaa belt, western Finland, has been known as an antimony province since the 1960s (Pääkkönen, 1966). Subsequently, it has become evident that the antimony is closely associated with gold, and the Seinäjoki Au-Sb metallogenic area has been recently established (Eilu and Kärkkäinen, 2012). Around 10 orogenic gold occurrences have so far been discovered, but no mining activities have begun in the area.

The situation is different in the northwest part of the Savo belt, where the Laivakangas gold mine started gold production in 2011, adding a primary gold source to the belt's previous significant metal endowment from VMS and orogenic Ni-Cu deposits. At Laivakangas, gold occurs mainly as native grains in thin silicified shear zones and quartz veins cutting quartz diorite and intermediate to mafic metavolcanic rocks (Västi et al., 2012). The most notable Au occurrences in the southeast part of the Savo belt (or at the boundary between the Savo belt and ACF) is the Osikonmäki deposit (Fig. 2.2) hosted by quartz veins in a synkinematic tonalite (Kontoniemi, 1998).

According to Eilu (2015), the timing of orogenic gold mineralization in southern Finland is bracketed somewhere in the range of around 1840–1770 Ma and is related to a continent–continent collision between Fennoscandia, Sarmatia, and Amazonia. Recently, Saalmann et al. (2010) determined isotopic compositions of secondary zircon rims and titanite from the Jokisivu deposit, obtaining ages of around 1802 ± 15 Ma and 1801 ± 18 Ma, respectively. These are similar in age (1807 ± 3 Ma) to the pegmatite dikes cutting the mineralized zone and coeval to the late- to postorogenic magmatism.

Also the Svecofennian postcollisional granitic magmatism encompasses some metallogenic events, though none of the deposits have so far been sufficiently rich to be economic. One of them is the molybdenite deposit in the Tepasto aplite granite (Fig. 2.3) in the Kittilä area (Front et al., 1989). This intrusion belongs to the 1800–1760 Ma, A-type granitic stocks and batholiths, the so-called Nattanen-type granites, which can be found in northern Finland and the Kola Peninsula and are manifestations of reactivation of the Archean crust (Heilimo et al., 2009). Another mineralization type related to the ~1790 Ma postorogenic magmatism, in western and southern Finland at least, is the spodumene-bearing Li pegmatites (Fig. 2.2), which are currently under active exploration and development in the Emmes area, western Finland (Fig. 2.2) (Alviola et al., 2001; Ahtola, 2012). These pegmatites are fractionation products of melts generated by melting of juvenile Svecofennian crust.

There are two further mineral deposits within the Svecofennian domain that deserve to be mentioned. The Korsnäs Pb-REE deposit in westernmost Finland (Fig. 2.2) was mined from 1961–1972 producing both Pb and REE. It occurs in ~1830 Ma, 20-m-thick carbonatite dike cutting Svecofennian mica gneisses (Isokangas, 1978). The dike contains galena and, as REE minerals, allanite, monazite, and apatite (Papunen and Lindsjö, 1972; Sarapää et al., 2015). The single notable uranium deposit in southern Finland, Palmottu, is associated with late-orogenic granites (Pohjolainen, 2015).

METALLOGENY OF THE ANOROGENIC MAGMATISM

In the Earth's history, the time period of around 1700–800 Ma is characterized by a distinct paucity of several mineral deposit types, such as banded iron formation or orogenic gold (Cawood and Hawkesworth, 2015). This time frame is coeval with the most significant global occurrence of anorogenic magmatic rocks. Ore formation was not controlled by a supercontinent cycle and related plate margin processes, but by intra-plate processes such as magmatism associated with mantle-plume activity. Most important mineralization types are represented by Fe-Ti-V deposits hosted by anorogenic gabbro-anorthosite complexes, and greisen systems related to A-type granites enriched in many elements such as Sn, W, Cu, Zn, Bi, Mo, U, and F (Pirajno, 2009).

In Finland, this stage is represented by rapakivi granites and related mafic intrusive rocks. Signs of mineralization are found in several places (Haapala and Rämö, 2015). Good examples are the hydrothermal greisen veins with Sn-Be-W-Zn-Pb mineralization associated with the Eurajoki and Kymi topaz granite stocks (Fig. 2.2). These stocks represent the latest volatile-rich crystallizing phases of the rapakivi granites (Rämö and Haapala, 2005; Haapala and Lukkari, 2005; Haapala and Rämö, 2015). Some of the greisen veins have turned out to be enriched in indium (Cook et al., 2001). Until now, no mining of anaorogenic magmatic rocks other than that related to dimension stone and semiprecious stone production has been carried out in Finland. In the recent metallogenic map, no metallogenic areas are defined in Finland that are directly related to rapakivi magmatism (Eilu, 2012a).

Nevertheless, one should not ignore the role of the anorogenic magmatism in ore formation as indicated by the Northern Ladoga Sn-Zn-Pb, U, Au, W metallogenic area, delineated by Korsakova (2012) on the northern shore of Lake Ladoga, Russian Karelia, which manifests a typical metal combination of rapakivi-related deposits. There are many promising prospects in this area associated with greisen veins and skarn rocks in country rocks of the 1540–1560 Ma Salmi rapakivi massif, including that at Pitkäranta, which was mined for Fe, Sn, and Zn from 1772–1920 (Haapala, 1995).

Anorogenic mafic magmas occur in Finland mainly as diabase dikes, which are generally barren with respect to mineral deposits. One exception is the small Petolahti Ni-Cu deposit (Fig. 2.2) located at the western coast of Finland, close to the Korsnäs Pb-REE deposit (Sipilä et al., 1985). Its resources were sufficient for mining in one year, 1973. The deposit is hosted by an olivine diabase dike, and Sipilä et al. (1985) regarded it as part of the Häme dike swarm, which has been dated at around 1665 Ma by Vaasjoki and Sakko (1989).

LATER MINERALIZATION EVENTS

Neoproterozoic or younger magmatism is scarce in Finland, being restricted to some occurrences of alkaline rocks, including kimberlites. Diamond-bearing kimberlites are known to occur in ancient stable cratons having deep, cool lithospheric mantle roots. This applies to the eastern part of the Fennoscandian Shield. Approximately 25 kimberlite pipes have been found in eastern Finland from Kaavi to Kuusamo, with some of them being diamondiferous (see O'Brien, 2015). They seem to form at least two age groups. Olivine lamproites and Group II kimberlites occurring in the Kuhmo-Lentiira region have yielded Ar-Ar pholopite ages of close to 1200 Ma (O'Brien et al., 2007), whereas Group I kimberlites from the Kaavi-Kuopio region have given U-Pb perovskite ages of around 600 Ma (O'Brien et al., 2005).

The ~365 Ma Sokli carbonatite in eastern Finnish Lapland (Fig. 2.3) is the westernmost igneous body of the Kola alkaline province. It is described by O'Brien and Lee (2015). A supergene phosphorus ore deposit enriched in Nb, Ta, Zr, REE, and U has been developed by weathering in the uppermost part of the carbonatite intrusion. There are also late carbonate veins in the fenite aureole that are prospective for REE (Sarapää et al., 2015).

ACKNOWLEDGMENT

Pasi Eilu and Wolfgang Maier are thanked for useful comments on the manuscript and Jukka-Pekka Ranta for assisting with the preparation of the figures.

REFERENCES

Äikäs, O., Sarikkola, R., 1987. Uranium in lower Proterozoic conglomerates of the Koli area, eastern Finland. In: Uranium Deposits in Proterozoic Quartz-Pebble Conglomerates—Report of the Working Group on Uranium Geology. IAEA TECDOC 427, 189–234.

Ahtola, T., 2012. Emmes Li. In: Eilu, P. (Ed.), Mineral Deposits and Metallogeny of Fennoscandia. Geological Survey of Finland Special Paper 53, 260–262.

Alapieti, T.T., Huhtelin, T.A., 2005. The Kemi intrusion and associated chromitite deposit. Geological Survey of Finland, Guide 51a, 13–32.

Alapieti, T.T., Kujanpää, J., Lahtinen, J.J., Papunen, H., 1989. The Kemi stratiform chromitite deposit, northern Finland. Economic Geology 84, 1057–1077.

Alviola, R., Mänttäri, I., Mäkitie, H., Vaasjoki, M., 2001. Svecofennian rare-element granitic pegmatites of the Ostrobothnia region, western Finland; their metamorphic environment and time of intrusion. Geological Survey of Finland, Special Paper 30, 9–29.

Amelin, Yu.V., Heaman, L.M., Semenov, V.S., 1995. U-Pb geochronology of layered mafic intrusions in the eastern Baltic Shield; implications for the timing and duration of Paleoproterozoic continental rifting. Precambrian Research 75, 31–46.

Andersen, J.C.Ø., Thalhammer, O.A.R., Schoenberg, R., 2006. Platinum-group element and Re-Os isotope variations of the high-grade Kilvenjärvi platinum-group element deposit, Portimo layered igneous complex. Finland. Economic Geology 101, 159–177.

Barley, M.E., Bekker, A., Krapez, B., 2005. Late Archean to Early Paleoproterozoic global tectonics, environmental change and the rise of atmospheric oxygen. Earth and Planetary Science Letters 238, 156–171.

Barnes, S.J., Makkonen, H.V., Dowling, S.E., et al., 2009. The 1.88 Ga Kotalahti and Vammala nickel belts, Finland: geochemistry of the mafic and ultramafic metavolcanic rocks. Bulletin of the Geological Society of Finland 81, 103–141.

Bekker, A., Barley, M.E., Fiorentini, M.L., et al., 2009. Atmospheric sulfur in Archean komatiite-hosted nickel deposits. Science 326, 1086–1089.

Bierlein, F.P., Groves, D.I., Cawood, P.A., 2009. Metallogeny of accretionary orogens—The connection between lithospheric processes and metal endowment. Ore Geology Reviews 36, 282–292.

Bleeker, W., Ernst, R.E., 2006. Short-lived mantle generated magmatic events and their dyke swarms: the key unlocking Earth's paleogeographic record back to 2.6 Ga. In: Hanski, E., Mertanen, S., Rämö, T., Vuollo, J. (Eds.), Dyke Swarms—Time Markers of Crustal Evolution. A.A. Balkema, Rotterdam, pp. 3–26.

Brownscombe, W., Ihlenfeld, C., Hartshorne, C., Coppard, J., Klatt, S., 2015. The Sakatti Cu-Ni-PGE sulphide deposit, northern Finland. In: Maier, W.D., O'Brien, H., Lahtinen, R. (Eds.), Mineral Deposits of Finland. Elsevier, Amsterdam. pp. 211–251.

Cawood, P.A., Hawkesworth, C.J., 2015. Temporal relations between mineral deposits and global tectonic cycles. In: Jenkin, G.R.T., Lusty, P.A.J., Mcdonald, I., et al. (Eds.), Ore Deposits in an Evolving Earth. Geological Society, London. Special Publication 393, 9–21.

Cook, N.J., Sundblad, K., Valkama, M., et al., 2001. Indium mineralisation in A-type granites in southeastern Finland: Insights into mineralogy and partitioning between coexisting minerals. Chemical Geology 284, 62–73.

Coppard, J., Klatt, S., Ihlenfeld, C., 2013. The Sakatti Ni-Cu-PGE deposit in northern Finland. In: Hanski, E., Maier, W. (Eds.), Excursion Guidebook FINRUS, Ni-Cr-PGE deposits of Finland and the Kola Peninsula, the 12th SGA Biennial Meeting. Mineral Deposit Research for a High-Tech World, Geological Survey of Sweden, Uppsala, pp. 10–13.

Eilu, P., 2015. Introduction and overview on Finnish gold deposits. In: Maier, W.D., O'Brien, H., Lahtinen, R. (Eds.), Mineral Deposits of Finland. Elsevier, Amsterdam. pp. 377–403.
Eilu, P. (Ed.), 2012a. Mineral deposits and metallogeny of Fennoscandia. Geological Survey of Finland, Special Paper 53, 401.
Eilu, P., 2012b. Gold mineralisation in southwestern Finland. Geological Survey of Finland, Special Paper 52, 11–22.
Eilu, P., 2012c. The Haveri copper-gold deposit: Genetic considerations. Geological Survey of Finland, Special Paper 52, 255–266.
Eilu, P., Bergman, T., Bjerkgård, T., et al., 2013. Metallic Mineral Deposit Map of the Fennoscandian Shield 1:2,000,000. Revised edition (comp.). Geological Survey of Finland, Espoo.
Eilu, P., Kärkkäinen, N., 2012. Seinäjoki Au-Sb. In: Eilu, P. (Ed.), Mineral Deposits and Metallogeny of Fennoscandia. Geological Survey of Finland. Special Paper 53, 248–249.
Eilu, P., Pankka, H., 2009. FINGOLD—A public database on gold deposits in Finland, Version 1.0. Geological Survey of Finland. Digital data product 4 (CDROM optical disc).
Eilu, P., Pankka, H., Keinänen, V., et al., 2007. Characteristics of gold mineralisation in the greenstone belts of northern Finland. Geological Survey of Finland. Special Paper 44, 57–106.
Eilu, P., Sorjonen-Ward, P., Nurmi, P., Niiranen, T., 2003. A review of gold mineralization styles in Finland. Economic Geology 98, 1329–1353.
Eilu, P., Tontti, M., 2012. Orijärvi Zn-Cu. In: Eilu, P. (Ed.), Mineral Deposits and Metallogeny of Fennoscandia. Geological Survey of Finland. Special Paper 53, 209–212.
Ekdahl, E., 1993. Early Proterozoic Karelian and Svecofennianformations and the evolution of the Raahe-Ladoga Ore Zone, based on the Pielavesi area, central Finland. Geological Survey of Finland, Bulletin 373, 137.
Front, K., Vaarma, M., Rantala, E., Luukkonen, A., 1989. Early Proterozoic Nattanen-type granite complexes in central Finnish Lapland: Rock types, geochemistry and mineralization. Geological Survey of Finland, Report of Investigation 85, 77.
Gaál, G., 1990. Tectonic styles of Early Proterozoic ore deposition in the Fennoscandian Shield. Precambrian Research 46, 83–114.
Gaál, G., Isohanni, M., 1979. Characteristics of igneous intrusions and various wall rocks in some Precambrian porphyry copper-molybdenum deposits in Pohjanmaa, Finland. Economic Geology 74, 1198–1210.
Gautneb, H., Ahtola, T., Lintinen, P., et al., 2013. Industrial Mineral Deposit Map of the Fennoscandian Shield 1:2,000,000 (comp.). Geological Survey of Finland, Espoo.
Gehör, S., Havola, M., 1988. The depositional environment of the early Proterozoic Tuomivaara iron-formation and associated metasediments, eastern Finland. Geological Survey of Finland, Special Paper 5, 109–133.
Goldfarb, R.J., Groves, D.I., Gardoll, S., 2001. Orogenic gold and geologic time: a global synthesis. Ore Geology Reviews 18, 1–75.
Gor'kovets, V.Y., Raevskaya, M.B., Belousov, E.F., Inina, K.A., 1981. Geology and metallogeny of the Kostomuksha Iron Deposit. Karelian Research Centre of the USSR Academy of Sciences, Petrozavodsk. pp. 143 (in Russian).
Grönholm, S., Kärkkäinen, N. (Eds.), 2012. Gold in Southern Finland: Results of GTK studies 1998–2011. Geological Survey of Finland, Special Paper 52, 276 p.
Groves, D.I., Bierlein, F.P., Meinert, L.D., Hitzman, M.W., 2010. Iron oxide copper-gold (IOCG) deposits through Earth history: Implications for origin, lithospheric setting, and distinction from other epigenetic iron oxide deposits. Economic Geology 105, 641–654.
Haapala, I., 1995. Metallogeny of rapakivi granites. Mineralogy and Petrology 54, 149–160.
Haapala, I., Lukkari, S., 2005. Petrological and geochemical evolution of the Kymi stock, a topaz granite cupola within the Wiborg rapakivi batholith, Finland. Lithos 80, 347–362.
Haapala, I., Rämö, O.T., 2015. Deposits related to granites. In: Maier, W.D., O'Brien, H., Lahtinen, R. (Eds.), Mineral Deposits of Finland. Elsevier, Amsterdam. pp. 531–551.
Häkli, T.A., 1963. Distribution of nickel between the silicate and sulphide phases in some basic intrusions in Finland. Bulletin de la Commission Géologique de Finlande 209, 54.

Häkli, T.A., 1971. Silicate nickel and its application to the exploration of nickel ores. Bulletin of the Geological Society of Finland 43, 247–263.

Halkoaho, T., Alapieti, T., Lahtinen, J., 1990. The Somppujärvi PGE reef in the Penikat layered intrusion, northern Finland. Mineralogy and Petrology 42, 39–55.

Halla, J., 2005. Late Archean high-Mg granitoids (sanukitoids) in the southern Karelian domain, eastern Finland: Pb and Nd isotopic constraints on crust-mantle interactions. Lithos 79, 161–178.

Hanski, E., 1997. The Nuttio serpentinite belt, central Lapland: An example of Paleoproterozoic ophiolitic mantle rocks in Finland. Ofioliti 22, 35–46.

Hanski, E., 2012. Evolution of the Palaeoproterozoic (2.50–1.95 Ga) nonorogenic magmatism in the eastern part of the Fennoscandian Shield. In: Melezhik, V., Prave, A., Hanski, E., et al. (Eds.), Reading the Archive of Earth's Oxygenation, Volume 1: The Palaeoproterozoic of Fennoscandia as context for the Fennoscandian Arctic Russia—Drilling Early Earth Project. Springer-Verlag, Berlin/Heidelberg, pp. 179–245.

Hanski, E., Huhma, H., 2005. Central Lapland Greenstone Belt. In: Lehtinen, M., Nurmi, P.A., Rämö, O.T. (Eds.), Precambrian Bedrock of Finland—Key to the Evolution of the Fennoscandian Shield. Elsevier, Amsterdam, pp. 139–194.

Hanski, E., Huhma, H., Suominen, I.M., Walker, R.J., 1997. Geochemical and isotopic (Os, Nd) study of the Keivitsa intrusion and its Cu-Ni deposit, northern Finland. In: Papunen, H. (Ed.), Mineral Deposits: Research and Exploration—Where Do They Meet? Proceedings of the Fourth Biennial SGA Meeting, Turku/Finland, August 11–13. A.A. Balkema, Rotterdam, pp. 435–438.

Hanski, E., Huhma, H., Rastas, P., Kamenetsky, V.S., 2001a. The Palaeoproterozoic komatiite-picrite association of Finnish Lapland. Journal of Petrology 42, 855–876.

Hanski, E.J., Melezhik, V.A., 2012. Litho- and chronostratigraphy of the Karelian formations. In: Melezhik, V., Prave, A., Hanski, E., et al. (Eds.), Reading the Archive of Earth's Oxygenation, Volume 1: The Palaeoproterozoic of Fennoscandia as Context for the Fennoscandian Arctic Russia—Drilling Early Earth Project. Springer-Verlag, Berlin/Heidelberg, pp. 39–110.

Hanski, E., Walker, R.J., Huhma, H., Suominen, I., 2001b. The Os and Nd isotopic systematics of the 2.44 Ga Akanvaara and Koitelainen mafic layered intrusions in northern Finland. Precambrian Research 109, 73–102.

Härkönen, I., 1984. The gold-bearing conglomerates of Kaarestunturi, central Finnish Lapland. In: Foster, R.P. (Ed.). Gold '82: The Geology, Geochemistry and Genesis of Gold Deposits, 1. Geological Society of Zimbabwe, Special Publication, pp. 239–247.

Heilimo, E., Halla, J., Lauri, L.S., et al., 2009. The Paleoproterozoic Nattanen-type granites in northern Finland and vicinity—a postcollisional oxidized A-type suite. Bulletin of the Geological Society of Finland 81, 7–38.

Hiltunen, A., 1982. The Precambrian geology and skarn iron ores of the Rautuvaara area, northern Finland. Geological Survey of Finland, Bulletin 318, 133.

Hietanen, A., 1975. Generation of potassium poor magmas in the northern Sierra Nevada and the Svecofennian in Finland. Journal Research U.S. Geological Survey 3, 631–645.

Himmi, R., Huhma, M., Häkli, T., 1979. Mineralogy and metal distribution in the copper-tungsten deposit at Ylöjärvi, southwest Finland. Economic Geology 74, 1183–1197.

Hitzman, M.W., Selley, D., Bull, S., 2010. Formation of sedimentary rock-hosted stratiform copper deposits through Earth history. Economic Geology 105, 627–639.

Holland, H.D., 2006. The oxygenation of the atmosphere and oceans. Philosophical Transactions of the Royal Society. Biological Sciences 361, 903–915.

Hölttä, P., Heilimo, E., Huhma, H., et al., 2012a. Archaean complexes of the Karelia Province in Finland. Geological Survey of Finland, Special Paper 54, 9–20.

Hölttä, P., Heilimo, E., Huhma, H., et al., 2012b. The Archaean of the Karelia Province in Finland. Geological Survey of Finland, Special Paper 54, 21–73.

Huhma, H., 1986. Sm-Nd, U-Pb and Pb-Pb isotopic evidence for the origin of the early Proterozoic Svecokarelian crust in Finland. Geological Survey of Finland, Bulletin 337, 48.

Huhma, H., Kontinen, A., Mikkola, P., et al., 2012b. Nd isotopic evidence for Archaean crustal growth in Finland. Geological Survey of Finland, Special Paper 54, 176–213.

Huhma, H., Mänttäri, I., Peltonen, P., et al., 2012a. The age of the Archean schist belts in Finland. Geological Survey of Finland, Special Paper 54, 74–175.

Huhma, H., O'Brien, H., Lahaye, Y., Mänttäri, I., 2011. Isotope geology and Fennoscandian lithosphere evolution. Geological Survey of Finland, Special Paper 49, 35–48.

Huhtelin, T., 2015. The Kemi Cr deposit. In: Maier, W.D., O'Brien, H., Lahtinen, R. (Eds.), Mineral Deposits of Finland. Elsevier, Amsterdam. pp. 165–177.

Iljina, M., 2005. Portimo layered Igneous complex. In: Alapieti, T.T., Kärki, A.J. (Eds.), Field Trip Guidebook: Early Proterozoic (2.5–2.4 Ga) Tornio-Näränkävaara Layered Intrusion Belt and Related Chrome and Platinum-Group Element Mineralization. Geological Survey of Finland, Northern Finland, pp. 77–100. Guide 51a.

Iljina, M., Hanski, E., 2005. Layered mafic intrusions of the Tornio-Näränkävaara belt. In: Lehtinen, M., Nurmi, P.A., Rämö, O.T. (Eds.), Precambrian Bedrock of Finland—Key to the Evolution of the Fennoscandian Shield. Elsevier, Amsterdam, pp. 103–138.

Iljina, M., Maier, W.D., Karinen, T., 2015. PGE-(Cu-Ni) deposits of the Tornio-Näränkavaara belt of intrusions (Portimo, Penikat and Koillismaa). In: Maier, W.D., O'Brien, H., Lahtinen, R. (Eds.), Mineral Deposits of Finland. Elsevier, Amsterdam. pp. 133–162.

Isohanni, M., 1985. The Oravainen nickel occurrence in western Finland. Geological Survey of Finland, Bulletin 333, 189–210.

Isokangas, P., 1978. Finland. In: Bowie, S.H.U., Kvalheim, A., Haslam, H.W. (Eds.), Mineral Deposits of Europe, Volume 1—Northwest Europe. The Institution of Mining and Metallurgy and the Mineralogical Society, London, pp. 39–92.

Kähkönen, Y., 2005. Svecofennian supracrustal rocks. In: Lehtinen, M., Nurmi, P.A., Rämö, O.T. (Eds.), Precambrian Geology of Finland—Key to the Evolution of the Fennoscandian Shield. Elsevier, Amsterdam, pp. 343–406.

Kahma, A., 1979. The main metallogenic features of Finland. Geological Survey of Finland, Bulletin 265, 28.

Karhu, J.A., 2005. Paleoproterozoic carbon isotope excursion. In: Lehtinen, M., Nurmi, P.A., Rämö, O.T. (Eds.), Precambrian Geology of Finland—Key to the Evolution of the Fennoscandian Shield. Elsevier, Amsterdam, pp. 669–680.

Karinen, T., Hanski, E., Taipale, A., 2015. The Mustavaara Fe-Ti-V oxide deposit. In: Maier, W.D., O'Brien, H., Lahtinen, R. (Eds.), Mineral Deposits of Finland. Elsevier, Amsterdam. pp. 179–192.

Kärkkäinen, N., 1999. The age of the Koivusaarenneva ilmenite gabbro, western Finland. Geological Survey of Finland, Special Paper 27, 35–37.

Kärkkäinen, N., Bornhorst, T.J., 2003. The Svecofennian gabbro-hosted Koivusaarenneva magmatic ilmenite deposit, Kälviä, Finland. Mineralium Deposita 38, 169–184.

Karvinen, A., Kojonen, K., Johanson, B., 1988. Geology and mineralogy of the Karhujupukka Ti-Fe deposit in Kolari, northern Finland. Geological Survey of Finland, Special Paper 10, 95–99.

Kerrich, R., Goldfarb, R.J., Richards, J.P., 2005. Metallogenic provinces in an evolving geodynamic framework. Economic Geology 100th Anniversary Volume 1097–1136.

Kinnunen, A., 2008. A Palaeoproterozoic high-sulphidation epithermal gold deposit at Orivesi, southern Finland. Ph.D. thesis, Acta Universitatis Ouluensis, Series A. Scientiae Rerum Naturalium 507, 183.

Kohonen, J., Rämö, O.T., 2005. Sedimentary rocks, diabases, and late cratonic evolution. In: Lehtinen, M., Nurmi, P.A., Rämö, O.T. (Eds.), Precambrian Geology of Finland—Key to the Evolution of the Fennoscandian Shield. Elsevier, Amsterdam, pp. 563–604.

Konnunaho, J., Halkoaho, T., Hanski, E., Törmänen, T., 2015. Komatiite-hosted Ni-Cu-PGE deposits in Finland. In: Maier, W.D., O'Brien, H., Lahtinen, R. (Eds.), Mineral Deposits of Finland. Elsevier, Amsterdam. pp. 93–127.

Konnunaho, J., Hanski, E.H., Bekker, A., et al., 2013. The Archean komatiite-hosted, PGE-bearing Ni-Cu sulfide deposit at Vaara, eastern Finland: Evidence for assimilation of external sulfur and post-depositional desulfurization. Mineralium Deposita 48, 967–989.

Kontinen, A., 1987. An early Proterozoic ophiolite—the Jormua mafic-ultramafic complex, northeastern Finland. Precambrian Research 35, 313–341.

Kontinen, A., Hanski, E., 2015. The Talvivaara black shale-hosted Ni-Zn-Cu-Co deposit in eastern Finland. In: Maier, W.D., O'Brien, H., Lahtinen, R. (Eds.), Mineral Deposits of Finland. Elsevier, Amsterdam. pp. 557–607.

Kontinen, A., Käpyaho, A., Huhma, H., Karhu, et al., 2007. Nurmes paragneisses in eastern Finland, Karelian craton: Provenance, tectonic setting and implications for Neoarchaean craton correlation. Precambrian Research 152, 119–148.

Kontinen, A., Paavola, J., Lukkarinen, H., 1992. K-Ar ages of hornblende and biotite from Late Archaean rocks of eastern Finland—interpretation and discussion of tectonic implications. Geological Survey of Finland, Bulletin 365, 31.

Kontoniemi, O., 1998. Geological setting and characteristics of the Palaeoproterozoic tonalite-hosted Osikonmaki gold deposit, southeastern Finland. Geological Survey of Finland, Special Paper 25, 39–80.

Kopperoinen, T., Tuokko, I., 1988. The Ala-Luoma and Taivaljärvi Zn-Pb-Ag-Au deposits, eastern Finland. Geological Survey of Finland, Special Paper 4, 131–144.

Korkalo, T., 2006. Gold and copper deposits in central Lapland, northern Finland, with special reference to their exploration and exploitation. Acta Universita Ouluensis, Series A. Scientiae Rerum Naturalium 461, 122.

Korkiakoski, E., 1992. Geology and geochemistry of the metakomatiite-hosted Pahtavaara gold deposit in Sodankylä, northern Finland, with emphasis on hydrothermal alteration. Geological Survey of Finland, Bulletin 360, 96.

Korsakova, M., 2012. Northern Ladoga Sn-Zn-Pb, U, Au, W. In: Eilu, P. (Ed.), Mineral Deposits and Metallogeny of Fennoscandia. Geological Survey of Finland, Special Paper 53, 389–390.

Kuivasaari, T., Torppa, A., Äikäs, O., Eilu, P., 2012. Otanmäki V-Ti-Fe. In: Eilu, P. (Ed.), Mineral Deposits and Metallogeny of Fennoscandia. Geological Survey of Finland, Special Paper 53, 272–275.

Laajoki, K., 2005. Karelian supracrustal rocks. In: Lehtinen, M., Nurmi, P.A., Rämö, O.T. (Eds.), Precambrian Geology of Finland—Key to the Evolution of the Fennoscandian Shield. Elsevier, Amsterdam, pp. 279–342.

Laajoki, K., Saikkonen, R., 1977. On the geology and geochemistry of the Precambrian iron formations in Väyrylänkylä, South Puolanka area, Finland. Geological Survey of Finland, Bulletin 292, 137.

Lahtinen, R., Huhma, H., 1997. Isotopic and geochemical constraints on the evolution of the 1.93–1.79 Ga Svecofennia crust and mantle in Finland. Precambrian Research 82, 13–34.

Lahtinen, R., Huhma, H., Kontinen, A., et al., 2010. New constraints for the source characteristics, deposition and age of the 2.1–1.9 Ga metasedimentary cover at the western margin of the Karelian Province. Precambrian Research 176, 77–93.

Lahtinen, R., Korja, A., Nironen, M., Heikkinen, P., 2009. Palaeoproterozoic accretionary processes in Fennoscandia. In: Cawood, P.A., Kröner, A. (Eds.), Earth Accretionary Systems in Space and Time. The Geological Society, Special Publication. 318, 237–256.

Lahtinen, R., Korja, A., Nironen, M., 2005. Palaeoproterozoic tectonic evolution. In: Lehtinen, M., Nurmi, P.A., Rämö, O.T. (Eds.), Precambrian Geology of Finland—Key to the Evolution of the Fennoscandian Shield. Elsevier, Amsterdam, pp. 481–532.

Lamberg, P., 2005. From genetic concepts to practice—petrochemical identification of Ni-Cu mineralised intrusions and localization of the ore. Geological Survey of Finland, Bulletin 402, 266.

Latvalahti, U., 1979. Cu-Zn-Pb ores in the Aijala-Orijarvi area, southwestern Finland. Economic Geology 74, 1035–1059.

Lauri, L.S., Andersen, T., Hölttä, P., et al., 2011. Evolution of the Archaean Karelian Province in the Fennoscandian Shield in the light of U–Pb zircon ages and Sm–Nd and Lu–Hf isotope systematics. Journal of the Geological Society 168, 201–218.

Lauri, L.S., Rämö, O.T., Huhma, H., et al., 2006. Petrogenesis of silicic magmatism related to the ~2.44 Ga rifting of Archean crust in Koillismaa, eastern Finland, Lithos 86, 137–166.

Laznicka, P., 2010. Giant Metallic Deposits. Future Sources of Industrial Metals. Springer, Berlin. p. 950.

Lehtinen, M., 2015. Industrial minerals and rocks in Finland. In: Maier, W.D., O'Brien, H., Lahtinen, R. (Eds.), Mineral Deposits of Finland. Elsevier, Amsterdam. pp. 685–705.

Lehtinen, M., Nurmi, P.A., Rämö, O.T. (Eds.), 2005. Precambrian Bedrock of Finland—Key to the Evolution of the Fennoscandian Shield. Elsevier, Amsterdam, p. 736.

Lehto, T., Niiniskorpi, V., 1977. The iron formations of northern and eastern Finland. Geological Survey of Finland, Report of Investigation 22, 49 (in Finnish with English summary).

Liipo, J., Laajoki, K., 1991. Mineralogy, geochemistry and metamorphism of the early Proterozoic Vähäjoki iron ores, northern Finland. Bulletin of the Geological Society of Finland 63, 69–85.

Lindborg, T., Papunen, H., Parkkinen, J., Tuokko, I., 2015. The Taivaljärvi (Silver Mine) Ag-Au-Pb-Zn deposit in Archaean greenstone belt, eastern Finland. In: Maier, W.D., O'Brien, H., Lahtinen, R. (Eds.), Mineral Deposits of Finland. Elsevier, Amsterdam. pp. 633–655.

Lindholm, O., Anttonen, R., 1980. Geology of the Otanmäki Mine. In: Häkli, T.A. (Ed.), Precambrian Ores of Finland: Guide to Excursions 078 A + C, Part 2 (Finland). Proceedings of the 26th International Geological Congress. Geological Survey of Finland, Espoo, Paris, pp. 25–33.

Loukola-Ruskeeniemi, K., Gaál, G., Karppanen, T., 1991. Geochemistry, structure and genesis of the Hammaslahti copper mine-explorational tools for a sediment-hosted massive sulphide deposit. Geological Survey of Finland, Special Paper 12, 101–106.

Loukola-Ruskeeniemi, K., Lahtinen, H., 2012. Multiphase evolution in the black-shale-hosted Ni-Cu-Zn-Co deposit at Talvivaara, Finland. Ore Geology Reviews 52, 85–99.

Luukkonen, A., 1994. Main geochemical features, metallogeny and hydrothermal alteration phenomena of certain gold and gold-tin-tungsten prospects in southern Finland. Geological Survey of Finland, Bulletin 377, 153.

Maier, W.D., Peltonen, P., Halkoaho, T., Hanski, E., 2013. Geochemistry of komatiites from the Tipasjärvi, Kuhmo, Suomussalmi, Ilomantsi and Tulppio greenstone belts, Finland: Implications for tectonic setting and Ni sulphide prospectivity. Precambrian Research 228, 63–84.

Mäkelä, U., 1989. Geological and geochemical environments of Precambrian sulphide deposits in southwestern Finland. Annales Academiae Scientiarum Fennicae, Series AIII. Geologica-Geographica 151, 102.

Mäki, T., Osorio, M.I., Kousa, J., Luukas, J., 2015. The Pyhäsalmi Mine. In: Maier, W.D., O'Brien, H., Lahtinen, R. (Eds.), Mineral Deposits of Finland. Elsevier, Amsterdam. pp. 507–528.

Mäki, T., Puustjärvi, H., 2003. The Pyhäsalmi massive Zn-Cu-pyrite deposit, middle Finland—a Palaeoproterozoic VMS-class "giant.". In: Kelly, J.G., Andrew, C.J., Ashton, J.H., et al. (Eds.). Europe's Major Base Metal Deposits. The Irish Association for Economic Geology, Dublin, pp. 87–100.

Mäkinen, J., 1987. Geochemical characteristics of Svecokarelidic mafic-ultramafic intrusions associated with Ni-Cu occurrences in Finland. Geological Survey of Finland, Bulletin 342, 109.

Makkonen, H., 2015. Ni deposits of the Vammala and Kotalahti belt. In: Maier, W.D., O'Brien, H., Lahtinen, R. (Eds.), Mineral Deposits of Finland. Elsevier, Amsterdam, pp. 253–285.

Makkonen, H.V., 1996. 1.9 Ga tholeiitic magmatism and related Ni-Cu deposition in the Juva area, SE Finland. Geological Survey of Finland, Bulletin 386, 101.

Makkonen, H., Huhma, H., 2007. Sm-Nd data for mafic-ultramafic intrusions in the Svecofennian (1.88 Ga) Kotalahti Nickel Belt, Finland–implications for crustal contamination at the Archaean/Proterozoic boundary. Bulletin of the Geological Society of Finland 79, 175–201.

Makkonen, H.V., Mäkinen, J., Kontoniemi, O., 2008. Geochemical discrimination between barren and mineralized intrusions in the Svecofennian (1.9 Ga) Kotalahti Nickel Belt, Finland. Ore Geology Reviews 33, 101–114.

Mänttäri, I., Hölttä, P., 2002. U–Pb dating of zircons and monazites from Archean granulites in Varpaisjärvi, Central Finland: Evidence for multiple metamorphism and Neoarchean terrane accretion. Precambrian Research 118, 101–131.

Marmo, J.S., Ojakangas, R.W., 1984. Lower Proterozoic glaciogenic deposits, eastern Finland. Geological Society of America Bulletin 95, 1055–1062.

Melezhik, V.A., Fallick, A.E., Brasier, A.T., Salminen, P.E., 2013a. Lomagundi-Jatuli excursion as seen from the Fennoscandian Shield record. In: Melezhik, V., Kump, L., Fallick, A., et al. (Eds.), Reading the Archive of Earth's Oxygenation, Volume 3: Global Events and the Fennoscandian Arctic Russia—Drilling Early Earth Project. Springer-Verlag, Berlin/Heidelberg, pp. 1117–1123.

Melezhik, V.A., Fallick, A.F., Filippov, M.M., et al., 2013b. Giant Palaeoproterozoic petrified oil field in the Onega Basin. In: Melezhik, V., Kump, L., Fallick, A., et al. (Eds.), Reading the Archive of Earth's Oxygenation, Volume 3: Global Events and the Fennoscandian Arctic Russia—Drilling Early Earth Project. Springer-Verlag, Berlin/Heidelberg, pp. 1202–1212.

Melezhik, V.A., Kump, L.R., Hanski, E., et al., 2012. Tectonic evolution and major global earth-surface palaeoenvironmental events in the Palaeoproterozoic. In: Melezhik, V.A., Prave, A.R., Hanski, E.J., et al. (Eds.), Reading the Archive of Earth's Oxygenation, Volume 1: The Palaeoproterozoic of Fennoscandia as Context for the Fennoscandian Arctic Russia—Drilling Early Earth Project. Springer-Verlag, Berlin/Heidelberg, pp. 3–21.

Mikkola, P., Huhma, H., Heilimo, E., Whitehouse, M., 2011. Archean crustal evolution of the Suomussalmi district as part of the Kianta complex, Karelia: Constraints from geochemistry and isotopes of granitoids. Lithos 125, 287–307.

Moilanen, M., Peltonen, P., 2015. Hannukainen Fe-Cu-Au deposit, western Finnish Lapland: Deposit model updated. In: Maier, W.D., O'Brien, H., Lahtinen, R. (Eds.), Mineral Deposits of Finland. Elsevier, Amsterdam. pp. 485–504.

Morozov, A.F., Hakhaev, B.N., Petrov, O.V., et al., 2010. Rock-salts in Palaeoproterozoic strata of the Onega depression of Karelia (based on data from the Onega parametric drillhole). Transactions of the Academy of Sciences 435, 230–233.

Müller, S.G., Krapež, B., Barley, M.E., Fletcher, I.R., 2005. Giant iron-ore deposits of the Hamersley province related to the breakup of Paleoproterozoic Australia: New insights from in situ SHRIMP dating of baddeleyite from mafic intrusions. Geology 33, 577–580.

Muntean, J.L., Cline, J.S., Simon, A.C., Longo, A.A., 2011. Magmatic–hydrothermal origin of Nevada's Carlin-type gold deposits. Nature Geoscience 4, 122–127.

Mutanen, T., 1997. Geology and ore petrology of the Akanvaara and Koitelainen mafic layered intrusions and the Keivitsa-Satovaara layered complex, northern Finland. Geological Survey of Finland, Bulletin 395, 233.

Mutanen, T., Huhma, H., 2003. The 3.5 Ga Siurua trondhjemite gneiss in the Archaean Pudasjärvi Granulite Belt, northern Finland. Bulletin of the Geological Society of Finland 75, 51–68.

Naldrett, A.J., 2004. Magmatic Sulfide Deposits: Geology, Geochemistry and Exploration. Springer, Berlin. 728.

Niiranen, T., Hanski, E., Eilu, P., 2003. General geology, alteration, and iron deposits in the Palaeoproterozoic Misi region, northern Finland. Bulletin of the Geological Society of Finland 75, 69–92.

Niiranen, T., Mänttäri, I., Poutiainen, M., et al., 2005. Genesis of the early Proterozoic iron skarns in Misi region, northern Finland. Mineralium Deposita 40, 192–217.

Niiranen, T., Poutiainen, M., Mänttäri, I., 2007. Geology, geochemistry, fluid inclusion characteristics, and U-Pb age studies on iron oxide–Cu-Au deposits in the Kolari region, northern Finland. Ore Geology Reviews 30, 75–105.

Nironen, M., 2005. Proterozoic orogenic granitoid rocks. In: Lehtinen, M., Nurmi, P.A., Rämö, O.T. (Eds.), Precambrian Geology of Finland—Key to the Evolution of the Fennoscandian Shield. Elsevier, Amsterdam, pp. 443–480.

Nurmi, P.A., Front, K., Lampio, E., Nironen, M., 1984. Svecokarelian porphyry-type molybdenum and copper occurrences in southem Finland: Their granitoid host rocks and lithogeochemical exploration. Geogical Survey of Finland, Report of Investigation 67, 88 (in Finnish with English summary).

Nurmi, P., Sorjonen-Ward, P. (Eds.), 1993. Geological development, gold mineralization and exploration methods in the Late Archean Hattu schist belt, Ilomantsi, eastern Finland. Geological Survey of Finland, p. 386. Special Paper 17.

Nurmi, P.A., Sorjonen-Ward, P., Damstén, M., 1993. Geological setting, characteristics and exploration history of mesothermal gold occurrences in the late Archean Hattu schist belt, Ilomantsi, eastern Finland. Geological Survey of Finland, Special Paper 17, 193–223.

O'Brien, H.E., 2015. Kimberlite-hosted diamond deposits. In: Maier, W.D., O'Brien, H., Lahtinen, R. (Eds.), Mineral Deposits of Finland. Elsevier, Amsterdam. pp. 345–366.

O'Brien, H.E., Heilimo, E., 2015. The Sillinjärvi carbonatite. In: Maier, W.D., O'Brien, H., Lahtinen, R. (Eds.), Mineral Deposits of Finland. Elsevier, Amsterdam. pp. 327–342.

O'Brien, H.E., Lee, M.J., 2015. The Sokli carbonatite. In: Maier, W.D., O'Brien, H., Lahtinen, R. (Eds.), Mineral deposits of Finland. Elsevier, Amsterdam. pp. 305–323.

O'Brien, H.E., Peltonen, P., Vartiainen, H., 2005. Kimberlites, carbonatites, amd alkaline rocks. In: Lehtinen, M., Nurmi, P.A., Rämö, O.T. (Eds.), Precambrian Geology of Finland—Key to the Evolution of the Fennoscandian Shield. Elsevier, Amsterdam, pp. 605–644.

O'Brien, H.O., Phillips, D., Spencer, R., 2007. Isotopic ages of Lentiira-Kuhmo-Kostomuksha olivine lamproite—Group II kimberlites. Bulletin of the Geological Society of Finland 79, 203–215.

Paakkola, J., 1971. The volcanic complex and associated manganiferous iron formation of the Porkonen–Pahtavaara area in Finnish Lapland. Geological Survey of Finland, Bulletin 247, 83.

Pääkkönen, V., 1966. On the geology and mineralogy of the occurrence of native antimony at Seinäjoki, Finland. Bulletin de la Commission Géologique de Finlande 225, 71.

Papunen, H., 1987. Outokumpu-type ores. Geological Survey of Finland, Special Paper 1, 41–50.

Papunen, H., Gorbunov, G.I. (Eds.), 1985. Nickel–copper deposits of the Baltic Shield and Scandinavian Caledonides. Geological Survey of Finland, p. 394. Bulletin 333.

Papunen, H., Halkoaho, T., Luukkonen, E., 2009. Archean evolution of the Tipasjärvi-Kuhmo-Suomussalmi greenstone complex, Finland. Geological Survey of Finland, Bulletin 403, 68.

Papunen, H., Kopperoinen, T., Tuokko, I., 1989. The Taivaljärvi Ag-Zn deposit in the Archean greenstone belt, eastern Finland. Economic Geology 84, 1262–1276.

Papunen, H., Lindsjö, O., 1972. Apatite, monazite and allanite: three rare earth minerals from Korsnäs, Finland. Bulletin of the Geological Society of Finland 44, 123–129.

Patison, N.L., 2007. Structural controls on gold mineralisation in the Central Lapland Greenstone Belt. Geological Survey of Finland, Special Paper 44, 107–124.

Patison, N.L., Välimaa, J., Koppström, K., Kortelainen, V.J., 2015. Suurikuusikko gold deposit (Kittilä Mine), northern Finland. In: Maier, W.D., O'Brien, H., Lahtinen, R. (Eds.), Mineral Deposits of Finland. Elsevier, Amsterdam. pp. 411–430.

Peltola, E., 1978. Origin of Precambrian copper sulfides of the Outokumpu district, Finland. Economic Geology 73, 461–477.

Peltonen, P., 1995. Magma-country rock interaction and the genesis of Ni-Cu deposits in the Vammala Nickel Belt, SW Finland. Mineralogy and Petrology 52, 1–24.

Peltonen, P., 2005a. Ophiolites. In: Lehtinen, M., Nurmi, P.A., Rämö, O.T. (Eds.), Precambrian Geology of Finland—Key to the Evolution of the Fennoscandian Shield. Elsevier, Amsterdam, pp. 237–278.

Peltonen, P., 2005b. Svecofennian mafic–ultramafic intrusions. In: Lehtinen, M., Nurmi, P.A., Rämö, O.T. (Eds.), Precambrian Geology of Finland—Key to the Evolution of the Fennoscandian Shield. Elsevier, Amsterdam, pp. 407–442.

Peltonen, P., Kontinen, A., Huhma, H., Kuronen, U., 2008. Outokumpu revisited: new mineral deposit model for the mantle peridotite-associated Cu-Co-Zn-Ni-Ag-Au sulphide deposits. Ore Geology Reviews 33, 559–617.

Piirainen, T., 1968. Die Petrologie und die Uranlagerstätten des Koli-Kaltimogebiets im Finnischen Nordkarelien. Geological Survey of Finland, Bulletin 237, 99.

Pirajno, F., 2009. Hydrothermal Processes and Mineral Systems. Springer, Berlin. 1250.

Pohjolainen, E., 2015. Uranium deposits of Finland. In: Maier, W.D., O'Brien, H., Lahtinen, R. (Eds.), Mineral Deposits of Finland. Elsevier, Amsterdam. pp. 659–681.

Poutiainen, M., Grönholm, P., 1996. Hydrothermal fluid evolution of the Palaeoproterozoic Kutemajärvi gold-telluride deposit, SW Finland. Economic Geology 91, 1335–1353.

Puchtel, I.S., Hofmann, A.W., Mezger, K., et al., 1998. Oceanic plateau model for continental crustal growth in the Archaean: A case study from the Kostomuksha greenstone belt, NW Baltic Shield. Earth and Planetary Science Letters 155, 57–74.

Rämö, O.T., Haapala, I., 1995. One hundred years of rapakivi granite. Mineralogy and Petrology 52, 129–185.

Rämö, O.T., Haapala, I., 2005. Rapakivi granites. In: Lehtinen, M., Nurmi, P.A., Rämö, O.T. (Eds.), Precambrian Geology of Finland—Key to the Evolution of the Fennoscandian Shield. Elsevier, Amsterdam, pp. 533–562.

Robb, L., 2005. Introduction to Ore-Forming Processes. Blackwell Publishing, Oxford. 373.

Roberts, M.D., Oliver, N.H.S., Lahtinen, R., 2004. Geology, lithogeochemistry and paleotectonic setting of the host sequence to the Kangasjärvi Zn-Cu deposit, central Finland: Implications for volcanogenic massive sulphide exploration in the Vihanti–Pyhäsalmi district. Bulletin of the Geological Society of Finland 76, 31–62.

Rouhunkoski, P., Isokangas, P., 1974. The copper-gold vein deposit of Kivimaa at Tervola, N-Finland. Bulletin of the Geological Society of Finland 46, 29–35.

Rutland, R.W.R., Williams, I.S., Korsman, K., 2004. Pre-1.91 Ga deformation and metamorphism in the Palaeoproterozoic Vammala migmatite belt, southern Finland, and implications for Svecofennian tectonics. Bulletin of the Geological Society of Finland 76, 93–140.

Saalmann, K., Mänttäri, I., Peltonen, P., et al., 2010. Geochronology and structural relationships of mesothermal gold mineralization in the Palaeoproterozoic Jokisivu prospect, southern Finland. Geological Magazine 147, 551–569.

Santaguida, F., Luolavirta, K., Lappalainen, M., et al., 2015. The Kevitsa Ni-Cu-PGE deposit, Central Lapland Greenstone Belt, Finland. In: Maier, W.D., O'Brien, H., Lahtinen, R. (Eds.), Mineral Deposits of Finland. Elsevier, Amsterdam. pp. 195–209.

Sarapää, O., Kärkkäinen, N., Ahtola, T., Al-Ani, T., 2015. Hi-tech metals in Finland. In: Maier, W.D., O'Brien, H., Lahtinen, R. (Eds.), Mineral Deposits of Finland. Elsevier, Amsterdam. pp. 613–629.

Simmons, S.F., White, N.C., John, D.A., 2005. Geological characteristics of epithermal precious and base metal deposits. Economic Geology 100th Anniversary Volume 485–522.

Sipilä, P., Ervamaa, P., Papunen, H., 1985. The Petolahti nickel-copper occurrence. Geological Survey of Finland, Bulletin 333, 293–303.

Sorjonen-Ward, P., 2012. Huhus Fe. In: Eilu, P. (Ed.), Mineral Deposits and Metallogeny of Fennoscandia. Geological Survey of Finland Special Paper 53, 281–284.

Sorjonen-Ward, P., Luukkonen, E.J., 2005. Archean rocks. In: Lehtinen, M., Nurmi, P.A., Rämö, O.T. (Eds.), Precambrian Geology of Finland—Key to the Evolution of the Fennoscandian Shield. Elsevier, Amsterdam, pp. 19–99.

Sorjonen-Ward, P., Hartikainen, A., Liikanen, J., et al., 2015. Exploration targeting and geological context of gold mineralization in the Neoarchean Ilomantsi greenstone belt, eastern Finland. In: Maier, W.D., O'Brien, H., Lahtinen, R. (Eds.), Mineral Deposits of Finland. Elsevier, Amsterdam. pp. 435–464.

Strauss, H., Melezhik, V.A., Reuschel, M., et al., 2013. Abundant marine calcium sulphates: Radical change of seawater sulphate reservoir and sulphur cycle. In: Melezhik, V., Kump, L., Fallick, A., et al. (Eds.), Reading the Archive of Earth's Oxygenation, Volume 3: Global Events and the Fennoscandian Arctic Russia—Drilling Early Earth Project. Springer-Verlag, Berlin/Heidelberg, pp. 1169–1194.

Sundblad, K., 2003. Metallogeny of gold in the Precambrian of northern Europe. Economic Geology 98, 1271–1309.

Trendall, A.F., Blockley, J.G., 2004. Precambrian iron-formation. In: Eriksson, P.G., Altermann, W., Nelson, D.R., et al. (Eds.), The Precambrian Earth: Tempos and Events. Elsevier, Amsterdam, pp. 403–420.

Tsuru, A., Walker, R., Kontinen, A., et al., 2000. Re-Os isotopic systematics of the 1.95 Ga Jormua ophiolite complex, northeastern Finland. Chemical Geology 164, 123–141.

Vaasjoki, M., Huhma, H., Lahtinen, R., Vestin, J., 2003. Sources of Svecofennian granitoids in the light of ion probe U-Pb measurements on their zircons. Precambrian Research 121, 251–262.

Vaasjoki, M., Korsman, K., Koistinen, T., 2005. Overview. In: Lehtinen, M., Nurmi, P.A., Rämö, O.T. (Eds.), The Precambrian Geology of Finland—Key to the Evolution of the Fennoscandian Shield. Elsevier, Amsterdam, pp. 1–18.

Vaasjoki, M., Sakko, M., 1989. The radiometric age of the Virmaila diabase dyke: Evidence for 20 Ma of continental rifting in Padasjoki, southern Finland. Geological Survey of Finland, Special Paper 10, 43–44

Vanhanen, E., 2001. Geology, mineralogy and geochemistry of the Fe-Co-Au-(U) deposits in the Paleoproterozoic Kuusamo schist belt, northeastern Finland. Geological Survey of Finland, Bulletin 399, 229.

Vanhanen, E., Cook, N.D.J., Hudson, M.R., et al., 2015. Rompas prospect, Peräpohja schist belt, Northern Finland. In: Maier, W.D., O'Brien, H., Lahtinen, R. (Eds.), Mineral Deposits of Finland. Elsevier, Amsterdam. pp. 467–484.

Västi, K., 2012. Vihanti–Pyhäsalmi Zn-Cu. In: Eilu, P. (Ed.), Mineral Deposits and Metallogeny of Fennoscandia. Geological Survey of Finland, Special Paper 53, 272–275.

Västi, K., Nikander, J., Kontoniemi, O., Eilu, P., 2012. Laivakangas Au. In: Eilu, P. (Ed.), Mineral Deposits and Metallogeny of Fennoscandia. Geological Survey of Finland, Special Paper 53, 267–269.

Vuollo, J., Huhma, H., 2005. Paleoproterozoic mafic dikes in NE Finland. In: Lehtinen, M., Nurmi, P.A., Rämö, O.T. (Eds.), Precambrian Geology of Finland—Key to the Evolution of the Fennoscandian Shield. Elsevier, Amsterdam, pp. 195–278.

Vuollo, J., Liipo, J., Nykänen, V., et al., 1995. An early Proterozoic podiform chromitite in the Outokumpu ophiolite complex. Economic Geology 90, 445–452.

Wanhainen, C., Broman, C., Martinsson, O., Magnor, B., 2012. Modification of a Palaeoproterozoic porphyry-like system: Integration of structural, geochemical, petrographic, and fluid inclusion data from the Aitik Cu-Au-Ag deposit, northern Sweden. Ore Geology Reviews 48, 306–331.

Weihed, P., Arndt, N., Billström, K., et al., 2005. Precambrian geodynamics and ore formation: The Fennoscandian Shield. Ore Geology Reviews 27, 273–322.

Williams, P.J., Barton, M.D., Johnson, D.A., et al., 2005. Iron oxide copper-gold deposits: Geology, space-time distribution, and possible modes of origin. Economic Geology 100th Anniversary Volume 371–405.

Williams, I.S., Rutland, R.W., Kousa, J., 2008. A regional 1.92 Ga tectonothermal episode in Ostrobothnia, Finland: Implications for models of Svecofennian accretion. Precambrian Research 165, 15–36.

Wilkinson, B.H., Kesler, S.E., 2007. Tectonism and exhumation in convergent margin orogens: insights from ore deposits. Journal of Geology 115, 611–627.

Yang, S.-H., Maier, W.D., Hanski, E., et al., 2013. Origin of ultra-nickeliferous olivine in the Kevitsa Ni-Cu-PGE-mineralized intrusion, Lapland, Finland. Contributions to Mineralogy and Petrology 166, 81–95.

PART

MAGMATIC NI-CU-PGE-CR-V DEPOSITS 3

GEOLOGY AND PETROGENESIS OF MAGMATIC NI-CU-PGE-CR-V DEPOSITS: AN INTRODUCTION AND OVERVIEW 3.1

W.D. Maier

ABSTRACT

Magmatic Ni-Cu-PGE-Cr-V Deposits formed in a variety of geologic environments and periods. PGE-Cr-V deposits predominantly occur in large layered intrusions emplaced during the late Archean and early Proterozoic into stabilized cratonic lithosphere. The magmas ascend through translithospheric sutures characterized by limited extension and rifting. The laterally extensive ore layers (so-called reefs) formed through hydrodynamic phase sorting when the central portions of large, incompletely solidified magma chambers subsided due to crustal loading. The bulk of global PGE-Cr-V resources occur in the largest layered intrusions, namely the Bushveld complex of South Africa, the Great Dyke of Zimbabwe, and the Stillwater complex of the USA, but significant deposits additionally occur in Finland, namely in the Kemi, Portimo, and Koillismaa intrusions. Due to the large size (tens of kilometers) and limited complexity of the deposits and their host intrusions, they are relatively easy to locate and delineate. As a result, the search space is relatively mature and few new discoveries have been made in the last few decades. The parental magmas to most intrusions are likely derived from the asthenosphere, followed by contamination with crust. Some intrusions, notably the Bushveld complex, may have crystallized from magma sourced in the subcontinental lithospheric mantle, as suggested, for example, by Os isotopic data.

Mineral Deposits of Finland. http://dx.doi.org/10.1016/B978-0-12-410438-9.00003-0

In contrast to PGE-Cr-V deposits, magmatic Ni-Cu sulfide deposits formed throughout geologic time, and there are many deposits that are presently mined, including the Kevitsa deposit in Finland. The ores formed under highly dynamic magma emplacement conditions within lava channels or feeder conduits. The deposits are preferentially located near craton margins where mantle magmas could ascend through translithospheric structures. Magma flow was focused and locally enhanced by shifting compressive-extensional tectonic regimes. Abundant S-rich crustal rocks provided an external S source that is required for the majority of deposits. The igneous bodies hosting the deposits tend to be irregular and small, 10–100s m in width and height, and are difficult to locate. As a result, the search space remains relatively immature, as indicated by the fact that significant discoveries continue to be made (e.g., Sakatti in Lapland).

Keywords: magmatic PGE-Cr-V-Ni-Cu ore deposits; Finland; layered intrusions; magma conduits; crystal mush

INTRODUCTION

Magmatic deposits host the bulk of global resources for PGE-Cr-V and a significant proportion for Ni and Cu. Fennoscandia has traditionally been an important source for these commodities, starting with mining of Ni in Norway in the 1800s. Finland now hosts the bulk of PGE-Cr-V and Ni resources in the European Union, mostly within magmatic deposits. The platinum-group element (PGE) deposits in the Finnish layered intrusions of the Tornio-Näränkävaara belt (Portimo, Penikat, Koillismaa) share many geological features with the world's main PGE reef-type deposits (Alapieti and Lahtinen, 2002; Iljina et al., 2015) and have been known to host important PGE mineralization since their discovery in the late 1980s. In 2002, the total resources were ~14 Moz Pt + Pd (Naldrett, 2009), mainly within the Portimo intrusion. The Kemi intrusion contains one of the world's largest Cr deposits, whereas the Koillismaa, Koitelainen, and Akanvaara intrusions contain significant V deposits, namely at Mustavaara. Magmatic Ni and Cu have traditionally been one of Finland's main mining products, notably in the Kotalahti and Vammala Ni-Cu belts, mined almost continuously since the 1920s, and the Kevitsa deposit that has been mined since 2012. The recent discovery of the Sakatti deposit (Brownscombe et al., 2015) suggests that there remains significant discovery potential for further large deposits in northern Finland. This chapter, constituting an updated version of Maier and Groves (2011), provides a summary of the geology and petrogenesis of magmatic PGE-Cr-V-Ni-Cu deposits globally, with special emphasis on the Finnish examples.

GEOLOGY OF DEPOSITS

Deposits of PGE, Cr, and V form stratiform or stratabound horizons within layered, differentiated, mafic–ultramafic intrusions. There is a clear tendency of the richest and most continuous deposits to occur in the largest intrusions; the Bushveld, Great Dyke, and Stillwater complexes contain ~90% of global Pt, and ~70% of global Pd, Cr, and V resources (Crowson, 2001; Naldrett, 2009). Some of the PGE, Cr, and V reefs in these intrusions can be laterally correlated over >100 km. In contrast, smaller intrusions tend to have less continuous, more irregular, and lower grade reefs: for example, the PGE reefs of the Munni Munni and Panton intrusions in Western Australia, the Stella intrusion of South Africa, and the Finnish layered intrusions (Alapieti and Lahtinen, 2002; Iljina et al., 2015); the Cr seams of the Panton intrusion, Australia (Hoatson and Blake, 2000); or the V reefs of the Mambula, Kaffirskraal, and Rooiwater intrusions, South Africa (Reynolds, 1979). The thicknesses of the PGE-Cr-V reefs can vary between a few centimeters to tens of meters. In some cases, there is evidence that the thickness of the reefs varies systematically from the margin to the center of intrusions. For example, the chromitite layer in the Kemi intrusion increases in thickness from a few cm at the margin to >70 m in the center (Alapieti et al., 1989; Huhtelin et al., 2015), and an increase in thickness is also seen at the Sopcha chromitite within the Monchegorsk pluton of the Kola Peninsula (Chaschin et al., 1999). There are indications that the thickness of the Merensky Reef also thickens toward the center of the Bushveld complex (Maier et al., 2013, and references therein).

The stratigraphic level of, and host rocks to, the different types of reefs vary considerably. Chromitite layers tend to occur in the ultramafic basal portion of layered intrusions, and V reefs in the mafic upper portions, whereas PGE reefs can be located within mafic–ultramafic rocks of a wide range of compositions, including dunite, harzburgite, pyroxenite, norite-gabbronorite-gabbro, troctolite, anorthosite, chromitite, and magnetite; that is, no specific rock type has a markedly enhanced PGE prospectivity. However, the best PGE reefs are generally located in the ultramafic–mafic transition intervals of the intrusions (Fig. 3.1.1). Maier et al. (2013) distinguished seven types of PGE deposits in layered intrusions:

(1) Contact reefs at the base and sidewall of intrusions
(2) PGE reefs in the peridotitic and pyroxenitic lower portions of layered intrusions

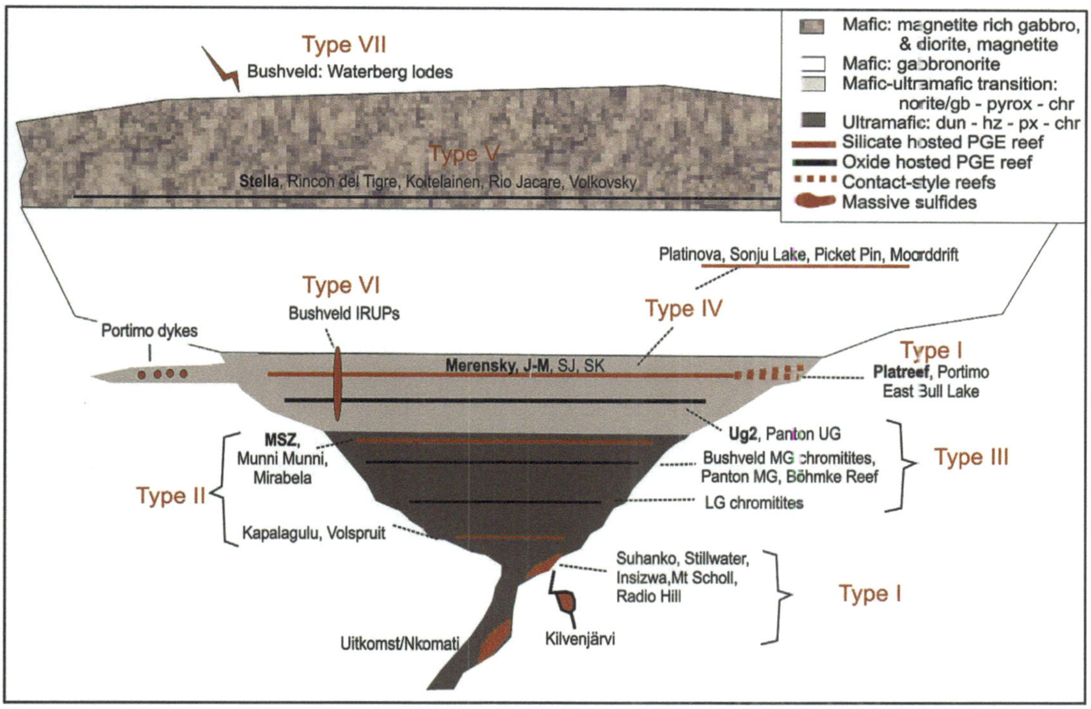

FIGURE 3.1.1 Schematic diagram showing occurrence of PGE mineralization in mafic–ultramafic intrusions and their feeder conduits.

JM = J-M reef of Stillwater complex, SJ = Sompujärvi reef of Penikat intrusion, SK = Siika Kämä reef of Portimo intrusion, LG-MG-UG = lower-middle-upper group chromitites of Bushveld, MSZ = main sulfide zone of Great Dyke.

Source: *Modified from Maier (2005).*

(3) PGE-enriched chromitite layers
(4) Silicate-hosted PGE reefs in interlayered mafic–ultramafic rocks, commonly within the central portions of layered intrusions
(5) PGE reefs in the magnetite-enriched upper portions of layered intrusions
(6) PGE-mineralized transgressive Fe-rich ultramafic pipes
(7) Vein-hosted PGE deposits in the roof (and floor) of the intrusions

Most reef rocks are medium-grained, but pegmatoidal textures occur in some portions of the Merensky and J-M reefs. The country rocks to many PGE enriched intrusions are granitic (Great Dyke, Penikat-Portimo-Koillismaa), but felsic volcanics (Koillismaa, Akanvaara), quartzite, dolomite, and ironstone (Bushveld) also occur. Where sulfide-rich pelitic rocks form the floor of layered intrusions (e.g., in parts of its northern lobe of the Bushveld complex hosting the Platreef), the PGE mineralization tends to be more sulfide rich and of lower grade, suggesting that significant addition of external S to the magma is a negative factor.

The mineralogy of the chromitite and titanomagnetite layers is relatively simple. In some cases, the rocks are monomineralic oxide layers, but they may also contain significant intercumulus silicates, namely pyroxenes, amphibole, plagioclase, or their alteration products. The mineralogy of the PGE reefs is more variable. The PGE mineralization is mostly hosted in sulfides (pyrrhotite-pentlandite-chalcopyrite), except for many PGE-bearing chromitite and magnetite layers where the PGE are hosted predominantly by platinum-group minerals and alloys (Cawthorn et al., 2005). In all types of PGE reefs, the sulfide contents tend to be relatively low (<1–3%) but the metal tenors of the sulfides are generally relatively high, often in excess of 1000 ppm PGE. In some metamorphosed reefs, there are no visible sulfides (e.g., in many of the Finnish intrusions), and in these cases, location of the reefs requires assaying of profiles essentially covering the entire intrusion.

Magmatic sulfide deposits in which Ni and Cu are the primary products occur within ultramafic–mafic rocks, including komatiite (i.e., many deposits in Western Australia and the Abitibi belt, Canada), dunite (Jinchuan, China), harzburgite (Kabanga, Tanzania), pyroxenite (Santa Rita, Brazil; Selebi-Phikwe, Botswana), olivine gabbro (Noril'sk, Russia), gabbronorite (Sudbury, Canada; Nebo-Babel, Australia; Phoenix, Botswana), and troctolite (Voisey's Bay, Canada) as summarized by Barnes and Lightfoot (2005). In many cases, the rocks are varied textured, that is, they have variable grain sizes, grain morphologies, and color indexes and they contain abundant xenoliths of the country rocks or autoliths from other, related intrusive phases. However, the ranges of lithologies and chemical compositions are more restricted than in PGE deposits, with differentiated, low-MgO rocks (e.g., ferro-gabbronorites, anorthosites, magnetites) being rare hosts to ore. The igneous host bodies are mostly relatively small (tens to hundreds of meters, rarely a few kilometers in diameter) and irregular in shape, comprising lava channels, dikes, sills, and chonoliths, all of which are generally interpreted as magma feeder conduits (Naldrett, 1997). The sulfides may occur near the base, the top, or within the bodies, reflecting the complex intrusive architecture of the host bodies, or they may be dislocated from the bodies, particularly in tectonized settings (e.g., in Archean greenstone belts; Lesher and Barnes, 2009, or the Finnish Kotalahti and Vammala Ni belts, Makkonen et al., 2015). The largest Ni-Cu deposit on Earth, at Sudbury, is anomalous in many regards, in that the sulfides are located at the base and in the footwall of a giant layered complex that formed in response to a meteorite impact (Naldrett, 1997).

Sulfide contents in the Ni-Cu deposits are ~5–100%, and many deposits are characterized by extensive halos of low-grade disseminated sulfides (e.g., Kabanga, Sudbury, Noril'sk, Pechenga, Jinchuan).

Thus, the amount of sulfides in Ni-Cu deposits is far larger than in PGE deposits. The main sulfide minerals are, however, similar, consisting of pyrrhotite, pentlandite (which may be dominant in komatiitic ores), and chalcopyrite (which may be dominant in gabbronoritic ores and in vein-hosted ores that crystallized from fractionated sulfide liquids). Metal tenors of the sulfides range between <1 and ≤20% Ni and Cu each, whereas PGE tenors of the sulfides are much lower than in the PGE deposits, between <1 and approximately 50 ppm (Noril'sk-Talnakh).

LOCATION OF DEPOSITS

All significant PGE-Cr deposits and most V deposits are located within stabilized cratons, and many are in the central segments of the cratons (Groves et al., 1987; Maier and Groves, 2011). It has been suggested that many of the Fennoscandian layered intrusions, including Penikat, Koillismaa, Panski, and Fedorova Tundra, are located near intracratonic rifts (Mitrofanov et al., 2012) which may constitute reactivated Archean suture zones along which the PGE bearing magmas ascended. The Bushveld complex was postulated to have been emplaced along the Thabazimbi-Murchison Lineament (Silver et al., 2004). However, there is no evidence for significant extension or flood basalt magmatism associated with most intrusions. This could indicate that any rifting that may have occurred was aborted. Under such conditions, the magmas could pond within the crust rather than erupt. This would have favored the emplacement and inflation of sills and ultimately the formation of large layered intrusions.

An exception to the pattern discussed appears to be the Giles complex, located within the Musgrave Province of central Australia. The complex hosts significant PGE and V reefs in layered intrusions emplaced in an extracratonic environment (Wingellina Hills intrusion and Jameson Range, Maier et al., 2014). The Musgrave Province was lodged between three cratons for >1.4 Ga (Smithies et al., 2011), thereby providing an exceptionally stable, far-field compressive tectonic setting favoring emplacement of thick sills.

Some cratons (e.g., Kaapvaal, Superior, Karelia, Zimbabwe) show a distinct relative enrichment in PGE deposits, whereas other cratons (e.g., Yilgarn) appear to be PGE-poor. The reasons for this pattern remain unresolved, but Archean melt depletion and refertilization of the protocratonic nuclei may play an important role (Maier and Groves, 2011).

Ni-Cu deposits also show a strong spatial association with Archean cratons (Fig. 3.1.2). Particularly well mineralized cratons include the Yilgarn, Superior, Zimbabwe, Karelia, and Kola cratons. Some cratons appear to be unmineralized (e.g., West Africa, Congo, Volga-Uralia, and Rio de la Plata), possibly due to poor exposure and underexploration. Several, mostly smaller, Ni-Cu deposits occur distal to cratons, particularly in orogenic belts (Aguablanca in Spain, Kalatongke in northwest China, deposits in the Caledonides of Norway, and the Appalachians of North America), but in some of these the presence of hidden Archean cratonic blocks has been postulated (Begg et al., 2010).

In marked contrast to PGE deposits, the Ni-Cu deposits are concentrated near the outer margins of the cratons (Groves et al., 2005; Kerrich et al., 2005; Begg et al., 2010). Examples from the Karelia-Kola craton include Pechenga, Kevitsa-Sakatti, the Kotalahti belt, and several other deposits shown in Fig. 3.1.2. Archean komatiite-hosted Ni-Cu deposits form an apparent exception to this spatial pattern, that is, they tend to be situated within, rather than at the margins of cratons, for example in the Karelian (Fig. 3.1.2), Yilgarn, Zimbabwe, and Superior cratons (Lesher and Barnes, 2009). However, the

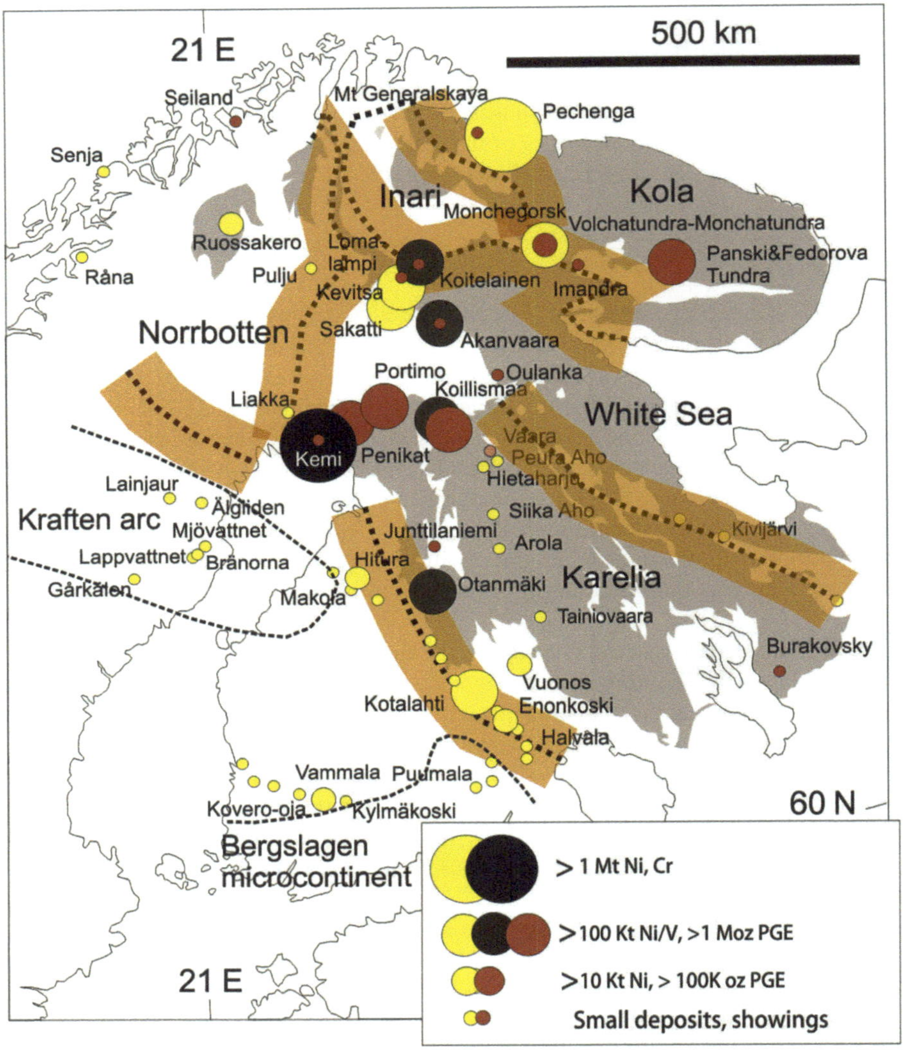

FIGURE 3.1.2 Location of PGE-Cr-V and Ni-Cu deposits in north-eastern Fennoscandia.

The distribution of exposed Archean crust (gray shade) is shown after Bleeker (2003). Craton margins are shown as stippled lines (after Lahtinen et al., 2005). Thick orange bars denote 100-km corridors centered on craton boundaries. Thin stippled lines denote crustal blocks with possible cratonic roots.

Source: *Modified after Maier and Groves (2011).*

greenstone belts that host the deposits represent the Archean sutures of protocratonic nuclei (Cassidy et al., 2006, Said et al., 2010), and therefore the craton-margin association remains valid.

All large deposits globally are preferentially associated with major lineaments (Begg et al., 2010), interpreted to be translithospheric faults, rift zones, or major shear zones through which large amounts of mantle-derived magma can ascend. Intriguingly, there are few Ni-Cu deposits in the refertilized/orogenically overprinted peripheral portions of cratons; exceptions include smaller deposits within southern Finland (Fig. 3.1.2) and Namaqualand in South Africa.

DISTRIBUTION OF DEPOSITS IN TIME

Both PGE-Cr-V and Ni-Cu deposits occur throughout most of geological time, but are concentrated in distinct periods (Fig. 3.1.3). All large deposits of PGE-Cr (Bushveld, Great Dyke, Stillwater, Lac des Iles in Canada, Stella in South Africa, Kemi), and many high-grade V deposits (Bushveld, Windimurra in Australia, Rio Jacare in Brazil) are older than ~1.8 Ga and younger than ~3 Ga; that is, no significant PGE deposits are yet known from the exposed early Archean rocks of the Kaapvaal, Slave, Pilbara, and Greenland cratons.

Nickel-Cu deposits occur over a wider geologic timespan than PGE-Cr deposits, with significant deposits occurring at ~2.7 Ga (komatiites of the Yilgarn and Zimbabwe cratons and the Abitibi greenstone belt, intrusions of the Tati belt of Botswana), ~2.0–1.8 Ga (Pechenga-Russia, Thompson, Raglan and Sudbury-Canada, Kevitsa, Finnish Ni belt), 1.4–1.3 Ga (Kabanga, Voisey's Bay, Nova-Australia), 1.1 Ga (Duluth-USA, Nebo-Babel), 0.82 Ga (Jinchuan), and 0.3–0.25 Ga (Noril'sk as well as several deposits in northwest China). There is a clear trend of Archean and Proterozoic deposits being more Ni-rich than the Phanerozoic deposits, largely because the former are hosted predominantly by komatiites (see Naldrett, 2010, for a discussion). Furthermore, there appears to be a trend of the younger (<2 Ga) deposits and their host rocks to contain more sulfide. Most of the komatiite-hosted deposits consist of either relatively small accumulations of massive high-grade sulfides, or of large masses of relatively S-poor rocks. In contrast, the younger deposits at, for example, Sudbury, Thompson, Pechenga, Kabanga, Duluth, and Noril'sk, are characterized by immense amounts of both economic and uneconomic magmatic sulfides, and their host rocks are also very S rich.

Both PGE-Cr-V and Ni-Cu deposits show a good correlation with rates of formation of juvenile crust (Fig. 3.1.3), representing supercontinent amalgamation and breakup (Groves et al., 2005). Deposits cluster at ~2.75–2.60 Ga during Kenorland amalgamation, 2.05–1.8 Ga during Columbia amalgamation, 1.4–1.3 Ga during Columbia breakup, ~1.1 Ga during Rodinia amalgamation, 0.45 Ga during Pangaea amalgamation, and ~0.25 Ga during Pangea breakup. Deposits whose age overlaps with supercontinent amalgamation include Stillwater, Great Dyke (Kenorland), Bushveld, Santa Rita, Panton, Pechenga, Thompson, Sudbury, Kevitsa, the Finnish Ni belt (Columbia), Duluth, and Nebo Babel (Rodinia). Deposits that formed during supercontinent breakup include Kabanga and Voisey's Bay (Columbia), and Noril'sk (Pangea). Whether the ~2.45–2.5 Ga PGE mineralized layered intrusions of the Karelia-Kola and southern Superior cratons formed during supercontinent amalgamation or breakup remains a matter of debate. Deposits tend to be rare in periods of low crust production, represented by supercontinent stability.

Whereas in some cratons, magmatic Ni-Cu-PGE-Cr-V deposits formed only in specific periods, for example, the Yilgarn komatiite-hosted Ni-Cu deposits at ~2.9–2.7 Ga, other cratons are characterized by multiple mineralization events. On the Fennoscandian Shield, PGE-Cr-V-Ni-Cu deposits were emplaced at 2.7–2.8 Ga (komatiites), 2.45 Ga (layered intrusions including Kemi, Penikat, Portimo, Koillismaa, Monchegorsk), 2.05 Ga (Kevitsa, Otanmäki), and 1.87–1.97 Ga (Pechenga, Finnish Ni belt), as summarized in Lehtinen et al. (2005). Nickel-Cu-PGE-Cr-V rich events on the Superior craton (Canada-USA) include the 2.7–2.9 Ga komatiites and associated intrusions (e.g., in the Abitibi and McFaulds Lake belts), the ~2.7–2.8 Ga Big Trout Lake and Highbank Lake layered intrusions, the 2.45 Ga East Bull Lake suite, the 1.9 Ga circum-Superior Raglan and Thompson belts, the 1.85 Ga Sudbury event, and the 1.1 Ga Duluth event.

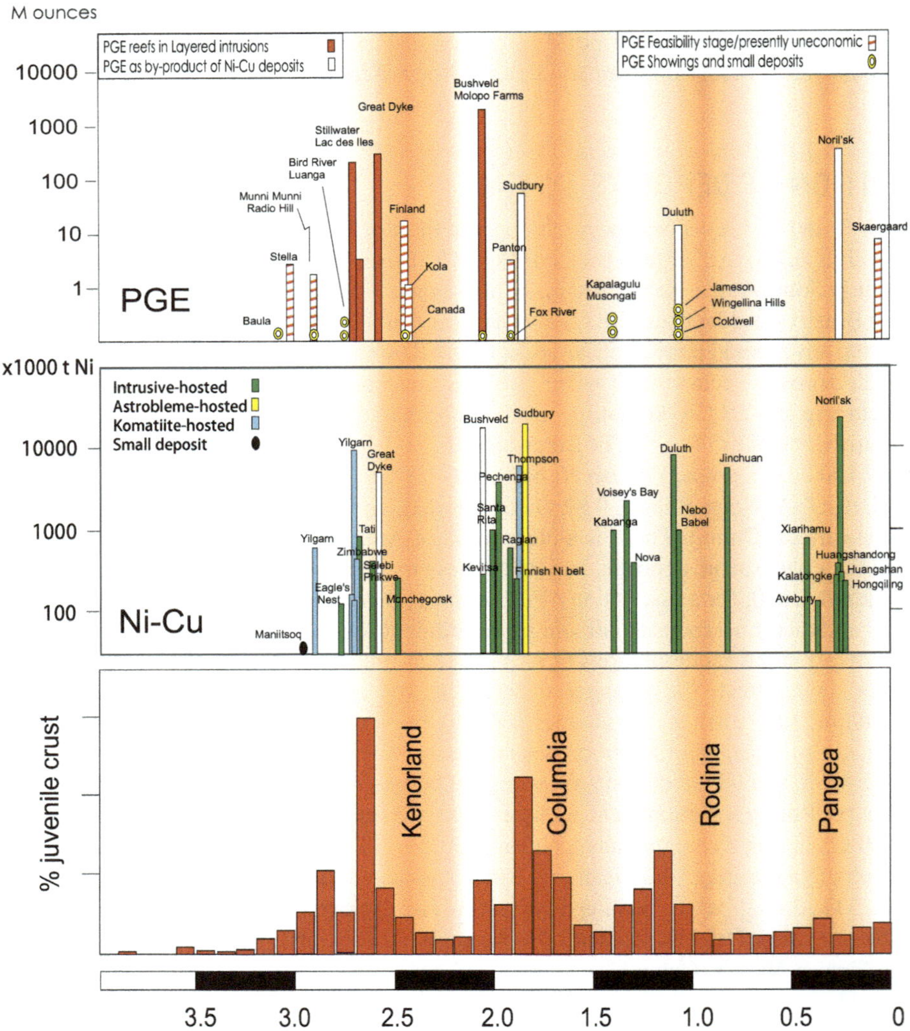

FIGURE 3.1.3 Secular distribution of PGE deposits. PGE deposits are concentrated in the late Archean and early Proterozoic.

Note: Correlation of most large deposits with periods of supercontinent amalgamation and breakup, and with juvenile crust production rate (from Condie, 2001). Approximate ages for supercontinents are: Kenorland 2.8–2.5 Ga (amalgamation) and 2.45–2.1 Ga (breakup), Columbia 2.1–1.8 and 1.6–1.3 Ga, Rodinia 1.3–1.1 and 0.85–0.6 Ga, and Pangea 0.6–0.3 and 0.2–0.06 Ga (from Goldfarb et al., 2010).

Source: *Modified from Maier and Groves (2011).*

FORMATION OF MAGMATIC PGE-Cr-V-Ni-Cu DEPOSITS

MAGMA SOURCE—DEGREE OF PARTIAL MELTING

Magmas parental to magmatic PGE-Cr-Ni deposits are derived from the mantle, as crustal rocks tend to be poor in PGE, Cr, and Ni (e.g., 1 ppb Pd in average crust, vs. 4–7 ppb in primitive mantle, 100 ppm Cr and Ni in average crust vs. 2000 ppm in mantle; Taylor and McLennan, 1985; Becker et al., 2006). Copper and V levels of the crust are higher, therefore, in theory, Cu and V of these deposits could be derived from the crust. The association of most V deposits with layered intrusions suggests a mantle derivation, but for certain magmatic Cu deposits a model of crustal derivation of the metal has been proposed (Maier, 2000).

For mantle magmas to be fertile in terms of the chalcophile metals (PGE, Ni, Cu), the degree of mantle melting needs to be relatively large. Nickel in the mantle is predominantly hosted by olivine, thus the Ni content of the magmas increases proportionally with the olivine content of the melted mantle and the degree of melting (Fig. 3.1.4). Basalts have a few hundred ppm Ni, whereas komatiites have >1000 ppm Ni. The PGE and Cu in the mantle are predominantly hosted by sulfides (Alard et al., 2000). For the magma to be PGE and Cu rich, the mantle sulfides need to be dissolved in the partial melt. Due to relatively low S solubilities of most basaltic-komatiitic magmas, dissolution of the sulfides is normally believed to require ~20% melting (Fig. 3.1.4) (see references in Naldrett, 2004). In oxidized mantle a smaller degree of melting may be required to dissolve the sulfides because S solubility in oxidized magmas is much higher (Jugo et al., 2005). This may lead to moderate PGE enrichment in some small degree melts derived from oxidized, metasomatized mantle, for example, meimechites (Mungall et al., 2006).

The preceding theoretical considerations are broadly consistent with the fact that most PGE and Ni deposits tend to be associated with rocks that crystallized from magmas with >8–10% MgO. In exceptional cases, PGE-rich sulfides can precipitate from less magnesian magmas if these are differentiates from more magnesian parental magmas in which S saturation has been delayed, as discussed in the following.

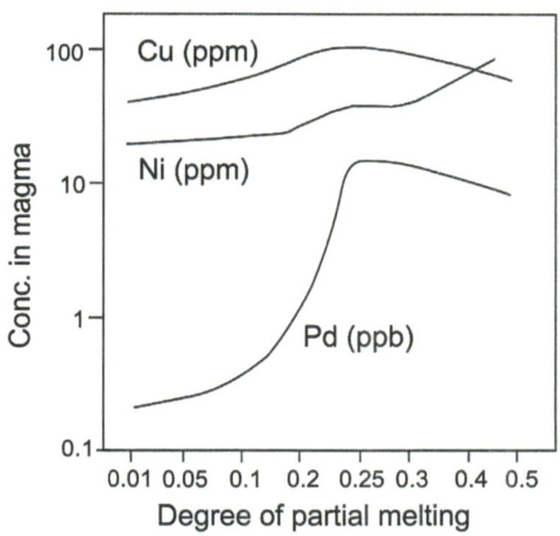

FIGURE 3.1.4 Concentration of Pd, Cu, and Ni in partial melts of mantle.

Source: *Modified after Barnes and Maier (1999).*

Chromium in the mantle is mainly hosted by orthopyroxene and chromite. Chromite may be a refractory phase during melting, as reflected by the increase in Cr contents from basalts to komatiites, but relatively Cr-rich magmas can also be generated by low degree partial melting of volatile-rich mantle (e.g., kimberlites). As chromite deposits occur associated with both komatiites (McFaulds Lake) and basalts (many deposits), the degree of melting is apparently less important than the concentration mechanism during crystallization.

Vanadium behaves as an incompatible element during melting of most mantle rocks and fractional crystallization prior to the onset of magnetite crystallization. Vanadium contents of basalts are thus higher than those of komatiites, and V deposits tend to be found in the differentiated portions of layered intrusions. Thus, fractional crystallization is more important in the generation of magmatic V deposits than the nature of the mantle and the degree of partial melting.

COMPOSITION OF THE MANTLE SOURCE

Primary PGE-Cr-Ni-rich magmas could be derived from two contrasting mantle sources, namely the convecting (asthenospheric) mantle or the lithospheric mantle (Fig. 3.1.5). In addition to GpII kimberlites and lamproites that are normally considered to be derived from the sub-continental lithospheric mantle (SCLM) (Lambert et al., 1995; Gurney et al., 2005), most mantle magmas are generally interpreted to be derived from the asthenospheric mantle. Some continental flood basalts that are highly enriched in crustal components, for example, Ventersdorp (Marsh et al., 1992), have also been proposed to be SCLM derived, as have some alkali basalts (Francis and Ludden, 1990).

The location of most large PGE deposits within cratons suggests that an SCLM control on magma generation is a possibility (Maier and Groves, 2011). However, the SCLM has been shown to be

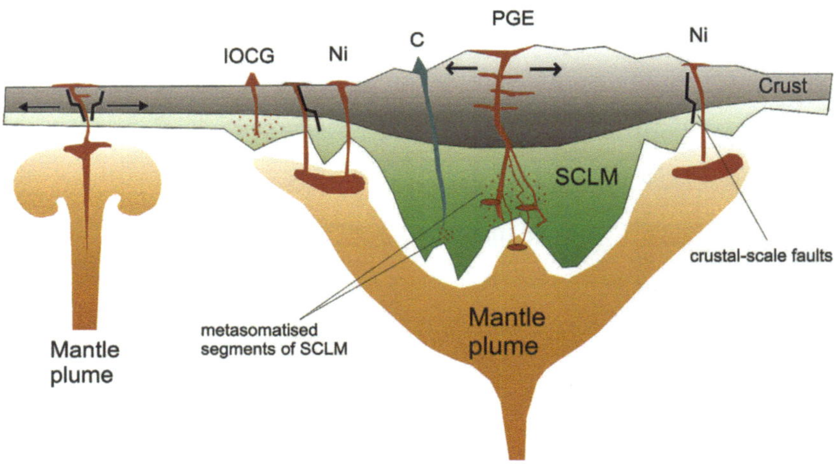

FIGURE 3.1.5 Sketch diagram showing possible mantle sources to PGE deposits.

When a mantle plume impacts on the base of a composite craton and undergoes limited adiabatic melting, melts infiltrate the SCLM. The heat of the plume and infiltrating plume melts cause melting of metasomatized domains of the SCLM. Melts of the SCLM and the plume mix and ascend along translithospheric suture zones through the SCLM and crust to form layered intrusions.

relatively PGE depleted (e.g., Pearson et al., 2004, Maier et al., 2012; for an exception, see McInnes et al., 1999), in response to Archean and early Proterozoic melting events. It thus seems counterintuitive to explain the magmas to the world's largest PGE deposits by melting of the most PGE-depleted mantle known. On the other hand, an SCLM component in the magmas is consistent with the relative Pt enrichment of many SCLM xenoliths (Maier and Barnes, 2004; Pearson et al., 2004) that matches the Pt enrichment of the Bushveld complex (Maier et al., 2013). Barnes et al. (2010) proposed that magmas derived from the sublithospheric mantle could infiltrate the SCLM and preferentially extract Pt from relatively PGE-poor lithologies, essentially representing a type of zone melting.

Metasomatic introduction of volatiles may be a key requirement because this could cause refractory Pt alloys to become fusible in some portions of the SCLM. The main alternative mantle reservoir is the convecting mantle. In this case, the high crustal component of many of the magmas has to be modeled via significant contamination with upper crust, but because most crust is PGE poor, the model implies that the PGE contents of the mantle magmas were originally even higher than those recorded. As the 19 ppb Pt found in the Bushveld magmas constitutes the upper limit of the terrestrial data field of basaltic-komatiitic magmas, the putative primary (precontamination) magma would have been unusually PGE rich, unless the assimilated crust contained significant PGE, such as black shale (Luokala-Ruskeeniemi, 1996; Wille et al., 2007). Another possibility is that the PGE-rich magmas hosting PGE deposits are derived from anomalously Pt-enriched mantle, perhaps resulting from heterogenous mixing-in of Pt-rich late veneer (Pt/Pd ~1.8; Palme and Jones, 2005), or containing a component of the Earth's core.

Many Ni-Cu deposits, particularly those hosted by komatiites, crystallize from primitive magmas that have flat mantle-normalized trace-element patterns. These magmas are unlikely to be derived from metasomatized SCLM and are best explained as large degree partial melts from the sublithospheric convecting mantle. Some Ni-Cu deposits contain pervasive crustal components and an unequivocal distinction between synemplacement upper-crustal contamination and SCLM derivation is often not possible.

MAGMA ASCENT

It has been suggested that komatiitic and magnesian basaltic magmas cannot ascend through relatively light upper crust unless the magma pathways extend to the base of the lithosphere (Naldrett, 2010). Such translithospheric pathways may be generated or reactivated during transpression within far-field compressional regimes, for example in volcanic arcs such as the Andes (Kerrich et al., 2005; Groves et al., 2010). This could result in rapid magma ascent, thereby delaying differentiation and loss of metals to segregating olivine and sulfide. Intracratonic magma pathways may exploit long-lived suture zones along which the earliest cratonic nuclei were amalgamated. Where extension is significant, rifting and cratonic-continental breakup typically characterized by flood basalt magmatism may result. Where extension is minor, rifting can be aborted, providing ideal conditions for the emplacement of large, layered intrusions. In the case of the Bushveld magmas, the transpressional Thabazimbi-Murchison Lineament (TML) is considered to have acted as a magma ascent route, following collision between the Zimbabwe and Kaapvaal cratons (Silver et al., 2004). Other intrusions that show evidence for structural control of magma ascent include the Great Dyke, the Jimberlana intrusion of Australia, and the Monts de Cristal intrusion of Gabon which show extreme length-to-width ratios.

Crustal-scale lineaments are particularly common along craton margins because deformational strain is focused in these environments (Begg et al., 2010). Kerrich et al. (2000, 2005) suggest that mantle plumes may be channeled to irregularities along lithospheric boundaries below craton margins. Together, these factors explain the occurrence of many mafic–ultramafic belts and Ni-Cu deposits there.

SULFUR SATURATION HISTORY AND SULFIDE CONCENTRATION

To form PGE-Ni-Cu deposits, the mantle magmas need to reach S saturation during emplacement. Primitive magmas initially are normally sulfide undersaturated during emplacement, because of the inverse relationship between S solubility and pressure (e.g., Mavrogenes and O'Neill, 1999). Fractional crystallization increases the S content of the residual magma because S is not incorporated in silicate minerals. This will ultimately result in sulfide melt saturation of the magma. For the parent magmas of many PGE deposits, modeling indicates that about 20–40% fractionation would cause sulfide melt saturation (Fig. 3.1.6; Li and Ripley, 2005, Ripley and Li, 2013), consistent with the stratigraphic location of most PGE reefs near the transition from the ultramafic to the mafic portions of intrusions. However, as most PGE reefs have sulfide contents above the cotectic ratio of ~0.5%, some concentration of sulfide is additionally required, perhaps during segregation of sulfide liquid through the magma or percolation of the sulfide liquid through the cumulate mush. In view of the low S content of Bushveld magmas, Maier et al. (2013) argued that mixing of resident and replenishing magma, one of the most popular models to explain PGE reefs, cannot be important in the formation of the deposits. Addition of external S is equally unlikely to have triggered reef formation because of the small amount of sulfides in most reefs, the mantle-like S isotopic compositions, and the high metal tenors. Instead, too much added S may result in abundant sulfide that dilutes the PGE thereby causing low-tenor reefs.

In sharp contrast to PGE deposits, most economic Ni-Cu deposits appear to have assimilated external S, as suggested by S isotopic data. The model is consistent with the high ratio of sulfide to silicate in many Ni-Cu deposits. Potential crustal rocks that may provide S to the magma include black shales (Pechenga, Barnes et al., 2001; Kabanga, Maier et al., 2011), paragneiss (Voisey's Bay; Li and Naldrett, 1999), Banded Iron Formation (BIF) (Platreef, Holwell et al., 2009), felsic volcanic rocks and sulfidic cherts (komatiite-hosted ores, Bekker et al., 2009), and evaporites (Noril'sk, Naldrett, 1997). In a few deposits, sulfur isotopic data cluster tightly around mantle values (Phoenix and Selkirk, Maier et al., 2008; Nebo Babel, Seat et al., 2009). This could indicate that concentration of mantle-derived S can be extremely efficient in some cases, or that S was added from an external source that had mantle-like S isotopic signatures.

Some magmatic ore deposits have unusually high metal tenors, for example, Noril'sk (Naldrett, 1997, Barnes and Lightfoot, 2005), Kevitsa (Mutanen, 1997; Yang et al., 2013), and Santa Rita (Barnes et al., 2011). It has been suggested that some of these deposits could have formed by assimilation of magmatic proto-ores ("cannibalization"), either derived from earlier magma fluxes of the same event, or from unrelated earlier magmatic sulfide deposits (Mutanen, 1997). For example, the Kevitsa Ni-rich disseminated sulfide ores have tenors of 50–70% Ni and tens of ppm PGE at low Cu contents (<3%). Yang et al. (2013) suggested that this could be due to assimilation of komatiitic proto-ores that were Ni- and PGE-rich, but Cu-poor.

Another potential indicator for cannibalization is isotopic decoupling, for example at Kabanga where some of the sulfide ores have extremely heavy S isotopic signatures ($\delta^{34}S$ up to +23) at mantle-like O isotopic signatures. Maier et al. (2011) proposed a multistage model whereby early picritic magma surges assimilated sulfide-bearing crustal rocks resulting in segregation of sulfide liquids that have heavy S isotopic signatures. Subsequent magma surges that used the same conduits were less

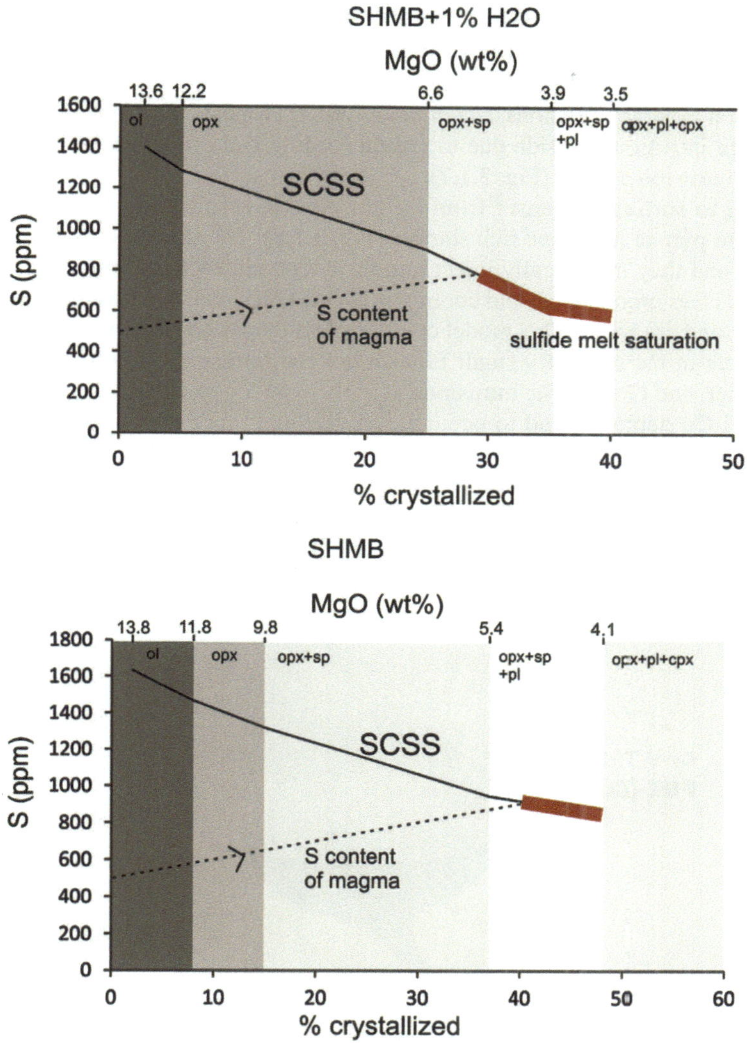

FIGURE 3.1.6 Modeling of S content at S saturation (SCSS) in hydrous and dry Bushveld B1 magma.

Source: *This uses data on the S content of Bushveld magmas from Barnes et al. (2010), the equation to calculate SCSS of Li and Ripley (2005), and thermodynamic modeling software PELE (Boudreau, 1999).*

contaminated with crust, perhaps due to lining of the conduits by the early magma surges. The late surges progressively flushed the early semiconsolidated silicate slurries out of the conduits, and then cannibalized the isotopically heavy sulfides that had accumulated from the early surges along the base of the conduits. After cooling, the later magma surges crystallized to form sulfide mineralized harzburgites with mantle-like O isotopic signatures and crustal S isotopic signatures.

MAGMA EMPLACEMENT AND CRYSTALLIZATION

Many PGE-Cr-V bearing layered intrusions are characterized by remarkably regular and laterally continuous layering. This suggests that the magmas were emplaced in a stable tectonic setting, which favors the formation of large intrusions. Maier et al. (2013) proposed that PGE-Cr-V reefs form when the centers of large intrusions subside due to crustal loading and semiconsolidated cumulate slurries slump toward the intrusion centers (Fig. 3.1.7).

This may lead to sorting of denser from lighter phases, resulting in the formation of sulfide-chromite-magnetite-pyroxene-olivine rich slurries (Fig. 3.1.8). The slurries crystallize to form distinct ultramafic layers, and they may locally inject into their footwall rocks to form sills. In smaller intrusions, subsidence is less pronounced, and cooling is faster. Thus, reefs are less regular, less continuous, have lower tenors, and are wider. This model could explain the less continuous nature and lower grade of PGE and V reefs in the relatively small Finnish layered intrusions, compared to the much larger Bushveld, Stillwater, and Great Dyke intrusions.

In contrast, Ni-Cu deposits tend to occur in smaller and more irregular, dike, sill, and chonolithic intrusions. Many of these intrusions are characterized by abundant xenoliths, compositional reversals, and irregular layering, interpreted to reflect dynamic emplacement settings such as magma conduits and feeders in which sulfides may be upgraded during entrainment and/or cannibalization by later magma pulses (Naldrett, 1997). A relatively high abundance of magma

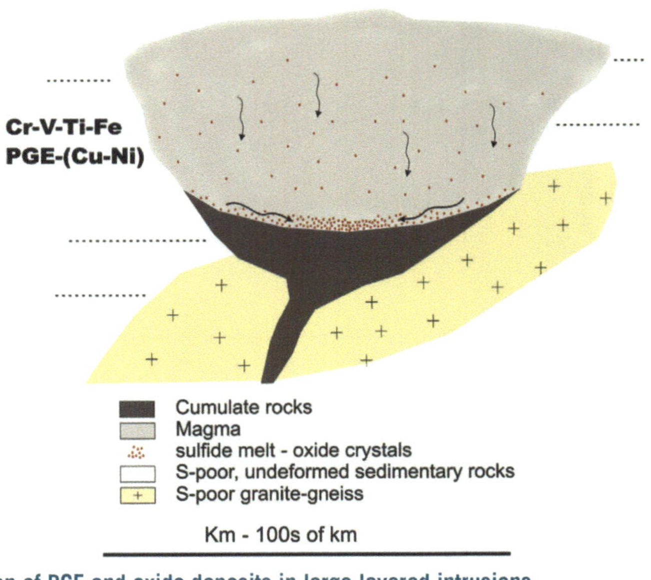

FIGURE 3.1.7 Formation of PGE and oxide deposits in large layered intrusions.

After saturation of the magma in oxide or sulfide liquid, the oxide/sulfide segregates to the top of the cumulate pile. It is further concentrated when the central portion of the magma chamber subsides due to magma loading, resulting in slumping of semiconsolidated cumulate mushes at the top of the cumulate pile and hydrodynamic sorting of dense phases.

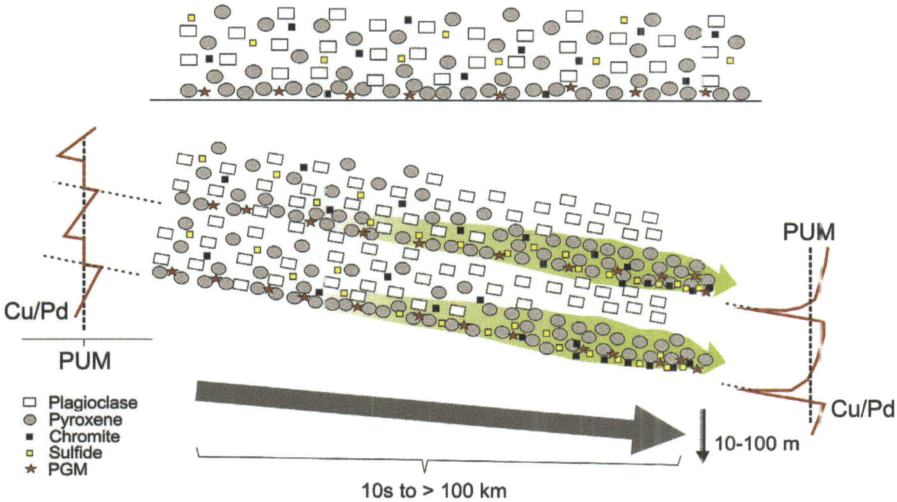

FIGURE 3.1.8 Schematic diagram illustrating formation of ultramafic and mafic layers during subsidence of intrusion.

At the top is a semiconsolidated protocumulate deposited by a fresh influx of magma prior to subsidence of the chamber, precipitating first pyroxene, chromite, and PGM, and later also plagioclase and sulfide. Cu/Pd is initially at mantle level, but falls below mantle once sulfide precipitates, and then progressively increases during Raleigh fractionation of sulfide. The lower part of the panel shows how cumulate layers are progressively sorted during subsidence, resulting in pronounced layering of pyroxenite, norite, and anorthosite in a down-dip direction. Sulfides percolate toward the bottom of the ultramafic layer (where they may absorb early formed PGM), thereby also shifting the reversal in Cu/Pd toward the base of the layer, and making reversal more pronounced.

Source: *Modified from Maier et al. (2013).*

conduits may be expected at tectonically active craton margins (Begg et al., 2010). Furthermore, ascent of magma entraining a load of dense sulfide liquids may be facilitated in active tectonic regimes where faulting can lead to repeated opening and closure of conduits thereby focusing and enhancing magma flux. Occasionally, this may even lead to seismically induced filter pressing and pumping of dense sulfide liquids into overlying parts of the intrusion or into fractures leading away from the host intrusion.

SUMMARY

The formation of magmatic PGE-Cr-V and Ni-Cu deposits is controlled by several factors, including geologic age as well as the architecture and composition of the lithosphere. Groves et al. (2005) suggested that the concentration of PGE deposits within cratons could be partly a preservational effect. The harzburgitic SCLM underlying the cratons is relatively buoyant due to the extraction of dense Fe-rich components (Griffin et al., 2008) and thus cannot be reprocessed. Furthermore, it is capable of bearing the weight of thousands of km^3 of heavy basic magma, and thus large intrusions in the cratonic crust are shielded from tectonic deformation and dismemberment. Another positive factor is the relatively stable cratonic environment, allowing the formation of large sill-like bodies and, during central subsidence of intrusions and late-magmatic mobilization of crystal slurries, that of laterally extensive reefs. This model is consistent with the occurrence of the best deposits in the largest, most slowly cooling intrusions.

For the formation of Ni-Cu deposits, a more dynamic environment is required, where mantle plumes are channeled toward lithospheric boundaries (Kerrich et al., 2000), large volumes of primitive magma can interact with S-rich crust, and sulfide-bearing magmas can ascend through abundant dynamic magma conduits resulting in flow-dynamic concentration of the sulfides. Such environments occur predominantly at craton margins (Groves et al., 2005; Begg et al., 2010). In addition, some Ni-Cu deposits may be transposed to craton margins during seafloor spreading (Kambalda in Australia, Raglan in Canada), particularly if the mantle plume impacts near craton margins. Deposits tend to be rare along the margins of refertilized cratons, possibly because of destruction of deposits and/or S-rich crustal rocks through tectonism, magmatism, and metamorphism. Furthermore, lineaments along such modified cratons could be less long-lived.

Secular trends are observed in both types of deposits. PGE deposits are concentrated in the Archean and the early Proterozoic, possibly because crustal S was less abundant, and because production of PGE-rich magmas may require large degrees of mantle melting. The paucity of PGE deposits in the early Archean may be due to poor preservation, whereas the Phanerozoic lacked sufficient heat flux to trigger large degree mantle melting. In contrast, in the late Archean and early Proterozoic, a hotter Earth resulted in a greater incidence of large mantle plumes (Condie, 2001).

A relatively high heat flux and large degree mantle melting are also responsible for the concentration of particularly Ni-rich deposits in the Archean (Naldrett, 2010), whereas younger deposits tend to have important Cu credits. As in the case of PGE deposits, early Archean Ni-Cu deposits are rare, possibly because of lack of exposure. In contrast, the largest accumulations of Ni-Cu sulfides are younger than ~2 Ga. It is possible that this reflects a secular increase in crustal S, consistent with the occurrence of abundant Ni deposits at the peak in crustal S inventory (Canfield, 2004). Of further note is the paucity of Ni deposits in the Neoproterozoic, coincident with the minimum in crustal S inventory.

The correlation of some important PGE and Ni-Cu deposits with the initial stages of supercontinent breakup (Kabanga, Voisey's Bay, Jinchuan, Noril'sk-Talnakh) can be understood in terms of mantle-plume activity and resulting crustal extension that facilitated magma generation and ascent. The clustering of deposits during the late stages of supercontinent amalgamation is more difficult to understand as one would assume that continent amalgamation is mainly driven by offshore mantle-plume activity. Whereas this may lead to the formation of oceanic Ni-Cu deposits which could be transposed to craton margins due to plate tectonics, the formation of large intracontinental layered intrusions at these times requires a different explanation. One possibility is that subduction rollback causes temporary lithospheric extension. This could aid complex Ni-Cu systems in back arcs, and it could lead to emplacement of intracratonic intrusions in failed rift settings where extension was limited. A further possibility is that enhanced metasomatism of the SCLM roots in response to supercontinent amalgamation triggered pervasive SCLM delamination, facilitating the establishment of mantle plumes beneath craton interiors.

REFERENCES

Alapieti, T.T., Lahtinen, J.J., 2002. Platinum-group element mineralization in layered intrusions of northern Finland and the Kola Peninsula, Russia. In: Cabri, L.J. (Ed.), The Geology, Geochemistry, Mineralogy and Mineral Beneficiation of Platinum-Group Elements, 54. Canadian Institute of Mining, Metallurgy and Petroleum, pp. 507–546.

Alapieti, T.T., Kujanpää, J., Lahtinen, J.J., Papunen, H., 1989. The Kemi stratiform chromitite deposit, northern Finland. Economic Geology 84, 1057–1077.

Alard, O., Griffin, W.L., Lorand, J.-P., et al., 2000. Non-chondritic distribution of the highly siderophile elements in mantle sulfides. Nature 407, 891–894.

Barnes, S.-J., Maier, W.D., 1999. The fractionation of Ni, Cu and the noble metals in silicate and sulphide liquids. In: Keays, R.R., Lesher, C.M., Lightfoot, P.C., Farrow, C.E.G. (Eds.), Dynamic Processes in Magmatic Ore Deposits and Their Application to Mineral Exploration, 13. Geological Association of Canada, Short Course Notes, pp. 69–106.

Barnes, S.-J., Lightfoot, P., 2005. Formation of magmatic nickel sulfide ore deposits and processes affecting the Cu and PGE contents. Economic Geology 100th Anniversary Volume, 179–214 .

Barnes, S.-J., Maier, W.D., Curl, E., 2010. Composition of the marginal rocks and sills of the Rustenburg Layered Suite, Bushveld complex, South Africa: Implications for the formation of the platinum-group element deposits. Economic Geology 105, 1491–1511.

Barnes, S.-J., Melezhik, V.A., Sokolov, S.V., 2001. The composition and mode of formation of the Pechenga nickel deposits, Kola Peninsula, northwestern Russia. Canadian Mineralogist 39, 447–471.

Barnes, Steven, J., Osborne, G.R., et al., 2011. The Santa Rita nickel sulfide deposit in the Fazenda Mirabela Intrusion, Bahia, Brazil: Geology and sulfide geochemistry. Economic Geology 106, 1083–1110.

Becker, H., Horan, M.F., Walker, R.J., et al., 2006. Highly siderophile element composition of the Earth's primitive mantle: Constraints from new data on peridotite massifs and xenoliths. Geochimica Cosmochimica Acta 70, 4528–4550.

Bekker, A., Barley, M.E., Fiorentini, M.L., et al., 2009. Atmospheric sulfur in Archean Komatiite-hosted nickel deposits. Science 326, 1986–1089.

Begg, G.C., Hronsky, J.M.A., Arndt, N.T., et al., 2010. Lithospheric, Cratonic and geodynamic setting of Ni-Cu-PGE sulfide deposits. Economic Geology 105, 1057–1070.

Bleeker, W., 2003. The late Archean record: a puzzle in ca 35 pieces. Lithos 71, 99–134.

Boudreau, A.E., 1999. PELE—A version of the MELTS software program for the PC platform. Computers & Geosciences 25, 201–203.

Brownscombe, W., Ihlenfeld, C., Coppard, J., et al., 2015. Sakatti CU-NI-PGE sulfide deposit in Northern Finland. Mineral Deposits of Finland. Elsevier, Amsterdam. pp. 211-251.

Canfield, D.E., 2004. The evolution of the Earth surface sulfur reservoir. American Journal of Science 304, 839–861.

Cassidy, K.F., Champion, D.C., Krapez, B., et al., 2006. A revised geological framework for the Yilgarn craton, Western Australia. Geological Survey of Western Australia, Record 8, 8.

Cawthorn, R.G., Barnes, S.J., Ballhaus, C., Malitch, K.N., 2005. Platinum-group element, chromium and vanadium deposits in mafic and ultramafic rocks. Economic Geology 100th Anniversary Volume, 215–250.

Chaschin, V.V., Galkin, A.S., Ozeryansky, V.V., 1999. The Sopcheozero chromite deposit and its PGE content, Monchepluton (Kola Peninsula, Russia). Geology of Ore Deposits 41, 507–513 (in Russian).

Condie, K.C., 2001. Mantle Plumes and Their Record in Earth History. Cambridge University Press, Oxford. p. 306.

Crowson, P., 2001. Minerals Handbook 2000–2001. Mining Journal Books, Kent. U.K. p. 486.

Francis D., Ludden J., (1990) The mantle source for olivine nephelinite, basanite, and alkaline olivine basalt 698 at Fort Selkirk, Yukon, Canada. J Petrol 31, 371–400.

Goldfarb, R.J., Bradley, D., Leach, D.L., 2010. Secular variation in economic geology. Economic Geology 105, 459–465.
Griffin, W.L., O'Reilly, S.Y., Afonso, J.C., Begg, G.C., 2008. The composition and evolution of lithospheric mantle: a re-evaluation and its tectonic implications. Journal of Petrology 50, 1185–1204.
Groves, D.I., Ho, D.E., Rock, N.M.S., et al., 1987. Archean cratons, diamond and platinum: Evidence for coupled long-lived crust-mantle systems. Geology 15, 801–805.
Groves, D.I., Vielreicher, R.M., Goldfarb, R.J., Condie, K.C., 2005. Controls on the heterogenous distribution of mineral deposits through time. Geological Society London Special Publications 248, 71–101.
Groves, D.I., Bierlein, F.P., Meinert, L.D., Hitzman, M.W., 2010. Iron oxide copper-gold (IOCG) deposits through Earth history: Implications for origin, lithospheric setting, and distinction from other epigenetic iron oxide deposits. Economic Geology 105, 641–654.
Gurney, J.J., Helmstaedt, H.H., LeRoex, A.P., et al., 2005. Diamonds: Crustal distribution and formation processes in time and space and an integrated deposit model. Economic Geology 100th Anniversary Volume, 143–178.
Hoatson, D.M., Blake, D.H., 2000. Geology and economic potential of the Palaeoproterozoic layered mafic–ultramafic intrusions in the East Kimberley, Western Australia. Australian Geological Survey Organization, Bulletin 246, 469.
Holwell, D.A., Boyce, A.J., McDonald, I., 2009. Sulfur isotope variations within the Platreef Ni-Cu-PGE deposit: Genetic implications for the origin of sulfide mineralization. Economic Geology 102, 1091–1110.
Huhtelin, T., 2015. The Kemi Cr deposit. In: Maier, W.D., O'Brien, H., Lahtinen, R. (Eds.), Mineral Deposits of Finland. Elsevier, Amsterdam. pp. 165–177.
Iljina, M., Maier, W.D., Karinen, T., Halkoaho, T., 2014. PGE deposits of the Tornio-Näränkavaara belt of intrusions (Portimo, Penikat and Koillismaa). In: Maier, W.D., Lahtinen, R., O'Brien, H. (Eds.), Mineral Deposits of Finland p......
Jugo, P.J., Luth, R.W., Richards, J.P., 2005. An experimental study of the sulfur content in basaltic melts saturated with immiscible sulfide or sulfate liquids at 1300 C and 1.0 GPa. Journal of Petrology 46, 783–798.
Kerrich, R., Goldfarb, R.J., Groves, D.I., Garwin, S., 2000. The geodynamics of world-class gold deposits: characteristics, space-time distribution, and origins. Reviews in Economic Geology 13, 501–552.
Kerrich, R., Goldfarb, R.J., Richards, J.P., 2005. Metallogenic provinces in an evolving geodynamic framework. In: Hedenquist, J.W., Thompson, J.F.H., Goldfarb, R.J., Richards, J.P. (Eds.), Economic Geology 100th Anniversary Volume, pp. 1097–1136.
Lahtinen, R., Korja, A., Nironen, M., 2005. Paleoproterozoic tectonic evolution. In: Lehtinen, M., Nurmi, P.A., Rämö, O.T. (Eds.), Precambrian Geology of Finland—Key to the Evolution of the Fennoscandian Shield. Elsevier, Amsterdam, p. 736.
Lambert, D.D., Shirey, S.B., Berman, S.C., 1995. Proterozoic lithospheric mantle source for the Prairie Creek lamproites: Re-Os and Sm-Nd isotopic evidence. Geology 23, 273–276.
Lehtinen, M., Nurmi, P.A., Rämö, O.T., 2005. Precambrian Geology of Finland—Key to the Evolution of the Fennoscandian Shield. Developments in Precambrian Geology, 14. Elsevier, Amsterdam. 736.
Lesher, C.M., Barnes, S.J., 2009. Komatiite-associated Ni-Cu-(PGE) deposits, 27-120. In: Li, C., Ripley, E.M. (Eds.), New Developments in Magmatic Ni-Cu and PGE Deposits. Geological Publishing House, Beijing, p. 290.
Li, C., Naldrett, A.J., 1999. Geology and petrology of the Voisey's Bay intrusion: Reaction of olivine with sulfide and silicate liquids. Lithos 47, 1–31.
Li, C., Ripley, E.M., 2005. Empirical equations to predict the sulfur content of mafic magmas at sulfide saturation and applications to magmatic sulfide deposits. Mineralium Deposita 40, 218–230.
Luokola-Ruskeeniemi, K., Heino, T., 1996. Geochemistry and genesis of the black-shale hosted Ni-Cu-Zn deposit at Talvivaara, Finland. Economic Geology 91, 80–110.
Maier, W.D., 2000. Concentrations of platinum-group elements in Cu-sulphide ores at Carolusberg and East Okiep, Namaqualand, South Africa. Mineralium Deposita 35, 422–429.

Maier, W.D., 2005. Platinum-group element (PGE) deposits and occurrences: mineralization styles, genetic concepts, and exploration criteria. Journal of African Earth Sciences 41, 165–191.

Maier, W.D., Barnes, S.-J., 2004. Pt/Pd and Pd/Ir ratios in mantle-derived magmas: A possible role for mantle metasomatism. South African Journal of Geology 107, 333–340.

Maier, W.D., Groves, D.I., 2011. Temporal and spatial controls on the formation of magmatic PGE and Ni-Cu deposits. Mineralium Deposita 46, 841–857.

Maier, W.D., Barnes, S.-J., Chinyepi, G., et al., 2008. The petrogenesis of magmatic Ni-Cu-(PGE) deposits in NE Botswana: Compositional data from Phoenix, Selkirk, Tekwane, and intrusions in the Selebi-Phikwe area (Phikwe, Dikoloti, Lentswe, Phokoje). Mineralium Deposita 43, 37–60.

Maier, W.D., Barnes, S.-J., Ripley, E.M., 2011. The Kabanga Ni sulfide deposits, Tanzania: A review of ore forming processes. Economical Geology Special Paper, Reviews in Economic Geology 17, 217–234.

Maier, W.D., Peltonen, P., McDonald, I., et al., 2012. The platinum-group element budget of the Kaapvaal and Karelian sub-continental lithospheric mantle: Implications for mantle evolution. Chemical Geology 302-303, 119–135.

Maier, W.D., Barnes, S.-J., Groves, D.I., 2013. The Bushveld complex, South Africa: Formation of platinum–palladium, chrome- and vanadium-rich layers via hydrodynamic sorting of a mobilized cumulate slurry in a large, relatively slowly cooling, subsiding magma chamber. Mineralium Deposita 48, 1–56.

Maier, W.D., Howard, H.M., Smithies, R.H., et al., 2014. Mafic–ultramafic intrusions of the Giles Event, Western Australia: Petrogenesis and prospectivity for magmatic ore deposits. GSWA Report 134, 82.

Makkonen, H., 2015. Ni deposits of the Vammala and Kotalahti belt. In: Maier, W.D., O'Brien, H., Lahtinen, R. (Eds.), Mineral Deposits of Finland. Elsevier, Amsterdam. pp. 253–285.

Marsh, J.S., Bowen, M.P., Rogers, N.W., Bowen, T.B., 1992. Petrogenesis of Late Archean flood-type basic lavas from the Klipriviersberg group, Ventersdorp Supergroup, South Africa. Journal of Petrology 33, 817–847.

Mavrogenes, J.A., O'Neill, H.StC., 1999. The relative effects of pressure, temperature and oxygen fugacity on the solubility of sulfide melts in mafic magmas. Geochimica Cosmochimica Acta 63, 1173–1180.

McInnes, B.I.A., McBride, J.S., Evans, N.J., et al., 1999. Osmium isotope constraints on metal recycling in subduction zones. Science 286, 512–516.

Mitrofanov, F.P., Malitch, K.N., Bayanova, T.B., et al., 2012. Comparison of East-Scandinavian and Norilsk large plume mafic igneous provinces of PGE ores. Proceedings of MSTU 15, 380–394.

Mungall, J.E., Hanley, J.J., Arndt, N.T., Debecdelievre, A., 2006. Evidence from meimechites and other low-degree mantle melts for redox controls on mantle-crust fractionation of platinum-group elements. PNAS 103, 12695–12700.

Mutanen, T., 1997. Geology and petrology of the Akanvaara and Koitelainen mafic layered intrusions and Keivitsa-Satovaara layered complex, northern Finland. Geological Survey of Finland, Bulletin 395, 233.

Naldrett, A.J., 1997. Key factors in the genesis of Noril'sk, Sudbury, Jinchuan, Voisey's Bay and other world-class Ni-Cu-PGE deposits: implications for exploration. Australian Journal of Earth Sciences 44, 283–315.

Naldrett, A.J., 2004. Magmatic Sulfide Deposits. Springer, Heidelberg, p. 728.

Naldrett, A.J., 2009. Fundamentals of Magmatic sulfide deposits, 1-26. In: Li, C., Ripley, E.M. (Eds.), New Developments in Magmatic Ni-Cu and PGE Deposits. Geological Publishing House, Beijing, p. 290.

Naldrett, A.J., 2010. Secular variation of magmatic sulfide deposits and their source magmas. Economic Geology 105, 669–688.

Palme, H., Jones, A., 2005. Solar system abundances of the elements, 41–61. In: Davis, A.M. (Ed.), Meteorites, Comets and Planets, Treatise on Geochemsitry, V1. Elsevier, Amsterdam, p. 737.

Pearson, D.G., Irvine, G.J., Ionov, D.A., et al., 2004. Re-Os isotope systematics and platinum-group element fractionation during mantle melt extraction: a study of mass if and xenolith peridotite suites. Chemical Geology 208, 29–59.

Reynolds, I.M., 1979. Vandium bearing titaniferous iron ores from the Rooiwater, Usushwana, Mambula, Kaffirskraal and Trompsburg igneous complexes. National Institute for Metallurgy, Randburg. Rep 2017, p. 61.

Ripley, E., Li, C., 2013. Sulfide saturation in mafic magmas: Is external sulfur required for magmatic Ni-Cu-(PGE) ore genesis? Economic Geology 108, 45–58.

Said, N., Kerrich, R., Groves, D.I., 2010. Geochemical systematic of basalts of the Lower Basalt Unit, 2.7 Ga Kambalda Sequence, Yilgarn craton, Australia: Plume impingement at a rifted craton margin. Lithos 115, 82–100.

Seat, Z., Beresford, S., Grguric, B.A., Gee, M.A.M., and Grassineau, N.V., 2009. Reevaluation of the role of external sulfur addition in the genesis of Ni-Cu-PGE deposits: evidence from the Nebo-Babel Ni-Cu-PGE deposit, West Musgrave, Western Australia: Econ Geol 104, 521–538.

Silver, P., Fouch, M.J., Gao, S.S., et al., 2004. Seismic anisotropy, mantle fabric, and the magmatic evolution of Precambrian southern Africa. South African Journal of Geology 107, 45–58.

Smithies, R.H., Howard, H.M., Evins, P.M., et al., 2011. Mesoproterozoic high temperature granite magmatism, crust-mantle interaction and the intracontinental evolution of the Musgrave Province. Journal of Petrology.

Taylor, S.R., McLennan, S.M., 1985. The continental crust: Its composition and evolution. Blackwell, Oxford. p. 312.

Wille, M., Kramers, J.D., Nägler, T.F., et al., 2007. Evidence for a gradual rise of oxygen between 2.6 and 2.5 Ga from Mo isotopes and Re-PGE signatures in shales. Geochimica et Cosmochimica Acta 71, 2417–2435.

Yang, S., Maier, W.D., Hanski, E., et al., 2013. Origin of ultra-nickeliferous olivine in the Kevitsa Ni-Cu-PGE mineralized intrusion, Lapland, Finland. Contributions to Mineralogy and Petrology 166, 81–95.

CHAPTER

KOMATIITE-HOSTED NI-CU-PGE DEPOSITS IN FINLAND 3.2

J. Konnunaho, T. Halkoaho, E. Hanski, T. Törmänen

ABSTRACT

Eight economically promising komatiite-hosted sulfide deposits have so far been found in eastern and northern Finland. They are related both to Archean (Tainiovaara, Hietaharju, Peura-aho, Vaara, Ruossakero, and Sarvisoaivi) and Paleoproterozoic (Hotinvaara and Lomalampi) komatiitic to komatiitic basalt magmatism. The deposits are mostly associated with the Al-undepleted komatiite type (AUK) (Al_2O_3/TiO_2 = 15-20), but Ti-depleted komatiites (Al_2O_3/TiO_2 > ~20) are predominant in the Hotinvaara, Ruossakero, and Sarvisoaivi areas. Disseminated Fe-Ni-Cu sulfides (commonly <5 wt% S) and minor massive sulfide accumulations as well as remobilized massive veins are associated with thick, MgO-rich and Cr-poor metacumulates. In some cases, sulfide mineralization is partly hosted in associated schist, as in the Peura-aho deposit. Most of the deposits were modified by postmagmatic processes to various degrees, resulting in a notable increase in metal contents of the sulfide fraction in some of the deposits, notably the Vaara deposit.

The deposits are divided into two main groups based on their metal content: (1) deposits that are enriched in PGE (Pd + Pt > 500 ppb) and Cu (Ni/Cu < 13) (Vaara, Hietaharju, Peura-aho, and Lomalampi), and (2) deposits that are enriched in Ni (Ni/Cu > 15) and have low PGE contents (Sarvisoaivi, Ruossakero, and Hotinvaara). The Tainiovaara deposit has an intermediate composition between these two groups. Lomalampi is a unique deposit, because it has relatively high levels of PGE compared to base metals (i.e., it is a PGE-(Ni-Cu) deposit)—and a Pt/Pd ratio of around 2, as opposed to most other komatiite-hosted Ni-Cu-(PGE) deposits globally that have Pt/Pd around unity. The Lomalampi deposit contains isotopically heavy sulfur with $\delta^{34}S$ of +10 to +15‰, which is consistent with the presence of a large proportion of crustal sulfur in this deposit. Most of the Finnish Ni-Cu-(PGE) deposits have $\delta^{34}S$ close to the mantle range (±2 per mil). Multiple isotope analyses of the Archean Vaara and Hietaharju deposits have revealed a considerable mass-independent fractionation of sulfur isotopes in both the ores and their country rocks, demonstrating a significant role of external sulfur assimilation in ore formation. Although no significant economic deposits have so far been found, the number of known mineralizations and the high PGE contents in some of them indicate that komatiites in Finland still provide potential targets for future exploration work.

Keywords: Komatiites; komatiite-hosted Ni-Cu-PGE deposits; PGE; sulfides; sulfur isotopes; alteration; greenstone belt; Suomussalmi greenstone belt; Central Lapland Greenstone Belt; Finland.

INTRODUCTION

Recognition of primary, high-temperature ultramafic lavas, called komatiites, in the late 1960s (Viljoen and Viljoen, 1982) started a new era of nickel exploration. In favorable environments and under appropriate dynamic physical conditions, ultramafic lavas can generate profitable nickel-copper sulfide deposits (e.g., Naldrett, 1966; Woodall and Travis, 1970; Ross and Hopkins, 1975). In 2006, approximately 18% of global nickel sulfide reserves were estimated to be associated with komatiites (Hronsky and Schodde, 2006). The komatiitic ultramafic lavas and komatiitic basalts and, in particular, associated olivine-rich cumulates in lava channels are potential host rocks for nickel deposits. In addition, sills and feeder dikes related to komatiitic magmatism are also potential environments for Ni-Cu-PGE ores.

Two main types of nickel deposits occur in komatiitic systems. Type I consists of accumulations of massive and/or matrix ore at the base of komatiitic lava channels. The nickel grades in massive sulfide ores vary between 2 and 20 wt%. Type II is characterized by sulfide disseminations within, rather than at the base of, the lava channel cumulates. In this deposit type, the whole-rock Ni contents are generally less than 1 wt%. Komatiite-related nickel deposits can be transitional between these two end-member types and deposits containing both types of mineralization have been described. In some of the deposits, the metals have been mobilized during postmagmatic tectono-metamorphic processes (type IV–V) (e.g., Lesher and Keays, 2002; Lesher and Barnes, 2009).

In Finland, no significant economic komatiite-hosted Ni-Cu-PGE deposits have been discovered yet. The small Tainiovaara deposit is the only one that has so far been mined, in 1989. Nevertheless, several small, low-grade sulfide deposits occur in Archean and Paleoproterozoic greenstone belts in eastern and northern Finland (Fig. 3.2.1 and Table 3.2.1).

NATURE OF KOMATIITE-HOSTED NI-CU-PGE DEPOSITS

Most komatiitic magmas are thought to have resulted from an extensive degree of partial melting of depleted mantle peridotite (e.g., Herzberg, 1992). As a consequence of the high degree of melting (>30%), these magmas are sulfur-undersaturated after segregation from their mantle source and carry elevated contents of chalcophile elements (Ni, Cu, and PGE). Critical factors controlling the genesis of magmatic Ni-Cu-PGE deposits include the formation of an immiscible sulfide liquid during magma ascent through the Earth's crust or after its eruption on the Earth's surface. To achieve high Ni tenors, the separation of the immiscible sulfide liquid has to take place before a significant amount of mafic silicates, particularly olivine, has crystallized from the magma. Additionally, because komatiitic magmas are commonly strongly S-undersaturated when emplaced in the upper crust, it is generally believed that the generation of Ni sulfide ores requires addition of external S to the magma.

This explains why Ni-Cu-PGE deposits are usually found in dynamic magma channel environments where thermo-mechanical erosion of the host rocks is believed to have been most efficient (e.g., Lesher and Groves, 1986; Keays, 1995; Lesher and Arndt, 1995; Lesher et al., 2001; Sproule et al., 2005; Barnes, 2007; Arndt et al., 2008). A dynamic magma environment also allows the segregated sulfides to be equilibrated with a large amount of silicate magma (cf. Campbell and Naldrett, 1979), thereby efficiently sequestering chalcophile metals from the magma. Recent studies of Bekker et al. (2009), Fiorentini et al. (2012a, b), and Konnunaho et al. (2013) have shown that coupled $\delta^{34}S$ and $\delta^{33}S$ measurements provide a powerful tool to constrain sulfur sources for Archean komatiite-hosted Ni-Cu deposits even in situations where $\delta^{34}S$ values are nondiagnostic.

The concentration of chalcophile elements in mineralized and unmineralized komatiitic rocks can be affected by primary magmatic processes (e.g., degree of mantle melting, composition of mantle, varying R factor, crustal contamination) and/or secondary processes. In general, the chalcophile

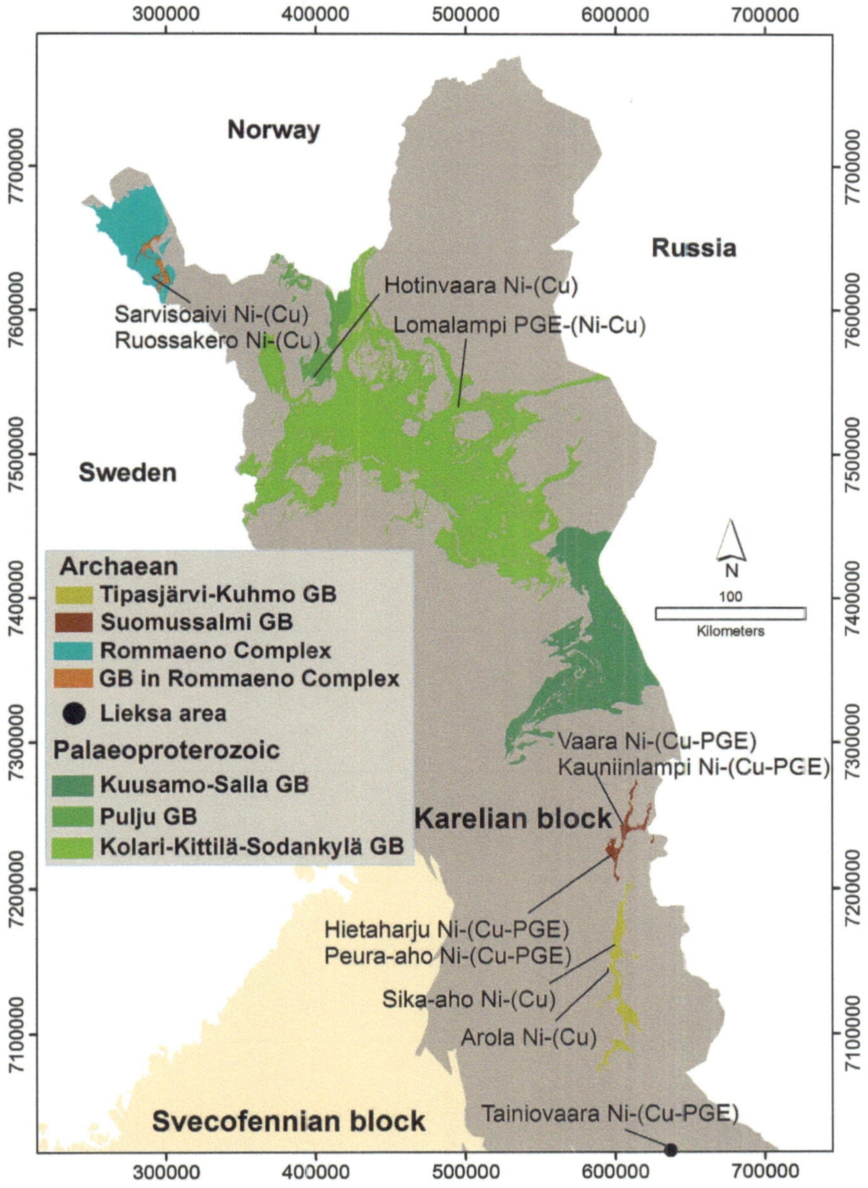

FIGURE 3.2.1 Archaean and Paleoproterozoic greenstone belts (GB) in northern and eastern Finland and associated komatiite-hosted Ni-Cu-PGE mineralizations.

Table 3.2.1 Classification of Finnish komatiite hosted Ni-Cu-PGE deposits (GB = greenstonen belt, AUK = aluminum undepleted komatiite, CLGB = Central Lapland greenstone belt)

						Mineralisation type			Ratios			
Deposit name	**Area**	**Age**	**Magma Type**	**Host rock**	**Deposit type**	**I**	**II**	**IV**	**Al/Ti**	**Ni/Cu**	**Pt/Pd**	**Pd/Ir**
Vaara	Suomussalmi GB	Archaean	AUK	oOC to oMC	Ni-(Cu-PGE)		x		19	13	0.4	28
Hietaharju	Suomussalmi GB	Archaean	AUK	oOC to oMC	Ni-(Cu-PGE)	(x)	x		15	4	0.5	22
Peura-aho	Suomussalmi GB	Archaean	AUK	oOC to oMC (+ surrounding schists)	Ni-(Cu-PGE)	(x)	x	x	15	3	0.5	14
Tainiovaara	Ultramafic body within Archean basement	Archaean	AUK	oMC	Ni-(Cu-PGE)	(x)	x		23	19	0.4	10
Lomalampi	Kolari-Kittilä-Sodankylä GB (CLGB)	Palaeoproterozoic	AUK	oOC to oMC	PGE-(Ni-Cu)		x		15	4	2	15
Hotinvaara	Pulju GB (CLGB)	Palaeoproterozoic	AUK	oMC to oAC	Ni-(Cu)	(x)	x	(x)	34	36	0.6	10
Ruossakero	Rommaeno Complex	Archaean	Ti-depleted	oMC to oAC	Ni-(Cu)		x		46	55	0.4	15
Sarvisoaivi	Rommaeno Complex	Archaean	Ti-depleted	oMC to oAC	Ni-(Cu)		x		37	15	0.2	47

oOC = olivine orthocumulate
oMC = olivine mesocumulate
oAC = olivine adcumulate
Deposit type ( ) = by product
Mineralisation type ( ) = minor type

element concentrations in the sulfide fraction vary from area to area and deposit to deposit, although they are commonly higher in disseminated ores than in massive ores (Lesher and Barnes, 2008; Fiorentini et al., 2010); it seems that disseminated sulfide deposits were formed at higher R factors than massive sulfide deposits (e.g., Lesher and Barnes, 2008). After crystallization, magmatic sulfides may become further enriched in precious metals if the host igneous body is subjected to strong hydrothermal alteration (e.g., oxidation, serpentinization, carbonatization, or desulfurization) or supergene alteration. This concerns particularly disseminated deposits, which may become markedly upgraded in their metal tenors (e.g., Stone et al., 2004; Barnes et al., 2009; Lesher and Barnes, 2009; Konnunaho et al., 2013). Secondary enrichment can explain the notable PGE (Pd + Pt) enrichment in some deposits in Finland. On the other hand, high metal tenors could also be due to PGE-rich mantle sources. There are still many open questions related to the noncoherent behavior of PGE in komatiites, for example, the origin of the different Pt/Pd in mineralized and unmineralized komatiites. In komatiite-hosted Ni-Cu-PGE deposits, Pd is commonly enriched over Pt by a factor of around 1.5–2, whereas most unmineralized komatiites show Pt/Pd close to unity (Lesher and Keays, 2002; Fiorentini et al., 2010). Barnes et al. (2012) discussed potential explanations for the high Pd/Pt in ores and concluded that this can be a primary magmatic signal in addition to postmagmatic alteration.

LOCATION AND CLASSIFICATION OF KOMATIITE-HOSTED NI-CU-(PGE) DEPOSITS IN FINLAND

Komatiite-hosted Ni-Cu-PGE deposits are found both in Archean and Paleoproterozoic greenstone belts in Finland (Fig. 3.2.1 and Table 3.2.1). In this chapter we will first describe several Archean deposits, including three major deposits within the Kuhmo-Suomussalmi greenstone belt (Vaara, Hietaharju, and Peura-aho), the Tainiovaara deposit located in a greenstone relict in eastern Finland, and the Ruossakero and Sarvisoaivi deposits in the Enontekiö-Käsivarsi area of northwestern Finland. In addition, some smaller Archean showings are also mentioned. This will be followed by a description of two notable Paleoproterozoic deposits, namely the Hotinvaara and Lomalampi deposits in the Central Lapland Greenstone Belt.

The Finnish Ni-Cu-PGE deposits belong mainly to the type II (disseminated type) of deposits, but some deposits belong to the type I (massive type) and type IV (hydrothermally, metamorphically mobilized type) deposits (Table 3.2.1) (cf. Barnes, 2006; Lesher and Barnes, 2009). Some deposits represent a mixture of types II, I and/or IV. All eight main deposits are associated with the MgO-rich and Cr-poor cumulate portion of komatiitic or komatiitic basalt units and they mostly consist of disseminated sulfides (e.g., Vaara, Lomalampi, and Ruossakero) and minor massive to semimassive sulfides (e.g., Tainiovaara, Hietaharju, and Peura-aho). Some of the deposits were modified by postmagmatic processes to various degrees (e.g., Vaara). In some deposits, Ni-Cu sulfides may be partly associated with the country rocks, including the Peura-aho deposit and the Sika-aho and Arola showings (Fig. 3.2.1).

In terms of their PGE content, the Finnish deposits can be divided into PGE-enriched and PGE-poor types. They have been classified as Ni-(Cu-PGE) and Ni-(Cu) deposits in this work, respectively. The PGE-enriched deposits are found in both age groups and are either S-poor (e.g., Lomalampi) or S-rich (e.g., Hietaharju and Peura-aho). The PGE-enriched deposits also tend to be enriched in Cu compared to the PGE-poor deposits (e.g., Sarvisoaivi, Ruossakero, and Hotinvaara).

NI-CU-PGE DEPOSITS OF THE SUOMUSSALMI GREENSTONE BELT

GEOLOGICAL SETTING

The Archean Suomussalmi greenstone belt represents the northernmost segment of the ~220-km-long and 10-km-wide, north–south trending Kuhmo-Suomussalmi greenstone belt complex (Fig. 3.2.1). A recent review of the complex was published by Papunen et al. (2009). The whole belt is characterized by the presence of komatiitic volcanic rocks, which show in some places well-preserved primary volcanic structures (Hanski, 1980; Tourpin et al., 1991; Gruau et al., 1992; Maier et al., 2013).

In the Suomussalmi belt, four lithostratigraphic formations have been defined. From the oldest to the youngest, these include the Luoma, Mesa-aho, Tervonen, Saarikylä, and Huutoniemi formations. The Luoma formation consists predominantly of andesitic to dacitic volcanic rocks, but also includes mafic and felsic volcanic rocks, quartz porphyry dikes, and banded amphibolites (Papunen et al., 2009). The U-Pb zircon data show that the age of the felsic volcanic rocks of the Luoma formation is ~2.95 Ga (Huhma et al., 2012). Most of the greenstone belt consists of tholeiitic basalts with Banded Iron Formation (BIF) interlayers belonging to the Tervonen formation. Stratigraphically above the Tervonen formation is the Saarikylä formation, which contains felsic metavolcanic rocks and graphitic black schists overlain by Cr-rich basalts and komatiitic rocks. The latter formation includes komatiitic olivine and olivine-pyroxene cumulates and lava flows, and komatiitic basalts (Halkoaho et al., 2000a,b; Luukkonen et al., 2002; Papunen et al., 2009).

The Vaara, Kauniinlampi, Hietaharju, and Peura-aho Ni-(Cu-PGE) deposits occur in this formation. The uppermost part of the stratigraphic sequence comprises felsic to intermediate metavolcanic rocks and graphite-bearing metasediments of the Huutoniemi formation. A felsic volcanic rock from the eastern branch, correlative with the Huutoniemi formation, has given a U-Pb zircon age of ~2.82 Ga (Huhma et al., 2012). From this, it can be concluded that the geochronological resolution currently available does not provide tight constraints on the depositional ages of the Mesa-aho, Tervonen, Saarikylä, and Huutoniemi formations. In analogy with other komatiitic rocks of the Tipasjärvi-Kuhmo-Suomussalmi greenstone complex, the komatiite-bearing Saarikylä formation is thought to have been formed at ~2.82 Ga (cf. Huhma et al., 2012).

GEOLOGY AND KOMATIITES OF THE VAARA REGION

The Vaara extrusive body is one of five ultramafic lenses that form a 15-km-long, north–south trending chain of bodies in the Saarikylä area, easily recognizable on magnetic maps. On the surface, the Vaara ultramafic lens is approximately 1 km long and 400 m wide. It is truncated by numerous northwest–southeast trending faults and is folded in a complicated manner (Konnunaho et al., 2013, see their Fig. 2). According to the current geological interpretation (Luukkonen et al., 2002; Papunen et al., 2009), the direction of the stratigraphic top is to the east. In the immediate footwall of the Vaara body there are phyllites, black schists, and sulfide-bearing sericite schists, likely belonging to the Mesa-aho formation (Luukkonen et al., 2002). Further to the west, felsic volcanic rocks of the Luoma formation occur. Due to folding, there also are felsic volcanic rocks intervened with cumulate rocks in the Vaara area, and these could be part of the Saarikylä formation. The phyllite-black schist association to the east of the Vaara body belongs to the Huutoniemi formation. The tholeiitic basalts on the southern and northern flanks of the Vaara body are assigned to the Tervonen formation (Konnunaho et al., 2013, see their Fig. 2).

At Vaara and elsewhere in the Tipasjärvi-Kuhmo-Suomussalmi greenstone belt complex, the komatiitic lavas have been pervasively metamorphosed to (1) tremolite-chlorite (± serpentine, talc, carbonate, albite) rocks, representing metamorphosed massive flow lobes and/or upper parts (A zone) of differentiated komatiite lava flows, and (2) serpentinites (±talc, tremolite, carbonate chlorite) representing metamorphosed massive lava flows and/or lower olivine cumulate zones (B zone) of differentiated lava flows. At Vaara, serpentinites are more abundant than tremolite-chlorite rocks. Commonly, tremolite-chlorite rocks change gradually to serpentinites, representing a transition from the A to B zone in the layered lava flows.

Although the rocks in the Vaara area have been metamorphosed at upper greenschist to middle amphibolite facies conditions and have undergone pervasive alteration, primary igneous textures can still be recognized in places, particularly among olivine cumulates. Magmatic silicates are totally altered, but chrome spinel is locally well preserved. In the northern part of the Vaara ultramafic lens, some thin lava flows altered to tremolite-chlorite rocks still preserve randomly oriented spinifex textures. Olivine cumulates are mainly homogenous, massive, and fine- to medium-grained rocks. The primary textures in these rocks include: (1) unimodal polyhedral cumulate textures, (2) bimodal polyhedral cumulate textures (equant olivine grains together with scattered olivine plates), and (3) harrisitic textures. Accessory opaque minerals include chromite, magnetite, ilmenite, and Fe-Ni-Cu sulfides.

VAARA NI-(CU-PGE) DEPOSIT

The Vaara Ni-(Cu-PGE) deposit was discovered in 1998 by the Geological Survey of Finland (GTK). The deposit is composed exclusively of disseminated Fe-Ni-Cu sulfides that are hosted by komatiitic olivine cumulates (Figs. 3.2.2, 3.2.3, Table 3.2.1). It belongs to the type II komatiitic deposits (i.e., stratabound internal), comprising disseminated sulfides mainly in the central part of olivine cumulate bodies. The deposit is north–south trending and has been delineated by drilling for a strike length of ~450 m and to a depth of 50–180 m below the surface (Halkoaho et al., 2000a). The deposit consists of three separate mineralized sulfide horizons (Figs. 3.2.2. and 3.2.3), the thickness of which varies from ~2–3 m (at the northernmost end) to ~50 m (at the southernmost end). As shown by the drillcore profile in Fig. 3.2.3, the mineralization is associated with relatively Cr-poor olivine cumulates. The indicated reserves of the deposit are 2.62 Mt of ore at 0.49 wt% Ni, 0.04 wt% Cu, 0.01 wt% Co, 0.28 ppm Pd, and 0.11 ppm Pt (Altona Mining, 2012).

Disseminated sulfides (Ø 0.1 to 1 mm) occur mainly in the interstitial space between former olivine crystals, and the original shapes of the sulfide blebs are well preserved. Of the sulfide minerals, millerite (50–75 vol%), pyrite (15–35 vol%), and chalcopyrite (~10 vol%) are the most abundant. Ni-rich pentlandite (41 wt% Ni) and violarite (38 wt% Ni) are less abundant, but can be locally important (Fig. 3.2.4 (A)) (Halkoaho et al., 2000a; Konnunaho et al., 2013). The volume of magnetite within sulfide blebs varies from 40–80 vol%. The quantity of magnetite at Vaara is higher than in most other komatiite-related Ni-Cu deposits (e.g., Heath et al., 2001). However, millerite-pyrite-magnetite blebs similar to those occurring at Vaara, with the magnetite content reaching up to 50 vol%, have been described from disseminated ores in the Otter Shoot and Black Swan deposits in Western Australia (Keele and Nickel, 1974; Barnes et al., 2009). Extensive replacement of interstitial sulfides by magnetite (Fig. 3.2.4(A)) and the presence of millerite- and violarite-bearing, pyrrhotite-free sulfide assemblages indicate postmagmatic, low-temperature hydrothermal oxidation of the primary magmatic pyrrhotite-pentlandite-chalcopyrite assemblages and associated sulfur loss. This process has led to a significant upgrading of the original metal tenors of the Vaara deposit (Konnunaho et al., 2013).

Platinum-group and tellurium minerals are mainly associated with Ni-bearing sulfides, chalcopyrite, and pyrrhotite, and occur as small grains (<10 μm) inside or at the edge of sulfide grains. In some

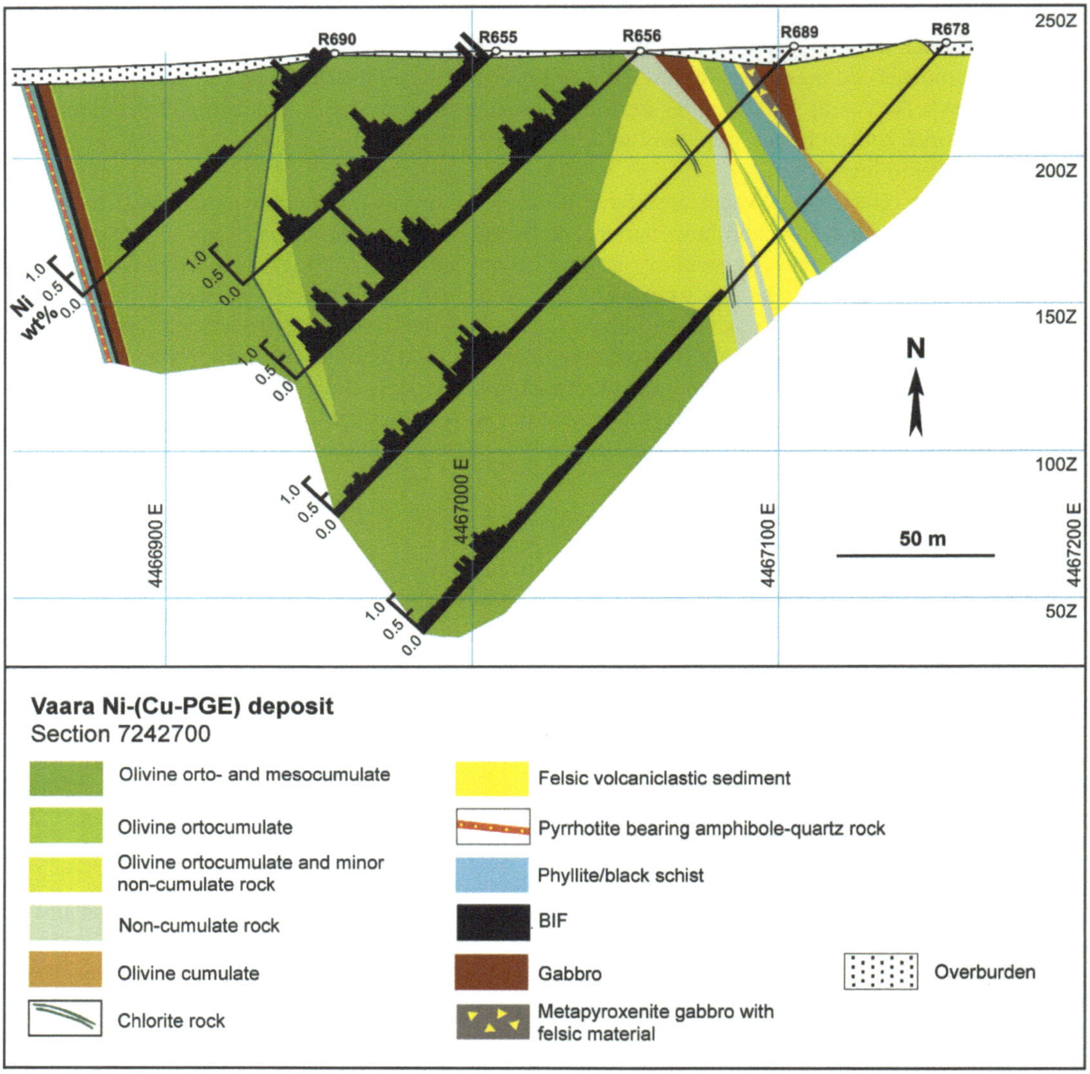

FIGURE 3.2.2 Vertical section of the Vaara ultramafic body, showing variation of whole-rock nickel content in a drilling profile.

Source: *Modified after Luukkonen et al. (2002) and Konnunaho et al. (2013).*

cases, PGE and tellurium minerals have also been found within silicates (notably serpentinized olivine) and oxides (ilmenite and chromite). The most common Pt-bearing mineral is sperrylite $PtAs_2$. Several different Pd-bearing minerals have also been identified (Konnunaho et al., 2013).

During the exploration within the area, GTK also found two small Ni-(Cu-PGE) occurrences called Kauniinlampi North and Kauniinlampi South (Fig. 3.2.1). These occurrences are associated

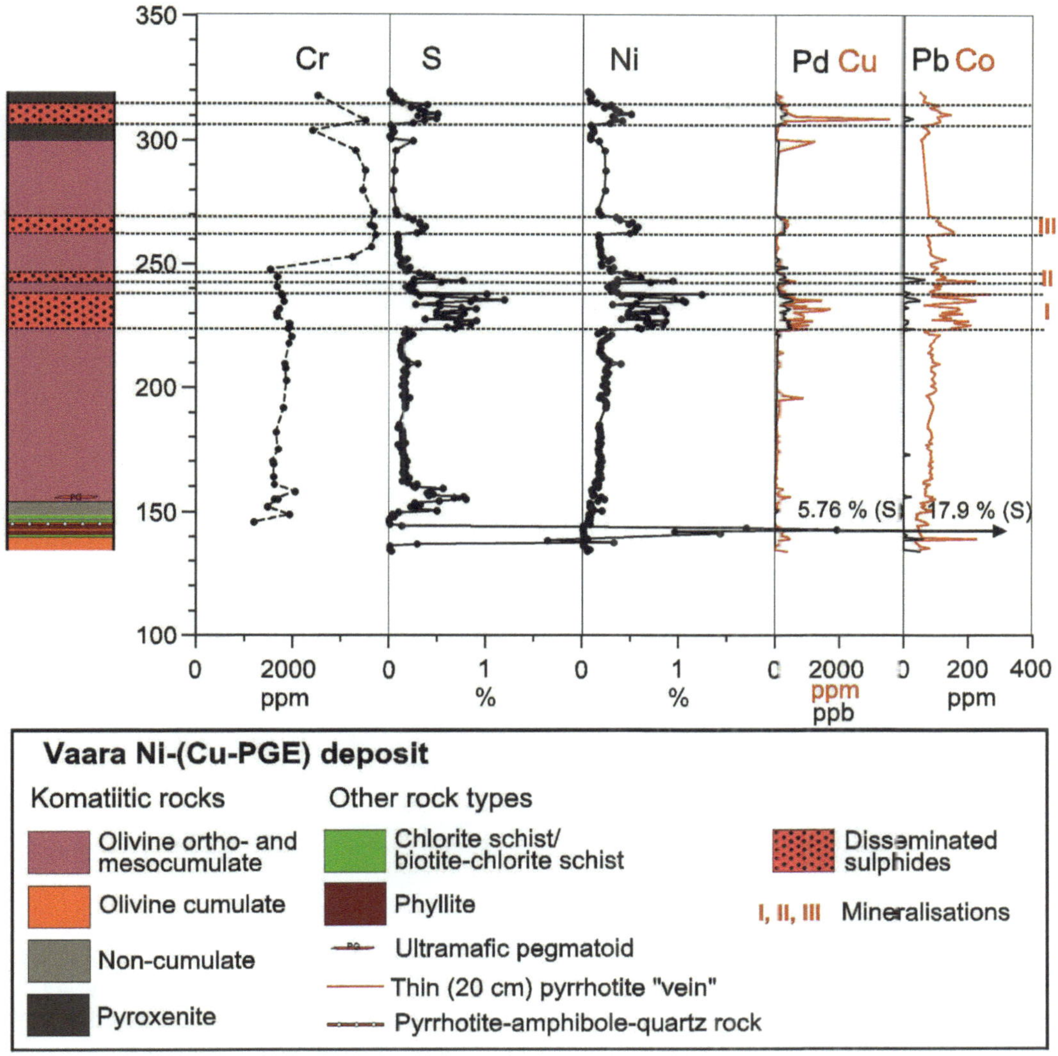

FIGURE 3.2.3 Variation in Cr, S, and chalcophile element concentrations across the Vaara deposit.

Source: *Modified after Luukkonen et al. (2002) and Konnunaho et al. (2013).*

with the Kauniinlampi cumulate body in the Saarikylä area (a few kilometers north of the Vaara cumulate body), which is cut by a shear zone. PGE-enriched sulfides have been remobilized into this zone at Kauniinlampi North. Kauniinlampi South consists of Ni-poor disseminated iron sulfides in the cumulate body. These two occurrences are not economically important, but they provide further evidence for operation of ore-forming processes in the Saarikylä area (Halkoaho et al., 2000a).

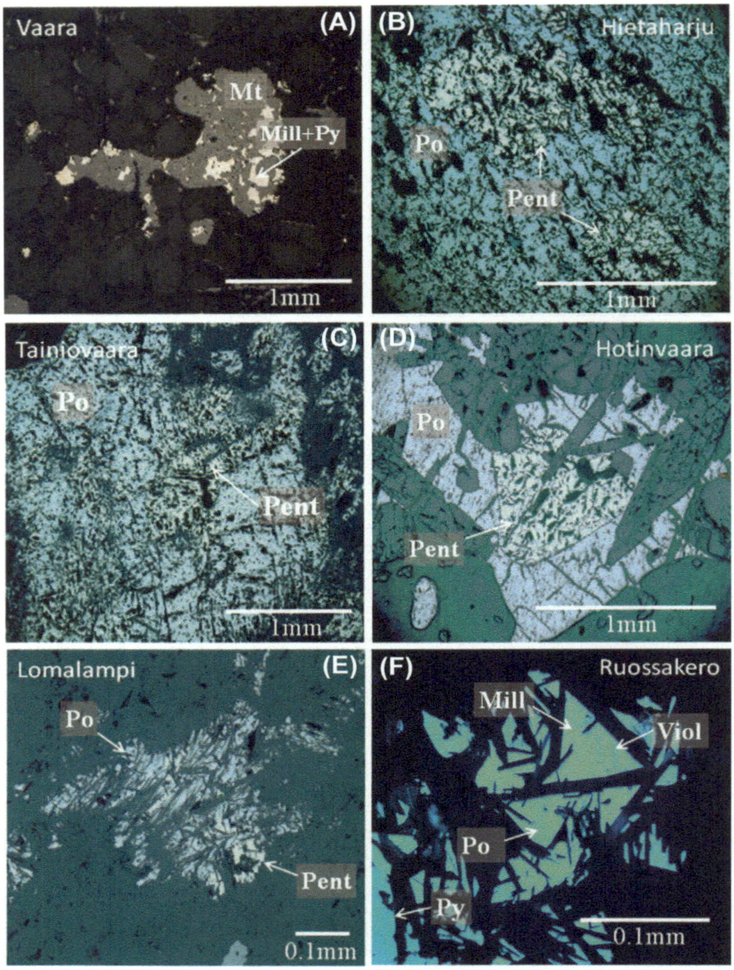

FIGURE 3.2.4 Photomicrographs in reflected light of sulfide aggregates in olivine cumulates from (A) Vaara (disseminated sulfide-magnetite aggregates), (B) Hietaharju (massive sulfides), (C) Tainiovaara (massive sulfides), (D) Hotinvaara (coarse grained sulfide blebs), (E) Lomalampi (disseminated sulfides), and (F) Ruossakero (disseminated sulfides).

Note the extensive replacement of sulfides by magnetite in the Vaara deposit. Abbreviations: Py = pyrite, Mill = millerite, Viol = violarite, Pent = pentlandite, Po = pyrrhotite, Mt = magnetite

GEOLOGY AND KOMATIITES OF THE HIETAHARJU AND PEURA-AHO AREAS

The komatiite-hosted Ni-Cu-PGE sulfide deposits of Peura-aho and Hietaharju are located in the Kiannanniemi area (Fig. 3.2.1 and Table 3.2.1), approximately 13 and 18 km southwest of the Vaara deposit, respectively. The Ni-Cu-PGE deposits occur in metamorphosed olivine-pyroxene cumulates of komatiitic basalt and surrounding schists belonging to the Saarikylä formation. The lithological unit containing felsic and mafic metavolcanic rocks, komatiitic basalts, and interbedded sulfur-bearing metasedimentary rocks (phyllites, black schists, and felsic schists) ranges from 100–150 m in thickness. Komatiitic lava flows and related olivine cumulates also occur in the Kiannanniemi area.

Fine- to medium-grained tremolite-chlorite-(±serpentine) rocks represent komatiitic noncumulates to orthocumulates or pyroxenitic cumulate portions. Komatiitic basalts are mainly massive hornblende-plagioclase rocks. Primary flow structures have been almost completely obliterated, but some well-preserved pillow and variolitic structures were found in outcrops of komatiitic basalts. Komatiitic olivine-pyroxene cumulates were metamorphosed to fine- to medium-grained serpentine-talc-carbonate-chlorite rocks. Talc-carbonate rocks (i.e., soapstones) are common, especially in the Hietaharju area. In some cases, cumulate textures occur in less altered and sheared orthocumulates and pyroxenites.

HIETAHARJU NI-(CU-PGE) DEPOSIT

The Hietaharju Ni-Cu-PGE deposit was found in the early 1960s by Outokumpu Oy. The deposit is mainly composed of disseminated Fe-Ni-Cu sulfides, but massive to semimassive veins and lenses as well as breccias are also present. The deposit is hosted by olivine-pyroxene cumulates of komatiitic basalts (Fig. 3.2.5). It belongs to the type II komatiitic deposits (i.e., stratabound internal), comprising disseminated sulfides mainly in the central part of olivine cumulate bodies. However, Hietaharju also contains some features of type I komatiitic deposits (Lesher and Keays, 2002) (Table 3.2.1). The host unit is approximately 100 m thick and 1 km long. The deposit is north–south trending and has a strike length of ~200 m, a width of 50 m, and a depth of at least 200 m (GTK Mineral Deposit Database, 2014). The deposit consists of several subparallel mineralized horizons or lenses (Fig. 3.2.5) whose thicknesses vary from ~0.5–10 m. The estimated reserves of the Hietaharju deposit are 0.85 Mt of ore at 0.85 wt% Ni, 0.44 wt% Cu, 0.06 wt% Co, 1.25 ppm Pd, and 0.53 ppm Pt (Altona Mining, 2012).

Pyrrhotite, pentlandite, and chalcopyrite are the most abundant sulfide minerals at Hietaharju. Based on Kojonen (1981) and Kurki and Papunen (1985), the ores can be divided into three main types: (1) finely disseminated sulfides and thin veinlets, (2) massive (Fig. 3.2.4(B)), granular sulfides, and (3) brecciated sulfides. Small amounts of sphalerite, cubanite, and mackinawite occur as inclusions in chalcophyrite and pyrite. Marcasite and violarite are common alteration phases. Arsenides (e.g., gersdorffite and cobaltite) are associated with disseminated sulfides in talc-carbonate rocks, and gersdorffite occurs as one of the main ore minerals in the most intensively carbonated parts of the deposit (Kojonen, 1981; Halkoaho and Papunen, 1998).

Gold, tellurides (i.e., Pd- and Bi-bearing tellurides), sperrylite, and other PGMs (platinum group minerals) occur mainly as inclusions in other ore minerals, but they are also found within silicates and oxides. The oxide minerals magnetite, ilmenite, and zoned chromite are common in all rock types, occurring as fine disseminations through the ultramafic body. All ore types contain disseminated magnetite, but in contrast to the Vaara deposit, sulfide replacement by magnetite is rare (Figs. 3.2.4(A,B)). Pyrrhotite, pentlandite, and chalcopyrite represent the primary magmatic sulfide assemblage, which was modified by postmagmatic processes (e.g., formation of Ni-Co arsenides). Massive and brecciated sulfides are probably the result of postmagmatic sulfide mobilization, but some massive lenses at the contact between the ultramafic body and metasedimentary rocks could represent reworked primary massive sulfide accumulations. Having relatively high Pd and Pt concentrations, the Hietaharju deposit belongs to the PGE-rich group of the komatiite-hosted Ni-Cu-PGE deposits in Finland (refer to Table 3.2.1).

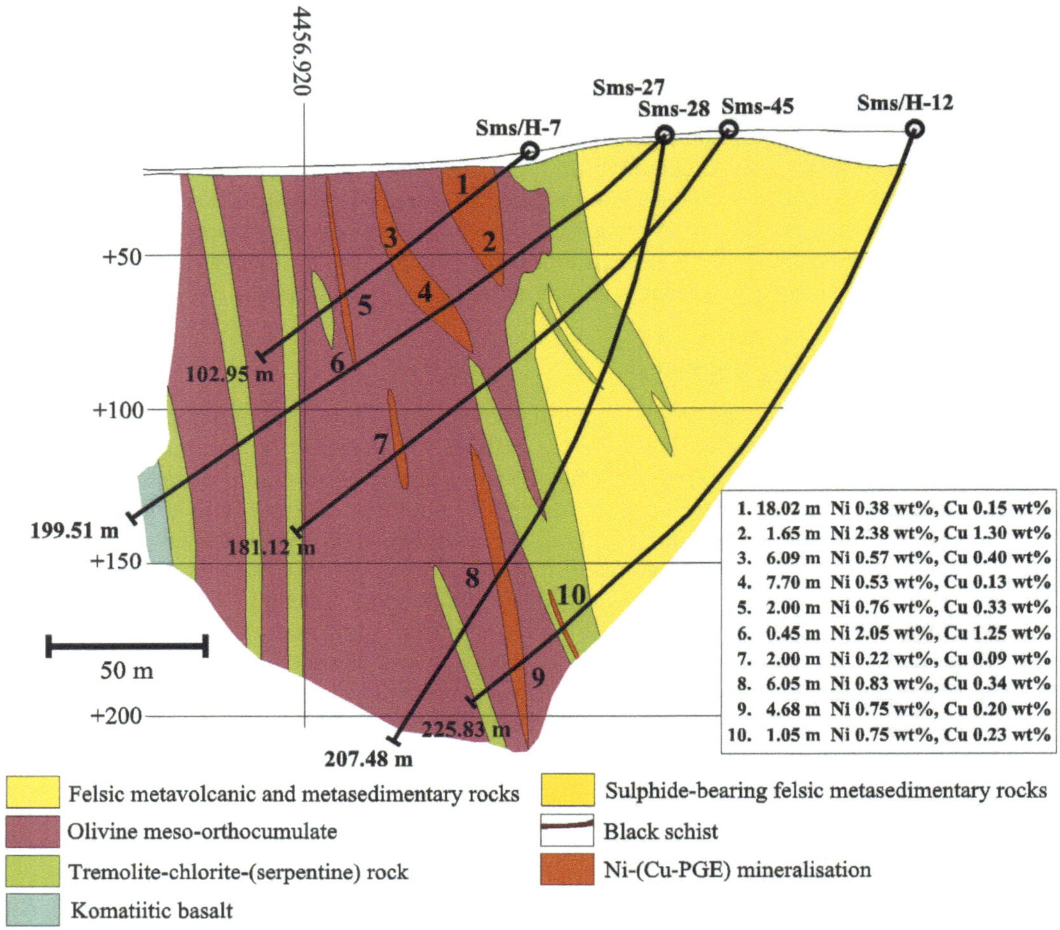

FIGURE 3.2.5 Vertical section of the Hietaharju nickel deposit showing locations of drillcores and their Ni and Cu contents.

Source: *Modified after Kojonen (1981) and Kurki and Papunen (1985). The table shows core lengths and average Ni and Cu contents of intersected sulfide-bearing zones.*

PEURA-AHO NI-(CU-PGE) DEPOSIT

The Peura-aho Ni-Cu-PGE deposit was found during the same exploration activities as the Hietaharju deposit. The Peura-aho area is characterized by an anticlinal structure with the fold axis dipping 65–70° to the east (Kojonen 1981; Kurki and Papunen, 1985) (Fig. 3.2.6). The Peura-aho Ni-Cu-PGE deposit consists of five sulfide-bearing lenses (A–E), whose length varies from ~30–250 m and thickness from ~3–10 m (Kojonen, 1981; Kurki and Papunen, 1985) (Fig. 3.2.6). The ore deposit is approximately 200 m long and 50 m wide on the surface and extends to a depth of at least 150 m (GTK Mineral Deposit Database, 2014). The main mineralized lenses are termed the A and B mineralizations. The disseminated Fe-Ni-Cu sulfides of mineralization A are located in the central part of an olivine ortho- to

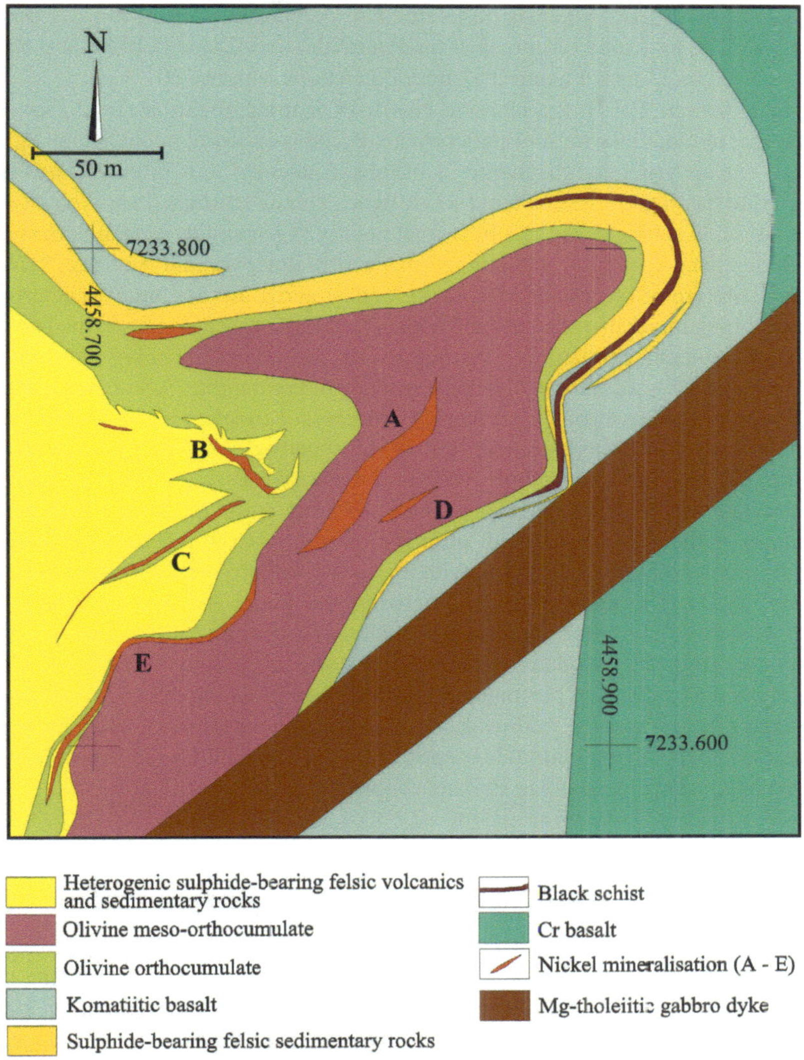

FIGURE 3.2.6 Geological map of the Peura-aho area.

Source: *Modified after Kojonen (1981) and Kurki and Papunen (1985).*

mesocumulate body (serpentinite), whereas massive sulfides of mineralization B are associated with tremolite-chlorite-(± serpentine) rocks, chlorite schists, and/or felsic metavolcanic/metasedimentary rocks. The sulfide disseminations in mineralizations C, D, and E are relatively weak. The disseminated sulfides in the C mineralization occur in tremolite-chlorite-(±serpentine) rocks stratigraphically below the serpentinite lens. Mineralization D occurs in the middle part of the small serpentinite lens and mineralization E in the basal part of a small serpentinite lens between the serpentinite and tremolite-chlorite-(±serpentine) rock or quartz-feldspar schist (Fig. 3.2.6). Mineralizations A, C, D, and E belong to the type II komatiitic deposit, whereas mineralization B represents the type IV (Lesher and Keays, 2002) (Table 3.2.1). The indicated resources of the Peura-aho deposit are 0.40 Mt at 0.63 wt% Ni, 0.29 wt% Cu, 0.04 wt% Co, 0.28 ppm Pt, and 0.62 ppm Pd (Altona Mining, 2012).

According to Kojonen (1981), the massive Ni-Cu-PGE mineralization (B in Fig. 3.2.6) consists mostly of pyrrhotite, pyrite, violarite, and chalcopyrite. Pyrite is euhedral or subhedral and contains marcasite, pyrrhotite, chalcopyrite, and magnetite inclusions. Chalcopyrite is often twinned and consists of lamellae. Pentlandite has mainly been altered to violarite. The disseminated and vein-style Ni-Cu-PGE mineralizations (A, C, D, and E in Fig. 3.2.6) consist mainly of pyrrhotite, pentlandite, and chalcopyrite. The oxides include zoned chromite, fine-grained magnetite, ilmenite, and hematite. Marcasite, mackinawite, and sphalerite are minor phases. Marcasite and violarite are common weathering products of sulfides. Platinum-group minerals are associated mainly with sulfides, analogous to the Hietaharju deposit.

The origin of the sulfides can largely be explained by alteration of primary pyrrhotite, pentlandite, and chalcopyrite. Violarite is a common alteration product of pentlandite. According to Kojonen (1981), it formed at a low temperature through weathering processes. Kojonen (1981) divided the pyrite in the Peura-aho mineralization into three generations: (1) Co-bearing euhedral pyrite in the disseminated A mineralization, (2) Ni-bearing euhedral pyrite in the massive B mineralization, and (3) secondary pyrite formed through weathering of pyrrhotite. The Co-bearing pyrite might be a high temperature mineral. The Ni-bearing pyrite is associated with sulfurization of Fe and Ni released during serpentinization. The secondary pyrite is normally intergrown with marcasite, which formed through weathering of pyrrhotite (Kojonen, 1981). The main difference between the Peura-aho and Hietaharju Ni-(Cu-PGE) mineralizations is that, in the latter, the talc-carbonate rocks contain Ni-Co arsenides. According to Kojonen (1981), this suggests introduction of arsenic by alteration fluids. Massive sulfide mineralizations probably formed as a result of postmagmatic sulfide mobilization, but some massive lenses could represent reworked primary massive sulfide accumulations, analogous to the Hietaharju deposit. Similarly to the Hietaharju deposit, the Peura-aho deposit belongs to the PGE- and Cu-enriched group of the komatiite-hosted Ni-Cu-PGE deposits in Finland (Table 3.2.1).

THE TAINIOVAARA NI-CU-PGE DEPOSIT, EASTERN FINLAND

The Tainiovaara deposit is located ~7 km northeast of the town of Lieksa. A small Archean serpentinite body and associated amphibolites form a greenstone belt relict within Archean tonalite gneisses of the Lieksa complex (GTK, DigiKP, 2014) (Figs. 3.2.1 and 3.2.7). An approximately 180-m-long and 80-m-wide serpentinite lens dips 70° to the west, with its longitudinal axis plunging approximately 35° to the northwest (Pekkarinen, 1980; Vanne, 1981) (Fig. 3.2.7). The Tainiovaara deposit occurs in the central part of the serpentinite lens, consisting mainly of altered olivine mesocumulates (Fig. 3.2.7). During the exploration activities by GTK, several other similar serpentinite bodies were found in the area (Halkoaho and Niskanen, 2004).

The Tainiovaara deposit is approximately 130 m long and 25 m wide and extends to a depth of at least 50 m (Vanne, 1981). It consists mainly of disseminated sulfides (type II komatiitic deposit), but there are also small massive to semimassive and net-textured sulfide accumulations (type I) at the bottom of the serpentinite lens (Vanne, 1981; Papunen, 1989) (Fig. 3.2.7). The major ore minerals are pyrrhotite and pentlandite, and chalcopyrite and pyrite form minor phases. Magnetite, zoned chromite, and ilmenite occur in disseminated form throughout the ore body. The main gangue minerals are serpentine, talc, chlorite, carbonate, and tremolite.

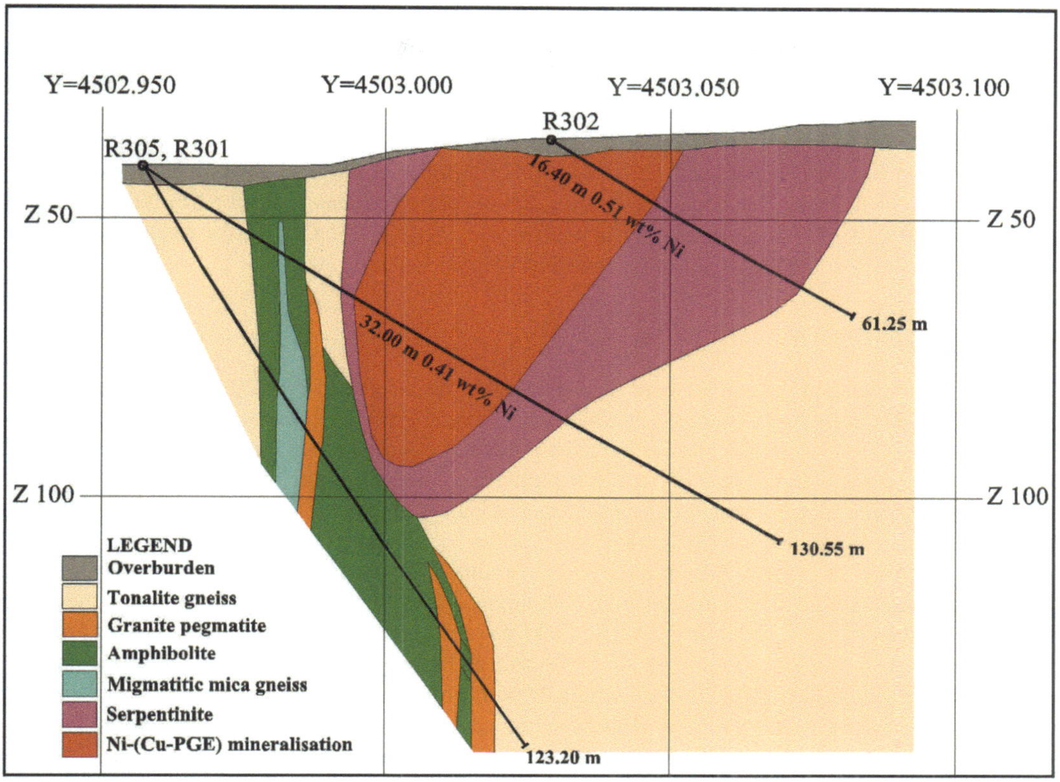

FIGURE 3.2.7 Vertical section of the Tainiovaara nickel deposit, showing locations of drillcores and their Ni content.

Source: *Modified after unpublished figure prepared by L. Pekkarinen.*

The estimated resources of the Tainiovaara deposit are 0.45 Mt at 0.5 wt% Ni, 0.03 wt% Cu, and 0.01 wt% Co (Vanne, 1981). The deposit was mined in 1989 by Outokumpu Oy. The total production was approximately 20,000 t at 1.4 wt% Ni and 0.12 wt% Cu (Puustinen et al., 1995). The deposit still contains 0.43 Mt of unexploited nickel ore. The Tainiovaara deposit is clearly magmatic in origin, as indicated by the magmatic sulfide assemblages, but was modified by postmagmatic processes. This is indicated by minor to moderate sulfide replacement by magnetite, especially in disseminated ore (Fig. 3.2.4(C)). The deposit is enriched in Cu and PGE, but not as much as some of the other Finnish PGE-enriched deposits (e.g., Hietaharju, Peura-aho, Lomalampi, and Vaara) (Table 3.2.1).

NI-CU-PGE DEPOSITS OF THE CENTRAL LAPLAND GREENSTONE BELT

GENERAL GEOLOGICAL SETTING

The Paleoproterozoic Central Lapland Greenstone Belt (CLGB) is one of the largest greenstone belts in the Fennoscandian Shield. The CLGB is generally subdivided into three subdomains (Fig. 3.2.1): (1) Pulju belt, (2) Kolari-Kittilä-Sodankylä belt, and (3) Kuusamo-Salla belt. The volcanic-sedimentary sequence of the CLGB is assigned to seven lithostratigraphic groups deposited on the Archean basement. From the oldest to the youngest these are: (1) Salla, (2) Vuojärvi, (3) Kuusamo, (4) Sodankylä, (5) Savukoski, (6) Kittilä, and (7) Kumpu. These groups have been subdivided into several formations by Lehtonen et al. (1998), Hanski et al. (2011), and Hanski and Huhma (2005). The oldest metavolcanic rocks (Salla group) erupted ~2.45 Ga ago, whereas the youngest metasedimentary rocks of the Kumpu group are younger than ~1.88 Ga (Hanski and Huhma, 2005); thus the geological evolution of the sequence spanned more than 0.5 Ga. The komatiite-bearing Savukoski group has yielded a Sm-Nd age of ~2.06 Ga (Hanski et al., 2011). For a more comprehensive description of the CLGB, the reader is referred to Lehtonen et al. (1998), Hanski et al (2011), and Hanski and Huhma (2005).

THE PULJU BELT

The Paleoproterozoic supracrustal rocks of the Pulju belt cover an area of 10 × 20 km in the northwestern part of the CLGB. The belt can be traced into Norway where it joins the Karasjok greenstone belt (Barnes and Often, 1990) (Fig. 3.2.1). In its lower part, the Pulju belt consists of a metasedimentary unit (quartzites and biotite-hornblende gneisses) and minor mafic metavolcanic rocks (Sietkuoja formation) of the Sodankylä group. The metavolcanic and metasedimentary units in the middle part (Mertavaara formation) of the sequence are associated with komatiitic rocks of the Savukoski group. MgO-rich olivine cumulates are rare in the CLGC, but in the Pulju belt, they are common. These cumulate bodies host the Hotinvaara Ni-(Cu) deposit and some other minor Ni-(Cu) showings. Sulfur-rich metasedimentary rocks (metacherts and calc-silicate rocks) and felsic metavolcanic rocks are among the lithological components of the Mertavaara formation. Komatiites are interbedded with sulfide-bearing metasedimentary rocks and metavolcanic rocks. The metasedimentary unit (paraschists with graphite-bearing interlayers) of the Vittaselkä formation (Savukoski group) forms the uppermost part of the stratigraphical succession in the Pulju belt (Inkinen et al., 1984; Papunen, 1998; DigiKp, 2014) (Figs. 3.2.8 and 3.2.9).

The komatiitic rocks of the Pulju belt were subdivided into two groups (Papunen, 1998): (1) nondifferentiated komatiitic lava flows (i.e., tremolite-chlorite rocks) without significant cumulate portions, and (2) differentiated komatiitic lava flows with extensive cumulate bodies (i.e., tremolite-chlorite-serpentine rocks to serpentinites and olivine rocks). Nondifferentiated komatiitic lava flows occur as independent layers together with mafic metavolcanic rocks of the Mertavaara formation. These rocks are characterized by well-preserved primary structures including volcanic breccias, pillows, and tuffogenic layering. They have been correlated with similar komatiites in the Sattasvaara formation of the Savukoski group (Lehtonen et al., 1998) and the Karasjok greenstone belt (Barnes and Often, 1990). Deviating from the stratigraphic position of the nondifferentiated lava flows, differentiated komatiitic lava flows occur in association with S-bearing metasediments and calc-silicate rocks occurring in the lower parts of the Mertavaara formation. Differentiated lava flows are typically coarse-grained and less foliated than nondifferentiated lava flows. Primary magmatic textures have not been recognized in differentiated lava flows. The gradual change from tremolite-chlorite-serpentine rocks to pure serpentinites indicates internal differentiation of flow units into A and B zones. In some places, tremolite-chlorite rocks occur as interbeds within sulfide-bearing metasediments and irregular masses within cumulates (Papunen, 1998).

Komatiites and associated supracrustal rocks were folded and sheared in at least four deformation phases and affected by hydrothermal alteration in several stages (Papunen, 1998). Relicts of an olivine spinifex texture were discovered in one drillcore in the Hotinvaara area (Papunen, 1998). The olivine

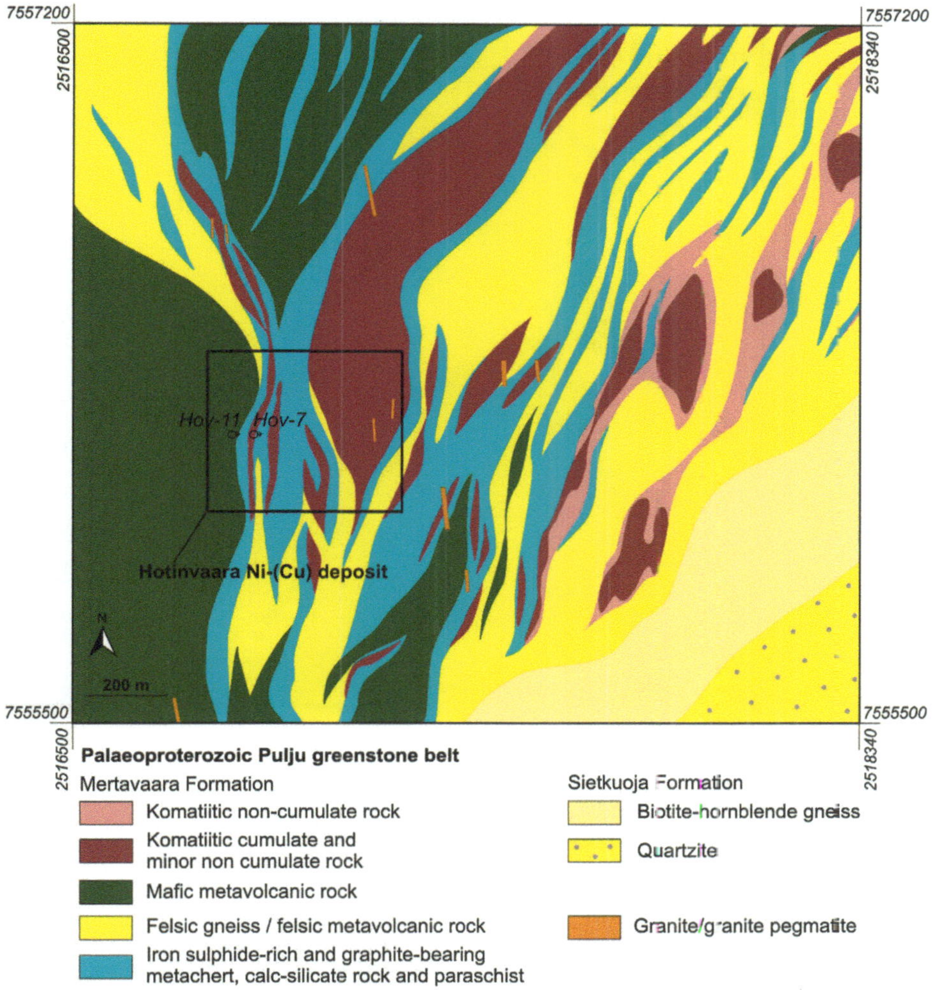

FIGURE 3.2.8 Geological map of the Hotinvaara area and location of the drilling profiles shown in Fig. 3.2.9.

Source: *Modified after Inkinen et al. (1984).*

cumulates are very heterogeneous, medium- to coarse-grained rocks, in which primary magmatic minerals and textures are not preserved. The cumulate portion consists of various serpentine-chlorite-tremolite rocks (±carbonate-talc) to almost pure olivine rocks (i.e., metadunites and metaperidotites). The metaperidotites contain metamorphic olivine, phlogopite, and pyroxenes. Accessory opaque minerals include chromite, magnetite, ilmenite, and Fe-Ni-Cu sulfides. Some chromite grains with an irregular form and without typical magnetite rims are also interpreted to be of metamorphic origin. Magnetite occurs as a fine-grained dissemination and dust, or forms crosscutting veinlets. Some magnetite was produced by oxidation of sulfides (Papunen, 1998).

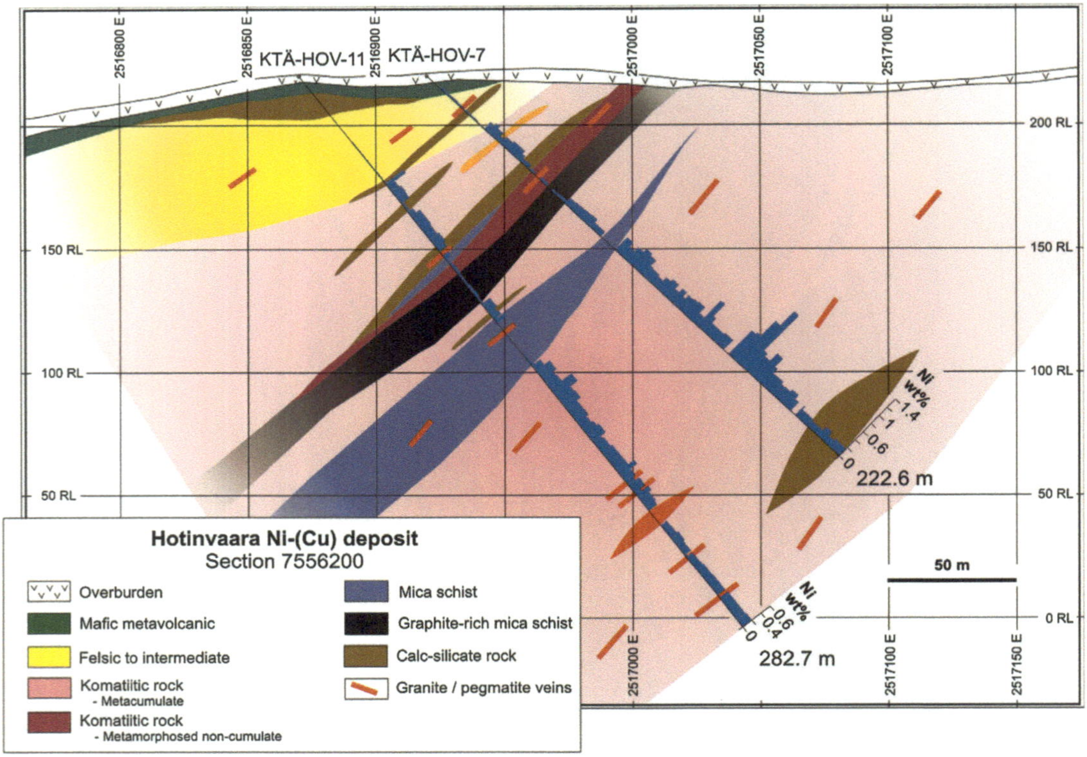

FIGURE 3.2.9 Vertical section of the Hotinvaara ultramafic body, showing locations of drillcores and their Ni content (see Fig. 3.2.8 for the location of the section).

HOTINVAARA NI-(CU) DEPOSIT

The Hotinvaara Ni-(Cu) mineralization was discovered as a result of exploration carried out by Outokumpu Oy in the early 1980s (Inkinen et al., 1984). Exploration activities of the company were mainly focused on an approximately 6-km-long and 1.3-km-wide zone in the Hotinvaara and Mertavaara areas where the komatiitic cumulates are most abundant. There are also other Ni-(Cu) occurrences (e.g., Mertavaara and Siettelöjoki) in the Pulju belt, but they are economically insignificant (Inkinen et al., 1984).

The Hotinvaara deposit is composed mainly of disseminated Fe-Ni-Cu sulfides and belongs to the type II komatiitic mineralization, but massive to semimassive sulfides (type I) have also been found in some drillcores. The Ni-(Cu) mineralization is hosted by strongly metamorphosed komatiitic olivine cumulates associated with differentiated komatiitic lava flows (Figs. 3.2.8, 3.2.9, and Table 3.2.1). In places, Ni-bearing iron sulfides associated with surrounding schists represent the type IV mineralization. The olivine cumulate body that hosts the mineralization is approximately 1.6 km × 1 km in size. Ni-(Cu) mineralization is roughly north–east trending and has been followed by drilling in a zone i.e., ~200 m long along strike, ~200 m wide, and ~200 m deep below the surface. The disseminated mineralization occurs in several subzones without any sharp contacts (Fig. 3.2.9). Massive to semimassive sulfides occur at the basal contact of the cumulate pile or close to the contact between the cumulates and intervening sediments. The preliminary feasibility study of the Hotinvaara deposit performed by Outokumpu Oy indicated Ni resources of 0.3 Mt ore at 0.66 wt% Ni (Inkinen et al., 1984).

Hexagonal pyrrhotite and pentlandite are the most abundant sulfide minerals at Hotinvaara. Pyrrhotite occurs as coarse and roundish grains or as a fine-grained dissemination. Monoclinic pyrrhotite occurs at the margins of, and in cracks within, hexagonal pyrrhotite grains. Pentlandite is associated with pyrrhotite and often occurs as individual coarse grains or as fine-grained dissemination and flames within pyrrhotite. Some of the sulfides, especially coarse-grained pentlandite and pyrrhotite, were formed by recrystallization of primary sulfides. Chalcopyrite, mackinawite, gersdorffite, troilite, and vallerite are accessory sulfides, and most of the Ni-rich sulfides are alteration products of pentlandite. Molybdenite and sphalerite have also been identified in mineralized samples (Sotka, 1984, 1986).

The chalcopyrite-poor and pyrrhotite-pentlandite-rich sulfide assemblage of the Hotinvaara mineralization is mainly a result of hydrothermal alteration, metamorphic recrystallization, and postmagmatic oxidation. Some disseminated and massive sulfides might be magmatic in origin (Papunen, 1998), but this question has not been investigated sufficiently. Only minor sulfide replacement by magnetite occurs in the Hotinvaara deposit (Fig. 3.2.4(D)). Platinum-group minerals have not been observed. Measured PGE abundances are very low and hence the deposit belongs to the PGE-poor group of the komatiite-hosted Ni-Cu-PGE deposits in Finland (Table 3.2.1).

GEOLOGICAL SETTING OF THE LOMALAMPI AREA

The Lomalampi deposit is located in the northern part of the Kittilä-Sodankylä belt (Fig. 3.2.1). In the Lomalampi area, the bedrock is mostly composed of metasedimentary rocks belonging to the Matarakoski formation (Savukoski group) and various types of ultramafic (komatiitic to komatiitic basalts) volcanic and cumulate rocks of the Peurasuvanto formation (Savukoski group). Minor olivine cumulates are associated with thin (~5–10 m thick) volcanic flows, but mostly they occur as sheet-like cumulate bodies up to several tens of meters in thickness that can be traced for 500–1500 m along strike.

KOMATIITES OF THE LOMALAMPI AREA

Based on drillcore data, komatiitic volcanic rocks are widespread in the Lomalampi area. Most common are fragmental rocks (hyaloclastites, tuffs, breccias) and relatively thin (3–10 m) flows. Macroscopically, they are mostly fine-grained, massive rocks composed of amphibole and chlorite with biotite present in more strongly altered rocks. Hyaloclastites and some other fragmental komatiites have preserved their original textures, but generally, primary textures of the volcanic rocks are destroyed or obscured by alteration and shearing.

Four significant massive komatiitic cumulate bodies occur in the Lomalampi area. The rocks are composed of serpentine-chlorite ± talc ± amphibole, with locally developed talc ± carbonate rocks. Original cumulus olivine is replaced by serpentine, while the interstitial material was altered to chlorite ± amphibole. Based on locally well-preserved textures, the cumulate rocks were originally olivine orthocumulates, which locally grade into mesocumulates. Olivine pseudomorphs show rounded to elongate habits with the average grain size varying between 0.5 and 1.0 mm. Poikilitic textures are locally preserved where original pyroxene oikocrysts are replaced by amphibole, and olivine inclusions by serpentine.

LOMALAMPI PGE-(NI-CU) DEPOSIT

The Lomalampi PGE-Ni-Cu deposit was discovered in 2004 by the Geological Survey of Finland. It is composed mainly of disseminated sulfides and classified as a type II komatiitic deposit (i.e., stratabound internal). Geochemically, it is an unusual type of mineralization, being enriched in Pt and Pd but containing relatively low concentrations of base metals (Ni, Cu, Co) (Table 3.2.1). The mineralization is hosted by a 30–65-m-thick, northeast–southwest trending olivine cumulate body, and it has been traced by drilling for ~550 m along the strike and to a maximum depth of 130 m. The deposit typically consists of one to three, several-meter-thick zones with 0.5–2.0 ppm Pt + Pd within a wider zone (up to 40 m) of weakly mineralized rock (0.1–0.5 ppm Pt + Pd) (Figs. 3.2.10 and 3.2.11) (Törmänen et al., 2010; Törmänen et al., *in preparation*). The mineralized zone usually occurs in the lower or middle part of the host cumulate, but it can also occur at the upper contact of the cumulate (Figs. 3.2.10 and 3.2.11). The preliminary mineral resource estimate published by GTK contains (at 0.3 ppm Pt cutoff) 1.05 Mt at 0.21 wt% Ni (including some silicate bound Ni), 0.078 wt% Cu, 0.42 ppm Pt, 0.19 ppm Pd, and 0.1 ppm Au (Koistinen and Heikura, 2010; Törmänen et al., 2010).

Disseminated sulfides occur as lobate to irregular grains and grain aggregates <0.1–0.5 mm in size, located in the interstitial space between altered olivine grains. In more sulfide-rich samples, recrystallized sulfide blebs can be up to 5 mm in size. Pyrrhotite is by far the most abundant sulfide phase, whereas pentlandite and chalcopyrite are subordinate (Fig. 3.2.4(E)). Arsenic-bearing samples contain

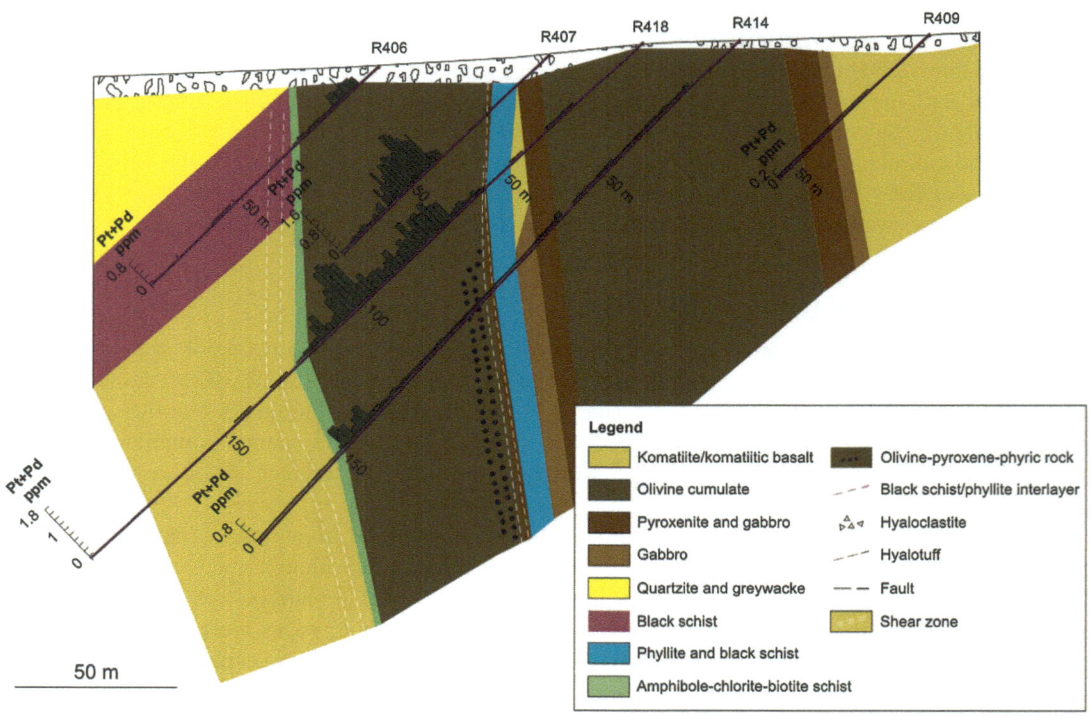

FIGURE 3.2.10 Vertical section of the Lomalampi deposit showing the mineralized and barren cumulate bodies and Pt + Pd contents in drillcores.

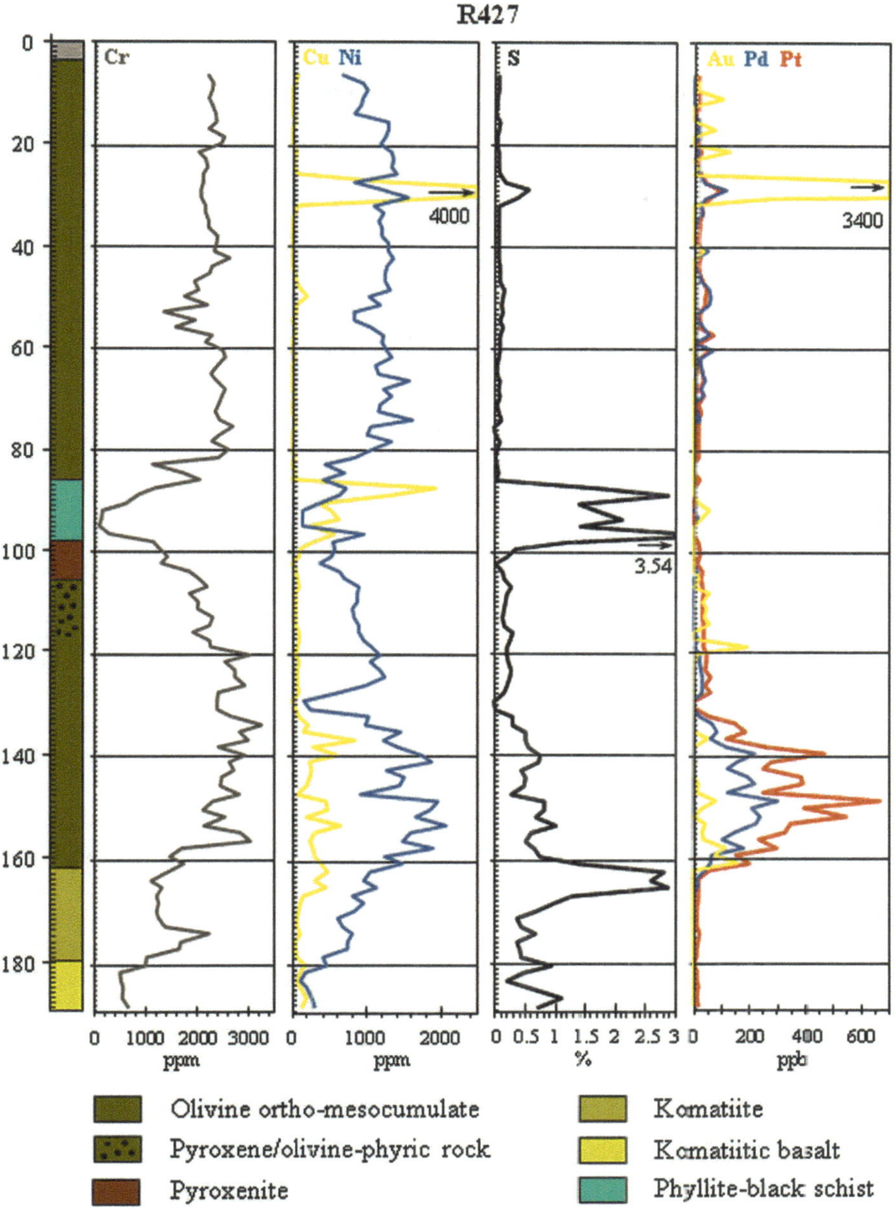

FIGURE 3.2.11 Variation of Cr, Ni, Cu, S, Au, Pd, and Pt contents across the barren and mineralized cumulate bodies in drillcore R427 from the Lomalampi deposit.

some sulfarsenides and nickeline grains. Other trace phases include pyrite, millerite, sphalerite, galena, and molybdenite.

Platinum-group minerals occur as small (<10 μm) grains associated with secondary silicates, base metal sulfides (mostly pyrrhotite and pentlandite), MeAsS-MeAs phases, oxides (magnetite, chromite), and carbonates. The only significant Pt-bearing phase is sperrylite, which occurs mostly with silicates (80%), and in lesser amounts in sulfides (8%) and oxides (8%). Palladium occurs as Pd- and Bi-bearing melonite—(Ni, Pd)(Te, Bi)$_2$—and as two unnamed Pd-Ni-Te-Sb ± Bi phases. The main host phases for Pd-minerals are silicates (~50%) and sulfides (39%), but the situation varies greatly between individual samples; the percentage of Pd-phases hosted by sulfides varies from 18–72 (Törmänen et al., *in preparation*).

The Lomalampi deposit belongs to the PGE-rich group of komatiite-hosted Ni-Cu-PGE deposit in Finland (i.e., it is a PGE-(Ni-Cu) deposit) (Table 3.2.1). It is magmatic in origin, but the high PGE content coupled with the low Ni-Cu-Co content is an unusual feature for komatiite-hosted Ni-Cu-PGE deposits worldwide.

NI-(CU) DEPOSITS OF THE ENONTEKIÖ-KÄSIVARSI AREA

The Archean Ropi terrane (referred to as Rommaeno complex) is located in the northwestern corner of Finland (Fig. 3.2.1) and northern Sweden. Reviews of the general geology of the area have been published by Bergman et al. (2001) and Sorjonen-Ward and Luukkonen (2005). The terrane is separated from the Karelian domain by the Karesuando-Arjeblog deformation zone and several Svecofennian granitoids (Bergman et al., 2001; Sorjonen-Ward and Luukkonen, 2005). On the western side, it is covered by the Caledonides. The Ropi terrane consists mainly of Archean supracrustal rocks surrounded by gneissose granitoids. Bodies of Archean felsic intrusive rocks also occur in the area (Bergman et al., 2001). Archean rocks are locally covered unconformably by Paleoproterozoic supracrustal rocks (e.g., Lätäseno group). The unconformity between the Archean and Paleoproterozoic rocks served as a zone where 2.5–2.4 Ga mafic intrusions (e.g., Tsohkoaivi and Kelottijärvi intrusions) were injected. The Archean supracrustal belts are characterized by the presence of komatiitic rocks, but so far only few studies of the Archean rocks from the area have been published (e.g., Heggie et al., 2013).

In recent geological interpretations by the Geological Survey of Finland (DigiKP, 2014), the Archean Rommaeno complex was divided into four formations (from oldest to youngest): Vuoskujoki, Ruossakero, Aatsakuru, and Pailuajärvi. These formations consist of metasedimentary and metavolcanic schists and belong to the Ropi group. The Vuoskujoki formation consists predominantly of sulfide-bearing, felsic to intermediate metatuffs with a U-Pb zircon age of ~2.93 Ga (in the felsic rocks) (GTK, unpublished data). This age is close to the U-Pb zircon age (2.95 Ga) of felsic metavolcanites from the Luoma formation in the Suomussalmi greenstone belt (Huhma et al., 2012). The Ruossakero formation consists of mafic and ultramafic metavolcanic rocks (komatiites) with BIF interlayers in the former (Fig. 3.2.12). It includes thick komatiitic olivine and olivine-pyroxene cumulates and thin lava flows, and minor amounts of komatiitic basalts. The Ruossakero and Sarvisoaivi Ni-(Cu) deposits occur within this formation.

The Aatsakuru formation consists of reworked sulfide-bearing felsic to intermediate metavolcanic rocks. U-Pb zircon data show that the age of these rocks is ~2.76 Ga (GTK, unpublished data). The uppermost part of the stratigraphic sequence comprises metasedimentary rocks such as sericite and fuchsite quartzites, and conglomerates of the Pailuajärvi formation. There is probably a hiatus between the Ruossakero and Aatsakuru formations.

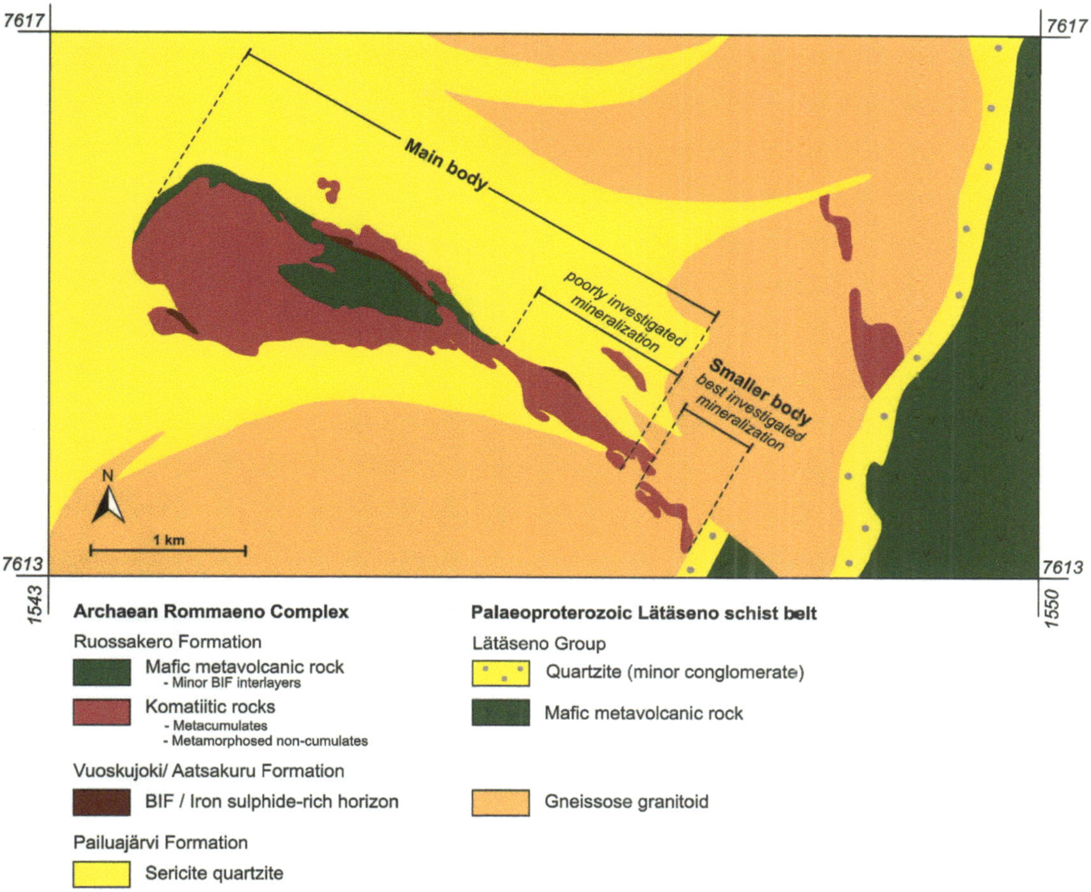

FIGURE 3.2.12 Geological map of the Ruossakero area.

Source: *Modified after Isomaa (1988).*

GEOLOGY AND KOMATIITES OF THE RUOSSAKERO AND SARVISOAIVI AREAS

The Ruossakero komatiitic body is one of the largest ultramafic lenses in the Archean Rommaeno complex (Figs.3.2.1 and 3.2.12). On the surface, the lens is ~7 km long and 500 m to 3 km wide. The whole ultramafic body and adjacent supracrustal rocks are folded in a complicated manner and there is a major shear zone at the northern contact of the ultramafic body. Isomaa (1988) divided the lens into two parts: (1) the main body and (2) the smaller body. According to the geological interpretation of the area, the main body is surrounded by Archean metasedimentary and metavolcanic rocks and the smaller body is located within Archean granitoids close to the contact of the Paleoproterozoic Lätäseno schist belt. The mafic metavolcanic rocks of the Ruossakero formation are associated with the Ruossakero main body and these include sulfide-bearing schist layers, which might be equivalents to the BIF interlayers or sulfide-bearing volcanic rocks of the Vuoskujoki or Aatsakuru formations. Minor amounts of

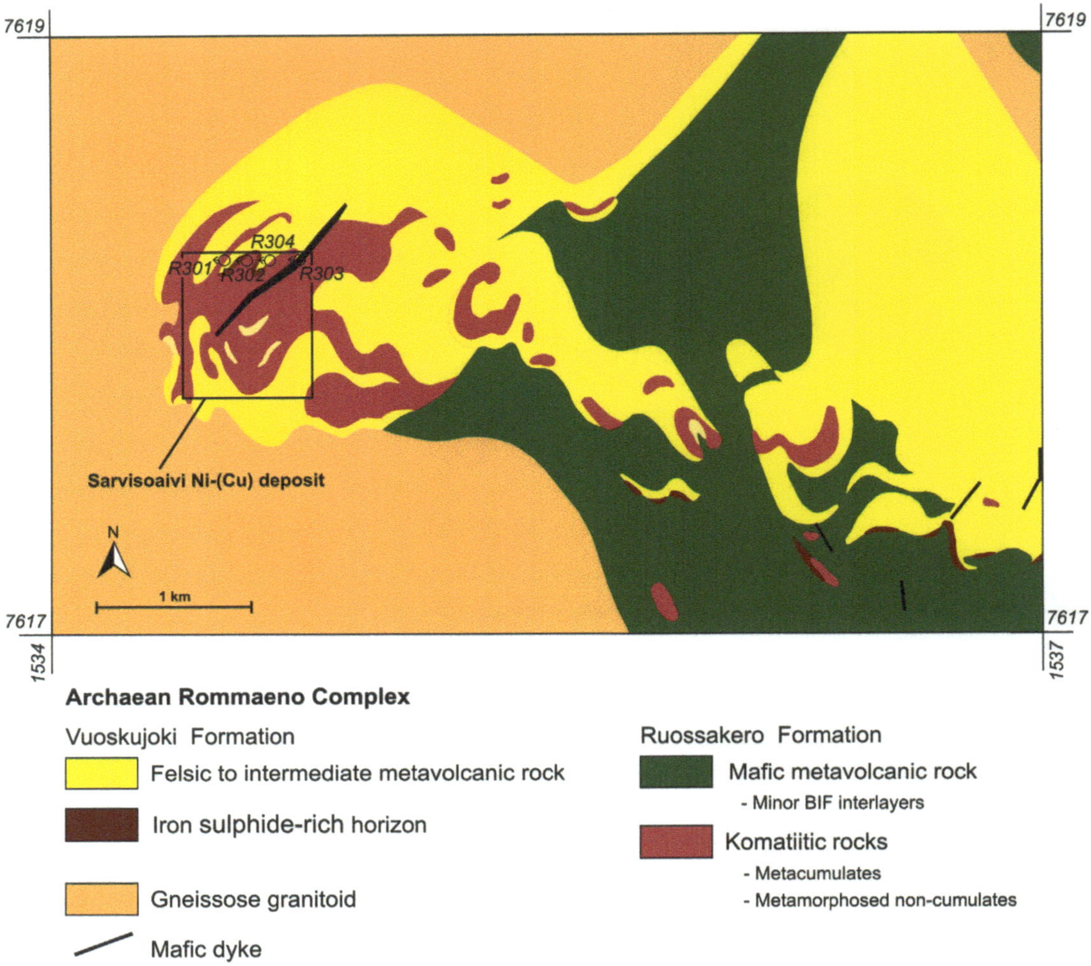

FIGURE 3.2.13 Geological map of the Sarvisoaivi area and location of the drilling profiles shown in Fig. 3.2.14.

Source: *Modified after Isomaa (1982).*

komatiitic lavas (tremolite-chlorite rocks) are also associated with the Ruossakero main and smaller cumulate bodies.

The Sarvisoaivi komatiitic body is located ~9 km west of the Ruossakero komatiitic body in the western corner of the Archean greenstone belt within the Rommaeno complex (Figs. 3.2.1 and 3.2.13). Komatiitic bodies in the Sarvisoaivi area can be divided into the main mineralized body and smaller unmineralized bodies on the eastern side of the main body (Fig. 3.2.13). The main body is complexly folded and cut by several shear zones (Isomaa, 1982).

The Sarvisoaivi komatiitic bodies are surrounded by Archean felsic to intermediate metavolcanic rocks of the Vuoskujoki formation and mafic metavolcanic rocks of the Ruossakero formation.

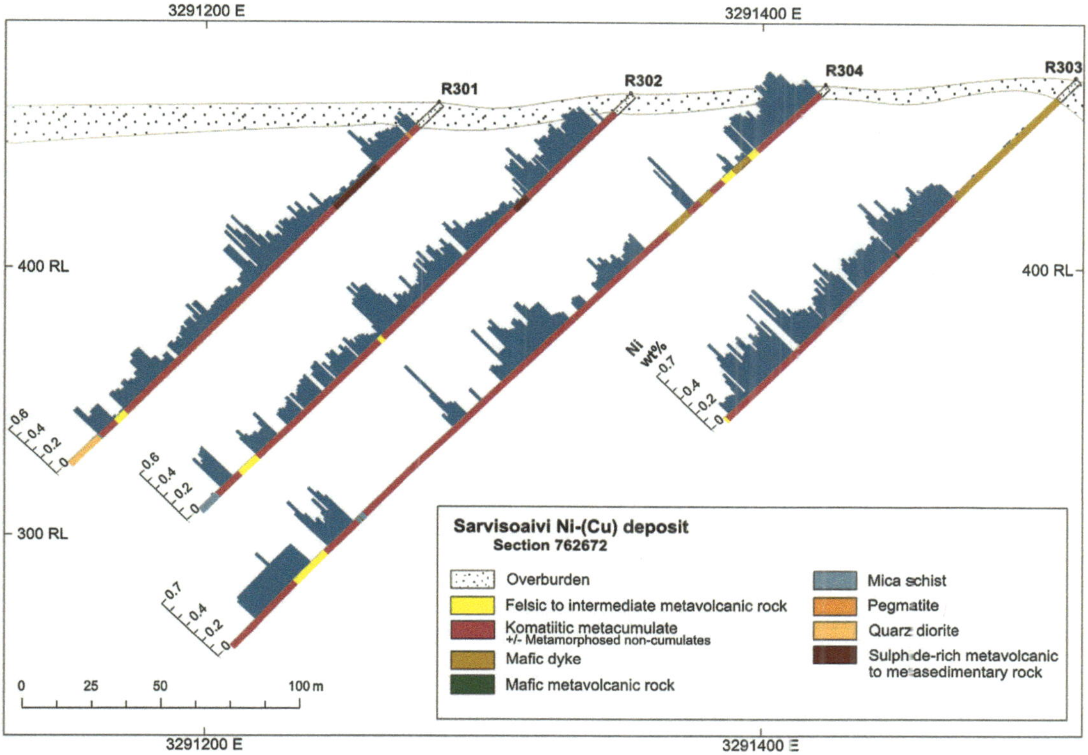

FIGURE 3.2.14 Vertical section of the Sarvisoaivi ultramafic body showing whole-rock Ni contents in drillcores (see Fig. 3.2.13 for the location of the section).

The metavolcanic rocks of the Vuoskujoki formation are rich in iron sulfides. Minor amounts of metasedimentary rocks (e.g., quartzites) of the Pailuajärvi formation also occur in the area. Thin zones of tremolite-chlorite rock are associated with the Sarvisoaivi ultramafic bodies. In the eastern part of the Sarvisoaivi area, they also occur as small lenses within metavolcanic and metasedimentary rocks (Fig. 3.2.13).

In the Ruossakero and Sarvisoaivi areas, komatiitic rocks consist of homogeneous, massive, and medium- to coarse-grained metamorphosed peridotites and dunites, as well as minor pyroxenites within dunites. The rocks were metamorphosed at amphibolite facies conditions, and the primary silicates have been replaced mainly by serpentine as well as chlorite, amphibole, talc, and carbonate. They commonly contain olivine and pyroxene of metamorphic origin. Despite the pervasive alteration, primary cumulus textures can still be recognized locally in ultramafic cumulates. Some ultramafic cumulates display a serpentine- and talc-pseudomorphed texture that resembles a spinifex texture, but it is metamorphic in origin. Snoken and Calk (1978) have called this kind of texture *jackstraw texture*.

The presence of metamorphic olivine and pyroxene is a characteristic feature in the Sarvisoaivi and Ruossakero cumulates. Large olivine grains are often zoned (Isomaa, 1982). In some places,

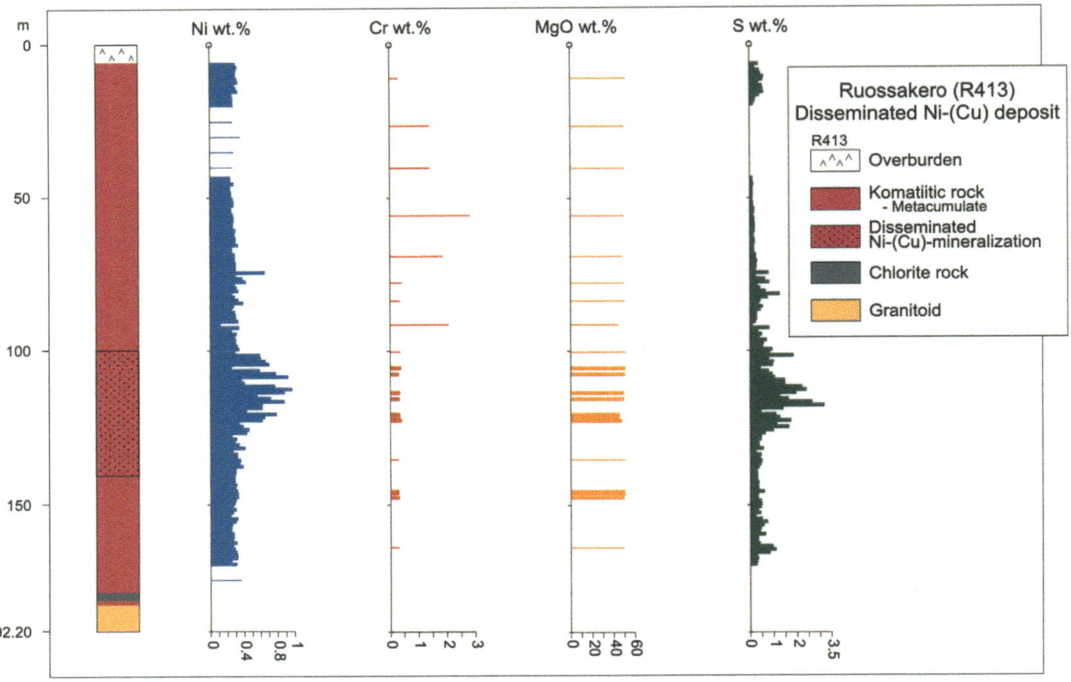

FIGURE 3.2.15 Variation of Ni, Cr, MgO, and S across the Ruossakero mineralized cumulate body in drillcore R413.

metamorphic pyroxene occurs as large (Ø >1 cm) grains poikilitically enclosing metamorphic olivine and other metamorphic silicates. Accessory opaque minerals include chromite, magnetite, ilmenite, and Fe-Ni-Cu sulfides. Thin zones of tremolite-chlorite rock are associated with the contact zone of the Ruossakero and Sarvisoaivi ultramafic bodies. In the Sarvisoaivi area, they also occur as small lenses within metavolcanic rocks. The komatiites that represent noncumulates contain variable proportions of amphibole (tremolite), chlorite, serpentine, and metamorphic olivine.

RUOSSAKERO AND SARVISOAIVI NI-(CU) DEPOSITS

The Sarvisoaivi deposit was discovered during a drilling program of the Geological Survey of Finland in 1978 and the Ruossakero deposit was discovered in 1980. Both deposits are composed of disseminated Fe-Ni-Cu sulfides hosted by metamorphosed Cr-poor komatiitic olivine cumulates (Figs. 3.2.14 and 3.2.15). The deposits belong to the type II komatiitic mineralization of Lesher and Keays (2002) (i.e., stratabound internal), comprising disseminated sulfides mainly in the central part of an olivine cumulate body (Figs. 3.2.14 and 3.2.15). Disseminated sulfide mineralization has been found in several distinct subzones in both deposits (Korhonen, 1981; Isomaa, 1982, 1986, 1988).

At Ruossakero, the smaller komatiitic body hosts the most thoroughly investigated mineralization. It is roughly west–east trending, extending for ~340 m along the strike. It is 100 m wide and 50–100 m

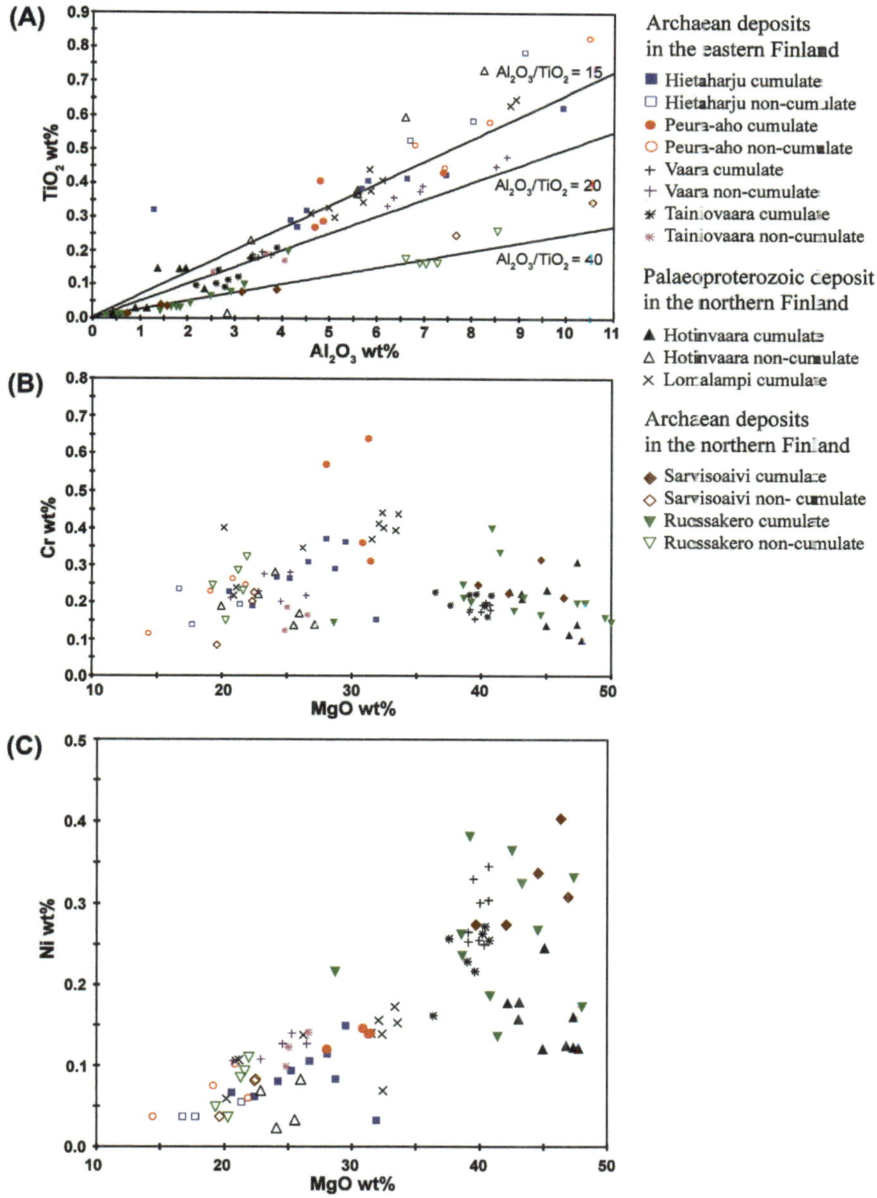

FIGURE 3.2.16 Variation diagrams for (A) TiO_2 versus Al_2O_3, (B) Cr vs. MgO, and (C) Ni vs. MgO for unmineralized komatiitic cumulate and noncumulate rocks.

Source: *From the Hietaharju, Hotinvaara, Lomalampi, Peura-aho, Ruossakero, Sarvisoaivi, Tainiovaara, and Vaara deposits.*

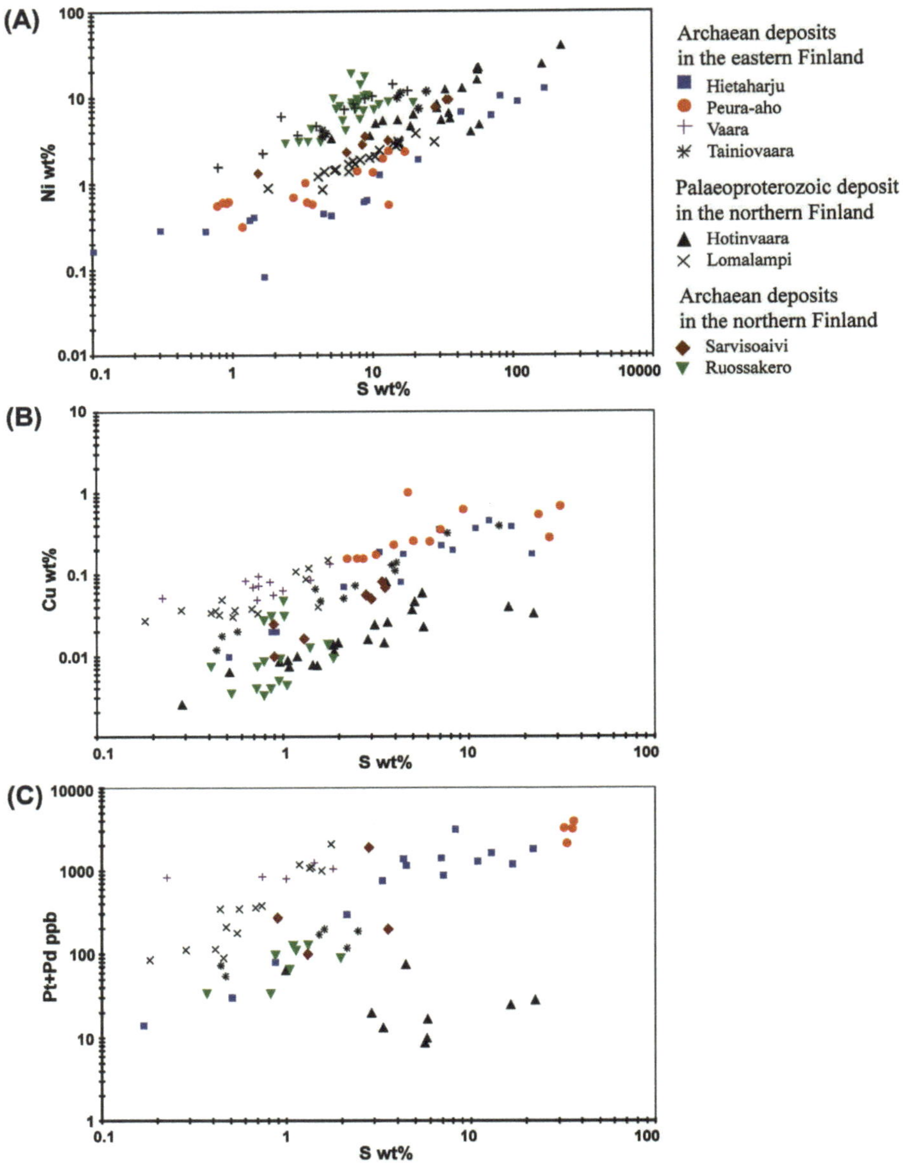

FIGURE 3.2.17 Plots of (A) Ni vs. S, (B) Cu vs. S, and (C) Pt + Pd vs. S for mineralized komatiitic cumulate samples.

Source: *From the Hietaharju, Hotinvaara, Lomalampi, Peura-aho, Ruossakero, Sarvisoaivi, Tainiovaara, and Vaara deposits.*

deep. The eastern part of the main komatiitic body hosts poorly investigated mineralization, which has a surface extent of ~1300 m, is 100–300 m wide, and extends to a depth of more than 400 m below the surface (Fig. 3.2.12). The preliminary feasibility study of the Ruossakero deposit performed by Outokumpu Oy suggests inferred resources of 4.2 Mt ore at 0.52 wt% Ni (0.5 wt% Ni cutoff) or 35.6 Mt ore at 0.33 wt% Ni (0.3 wt% Ni cutoff) (Lahtinen, 1996). The mineralized main body at Sarvisoaivi is rounded in shape, has a size of ~700 m × 700 m on the surface and reaches a depth of 200–300 m. No reliable resource estimates have so far been generated for the Sarvisoaivi deposit.

Despite the overall similarity of their host rocks and tectono-metamorphic history, the two ore deposits differ in terms of their sulfide mineralogy. At Ruossakero, the sulfide assemblage is dominated by pyrite, millerite, and pyrrhotite, whereas pyrrhotite and pentlandite are predominant at Sarvisoaivi. At Ruossakero, the mode of occurrence of pyrite varies from a fine-grained dissemination to 5-mm-sized, euhedral grains. Millerite occurs as rounded or elongate crystals (Ø 0.1–0.5 mm). In some cases, rounded millerite crystals are replaced by violarite. Chalcopyrite and pentlandite have also been occasionally found among other sulfide minerals. The Ruossakero deposit is similar to the Vaara deposit (Konnunaho et al., 2013) in terms of the presence of a millerite- and violarite-bearing sulfide assemblage. However, the former does not show such a high degree of replacement of sulfides by secondary magnetite as is the case at Vaara (refer to Figs. 3.2.4(A,F)).

The Sarvisoaivi mineralized body contains pentlandite and pyrrhotite as a fine- to medium-grained dissemination. Pentlandite also forms flames within pyrrhotite grains. Pyrite is found as cataclastic grains (Ø < 1 mm). Fractures cutting pyrite grains are often filled by pyrrhotite (Korhonen, 1981). Minor chalcopyrite is associated with other sulfides. Ni-rich sulfides (heazlewoodite, millerite, and violarite) occur as alteration products of Ni-bearing sulfides in the most altered and weathered ultramafic rocks of the deposit (Korhonen, 1981). Some thin (10–30 cm), massive pentlandite-pyrrhotite veins have also been observed. The presence of pyrrhotite, pentlandite, chalcopyrite, and pyrite at Sarvisoaivi indicates a primary magmatic nature of the sulfide assemblage.

Platinum-group minerals have not been found in the Ruossakero and Sarvisoaivi deposits, and measured PGE abundances are very low. Thus both deposits belong to the PGE-poor group of komatiite-hosted Ni-(Cu) deposits in Finland (Table 3.2.1).

GEOCHEMISTRY OF THE FINNISH KOMATIITE-HOSTED NI-CU-PGE DEPOSITS

WHOLE-ROCK GEOCHEMISTRY

Most of the komatiite-hosted Ni-Cu-PGE deposits in Finland are associated with the Al-undepleted komatiite type (see Table 3.2.1), showing moderate Al_2O_3/TiO_2 (15–22). Undifferentiated lava flows at Hietaharju, Peura-aho, Lomalampi, and Hotinvaara have subchondritic Al_2O_3/TiO_2 ratios (~15). Komatiites in the Enontekiö area and differentiated komatiitic lava flows with extensive cumulate bodies in the Hotinvaara area exhibit higher Al_2O_3/TiO_2 (20-65), whereas komatiites of the Tainiovaara area represent intermediate Al_2O_3/TiO_2 (20–30) (Fig. 3.2.16(A)).

The Hietaharju and Peura-aho deposits occur in the cumulate portions of lava flows generated from komatiitic basalt. The MgO content of the cumulates varies from 20–31 wt%. The MgO content of the noncumulates ranges between 14 and 21 wt%. The Lomalampi, Vaara, and Tainiovaara deposits are associated with cumulates of low-MgO komatiitic flow units. The MgO content of the cumulates is

20–41 wt%, with relatively high MgO contents in the Vaara and Tainiovaara cumulates (~36–40 wt%). The MgO content of noncumulates ranges from 20–27 wt%. Komatiitic cumulates of the Sarvisoaivi, Ruossakero, and Hotinvaara deposits have the highest MgO contents (40–50 wt%) and noncumulate rocks display MgO contents of 19–29 wt% (Figs. 3.2.16(B,C)). The absence of preserved spinifex-textured rocks, flow-top units, and magmatic silicate minerals (e.g., olivine) have handicapped the precise estimation of the MgO contents of the komatiitic parental melts of the Ni-Cu deposits in Finland.

Cumulates of komatiitic basalt in the Hietaharju and Peura-aho areas have Cr contents of 0.2–0.65 wt%, while the Cr content in the Lomalampi cumulates varies between 0.2 and 0.45 wt%. These komatiites show a positive correlation between MgO and Cr, following a cotectic olivine-chromite cumulate trend; thus, they were derived from Cr-saturated magmas (Fig. 3.2.16(B)). The Cr contents in cumulates of low-Mg komatiites (Vaara and Tainiovaara) and komatiites (Sarvisoaivi, Ruossakero, and Hotinvaara) vary between 0.2 and 0.4 wt%. These komatiites show a weakly negative or weakly positive correlation (Tainiovaara) between Cr and MgO. The corresponding cumulates were derived from Cr-undersaturated magmas with their compositional trend following an olivine-liquid mixing trend (cf. Barnes and Fiorentini, 2012 see their Fig. 8A).

Most of the studied areas (e.g., Sarvisoaivi, Ruossakero, Vaara, and Hotinvaara) comprise cumulates that have Cr contents clearly outside of the previously described trend (i.e., MgO-rich cumulates that plot between the cotectic olivine-chromite cumulate line and the olivine-liquid mixing line). This is common in mineralized systems, but not in komatiitic basalt systems, as discussed by Barnes and Brand (1999) and Barnes and Fiorentini (2012). However, most of the Ni-Cu-PGE-mineralized rocks comprise Mg-rich cumulates plotting close to the olivine-liquid mixing line (e.g., Barnes and Fiorentini, 2012; see Fig. 3.2.8(A)), as in the Vaara deposit (Konnunaho et al., 2013). This is also the case with the Sarvisoaivi, Ruossakero, and Vaara deposits, which are associated with Cr-poor cumulates (Figs. 3.2.3, 3.2.15, and 3.2.16(B)).

Ni contents show a positive correlation with MgO, reflecting Ni control by olivine in unmineralized samples (S < 0.2 wt%) at all studied localities (Fig. 3.2.16(C)). In the komatiitic basalts (Hietaharju and Peura-aho), Ni contents (S < 0.2 wt%) vary from 0.04–0.15 wt%, with high values occurring in cumulates. Unmineralized low-Mg komatiites in the Lomalampi area show Ni contents between 0.06 and 0.17 wt%, while in the Tainiovaara and Vaara areas, Ni contents of unmineralized rocks (S < 0.2 wt%) vary from 0.1–0.3 wt%. The scatter in Ni is relatively pronounced in the Sarvisoaivi, Ruossakero, and Hotinvaara komatiites. In the Hotinvaara area, Ni contents vary between 0.02 and 0.25 wt% and in the Ruossakero and Sarvisoaivi areas, they fall in the range of 0.03–0.4 wt%, and are highest in cumulates. The Tainiovaara and Vaara samples are slightly enriched in Ni compared to samples from the other localities. The Hotinvaara cumulates are clearly depleted in Ni, and a similar depletion has also been observed in some low-S (<0.2 wt%) Ruossakero samples. The Hotinvaara cumulates consist of metamorphic silicates (olivine and pyroxenes) and they have apparently partly lost part of their primary Ni content. The same feature has been found in some cumulates at Ruossakero (Fig. 3.2.16(C)).

BASE METAL AND PGE GEOCHEMISTRY

The studied deposits consist mainly of disseminated sulfides. The S contents of the deposits are commonly less than 5 wt%, except for less-abundant densely disseminated samples that may contain up to 5–10 wt% sulfur. Sulfur-rich samples (S >15 wt%) represent massive to semimassive ores from the Hietaharju, Peura-aho, Hotinvaara, and Tainiovaara deposits (Figs. 3.2.17(A,B,C)). Base metals (Ni, Cu) generally show a positive correlation with sulfur in mineralized komatiitic rocks in all deposits, except at Ruossakero and Hotinvaara, where there is no clear correlation between S and Ni or there are sample populations with different correlations. This feature has also been observed between S and Cu in Ruossakero (Figs. 3.2.17(A,B)) and is likely related to different ore types within the deposit (i.e., variety of sulfide mineralogy and host rock types). Platinum and palladium show a positive correlation with S in most of the deposits, except for the Hotinvaara deposit (Fig. 3.2.17(C)).

Even though nickel is the most important base metal in all deposits, Figs. 3.2.17(B) and 3.2.18(A) clearly show that the Lomalampi, Vaara, Peura-aho, and Hietaharju deposits are enriched in Cu compared to the other deposits. Ruossakero, Sarvisoaivi, and Hotinvaara are low in Cu although there are some moderately Cu-enriched parts in the deposits. The Tainiovaara deposit represents intermediate compositions between these groups. The Ni/Cu ratio in the Cu-enriched deposits is ~3–13, whereas in the Cu-depleted deposit, the ratio is ~15–36 and at Tainiovaara, it is 19 (Table 3.2.1). The Cu-enriched deposits are also clearly enriched in PGE, especially Pd and Pt, compared to the other deposits (Figs. 3.2.17(B,C) and 3.2.18(A,B)). Low Ni/Cu (<6) is typical for Ni-Cu-(PGE) deposits in the Raglan and Delta Horizon areas of the Cape Smith belt, Canada, whereas higher Ni/Cu ratios (>13) are typical for other komatiite-hosted Ni-Cu-(PGE) deposits around the world (e.g., Naldrett, 2004; Lesher, 2007).

Figure 3.2.17(A) shows that some of the S-poor (<2 wt% S) disseminated deposits (Vaara, Tainiovaara, Ruossakero) are relatively rich in Ni (Ni/S > 0.7). The composition of flotation test concentrates demonstrate that the Ni content of the sulfide fraction in the Vaara deposit is extremely high, reaching 38 wt%, which has been explained by postmagmatic oxidation of sulfides (Konnunaho et al., 2013). Based on bromine-methanol leaching of representative mineralized samples, Ni-rich sulfides are also present in the Ruossakero (ave. Ni ~23 wt%), Tainiovaara (ave. Ni ~20 wt%), and Sarvisoaivi (ave. Ni ~16 wt%) deposits. The most Ni-poor sulfides are found in the Hietaharju (~3 wt% Ni on average), Peura-aho, and Lomalampi (~6 wt% Ni) deposits, which are also relatively rich in Cu (>1 wt%), analogous to the Vaara deposit (Fig. 3.2.18(A)). In some deposits, the scatter of whole-rock Ni is very large, preventing precise estimation of the Ni tenor (Hotinvaara Ni ~3–14 wt%, Tainiovaara Ni ~13–26 wt%). The metal content of the sulfide fraction at Peura-aho and Tainiovaara was estimated using whole-rock ICP-OES compositions from which silicate-bound Ni was subtracted.

The Vaara and Ruossakero deposits consist of Ni-rich sulfide (millerite, violarite) disseminations, whereas the Hotinvaara, Sarvisoaivi, and Tainiovaara deposits consist of Ni-rich pentlandite disseminations. Nickel-rich sulfides (e.g., violarite and millerite) are also found in the Sarvisoaivi deposits. The Cu-enriched deposits (Hietaharju, Peura-aho, Vaara, and Lomalampi) contain a slightly higher amount of chalcopyrite than the other deposits. The Cu content of the sulfide fraction ranges from ~1–4 wt% in the Cu-enriched deposits, and is highest in the Vaara (~4 wt%) and Lomalampi deposits (~3 wt%), whereas in the other deposits, the Cu tenor is less than 1 wt% (Fig. 3.2.18(A)).

In the Finnish komatiite-hosted Ni-Cu-PGE deposits, the average Pd and Pt contents vary in the range of 170–2160 and 300–1080 ppb, respectively. In disseminated sulfide samples, Pd ranges from 170–760 ppb and Pt from 305–408 ppb. The highest average Pd + Pt concentrations occur in the Peura-aho (3240 ppb), Hietaharju (1080 ppb), Vaara (960 ppb), and Lomalampi (580 ppb) deposits, all of which represent PGE-enriched deposits. The sum of Pt and Pd in the other deposits averages less than 190 ppb (Fig. 3.2.17(C)). Figure 3.2.18(C) shows Pt/Pd and Pd/Ir ratios in the Finnish deposits. Average Pt/Pd is ~0.4 in all deposits except Lomalampi, where the average ratio is significantly higher, at approximately 2.0. Typically, komatiite-hosted Ni-Cu-PGE deposits around the world are enriched in Pd over Pt, and their Pt/Pd ratio is similar to that in most Finnish deposits (e.g., Lesher and Keays, 2002; Fiorentini et al., 2010), but the Lomalampi

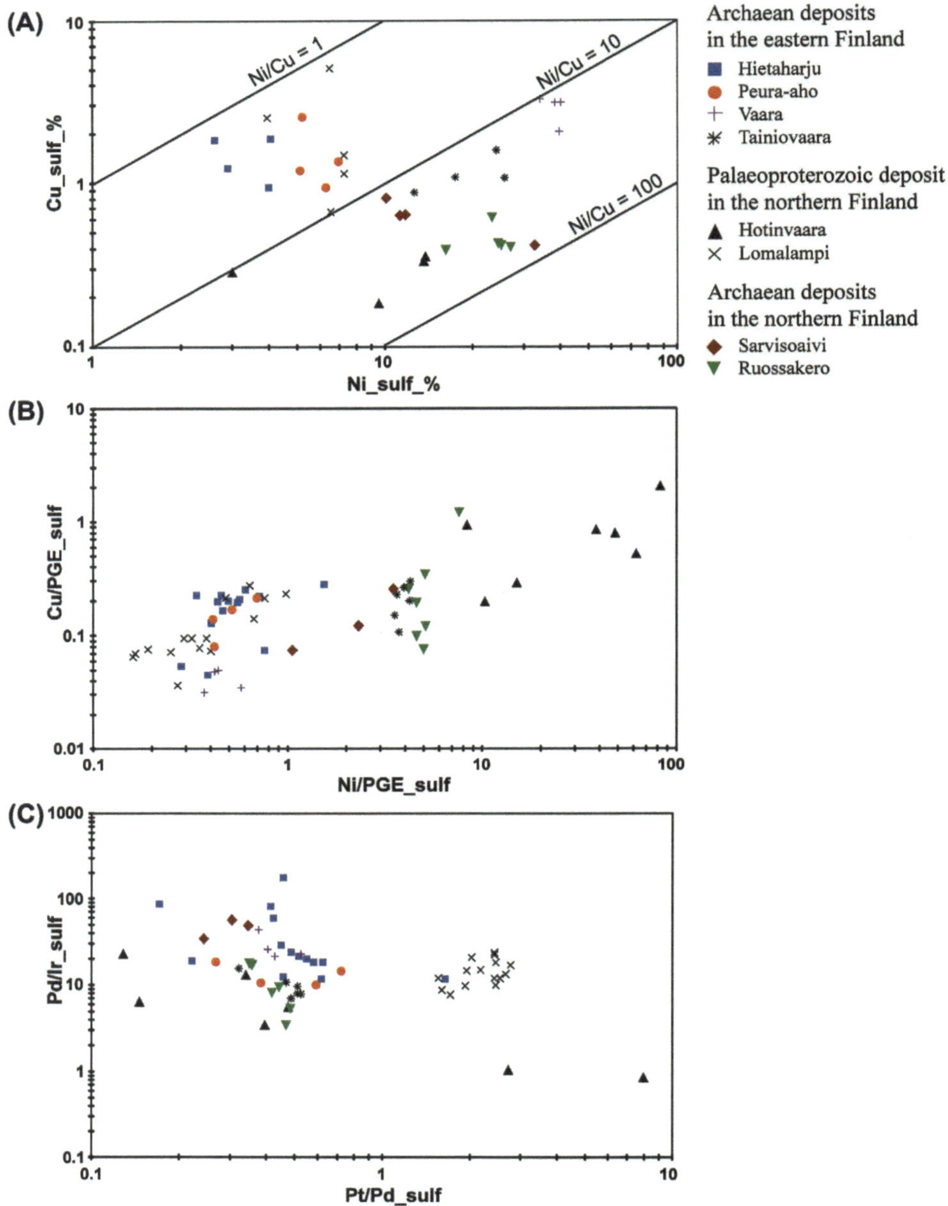

FIGURE 3.2.18 Plots of (A) Cu vs. Ni, (B) Cu/PGE vs. Ni/PGE, and (C) Pd/Ir vs. Pt/Pd for mineralized komatiitic cumulate samples.

Source: *From the Hietaharju, Hotinvaara, Lomalampi, Peura-aho, Ruossakero, Sarvisoaivi, Tainiovaara, and Vaara deposits. In (A), Cu and Ni contents in the sulfide fraction was estimated based on flotation test concentrates for Vaara, bromine-methanol leaching for Hietaharju, Hotinvaara, Lomalampi, Ruossakero, and Sarvisoaivi, and aqua regia leaching for Peura-aho and Tainiovaara with a silicate Ni correction in the two last cases.*

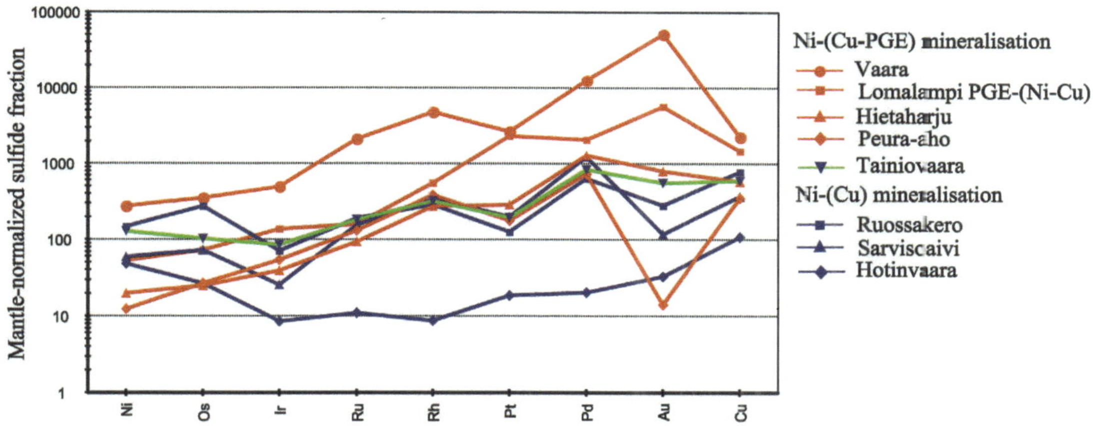

FIGURE 3.2.19 Primitive mantle-normalized chalcophile element patterns (metal contents normalized to 100% sulfide).

Source: *From the Hietaharju, Hotinvaara, Lomalampi, Peura-aho, Ruossakero, Sarvisoaivi, Tainiovaara, and Vaara mineralizations. Normalization values are from Palme and O'Neill (2004).*

deposit has a uniquely high Pt/Pd ratio (Fig. 3.2.18(C)). The average Pd/Ir in all Finnish deposits is ~20. The Hietaharju, Vaara, and Sarvisoaivi deposits have slightly higher Pd/Ir (29-47) than the other deposits, exhibiting more fractionated PGE patterns (Figs. 3.2.18(C), 3.2.19, and Table 3.2.1). Figure 3.2.19 shows that the Hotinvaara deposit has a "U-shaped" and unfractionated metal pattern and all PGE (except Os) are depleted compared to the other deposits.

PGM occur in most deposits and are mainly associated with base metal sulfides or, to a lesser degree, with silicates. Sperrylite is the most abundant Pt mineral in all deposits, but Pd is commonly associated with several Te-Bi-Sb-bearing phases.

S ISOTOPES

The sulfur isotope composition ($\delta^{34}S$) of uncontaminated mantle-derived magmas is around −2 to +2‰ (Ripley and Li, 2003) and in komatiite-hosted Ni-Cu-(PGE) deposits $\delta^{34}S$ is commonly close to zero or moderately positive (e.g., Ripley, 1999; Lesher and Keays, 2002). Determinations of $\delta^{34}S$ from Finnish komatiite-hosted Ni-Cu-(PGE) deposits show variable values between −0.7 to +15.0‰ (Fig. 3.2.20). In the Lomalampi deposit, the measured $\delta^{34}S$ values are the highest, varying from +9.8–15.0‰ (Törmänen et al., 2013, *in preparation*), while the associated black schists have an even heavier sulfur isotope composition (+17.2‰ and +24.4‰). These data imply a considerable crustal component in the Lomalampi sulfides. In the Vaara and Hietaharju deposits, $\delta^{34}S$ is between −0.7 and +2.7 ‰ (Konnunaho et al., 2013; unpublished data) and in sulfide-bearing country rocks, $\delta^{34}S$ ranges from −1.8 to +4.5 ‰ at Vaara and from −0.26–3.44 ‰ at Hietaharju (Fig. 3.2.20). Thus, these two deposits show $\delta^{34}S$ values that overlap with those of the country rocks, but are not clearly distinguishable from the mantle-sulfide composition. However, the available multi-isotope analyses have revealed considerable mass-independent fractionation of sulfur isotopes in both deposits and their country rocks, demonstrating a significant role of external sulfur assimilation in ore formation (Konnunaho et al., 2013; unpublished data), analogously with some other Archean deposits in Canada and Australia (e.g., Bekker et al., 2009; Fiorentini et al., 2012a).

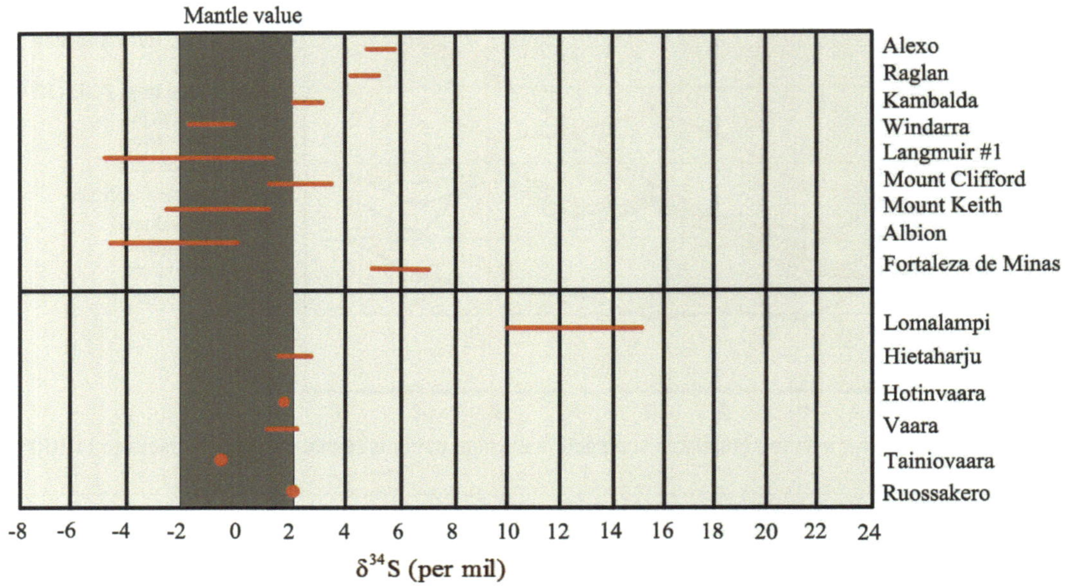

FIGURE 3.2.20 Sulfur isotope compositions ($\delta^{34}S$) of Finnish komatiite-hosted Ni-Cu-PGE deposits and other komatiite-hosted Ni-Cu-PGE deposits globally.

Source: *Reference data from Fiorentini et al. (2012a) and Choudhuri et al. (1997).*

SUMMARY

Several komatiite-hosted Ni-Cu-PGE deposits have been discovered in Archean and Paleoproterozoic greenstone belts in the eastern and northern parts of Finland (Fig. 3.2.1 and Table 3.2.1). Economically, the known deposits are small in size and dispersed, and the concentrations of nickel are low (<0.5 wt%). Part of the Tainiovaara deposit was mined in the 1980s, but the other deposits remain unexploited. Parental magmas of the deposits varied from komatiitic basalt (at Hietaharju and Peura-aho) to low-Mg komatiite (at Vaara, Lomalampi, and Tainiovaara) and komatiite (Hotinvaara, Ruossakero, and Sarvisoaivi). Most of the deposits formed from Al-undepleted magma types, but in the case of Ruossakero, Sarvisoaivi, and Hotinvaara, the magma was Ti-depleted komatiite (Table 3.2.1). Most of the ultramafic bodies hosting the deposits are extrusive in origin, but the Hotinvaara, Ruossakero, Sarvisoaivi, and Lomalampi metacumulate bodies might represent sills.

All komatiite-hosted Ni-Cu-PGE deposits in Finland belong to the type II komatiitic mineralization of Lesher and Keays (2002), and in most cases sulfides are associated with cumulates having low Cr contents (e.g., Vaara, Ruossakero, Tainiovaara, and Sarvisoaivi). Some of the deposits contain small massive or semimassive sulfide accumulations (e.g., Tainiovaara, Hietaharju, and Peura-aho). The deposits are magmatic in origin, but they have been modified to various degrees by postmagmatic metamorphic processes. Some sulfides (e.g., Peura-aho) can also occur within associated country rocks close to the komatiitic host rocks due to remobilization of sulfides.

Based on their chalcophile element concentrations, the komatiite-hosted Ni-Cu-PGE deposits can be divided into two groups:

(1) Ni-Cu-PGE (Hietaharju, Peura-aho, Vaara, and Lomalampi) deposits with elevated Pd + Pt and Cu levels (Pd + Pt > 500 ppb and Ni/Cu ~3–13)
(2) Ni-(Cu) deposits (Ruossakero, Sarvisoaivi, and Hotinvaara) with low PGE and Cu concentrations (Pd + Pt < 500 ppb and Ni/Cu ~15–36)

The Tainiovaara deposit has an intermediate composition between these two groups. Lomalampi is a unique deposit because it has an exceptionally high PGE/(Ni + Cu) ratio (i.e., PGE-(Ni-Cu) deposit) (Figs. 3.2.16 through Fig. 3.2.18 (B)). Most of the Finnish komatiite-hosted Ni-Cu-PGE deposits show mantle-like $\delta^{34}S$ values (Fig. 3.2.20), but the Paleoproterozoic Lomalampi deposit contains heavier sulfur, with $\delta^{34}S$ from +9.8–15 per mil. Available multiple S isotope data ($\delta^{34}S$ and $\Delta^{33}S$) from some Archean deposits show that contamination with S-bearing substrate has played an important role in the genesis of the sulfides (e.g., Vaara; Konnunaho et al., 2013).

Most of the Finnish komatiite-hosted Ni-Cu-PGE deposits are geochemically similar to komatiitic deposits globally. Some of the Finnish deposits are highly enriched in PGE (Pd + Pt) and, thus, generated particular economic and scientific interest. The Lomalampi deposit is unique among the disseminated deposits in showing abnormally high Pt/Pd (~2.0) and (Pt + Pd)/(Ni + Cu) ratios.

The Cr content of olivine cumulates can be used to identify the most favorable rocks to host sulfide mineralization in komatiitic magmatic systems; for example, at Vaara (Fig. 3.2.3) and Ruossakero (Fig. 3.2.15). Nickel depletion or enrichment in olivine and their host rocks has been used in Ni exploration in environments containing mafic–ultramafic rocks (e.g., Häkli, 1971; Duke and Naldrett, 1978; Makkonen, 1996; Lesher et al., 2001) but other studies have questioned whether mineralized komatiite environments show detectable Ni depletion haloes (Barnes and Fiorentini, 2012). However, the lack of magmatic silicate minerals in the Finnish komatiites precludes the use of this method. One potential method is to look for signs of anomalous PGE depletion in barren komatiitic magma suites, as this may indicate sulfide saturation elsewhere (Fiorentini et al., 2010; Fiorentini et al., 2011; Heggie et al., 2013; Maier et al., 2013). In any case, there is no single geochemical detector for indicating the presence of komatiite-hosted Ni-Cu-PGE deposits (Lesher and Barnes, 2009). In the view of the present authors, the discovered mineralizations show that Archean and Paleoproterozoic Finnish komatiites show considerable exploration potential for Ni-Cu-PGE ores.

ACKNOWLEDGMENTS

Geochemical data for this study were obtained from the Geological Survey of Finland. Part of the base metal and PGE analyses of the Hietaharju and Peura-aho deposits were obtained from Altona Mining Limited. We thank Viena Arvola and Pertti Telkkälä from the Geological Survey of Finland for preparing some of the figures. We are grateful to the following people for their help and discussions during this study: Jarmo Vesanto, Sanna Juurela, and Jarmo Lahtinen. The valuable comments on the manuscript from Steve Barnes and Wolfgang Maier are gratefully acknowledged.

REFERENCES

Altona Mining, 2012. Annual report, p. 108.

Arndt, N., Lesher, C.M., Barnes, S.J., 2008. Komatiites. Cambridge University Press. p. 467.

Barnes, S.-J., Often, M., 1990. Ti-rich komatiites from northern Norway. Contributions to Mineralogy and Petrology 105, 42–54.

Barnes, S.J., 2006. Komatiite-hosted nickel sulphide deposits: Geology, geochemistry, and genesis. Society of Economic Geologists, Special Publication 13, 51–118.

Barnes, S.J., 2007. Cotectic precipitation of olivine and sulphide liquid from komatiite magma, and the origin of komatiite hosted disseminated nickel sulphide mineralization at Mt. Keith and Yakabindie, Western Australia. Economic Geology 102, 299–304.

Barnes, S.J., Brand, N.W., 1999. The distribution of Cr, Ni and chromite in komatiites, and application to exploration for komatiite-hosted nickel sulphide deposits. Economic Geology 94, 129–132.

Barnes, S.J., Fiorentini, M.L., 2012. Komatiite magmas and sulphide nickel deposits: A comparison of variably endowed Archaean Terranes. Economic Geology 107, 755–780.

Barnes, S.J., Wells, M.A., Verrall, M.R., 2009. Effects of magmatic processes, serpentinization, and talc-carbonate alteration on sulphide mineralogy and ore textures in the Black Swan disseminated nickel sulphide deposit, Yilgarn Craton. Economic Geology 104, 539–562.

Barnes, S.J., Fiorentini, M.L., Fardon, M.C., 2012. Platinum group element and nickel sulphide ore tenors of the mount Keith nickel deposit, Yilgarn Craton, Australia. Mineralium Deposita 47, 129–150.

Bekker, A., Barley, M.E., Fiorentini, M.L., et al., 2009. Atmospheric sulphur in Archaean komatiite-hosted nickel deposits. Science 326, 1086–1089.

Bergman, S., Kübler, L., Martinsson, O., 2001. Description of regional geological and geophysical maps of northern Norrbotten County (east of the Caledonian orogen). Sveriges geologiska Undersökning Ba 56, p. 104.

Campbell, I.H., Naldrett, A.J., 1979. The influence of silicate: sulphide ratios on the geochemistry of magmatic sulphides. Economic Geology 74, 1503–1506.

Choudhuri, A., Iyer, S.S., Krouse, H.R., 1997. Sulfur isotopes in komatiite-hosted Ni-Cu sulphide deposits from the Morro do Ferro greenstone belt, Southeastern Brazil. International Geology Review 39, 230–238.

DigiKP, Bedrock of Finland, 2014. Digital map database [Electronic resource]. Espoo: Geological Survey of Finland [referred to 29.1.2014], Version 1.0. Available at www.geo.fi/en/bedrock.html.

Duke, J.M., Naldrett, A.J., 1978. A numerical model of the fractionation of olivine and molten sulphide from komatiitic magma. Earth and Planetary Science Letters 39, 255–266.

Fiorentini, M.L., Barnes, S.J., Lesher, C.M., et al., 2010. Platinum group element geochemistry of mineralised and nonmineralised komatiites and basalts. Economic Geology 105, 795–823.

Fiorentini, M.L., Barnes, S.T., Maier, W.D., et al., 2011. Global variability in the platinum-group element contents of komatiites. Journal of Petrology 52, 83–112.

Fiorentini, M.L., Bekker, A., Rouxel, O., et al., 2012b. Multiple sulphur and iron isotope composition of magmatic Ni-Cu-(PGE) sulphide mineralisation from eastern Botswana. Economic Geology 107, 105–116.

Fiorentini, M., Beresford, S., Barley, M., et al., 2012a. District to camp controls on the genesis of komatiite-hosted nickel sulphide deposits, Agnew-Wiluna greenstone belt, Western Australia: Insights from the multiple sulphur isotopes. Economic Geology 107, 781–796.

Gruau, G., Tourpin, S., Fourcade, S., Blais, S., 1992. Loss of isotopic (Nd, O) and chemical (REE) memory during metamorphism of komatiites: new evidence from eastern Finland. Contributions to Mineralogy and Petrology 112, 66–82.

GTK, Mineral Deposit Database of Finland, 2014. Digital map database (DigiKP) [Electronic resource]. Espoo: Geological Survey of Finland [referred to 3.2.2014]. Available at http://gtkdata.gtk.fi/MDaE/index.html.

Häkli, T.A., 1971. Silicate nickel and its application to the exploration of nickel ores. Bulletin of the Geological Society of Finland 43, 247–263.

Halkoaho, T., Papunen, H., 1998. Geology and mineral deposits of the Kiannanniemi area, Suomussalmi. Integrated technologies for mineral exploration pilot project for nickel ore deposits. Brite-EuRam BE-1117 GeoNickel, Task 1.2 Mineralogy and modeling of Ni sulphide deposits in komatiitic/picritic extrusives 15 Technical Report 6.4.

Halkoaho, T., Koistinen, E., Luukkonen, E., et al., 2000a. The Vaara–Kauniinlampi komatiitic nickel prospects in Suomussalmi, eastern Finland (register numbers of claims 5376/1, 5376/2, 5789/1, 5789/2, 6171/1, 6271/2, 6273/1, 6676/1, 6770/1, and 6938/1). Report CM 06/4513-4514/2000/1, Geological Survey of Finland 37 Espoo.

Halkoaho, T., Liimatainen, J., Papunen, H., Välimaa, J., 2000b. Exceptional Cr-rich basalts in the komatiitic volcanic association of the Archaean Kuhmo greenstone belt, eastern Finland. Mineralogy and Petrology 70, 105–120.

Halkoaho, T., Niskanen, M., 2004. A Research report of nickel studies concerning the claim of Jamali 1 (register number of claim 7626/1) in the Lieksa during years 2002–2003. Archive report M 06/4314/2004/1/10, Geological Survey of Finland p. 14.

Hanski, E.J., 1980. Komatiitic and tholeiitic metavolcanics of the Kellojärvi group in the Siivikkovaara area of the Archaean Kuhmo greenstone belt, eastern Finland. Bulletin of Geological Society of Finland 52, 67–100.

Hanski, E., Huhma, H., 2005. Central Lapland Greenstone Belt. In: Lehtinen, M., Nurmi, P.A., Rämö, O.T. (Eds.), Precambrian Geology of Finland—Key to the Evolution of the Fennoscandian Shield. Elsevier, Amsterdam, pp. 141–194.

Hanski, E., Huhma, H., Rastas, P., Kamenetsky, V.S., 2011. The Palaeoproterozoic komatiite-picrite association of Finnish Lapland. Journal of Petrology 42, 855–876.

Heath, C.J., Lahaye, Y., Stone, W.E., Lambert, D.D., 2001. Origin of variations in nickel tenor along the strike of the Edwards Lode nickel sulphide ore body, Kambalda, Western Australia. Canadian Mineralogist 39, 655–671.

Heggie, G.J., Barnes, S.J., Fiorentini, M.L., 2013. Application of lithogeochemistry in the assessment of nickel-sulphide potential in komatiite belts from northern Finland and Norway. Bulletin of the Geological Society of Finland 85, 107–126.

Herzberg, C., 1992. Depth and degree of melting of komatiites. Journal of Geophysical Research 97, 4521–4540.

Hronsky, J.M., Schodde, R.C., 2006. Nickel exploration history of the Yilgarn Craton: from the nickel boom to today. Society of Economic Geology. Special Publications 13, 1–11.

Huhma, H., Mänttäri, I., Peltonen, P., et al., 2012. The age of the Archaean greenstone belts in Finland. Geological Survey of Finland,. Special Paper 54, 74–175.

Inkinen, O., Ilvonen, E., Pelkonen, R., 1984. Report of investigations of the Pulju schist belt and Hotinvaara area in 1982–1984. Outokumpu Oy, Lapin Malmi Exploration, Report 001/2742/OI, EI, RP/84/21, p. 54. (in Finnish).

Isomaa, J., 1982. Report of investigations on the nickel mineralisation in the Sarvisoiaivi area, Enontekiö, in 1978–1980. Geological Survey of Finland, Report M19/1832/-86/1/10, p. 41. (in Finnish).

Isomaa, J., 1986. Report of investigations on the nickel mineralisation in the Sarvisoaivi area, Enontekiö, in 1980–1983. Geological Survey of Finland, Report M19/1832/-86/1/10. p. 4. (in Finnish).

Isomaa, J., 1988. Report of investigations on the nickel mineralisation in the Ruossakero area, Enontekiö. Geological Survey of Finland, Report M19/1834/-88/1/10. p. 27. (in Finnish).

Keays, R.R., 1995. The role of komatiitic and picritic magmatism and S-saturation in the formation of ore deposits. Lithos 34, 1–18.

Keele, R.A., Nickel, E.H., 1974. The geology of a primary millerite-bearing sulphide assemblage and supergene alteration at the Otter shoot, Kambalda, Western Australia. Economic Geology 69, 1102–1117.

Koistinen, E., Heikura, P., 2010. Mineral resource assessment and 3D modelling of the Lomalampi deposit, Sodankylä Finland. Report, M19/3723/2010/50, Geological Survey of Finland, Espoo, p. 58.

Kojonen, K.K., 1981. Geology, geochemistry and mineralogy of two Archaean nickel-copper deposits in Suomussalmi, eastern Finland. Geological Survey of Finland Bulletin 315, p. 54.

Konnunaho, J.P., Hanski, E.J., Bekker, A., et al., 2013. The Archaean komatiite-hosted, PGE-bearing Ni-Cu sulphide deposit at Vaara, eastern Finland: evidence for assimilation of external sulphur and post-depositional desulphurization. Mineralium Deposita 48, 967–989.

Korhonen, P., 1981. Petrology, mineralogy and geochemistry of the Sarvisoaivi nickel mineralisation in Enontekiö. M.Sc. Thesis, Department of Geology and Mineralogy. University of Helsinki. p. 65. (in Finnish).

Kurki, J., Papunen, H., 1985. Geology and nickel-copper deposits of the Kianta area, Suomussalmi. In: Papunen, H., Gorbunov, G.I. (Eds.), Nickel-Copper Deposits of the Baltic Shield and Scandinavian Caledonides. Geological Survey of Finland Bulletin 333, pp. 155–161.

Lahtinen, J., 1996. Nickel mineral resource estimate of the Ruossakero nickel deposit, Enontekiö, northwestern Finland. Lapin Malmi Oy. Report 035/1834/JJL/96. p. 21.

Lehtonen, M., Airo, M.-L., Eilu, P., et al., 1998. The stratigrapy, petrology and geochemistry of the Kittilä greenstone area, northern Finland. Geological Survey of Finland, Report of Investigations 140, p. 144. (in Finnish with English summary).

Lesher, C.M., 2007. Ni-Cu-(PGE) deposits in the Raglan area, Cape Smith belt, New Quebec. In: Goodfellow, W.D. (Ed.), Mineral Deposits of Canada: A Synthesis of Major Deposit Types, District Metallogeny, the Evolution of Geological Provinces and Exploration Methods. Geological Association of Canada. Special Publication 5, pp. 351–387.

Lesher, C.M., Arndt, N.T., 1995. REE and Nd isotope geochemistry, petrogenesis and volcanic evolution of contaminated komatiites at Kambalda, Western Australia. Lithos 34, 127–157.

Lesher, C.M., Barnes, S.J., 2008. Komatiite-associated Ni-Cu-PGE deposits. In: Arndt, N. (Ed.), Komatiite. Cambridge University Press, Cambridge, pp. 295–326.

Lesher, C.M., Barnes, S.J., 2009. Komatiite-associated Ni-Cu-(PGE) deposits. In: Li, C., Ripley, E.M. (Eds.), New Developments in Magmatic Ni-Cu and PGE Deposits. Geological Publishing House, Beijing, pp. 27–120.

Lesher, C.M., Burnham, O.M., Keays, R.R., et al., 2001. Trace-element geochemistry and petrogenesis of barren and mineralised komatiites associated with magmatic Ni-Cu-(PGE) sulphide deposits. Canadian Mineralogist 39, 673–696.

Lesher, C.M., Groves, D.I., 1986. Controls on the formation of komatiite-associated nickel-copper sulphide deposits. In: Friedrich, G.H., et al. (Ed.), Proceedings of the 27th International Geological Congress. Springer-Verlag, Berlin, pp. 43–62.

Lesher, C.M., Keays, R.R., 2002. Komatiite-associated Ni-Cu-(PGE) deposits: geology, mineralogy, geochemistry and genesis. In: Louis, J.C. (Ed.), The Geology, Geochemistry, Mineralogy and Mineral Beneficiation of Platinum-Group Elements. Canadian Institute of Mining and Metallurgy, Special Volume 54, pp. 579–619.

Luukkonen, E., Halkoaho, T., Hartikainen, A., et al., 2002. Report of activities in 1992–2001 of the project Archaean Areas in Eastern Finland (12201 ja 210 5000) in the areas of Suomussalmi, Hyrynsalmi, Kuhmo, Nurmes, Rautavaara, Valtimo, Lieksa, Ilomantsi, Kiihtelysvaara, Eno, Kontiolahti, Tohmajärvi, and Tuupovaara. Geological Survey of Finland, Archive Report M19/4513/2002/1, p. 265. (in Finnish).

Maier, W.D., Peltonen, P., Halkoaho, T., Hanski, E., 2013. Geochemistry of komatiites from the Tipasjärvi, Kuhmo, Suomussalmi, Ilomantsi and Tulppio greenstone belts, Finland: Implications for tectonic setting and Ni sulphide prospectivity. Precambrian Research 228, 63–84.

Makkonen, H., 1996. 1.9 Ga tholeiitic magmatism and related Ni-Cu deposition in the Juva area, SE Finland. Geological Survey of Finland Bulletin 386, 101.

Naldrett, A.J., 2004. Magmatic Sulphide Deposits: Geology, Geochemistry and Exploration. Springer-Verlag, Berlin/Heidelberg, p. 727.

Naldrett, A.J., 1966. The role of sulphurization in the genesis of iron-nickel sulphide deposits of the Porcupine district, Ontario. Canadian Institute of Mining and Metallurgy Transactions 69, 147–155.

Palme, H., O'Neill, H.St.C., 2004. Cosmochemical estimates of mantle composition. In: Carlson, R.W. (Ed.), The Mantle and Core. Treatise on Geochemistry Vol. 2, p. 1–38.

Papunen, H., 1998. Geology and ultramafic rocks of the Palaeoproterozoic Pulju greenstone belt, Western Lapland. Integrated technologies for mineral exploration pilot project for nickel ore deposits, Brite-EuRam BE-1117 GeoNickel, Task 1.2 Mineralogy and modeling of Ni sulphide deposits in komatiitic/picritic extrusives. Technical Report 6.5, p. 41.

Papunen, H., 1989. Platinum-group elements in metamorphosed Ni-Cu deposits in Finland. In: Prendergast, M.D., Jones, M.J. (Eds.), Magmatic Sulphides—the Zimbabwe Volume. The Institution of Mining and Metallurgy, London, pp. 165–176.

Papunen, H., Halkoaho, T., Luukkonen, E., 2009. Archaean Evolution of the Tipasjärvi-Kuhmo-Suomussalmi greenstone complex, Finland. Geological Survey of Finland Bulletin 403, p. 68.

Pekkarinen, L., 1980. Ni-deposit of the Lieksa Tainiovaara. Geologi 7, p. 92. (in Finnish).

Puustinen, K., Saltikoff, B., Tontti, M., 1995. Distribution and metallogenic types of nickel deposits in Finland. Geological Survey of Finland, Report of Investigations 132, p. 38.

Ripley, M.E., 1999. Systematics of sulphur and oxygen isotopes in mafic igneous rocks and related Cu-Ni-PGE mineralization. In: Keays RR, Lesher RR, Lightfoot PC and Farrow CEG (Eds.), Dynamic processes in magmatic ore deposits and their application in mineral exploration. Geol Assoc Can, Short Course Notes Vol.13, pp. 133–158.

Ripley, E.M., Li, C., 2003. Sulphur isotope exchange and metal enrichment in the formation of magmatic Cu-Ni-(PGE) deposits. Economic Geology 98, 635–641.

Ross, J.R., Hopkins, G.M.F., 1975. The nickel sulphide deposits of Kambalda, Western Australia. In: Knight, C. (Ed.), Economic Geology of Australia and Papua New Guinea Volume 1. Australian Institute of Mining and Metallurgy, Metals, Melbourne, Monograph 5, 100–121.

Snoke, A.W., Calk, L.C., 1978. Jackstraw-textured talc-olivine rocks, Preston Peak area, Klamath Mountains, California. Geological Society of America Bulletin 89, 223–230.

Sorjonen-Ward, P., Luukkonen, E.J., 2005. Archaean rocks. In: Lehtinen, M., Nurmi, P.A., Rämö, O.T. (Eds.), Precambrian Geology of Finland—Key to the Evolution of the Fennoscandian Shield. Elsevier, Amsterdam, pp. 19–99.

Sotka, P., 1986. Hotinvaara, Kittilä: thin section study of the Ni mineralisation. Outokumpu Oy, Unpublished Report 070/Hotinvaara/PMS/1986/21, p. 2. (in Finnish).

Sotka, P., 1984. Hotinvaara Kittilä: ore mineralogy of the Ni mineralisation. Outokumpu Oy, Unpublished Report 070/274204, Hotinvaara, Ni-miner./PMS/1984/21, p. 4. (in Finnish).

Sproule, R.A., Lesher, C.M., Houlé, M., et al., 2005. Chalcophile element geochemistry and metallogenesis of komatiitic rocks in the Abitibi greenstone belt, Canada. Economic Geology 100, 1169–1190.

Stone, W.E., Heydari, M., Seat, Z., 2004. Nickel tenor variations between Archaean komatiite-associated nickel sulphide deposits, Kambalda ore field, Western Australia: The metamorphic modification model revisited. Mineralogy and Petrology 82, 295–316.

Törmänen, T., Heikura, P., Salmirinne, H., 2010. The komatiite-hosted Lomalampi PGE-Ni-Cu-Au deposit, northern Finland. Report M19/3723/2010/52, Geological Survey of Finland, Espoo, p. 39.

Törmänen, T., Konnunaho., J., Hanski, E., et al., (in preparation) A unique Palaeoproterozoic komatiite-related PGE mineralisation in the Central Lapland Greenstone Belt, northern Finland.

Tourpin, S., Gruau, G., Blais, S., Fourcade, S., 1991. Resetting of REE, and Nd and Sr isotopes during carbonitization of a komatiite from Finland. Chemical Geology 90, 15–29.

Vanne, J., 1981. A research report of nickel studies concerning the claim of Tainiovaara 1 (register number of claim 2538/1). Archive Report M 06/4332/81/1/10, Geological Survey of Finland, p. 9. (in Finnish).

Viljoen, R.P., Viljoen, M.J., 1982. Komatiites—An historical review. In: Arndt, N.T., Nisbet, E.G. (Eds.), Komatiites. George Allen and Unwin, London, pp. 5–18.

Woodall, R., Travis, G.A., 1970. The Kambalda nickel deposits. Ninth Commonwealth Mining and Metallurgy Congress, London Vol. 2, 517–533.

CHAPTER

PGE-(CU-NI) DEPOSITS OF THE TORNIO-NÄRÄNKÄVAARA BELT OF INTRUSIONS (PORTIMO, PENIKAT, AND KOILLISMAA)

3.3

M. Iljina, W.D. Maier, T. Karinen

ABSTRACT

The layered intrusions of the Tornio-Näränkävaara belt in north-central Finland host Europe's most significant platinum-group element (PGE) deposits. At the most advanced stage of exploration are contact-type deposits of the Portimo complex (i.e., those at Ahmavaara and Konttijärvi), where the total PGE-Ni-Cu resources are ~263 million tons. The contact-type Haukiaho and Kaukua deposits of the Koillismaa intrusion are also reported to contain significant PGE-Ni-Cu resources, amounting to approximately 46.8 Mt. In addition to the contact-type deposits, the intrusions host several other styles of PGE mineralization, including internal PGE reefs (most significant being the SJ and PV Reefs of the Penikat intrusion and the SK Reef of the Portimo complex), and offset-type mineralization below the Narkaus intrusion in the Portimo complex. We present key geological features of various deposits and place them into a stratigraphic context with their host intrusions. We also refer to an exploration model that was successfully used in the discovery of these metal enrichments.

Keywords: Platinum-Group Elements; nickel mineralization; copper mineralization; Portimo; Suhanko; Penikat; Koillismaa; PGE reef; contact-type deposit.

INTRODUCTION

Northern Fennoscandia contains approximately two dozen Paleoproterozoic layered mafic–ultramafic intrusions (Fig. 3.3.1) (Alapieti et al., 1990). Based on U-Pb zircon and Sm-Nd whole-rock isotope data, the mean age for the Finnish intrusions is 2440 Ma (Huhma et al., 1990, 2011). These intrusions belong to bimodal igneous activity, which resulted in mafic and felsic, plutonic and extrusive formations not only in Finland but also in northwest Russia. The igneous activity was related to mantle plume-generated rifting of Archean continent (Amelin et al., 1996; Hanski et al., 2001) and many of the formations form long linear belts following intracratonic rift zones (Fig. 3.3.1). In Finland, the 2.44 Ga intrusions are concentrated into two zones: one, as represented by the Koitelainen and Akanvaara intrusions (Mutanen, 1997; Maier, 2015) is located in central Lapland and the other one forms the approximately 300-km-long, west–east trending Tornio-Näränkävaara belt (TNB) that starts from the

Mineral Deposits of Finland. http://dx.doi.org/10.1016/B978-0-12-410438-9.00005-4

Finnish-Swedish border and extends into Russia. Based on the similarities in their stratigraphy, some of the TNB igneous bodies represent dismembered fragments of larger intrusions.

The intrusions of the western part of the Tornio-Näränkävaara belt have Archean basement complex on their footwall side and the hanging wall rocks are younger Paleoproterozoic subracrustal sequences

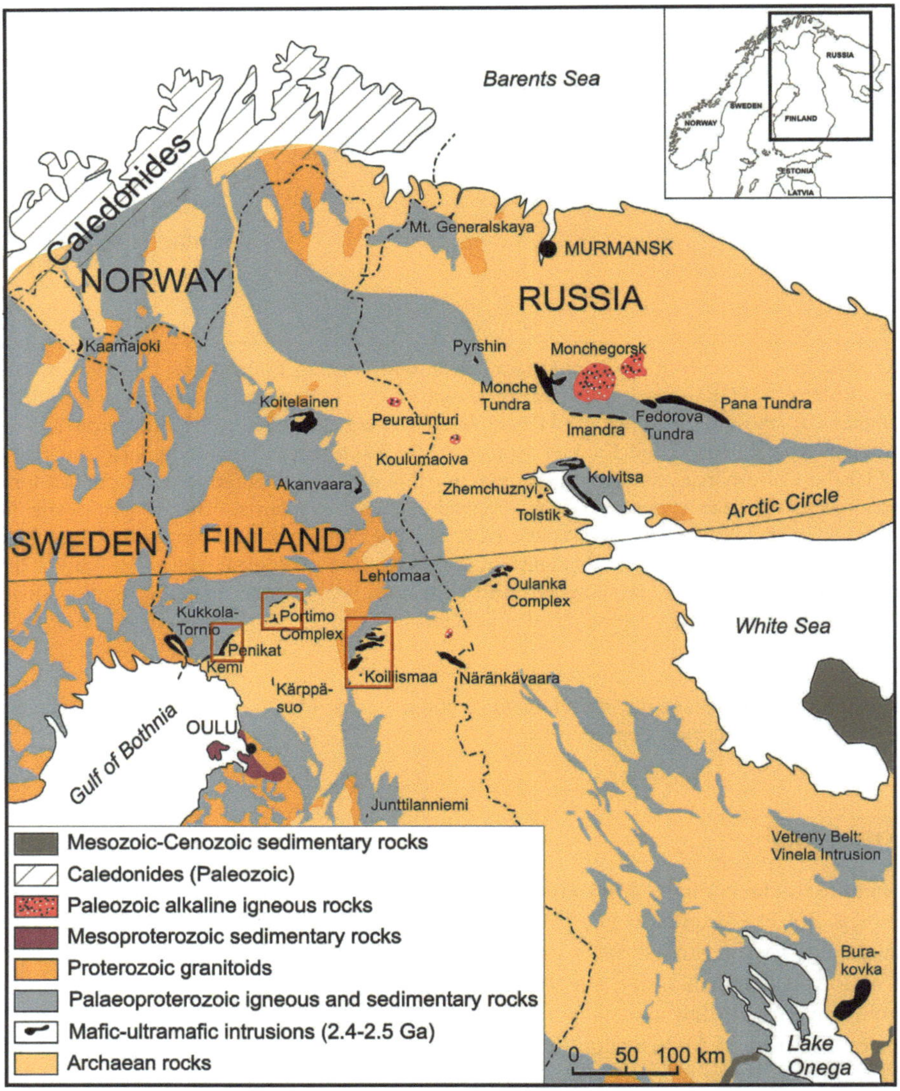

FIGURE 3.3.1 Generalized geological map of the northeastern Fennoscandian Shield, with Paleoproterozoic layered intrusions.

The intrusions mentioned in the text are highlighted.

Source: *Modified from Karinen (2010).*

deposited on partly eroded and block-faulted intrusions, referring to a rather shallow intrusion depth and rapid uplift. In the east, the Koillismaa intrusion has its Paleoproterozoic volcanic roof rocks partly preserved (Lauri et al., 2003). The easternmost body, the Näränkävaara intrusion, was emplaced totally into Archean basement rocks. Magmatic layering of the intrusions dips toward the overlying subra-crustal rocks, and gravity and seismic surveys indicate intrusions to plunge underneath Proterozoic rocks. Modal mineralogy of intrusions varies from ultramafic dominated (Tornio and Näränkävaara) to intrusions composed almost exclusively of mafic cumulates (Koillismaa). A more detailed description of the Tornio-Näränkävaara belt can be found in Iljina and Hanski (2005).

Many of the ~2.44–2.50 Ga Paleoproterozoic layered intrusions are platinum-group element (PGE) and Ni-Cu mineralized, most notably Portimo, Penikat, and Koillismaa in the Tornio-Näränkävaara belt, and Monchegorsk, Fedorova Tundra, and Pana Tundra in the Kola Peninsula, Russia, although none of them is currently exploited for PGE (several deposits hosted by the Monchegorsk intrusion were mined for Ni and Cu(-PGE) from 1935–1975). In the present chapter, the geology and PGE mineralization styles in intrusions of the Tornio-Näränkävaara belt are described. This belt is of considerable economic and academic interest because it hosts several forms of PGE enrichment, including reef-, contact-, and offset-type mineralization.

THE PORTIMO COMPLEX AND ITS MINERALIZATION

STRUCTURAL UNITS AND STRATIGRAPHIC SECTIONS

The Portimo complex (Fig. 3.3.2) is composed of four principal intrusive blocks/structural units (Alapieti et al., 1989; Iljina, 1994, 2005; Iljina and Hanski, 2005), including the Narkaus, Suhanko, and Konttijärvi intrusions. These blocks represent tectonically dismembered bodies of one or two originally larger intrusions. The Portimo complex was deformed and metamorphosed during several tectonic events. During these events, almost all primary magmatic minerals were altered to secondary minerals. In the mafic

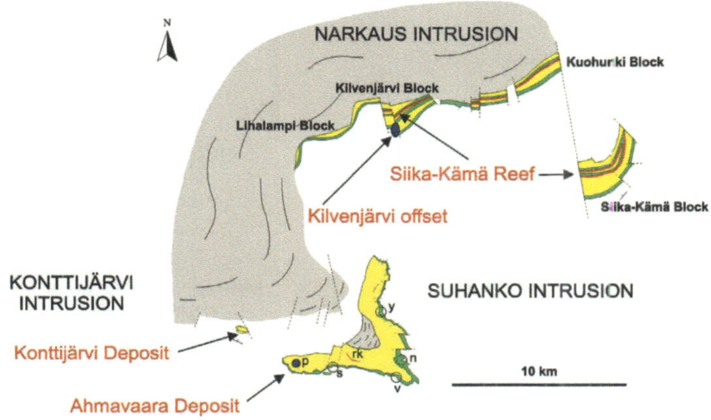

FIGURE 3.3.2 General geological map and principal PGE occurrences of the Portimo layered igneous complex.

rk = Rytikangas PGE Reef, p = clinopyroxene pegmatite pipe; the massive sulfide deposits in the Suhanko intrusion are also shown: s = Suhanko proper, v = Vaaralampi, n = Niittylampi, and y = Yli-Portimojärvi.

Source: *Modified after Iljina, 1994.*

rocks, the metamorphic assemblage consists of amphiboles and plagioclase. Polyphase metamorphism, probably comprising several prograde and retrograde phases (Iljina, 1994), is expressed by greenish hornblende rims around colorless amphiboles and the subsequent development of colorless amphibole around greenish amphibole. In spite of the metamorphic alteration, it is usually possible to recognize the original cumulus textures based on the pseudomorphs of the original minerals. Clinopyroxene and orthopyroxene relicts in the similar, but less altered Penikat intrusion are augitic and bronzitic in composition, respectively. Therefore, where it is possible to distinguish two different pyroxenes in the Portimo rocks on the grounds of the habitus of their pseudomorphs, the clinopyroxene has been generally referred to as augite and the orthopyroxene as bronzite. These terms are also used in the rock nomenclature. Where the nature of the primary minerals is unclear, less-specific rock names, such as pyroxenite and peridotite, are used.

Each individual block of the Portimo complex contains a basal marginal series of variable thickness and an overlying layered series (Fig. 3.3.3). In all cases, the footwall consists of Archean granite-gneiss, or locally also quartzite and mafic volcanic rocks of Archean greenstone belts. The marginal series of the Narkaus intrusion differs from those in the Suhanko and Konttijärvi intrusions in its thickness and prevailing rock types. The Narkaus marginal series generally varies from 10–20 m in thickness, whereas the Suhanko and Konttijärvi marginal series reach a thickness of 40–80 m and 100–150 m, respectively. The

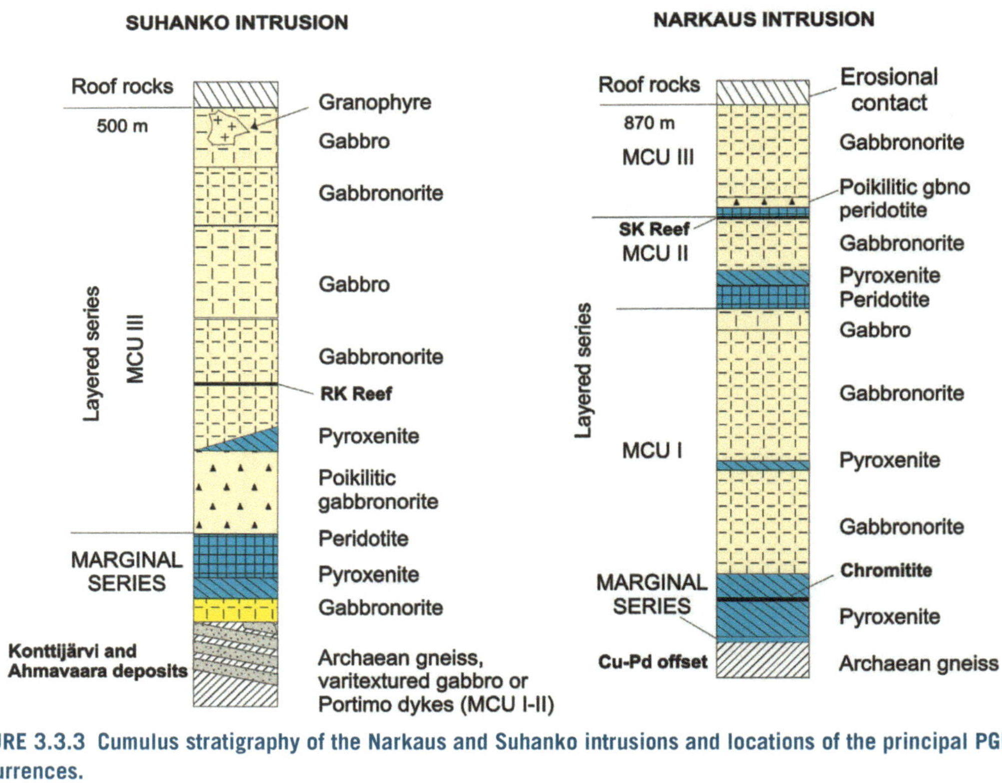

FIGURE 3.3.3 Cumulus stratigraphy of the Narkaus and Suhanko intrusions and locations of the principal PGE occurrences.

MCU = megacyclic unit.

Source: *Modified after Iljina (1994).*

Narkaus marginal series is mainly composed of pyroxenite, with some relatively feldspathic varieties in its lower parts, whereas the Suhanko and Konttijärvi marginal series contain thick, olivine-rich cumulates (lherzolite) in their upper portions, which are underlain by pyroxenite and then gabbronorite (Fig. 3.3.3).

Fine-grained, granular-textured gabbronoritic bodies or fragments, ranging in thickness from a few centimeters to a few tens of meters and reaching several hundred meters in length, occur in many places in the Suhanko marginal series (Figs. 3.3.4B and 3.3.5). The chemical composition of these rocks seems to vary along the strike. For example, at Ahmavaara, they have a distinctly higher Cr content and slightly higher MgO content than in the southeastern portion of the Suhanko intrusion; for example, in the Niittylampi area (refer to Fig. 3.3.7). The Ahmavaara fragments resemble chemically the Portimo dikes located below the Konttijärvi and Ahmavaara marginal series, whereas the chemical composition of the Niittylampi fragments is similar to the mean composition of the Suhanko intrusion (see Table 3.3.4 later in chapter). The chemical features and the mode of occurrence of these fine-grained gabbronoritic rocks have led to the idea that they are autoliths and represent chilled margin rocks that were disrupted and entrained by subsequent magma pulses (Iljina, 1994).

At the base of the layered series of the Suhanko intrusion are gabbronoritic orthocumulates (with poikilitic augite), which also contain some pyroxenitic interlayers up to a few meters in thickness. The poikilitic gabbronorite is separated from the overlying, rather monotonous gabbronoritic adcumulates by a few-meters-thick pyroxene-rich layer (except at Konttijärvi and Ahmavaara, see the following heading). Approximately midway in the stratigraphy of the layered series, bronzite disappears as a cumulus mineral, but returns higher up in the sequence. Four poikilitic anorthosite layers also occur in the upper Suhanko layered series. Granophyric material is found as discontinuous patches and cross-cutting dikes in both the upper Suhanko and upper Konttijärvi layered series.

In the Narkaus intrusion, the layered series contains thick peridotite layers that are absent at Suhanko and Konttijärvi. The peridotite layers have been interpreted to represent compositional reversals resembling those in the Penikat intrusion which enables the subdivision of the Narkaus layered series into three megacyclic units (MCU I–III, Fig. 3.3.3). The lowermost unit (MCU I) commences with a thick (~80 m) orthopyroxenite sequence that hosts a massive, 30-cm-thick chromitite in its middle part. The rest of MCU I comprises mainly gabbronoritic adcumulates. MCU II is found only in the Kilvenjärvi block and fades away eastward. It consists of a basal peridotite overlain by pyroxenite and gabbronorite. MCU III commences with a peridotite layer overlain by gabbronorite orthocumulates (plagioclase-bronzite cumulates with poikilitic augite) followed by gabbronorite adcumulates.

Mafic and ultramafic dikes, known as the Portimo dikes, occur in the basement below the Konttijärvi intrusion and in the Ahmavaara area of the Suhanko intrusion (Figs. 3.3.3, 3.3.6, and 3.3.11). They have also been detected as fragments in the Konttijärvi marginal series (Fig. 3.3.4D). The dikes have not been dated and their association to the main intrusions is based on geochemical and geologic observations, as discussed in the following. The dikes strike subparallel to the basal contact of the intrusions and merge locally with them.

Special stratigraphic features of the Konttijärvi and Ahmavaara areas

The highest grade of PGE mineralization within the Portimo complex occurs in the Konttijärvi intrusion and in the Ahmavaara area at the western end of the Suhanko intrusion. The geology and stratigraphy of these areas is therefore discussed in more detail in the following.

The lowermost homogenous gabbronoritic cumulates of the Konttijärvi marginal series are underlain by the varitextured gabbro zone (also called the *transition* or *mixing zone*), which constitutes a heterogeneous mixture of pyroxene cumulates that have been infiltrated by variable proportions of felsic melts

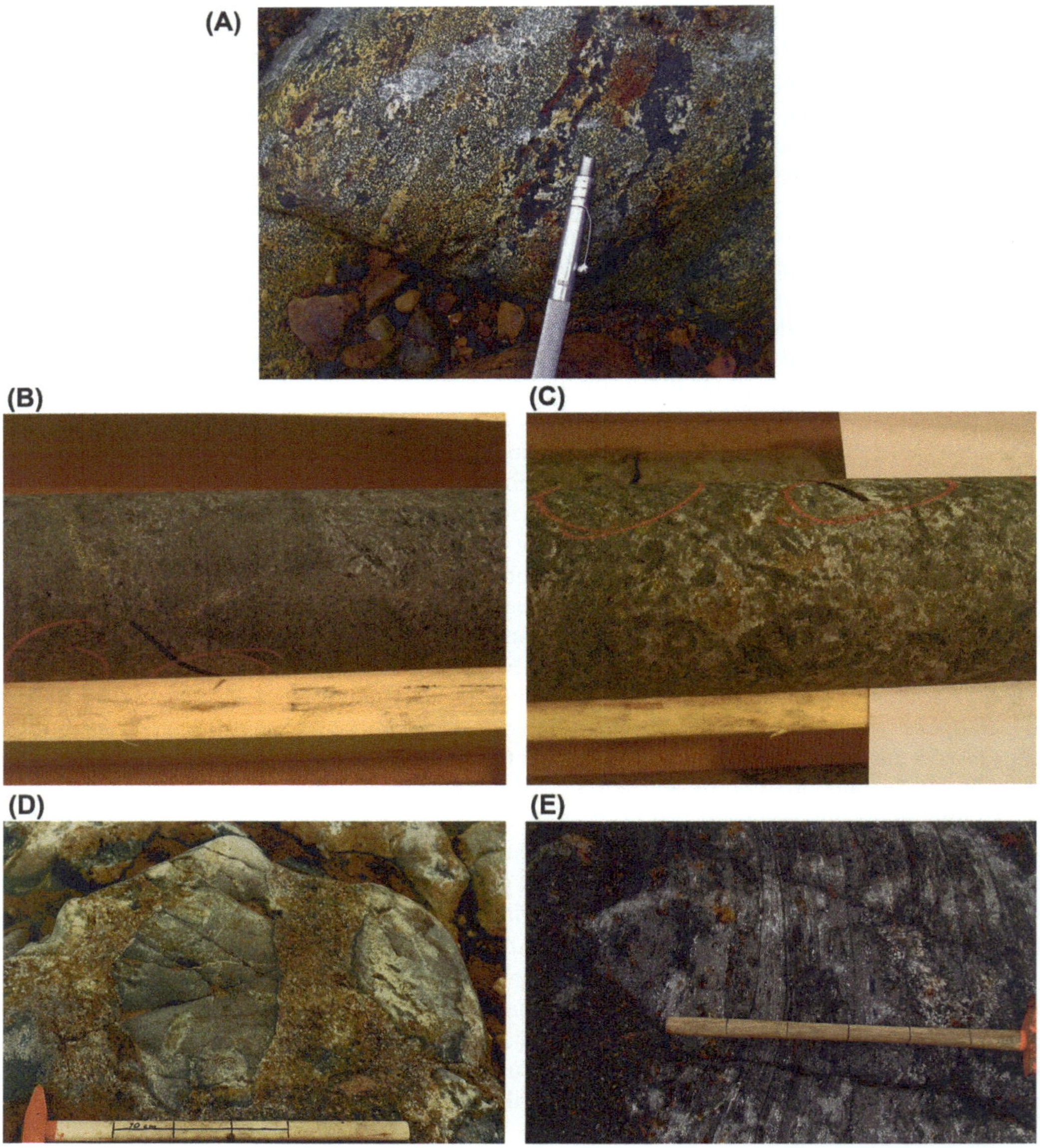

FIGURE 3.3.4 (A) Varitextured gabbro at Konttijärvi. (B) Fine-grained and sharp-edged chilled margin fragments in the Ahmavaara drill core. For the structural position, see Fig. 3.3.5. (C) Cresscumulate with sulfides in the lower part of the Ahmavaara marginal series. (D) Large blocks of the Portimo dikes in the olivine cumulate (darker) of the Konttijärvi marginal series. (E) Banded gabbro, varitextured zone, Konttijärvi.

Source: *Photos by M. Iljina.*

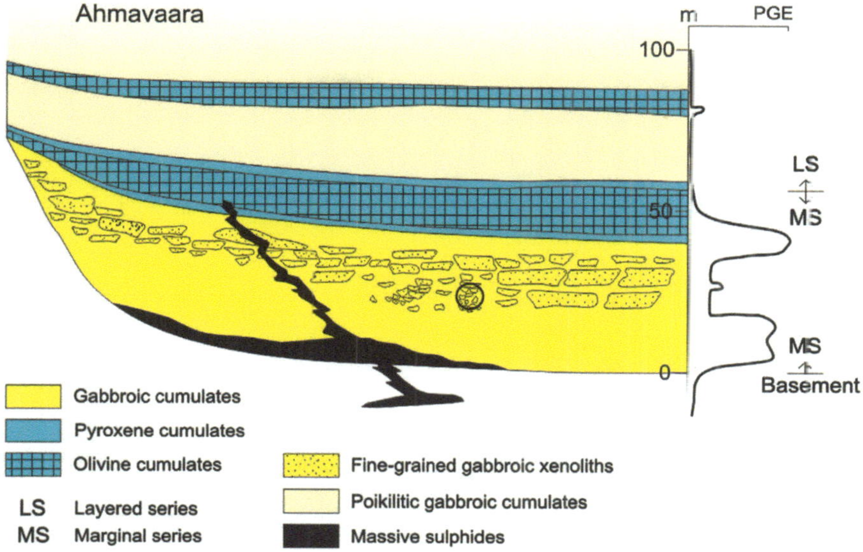

FIGURE 3.3.5 Schematic cross section of the Ahmavaara marginal series and lowermost layered series (Suhanko intrusion).

The circle shows the location of the photograph in Fig. 3.3.4B.

derived from the footwall rocks. These rocks are referred to as *hybrid gabbro* and *banded gabbro* (Figs. 3.3.4, 3.3.6, and 3.3.8). The hybrid gabbro is characterized by fine- to medium-grained rocks but also contains partially digested felsic fragments derived from the footwall. Further away from the intrusion, the hybrid gabbro turns into banded gabbro (Fig. 3.3.4E). This banding is probably due to flow of a heterogeneous mixture of mafic magma and anatectic felsic melts. Contacts between the homogenous gabbro, hybrid gabbro, and banded gabbro are diffuse, but the hybrid and banded gabbros together form a distinct mappable unit that also contains nonassimilated basement gneiss blocks up to several tens of meters in size. The varitextured gabbro zone appears to be a result of the mechanical and metasomatic mixing of melted Archean gneiss and mafic magma, indicating dynamic intrusion of the mafic magma.

The cumulus sequences of the lower layered series of the Konttijärvi and Ahmavaara areas bear a strong resemblance. The pyroxenite occurring between the lowermost poikilitic gabbronorite of the layered series and the overlying gabbronoritic adcumulate attains a thickness of several tens of meters in both sections. It measures only a tenth of the thickness elsewhere in the Suhanko intrusion.

THREE-DIMENSIONAL STRUCTURE OF THE PORTIMO COMPLEX

The surface exposure and geological cross sections of the Narkaus intrusion indicate a sill-like body (Iljina, 1994; Iljina and Salmirinne, 2011) dipping at a moderate angle beneath the sedimentary roof rocks, down to a depth of ~700 m, in a similar way as the Penikat intrusion does (Lerssi, 1990). In contrast, the Suhanko intrusion has a more lopolithic shape, which is also reflected in the variability of the stratigraphic sequences along surface profiles and the location of mineral deposits. Interpretations of

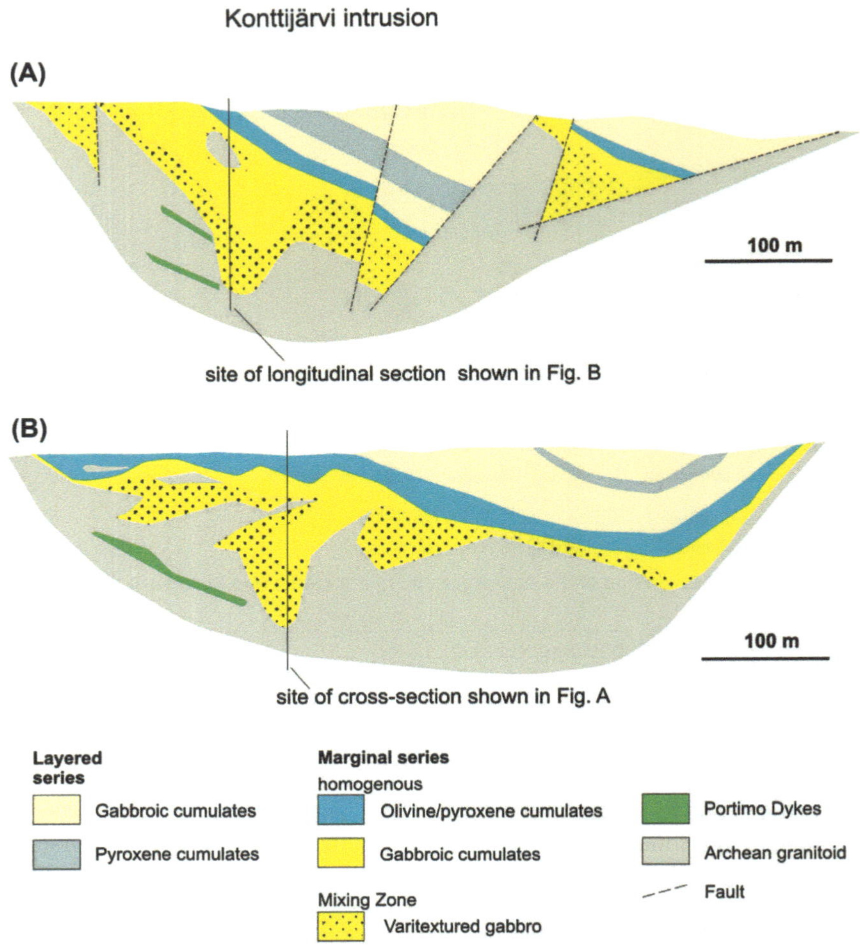

FIGURE 3.3.6 Cross section (A) and longitudinal section (B) of the Konttijärvi intrusion.

Source: *Modified after Iljina and Hanski (2005).*

gravimetric measurements (Pernu et al., 1986) and seismic reflection surveys (Iljina and Salmirinne, 2011) have revealed the three-dimensional structure of the Suhanko intrusion. The magmatic body has an estimated present-day volume of about 13.9 km^3. Figure 3.3.7 presents horizontal sections through the intrusion at variable depths. The most significant feature is that the central and northern parts of the Suhanko intrusion plunge to the northwest at a low angle, reaching a depth of about 1 km. The lower contact of the intrusion is *transgressive*; that is, the base of the southeastern tip is in contact with relatively higher levels of the Archean basement complex than the remainder of the intrusion. In view of its cumulus stratigraphy and chemistry, the Ahmavaara section is interpreted to represent the stratigraphically lowest part of the Suhanko intrusion, partly because it contains the highest proportion of ultramafic cumulates. In contrast, the southeastern tip of the intrusion represents a relatively elevated stratigraphic level.

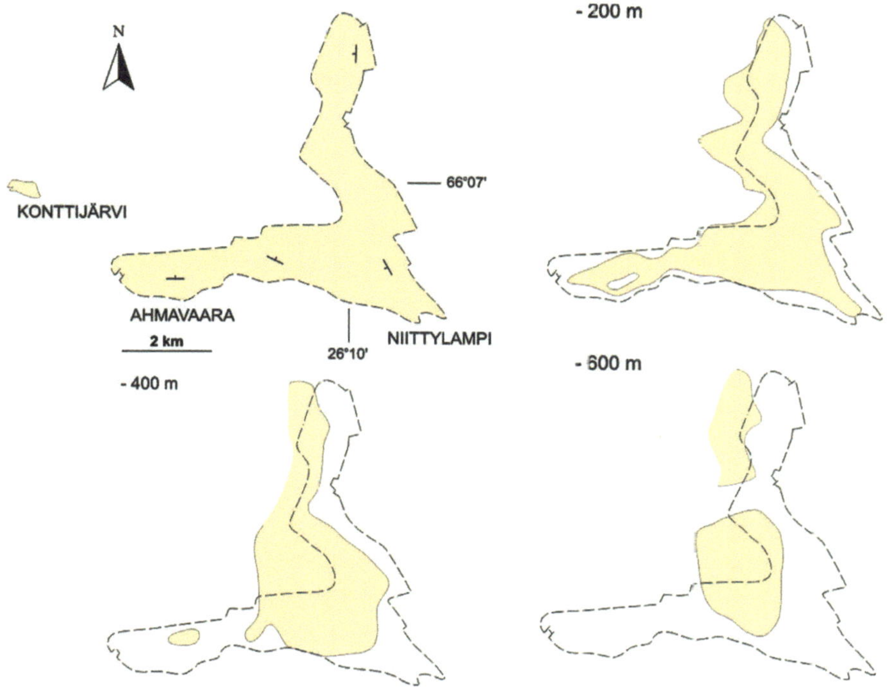

FIGURE 3.3.7 Suhanko intrusion: plan view and horizontal cross sections at depths of 200, 400, and 600 m.

Source: *Modified after Pernu et al. (1986).*

CU-NI-PGE MINERALIZATION IN THE PORTIMO COMPLEX

Of the Finnish layered intrusions, the Portimo complex contains the most diverse types of PGE mineralization, including internal PGE reefs, contact-type PGE reefs, and so-called offset deposits (refer to Figs. 3.3.8–3.3.11). Based on Iljina (1994), the following are the principal mineralization types.

Contact-type deposits:

- PGE-bearing Cu-Ni-PGE sulfide disseminations in the marginal series of the Konttijärvi and Suhanko intrusions (see Figs. 3.3.8 and 3.3.9).
- Predominantly massive pyrrhotite deposits located close to the basal contact of the Suhanko and Narkaus intrusions (Fig. 3.3.2).
- Offset-type Cu-PGE mineralization below the Narkaus intrusion (Fig. 3.3.10).

Reef-type deposits:

- The Rytikangas PGE Reef in the layered series of the Suhanko intrusion (Figs. 3.3.11 and 3.3.20).
- The Siika-Kämä PGE Reef in the layered series of the Narkaus intrusion (Figs. 3.3.11 and 3.3.20).

Four other PGE enrichment types are depicted in Fig. 3.3.11: (1) PGE enrichment in the Portimo dikes below the Konttijärvi and Ahmavaara marginal series, (2) PGE enrichment near the roof of the Suhanko intrusion, mostly associated with pegmatites, (3) a Pt-enriched clinopyroxenitic pegmatite

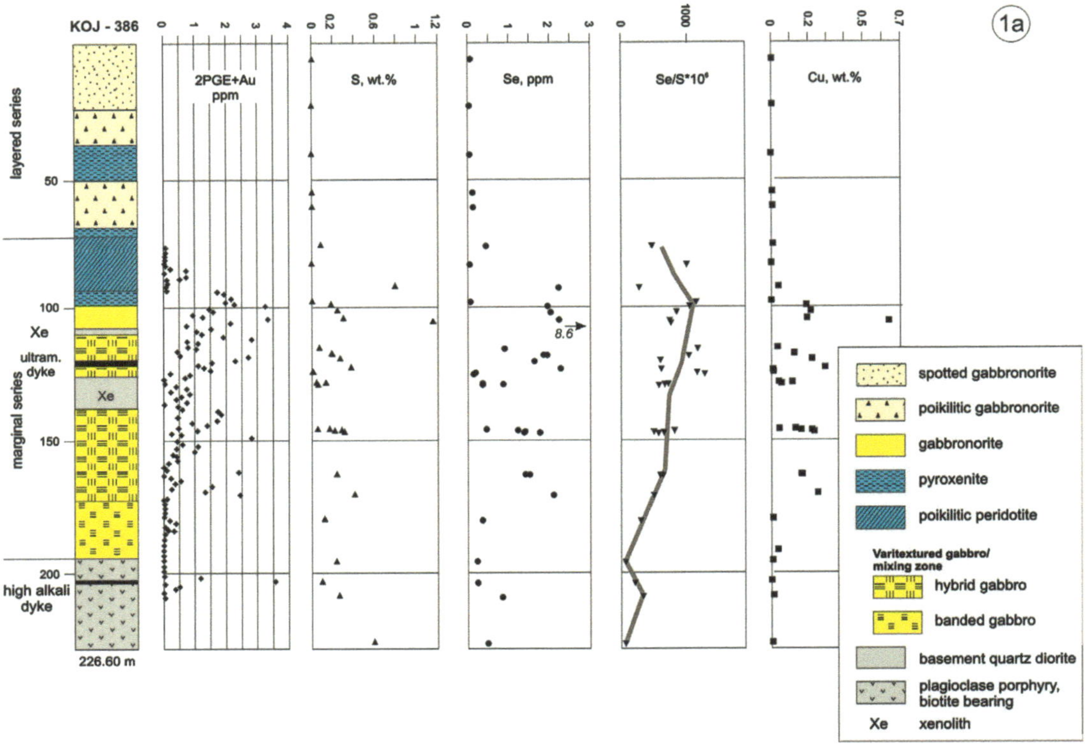

FIGURE 3.3.8 Stratigraphic sequence of the Konttijärvi marginal series.

This shows variations in bulk Pt + Pd + Au, S, Se, Se/S, and Cu. For structural position, see circle 1a in Fig. 3.3.11.

Source: *Modified after Iljina (2005).*

pipe in the western limb of the Suhanko intrusion, and (4) chromite and silicate-associated PGE enrichments in the lower parts of the Narkaus intrusion and MCU II. Fig. 3.3.11 shows the structural model for the Portimo complex and the positions of the deposits referred to earlier. Table 3.3.4 shows grade and tonnage information of various deposits in the Portimo complex.

Disseminated contact-type Cu-Ni-PGE mineralization in the Suhanko and Konttijärvi intrusions

Disseminated PGE-bearing base-metal-sulfide mineralized zones, normally 10–30 m thick, occur practically throughout the marginal series of the Suhanko and Konttijärvi intrusions (Figs. 3.3.8, 3.3.9, and 3.3.11 later). The distribution of the mineralization is erratic; it generally extends from the peridotitic layer of the marginal series downward for some 30 m into the granitic basement. The PGE contents vary from only weakly anomalous values (100 s of ppb) to 2 ppm in much of the marginal series of the Suhanko intrusion, but rise to >10 ppm in several, ~1-m-long drill core intervals in the Konttijärvi and Ahmavaara areas.

Fig. 3.3.8 depicts the variation in the copper and precious metal contents and Se/S ratio in one drill hole across the Konttijärvi marginal series. The whole-rock PGE contents show a good positive

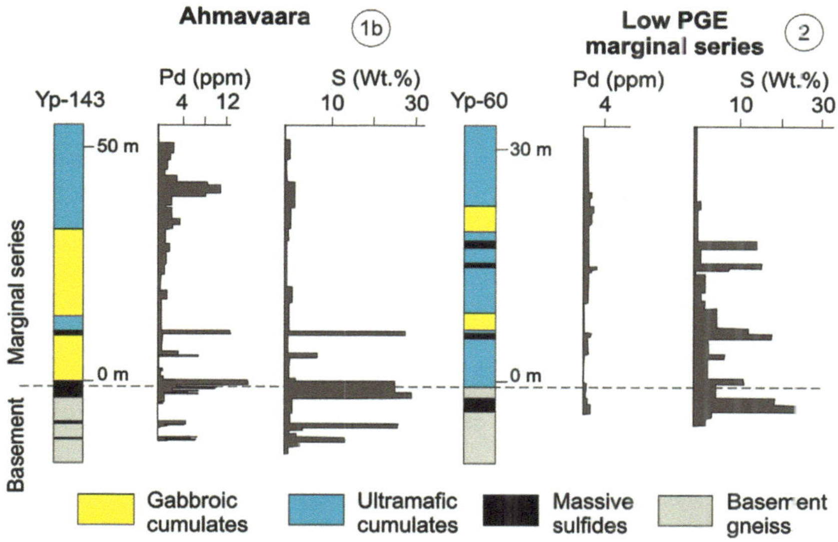

FIGURE 3.3.9 Comparison of the Ahmavaara deposit and one low-grade (Suhanko proper, Fig. 3.3.2), disseminated and massive sulfide deposit of the Suhanko intrusion (Fig. 3.3.2).

For the structural position, see circles 1b and 2 in Fig. 3.3.11.

Source: *Modified after Iljina et al. (1992).*

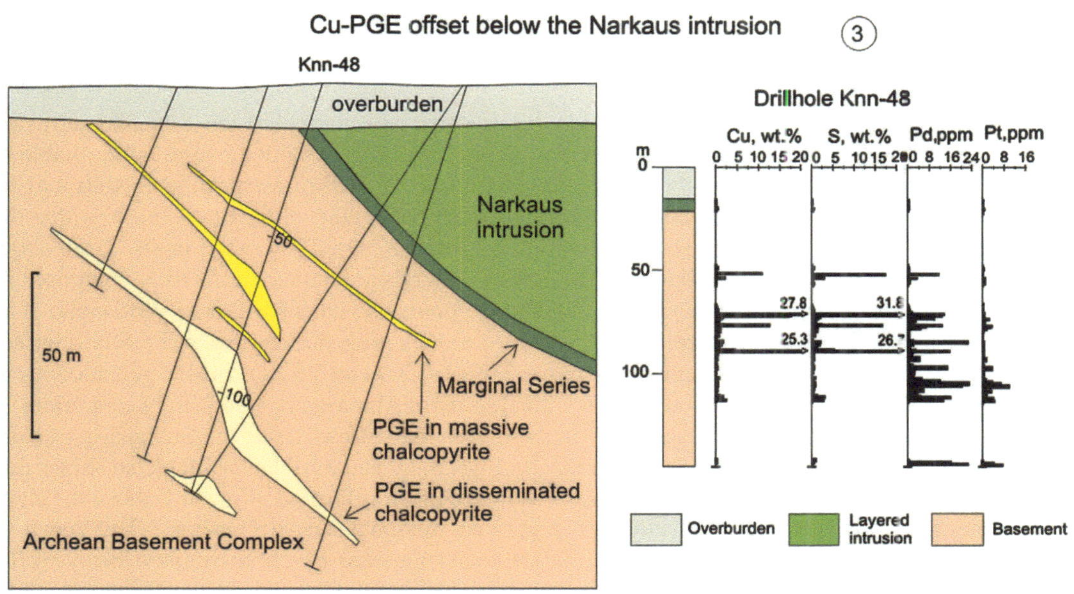

FIGURE 3.3.10 Offset Cu-Pd deposit at Kilvenjärvi below the Narkaus intrusion.

For the structural position, see circle 3 in Fig. 3.3.11.

Source: *Metal data from Huhtelin et al., 1989.*

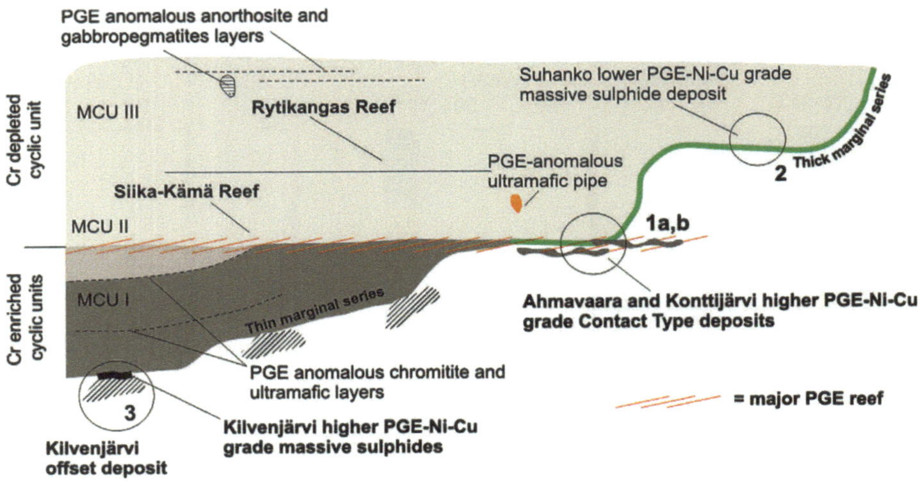

FIGURE 3.3.11 Schematic presentation of the locations of the various PGE enrichments encountered in the Portimo complex.

The circled numbers refer to Figs. 3.3.8–3.3.10 in this chapter.

Source: *Modified after Iljina (1994).*

correlation with Cu, Se, and Se/S. As a whole, half of the metal content of the entire Konttijärvi deposit is hosted by varitextured gabbro and basement rocks located below the homogenous marginal series proper.

Massive sulfide mineralization

Massive sulfide mineralization is characteristic for the marginal series of the Suhanko intrusion (Fig. 3.3.2). The mineralization occurs in the form of dikes and also plate-like bodies conformable to the layering, and generally varies in thickness from 20 cm to 20 m. The massive sulfide bodies may be located within felsic basement gneiss up to 30 m below the basal contact of the intrusion, or within the intrusion, up to 20 m above the basal contact (Figs. 3.3.5 and 3.3.9). They range in size from about 1 million tons to more than 10 million tons. The sulfide minerals are almost exclusively composed of pyrrhotite, except in the Ahmavaara deposit, which also contains chalcopyrite and pentlandite. The massive pyrrhotite deposits show relatively low PGE values with the maximum Pt + Pd content normally reaching a few ppm (exemplified by circle 2, Figs. 3.3.9 and 3.3.11). However, analogous to the disseminated sulfide mineralization in the marginal series of Ahmavaara, the PGE concentrations in the Ahmavaara massive sulfides are generally much higher than elsewhere in the Suhanko intrusion, attaining a level of 20 ppm (exemplified by circle 1b, Figs. 3.3.9 and 3.3.11). Drilling also shows that the relatively PGE-poor Suhanko (proper) massive sulfide deposit is located physically above the Rytikangas Reef due to the transgressive nature of the Suhanko marginal series (refer Fig. 3.3.12 later).

A less well-known massive sulfide deposit is located at the base of the Kilvenjärvi block of the Narkaus intrusion, above the Kilvenjärvi offset deposit. The base and precious metal tenors of this deposit resemble those of the Ahmavaara deposit in being distinctly higher than in the low-PGE grade deposits of the Suhanko intrusion.

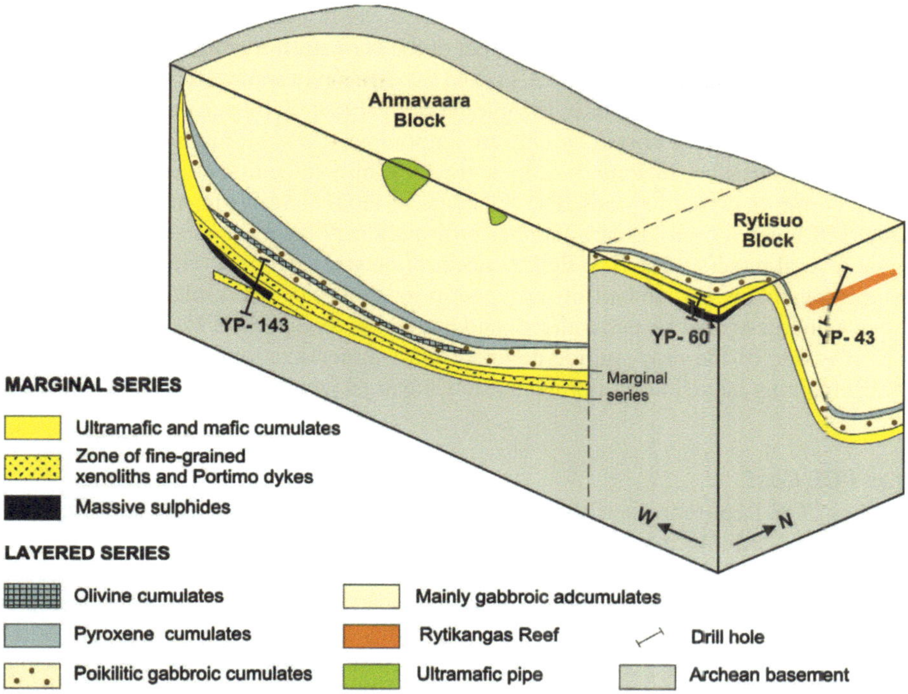

FIGURE 3.3.12 Perspective section from the western lobe of the Suhanko intrusion, showing the structural relationships between the Portimo dikes, Ahmavaara, and Suhanko proper massive sulfide deposits and Rytikangas Reef.

Drill hole sections YP-60 and YP-143 refer to Fig. 3.3.9.

Modified after Iljina et al. (1992).

Offset-type Cu-Pd mineralization

Copper-Pd mineralization is sporadically distributed in the basement gneisses and granites below the entire length of the Narkaus intrusion. It is usually referred to as *offset-type* mineralization (Huhtelin et al., 1989; Iljina, 1994; Andersen et al., 2006). The largest and best-known deposit is situated below the Kilvenjärvi block (Figs. 3.3.2, 3.3.10, and 3.3.11). This deposit is composed of a cluster of closely grouped ore bodies and is located in and near an N-trending major fault zone, some tens of meters wide, against which the Kilvenjärvi block terminates. The offset mineralization represents the richest PGE deposit type in the Portimo area, with whole-rock Pt + Pd contents reaching 100 ppm. The offset mineralization is predominantly a Pd deposit, as it has a much higher Pd/Pt ratio than the other Portimo deposits (or any other PGE deposit in the Tornio-Näränkävaara belt), whereas the Os, ir, Ru, and Rh concentrations are extremely low. The mineralization may comprise disseminated "clouds" of sulfide and PGM, massive sulfide veins or bodies, and sulfide-bearing granite breccias. The proportions of base metal sulfides and PGM are highly variable. The massive sulfide bodies are always rich in PGE, but some samples containing almost no visible sulfides can also carry several tens of ppm of Pd. In general terms, the more sulfide-rich occurrences are situated closer to the basal contact of the intrusion and those poorer in sulfides are encountered a greater distance below the intrusion (Fig. 3.3.10).

Siika-Kämä PGE Reef

The laterally persistent Siika-Kämä (SK) PGE Reef of the Narkaus intrusion is most commonly located at the base of MCU III (Figs. 3.3.2, 3.3.3, 3.3.11, and 3.3.20), but in some cases, it may occur somewhat below the basal contact of MCU III, or in the middle of the basal olivine cumulate layer (peridotite) of MCU III. The reef is normally hosted by chlorite-amphibole schist, similarly to the Sompujärvi PGE Reef in the Penikat intrusion. In some locations, the PGE mineralization is associated with thin chromite seams or chromite disseminations. The thickness of the reef varies from <1 m to several meters, and in many drill holes it consists of a number of mineralized layers separated by PGE-poor layers, which can be several meters thick. The PGE concentration varies from several hundred ppb to tens of ppm. Gabbroic pegmatites, abundant in the uppermost gabbroic, adcumulate tens of meters below MCU III, can sometimes also be mineralized, and contain several ppm of Pd and Pt. The Siika-Kämä mineralization is one of the most sulfide-deficient PGE mineralizations in the Portimo complex, in some places containing no visible sulfides and rarely having a whole-rock sulfur content of higher than 1 wt%.

Rytikangas PGE Reef

The Rytikangas PGE Reef represents the main PGE occurrence in the layered series of the Suhanko intrusion (Figs. 3.3.3, 3.3.11, and 3.3.12). It is located in the central portion of the cumulate sequence, about 170 m above the base of the western limb of the intrusion. The reef is verified by drilling to be continuous over the strike length of 1.5 km. It is hosted by anorthosites and gabbronorites (plagioclase-bronzite-augite orthocumulates with poikilitic augite). This cumulate series overlies a 70-m sequence of monotonous gabbronorite adcumulates and underlies 10 m of homogenous gabbronorite mesocumulates, in which augite has a nonpoikilitic intercumulus habit. The orthocumulate layer hosting the reef varies in thickness from 30 cm to 10 m. The thickness of the reef itself is 30–50 cm, and it typically occurs in the uppermost part of the poikilitic orthocumulate layer. The cumulus stratigraphy and compositional variation in major and trace elements across the Rytikangas Reef resembles those of the Ala-Penikka Reef in the Penikat intrusion.

Composition of the sulfide fraction and PGE characteristics

The concentrations of sulfur and base and noble metals in representative samples, and their values as recalculated to 100% sulfides, are presented in Table 3.3.1. Average concentrations for a large number of samples from each deposit, also recalculated to 100% sulfides, are also given in Table 3.3.1. A graphical presentation of the base and precious metal concentrations in 100% sulfides for massive sulfide deposits is shown in Fig. 3.3.13.

The massive sulfide deposits at the base of the Suhanko intrusion, with the exception of Ahmavaara, proved to be poor in nickel and copper, having Ni and Cu tenors from 0.48–2.2 wt% and from 0.37–2.4 wt%, respectively. Conversely, the Ahmavaara massive sulfide deposit has a markedly higher nickel content, with 2.7 wt% Ni in the sulfide fraction (Table 3.3.1).

Pd/Pt, Pd/Ir, and (Pt + Pd)/(Os + Ir + Ru) ratios of the deposit types are presented in Table 3.3.2. Most deposit types are characterized by a predominance of palladium over platinum, with only the Siika-Kämä Reef and the chromite and silicate-associated PGE enrichments of MCU II locally showing Pt/Pd ratios above unity. The Konttijärvi PGE-rich marginal series, the Ahmavaara massive sulfide deposit, and the Rytikangas and Siika-Kämä PGE Reefs have similar Pd/Ir and (Pt + Pd)/(Os + Ir + Ru) ratios.

Table 3.3.1 A: Average whole-rock Ni, Cu, S, PGE, and Au concentrations for selected type samples
B: Concentrations in the type samples, recalculated to 100% sulfide
C: Metal concentrations in a large number of samples, recalculated to 100% sulfide

		Ni(wt%)	Cu	S	Os-(ppb)	Ir	Ru	Rh	Pt	Pd	Au
Portimo dikes and sulfide disseminated marginal series											
Portimo dikes	A	0.025	0.093	0.183	3.0	0.5	-	2.0	510	2200	47
n = 1	B	5.0	18.5	36.5	598	99.7		399	101 700	438 700	9370
Konttijärvi	A	0.056	0.239	0.323	8.6	19.0	14.2	92	1 300	4 070	240
marginal series	B	6.1	25.9	35.0	930	2 060	1 540	9 970	140 800	440 900	26 000
n = 5	C	5.4	14.4	36.7							
Massive sulfide deposits of marginal series											
Ahmavaara	A	2.00	0.719	25.8	20	50	44	357	1 510	11 030	104
n = 3	B	3.0	1.1	39	30	76	67	540	2 280	16 700	160
	C	2.7	2.4	37					2 120	15 200	
Suhanko	A	0.919	0.807	20.7	30	64	64	222	207	1 230	7.3
n = 3	B	1.7	1.5	39.1	57	120	120	420	390	2 320	14
	C	1.5	2.4	37							
Vaaralampi	A	0.284	0.143	23.3	25	8.5	89	24		485	5.0
n = 2	B	0.48	0.24	39.5	42	14	151	41		320	8.5
	C	0.94	0.63	37							
Niittylampi,	A	1.67	0.305	32.7	23	79	36	550	136	835	19
n = 2	B	2.0	0.37	39.2	28	95	43	660	160	1 000	23
	C	2.2	1.6	37					870	2 190	
Yli-Portimojärvi	A	0.456	0.575	24.9		111		199	171	930	7.0
n = 2	B	0.72	0.91	39.4		180		310	270	1 470	11
Kilvenjärvi	A	2.92	0.6	20.1	15	27	60	207	625	3 800	85
n = 2	B	5.6	1.15	38.6	29	52	115	397	1 200	7 280	163
Layered series											
Rytikangas reef	A	0.063	0.249	0.291	28	32	24	171	1 630	7 240	207
n = 5	B	7.4	29.3	34.2	3 290	3 760	2 820	20 100	191 600	850 900	24 300
	C	6.4	38.7	33.0							
Siika-Kämä reef	A	0.080	0.360	0.454	32	65	47	330	2 850	9 980	337
n = 4	B	6.2	27.7	34.9	2 460	5 000	3 610	25 370	219 000	766 900	25 910
Mineralized upper	A	0.035	0.231	0.492	23	27	76	53	605	1270	40
Suhanko layered series *n* = 2	B	2.7	17.9	38.2	1 790	2 100	5 900	4 120	47 000	98 600	3 100

n = number of samples
Source: *Data from Iljina (1994) and references therein.*

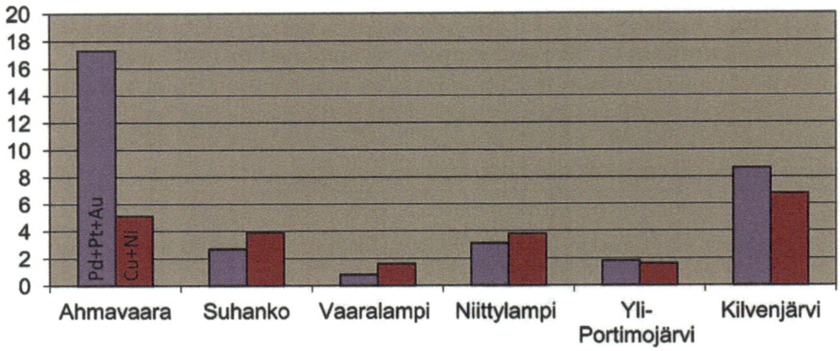

FIGURE 3.3.13 Nickel, copper, and precious metal concentrations recalculated in 100% in various massive sulfide deposits of the Portimo complex.

Base metals in wt% and precious metals in ppm.

The Portimo dikes have markedly higher Pd/Ir of ~3840 and (Pt + Pd)/(Os + Ir + Ru) of >267, respectively, whereas the PGE-poor massive sulfide deposits have lower Pd/Ir and (Pt + Pd)/(Os + Ir + Ru).

Chondrite-normalized metal patterns are presented in Fig. 3.3.14. The Konttijärvi and Ahmavaara disseminated, PGE-rich marginal series, the Ahmavaara and Kilvenjärvi massive sulfides, and the Rytikangas and Siika-Kämä Reefs have very similar metal patterns, all showing a steep positive slope and a negative Ru anomaly. In contrast, the PGE-poor massive pyrrhotite deposits have flatter metal patterns, with negative Pt anomalies instead of negative Ru anomalies (Fig. 3.3.14). The Kilvenjärvi offset deposit shows the most fractionated pattern with a strong relative depletion in Os, Ir, Ru, and Rh.

Mineralogy of base metal sulfides

The most common base metal sulfides are pyrrhotite, chalcopyrite, and pentlandite, occurring in variable proportions. The massive sulfide deposits at the base of the Suhanko and Narkaus intrusions are deficient in pentlandite and chalcopyrite, except in the Ahmavaara and Kilvenjärvi deposits. In the Rytikangas PGE Reef, Fe-deficient copper sulfides, (i.e., bornite and chalcocite), dominate over chalcopyrite and Fe(-Ni) sulfides. Among the nickel-bearing sulfides, millerite (NiS) is estimated to contain half of the sulfidic nickel in the Rytikangas Reef.

An indication of low-temperature sulfide mineralization is noted in the basement granitoids below the Konttijärvi marginal series, where pyrite grains were found to contain polymineralic chalcopyrite, pyrrhotite, and pentlandite inclusions. A peculiar sulfide aggregate composed almost in equal proportions of pentlandite, galena, sphalerite, and pyrite is also common in the mineralized basement below the Konttijärvi marginal series.

Even a lower crystallization temperature is indicated by the secondary sulfide assemblage of the offset mineralization, which is composed of chalcopyrite, polydymite ($NiNi_2S_4$), melnikovite-pyrite, and bravoite. Bravoite is an intermediate member (6.8–19.8 atomic % Ni) of the solid-solution series of vaesite (NiS_2) and pyrite (FeS_2). This atypical sulfide mineralogy of the Kilvenjärvi offset deposit occurs in deep drill holes, suggesting that surface alteration played no role in mineralization.

Table 3.3.2 Average Pd/Pt, Pd/Ir, and (Pt + Pd)/(Os + Ir + Ru) ratios in the mineralized zones (range in parenthesis)

	N	Pd/Pt	Pd/Ir	(Pt+Pd)/ (Os+Ir+Ru)
Offset, Portimo dikes, and Konttijärvi marginal series				
Offset	1	6.7	>12 900	>687
Portimo dikes	2	3.4	3 840	>267
		(2.6–4.3)	(3 270–4 400)	
Konttijärvi marginal series	5	3.0	212	141
		(1.4–4.0)	(138–429)	(67–300)
Massive sulfide deposits				
Ahmavaara	3	8.0	243	116
		(6.2–10.3)	(114–335)	(54–175)
Suhanko	3	6.6	20.6	10.5
		(4.5–9.3)	(15.2–24.3)	(6.6–14.2)
Vaaralampi	1	3.6	42.7	4.2
Niittylampi	2	7.6	10.7	7.1
		(4.3–10.9)	(9.7–11.7)	(6.8–7.3)
Yli-Portimojärvi	1	5.3	6.3	2.3
Kilvenjärvi	2	6.1	146	43.8
		(5.6–6.6)	(123–170)	(42.7–44.9)
Layered series				
Rytikangas reef	5	4.4	270	132
		(3.7–5.1)	(147–382)	(77–184)
Siika-Kämä reef	8	1.7	83.6	52.2
		(0.8–3.0)	(32.1–132)	(41.7–93.3)
Mineralized upper	2	2.1	46.6	14.9
Suhanko layered series		(2.0–2.2)	(44.5–48.6)	(14.4–15.4)
Chromitite layer, MCU I	1	3.8	17.1	5.6
Chromite and silicate-associated MCU II	3	0.9	49.8	43.1

N = number of samples
Source: *Data from Iljina (1994) and references therein.*

Platinum-group minerals

The platinum-group mineralogy is dominated by sperrylite ($PtAs_2$) and various Pd-Sb-As and Pd-Te-Bi phases, while PGE sulfides are rare and alloys are absent. The platinum-group sulfarsenides are represented by hollingworthite (RhAsS) and platarsite (PtAsS), which were also found to be the only carriers of osmium, iridium, ruthenium, and rhodium. Sn-bearing Pd minerals and pure Pd-Sn minerals have been identified in the Konttijärvi and Ahmavaara marginal series, while the high whole-rock Se/S ratio of the Portimo dikes is reflected also by the presence of PGE-Se mineral phases. Other rare PGM phases include a Pd-Ni-As mineral and Pd-Ag-Te phases found in the Rytikangas Reef.

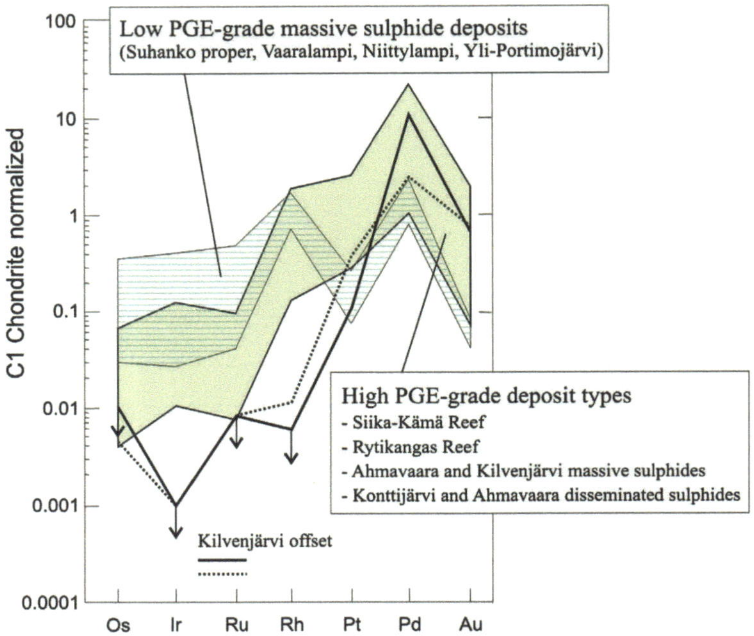

FIGURE 3.3.14 Chondrite-normalized PGE and Au patterns of the Kilvenjärvi offset mineralization, the Siika-Kämä and Rytikangas PGE Reefs, Konttijärvi and Ahmavaara high-grade disseminated and massive sulfide PGE deposits, and lower-grade massive sulfide PGE deposits of the Suhanko marginal series.

Sources: *Chondrite values are from Naldrett (1981). Modified after Iljina (1994).*

An atypical platinum-group mineralogy is observed locally in the Kilvenjärvi offset deposit. In view of the very high whole-rock PGE contents (up to several tens of ppm), large PGM grains are scarcer than expected. Instead, polydymite grains are found to host a large number of tiny (≤1 μm) Pd mineral grains containing the bulk of the Pd in the rock.

Mineral resources of contact-type deposits

Classified mineral resources of the Portimo contact type deposits are presented in Table 3.3.3. The total resources (Puritch et al., 2007) are ~263 million tons including measured, indicated, and inferred resources. These resources result in about 10 million ounces of platinum, palladium, and gold combined.

CONCLUDING REMARKS

- Of the Finnish layered intrusions, the Portimo complex is exceptional in hosting a variety of PGE mineralization types.
- The lower contact of the Suhanko intrusion is *transgressive* relative to the Archean basement; the southeastern tip of the intrusion is in contact with stratigraphically higher levels of the Archean basement complex than the remainder of the intrusion.
- In the contact-type deposits of the Portimo complex, the Cu, Ni, and precious metal concentrations in the whole-rock samples and the sulfide fraction are highest in the western portion of the complex (Konttijärvi and Ahmavaara areas).

Table 3.3.3 Classified mineral resources and metal grade information for contact-type PGE deposits

Intrusion/ complex	Tons (Mt)	Pd (ppm)	Pt (ppm)	Au (ppm)	Ni (wt%)	Cu (wt%)
Portimo Ahmavaara						
Measured	38.197	0.707	0.141	0.086	0.061	0.161
Indicated	114.816	0.851	0.175	0.104	0.071	0.181
Inferred	34.757	0.867	0.187	0.103	0.070	0.170
Total	**187.770**					
Konttijärvi						
Measured	26.031	1.064	0.307	0.081	0.059	0.100
Indicated	37.571	0.902	0.255	0.071	0.043	0.095
Inferred	11.638	0.867	0.241	0.062	0.027	0.097
Total	**75.240**					
Koillismaa Kaukua						
Indicated	10.4	0.73	0.26	0.08	0.1	0.15
Inferred	13.2	0.63	0.22	0.06	0.1	0.13
Total	**23.6**					
Haukiaho						
Inferred	23.2	0.31	0.12	0.10	0.14	0.21
Total	**23.2**					

Note: *Calculated using cutoff grades of 1.0 g/t* Pd_{Eq} *for Portimo deposits and 0.1 g/t Pd for Koillismaa deposits.*
Sources: *Portimo deposits from Puritch et al. (2007), and Koillismaa deposits from the Finore Mining Inc. press release of September 19, 2013.*

- The relative proportion of ultramafic cumulates within the Portimo complex is highest at Konttijärvi.
- Based on its cumulus stratigraphy and chemistry, the Ahmavaara section is interpreted to represent the stratigraphically lowest part of the Suhanko intrusion.
- Megacyclic units form two groups in terms of their chromium content, one being relatively rich and the other relatively depleted. The former is represented by MCU I and II of the Narkaus intrusion, as well as the Portimo dikes, and the latter is represented by MCU III and the Suhanko and Konttijärvi intrusive bodies.
- The Portimo dikes have so far only been identified below the Konttijärvi and Ahmavaara marginal series.
- Autoliths interpreted to represent the chilled margin of the intrusion become less Cr- and MgO-rich from the west (Ahmavaara) to the east (Niittylampi) within the Suhanko intrusion, representing stratigraphically progressively higher levels. This trend is reminiscent of the change from B1 to B3 type sills in the floor rocks of the lower to main zone of the Bushveld complex.

REEF-TYPE PGE MINERALIZATION IN THE PENIKAT INTRUSION

The Penikat intrusion is located approximately 30 km to the northeast of the town of Kemi and 30 km to the southwest of the Portimo complex. It is 23 km long and 1.5–3.5 km wide (Halkoaho, 1993; Iljina and Hanski, 2005). The magmatic body has an intrusive contact to K-rich granite in its footwall, but the uppermost rocks were eroded away before the deposition of the overlying sedimentary rocks.

The basal contact of the intrusion is sharp, defined by a chilled margin, which is relatively biotite-rich and thus believed to be contaminated. This model is consistent with the relatively evolved composition of the rocks. Two samples analyzed by Halkoaho (1993) gave 6.5–8 wt% MgO, 0.3–0.6 wt% TiO_2, 440–470 ppm Cr, 90–100 ppm Ni, and 12–14 ppb Pd. The chilled margin is overlain by approximately 10–20 m of marginal series rocks consisting of subophiticgabbro (18 wt% MgO; Halkoaho, 1993), as well as gabbronoritic and orthopyroxenitic cumulates, which may be injected by rheomorphic granite veins.

The layered series is close to 3000 m thick and can be subdivided into five megacyclic units (MCU I–V). MCU I is between 270 and 410 m thick and contains a thin basal chromitite overlain by chromite-bearing orthopyroxenite and then gabbronorite. MCU II is 160–230 m in thickness and shows more variability than MCU I. It consists of basal websterite, overlain by lherzolite with thin chromitite stringers; websterite containing interlayers of gabbronorite and lherzolite; and finally, two chromite-bearing gabbronorite layers separated by websterite. MCU III is 75–330 m thick and consists of basal chromite-bearing websterite that may host a lherzolite layer. This is overlain by gabbronorite adcumulates and, locally, by poikilitic gabbronorite associated with gabbronoritic pegmatoids and disseminations, schlieren, and layers of chromite. MCU IV is 760–1110 m thick and hosts the bulk of the PGE mineralization in the form of three reefs. The first one, named the Sompujärvi or SJ Reef, occurs near the base of the unit. The second reef is termed the Ala-Penikka or AP Reef and occurs in the center of the unit, whereas the third reef, termed the Pasivaara or PV Reef, is located at the top of the unit. The base of MCU IV is transgressive and shows a 10–20-m-thick ultramafic layer consisting of a thin basal chlorite schist that can be highly enriched in chromite. This is overlain by a 1-m-thick layer of orthopyroxenite, a thick olivine cumulate layer containing intercumulus ortho- and clinopyroxene, and finally, a further orthopyroxenite. Next follow gabbronoritic cumulates, in which clinopyroxene has a poikilitic intercumulus status, followed by gabbronorite where clinopyroxene has a cumulus habit. In places, pronounced rhythmic layering is seen, including horizons of orthopyroxenite, gabbronorite, and anorthosite. Some 200–300 m above the base of the MCU IV is the AP Reef, hosted by mottled anorthosite and gabbronorite. At the top of MCU IV is a 40–60 m thick, highly complex zone consisting of gabbronorite, anorthosite, pegmatoid, and what could represent a magmatic breccia of anorthosite and gabbronorite (Huhtelin et al., 1990). This zone hosts the PV Reef. MCU V has a highly variable thickness (0–900 m), likely due to erosion. At its base is an orthopyroxenite overlain by norite. Toward the top, there also occurs some gabbro and magnetite-bearing leucogabbro, anorthosite, and plagioclase-quartz-biotite rock resembling the granophyre in the roof of the Koillismaa intrusion (Iljina and Hanski, 2005).

The compositional variation through the intrusions is summarized in Fig. 3.3.15. As a whole, the intrusion is relatively mafic, with ultramafic rocks making up <10% of the stratigraphy. This is a major difference compared to the Kemi intrusion, located immediately to the southwest, together with a relatively lower proportion of chromite, but much higher PGE grades at Penikat. The Penikat megacyclic units are well delineated by a progressive decrease in Cr/V from the base to the top. The

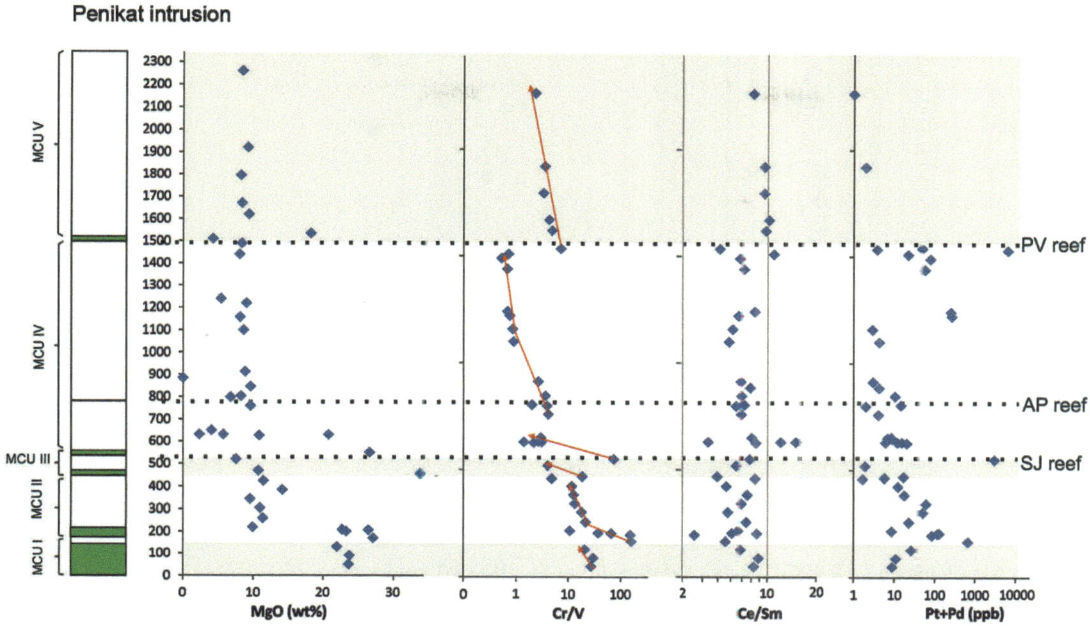

FIGURE 3.3.15 Litho- and chemostratigraphy of the Penikat intrusion.

Source: *Based on Halkoaho, 1993, and unpublished data of W.D. Maier.*

PGE-enriched zones of the SJ and PV Reefs are located at compositional and lithological reversals, represented by high Cr/V and the reappearance of orthopyroxene in the liquidus, analogously to many other PGE mineralized layered intrusions. However, in contrast to the Bushveld complex, for example, incompatible trace element ratios give no strong evidence for mixing between compositionally distinct magma pulses at Penikat, except for an increase in Ce/Sm above the SJ Reef, and a minor increase in Ce/Sm at the level of the PV Reef. Thus, if there was significant magma replenishment in the Penikat intrusion, the involved magma had a broadly similar chemical character in terms of elements discussed above. The compositional and lithological variation across the reefs is described in detail in Halkoaho (1993) and Halkoaho et al. (1990a,b).

PGE REEFS

The PGE in the SJ Reef are mostly concentrated in the basal portion of MCU IV, but can locally occur in the overlying peridotite or the underlying gabbronorite (see Halkoaho, 1990a). The PGE can be associated with sulfide enrichments of up to ~1 wt% (base metal-type) or with S-poor silicate rocks and chromitites, to the effect that the mineralization is invisible to the naked eye. The SJ Reef has been correlated with the Siika-Kämä Reef of the Portimo complex (see earlier and Iljina and Hanski, 2005). It locally contains grades of up to >100 ppm PGE over several dm. However, the average grade is much lower, and the thickness is usually ~1 m. In general, all three Penikat reefs show much more variation in thickness and lateral consistency than the Bushveld reefs.

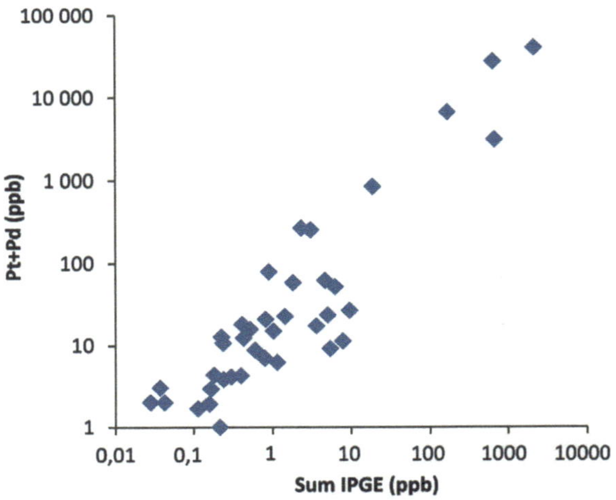

FIGURE 3.3.16 Pt + Pd vs. total IPGE + Rh diagram for SJ, AP, and PV Reefs.

The Ala-Penikka (AP) PGE Reefs (Halkoaho et al., 1990b) are located 250–450 m above the base of MCU IV. They are developed along most of the strike length of the intrusion and have been correlated with the Rytikangas Reef of the Portimo intrusion (Iljina and Hanski, 2005). The API Reef is normally 20–40 cm thick and contains erratic PGE mineralization hosted by gabbronorite and anorthosite. In the so-called depression structures (which are likely equivalents to the Bushveld potholes), the reef can be up to 20 m thick. Laterally less continuous APII Reef, some tens of meters above API, strongly resembles the API Reef in terms of host rock types, but contains less base metal sulfides. PGE grades in API may locally reach several tens of ppm, but are mostly much lower. The sulfide fraction shows the following element ratios: Cu/Ni 1.4–2.7, Pd/Pt 3.7, and Pd/Ir 170–240. Primitive mantle-normalized metal patterns of the AP reefs are more fractionated than those of the SJ Reef.

The Pasivaara (PV) Reef (Huhtelin et al., 1990) is located ~700–1000 m above the AP Reefs. Like the other reefs, it is broadly continuous along the entire strike of the intrusion, with the exception of the Keski-Penikka block where it has been eroded away. The PGE mineralization is erratically distributed within the reef. The best grades are usually found within the anorthosite near the top of MCU IV, or up to 30 m below the peak mineralization. The average thickness of the reef is 1 m with element ratios Cu/Ni 1, Pt/Pd 2 (the highest in the Penikat intrusion), and Pd/Ir 28. The primitive mantle-normalized metal patterns resemble those of the SJ base metal-type mineralization and the Merensky Reef, and are less fractionated than those of the AP Reefs.

There is generally a good correlation between PPGE and IPGE contents in the Penikat intrusion (Fig. 3.3.16), suggesting that the original PGE concentration mechanism was magmatic, possibly followed by some mobility of S.

PLATINUM-GROUP MINERALS

The platinum-group mineralogy of the SJ Reef is characterized by Pt-As, Pd-Sb-As, Pd-Te-Bi, and Rh-As-S phases resembling the mineralogy of various Portimo PGE enrichments. In chromite-bearing samples, platinum and palladium sulfides are relatively abundant. Alloys are exclusively found in chromite-bearing samples of the SJ Reef (Halkoaho et al., 1990b). The mineralogy of the AP and PV Reefs resembles that of the SJ Reef (Huhtelin et al., 1990).

PARENTAL MAGMA

The parental magma composition may be represented by the Loljunmaa dike, to the east of the Penikat intrusion (Alapieti et al., 1990). The dike has 11.2 wt% MgO, 0.86 wt% TiO_2, 990 ppm Cr, 350 ppm Ni, and 420 ppm S, showing a composition that is broadly similar to that of the Bushveld B1 type magma (Table 3.3.4). However, it is less enriched in incompatible trace elements and has half the PGE contents of Bushveld B1 (9 ppb Pt and 6 Pd vs. 20 ppb Pt and 13 ppb Pd; Barnes et al., 2010). Notably, the Loljunmaa dike has Ce/Sm identical to the ratio in Penikat cumulates, which is consistent with it being parental to the intrusion. The average composition of 2.45 Ga siliceous high-magnesium basalt dikes in northern Finland (Guo and Maier, 2013) is 11.21 wt% MgO, 0.62 wt% TiO_2, 960 ppm Cr, 297 ppm Ni, 6.8 ppb Pd, 6.9 ppb Pt, and Ce/Sm 8, showing an overlap with the Loljunmaa dike composition.

CONCLUDING REMARKS

The Penikat intrusion contains several PGE-enriched zones or reefs, analogously to many other PGE mineralized layered intrusions. However, the Penikat reefs are of a lower lateral persistence than, for example, those in the Bushveld complex or the Stillwater complex. It is possible that this reflects a faster cooling rate of the relatively small Penikat intrusion. A notable difference between the Bushveld and Portimo complexes is the absence of a mineralized marginal zone at Penikat. In the Bushveld and Portimo complexes, it has been proposed that the best contact-type deposits are developed where the internal reefs abut against the floor (Iljina, 1994; Iljina and Lee, 2005; Maier et al., 2013). Such structural sections are absent at the present erosional level of the Penikat intrusion.

CONTACT-TYPE CU-NI-PGE MINERALIZATION OF THE KOILLISMAA INTRUSION

The Koillismaa intrusion of the Koillismaa-Näränkävaara complex hosts two principal types of orthomagmatic metallic mineralization: Cu-Ni-PGE-bearing sulfide occurrences and a vanadium deposit hosted by magnetite gabbro (Karinen et al., 2015).

The marginal series of the Koillismaa intrusion is mineralized with Cu, Ni, and PGE along its entire strike length of approximately 100 km. Here we focus on two sections that have been investigated in detail (i.e., Kaukua and Haukiaho). On average, the chalcophile element contents of the Koillismaa marginal series are low, reaching 0.2–0.4 wt% Cu and 0.2–0.3 wt% Ni in 1-m-long drill core samples, but may occasionally exceed 1.0 wt% in both elements. The principal sulfide minerals are pyrrhotite, chalcopyrite, pentlandite, and, in places, pyrite. Variations in precious metal content broadly correlate with those of the base metal sulfides and in many samples, the grade is on the order of 0.5–1.0 ppm of combined Pt + Pd + Au. Tellurides, bismuthides, and antimonides dominate the Pd mineralogy, while sperrylite is the dominant Pt mineral (Alapieti, 1982; Kojonen and Iljina, 2001; Iljina and Hanski, 2005; Karinen 2010).

Table 3.3.4 Chemical compositions for the evaluation of the parental magmas

wt%	1	2	3	4	5	6	7	8	9	10	11	12
SiO_2	54.0	51.4	52.0	53.5	53.4	53.4	50.7	52.7	52.2	51.3	56.1	51.7
TiO_2	0.16	0.40	0.68	0.21	0.24	0.30	0.22	0.33	0.15	0.12	0.34	0.19
Al_2O_3	3.2	8.0	12.3	13.4	12.2	9.8	14.2	16.9	16.1	17.1	11.5	16.8
FeO_{tot}	7.9	12.0	10.2	9.7	9.9	10.5	7.1	7.3	7.4	5.9	9.5	7.1
MnO	0.2	0.3	0.18	0.2	0.2	0.2	0.1	0.2	0.2	0.2	0.2	0.2
MgO	23.3	13.8	13.0	12.6	13.7	15.1	14.7	9.0	9.8	10.6	13.0	7.9
CaO	9.9	11.0	9.26	8.0	8.0	9.2	11.6	10.7	10.9	12.2	6.7	11.8
Na_2O	0.11	0.62	1.74	2.14	2.05	1.29	1.22	2.38	2.84	2.26	1.68	2.31
K_2O	0.01	0.45	0.71	0.10	0.19	0.11	0.22	0.57	0.42	0.18	0.80	0.24
P_2O_5	0.03	0.07	0.10	0.01	0.01	0.02	0.02	0.04	0.01	0.01	0.07	0.02
ppm												
V	50	230		127	126	157			150	150	167	130
Cr	1800	2300	1180	1137	951	1000	3330	280	290	380	1240	550
Ni	610	250	362	375	440	318	270	280	160	190	295	130
Zn	85	160		83	81	79			60	58	78	56
Sc	23	37		28	29	30			33	35		31
Sr	10	20	200	382	308	157	150	210	340	500	160	330
Y	4.8	7.0		4.1	4.9	6.2			2.5	1.0	13	11
Zr	20	40	70	10	10	15	25	47	30	1.0	77	26
Nb	1.9	2.0		<0.2	0.34	0.22			2.7	1.9	3.6	
Ba	53	47	210	68	108	63	75	180	110	52		
La		10.9		1.83	3.81	2.51			2.50	1.10	14.8	3.80
Ce		25.0		3.37	7.44	5.83			7.00	2.00	29.5	8.30
Nd		12.0		1.76	3.95	4.04			4.00	1.00	12.1	
Sm		2.50		0.26	0.53	0.77			0.75	0.37	2.70	0.86
Eu		0.61		0.37	0.52	0.38			0.32	0.26	0.72	0.52

Continued

Table 3.3.4 Chemical compositions for the evaluation of the parental magmas—cont'd

wt%	1	2	3	4	5	6	7	8	9	10	11	12
Tb		0.30		<0.1	0.13	0.19			0.20	<0.10	0.30	
Yb		1.08		0.58	0.55	0.48			0.59	0.30	1.16	0.64
Lu		0.16		0.10	0.1	<0.10			0.09	0.04	0.15	0.07
U	0.80	1.20		<0.2	<0.2	<0.2			<0.10	<0.10		

Cr-richer group
1. Westerite Portimo dike, Konttijärvi. Iljina (1994).
2. Westerite Portimo dike, Ahmavaara. Iljina (1994).
3. Loljunmaa dike. (Alapieti et al., 1990). Iljina (2005).
4. Ahmavaara autolith, YP-517/109.9 m. Iljina (2005).
5. Ahmavaara autolith, YP-517/115.5 m. Iljina (2005).
6. Ahmavaara autolith, YP-517/126.4 m. Iljina (2005).
7. Weighted average of Narkaus MCU I. Alapieti et al. (1990).
Cr-poorer group
8. Weighted average of the Suhanko intrusion (MCU III), Alapieti et al., (1990).
9. Niittylampi autolith. Iljina (1994).
10. Niittylampi autolith. Iljina (1994).
Bushveld
11. Bushveld B1 Sill. Sharpe and Hulbert (1985).
12. Fine-grained marginal rock adjacent to the Main Zone/Bushveld complex. Hatton and Sharpe (1989).

HAUKIAHO AND KAUKUA DEPOSITS

Figure 3.3.17 depicts a typical cross section from Haukiaho. The sulfide-bearing gabbroic zone, about 30 m in thickness, is heterogeneous and contains albite-quartz veins and basement inclusions. The dip of the magmatic layering in the deposit is 55–85° to the north. Sulfides comprise 1–5 vol% of the rock and occur mostly as a blebby dissemination in the interstices of silicate grains. The lowermost part of the marginal series includes a light-colored albite and quartz-rich rock, which grades into granitoidal rock of the basement complex. The albite and quartz-rich rock is considered to represent metasomatically altered portions of the basement complex.

The mineralized Kaukua marginal series, which dips ~35° to the south, represents the thickest part of the Koillismaa marginal series (Fig. 3.3.18), but it is possible that layer-parallel shearing has duplicated the mineralization. In addition to the contact-type mineralization, a low-grade reef-type mineralization is found just above the marginal series. Locally these two mineralization types merge together. Relatively feldspathic cumulates are found near the basal contact of the intrusion, whereas the upper part of the marginal series is composed exclusively of ultramafic cumulates.

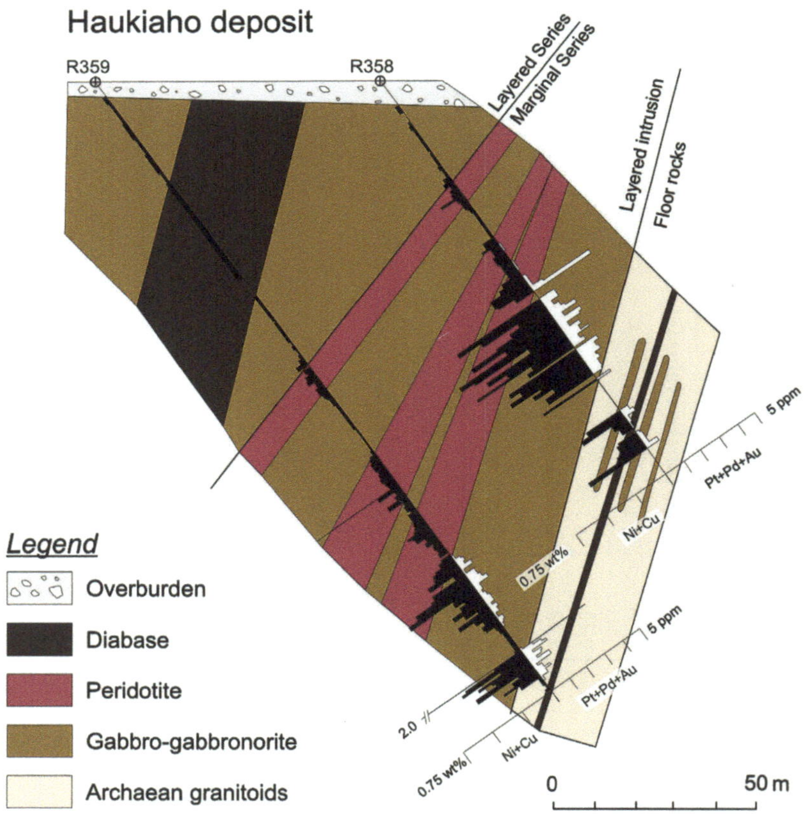

FIGURE 3.3.17 Cross section of the Haukiaho deposit along drill hole profile R358-R359, as seen from the west.

Source: *Modified from Iljina and Hanski (2005).*

In general, the Kaukua marginal series is more heterogeneous than that at Haukiaho. The sequence includes multiple thin layers of more magnesian cumulates and zones of gabbroic rocks containing mafic autoliths and minor amounts of felsic xenoliths derived from the basement. It is therefore difficult to map coherent units of mafic and ultramafic cumulates in the Kaukua deposit, and only the stratigraphically highest layer of peridotite is marked in Fig. 3.3.18. In contrast to Haukiaho, the marginal series in the Kaukua area is mineralized throughout, comprising 1–5 vol% of finely disseminated sulfides. The bottom part of the Kaukua section contains the so-called mixed zone, which is composed of basement rocks with thin mafic cumulate intercalations or xenoliths. The abundance of the mafic layers or fragments decreases downward until the rock consists entirely of basement granitoid.

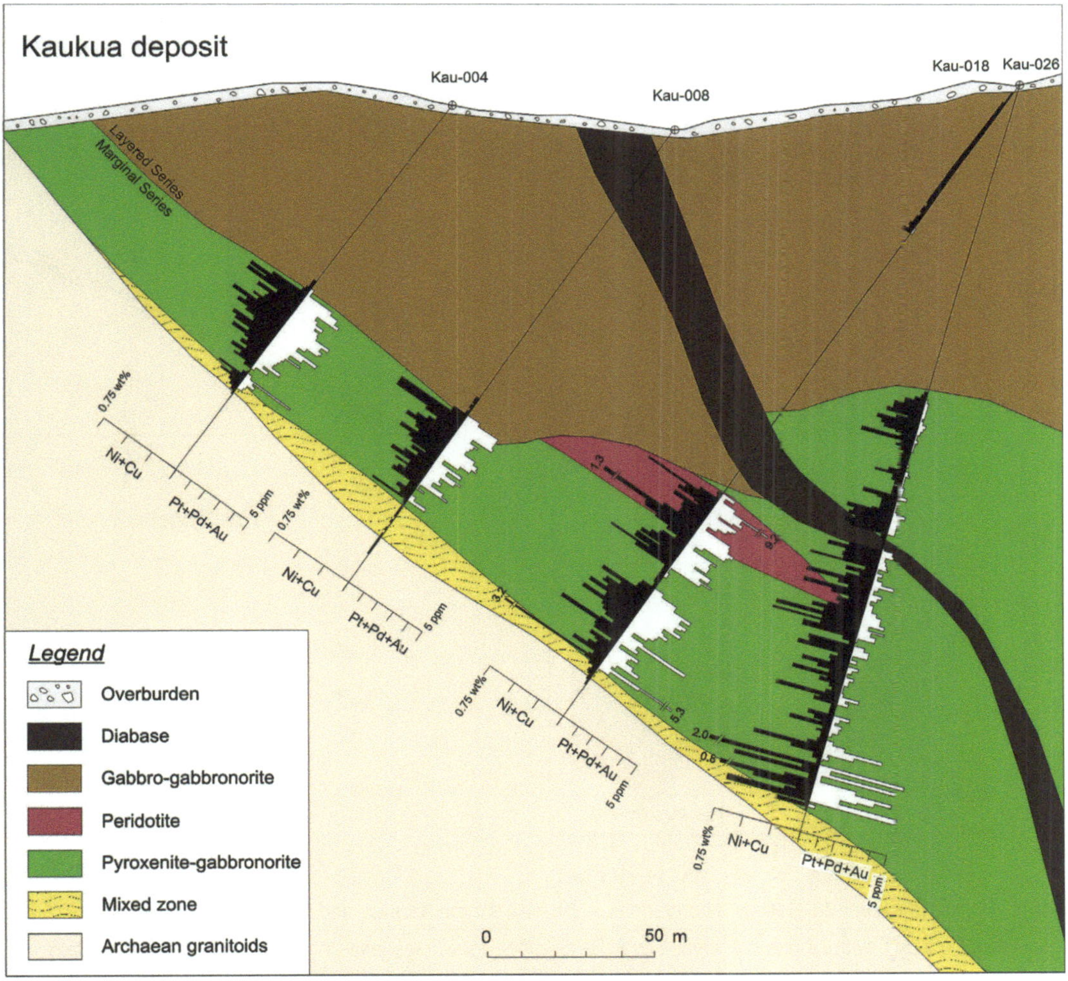

FIGURE 3.3.18 Cross section of the Kaukua deposit along Section 3553700E, as seen from the west.

Figure provided by Finore Mining Inc.

The current mineral resources for Kaukua are 23.6 Mt (refer to Table 3.3.4) of ore containing 1.07 ppm and 0.91 ppm Pd + Pt + Au in indicated and inferred categories, respectively. For the Haukiaho deposit, the inferred mineral resources are 23.2 Mt ore grading 0.53 ppm of combined Pt + Pd + Au, 0.21 wt% Cu, and 0.14 wt% Ni.

Both the Kaukua and Haukiaho deposits show similar Cu/Ni (~2.0) and Pd/Pt (~2.8), but it is noteworthy that Kaukua has higher values of PGE + Au whereas Haukiaho is relatively more enriched in base metals (Figs. 3.3.19A and B and Table 3.3.3). Based on sulfide-specific assays, the sulfide fraction from Kaukua contains 4–6 wt% Ni, 7–17 wt% Cu, and 80–100 ppm combined Au + Pd + Pt. At Haukiaho, the sulfide fraction contains 4–8 wt% Ni, 12–35 wt% Cu, and 40 ppm combined Au + Pd + Pt (Iljina et al., 2012). Both deposits display similar chondrite-normalized chalcophile element patterns (Fig. 3.3.19C). This indicates a similar parental magma, and thus the higher PGE + Au grades at Kaukua may be related to a higher R factor, reflecting more dynamic magma emplacement conditions as also evidenced by repeated ultramafic and mafic layers and higher abundance of rock fragments.

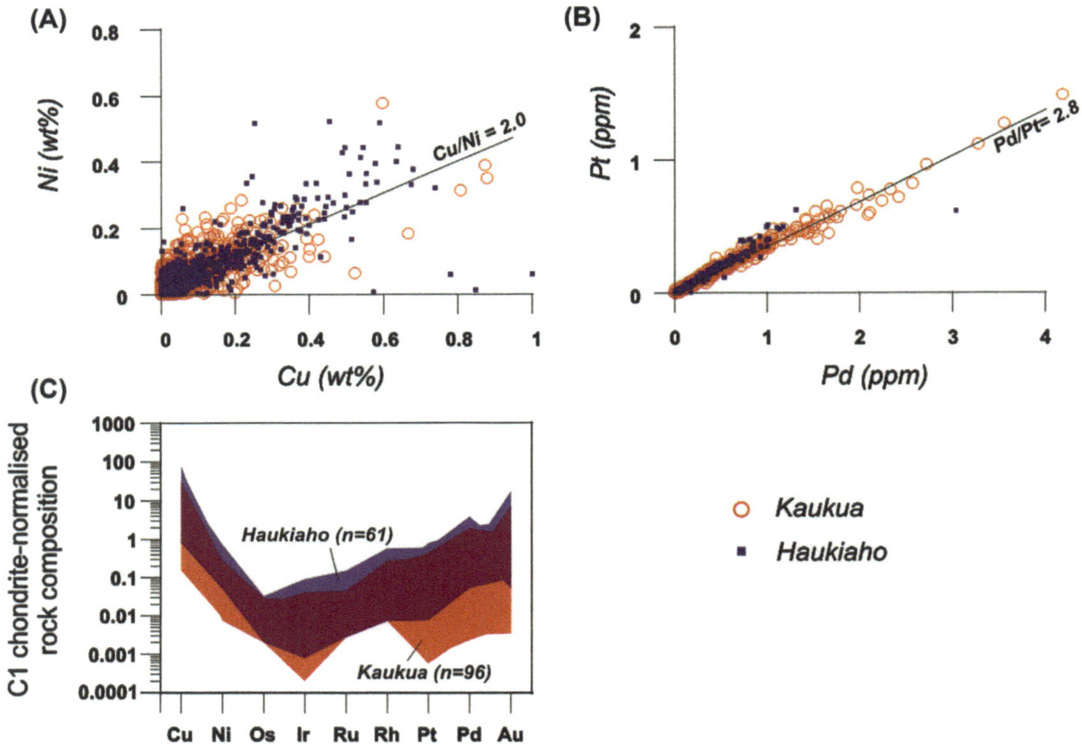

FIGURE 3.3.19 Chalcophile element abundances in rocks from the Kaukua and Haukiaho deposits.

(A) Ni versus Cu plot, with samples analyzed by ICP-AES after aqua regia leaching; thus compositions contain sulfidic and some silicate nickel. (B) Pt versus Pd plot with metals measured by ICP-AES after fire-assays preconcentration. (C) Chondrite-normalized metal patterns for rocks from the two deposits (samples analyzed by ICP-MS).

Sources: *Chondrite values are from Barnes and Maier (1999). Data provide by Finore Mining Inc., 2013.*

COMPOSITIONAL DIFFERENCES BETWEEN CYCLIC UNITS AND IMPLICATIONS FOR EXPLORATION

During the PGE exploration in the Portimo and Penikat intrusions during the 1980s, it was recognized that the cumulates of various megacyclic units formed two contrasting groups in terms of their Cr contents (Table 3.3.4). The lower portions of the intrusions were found to be composed of relatively Cr-enriched cyclic units, whereas the upper portions are composed of relatively Cr-depleted units. This was attributed to differences in the parental magma composition of the cyclic units (Lahtinen et al., 1989; Alapieti et al., 1990; Saini-Eidukat et al., 1997).

The stratigraphically lowermost significant PGE reef (the SJ Reef in the Penikat intrusion and the SK Reef in the Portimo intrusion) was found to be located in the transition zone between the uppermost Cr-enriched and the lowermost Cr-depleted cyclic unit (Fig. 3.3.20). Similarly, the Konttijärvi and Ahmavaara PGE-enriched marginal series are located in the same transitional interval, as indicated by the observation that the immediately underlying Portimo dikes belong to the relatively Cr-enriched group. The Rytikangas (RK) and Ala-Penikka (AP) Reefs are found stratigraphically above the transition level. This model, as first published by Lahtinen et al. (1989), was developed in a collaborative research project between the Oulu University and the exploration department of Outokumpu Oy and was successfully applied to PGE exploration in the 1980s when PGE enrichments in the Portimo and Penikat areas were discovered.

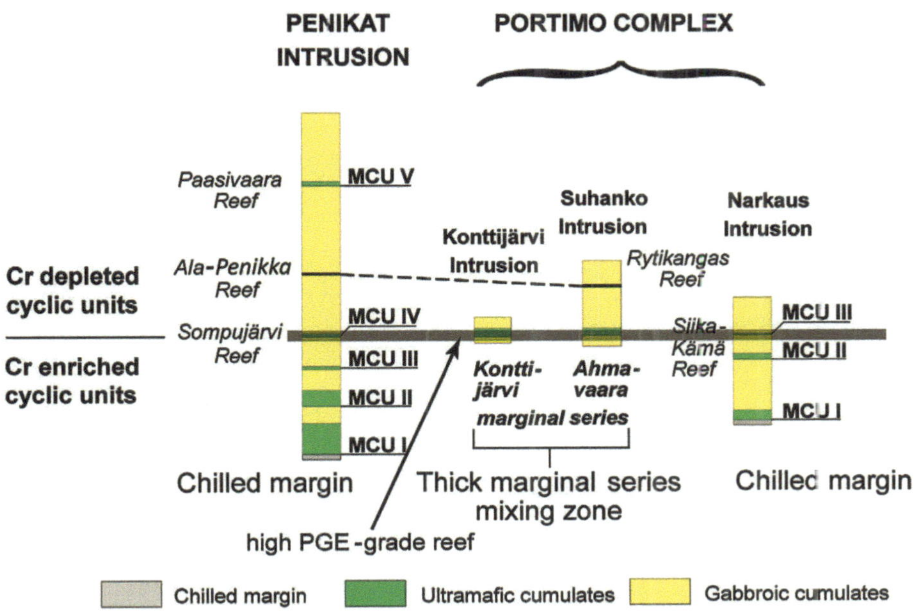

FIGURE 3.3.20 Stratigraphical correlation of the Portimo and Penikat complexes.

For discussion, see text.

Source: *Modified after Lahtinen et al. (1989).*

ACKNOWLEDGMENTS

We would like to express our appreciation to Finore Mining Inc. for providing analytical data for this chapter. Credits are also extended to Dr. Tapio Halkoaho, Geological Survey of Finland, and Turkka Rekola, former exploration geologist of Finore Mining Inc., for their contributions to the Penikat and Koillismaa sections of this paper, respectively.

REFERENCES

Alapieti, T., 1982. The Koillismaa layered igneous complex, Finland—its structure, mineralogy and geochemistry, with emphasis on the distribution of chromium. Geological Survey of Finland, Bulletin 319, p. 116.

Alapieti, T., Lahtinen, J., Huhma, H., et al., 1989. Platinum-group element-bearing Cu-Ni mineralization in the marginal series of the early Proterozoic Suhanko-Konttijärvi layered intrusion, Northern Finland. In: Prendergast, M., Jones, M. (Eds.), Magmatic Sulphides–Zimbabwe Volume. Institute of Mining Metallurgy, London, pp. 177–187.

Alapieti, T., Filen, B., Lahtinen, J., et al., 1990. Early Proterozoic layered intrusions in the northeastern part of the Fennoscandian Shield. Mineralogy and Petrology 42, 1–22.

Amelin, Y.V., and Semenov, V.S., 1996. Nd and Sr isotopic geochemistry of mafic layered intrusions in the eastern Baltic shield: implications for the evolution of Paleoproterozoic continental mafic magmas. Contrib. Mineral Petrol 124, 255–272.

Andersen, J.C.Ø., Thalhammer, O., Schoenberg, R., 2006. Platinum-group element and Re-Os isotope variations of the high-grade Kilvenjärvi platinum-group element deposit, Portimo layered igneous complex. Finland. Economic Geology 101, 159–177.

Barnes, S.-J., Maier, W.D., 1999. The fractionation of Ni, Cu and the noble metals in silicate and sulfide liquids. In: Keays, R.R., Lesher, C.M., Lightfoot, P.C., Farrow, C.E.G. (Eds.), Magmatic Ore Deposits and Their Application in Mineral Exploration. Geological Association of Canada, pp. 69–106. Short Course Volume 13.

Barnes, S.-J., Maier, W.D., Curl, E., 2010. Composition of the marginal rocks and sills of the Rustenburg Layered Suite, Bushveld complex, South Africa: Implications for the formation of the platinum-group element deposits. Economic Geology 105, 1481–1511.

Finore Mining Inc, 2013. Finore reports an increased mineral resource estimate for the Lantinen-Koillismaa project. Finland. September 19, 2013, press release.

Guo, F.-F. and Maier, W., 2013. Geochemistry of ~2.45 Ga mafic dykes in Northern Finland: constraints on the origin of PGE mineralization in coeval layered intrusions. Abstract, 12th SGA biennial meeting, Uppsala, Sweden.

Halkoaho, T., 1993. The Sompujärvi and Ala-Penikka PGE reefs of the Penikat layered intrusion, Northern Finland—Implications for PGE-reef forming processes. Acta Universitatis Ouluensis, Series A, Scientiae Rerum Naturalium 249, 122.

Halkoaho, T., Alapieti, T., Lahtinen, J., 1990a. The Sompujärvi PGE reef in the Penikat layered intrusion, Northern Finland. Mineralogy and Petrology 42, 39–55.

Halkoaho, T., Alapieti, T., Lahtinen, J., 1990b. The Ala-Penikka PGE reefs in the Penikat layered intrusion, Northern Finland. Mineralogy and Petrology 42, 23–38.

Hanski, E., Walker, R.J., Huhma, H., Suominen, I., 2001. The Os and Nd isotopic systematics of the 2.44 Ga Akanvaara and Koitelainen mafic layered intrusions in northern Finland. Precambrian Research 109, 73–102.

Huhma, H., Cliff, R., Perttunen, V., Sakko, M., 1990. Sm-Nd and Pb isotopic study of mafic rocks associated with early Proterozoic continental rifting: The Peräpohja schist belt in Northern Finland. Contributions to Mineralogy and Petrology 104, 369–379.

Huhma, H., O'Brien, H., Lahaye, Y., Mänttäri, I., 2011. Isotope geology and Fennoscandian lithosphere evolution. Geological Survey of Finland, Special Paper 49, 35–48.

Huhtelin, T., Lahtinen, J., Alapieti, T., et al., 1989. The Narkaus intrusion and related PGE and sulphide mineralization. In: Alapieti, T. (Ed.), Proceedings of the 5th International Platinum Symposium. Guide to the Post-symposium Field Trip. Geological Survey of Finland. Guide 29, 145–161.
Huhtelin, T., Alapieti, T., Lahtinen, J., 1990. The Pasivaara PGE reefs in the Penikat layered intrusion, northern Finland. Mineralogy and Petrology 42, 57–70.
Hatton, C., Sharpe, M., 1989. Significance and origin of boninite-like rocks associated with the Bushveld complex. In: Crawford, A. (Ed.), Boninites. University Press, Cambridge, U.K, pp. 174–207.
Iljina, M., 1994. The Portimo layered igneous complex with emphasis on diverge sulphide and platinum-group element deposits. Acta Universitatis Ouluensis, Series A, Scientiae Rerum Naturalium 258, 158.
Iljina, M., 2005. Portimo layered igneous complex. In: Alapieti, T., Kärki, A. (Eds.), Proceedings of the 10th International Platinum Symposium, Field Trip Guidebook. Geological Survey Finland. Guide 51a, 77–100.
Iljina, M., Hanski, E., 2005. Layered mafic intrusions of the Tornio–Näränkävaara belt. In: Lehtinen, M., Nurmi, P.A., Rämö, O.T. (Eds.), Precambrian geology of Finland—Key to the Evolution of the Fennoscandian Shield. Elsevier, Amsterdam, pp. 101–138.
Iljina, M., Lee, C.A., 2005. PGE deposits in the marginal series of layered intrusions. In: Mungall, J.E. (Ed.), Exploration for Platinum-Group Element Deposits. Mineralogical Association of Canada. Short Course vol. 35, 75–96.
Iljina, M., Salmirinne, H., 2011. Suhanko seismic reflection profile and integrated geological-geophysical model of the Portimo area. Geological Survey of Finland, Report of Investigation 189, 20.
Iljina, M., Alapieti, T., McElduff, B., 1992. Platinum-group element mineralization in the Suhanko-Konttijärvi intrusion. Australian Journal of Earth Sciences 39, Finland, 303–313.
Iljina, M., Duke, C., Hinzer, J., 2012. A Technical Review of the Läntinen Koillismaa Project. Finland for Finore Mining Inc NI 43–101 Technical Report.
Karinen, T., Hanski, E., Taipale, A., 2015. The Mustavaara Fe-Ti-V oxide deposit. In: Maier, W.D., O'Brien, H. Lahtinen, R. (Eds.), Mineral Deposits of Finland. Elsevier, Amsterdam, pp. 179–192.
Karinen, T., 2010. The Koillismaa intrusion, northeastern Finland—Evidence for PGE reef forming processes in the layered series. Geological Survey of Finland, Bulletin. 404, p. 176.
Kojonen, K., Iljina, M., 2001. Platinum-group minerals in the Early Proterozoic Kuusijärvi marginal series, Koillismaa layered Igneous complex, northeastern Finland. In: Piestrzynski, A. (Ed.), Proceedings of the Sixth Biennial SGA-SEG meeting. Balkema Publishers, Lisse, pp. 653–656.
Lahtinen, J., Alapieti, T., Halkoaho, T., et al., 1989. PGE mineralization in the Tornio-Näränkävaara layered intrusion belt. In: Alapieti, T. (Ed.), Proceedings of the 5th International Platinum Symposium. Guide to the Post-Symposium Field Trip. Geological Survey of Finland, Guide. 29, 43–58.
Lauri, L.S., Karinen, T., Räsänen, J., 2003. The earliest Paleoproterozoic supracrustal rocks in Koillismaa, northern Finland—their petrographic and geochemical characteristics and lithostratigraphy. Bulletin of the Geological Society of Finland 75, 29–50.
Lerssi, J., 1990. Geophysical investigations. In: Alapieti, T., Halkoaho, T., Huhtelin, T., et al. (Eds.), Final Report of the Peräpohja Platinum Project. Unpublished report of PGE Research Project. University of Oulu, Report. 3, 122–149 (in Finnish).
Maier, W.D., Barnes, S.-J., Groves, D.I., 2013. The Bushveld complex, South Africa: Formation of platinum-palladium, chrome and vanadium-rich layers via hydrodynamic sorting of a mobilized cumulate slurry in a large, relatively slowly cooling, subsiding magma chamber. Mineralium Deposita 48, 1–56.
Maier, W.D., 2015. Geology and petrogenesis of magmatic Ni-Cu-PGE-Cr-V deposists: an introduction and overview. In: Maier, W.D., O'Brien, H. Lahtinen, R. (Eds.), Mineral Deposits of Finland. Elsevier, Amsterdam, pp. 73–88.
Mutanen, T., 1997. Geology and ore petrology of the Akanvaara and Koitelainen mafic layered intrusions and the Keivitsa Satovaara layered complex, northern Finland. Geological Survey of Finland, Bulletin 395, p. 233.
Naldrett, A., 1981. Platinum-group element deposits. In: Cabri, L. (Ed.), Platinum-Group Elements: Mineralogy, Geology, Recovery. Canadian Institue of Mining and Metallurgy. Special Volume 23, 197–231.

Pernu, T., Keränen, T., Lerssi, J., Hjelt, S.-E., 1986. The geophysical structure of the Suhanko layered intrusion. Unpublished report of Geophysical Research Project. University of Oulu. p. 44 (in Finnish).
Puritch, E., Ewert, W., Brown, F.H., et al., 2007. Technical report, mineral resource estimate, and preliminary economic assessment (scoping study) of the Suhanko project northern Finland. For North American Palladium Ltd by Aker Kvaerner, P&E Mining Consultants Inc. F.H. Brown. NI-43–101 and 43-101F1 Technical Report No 135.
Saini Eidukat, B., Alapieti, T., Thalhammer, O.A.R., Iljina, M., 1997. Siliceous, high magnesian parental magma compositions of the PGE rich early Paleoproterozoic layered intrusion belt of northern Finland. In: Rongfu, P. (Ed.), Proceedings of the 30th International Geological Congress vol. 9, 177–197.
Sharpe, M., Hulbert, L., 1985. Ultramafic sills beneath the eastern Bushveld complex: Mobilized suspensions of early lower zone cumulates in a parental magma with boninitic affinities. Economic Geology 80, 849–871.

CHAPTER

THE KEMI CHROMITE DEPOSIT

3.4

T. Huhtelin

ABSTRACT

The Kemi stratiform chromitite deposit, hosted by an Early Proterozoic (2.44 Ga) layered intrusion, is located northeast of the town of Kemi on the coast of the Gulf of Bothnia. The chrome ore has been exploited since 1968 by the Outokumpu company, operating the only chrome mine within the European Union. The present surface section of the Kemi intrusion is lenticular in shape, being about 15 km long in a southwest–northeast direction and 0.2–2 km wide. The intrusion comprises an ultramafic lower part and a gabbroic upper part. The intrusion was tilted by tectonic movements during the Svecokarelidic orogeny to form a body dipping about 70° to the northwest. The chromitite layer, broken into several blocks, extends along the entire length of the basal part of the Kemi layered intrusion, but the thickness of the layer varies from a few centimeters at both ends of the intrusion to well over 100 m in the central part, with the average thickness being approximately 40 m in the mining area.

During the Svecokarelidic orogeny, the area of the Kemi intrusion underwent lower-amphibolite facies metamorphism. The original magmatic silicates have been completely altered to secondary minerals in the lower and upper parts of the intrusion, whereas those in the center have been preserved. Alteration had a negative effect on the grade of the chrome ore. Chromite grains have nevertheless preserved their original composition in their cores, even though the silicates of the same rock have been completely altered. The altered rocks have preserved their cumulate textures fairly well, and, despite alteration, many of the primary minerals can still be recognized by means of their pseudomorphs (Alapieti et al., 1989).

The Kemi chromite mineralization was discovered in 1959 and mining commenced in 1968. The mine's current proven ore reserves total 50.1 Mt with an average grade of about 26.0% Cr_2O_3 and a chrome–iron ratio of 1.6–1.7. Additional mineral resources, which have been estimated to a depth of 1 km, comprise 97.8 Mt (Outokumpu, 2014). According to a geophysical seismic reflection survey by the Geological Survey of Finland, it is possible that the chromitite layer continues to the depth of 2–2.5 km or even deeper, the Kemi layered intrusion possibly extending down to about 3–4 km (Kukkonen et al., 2009).

In 1968, the first closed furnace was completed at Tornio and ferrochrome production started. The ferrochrome works was expanded by two more furnaces in 1985 and 2012. The construction process of the stainless steel works was completed in 1976, and the hot rolling mill was inaugurated in 1988. This integrated and technologically advanced production chain from chrome ore to stainless steel coils and plates in the Kemi-Tornio area is unique in the world today.

Keywords: Kemi Mine; chromite; chromitite; layered intrusion; ferrochrome.

INTRODUCTION

The Kemi layered intrusion is economically the most significant amongst the Fennoscandian layered intrusions, and the only one presently mined. The Kemi Mine is the only operating chrome mine within the European Union. The layered intrusion hosting the Kemi chromite deposit extends some 15 km northeast of

Mineral Deposits of Finland. http://dx.doi.org/10.1016/B978-0-12-410438-9.00006-6

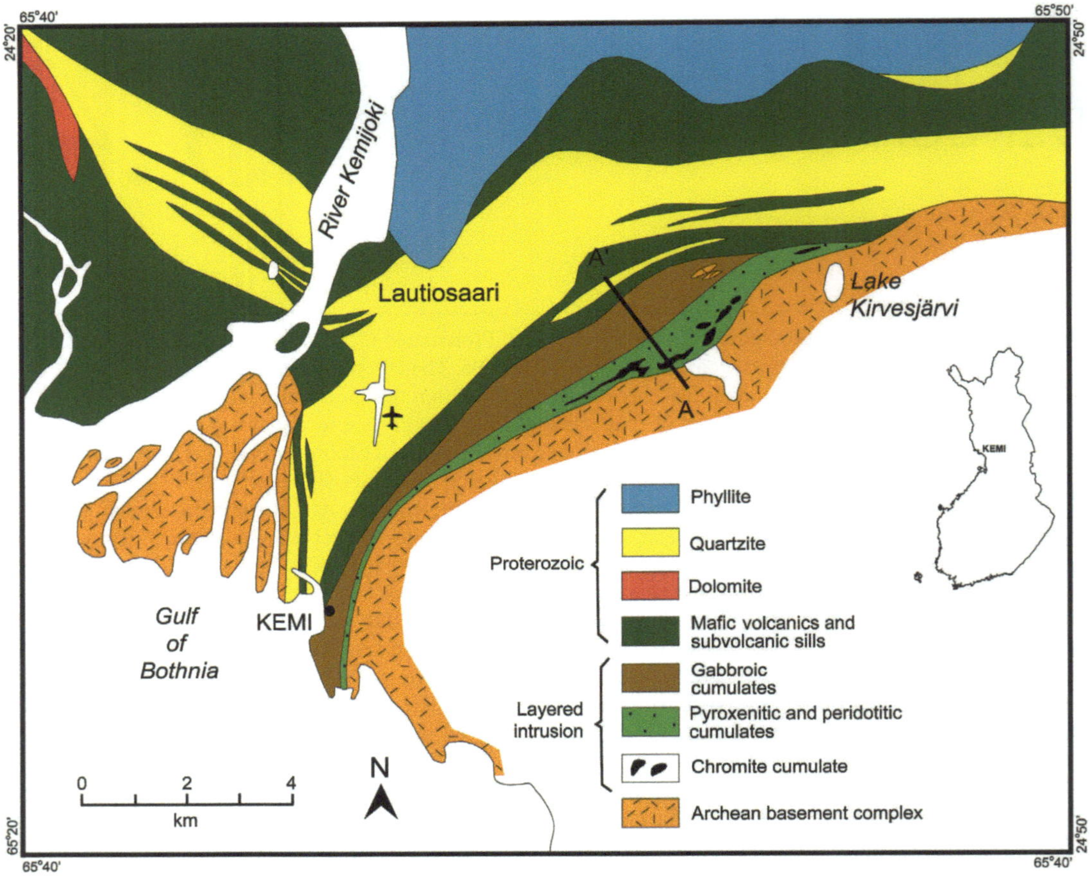

FIGURE 3.4.1 Generalized geological map of the Kemi region.

The section depicted in Fig. 3.4.2 is indicated by the line A–A′.

Source: *After Alapieti et al. (1989).*

Kemi, a town on the coast of the Gulf of Bothnia in northern Finland (Fig. 3.4.1). The chromitite layer can be traced for the entire length of the intrusion, varying in thickness from a few centimeters to well over 100 m in the central part of the intrusion. The U-Pb zircon data yield an age of 2.44 Ga for the Kemi intrusion (Patchett et al., 1981), and whole rock Pb-Pb data define an age of 2.44 ± 0.16 Ga (Manhes et al., 1980).

SHAPE AND LAYERING OF THE INTRUSION

The present surface section of the Kemi intrusion is lenticular in shape, being about 15 km long and 0.2–2 km wide. It represents a cross section of an originally funnel-shaped intrusion which was tilted by tectonic movements during the Svecokarelidic orogeny to form a body dipping about 70° to the northwest (Alapieti et al., 1989). The total length of the intrusion is interpreted to be about 22 km, as

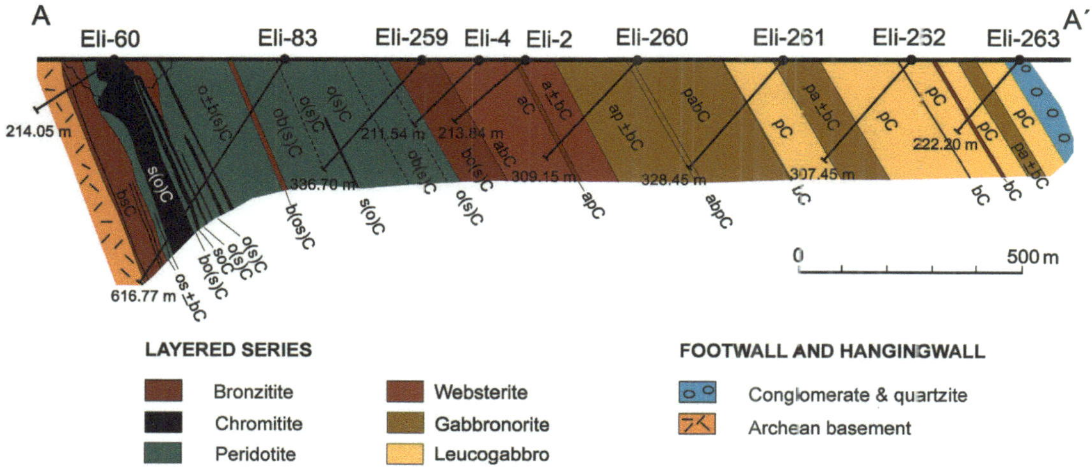

FIGURE 3.4.2 Cross section of the Kemi intrusion based on the drilled data along profile A–A′ indicated in Fig. 3.4.1. For abbreviations, see Table 3.4.1.

Source: *Modified from Alapieti et al. (1989).*

the intrusion extends, according to aeromagnetic data, about 7 km further southwest under the sea. A chain of small islands offshore from Kemi, the biggest of which is 2.5 km long, comprises gabbroic cumulates. According to a geophysical seismic reflection survey by the Geological Survey of Finland, the layered intrusion extends to the depth of about 3 km, possibly even to 4 km (Kukkonen et al., 2009).

The intrusion comprises a lower ultramafic part and an upper gabbroic part. The individual cumulate layers are thickest in the central part of the intrusion and become thinner toward the ends of the present surface section (Figs. 3.4.1 and 3.4.2). In the center of the intrusion, the approximate thickness of the ultramafic part is 1000 m, and the thickness of the mafic part is 900 m. Underground inclined tunneling in the footwall granite has proved that a feeder conduit to the magma chamber was located below the thickest part of the intrusion, as originally suggested by Alapieti et al. (1989). The feeder dike, about 20 m thick, was encountered 260 m below the present surface. It comprises fine-grained, uralitized rock types close to the contacts with the country rocks, grading to coarser grained rocks inward. The rocks mainly consist of chlorite schists and talc-carbonate rocks. The center of the dike is composed of an almost vertical chromitite reaching a thickness of several meters. The feeder dike is also exposed within basement granite (Fig. 3.4.3) forming the southeastern wall of the central open pit (Kemppainen, 2002). Notably, chromite grains from the dike show the most Cr-enriched compositions in the Kemi intrusion (Alapieti and Huhtelin, 2005).

The footwall of the Kemi intrusion consists of late Archean granitoids. These comprise albite or potassium-feldspar granites, and in places gneisses and amphibolites. The hanging-wall rocks comprise younger mafic volcanic rocks or subvolcanic sills yielding a U-Pb zircon age of 2.15 Ga (Sakko, 1971) and a polymictic conglomerate of unknown but younger age than the intrusion. This indicates that the present upper contact is erosional, implying that the original roof rocks and the uppermost cumulates of the layered series, together with possible granophyre, have been obliterated by erosion. The feeder dikes of the subvolcanic sills, referred to as albite diabase, intersect the Kemi intrusion (Alapieti et al., 1989).

FIGURE 3.4.3 "Feeder dike" within basement granite on the uppermost part of the main open pit looking southeast.

The eastern margin of the dike is characterized by a fault. Hatched lines highlight the margins of the dike.

Photo by Marko Matinlassi.

Table 3.4.1 Rock-type nomenclature

Conventional rock name	Cumulus mineral assemblage	Cumulate abbreviation
Peridotite	Olivine + bronzite + chrome spinel	o + b + sC
Chromitite	Chrome spinel (+ minor olivine)	s(o)C
Bronzitite and olivine-bronzitite	Bronzite + olivine (+ minor chrome spinel)	bo(s)C
Websterite and clinopyroxenite	Augite + bronzite	a + bC
Gabbronorite and gabbro	Plagioclase + augite + bronzite	pa + bC
Leucogabbro and anorthosite	Plagioclase	pC

Source: *After Alapieti et al. (1989).*

The area of the Kemi intrusion underwent lower-amphibolite facies metamorphism during the Svecokarelidic orogeny, probably during 1.9–1.8 Ga. The original magmatic silicates have been completely altered to secondary minerals in the lower and upper parts of the intrusion, whereas those in the middle are relatively unaltered. Alteration affected the Cr_2O_3 content of chromite crystals with iron replacing chrome on the outer rims and along cracks of the crystals (Huovinen, 2007). Many chromites have nevertheless preserved their original composition in their cores, even though the silicates of the same rock have been completely altered. The altered rocks have preserved their cumulate textures fairly well and, despite alteration, many of the primary minerals can still be recognized by means of pseudomorphs, thus enabling the cumulate sequences to be determined (Alapieti et al., 1989). Rock-type nomenclature and cumulus mineral assemblage are depicted in Table 3.4.1. Although all rocks are metamorphosed, the prefix "meta" has been omitted for simplicity.

THE BASAL CONTACT SERIES

The basal contact series in the Kemi intrusion is highly altered. Its thickness varies between 10 and 40 meters. It is mainly represented by a mylonitic talc-chlorite-carbonate schist, and in places by a talc-carbonate rock, in contact with the basement granitoid. During the development of the underground mine from level 400 downward, an approximately 10-m-thick gabbro has also been encountered in some of the tunnels intersecting the basal contact. This gabbro does not form a continuous layer and its contacts are commonly quite irregular. It is mainly encountered within the mylonite and the talc-carbonate rock. The grain size of the gabbro is fine, close to the basement granitoid, becoming medium to coarse grained further away from the contact. In a tunnel at level 450, where the basal contact series is quite well preserved, a pyroxenite is also encountered, overlaying the gabbro. The MgO content of the rocks in the relict marginal series increases from the contact upward, from 7–14% MgO in gabbros to about 18–22% in the pyroxenites and mylonites, the average being 23–25% MgO in the altered ultramafic rocks of the layered series (Posio, 2008). It seems possible, in accordance with Alapieti et al. (1989), that the basal contact series may have been partially obliterated by erosion during the magmatic stage, as suggested by the sporadic appearance of the gabbro.

THE LAYERED SERIES

The sequence of tectonic mylonite, talc-carbonate rock and sporadic gabbro is overlain by a markedly altered sequence 50–100 m thick. The lower part of this sequence is composed of a bronzite-chromite cumulate, containing gneissic xenoliths from the underlying basement complex (Alapieti et al., 1989). The upper part of the altered sequence encountered between the basal contact and the main chromitite comprises an olivine-chromite ± bronzite cumulate, with chromitite interlayers from 0.5–1.5 m thick (Fig. 3.4.2). In addition to chromitite interlayers, varying amounts of disseminated chromite and roundish lumps of chromitite, whose diameter varies from a few centimeters up to several meters, can be encountered in this sequence (see Fig. 3.4.9 later in chapter). In places, where the smaller roundish lumps of chromitite are abundant, the texture can even be said to resemble a conglomerate.

The sequence just described is overlain by the main chromitite unit (Fig. 3.4.2), which in the central part of the intrusion is composed of two layers separated by a relatively more silicate-rich rock (see Fig. 3.4.5 later). The average thickness of the main chromitite unit in the mining area is 40 m. Its cumulus minerals are chromite and olivine, and the intercumulus minerals comprise poikilitic bronzite and, to a lesser extent, augite. The proportion of cumulus olivine in relation to chromite is comparatively low in the upper part of the chromitite layer. However, a thin but quite continuous layer comprising cumulus olivine is encountered within the upper part of the unit, referred to as the "spotted type" in mine nomenclature (see Figs. 3.4.7(E) and (F) later). Bronzite locally occurs as the cumulus phase in the more silicate-rich interlayer, which is typified by accumulations of small chromite grains around the larger cumulus silicate minerals. The lower part of the main chromitite unit is typically nonlayered, is partly disseminated, and contains abundant ultramafic inclusions (Alapieti et al., 1989). The chromite dissemination locally forms net-like textures, with chromite crystals surrounding large pseudomorphs of olivine.

The main chromitite unit is overlain by about 550 m of peridotitic cumulates (Fig. 3.4.2) with olivine, chromite, and occasional bronzite as the cumulus minerals. This thick cumulate sequence contains about 15 chromite-rich interlayers varying in thickness from 5 cm to 2.5 m; the uppermost layer is located about 370 m above the main chromitite layer (Alapieti et al., 1989). Six of the thickest chromite-rich interlayers, including the uppermost one, are shown in Fig. 3.4.2. Three pyroxenite interlayers, approximately 10–30 m thick, occur in the lower part of the peridotitic sequence. The uppermost pyroxenite is situated between sequences displaying well-developed rhythmic layering that are approximately 30–50 m thick. The rhythmic intervals are composed of alternating layers of olivine-(chromite) cumulates and olivine-bronzite-(chromite) cumulates; augite is the main intercumulus mineral in both rock types. A similar type of rhythmic sequence occurs in the third megacyclic unit of the Penikat layered intrusion, situated some 10 km northeast of the Kemi intrusion (Huhtelin et al., 1989). Bronzite becomes the dominant cumulus mineral about 700 m above the basal contact of the intrusion, with olivine and chromite as additional cumulus phases. After another 100 m, augite becomes the dominant cumulus mineral alongside bronzite, whereas olivine and chromite disappear. In places, the rocks are clinopyroxenites (Alapieti et al., 1989).

At about 1000 m above the base of the intrusion, plagioclase becomes the cumulus phase alongside augite and bronzite (Fig. 3.4.2). These gabbronoritic cumulates continue for about 800 m upward to the contact with the hanging wall. In the upper part of the sequence, augite occurs as the intercumulus phase, and there is little or no Ca-poor pyroxene. In conventional terms, these rocks are therefore leucogabbros or anorthosites (Alapieti et al., 1989).

THE CHROMITE MINERALIZATION

The main chromitite layer that parallels the basal contact zone of the Kemi intrusion is known to occur over the whole length of the complex, extending for 15 km from the town of Kemi to the northeast. In the central part of the intrusion, where the complex is at its thickest, the chromitite layer has an average width of 40 m, locally reaching a thickness of well over 100 m (Figs. 3.4.4, 3.4.5, and 3.4.6). This thickened part of the chromitite layer occurs along a strike of about 4 km. Toward both ends of the intrusion, the thickness of the chromitite layer averages some tens of centimeters. The chromite-rich unit has an average dip of 70° to the northwest (Alapieti et al., 1989).

The Cr/Fe ratio of the chromitite decreases along the strike, from 1.6–1.7 in the central part of the intrusion to less than 1.0 at the margins. The upper contact of the chromitite unit lies stratigraphically 100–150 m above the basal contact of the complex, but its position has been changed by several strike-slip faults (Fig. 3.4.4). The upper part of the main chromitite unit is layered in structure; the hanging wall contact of the ore is sharp, whereas the lower part is nonlayered and partly brecciated, characterized by gradually diminishing chromite dissemination toward the bottom of the intrusion and accompanied by irregular ore lumps that can reach the size of tens of meters (Figs. 3.4.5 and 3.4.6). The lower contact of the ore body is, therefore, quite complex (Alapieti and Huhtelin, 2005). The chromitite unit contains abundant barren ultramafic inclusions, especially in its lower part. Gneissic or granitic xenoliths have also been encountered in places within the main chromitite layer.

The chromitite unit is cut into several distinct ore bodies by numerous faults, and these bodies are treated, to a certain degree, as separate units for the purposes of mining, beneficiation, and metallurgy. The reason for this is that the separate ore bodies, due to variation in the degree of metamorphism, host different secondary gangue minerals in the interstices of chromite grains, which has an effect on hardness and specific gravity of the ore. The ore bodies are classified to four main mineralogical ore types: (1) tremolite-chlorite, (2) serpentine, (3) talc-carbonate, and (4) chlorite-based ores. Tremolite-chlorite and serpentine-based ores are dominant in the western part of the mining area, whereas talc-carbonate and chlorite-based ores are typical in the central and eastern parts of the mining area (Table 3.4.2). Typical cumulus textures and mineralogical ore types encountered in the main chromitite unit are

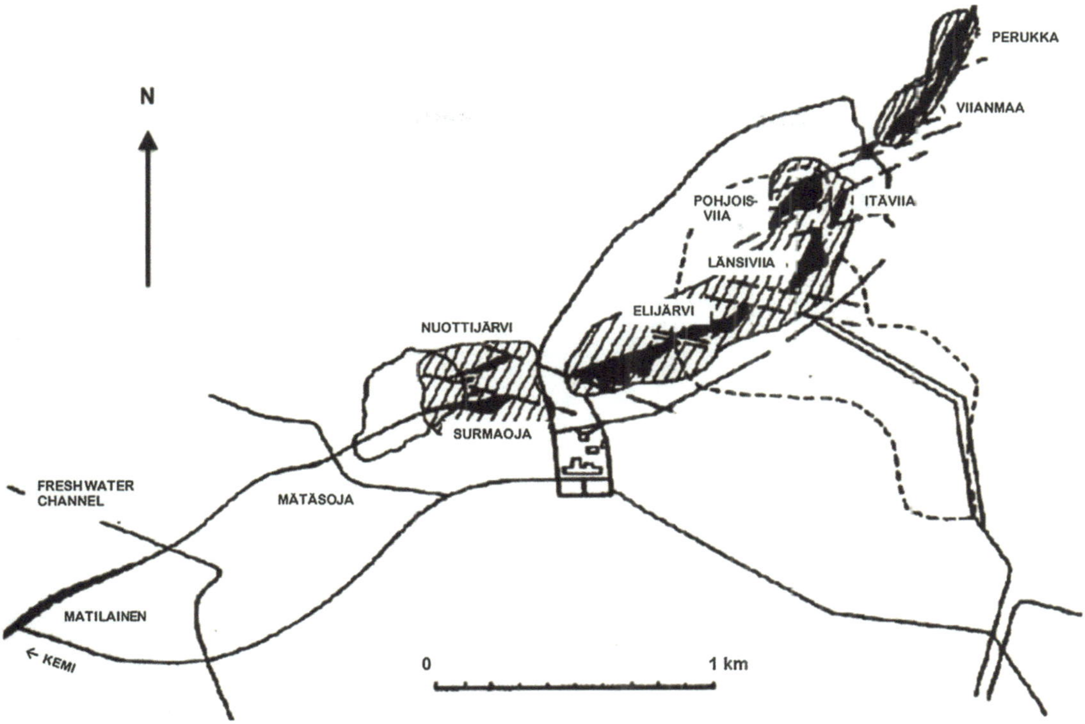

FIGURE 3.4.4 Surface plan of the mining area and the main chromitite unit faulted to several distinct ore bodies.

Lined areas denote open pits.

Source: *Modified from Alapieti et al. (1989).*

depicted in Fig. 3.4.7. Variation in hardness of the ore has an effect on the degree of liberation of chromite in grinding. Also, the chromite grains are characterized by extensive microcracking and fragmentation in certain ore bodies (Leinonen, 1998), thus reducing the size of purely ground chromite in concentration (Fig. 3.4.8).

FROM CHROME ORE TO STAINLESS STEEL

MINE HISTORY AND RESOURCE

The Kemi chromite mineralization was discovered in 1959, when a freshwater channel was being excavated in an area covered with woods and bogs. Mr. Martti Matilainen, a local diver who was interested in ore prospecting, discovered the first chromite-bearing blocks. Subsequent exploration by the Geological Survey of Finland soon led to the discovery of a thick chromite-rich layer. From 1960 onward, the exploration was conducted by Outokumpu under contract from the government of Finland. The prospecting period and the general geology of the Kemi chromite deposit were first described by Kahma et al. (1962). The decision to exploit the deposit was made in 1964, by which time 30 million metric tons of chromite ore had been located in the basal part of the intrusion (Alapieti et al., 1989).

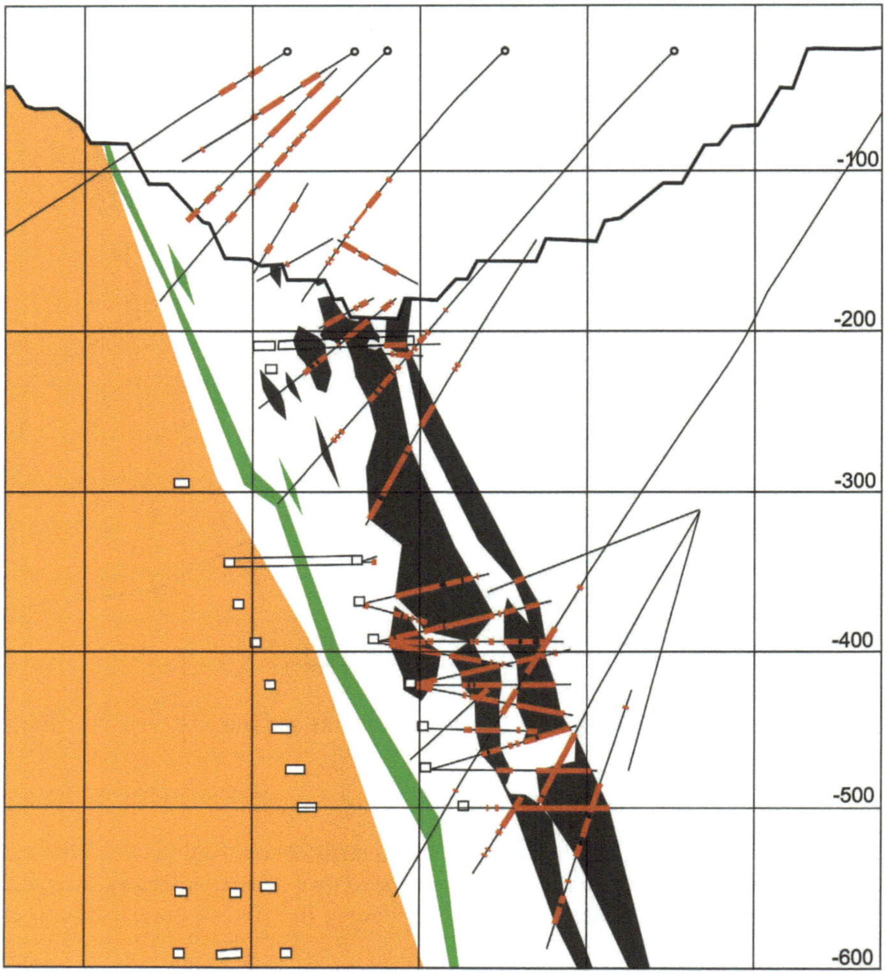

FIGURE 3.4.5 Typical cross section of the lower part of the Kemi layered intrusion, Elijärvi ore body.

Gray color indicates the main chromitite layer; mylonite is represented by green; and light brown color denotes Archean basement (grid is 100 m × 100 m).

The mine's current proved ore reserves total 50.1 Mt. In addition, it is estimated that there are 97.8 Mt of mineral resources (Outokumpu, 2014). The mineral resources have been estimated to a depth of 1 km below the present underground mine. The average Cr_2O_3 content of the ore is about 26% and its average chrome–iron ratio is about 1.7.

MINING AND PROCESSING

Full-scale open pit mining commenced in 1968, and was carried out until December 2005, when the main open pit had reached a depth of approximately 200 m. Construction of the underground mine

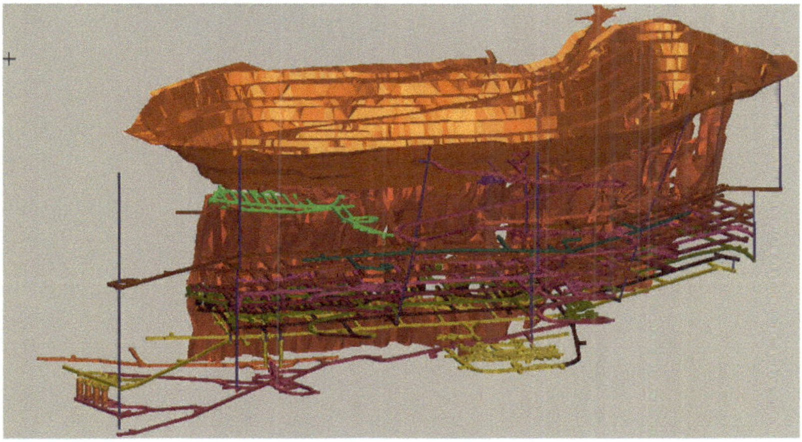

FIGURE 3.4.6 3D model of the Kemi Mine below the central open pit, which is shown as a brown color.

The red color indicates the almost vertical main chromitite layer.

Table 3.4.2 Average grades and Cr/Fe ratios of the ore bodies

Ore body	Cr_2O_3%	Cr/Fe	Mineralogical ore type
Matilainen	26.8	1.43	Talc-carbonate
Nuottijärvi	29.5	1.53	Talc-carbonate/chlorite
Surmaoja	30.0	1.60	Tremolite-chlorite/ serpentine/talc-carbonate
Elijärvi	29.6	1.59	Tremolite-chlorite/ serpentine/talc-carbonate
Länsiviia	29.2	1.56	Talc-carbonate
Pohjoisviia	29.3	1.65	Talc-carbonate/chlorite
Viianmaa	29.1	1.58	Talc-carbonate/serpentine
Perukka	28.3	1.56	Serpentine

Note: *Also shown are the dominant mineralogical ore types in each body.*
Source: *Modified from Alapieti et al. (1989).*

began in 1999, and it was officially opened in September 2003. The stoping method is bench cut and fill. Stoping commences at level 500 and progresses upward toward the bottom of the open pit. The sublevel caving method has also been used on a limited scale below the western part of the main open pit (Fig. 3.4.6). Annual production has amounted to around 1.2 Mt of ore, but the mill capacity will be 2.7 Mt of ore from 2015 onward due to recent expansion of the ferrochrome production capacity.

The portion of the chromitite layer being mined at present extends from the Elijärvi ore body in the west to the Pohjois-Viia ore body in the east (Figs. 3.4.4 and 3.4.6) situated below the main open pit. The length of the chromitite unit under mining is about 1.5 km.

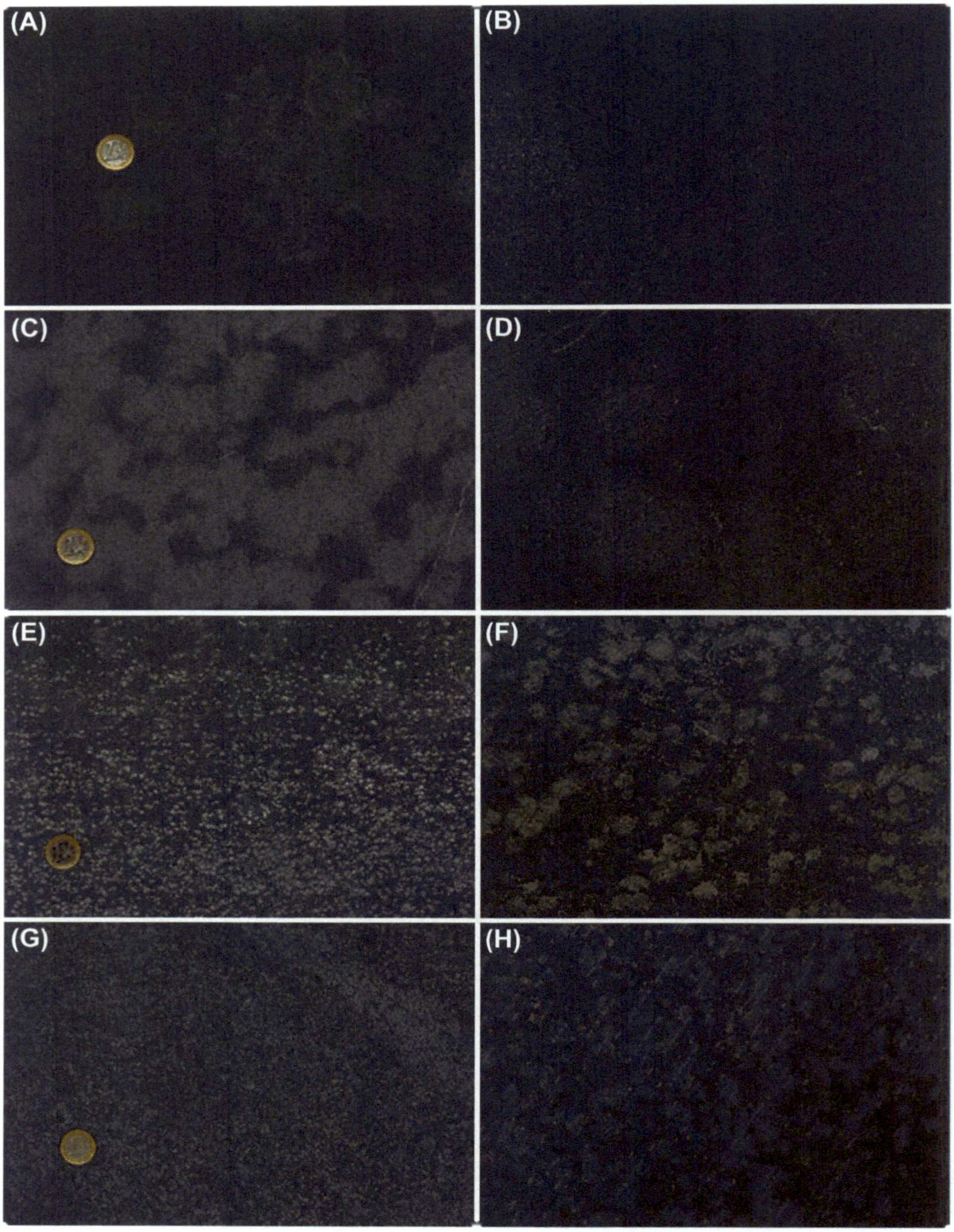

FIGURE 3.4.7 Mineralogical ore types and textures of the main chromitite layer.

Basic picture on the left with 1 € coin as a scale bar (diameter 23.25 mm); enlargement on the right. (A) and (B): tremolite-kaemmererite based ore with tremolite oikocrysts from the upper part of the main chromitite layer. (C) and (D): talc-carbonate based, high-grade "gray-spotted type" with lighter, roundish spots containing very fine-grained chromite and a matrix of medium-grained chromite crystals. (E) and (F): talc-carbonate based "spotted type" with pseudomorphs of cumulus olivine. (G) and (H): serpentine-based disseminated type from the footwall of the main chromitite layer.

Photos by Jari Mäntylä.

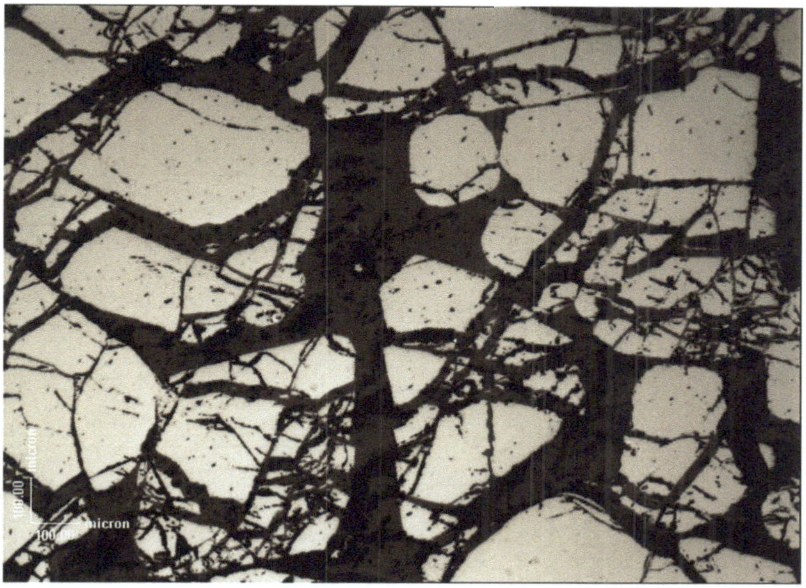

FIGURE 3.4.8 Microscope image of a polished surface in reflected light depicting typical fracture patterns of the chromite crystals.

Scale bar is 100 μm.

FIGURE 3.4.9 Rounded to subrounded lumps of chromitite clasts ("pebbles") within a talc-carbonate altered ultramafic rock with disseminated chromite.

Diameter of the drill core is 30 mm.

Photo by Jari Mäntylä.

The ore feed is concentrated into upgraded lumpy ore and fine concentrate. Concentrating is based on different densities of minerals and is virtually chemical free. In the first stage of the process, the ore is crushed and screened to a diameter of 12–100 mm. After crushing, ore lumps are processed by heavy-medium separation. In this process, upgraded lumpy ore and lumpy waste are separated from the ore. Lumpy waste is used underground in backfilling. Further processing takes place in the concentrating plant where the ore is first ground in a rod mill. The fine concentrate is then produced by gravity separation using spirals (Figs. 3.4.10 and 3.4.11). Annual production of lumpy ore and fine concentrate is about 400,000 t and 850,000 t, respectively.

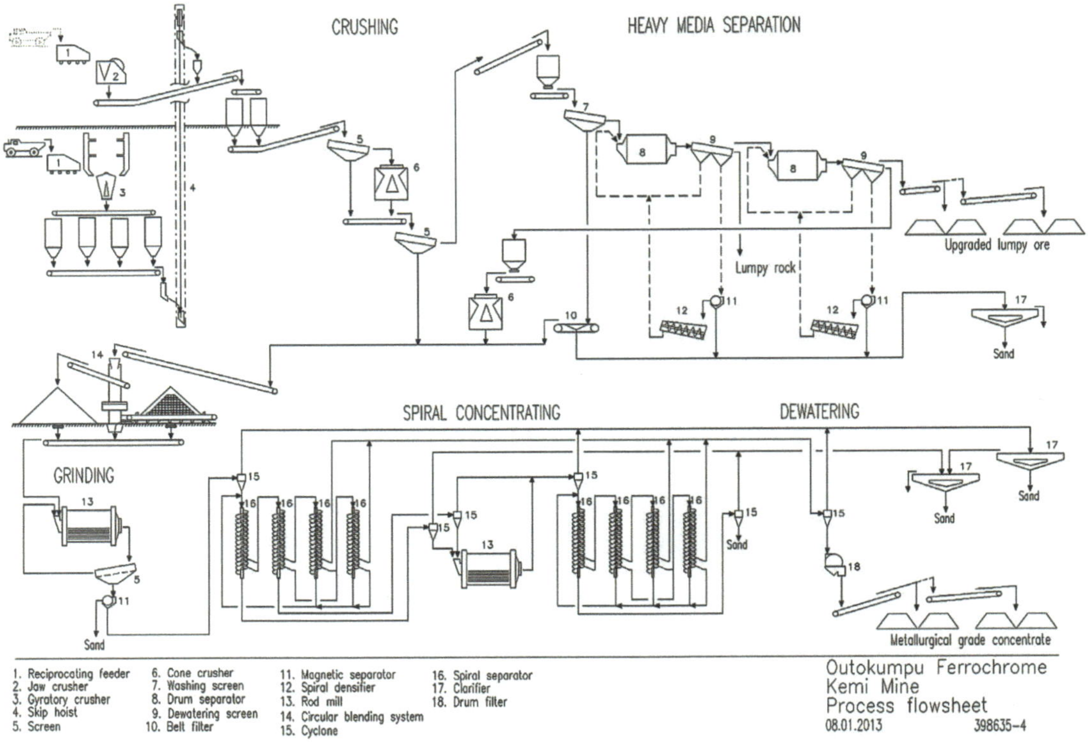

FIGURE 3.4.10 Process flowsheet of the Kemi Mine.

FIGURE 3.4.11 Aerial photograph of the Kemi Mine looking north.

FERROCHROME PLANT AND THE STEEL WORKS

The ferrochrome plant and the steel works were built close to the harbor of Tornio, a town situated some 25 km from the Kemi Mine. Production of ferrochrome commenced in 1968. The ferrochrome works was expanded with the second smelting furnace in 1985. The third closed furnace, which is the biggest in the world today, became operational in late 2012. The construction process of the stainless steel works at Tornio was completed in 1976. The unique, integrated production chain from chrome ore to stainless steel coils and plates was completed in 1988, when the hot rolling mill was inaugurated. Further investments in the 1990s and during recent years have increased the annual production capacity of Tornio Works to 1.7 Mt of rolled products.

SUMMARY

The Kemi chromite deposit is hosted by the originally funnel-shaped 2.4 Ga Kemi layered intrusion, which comprises a lower ultramafic part and an upper gabbroic part. The thickness of the cumulate sequence is about 1.9 km in the central part of the intrusion, and the total length of the intrusion is about 22 km. The thickness of the main chromitite layer averages 40 m in the central part of the intrusion, but decreases rapidly toward the margins. The upper contact of the chromitite unit lies stratigraphically 100–150 m above the basal contact of the complex, but its position has been changed by several strike-slip faults.

The Kemi intrusion was tilted by tectonic movements during the Svecokarelidic orogeny to form a body dipping about 70° to the northwest, probably at around 1.9–1.8 Ga. According to a geophysical seismic reflection survey, the intrusion extends to a depth of about 3 km, possibly even to 4 km. Due to metamorphism, the original magmatic silicates have been completely altered to secondary minerals, especially in the lower and upper parts of the intrusion. Alteration also affected the Cr_2O_3 content of chromite crystals with iron replacing chrome. The average chromium content of the ore feed is, therefore, only about 26%. Concentrating is based on different densities of minerals and is virtually chemical free. The mine's current ore reserves comprise 50.1 Mt. In addition, there are 97.8 Mt of mineral resources, estimated to a depth of 1 km.

The Kemi chrome mine is a fine example of the exploitation of a low-grade ore body, distinctly lower in grade than the stratiform chrome deposits of the Bushveld complex or Great Dyke in southern Africa, which grade between 40–48 % Cr_2O_3. The success of the operation is due to the convenient location of the deposit relative to existing infrastructure, combined with advanced mineral processing and ferrochrome production technology.

REFERENCES

Alapieti, T.T., Huhtelin, T.A., 2005. The Kemi intrusion and associated chromitite deposit. In: Alapieti, T.T. (Ed.), Proceedings of the 10th International Platinum Symposium. Field Trip Guidebook. Geological Survey of Finland, Guide 51a, pp. 13–32.

Alapieti, T.T., Kujanpää, J., Lahtinen, J.J., Papunen, H., 1989. The Kemi stratiform chromitite deposit, northern Finland. Economic Geology 84, 1057–1077.

Huhtelin, T.A., Alapieti, T.T., Lahtinen, J.J., 1989. Megacyclic units I, II and III in the Penikat layered intrusion. In: Alapieti, T.T. (Ed.), Proceedings of the 5th International Platinum Symposium. Guide to the post-symposium field trip, Geological Survey of Finland, Guide 29, pp. 59–69.

Huovinen, I., 2007. Kemin kromitiitin Cr/Fe-suhde kromiitin raekoon funktiona Pohjois-Viian, Elijärven ja Elijärven E-malmioissa. Unpublished M.Sc. thesis, University of Oulu, p. 104. (in Finnish)

Kahma, A., Siikarla, T., Veltheim, V., et al., 1962. On the prospecting and the geology of the Kemi chromite deposit, Finland. A preliminary report. Commission Géologique de Finlande, Bulletin 194, p. 91.

Kemppainen, K., 2002. Kemin kerrosintruusion oletettu tulokanava ja sen vaikutus maanalaiseen kaivostoimintaan. Unpublished M.Sc. thesis, University of Oulu, p. 92. (in Finnish)

Kukkonen, I., Heikkinen, P., Elo, S., et al., 2009. HIRE Working Group of the Geological Survey of Finland. HIRE Seismic Reflection Survey in the Kemi Cr Mining Area, Northern Finland. Geological Survey of Finland Report Q23/2009/64.

Leinonen, O., 1998. Use of chromite microstructure image analysis to estimate concentration characteristics in the Kemi chrome ore. Acta Universitatis Ouluensis Series A, Scientiae Rerum Naturalium 305.

Manhes, G., Allegre, C.J., Dupre, B., Hamelin, B., 1980. Lead isotope study of basic-ultrabasic layered complexes: speculation about the age of the earth and primitive mantle characteristics. Earth Planet Science Letters 47, 370–382.

Patchett, P., Kouvo, O., Hedge, C.E., Tatsumoto, M., 1981. Evolution of continental crust and mantle heterogeneity: evidence from Hf isotopes. Contributions to Mineralogy and Petrology 78, 279–297.

Posio, M.A., 2008. Kemin kerrosintruusion reunasarja. Unpublished M.Sc. thesis, University of Oulu, p. 121. (in Finnish)

Sakko, M., 1971. Varhais-karjalaisten metadiabaasien zirkoni-ikiä. With English summary: Radiometric zircon ages on the early Karelian metadiabases, Geologi, Helsinki 23, 117–118.

CHAPTER

THE MUSTAVAARA FE-TI-V OXIDE DEPOSIT

3.5

T. Karinen, E. Hanski, A. Taipale

ABSTRACT

The Koillismaa mafic layered intrusion in northeastern Finland is part of the ~2440 Ma Tornio-Näränkävaara intrusion belt. After its emplacement between the Archean basement and the overlying Paleoproterozoic sedimentary-volcanic cover and its solidification, the intrusion was faulted and fragmented into several blocks. One of these is the 4-km-wide and 20-km-long Porttivaara block. It has a vertical stratigraphic thickness of ~2500 m and can be divided from the base upwards into the marginal series and the layered series, the latter containing the lower, middle, and upper zone. The Mustavaara Fe-Ti-V oxide deposit occurs within a magnetite gabbro unit in the lower part of the upper zone. This deposit represents the only economic Fe-Ti-V oxide deposit in Finland related to the ~2440 Ma magmatism, containing approximately 100 million tons of ore reserves with 14 wt% of vanadiferous ilmenomagnetite.

The ore of the Mustavaara deposit is composed of equigranular, small- to medium-grained plagioclase-augite-ilmenomagnetite adcumulate with a very sharp lower contact and gradual upper contact. The deposit includes three different conformable ilmenomagnetite-rich ore layers, known as the lower, middle, and upper ore layers, with a total thickness of approximately 80 m. The highest ilmenomagnetite grades (15–35 wt% ilmenomagnetite) are found in the lower and upper ore layers. All three ore layers are continuous along the strike of the Porttivaara block. However, outwards from the Mustavaara deposit area, they show decreasing metal grades and vanadium contents of ilmenomagnetite.

The V_2O_3 content of magnetite is highest (1.7 wt%) in the lower and upper ore layers, whereas in the other ore layers the content is lower (1.5 wt%). The FeO^{TOT} content of magnetite is constant, at 90 wt%, and the TiO_2 content is generally low, at <1.0 wt%. The average chemical composition of the ilmenomagnetite-rich concentrate is the following: 54.2 wt% Fe_2O_3, 30.4 wt% FeO (62.3 wt% Fe), 7.5 wt% TiO_2 (4.6 wt% Ti), and 0.91 wt% V. Compositional and textural data indicate that the deposit formed from a relatively differentiated tholeiitic magma under oxygen fugacity conditions around the NiO buffer, favoring strong enrichment of vanadium in magnetite.

Keywords: layered intrusion; oxide ore; Fe; Ti; V; ilmenomagnetite.

INTRODUCTION

Vanadium-bearing magnetite of igneous origin is the main source of vanadium globally, currently accounting for approximately 85% of global V_2O_5 production. Although global vanadium resources are large, exceeding 63 million tons, the majority of them are located in three countries: China, Russia, and South Africa (U.S. Geological Survey, 2013). Therefore, vanadium is regarded as one of the strategic metals that have many important industrial applications. The Bushveld complex of South Africa hosts

Mineral Deposits of Finland. http://dx.doi.org/10.1016/B978-0-12-410438-9.00007-8

the largest vanadium reserves and presently accounts for about 40% of global vanadium production (Yager, 2012). From the 1950s to the 1980s, Finland also had a significant role as a vanadium producer, based on two mines, Otanmäki and Mustavaara. A decline in the vanadium price led to the closure of these mines, but both of them are currently being evaluated to be reopened.

The Mustavaara Fe-Ti-V oxide deposit is hosted by the Koillismaa intrusion, which forms part of the ~2440 Ma Tornio-Näränkävaara intrusion belt running across Finland from the Finnish-Swedish border to the Finnish-Russian border (Iljina and Hanski, 2005). The Mustavaara ore consists of an ilmenomagnetite-rich layer within the upper part of the Koillismaa intrusion (Juopperi, 1977). In this chapter, we give an account of the exploration history of the Mustavaara deposit and current estimates of ore reserves; describe the structure, geochemistry, and mineralogy of the ore layers; and propose a petrogenetic model for the deposit.

HISTORY OF THE MUSTAVAARA MINE

The history of the Mustavaara mine dates back to the summer of 1957 when forestry manager Antti Oikarinen observed compass interferences in the Porttivaara-Haukivaara area. This encouraged him to send oxide-rich samples from the area to different mining companies. The first assayed vanadium grades were not encouraging enough for the Otanmäki Company to conduct further investigations in the area. Nevertheless, Oikarinen continued prospecting and recognized that in the Mustavaara area, the samples had higher magnetite contents. This led the Otanmäki Company to launch an exploration project in that area. The decision was also facilitated by the aeromagnetic anomalies that were detected in the area in the first national airborne geophysical mapping program (high-altitude survey program in 1951–1972) carried out by the Geological Survey of Finland (GTK). These measurements showed that the samples represented a coherent layer rather than scattered anomalies in the Mustavaara area. Over the next 10 years, the Otanmäki Company performed exploration in the area using ground magnetic surveys, diamond drilling, and outcrop mapping. Eventually, in 1967, this work led to the discovery of a V-bearing magnetite gabbro layer in the upper part of the Porttivaara block of the Koillismaa intrusion (Isokangas, 1957; Juopperi, 1977; Markkula, 1980).

The decision to open a mine in the Mustavaara area was made by the Rautaruukki Steel Company in 1971 (the Otanmäki Company was merged with Rautaruukki in 1968). Open pit mining began in 1976 and terminated in 1985 due to the low vanadium price at the time (Juopperi, 1977; Puustinen, 2003). The ore reserves of the Mustavaara Mine were 38 Mt at 16.8 wt% ilmenomagnetite concentrate, with a cutoff value of 11.9 wt% (Paarma, 1971). These reserves were estimated to 100 m below the topographic surface. During its operation, the mine produced 13.45 Mt of ore and 1.97 Mt of ilmenomagnetite-rich concentrate averaging 0.91 wt% V. The annual production was approximately 240,000 t of concentrate and 2500–3000 t of V_2O_5 (1400–1700 t V) accounting for some 6–9% of the global production of vanadium at that time.

The mine area has recently been reevaluated for additional ore potential by Mustavaara Mine Ltd. A drilling program by the company has outlined a down-dip continuation of the magnetite gabbro. A new ore reserve estimate has been calculated to a depth of 250 m below the topographic surface. The current reserves are 99 Mt of ore grading 14.0 wt% ilmenomagnetite with vanadium contents of 0.91 wt%. The reserves include 64 Mt in proven and 35 Mt in the probable class. These estimates were calculated using an ilmenomagnetite cutoff value of 8.0 wt% (Mustavaaran Kaivos Oy, 2013).

GEOLOGICAL SETTING OF THE MUSTAVAARA DEPOSIT

The Koillismaa intrusion comprises the western part of the Koillismaa-Näränkävaara complex, whereas the eastern part of the complex forms the Näränkävaara intrusion (Fig. 3.5.1). The two intrusions are linked by a strong positive gravity anomaly interpreted to possibly represent a nonexposed dike.

The Koillismaa intrusion consists of several distinct blocks believed to have originally formed a single, sheet-like layered intrusion that was later dismembered by tectonic movements (Karinen, 2010). The magma that formed the Koillismaa intrusion was emplaced close to the boundary between the Archean basement complex and the overlying supracrustal rocks of the Paleoproterozoic Kuusamo schist belt Based on analyses on chilled margins, Karinen (2010) proposed a high-alumina tholeiitic basalt as the parental magma composition for the Koillismaa intrusion. In contrast, the Näränkävaara intrusion was emplaced into granitoids of the Archean basement, and its parental magma was likely more magnesian (Alapieti, 1982).

The stratigraphy of the Koillismaa intrusion is most complete in the Porttivaara block, which hosts the Mustavaara ore deposit. The cumulate succession has a thickness of ~2500 m. Based on the appearance of major cumulus minerals, it is divided into several subunits (Fig. 3.5.2). The marginal series (MS) is up to 200 m thick, and contains gabbroic cumulates at the base, which are overlain by progressively more

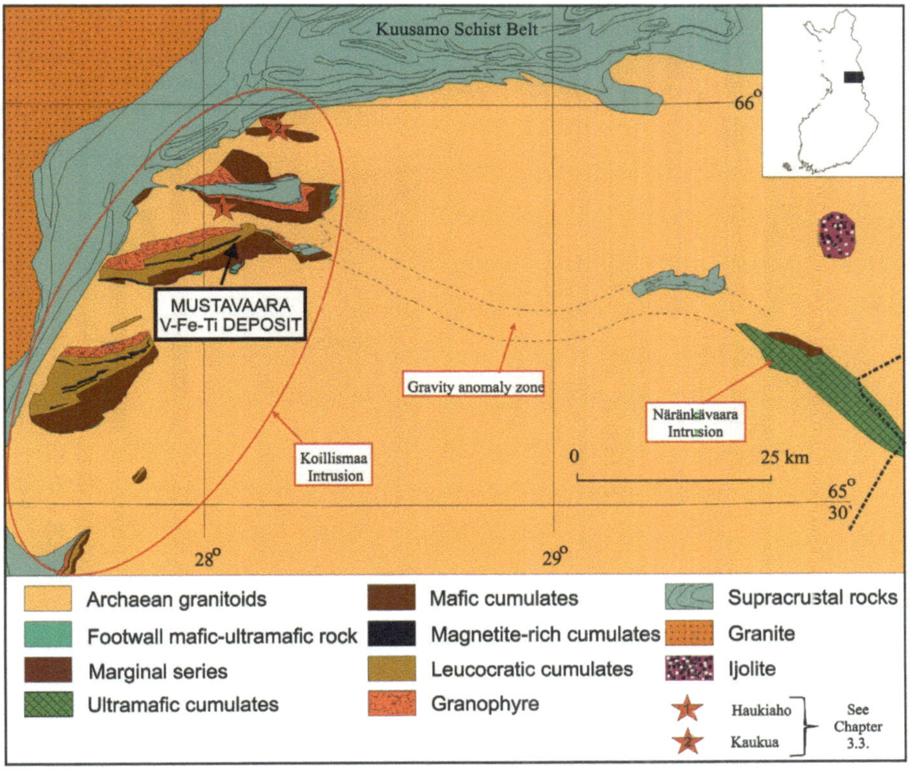

FIGURE 3.5.1 Generalized geological map of the Koillismaa-Näränkävaara layered complex.

Source: *Modified from Karinen (2010).*

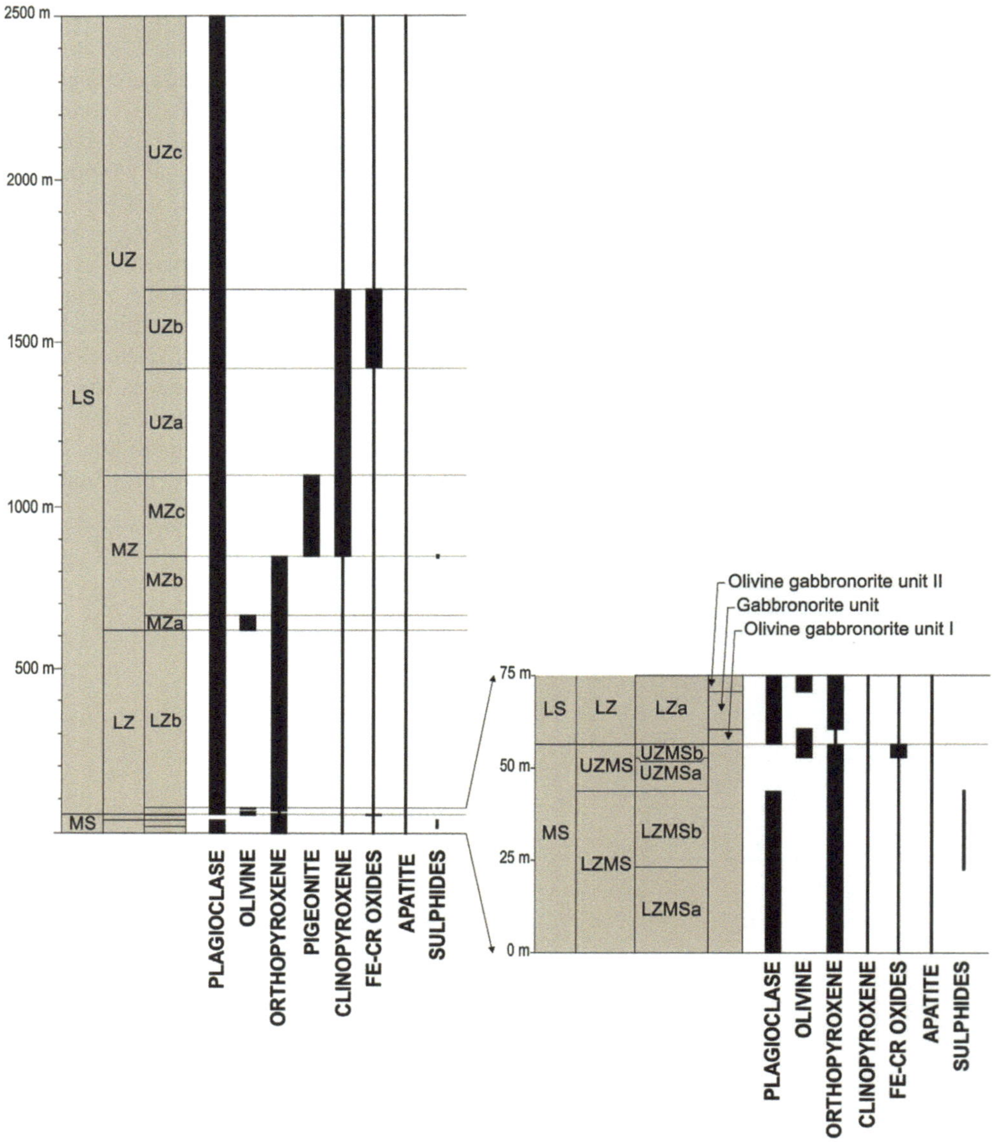

FIGURE 3.5.2 Stratigraphic section and cumulus stratigraphy in the profile across the Porttivaara block.

Thick vertical lines indicate the presence of cumulus minerals and thin vertical lines indicate intercumulus minerals.

Source: *Modified from Karinen (2010).*

magnesian rocks, including pyroxenites and peridotites. The rocks contain Ni-Cu-PGE-enriched sulfides. The layered series (LS) comprises three subzones. The lower zone (LZ) is composed of olivine gabbronorite in its lower part (LZa) and gabbronorite in its upper part (LZb). The middle zone (MZ) begins with a thin layer of olivine gabbronorite (MZa), which is overlain by a thick sequence of gabbronorite (MZb). The middle part of the middle zone is marked by the first appearance of cumulus augite and inverted cumulus pigeonite. This level, which is defined as the boundary between subzones MZb and MZc, contains a subeconomic PGE-Cu reef (up to 1 ppm Pt + Pd + Au over a few meters), known as the Rometölväs Reef. The upper zone (UZ) is characterized by the presence of Fe-Ti-V-rich magnetite gabbro (UZb), which is the host rock of the Mustavaara deposit. The magnetite gabbro is sandwiched between anorthosites and leucogabbros (UZa and UZc, respectively). The lower and upper chilled margins of the intrusion are locally exposed, with the latter separating the cumulates from a thick and homogenous layer of granophyre (Karinen, 2010). The genesis of the granophyre has long been controversial, but it is now regarded as representing preintrusion volcanic rocks (Lauri et al., 2003).

GEOLOGY OF THE MUSTAVAARA ORE DEPOSIT

The Mustavaara deposit is located in the eastern part of the magnetite gabbro layer (UZb) that can be traced throughout the Porttivaara block ("magnetite-rich cumulates" in Fig. 3.5.1). The Porttivaara block is one of the largest blocks of the intrusion, and it has been gravimetrically modeled to extend to a depth of ~2000 m. It is 20 km long and 4 km thick with an exposed area of ~80 km^2. The magnetite gabbro is thicker here than in the other blocks of the Koillismaa intrusion, reaching a thickness of 200 m over a strike of 19 km. It strikes nearly east–west and dips 35–45° to the north (Juopperi, 1977; Ruotsalainen, 1977; Piirainen et al., 1978; Alapieti, 1982; Karinen, 2010).

The Mustavaara deposit comprises an ~80-m-thick succession of ilmenomagnetite-rich sublayers in the lower part of the magnetite gabbro layer. The dip of the ore layer in the area of the deposit is 40° to the north, but steepens to 60° east of the old open pit, where the layer also becomes narrower, at only 20 m thick. Depending on the amount of ilmenomagnetite, the deposit is divided into distinct sublayers comprising from the bottom upward: the lower ore layer (LOL; 5 m thick; 20–35 wt% ilmenomagnetite), the middle ore layer (MOL; 15–50 m thick; 10–15 wt% ilmenomagnetite), the upper ore layer (UOL; 10–40 m thick; 15–25 wt% ilmenomagnetite), and the disseminated ore layer (DOL; <10 wt% ilmenomagnetite) (Figs. 3.5.3 and 3.5.4). These sublayers extend throughout the Porttivaara block as established by the diamond drilling of the Rautaruukki Company. These drillings revealed that the sublayers are continuous, but their ilmenomagnetite and vanadium grades become lower toward the west of the deposit area (Markkula, 1980). In the area of the Mustavaara deposit, the magnetite gabbro contains an average of 14.7 wt% ilmenomagnetite and 0.88 wt% V, whereas outside the mine area, in the Haukivaara-Porttivaara area, the average grades are lower, at 9.1 wt% ilmenomagnetite and 0.77 wt% V, respectively (Fig. 3.5.5(A) and (B)).

The magnetite gabbro of the Mustavaara deposit is an equigranular, small- to medium-grained plagioclase-augite-ilmenomagnetite adcumulate. Plagioclase occurs typically as idiomorphic grains 1–5 mm in size, with their preferred orientation giving the rock a distinct igneous lamination. Augite has undergone adcumulus growth and is 0.5–1 mm in size and completely altered to amphibole group minerals. Ilmenomagnetite occurs as interstitial fillings between the silicate minerals throughout the lower part of the deposit, but can also locally display an idiomorphic habit in the uppermost parts of the magnetite gabbro layer, especially in the DOL. The footwall contact of the deposit is sharp, but toward the hanging wall, the

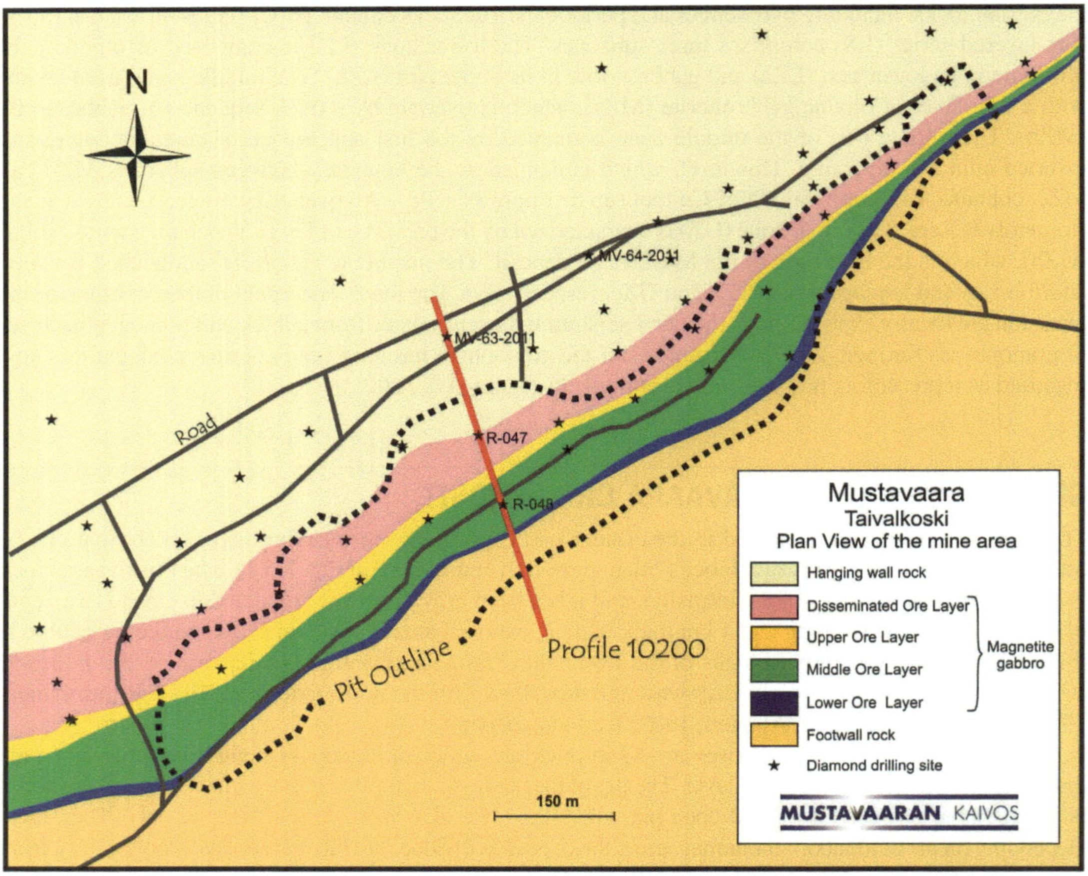

FIGURE 3.5.3 Plan view of the Mustavaara open pit area showing the ore layers projected to surface.

magnetite gabbro is characterized by an increasing amount of leucocratic fragments or interlayers. A typical feature is that wherever any of the sublayers include these fragments or interlayers, the sublayers thicken in comparison to the areas free of fragments or interlayers (Fig. 3.5.4). In some drill cores, at a level of a few tens of meters above the DOL, the hanging wall includes a thin, <0.5-m-thick section of semimassive to massive oxide, which has been named "the Ore" (for the location of the ore in a diamond drill core section; see Figs. 3.5.7 and 3.5.8 later). The hanging and footwall rocks are composed of light-colored anorthosite and leucogabbro that typically display a much coarser grain size (>5 mm for plagioclase) than the magnetite gabbro, which makes these rocks distinguishable from oxide-bearing rocks.

In general, ilmenomagnetite is an Fe-, V-, and Ti-rich oxide that originally crystallized as titanomagnetite (magnetite-ulvöspinel solid solution), but was oxidized during the subsolidus stage to form composite grains of fine ilmenite exsolutions in a V-bearing magnetite host (Buddington and Lindsley, 1964). Juopperi (1977) described three exsolution types in ilmenomagnetites of the Mustavaara deposit: (1) a coarse type exhibiting homogenous and granular lamellae, (2) long, thin lamellae occurring parallel to the {111} plane of the magnetite host, and (3) small, splinter-like lamellae parallel to

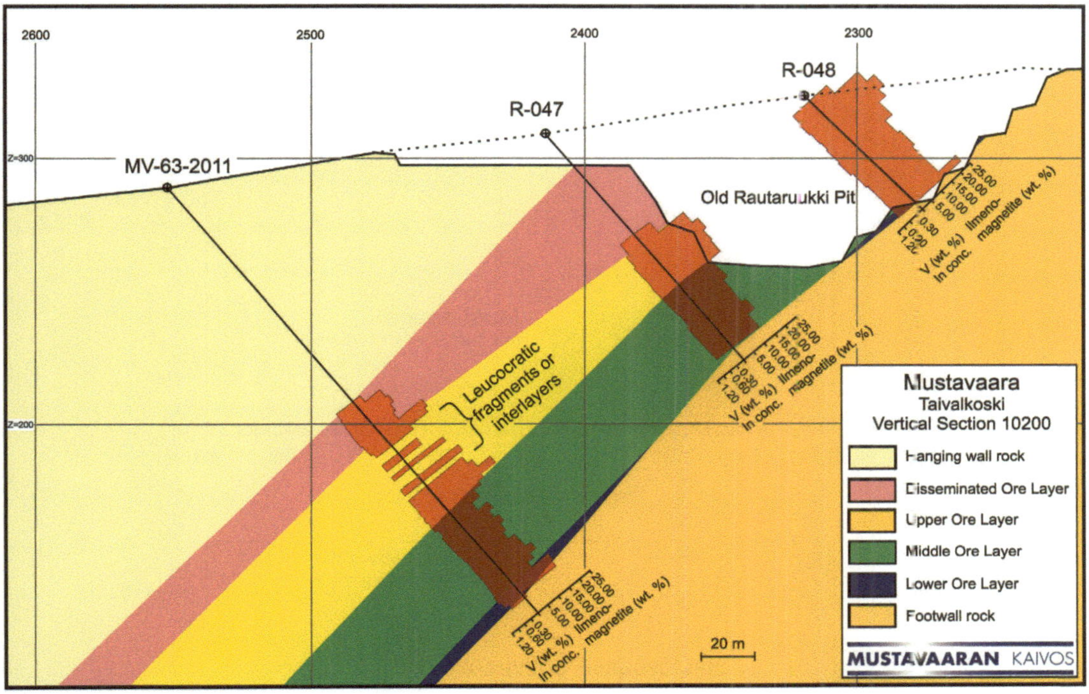

FIGURE 3.5.4 Vertical cross section of the ore of the Mustavaara deposit along profile 10200.

See location of the profile in Fig. 3.5.3.

the {111} plane, which occur between the lamellae of the second type but do not touch them. Juopperi (1977) noticed that the sublayers of the deposit display different exsolution textures in oxide grains. In the oxide grains of the LOL, the {111} lamellae are narrow (<5 μm in width), and between them the magnetite host includes abundant small, splinter-like lamellae (Fig. 3.5.6(A)). In the MOL, UOL, and DOL, exsolutions parallel to the {111} plane are broader (10–60 μm) and the splinter-like lamellae are less abundant. In these sublayers, the coarse, granular lamellae are more common than in the LOL (Figs. 3.5.6(B), (C), and (D)). The exsolution textures in the massive-semimassive Ore are comparable to the textures in the LOL.

Karinen (2010) calculated the following composition for the primary titanomagnetite: 42.0 wt% Fe_2O_3, 44.0 wt% FeO, and 14.0 wt% TiO_2 ($Mt_{60}Usp_{40}$), corresponding to oxygen fO_2 conditions that match closely the conditions of extreme enrichments of vanadium in the experimental study of Toplis and Corgne (2002). The authors noticed that during fractional crystallization of basaltic melts, V-rich magnetite crystallizes in a relatively narrow range of fO_2 conditions, between NNO and NNO-1.5. In their XANES spectroscopy study of V-bearing magnetite grains from layered intrusions, Balan et al. (2006) measured one magnetite sample from Mustavaara that yielded a V^{4+}/V^{3+} ratio of 0.18. This is similar to values obtained from the Bushveld and Skaergaard V-rich magnetite and consistent with fO_2 conditions around the NNO buffer.[1]

[1]NNO is nickel-nickel oxide buffer.

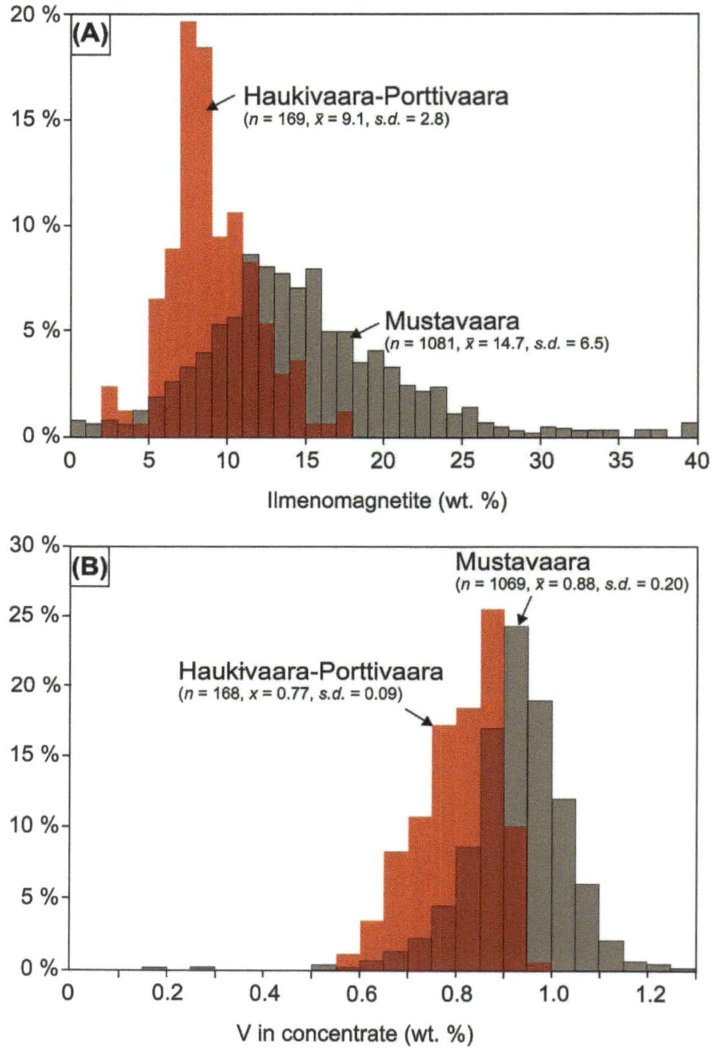

FIGURE 3.5.5 Frequency distribution of ilmenomagnetite grades (A) and V content of the concentrate (B) of the magnetite gabbro samples collected at the Mustavaara and Haukivaara-Porttivaara areas of the Porttivaara block, Koillismaa intrusion.

The amount of ilmenomagnetite was determined with the Dings-Davis tube (DDT) method.

Source: *Geochemical analyses are compiled from the data of Rautaruukki Oy and Mustavaara Mine Ltd. Geochemical analysis of vanadium by Mustavaara Mine Ltd. was made with ICP-OES at the geochemical laboratory of Labtium Oy, and these analyses correlate well with the old Rautaruukki data.*

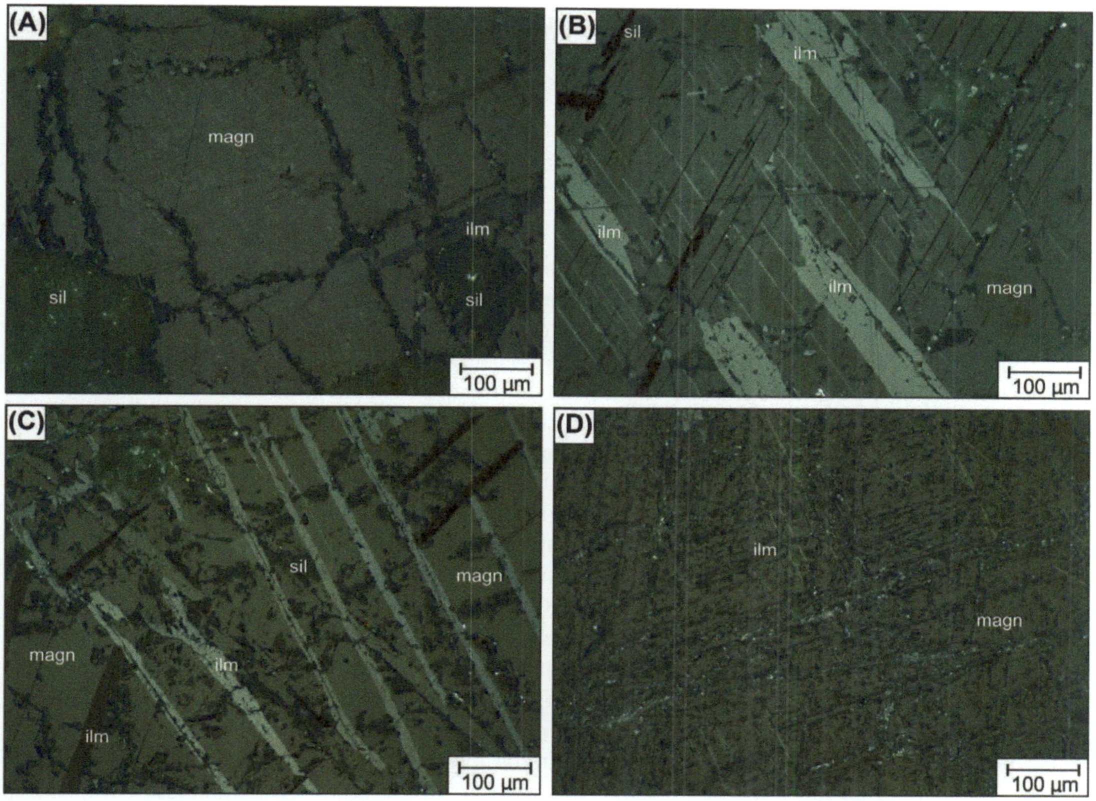

FIGURE 3.5.6 Photomicrographs of exsolution textures in different sublayers of the Mustavaara deposit.

(A) Lower ore layer showing sparse {111} lamellae. Between these lamellae the magnetite host includes abundant small, splinter-like lamellae (sample MV-64-2011, 160.80 m, reflected light). (B) Middle ore layer (sample MV-64-2011, 158.30 m, reflected light). (C) Upper ore layer (sample MV-64-2011, 125.25 m, reflected light). (D) Disseminated ore layer (sample MV-64-2011, 109.19 m, reflected light). In the MOL, UOL, and DOL ((B), (C), and (D)), exsolutions parallel to the {111} plane are broader and more abundant and the splinter-like lamellae are less abundant. Abbreviations: sil = silicate, magn = magnetite, ilm = ilmenite.

ANALYTICAL METHODS

Mineral analyses were performed with the wavelength dispersive spectrometer of a JEOL JXA-8200 electron microprobe at the Center of Microscopy and Nanotechnology, University of Oulu, and were part of the M.Sc. thesis work of Taipale (2013). The analyses were performed with a beam diameter of 1 μm on plagioclase grains and magnetite in host ilmenomagnetite grains in 15 polished thin sections collected from the drill core. For whole-rock analyses, the drill core was sampled at up to 3 m intervals paying attention to the contacts of different lithologies and ore subunits. The samples were analyzed at Labtium Oy, Rovaniemi. The samples were pulverized with a carbon steel mill (>90% <100 μm grain size). Prior to geochemical analyses, the pulverized samples were also processed electromagnetically with the Dings-Davis tube (DDT) method to obtain concentrate powders. The pretreatment method before analysis was sodium peroxide fusion of 0.2 g of the pulverized sample. The analyses were performed with inductively coupled plasma optical emission spectrometry (ICP-OES), and the results represent total analyses of the samples.

GEOCHEMISTRY OF THE ORE

The following description of the compositional variation in the Mustavaara deposit is based on diamond drill core MV-64-2011, which was obtained during the drilling campaign of the Mustavaara Mine Ltd. in late 2011. In the core, the LOL-UOL succession is continuous, but the DOL is in two parts due to a hanging wall fragment or interlayer. The drill core includes a <1-m-thick section of the Ore located in the hanging wall, ~70 m above the upper contact of UOL. The LOL-UOL succession forms a 48-m-thick layer, wherein the LOL is ~3 m, MOL 27 m, and UOL 18 m thick.

Fig. 3.5.7 shows a section of drill core MV-64-2011 with compositional variations of magnetite and plagioclase. The Fe content of the magnetite phase in ilmenomagnetite grains does not display a distinct variation along the drill core and averages 90.0 wt% FeO^{TOT}. The V_2O_3 content of magnetite is highest in the LOL and UOL, ~1.7 wt%, whereas in the under- and overlying sublayers, the content is lower, ~1.5 wt% V_2O_3. There is no systematic variation in the vanadium content of magnetite along the length of the core. The titanium content of magnetite is generally low, due to complete exsolution, as evidenced by microscopic studies. The elevated TiO_2 content in the magnetite in the upper part of the diamond drill core, close to the semimassive-massive oxide layer/dike, is likely due to very narrow and frequent splinter-like ilmenite lamellae. The magnetite grains richest in chromium are in the LOL, where magnetite may contain up to 0.15 wt% Cr_2O_3. Below the magnetite gabbro, the compositional variation of plagioclase displays a normal trend of differentiation with a decrease of the anorthite content from An_{60} to An_{50} with stratigraphic height. In the lower part of the deposit, the composition of plagioclase is more calcic, at $An_{60\text{-}70}$. Above this reversal in the core, the plagioclase composition shows a normal differentiation trend toward the UOL, where the most evolved plagioclase composition is reached (An_{50}). In the hanging wall, plagioclase becomes more calcic with height, reaching An_{55}.

Whole-rock compositions of the sublayers of the deposit are presented in Table 3.5.1 and in Fig. 3.5.8. The whole-rock V and Ti contents are mainly controlled by the amount of ilmenomagnetite, whereas the Fe content is also affected by the occurrence of iron-bearing silicates. The average chemical composition of the concentrate is 54.2 wt% Fe_2O_3, 30.4 wt% FeO (62.3 wt% Fe), 7.5 wt% TiO_2 (4.6 wt% Ti), and 0.91 wt% V. In concentrate analyses, the abundance of Ti displays a negative correlation with that of V (Fig. 3.5.9), which reflects the composition and distribution of magnetite host and ilmenite exsolutions in ilmenomagnetite grains. The concentrates of the Ore and LOL appear higher in their average Ti contents than the concentrates of other sublayers. This is because of the relatively higher abundance of small, splinter-like ilmenite lamellae in the ilmenomagnetite grains in these sublayers. It is impossible to extract such ilmenite lamellae completely in the concentration process such as the DDT method.

GENESIS OF THE MUSTAVAARA ORE DEPOSIT

Genetic models explaining orthomagmatic Fe-Ti-V oxide deposits usually involve late-stage magma differentiation in accordance to the Fenner crystallization trend. However, the models differ from each other with respect to the factors that control the deposition of the ore. These include gravity settling of oxide

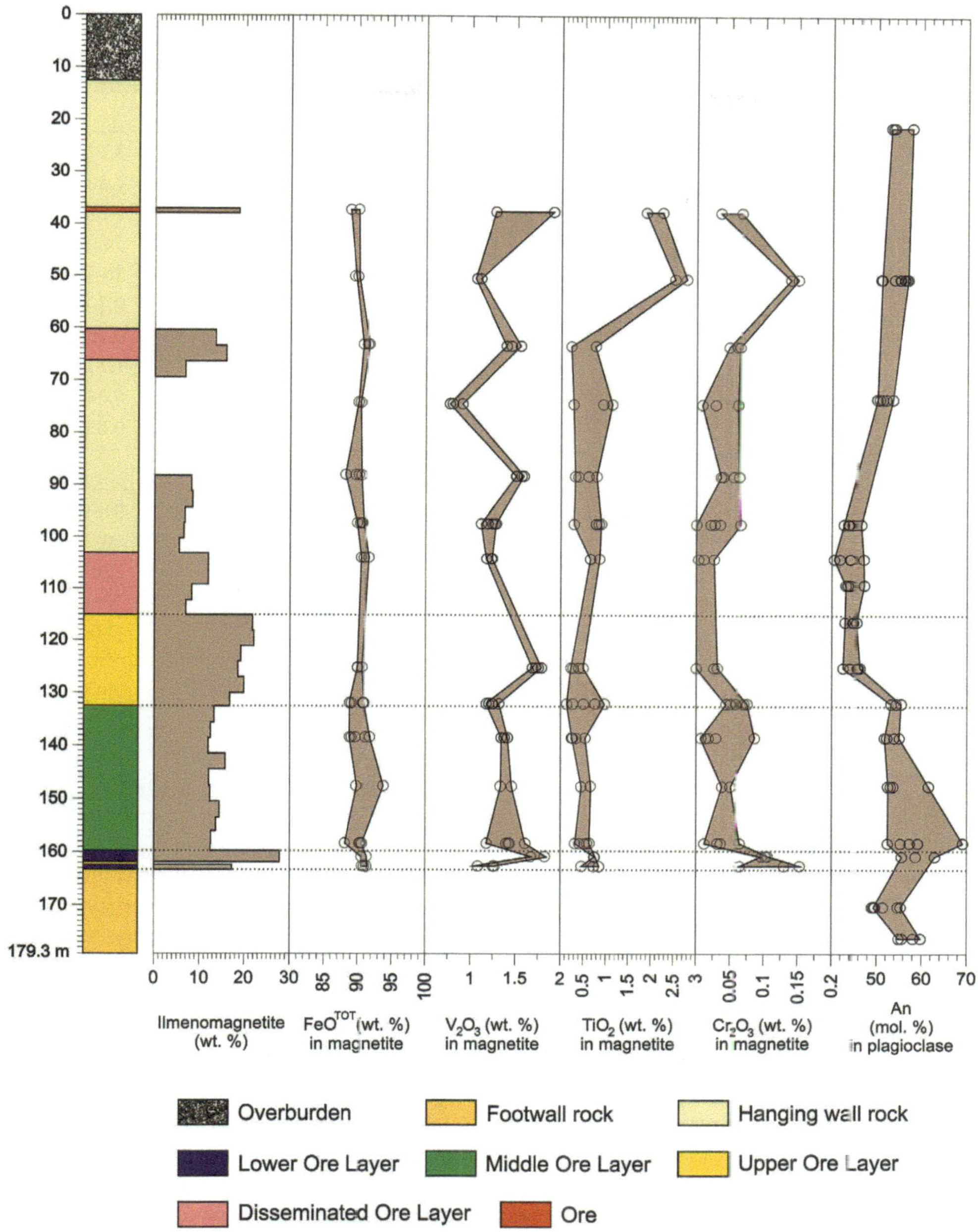

FIGURE 3.5.7 Section of drill core MV-64-2011 at the Mustavaara deposit.

This shows the variation in the amount of ilmenomagnetite, the FeO, V_2O_3, TiO_2, and Cr_2O_3 contents of the magnetite host of ilmenomagnetite grains, and An contents of plagioclase (anorthite end-member). The amount of ilmenomagnetite was determined using a Dings-Davis tube (DDT). See Fig. 3.5.3 for the location of the diamond drilling site.

Table 3.5.1 Composition of different sublayers of the Mustavaara deposit

Sample	1	2	3	4	5
wt %					
SiO_2	34.99	42.72	38.99	44.76	3.12
TiO_2	2.79	2.00	2.11	2.06	7.50
Al_2O_3	10.44	12.27	12.49	14.69	1.16
Fe_2O_3	19.81	11.63	16.49	8.73	54.20
FeO	15.61	11.08	12.85	9.56	30.40
MnO	0.21	0.17	0.17	0.21	0.30
MgO	3.17	5.18	4.28	5.90	0.75
CaO	7.31	9.21	8.61	10.88	1.17
Na_2O	2.32	2.48	2.47	2.53	0.04
K_2O	0.89	0.35	0.34	0.30	0.03
ppm					
V	3800	2200	2600	1500	9180
Cr	130	30	20	20	60
Ni	290	140	110	90	220
Cu	610	760	780	320	70
Co	130	100	130	180	180

1= lower ore layer
2= middle ore layer
3= upper ore layer
4= disseminated ore layer
5= ilmenomagnetite concentrate
Source: *Data of* **1–4** *collected from Juopperi (1977), and* **5** *from the annual report 1982 of Rautaruukki Company (Rautaruukki Oy, 1982).*

grains either by crystal settling (Wager and Brown, 1967; Pang et al., 2008) or density flows (Klemm et al., 1985; Charlier et al., 2006; Maier et al., 2013), magma addition and/or mixing (Harney et al., 1990), changes in fO_2 and/or total pressure in the magma chamber (Cawthorn and McCarthy, 1980; Klemm et al., 1982; Reynolds, 1985), and crystallization from an immiscible Fe-Ti-rich liquid accumulated from silicate magma (Buddington et al., 1955; Lister, 1966; Naslund, 1983; Zhou et al., 2005).

In the case of the Mustavaara deposit, Juopperi (1977) favored the liquation hypothesis largely on the basis of the compositional variation of augite with height in the deposit. Juopperi (1977) showed that the Mg/($Mg+Fe^{2+}$) ratio of augite decreases systematically from 0.70 to 0.60 from the LOL to DOL and postulated that if the deposit had been formed by crystal fractionation, the precipitation of titanomagnetite should have caused a decrease in Fe content and thus an increase in Mg/Fe in the residual magma and any augite crystallizing from it. However, the liquation model may produce the same effect on the augite composition. Another problem with the liquation hypothesis is that the basal contact of the deposit is commonly sharp and planar. It is likely that a dense oxide liquid would have percolated into the semiconsolidated footwall cumulates or into fractures of the footwall rock to form oxide-rich veins, but this is not observed in the Mustavaara deposit. Therefore, it is likely that the Mustavaara ore was formed by another process, for instance, by relatively effective gravitative concentration of oxide crystals, or by sorting of a magnetite slurry.

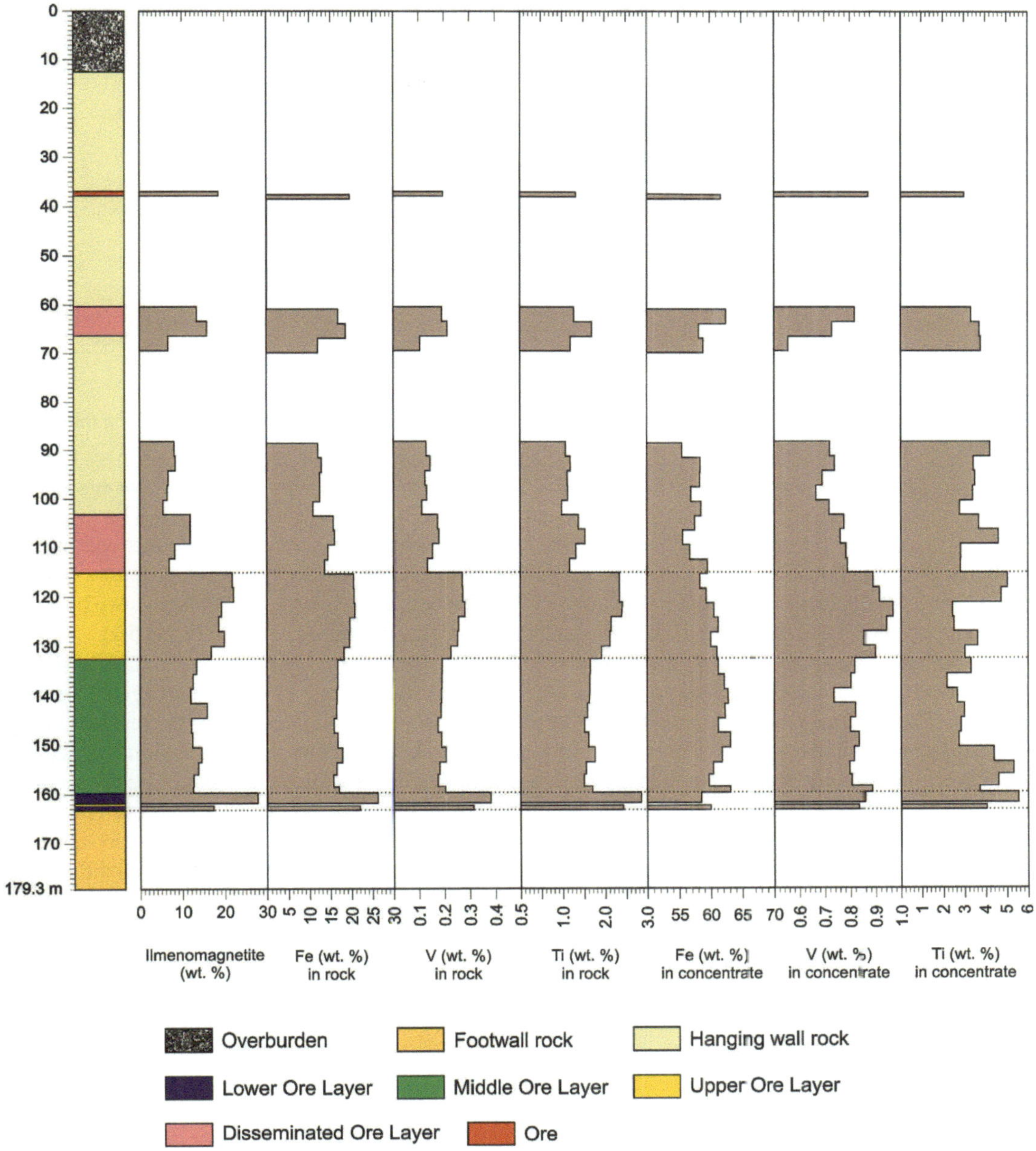

FIGURE 3.5.8 Section of drill core MV-64-2011 at the Mustavaara deposit.

This shows the variation in the amount of ilmenomagnetite, and Fe, V, and Ti contents of whole rock and concentrate samples. The amount of ilmenomagnetite was determined using a Dings-Davis tube (DDT). See Fig. 3.5.3 for the location of the diamond drilling site.

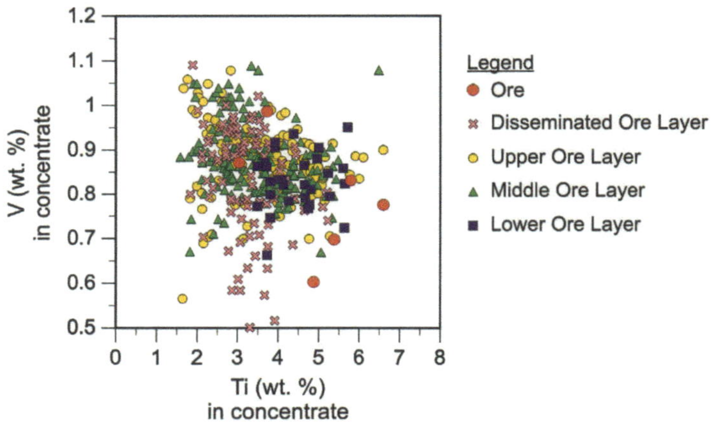

FIGURE 3.5.9 TiO_2 versus V diagram compiled for the concentrate analyses of samples collected from the sub-layers of the Mustavaara deposit.

Source: *Data compiled from the analyses of the samples of the 2011 diamond drill core campaign of the Mustavaara Mine Ltd.*

In this chapter it is argued that the genesis of the Mustavaara deposit is consistent with fractional crystallization rather than liquation. First, the modal proportion of magnetite at Mustavaara is 5–30 wt%, which is in the range of the cotectic ratio (Toplis and Carrol, 1996). Second, the compositional trends shown by augite (see Juopperi, 1977) and plagioclase are consistent with fractional crystallization (See Fig. 3.5.7).

The vanadium content of magnetite depends on the concentration of vanadium in the magma and the mineral-melt partition coefficient of vanadium. During the crystallization of Fe-Ti-rich magmas, the proportions of crystallizing silicates and oxides control the concentration of vanadium in the residual liquid. Crystallization of silicates usually causes the residual liquid to become enriched in vanadium, but because vanadium is a strongly compatible element within the magnetite structure, the residual liquid becomes rapidly depleted in vanadium after the commencement of magnetite precipitation. This has been recorded, for example, in the magnetite-rich layers in the upper zone of the Bushveld complex, where the vanadium content in the oxides in these layers decreases systematically from 2.5 to 0.2 wt% V_2O_3 across the 1500 m height of the upper zone (Klemm et al., 1985; Reynolds, 1985; Ashwal et al., 2005).

In contrast, at the Mustavaara deposit the oxide phase remains relatively V-rich across the magnetite gabbro (Fig. 3.5.7). Of further note is that the vanadium concentration of the magnetite gabbro unit and the ilmenomagnetite grains varies along the strike, decreasing from the deposit area toward the west (Fig. 3.5.5). This could potentially indicate that magnetite gabbro in the deposit area crystallized from relatively less evolved magma, and/or that there were fluctuations in the oxygen or total pressure during the crystallization of ilmenomagnetite.

ACKNOWLEDGMENTS

The first author wishes to thank Aarre Juopperi for stimulating discussion on the genesis of the Mustavaara deposit. We would also like to express our appreciation to Mustavaara Mine Ltd. for providing sample material and analytical data for this section.

REFERENCES

Alapieti, T., 1982. The Koillismaa layered igneous complex, Finland—its structure, mineralogy and geochemistry, with emphasis on the distribution of chromium. Geological Survey of Finland Bulletin 319, 116.

Ashwal, L.D., Webb, S.J., Knoper, M.W., 2005. Magmatic stratigraphy in the Bushveld Northern Lobe: continuous geophysical and mineralogical data from the 2950 m Bellevue drillcore. South African Journal of Geology 108, 199–232.

Balan, E., De Villiers, J.P.R., Eeckhout, S.G., et al., 2006. The oxidation state of vanadium in titanomagnetite from layered basic intrusions. American Mineralogist 91, 953–956.

Buddington, A.F., Flahey, Y., Vlisidis, A., 1955. Thermometric and petrogenetic significance of titaniferous magnetite. American Journal of Science 253, 497–532.

Buddington, A.F., Lindsley, D.H., 1964. Iron-titanium oxide minerals and synthetic equivalents. Journal of Petrology 5, 310–357.

Cawthorn, R.G., McCarthy, T.S., 1980. Variations in the Cr content of magnetite from the upper zone of the Bushveld complex—evidence for heterogeneity and convection currents in magma chambers. Earth and Planetary Science Letters 46, 335–343.

Charlier, B., Duchesne, J.-C., Van der Auwera, J., 2006. Magma chamber processes in the Tellnes ilmenite deposit (Rogaland Anorthosite Province, SW Norway) and the formation of Fe-Ti ores in massif-type anorthosites. Chemical Geology 234, 264–290.

Harney, D.M.W., Merkle, R.K.W., Von Gruenewaldt, G., 1990. Platinum-group element behavior in the lower part of the upper zone, eastern Bushveld complex—implications for the formation of the main magnetite layer. Economic Geology 85, 1777–1789.

Iljina, M., Hanski, E., 2005. Layered mafic intrusions of the Tornio-Näränkävaara belt. In: Lehtinen, M., Nurmi, P., Rämö, T. (Eds.), Precambrian Bedrock of Finland—Key to the Evolution of the Fennoscandian Shield. Elsevier, Amsterdam, pp. 103–138.

Isokangas, P., 1957. Iron ore prospects in the Taivalkoski area. Outokumpu Oy, Mineral Exploration Department, Vihanti. Report of Activities, p. 2. (in Finnish).

Juopperi, A., 1977. The magnetite gabbro and related Mustavaara vanadium ore deposit in the Porttivaara layered intrusion, north-eastern Finland. Geological Survey of Finland, Bulletin 288, 68.

Karinen, T., 2010. The Koillismaa intrusion, northeastern Finland—Evidence for PGE reef forming processes in the layered series. Geological Survey of Finland, Bulletin 404, 176.

Klemm, D.D., Snethlage, R., Dehm, R.M., et al., 1982. The formation of chromite and titanomagnetite deposits within the Bushveld igneous complex. In: Amstutz, G.C., El Goresy, A., Frenzel, G., et al. (Eds.), Ore Genesis, The State of the Art. Spriger-Verlag, Berlin-Heidelberg-New York, pp. 351–370.

Klemm, D.D., Henckel, J., Dehm, R., Von Gruenewaldt, G., 1985. The geochemistry of titanomagnetite in magnetite layers and their host rocks of the eastern Bushveld complex. Economic Geology 80, 1075–1088.

Lauri, L.S., Karinen, T., Räsänen, J., 2003. The earliest Paleoproterozoic supracrustal rocks in Koillismaa, northern Finland—their petrographic and geochemical characteristics and lithostratigraphy. Bulletin of the Geological Society of Finland 75, 29–50.

Lister, G.F., 1966. The composition and origin of selected iron-titanium deposits. Economic Geology 61, 275–310.

Maier, W.D., Barnes, S.-J., Groves, D.I., 2013. The Bushveld complex, South Africa: formation of platinum-palladium, chrome- and vanadium-rich layers via hydrodynamic sorting and mobilized cumulate slurry in a large, relatively slow cooling, subsiding magma chamber. Mineralium Deposita 48, 1–56.

Markkula, H., 1980. Vanadium investigations in the Mustavaara-Porttivaara area in 1980. Rautaruukki Oy, Mineral Exploration. Unpublished Report Ro 13/80. p. 6. (in Finnish).

Mustavaaran Kaivos Oy, 2013. Press release 19.06.2013. Mustavaara Ore Reserve update (www.mustavaarankaivos.com/en/events/2013/mustavaara-ore-reserve-update).

Naslund, H.R., 1983. The effect of oxygen fugacity on liquid immiscibility in iron-bearing silicate melts. American Journal of Science 283, 1034–1059.

Paarma, H., 1971. Ore reserve estimates of the Mustavaara ore. Rautaruukki Oy, Mineral Exploration. Unpublished Report Ou 8/71. p. 7. (in Finnish).

Pang, K.N., Zhou, M.-F., Lindsley, D., et al., 2008. Origin of Fe-Ti oxide ores in mafic intrusions: Evidence from the Panzhihua intrusion, SW China. Journal of Petrology 49, 295–313.

Piirainen, T., Hugg, R., Aario, R., et al., 1978. Final Report of the Koillismaa Research Project in 1976. Geological Survey of Finland, Report of Investigation 18, 51. (in Finnish).

Puustinen, K., 2003. Mining industry and production of mineral raw materials in 1530–2001. Historical review in the light of production numbers. Geological Survey of Finland, Report M 10.1/2003/3. p. 578. (in Finnish).

Rautaruukki Oy, 1982. Annual Report, p. 57. (in Finnish).

Reynolds, I.M., 1985. The nature and origin of titaniferous magnetite-rich layers in the upper zone of the Bushveld complex: A review and synthesis. Economic Geology 80, 1089–1108.

Ruotsalainen, A., 1977. Geophysical interpretation of the structure of the Koillismaa intrusions. M.Sc. Thesis, Department of Geology, University of Oulu, p. 64. (in Finnish).

Taipale, A., 2013. Composition of magnetite in gabbros of the Mustavaara Fe-Ti-V deposit. M.Sc. Thesis, Department of Geosciences, University of Oulu, p. 55.

Toplis, M.J., Carrol, M.R., 1996. Differentiation of ferro-basaltic magmas under conditions open and closed to oxygen: Implications for the Skaergaard intrusion and other natural systems. Journal of Petrology 37, 837–858.

Toplis, M.J., Corgne, A., 2002. An experimental study of partitioning between magnetite, clinopyroxene, and iron-bearing silicate liquids with particular emphasis on vanadium. Contributions to Mineralogy and Petrology 144, 22–37.

U.S. Geological Survey, 2013. Minerals Commodity Summaries. p. 198 http://minerals.usgs.gov/minerals/pubs/mcs/2013/mcs2013.pdf.

Wager, L., Brown, G., 1967. Layered Igneous Rocks. W. H. Freeman and Company, San Francisco. p. 588.

Yager, T.R., 2012. The mineral industry of South Africa. In: Minerals Yearbook. Area Reports: International 2010, Volume III: Africa and the Middle East. U.S. Department of the Interior, U.S. Geological Survey.

Zhou, M.-F., Robinson, P.T., Lesher, C.M., et al., 2005. Geochemistry, petrogenesis and metallogenesis of the Panzhihua gabbroic layered intrusion and associated Fe-Ti-V oxide deposit, Sichuan province, SW China. Journal of Petrology 46, 2253–2280.

CHAPTER

THE KEVITSA NI-CU-PGE DEPOSIT IN THE CENTRAL LAPLAND GREENSTONE BELT IN FINLAND

3.6

F. Santaguida, K. Luolavirta, M. Lappalainen, J. Ylinen, T. Voipio, S. Jones

ABSTRACT

The Kevitsa Mine is a Ni-Cu-PGE magmatic-type deposit hosted within a composite ultramafic intrusion within the Lapland Greenstone Belt. Discovered in 1987 by GTK, mine production began in 2012 with an expected life of over 20 years. Mineralization is predominantly disseminated pyrrhotite, pentlandite, and chalcopyrite with an irregular shape that roughly conforms to magmatic layering within the deposit area. The host rock is an olivine-pyroxenite containing both clinopyroxene and orthopyroxene. Amphibole and serpentine-chlorite alteration is prevalent throughout the intrusion and obscures primary relationships, but overall does not impact the distribution of metals at the deposit-scale. The disseminated style, high metal tenor, and Cu endowment over Ni is unique compared other magmatic deposits globally.

Keywords: magmatic; nickel; copper; Kevitsa; Lapland; olivine pyroxenite; disseminated.

INTRODUCTION

The Kevitsa mine is Europe's most recent nickel mine. Production commenced in 2012, with annual output in the range of 17,000–19,000 t of copper, 9000–10,000 t of nickel, 12,000–13,000 oz of gold, and 22,000–24,000 oz each of platinum and palladium. Plans for expansion will accelerate mining, but the approximate life of the mine will be more than 20 years.

Compared to most other magmatic Ni-Cu-PGE deposits globally, the Kevitsa deposit is unique. The sulfide mineralization is mostly disseminated, the ore is enriched in Cu relative to Ni, and the tenor is consistently high. While previous work (Hanski et al., 1997; Mutanen, 1997; Gervilla and Kojonen, 2002) laid the foundation to the present petrogenetic model for Kevitsa, this chapter presents a description of the deposit and its host rocks based on drill core analysis, lithogeochemical interpretation, and several geophysical surveys conducted over the last decade. Understanding of ore formation and intrusion architecture has advanced most recently from pit mapping and modeling grade control data during mining. Further insight is expected from two Ph.D. dissertations, by K. Luolavirta at Oulu University in Finland and M. LeVaillant at the University of Western Australia, that are in progress at the time of writing this chapter.

Mineral Deposits of Finland. http://dx.doi.org/10.1016/B978-0-12-410438-9.00008-X

REGIONAL GEOLOGY

The Kevitsa deposit is located within the Central Lapland Greenstone Belt (CLGB), a Paleoproterozoic volcano-sedimentary sequence. Ultramafic intrusive and volcanic rocks are widespread and occur at several stratigraphic levels within the CLGB (Hanski et al., 2001; Hanski and Huhma, 2005). The ultramafic intrusion hosting the Kevitsa deposit is a composite olivine pyroxenite to gabbro complex dated at 2058 ± 4 Ma (Mutanen and Huhma, 2001). At the surface, the intrusion has an arcuate shape (Fig. 3.6.1). Drilling has established a thickness of more than 1.5 km, with an increasingly complex geometry at depth that in part reflects regional deformation effects as well as the original magmatic emplacement relationships.

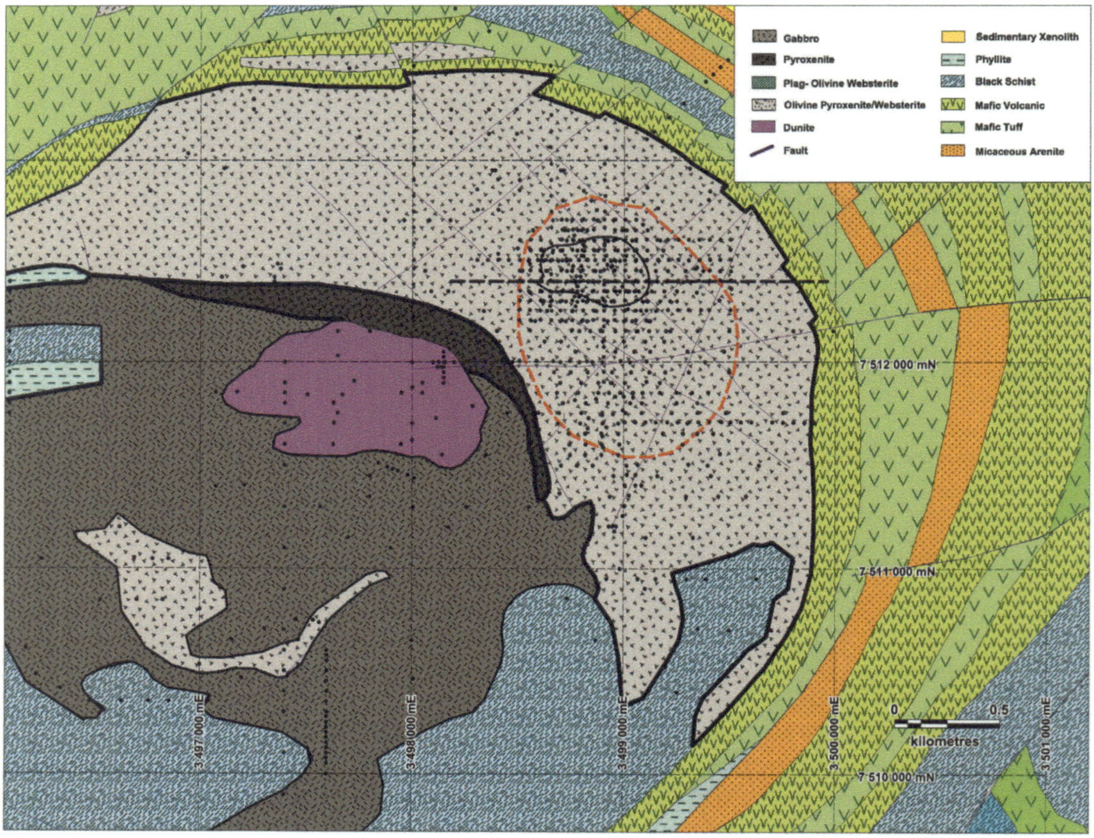

FIGURE 3.6.1 Bedrock geology of the Kevitsa intrusion and surrounding area.

This is based on surface mapping, diamond drill holes, and geophysical interpretation. The outline of the final open pit is indicated by the red line and roughly shows the location and extent of Ni-Cu mineralization. Diamond drill holes are shown by the black dots. Coordinate numbers are Finnish KKJ projection (Zone 3). The black dashed line shows the location of the east–west section shown in Fig. 3.6.7.

EXPLORATION AND DEVELOPMENT HISTORY

The Kevitsa Ni-Cu-PGE deposit was discovered in 1987 by the Geological Survey of Finland (Table 3.6.1) by drilling a glacial till geochemical anomaly following previous findings of disseminated pyrrhotite in outcropping peridotite. Early metallurgical tests returned uneconomic recoveries of Ni and Cu. Several subsequent iterations of mineral resource modeling did not improve the economic outlook of Kevitsa. Prior to the mine development decision in 2009, First Quantum Minerals Ltd. (FQML) initiated a comprehensive drilling campaign to confirm previous results and to increase the confidence level of the previous resource model, which was calculated in September 2008. In late 2009, FQML released its first NI 43-101 compliant report regarding the Kevitsa Ni-Cu-PGE project. On the basis of the 2009 mineral resources, an updated mineral reserve for Kevitsa was disclosed. The new reserve, together with improved metal prices, plus other technical factors, encouraged a board decision to start development of the Kevitsa Ni-Cu-PGE project.

Between October 2009 and August 2010, FQML completed a second major resource definition drilling campaign following a successful near-mine exploration program. The exploration program resulted in a new geological and structural model that led to drilling of geophysical and conceptual targets. The second definition drilling campaign focused on the previously unknown southern extensions of the Kevitsa deposit as well as upgrading the inferred resources from 2009 modeling to measured and indicated classes. In March 2011, FQML published its second NI 43-101 compliant technical report showing a significant increase in the mineral resource and reserves compared to the 2009 calculations (Lappalainen and White, 2010). As a positive consequence, FQML is actively looking for possibilities to expand the planned annual production. The current measured and indicated mineral resource for the Kevitsa Ni-Cu-PGE deposit is 240 Mt at 0.30% Ni, 0.41% Cu, 0.21 gpt Pt, 0.15 gpt Pd, and 0.11 gpt Au. Additionally, an inferred resource of 34.7 Mt with comparable grade is calculated.

Table 3.6.1 Summary of the exploration history of the Kevitsa Ni-Cu-PGE deposit

Years	Owner	Description
1960s	GTK	Mapping of outcrops and river boulders
1970s	Outokumpu Ltd.	Reconnaissance regional exploration
1984	GTK	Initial diamond drilling; discovery of "false ores"
1984–1987	GTK	Basal till geochemistry and ground geophysical surveys (magnetic, gravity, electromagnetic)
1987	GTK	Diamond drilling and discovery of Ni-Cu mineralization
1990	GTK	Diamond drilling (exploration)
1992–1993	GTK	Resource definition drilling and trenching program, resource modeling
1994	GTK	Airborne geophysical survey
1997	GTK	Publication of Mutanen Ph.D. thesis
1996–1998	Outokumpu Ltd.	Till geochemistry, diamond drilling, and metallurgical test work
2000–2008	Scandinavian Minerals	Acquisition; resource and exploration diamond drilling
2008	First Quantum Minerals Ltd.	Acquisition
2008–present	First Quantum Minerals Ltd. (Kevitsa Mining Oy)	Resource and exploration diamond drilling; 3D seismic and magnetotelluric geophysical surveys

MINE GEOLOGY

MINERALIZATION

Due to the disseminated style of the sulfide mineralization, the ore body has an irregular shape. For reserve and resource estimation, several distinct ore domains were defined to coincide with general lithological layering in the deposit area (Fig. 3.6.2). Mineralization is concentrated in the center of the intrusion and not along the basal contact, unlike many other Ni-Cu magmatic ore bodies.

Pentlandite and chalcopyrite are the dominant ore minerals occurring together with pyrrhotite and magnetite (Fig. 3.6.3). Magmatic textures are commonly preserved, with sulfides and magnetite occurring between cumulate olivine and pyroxene grains. In places, the sulfide mineralization is net-textured, but such zones cannot be defined beyond one or two diamond drill holes. In the net-textured zones, pyrrhotite is relatively abundant, thus tenor tends to be relatively low. Mutanen (1997) identified four ore types based largely on Ni tenor: *regular ore*, *transitional ore*, *Ni-PGE ore*, and *false ore*. Subsequent drilling and ore resource modeling demonstrated that transitional ore was not truly a distinct type, but represents a lower grade of Ni-PGE ore (Lappalainen and White, 2010).

In the ore reserves and resources, regular ore has an Ni tenor of 4–7% and combined PGE contents of less than 1 g/t. The regular ore constitutes approximately 95% of the resource, although mining has exposed small pods of Ni-PGE ore not known from the resource drilling.

Ni-PGE ore is distinguished by high Ni tenor of greater than 10 and the presence of millerite and heazlewoodite as additional Ni mineral phases. Pyrite is also present and appears contemporaneous with millerite. The host rocks to the Ni-PGE ore are similar to regular ore, but generally contain more clinopyroxene than orthopyroxene. The Ni content within olivine is unusually high, up to 14,000 ppm (Hanski et al., 1997; Määttä, 2012; Yang et al., 2013). The Ni-PGE ore zones are discordant to the regular ore (see Fig. 3.6.3); thus, they are generally considered to have formed separately, but their origin remains debated (Mutanen, 1997; Hanski et al., 1997; Yang et al., 2013).

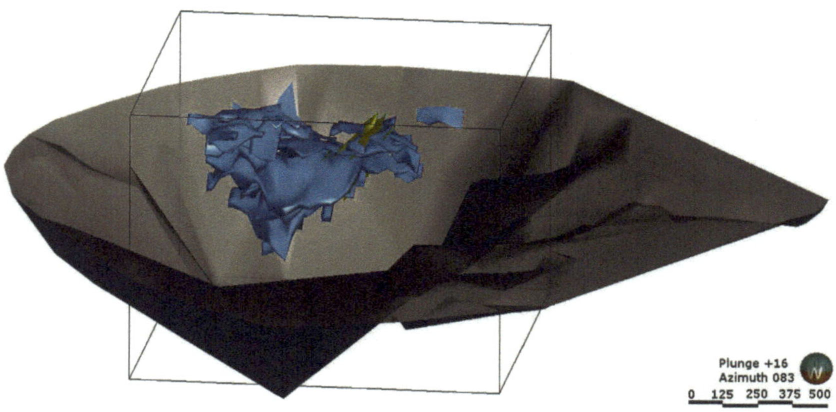

FIGURE 3.6.2 3D model of the Ni-Cu-Pt-Pd ore domains at Kevitsa.

Regular ore type is in blue and Ni-PGE ore type is in yellow. The basal intrusive contact with the country rocks is shown in gray. The view is looking eastward from above. Ore domains in the southern portion of the deposit appear to abruptly terminate, but this is due to a lack of drilling in this area.

Source: *Ore domain solids as published in the NI43-101 report (2011).*

The variability between Ni, Cu, Co, Pt, Pd, and Au within the regular ore is relatively low, with copper contents generally twice as high as Ni (Fig. 3.6.4). Cobalt typically correlates well with Ni and is hosted within pentlandite and pyrrhotite, rather than as a discrete sulfide mineral phase. Elevated Pt-Pd-Au occurs with both Cu-rich and Ni-rich mineralization, but the highest grades occur in Ni-PGE ore. In general, Pt:Pd ratios throughout the deposit are quite consistent at around 4:3, but approach 1:1 in Ni-rich domains and specifically within the Ni-PGE ore. High Au values, greater than 0.2 g/t, occur locally and coincide with Cu-rich ores.

Aside from the different types of ore, Ni-Cu variability can be high at the local scale (20 m × 20 m), but general zoning is also apparent at the deposit-scale (Fig. 3.6.5). Both Ni-rich and Cu-rich types of mineralization occur as discrete zones. Nickel-rich mineralization is particularly prominent at depth and in the southern portion of the ore body. Copper-rich zones typically occur in the central part of the ore body.

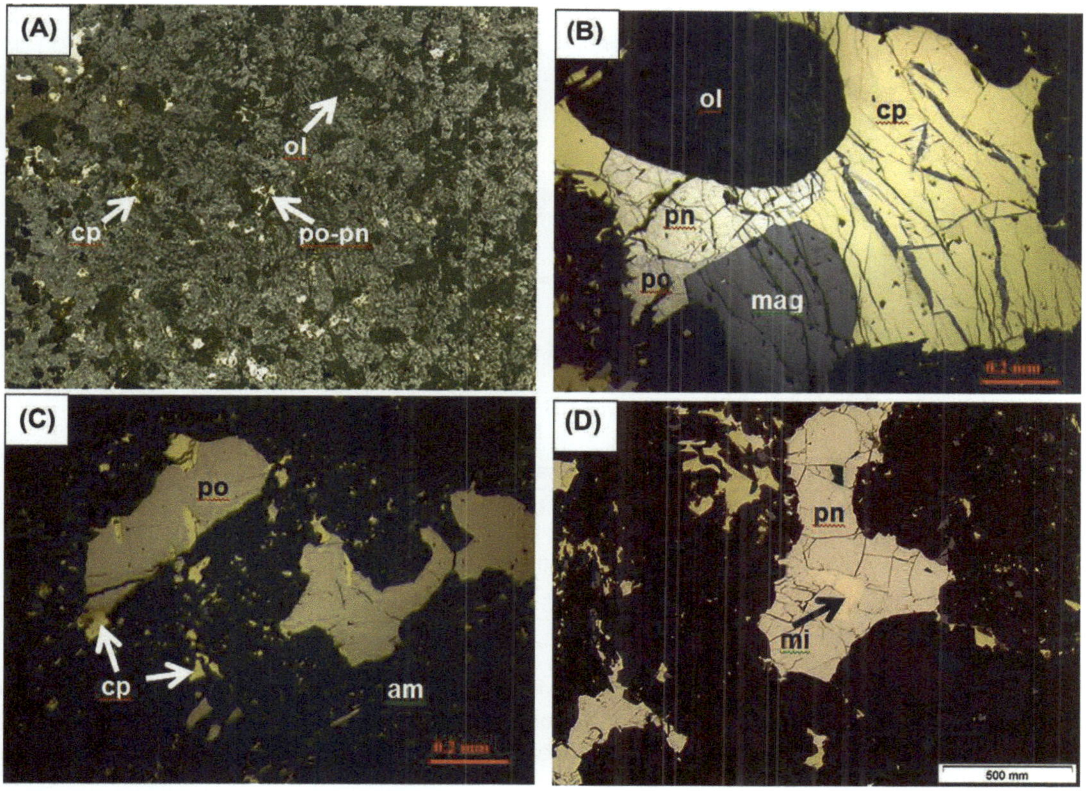

FIGURE 3.6.3 Typical textures of Kevitsa Ni-Cu-PGE mineralization.

(A) Polished hand specimen of disseminated mineralization (width of photo is 3 cm). (B) Interstitial texture of ore minerals in relatively fresh olivine pyroxenite. (C) Ore minerals in intensely amphibole-altered rock. Fine sulfide grains are hosted within tremolite. (D) Millerite within core of pentlandite grain in Ni-PGE ore. Abbreviations: po = pyrrhotite, cp = chalcopyrite, pn = pentlandite, mag = magnetite, mi = millerite, ol = olivine, am = amphibole.

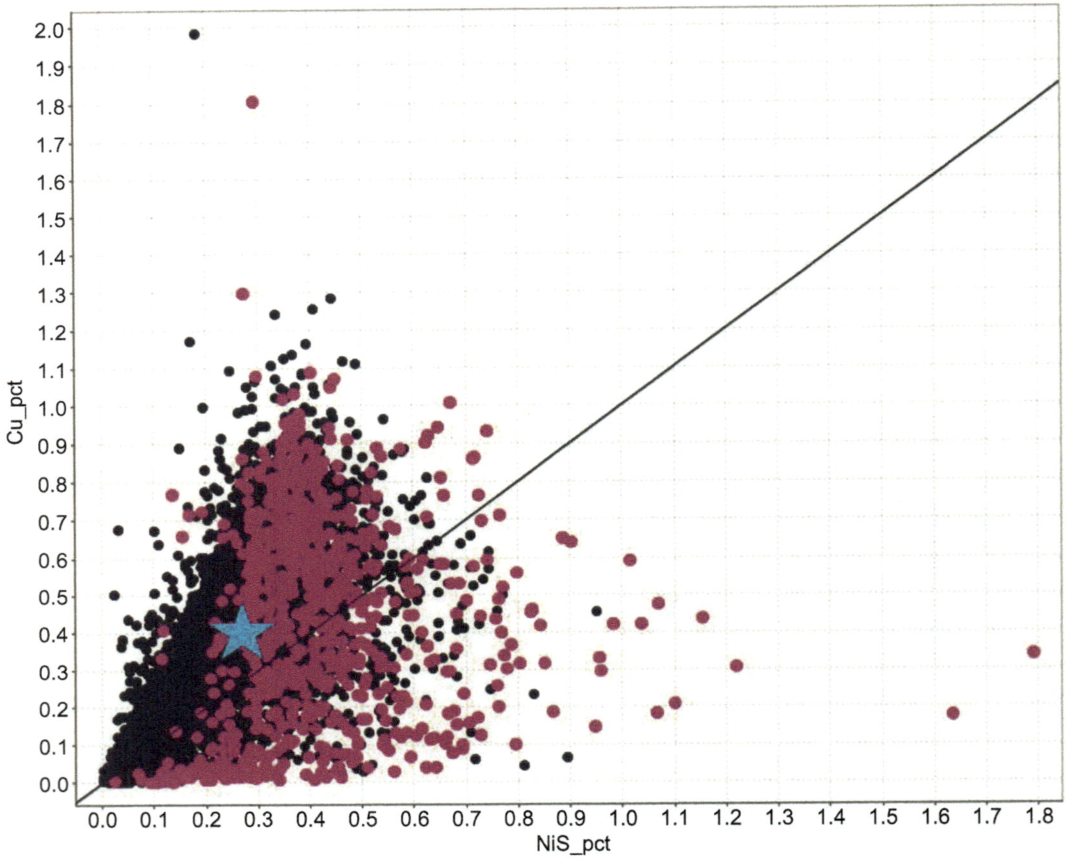

FIGURE 3.6.4 Grade distribution of metals shown by assay data contained in the resource model.

For this, more than 15,400 metals were analyzed. The blue star indicates the mean for this dataset. The 1:1 line cuts Cu-rich from Ni-rich ore. Nickel represents Ni in sulfide from ammonium citrate digestion that does not dissolve silicate minerals (see Lappalainen and White, 2010, for a description of the method and data quality assurance and control). Samples with Pt + Pd + Au > 600 ppm are highlighted by cyan color representing "high grade."

Small pods of Ni-Cu-PGE disseminated mineralization also occur outside of the Kevitsa deposit, but within the intrusion. These have not yet been fully defined or well understood, but their existence suggests that there is potential for economic mineralization beyond the presently identified deposit.

As in the case of many other magmatic Ni-Cu-PGE deposits, the number of platinum-group minerals (PGM) is extensive. They range in size from 100 to a few microns (Gervilla and Kojonen, 2002). The dominant minerals are Pt-Pd bismuth tellurides and sperrylite ($PtAs_2$). Braggite (Pt-PdS) is also common, but typically small in size and not as volumetrically significant. The proportions of the PGMs and their grain size vary at the meter scale within single drill holes, but mineral zoning is

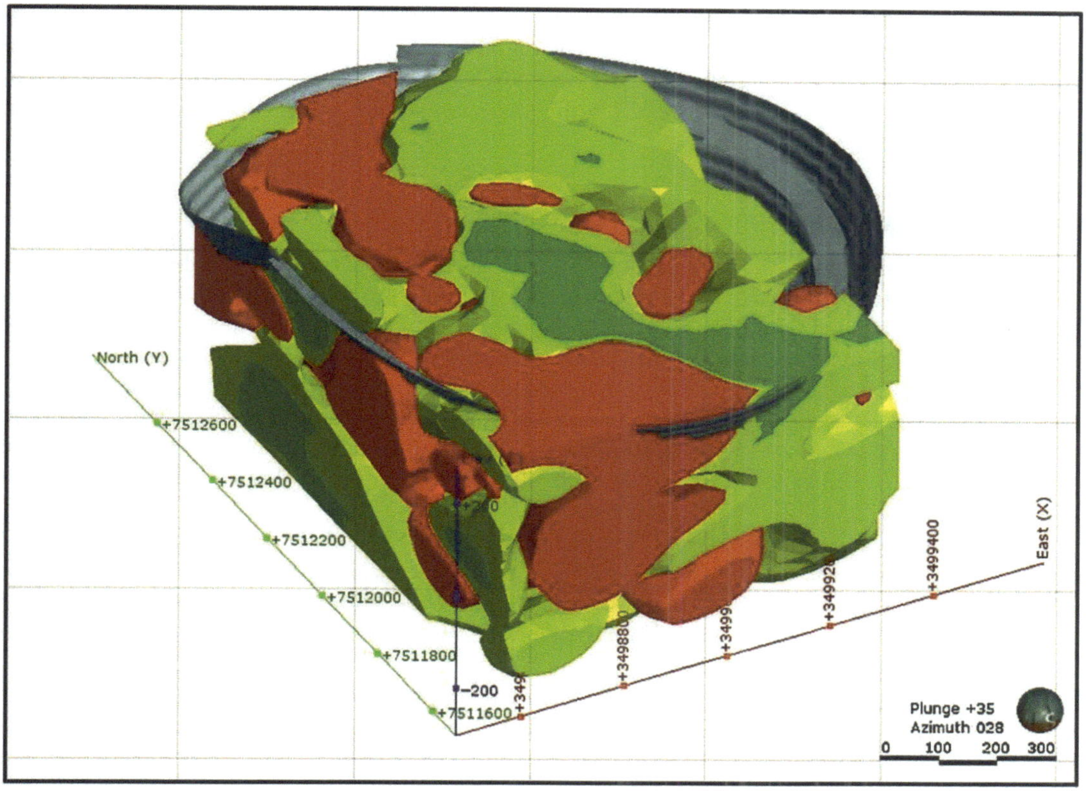

FIGURE 3.6.5 Ni and Cu variability in the deposit area.

View toward northwest. Modeling is done using the diamond drill hole assay database. Ni:Cu ratios represented by solid colors: red is >1.5; light green is 1.5–0.5 and dark green is <0.5. Gaps in the model indicate waste. Large, broad solids at depth and in the western part of the deposit are considered interpretive (inferred) due to the wide drill hole spacing lying outside of the resource area. The planned final open pit is shown in gray. The elongated Ni-rich and Cu-rich zones in the central part of the deposit reflect, in part, sulfides concentrated within veins.

not apparent at the deposit scale. The size distribution of the PGM is currently not well constrained. Silver and Au tellurides are both present, and rare native gold also occurs.

False ore consists entirely of pyrrhotite, with only rare chalcopyrite and pentlandite; it is defined by relatively low Ni tenor of 2–3%. False ore is easily recognized as it forms good electromagnetic conductors due to the high interconnection of the disseminated sulfides, thus it is well mapped by geophysical methods. South of the deposit, but still within the Kevitsa intrusion, a horizon of false ore contains net textured sulfides with >10 pyrrhotite over a width of several tens of meters and >200 m strike length. To the west of the deposit area, intersections of Ni-poor semimassive sulfides have been drilled at depth (>1000 m), immediately above the base of the intrusion. These are called *contact ore*. Despite the low tenor, both pentlandite and chalcopyrite are present.

Sulfides also occur within veins crosscutting the ore body. These have previously been considered unimportant, but mining has exposed several quartz-carbonate veins with a width of more than 1 m that have massive sulfide selvages enriched in Cu and locally Ni. The Ni-Cu-PGE content in the veins does not greatly impact the resource at Kevitsa, but the presence of vein sulfides does indicate that metal mobilization has occurred.

ROCK TYPES AND STRATIGRAPHY

The Kevitsa intrusion consists of an ultramafic lower part (approximately 1 km thick) overlain by gabbroic rocks. A large lherzolite body occurs in the central part of the intrusion (see Fig. 3.6.1), but is not spatially associated with the ore deposit. Compositional variations within the lower ultramafic portion are minor, but discrete lithological units can nevertheless be mapped. Layering is locally developed, particularly within the deposit, but in general, the contacts between rock types are diffuse. Alteration of pyroxene and olivine is intense in places, making primary rock types difficult to recognize and further complicating stratigraphic correlation. Images of the main rock types are shown in Fig. 3.6.6. The distribution and stratigraphic sequence of the rock types described here are shown in Fig. 3.6.7.

Olivine websterite is the dominant rock type and host rock for the sulfide mineralization, defined locally as containing more than 5% orthopyroxene. Olivine occurs as discrete grains or clusters a few millimeters in size. The rock has a poikilitic (heteradcumulate) texture, with orthopyroxene-forming oikocrysts. Typical accessory minerals include plagioclase, magnetite, sulfides, and apatite. Hornblende and phlogopite also occur locally.

Olivine pyroxenite resembles the olivine websterite in terms of texture, but is devoid of orthopyroxene (<5%). Because pyroxene is susceptible to overprinting by amphibole, it is difficult to distinguish these two rock types; however, the olivine clinopyroxenites can be clearly identified in thin section and, in general, are more prevalent outside of the ore body.

Plagioclase-bearing (olivine) websterite occurs as discontinuous zones within the olivine websterite/clinopyroxenite. The plagioclase-bearing (ol) websterites show orthocumulate textures and contain visible plagioclase (>10%) as an intercumulus phase. Orthopyroxene oikocrysts are also more abundant (15–25%) than in typical olivine websterite. Olivine is absent or rare (<15%). Contacts with the olivine websterite/clinopyroxenite are mainly diffuse, but can be locally quite sharp. In places, the plagioclase-bearing (olivine) websterite forms marker horizons characterized by magmatic layering, but in most cases, the layers are discontinuous and cannot be traced beyond a few hundred meters (see Fig. 3.6.7). Overall, these rocks are weakly mineralized and not found outside of the mineralized area.

Pyroxenite, with <5% olivine, forms the uppermost ultramafic cumulate unit below the gabbroic rocks, outside the mineralized area. The transition between olivine websterite and pyroxenite is highly gradational.

The marginal rocks of the intrusion are composed of pyroxenite (± minor olivine) and gabbro. Mutanen (1997) considered these rocks to be "microgabbros." The contact between olivine websterite and the marginal rocks is gradational. In places, distinct layering or banding between pyroxene-rich and plagioclase-rich rocks is seen, but most commonly the marginal rocks are varitextured. The marginal rocks vary in thickness from a few meters to more than 50 m. In places, the marginal rocks are absent and faulting is inferred. Where the marginal rocks are sulfide mineralized, they form so-called contact ore, dominated by pyrrhotite, and thus are uneconomic. Fragments of country rock are

FIGURE 3.6.6 Kevitsa rock types.

(A) Unaltered olivine websterite. Orthopyroxene (opx) stands out as brownish oikocrysts. Plagioclase (pla) is interstitial to olivine (ol) and clinopyroxene (cpx). Opaque minerals are magnetite and sulfides. Scanned thin section image, width of the photo ~1.5 cm. (B) Ore-bearing (opaque intercumulus sulfides + magnetite) olivine clinopyroxenite, with clinopyroxene altered to amphibole; olivine (ol) is preserved. Scanned thin section image, width of the photo ~1.5 cm. (C) Plagioclase-bearing, olivine websterite. Clinopyroxene replaced by amphibole and magnetite. Scanned thick section image, width of the photo ~1.5 cm. (D) Varitextured pyroxenite of the marginal zone, weakly altered by amphibole, width of photo is 5 cm. (E) Lherzolite-occurring as an angular fragment in olivine websterite (F) Mixed olivine websterite (light coloured) and lherzolite (dark coloured). (G) Magmatic breccia of lherzolite fragments in olivine websterite (autolith?). (H) Gabbro containing intercumulus orthopyroxene (op) and plagioclase (pla). Clinopyroxene (cpx) occur as phenocrysts. (width of photo is 2 cm). (I) Banded metasedimentary xenolith (width of photo is 5 cm).

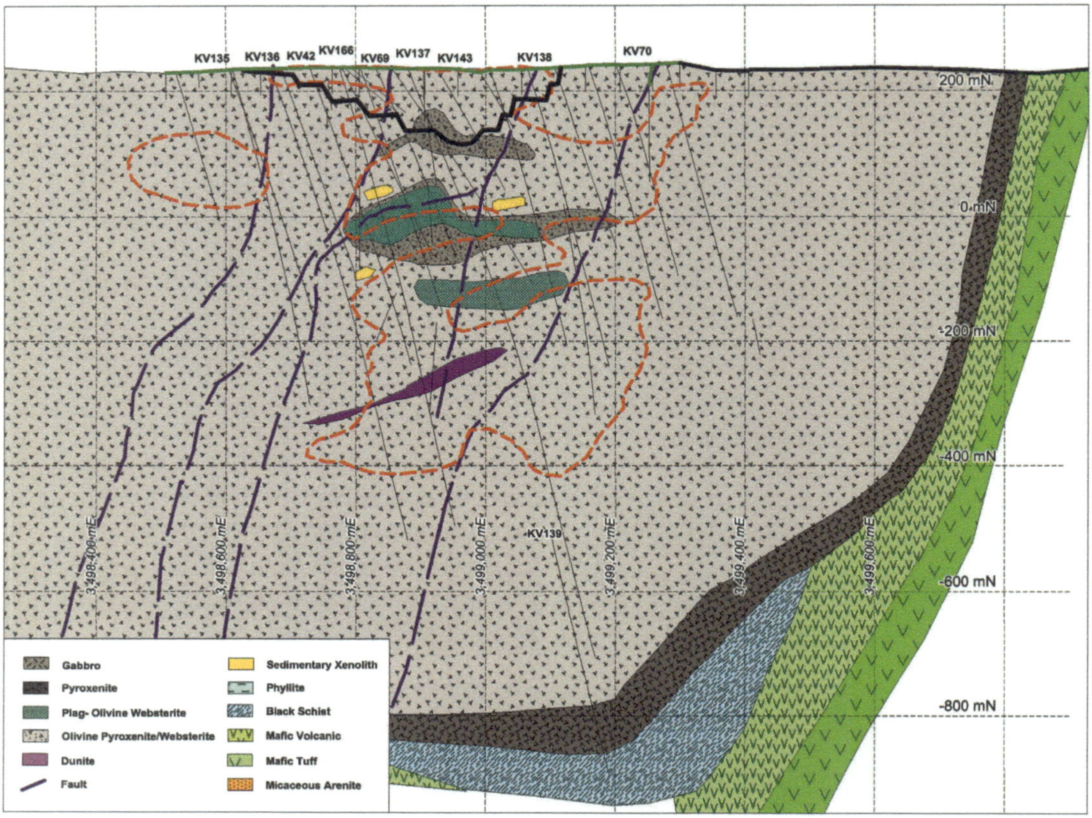

FIGURE 3.6.7 East–west section through the Kevitsa intrusion defined by drilling and geophysical data interpretation.

Deep diamond drill holes are labeled (KV series). The thick black line indicates the Stage 1 pit outline. The shape of the Ni-Cu-PGE mineralization is shown by the red dashed line and in part reflects the lateral variation of pyroxene-rich to olivine-rich layering within the immediate host rocks. This layering is not continuous outside of the mineralized zone.

also common within the marginal sequence. The immediate country rocks to the intrusion consist of mafic volcanic flows and epiclastic rocks, as well as micaceous phyllites and carbonaceous schists. In many places, the contact is sharp and intact, although faulting is prevalent at the southern margin of the intrusion.

The intrusion contains a number of distinct olivine-rich bodies and lenses that contain >50% olivine. They are of lherzolitic to wehrlitic composition, but have been collectively termed *dunite* in Kevitsa mine terminology. The rocks are intensely serpentinized, particularly near the surface. A large lherzolite body occurs in the central portion of the intrusion, but shows no spatial relationship with the mineralization (refer to Fig. 3.6.1). Lherzolite has also been intersected by drilling below the deposit; whether this lherzolite is related to the previously mentioned lherzolite body is currently unknown. Lherzolite clasts occur throughout the mineralized zone, but their origin remains contentious (Mutanen, 1997; Yang et al., 2013). The clasts are highly variable in size, ranging from centimeters to traceable zones roughly tens of meters in thickness (see Fig. 3.6.7). Cumulate texture of olivine is locally preserved, although most clasts are foliated along serpentinized planes. The clasts may occur as discrete, rounded fragments, or, in places, lherzolite is intermingled with olivine websterite. Pyrrhotite is common within lherzolite, whereas pentlandite and chalcopyrite only occur locally. In general, the lherzolite clasts host the same sulfide assemblage as their surrounding olivine websterite/pyroxenite.

Gabbroic rocks occur on top of the ultramafic cumulates. They are particularly prominent in the southwestern portion of the intrusion (see Fig. 3.6.1). Plagioclase is the dominant mineral along with clinopyroxene and accessory olivine. Modally, the rocks are gabbros, olivine gabbros, and gabbronorites. Apatite, magnetite, and ilmenite are common accessory phases. Magnetite-rich hornblende gabbro is prominent along the southern portion of the intrusion. Drilling has shown that the gabbroic rocks form a relatively thin unit (<500 m) overlying the thick ultramafic portion of the intrusion. Copper-Au mineralization occurs at two prospects along the gabbroic margins of the intrusion, but overall, sulfide minerals are rare and consist mostly of pyrite.

Xenoliths of hornfelsed pelitic sediments and mafic volcanics are common throughout the intrusion, but are particularly concentrated within the deposit area where they are spatially associated with lherzolite clasts. Xenoliths are concentrated in discrete zones that measure several meters in thickness and extend for several hundreds of meters in a north–south direction. Most xenoliths are pervasively altered to phlogopite.

Numerous dikes crosscut the intrusion and the mineralization. Most are olivine gabbroic in composition. The coarse-grained dikes rarely contain sulfide minerals, although veins containing pyrrhotite-chalcopyrite ± pentlandite may form locally along the margins. Fine-grained gabbroic dikes cut the mineralized ultramafic rocks and often contain pyrrhotite-chalcopyrite ± pentlandite. These are altered to a chlorite-actinolite-magnetite assemblage. Felsic dikes consisting of feldspar, quartz, and minor amounts of mafic minerals also occur. Dikes are rarely traceable beyond a single drill hole, thus their orientations are not established.

Granophyre occurring along the southern margin of the intrusion has been described by Mutanen (1997). Despite extensive drilling, these rocks have not been encountered in the present exploration and mining operation. Instead, several tens of meters of albitized gabbroic rocks and dikes have been intersected in some drill holes along the southern margin of the intrusion. These rocks do not host mineralization, thus they have not been intensely studied nor are they shown on the most current geological maps (see Fig. 3.6.1).

Lithogeochemistry has been useful for discriminating between rock types in the Kevitsa intrusion (Fig. 3.6.8). Multielement ICP analyses have been done on selected drill holes to improve correlation within the lithological units. Specifically, Al/Cr ratios have been found to be reliable proxies to the presence of plagioclase even where alteration is prominent, allowing plagioclase-bearing (ol) websterites to be more confidently recognized. Olivine-rich rocks have an inherently low Al/Cr ratio compared to pyroxenite. Notably, high Cr typically reflects the presence of clinopyroxene, although Cr may locally be hosted by magnetite. High Al/Cr is characteristic of fine-grained gabbroic dikes, and these are easily recognized during logging. Magnesium correlates well with Cr and is low within the plagioclase-bearing rocks, corresponding to reduced olivine content rather than alteration.

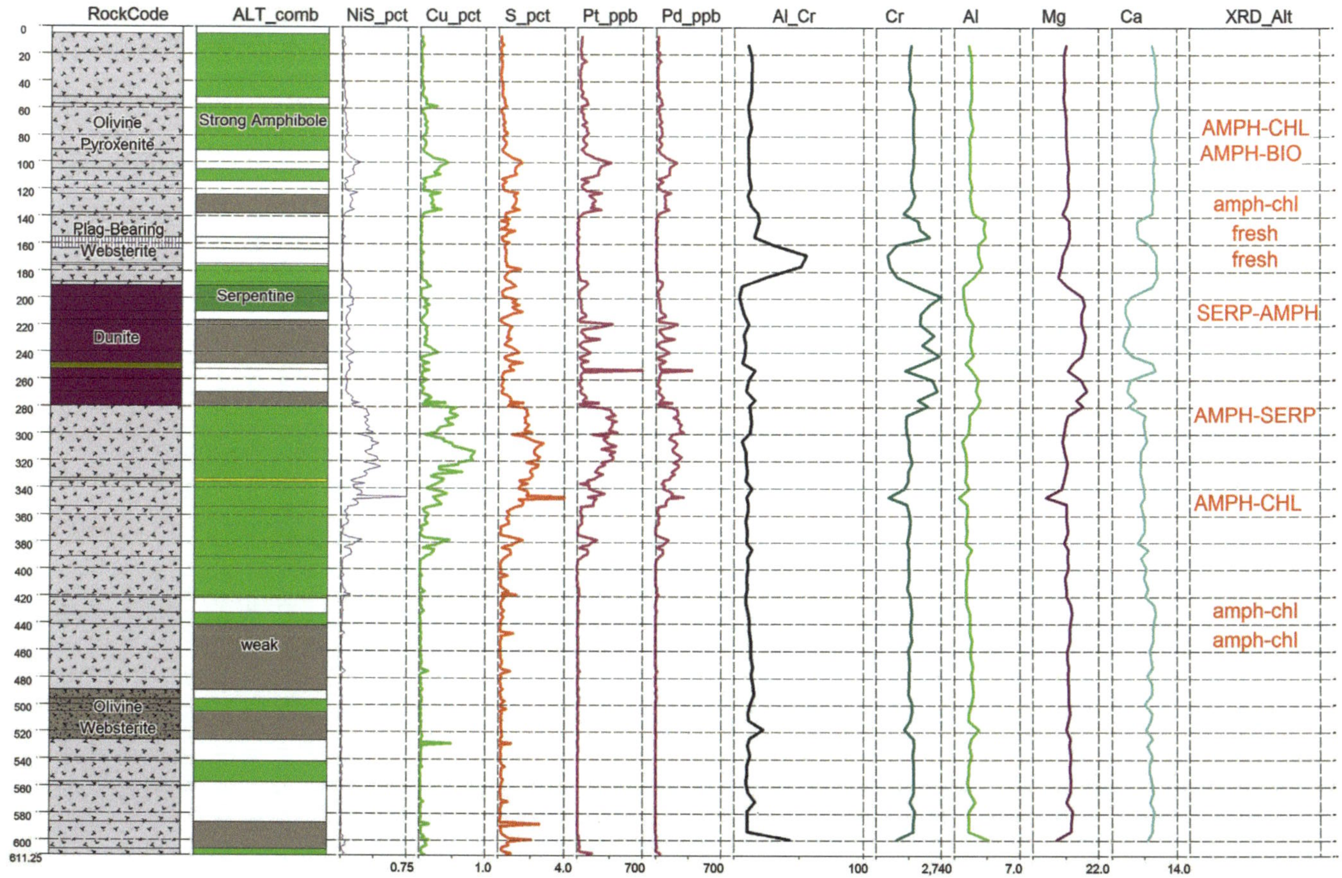

FIGURE 3.6.8 Representative diamond drill hole (KV157) through the Kevitsa deposit showing relationships of lithology, alteration mineral assemblages, and lithogeochemical data.

The rock code legend is the same as in Fig. 3.6.1. Alteration column abbreviations: AMPH = amphibole-chlorite-serpentine assemblage, CARB = carbonate mineral assemblage. XRD_Alt column abbreviations: AMPH = amphibole, SERP = serpentine, CHL = chlorite, BIO = biotite/phlogopite. Capital letters denote abundances >30%. Lowercase letters denote abundances >15%. Rock is considered unaltered where alteration minerals <15%.

HYDROTHERMAL ALTERATION

Hydrothermal alteration is notable throughout the Kevitsa deposit and its host intrusion (Fig. 3.6.9). Many of the rocks hosting mineralization have been overprinted by amphibole. Clinopyroxene is most susceptible to replacement, but, in places, olivine and orthopyroxene have also been completely altered to amphibole, chlorite, and serpentine. The distribution of the alteration is difficult to assess, but the most intense zones appear to be directly associated with relatively late mafic dikes and veins. A systematic study of the amphibole minerals has not been done, but microprobe analyses have typically returned tremolite-actinolite compositions. In general, the amphibole overprint does not affect Ni-Cu-PGE grades, but individual platinum-group minerals have been found enclosed in tremolite, and textures of sulfides reflect remobilization at a microscale. X-ray diffraction analyses have been done in places to

FIGURE 3.6.9 Alteration textures and minerals.

(A) Thin section scan of amphibole-altered olivine pyroxenite. Light to dark gray areas consist entirely of amphibole. Dark brown mineral is hornblende. Olivine is locally preserved. (B) Magnetite veins crosscutting amphibole and sulfide minerals. (C) Veins with massive sulfide (core from separate drill holes). (D) Falso colour Scanning Electron Microprobe image of vein mineralization.

further characterize the amphibole alteration (see Fig. 3.6.8). The Al/Cr ratios are not affected by intense amphibole alteration and confirm previous assessments that the hydrothermal overprint is generally isochemical.

Magnetite also occurs as a secondary mineral, clearly seen as thin, millimeter-sized veins throughout the deposit area. An association to the amphibole alteration is evident in places where anomalous down-hole magnetic susceptibility readings have been logged within intense alteration zones, but this relationship is not consistent and requires more investigation. Chlorite and serpentine replace amphibole and olivine, respectively. This is largely attributed to regional metamorphism. Neither mineral is ubiquitous, but is locally concentrated where gabbroic dikes are present, suggesting an association with permeable structures.

Veining occurs throughout the deposit. Most commonly, veins consist of dolomite, actinolite, biotite, serpentine, talc, and, locally, quartz. Pyrrhotite is ubiquitous within veins, but chalcopyrite and pentlandite can also occur locally. The veins range in thickness from less than 1 cm to more than 1 m. In the open pit, individual veins have been traced for more than 100 m. Some sulfide-bearing veins are enriched in Ni, Cu, Pt, and Pd, suggesting a magmatic origin, but more detailed studies are required to investigate this.

STRUCTURE

Overall, faulting does not offset the ore body over significant distances (Standing et al., 2009). Most faults and fractures are steeply dipping and displacement is typically <10 m; however due to the lack of many stratigraphic marker horizons, exact displacement is not well constrained. Fractures have been developed close to the surface (<50 m below the bedrock interface), enhancing weathering and causing minor oxidation of the sulfide minerals. Regional northeast–southwest trending structures that transect the intrusion and offset the country rocks are indicated in magnetic surveys (see Fig. 3.6.1).

Steeply dipping, north–south trending faults are exposed in the open pit. Veins containing massive pyrrhotite, chalcopyrite, and, locally, pentlandite are concentrated along one single fault zone in the central part of the mine, resulting in increased Ni-Cu grades (refer to Fig. 3.6.5). Based on the orientation of the ore body, these north–south trending structures appear to dip westward, but do not demonstrate significant offset. Other east–west trending structures are also exposed in the open pit where they are defined by calcite veining. Because they do not contain sulfides, these structures may be part of the fault set developed during later regional deformation.

COMPARISON TO OTHER MAGMATIC NI-CU-PGE DEPOSITS

Most magmatic sulfide deposits in northern Finland are considered to be “reef-type” (Eckstrand and Hulbert, 2007) or “PGE-type” deposits (Naldrett et al., 2011). Examples are Ahmavaara and Suhanko in the Portimo complex (Lahtinen et al., 1989; Iljina et al., 2015) and Haukiaho in the Koillismaa complex (Iljina et al., 2015). The Ni-Cu sulfide deposits within the 1.89–1.87 Ga Svecofennian mafic–ultramafic intrusions in southern and central Finland show certain petrogenetic commonalities to Kevitsa, but their tectonic setting is very different, with Kevitsa being emplaced during rifting in an intracontinental setting, whereas the Svecofennian deposits have been ascribed to arc magmatism (Peltonen, 2005).

Among Ni-Cu sulfide deposits globally, Kevitsa has relatively high Cu/Ni ratios above unity. Few other major deposits are Cu-enriched, however, some of the most notable are the deposits in the Noril'sk district Russia (Naldrett et al., 2011) and Okiep South Africa (Maier et al., 2013). Many magmatic systems contain Cu-rich portions and several of these have been shown to be influenced by Cl-rich hydromagmatic fluids, for example, the Sudbury Canada footwall deposits (Farrow and Watkinson, 1997). While Cl-bearing minerals such as hornblende, biotite, and apatite are present at Kevitsa, their paucity suggests that the Cu enrichment is of magmatic origin.

Burrows and Lesher (2012) provide a review of Cu-rich magmatic Ni-Cu-PGE deposits highlighting a number of processes as part of intrusion emplacement to generate this style of mineralization. Distinct Ni and Cu rich portions in magmatic ore deposits can be generated by fractionation of sulfide melt during magma emplacement (Naldrett, 2004; Naldrett et al., 2011). Conspicuously, the Sakatti deposit near Kevitsa also contains Cu-enriched mineralization in addition to Ni-rich ores considered to be generated by this process (Coppard, 2011; Brownscombe et al., 2015). This model could suggest some potential for discovery of Ni-rich sulfides in the Kevitsa intrusion. A better understanding of the magmatic architecture and the emplacement history of the Kevitsa intrusion would clearly be beneficial to define further exploration targets.

ACKNOWLEDGMENTS

Steve Beresford and Petri Peltonen have both greatly encouraged the writing of this chapter and influenced the ideas on the stratigraphy and petrogenesis of the Kevitsa deposit. Mike Christie continually supported geological work at Kevitsa that led to the expansion of the ore resources. Petrologic work led by Hugh O'Brien at GTK, especially on the platinum-group minerals, has been crucial in determining their representative distribution. Some of his micro-photographs have been used for this chapter. Dr. T. Mutanen is thanked for conducting much of the early work on Kevitsa, some of which remains unpublished. Supervision of graduate students working at Kevitsa by Wolf Maier, Eero Hanski, Marco Fiorentini, and Stephen Barnes has led to many fruitful discussions concerning rock types and ore genesis. Wolf Maier also greatly improved the writing and clarity of this chapter.

REFERENCES

Brownscombe, W., Ihlenfeld, C., Hartshorne, C., Coppard, J., Klatt, S., siikaluoma, J., herrington, R.J., 2015. The Sakatti Cu-Ni-PGE sulphide deposit, northern Finland. In: Maier, W.D., O'Brien, H., Lahtinen, R. (Eds.), Mineral Deposits of Finland. Elsevier Amsterdam. pp. 211–251.

Burrows, D.R., Lesher, C.M., 2012. Copper-rich magmatic Ni-Cu-PGE deposits. Society of Economic Geologists, Inc. Special Publication 16, 515–552.

Coppard, J., 2011. Sakatti Discovery, Northern Finland. Fennoscandian Exploration and Mining Conference. Levi, Finland.

Eckstrand, O.R., Hulbert, L.J., 2007. Magmatic nickel-copper-platinum group element deposits. In: Goodfellow, W.D. (Ed.), Mineral Deposits of Canada: A Synthesis of Major Deposit Types, District Metallogeny, the Evolution of Geological Provinces, and Exploration Methods: Geological Association of Canada, Mineral Deposits Division, 5. Special Publication, pp. 205–222.

Farrow, C.E.G., Watkinson, D.H., 1997. Diversity of precious metal mineralization in footwall Cu-Ni-PGE deposits. Canadian Mineralogist, 35,Implications for hydrothermal models of formation, Sudbury, Ontario. 817–839.

Gervilla, F., Kojonen, K., 2002. The platinum-group minerals in the upper section of the Keivitsansarvi Ni-Cu-PGE deposit, northern Finland. Canadian Mineralogist 40, 377–394.

Hanski, E., Huhma, H., Rastas, P., Kamenetsky, V., 2001. The Paleoproterozoic komatiite-picrite association of Finnish Lapland. Journal of Petrology 42, 855–876.

Hanski, E., Huhma, H., Suominen, I.M., Walker, R.J., 1997. Geochemical and isotopic (Os, Nd) study of the Keivitsa intrusion and its Cu-Ni deposit, northern Finland. Turku, Finland, August 11–13. In: Papunen, H. (Ed.), Mineral Deposits: Research and Exploration—Where Do They Meet? Proceedings of the 4th Biennial SGA Meeting. A.A. Balkema, Rotterdam, pp. 435–438.

Hanski, E., Huhma, H., 2005. Central Lapland Greenstone Belt. In: Lehtinen, M., Nurmi, P.A., Ramo, O.T. (Eds.), Precambrian Geology of Finland—Key to the Evolution of the Fennoscandian Shield. Elsevier, Amsterdam, pp. 139–194.

Iljina, M., Maier, W.D., Karinen, T., 2015. PGE-(Cu-Ni) deposits of the Tornio-Näränkavaara belt of intrusions (Portimo, Penikat and Koillismaa). Mineral Deposits of Finland. Elsevier, Amsterdam.

Lahtinen, J.J., Alapieti, T.T., Halkoaho, T.A.A., et al., 1989. PGE mineralization in the Tornio–Narankavaara layered intrusion belt. Geological Survey of Finland. Guide 29, 43–58.

Lappalainen, M., White, G., 2010. 43-101 Technical Report on Mineral Resources of the Kevitsa Ni-Cu-PGE Deposit, Finland, p. 307

Maier, W.D., Andreoli, M.A.G., Groves, D.I., Barnes, S.-J., 2013. Petrogenesis of Cu-Ni sulfide ores from O'Okiep, Kliprand, Namaqualand, South Africa: constraints from chalcophile metal contents. S Afr J Geol 115, 499–514.

Määttä, S. 2012. Origin of Ni-rich sulfide ore in the Kevitsa intrusion, northern Finland. Unpublished M.Sc. Thesis, University of Oulu, Finland, p. 97.

Mutanen, T., 1997. The geology and petrology of the Akanvaara and Koitelainen layered mafic intrusions and the Keivitsa-Satovaara layered complex, northern Finland. Geological Survey of Finland Bulletin 395, p. 233.

Mutanen, T., Huhma, H., 2001. U-Pb geochronology of the Koitelainen, Akanvaara and Keivitsa layered intrusions and related rocks. In: Vaasjoki, M. (Ed.), Geological Survey of Finland Vol. 33. Special Paper, pp. 229–246.

Naldrett, A.J., 2004. Magmatic sulfide deposits: geology, geochemistry and exploration. Springer-Verlag. 728.

Naldrett, A.J., 2011. Fundamentals of magmatic sulfide deposits. Reviews in Economic Geology 17, 1–50.

Standing, J., De Luca, K., Outhwaite, M., et al., 2009. Report and recommendations from the Kevitsa campaign Finland. Unpublished confidential report for First Quantum Minerals Ltd. 118 pages.

Yang, S.H., Maier, W., Hanski, E., et al., 2013. Origin of ultra-nickeliferous olivine in the Kevitsa Ni–Cu–PGE-mineralized intrusion, northern Finland. Contributions to Mineralogy and Petrology 166 (1), 81–95.

CHAPTER

THE SAKATTI CU-NI-PGE SULFIDE DEPOSIT IN NORTHERN FINLAND

3.7

W. Brownscombe, C. Ihlenfeld, J. Coppard, C. Hartshorne, S. Klatt, J.K. Siikaluoma, R.J. Herrington

ABSTRACT

The Sakatti Cu-Ni-PGE sulfide deposit was discovered by Anglo American in northern Finland in 2009. It consists of both disseminated and massive sulfides, hosted by a large olivine cumulate body, surrounded by volcanic rocks that make up the footwall and part of the hanging wall. A polymict breccia constitutes the majority of the hanging wall. Two smaller satellite olivine cumulate bodies also host mineralization. The sulfide mineralization is generally Cu-dominated, especially in the shallower part of the deposit, with a transition to more Ni-rich and Ni-dominated mineralization in its deeper part. Disseminated mineralization is only found within the olivine cumulate bodies, whereas massive sulfides extend up to 150 m into the footwall, where they rapidly evolve toward very low Ni/Cu ratios.

Keywords: Sakatti; Ni-Cu-PGE; magmatic sulfide; ultramafic; nickel; olivine cumulate.

INTRODUCTION

Sakatti is one of only a handful of significant Ni-Cu sulfide deposits discovered globally in the past decade, and is one of the most significant discoveries in Finland for more than a generation. The deposit was discovered during a regional Ni sulfide exploration program initiated by Anglo American in Fennoscandia in 2002 under the leadership of Jim Coppard.

In this chapter, the main features of the Sakatti deposit are presented and the current level of understanding is briefly explored. The information presented extends up to the time of writing but inevitably, with a relatively recent discovery, there is significant uncertainty and also a large amount of evidence and information still to be discovered. The deposit is located within the Central Lapland Greenstone Belt (CLGB), which extends across Lapland from northern Norway to the Finnish–Russian border (Fig. 3.7.1). The CLGB contains numerous ultramafic intrusive and extrusive bodies and, following the Sakatti discovery, can be viewed as being highly prospective for Ni-Cu sulfide deposits.

DISCOVERY HISTORY

Nickel sulfide exploration across Fennoscandia by Anglo American began in 2002. Initial tenure applications in the region were made in 2003 over mapped mafic–ultramafic sills intruded within the outlines of the Central Lapland Greenstone Belt (CLGB). The high-quality Aeromagnetic and Frequency

Mineral Deposits of Finland. http://dx.doi.org/10.1016/B978-0-12-410438-9.00009-1

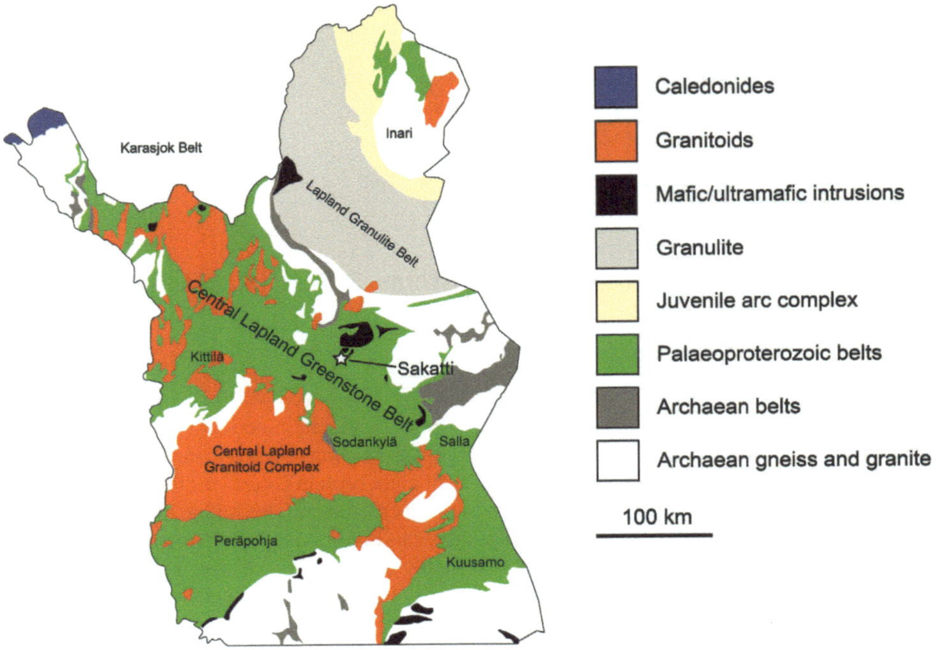

FIGURE 3.7.1 Regional map of the CLGB showing the location of the Sakatti deposit.

Source: *After Hanski and Huhma (2005).*

Electromagnetic (AMFEM) dataset (200 m line spacing, 30 m altitude) of the Geological Survey of Finland (GTK) covering the whole region was the primary base dataset used in initial target generation. Combined with the public domain digital geological maps and geochemical data from GTK, the AMFEM data enabled an initial prioritization of targets, with Sakatti chosen as target number eight (MOS8) on the first day of detailed targeting. At that time, the base datasets used had been in the public domain for more than a decade.

The initial targets were subsequently followed up by ground geophysical methods and soil geochemistry. An infill GTK AMFEM survey was later carried out to provide 100-m line spacing. This was done to better define the priority targets and to generate additional targets within the region. With processing of the wing-tip sensors, this produced an effective 75 m line spacing coverage of the area of exploration interest. The exploration methodologies were continuously adapted and improved until an effective system had been developed. It was apparent from geochemical orientation surveys conducted during the first field season that surface soil geochemistry was ineffective due to a number of reasons, including complex ice flow directions, multiple till layers, and airborne pollution from the Norilsk Nickel smelter at Nikel in Russia. Base of till (BOT) geochemistry, which involves the collection of till samples at the till–bedrock interface, proved to be the most effective geochemical tool.

Initial drill testing of Sakatti started in the late spring of 2006 when three diamond drill holes (DDH) were drilled at the eastern edge of the magnetic anomaly. Borehole DDH 06MOS08003 returned a number

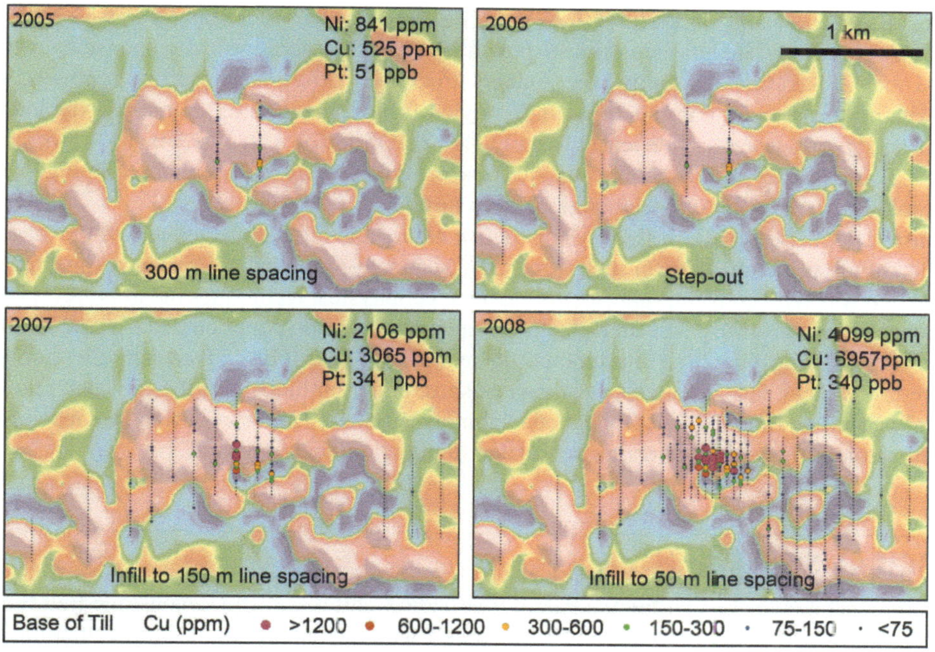

FIGURE 3.7.2 Base-of-till geochemistry undertaken over the Sakatti deposit prior to discovery, overlain on an airborne magnetic map (reduced to pole).

The base-of-till anomaly was well defined by 2008 and was approximately 150 m × 150 m. Given values refer to the highest Cu sample in each year.

of short significant intervals of Cu-PGE[1]-Au mineralization but with only background Ni. Importantly, the DDH was abandoned within the olivine cumulate due to the beginning of the spring thaw, when ground conditions became unstable. A year-end review of the 2006 field program by Anglo American's internal and external nickel commodity experts severely downgraded the Sakatti target and all work on it ceased.

An exploration review in 2007 within the Finland office team highlighted once again the potential of Sakatti and a detailed BOT grid was undertaken (Fig. 3.7.2) with resources diverted from other targets and projects within Fennoscandia. This detailed BOT geochemistry program returned highly significant coherent and coincident Ni, Cu, and PGE anomalies with values up to 6957 ppm Cu, 5575 ppm Ni, and 739 ppb Pt over the sub-cropping portion of the Sakatti deposit (Fig. 3.7.2).

Funding to continue the program was secured from senior management in London, and a second drill phase commenced in the winter of 2008, when the first important disseminated and minor vein-related mineralization was intersected in DDH 08MOS08007. This phase of exploration highlighted the economic potential of Sakatti. The official discovery borehole at Sakatti was DDH 09MOS08013, drilled in 2009. It intersected 110.00 m at 1.3 wt% Cu, 0.2 wt% Ni, 0.5 g/t Pt, 0.3 g/t Pd, and 0.4 g/t Au. This DDH proved that Sakatti is a significant mineralized system.

[1] PGE = platinum-group elements (Pt, Pd, Rh, Ir, Os, and Ru).

The GTK AMFEM and Geotech VTEM (Versatile Time Domain Electromagnetic) airborne survey systems flown over the Sakatti area do not show a recognizable geophysical response over the main cumulate body at Sakatti, but highlight the smaller, structurally offset, northeast and southwest satellite bodies. The northeast body consists of a similar style of mineralization to the main body and drilling has intersected two massive sulfide lenses (e.g., DDH 12MOS08079 with 39.95 m at 3.40 wt% Cu, 3.54 wt% Ni, 1.81 g/t Pt, 2.09 g/t Pd, and 0.45 g/t Au). The southwest body includes sub-cropping, weathered massive sulfides, and deeper, unweathered massive sulfides.

Due to the depth (>350 m), the highly conductive nature and interconnectedness of the massive and vein-style sulfide mineralization in the main body, very late-time channel responses are required to detect the deposit. With conventional electromagnetic geophysical techniques the noise level overpowers the response in these very late-time channels. In contrast, Anglo American's in-house low-temperature superconducting quantum interference device (LTSQUID) TEM (Transient Electromagnetic) system returned clean late-time responses over Sakatti, and with hindsight, the modeled results were remarkably accurate. In addition, the magnetic anomaly at Sakatti is now known to be a response of the serpentinization of the peridotite and is not related to the mineralization itself.

GEOLOGICAL SETTING

The Sakatti deposit is located within the Paleoproterozoic Central Lapland Greenstone Belt (CLGB), a complex succession of sedimentary and volcanic rocks, with the latter ranging from komatiitic to rhyolitic in composition. The evolution of the CLGB spans about 600 Ma starting with andesitic volcanism at ~2.5 Ga and ending with deposition of molasse-type coarse clastic sediments at <1.9 Ga. The belt is also host to several mafic–ultramafic intrusions, including the 2439 ± 3 Ma Koitelainen layered intrusion and the 2058 ± 4 Ma Kevitsa intrusion (Mutanen and Huhma, 2001). The geology of this area is exceptionally well described in publications by GTK. A comprehensive summary of this work is given in Hanski and Huhma (2005).

The stratigraphic level at which Sakatti was located within the CLGB has not yet been resolved, as the units so far intersected do not correspond conclusively to any particular part of the CLGB succession. On a regional scale, the deposit is surrounded by pelitic metasediments of the Matarakoski formation and quartzites of the Sodankylä group, which were deposited between ~2.3 and 2.06 Ga. However, neither of these rock formations has been identified in drill core at Sakatti. It is interesting to note in this context that the Ni-Cu ore-bearing Kevitsa intrusion, which is located approximately 15–20 km northeast of Sakatti, was emplaced in mica schists and black schists of the Matarakoski formation.

DEPOSIT OVERVIEW

Based on current understanding, the Sakatti deposit consists of three spatially distinct mineralized bodies of olivine cumulate named "main body," "northeast body," and "southwest body" (Fig. 3.7.3). Mineralization in each of the three is hosted within or at a basal contact of the olivine cumulate. In hand specimens, there are no discernible petrological differences between these three bodies.

The major host and wall rock units of the deposit comprise the *peridotite unit, aphanitic unit, mafic suite, breccia unit* and *volcaniclastic unit*. These units are defined in the following and their relationship is illustrated later in Figs. 3.7.5, 3.7.6, and 3.7.7. The composition and location of these units are

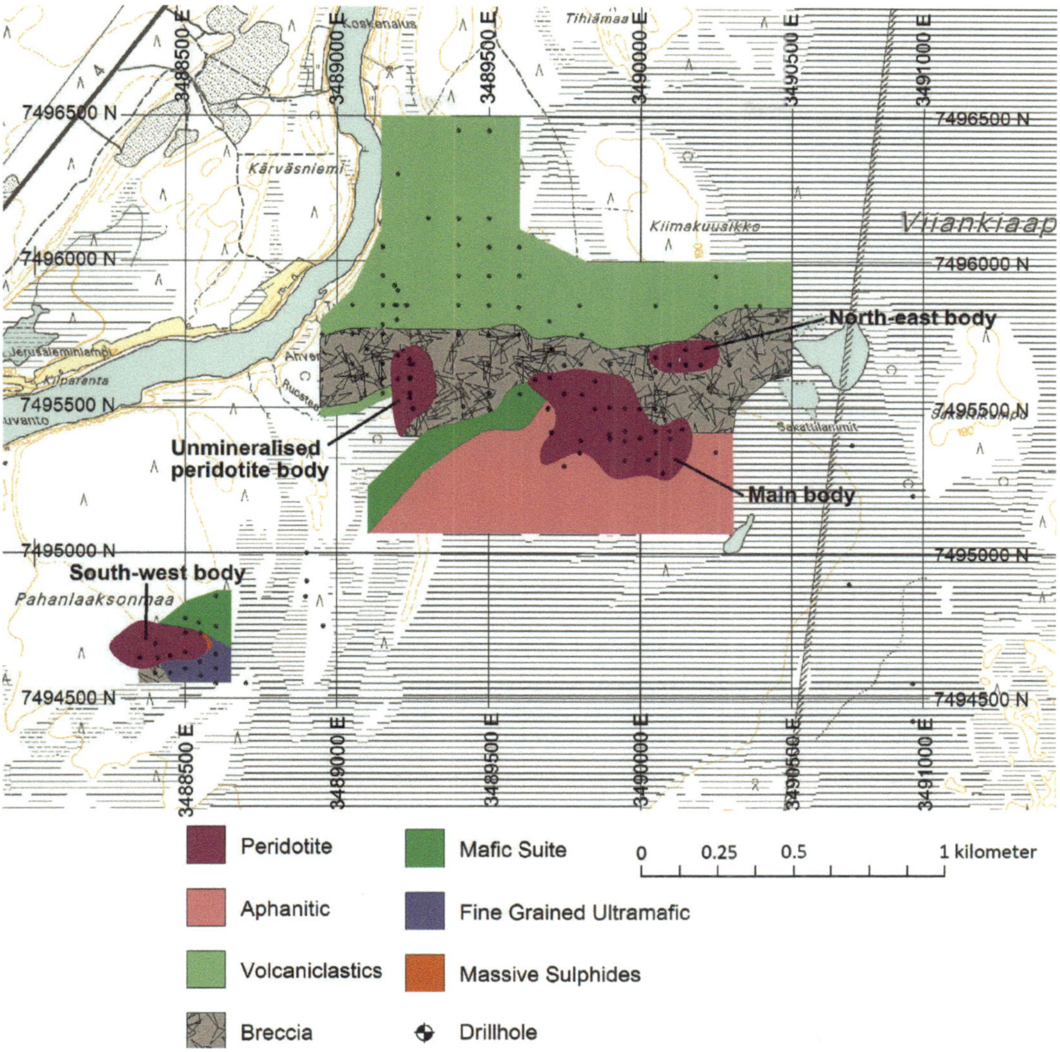

FIGURE 3.7.3 Plan map of the Sakatti deposit showing the interpreted geology and drill hole locations.

The horizontal lines indicate wetland.

briefly described in this section and their mineralogical and chemical compositions are discussed in more detail in the "Petrology and geochemistry" section later in the chapter.

KEY LITHOLOGIES

Peridotite unit

This is an olivine cumulate with variable oikocrystic pyroxene content and rare minor interstitial plagioclase. The peridotite unit is the principal constituent of the main cumulate body, which itself

FIGURE 3.7.4 Photographs of drill core from the Sakatti deposit.

(A,B) Typical peridotite from the main cumulate body; (C,D) typical samples of the aphanitic unit in contact with the cumulate body; (E) aphanitic unit intruded by mineralized peridotite of the main body; (F) aphanitic unit that is located more than 100 m from the cumulate body.

can be more than 400-m thick. Most of the unit is pervasively serpentinized, and should technically be termed a serpentinite. However, the cumulate texture is preserved (Fig. 3.7.4(A) and (B)), and therefore the rock can be considered in terms of its protolith. Textures within the main body range from adcumulate to orthocumulate with the groundmass typically also composed of serpentine with minor talc.

The *altered ultramafic* is a logging unit interpreted to be a talc-carbonate alteration product of the peridotite. This unit is invariably present where the peridotite is in direct contact with the overlying breccia unit. The *dunite* is another logging subunit of the peridotite. It is an olivine adcumulate marked by an almost complete lack of serpentinization and forms a continuous, discrete package at the base of the peridotite in the northwestern part of the main body. Primarily in the upper part of the peridotite unit, coarse, *pegmatoidal gabbro* may occur. These rocks display sharp contacts with peridotite and comprise intersections of between 50 cm and 15 m.

Aphanitic unit

The aphanitic unit, so named because of its grain size and likely volcanic origin, forms the hanging wall, footwall, and sidewall to much of the main cumulate body, notably along the southern edge and the far western side. Referring to a unit as aphanitic is clearly problematic as it is not an adequate rock name. This name has endured because the unit has proved difficult to classify. The mineralogical and textural features of this unit all point toward a volcanic origin.

The rock is classified as a plagioclase-rich picrite with 10–15% small (<0.5 mm) phenocrysts of olivine and minor plagioclase (up to 1 mm) in a fine-grained groundmass of plagioclase, pyroxene, and minor olivine. Where it is within 50 m of the contact with the peridotite unit, the aphanitic unit exhibits an unusual texture containing injections of the peridotite unit (Fig. 3.7.4(C), (D), and (E)). Further from the contact, this texture is not present and the rocks merely show serpentine veining (Fig. 3.7.4(F)).

The *footwall mafic rock* is a logging unit referring to what is interpreted as an alteration product of the aphanitic unit. The rock consists mainly of chlorite, phlogopite, and talc, accompanied by carbonate veins and fracture fill.

Mafic suite

In addition to the aphanitic unit, the hanging wall of the Sakatti deposit contains several other lithological units, including the mafic suite. It is present in the southwestern part of the deposit where it occurs between the peridotite or aphanitic units and the breccia unit. It comprises three separate logging units: the *mafic volcanic rock, the scapolite-mica rock* and the *hanging-wall gabbro*. The mafic volcanic rock is a strongly chlorite-amphibole-altered lithology that, when in close proximity to the breccia unit, has undergone in situ brecciation and precipitation of matrix and vein calcite. The scapolite-mica rock is a strongly foliated, almost schistose rock with a biotite matrix hosting scapolite porphyroblasts. The hanging-wall gabbro comprises a series of gabbroic sills that intrude the mafic volcanic and scapolite-mica rocks but do not intrude the main cumulate body. At the contact between the scapolite-mica rock and the aphanitic unit, a cryptocrystalline *serpentinite unit* is frequently present.

Breccia unit

The breccia unit is a 100–300-m-thick hematite-dolomite-albite-talc altered, and exceptionally heterogeneous, polymict breccia package that lies stratigraphically above the main cumulate body. Various zones can be differentiated within the breccia unit, including zones with predominant albite or carbonate (calcite/dolomite) alteration, as well as polymict zones where rounded to angular clasts of talc, chlorite, and quartz typically occur in a calcite matrix.

Volcaniclastic unit

The volcaniclastic unit is the stratigraphically uppermost unit in the hanging wall of the Sakatti deposit. It is a phyllite with biotite porphyroblasts and a crenulation cleavage throughout. It is interpreted as a metamorphosed volcano-sedimentary package. The thickness of this unit is at least 600 m.

Footwall units

In the eastern part of the deposit, where the aphanitic unit forms the primary footwall below the peridotite unit, a clay-rich zone, interpreted as a fault structure, is present beneath the aphanitic unit. In the west, this fault structure occurs directly at the base of the peridotite unit and has often necessitated the cessation of drilling. Beneath this fault, a strongly laminated carbonate-rich metasediment is present. No sulfides have been observed in this metasediment.

MORPHOLOGY OF THE SAKATTI DEPOSIT

Main cumulate body

The northwest-plunging main cumulate body is by far the largest of the three cumulate bodies (see Fig. 3.7.3), with a currently delineated extent of more than 1100 m down-plunge from surface to a depth of 1220 m below surface, and a maximum thickness of more than 400 m. The cumulate body is roughly tubular, yet irregular in shape, with a shallower plunge near surface in the east and at depth in the north and west. These two shallowly plunging parts are connected by a more steeply plunging north-northwest to south-southeast orientated part of the body (Fig. 3.7.5).

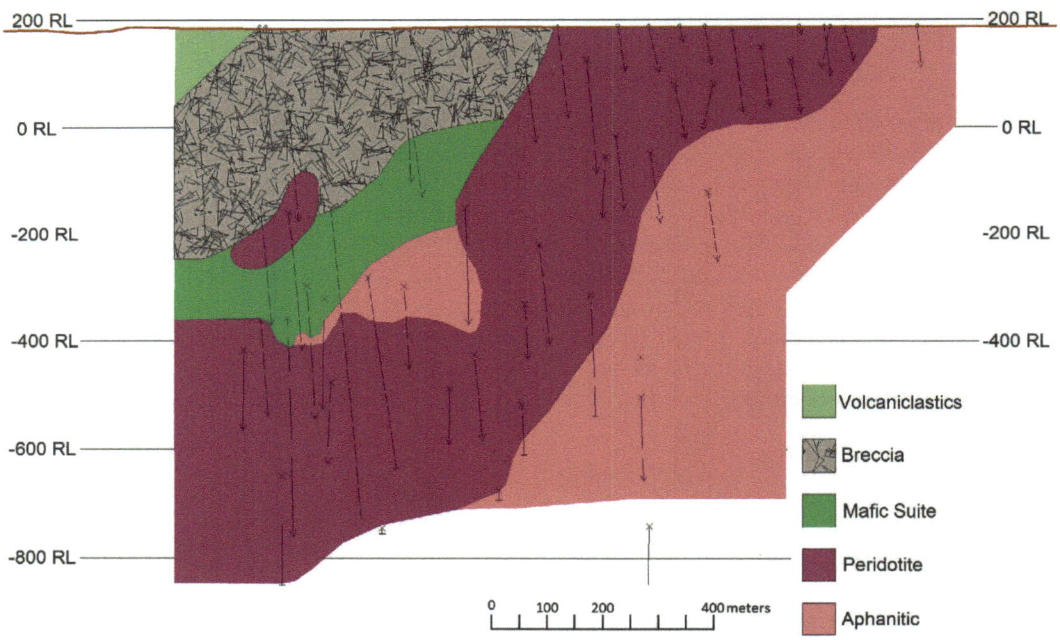

FIGURE 3.7.5 West-northwest to east-southeast angled section, ±25 m clipping, through the main body of the Sakatti deposit at 110° through center point 3489600E 7495600N.

This shows the changing geometry of the cumulate body and the crosscutting relationship of the cumulate with both the mafic suite and the aphanitic unit. Boreholes intersecting the 50 m wide section are shown as lines, with cross and arrows indicating where they enter and leave the section or crossbars where they terminate within the section.

Broadly speaking, the main cumulate body occurs stratigraphically below the breccia unit but above the aphanitic unit; however, the exact stratigraphic setting is locally more complex. The geology in the eastern portion of the main body is relatively simple (Fig. 3.7.6). The peridotite subcrops beneath the glacial till cover, and a mixture of aphanitic and altered aphanitic rocks form the southern sidewall and basal contact to the cumulate body.

With increasing depth further to the west, an extensive package of nonmineralized hanging wall rocks is encountered before reaching the cumulate body (Fig. 3.7.7). In this western part of the deposit,

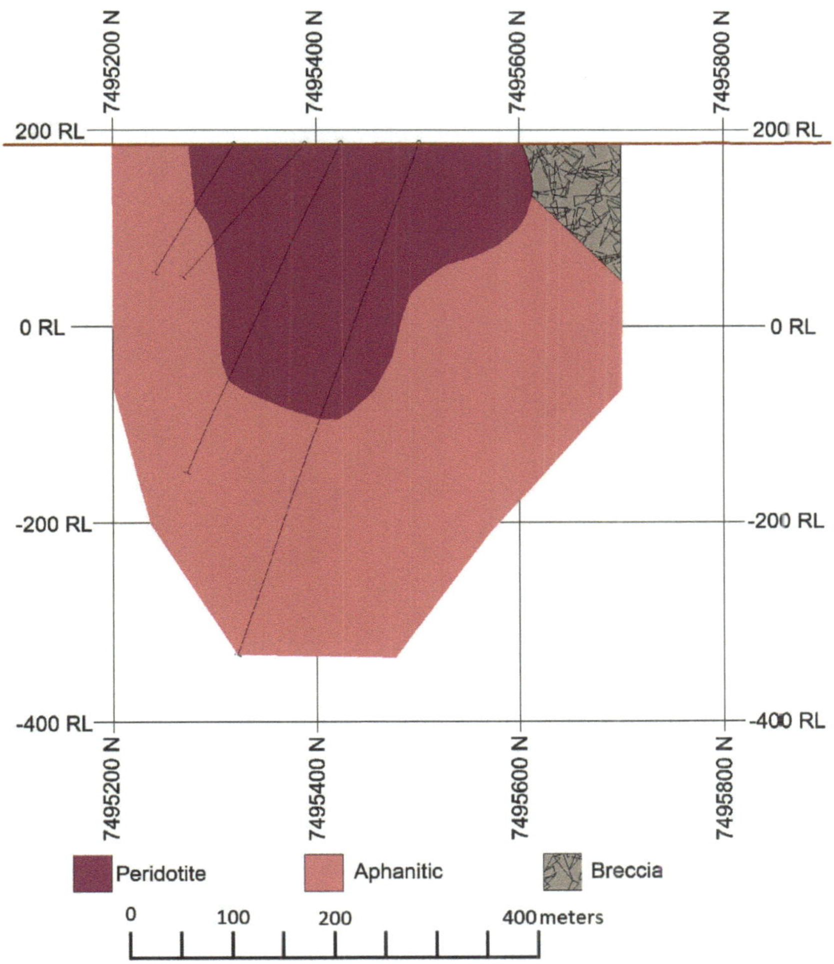

FIGURE 3.7.6 South to north cross section, ±25 m clipping, through the main body of the Sakatti deposit on 3489950E.

This shows the aphanitic footwall surrounding the main cumulate body and a small portion of the hanging-wall breccia. Boreholes intersecting the 50 m wide section are shown as lines, with cross and arrows indicating where they enter and leave the section or crossbars where they terminate within the section.

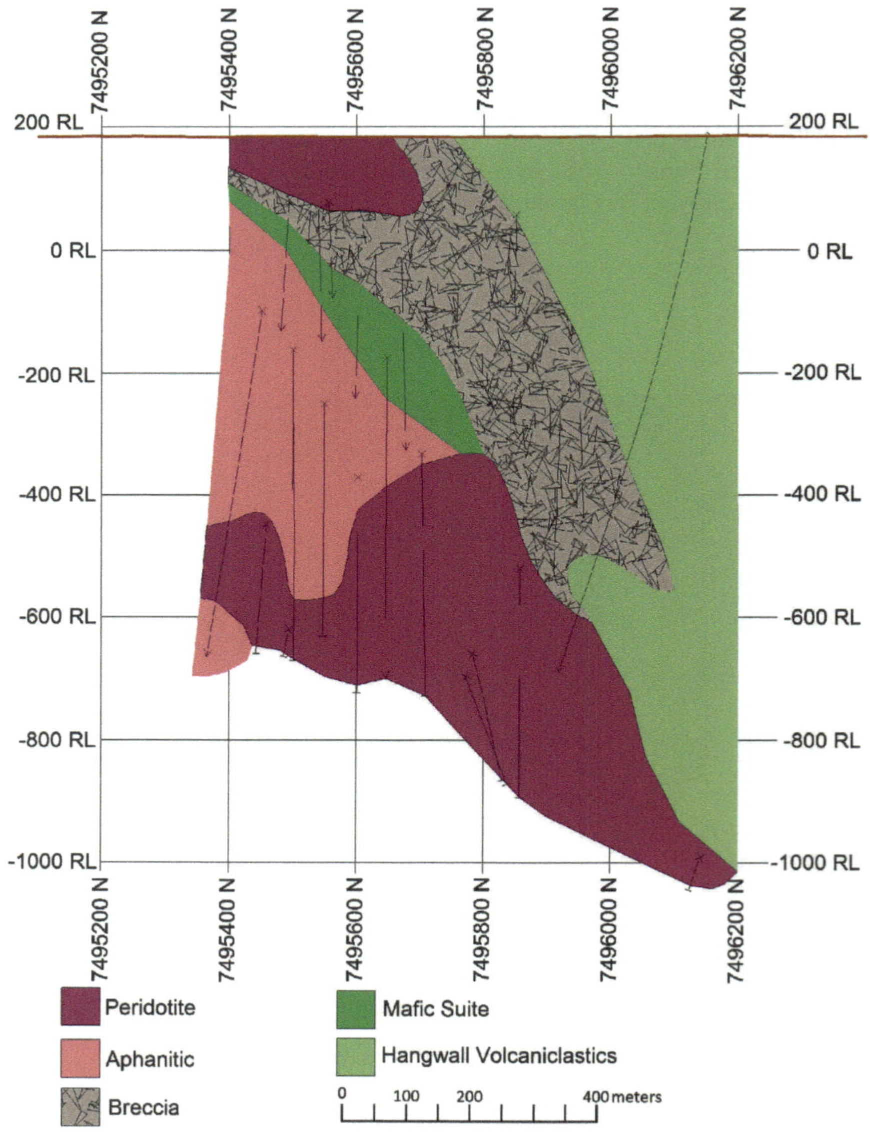

FIGURE 3.7.7 South to north cross section, ±25 m clipping, through the main body of the Sakatti deposit on 3489300E.

The hanging wall contains volcaniclastics, breccia, the mafic suite, and an unmineralized peridotite. The aphanitic unit is present both above and below the peridotite unit. The base of the cross section is a faulted contact. Boreholes intersecting the 50 m wide section are shown as lines, with cross and arrows indicating where they enter and leave the section or crossbars where they terminate within the section.

the aphanitic unit constitutes both the hanging wall and part of the footwall of the cumulate body (the remainder of the footwall is a fault structure).

The upper contact of the aphanitic unit with the remainder of the hanging wall is always marked by a cryptocrystalline serpentinite unit. Stratigraphically above the aphanitic unit and the serpentinite is the northwest-dipping mafic suite. Above this package is the breccia unit, which also dips to the northwest and appears to be concordant with the underlying mafic suite. Toward the north, the package of aphanitic unit, serpentinite, and mafic suite pinches out and the cumulate body is in direct contact with the breccia unit. Further to the north, the breccia unit itself pinches out to leave the volcaniclastic unit in contact with the cumulate body. An apparently isolated nonmineralized peridotite cumulate body subcrops at surface within the breccia unit in this western portion (see Fig. 3.7.7), without any established links to the main cumulate body.

Northeast cumulate body

Based on the evidence of current drilling, the northeast cumulate body is smaller than the main cumulate body and more cylindrical, albeit with an elongate tail to depth, giving it an inverted teardrop-like shape (Fig. 3.7.8). It has a west–east strike and gently plunges to the east. It occurs at a different stratigraphic level to the main cumulate body, as it is bound by the volcaniclastic unit to the north and the breccia unit to the south and at depth.

Talc-carbonate-altered ultramafic rocks are also more common when compared to the main cumulate body, particularly to the east. In the west, a 20–30-m-thick cryptocrystalline serpentinite body occurs within the peridotite, almost perpendicular to the contact between the cumulate body and the volcaniclastic unit. Only one drill hole has extended from the northeast cumulate body toward the main cumulate body and this hole intersected a fault at depth.

Southwest cumulate body

The southwest cumulate body is ovoid in shape, with a central sub-cropping portion and a gently dipping eastern edge. The western edge is untested by drilling. Scapolite-mica rock occurs to the north of the cumulate body, and a breccia occurs at depth and to the south. At surface in the southeast is a fine-grained, amphibole-rich, ultramafic rock that has not been intercepted elsewhere within the Sakatti area.

MINERALIZATION

Mineralization at Sakatti consists of disseminated, vein, semi-massive, and massive sulfides (Fig. 3.7.9). Vein, semi-massive, and massive mineralization is found mostly within the olivine cumulate bodies but can extend into the aphanitic footwall and hanging wall. In contrast, the disseminated mineralization is only found within the olivine cumulate bodies. Not all of the main cumulate body is mineralized; within the relatively thick central and eastern portions of the body, typically only the bottom half hosts mineralization. In the far west and north where the cumulate body is relatively thin, mineralization occurs throughout the entire cumulate package.

Main body

The massive sulfide mineralization is present within the peridotite and also extends into the aphanitic footwall and sidewall of the peridotite for up to approximately 150 m. There are several different lenses of massive sulfides, at least two of which can be correlated between drill holes. The lenses are thickest

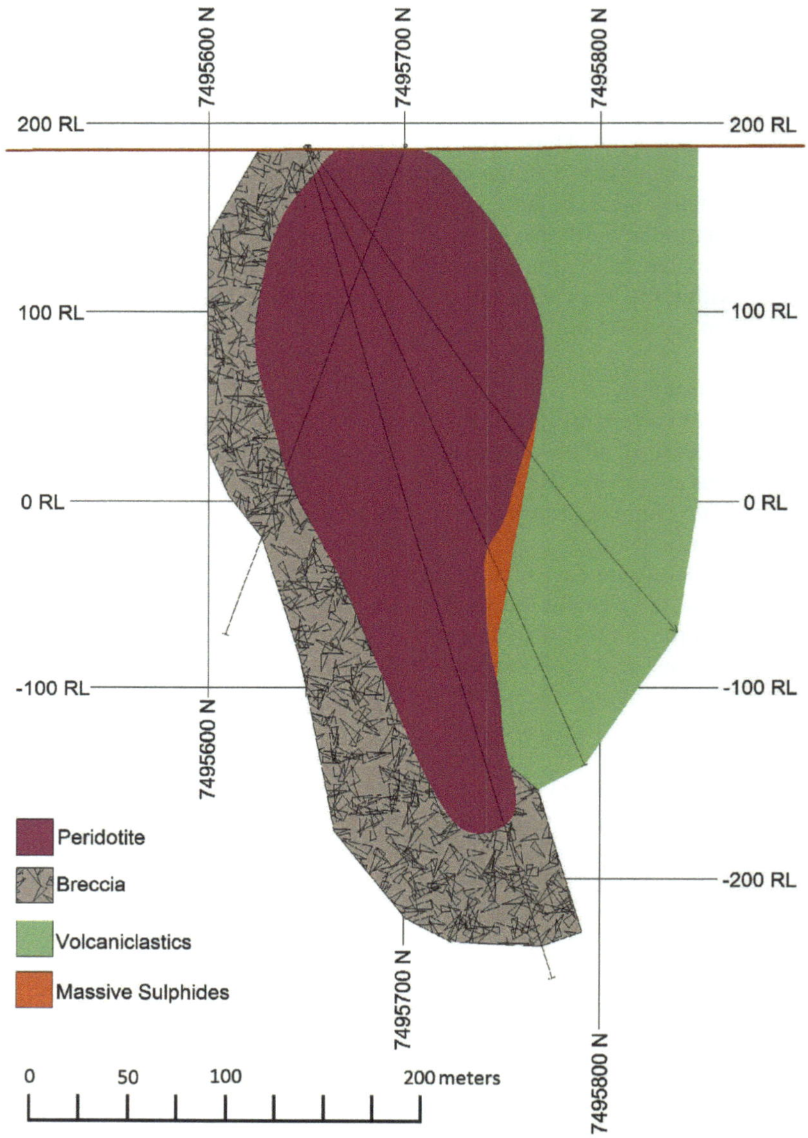

FIGURE 3.7.8 South to north cross section through the northeast body of the Sakatti deposit on 3490150E.

Peridotite occurs between breccia and volcaniclastics. Boreholes intersecting the 50 m wide section are shown as lines, with cross and arrows indicating where they enter and leave the section or crossbars where they terminate within the section. Solid lines represent boreholes.

FIGURE 3.7.9 Photographs showing Ni-Cu sulfide mineralization in drill core from the Sakatti deposit.

(A) Chalcopyrite-dominated massive sulfide mineralization with streaks of pyrrhotite and minor pentlandite and magnetite; (B) contact of pyrrhotite-pentlandite-pyrite massive sulfide mineralization with olivine cumulate; (C) pyrrhotite-pentlandite massive sulfides with minor chalcopyrite and euhedral magnetite; (D) massive sulfides showing segregation of pyrrhotite-pentlandite and chalcopyrite; (E) pyrite style of mineralization characterized by large pyrite grains typical of, but not exclusive to, the northeast body mineralization; (F) typical chalcopyrite-dominated disseminated/interstitial mineralization; (G) chalcopyrite- and pyrite-rich vein-style of mineralization; (H) semi-massive mineralization hosted by pegmatoidal gabbro.

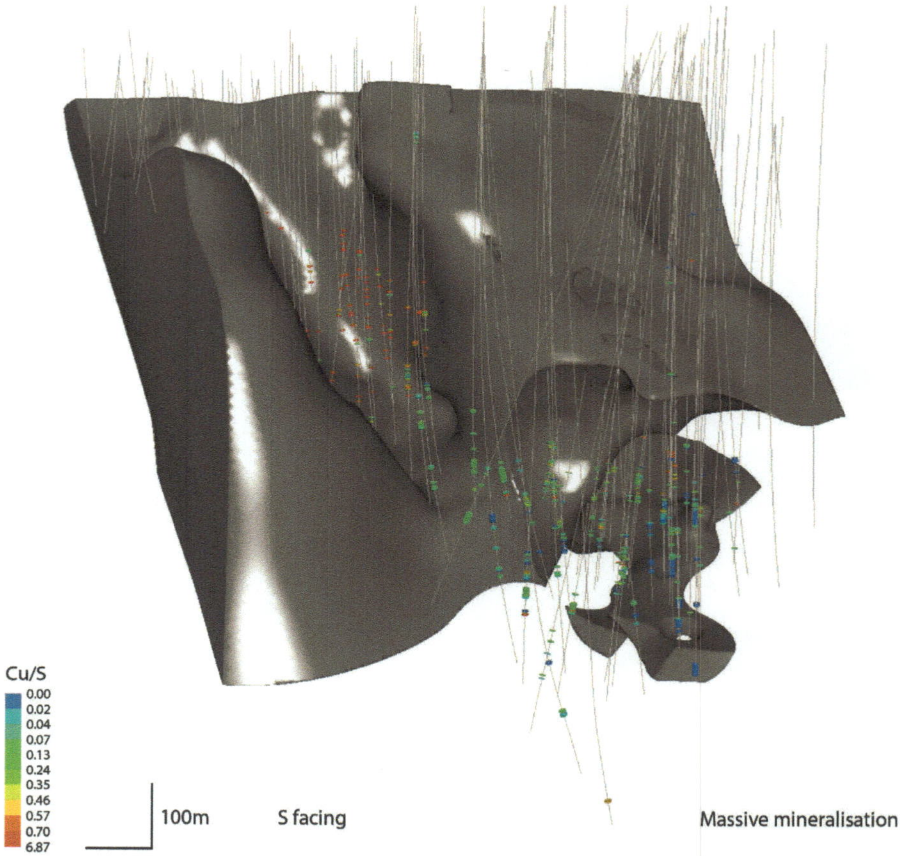

FIGURE 3.7.10 3D diagram showing Cu-S ratios for intersections of massive sulfides in the Sakatti deposit.

The upper contact of the aphanitic unit is shown in gray and the peridotite unit has been removed. Solid lines represent boreholes.

(up to 25 m) in the center of the deposit and thin out to as little as 0.5 m toward both the northwest and southeast.

The massive and veined sulfide mineralization within the main cumulate body shows a distinct fractionation in Ni-Cu ratios. In the west and northwest (i.e., in the deeper portions of the deposit) massive sulfides are composed of pyrrhotite-pentlandite-chalcopyrite with overall Ni-Cu ratios greater than one. The composition of the massive sulfides evolves up-plunge to become increasingly chalcopyrite-dominated (Fig. 3.7.10).

Massive sulfide lenses extending from the cumulate body into the aphanitic unit show decreasing thickness and Ni-Cu ratios with distance from the cumulate body, but these trends occur over a much shorter distance than those observed up-plunge within the cumulate body. At the terminal extremities of the massive sulfide lenses, both within the cumulate body and the aphanitic unit, there is a marked increase in precious metal tenors, in particular that of Pt.

An additional style of mineralization is typified by coarse, pegmatoidal gabbroic silicates with semi-massive sulfides (Fig. 3.7.9(H)). The coarse, pegmatoidal gabbroic subunit occurs both with and without associated sulfide mineralization.

Vein sulfides are present to a lesser extent throughout the main peridotite cumulate body, but are most abundant in its shallow southeastern part where the massive sulfide lenses show relatively low Ni-Cu ratios and reduced thickness. The sulfide veins are mostly relatively thin with a broad range of orientations and contain predominantly chalcopyrite. As such, they are judged to be distinct from the massive sulfides. The veins are generally 5–20 cm thick, but can be up to 50 cm in thickness. The use of the term "vein" is not intended to imply an epigenetic hydrothermal origin; rather, the veins are interpreted to represent fractionated apophyses of the main massive sulfide lenses.

Disseminated sulfides or the interstitial mineralization, as it is commonly called, occur predominantly in the central, steeply plunging and the southeast, shallowly plunging parts of the main cumulate body. In terms of composition, the mineralization is generally chalcopyrite-dominated with only minor pyrrhotite, pentlandite, and pyrite. The occurrence of this style of mineralization is spatially independent of the other styles. In many places the disseminated mineralization is cut by massive sulfide veins of varying composition and size. However, there are numerous examples where disseminated sulfides show no spatial association with massive sulfides, either vertically or laterally. This is especially the case at depth in the northwestern part of the body.

The disseminated sulfides are generally chalcopyrite-dominated but the Cu-S ratio shows considerable variation, being relatively Cu-poor in the deeper northwestern part of the main body and relatively Cu-rich in the shallow southeastern part (Fig. 3.7.11). As discussed in greater detail later in this chapter, minor interstitial chalcopyrite also occurs in small peridotitic intrusions within the hanging wall and footwall aphanitic unit, although this is restricted to within 20 m of the contact with the cumulate body.

Northeast body

Mineralization in the northeast body is almost entirely limited to massive sulfides at the basal contact between the cumulate body and the volcaniclastic package. Disseminated mineralization is of minor significance.

The metal tenors of the massive sulfides are very similar to those of massive sulfides at depth in the main cumulate body; Ni contents are roughly equal to or greater than Cu contents, and Pt and Pd are present in a ratio of 2:1, although both Pt and Pd are enriched compared to massive sulfides in the main cumulate body. One notable difference when compared to the main body is the abundance of coarse-grained pyrite and the lack of pyrrhotite.

Southwest body

Mineralization in the southwest cumulate body occurs in a variety of forms. On the eastern side of the body, mineralization is massive and semi-massive and located within chlorite gouge at the upper contact of the cumulate. Moving further west, massive sulfides and semi-massive sulfides, as well as clasts of massive sulfide, are found within the body. All mineralization in the southwest body is Ni-dominated with a Ni-Cu ratio of 2.

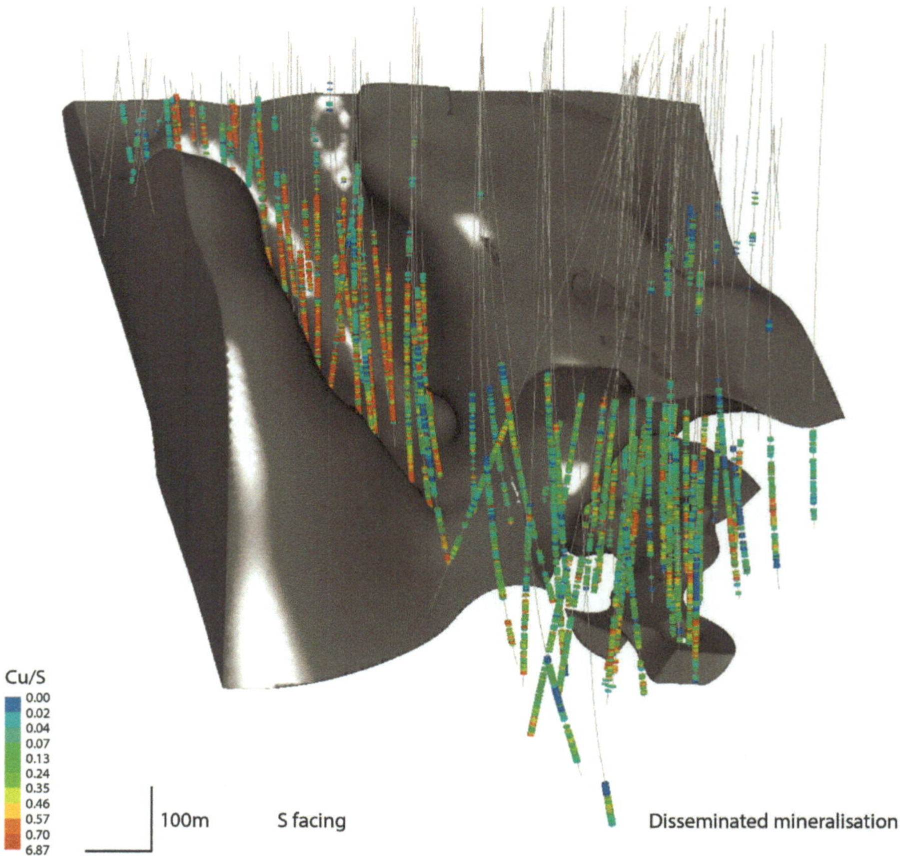

FIGURE 3.7.11 3D diagram showing Cu-S ratios for intersections of disseminated sulfides in the Sakatti deposit.

The upper contact of the aphanitic unit is shown in gray and the peridotite unit (where mineralization occurs) has been removed.

PETROLOGY AND GEOCHEMISTRY

Investigation of the lithologies of the Sakatti deposit beyond hand specimen identification has involved a variety of techniques, relying primarily on conventional thin section microscopy and whole-rock geochemistry. In addition, the mineral chemistry of preserved magmatic minerals has been extensively investigated to constrain the petrogenesis of the deposit.

PERIDOTITE UNIT

Petrology

Due to pervasive serpentinization, most samples do not contain primary magmatic minerals (Fig. 3.7.12(A) and (B)). However, in some samples, primary magmatic minerals are preserved. The main cumulus phases are olivine (typically 0.5–2 mm in width) and chromite (typically 0.1 mm).

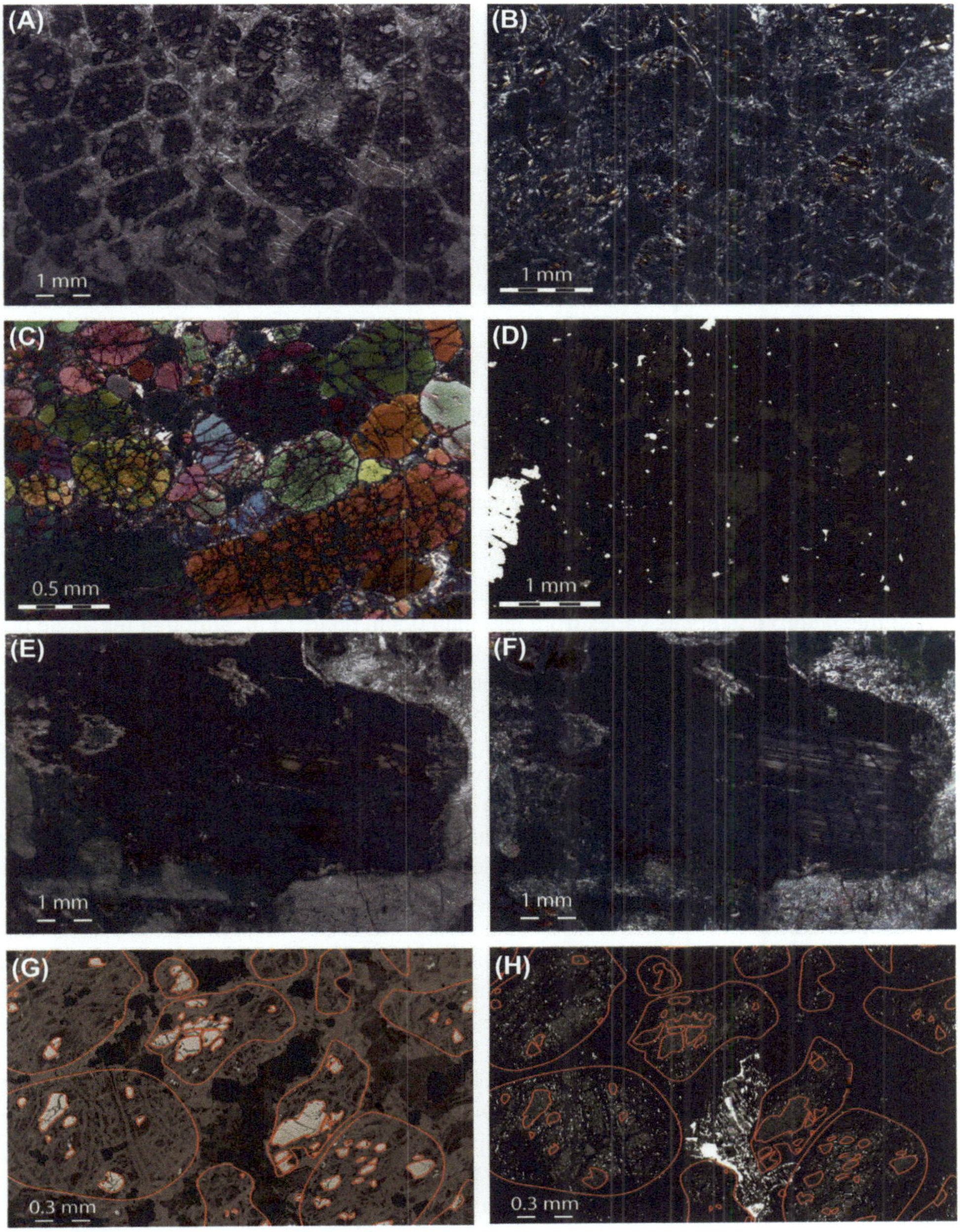

FIGURE 3.7.12 Images of thin sections from the peridotite unit.

(A) Serpentinized orthocumulate from the Sakatti main body, plane-polarized light; (B) serpentinized mesocumulate, crossed polarizers; (C) weakly altered olivine adcumulate from the dunite subunit, crossed polarizers; (D) backscattered electron image showing pargasite, enstatite, and diopside intercumulus and partially serpentinized olivine cumulate; (E) black plagioclase from the pegmatoidal gabbro subunit, plane polarized light; (F) black plagioclase from the pegmatoidal gabbro subunit, crossed polarizers; (G) wavelength-dispersive spectroscopy (WDS) chemistry map showing relative MgO contents of partially serpentinized olivine cumulate; cumulus olivine shapes have been outlined as well as their contained preserved olivine; (H) WDS chemistry map showing relative NiO contents of Fig. 3.7.12(G) showing variable Ni in preserved olivine as well as immobility of Ni during serpentinization.

Clinopyroxene is the most common intercumulus mineral, occurring as oikocrysts up to 2 cm across that are frequently altered to tremolite. Orthopyroxene also forms oikocrysts, but is less abundant.

The rock varies from orthocumulate to adcumulate on the 10^{-2} to 10^{2} m scale. Orthocumulate is more abundant toward the top of the succession, although there is local variability throughout. The dunite subunit is an adcumulate situated at the base of the intrusion. Rather than being distinct on a textural or primary mineralogical basis, it is different from the rest of the cumulate primarily because of an almost complete preservation of magmatic olivine (Fig. 3.7.12(C)). Possibly, the lack of intercumulus melt during crystallization has limited later serpentinization.

There are multiple occurrences within the peridotite package where olivine cumulate is absent and the rock is a coarse pyroxenite. These drill core intersections are generally less than 15 cm wide and are composed of up to 3-cm-sized diopside grains, as well as minor enstatite. These intersections could represent stratigraphically important levels, such as the tops of separate pulses of magmatism; however, to date, they do not correlate with the identified whole rock and mineral chemistry changes presented later in this section.

Pegmatoidal gabbro is similar to pyroxenite in that it contains very coarse pyroxene; pyroxenite could simply be a less evolved form of this subunit. The pegmatoidal gabbro is particularly prevalent in the central and eastern portions of the main cumulate body, and also toward its top. The down-hole thickness of the gabbroic intervals varies from less than 0.5 m up to 15 m, although the angle between the contacts and the core axis is highly variable and thus the true thickness remains unclear.

The pegmatoidal gabbro contains large euhedral plagioclase grains in addition to pyroxene and, possibly secondary, amphibole. Unusually, the plagioclase is black in color while the pyroxene is lighter gray. The rocks have sharp contacts to the surrounding cumulate and are frequently associated with mineralization (see Fig. 3.7.9(H)), either in the form of interstitial sulfides within the gabbro or as massive sulfides adjacent to the gabbro. It has not been possible to connect the coarse grained pyroxenites or the pegmatoidal gabbros laterally between boreholes, and their stratigraphic position does not appear to define chemical reversals. They are thus interpreted as late-stage dikes of more evolved melt, coeval with late-stage mineralization processes.

Mineral compositions

The composition of the main minerals was determined by electron microprobe at the Natural History Museum, London, using a Cameca SX 100 instrument.

Olivine

The olivine in the peridotites is characterized by Mg# (Mg# = atomic Mg/(Mg + Fe)) between 0.85–0.91 and Ni contents clustering between 3000–3500 ppm (Fig. 3.7.13). Compared to olivine from most other mafic–ultramafic intrusions with comparable Mg#, the Ni contents are relatively high, inconsistent with crystallization from a melt that was S saturated (Brenan and Caciagli, 2000).

In general, the Ni content of olivine does not vary significantly within the intrusion. However, some samples contain highly variable olivine Ni compositions, including a number of grains with a relatively low Ni content, similar to that of olivine in the aphanitic unit (Fig. 3.7.13). These crystals are relatively small in size (<20 μm) and are interpreted as entrained olivines derived from the footwall rocks.

In samples with significant disseminated mineralization, olivine may be zoned with decreasing Ni content toward grain margins (see Fig. 3.7.12(H)), possibly due to diffusive loss of Ni from olivine to surrounding sulfides or Ni-depleted silicate melt. However, zoning as a result of olivine growth in progressively Ni-depleted melt cannot be ruled out. Notably, Ni appears to be immobile during pervasive serpentinization. This is evident in an Ni distribution map of peridotite (see Fig. 3.7.12(H)), showing similar Ni contents in preserved olivine and serpentine.

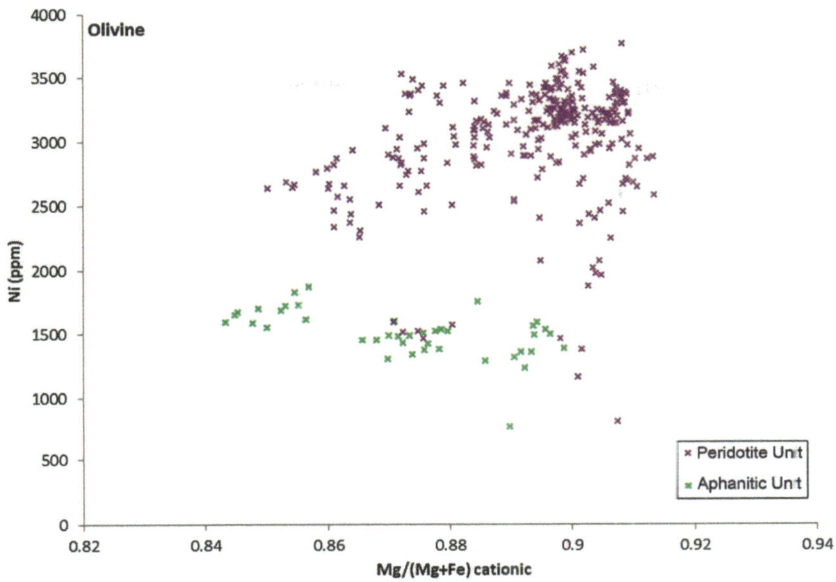

FIGURE 3.7.13 Olivine chemistry from the main cumulate body in the Sakatti deposit and the aphanitic footwall.

WDS (Wavelength Dispersive Spectra) data acquired on a Cameca SX100 EMPA (Electron Probe Micro Analyzer) at the Natural History Museum, London.

Overall, the olivine compositions are broadly comparable with "main ore" olivine chemistry reported from the Kevitsa deposit (Mutanen, 1997). At Sakatti, however, no olivine has been identified that approaches the unusual ultranickeliferous (up to 14,000 ppm) composition found in the "Ni-PGE ore" at Kevitsa (Yang et al., 2013).

Pyroxene

Pyroxene in the peridotite unit is invariably oikocrystic, surrounding and encompassing smaller cumulus olivine. Oikocrysts may be 1–2 cm wide. The pyroxene is frequently altered to tremolite-actinolite. In some cases, unaltered oikocrysts contain preserved olivine whereas surrounding olivine has been serpentinized. Enstatite oikocrysts have Mg# of 0.80–0.90 and Ni contents in the range of 400–1000 ppm, whereas diopside oikocrysts have Mg# of 0.85–0.92 and Ni contents of 200–600 ppm. Nickel contents correlate positively between ortho- and clinopyroxene in the same sample, but neither correlate with olivine Ni contents.

Amphibole

Chemical analyses reveal two distinct populations of amphibole. Amphiboles of the tremolite-actinolite series are the predominant type and are texturally identifiable as alteration products of pyroxene. In addition, there is a significant population of pargasite that is interpreted as a primary magmatic phase, based largely on texture (see Fig. 3.7.12(D)).

The presence of a magmatic hydrous phase has important implications for the composition of the melt (e.g., Fiorentini et al., 2008) and also raises the possibility of autoserpentinization associated with late-stage magmatic processes.

Plagioclase

Plagioclase is rarely present in the peridotite unit, except as an intercumulus phase near the contact with the aphanitic unit. This is interpreted to be a result of assimilation of rocks of the aphanitic unit, given the nature of the contact between the two units (see "Aphanitic Unit" section below). Plagioclase is also present in the pegmatoidal gabbroic rocks where it is black in hand specimens (see Fig. 3.7.12(E) and (F)). In both cases, the plagioclase is anorthite, although the Na:Ca ratio is variable.

Chromite

Chromite grains are euhedral and form the only other cumulus phase besides olivine. Most cumulus chromite has been altered to chrome-bearing magnetite. However, the grains frequently contain cores of chromite that potentially reflect the original composition. These cores have Cr_2O_3 contents between 27–37 wt% and Ni contents of 1000–2500 ppm. The presence of cumulus chromite suggests that the parental melt had an MgO content below 24 wt%, as otherwise chromite would not have been on the liquidus (Barnes, 1998). See details in Table 3.7.1.

Whole-rock chemistry

The whole-rock geochemical data for the majority of samples from the peridotite unit at Sakatti indicate primary control by accumulation of cumulus olivine (Fig. 3.7.14).

The MgO content of the parental melt from which the peridotite had formed can be estimated at approximately 19% using whole rock MgO and FeO contents, as well as olivine compositions (Bickle, 1982; Nisbet et al., 1993; Arndt, 2008; Fig. 3.7.15). Note that this estimate relies on a number of assumptions and should be treated with some caution (Arndt, 2008).

Identifying layering or separating different pulses of magmatism within the deposit has not been conclusive from simple drill core observations as the cumulate rocks are texturally relatively homogenous. Whole-rock chemical data have been used to delineate distinct pulses within the cumulate body (Fig. 3.7.16). The ratio (Mg + Fe)/Si is essentially controlled by modal proportions of olivine to pyroxene. In each drill core for which whole-rock geochemistry data are available it is possible to identify at least three separate magma pulses (Table 3.7.2).

Mineral chemistry data broadly confirm the separate layers seen in the whole-rock geochemistry, although sampling density is suboptimal due to the fact that most samples are pervasively serpentinized and therefore do not contain preserved magmatic minerals.

Trace element ratios are relatively uniform throughout the main cumulate body, both laterally and vertically, but tend to show some variations consistent with a relatively more contaminated signature (i.e., lower Nb-La, Nb/Th, and higher La-Sm, Th-Yb) in the upper part of the body (Fig.3.7.20).

The heavy rare earth elements (HREE) are relatively unfractionated, as indicated by primitive mantle-normalized Gd/Yb around 1.5. This suggests that garnet was not a residual phase during mantle melting, consistent with magma generation at relatively shallow depths of <90 km. Moderate crustal contamination of the mantle-derived melt prior to its emplacement is suggested by a number of trace element ratios (all normalized to primitive mantle values of McDonough and Sun, 1995), including Nb-La (0.15–0.45), Nb-Th (0.10–0.25), La-Sm (1.6–3), and Th-Yb (5–8).

Due to the moderately to highly incompatible nature of the lithophile trace elements, the information provided by these elements reflects primarily the composition of the intercumulus melt and may not necessarily be representative of the melt from which the cumulus phases (e.g., olivine) and sulfide mineralization formed (see "Mineralization" section below).

Table 3.7.1 WDS data showing representative mineral compositions of the main magmatic phases and their common alteration products in the main Sakatti cumulate body

	Olivine	Serpentine	Diopside	Tremolite	Enstatite	Pargasite	Plagioclase	Chromite	Magnetite
Na_2O	<0.05	<0.05	0.35	0.34	<0.05	2.37	2.26	na	na
MgO	48.73	36.06	17.47	19	31.88	19.02	0.05	5.52	1.45
Al_2O_3	<0.04	0.22	3.52	2.2	1.43	10.46	32.85	12.96	1.29
SiO_2	40.34	40.56	51.87	51.47	56.37	46.56	47.8	<0.03	<0.03
K_2O	<0.05	<0.05	<0.05	<0.05	<0.05	0.57	0.07	na	na
CaO	0.03	0.11	20.83	20.54	2.26	11.65	16.33	<0.02	<0.02
TiO_2	0.04	0.03	0.5	0.24	0.14	1.18	<0.02	0.35	0.43
Cr_2O_3	<0.04	<0.04	1.04	1.26	0.54	1.81	<0.04	33.04	11.63
V_2O_3	<0.04	<0.04	<0.04	<0.04	<0.04	0.06	<0.04	0.14	0.17
MnO	0.13	0.27	0.13	0.11	0.17	0.03	<0.03	0.61	0.23
FeO	10.22	8.4	4.36	3.5	7.51	4.61	0.51	43.6	78.82
CoO	<0.06	<0.06	<0.06	<0.06	<0.06	<0.06	<0.06	<0.06	<0.06
NiO	0.42	0.44	0.06	0.13	0.12	0.09	<0.02	0.07	0.14
Totals	100	86.12	100.2	98.82	100.5	98.41	99.87	96.34	94.19

na = not analyzed

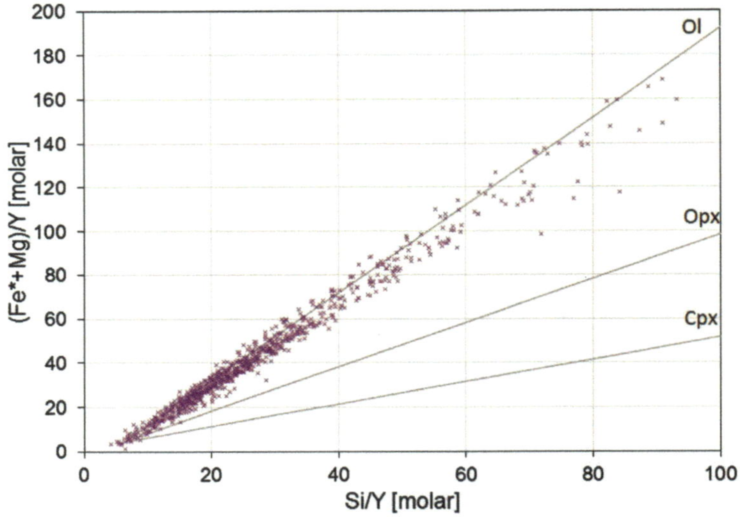

FIGURE 3.7.14 Whole-rock geochemistry plot showing Fe* + Mg versus Si normalized to Y for the peridotite unit in molar proportions.

Olivine (Ol), orthopyroxene (Opx), and clinopyroxene (Cpx) lines are calculated using WDS mineral data. All data recalculated volatile free; Fe* = Fe corrected to remove sulfide Fe.

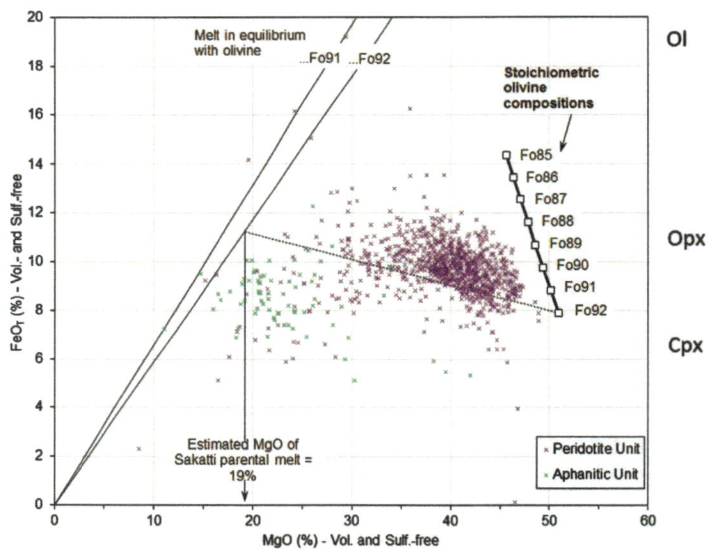

FIGURE 3.7.15 Binary variation plot of whole rock FeO_T versus MgO in peridotite and aphanitic units.

Data have been normalized to volatile and sulfide-free compositions. Melt-olivine equilibrium lines have been calculated assuming $KD_{(FeO/MgO_{Ol})/(FeO/MgO_L)} = 0.3$. *Arndt (2008).*

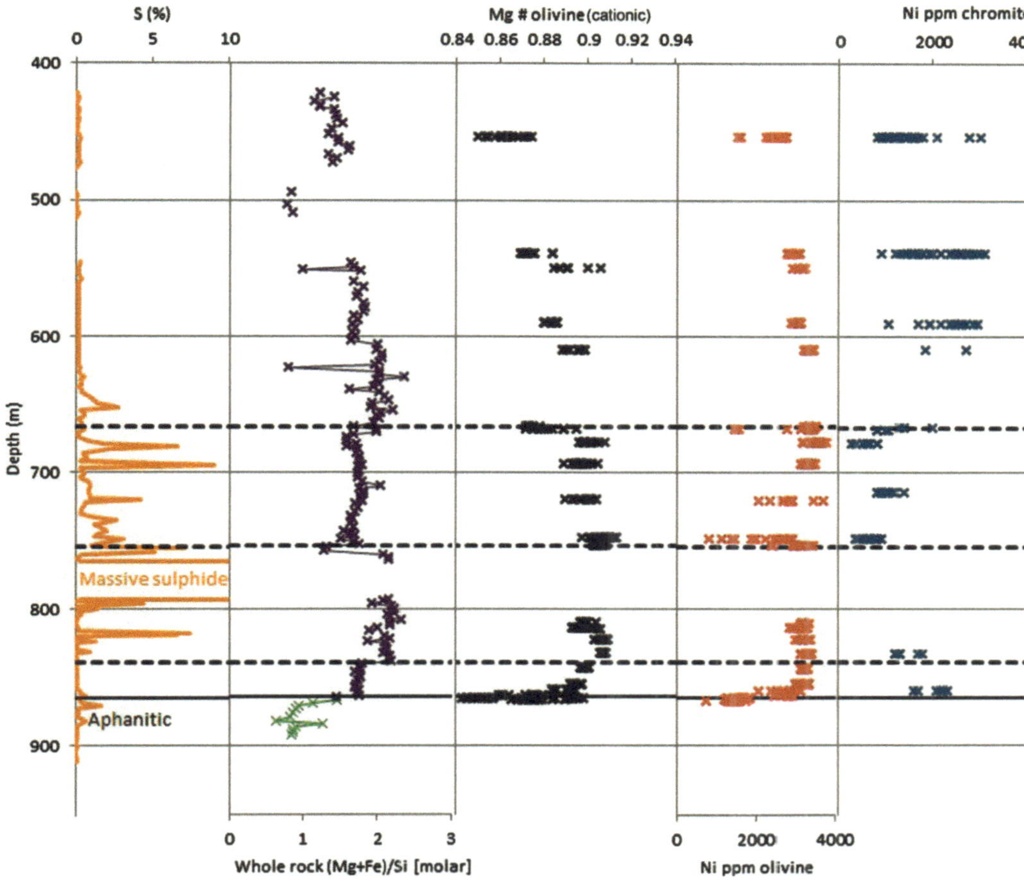

FIGURE 3.7.16 Down-hole plot showing whole-rock geochemistry and mineral chemistry data.

Plotted from left to right are: whole-rock S%, whole-rock (Mg + Fe/Si (volatile and sulfide corrected), WDS data for Mg/(Mg + Fe) in olivine, Ni ppm in olivine, and Ni ppm in chromite. Layers can be identified in the whole-rock geochemistry and these correlate broadly with subtle changes in mineral chemistry.

APHANITIC UNIT

Petrology

The unit is composed of a groundmass of fine-grained (20–50 μm) interlocking plagioclase and pyroxene in roughly equivalent proportions (~35–45% of groundmass each) and a smaller amount of fine-grained olivine (~15–20% of groundmass).

Phenocrysts of olivine and rare plagioclase make up about 10% of the rock, although the olivine phenocrysts are generally less than 1 mm in size and difficult to see in hand specimens. Olivine phenocrysts form elongate, narrow grains, with the longest one observed being a 4 mm × 0.2 mm single crystal (Fig. 3.7.17(A) and (C)). The cores of phenocrysts are frequently hollow. These are interpreted as extrusive olivine textures, which are compatible with the fine-grained nature of the groundmass. The

Table 3.7.2 Whole-rock geochemistry for major units of the Sakatti deposit

		Peridotite										
SiO_2	%	46.96	43.85	41.8	43.42	39.79	47.42	43.25	40.48	42.59	43.52	40.91
TiO_2	%	0.22	0.19	0.25	0.17	0.08	0.36	0.15	0.14	0.18	0.33	0.2
Al_2O_3	%	4.39	2.21	3.15	2.2	1.16	4.57	2.05	1.66	2.56	2.54	2.42
Fe_2O_3	%	1.61	1.59	1.69	1.45	1.32	1.4	1.54	1.32	1.48	1.6	1.76
FeO	%	9.67	9.54	10.15	8.72	7.94	8.39	9.24	7.91	8.88	9.59	10.54
MnO	%	0.11	0.16	0.17	0.18	0.18	0.22	0.14	0.13	0.18	0.16	0.13
MgO	%	29.21	39.76	37.88	41.53	46.12	30.74	40.09	46.26	41.55	38.74	40.5
CaO	%	6.48	0.9	1.98	0.76	1.48	5.22	1.03	0.57	0.8	1.61	1.21
Na_2O	%	0.11	0.07	0.24	<0.01	<0.01	0.49	<0.01	<0.01	<0.01	0.05	<0.01
K_2O	%	0.02	0.09	0.21	0.05	0.01	0.21	<0.01	0.04	0.12	0.08	<0.01
P_2O_5	%	<0.01	0.01	0.05	0.01	0.04	0.03	<0.01	0.02	0.04	0.05	0.03
LOI	%	8.5	13.1	11.9	14.8	15	9.3	14.4	15.3	14.1	12.1	13.4
C	%	0.26	0.13	0.15	0.13	0.55	0.11	0.21	0.13	0.18	0.1	0.24
S	%	0.5	0.2	0.9	0.43	0.13	0.23	0.66	0.11	0.14	0.53	1.23
Se	ppm	2.1	1.1	5.5	0.5	0.6	0.4	2.3	0.1	0.7	3.9	9.6
Ni	ppm	2645	2843	2625	2745	3077	1544	2914	2786	2497	2588	2574
Cu	ppm	172	376	5337	30	94	58	303	10	18	2021	7265
Co	ppm	109.2	130.7	131.3	125.1	131.5	93.7	133.7	117.4	131.2	111	182.5
Cr	ppm	3808	7469	7317	6212	9348	3738	11826	7023	8139	6228	3906
V	ppm	102	66	78	57	37	128	39	51	75	85	87
Sc	ppm	22	9	12	10	6	21	8	10	12	13	10
Zn	ppm	9.1	24.5	34.3	37.1	8.7	83.7	22.9	6.3	31.4	31.9	45.9
Rb	ppm	0.7	4	6.3	2.5	1.2	5.5	1.1	1.8	4.7	2.9	1.2
Sr	ppm	11.5	7.4	13.8	5.9	7.4	47.6	3.8	4.5	8	7.3	5
Ba	ppm	7	42	74	6	33	359	2	42	87	8	15

Th	ppm	0.3	0.5	0.6	0.5	<0.2	0.8	<0.2	0.2	0.2	0.6	0.2
U	ppm	<0.1	<0.1	0.1	0.1	<0.1	0.2	<0.1	<0.1	<0.1	<0.1	<0.1
Nb	ppm	0.4	0.1	<0.1	0.6	<0.1	0.9	<0.1	<0.1	0.4	1.6	0.7
Zr	ppm	12	13.6	22.9	12.1	5.5	22.6	15.6	10.4	18.3	30.9	14.8
Hf	ppm	0.4	0.5	0.6	0.2	0.2	0.8	0.4	0.2	0.5	0.8	0.3
Y	ppm	4	2.2	3.9	2.4	0.8	5.1	1.4	1.4	3.2	4.2	3.3
La	ppm	1.5	1.3	2.4	1	0.7	2.2	0.4	0.7	1.2	1.1	1.7
Ce	ppm	2.5	3.4	5	1.9	1.2	6.3	0.7	1.7	3.1	2.9	3.5
Pr	ppm	0.28	0.39	0.67	0.31	0.14	0.75	0.11	0.23	0.4	0.48	0.5
Nd	ppm	1.7	2.2	1.7	1.4	1.1	3.5	0.6	1.2	1.8	2.2	2.3
Sm	ppm	0.42	0.47	0.81	0.29	0.14	0.8	0.17	0.37	0.27	0.63	0.5
Eu	ppm	0.12	0.13	0.18	0.12	0.05	0.3	0.05	0.11	0.13	0.16	0.13
Gd	ppm	0.75	0.44	0.66	0.48	0.24	1.06	0.27	0.38	0.6	0.76	0.71
Tb	ppm	0.12	0.08	0.12	0.07	0.04	0.18	0.05	0.07	0.08	0.14	0.12
Dy	ppm	0.63	0.48	0.77	0.44	0.18	1.11	0.27	0.32	0.63	0.76	0.57
Ho	ppm	0.14	0.09	0.12	0.08	0.04	0.26	0.06	0.06	0.14	0.15	0.14
Er	ppm	0.42	0.27	0.41	0.32	0.08	0.7	0.17	0.18	0.27	0.42	0.34
Tm	ppm	0.06	0.04	0.06	0.04	0.02	0.08	0.02	0.02	0.05	0.07	0.05
Yb	ppm	0.42	0.26	0.31	0.29	0.11	0.6	0.19	0.2	0.29	0.37	0.4
Lu	ppm	0.06	0.06	0.06	0.04	0.01	0.09	0.02	0.04	0.04	0.06	0.06

Continued

Table 3.7.2 Whole-rock geochemistry for major units of the Sakatti deposit—cont'd

		Dunite (logging unit)		Altered Ultramafic			Aphanitic	
SiO_2	%	41.36	42.09	46.98	41.48	40.67	50.01	49.53
TiO_2	%	0.1	0.11	0.32	0.17	0.36	0.82	0.48
Al_2O_3	%	1.19	1.38	5.38	2.17	6.63	9.5	9.62
Fe_2O_3	%	1.4	1.5	1.42	1.27	0.9	1.34	1.37
FeO	%	8.39	8.98	8.52	7.6	5.37	8.02	8.21
MnO	%	0.13	0.14	0.18	0.14	0.09	0.13	0.14
MgO	%	44.51	42.92	29.53	37.85	26.62	20.7	20.51
CaO	%	0.79	0.95	6.35	7.9	16.84	6.65	7.39
Na_2O	%	0.07	0.05	0.18	0.04	1.59	2.25	2.08
K_2O	%	0.13	0.04	0.06	0.12	0.07	0.09	0.18
P_2O_5	%	0.02	<0.01	0.01	0.01	0.25	0.02	0.02
LOI	%	7.8	4.9	8.9	21.4	23.9	2.5	2.5
C	%	0.08	0.09	0.25	5.28	5.3	0.15	0.13
S	%	0.19	0.29	0.27	0.27	<0.02	0.06	0.12
Se	ppm	0.9	1.5	0.1	1.3	0.1	<0.1	0.2
Ni	ppm	3088	2795	1630	2990	1406	676	668
Cu	ppm	74	1443	85	923	41	41	47
Co	ppm	144.8	128.8	90.8	92.5	47	67.8	71.1
Cr	ppm	9478	7943	4739	4327	2643	1875	1947
V	ppm	54	41	120	71	112	188	140
Sc	ppm	6	7	18	10	13	29	32
Zn	ppm	18.7	19.4	27.6	3	6.5	11.8	17.9
Rb	ppm	5.9	2.8	2.2	5	3.7	2.4	4.8
Sr	ppm	3.6	8.8	16.8	36.6	83.3	96.7	91.8
Ba	ppm	13	21	18	4	46	74	62
Th	ppm	<0.2	<0.2	0.7	<0.2	2.2	0.9	0.4
U	ppm	<0.1	<0.1	0.2	<0.1	2.1	<0.1	<0.1
Nb	ppm	14.6	0.2	0.7	1.2	2.6	1.6	1.1
Zr	ppm	6.2	7	18.6	13.2	52.4	57.9	25.9

Hf	ppm	<0.1	0.2	0.6	0.4	1.5	1.6	0.9
Y	ppm	1.7	1.4	5.7	3.5	11.4	9.3	7.2
La	ppm	0.9	0.7	1.2	1.9	13.6	2.8	2.1
Ce	ppm	1.4	1.6	3.2	2.7	30.7	7	4.4
Pr	ppm	0.19	0.13	0.5	0.32	3.11	1	0.74
Nd	ppm	0.8	0.9	2.8	1.7	11.5	5.4	3.5
Sm	ppm	0.2	0.17	0.72	0.4	2.29	1.39	0.97
Eu	ppm	0.08	<0.02	0.21	0.05	0.6	0.38	0.34
Gd	ppm	0.24	0.2	1.01	0.54	2.08	1.79	1.32
Tb	ppm	0.06	<0.01	0.18	0.09	0.36	0.32	0.23
Dy	ppm	0.22	0.21	0.89	0.64	1.93	1.76	1.4
Ho	ppm	0.06	<0.02	0.22	0.15	0.49	0.38	0.27
Er	ppm	0.18	0.12	0.62	0.44	1.02	0.97	0.81
Tm	ppm	0.02	<0.01	0.08	0.05	0.16	0.13	0.13
Yb	ppm	0.13	0.11	0.57	0.38	1.2	0.91	0.8
Lu	ppm	0.02	<0.01	0.07	0.05	0.17	0.13	0.09

Note: FeO and Fe_2O_3 concentrations were calculated from Fe_2O_{3T} using a molar Fe_2O_3/FeO ratio of 0.15, typical of mantle-derived mafic–ultramafic magmas. All data are corrected to anhydrous compositions.

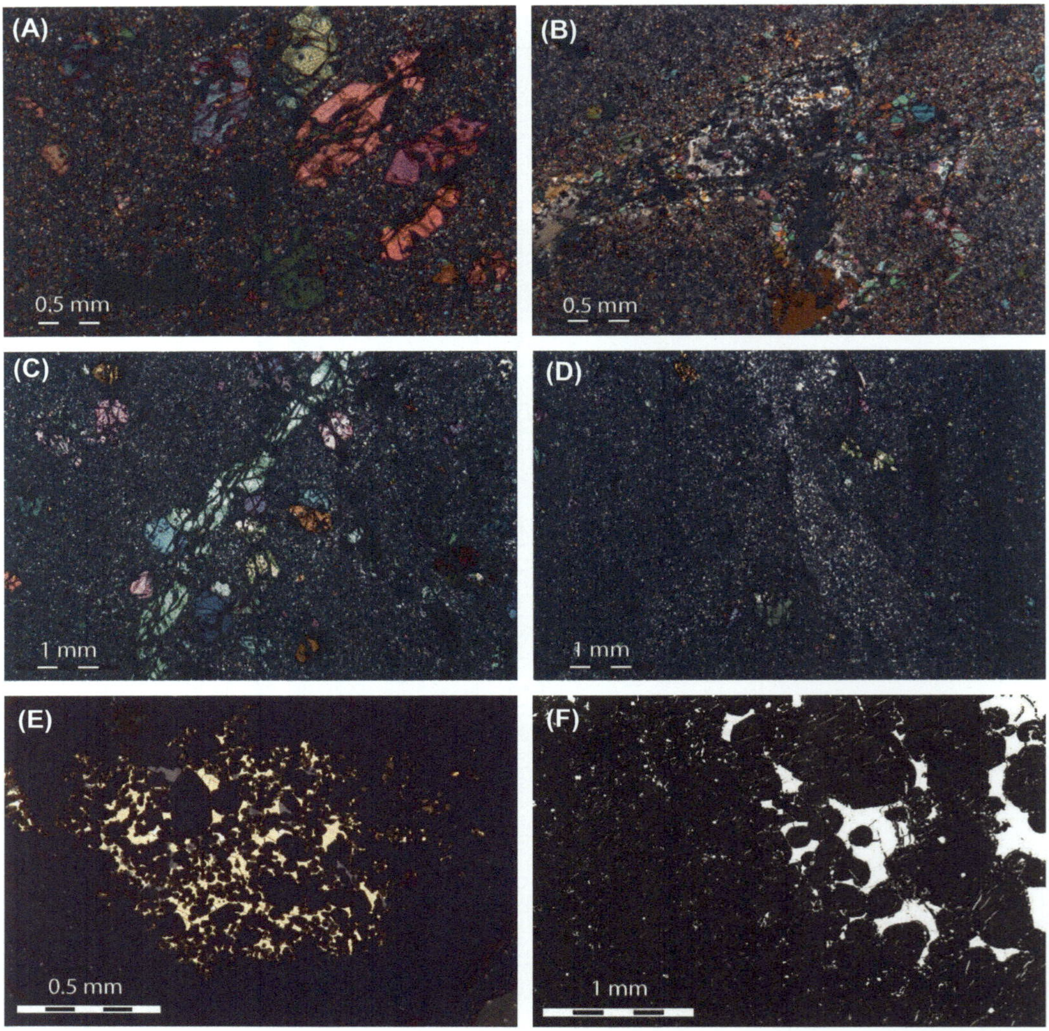

FIGURE 3.7.17 Images of thin sections from the aphanitic unit.

(A) Euhedral olivine in fine-grained pyroxene and plagioclase groundmass, typical of the aphanitic unit; (B) large plagioclase and olivine phenocrysts and serpentine veining; (C) elongate individual crystal of olivine and numerous smaller phenocrysts; (D) domaining of orthopyroxene- and clinopyroxene-dominated groundmass; (E) interstitial mineralization and olivine cumulate intruded into the aphanitic unit; (F) backscattered electron (BSE) image in Fig. 3.7.17(E) showing the difference between large cumulus olivines from the main cumulate body and phenocrystic olivine from the aphanitic unit.

chemistry of these olivines is similar to that of olivines in the peridotite unit, with Mg# of around 0.84–0.90. However, the Ni contents of the olivines in the aphanitic unit are distinctly lower (Ni 1800–1200 ppm) and do not exhibit the positive correlation with Mg# shown by the olivines of the peridotite unit (see Fig. 3.7.13).

Table 3.7.3 EDS (Energy Dispersive Spectra) data showing representative major element chemical composition of mineral phases in the Aphanitic Unit and their approximate proportions.

	Olivine (phenocryst)	Plagioclase (phenocryst)	Enstatite	Diopside	Olivine	Plagioclase	Ilmenite
Proportion of each	0.15	0.05	0.15	0.2	0.15	0.29	0.01
NaO	nd	5.4	nd	0.7	nd	7.3	nd
MgO	42.2	nd	31	16	42.9	nd	3.6
Al_2O_3	nd	27.6	0.7	2.1	nd	25.5	nd
SiO_2	39.4	54.7	56.9	52.8	40	59.2	nd
CaO	nd	10.8	1.2	21.5	nd	7.7	nd
TiO_2	nd	nd	nd	0.8	nd	nd	52.9
Cr_2O_3	nd	nd	nd	0.7	nd	nd	nd
MnO	nd	nd	nd	nd	nd	nd	0.6
FeO	18.5	0.7	11.8	5.4	17.2	0.5	42.9
Total	100.1	99.2	101.6	100	100.2	100.3	100.1

nd = not detected
Source: *Data acquired on the Zeiss EVO 15 LS at the Natural History Museum, London.*

Plagioclase is present as small crystals in the groundmass and as larger crystals (up to 3 mm in length) containing numerous inclusions of olivine but not pyroxene (Fig. 3.7.17(B)). These larger crystals were originally thought to be of hydrothermal origin; however, analysis revealed them to be more Ca-rich than their groundmass equivalents (Table 3.7.3). Thus, these plagioclase grains are interpreted as phenocrysts. Pyroxene is present only as a groundmass phase. As in the peridotite unit, orthopyroxene and clinopyroxene are present as enstatite and diopside respectively. These two pyroxenes are present in distinct domains, so that part of a sample is enstatite-dominated, whereas other parts are dominated by diopside (see Fig. 3.7.12(D)).

Texturally, the rock exhibits considerable variability and domaining of minerals at the microscopic and hand specimen scale, which is interpreted as autobrecciation. Unaltered samples of this rock appear relatively homogenous, with minor serpentine veining being the only discernible texture. However, in more altered samples, separate domains become more obvious. Finer, moderately brecciated domains alternate with coarser, more homogenous layers. These are interpreted as flow-tops and cumulate portions, respectively. The relatively thin cumulate portions (<20 m) are distinct from the peridotites of the main cumulate body in that the former are finer-grained, much less serpentinized, and have exclusively orthocumulate textures.

Alteration

One of the most striking characteristics of the aphanitic unit is that it is, in general, remarkably unaltered and composed almost entirely of fine-grained primary magmatic minerals. Small black serpentine veinlets are concentrated around lineations of phenocrysts, and the long axes of olivine crystals appear to be particularly susceptible.

In some occurrences, the aphanitic unit can be more pervasively altered. This does not appear to be spatially related to the peridotite unit. Instead, strongly altered segments can occur sporadically within the aphanitic unit, but particularly so in the vicinity of its contact with the hanging wall lithologies.

Whole-rock chemistry

As this unit is relatively unaltered, the whole-rock chemistry is particularly reliable in reflecting the original magmatic composition (see Table 3.7.2). The MgO contents of the rocks typically range between 19 and 22 wt% but may reach up to 30 wt%.

Such a high MgO content could, at first, appear to be in disagreement with petrological results, which show the unit to contain up to 35% plagioclase. However, the mineral chemistry (see Table 3.7.3) and the approximate proportions of each mineral from petrological observations were found to be in good agreement with the whole-rock geochemistry. The typical Na content of 1.5–2.5 wt% in the whole rock is reflective of the plagioclase content and is not related to alteration.

The observed variations in whole-rock major element geochemistry within the aphanitic unit are not consistent with simple addition and subtraction of olivine, but rather suggest the presence of plagioclase and olivine phenocrysts (Figs. 3.7.18 and 3.7.19).

Classification of the aphanitic unit

The rocks cannot be readily defined as ultramafic because they contain more than 10% felsic minerals; nor can they be defined as ultrabasic because they have more than 45 wt% SiO_2 (Le Maitre et al., 2002). High MgO content of more than 18 wt% precludes classification as a basalt. The lack of alteration and the high plagioclase content suggest that the Na values reflect the original magmatic values.

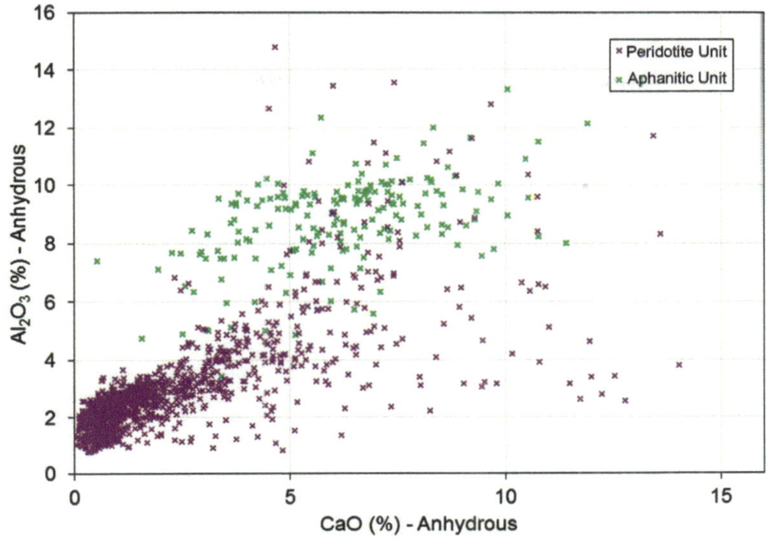

FIGURE 3.7.18 Binary variation diagram of whole-rock CaO versus Al_2O_3 for the peridotite and the aphanitic units.

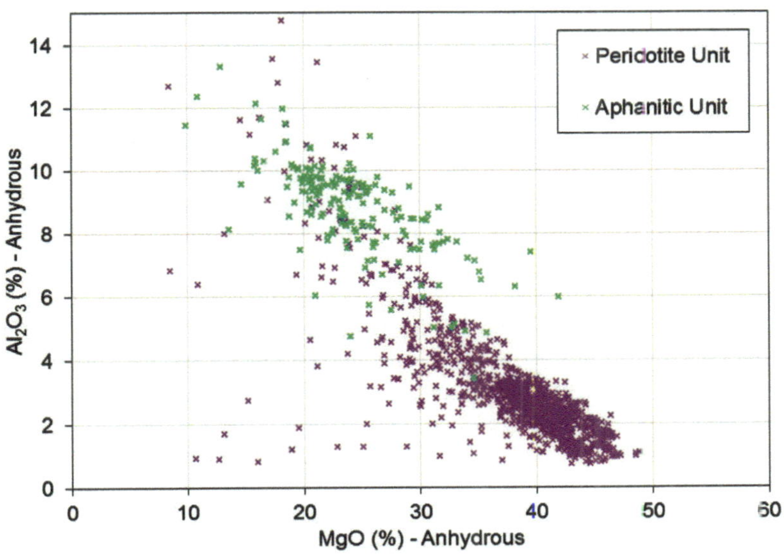

FIGURE 3.7.19 Binary variation diagram of whole-rock MgO versus Al_2O_3 showing both the peridotite and the aphanitic units.

The trend in the peridotite unit is consistent with olivine accumulation/dilution while the aphanitic unit shows a higher Al trend. This is potentially due to the presence of both olivine and plagioclase phenocrysts together.

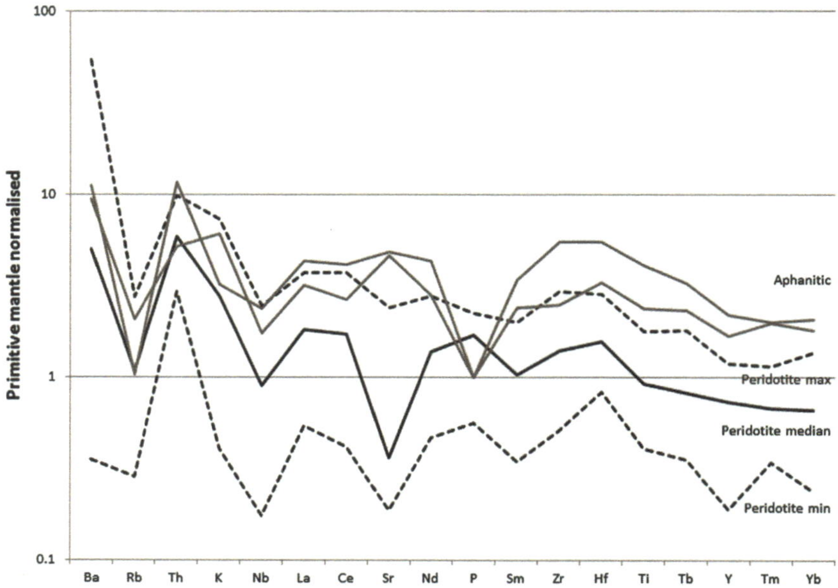

FIGURE 3.7.20 Trace element concentrations in representative samples of the peridotite and the aphanitic units.

Elements are arranged in order of compatibility and normalized to primitive mantle. *Normalization factors from McDonough and Sun (1995).*

According to Arndt (2008), the term komatiite should "be reserved solely for lavas with characteristic spinifex-textured olivines, or lavas that can be related directly, using field or petrological criteria, to lavas of this texture." Olivine spinifex texture is clearly identifiable in a 2-m section of one drill hole at the contact between the aphanitic unit and the peridotite unit. However, at present this is the only documented occurrence of olivine spinifex texture that has been observed in 155 drill holes and 99,388 m of diamond drilling.

For the purposes of this chapter it is deemed that this single occurrence of spinifex, lacking clear field relationships, is an insufficient basis on which to classify this unit as komatiitic. Furthermore, the high plagioclase content of this rock, meaning it is mafic rather than ultramafic, would make it an atypical komatiite. It is therefore termed a "plagioclase-rich picrite" on the basis of its high MgO content (Fig. 3.7.21). However, it is acknowledged that with the discovery of more olivine spinifex textured rocks, with clear field relationships, this definition could well be revised to a komatiite.

Nature of the contact between the peridotite and the aphanitic units

The contact between the peridotite unit and the aphanitic unit is generally diffuse, with the peridotite cumulate rock becoming more plagioclase- and pyroxene-rich (up to 30% pyroxene and 5% plagioclase) over an interval of 20–50 m approaching the contact. Despite the differing intercumulus material, the rock still remains identifiable as a cumulate and the Ni content of olivine does not change.

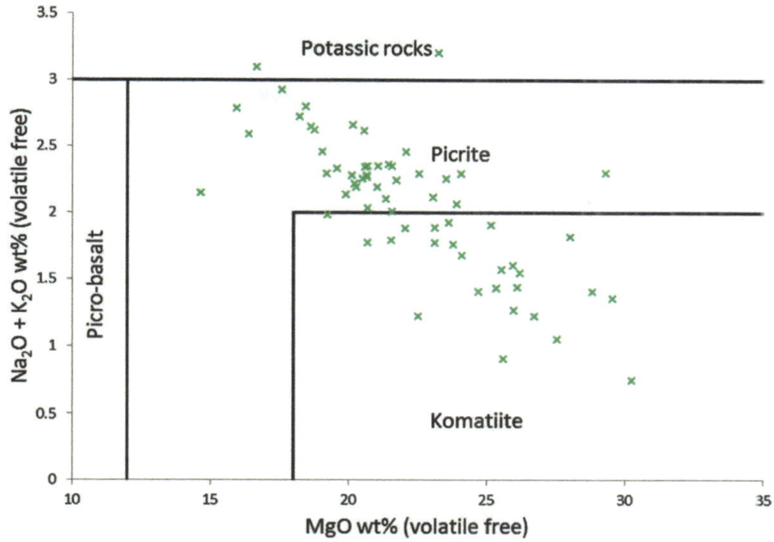

FIGURE 3.7.21 Whole-rock geochemistry of the aphanitic unit plotted according to the International Union of Geological Sciences (IUGS) komatiite-picrite definition.

All values are anhydrous.

Source: *From Le Maitre et al. (2002).*

In close proximity to the peridotite cumulate, the aphanitic unit contains 0.5–5 cm wide fingers or veinlets of peridotite. Within these fingers, cumulus euhedral olivine grains of up to 2 mm are distinguishable, identical to those in the peridotite unit (see Fig. 3.7.17(E) and (F)). In some cases, interstitial disseminated sulfide mineralization is present within the fingers (refer to Figs. 3.7.4(E) and 3.7.17(E) and (F)).

Microscopic analysis reveals that while the fingers are often relatively serpentinized, the surrounding aphanitic unit is usually relatively unaltered, retaining the fine-grained texture of plagioclase and pyroxene. Analysis of the mineral chemistry confirms the distinct nature of the two rock types as the olivine grains in the fingers exhibit a high Ni content while the olivines in the aphanitic unit have the low Ni content characteristic of this unit. It is also noted that generally the contacts between the fingers and the aphanitic unit are rich in large plagioclase crystals containing inclusions of olivine but no pyroxene. The origin of this texture is considered further in the "Discussion" section.

MINERALIZATION

Massive sulfides

The massive sulfides are characterized by a mixture of chalcopyrite and pyrrhotite-pentlandite (Fig. 3.7.22(A)). Pyrrhotite frequently carries up to 1 wt% Ni and contains small pentlandite "flames" orientated perpendicular to crystal boundaries (Fig. 3.7.22(G)). Pentlandite is often partly altered to violarite or less commonly millerite, particularly in samples where pyrite is present. Euhedral magnetite (typically 2–5 mm) is present throughout the massive sulfides and is interpreted to have crystallized from the sulfide liquid.

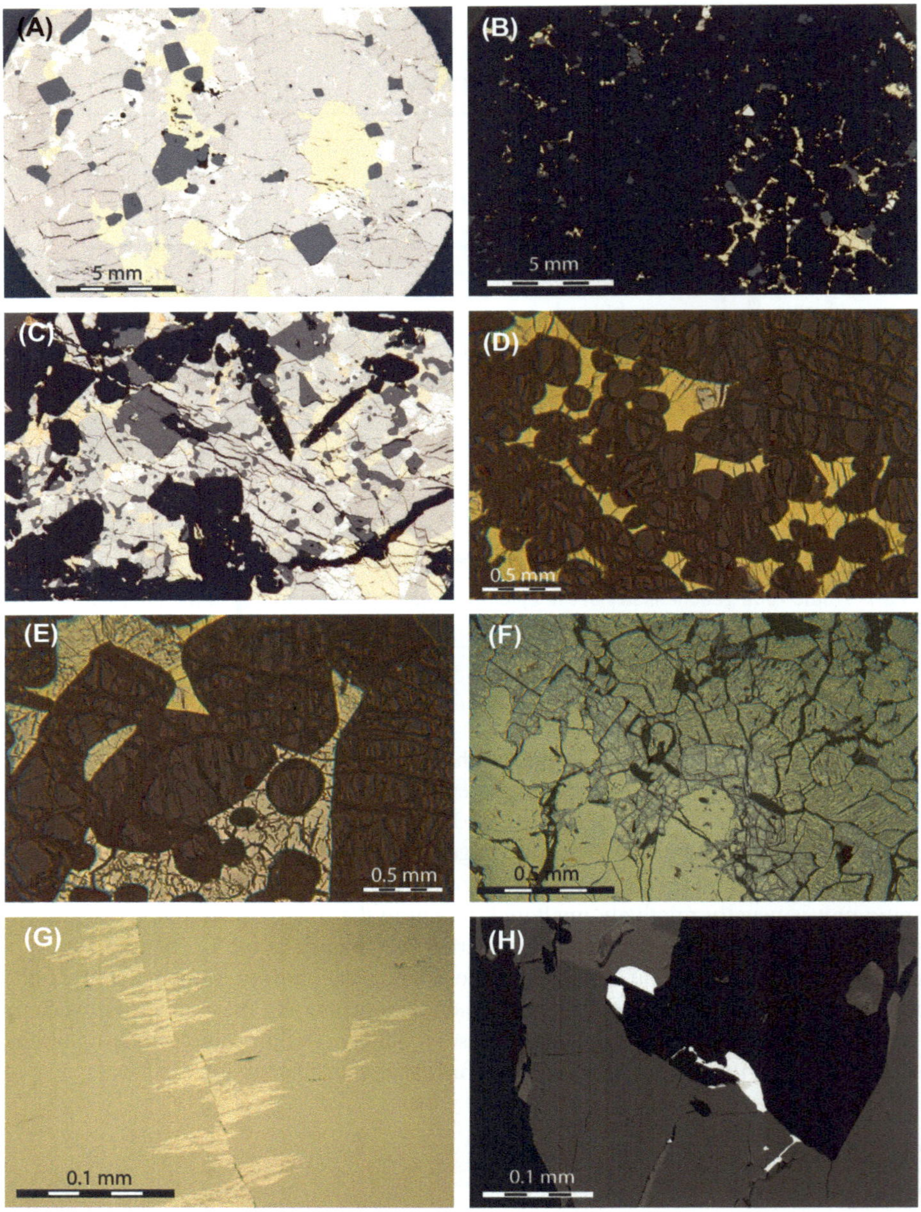

FIGURE 3.7.22 Polished block images of mineralization.

(A) Typical pyrrhotite-pentlandite massive sulfides with euhedral magnetite; (B) typical chalcopyrite-dominated disseminated sulfides in olivine cumulate; (C) semi-massive mineralization with large laths of black plagioclase and amphibole/pyroxene; (D) typical chalcopyrite-dominated disseminated mineralization interstitial to olivine cumulate; (E) atypical pyrrhotite-pentlandite disseminated mineralization in olivine cumulate; (F) pyrite-dominated mineralization containing (*left to right*) clean cobaltiferous pyrite, pentlandite, and violarite- and Ni-rich pyrite with magnetite lamellae; (G) pentlandite flames in pyrrhotite; (H) merenskyite (white) at a sulfide-silicate grain boundary.

PGE minerals of the merenskyite-moncheite-melonite series are present as small inclusions (1–100 μm), always associated with sulfide and primarily on grain boundaries of sulfide with either magnetite, silicate, or other sulfide (Fig. 3.7.22(H)). They have not been identified as being preferentially associated with any particular sulfide phase.

Pyrite is characteristic of the mineralization in the northeastern body, but it is also present in the main body, frequently toward the base of massive sulfide lenses. Pyrite is typically present as semi-spherical patches within other sulfides. This type of pyrite is frequently Co-rich (up to 1 wt.%) and Ni-poor. Where this pyrite is dominant, the interstitial spaces between the pyrite patches usually comprise Ni phases and Ni-rich pyrite with exsolved magnetite lamellae within it (Fig. 3.7.22(F)). The Ni phases comprise pentlandite, violarite, and minor millerite.

Vein sulfides

Sulfide veins are generally narrower than massive sulfide lenses and have variable orientations. The veins consist predominantly of chalcopyrite with minor pentlandite and pyrrhotite. Euhedral magnetite tends to be absent in this style of mineralization, distinguishing it from the massive sulfides. Platinum and Pd tenors are high and merenskyite-moncheite-melonite series tellurides are the only PGE minerals.

Disseminated sulfides

Disseminated sulfide mineralization occurs almost exclusively in the cumulate bodies, but can also be present in insignificant quantities in peridotite fingers within the aphanitic unit. Although the disseminated sulfides generally occur in the same portion of the cumulate body as the massive sulfides, they are not directly spatially related to the latter and are frequently more abundant distal from the massive sulfides. They are usually present as isolated, rounded blebs located interstitial to the olivines.

Chalcopyrite is invariably the dominant sulfide mineral in the disseminated mineralization. Magnetite lamellae are commonly present within chalcopyrite. The disseminated mineralization shares the same PGE mineralogy as the massive sulfides, with all PGE minerals being tellurides of the merenskyite-moncheite-melonite series.

Geochemistry

The distribution of Ni-Cu ratios within the deposit is not uniform. Massive sulfides show a shift in Ni-Cu ratios from relatively Ni-rich deep portions of the body in the northwest to relatively Cu-rich shallower portions in the southeast (Fig. 3.7.10). Vein and disseminated sulfides are generally more Cu-rich than massive sulfides. The Cu tenor of disseminated sulfides shifts from being relatively low in the northwest to relatively high in the southeast (Fig. 3.7.11). In the disseminated ore type, Pt and Pd occur roughly in the ratio of 2:1, which is not far from the average Pt-Pd ratio (1.8) measured for disseminated sulfides in the Kevitsa ore deposit (Mutanen, 1997). (See Table 3.7.4.)

The concentrations of the iridium-group PGEs (IPGE: Os, Ir, Ru) and palladium-group PGEs (PPGE: Rh, Pt, Pd) show variable primitive mantle-normalized patterns (Fig. 3.7.23) consistent with sulfide liquid fractionation in the massive sulfides (e.g., Ebel and Naldrett, 1996). Cu-rich sulfides are enriched in Pt and Pd and depleted in Os, Ir, Ru relative to Ni-rich sulfides. The disseminated sulfide samples show the same enrichment and depletion pattern as the Cu-rich massive sulfide (Fig. 3.7.23). Both the Ni-Cu and IPGE/PPGE ratios indicate that the Sakatti deposit has experienced considerable fractionation of the sulfide liquid.

Table 3.7.4 Whole-rock geochemistry showing examples of PGE, Ni, Cu, and Au values from mineralization in the Sakatti deposit

		Disseminated				Vein style			Massive (Cu-rich)				Massive (Ni-rich)			
Ni	%	0.22	0.04	0.20	0.20	0.05	0.5	1.21	0.6	1.93	1.70	4.22	4.9	4.76	8.16	9.46
Cu	%	3.22	0.6	0.20	0.26	0.05	6.95	18.44	28.4	6.55	2.60	5.84	2.24	1.92	2.23	2.34
S	%	3.78	1.59	0.43	1.87	0.10	10.13	28.55	30.84	16.68	43.46	32.26	34.02	33.51	35.16	23.69
Os	ppb	3	2	2	2	2	0.5	2	1	0.5	23	9	16	22	22	6
Ir	ppb	1	1	2	2	2	1	0.5	0.5	0.5	24	7	27	36	33	7
Ru	ppb	7	4	7	5	7	2	17	14	3	10	8	10	15	12	7
Rh	ppb	11	2	2	2	3	10	227	121	13	52	8	59	100	99	14
Pt	ppb	4941	2666	3136	2387	1124	15273	2443	1503	19618	2653	2469	1399	1935	472	1255
Pd	ppb	1644	1628	1307	1445	126	2064	26	1025	4010	1418	6095	1626	3536	1904	4198
Au	ppb	365	33	66	96	178	3119	2397	999	911	15	489	118	39	325	954

Note that grades across the deposit are variable, and while the reported values are typical, they are not meant to be representative of the whole deposit.

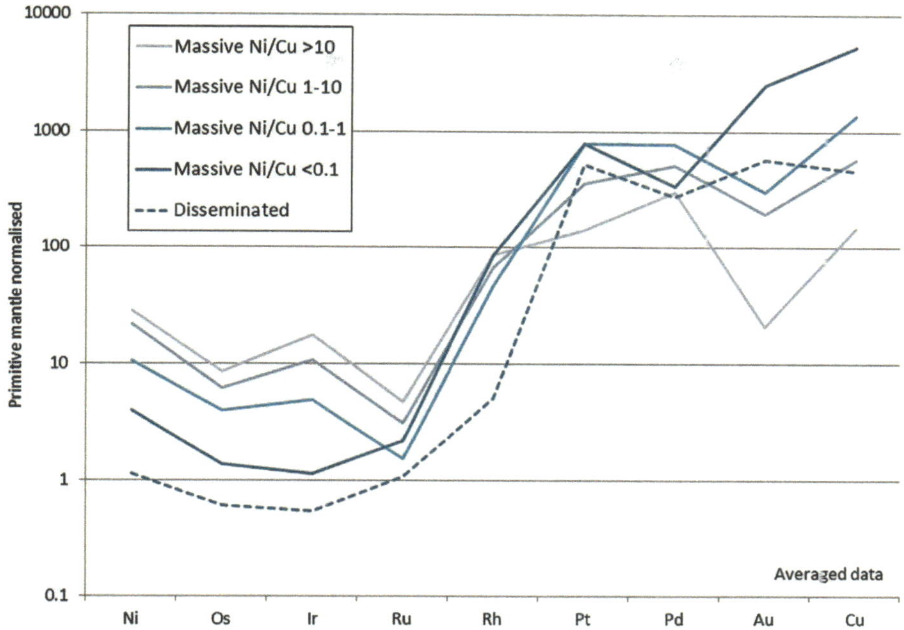

FIGURE 3.7.23 Whole-rock PGE, Ni, Cu, and Au data normalized to primitive mantle.

Massive sulfide data have been grouped according to Ni-Cu ratio. Vein sulfides have not been separated from massive sulfides. Note that the data are not normalized to 100% sulfide. *Normalization factors from McDonough and Sun (1995).*

S isotopes

The sulfides from the Sakatti deposit have a relatively uniform $\delta^{34}S$ (Vienna Canyon Diablo Troilite or VCDT) averaging +3‰ (Brownscombe et al., 2013). While sulfide-bearing sediments have not been intersected at Sakatti, samples of black schist from the Matarakoski formation in the surrounding area exhibit a wide spread of $\delta^{34}S$ (VCDT) signatures (Fig. 3.7.24). This is consistent with previous studies on the Matarakoski formation (Grinenko et al., 2003). The fact that the Sakatti data are so homogenous and plot relatively close to mantle values suggests that S saturation in the deposit is unlikely to have been caused by localized, in situ assimilation of the Matarakoski formation sediments.

Consistent with the S isotope results, S-Se ratios within the deposit (1500–4000, 10th to 90th percentile) are also in the range of typical mantle values, and likewise do not provide any evidence for assimilation of crustal S. The lower end of the range of S-Se ratios tends to be associated with Cu-rich sulfide, consistent with the experimentally derived partitioning behavior of S and Se in fractionating sulfide (Helmy et al., 2010).

If assimilation of S within the crust had occurred, the assimilant must have had $\delta^{34}S$ and S-Se signatures close to mantle values, and assimilation must have occurred prior to final emplacement allowing for high *R* factors and complete homogenization of the resultant disseminated and massive sulfides.

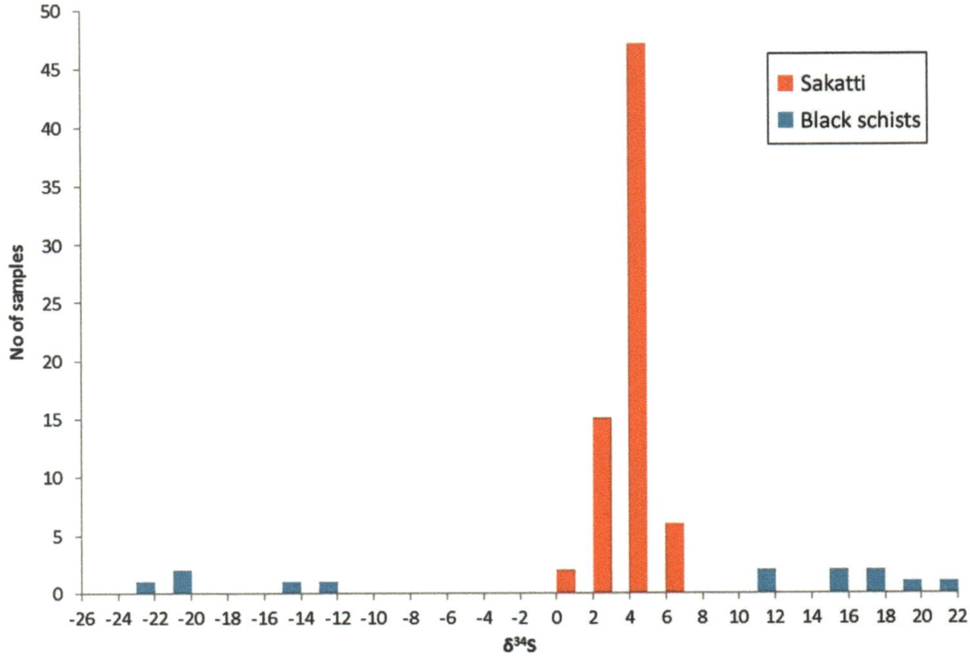

FIGURE 3.7.24 Histogram showing $\delta^{34}S$ values of sulfides from the Sakatti deposit and the regional Matarakoski formation black schists.

Analysis was undertaken at the Scottish Universities Environmental Research Centre, Glasgow.

Source: *From Brownscombe et al. (2013).*

Anhydrite is present within the intrusion, sometimes associated with pyrite. The pyrite has typical Sakatti $\delta^{34}S$ values of +3‰ whereas the anhydrite has $\delta^{34}S$ values of +10‰. Although it is theoretically possible that the anhydrite is externally derived, the fact that sulfides in close proximity do not show elevated $\delta^{34}S$ values is inconsistent with this hypothesis. Instead, the anhydrite is interpreted to be an oxidation product of magmatic sulfides.

Platinum-group element minerals

It is estimated that between 30 and 50% of the measured Pd in the massive sulfides resides in solid solution within pentlandite, based on LA-ICP-MS (Laser Ablation Inductively Coupled Plasma Mass Spectrometry) analysis and mass balance on eight samples from two drill holes. The remainder of the Pd and all of the Pt is located within telluride minerals of the moncheite-merenskyite-melonite series and minor michenerite. A wide range of Pt-Pd-Ni compositions is observed, which is unusual for this series (Fig. 3.7.25). By contrast, in the majority of known occurrences, these tellurides are restricted to either a Pt-Pd trend or a Pd-Ni trend (Helmy et al., 2007). Departure from these two trends, as well as the overall dominance of telluride phases, is shared by the Kevitsa deposits (Gervilla and Kojonen, 2002).

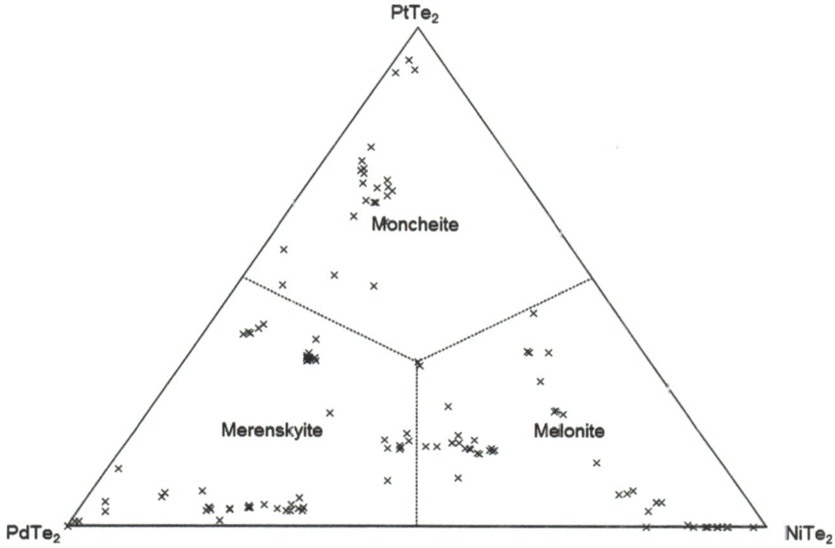

FIGURE 3.7.25 Chemistry of the PGE phases at the Sakatti deposit in atomic proportions.

Source: *Measured by WDS using the Cameca SX 100 at the Natural History Museum, London.*

DISCUSSION

INTRUSIVE VERSUS EXTRUSIVE

The Sakatti main cumulate body has been interpreted as a shallow level, conduit-like intrusion (Brownscombe et al., 2013) or as a cumulate portion of the picritic-komatiitic lava flows forming the aphanitic unit (T. Halkoaho, personal communication, 2014). While it is true that the Sakatti main cumulate body is hosted by volcanic rocks, and ultrabasic lava successions are known to have associated peridotitic to dunitic cumulate bodies, there are several lines of evidence that suggest that Sakatti is not a lava channel cumulate but an intrusion into a lava flow field:

- The tubular shape of the cumulate body (see Fig. 3.7.10).
- The peridotite has an apparent intrusive contact with the aphanitic unit both above and below the cumulate body, with finger-like intrusions of cumulate peridotite into the aphanitic footwall and hanging wall.
- The cumulate body is texturally and chemically different from the aphanitic unit such as the abundant plagioclase and low Ni contents of olivine in the latter. The aphanitic unit itself contains minor interlayered cumulate portions, that are distinct from the cumulate unit.
- The cumulate body is large and homogenous. It is more than 400-m thick in sections, with no clear evidence of flow tops or chilled margins.

It is suggested that the aphanitic unit and the main peridotite cumulate body are probably related to the same magmatic event and the same magmatic feeder system, but that the main cumulate body was formed slightly later as a subvolcanic intrusion.

NATURE OF CONTACT WITH THE APHANITIC UNIT

The contact between the peridotite unit and the aphanitic unit in both the hanging wall and the footwall consists of fingers of peridotite injecting into the lava.

This pattern of microintrusions suggests that the peridotite exploited a preexisting texture or weakness in the aphanitic unit. Petrographic examination of samples of the aphanitic unit reveal a network of fine serpentine veins that are concentrated along olivine and plagioclase phenocrysts.

It is proposed that the serpentine veining occurred prior to emplacement of the main cumulate body, or was potentially associated with it. The zones of hydrated microveins along with plagioclase phenocrysts would have a lower solidus than the rest of the aphanitic rock, resulting in preferential melting during emplacement of the peridotite and local injection of the latter.

DECOUPLING OF THE COMPOSITION OF OLIVINE AND SULFIDE MINERALIZATION

Olivine within the main cumulate body contains relatively high concentrations of Ni, which indicates crystallization from a melt that had not been depleted in Ni in response to segregating sulfide melt. Furthermore considering the primitive nature of the host rocks, the mineralization is, overall, unusually Cu-rich. These observations imply that the olivine was not in equilibrium with the sulfide melt. The current silicate host cannot, therefore, be considered as having crystallized from magma parental to the sulfides.

It is proposed that the sulfides were initially deposited by earlier magmatic activity "upstream" in the Sakatti conduit system and were subsequently remobilized by an olivine-charged silicate magma to be deposited in the current location. In addition, sulfide liquid was mobilized during or after in situ sulfide melt fractionation to form distinct veins of massive sulfides. The veins are relatively Cu rich because Cu-rich residual liquid has a lower solidus than the silicate melt. This model is consistent with the sharp contacts of the massive sulfide lenses and veins and their intrusion into the footwall where they continue to undergo fractionation.

SUMMARY

The Sakatti Cu-Ni-PGE deposit is a Cu-rich magmatic Ni sulphide deposit hosted by an olivine cumulate.

- The footwall and in some places hanging wall rock is a high Mg volcanic succession.
- The cumulus olivine contains high Ni values, up to 3700 ppm.
- Disseminated sulphide mineralisation is Cu-dominated and shows evolved PGE patterns.
- Massive sulphide mineralisation can be either Ni- or Cu-dominated, shows clear spatial zonation in Ni/Cu ratios, and indicates typical sulphide fractionation.

The deposit is interpreted as a conduit-like intrusion through the volcanic succession. Sulphide formation is inferred to have occurred at an earlier stage and undergone a degree of fractionation before being remobilised by the current silicate host.

ACKNOWLEDGMENTS

The Anglo American Finland Exploration team has been a tight-knit unit and has seen discovery of this deposit through to the current level of understanding. All of the work presented builds on the work of the team, which includes, but is not limited to, geologists Sebastian Stelter, Ryan Preece, Klara Collis, Peter Dodds, Louise Wright, and Catherine Reynolds.

Special thanks go to Brian Williams and Denis Fitzpatrick who were crucial in the discovery of Sakatti. In addition, Johanna Alitalo and Jorgen Ylitalo were part of the team during the discovery period, and deserve credit. Furthermore, without the belief and support of Owen Bavinton and Graham Brown, Sakatti would not have been discovered by Anglo American.

Thanks also go to the current Sakatti project manager Jukka Jokela and to Tapio Halkoaho (GTK) for stimulating discussions.

Anglo American is thanked for the permission to publish this chapter.

Thanks are due to Jamie Wilkinson of the Natural History Museum, London, who was a supervisor of the senior author's (W. Brownscombe) Ph.D. project from which this chapter arose. Thanks are also due to Anton Kearsley and John Spratt of the Natural History Museum, London, for their assistance with SEM and EPMA techniques.

REFERENCES

Arndt, N.T. (Ed.), 2008. Komatiite, second ed, Cambridge University Press, Cambridge, p. 467.

Barnes, S.J., 1998. Chromite in komatiites: 1. magmatic controls on crystallisation and composition. Journal of Petrology 39, 1689–1720.

Bickle, M.J., 1982. The magnesium contents of komatiitic liquids. In: Arndt, N.T., Nisbet, E.G. (Eds.), Komatiites. George Allen and Unwin, London, pp. 479–494.

Brownscombe, W., Herrington, R.J., Wilkinson, J.J., et al., 2013. Geochemistry of the Sakatti magmatic Cu-Ni-PGE deposit, northern Finland. In: Proceedings of the 12th Biennial SGA Meeting on Mineral Deposit Research for a High-Tech World, August 12–15, Uppsala. Geological Survey of Sweden, Uppsala, pp. 956–959.

Brenan, J.M., Caciagli, N.C., 2000. Fe-Ni exchange between olivine and sulphide liquid: implications for oxygen barometry in sulphide-saturated magmas. Geochimica et Cosmochimica Acta 64, 307–320.

Ebel, D.S., Naldrett, A.J., 1996. Fractional crystallization of sulfide ore liquids at high temperature. Economic Geology 91, 607–621.

Fiorentini, M.L., Beresford, S.W., Deloule, E., et al., 2008. The role of mantle-derived volatiles in the petrogenesis of Palaeoproterozoic ferropicrites in the Pechenga greenstone belt, northwestern Russia: Insights from in-situ microbeam and nanobeam analysis of hydromagmatic amphibole. Earth and Planetary Science Letters 268, 2–14.

Gervilla, F., Kojonen, K., 2002. The platinum-group minerals in the upper section of the Keivitsansarvi Ni-Cu-PGE deposit, northern Finland. Canadian Mineralogist 40, 377–394.

Grinenko, L.N., Hanski, E., Grinenko, V.A., 2003. Formation conditions of the Keivitsa Cu-Ni deposit, northern Finland: Evidence from S and C isotopes. Geochemistry International 41 (2), 154–167.

Hanski, E., Huhma, H., 2005. Central Lapland Greenstone Belt. In: Lehtinen, M., Nurmi, P.A., Rämö, O.T. (Eds.), The Precambrian Geology of Finland—Key to the Evolution of the Fennoscandian Shield. Elsevier, Amsterdam, pp. 139–193.

Helmy, H.M., Ballhaus, C., Berndt, J., et al., 2007. Formation of Pt, Pd and Ni tellurides: experiments in sulphide-telluride systems. Contributions to Mineralogy and Petrology 153, 577–591.

Helmy, H.M., Ballhaus, C., Wohlgemuth-Ueberwasser, C., et al., 2010. Partitioning of Se, As, Sb, Te and Bi between monosulfide solid solution and sulfide melt—Application to magmatic sulfide deposits. Geochimica et Cosmochimica Acta 74, 6174–6179.

Le Maitre, R.W., Streckeisen, A., Zanettin, B., et al. (Eds), 2002. In: Igneous Rocks. A Classification and Glossary of Terms. Recommendations of the International Union of Geological Sciences Subcommission on the Systematics of Igneous Rocks. Cambridge University Press, Cambridge, p. 236.

McDonough, W.F., Sun, S.-S., 1995. The composition of the Earth. Chemical Geology 120, 223–253.

Mutanen, T., 1997. Geology and ore petrology of the Akanvaara and Koitelainen mafic layered intrusions and the Keivitsa-Satovaara layered complex, northern Finland. Geological Survey of Finland, Bulletin. 395, p. 233.

Mutanen, T., Huhma, H., 2001. U-Pb geochronology of the Koitelainen, Akanvaara and Keivitsa mafic layered intrusions and related rocks. Geological Survey of Finland. Special Paper 33, 229–246.

Nisbet, E.G., Cheadle, M.J., Arndt, N.T., Bickle, M.J., 1993. Constraining the potential temperature of the Archaean mantle: a review of the evidence from komatiites. Lithos 30, 291–307.

Roeder, P.L., Emslie, R.F., 1970. Olivine-liquid equilibrium. Contributions to Mineralogy and Petrology 29, 275–289.

Yang, S.-H., Maier, W.D., Hanski, E.J., et al., 2013. Origin of ultra-nickeliferous olivine in the Kevitsa Ni-Cu-PGE-mineralised intrusion, northern Finland. Contributions to Mineralogy and Petrology 166, 81–95.

CHAPTER

NICKEL DEPOSITS OF THE 1.88 GA KOTALAHTI AND VAMMALA BELTS

3.8

H.V. Makkonen

ABSTRACT

The Kotalahti and Vammala belts in central and southern Finland, respectively, contain most of the 1.88 Ga intrusive Svecofennian nickel deposits. The Kotalahti belt is located close to the Archean craton margin, while the Vammala belt occurs further to the southwest within the Svecofennian domain. The total ore production from the 10 Svecofennian nickel mines in Finland has been about 45 Mt at 0.7% Ni and the total premining resource of all deposits known to date is about 73 Mt at 0.6% Ni. The belts are characterized by amphibolite facies to granulite facies metamorphic grade and abundant schollen- and schlieren-migmatites. The intrusions were emplaced during peak deformation and metamorphism. This resulted in highly variable settings of the intrusions and, in many cases, their dismemberment. The weakly differentiated, dominantly ultramafic Vammala-type intrusions consist almost entirely of olivine cumulates and represent magma conduits. The more strongly differentiated, mafic and mafic–ultramafic, Kotalahti-type intrusions consist of olivine cumulates, pyroxene cumulates, and plagioclase-bearing cumulates. In both belts, the parental magma was basaltic with MgO contents mostly around 10–12 wt%. Sulfide segregation took place via crustal contamination when sulfur was added to the magma from the wall rocks. The mass ratio of silicate melt/sulfide melt (R factor) varied mostly between 200 and 1300 and, consequently, a wide range in nickel tenor is observed.

Keywords: Finland; Svecofennian; nickel; basalt; contamination; migmatites; olivine.

INTRODUCTION

The nickel-(copper) sulfide deposits hosted by the Svecofennian 1.88 Ga mafic–ultramafic intrusions have played a major role in Finnish nickel mining history. Altogether, 10 deposits have been mined, beginning in 1941 with the Makola deposit and extending to 2013 at the Hitura mine, which has become the largest nickel deposit of this type in Finland (the total amount of hoisted ore is about 16.5 Mt at 0.6% Ni and 0.2% Cu). The total ore production of the Svecofennian nickel mines in Finland stands at about 45 Mt at 0.7% Ni. The total premining resource of all the deposits known to date is about 73 Mt at 0.6% Ni, 0.3% Cu, and 0.03% Co (Tables 3.8.1 and 3.8.2). The current estimate for undiscovered resources down to a depth of 1 km (50% probability) is at least 480,000 t Ni, 200,000 t Cu, and 23,000 t Co (Rasilainen et al., 2012), representing a similar amount of metals as the total premining resources.

Owing to the synorogenic timing of the magmatism, the intrusions have a complicated tectonomagmatic history. This makes the Svecofennian intrusions quite distinct when compared to anorogenic nickel-sulfide-bearing intrusions such as Sudbury (Canada) or Norilsk (Russia). However,

Mineral Deposits of Finland. http://dx.doi.org/10.1016/B978-0-12-410438-9.00010-8

Table 3.8.1 Svecofennian nickel mines, total hoisted ore and average metal grades, in Finland 1941–2013

Mine	Mt	Ni (%)	Cu (%)	Ni (met. t)
Hitura	16.50	0.60	0.20	99,000
Kotalahti	12.36	0.66	0.26	81,576
Vammala	7.57	0.68	0.42	51,476
Enonkoski	6.71	0.76	0.22	50,996
Kylmäkoski	0.69	0.36	0.27	2484
Telkkälä	0.61	1.29	0.33	7869
Makola	0.41	0.81	0.43	3321
Hälvälä	0.25	1.41	0.35	3525
Särkiniemi	0.12	0.92	0.44	1104
Puumala	0.02	0.67	0.24	134
Total	**45.24**	**0.67**	**0.26**	**301,485**

deposits that resemble the Finnish ones in terms of tectonic setting are known from the Lappvattnet nickel belt in Sweden (Weihed et al., 1992); the Sveconorwegian (Norway) and Grenville (Canada) belts (Boyd and Mathieson, 1979; Boyd et al., 1988; Lamberg, 2005); the Variscan orogeny of southwestern Spain (Aguablanca; e.g., Piña et al., 2010); the circum-Superior belt of Canada, which contains the Thompson nickel belt (1.88 Ga; Hulbert et al., 2005), Fox River belt (1.88 Ga; Heaman et al., 1986), and Lynn Lake belt (1.87 Ga; Turek et al., 2000); and the mafic–ultramafic intrusions of the Halls Creek orogen in the East Kimberley, Western Australia (Hoatson and Blake, 2000).

The Svecofennian nickel-bearing intrusions of Finland have been comprehensively studied since the 1960s, by academic institutions and exploration geologists of the Outokumpu Company and the Geological Survey of Finland (Häkli, 1963, 1968, 1970, 1971; Papunen, 1970, 1974, 1980, 1985, 1986, 1989, 2003, 2005; Gaál, 1972, 1980; Häkli et al., 1975, 1976, 1979; Papunen et al., 1979; Papunen and Mäkelä, 1980; Grundström, 1980, 1985; Papunen and Koskinen, 1985; Papunen and Vorma, 1985; Papunen and Gorbunov, 1985; Mäkinen, 1987; Isomäki, 1994; Peltonen, 1995a, 1995b, 2005; Papunen and Penttilä, 1996; Makkonen, 1996, 2005; Mäkinen and Makkonen, 2004; Lamberg, 2005; Makkonen and Huhma, 2007; Makkonen et al., 2008, 2010; and Barnes et al., 2009).

Peltonen (2005) summarized the age data for the Svecofennian mafic–ultramafic intrusions. Since then, only one new age determination has been made. The Keskimmäinen gabbro-pyroxenite intrusion in the Siilinjärvi area yielded a U-Pb zircon age of 1879 ± 6 Ma (in situ laser ablation mass spectrometer; Kalliomäki, 2013). Most of the mafic–ultramafic intrusions record ages between 1875 and 1885 Ma, and all Ni-bearing intrusions belong to the 1880 Ma age group. Notably, this is the same age as that for the Paleoproterozoic nickel belts of Thompson, Cape Smith, Fox River, and Lynn Lake in Canada.

AREAL DISTRIBUTION OF DEPOSITS

The Svecofennian 1.89–1.87 Ga mafic–ultramafic intrusions occur throughout the Svecofennian of central and southern Finland, but most of the nickel-bearing intrusions occur within the Kotalahti and Vammala nickel belts to the south and east of the Central Finland Granitoid Complex (Fig. 3.8.1).

Table 3.8.2 Svecofennian nickel deposits in Finland, estimated premining ore tonnages and average metal grades

Deposit	Status	Tonnage	Ni (%)	Cu (%)	Co (%)
Hitura	Mine	19,300,000	0.61	0.21	0.020
Kotalahti	Mine	13,300,000	0.66	0.26	0.030
Vammala	Mine	9,200,000	0.64	0.40	0.040
Laukunkangas	Mine	7,900,000	0.72	0.20	0.030
Ruimu	Deposit	3,500,000	0.32	0.28	0.040
Sääksjärvi	Deposit	3,500,000	0.24	0.33	0.030
Niinimäki	Deposit	2,700,000	0.40	0.14	0.018
Sahakoski	Deposit	1,600,000	0.65	0.19	0.030
Rytky	Deposit	1,540,000	0.71	0.29	0.030
Oravainen	Deposit	1,300,000	0.95	0.16	0.030
Ekojoki	Deposit	1,140,000	0.53	0.42	0.024
Kovero-oja	Mine	1,100,000	0.40	0.33	0.017
Makola	Mine	1,000,000	0.80	0.45	0.050
Hyvelä	Deposit	807,000	0.52	0.26	0.034
Kylmäkoski	Mine	690,000	0.36	0.27	0.010
Telkkälä	Mine	605,000	1.29	0.33	0.050
Makkola	Deposit	526,000	0.52	0.18	0.034
Mäntymäki	Deposit	466,000	0.73	0.20	0.010
Hälvälä	Mine	448,000	1.50	0.36	0.075
Rausenkulma	Deposit	375,000	0.36	0.49	0.023
Särkiniemi	Mine	292,000	0.91	0.53	0.063
Liakka	Deposit	250,000	0.37	0.78	0.020
Ilmolahti	Deposit	210,000	0.37	0.28	0.040
Sarkalahti	Deposit	190,000	1.02	0.33	0.027
Tevanniemi	Deposit	182,000	0.63	0.15	0.030
Hanhisalo	Deposit	143,000	0.61	0.20	nd
Törmälä	Deposit	116,000	0.60	0.33	0.030
Mäkisalo	Deposit	104,000	0.43	0.28	0.030
Niinikoski	Deposit	83,000	0.43	0.13	0.045
Rietsalo	Deposit	56,000	0.53	0.53	0.015
Heiskalanmäki	Deposit	55,000	0.55	0.25	0.015
Kekonen	Deposit	50,000	0.54	0.21	0.019
Härmäniemi	Deposit	37,000	0.84	0.24	0.040
Vehmasjoki	Deposit	36,000	0.94	0.69	0.060
Kitula	Mine	34,600	0.87	0.24	0.067
Pihlajasalo	Deposit	20,000	0.84	0.17	0.020
Total		**72,855,600**	**0.61**	**0.27**	**0.029**

Source: *Modified from Rasilainen et al. (2012, Table 1) and references therein. Tevanniemi after Eeronheimo and Pietilä (1988), Hälvälä after Eeronheimo (1985), and Hanhisalo after Kontoniemi and Forss (1998).*

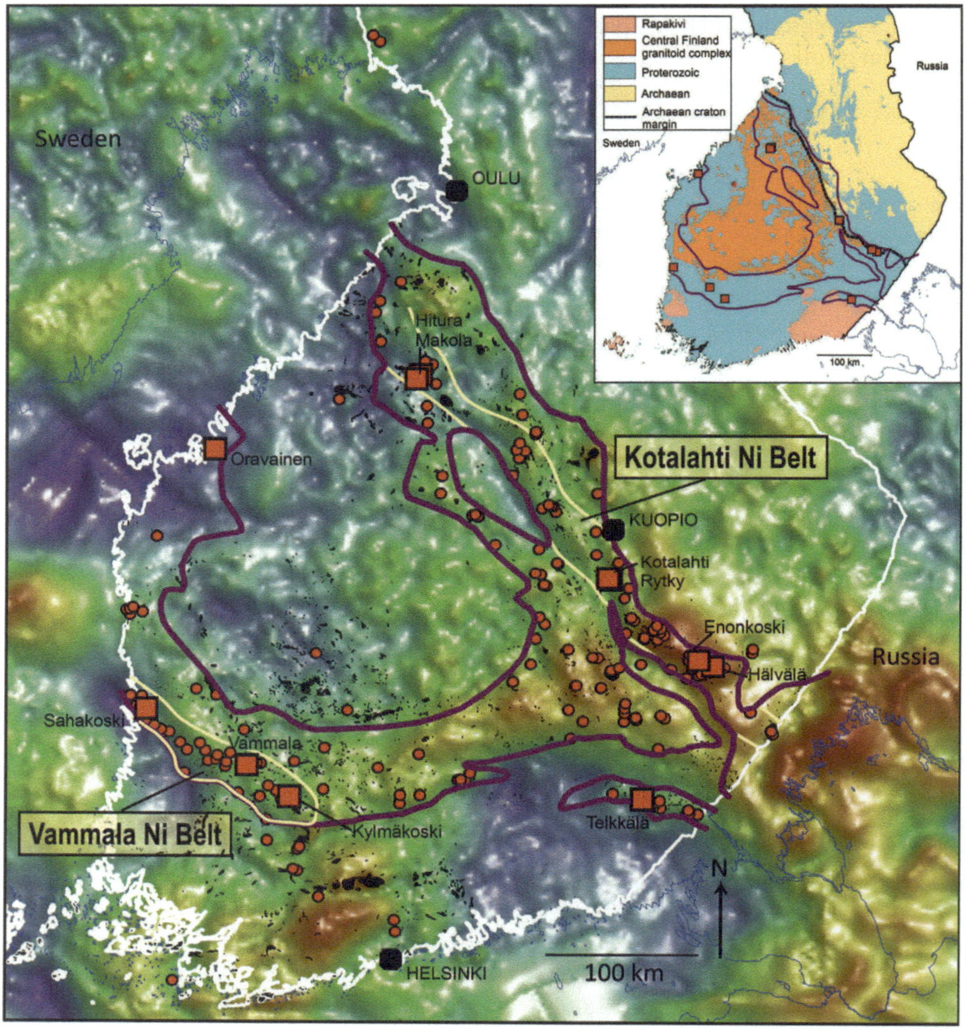

FIGURE 3.8.1 Gravimetric map showing the main domains of the Svecofennian Ni deposits and prospects in central and southern Finland (purple lines).

Nickel deposits are concentrated in the Kotalahti and Vammala Ni belts (highlighted by light yellow lines). The most important Ni deposits are indicated by red squares. The definition for the main Ni domains has been made on the basis of deposit distribution, lithology, regional metamorphic grade, gravity, and related tract definition by Rasilainen et al. (2012). Svecofennian mafic and ultramafic intrusions are shown in black. Gravimetric map is a hill-shaded (from northeast) Bouguer anomaly map with maximum intensity shown in red and minimum intensity in blue; map by Jouni Lerssi, GTK. Inset shows the main Ni domains superimposed on a simplified geological map, based on the GTK bedrock map database (Bedrock of Finland–DigiKP).

Peltonen (2005) classified the Svecofennian mafic–ultramafic intrusions into three groups on the basis of their geotectonic setting. Only the Group I hosts nickel deposits, and these were divided into (1) intrusions located close to the craton margin (Group Ia, referred to as the Kotalahti nickel belt) and (2) intrusions of the Tampere and Pirkanmaa belts (Group Ib, referred to as the Vammala nickel belt). Originally, the Kotalahti nickel belt was defined as a very narrow belt located at the craton margin (Gaál, 1972), but subsequently, nickel deposits located further to the southwest of the craton margin were included in the Kotalahti belt (e.g. Mäkinen and Makkonen, 2004; Makkonen, 2005). The Telkkälä nickel belt is a separate belt north of the Mesoproterozoic Viborg rapakivi massif. The Liakka nickel deposit within the Haaparanta suite north of Oulu constitutes a distinct Svecofennian deposit (Fig. 3.8.1).

The host rocks to the Svecofennian nickel-bearing intrusions are of a highly variable nature. The Vammala nickel belt occurs in the Pirkanmaa migmatite suite, which continues eastward to the Juva area in southern Savo province where several further nickel deposits occur. The Juva area, in turn, belongs partly to the Häme migmatite suite. East of the Juva area is the Haukivesi complex, which hosts the nickel deposits of the Kotalahti nickel belt. Thus, in the southeast corner of the Central Finland Granitoid Complex several lithological suites meet, giving rise to local overthrusting, with all of them containing Svecofennian nickel deposits (Fig. 3.8.1). This suggests that the exposed nickel deposits probably represent different crustal levels.

The distribution of the Svecofennian metapicrites can be used as a guide to determine the areal distribution of the Svecofennian nickel-bearing intrusions. Both lithologies host nickel mineralization, they occur together (e.g., in the southern Savo area), and are thought to represent the same magmatic event (Makkonen, 1996; Makkonen and Huhma, 2007; Barnes et al., 2009). The metapicrites are found all around the Central Finland Granitoid Complex, similar to the nickel-bearing intrusions. An analogy is found within the Thompson nickel belt, Canada, where the Opswagen group hosts nickel-bearing ultramafic sills and comagmatic mafic to ultramafic volcanic rocks (picrites) (Bleeker, 1990).

Most of the nickel deposits related to the Svecofennian Group I intrusions occur within zones characterized by amphibolite facies to granulite facies metamorphic grade, schollen and schlieren migmatite-textured sulfide- and graphite-bearing metasedimentary rocks, abundant mafic–ultramafic intrusions, and a high regional gravity signature. Gravity maps suggest the presence of more voluminous intrusions, notably to the east of the Vammala belt (Fig. 3.8.1).

MODE OF OCCURRENCE

The Svecofennian nickel-bearing mafic and ultramafic intrusions are mainly found within migmatitic mica gneisses, which often include graphite-bearing layers. However, in the Kotalahti belt, and especially in the Kotalahti area, some intrusions occur within Archean gneisses or within rocks of the craton margin sequence composed of Paleoproterozoic quartzites, limestones, calc-silicate rocks, black schists, and diopside-bearing banded amphibolites.

The most typical and easily recognized migmatite within the nickel belts is the schollen migmatite (Fig. 3.8.2), but schlieren migmatites are also common. The migmatite belts are considered as the most prospective areas for 1.88 Ga nickel deposits in Finland and they have thus been mapped in detail (e.g., Aarnisalo, 1988; Jokela, 1994). A genetic relationship between the migmatites and nickel-bearing intrusions has been suggested by Korsman et al. (1999), who proposed that the low-P/high-T metamorphism and the generation of tonalite-trondhjemite migmatites in the Svecofennian crust were caused by extensive

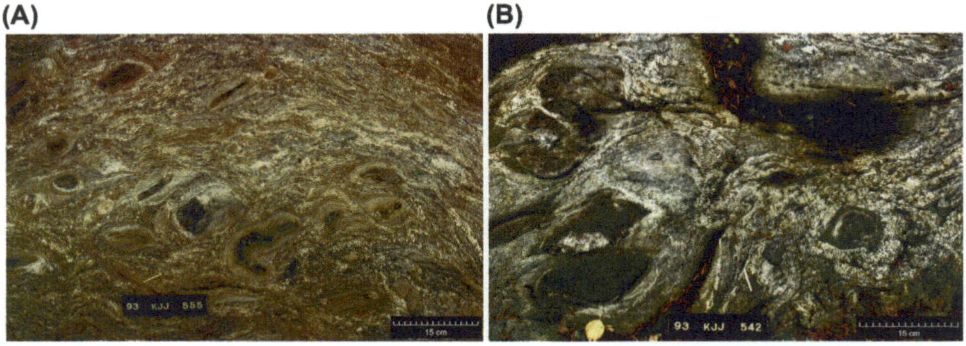

FIGURE 3.8.2 Migmatites from the Kotalahti Ni belt. (A) Schollen migmatite with calc-silicate rock and psammitic fragments (from Jokela, 1994, Fig. 30), (B) Gabbro fragments within schlieren-schollen migmatite (from Jokela, 1994, Fig. 42).

magma under/intraplating during and soon after subduction and crustal thickening (1885 Ma). According to the study of Lyubetskaya and Ague (2010) on migmatites in Scotland, the peak temperatures in crustal rocks around basaltic intrusions may be increased by more than 100 °C. In Scotland, the emplacement of mafic intrusions at midcrustal depths of 15–35 km produced minerals of low-pressure/high-temperature metamorphic grade such as cordierite-biotite-K-feldspar. Makkonen (2005) suggested that the heat of the mafic magma, together with the latent heat produced by the crystallization of the magma, enabled migmatite neosome formation within the Svecofennian nickel belts. In extensional environments, neosome formation was promoted by pressure release and the melt concentrated as tonalitic bodies.

Two different variants of Svecofennian nickel-bearing intrusions can be distinguished (e.g. Mäkinen, 1987; Lamberg, 2005; Peltonen, 2005). The weakly differentiated, dominantly ultramafic Vammala-type intrusions consist almost entirely of olivine cumulates and occur as small boudinaged lenses or pipes with a diameter of 100–1000 m within polydeformed paragneisses. The strongly differentiated, mafic and mafic–ultramafic, Kotalahti-type intrusions consist of olivine cumulates, pyroxene cumulates, and plagioclase-bearing cumulates, and are commonly up to several kilometers (km) long and a few hundred meters wide at surface. Magmatic layering is locally visible, although the original layered structure is commonly obscured by polyphase deformation.

The intrusions were emplaced during peak deformation and metamorphism (e.g., Marshall et al., 1995; Koistinen et al., 1996; Kilpeläinen, 1998; Mäkinen and Makkonen, 2004; Makkonen, 2005; Peltonen, 2005). The country rocks surrounding the intrusions were in most cases extensively metamorphosed and deformed during the early stage of the Svecofennian orogeny (Gaál, 1980; Koistinen, 1981; Kilpeläinen, 1998; Mäkinen and Makkonen, 2004). Overthrusting and faulting resulted in fragmentation of both the intrusions and the country rocks. Consequently, the shape of the intrusions in surface plan view varies widely, depending on the structural history and the depth of exposure. The dimensions of the intrusions also vary, from decimeter-scale schollen-migmatite fragments to more than 10-km-long intrusive bodies. The intrusions often form oval-shaped bodies in surface plan section and may consist of gabbro only, peridotite only, or gabbro-peridotite.

The intrusions show D2 structures and they seem to be elongated along the F2 fold axis. Excluding the sill-like intrusions, the cross section perpendicular to the F2 fold axis is often roundish. Thus, the form of the surface section varies according to the dip of the F2 fold axis; in areas with steep F2 fold axes, roundish and oval-shaped sections dominate, while in areas with gentle F2 fold axes, elongated sections are typical. Because D2 structures were formed subhorizontal (e.g., Koistinen, 1981), intrusions emplaced along F2 were also originally subhorizontal. Gravitative fractionation was therefore probable during magma flow in the chamber and, as a result, ultramafic cumulates and sulfides were concentrated at the base of the intrusions. When these ore-bearing intrusions were tilted with the F2 fold axis, the result in the surface section is an elongated/oval body with the nickel ore and ultramafic cumulates, and gabbros and more felsic differentiates, at opposing ends of the body. This observation is important from the exploration point of view. In most of the studied intrusions within the Kotalahti belt, it has been possible to recognize the stratigraphic footwall of the intrusions as the most prospective locality for nickel ore (e.g., Forss et al., 1999; Makkonen et al., 2003). However, in the case of the weakly differentiated, ultramafic Vammala-type intrusions, it is not possible to define the facing of the intrusion. In fact, some of the intrusions are concentric, suggesting they represent feeder conduits (Peltonen, 1995b; Lamberg, 2005).

In theory, the shallow intrusions would be expected to have a more sill-like form than those crystallized at a greater depth, as is seen in southern Savo province where gabbro-diorite intrusions form sills that were later folded upright. These intrusions can be interpreted to represent the most fractionated and high-level melts, whereas the peridotite and peridotite-gabbro bodies occur as more rounded and less elongated bodies (c.f. Makkonen, 1996). Tuisku and Makkonen (1999) estimated, on the basis of the chemistry of orthopyroxene-clinoamphibole-spinel coronas located between plagioclase and olivine, that the crystallization depth for the Saarijärvi intrusion in southern Savo was 17–20 km (900 °C, 5–6 kbar). In comparison, a similar study by Trudu and Hoatson (2000) of the East Kimberley intrusions resulted in a range from 2.4–6.7 ± 0.6 kb, which equates to upper-middle crustal depths of around 8–23 km. Fig. 3.8.3 summarizes the structural history of the Svecofennian intrusions.

GEOTECTONIC SETTING

According to Eilu and Lahtinen (2013), the Svecofennian nickel deposits (as well as the orogenic gold and Ti-V-Fe deposits) formed during collisions between arcs and microcontinents, and incipient amalgamation of Fennoscandia and Laurentia. The mafic magma intruded into tensional structures above the subduction zone (Nironen, 1997; Peltonen, 2005), and the rocks show evidence for syncrystallization deformation and assimilation of country rocks. The intrusions of the Vammala belt in southern Finland represent conduits of arc basalts or were emplaced within an accretionary wedge (Lahtinen, 1994; Peltonen, 1995b), whereas the intrusions of the Kotalahti belt along the Archean craton margin were emplaced along major transtensional shear zones during the Svecofennian arc–Archean craton collision (Peltonen, 2005).

In terms of their geochemical signatures, the Svecofennian intrusions and comagmatic volcanic rocks in southeast Finland conform to a back-arc environment. The southeast Finnish metatholeiites representing the parental magma to the Svecofennian nickel-bearing intrusions occur together with limestones, cherts, and iron formations and commonly exhibit pillow structures. These features, together with the fact that the tholeiitic magmas have an enriched mid-ocean ridge basalt (EMORB) affinity, suggest a cratonic margin or marginal basin environment; for example, a back-arc setting (Makkonen, 1996; Makkonen and Huhma, 2007).

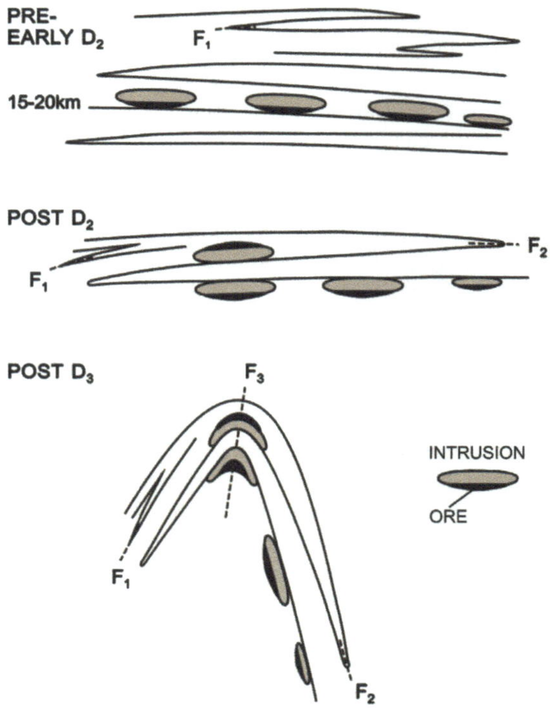

FIGURE 3.8.3 Simplified structural model for the intrusions.

Note that individual intrusions may represent boudinaged fragments of originally larger sill-like intrusions.

Comparison can be made with the circum-Superior belt in Canada. The tectonic evolution of the Thompson nickel belt probably involved an early stage of rifting during the onset of the Ospwagan group sedimentation and a later stage of peak deformation and mafic–ultramafic volcanism (Zwanzig, 2005). Subsequent emplacement of the mineralized ultramafic sills and mafic dikes possibly occurred in a continental back-arc environment and was associated with felsic calc-alkaline magmatism (Percival et al., 2004, 2005; Zwanzig et al., 2007, and references therein). The change from stable platform deposits to syntectonic filling and emplacement of mafic intrusions in the Ospwagan group is attributed to the convergence between the Reindeer Zone and the Superior Province at 1891–1885 Ma (Machado et al., 2011).

According to Heaman et al. (2009), the Molson magmatic event (1.88 Ga) in the northwest Superior craton was relatively short-lived (<8 Ma), and consists of a dike swarm emplaced parallel to the craton margin (Molson dike swarm), mafic–ultramafic volcanic sequences in both the Thompson nickel and Fox River belts, and sill complexes. Synchronous and geochemically similar mafic/ultramafic magmatism occurs in the Cape Smith belt and New Québec orogen, indicating that 1.88 Ga magmatism occurred for more than 3000 km along the previously rifted margin of the Superior craton. On the basis of the MORB-like affinity of the magmas, the Molson magmatic event possibly represents a series of back-arc basins, local transtensional pull-apart basins, or large-scale passive upwelling of asthenospheric mantle and concentration of magmatism along an already thinned and rifted margin of the Superior craton.

In summary, the craton margin in both the Fennoscandian and Canadian shields was intruded at ~1.88 Ga by mafic–ultramafic mantle-derived magma, resulting in the formation of nickel-copper sulfide deposits. The geotectonic setting was likely one of a marginal (or possibly back-arc) basin. The event is related to a major crustal growth period at ~1.9 Ga, interpreted to mark the amalgamation of the Columbia supercontinent (Groves and Bierlein, 2007).

NICKEL ORES

The parental magma for the 1.88 Ga intrusions in Finland contained up to 15 wt% MgO, but more usually around 10–12 wt%. Depending on the amount of sulfur available during sulfide segregation (i.e., the R factor) and on the state of fractionation of the silicate melt, the nickel content of the sulfide fraction is between 2–14 wt% and Ni/Cu between 1–4. The R factor varied mostly between 200 and 1300 (e.g., Makkonen et al., 2003). The nickel content of the ore is dependent on the amount of sulfides. In disseminated ore types, the average nickel grade is usually ≤1 wt%. In massive ores the nickel content can be up to 7 wt%.

Tables 3.8.1 and 3.8.2 give the average metal contents of the Svecofennian nickel deposits. Only in a few deposits is the average nickel grade higher than 1%, and all deposits are low in nickel when compared to nickel deposits globally. The average nickel grade calculated for the deposits in Table 3.8.2 is 0.61 wt%.

The nickel ores are usually located in the stratigraphically lowermost portions of the intrusions. In addition, the ores are often zoned, with the most massive ore being located at the basal contact. The massive ore can be up to a few meters thick, and it often forms breccias with the wall rock. The disseminated ores are the most common, and they too show compositional zoning, in that Ni/Co and Ni tenors of the sulfides increase toward the stratigraphic base (e.g., Mäkinen and Makkonen, 2004). With increasing amount of sulfide, disseminated ore turns into net-textured ore in which the sulfides form an interconnected network enclosing silicate minerals. In some peridotitic intrusions, a special nodular or orbicular texture is present, for example, at Kylmäkoski, Ekojoki, and Vammala (Papunen, 1980; Lamberg, 2005). Svecofennian Ni deposits often include so-called *offset ores*, which are hosted by the wall rocks (mica gneiss, black schist), likely in the stratigraphic floor of the bodies, and tend to be separated from the main intrusion by about 50–200 m. Typical examples are the Jussi ore body at Kotalahti (Papunen and Koskinen, 1985) and the Leo ore body at Laukunkangas (Grundström, 1985). At Telkkälä, the "B" ore consists of a massive sulfide dike within mica gneiss (Eeronheimo and Pietilä, 1988a,b). At Stormi, the sulfides extend for 10 m into the mica gneiss (Häkli and Vormisto, 1985). In the small Ni deposits of the Juva area, the offset ores are thin but similar in mode of occurrence to those mentioned earlier.

Platinum-group element (PGE) contents of the Svecofennian intrusions are typically low, and mantle- or chondrite-normalized spidergrams usually show a negative Pt anomaly (Papunen, 1986, 1989; Makkonen and Halkoaho, 2007) (see Fig. 3.8.4) At Hitura, the nickel ore contains subeconomic values of PGE, about 0.2 ppm Pt + Pd on average. In some other deposits that are not mined, elevated average values are observed, for example, Ekojoki, Uudiskorhola (Papunen, 1989), Tiemasoja (Makkonen and Forss, 2003), and Vehmasjoki (Makkonen and Forss, 1999). Papunen (1989) attributed the negative Pt anomalies to redistribution of Pt during metamorphism and alteration of the host rocks. According to

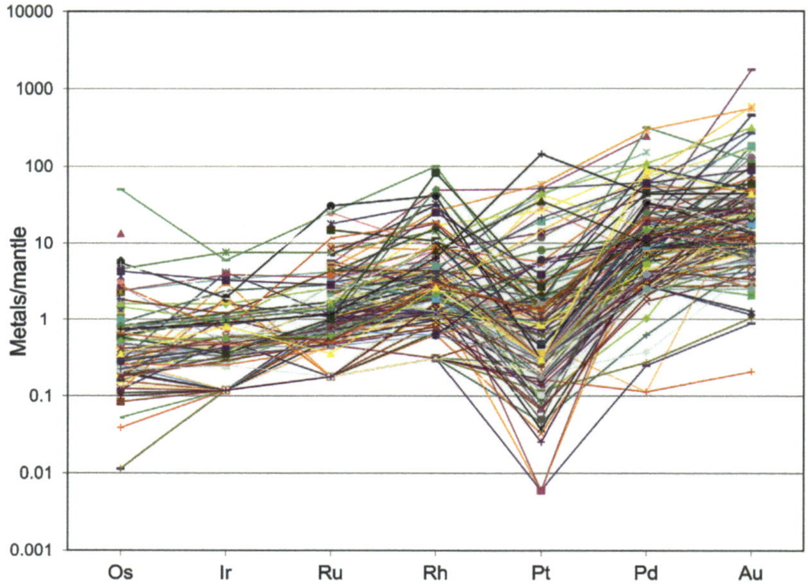

FIGURE 3.8.4 Mantle-normalized multielement diagram for noble metals in Svecofennian nickel-copper deposits.

Number of samples =147. Platinum values of 0.5 by detection limit were used for samples below the detection limit.

Source: *Data from Papunen (1989), Lamberg (2005), and Makkonen and Halkoaho (2007). Mantle values after Barnes (2006).*

Barnes (2006), massive ores are typically enriched in the Ir platinum-group elements (IPGE; Os, Ir, Ru) and Rh relative to Pt and Pd, and have negative Pt anomalies, whereas matrix and disseminated ores are enriched in Pt and Pd. This is explained by the concentration of IPGE into early crystallizing nickel-rich pyrrhotite or monosulfide solid solution (MSS). Fractionated, IPGE-depleted sulfide liquid becomes progressively concentrated within the overlying matrix ore. The PGE data on the Svecofennian deposits are consistent with this model: there is an overall positive correlation between the sulfur content and the magnitude of the negative Pt anomaly as noted also by Papunen (1989). However, other factors may also have caused the negative Pt anomaly, as indicated by the scatter in Fig. 3.8.5. A negative Pt anomaly has also been observed in the massive, semimassive, and overlying net-textured nickel ores at Jinchuan (Tang et al., 2009) and in nickel ores of the Thompson belt (Bleeker, 1990; Layton-Matthews et al., 2011).

A large amount of analytical data has been collected from the Svecofennian nickel ores (e.g., Makkonen et al., 2003; Lamberg, 2005) making it possible to evaluate the distribution of nickel between different phases. Preliminary estimates for the $D_{Ni}^{sul/sil}$ values in 14 nickel-bearing intrusions that have sufficient data on both the nickel ore and parental magma, vary between 310 and 1080, for an MgO content in the parental magma of between 7.3 and 10.4 wt%. A negative correlation with the MgO content of the parental magma is observed, consistent with the empirical and experimental results (c.f. data compilations by Lamberg, 2005, and Naldrett, 2011).

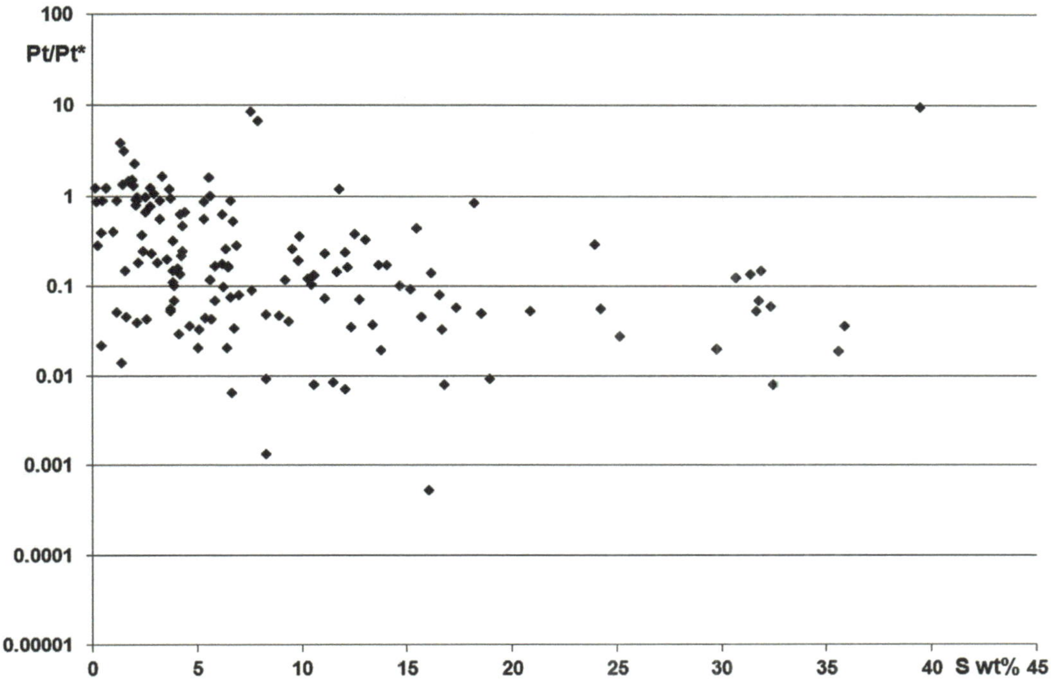

FIGURE 3.8.5 Diagram of platinum anomaly (Pt/Pt*) versus sulfur in Svecofennian nickel deposits.

Pt* = (Pdmn × Rhmn)$^{0.5}$. Number of samples = 147. Platinum values of 0.5 by detection limit were used for samples below the detection limit.

Source: *Data from Papunen (1989), Makkonen (1996), Makkonen and Forss (2003), Lamberg (2005), and Makkonen and Halkoaho (2007). Mantle values after Barnes (2006).*

ORE MINERALOGY

The ore mineralogy of many Svecofennian nickel deposits was comprehensively studied by Papunen during the 1970s and 1980s. (e.g., Papunen, 1970; 1974; 1980; Papunen et al., 1979). Because the deposits formed via concentration of immiscible sulfide liquid that segregated from basaltic magma, followed by fractionation of MSS, the ore mineral assemblage typically consists of pyrrhotite-pentlandite-chalcopyrite. More detailed data on the ore mineralogy is given in the chapters describing the type intrusions (p. 264-274).

In addition to magmatic processes producing the present ore mineral assemblage, one must consider the effect of metamorphism. According to Tomkins et al. (2007), massive Ni-Cu-PGE deposits undergo insignificant melting under common crustal metamorphic conditions. Only the highly fractionated low-temperature melts (crystallized as arsenopyrite, gersdorffite, cobaltite, nickeline, maucherite, löllingite, westerveldite, safflorite, chalcopyrite, and cubanite) may be subject to remelting, and thus mobilization, during amphibolite- to granulite-facies metamorphism. In the Svecofennian nickel deposits, small-scale metamorphic remobilization of sulfides can be traced along lithological contacts and shear zones, but only for a maximum distance of a few meters (Papunen, 2003; present author).

Chalcopyrite veins of various thicknesses are common within and near the contact of the Svecofennian nickel deposits, especially within areas characterized by upper-amphibolite to granulite-facies metamorphism. This suggests mobilization of magmatic chalcopyrite. Pentlandite may also occur as veins or bands and crack fillings (e.g., Makkonen, 1996; Kojonen, 1999; Kojonen et al., 2002), whereby it is possible that in some deposits, some of the primary pentlandite was mobilized during high-temperature metamorphism. Alternating pentlandite- and pyrrhotite-rich bands in nickel ores have been attributed to breakdown of metamorphic MSS within a stress field (McQueen, 1987).

The fact that the Svecofennian intrusions were emplaced during peak metamorphism, that is, into a hot environment of upper-amphibolite to granulite-facies metamorphism, makes it difficult to distinguish between late magmatic phases and phases crystallized from metamorphic MSS.

TYPE DEPOSITS

As noted earlier, the intrusions of the Vammala and Kotalahti belts show some systematic geological and compositional differences. In the following section, the Kotalahti and Rytky deposits are described as examples of the Kotalahti type deposits and, although geographically separate from the Vammala belt, the Hitura deposit as an example of Vammala-type deposits. Of the remaining economically important deposits (refer to Table 3.8.1), the Enonkoski and Vammala deposits have been described in detail by Lamberg (2005).

KOTALAHTI AND RYTKY

Most of the mafic–ultramafic Ni-mineralized intrusions in the Kotalahti Ni belt are rather small, with only a few being up to 10 km in diameter at the present erosion level. Typically, the maximum horizontal dimension is less than 2 km. The intrusions are differentiated from ultramafic to gabbroic (locally to dioritic and quartz dioritic) rocks. Intrusions composed solely of ultramafic rocks are rare. Intrusions are mainly found in areas of relatively high metamorphic grade, suggesting a relatively deep crustal section. These areas are commonly located within local gravimetric highs, in part because of the abundance of mafic–ultramafic rocks, but also because of the abundance of relatively dense minerals in the country rocks.

The Kotalahti and Rytky intrusions are located within the Kotalahti dome, which is composed of Archean gneiss surrounded by a Paleoproterozoic craton-margin supracrustal sequence of quartzites, limestones, calc-silicate rocks, black schists, and banded diopside amphibolites. The Särkiniemi intrusion lies within mica gneiss, higher in the stratigraphy than Kotalahti and Rytky (Fig. 3.8.6). Metamorphism reached the amphibolite facies, causing gneissose and migmatitic textures in the supracrustal rocks.

The Kotalahti deposit has been of high importance for the Finnish nickel industry. The discovery of the deposit in 1954 and the opening of the mine in 1959 led to the establishment of a nickel smelter at Harjavalta. The smelter has been in operation since 1959 and has a capacity of 50,000 tons of Ni per annum. In total, 12.35 Mt of ore at 0.66% Ni, 0.26% Cu, and 0.03% Co was mined at Kotalahti from 1959–1987 (Puustinen et al., 1995).

The Kotalahti intrusion has been dated at 1883 ± 6 Ma (Gaál, 1980). It forms a subvertical sheet with a length of approximately 1.3 km and a maximum width of 200 m. The southernmost intrusive segment extends to a depth of more than 1000 m (Papunen, 2003). The wall rocks of the intrusion consist mainly of Archean gneiss (Fig. 3.8.6).

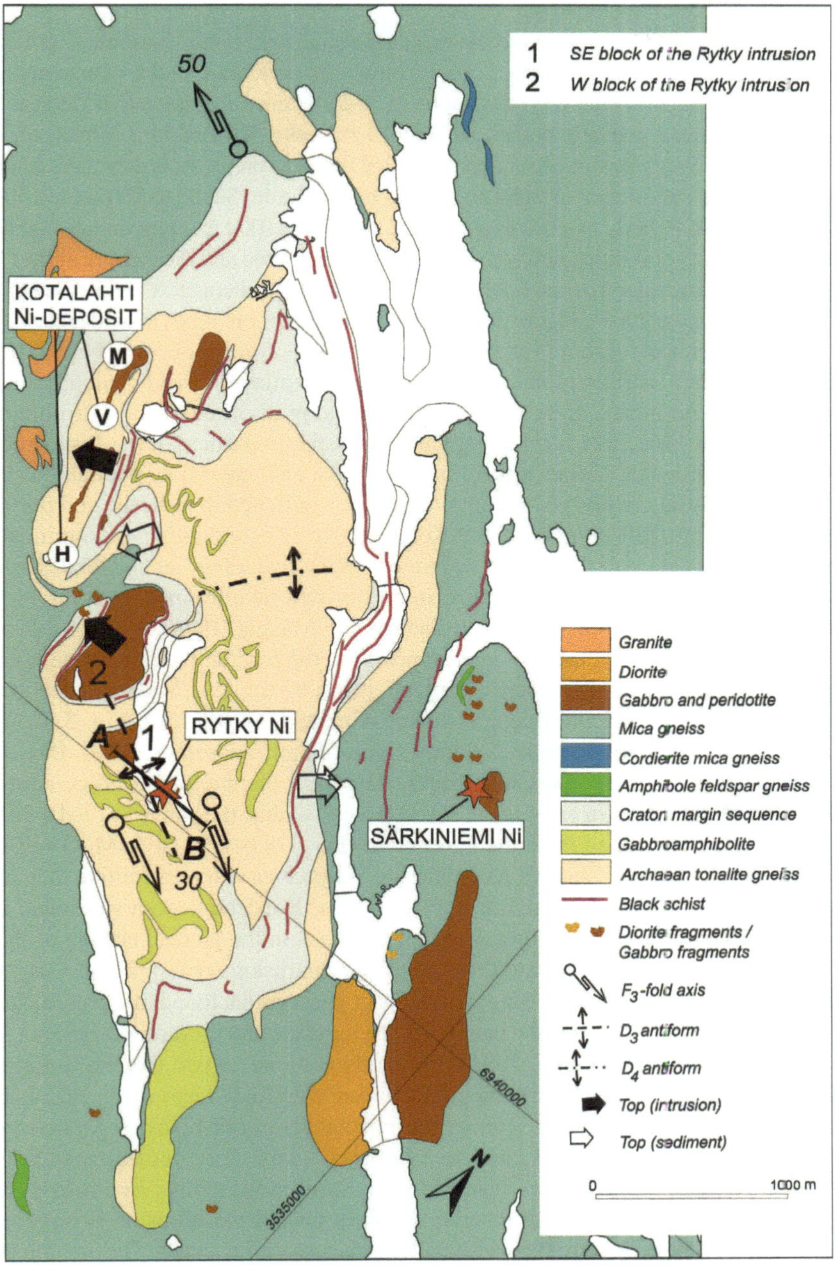

FIGURE 3.8.6 Geological map of the Kotalahti dome, showing location of Huuhtijärvi (H) Välimalmi (V) and Mertakoski (M) ore bodies.

The Rytky and Särkiniemi Ni deposits are indicated by red stars. The line AB represents the location of the cross section shown later in Fig. 3.8.11

Source: *Modified from Mäkinen and Makkonen (2004).*

The intrusion consists of olivine- and olivine-enstatite cumulates, orthopyroxenites, poikilitic gabbros, ophitic gabbronorites, and diorites (Papunen, 2003). Among the peridotitic rocks, coarse-grained lherzolite forms the stratigraphically lowest unit. It is overlain by medium-grained lherzolite (Mäkinen and Makkonen, 2004). From north to south, the ore bodies include Mertakoski, Välimalmio, Vehka, and Huuhtijärvi (Fig. 3.8.7). A separate, mostly massive offset deposit, called the Jussi ore body, is present as a subvertical slab hosted by black-schist and calc-silicate wall rock (but also by peridotite; Seppä, 2009) some 150 m east of the Vehka ore body. The Jussi ore body extends to a depth of more than 1000 m. Disseminated sulfides are common in ultramafic rocks and poikilitic gabbros, whereas ophitic gabbros and diorites are almost barren. Breccia-type sulfides occur as irregular masses, commonly along lithological contacts in the relatively thin central part of the intrusion (Papunen, 2003). Figure 3.8.8 shows examples of ore types at Kotalahti.

According to Papunen and Vorma (1985), the Kotalahti ore averages 0.70% Ni, 0.27% Cu, 0.03% Co, and 4.00% S. The main sulfide minerals are pyrrhotite, pentlandite, and chalcopyrite; in the Jussi ore body, pyrite is also significant. In the disseminated ores, pyrrhotite occurs as troilite or hexagonal pyrrhotite, whereas in the breccia ores (vein network ores) and, in particular, the Jussi ore body, monoclinic pyrrhotite predominates. Gersdorffite, mackinawite, and argentian pentlandite are common accessory sulfides. The Jussi ore body also contains portions enriched in millerite and bornite (Papunen and Koskinen, 1985). The PGE content of the Kotalahti ore is economically insignificant. The PGE patterns show a distinct negative Pt anomaly, as in most of the Svecofennian nickel deposits (Papunen, 1989; Makkonen and Halkoaho, 2007). The $\delta^{34}S$ in ores hosted by ultramafic rocks is +1.4 to +2.6‰, similar to the Jussi ore body (+1.4 to +2.8‰) (Papunen and Mäkelä, 1980).

In terms of its geology, host rocks, and ore composition, the Kotalahti deposit resembles the neighboring Rytky deposit. At Kotalahti, forsterite contents reach 89 mole% (Mäkinen, 1987), which indicates an MgO content of the parental magma of approximately 17 wt% (c.f. Fig. 3.8.15b). A similar estimation can be made from the whole-rock composition (data from Seppä, 2009). Based on the TiO_2 and Al_2O_3 versus MgO plots of samples with MgO > 10 wt%, the maximum Fo content is 88.9. The MgO content of the chilled margin samples at Rytky varies within the range 9.17–10.50 wt%, which corresponds to the value calculated from the Fo content (Fo = 78.6–84.7 m%; MgO = 9.0–11.2 wt%) (Mäkinen and Makkonen, 2004). Thus, at Kotalahti the magma was more primitive than at Rytky.

In the study by Seppä (2009), the peridotitic host rock of the Kotalahti deposit was found to be layered, analogous to the Rytky intrusion. Layering is seen, for example, in the gradual change of the whole-rock Mg-number, MgO, and Ni/Co values toward the eastern basal intrusion contact. In the Jussi ore body, layered peridotite was found as the host rock. Magmatic contact phenomena, found in places, include the occurrence of fine-grained amphibolitic rock at the margins of the intrusion and the disappearance of S2 schistosity in the wall rock (Mäkinen and Makkonen, 2004). These observations suggest that the Kotalahti and Rytky intrusions originally formed subhorizontal bodies, which were later folded into a subvertical orientation.

The surface section (0.5 × 1 km) of the Rytky intrusion comprises two blocks separated by Proterozoic supracrustal rocks and Archean tonalite gneisses (see Fig. 3.8.6). The southeast block is surrounded mainly by Archean rocks, whereas the northwest block is surrounded mainly by Proterozoic rocks. Primarily, the intrusion is funnel-shaped and layered. A minor part of the intrusion is represented by sills located stratigraphically below the intrusion, within the Archean tonalite gneiss. The magma was emplaced during D2 and detached fragments of the intrusion are found in the surrounding gneisses. During D3 and D4, the magmatic layering and the ores were folded into a subvertical position (c.f. Fig. 3.8.9) (Mäkinen and Makkonen, 2004).

The following stratigraphic sequence is proposed, from base to top: (1) coarse-grained lherzolites and websterites/melagabbros, (2) medium-grained lherzolites, websterites, and gabbronorites, and

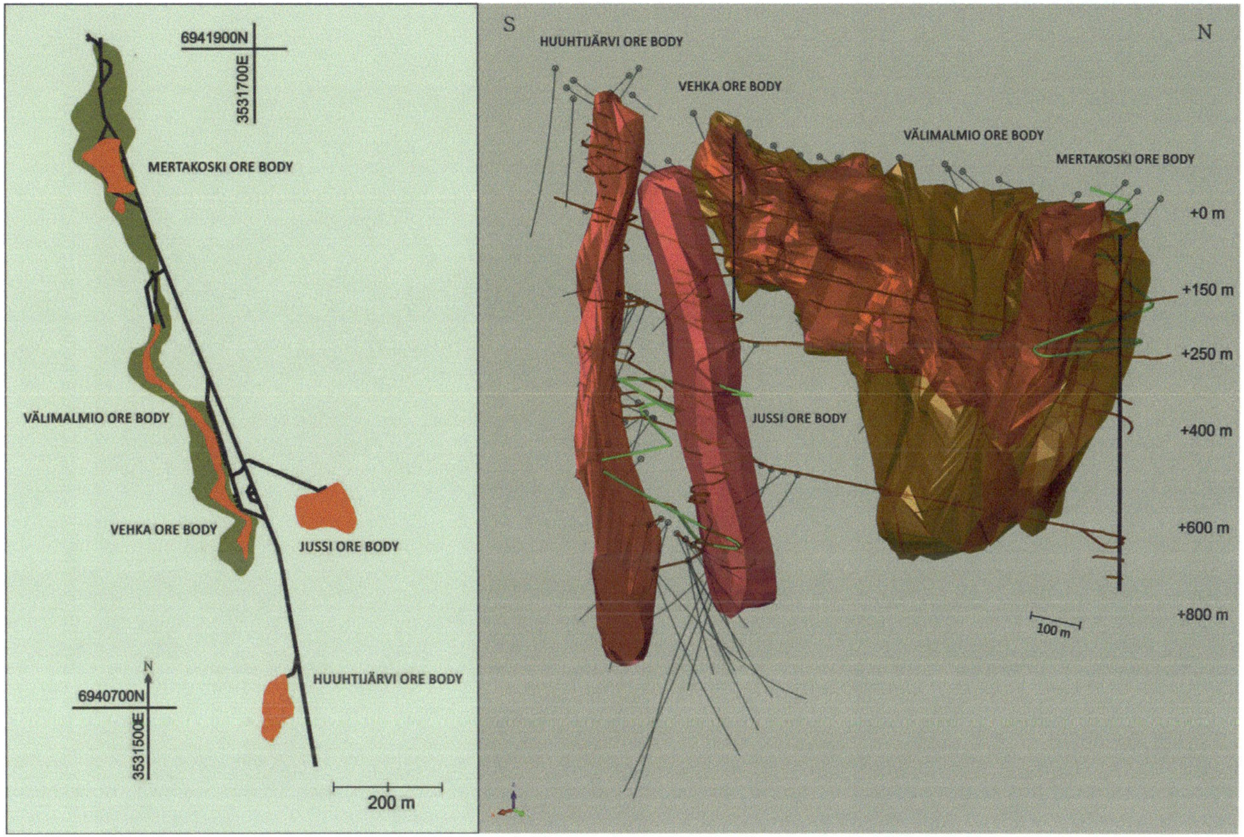

FIGURE 3.8.7 Plan section (level 250 m) and 3D bird's-eye view from northeast of the Kotalahti deposit.

Ore bodies are shown in red and the host intrusion in brown. Some tunnels and boreholes are also shown.

Source: *Figures generated by Janne Hokka, GTK.*

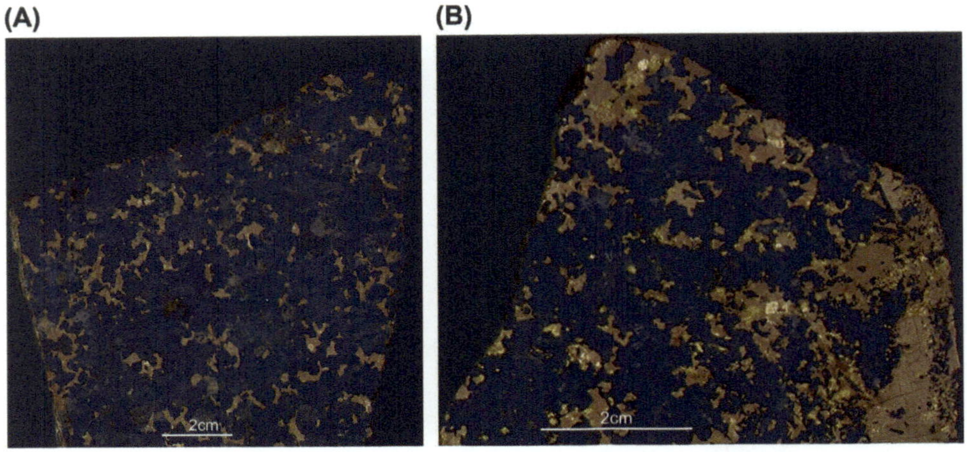

FIGURE 3.8.8 Scanned sections of ore samples from Kotalahti.

(A) Pyrrhotite-pentlandite dissemination in coarse-grained lherzolite. (B) Disseminated-breccia ore in metapyroxenite, showing abundant dull pyrrhotite, yellow chalcopyrite veinlets and disseminations, and bright pentlandite grains.

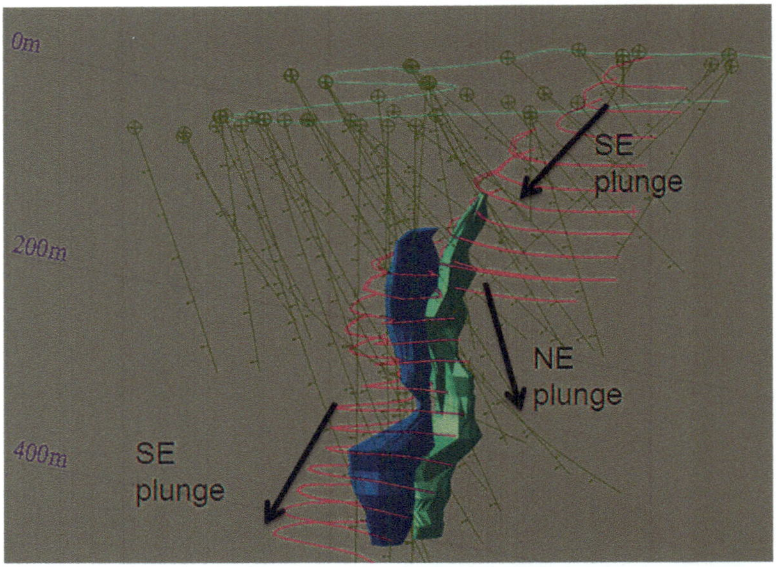

FIGURE 3.8.9 3D bird's-eye view from the southeast depicting the complex structural control of the Rytky deposit.

The ore hosted by coarse-grained lherzolite is shown in dark blue and the ore hosted by medium-grained lherzolite is shown in light blue. The contact of the intrusion is marked by red lines.

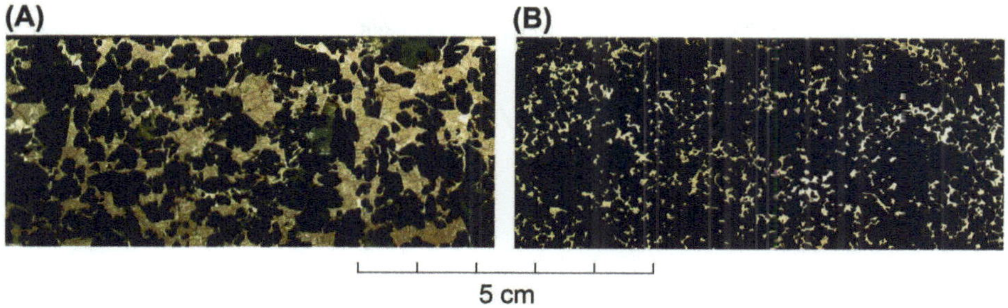

FIGURE 3.8.10 Scanned sections of the main ore types at Rytky.

(A) Matrix ore in coarse-grained lherzolite. (B) Disseminated ore in medium-grained lherzolite.

(3) subophitic gabbros. The first rock suite is up to 35-m thick and found at the margin of the intrusion. The coarse-grained lherzolite also occurs as sills separate from the main intrusion. The most typical feature of this rock type is the occurrence of euhedral to subhedral elongated olivine grains, or medium- to coarse-grained aggregates (<14 mm; c.f. Fig. 3.8.10), showing a mesocumulate texture. The medium-grained lherzolite is located above the coarse-grained lherzolite, but has been observed to crosscut it. The contact between the coarse-grained and the medium-grained lherzolite is sharp. The thickest and most uniform part of the medium-grained lherzolite occurs in the southeast part of the intrusion and indicates a decrease in the amount of olivine and sulfides toward the top. The transition from lherzolites to websterites is indicated by a zone of lherzolite/websterite interlayering. Most of the intrusion is composed of the second phase of rocks (i.e., medium-grained lherzolites, websterites, and gabbronorites). Subophitic and granular textured gabbros with varying grain sizes form a relatively minor portion of the intrusion and they crosscut the websterites. Fragments of a chilled margin are encountered in outcrops near the western edge of the intrusion.

The chemostratigraphy of the intrusion and the nickel ore is depicted in Fig. 3.8.11 The Ni/Co ratio of the sulfide fraction is highest in the ore hosted by the coarse-grained lherzolite, at the base of the cumulate pile. Aluminum and Ti contents increase toward the top of the intrusion.

The ore types in the Rytky deposit comprise fine-grained disseminations, coarse-grained disseminations, matrix (net-textured) ore, and massive ore (Fig. 3.8.10). The major sulfide ore minerals are pyrrhotite, pentlandite, and chalcopyrite. Cubanite is occasionally observed in the form of lamellae in chalcopyrite, cobaltite may form euhedral grains in the sulfides, and mackinawite inclusions occur in chalcopyrite and pentlandite.

Electron microprobe analyses of pyrrhotite (N = 572) mainly show a hexagonal NC (57.5%), NC + 4C (21%), or troilite composition (8.7%). Only a minor proportion of the analyses (4.7%) indicate a monoclinic 4C composition. The average Fe content of all pyrrhotite samples is 47.75 wt%, corresponding to the hexagonal NC type. The main carrier of nickel is pentlandite, which occurs in subhedral to euhedral grains, up to several millimeters in diameter, forming bands between pyrrhotite grains having a grain size of 75–200 microns, and in flame-like inclusions within pyrrhotite, usually 10–30 microns long and 5–10 microns wide (Kojonen et al., 2002).

HITURA

Hitura is the largest Svecofennian nickel deposit in Finland, with total resources of 19.3 Mt at 0.61% Ni, 0.21% Cu, and 0.02% Co (Rasilainen et al., 2012). Mining started in 1970 and was continued until

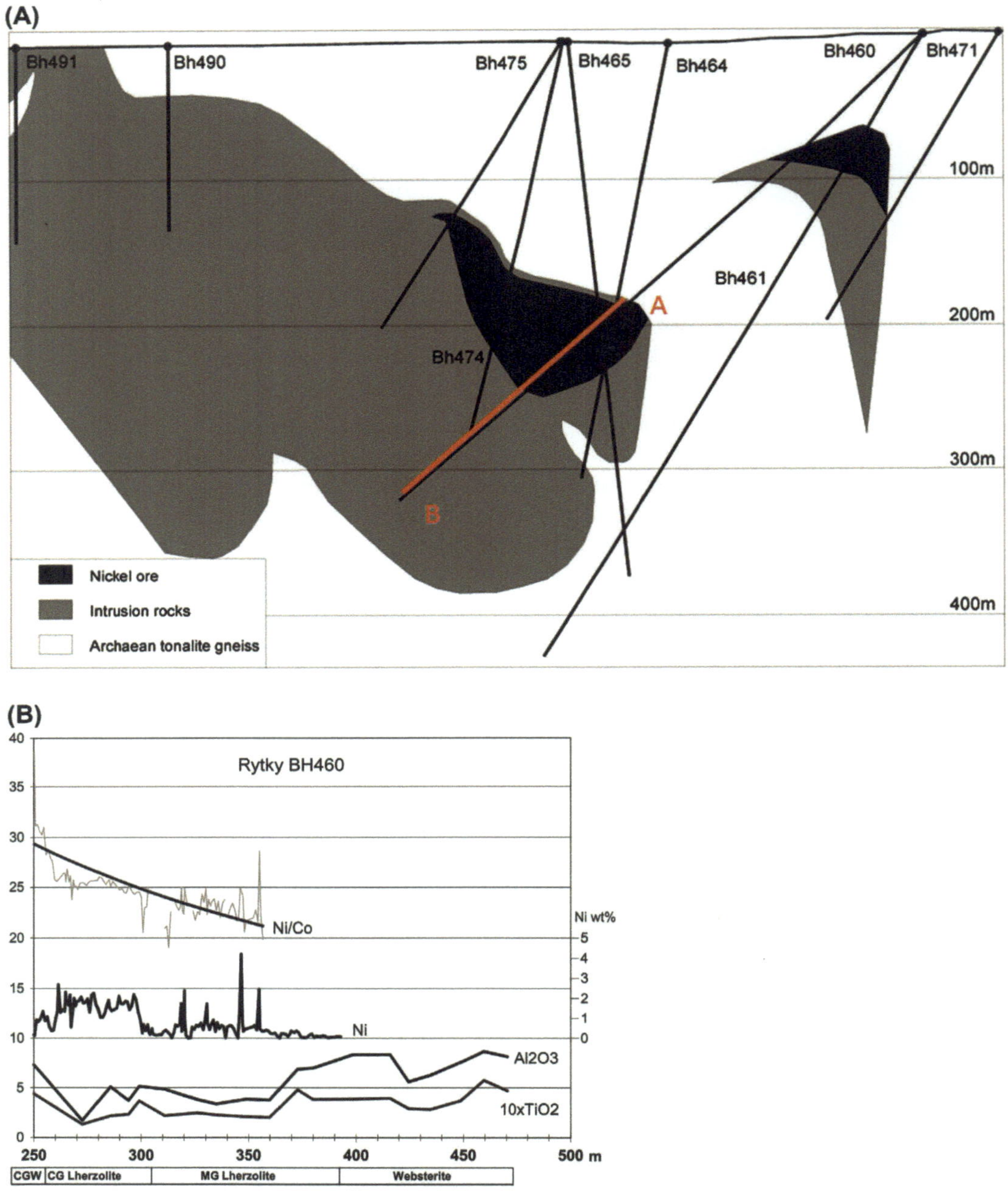

FIGURE 3.8.11 (A) Cross section through the Rytky nickel deposit (along line AB in Fig. 3.8.6) and (B) related chemostratigraphy. (along line AB in the cross section).

Ni/Co ratio (with trend line) is shown for samples where the nickel content is >0.4 wt%. CGW = coarse-grained websterite. Units for TiO_2, Al_2O_3 (in wt%), and Ni/Co are given at the left vertical axis.

Source: *Modified from Mäkinen and Makkonen (2004).*

2013 with intermittent breaks. A total of 16.5 Mt ore at 0.6% Ni was mined. Although geographically within the Kotalahti nickel belt, the Hitura deposit is geologically similar to the intrusions of the Vammala belt in that it contains weakly differentiated olivine cumulates and is strongly serpentinized.

According to Isohanni et al. (1985), the main rock types around Hitura (i.e., in the Nivala area) are intensely metamorphosed metasediments, intervening metavolcanics and extensive felsic intrusives. The stratigraphy of the lower group of the supracrustals consists of migmatitic mica gneisses and interlayers of graphite- and sulfide-bearing gneisses and amphibolites. The migmatitic mica gneiss forms the country rock to the ultramafic intrusions at Hitura, Makola, and many other locations. The least migmatized country rocks consist of medium-grained banded metagraywacke. The bulk of the mica gneiss has undergone intense migmatization and now consists of veined gneiss. Based on geophysical data, the amphibolites constitute long and narrow ribbon-like zones that often occur either at the contact between plutonic rocks and migmatitic mica gneiss or as interlayers in the migmatitic and gneissic metasediments. The amphibolites are banded or striped in structure. The well-preserved portions suggest that the amphibolites derive from pyroclastic lavas and dikes. The upper group of supracrustals consists of two volcanic suites between which there is a sequence of sedimentary rocks composed mainly of graywackes.

The Hitura ultramafic complex consists of three separate, closely-spaced serpentinite massifs surrounded by migmatized mica gneiss (Figs. 3.8.12 and 3.8.13) The metamorphic grade in the Hitura area

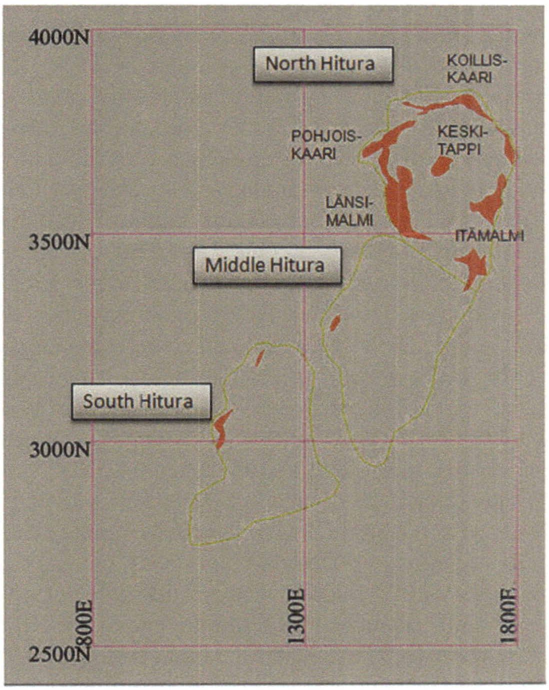

FIGURE 3.8.12 Plan view at the 200-m level of the Hitura nickel deposit.

The margins of the intrusion are shown by a brown line and the ore is shown in red. Map coordinates are shown in meters.

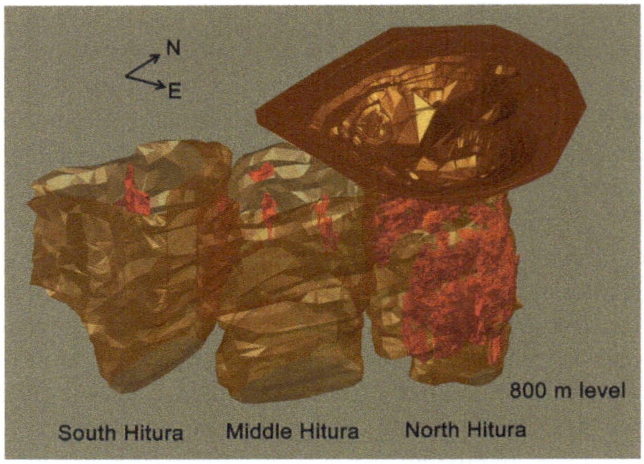

FIGURE 3.8.13 3D bird's-eye view from southeast of the Hitura deposit.

This shows the three serpentinite bodies (in brown) and the ores (in red). Open pit is also shown.

is at upper-amphibolite facies, which peaked during D2 and produced various types of migmatites, including schollen migmatite typical to all Svecofennian nickel prospects. In addition to migmatized gneisses, a belt of sulfide- and graphite-bearing schists and mylonitic rocks occurs in the vicinity of the Hitura body. The horizontal extent of the ultramafic complex is 0.3 × 1.3 km. The deepest drilling intersections reach a depth of about 800 m. Geophysical surveys indicate that the intrusion continues to a depth of, at least, 1000 m below the surface. The core of the complex is serpentinite and the marginal zones consist of amphibole-rich ultramafic rocks, originally representing olivine adcumulates and olivine orthocumulates or olivine-orthopyroxene cumulates (Papunen and Penttilä, 1996). Crosscutting pegmatitic dikes, with an age of 1877 ± 2 Ma, are not uncommon (Isohanni et al., 1985).

The contacts of the Hitura complex against gneissic wall rocks are commonly tectonic. The contact zone is characterized by dislocated mafic blocks, erratic wall-rock inclusions, and, locally, by massive sulfide lumps in a soft talc-rich matrix, indicating late-tectonic movement and faults. The gneiss near the contact ("contact gneiss") is typically homogenous and contains small garnet crystals and large pale dots of feldspars and quartz. Pyrrhotite disseminations are common in the gneisses near the serpentinite body. Some small serpentinite tongues in mica gneisses, partly nickel mineralized, are found at the western side of the Hitura massif. Shear zones with narrow tongues of mica gneiss separate the three ultramafic massifs (Papunen and Penttilä, 1996; Meriläinen et al., 2008).

Several nickel ore bodies are situated at the contact zones and in the core of the North Hitura serpentinite massif. From west to east, the ore bodies in the contact zones are Länsimalmi, Pohjoiskaari, Koilliskaari, and Itämalmi. In the center of the North Hitura body, there is the Keskitappi (central core) ore body (Fig. 3.8.12). Low-grade nickel mineralization occurs in the intrusive rocks between the core zone and the contacts. The mineral resources of the Middle and South Hitura massifs are not known (Meriläinen et al., 2008). According to Papunen and Penttilä (1996), three main ore types can be recognized in the North Hitura ore body: (1) scattered fine-grained sulfides disseminated in the serpentinite core zone, (2) medium- to coarse-grained sulfide disseminations in the marginal serpentinite-amphibole rock, (3) high-grade interstitial disseminated sulfides and massive accumulations in the amphibole rock locally extending into the gneissic wall rock (see Fig. 3.8.14).

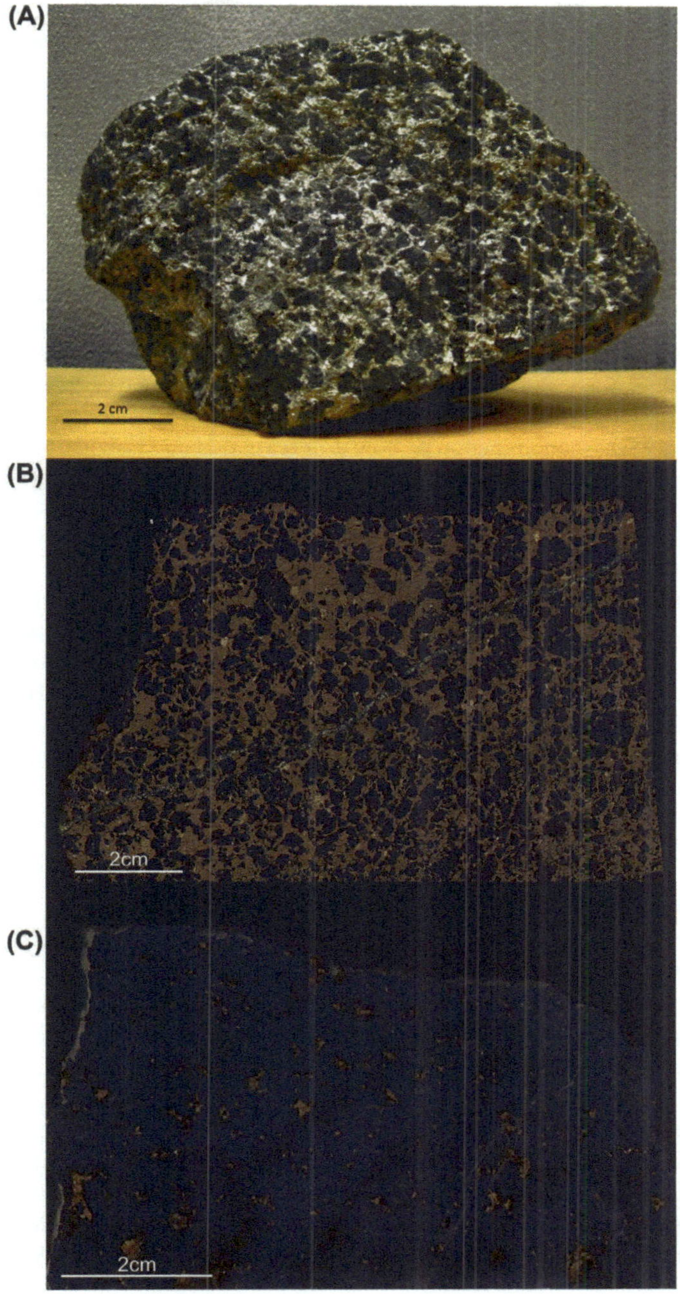

FIGURE 3.8.14 Ore samples from North Hitura.

(A) Net-textured ore at level 550 m, photo Hannu Makkonen; (B) net-textured ore in serpentinite (primarily olivine cumulate) at level 350 m, scanned section; (C) disseminated ore in serpentinite, scanned section.

The main ore minerals at Hitura are pyrrhotite and pentlandite, but in places, valleriite, mackinawite, chalcopyrite, and cubanite are abundant. Pentlandite is the main nickel-bearing mineral but mackinawite containing up to 6 wt% Ni is locally also important. Copper is mainly hosted in valleriite, but locally by chalcopyrite and cubanite. Many accessory minerals have been identified, including pyrite, violarite, maucherite, niccolite, gersdorffite, and millerite. Pyrite occurs only in joints associated with carbonates. Platinum-group minerals, such as sperrylite, michenerite, irarsite, iridarsenite, and hollingworthite, have also been detected and the ore averages around 0.2 ppm Pt + Pd. Platinum contents as high as 600 ppm have been found in a sperrylite-rich sulfide concentration in chlorite-amphibole rock at the intrusion margin (Häkli et al., 1976; Papunen and Penttilä, 1996; Meriläinen et al., 2008). A new mineral species, tarkianite, with a general formula $(Cu,Fe,Co,Ni)(Re,Mo,Os)_4S_8$, was discovered in sulfide concentrates. The average Re content in the concentrate is 20 ppb (Kojonen et al., 2004).

The internal stratigraphy of the northern ultramafic body is reflected in increases in whole-rock MgO, normative olivine, forsterite and Ni/Co of the sulfide fraction; and decreases in whole-rock SiO_2 content toward the contact of the intrusion (Suikkanen, 2011). A possible mechanism for generating this chemostratigraphy was a combination of flow differentiation and gravitational settling. Calculated and analyzed nickel and forsterite contents of olivine indicate that the mineral crystallized under sulfur-saturated conditions. The North and Middle Hitura bodies have similar nickel contents of olivine whereas the South Hitura body has distinctly lower Ni contents, suggesting more depleted parent melts (Makkonen et al., 2011).

The parental magma at Hitura was basaltic. Estimates of the MgO content of the magma vary from about 11 wt% (Papunen and Penttilä, 1996) to 12.5 wt% (Suikkanen, 2011). According to Papunen and Penttilä (1996), the shape of the Hitura complex implies that the melt fractionated in a tubular body, probably a feeder channel to the overlying volcanic rocks. Alternatively, the Hitura complex can be interpreted as a tubular or sill-like body, which was subsequently folded (mainly in D3). Consequently, the richest nickel sulfide ore is found in the marginal zone of the ultramafic complex, which represents the stratigraphic footwall of the body (Makkonen et al., 2011).

PARENTAL MAGMA AND CRUSTAL CONTAMINATION

On the basis of whole rock and olivine composition, the parental magma to the Svecofennian nickel-bearing intrusions was tholeiitic basalt with an MgO content of mostly 10–12 wt% (e.g., Peltonen, 1995a; Makkonen, 1996) and trace element contents similar to EMORB (Makkonen, 1996; Lamberg, 2005; Makkonen and Huhma, 2007). Ultramafic metapicrites and adjacent metatholeiites have been proposed to represent the extrusive counterparts of the intrusions (Makkonen, 1996; Makkonen and Huhma, 2007). Hill et al. (2005) and Barnes et al. (2009) concluded that the parent magmas to the intrusions and the extrusive picrites had a common mantle source, but the magmas ascended through different routes into the upper crust. The intrusive phases were more contaminated by crustal material than the extrusive phases. Initial epsilon Nd (1.9 Ga) values around +4 have been determined for the metapicrites, suggesting a depleted mantle source. The CaO-MgO-Al_2O_3 (CMA) plot (Fig. 3.8.15a) indicates the distinct differentiation series of the clinopyroxene-dominated Vammala-type and the orthopyroxene-plagioclase-dominated Kotalahti-type. The Al_2O_3-FeOtot-MgO (AFM) plot (Fig. 3.8.15b) highlights the tholeiitic trend of the rocks. The MgO/FeO ratio of the most MgO-rich cumulates correlates with the MgO content of the parental magma.

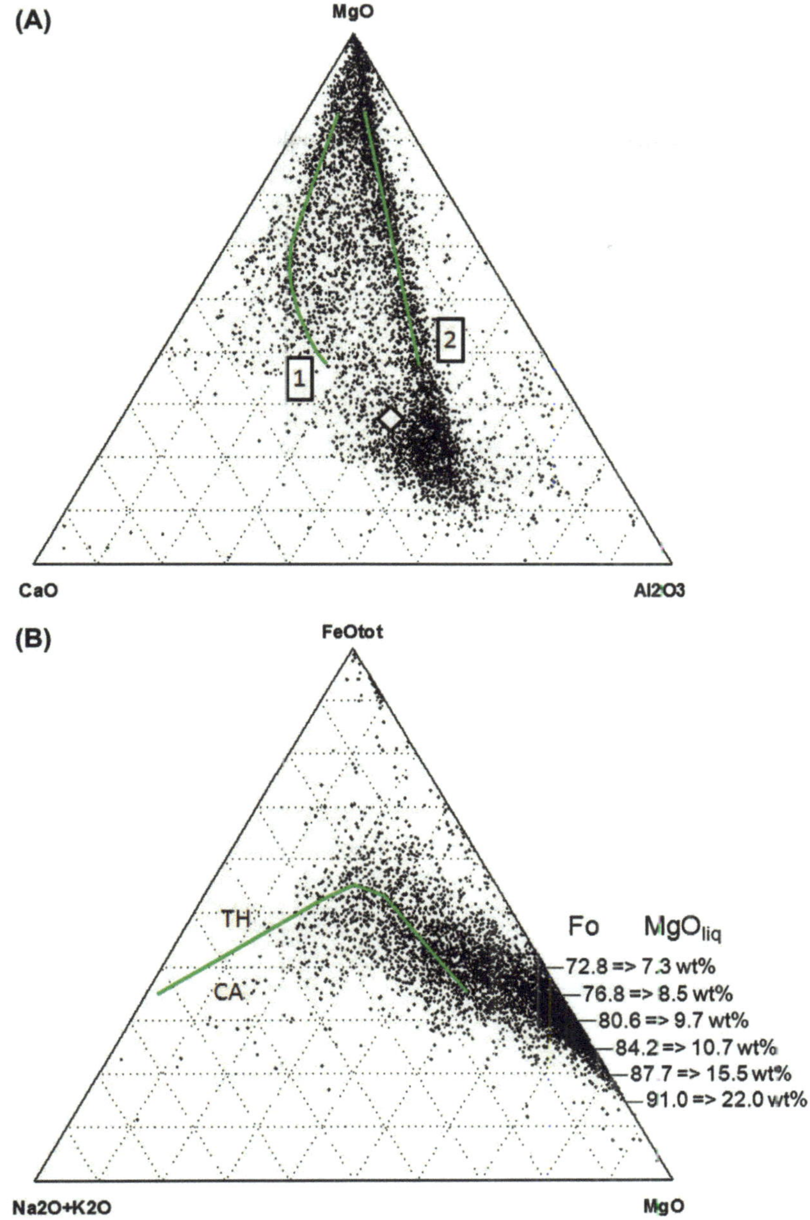

FIGURE 3.8.15 CMA (A) and AFM (B) ternary diagrams for the Svecofennian mafic–ultramafic intrusions.

In the CMA diagram, the fractionation trends of the Vammala-type (1) and the Kotalahti-type (2), toward plagioclase are shown by green lines. The proposed parental magma composition of Makkonen and Huhma (2007) is marked by a diamond. In the AFM diagram, the boundary line between the tholeiitic (TH) and calc-alkaline (CA) compositions is drawn after Irvine and Baragar (1971). The MgO/FeO ratio for the parental magma was calculated using K_D value of 0.33. The MgO content was then calculated from the correlation of the MgO and Mg number in the comagmatic tholeiites-picrites (Makkonen and Huhma, 2007; Barnes et al., 2009) as follows: Mg-number = 0.2189 × ln MgO (wt%) + 0.0981, and for the three lowest Fo contents based on the equation: $\log MgO_{Liq}$ (wt%) = (logFo-log (252.7833 − 0.7825 × Fo))/0.4602 + 1.795 (Makkonen, 1996). Number of samples = 6192.

Source: *Data from Mäkinen (1987), Makkonen et al. (2003), and Lamberg (2005).*

Crustal contamination of the mafic magma has been indicated in both the Vammala (Peltonen, 1995a; Barnes et al., 2009) and Kotalahti belts (Makkonen, 1996; Mäkinen and Makkonen, 2004; Lamberg, 2005; Makkonen et al., 2008; Makkonen and Huhma, 2007; Barnes et al., 2009). Evidence for wall rock/country rock assimilation is seen in many outcrops, in the form of xenoliths of metasediments enclosed in the mafic–ultramafic intrusive rocks.

In the Vammala Ni province, black schist was an important contaminant of the magma. The circulation of H_2S-bearing C-O-H-S fluids enabled selective transfer of large quantities of S and Zn from black schist into the cooling magma (Peltonen, 1995a). At Enonkoski, sulfide saturation was caused by black schist contamination (Lamberg, 2005). Graphite is ubiquitous in the intrusions and its abundance correlates broadly with that of sulfides. Hybrid rocks and remnants of graphite schist are associated with anomalously high contents of V, Mo, Mn, and Zn in the ore and its surroundings. In the Kotalahti belt, crustal assimilation of 5–40% by mass has been suggested (Makkonen, 1996; Makkonen and Huhma, 2007), similar to the estimates of Mungall (2007) in the 1.88 Ga Expo Intrusive Suite of the Cape Smith fold belt, Canada.

In Fig. 3.8.16 the Sm-Nd data for the Svecofennian intrusions and related picrite/tholeiite suites are compared to those of Svecofennian metasediments and Archean felsic rocks in Finland. Mixing between the volcanics, metasediments, and Archean rocks can produce the composition of the intrusive rocks.

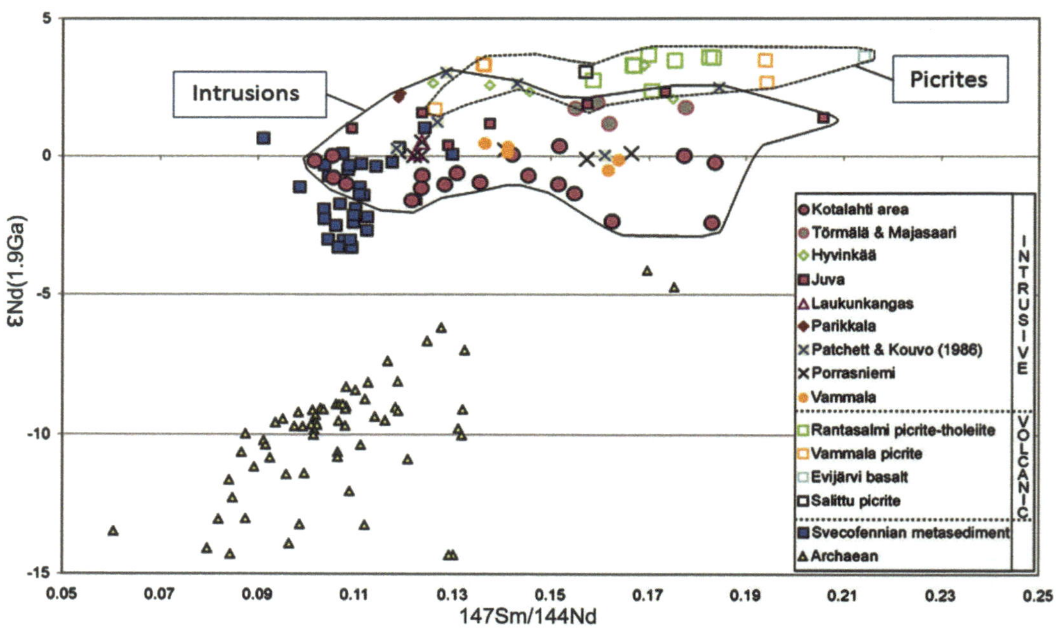

FIGURE 3.8.16 Sm-Nd isotope data for Svecofennian mafic–ultramafic intrusions, as well as for metapicrites and metabasalts, representing the proposed parental magma, the Svecofennian metasediments, and Archean felsic rocks in Finland.

εNd (1.9 Ga) in intrusions ranges from −2.4 to +3.1 (±0.4), in picrites from +1.7 to +3.7 (±0.4).

Source: *Modified from Makkonen and Huhma (2007).*

To obtain a negative εNd value for the intrusions, bulk contamination of about 10% of Archean gneiss and about 30% of Svecofennian metasediment is required. The composition of the intrusive rocks is controlled by the type of the country rock, as indicated by the fact that intrusions adjacent to Archean rocks have the lowest (negative) εNd values (Figs. 3.8.16 and 3.8.17) (Makkonen and Huhma, 2007).

Peltonen (1995a) found that in the Vammala area, the ore-bearing intrusions have higher S/Se than barren intrusions, consistent with country rock assimilation. The results of a similar study in the Kotalahti area are less clear. The S/Se of most of the nickel ores is much higher than the mantle value (28 samples, average S/Se = 9000, ranging from 2430–14,650—data by the author; mantle S/Se = 3333—McDonough and Sun, 1995), suggesting that external sulfur was added to the magma. However, the S/Se in the black schists and mica gneisses of the Kotalahti area is in the range of the mantle (6 samples, average S/Se = 4230, ranging from 2440–5500—data by the author) and in the B1 and B2 boundary zones of the Kotalahti belt, with S/Se averages of 10,500 and 5100, respectively (Lahtinen, 2000).

Consequently, assimilation of the metasediments cannot explain the high S/Se in some of the ores by a simple bulk contamination process. Queffurus and Barnes (2015) have shown that S/Se ratios are controlled by several processes, including addition of S from sedimentary rocks, variations in the R factor during sulfide liquid segregation, depletion of the silicate magma in Se by early segregation of the sulfide liquid, and moderate incompatibility of Se into the first sulfide minerals to crystallize from the MSS, resulting in a change in S/Se ratio between the Fe-rich and Cu-rich zones of magmatic sulfide ores. The overall low PGE values for the Svecofennian deposits are consistent with early sulfide segregation, but more work is required to constrain the processes controlling the S/Se ratio in the Svecofennian nickel deposits.

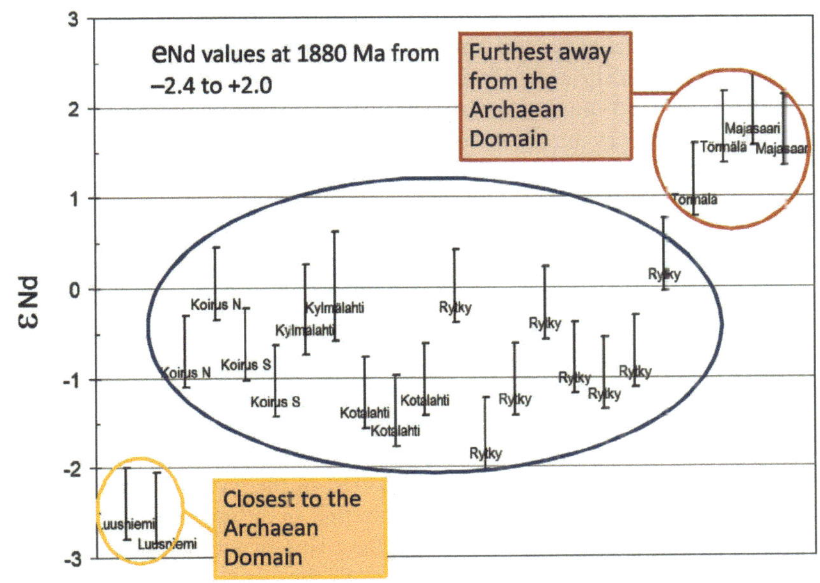

FIGURE 3.8.17 εNd (1880 Ma) values for intrusions from the Kotalahti nickel belt.

These are grouped according to proximity relative to the Archean crust. Intrusions within Kotalahti area are within the blue ellipse.

Source: *Modified from Makkonen and Huhma (2007).*

ORE MODELS

Because of their economic importance, numerous Svecofennian nickel deposits have been studied in great detail, particularly with regard to their sulfide mineralization. These data provide a good basis for ore petrological studies (e.g., Lamberg, 2005).

Peltonen (1995a) proposed that the intrusions of the Vammala belt were emplaced above an active subduction zone during progressive regional metamorphism and deformation. During their ascent to higher crustal levels, some of the magma intruded sulfidic black schist layers. Assimilation of external sulfur triggered segregation of an immiscible sulfide melt, forming the nickel deposits at the stratigraphic base of the intrusions. The residual barren magmas ascended to higher crustal levels and formed the unmineralized intrusions.

Papunen (2003) proposed a slightly different model for synorogenic differentiated intrusions and related ores (e.g., Kotalahti) whereby the magma was contaminated at an early stage, followed by fractionation and segregation of sulfide melt in a lower crustal magma chamber prior to final emplacement to the present position in the form of distinct magma pulses. Sulfides were injected separately into low-stress areas due to tectonic filter pressing, forming high-grade ores either in the mafic–ultramafic intrusions or as high-grade offset ores.

To explain the intrusions containing only ultramafic cumulates (e.g., at Vammala), it was proposed that the residual mafic magma efficiently escaped from the magma chamber to the surface (Papunen, 2005).

The offset ores have been economically important because their nickel content is higher than in other ore types (around 3–6% vs. 1%). The offset ores are thus important exploration targets. Häkli et al. (1975) proposed that at Telkkälä, the wall rocks of the intrusion were fractured during cooling of the intrusion, resulting in infiltration of sulfide liquid and the formation of breccia and massive ore. Papunen (2003) suggested that pulses of silicate and sulfide melt were injected upward and outward of the magma chambers during tectonism, with the sulfide melt forming the last pulse. A similar model was proposed by Piña et al. (2010) for the Aguablanca deposit in Spain. Mobilization and concentration of magmatic sulfides by deformation has been proposed for the origin of the Svecofennian offset ores (e.g., Peltonen, 2005). Arndt et al. (2013) argued that sulfide melts that had accumulated at relatively high levels in intrusions could percolate to the base of intrusions and into their footwall.

In the experience of the author, offset ores tend to occur in the stratigraphic footwall of the intrusions, usually where the intrusion is thickest and contains the greatest proportion of ultramafic cumulates and sulfides. This suggests that the offset ores formed during the primary magmatic stage. The sulfide melt concentrated at the bottom of the intrusion and invaded the wall rock along synintrusion fractures produced during deformation and/or by partial melting of the wall rock.

Fig. 3.8.18 shows a schematic model for the formation of the Svecofennian nickel ores. Mantle-derived tholeiitic magma rises along different paths resulting in different amounts of crustal assimilation. Some magmas reach the surface without undergoing significant assimilation. This is consistent with their initial εNd (1880 Ma) values of around +4, and by the fact that chalcophile element depletion (Ni, PGE) is relatively minor (Barnes et al., 2009). In contrast, most of the Svecofennian intrusions are strongly depleted in PGE. The initial εNd (1880 Ma) values for the intrusions show a wide range (from −2.4 to +3.1, ±0.4), which is attributed to variation in the amount and type of assimilated material (Makkonen, 1996; Makkonen and Huhma, 2007). Sulfide segregation took place within the magma

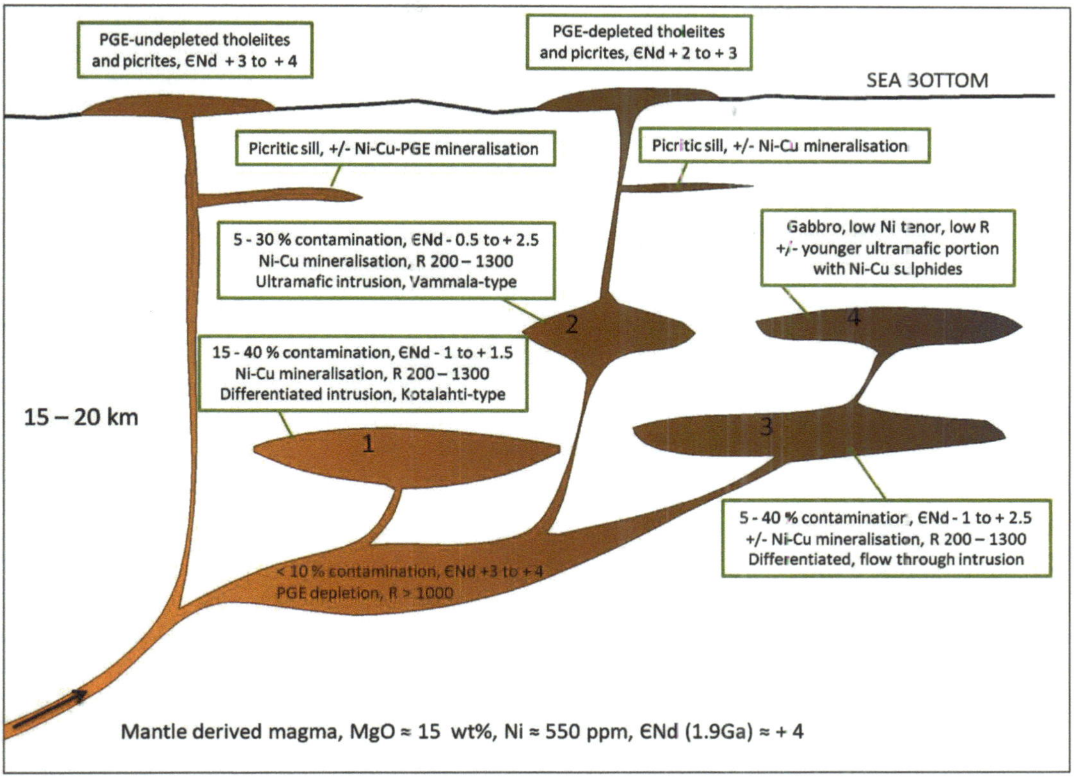

FIGURE 3.8.18 Schematic model for Svecofennian mafic–ultramafic magmatism and related nickel deposition.

Mantle-derived tholeiitic magma with MgO content of around 15 wt% intrudes sulfur-bearing sedimentary rocks at a crustal depth of 15–20 km. Magma ascends along different routes while fractionating olivine, so that the parental magma composition for each individual intrusion varies. Contamination by the sedimentary rocks is most effective in relatively large magma chambers. The amount and type of the assimilated sulfur-bearing sediment resulted in a large range of R values and, thus, Ni tenor as well as initial Nd isotope values (the Nd isotope values shown in the text boxes relate to cases where the contaminant is a Svecofennian sedimentary rock; in intrusions near the Archean rocks, εNd is lower).
Numbers 1–4 refer to (1) Kotalahti-type intrusions, where differentiation produced layered ultramafic to gabbroic rocks; (2) Vammala-type intrusions, representing ultramafic, weakly differentiated magma conduits; (3) ultramafic cumulate bodies from which the residual melt has been efficiently expelled upward; (4) gabbroic intrusions, with or without peridotite as a result of later pulses of olivine-bearing melt from lower magma chambers.

conduits and in somewhat larger chambers. Sulfide segregation was mainly caused by assimilation of sulfur-bearing sedimentary rocks. The level where this assimilation took place varies. If sulfur saturation was reached after significant olivine fractionation, the intrusion may contain Cu ± PGE rich mineralization; for example, in gabbros in the Juva area and in hornblendites in central and eastern Finland.

In the Kotalahti and Vammala belts, the largest intrusions are not known to contain significant Ni-Cu mineralization. Examples include Ylivieska, Alpua in Vihanti, and Saarela in Pielavesi. This is consistent with the model of Maier et al. (2001), which shows that economically important magmatic Ni-Cu sulfide deposits tend to occur in dynamic magma conduit systems rather than within large layered intrusions. The conduits are characterized by a highly dynamic flow environment characterized by multiple flows of magma. Sulfides are concentrated in the widened parts of the conduits, owing to a decrease in the flow velocity of the magma.

Many of the intrusions contain a gabbroic and an ultramafic portion separated by a sharp contact. The ultramafic phase often shows intrusive relationships relative to the gabbroic portion and thus appears to be younger, analogous to many Alaskan-type intrusions (Ripley, 2009). This relationship can be explained by the following model. The ascending magma fractionated in a staging chamber below the final level of emplacement. The relatively differentiated magma at the top of the staging chamber was the first to ascend further. Subsequently, olivine-rich slurries, with 15–20 wt% MgO, followed via the same route and intruded the gabbroic body. Flow differentiation may have led to further concentration of olivine and sulfide, forming sulfidic peridotites.

EXPLORATION

Successful exploration methods for Svecofennian nickel deposits include geophysics, lithogeochemistry, and location of sulfides in outcrops and glacial boulders. Results of till geochemistry are often obscured by the silicate nickel from the host rocks, but the use of ratios and other derivatives from the till geochemical data can be successful. Stratigraphic considerations are also important.

All economically important Svecofennian nickel deposits are hosted by ultramafic cumulates in the basal part of the intrusion. Intrusions with only gabbroic differentiates are thus considered to have low nickel ore potential.

GEOPHYSICS

Gravity maps can define potential target areas and locate the intrusions. Areal gravity highs represent relatively high metamorphic grade, suggesting a deeply exposed section of the crust (refer to Fig. 3.8.1). In targeting individual intrusions, one must remember that the serpentinites produce gravity lows.

The olivine-bearing cumulates show up well in magnetic maps. The ore itself is often detectable by magnetic ground measurements because magnetic (monoclinic) pyrrhotite is the predominant sulfide mineral. However, in peridotites, pyrrhotite is often hexagonal and less magnetic (antiferromagnetic). Distinguishing sulfide disseminations from magnetite in serpentinized olivine-bearing cumulates is difficult using geophysical methods. In some cases, remanent magnetism also complicates the interpretations. Once the nickel ore has been located, mise-a-la-masse (charged potential) is useful to trace the continuation of the ore. Usually, economic nickel contents require such a large amount of sulfides that the conductivity of the rock is high enough for the mise-a-la-masse method. Different kinds of magnetic and electro-magnetic borehole measurements also proved to be useful.

LITHOGEOCHEMISTRY

Since the 1960s, lithogeochemistry has been used in nickel exploration in Finland (Häkli, 1963, 1971). Around 20,000 rock samples were analyzed to determine the compositions of whole rocks, olivine, pyroxene, and amphibole during the Ni exploration of the Outokumpu Company, which began in 1960 (c.f. Lamberg, 2005). From the beginning of the 1980s, the Geological Survey of Finland carried out similar studies. The most important processes and features that can be constrained by lithogeochemical studies are (1) magma composition, (2) magma contamination, (3) chalcophile element depletion, and (4) internal structure of an intrusion.

The MgO content of the magma provides information on the state of differentiation of the magma. In sulfide undersaturated magmas, the Ni content correlates positively with MgO, thus the less evolved the magma is, the more potential exists for the formation of Ni-rich sulfide ores. The MgO content of the magma from which the intrusion crystallized can be estimated from the whole rock cumulate composition, from olivine composition, or from the composition of chilled margins of the intrusion.

Magma contamination by the country rocks may result in elevated contents of SiO_2 (which in turn may result in orthopyroxene being more abundant than clinopyroxene), Al_2O_3, CaO, incompatible and high field strength (HFS) elements, and rare earth elements (REE) (Fig. 3.8.19). Isotope and S/Se studies can also reveal magma contamination, but as noted earlier, interpretations can be complicated in S/Se studies.

Chalcophile element depletion has been widely used to evaluate prospectivity. Studies on the nickel content of olivine have been particularly abundant. Other useful methods include plots of whole rock nickel versus MgO or Mg-number. PGE depletion has also been tested (Barnes et al., 2009) but more studies are needed to prove its effectiveness. One rarely used but worthwhile method may be Tl depletion. Univalent thallium follows rubidium in magmatic differentiation but because it also has a chalcophile nature it is depleted in sulfur-saturated systems (McGoldrick et al., 1979). Plots of Tl versus Rb (also K and Na) can be used to study the depletion (Forss et al., 1999).

Olivine is commonly at least partly preserved in most of the primary olivine cumulates and, thus, can be analyzed by microprobe. Orthopyroxene can be used in a similar way to olivine (c.f. Lamberg, 2005). Fig. 3.8.20 shows the Ni versus Fo plot for olivines analyzed from the mafic–ultramafic intrusions in the Kotalahti belt. The majority of the samples plot in the sulfur-saturated field. The most prospective intrusions are those that show a steep trend from high Fo (>80 m%) and Ni (>1500 ppm) contents toward low Ni contents (≈500 ppm). This trend suggests that sulfide segregation took place within the intrusion, (i.e., in situ). In contrast, if an intrusion contains only nickel-depleted olivine, this suggests that fractionation and/or nickel depletion occurred prior to final emplacement, resulting in low ore potential.

In addition to the magmatic trend shown in the plot, at least two further factors affect the composition of olivine: (1) Reaction between olivine and intercumulus melt can lower both the primary forsterite and nickel contents. The process was referred to as "trapped liquid shift" by Barnes (1986) and can produce significant scatter in the nickel versus forsterite plot. (2) Reaction between sulfide and olivine ($NiSi_{0.5}O_2$ + FeS = NiS + $FeSi_{0.5}O_2$; Fleet et al., 1977) can produce a positive correlation between Ni and Fe contents in olivine, resulting in scatter or a negative correlation between the nickel and forsterite contents in olivine.

In cases where a representative set of chemical data for a range of intrusions is available, a fertility analysis combining a number of critical factors in nickel ore formation (Lamberg, 2005) can be a useful tool to discriminate between barren and ore-bearing intrusions.

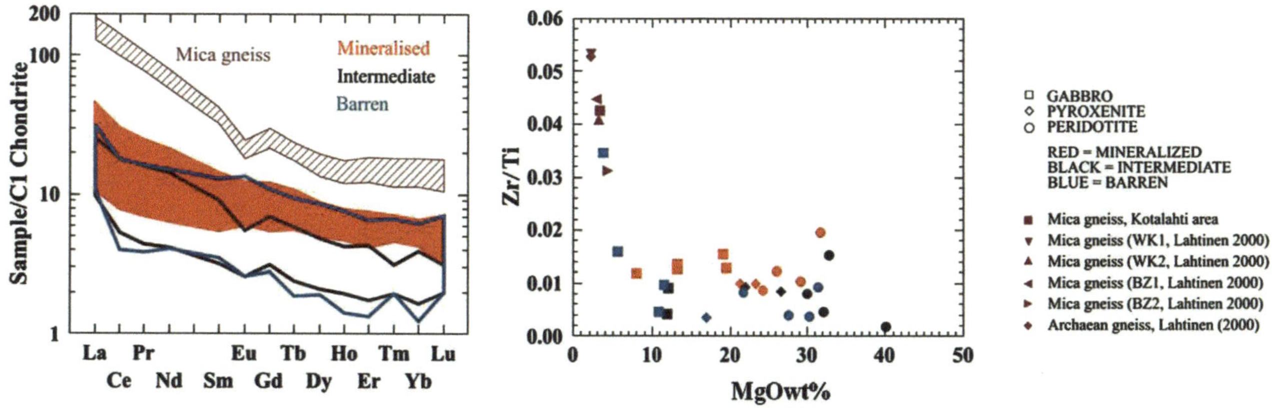

FIGURE 3.8.19 Diagrams showing the average compositions of mineralized (red), intermediate (black), and barren (blue) intrusions.

These are from the Kotalahti belt. In the REE diagram the composition ranges for peridotites only are shown. MgO as vol. free. total number of samples is 663. Compositions of the most probable contaminants are also given, namely mica gneiss (purple square and triangles) and Archean gneiss (purple diamond; data from Lahtinen, 2000).

Source: *Modified from Makkonen et al. (2008).*

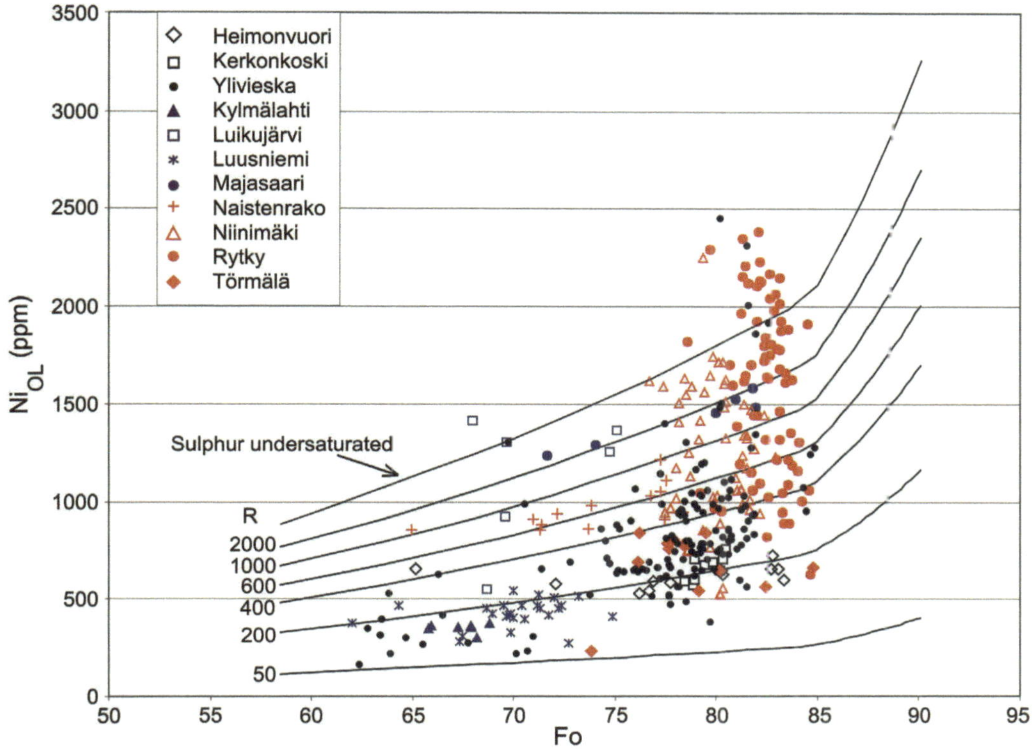

FIGURE 3.8.20 Diagram of nickel versus Fo contents in olivine for some intrusions of the Kotalahti belt.

Number of samples = 374. Olivines from strongly mineralized intrusions are shown in red, weakly mineralized intrusions in black, and barren intrusions in blue. The model curves show the composition of olivine crystallizing from parent tholeiite assuming different mass ratios between the silicate melt and sulfide melt (R). The curves are for different degrees of batch equilibrium crystallization. R values are calculated after Campbell and Naldrett (1979), using the following equations: logMgOLiq = (logFo − log(252.7833 − 0.7825 × Fo))/0.4602 + 1.795 and MgOLiq = $^{0.2588}\sqrt{(0.9554/(100/Fo - 0.67))}$ (Makkonen, 1996), NiLiq = 2.1257 × MgOLiq$^{2.05}$ (Makkonen et al., 2008), $D_{Ni}^{Ol/Liq} = e^{(4.961 - 1.266 \times \ln MgOLiq)}$ (Duke and Naldrett, 1978), $D_{Ni}^{sul/sil} = 350$ (Francis, 1990).

Source: *Reprinted from Makkonen et al. (2008, Fig. 3).*

SUMMARY

Most of the Svecofennian mafic–ultramafic intrusions record ages between 1875 and 1885 Ma, and all Ni-bearing intrusions belong to the 1880 Ma age group.

The total ore production of the 10 Svecofennian nickel mines in Finland is about 45 Mt at 0.67% Ni and 0.26% Cu, and the total premining resource of all the deposits known to date is about 73 Mt at 0.6% Ni and 0.3% Cu. From the mined deposits, only the four biggest, Hitura, Kotalahti, Vammala, and Enonkoski contained several million tons of ore.

Intrusions occur throughout the Svecofennian of central and southern Finland, but most of the nickel-bearing intrusions occur within the Kotalahti and Vammala nickel belts to the east and south of the Central Finland Granitoid Complex. The Kotalahti nickel belt extends close to the Archean craton margin. Most of the nickel-deposits occur within zones characterized by amphibolite facies to granulite facies metamorphic grade, schollen- and schlieren-migmatite-textured sulfide- and graphite-bearing metasedimentary rocks, abundant mafic–ultramafic intrusions, and a high regional gravity signature.

The intrusions were emplaced during peak deformation and metamorphism. Two different variants of Svecofennian nickel-bearing intrusions can be distinguished. The weakly differentiated, dominantly ultramafic Vammala-type intrusions consist almost entirely of olivine cumulates and occur within polydeformed paragneisses as small boudinaged lenses or pipes with a diameter of 100–1000 m. The differentiated, mafic and mafic–ultramafic, Kotalahti-type intrusions consist of olivine cumulates, pyroxene cumulates, and plagioclase-bearing cumulates, and are commonly up to several kilometers long and a few hundred meters wide at surface section. Magmatic layering is locally visible, although the layered structure is commonly obscured by polyphase deformation. The intrusions were emplaced during or before the main deformation phase of D2 and underwent further deformation during D3–D4, resulting in highly complex structures.

The parental magma for most intrusions contained 10–12 wt% MgO and 240–350 ppm Ni. The least evolved magma was derived from a depleted mantle source (initial εNd + 4) and probably contained around 15 wt% MgO and 550 ppm Ni as indicated by the high Fo contents up to 89 at Kotalahti. Geochemically, the magma was similar to EMORB. It intruded sulfur-bearing sedimentary rocks, resulting in crustal contamination of the mafic magma, indicated in both the Vammala and Kotalahti belts. In the Vammala Ni province, black schist was an important contaminant. In the Kotalahti area, contamination by both Svecofennian sediments and Archean gneiss took place. The distinct fractionation trends of the clinopyroxene-dominated Vammala type and the orthopyroxene-plagioclase-dominated Kotalahti type probably resulted from differences in the amount and type of assimilated crust. In the more contaminated Kotalahti-type intrusions, SiO_2 was added to the magma in abundance resulting in orthopyroxene crystallization. Based on Nd isotope data, the amount of the assimilated material varied between 10% and 40% by mass.

The metal contents of the parental magma control the composition of the related ores. Depending on the sulfur amount available during sulfide segregation (R factor) and on the fractionation of the silicate melt, the nickel content of the sulfide fraction varies between 2 and 14 wt% and Ni/Cu between 1 and 4. The nickel ore is located stratigraphically in the basal part of the intrusion. The ore is often zoned, with the most massive ore located at the basal contact. The PGE contents are typically low and PGE patterns usually show negative Pt anomalies in mantle- or chondrite-normalized spidergrams. The Svecofennian Ni deposits often contain offset ores, which are located 50–200 m below the main intrusion within mica gneiss and black schist. The ore mineral assemblage in the Svecofennian nickel-copper ores typically is pyrrhotite-pentlandite-chalcopyrite. The composition of pyrrhotite (monoclinal, hexagonal, troilite) varies between deposits and also between ore bodies in one deposit.

Sulfide segregation took place when sulfur was added to the magma from the wall rock in the magma conduit or in the final magma chamber. The ratio of silicate melt to sulfide melt (R factor) varied mostly between 200 and 1300 and, consequently, a wide range in nickel tenor is observed. Ultramafic, weakly differentiated Vammala-type intrusions can be interpreted as flow conduits. Kotalahti-type intrusions are more strongly differentiated and contain layered rocks, suggesting crystallization in a more quiescent magma chamber. Tholeiitic and picritic volcanic rocks spatially closely associated with the intrusions may represent the same magmatic event as the intrusions, but the magma probably ascended via different routes into the upper crust. Geochemical differences suggest that contamination was more significant for the intrusive phases. This is in accordance with the fact that nickel-copper mineralization is rare within the extrusive rocks.

ACKNOWLEDGMENTS

I wish to thank Tapio Halkoaho for comments on the first draft of the manuscript and useful discussions, Janne Hokka for the production of the Kotalahti 3D figures, Jukka Kousa and Jouni Luukas for useful discussions and help in using map data, Jouni Lerssi for the generation of the geophysical maps, and Riitta Turunen and Ritva Jokisaari for help in generating the diagrams. I am grateful to Wolf Maier for critical reading, valuable comments, and correcting the English, and to Petri Peltonen for constructive suggestions to improve the chapter. I also thank Belvedere Mining Oy for permission to publish the Hitura data and Altona Mining Ltd. for permission to publish the Rytky data.

REFERENCES

Aarnisalo, J., 1988. E-S Ni-vyöhykekartoitukset. Reports of Outokumpu Mining Company 230 020_11 Y3_88.

Arndt, N., Sobolev, S., Barnes, S., Robertson, J., 2013. Magma dynamics and the formation of magmatic sulphide deposits. In: Proceedings of the 12th Biennial SGA Meeting. Sweden, Uppsala, pp. 930–932. August 12–15.

Barnes, S.J., 1986. The effect of trapped liquid crystallisation on cumulus mineral compositions in layered intrusions. Contributions to Mineralogy and Petrology 93, 524–531.

Barnes, S.J., 2006. Komatiite-hosted nickel sulphide deposits: Geology, geochemistry and genesis. Society of Economic Geologists Special Publication 13, 51–118.

Barnes, S.J., Makkonen, H.V., Dowling, S.E., et al., 2009. The 1.88 Ga Kotalahti and Vammala nickel belts, Finland: geochemistry of the mafic and ultramafic metavolcanic rocks. Bulletin of the Geological Society of Finland 81 (2), 103–141.

Bleeker, W., 1990. Thompson area–general geology and ore deposits. Geology and Mineral deposits of the Flin Flon and Thompson Belts, Manitoba. Field Trip Guidebook, 8th IAGOD Symposium, 2165, Geological Survey of Canada, pp. 93–136. Open File.

Boyd, R., Mathieson, C.O., 1979. The nickel mineralisation of the Råna mafic intrusion, Nordland, Norway. Canadian Mineralogist 17, 287–298.

Boyd, R., Barnes, S.-J., Grönlie, A., 1988. Noble metal geochemistry of some Ni-Cu deposits in the Sveconorwegian and Caledonian orogens in Norway. In: Prichard, H.M., Potts, P.J., Bowles, J.F.W., Cribb, S.J. (Eds.), Geo-Platinum 87. Elsevier, Applied Science, London, pp. 144–158.

Campbell, I.H., Naldrett, A.J., 1979. The influence of silicate:sulphide ratios on the geochemistry of magmatic sulphides. Economic Geology 74, 1503–1505.

Duke, J.M., Naldrett, A.J., 1978. A numerical model of the fractionation of olivine and molten sulphide from komatiite magma. Earth and Planetary Science Letters 39, 255–266.

Eeronheimo, J., 1985. Kairaus-ja malmiarvioraportti, Kerimäki, Hälvälä. 030/4211 08/JJE/1985. Report of Outokumpu Oy, Exploration 10.

Eeronheimo, J., Pietilä, R., 1988a. Tutkimusraportti Savonlinna, Tevanniemi, NiCu. 001/421105B/JJE/RMP/1988. Report of Outokumpu Oy, Exploration 7.

Eeronheimo, J., Pietilä, R., 1988b. Tutkimus- ja malmiarvioraportti, Taipalsaari, Telkkälä. 001_313405D_88. Report of Outokumpu Oy, Exploration 12.

Eilu, P., Lahtinen, R., 2013. Fennoscandian metallogeny and supercontinent cycles. In: Proceedings of the 12th Biennial SGA Meeting. Sweden, Uppsala, pp. 1632–1634. August 12–15.

Fleet, M.E., MacRae, N.D., Herzberg, C.T., 1977. Partition of nickel between olivine and sulphide; a test for immiscible sulphide liquids. Contributions to Mineralogy and Petrology 65, 191–197.

Forss, H., Kontoniemi, O., Lempiäinen, R., et al., 1999. Ni-vyöhyke ja 1.9 Ga magmatismi –hankkeen (12204) toiminta vuosina 1992-1998 Tervo-Varkaus – alueella. Geological Survey of Finland 152 Archive Report M19/3241/99/1/10.

Francis, R.D., 1990. Sulfide globules in mid-ocean ridge basalts (MORB), and the effect of oxygen abundance in Fe-S-O liquids on the ability of those liquids to partition metals from MORB and komatiite magmas. Chemical Geology 85, 199–213.

Gaál, G., 1972. Tectonic control of some Ni-Cu deposits in Finland. In: Proceedings of the International Geological Congress, 24th session, Montreal, pp. 215–224 Section 4, Mineral Deposits.

Gaál, G., 1980. Geological setting and intrusion tectonics of the Kotalahti nickel-copper deposit, Finland. Bulletin of the Geological Society of Finland 52 (1), 101–128.

Groves, D.I., Bierlein, F.P., 2007. Geodynamic settings of mineral deposits. J. Geol. Soc 164, 19–30.

Grundström, L., 1980. The Laukunkangas nickel-copper occurrence in south-eastern Finland. Bulletin of the Geological Society of Finland 52 (1), 23–53.

Grundström, L., 1985. The Laukunkangas nickel-copper deposit. Geological Survey of Finland Bulletin 333, 240–256.

Häkli, A., 1963. Distribution of nickel between the silicate and sulphide phases in some basic intrusions in Finland. Bull. Comm. Geol. Finlande 209, 54.

Häkli, T.A., 1968. An attempt to apply the Makaopuhi nickel fractionation data to the temperature determination of a basic intrusive. Geochim. Cosmochim. Acta 32, 449–460.

Häkli, T.A., 1970. Factor analysis of the sulphide phase in mafic–ultramafic rocks in Finland. Bulletin of the Geological Society of Finland 42, 109–118.

Häkli, T.A., 1971. Silicate nickel and its application to the exploration of nickel ores. Bulletin of the Geological Society of Finland 43 (2), 247–263.

Häkli, T.A., Huhma, M., Viluksela, E., Vuorelainen, Y., 1975. A minor Ni-Cu-deposit at Telkkälä, SE-Finland. Bulletin of the Geological Society of Finland 47 (1–2), 55–70.

Häkli, T.A., Hänninen, E., Vuorelainen, Y., Papunen, H., 1976. Platinum-group minerals in the Hitura nickel deposit, Finland. Economic Geology 71 (7), 1206–1213.

Häkli, T.A., Vormisto, K., Hänninen, E., 1979. Vammala, a nickel deposit in layered ultramafite, southwest Finland. Economic Geology 74 (5), 1166–1182.

Häkli, T.A., Vormisto, K., 1985. The Vammala nickel deposit. In: Papunen, H., Gorbunov, G.I. (Eds.), Nickel-copper deposits of the Baltic Shield and Scandinavian Caledonides, 333. Geological Survey of Finland, pp. 273–286. Bulletin.

Heaman, L.M., Machado, N., Krogh, T.E., Weber, W., 1986. Precise U-Pb zircon ages for the Molson dyke swarm and the Fox River sill: Constraints for Early Proterozoic crustal evolution in northeastern Manitoba, Canada. Contributions to Mineralogy and Petrology 94, 82–89.

Heaman, L.M., Peck, D., Toope, K., 2009. Timing and geochemistry of 1.88 Ga Molson igneous events, Manitoba: Insights into the formation of a craton-scale magmatic and metallogenic province. Precambrian Research 172, 143–162.

Hill, R.E.T., Barnes, S.J., Dowling, S.E., et al., 2005. Chalcophile element distribution in mafic and ultramafic metavolcanic rocks of the Svecofennian (1. 9 Ga) Kotalahti and Vammala nickel belts, Finland—a test for a geochemical signature of subvolcanic magmatic ore forming processes. August 8–11, Oulu, Finland, extended abstracts. In: Proceedings of the 10th International Platinum Symposium: Platinum-group elements—from genesis to beneficiation and environmental impact. Geological Survey of Finland, Espoo, pp. 369–372.

Hoatson, D.M., Blake, D.H., 2000. Geology and economic potential of the Palaeoproterozoic layered mafic–ultramafic intrusions in the east Kimberley, Western Australia. AGSO Bulletin 246, 476.

Hulbert, L.J., Hamilton, M.A., Horan, M.F., Scoates, R.F.J., 2005. U-Pb zircon and Re-Os isotope geochronology of mineralised ultramafic intrusions and associated nickel ores from the Thompson nickel belt, Manitoba, Canada. Economic Geology 100, 29–41.

Irvine, T.N., Baragar, W.R.A., 1971. A guide to the chemical classification of the common volcanic rocks. Can. J. Earth Sci 8, 523–548.

Isohanni, M., Ohenoja, V., Papunen, H., 1985. Geology and nickel-copper ores of the Nivala area. Geological Survey of Finland Bulletin 333, 211–228.

Isomäki, O.-P., 1994. Hälvälän ja Telkkälän nikkeli-kuparikaivokset. Summary: The Hälvälä and Telkkälä nickel-copper mines. Geologi 46 (8), 112–114.

Jokela, J., 1994. Itä-Suomen Ni-vyöhykkeen kartoitus ja rakennetulkinta. Raportti kenttätöistä 1993–1994. Geological Survey of Finland 26 Archive Report, M19/3241/94/1/10.

Kalliomäki, H., 2013. Petrological, mineralogical and geochemical characteristics of the Keskimmäinen gabbroic intrusion, Eastern Finland. Unpublished Master's Thesis, University of Helsinki, Department of Geosciences and Geography, Division of Geology. p. 116.

Kilpeläinen, T., 1998. Evolution and 3D modelling of structural and metamorphic patterns of the Palaeoproterozoic crust in the Tampere-Vammala area, southern Finland. Geological Survey of Finland 124 Bulletin 397.

Koistinen, T., 1981. Structural evolution of an early Proterozoic strata-bound Cu-Co-Zn deposit, Outokumpu, Finland. Transactions of the Royal Society of Edinburgh: Earth sciences 72, 115–158.

Koistinen, T., Klein, V., Koppelmaa, H., et al., 1996. Paleoproterozoic Svecofennian orogenic belt in the surroundings of the Gulf of Finland. Explanation to the map of Precambrian basement of the Gulf of Finland and surrounding area 1:1 mill, 21, Geological Survey of Finland, pp. 21–57. Special Paper.

Kojonen, K., 1999. Rautalammin Törmälän Ni-Cu-malmiaiheen malmimineralogia. Geological Survey of Finland 7 Archive Report, M42.2/3223/99/1.

Kojonen, K., Johansson, B., Pakkanen, L., 2002. Leppävirran Rytkyn Ni-Cu sulfidimalmiaiheen malmimineraaleista. Geological Survey of Finland 17 Archive Report, M41/3241/2002/1.

Kojonen, K.K., Roberts, A.C., Isomäki, O.-P., et al., 2004. Tarkianite, (Cu,Fe)(Re,Mo)4S8, a new mineral species from the Hitura mine, Nivala, Finland. In: Platinum-group elements: petrology, geochemistry, mineralogy. The Canadian Mineralogist 42 (2), 539–544.

Kontoniemi, O., Forss, H., 1998. Tutkimustyöselostus Leppävirran kunnassa valtausalueilla Hanhi 1 (kaiv.rek.nro 5473/1) ja Hanhi 2 (kaiv.rek.nro 5771/1) suoritetuista nikkelimalmitutkimuksista vuosina 1993–1996. Geological Survey of Finland 13 Archive Report, M06/3241/98/1/10.

Korsman, K., Korja, T., Pajunen, M., Virransalo, P., 1999. The GGT/SVEKA transect: structure and evolution of the continental crust in the Paleoproterozoic Svecofennian orogen in Finland. International Geology Review 41 (4), 287–333.

Lahtinen, R., 1994. Crustal evolution of the Svecofennian and Karelian domains during 2.1–1.79 Ga, with special emphasis on the geochemistry and origin of 1.93–1.91 Ga gneissic tonalites and associated supracrustal rocks in the Rautalampi area, central Finland. Geological Survey of Finland 128 Bulletin 378.

Lahtinen, R., 2000. Archaean–Proterozoic transition: geochemistry, provenance and tectonic setting of metasedimentary rocks in central Fennoscandian Shield, Finland. Precambrian Research 104, 147–174.

Lamberg, P., 2005. From genetic concepts to practice: lithogeochemical identification of Ni-Cu mineralised intrusions and localisation of the ore. Geological Survey of Finland 264 Bulletin 402.

Layton-Matthews, D.M., Lesher, C.M., Liwanag, J., et al., 2011. Mineralogy, geochemistry, and genesis of komatiite-associated Ni-Cu-(PGE) mineralisation in the Thompson nickel belt, Manitoba. In: Ni-Cu, Magmatic, Deposits, P.G.E. (Eds.), Geology, Geochemistry, and Genesis, 17. Reviews in Economic Geology, pp. 123–143.

Lyubetskaya, T., Ague, J.J., 2010. Modeling metamorphism in collisional orogens intruded by magmas: I. Thermal evolution. American Journal of Science 310, 427–458.

Machado, N., Gapais, D., Potrel, A., et al., 2011. Chronology of transpression, magmatism, and sedimentation in the Thompson nickel belt (Manitoba, Canada) and timing of Trans-Hudson orogen–Superior Province collision. Canadian Journal of Earth Sciences 48 (2), 295–324.

Maier, W.D., Li, C., De Waal, S.A., 2001. Why are there no major Ni-Cu sulphide deposits in large layered mafic–ultramafic intrusions? Canadian Mineralogist 39, 547–556.

Mäkinen, J., 1987. Geochemical characteristics of Svecokarelidic mafic–ultramafic intrusions associated with Ni-Cu occurrences in Finland. Geological Survey of Finland 109 Bulletin 342.

Mäkinen, J., Makkonen, H.V., 2004. Petrology and structure of the Palaeoproterozoic (1.9 Ga) Rytky nickel sulphide deposit, Central Finland: a comparison with the Kotalahti nickel deposit. Mineralium Deposita 39, 405–421.

Makkonen, H.V., 1996. 1.9 Ga tholeiitic magmatism and related Ni-Cu deposition in the Juva area, SE Finland. Geological Survey of Finland 101 Bulletin 386.

Makkonen, H.V., 2005. Intrusion model for Svecofennian (1.9 Ga) mafic–ultramafic intrusions in Finland. In: Autio, S. (Ed.), Current Research 2003–2004. Geological Survey of Finland. Special Paper 38, 11–14. Electronic publication.

Makkonen, H., Forss, H., 1999. Tutkimustyöselostus Kuopion ja Karttulan kunnissa valtausalueilla Vehmasjoki 1 (kaiv. rek. nro 6329/1), Vehmasjoki 4 (kaiv. rek. nro 6371/1), Vehmasjoki 5 (kaiv. rek. nro 6502/1) ja Vehmasjoki 6 (kaiv. rek. nro 6502/2) suoritetuista nikkelimalmitutkimuksista vuosina, 1996–1998. Geological Survey of Finland 12 Archive Report, M06/3242/99/1/10.

Makkonen, H., Forss, H., 2003. Nikkelimalmitutkimukset Tiemasojan, Koirasaaren ja Pölkkyinniemen kohteissa Rantasalmen-Joroisten Tiemassaaressa vuosina 1992–1993. Geological Survey of Finland 18 Archive Report, M19/3234/2003/1/10.

Makkonen, H., Halkoaho, T., 2007. Whole rock analytical data (XRF, REE, PGE) for several Svecofennian (1. 9 Ga) and Archaean (2. 8 Ga) nickel deposits in eastern Finland. Geological Survey of Finland 49 Archive Report, M19/3241/2007/32.

Makkonen, H., Halkoaho, T., Tiainen, M., et al., 2010. FINNICKEL – a public database on nickel deposits in Finland. Version 1.1 [Electronic resource]. Digitaaliset tietotuotteet 11. Espoo: Geological Survey of Finland Optical disc (CD-ROM).

Makkonen, H.V., Huhma, H., 2007. Sm-Nd data for mafic–ultramafic intrusions in the Svecofennian (1. 88 Ga) Kotalahti nickel belt, Finland—implications for crustal contamination at the Archaean/Proterozoic boundary. Bulletin of the Geological Society of Finland 79 (2), 175–201.

Makkonen, H., Kontoniemi, O., Lempiäinen, R., et al., 2003. Raahe-Laatokka-vyöhyke, nikkelin ja kullan etsintä-hankkeen (2108001) toiminta vuosina, 1999–2003. Geological Survey of Finland 90 Archive Report, M10.4/2003/5/10.

Makkonen, H.V., Mäkinen, J., Kontoniemi, O., 2008. Geochemical discrimination between barren and mineralised intrusions in the Svecofennian (1. 9 Ga) Kotalahti nickel belt, Finland. In: Ore-forming processes associated with mafic and ultramafic rocks. Ore Geology Reviews 33 (1), 101–114.

Makkonen, H.V., Suikkanen, M., Isomäki, O.-P., Seppä, V.-M., 2011. Lithogeochemistry and geology of the Palaeoproterozoic (1.88 Ga) Hitura nickel deposit, western Finland. In: Proceedings of the 25th International Applied Geochemistry Symposium, August 22–26, Rovaniemi, Finland. Vuorimiesyhdistys. Sarja B 92–1, Espoo, pp. 119–120. Programme and Abstracts.

Marshall, B., Smith, J.V., Mancini, F., 1995. Emplacement and implications of peridotite-hosted leucocratic dykes, Vammala Mine, Finland. GFF 117 (4), 199–205.

McDonough, W.F., Sun, S.-S., 1995. The composition of the Earth. Chemical Geology 120, 223–254.

McGoldrick, P.J., Keays, R., Scott, B.B., 1979. Thallium: a sensitive indicator of rock/seawater interaction and on sulphur saturation of silicate melts. Geochim. Cosmochim. Acta 43, 1303–1311.

McQueen, K.G., 1987. Deformation and remobilization in some Western Australian nickel ores. Ore Geology Reviews 2, 269–286.

Meriläinen, M., Ekberg, M., Lovén, P., et al., 2008. Updated Reserve and Resource Estimate of the Hitura Nickel Mine in Central Finland. Belvedere Resources, National Instrument 43–101 65 Technical Report.

Mungall, J.E., 2007. Crustal contamination of picritic magmas during transport through dikes: the Expo Intrusive Suite, Cape Smith fold belt, New Quebec. Journal of Petrology 48, 1021–1039.

Naldrett, A.J., 2011. Fundamentals of magmatic sulphide deposits. In: Li, C., Ripley, E.M. (Eds.), Magmatic Ni-Cu and PGE Deposits: Geology, Geochemistry, and Genesis, Vol. 17. Society of Economic Geologists, pp. 1–50. Reviews in Economic Geology.

Nironen, M., 1997. The Svecofennian orogen: a tectonic model. Precambrian Research 86 (1–2), 21–44.

Papunen, H., 1970. Sulphide mineralogy of the Kotalahti and Hitura nickel-copper ores, Finland. Annales Academiae Scientiarum Fennicae, Series A3: Geologica-Geographica = Suomalaisen Tiedeakatemian Toimituksia 109 Sarja A3. Geologica-Geographica.

Papunen, H., 1974. The sulphide mineral assemblages of some Finnish Ni-Cu deposits. In: Problems of Ore Deposition. Fourth Symposium on the International Association on the Genesis of Ore Deposits, Varna. Publishing House of the Bulgarian Academy of Sciences, Sofia, pp. 311–319.

Papunen, H., 1980. The Kylmäkoski nickel-copper deposit in south-western Finland. Bulletin of the Geological Society of Finland 52 (1), 129–145.

Papunen, H., 1985. The Kylmäkoski nickel-copper deposit. In: Nickel-Copper Deposits of the Baltic Shield and Scandinavian Caledonides. Geological Survey of Finland. Bulletin 333, 264–273.

Papunen, H., 1986. Platinum-group elements in Svecokarelian nickel-copper deposits. Finland. Economic Geology 81 (5), 1236–1241.

Papunen, H., 1989. Platinum-group elements in metamorphosed Ni-Cu deposits in Finland. In: Magmatic Sulphides—The Zimbabwe Volume. The Institution of Mining and Metallurgy, London, pp. 165–176.

Papunen, H., 2003. Ni-Cu sulphide deposits in mafic–ultramafic orogenic intrusions – examples from the Svecofennian areas, Finland. In: Mineral Exploration and Sustainable Development: Proceedings of the Seventh Biennial SGA Meeting, Athens. Millpress, Rotterdam, pp. 551–554. August 24–28.

Papunen, H., 2005. Metallogeny of nickel in Fennoscandia. Short course notes on nickel metallogeny and exploration. Proceedings of the 5th Fennoscandian Exploration and Mining Conference Rovaniemi, p. 57.

Papunen, H., Gorbunov, G.I. (Eds.), 1985. Nickel-copper deposits of the Baltic Shield and Scandinavian Caledonides. Geological Survey of Finland, p. 394. Bulletin 333.

Papunen, H., Häkli, T.A., Idman, H., 1979. Geological, geochemical and mineralogical features of sulphide-bearing ultramafic rocks in Finland. The Canadian Mineralogist 17 (2), 217–232.

Papunen, H., Koskinen, J., 1985. Geology of the Kotalahti nickel-copper ore. In: Papunen, H., Gorbunov, G.I. (Eds.), Nickel-copper deposits of the Baltic Shield and Scandinavian Caledonides, 333. Geological Survey of Finland, pp. 228–240. Bulletin.

Papunen, H., Mäkelä, M., 1980. Sulphur isotopes in Finnish nickel-copper occurrences. Bulletin of the Geological Society of Finland 52, 55–66.

Papunen, H., Penttilä, V., 1996. Mineralogy and geology of the Serpentinite associated Hitura Ni-Cu deposit, Finland: Implications for beneficiation. Australasian Institute of Mining and Metallurgy 6/96 79–88.

Papunen, H., Vorma, A., 1985. Nickel deposits in Finland, a review. In: Papunen, H., Gorbunov, G.I. (Eds.), Nickel-copper deposits of the Baltic Shield and Scandinavian Caledonides, 333. Geological Survey of Finland, pp. 123–143. Bulletin.

Patchett, J., Kouvo, O., 1986. Origin of continental crust of 1.9–1.7 Ga age: Nd isotopes and U-Pb zircon ages in the Svecokarelian terrain of south Finland. Contributions to Mineralogy and Petrology 92 1–12.

Peltonen, P., 1995a. Magma-country rock interaction and the genesis of Ni-Cu deposits in the Vammala nickel belt, SW Finland. Mineralogy and Petrology 52, 1–24.

Peltonen, P., 1995b. Petrogenesis of ultramafic rocks in the Vammala nickel belt; implications for crustal evolution of the early Proterozoic Svecofennian arc terrane. Lithos 34, 253–274.

Peltonen, P., 2005. Mafic–ultramafic intrusions of the Svecofennian orogen. In: Lehtinen, M., Nurmi, P.A., Rämö, O.T. (Eds.), Precambrian of Finland—a Key to the Evolution of the Fennoscandian Shield. Elsevier, Amsterdam, pp. 413–447.

Percival, J.A., Whalen, J.B., Rayner, N., 2004. Pikwitonei—Snow Lake, Manitoba Transect (parts of NTS 63J, 63O, and 63P), Trans-Hudson orogen—Superior Margin Metallotect Project: initial geological, isotopic and SHRIMP U-Pb results. In: Report of Activities, Manitoba Industry, Economic Development and Mines. Manitoba Geological Survey, pp. 120–134.

Percival, J.A., Whalen, J.B., Rayner, N., 2005. Pikwitonei—Snow Lake Manitoba Transect (parts of NTS 63J, 63O, and 63P), Trans-Hudson orogen—Superior Margin Metallotect Project: new results and tectonic interpretation. In: Report of Activities, Manitoba Industry, Economic Development and Mines. Manitoba Geological Survey, pp. 69–91.

Piña, R., Romeo, I., Ortega, L., et al., 2010. Origin and emplacement of the Aguablanca magmatic Ni-Cu-(PGE) sulphide deposit, SW Iberia: A multidisciplinary approach. Bulletin of the Geological Society of America 122.5–6, 915–925.

Puustinen, K., Saltikoff, B., Tontti, M., 1995. Distribution and metallogenic types of nickel deposits in Finland. Geological Survey of Finland. Report of Investigation 132, p. 30.

Queffurus, M., Barnes, S-J., 2015. A review of sulfur to selenium ratios in magmatic nickel-copper and platinum-group element deposits. Ore Geology Reviews 69, 301–324.

Rasilainen, K., Eilu, P., Äikäs, O., et al., 2012. Quantitative mineral resource assessment of nickel, copper and cobalt in undiscovered Ni-Cu deposits in Finland. Geological Survey of Finland Report of Investigation 194, p. 514. Electronic publication.

Ripley, E.M., 2009. Magmatic sulphide mineralisation in Alaskan-type complexes. In: Li, C., Ripley, E.M. (Eds.), New Developments in Magmatic Ni-Cu and PGE Deposits. Geological Publishing House, Beijing, pp. 219–228.

Seppä, V-M., 2009. Kotalahden alueen nikkeliesiintymät. Unpublished Master's Thesis. University of Turku, Department of Geology, p. 77.

Suikkanen, M., 2011. Hituran intruusion geologia ja litogeokemia. Unpublished Master's Thesis. University of Oulu, Department of Geology, p. 89.

Tang, Z., Song, X.-Y., Su, S., 2009. Ni-Cu Deposits Related to High-Mg Basaltic Magma, Jinchuan, Western China. In: Li, C., Ripley, E.M. (Eds.), New Developments in Magmatic Ni-Cu and PGE Deposits, pp. 121–140.

Tomkins, A.G., Pattison, D.R.M., Frost, B.R., 2007. On the initiation of metamorphic sulphide anatexis. Journal of Petrology 48, 511–535.

Trudu, A.G., Hoatson, D.M., 2000. Depths of emplacement of the mafic–ultramafic intrusions. In: Hoatson, D.M., Blake, D.H. (Eds.), Geology and economic potential of the Palaeoproterozoic layered mafic–ultramafic intrusions in the east Kimberley, Western Australia, 246, pp. 201–219 AGSO Bulletin.

Tuisku, P., Makkonen, H.V., 1999. Spinel-bearing symplectites in Palaeoproterozoic ultramafic rocks from two different geological settings in Finland: thermobarometric and tectonic implications. GFF 121 (4), 293–300.

Turek, A., Woodhead, J., Zwanzig, H.V., 2000. U-Pb age of the gabbro and other plutons at Lynn Lake (part of NTS 64C). In: Report of Activities 2000, Manitoba Industry, Trade and Mines. Manitoba Geological Survey 97–104.

Weihed, P., Bergman, J., Bergström, U., 1992. Metallogeny and tectonic evolution of the Early Proterozoic Skellefte district, northern Sweden. Precambrian Research 58, 143–167.

Zwanzig, H.V., 2005. Geochemistry, Sm-Nd isotope data and age constraints of the Bah Lake assemblage, Thompson nickel belt and Kisseynew Domain margin: Relation to Thompson-type ultramafic bodies and a tectonic model (NTS63J, O and P). In: Report of Activities, Manitoba Industry, Economic Development and Mines, pp. 40–53.

Zwanzig, H.V., Macek, J.J., McGregor, C.R., 2007. Lithostratigraphy and geochemistry of the high-grade metasedimentary rocks in the Thompson nickel belt and adjacent Kisseynew Domain, Manitoba: Implications for nickel exploration. Economic Geology 102 (7), 1197–1216.

PART

DEPOSITS RELATED TO CARBONATITIES AND KIMBERLITES 4

INTRODUCTION TO CARBONATITE DEPOSITS OF FINLAND 4.1

H. O'Brien

ABSTRACT

Until the mid-1950s, the idea of an igneous rock made of carbonate was accepted by few geologists. Today, the validity of the idea of carbonatite magma is no longer questioned, because of direct observation of carbonatite lavas capping the Oldoinyo Lengai volcano; experimental work that has produced carbonatite melts during partial melting of peridotite; extensive geological mapping of plutonic complexes that found carbonatites in intrusive relationships with associated cumulate ultramafic rocks; and evidence of carbonatite melts and their ephemeral geochemical imprints in mantle peridotite xenoliths. Nevertheless, even though it is accepted that most carbonatites are generated in the mantle, the means by which they form, manage to ascend through the thick lithospheric mantle, and erupt at the Earth's surface are still open to considerable debate.

Carbonatites are not just of academic interest; their unusual chemistry means that they concentrate certain elements that are essential for modern society. Such elements include phosphorus, needed to help feed the growing world population, and rare earth elements, essential for many electronic appliances and magnets used, for example, in wind turbines. Niobium, nearly all of which is sourced from carbonatites, is needed to produce high-quality stainless steel.

Finland contains a number of carbonatites, with ages ranging from the Archean to the Devonian, and distributed across the country, from the southwestern corner to the very northeast. The approximately 2.0 Ga Laivajoki, Kortijärvi, and Petäikkö-Suvantovaara carbonatites are actively being tested for their mineral potential, while the group of rather small but mineralogically interesting 1.8 Ga Halpanen, Panjavaara, and Naantali carbonatites are,

Mineral Deposits of Finland. http://dx.doi.org/10.1016/B978-0-12-410438-9.00011-X

at least presently, mostly of academic interest. Two Finnish carbonatites, Sokli, apparently the world's largest, and Siilinjärvi, among the world's oldest, are of such geologic and economic importance as to deserve special attention. Hence, each is discussed in more detail (Chapters 4.2 and 4.3, respectively) following this introduction.

Keywords: carbonatite; Laivajoki; Kortijärvi; phosphorus; niobium; rare earth elements.

INTRODUCTION

Carbonatite is a unique igneous rock type, in that it formed predominantly of carbonate minerals (of which it must contain >50 modal % by definition) along with lesser silicates, phosphate minerals, and oxides. The term *carbonatite* was introduced by Brøgger in 1921 in his studies of the Fen complex in Norway (Brøgger, 1921), but the concept of a rock formed from carbonate-dominant magma was not generally accepted at that time. Further evidence for the existence of magmatic carbonatites came particularly from detailed studies of alkaline complexes in Northern Europe, for example at Afrikanda, Kola (Kupletsky, 1938); Alnö, Sweden (von Eckermann, 1948); and again at the Fen complex (Saether, 1957).

Strontium isotope work by Powell et al. (1962) provided some of the first strong geochemical evidence that carbonatites that came from the mantle were not recycled limestones. The discovery of extrusive carbonatite lavas at Oldoinyo Lengai in 1960 (reported by Dawson, 1962) proved without a doubt that carbonatite magmas exist. Only slightly later, experimental work (e.g., starting with Wyllie, 1965, as well as Wyllie, 1978) showed that small degree partial melts of carbonate-bearing peridotite are carbonatitic in character. The question as to whether these magmas exist has hence shifted to answering the question of how they form.

AGE AND OCCURRENCE

Carbonatites occur throughout the geologic record, but there is an apparent population maximum in post-Paleozoic times (greater than 50% of the 527 carbonatites in the Wooley and Kjarsgaard (2008a) compilation are <545 Ma). Although this may be a preservation phenomenon, with younger carbonatites preferentially surviving erosion, it may also reflect an increase in carbonatite activity of the Earth as it has aged. The latter idea is supported by the prevalence of carbonatites in Precambrian areas (Woolley and Kjarsgaard, 2008a) that are underpinned by remnants of the oldest mantle material on the planet, potentially arguing for the importance of aged mantle sources for carbonatites.

FORM AND ROCK TYPES

Carbonatites have been found in the form of flows, tuffs, ring dikes, linear dikes, and plutons—in fact, they occur in the same kind of bodies as silicate-dominated magmas. Extrusive carbonatites, although rare, exist in at least 49 localities (Woolley and Church, 2005), ranging from flows to tuffs, but they are almost all exclusively young. Due to erosional effects, the predominant form of carbonatite is as an

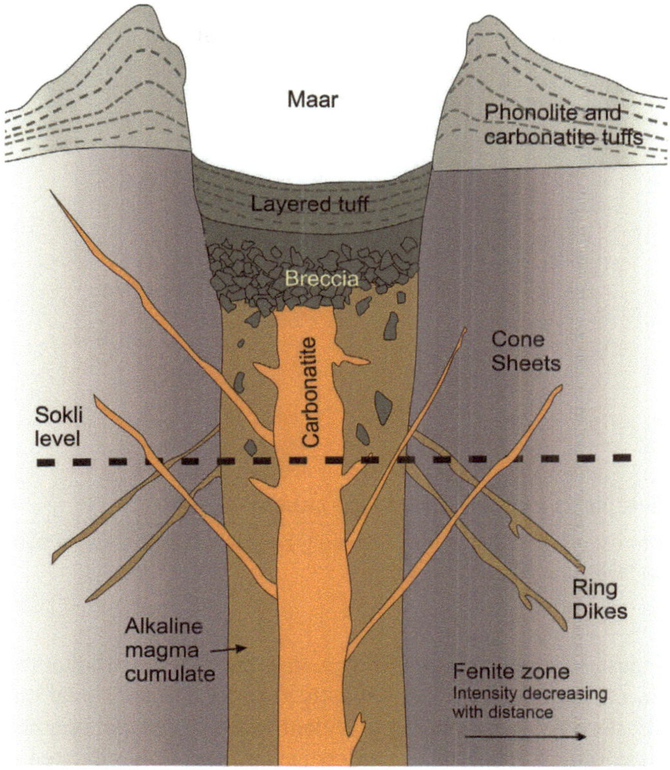

FIGURE 4.1.1 Cross section of a standard carbonatite, with Sokli erosion level estimated.

Source: *Diagram after Sage and Watkinson (1991).*

intrusion, exposed at levels several hundred meters below the preexisting volcanic edifice. Thus, the "standard" model for carbonatite bodies is in the form of a pluton, circular in plan, with outer rings of associated silicate rocks (magmas or cumulates), and a carbonatitic central plug (Fig. 4.1.1).

The Sokli carbonatite may fit this model, with obvious circular structures and cone sheets intruded into the surrounding fenite. In this case, much of the actual magma chamber must lie below this level. Other carbonatites that do not show this concentric shape possibly represent deeper levels below the volcanic edifice. Siilinjärvi may be one example where, due to the depth of erosion, the underlying magma chamber is exposed (Fig. 4.1.2).

Approximately one quarter of all carbonatites occur without other spatially associated igneous activity, in particular those that occur as sheet-like dikes. The remainder are spatially associated with some type of alkaline silicate igneous rock or rock suite. Woolley and Kjarsgaard (2008b) have distinguished seven such associations, with each providing general information on the type of mantle source for a particular carbonatite-containing complex. The list (with intrusive equivalents in parentheses) consists of nephelinite (ijolite), melilitite (melilitolite), basanite (alkali gabbro), phonolite (feldspathoidal syenite), trachyte (syenite), kimberlite, and ultramafic lamprophyre (aillikite, alnöite).

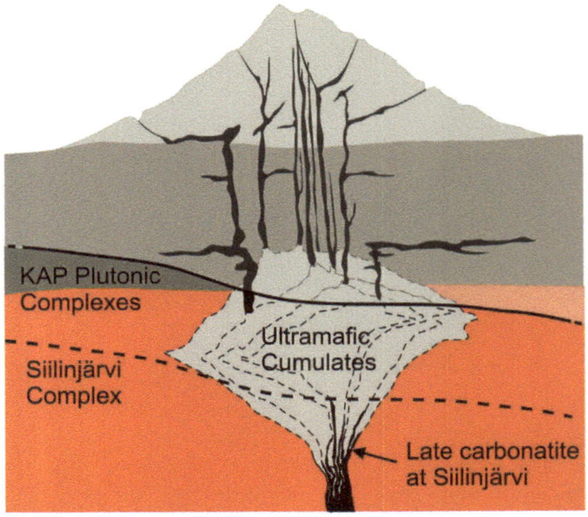

FIGURE 4.1.2 Subvolcanic plutonic complex and magma chamber storage.

Image was prepared with estimates of erosion level more typical in the Kola alkaline complexes in comparison to the estimated erosion level at Siilinjärvi.

Source: *After Nielsen (1994).*

The first three of these together represent the greatest number of carbonatite complexes, are more sodic than potassic in character, and are dominantly formed in extensional environments, with the Great Rift Valley of Eastern Africa providing abundant examples of carbonatite complexes from this tectonic setting. The last four associations, in contrast, are potassium-dominated systems, and although also found mostly in extensional zones, in many cases they occur where there has been prior collisional tectonics (e.g., subduction).

Proof of genetic relationships between the alkaline rocks typically hosting crosscutting younger carbonatite is not as common as one might expect given the numerous examples of spatial overlap. Besides the nearly one quarter of carbonatites without known associated silicate magmas, Gittins and Harmer (2003) make a strong case that spatial association may just as likely result from silicate magmas ascending via structural weaknesses to the surface, and carbonatite magmas, formed from a different mantle source (quite possibly deeper) taking advantage of the same structures. Isotopic evidence is equivocal on this question; for example, at Kovdor and Khibina the silicate and carbonatite magmas come from distinctly different isotope reservoirs, but at Vuorijärvi, both magma types come from isotopically identical isotope reservoirs (Brassinnes et al., 2003). Consequently, any genetic relationship of carbonatite and silicate magmas in an alkaline rock complex must be tested on a case-by-case basis.

MINERALOGY AND ORES

By definition, carbonatites consist of a volume greater than 50% carbonate minerals, mostly calcite-dominated, but dolomite carbonatites are typical as later stage carbonatites, and also appear to be more common in older cratonic areas (Harmer and Gittins, 1997). In an evolving carbonatite system, the final, most fractionated melts can form ferrocarbonatites with ankerite and/or siderite, or, in very rare cases, magnesite (e.g., Zaitsev et al., 2004). In many cases, these also contain otherwise rare carbonates, including Sr carbonates (strontianite, ancylite) and rare earth elements (REE) carbonates (especially bastnäsite (Ce, La)FCO_3). Nyerereite $(Na,K)_2Ca(CO_3)_2$ and gregoryite $(Na,K)_2CO_3$ are the two main components of Oldoinyo Lengai natrocarbonatite lavas. In addition to carbonates, the list of minerals that can be found in carbonatites, even omitting the less common minerals, is long (the most characteristic minerals are in bold):

- Phosphates—**apatite**, monazite
- Oxides—baddeleyite, zircon, hematite, **magnetite**, ilmenite, perovskite, pyrochlore
- Silicates—olivine, clinopyroxene (diopside to aegirine-augite), **phlogopite**, **amphibole**
- Sulfides—chalcopyrite, pyrite, pyrrhotite, galena
- Sulfates—barite
- Halides—fluorite, halite, sylvite

In terms of value, ore derived from carbonatites is dominantly targeted for the fertilizer industry. The phosphorous content of the ore is directly related to the high modal abundance of apatite generally found in carbonatites. Carbonatite-derived apatite is the most important source for hard-rock P_2O_5, although sediment-derived sources make up 75–80% of the 225 Mt global phosphate rock market (USGS Commodity Summaries, 2015). Apatite also contains significant contents of REE. It is not restricted to the carbonate-dominated portions of carbonatite complexes, but can also occur in high modal amounts in silicate rocks associated with the carbonatite. Siilinjärvi is a good example; the phlogopite rocks have, on average, only slightly lower grades of P_2O_5 than the carbonatites.

Rare earth elements (REE) are the second most important commodity by value derived from carbonatites, and carbonatites represent the source rocks of the majority of the roughly 135,000 ton/a global REE market. The Bayan Obo Fe-REE-Nb deposit is the world's largest REE resource at 40 Mt; there is lively debate on whether it is a carbonatite (e.g., Wu, 2008), but more recent studies (Yang et al., 2011) validate the carbonatite origin. Mount Weld is a huge REE deposit in lateritic soils above a carbonatite 3 km in diameter. The mine has recently begun production from a resource of 9.7 Mt ore at 11.7 wt% rare earth oxide (Lynas Corp.Ltd., 2014). The Mountain Pass mine located on the Proterozoic Sulfide Queen Carbonatite in California has been put back into production by Molycorp, with a projected annual production of 19,500 t of REE. Taken together, these large resources of REE in carbonatite suggest that elements will not be in short supply even over the long term.

The use of niobium in steelmaking continues to grow as the demand for increasingly higher grades of corrosion-resistant steel expands. Today, 99% of niobium mined globally comes from three carbonatites, Araxa and Catalao in Brazil, and Saint-Honoré in Quebec, Canada. In all cases, the main ore mineral is pyrochlore—that is, (Na,Ca)2Nb2O6(OH,F). A possible fourth world-class niobium project is presently being developed in the United States, specifically, in the Elk Creek carbonatite situated in southeastern Nebraska. Pyrochlore does not occur in significant quantities in many carbonatites, and requires considerable fractionation in the magmatic system before it becomes a liquidus phase.

Other important commodities derived from carbonatite complexes are fluorite, vermiculite (e.g., Phalaborwa), and magnetite (e.g., Kovdor). Sulfides are generally found only in the most evolved magmas within carbonatite complexes, and rarely in any significant quantity. The Phalaborwa carbonatite is the exception, representing the only carbonatite where sulfides are mined, and producing 60,000 t of Cu per year (Palabora Mining, 2013). Phalaborwa is unique in a number of ways, the first being mineralogical, in that sulfides in carbonatites generally demarcate evolved compositions, yet there are no accompanying enrichments of REE or Nb at Phalaborwa. The second is its isotopic composition (see the following).

GEOCHEMISTRY AND MANTLE SOURCES

Carbonatite magmas, in addition to the main components Ca, Mg, and CO_2, and minor components Si, Fe, and P, are also generally highly enriched in incompatible trace elements such as Ba, Sr, and REE including Y and Nb. Concentrations of these elements vary considerably, especially in more fractionated carbonatite magmas, in which these trace elements may reach wt% levels. However, some carbonatites show only modest REE enrichment (Hornig-Kjarsgaard, 1998), including a number of carbonatites from the Kola Alkaline Province (KAP) such as Sokli and Kovdor, as well as Siilinjärvi.

The isotopic compositions of carbonatites provide information on the ultimate source of the magmas that formed the rocks, and consequently on the sources of the essential elements such as P, Nb, and the REE, in which they are typically enriched. By far the bulk of the carbonatites plot in the "depleted mantle" field of the Sr-Nd isotope diagram (Fig. 4.1.3), suggesting a mantle source with long-term

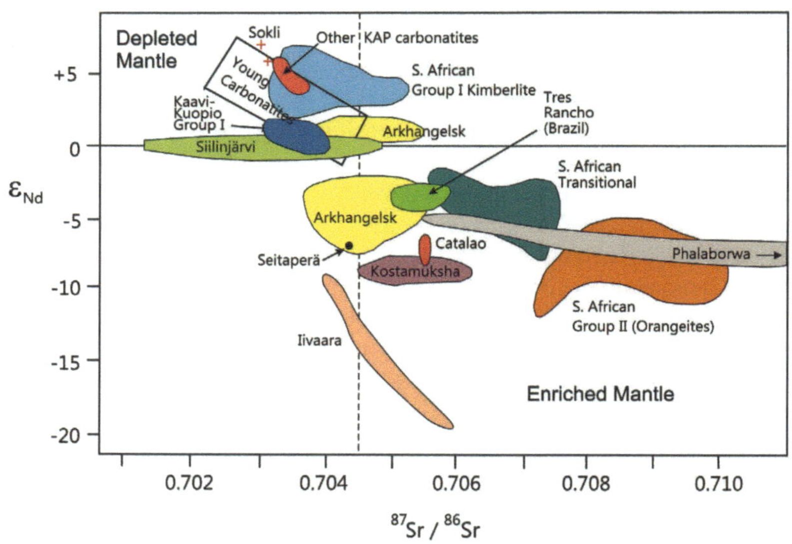

FIGURE 4.1.3 Plot of initial $^{87}Sr/^{86}Sr$ versus ε_{Nd} for selected kimberlites and carbonatites.

The young carbonatites box includes data from almost all carbonatites that are <200 Ma. Data from Sokli and other KAP carbonatites plot within or very near this box.

Source: *Data sources include Smith (1983), Harmer (1999), Mahotkin et al. (2000), Kramm (1993, 1994), Belyatsky et al. (1995), O'Brien and Tyni (1999), Cordeiro et al. (2010), and Eriksson (1989).*

incompatible element depletion (low Rb/Sr and light rare earth element (LREE)-depleted), despite the fact that the rocks are highly incompatible element-enriched. In this way, the bulk of carbonatites share many of the isotopic and trace element characteristics of kimberlites, and it is likely both magma types originate from partial melting of carbonated peridotites from below the lithosphere (see Bell and Simonetti, 2010).

Siilinjärvi and Sokli carbonatites fit this pattern, with Sr-Nd isotopic compositions similar to the Finnish kimberlites (Fig. 4.1.3). Incompatible element-rich magmas plotting in the depleted Sr-Nd field require sources enriched at the same time or shortly prior to magma formation. The unique characteristics of Phalaborwa, in terms of mineralogy (see earlier) and isotopic composition (Fig. 4.1.3), may indicate that this is an example of a lithospheric mantle-derived, carbonate-rich melt.

CARBONATITES IN FINLAND

The 10 known Finnish carbonatite occurrences are dispersed across the country and have a wide range of ages. Carbonatite intrusions occur in the very southwest of the country at Naantali, all the way to northeastern Finnish Lapland in Savukoski (Fig. 4.1.4). Siilinjärvi represents one of the oldest carbonatites on Earth at 2.6 Ga (probably the seventh oldest based on present data), while Sokli is Devonian in age, and represents one of the youngest intrusive bodies in Finland. Due to the fact that these two carbonatites are exceptional in terms of size and economic value, Sokli and Siilinjärvi are described in detail in Chapters 4.2 and 4.3, respectively, of this book. The remaining carbonatites in Finland are described briefly in the following sections, from older to younger.

LAIVAJOKI, KORTIJÄRVI, AND PETÄIKKÖ-SUVANTOVAARA CARBONATITES

The Laivajoki, Kortijärvi, and Petäikkö-Suvantovaara carbonatites are compositionally, temporally, and spatially associated, all occurring in the Koillismaa region of northeast Finland (Fig. 4.1.5). Laivajoki and Kortijärvi were discovered and drilled in 1971–1972 as a result of mineral exploration by Rautaruukki Oy. Beginning in 2010, the Geological Survey of Finland (GTK) has further investigated and evaluated this group of carbonatites, particularly with regard to their P, REE, and Zr contents (Lintinen, 2014). Diamond drilling has confirmed that Petäikkö-Suvantovaara is a carbonatite, as well as generating several new drill cores from Kortejärvi.

All three carbonatites occur as elongated bodies within the Hirvaskoski shear zone between the Kuhmo and the Pudasjärvi Archean blocks, with amphibolites as the main country rocks and lesser mica and quartz-feldspar gneisses (see Fig. 4.1.5). The carbonatite intrusions have highly elongated shapes concordant to the nearly vertical lithological layering of the country rocks, indicative of strong stretching during Proterozoic activation of the Hirvaskoski zone. Based on aeromagnetic data and limited drill results, Nykänen et al. (1997) estimated that the Kortejärvi deposit is about 30–60 m wide and 2 km long, while Laivajoki is about 20 m wide and 4 km long.

At Kortejärvi, the main rock type is bluish to white calcite carbonatite with calcite grains averaging 1.4 mm in size, but ranging up to 7.5 mm, and accessory minerals including dolomite (up to 10%), magnetite with ilmenite exsolution lamellae, tetraferriphlogopite, olivine, serpentine, tremolite, actinolite, apatite, allanite, and monazite. Significant amounts of slightly coarser yellowish dolomite

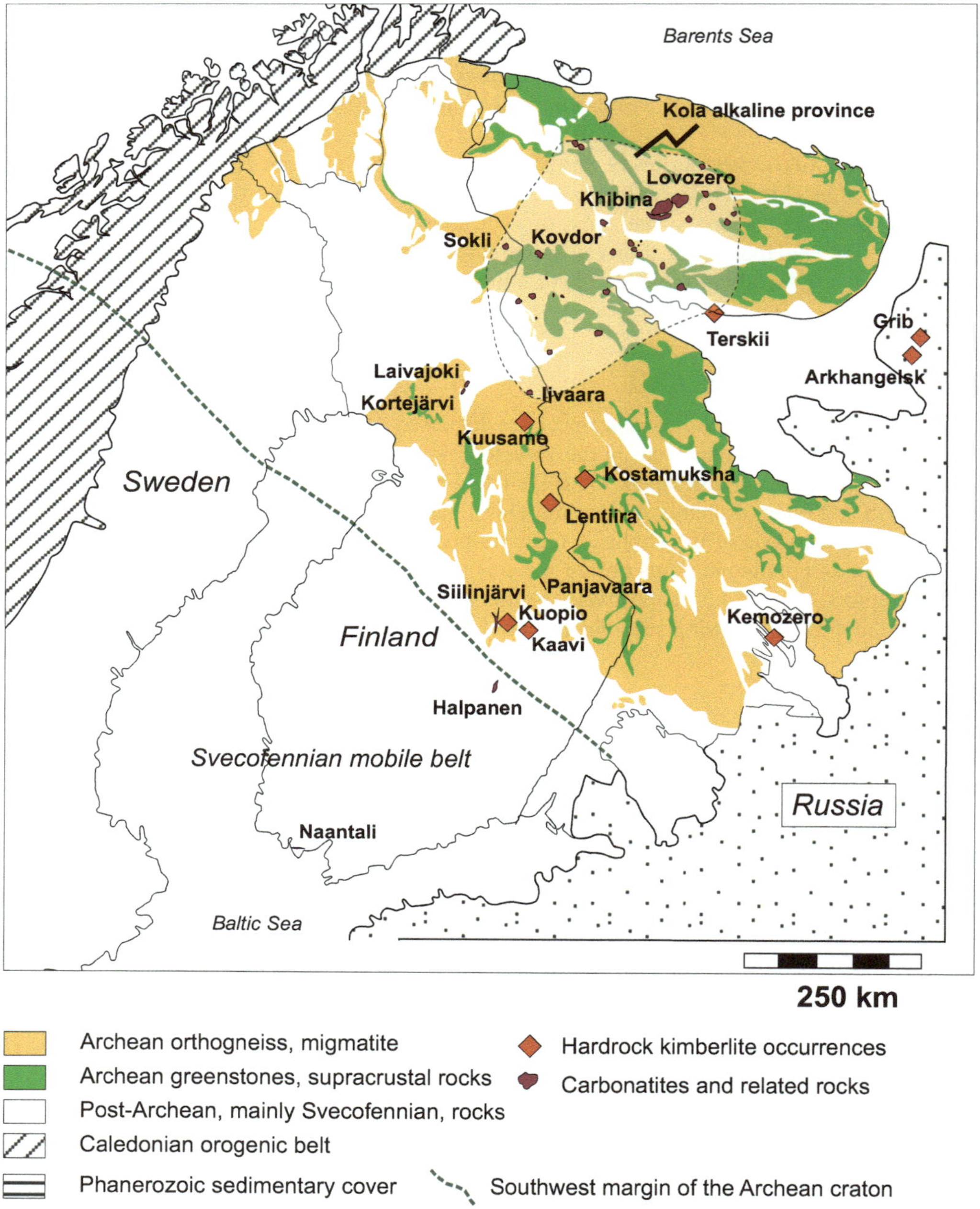

FIGURE 4.1.4 Location map of alkaline rocks in Finland and northwest Russia, with Devonian Kola alkaline province outlined.

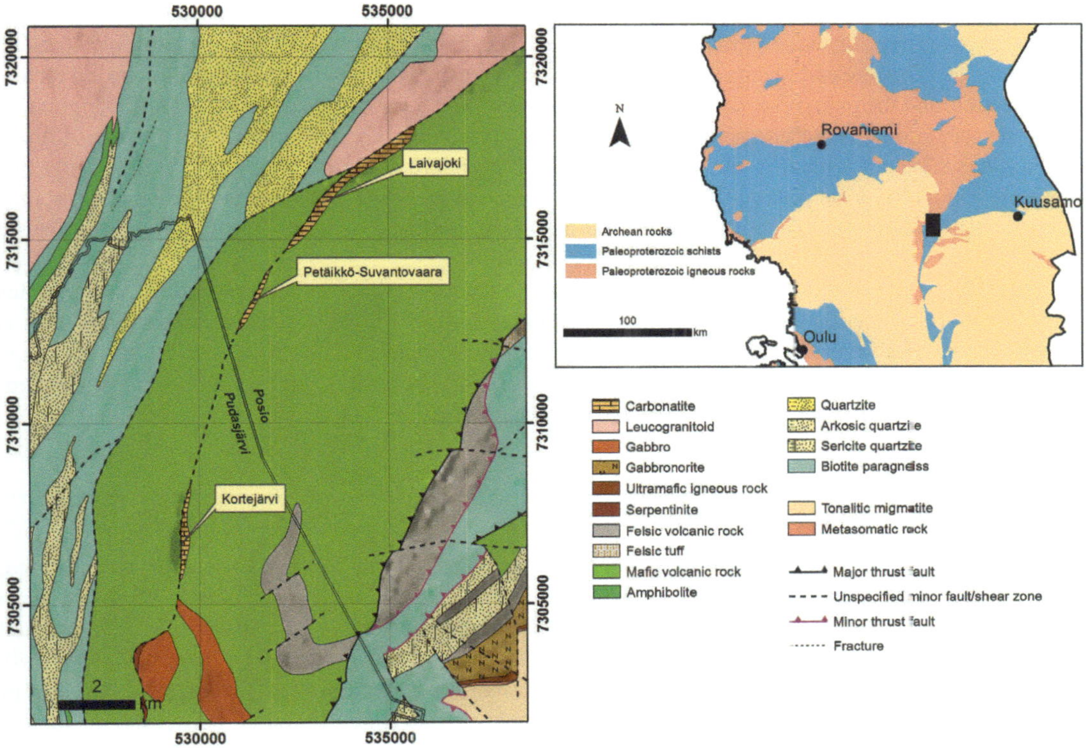

FIGURE 4.1.5 Location of the Kortejärvi, Petaikko-Suvantovaara, and Laivajoki carbonatites.

Aeromagnetic grayscale map can be seen in the background.

Source: *Geology from Bedrock of Finland–DigiKP database (accessed 1.30.2014). Contains data from the National Land Survey of Finland Topographic Database, 08/2012.*

carbonatite and dolomite-calcite carbonatite also occur, the former containing the most apatite-rich layers at Kortejärvi. Silicate-dominated rock types at Kortejärvi include:

- Dark green to brownish green glimmerite forming layers up to 4 m thick. The rocks are composed of phlogopite grains 0.4–1.8 mm in diameter with minor actinolite and edenite; accessory calcite, dolomite, and apatite; and rare diopside, zircon, allanite, sulfide, and magnetite.
- Olivine-magnetite rock layers less than 1 m thick, with unaltered or completely serpentinized olivine, iddingsite, bowlingite, magnetite-containing ilmenite exsolution lamellae, and accessory minerals including dolomite, tetraferriphlogopite, richterite, and zircon. These rocks differ from phoscorites (e.g., Phalaborwa in South Africa) in that there is no apatite.

Laivajoki is dominated by rocks called tremolite-rich carbonatite by Nykänen et al. (1997). They show considerable modal variation in carbonate and tremolite contents, but nevertheless average greater than 50% carbonates. Other minerals include magnetite (ranging from accessory to a main mineral), ilmenite, phlogopite, relatively abundant zircon, allanite, and a few sulfides. Apatite is rare, as in the Kortejärvi magnetite-rich rocks. Rarely, pyroxene remnants can be seen within the tremolite grains, perhaps suggesting this variety of carbonatite was originally pyroxene-rich. The main true carbonatite rock at Laivajoki is calcite carbonatite, but here the carbonatite is almost pure calcite with only occasional grains of dolomite, no olivine or serpentine, and the phlogopite does not show the reverse pleochroism indicative of tetraferriphlogopite.

Accessory minerals include apatite, actinolite, magnetite, ilmenite, tremolite, zircon, allanite, and a few sulfides. The other major rock type at Laivajoki is serpentine-talc-dolomite rock with the first two minerals being alteration products of olivine. In addition to dolomite, the rock contains small amounts of phlogopite, magnetite, ilmenite, sparse sulfide grains, and rare apatite. Thin layers of glimmerite similar to that at Kortejärvi also occur at this locality.

A single U-Pb zircon age of 2020 Ma is thought to represent the primary age of these carbonatites (personal communication by O. Kouvo in Vartiainen and Woolley, 1974) but the zircon population is heterogeneous and several younger zircon generations exist (Karhu et al., 2001). An earlier K-Ar age of 1875 Ma reported by Kresten et al. (1977) from Kortejärvi is probably a metamorphic age that represents the last major event along the Hirvaskoski shear zone.

1.8 Ga CARBONATITE DIKES

The Halpanen carbonatite dike occurs 12 km northeast of the city of Mikkeli in southeastern Finland, along a major, deep north–south fracture that also includes the Siilinjärvi carbonatite complex (refer to Fig. 4.1.4). It was recognized as a carbonatite by the exploration staff of Rautaruukki Oy and was first described by Puustinen (1986). According to geophysical data, the approximately 8 m wide, shallowly dipping (30–35°) dike-like body can be as much as 1.5 km long. It consists of fine-grained, massive or weakly banded calcite carbonatite, and has accessory apatite, magnetite, pyrite, barite, monazite, and fluorite, with the first two occurring in places as phenocrysts >1 cm in diameter. At the eastern contact of the dike within the main quarry there is a roughly 10-cm-wide apatite-rich zone that varies from 20–80% calcite to 20–80% apatite. Fenite alteration around the intrusion is limited to about 1 m, and is shown mostly by elevated SrO and BaO in the surrounding quartz-feldspar rocks. A precise zircon U-Pb age of 1792 ± 1 (2σ) was reported by Rukhlov and Bell (2010).

Other occurrences of thin carbonatite dikes also deserve mention, even if it is unlikely that they have potential as ore deposits. With a width of 3–60 cm, Panjavaara and Petäiskoski are two of the thicker dikes of a swarm of more than 50 dikes that occur near the town of Juuka, 180 km northeast of Halpanen (Tyni et al., 2003; Torppa and Karhu, 2007). These bastnäsite-bearing carbonate rocks contain up to several wt% of REE, Sr, and Ba, producing traceable geochemical anomalies in the overlying till (Tyni et al., 2003). An age estimate of ~1800 Ma based on conventional U-Pb on monazite was reported for the Panjavaara dikes by Torppa and Karhu (2007).

At the Naantali locality, near the city of Turku in southwest Finland, a swarm of dikes that also range in width from 2–60 cm, were intruded in an area of about 1 km × 2 km, parallel to an important crustal shear zone (Woodard, 2010). These dikes consist of calcite, fluorapatite, and accessory allanite, titanite, and minor bastnäsite, and are enriched in P, F, Sr, and REE. Rare zircon grains have provided a U-Pb age of 1796 ± 9 Ma (2σ) (Woodard and Hetherington, 2014).

In addition to all three carbonatite dike localities just mentioned being of the same or very similar age, they also have similar yet unusual carbon isotope signatures (relative to $\delta^{13}C$ PDB), at least for carbonatites: Halpanen −12.2 to −12.4‰; Panjavaara −15.5 to −16.4 ‰; Naantali −11.3 to −11.5‰ (Torppa and Karhu, 2007; Tyni et al., 2003; Woodard, 2010). All of these values are significantly lower than in mantle-derived carbonatites elsewhere on the planet, which means the magmas from which they formed were either derived from a unique crustal-derived carbon source in the mantle or they are crustal anatectic melts, as is suggested for the carbonate vein at Korsnäs (see Chapter 23, Subchapter 9.2).

SUMMARY

1. Carbonatites host important mineral deposits of phosphorus, rare earth elements (particularly light REE), and niobium. Many of these are "critical" metals, that is, those that are essential for modern high technology and whose supply is therefore vital to ensure.
2. Most carbonatites are associated with alkaline magma complexes, ranging from sodic (nephelinites) to potassic (phonolites and aillikites). The carbonatites of the Kola alkaline province (including Sokli) are related to potassic magmas.
3. The ~2.0 Ga carbonatites Laivajoki, Kortijärvi, and Petäikkö-Suvantovaara have a relatively strong metamorphic overprint, yet nevertheless some primary tetraferriphlogopite still exists. Evaluation for potential as phosphorus ore of these carbonatites is ongoing.
4. The ~1.8 Ga group of carbonatite dikes and dike swarms in Finland in some cases show extreme REE enrichment and also have unusual C and O isotope compositions that point to sources different from those of typical carbonatites.

ACKNOWLEDGMENTS

The author is grateful to Panu Lintinen of GTK for the use of the Laivajoki maps and to Harri Kutvonen for help with drafting the figures. R.E. Harmer and W.D. Maier provided helpful constructive criticism for which the author is grateful.

REFERENCES

Bell, K., Simonetti, A., 2010. Source of parental melts to carbonatites—critical isotopic constraints. Mineralogy and Petrology 98, 77–89.

Belyatsky, B.V., Savva, E.V., Nikitina, L.P., Levsky, L.K., 1995. Nd, Sr and Pb istotopes in lamproitic dykes of the eastern part of the Baltic Shield. Proceedings of the 9th Meeting of the Association of European Geology Society, Abstract, pp. 9–10.

Brassinnes, S., DeMaiffe, D., Balaganskaya, E., Downes, H., 2003. New mineralogical and geochemical data on the Vuojijarvi ultramafic alkaline and carbonatitic complex (Kola Region, NW Russia). Periodico di Mineralogia 72, 79–86.

Brøgger, W.C., 1921. Die Eruptivegesteine des Kristianiagebeites. IV. Das Fengebeit in Telemark, Norwegen. Norske Videnskaps-Selskabets Skripper I. Matchematisk-Naturviss Klasse 9, 1–408.

Cordeiro, P.F.O., Brod, J.A., Dantas, E.L., Barbosa, E.S.R., 2010. Mineral chemistry, isotope geochemistry and petrogenesis of niobium-rich rocks from the Catalão I carbonatite-phoscorite complex, Central Brazil. Lithos 118, 223–237.

Dawson, J.B., 1962. Sodium carbonate lavas from Oldoinyo Lengai, Tanganyika. Nature 195, 1075–1076.

Eriksson, S.C., 1989. Phalaborwa: a saga of magmatism, metasomatism and miscibility. In: Bell, K. (Ed.), Carbonatites: Genesis and Evolution. Unwin Hyman, London, pp. 221–254.

Gittins, J., Harmer, R.E., 2003. Myth and reality in the carbonatite-silicate rock "association.". Periodico di Mineralogia 72, 19–26.

Harmer, R.E., 1999. A common source for carbonatites, kimberlites and megacrysts? Proceedings of the 7th International Kimberlite Conference vol. 1, 332–340.

Harmer, R.E., Gittins, J., 1997. The origin of dolomitic carbonatites: field and experimental constraints. Journal of African Earth Sciences 25, 5–28.

Hornig-Kjarsgaard, I., 1998. Rare earth elements in sövitic carbonatites and their mineral phases. Journal of Petrology 39, 2105–2121.

Karhu, J.A., Mänttäri, I., Huhma, H., 2001. Radiometric ages and isotope systematics of some Finnish carbonatites (abstract). University of Oulu Research Terrae, Series A, No. 19, 8.

Kramm, U., 1993. Mantle components of carbonatites from the Kola Alkaline Province, Russian and Finland: a Nd-Sr study. European Journal of Mineral 5, 985–989.

Kramm, U., 1994. Isotope evidence for ijolite formation by fenitization: Sr-Nd data of ijolites from the type locality Iivaara. Finland Contributions Mineral Petrol. 115, 279–286.

Kresten, P., Printzlau, I., Rex, D., et al., 1977. New ages of carbonatitic and alkaline ultramafic rocks from Sweden and Finland. Geologiska Föreningens i Stockholm Förhandlingar 99s, 62–65.

Kupletsky, B.M., 1938. A pyroxenite intrusion at Afrikanda station in the Kola Peninsula. Trudy Petrographicheskogo Instituta AN SSSR 12, 71–88 (in Russian).

Lintinen, P., 2014. Preliminary results from new drillings and geochemical studies on the apatite deposits in the Kortejärvi and Petäikkö-Suvantovaara carbonatites, Pudasjärvi-Posio District, Northern Finland. Geological Survey of Finland. Report of Investigation 207, 100–103.

Lynas Corporation Ltd, 2014. Annual meeting presentation.

Mahotkin, I.L., Gibson, S.A., Thompson, R.N., et al., 2000. Late Devonian diamondiferous kimberlites and alkaline picrite (Proto-kimberlite?) magmatism in the Arkhangelsk Region, NW Russia. Journal of Petrol 41, 201–227.

Nielson, T.F.D., 1994. Alkaline dike swarms of the Gardiner complex and the origin of ultramafic alkaline complexes. Geochemistry International 31, 37–56.

Nykänen, J., Laajoki, K., Karhu, J., 1997. Geology and geochemistry of the early Proterozoic Kortejärvi and Laivajoki carbonatites, central Fennoscandian Shield, Finland. Bulletin of the Geological Society of Finland 69, 5–30.

O'Brien, H.E., Tyni, M., 1999. Mineralogy and Geochemistry of kimberlites and related rocks from Finland. In: Gurney, J.J., et al. (Ed.), Proceedings of the 7th International Kimberlite Conference, vol. 2, pp. 625–636.

Palabora Mining Company Limited, 2013. Annual Report.

Powell, J.L., Hurley, P.M., Fairbairn, H.W., 1962. Isotopic composition of strontium in carbonatites. Nature 196, 1085–1086.

Puustinen, K., 1986. Halpasen karbonatiitti Mikkelin mlk:ssa. Geologi 38, 1–5 (in Finnish with English summary).

Ruhklov, A.S., Bell, K., 2010. Geochronology of carbonatite from the Canadian and Baltic shields, and the Canadian Cordillera: clues to mantle evolution. Mineralogy and Petrology 98, 11–54.

Saether, E., 1957. The alkaline rock province of the Fen area in southern Norway. Det Konglige Norske Videnskabsselskaps Skrifter 1, 150p.

Sage, R.P., Watkinson, D.H., 1991. Alkalic rock carbonatite complexes of the Superior Structural Province, northern Ontario, Canada. Chronique de la Recherche Minière 504, 5–19.

Smith, C.B., 1983. Pb, Sr and Nd isotopic evidence for sources of southern African Cretaceous kimberlites. Nature 304, 51–54.

Torppa, O.A., Karhu, J.A., 2007. Ancient subduction recorded in the isotope characteristics of ~1.8 Ga Fennoscandian carbonatites. In: Abstracts of the 17th Annual V.M. Goldschmidt Conference, Cologne, p. A1032. August. Geochemica et Cosmochemica Acta 71.

Tyni, M., Puustinen, K., Karhu, J., Vaasjoki, M., 2003. The Petäiskoski carbonate veins at Juuka, Eastern Finland. Geological Survey of Finland, Special Paper 36, 13–16.

U.S. Geological Survey, 2015. Mineral commodity summaries 2015: U.S. Geological Survey. p. 196. http://dx.doi.org/10.3133/70140094.

Vartiainen, H., Woolley, A.R., 1974. The age of the Sokli carbonatite, Finland and some relationships of the North Atlantic alkaline igneous province. Bulletin of the Geological Society of Finland 46, 81–91.

von Eckermann, H., 1948. The alkaline district of Alnö Island. Sveriges Geologiska Undersökning Series C, No. 36, 176p.

Woodard, J., 2010. Genesis and emplacement of carbonatites and lamprophyres in the svecofennian domain. Annales Universitatis Turkuensis vol. 252, 184 A II (dissertation).

Woodard, J., Hetherington, C.J., 2014. Carbonatite in a post-collisional tectonic setting: Geochronology and emplacement conditions at Naantali, SW Finland. Precambrian Research 240, 94–107.

Woolley, A.R., Church, A.A., 2005. Extrusive carbonatites: A brief review. Lithos 85, 1–14.

Woolley, A.R., Kjarsgaard, B.A., 2008a. Carbonatite occurrences of the world: Map and Database. Geological Survey of Canada Open File 5796 (CD-ROM and 1 map).

Woolley, A.R., Kjarsgaard, B.A., 2008b. Paragenetic types of carbonatite as indicated by the diversity and relative abundances of associated silicate rocks: Evidence from a global database. The Canadian Mineralogist 46, 741–752.

Wu, C., 2008. Bayan Obo Controversy: Carbonatites versus Iron Oxide-Cu-Au-(REE-U). Resource Geology 58, 348–354.

Wyllie, P.J., 1965. Melting relationships in the system CaO-MgO-CO_2-H_2O, with petrological applications. Journal of Petrology 6, 101–123.

Wyllie, P.J., 1978. Silicate-carbonate systems with bearing on the origin and crystallization of carbonatites. Proc. 1st Int. Symp. on Carbonatites, Ministerio das Mina e Energia, Departamento Nacional da Produçao Mineral, Poços de Caldas. Minas Gerais, Brazil. pp. 61–78.

Yang, K.-F., Fan, H.-R., Santosh, M., et al., 2011. Mesoproterozoic carbonatitic magmatism in the Bayan Obo deposit, Inner Mongolia, north China—Constraints for the mechanism of super accumulation of rare earth elements. Ore Geology Reviews 40, 122–131.

Zaitsev, A.N., Sitnikova, M.A., Subbotin, V.V., et al., 2004. Sallanlatvi complex—rare example of magnesite and siderite carbonatites. In: Wall, F., Zaitsev, A.N. (Eds.), Phoscorites and Carbonatites from Mantle to Mine: The Key Example of the Kola Alkaline Province. The Mineralogical Society of Britain and Ireland, London, pp. 201–245.

CHAPTER

THE SOKLI CARBONATITE COMPLEX

4.2

H. O'Brien, E. Hyvönen

ABSTRACT

The Sokli carbonatite complex is one of 22 alkaline complexes comprising the Devonian Kola alkaline province (KAP). It represents the largest carbonatite intrusion of the province, and possibly the largest carbonatite on Earth. The Sokli complex shows many important similarities to other alkaline complexes of the KAP, especially Kovdor and Vuorijärvi, such as associated early ultramafic cumulate rocks, a well-developed phoscorite-carbonatite association and late carbonatites that represent carbonate liquid evolution to dolomite carbonatites to a final light rare earth element (LREE)-Sr-Ba dolomite carbonatite pulse. However, a critically important difference with Sokli is the vastly greater abundance of carbonatite relative to the associated cumulate ultramafic rocks.

At Sokli, a core zone of magmatic carbonatite is surrounded by a zone of precursor ultramafic rocks that have been almost completely transformed into carbonatites by outward migration of magmas and fluids from the core zone; the resulting rocks are termed *metasomatic carbonates*. Surrounding this is a highly heterogeneous outer transition zone of mixed rocks, composed of phlogopite-rich, carbonate-bearing varieties that have a very complex history and are derived variably from country rocks, fenites, ultramafic cumulates, and metasomatic carbonatites.

To push forward research on the Sokli complex, in 2008, Yara International ASA commissioned the Geological Survey of Finland to fly a high-density aerogeophysical program over the Sokli area, acquiring magnetic, electromagnetic, and radiometric data. Geophysical images derived from this program are used here to complement this review of the petrology and mineralogy of the Sokli complex. The aerogeophysical data allow further definition of the geology of the carbonate complex, as well as permit differentiation of the ore types of the weathered regolith covering the carbonatite.

Keywords: Sokli; carbonatite; phoscorite; aerogeophysical mapping; regolith; phosphate ore.

INTRODUCTION

The Sokli carbonatite complex is one of the 22 alkaline complexes that constitute the Kola alkaline province (KAP; Fig. 4.1.4) intruded during the Devonian, from 410–362 Ma, with most bodies being intruded in the relatively short period of 382–362 Ma (Kramm and Sindern, 2004). Rb/Sr dating of 5 Sokli samples, using biotite, carbonate, apatite, and whole rocks, give ages for Sokli ranging from 365 ± 3 Ma (Kramm et al., 1993), putting it at the younger end of the KAP.

The KAP comprises a range of alkaline complexes from the giant Khibiny and Lovozero nepheline syenite-dominated plutons, with Khibiny representing the largest alkaline complex in the world, covering 1327 km^2 (Ilyin, 1989), to the relatively small Sallanlatvi complex at 4.2 km^2 (Zaitsev et al., 2004). Fourteen of the KAP alkaline intrusions contain carbonatites, and Sokli, at 18 km^2 is the largest carbonatite among them, and possibly the largest in the world.

Mineral Deposits of Finland. http://dx.doi.org/10.1016/B978-0-12-410438-9.00012-1

The enormous volume of alkaline magmas of the KAP event produced a wide variety of rock types. Some are close to magma compositions, but most represent predominantly cumulates. The reader is referred to the volume edited by Wall and Zaitsev (2004) for a full review of mineral chemistry, rock types, magma fractionation paths, and relative ages.

Exposure at Sokli is relatively poor and the top of the body is heavily weathered, restricting all work on paragenetic relationships to drill-core observation and sampling. Moreover, there has been extensive overprinting of all older silicate cumulate rocks at Sokli by late voluminous carbonatite magmatism, as explained later. Based on these criteria, the best analogs for Sokli appear to be its two nearest KAP neighbors, the Kovdor and Vuorijärvi complexes, and reference to these intrusions is made throughout this chapter.

AEROGEOPHYSICAL SURVEY

As part of a research and ore evaluation program aimed at developing a better geological model for the Sokli complex, Yara International ASA commissioned the Geological Survey of Finland (GTK) to carry out an airborne survey on the Sokli area in 2008. Specifications of the survey are given in Table 4.2.1. Geophysical parameters measured were the Earth's magnetic field, the electromagnetic field (four-frequency system with frequencies of 0.9, 3, 12, and 25 kHz), and natural gamma radiation. The measurement data were interpolated into grids with a pixel size of 18.75 m × 18.75 m. The measurement systems and data processing are described in Hautaniemi et al. (2005). In 2012 the data were released and GTK received full rights to them.

The measured total magnetic induction (TMI) was reduced to the northern pole (RTP) to minimize remanence caused by the magnetite-bearing intrusion. The aeroelectromagnetic (AEM) data of higher frequencies (12 and 25 kHz) were used to delineate shallow subsurface conductors and lower frequencies to reveal deeper structures and magnetite-bearing conductors. The aeroradiometric data reflect surface mineralogy, and the approximate depth penetration is 0.5–1 m.

Changes in uranium (U), thorium (Th), and potassium (K) concentrations and concentration ratios reflect both lithological variations and degree of weathering. The concentration values of thorium, uranium, and potassium were converted to units of radiation (UR) and scaled similarly to keep them comparable. Water attenuates gamma-rays significantly and therefore wet areas appear as black on the radiometric images.

Table 4.2.1 Aerogeophysical survey specifications

Flight direction	East–West
Line spacing	75 m
Nominal flight altitude	30 m
Total line length	1842.4 km
Magnetometers	Two cesium magnetometers
Electromagnetic system	Frequencies: 912, 3005, 11,962, and 24,510 Hz
Radiometric	256-channel spectrometer; NaI (Tl) crystal volume 41 l

GENERAL DESCRIPTION OF THE SOKLI COMPLEX

The Sokli carbonatite complex is located in Finnish Lapland, with a center point near 67° 48′ 12.2″ north latitude, and 29° 19′ 2.3″ east longitude. It was discovered in 1967 during an iron ore prospecting campaign by Rautaruukki Oy in the area of Finland nearest to the Kovdor alkaline complex. Sokli is ellipsoidal in plan, with a major axis oriented northwest–southeast of about 6.2 km and a minor axis of 4.8 km. The total area of the complex is about 18 km^2. A comprehensive study of the complex was undertaken by Vartiainen and Paarma (1979) and Vartiainen (1980) to assess its ore potential.

The three-dimensional shape of Sokli is of a concentrically zoned, funnel-like body (Fig. 4.2.1) with a younger, plug-like carbonatite magmatic core measuring 2.5 km in width at the surface and 1 km in width at 5 km depth, as indicated by deep seismic soundings (Paarma et al., 1981). As described in Subchapter 4.1,

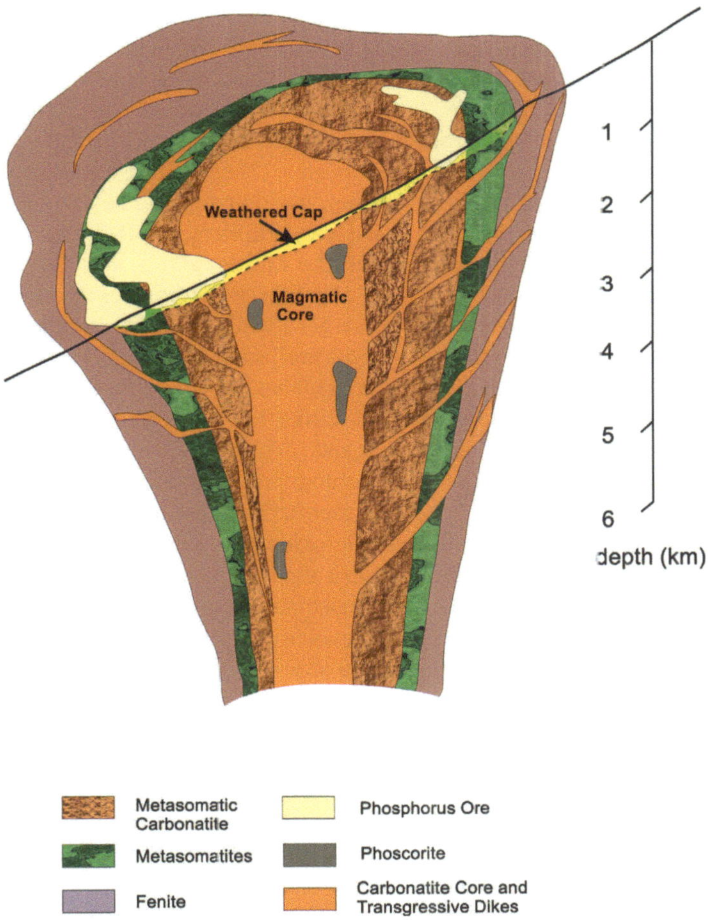

FIGURE 4.2.1 Schematic cross section through the Sokli carbonatite complex.

Source: *After O'Brien et al. (2005).*

the general model for carbonatite complexes around the world has carbonatite forming a distinctly younger intrusive body, typically as a central core, or at least the latest intrusive phase, of a zoned alkaline igneous complex (see Fig. 4.1.1). However, at Sokli, the younger magmatic core is not surrounded by neatly concentric layers of older, related cumulate rocks as in other KAP intrusions, but rather by a semicontinuous ring of metasomatically produced carbonated rocks. This in turn is followed outwards by an even more heterogeneous zone of carbonate and phlogopite-rich metasomatized ultramafic silicate rocks, amphibole rocks, pyroxene rocks, and blocks of fenite, all cemented together by metasomatic carbonate (Vartiainen, 1980).

ROCK TYPES

MAGMATIC CORE

Carbonatites with fully magmatic textures form an elliptical vertical pipe that is located relatively close to the center of the Sokli complex, although not exactly in the center. Within this inner pipe, carbonatite dominates, but there are many inclusions of metasomatic carbonatite, amphibole rocks, and mica-amphibole rocks ranging in size from centimeters to hundreds of meters. Crosscutting relationships led Vartiainen (1980) to suggest the following sequence of events for the Sokli central pipe: (1) emplacement of phoscorite following segregation from carbonatite magma, (2) main pulse of calcite carbonatite emplacement, (3) pneumatolitic-hydrothermal activity, (4) intrusion of dolomite and Fe-bearing carbonatites, and (5) intrusion of late dolomite veins containing also Sr and rare earth elements (REE) carbonates and barite.

Phoscorite is a forsterite-magnetite-apatite rock (Le Maitre, 2002), and has been described from only 21 localities globally (Krasnova et al., 2004). All of these occurrences are closely associated with carbonatites, providing strong evidence that they are genetically related to carbonatite. The textures of some of the phoscorites, described from Sokli and Vuorijärvi (Lapin and Vartiainen, 1983), provide strong evidence that liquid immiscibility has occurred in these systems. These observations include: (1) phoscorites with spherulitic structures, spherical to ellipsoidal aggregates of magnetite and forsterite with radial crystal habit, in a carbonate-rich matrix; and (2) phoscorites with orbicular textures, wherein ellipsoidal forsterite-magnetite, forsterite-tetraferriphlogopite-magnetite-calcite, or phlogopite-calcite-magnetite orbicules exist in a calcite ± magnetite matrix. The best interpretation of these phoscorite textures is that they represent conjugate immiscible oxide-phosphate-silicate magmas to carbonatite magmas, hence the similarity in the mineralogy and their major element compositions, with only a considerable variation in modal proportions of phases distinguishing the rocks.

Lee (2002) and Lee et al. (2004) conducted a detailed study of phoscorites and magmatic carbonatites at Sokli. Based on mineral chemistry data, he was able to show that the rocks formed through several immiscibility events, named P1C1 (phoscorite 1-carbonatite 1), P2C2, and P3C3. A thin section of a typical Sokli P1 is shown in Fig. 4.2.2. Lee was also able to show that phlogopite (Lee et al., 2003), oxide (Lee, 2005), and pyrochlore (Lee et al., 2006) compositions are all distinctive for each PC pair (see "Mineralogy" section, later). Moreover, each succeeding PC pair has minerals with increasingly evolved compositions. For example, phlogopite shows a systematic change from Ba-containing, slightly eastonitic phlogopite in P1C1 to steadily decreasing Al_2O_3 contents, leading to nearly pure Al-depleted tetraferriphlogopite in P3C3, indicative of crystallization from severely Al-depleted melts.

This evolutionary trend from normal phlogopite to tetraferriphlogopite is also seen at other phoscorite-bearing carbonatites of the Kola alkaline province, in particular at Kovdor (Kukarenko et al., 1965) and Vuorijärvi (Karchevsky and Moutte, 2004). In addition to overall magma evolution

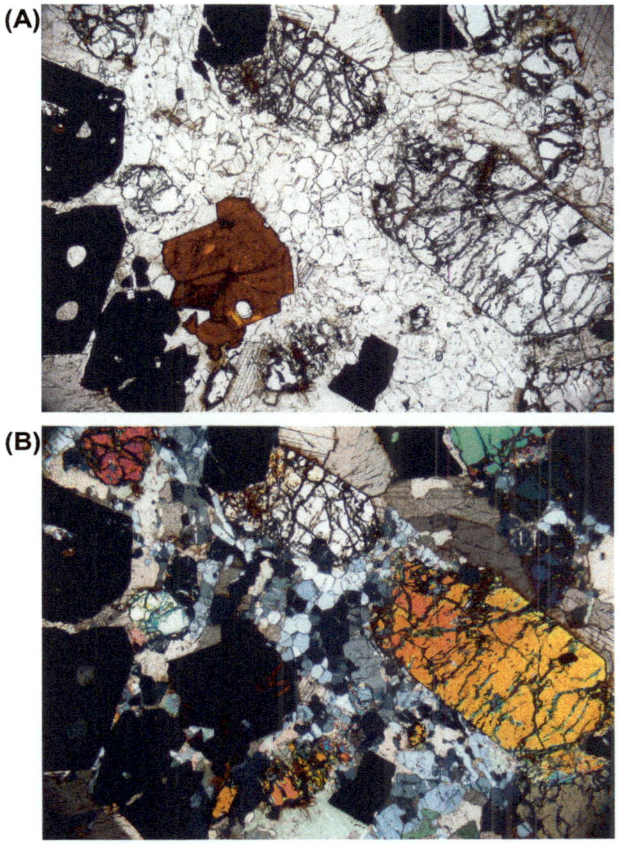

FIGURE 4.2.2 Photomicrographs of Sokli phoscorite (P1 phase) in thin section.

(A). Phoscorite composed of calcite, apatite, olivine, magnetite, and pyrochlore from magmatic carbonatite. Plane-polarized light. Width of field is 6.85 mm. (B). Same as A., polars crossed.

from P1C1 to P3C3, within each pair of immiscible magmas (crystals + liquid), the daughter liquids continued to evolve by fractionation, as evidenced by mineral core to rim zoning profiles (Lee, 2002). It appears that this immiscible process occurred three times on a large scale at Sokli, and, interestingly, despite the large differences in the size of the carbonatite systems, this is also the number of PC pairs reported at Kovdor and Vuorijärvi (Karchevsky and Moutte, 2004). The final product of this stepped process of magma immiscibility and fractionation at all three of these carbonatites led to a similar, highly evolved dolomite carbonatite (stage 5) containing a range of trace element rich minerals (Vartiainen and Vitikka, 1993).

In terms of its aeromagnetic response, the Sokli magmatic core is highly magnetic, heterogenous and fractured as seen in Fig. 4.2.3. The alteration zones surrounding the core are recognized by a decrease in magnetic intensity, forming a circular break that defines the edge of the magmatic core (Fig. 4.2.4).

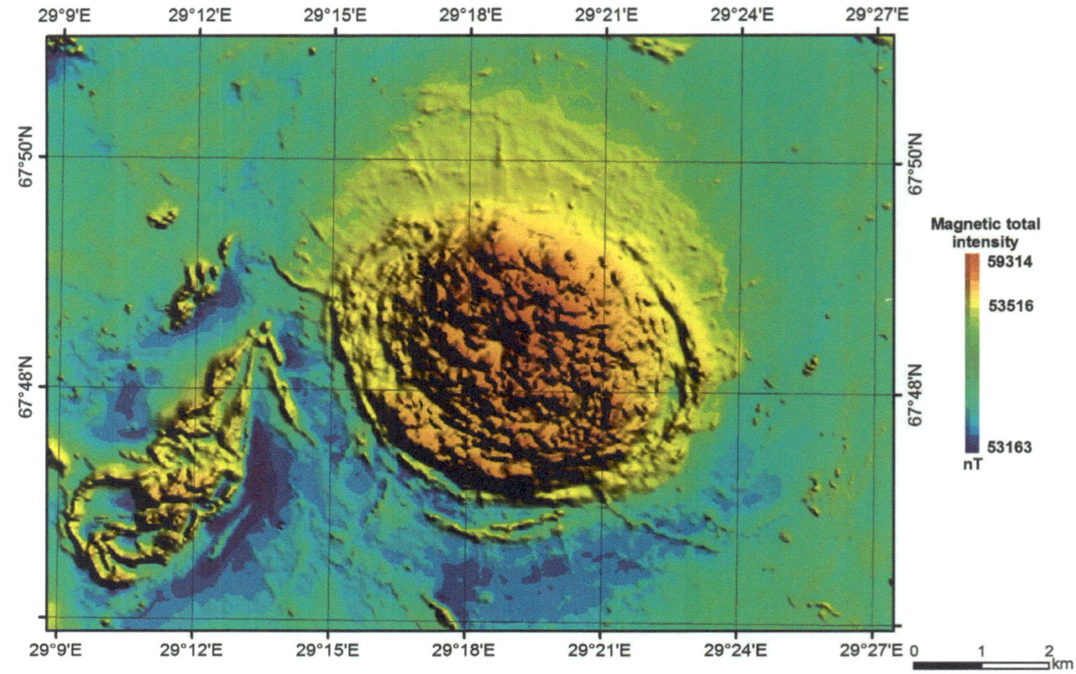

FIGURE 4.2.3 Shaded aeromagnetic TMI image of Sokli carbonatite complex and associated Tulppio olivinite.

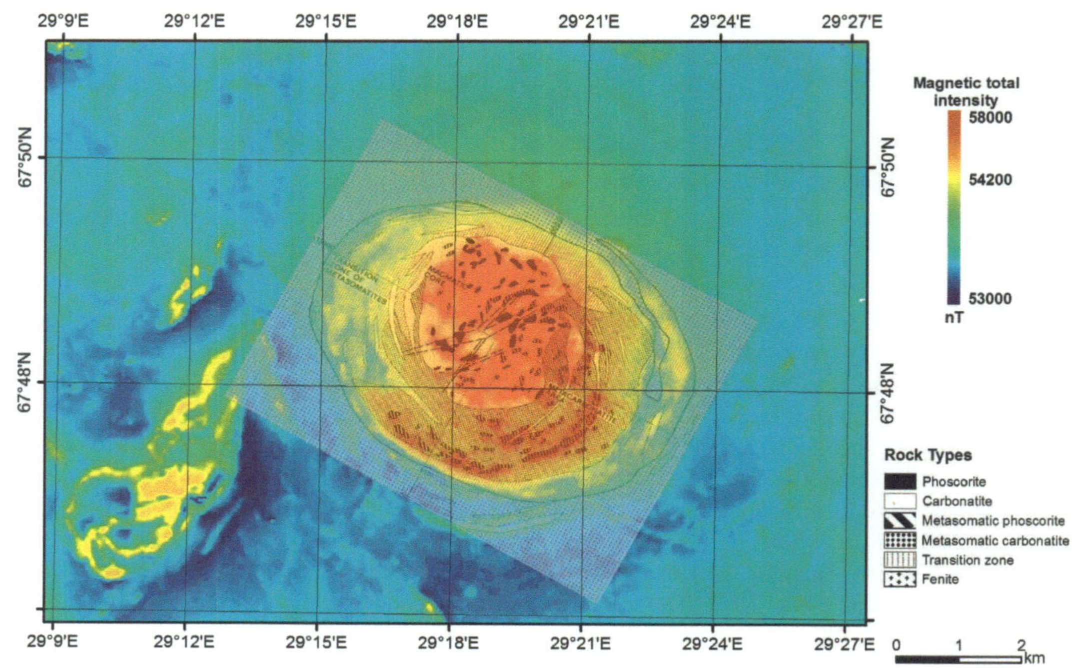

FIGURE 4.2.4 Geological map of the Sokli complex overlain on aerogeophysical TMI image.

Source: *Rotated and rectified from Figure 3 of Vartiainen (1980).*

METASOMATIC ROCKS

Based on petrographic evidence and gradational mineralogical contacts, Vartiainen (1980) was able to show that the metasomatic carbonatites (his "metacarbonatites") formed as a result of carbonitization of pyroxenites and coarse-grained olivinites (dunites with magnetite rather than chromite), the oldest rocks of the complex. The slightly carbonitized variants are massive, while the more strongly replaced ones are generally banded. Metasomatic silicate rocks formed from the ultramafic rocks of the original intrusion through complex replacement and substitution processes in a manner that, at its extreme, resulted in almost pure phlogopite rock. As intermediate products, there occurs a host of rock types containing variable amounts of amphibole, aegirine, and phlogopite as major minerals.

Although being just remnants, the few blocks of ultramafic rock within this zone allow analogies to be drawn between the precarbonatite rocks at Sokli and the more typical dominant rock types of the other Kola alkaline complexes. Especially relevant are the large volumes of ultramafic cumulates composed of olivine and clinopyroxene at Kovdor and Vuorijärvi. The geology of the Kovdor carbonatite complex is relatively well known, due to active mining of the carbonatite-phoscorite that has been ongoing for more than 50 years. Even though the rocks of the complex represent cumulates, they nevertheless record the sequence of magmas within the subvolcanic system now exposed.

Crosscutting relationships at Kovdor suggest the following sequence of emplacement: olivinite, phlogopite-diopside-olivine rock, clinopyroxenites, melilitolites, ijolite, and, finally, phoscorites, followed closely by carbonatite magmas (Krasnova and Kopylova, 1988; Verhulst et al., 2000; Krasnova et al., 2004). The cumulate sequence at Vuorijärvi is similar to Kovdor, but the volumes of each cumulate type are considerably different, with various types of clinopyroxenites constituting the vast majority of the Vuorijärvi complex, and olivinites occurring only as isolated blocks (xenoliths) within the pyroxenites.

These analogs suggest that there likely were large volumes of mafic mineral accumulations formed in subvolcanic magma chambers also at Sokli. Moreover, the olivinite mass at Tulppio, showing up as a distinctive 4 km × 2 km magnetic anomaly southwest of the Sokli complex proper (see Fig. 4.2.3), is likely a good example of what these cumulates were like, and clearly more studies of the Tulppio rocks are warranted. Summarizing these circumstantial pieces of evidence, it appears possible that prior to the emplacement of the main carbonatite, Sokli may have looked similar to the Vuorijärvi and/or Kovdor complexes.

Thus, what separates Sokli from the rest of the KAP intrusions is the amount of carbonatite magmatism that occurred. In the previously discussed Kovdor and Vuorijärvi cases, the volume of carbonatite is less than 10%, whereas at Sokli, magmatic carbonatite, or rocks that reacted with carbonatite magma or carbonate-rich fluid, represent nearly the entire intrusion. So thorough was this overprint in many cases that the precursor magmatic rocks are difficult, if not impossible, to identify, as these rocks are now composed of greater than 50% by volume of carbonate minerals.

The aeromagnetic response of the metasomatic zone is midway between that of the magmatic rocks and the transition zone in terms of intensity (see Fig. 4.2.3). Highly magnetic pods are more discrete, and far less abundant than in the magmatic core (see Fig. 4.2.4), but are still interpreted to represent magnetite-rich phoscorites (so-called metaphoscorites), as mapped by Vartiainen (1980).

TRANSITION ZONE

Composed of fragments of fenites and rocks formed of variable amounts of pyroxene, amphibole, and mica all enclosed in a matrix of metasomatic carbonatite, the *transition zone of metasomatites*, as termed by Vartiainen (1980), is clearly a very heterogeneous mixture of rock types with mineralogies and textures distinct from the remainder of the complex. Sequential potassic metasomatism turned initial ultramafic cumulates dominated by clinopyroxenites first into amphibole-rich, and finally phlogopite-rich, rock types (Vartiainen, 1980). Subsequent overprinting by later carbonate-dominated fluids or melts further transformed these original ultramafic rocks into mixed zones of silicate-dominated and carbonate-dominated rock types, now forming the "transition zone."

The fact that so much transformation of silicate to carbonate rock occurred distally to the magmatic core provides strong evidence that a considerable amount of carbonatite melt was, in fact, involved in producing the rocks, even of the transition zone. As we will show later (see "Ore deposits" section), this extensive overprinting was a very positive feature from an economic point of view, in that regolith ore deposits occur across the entire intrusion, derived from all three rock type zones, not just above the magmatic core.

The aeromagnetic response of the transition zone is markedly different from that of the two inner zones (see Fig. 4.2.3). Most importantly, there are no obvious highly magnetic pods as in the inner zones, suggesting no magnetite-rich phoscorites (see Fig. 4.2.4). In a number of places, sublinear anomalies, most likely representing dikes, appear to emanate from the two inner rock zones, and cut across the transitional zone while other anomalies seem to emanate from the edge of the outer transition zone. A circular zone of higher conductivity at AEM frequencies of 12 kHz and 25 kHz characterizes the rocks of the transition zone (Fig. 4.2.5). Because these conductors are recognized only from the high-frequency AEM data, they are shallow and mainly the result of weathering.

FENITES

Fenites surround the carbonatite-biotitite central core developed by sodium and potassium metasomatism of the surrounding granite gneiss, amphibolite, and hornblende schist. The Sokli fenite halo is developed up to 3 km from the carbonatite core and is manifested by the development of alkali feldspar, pyroxene (aegirine and aegirine-augite), alkali amphibole (arfvedsonite and eckermannite), and phlogopite, giving the rock a greenish cast (Vartiainen and Woolley, 1976). Levels of fenitization increase toward the carbonatite intrusion and in the zone of most intense fenitization proximal to the intrusion, potassic metasomatism is dominant, forming extensive phlogopitite and phlogopite-alkali amphibole rocks.

In terms of aeromagnetics, the fenites do not appear to have any noticeable effect on the country rock magnetic signature (see Fig. 4.2.3). Note that in this figure, the yellow region to the north of the Sokli complex is due to the dipole effect, and is unrelated to the fenite zone. Carbonatite ring dikes intruded into the fenite zone have highly magnetic signatures, similar to the rocks from the magmatic core.

AILLIKITE DIKES

Associated with many ultramafic alkaline complexes are ultramafic, phlogopite-rich, olivine- and carbonate-bearing dikes that are termed aillikite, a type of ultramafic lamprophyre (Rock et al., 1991; Tappe et al., 2005). They typically have low silica contents, and experimental evidence suggests that they represent very small partial melts of carbonated peridotite source rocks that compositionally can be gradational to Group I kimberlite (Dalton and Presnall, 1998).

Everywhere within the Sokli complex, and also up to several kilometers away from it, ultramafic lamprophyre dikes occur. They are generally dark rocks varying from reddish- to greenish-gray, occur

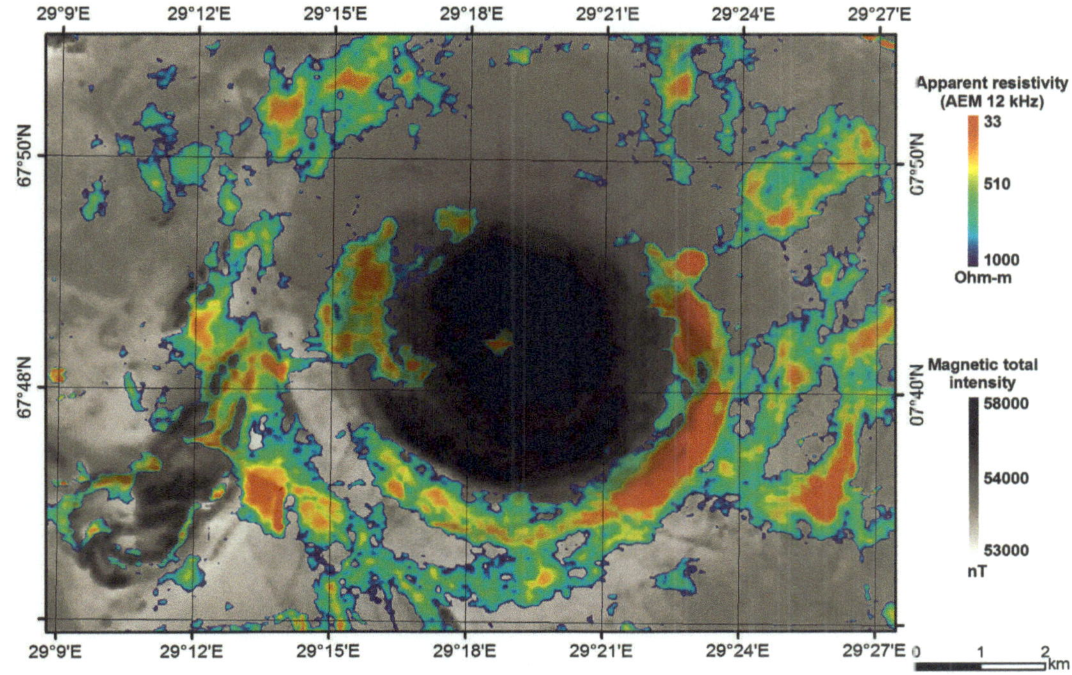

FIGURE 4.2.5 Aeroelectromagnetic map of the Sokli complex showing frequencies 12 kHz and 25 kHz overlain with the geological map of Vartiainen (1980).

These frequencies reveal shallow conductors, mainly as the result of weathering, and demarcate the transition-zone rocks rather well.

as swarms with different orientations, and, according to drilling results, average about half a meter in thickness. Vartiainen et al. (1978) divided the dikes into four subgroups based on texture: porphyritic, xenolithic (= autolithic), massive, and mica-rich. The last two groups dominate, while the first two groups occur mainly within the fenite zone and the bedrock area outside of the carbonatite proper. A typical example of a Sokli aillikite is shown in Fig. 4.2.6.

The mineralogy of the Sokli lamprophyres varies considerably. In the porphyritic group, the proportion of phenocrysts varies between 27 and 37%. Typically, olivine is strongly altered and is absent from the groundmass. The groundmass is very fine-grained (~0.1 mm) and crystallized in the following order: opaques, phlogopite-calcite, and richterite. Xenoliths in the xenolithic variety within the carbonatite complex include rock types of the complex itself and include phoscorite, carbonatite, and fenite.

Weak magnetic, dike-like anomalies radially surround the complex, and, as mentioned, seem to be both older and younger than the transition-zone rocks, based on crosscutting relationships. Lee (2002) also documented evidence that some lamprophyre dikes are affected by, and even cut by, carbonatites from the magmatic core. Hence, combining all observations described points to the existence of multiple generations of lamprophyre dikes, intruded at all stages of the development of the Sokli complex, from the very earliest stages to the very latest.

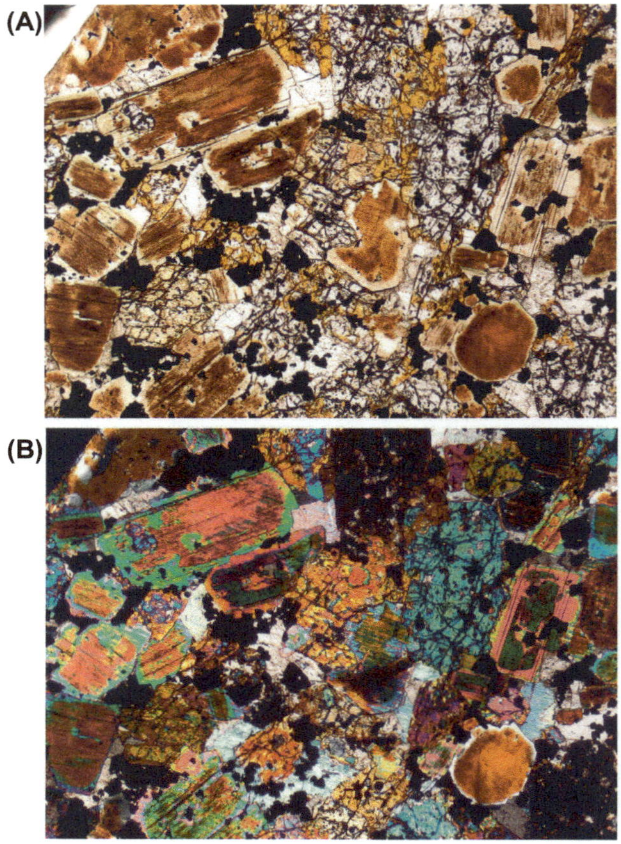

FIGURE 4.2.6 Photomicrographs of Sokli aillikite in thin section.

(A). Olivine is mostly fresh, but approximately 20% has been converted to alteration minerals, in this case predominantly iddingsite. The phlogopite shows systematic zoning from a cloudy core to clear subhedral mantles, to very late discontinuous tetraferriphlogopite rims (see large grain, middle-right). Calcite was late to crystallize, but is abundant. The remaining minerals include magnetite, apatite, and richterite. Plane-polarized light, width of field is 5.0 mm. (B). Same as A, polars crossed.

MINERALOGY

The following overview section on the mineralogy of the Sokli complex relies heavily on the comprehensive work undertaken on Sokli by H. Vartiainen (1980) and by further detailed studies made by Lee (2002, 2005) and Lee et al. (2003, 2006).

OLIVINE

Olivine is dominant in the Tulppio olivinite where it forms nearly monomineralic bodies up to several hundreds of meters in size, accompanied essentially only by magnetite. Within the Sokli complex, forsterite is also relatively important and occurs in blocks of metasomatic phoscorites, within the phoscorites proper of the magmatic core, and as phenocrysts in the aillikite dikes of the complex. In many cases, the existence of olivine in Sokli rocks can only be inferred from alteration minerals, clinohumite being the most typical alteration product.

Microprobe analyses (Vartiainen 1980; Lee, 2002) show that the olivine in the Tulppio olivinite and phenocrysts in the aillikites have similar compositions, with high Ni (0.2–0.45 wt%) and low Ca and Mn (<0.2 wt% both elements), whereas the olivines in the carbonatites and phoscorites are unsurprisingly just the opposite, with low Ni (<0.2 wt%) and high Ca and Mn (up to 0.46 and 1.26 wt%, respectively). This relationship holds, despite the fact that the carbonatite olivines have higher Fo content (Fo88–95) than the aillikite/Tulppio olivines (Fo83–92), presumably due to the relatively high mg# of the Sokli carbonatite magmas. Importantly, some of the aillikites also contain small olivine microphenocrysts that have Ni-poor, Mn-rich compositions (Lee, 2002) indicative of the fractionated and carbonate-rich nature of the matrix liquid composition, an environment much more similar to that in the carbonatite magmas.

CLINOPYROXENE

As we noted earlier, clinopyroxene was probably quite abundant during the early stages of development of the Sokli magmatic system, perhaps forming large cumulate bodies in a similar way to the KAP complexes of Kovdor and Vuorijärvi. However, with the much larger volume of carbonatite influx at Sokli, pyroxene was mostly transformed into other minerals, especially alkali amphibole and mica. Only small remnants of diopsidic pyroxene exist within blocks in the metasomatic carbonatites, or more rarely with the magmatic core, and these remain the only record of the original pyroxenes.

MICA

Compositions of mica range from slightly eastonite-bearing phlogopite (i.e., Al-Tschermak coupled substitution along the phlogopite-eastonite join) to nearly stoichiometric phlogopite to pure tetraferriphlogopite (Lee et al., 2003). Moreover, this variation is very systematic, with each successive PC stage (from P1C1 to P3C3) showing increasingly Al-depleted phlogopite compositions. Fluorine contents mimic this trend, changing from nearly pure OH-bearing phlogopite to a variety of fluorphlogopite in the D5 dolomite carbonatite.

Despite these overall trends, micas from each PC stage show significant variation in composition due to fractionation effects, and from stage to stage show a partial overlap in composition. This strongly implies a continuous progression of increasingly evolved liquids that formed the magmatic core, possibly all derived from a single parental liquid. Similar trends are seen in phlogopites at Kovdor and Vuorijärvi (Kukanrenko et al., 1965), again indicating similarities in magma compositions and crystallization processes among these KAP carbonatite complexes.

CARBONATES

The dominant carbonate minerals at Sokli are calcite and dolomite, with the ratio of the two being quite variable depending on the rock type considered (Vartiainen, 1980). For example, in the magmatic core rocks, the phoscorites contain a significantly higher ratio of dolomite to calcite (at a total CO_2 content mostly below 5 wt%), and vice versa for the conjugate carbonatites (with total CO_2 contents >30 wt%).

The metasomatically derived phoscorites and carbonatites follow a similar rule, with a few exceptions. The other important consistent carbonate mineral relationship is that the late stages 4 and 5 magmatic carbonatites shift almost completely to dolomite; these rocks crystallized from the last and most evolved liquids that formed Sokli. It is also within these late-stage evolved dolomite carbonatites that less common carbonate minerals become more abundant, for example, strontianite, ankerite, siderite, and several REE carbonate minerals are only reported from the late carbonatites.

FE-TI OXIDES

Oxides at Sokli include magnetite and ilmenite, with magnetite nearly ubiquitous throughout the complex. Magnetite is the main mineral in the phoscorites, ranging from 15–70 modal % and having grain sizes from 1–20 mm, with typically larger amounts in the younger phoscorites. Crystals also tend to be more euhedral in the younger phoscorites. Very similar to the progressive evolution evident in phlogopite compositions, the magnetite compositions also vary systematically from stage 1 to stage 5, with constantly decreasing mg# and decreasing concentrations of Mg, Al, Mn, and Ti due to depletion of these elements in the evolving melt. The end result of this process is the crystallization of nearly pure magnetite in the stage 5 carbonatites.

Ilmenite occurs mostly as exsolution features within the magnetite grains or as grains attached to magnetite grain boundaries. The composition of the ilmenite is very much in line with ilmenite from other carbonatites, with high MgO (up to 11.4 wt%) and diagnostically high MnO (up to 14.2%), the combination being the hallmark of carbonatite-related ilmenite (e.g., Mitchell, 1978). However, similar to magnetite, ilmenite compositions show a complete and consistent path toward Fe- enrichment, crystallizing as nearly pure $FeTiO_3$ in the latest magmatic stages (Lee, 2005). This decrease in MgO in the oxides occurs as the equilibrium carbonate phase shifts toward Mg-rich compositions, potentially explaining the MgO depletion in other minerals due to stronger partitioning of Mg into dolomite.

PYROCHLORE

Pyrochlore is the third mineral at Sokli providing evidence of a progressively fractionating system (Lee, 2006). Pyrochlore did not crystallize at Sokli until the magmas evolved to P2C2 compositions, but then it occurs in all later carbonatite magma types. Compositionally, the P2C2 pyrochlores are relatively U- and Ta-rich, but later P3C3 pyrochlore instead contains higher Ce and Th, while the pyrochlores from the latest carbonates are nearly end-member pyrochlore $(Ca, Na)_2Nb_2O_6F$. Further detail provided by Lee et al. (2006) shows that the pyrochlores within a given phoscorite have the same starting composition but a distinctly longer crystallization history than those from the conjugate carbonatite. This strongly suggests that the conjugate pairs formed by liquid immiscibility, with initially identical mineral compositions that then followed distinct fractionation trends after separation.

APATITE

Apatite occurs in all rock types of the Sokli carbonatite complex. Modal values of apatite range from up to 50% in some of the phoscorites, down to tens of percent in the late dolomite carbonatites, with respective grain sizes of millimeter to micrometers. Compositionally, the apatites are all fluorapatites (i.e., Cl is nearly nonexistent). As in many of the other minerals studied, the apatites show a progressive increase in elements such as F, Sr, Na, and light rare earth elements (LREE) from early P1C1 to late C5 carbonatite host rocks. However, such enrichments are even evident on the rims of some P2 phoscorite apatite grains, likely indicative of late-stage magmatic fluids migrating through the already crystallized carbonatite complex rocks.

OTHER MINERALS

Although baddeleyite was originally thought to be absent from the magmatic inner core carbonatites, as opposed to the metasomatized carbonatites where it occurs in trace amounts (Vartiainen, 1980), Lee (2002) reported small amounts of baddeleyite from magmatic core rocks. Rare zirconolite occurs in C2 carbonatite, P2, and P3 phoscorite and in the ultramafic lamprophyres, typically in contact with pyrochlore or baddeleyite, but also as tiny, late-crystallized grains.

GEOCHEMISTRY AND ISOTOPES

Given that the emphasis in this chapter is on petrology and mineralogy and its relevance to the Sokli geophysical signature, we do not discuss in detail the geochemistry of the complex. The average composition of magmatic carbonatite, metasomatic carbonatite, and metasomatic ultramafic rock are listed in Table 4.2.2. Of note are the following points: (1) the magmatic carbonatites are extremely depleted in Al_2O_3, yet contain significant K_2O, explaining the high modal abundance of tetraferriphlogopite in the system; (2) the average MgO content of the metasomatic carbonatites and the transitional zone rocks is roughly the same despite nearly three times the silica content in the latter. This might indicate

Table 4.2.2 Average chemical composition (wt%) of Sokli rocks

	Transition zone metasomatites	Metasomatic carbonatites	Magmatic carbonatites
SiO_2	29	10.5	2
TiO_2	1	1	0.1
Al_2O_3	6	2	0.4
Fe_2O_3	9.5	9	4.5
FeO	5	4	0.5
MnO	0.3	0.5	0.5
MgO	11	10	5
CaO	20	34	47
K_2O	3.5	1.5	0.5
P_2O_5	5	4	3.5
CO_2	9	23	34
S	0.7	0.5	0.5
Nb	0.01	0.02	0.15

Source: *Data from Vartiainen (1989).*

a significant magnesiocarbonatite component in the metasomatizing melts and fluids that affected the metasomatic carbonatite zone; and (3) as the carbonatite component decreases from the magmatic core to the transition zone, the average phosphorus content increases.

A very comprehensive geochemical dataset on the Sokli complex can be found in Lee (2002); it includes 225 whole-rock samples analyzed by X-ray fluorescence (XRF) and 21 repeat analyses using inductively coupled plasma mass spectrometry (ICP-MS). A number of important observations made by Lee are summarized in the following:

1. The types of magma evolution described by the mineral compositional variations, and summarized above, can also be traced in the whole-rock data.
2. The mineralogically described phoscorite-carbonatite conjugate pairs have parallel trace element patterns, albeit at significantly lower enrichment levels for the phoscorites of each pair.
3. Trace element data show that the calciocarbonatites at Sokli are very much average on a worldwide basis for most elements.
4. In terms of REE contents, however, the Sokli calciocarbonatites are relatively REE-poor, similar to contents in the Siilinjärvi carbonatites (see Subchapter 4.3).
5. A relatively clear compositional gap exists between calciocarbonatites and magnesiocarbonatites, implying that the former are predominantly cumulates without significant trapped evolved magnesiocarbonatite liquid.

The available isotope data comprise a few Sr and Nd analyses reported by Kramm (1993). The two data points plot in the depleted mantle field along with young carbonatites and near the Siilinjärvi composition, and implicate an Ocean Island Basalt (OIB)-type mantle source (see Subchapter 4.1 and figure 4.1.3). Primary carbonatite carbon and oxygen isotope data for Sokli, reported in Demeny et al. (2004), are all close to the values from the other KAP intrusions and well within the "mantle box" of Taylor et al. (1967). For the P1C1 through C3 stages, the values range from $\delta^{13}C = -3.0$ to $-4.2‰$, $\delta^{18}O = 7.1$ to $7.5‰$. However, P3, C4, and the metasomatic carbonatites show a well-defined array away from these mantle values toward $\delta^{13}C = -2.6‰$, $\delta^{18}O = 10‰$. Demeny et al. (2004) suggest that this trend could represent crystallization-induced Rayleigh fractionation, implying crystallization in a relatively shallow subvolcanic magma chamber.

GENESIS OF THE CARBONATITE

The fact that aillikite, a carbonate-rich ultramafic lamprophyre, formed dikes at Sokli that intruded before, during, and after the carbonatite, implies an important role for this magma type in the formation of the Sokli complex. Relatively low LREE enrichment, compared to other carbonatites in the world, appears to be a trademark of carbonatites derived from aillikite magmas (Hornig-Kjarsgaard, 1998). We suggest that the Sokli intrusion likely started as dominantly olivine and clinopyroxene cumulates, quite similar to what is seen at the Kovdor and Vuorijärvi complexes, from ascending and pooling aillikite magmas.

The Tulppio olivinite quite probably represents a remnant from that stage, but different interpretations have been presented (see Maier et al., 2013, who interpreted the Tulppio olivine rich rocks as being of komatiite affinity). Carbonatite magma, derived from an aillikite parental liquid, then intruded into the system, first forming a central core as in the ideal model shown earlier in Fig. 4.2.1. However, in the case of Sokli, an unprecedented volume of carbonatite magma continued to ascend into the system, causing pervasive carbonate overprinting of all previously formed rocks, including the magma chamber cumulates and the first intruded carbonatites and phoscorites. Subsequent erosion has brought these magma chamber rocks to the surface. This evidence suggests that the present level of erosion at Sokli is probably no more than a kilometer below the now-eroded volcanic edifice.

ORE DEPOSITS

WEATHERED CAP

All of the Sokli area is covered by weathered bedrock of variable thickness. It is likely that this thick layer of weathered material has been preserved due to the fact that Sokli is located in a glacial ice divide, the locus of least erosion under a large continental ice sheet.

The most important mineralogical changes associated with weathering are the partial dissolution of carbonates, local replacement of phlogopite by vermiculite, intense alteration of olivine, the partial replacement of magnetite by hematite (martitization), alteration of pyrochlore, and the total removal of sulfides.

ORE TYPES

The phosphorus ore developed from the carbonatite and underwent complex weathering, leaching, recrystallization, and lithification processes (Nuutilainen, 1973), driven by the prevailing tropical climatic conditions (Finland lay on the equator ~400 Ma ago). The end product is a reddish brown layer, averaging 26 m in thickness, which varies from solid rock to a soil in which the carbonate has been totally dissolved. Apatite, magnetite, hydrated mica, and patchy pyrochlore occur as restite minerals. Recrystallized phases include francolite (carbonate-fluorapatite), goethite, and manganese oxides (Vartiainen et al., 1990). Owing to this process, the P_2O_5 values have been elevated from the original 4–5 wt% to up to 30 wt%.

There are three main types of residual phosphorus ore at Sokli: lateritic phosphate ore, silicate-apatite ore, and apatite residue on the carbonatite ring dikes intruded into the region of fenites outside the carbonatite (Vartiainen, 1989). The lateritic ore is located above the magmatic carbonatite, while the silicate-apatite ore is located above the transition-zone rocks that contain significantly more silicate minerals (and thus silica; see Table 4.2.2).

GEOPHYSICAL MAPPING OF ORE TYPES

From the aeroradiometric data (Fig. 4.2.7), it is apparent that the carbonatite contains elevated levels of thorium and uranium relative to the surrounding country rocks, whereas it is depleted in potassium, consistent with the geochemistry of the rocks. In the overburden overlying the complex, thorium and uranium are adsorbed onto iron oxides. In detail, thorium concentrations in the overburden over the transition zone and well into the fenite zone are relatively higher than uranium. This is due to the greater extent of alteration of the overburden above the magmatic and metasomatic carbonatite zones, and the greater leaching of thorium downward. The niobium-rich mineral deposits have a high concentration of both uranium and thorium in the middle of the massif, whereas potassium is enriched along the outer edge of the massif.

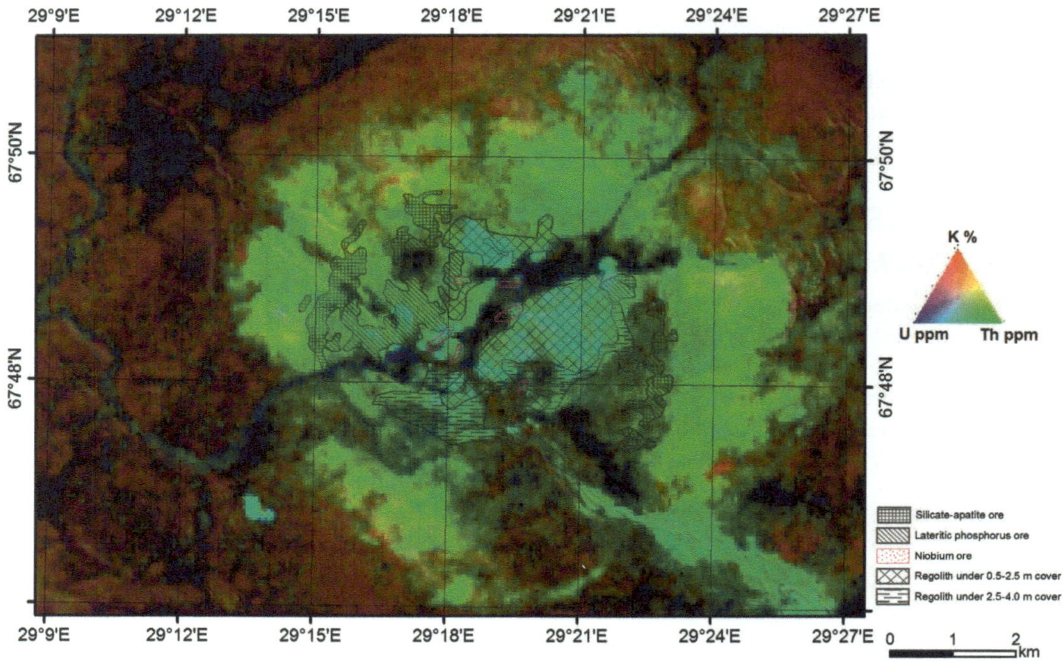

FIGURE 4.2.7 Aeroradiometric map of the Sokli complex overlain with the various surface ore types.

Water attenuates gamma rays significantly and, therefore, wet areas appear as black in the map.

The carbonatite contains abundant magnetite, which affects the negative response to the low-frequency AEM data. The reddish colors in Fig. 4.2.8(A) indicate rock units rich in magnetite, where the magnetic susceptibility is high and the electrical conductivity is low. This negative AEM response outlines the carbonatite core well. The innermost core also has a higher concentration of thorium and uranium so that the highest values are correlated with the lowest AEM-negative response (Fig. 4.2.7). The negative AEM response was roughly divided into two categories, which are thought to result from grain size variation of magnetite causing variation in the magnetic susceptibility (Fig. 4.2.8(A)). The silicate-apatite ore deposits are located in the surrounding weathered transition zone and they follow high-frequency conductivity anomalies represented with bluish colors in Fig. 4.2.8(B).

ORE RESERVE ESTIMATES

Vartiainen (1989) reported the ore reserves of the various residual phosphate ore types at Sokli to be approximately 110 Mt in total, at a P_2O_5 content of 16.5%. In the environmental impact statement prepared for Yara International ASA (Pöyry Environment Oy, 2009), the reserves reported are very similar to the earlier estimates, and are broken down by ore type in Table 4.2.3.

Ores based on the weathered crust of the carbonatite (regolith) were discussed by Vartiainen (1989) with an ore reserve estimate of ~200 Mt at a grade of 4.5% P_2O_5. The Pöryr EIS (Pöyry Environment

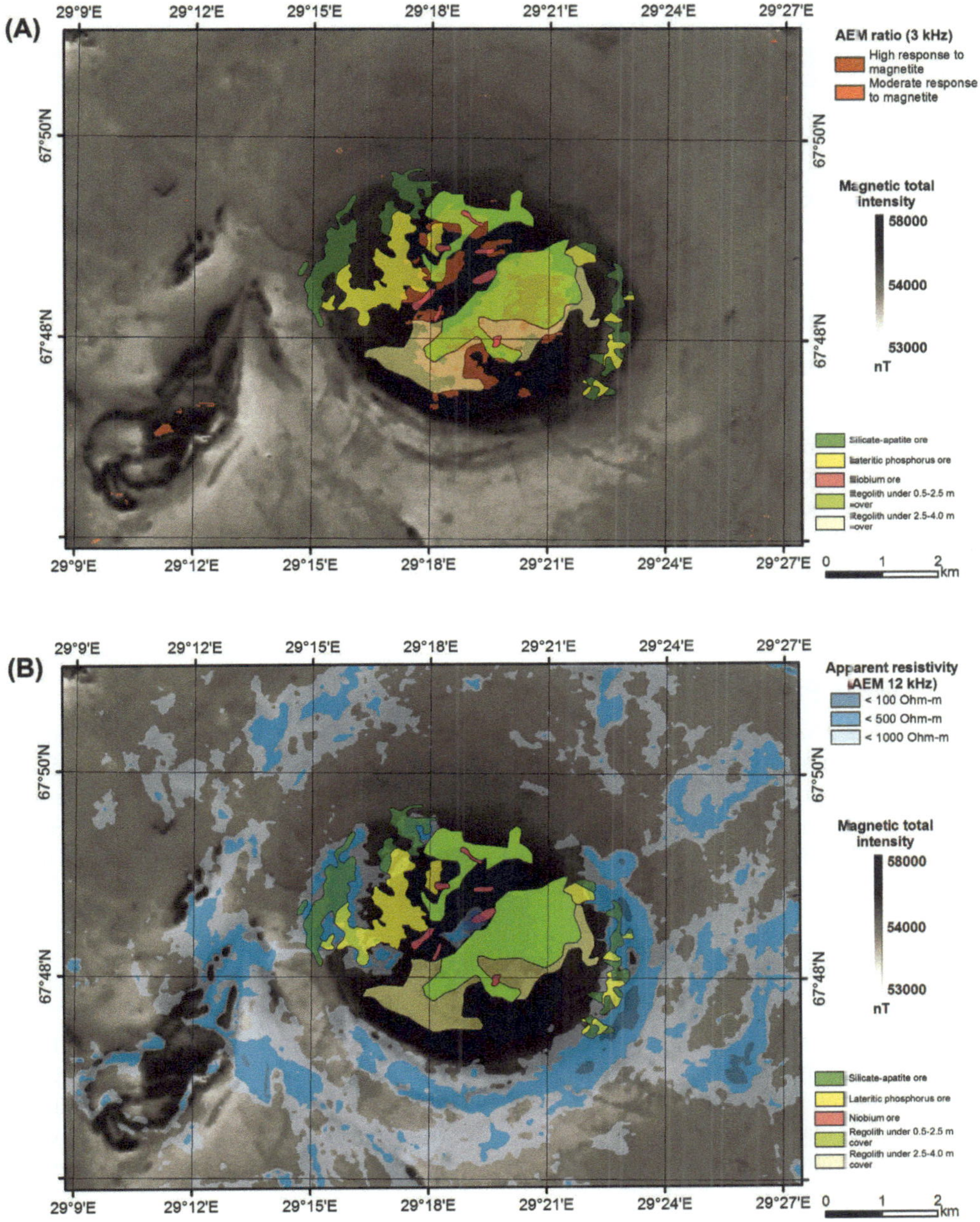

FIGURE 4.2.8 (A) Combined apparent resistivity and classified AEM ratio image compared to location of the various surface ore types. (B) Classified apparent resistivity compared to location of the various surface ore types.

Table 4.2.3 Sokli ore reserves by type

	Amount (Mt)	P_2O_5 wt%
Lateritic ore	36.7	18.7
Silicate-apatite ore	19.3	11.0
Other phosphorus ore	4.4	10.9
Ores in weathered rock	53.1	14.1
Subtotal/average	113.6	14.9
Regolith ores	75.2	5.6
Nb ores	1.9	6.7
Total/average	190.6	11.2

Source: *From Pöyry (2009), a JORC compliant report.*

Oy, 2009) puts the regolith ore reserves at 75 Mt by limiting the reserve to areas under the thinnest overburden, in the Kaulusrova area, with a grade of 5.6 wt% P_2O_5, and Nb content 0.2 wt% (Table 4.2.3). Other ores include the niobium ore, localized within several small areas in the carbonatite core area (see Fig. 4.2.8A) with a grade of 6.7 wt% P_2O_5, and Nb content 0.6 wt%. The grade of the unweathered carbonatite is on average 3.5% P_2O_5, with huge resources given the size of the complex. However, on a global scale, this is very low-grade ore.

ONGOING MINE PREPARATION

Yara International ASA is presently going through the processes necessary for establishing a phosphorous mine at Sokli. Commissioning an environmental impact assessment, completing further definition drilling, and spending a large effort negotiating with the government of Finland and the local population about acceptable plans for a new phosphorus mine are all activities currently underway or recently completed. Mine development plans are proceeding, and CTS Engtec reported completion of the definitive feasibility study for the mine in October 2013. As of this writing the mine decision is still pending.

SUMMARY

1. Sokli is a member of the Devonian Kola alkaline province. At the early magmatic stage it may have resembled the Kovdor and Vuorijärvi complexes in terms of rock types, zoning, and general layout. However, at Sokli, the later carbonatite magma overprinted the early silicate cumulates to such an extent that the latter are now rare, and studying them is difficult. The Tulppio olivinite, if proven to be related to the same magma system, may be the only real remnant from these early stages, and deserves further scrutiny.
2. Magnetic properties of the complex are distinct for each of the rock types described by Vartiainen (1980). Nevertheless, there has not been sufficient study of the metasomatic carbonatites, and particularly the transition-zone rocks, to fully illuminate the details of the processes by which they formed, including time sequencing and the geochemical mass balance. The

geophysical properties of these rocks can help map the distribution of rock types for these calculations.

3. Research by Lee (2002 and subsequent papers) concentrated on the magmatic carbonatites and showed quite convincingly how the phoscorite-carbonatite conjugate pairs were formed, documented both in terms of mineral and whole-rock chemistry. Valuable information remains to be gleaned from this extensive geochemical database.
4. Textures consistent with immiscibility, although rare, are known from a number of the KAP carbonatites, and although this process was almost certainly active at Sokli, modern methods of investigation have not been carried out on these spectacular samples.
5. Aillikite dikes were intruded throughout the development of the Sokli complex, and although it is difficult to prove, evidence including moderate REE enrichment, and the relatively high K_2O contents of the Sokli carbonatites point toward some type of aillikite magma as parental to the system.
6. The ore deposit is composed mostly of laterized materials at the top of the complex. The bulk of the ore is derived from weathering of the metasomatic carbonatite and the transition-zone rocks. It is an interesting observation that the rocks formed by metasomatism apparently have higher average P_2O_5 contents than the magmatic carbonatite core.
7. Geophysical mapping of ore deposit types is best accomplished with resistivity measurements at high frequency that show the strongest response to near surface, conductive features.
8. The Sokli ore reserves of "soft-rock" phosphate-rich materials are about 114 Mt with 15 wt% P_2O_5, with an additional approximately 75 Mt of weathered bedrock ore with about 5.6 wt% P_2O_5.

ACKNOWLEDGMENT

The authors would like to thank R.E. Harmer and W.D. Maier for their reviews of this section of Chapter 4, which markedly improved the presentation and content.

REFERENCES

Dalton, J.A., Presnall, D.C., 1998. The continuum of primary carbonatitic-kimberlitic melt compositions in equilibrium with lherzolite: Data from the system $CaO-MgO-Al_2O_3-SiO_2-CO_2$ at 6 GPa. J. Petrol. 39, 1953–1964.

Demeny, A., Sitnikova, M.A., Karchevsky, P.I., 2004. Stable C and O isotope compositions of carbonatite complexes of the Kola Alkaline Province: phoscorite-carbonatite relationships ans source compositions. In: Phoscorites and carbonatites from mantle to mine: the key example of the Kola Alkaline Province, F. Wall and A.N. Zaitsev, editors. Mineralogical Society Series 10, 407–431.

Hautaniemi, H., Kurimo, M., Multala, J., et al., 2005. The "three in one" aerogeophysical concept of GTK in 2004. In: Airo, M.-L. (Ed.), Aerogeophysics in Finland 1972–2004. Methods, System Characteristics and Applications, Geological Survey of Finland, Special Paper 39, pp. 21–74.

Hornig-Kjarsgaard, I., 1998. Rare earth elements in sövitic carbonatites and their mineral phases. Journal of Petrology 39, 2105–2121.

Ilyin, A.V., 1989. Apatite deposits in the Khibiny and Kovdor alkaline igneous complexes, Kola peninsula, northwestern USSR. Phosphate Rock Resources. In: Notholt, A.J.G., Sheldon, R.P., Davidson, D.F. (Eds.), Phosphate Deposits of the World, vol. 2. Cambridge University Press, Cambridge, U.K, pp. 394–397.

Karchevsky, P.I., Moutte, J., 2004. The phoscorite-carbonatite complex of Vuoriyarvi, northern Karelia. In: Wall, F., Zaitsev, A.N. (Eds.), Phoscorites and Carbonatites from Mantle to Mine: The Key Example of the Kola Alkaline Province. The Mineralogical Society of Britain and Ireland, London, pp. 163–199.

Kramm, U., 1993. Mantle components of carbonatites from the Kola alkaline province, Russian and Finland: a Nd-Sr study. Eur. J. Mineral 5, 985–989.

Kramm, U., Kogarko, L.N., Kononova, V.A., Vartiainen, H., 1993. The Kola alkaline province of the CIS and Finland: Precise Rb-Sr ages define 380–360 Ma age range for all magmatism. Lithos 30, 33–44.

Kramm, U., Sindern, S., 2004. Timing of Kola ultrabasic, alkaline and phoscorite-carbonatite magmatism. In: Phoscorites and carbonatites from mantle to mine: the key example of the Kola Alkaline Province, F. Wall and A.N. Zaitsev, editors. Mineralogical Society Series 10, 75–97.

Krasnova, N.I., Balaganskaya, E.G., Garcia, D., 2004. Kovdor—classic phosocorites and carbonatites. In: Wall, F., Zaitsev, A.N. (Eds.), Phoscorites and Carbonatites from Mantle to Mine: The Key Example of the Kola Alkaline Province. Mineralogical Society Series 10. The Mineralogical Society of Britain and Ireland, London, pp. 99–132.

Krasnova, N.I., Kopylova, L.N., 1988. The geological basis for mineral-technological mapping at the Kovdor ore deposit. International Geology Review 30, 307–319.

Kukharenko, A.A., Orlova, M.P., Bulakh, A.G., et al (Eds.), 1965. The Caledonian Ultramafic Alkaline Rocks and Carbonatites of the Kola Peninsula and Northern Karelia. Nedra Press, Moscow, p. 772. (in Russian).

Lapin, A.V., Vartiainen, H., 1983. Orbicular and spherulitic carbonatites from Sokli and Vuorijärvi. Lithos 16, 53–60.

Le Maitre, R.W. (Ed.), 2002. Igneous Rocks. A Classification and Glossary of Terms, 2nd edition. Cambridge University Press, Cambridge, U.K, p. 236.

Lee, M.J., 2002. Mineralogy, petrology and geochemistry of the phoscorite-carbonatite association of the Sokli alkaline complex, Finland. Dr. Sci. Thesis, Ecole des Mines de Saint Étienne p. 233.

Lee, M.J., Garcia, D., Moutte, J., Lee, J.I., 2003. Phlogopite and tetraferriphlogopite from phoscorite and carbonatite associations in the Sokli massif, Northern Finland. Geosciences Journal 7, 9–20.

Lee, M.J., Garcia, D., Moutte, J., et al., 2004. Carbonatites and phoscorites from the Sokli complex, Finland. In: Wall, F., Zaitsev, A.N. (Eds.), Phoscorites and Carbonatites from Mantle to Mine: The Key Example of the Kola Alkaline Province. The Mineralogical Society of Britain and Ireland, London, pp. 133–162.

Lee, M.J., 2005. Compositional variation of Fe-Ti oxides from the Sokli complex, northeastern Finland. Geosciences Journal 9, 1–13.

Lee, M.J., Lee, J.I., Garcia, D., et al., 2006. Pyrochlore chemistry from the Sokli phoscorite-carbonatite complex, Finland: Implications for the genesis of phoscorite and carbonatite association. Chemical Journal 40, 1–13.

Maier, W.D., Peltonen, P., Halkoaho, T., Hanski, E., 2013. Geochemistry of komatiites from the Tipasjärvi, Kuhmo, Suomussalmi, Ilomantsi and Tulppio greenstone belts, Finland: Implications for tectonic setting and Ni sulphide prospectivity. Precambrian Geology 228, 63–84.

Mitchell, R.H., 1978. Manganoan magnesian ilmenite and titanian clinohumite from the Jacupiranga carbonatite, Sao Paulo, Brazil. American Mineralogist 63, 544–547.

Nuutilainen, J., 1973. Soklin karbonatiittimassiivin geokemiallisista tutkimuksista. Geologi 25, 13–17.

O'Brien, H.E., Peltonen, P., Vartiainen, H., 2005. Kimberlites, carbonatites, and alkaline rocks. In: Lehtinen, M., Nurmi, P.A., Rämö, O.T. (Eds.), Precambrian Geology of Finland—Key to the Evolution of the Fennoscandian Shield. Elsevier, Amsterdam, pp. 101–138.

Pöyry Environment Oy, 2009. Sokli mine environmental impact statement. p. 86. Available at www.ymparisto.fi/download/noname/%7B9FF4A1D4-C572-4C7A-865D-143DD32F0B33%7D/42576

Rock, N.M.S., Bowes, D.R., Wright, A.E., 1991. Lamprophyres. Blackie and Son Ltd., Glasgow. p. 275.

Suppala, I., Oksama, M., Hongisto, H., 2005. GTK airborne EM system: Characteristics and interpretation guidelines. In: Airo, M.-L. (Ed.), Aerogeophysics in Finland 1972–2004. Methods, System Characteristics and Applications. Geological Survey of Finland, Special Paper 39, pp. 103–117.

Tappe, S., Foley, S.F., Jenner, G.A., Kjarsgaard, B.A., 2005. Integrating ultramafic lamprophyres into the IUGS classification of igneous rocks: Rationale and implications. Journal of Petrology 46, 1893–1900.
Taylor Jr., H.P., Frechen, J., Degens, E.T., 1967. Oxygen and carbon isotope studies of carbonatites from Laacher See District, West Germany and the Alnö District, Sweden. Geochimica et Cosmochimica Acta 31, 407–430.
Vartiainen, H., 1980. The petrography, mineralogy and petrochemistry of the Sokli carbonatite massif northern Finland. Geological Survey of Finland, Bulletin 313, p. 126.
Vartiainen, H., 1989. Phosphate deposits of the world. In: Notholt, A.J.G., Sheldon, R.P., Davidson, D.F. (Eds.), Phosphate Rock Resources, vol. 2. Cambridge University Press Cambridge, U.K. pp. 398–402.
Vartiainen, H., Kresten, P., Kafkas, Y., 1978. Alkaline lamprophyres from the Sokli complex, northern Finland. Geological Society of Finland. Bulletin 50, 59–68.
Vartiainen, H., Melnikov, I., Sullimov, B., 1990. The francolite ore deposits of Kovdor and Sokli. Proceedings of the Finnish-Soviet Symposium held in Helsinki. November 14–15. Research Report TKK-IGE A13, 7–14.
Vartiainen, H., Paarma, H., 1979. Geological characteristics of the Sokli carbonatite complex. Finland. Economic Geology 74, 1296–1306.
Vartiainen, H., Vitikka, A., 1993. The late dikes of the Sokli massif and their tectonic monitoring. Geochimia 8, 1241–1244 (in Russian).
Vartiainen, H., Woolley, A.R., 1976. The petrography, mineralogy and chemistry of the fenites of the Sokli carbonatite intrusion, Finland. Geological Survey of Finland, Bulletin 280, p. 87.
Verhulst, A., Balaganskaya, E., Kirnarsky, Y., Demaiffe, D., 2000. Petrological and geochemical (trace elements and Sr-Nd isotopes) characteristics of the Paleozoic Kovdor ultramafic, alkaline and carbonatite intrusion (Kola Peninsula, NW Russia). Lithos 51, 1–25.
Wall, F., Zaitsev, A.N., 2004. Phoscorites and carbonatites from mantle to mine: the key example of the Kola Alkaline Province, F. Wall and A.N. Zaitsev, editors. Mineralogical Society Series 10, 498 p.
Zaitsev, A.N., Sitnikova, M.A., Subbotin, V.V., et al., 2004. Sallanlatvi complex—rare example of magnesite and siderite carbonatites. In: Wall, F., Zaitsev, A.N. (Eds.), Phoscorites and Carbonatites from Mantle to Mine: The Key Example of the Kola Alkaline Province. The Mineralogical Society of Britain and Ireland, London, pp. 201–245.

CHAPTER

THE ARCHEAN SIILINJÄRVI CARBONATITE COMPLEX

4.3

H. O'Brien, E. Heilimo, P. Heino

ABSTRACT

The Siilinjärvi carbonatite complex is named after the nearby village of Siilinjärvi, located some 5 km west of the southern extension of the complex. It is one of the oldest carbonatites on Earth, and the oldest presently being mined. Due to its age, it has been metamorphosed, particularly during the 1.8 Ga Svecofennian event, and intruded by several generations of younger, mafic to intermediate dikes, and a tonalite-diorite intrusion. Despite this overprint, the bulk of the rocks of the complex show well-preserved igneous textures, primary magmatic compositions of constituent minerals, and retain most of their primary isotopic compositions, which point to a mantle derivation. As such, the carbonatite and associated rocks provide a rare opportunity to constrain the mineralogy and geochemistry of the Archean mantle. Dominant rock types in the complex include carbonatite, silicocarbonatite, carbonatite-glimmerite, and glimmerite (mainly tetraferriphlogopite) within the main intrusion, all of which are surrounded by a halo of metasomatically produced fenites. Primary magmatic saline aqueous fluids rich in alkali elements have been identified in fluid inclusions in zircon and apatite from the carbonatite, explaining the formation of the well-developed metasomatic fenite syenite halo. Mining for apatite as a source for phosphorus began in 1979 by Kemira Oy, but since 2007 the deposit has been under the ownership of Yara International ASA, presently producing about 11 Mt of ore per year combined from the main Särkijärvi pit in the south, and the satellite Saarinen pit in the north.

Keywords: carbonatite; glimmerite; tetraferriphlogopite; apatite; phosphorus ore; mine.

INTRODUCTION

The Siilinjärvi carbonatite complex is located in eastern Finland close to the city of Kuopio. It consists of a steeply dipping lenticular body roughly 16 km long with a maximum width of 1.5 km and a surface area of 14.7 km^2 intruded into basement gneiss (Fig. 4.3.1). It was discovered in 1950 after local mineral collectors sent samples of carbonatite to the Geological Survey of Finland (GTK). Exploration drilling began in 1958 by Lohjan Kalkkitehdas Oy and continued between 1964 and 1967 by Typpi Oy, and from 1967–1968 by Apatiitti Oy. After 1968, Kemira and its subsidiaries moved forward with laboratory and pilot plant work. An open pit mine for phosphorus ore was commissioned in 1979. The mine was sold to Yara Suomi Oy in 2007.

The main glimmerite-carbonatite intrusion within the Siilinjärvi complex occurs as a central tabular, up to 900 m wide, body of glimmerite and carbonatite running the length of the complex, surrounded by a fenite margin. Unlike many other carbonatite-bearing complexes that contain a sequence of phlogopite-rich rocks intruded by a core of carbonatite (c.f., Kovdor, Phalaborwa), at Siilinjärvi, the carbonatites and

Mineral Deposits of Finland. http://dx.doi.org/10.1016/B978-0-12-410438-9.00013-3

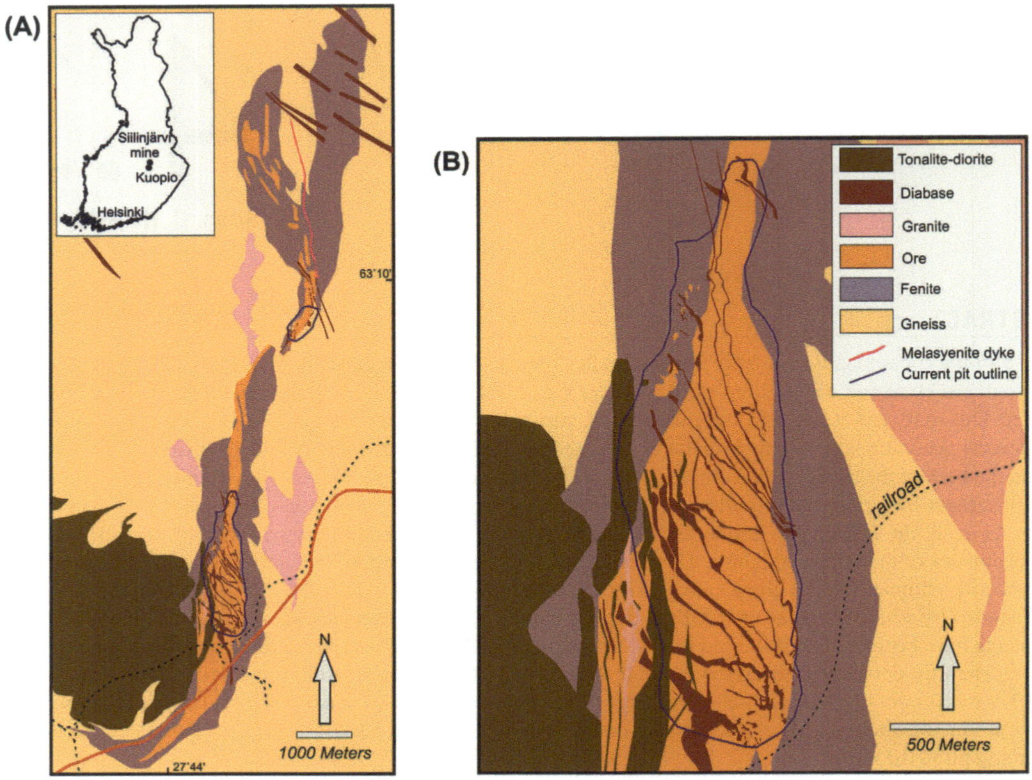

FIGURE 4.3.1 (A) Geological map of the Siilinjärvi carbonatite complex and (B) the area of the main pit at larger scale.

glimmerites are intimately mixed, varying from nearly pure glimmerites (tetraferriphlogopitites) to nearly pure carbonatites, with a well-developed subvertical to vertical lamination.

Although not strictly zoned, generally the volume of carbonatite is greatest near the center of the intrusion, which is cut by numerous subvertical carbonatite veins (Fig. 4.3.2). Glimmerites near the outer edges of the body can be nearly carbonate-free, yet still contain ore-grade amounts of apatite. Crosscutting relationships and xenoliths suggest that, at least at the present level of exposure, some of the fenites formed early because they occur as megaxenoliths within the magmatic glimmerite (Fig. 4.3.3).

ROCK TYPES

FENITES

Fenites surrounding the carbonatite-glimmerite central core developed as a result of sodium and potassic metasomatism of the surrounding granite gneiss country rocks. The main minerals in the fenites are microcline, amphibole, and pyroxene, but there exists a wide variety of fenite types

FIGURE 4.3.2 Photograph from the south wall of the main pit showing alternating glimmerite-rich and carbonatite-rich bands of the central portion of the main ore zone.

Source: *Photo by Esko Koistinen, GTK.*

FIGURE 4.3.3 Eastern wall of main mine pit, showing megablocks of fenitic syenite supported in a black biotite glimmerite matrix.

Biotite and chlorite have formed after primary tetraferriphlogopite in areas that have undergone significant shearing. Photograph shows about 90 m vertically of the 250 m high main pit wall.

including: pyroxene, amphibole, carbonate, quartz, aplitic, and quartz-aegirine varieties. Compositions of the fluids that produced these fenites have been determined from fluid inclusions within magmatic zircon and apatite (Poutiainen, 1995). Zircon crystals, which are thought to be more common in the amphibole-rich parts of the intrusion, contain two types of fluid inclusions trapped prior to emplacement of the carbonatites.

Type 1 fluid is an H_2O-CO_2 mixture with low salinity (1–4 wt% NaCl equivalent) whereas type 2 is of moderate salinity (7–18 wt% NaCl equivalent), alkali- and H_2O-rich. Thus, the development of H_2O- and alkali-rich late-stage fluids that formed the fenite halo was a direct consequence of the early crystallization of predominantly carbonate and apatite (Poutiainen, 1995). Ascent and hydrofracturing by the evolving H_2O-rich fluid may have facilitated the ascent of the parental magmas along deep crustal shears, with attendant fenitization along the path.

GLIMMERITE-CARBONATITE ROCK SERIES

The vast majority of the central ore body is made up of phlogopite-rich rocks ranging from almost pure glimmerite (dominantly tetraferriphlogopite, 0–10 modal% carbonates) to carbonate glimmerite (10–25 modal% carbonates) to silicocarbonatites (25–50 modal% carbonates). White to pinkish, medium-grained carbonatite (>50 modal% carbonates), typically with a greenish tinge reflecting the apatite grains it contains, represents roughly 1.5 vol% of the main intrusion (1.4 and 2.3 vol% at the Särkijärvi and Saarinen pits, respectively). Carbonatite occurs mostly as thin subvertical veins in glimmerite, with the veins being concentrated toward the center of the intrusion. In addition to carbonate and phlogopite, blue-green rocks with up to 50 modal% richterite can also be abundant in places (Puustinen, 1972).

Figure 4.3.4(A–D) shows textural examples of rock types from the glimmerite-carbonate series, including magmatic laminar texture, foliated texture due to shearing, less common pegmatoidal texture of mica, and an example of a zircon-rich sample. Even though all varieties of this magmatic pulse contain apatite, apatite is nonetheless only slightly more abundant in the glimmerite based on a new geochemical database (summarized in Table 4.3.1). The overall mode of the carbonatite-glimmerite portion of the complex, as indicated by the average composition of the Siilinjärvi ore, is 65% phlogopite (including tetraferriphlogopite), 19% carbonates (with a 4:1 calcite:dolomite ratio), 5% richterite, 10% apatite (equivalent to 4% P_2O_5 in the whole rock), and 1% accessory minerals comprising mainly magnetite and zircon.

MICA-BEARING DIKES

Crosscutting the surrounding bedrock, the fenite halo, and the central intrusive body is a 4-km-long, 20–30 m wide, north-trending mafic dike within the northern part of the complex (see Fig. 4.3.1) that was termed melasyenite by Puustinen (1971) and thought to be related to the same intrusive event as the carbonatite. The melasyenite is composed of alkali feldspar, biotite, alkaline amphibole, apatite, and magnetite. Suspected to be one of a series of dikes (H. Lukkarinen, personal communication, 2003), another dike discovered in 2005 (sample 14-PTP-05), taken from the same area, proved to be ultramafic in character. It is composed of minerals similar to those found in the Siilinjärvi carbonatite: phlogopite, alkali-amphibole, primary carbonate, apatite, magnetite, and titanite (Fig. 4.3.5(A)).

In a hand sample, a foliation is apparent, which in thin sections can be seen to have caused recrystallization of much of the mica. Nevertheless, remnant large phlogopite crystals with epitaxial alkali-amphibole overgrowths, along with primary carbonate segregations, can be observed in thin section

FIGURE 4.3.4 Examples of the Siilinjärvi glimmerite-carbonatite rock series.

(A) Laminated glimmerite-carbonatite typical of the main intrusion, primary magmatic texture. (B) Highly foliated carbonatite. This type of metamorphic foliation is not typical of most of the glimmerite-carbonatite rocks series, but can be found in the southern extension of the ore body. (C) Pegmatitic phlogopite book in carbonate matrix. (D) Zircon-rich sample of silicocarbonatite with no visible richterite, showing that zircon occurs also in relatively SiO_2-poor rock types.

Table 4.3.1 Siilinjärvi ore zone rocks, modal mineralogy, and calculated major element chemistry

	Ore[1]	Glimmerite	Carbonatite apatite containing	Carbonatite apatite poor	Lamprophyre dike[3]
Micas[2]	65.0	81.5	14.7	1.2	
Amphibole	5.0	4.5	0.6	0.2	
Calcite	15.0	1.6	61.2	86.8	
Dolomite	4.0	0.9	13.4	10.6	
Apatite	10.0	10.4	9.9	0.8	
Others	1.0	0.7	0.1	0.4	
wt%					
Na_2O	0.2	0.2	0.1	0.1	0.6
MgO	18.3	20.8	8.1	4.6	15.3
Al_2O_3	7.0	8.8	1.8	0.2	6.1
SiO_2	30.2	37.5	7.8	1.3	43.3
P_2O_5	4.2	4.1	4.5	0.5	0.3
K_2O	6.2	7.6	1.6	0.2	4.5
CaO	13.9	6.8	38.6	47.0	6.8
TiO_2	0.3	0.5	0.1	<0.1	2.9
MnO	0.1	0.0	0.1	0.2	0.2
Fe_2O_3	7.1	8.3	3.0	1.6	18.1

[1]*Average ore composition; there is significant variation from block to block.*
[2]*Mainly tetraferriphlogopite.*
[3]*Dike 14-PTP-05, Siilinjärvi potassic magma.*

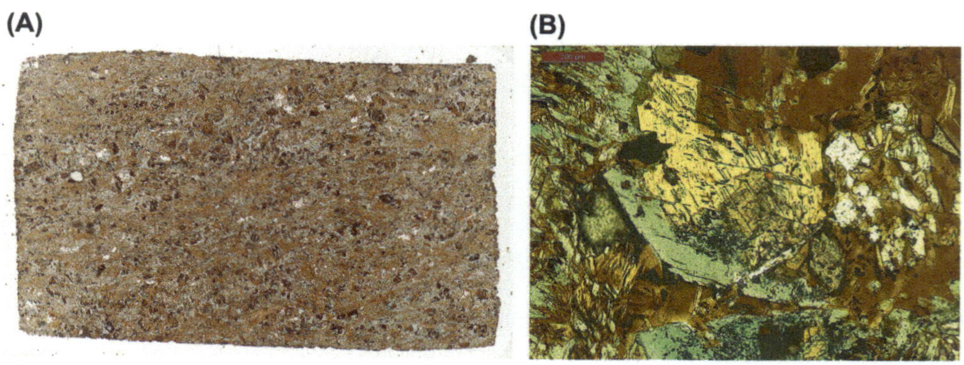

FIGURE 4.3.5 Thin section of ultramafic lamprophyre dike 14-PTP-05 from the northern part of the intrusion.

(A) Scan of thin section showing amphibole (blue) and mica (brown) with small euhedral clusters of calcite grains and opaque magnetite. Rock piece is 3 cm × 2 cm. (B) Photomicrograph of same, displaying epitaxial overgrowth of amphibole on mica. Plane polarized light.

(Fig. 4.3.5(B)). Further study is required to determine whether this is an amphibole-bearing aillikite (ultramafic lamprophyre) or some other type of lamprophyre. The magma from which this dike crystallized could be a potential candidate as a parental magma of the Siilinjärvi complex.

Even closer in appearance and mineralogy to the main carbonatite-glimmerite ore body are a number of 0.5-m to several-meters-thick potassic dikes that transect the fenites and in places form the matrix to fenite megablocks, which can be studied in exposures along the railroad track southeast of the main pit, in drill cores of the western margin of the main pit, and in the walls of the northern satellite pit.

These dikes are mineralogically similar to the main ore body, but importantly, they lack the laminar texture of the former, and instead show a much more chaotic distribution of carbonate-rich, amphibole rich, and phlogopite-rich patches. They lack foliation and in some cases are pegmatoidal (Fig. 4.3.6). These textures exclude their formation as fault breccias. Further study is required to determine whether

FIGURE 4.3.6 Glimmerite-carbonatite dike exposed in the railroad cut southeast of the main pit (see Fig. 4.3.1).

these dikes represent feeders to the main intrusion or apophyses derived from the main intrusion, but their importance lies in the clearly displayed relationship that at least some of the carbonatite component was unequivocally part of the same magmatic system as the glimmerite.

DIABASE DIKES

Crosscutting the entire Siilinjärvi complex are diabase dikes of basaltic composition with widths that range from centimeters to several meters (rarely up to 60 m), (refer to Fig. 4.3.1). The dikes show a strong northwest–southeast or north-northwest–south-southeast orientation. Where they are sufficiently wide they form non-ore rock that is stockpiled separately. To our knowledge, these dikes have not been studied in detail, but preliminary studies show that there are at least three varieties, including calcite-bearing diabase (with calcite probably derived from the carbonatite), sulfide-bearing diabase, and barren diabase (without calcite or sulfides). Sulfides in the diabases tend to be enriched in more highly sheared zones, especially in the northern Saarinen pit area.

STRUCTURES

Contacts between the country rocks and the main Siilinjärvi ore body are either primary magmatic or sheared. For example, contact between the ore and the surrounding fenite in the eastern part of the Särkijärvi pit is sharp and well preserved, whereas the southwest corner of the pit shows a mosaic of sheared blocks containing mingled tonalite-diorite, diabase, and fenite, as well as the carbonatite-glimmerite ore (see Fig. 4.3.1). In this chaotic zone, Paleoproterozoic diabase dikes, as well as apophyses of a younger tonalite-diorite intrusion, are intermingled and together transect the ore.

Overall, a roughly north–south striking, steep shearing is a common feature near the contacts of the deposit as well as within the carbonatite-glimmerite ore. The vertically laminar structure of carbonatite-glimmerite is "folded" at least twice. An older "folding" stage has produced open to tight fold structures with an almost horizontal north- and south-trending fold axis. Whether this represents a process during late consolidation of the cumulates or later deformation is not well understood, but flow folding is a common primary feature of carbonatite intrusions (R.E. Harmer, personal communication, 2015).

The second deformation stage comprises an induced, intense vertical north–south striking shearing that is also observed at the ore contacts. The shearing has produced small-scale left- and right-sided folds in the incompetent glimmerite ore. This younger folding event can be observed in both the ore and the Paleoproterozoic diabase dikes that crosscut the complex. These structural observations provide evidence of several deformation events that have affected the Siilinjärvi complex, perhaps earlier than, but certainly also during, the Svecofennian orogeny.

MINERALOGY

MICA

The dominant mica at Siilinjärvi is reddish-brown tetraferriphlogopite, much the same as at Sokli, described in Subchapter 4.2. This variety of mica crystallizes in magnesian alkaline magmas that have low Al, forcing ferric Fe into the tetrahedral site to fill the structure. A direct comparison of the micas from

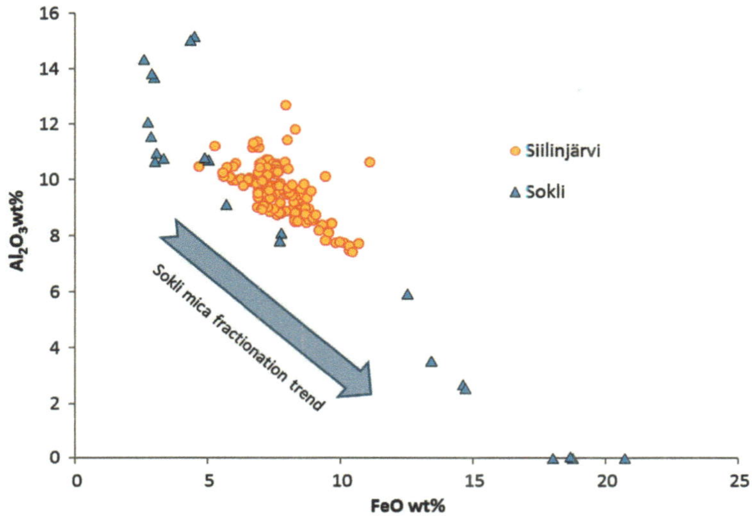

FIGURE 4.3.7 Mica compositions from Siilinjärvi compared to those from Sokli.

Siilinjärvi tetraferriphlogopites are restricted in composition and similar to those in the Sokli second series phoscorite-carbonatite conjugate pair (see Subchapter 4.2).

these two carbonatites (Fig. 4.3.7) shows that the Siilinjärvi tetraferriphlogopite is similar in composition to the Sokli cycle's two phoscorite-carbonatite tetraferriphlogopite compositions (see Section 4.2), but does not show the large compositional range to highly evolved varieties. The restricted Siilinjärvi mica compositional field implies a moderate level of fractionation of the source magma, and a large magmatic system that could produce enormous quantities of phlogopite of relatively constant composition.

In shear zones and contact zones where younger diabases have transected the carbonatite, tetraferriphlogopite has been converted to brown biotite-phlogopite and, in the most intensely affected zones, to black biotite and even chlorite. This is particularly true at the margins of the main intrusion, which appear to have absorbed considerable strain during regional metamorphism and acted as "great greasy shear plane(s)" (Puustinen, 1971).

CARBONATES

White to pink to gray calcite is the dominant carbonate mineral. Dolomite is nevertheless nearly ubiquitous, and commonly occurs as fine exsolution lamellae in calcite (refer to Fig. 9 in Puustinen, 1971; Puustinen, 1974). Microprobe analyses of 170 grains of calcite from a representative suite of main ore samples show that calcite contains a relatively uniform content of minor elements: 1.1 wt% SrO and 0.2 wt% MnO, along with <2 wt% MgO and <0.8 wt% FeO.

By running a scanning electron microscope (SEM)-based heavy mineral search on a set of thin sections from representative Siilinjärvi samples, tiny grains of strontianite, on the order of 100 μm wide, were discovered occurring as patches of small blocky grains within calcite (Al-Ani, 2013). This strontianite contains less than 1 wt% rare earth elements $(REE)_{tot}$, consistent with the low REE content of Siilinjärvi carbonatite in general.

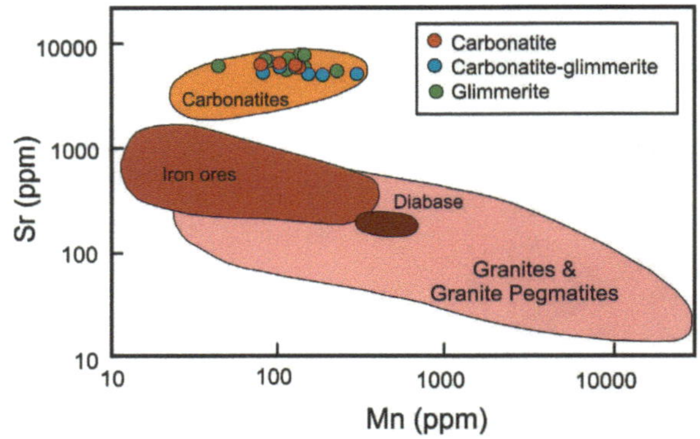

FIGURE 4.3.8 Sr-Mn discrimination diagram for apatite, with averages for about 340 Siilinjärvi ore apatite grains.

All data plotted in the field typical for carbonatite and cannot be differentiated based on host rock type.

Source: *Diagram after Belousova et al. (2002).*

APATITE

Light green to slightly gray apatite occurs in roughly equal amounts (~10 vol%) in ore rocks ranging from glimmerite to carbonatite, except for some examples of exceptionally apatite-rich glimmerites and carbonatites that have up to 30% apatite, and some 10–50 cm wide apatitite veins with up to 80 vol% apatite. Apatite forms rounded grains or euhedral hexagonal rods, typically several millimeters in diameter and several centimeters in length. All apatite at Siilinjärvi is fluoroapatite with between 2 and 4 wt% F, with an average for 160 grains of 3.2 wt% F and 0.75 wt% SrO. Despite the relatively low REE content of Siilinjärvi apatites (Hornig-Kjarsgaard, 1998), the Sr and Mn contents of the apatites are very typical of carbonatites (Fig. 4.3.8).

AMPHIBOLE

Blue-green richterite amphibole forms generally less than 15 vol% of the glimmerites and averages 5 vol% overall. However, Puustinen (1971, 1972) described carbonate-amphibole rocks that contain up to 50 vol% or more amphibole with richterite crystals up to several centimeters in length. In our experience, these rock types must be extremely rare. Microprobe analyses of 20 grains show uniform Al_2O_3 (<0.3%) and FeO/MgO, but variable K_2O (0.4–4.2 wt%) and Na_2O (1.0–6.3 wt%), indicating a range of amphibole types that require further study, particularly to determine their distribution within the ore body.

SERPENTINE

Serpentine is not a common mineral at Siilinjärvi; however, Puustinen (1971) reported a mica-bearing rock in the northern part of the Siilinjärvi complex that contains up to 80 vol% serpentine as pseudomorphs after olivine, along with apatite and magnetite. In this case, the rock is a serpentinized mica peridotite, or may represent a serpentinized phoscorite. Unlike other well-known carbonatites, including Sokli and Phalaborwa, phoscorite has not yet been described from the Siilinjärvi carbonatite.

OXIDES AND SULFIDES

Magnetite is the most common accessory mineral of both the carbonatites and the glimmerites, but generally represents less than 1 vol% of the ore. Pyrite, pyrrhotite, and lesser chalcopyrite represent the main sulfide minerals. They mostly constitute a small proportion of the ore, but can locally occur in massive form. Little research has been done on these minerals.

OTHER MINERALS

Relatively rare accessory minerals at Siilinjärvi include barite, strontianite, monazite, pyrochlore, zircon, baddeleyite, and ilmenite. Compositions of a number of these rare minerals from the Siilinjärvi carbonatite were reported by Al-Ani (2013) and are summarized here. Barite occurs as sparse, <50 µm inclusions in calcite, or as intergrowths with strontianite and contains 1–4 wt% SrO. Monazite occurs as <50 µm subhedral inclusions in calcite or apatite, or as more irregular grains along grain boundaries; grains up to 100 µm wide have been located. The grains analyzed so far show enrichments typical for carbonatite monazite with up to 23 wt% La_2O_3 and up to 40 wt% Ce_2O_3, but their relative rarity equates to low whole rock REE content.

Pyrochlore forms euhedral, typically 50–200-µm-wide grains as inclusions mostly in phlogopite. Few analyses exist, but these show typical compositions (77 wt% Nb_2O_5; 15 wt% CaO) with low contents of Ti, Ta, F, and REE. Zircon is rare in carbonatites, mostly because it requires a relatively high silica activity to become saturated in a melt. At Siilinjärvi, it occurs as euhedral grains ranging from tiny 100-µm-wide grains to megacrysts several centimeters in diameter, the latter typically containing large inclusions of baddeleyite. Allanite has been reported in thin section scans, but full microprobe analyses remain to be done.

AGE

There have been quite a number of studies concerning the age of the Siilinjärvi carbonatite complex (Puustinen, 1971; Basu and Puustinen, 1982; Karhu et al., 2001; Bayanova, 2006; Tichomirowa et al., 2006; Zozulya et al., 2007; Rukhlov and Bell, 2010; Tichomirowa et al., 2013). In short, the most precise data appear to be from U-Pb analyses of zircon, particularly a concordant zircon U-Pb of age 2610 ± 4 Ma (2σ; Fig. 4.3.9) measured by Olavi Kuovo (GTK unpublished report, 1984) on a large zircon megacryst. These U-Pb data indicate that Siilinjärvi is one of the oldest carbonatites in the world. However, K-Ar data (Puustinen, 1971) and Rb-Sr isochron data (Tichomirowa et al., 2006) for carbonatite and glimmerite samples with results of 1785–2030 Ma and 1754–2031 Ma, respectively, clearly show a Svecofennian orogenic overprint that is well documented throughout much of the Archean terranes of eastern Finland (Kontinen et al., 1992).

GEOCHEMISTRY AND ISOTOPES

The extreme mineralogic variability within a small volume of glimmerite-carbonatite of the main ore body translates into a large variability in the chemical composition of analyzed samples (Fig. 4.3.10). All ore samples essentially plot on a binary join between glimmerite and carbonatite. However, given the much greater volumes of glimmerite relative to carbonatites, the bulk composition of the Siilinjärvi

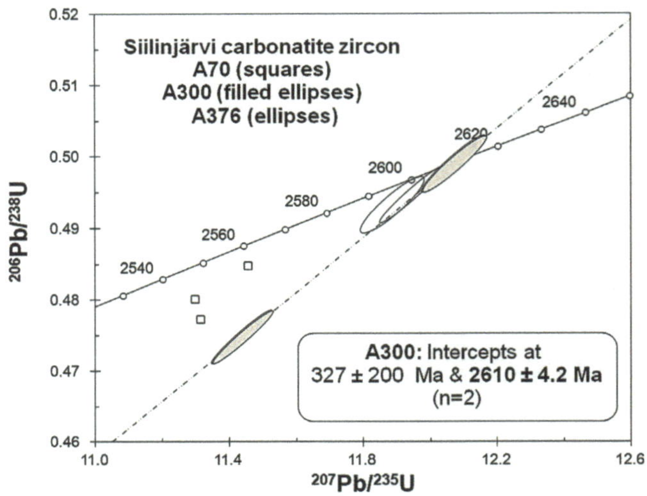

FIGURE 4.3.9 U-Pb concordia diagram for zircons from Siilinjärvi.

Source: *From original GTK report by O. Kouvo to H. Lukkarinen (1984).*

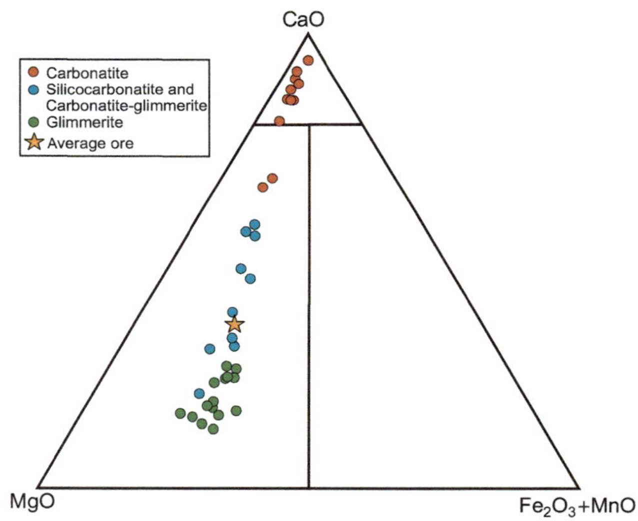

FIGURE 4.3.10 MgO-CaO-(FeO + MnO) diagram for 37 whole-rock analyses of the ore rocks and the average ore composition for comparison.

The two carbonatites with higher MgO contain 10–20 modal% dolomite.

main ore body differs only slightly from average glimmerite (see Table 4.3.1 and Fig. 4.3.10). Thus, even though the carbonatites are the most striking feature at Siilinjärvi, the average ore could be considered as a cumulate rock derived from a potassic melt that contained some carbonate.

One way to test the cumulate hypothesis is to compare the bulk ore composition to dike rocks identified by their mineralogy as potential parental magmas, such as potassic ultramafic dike rock

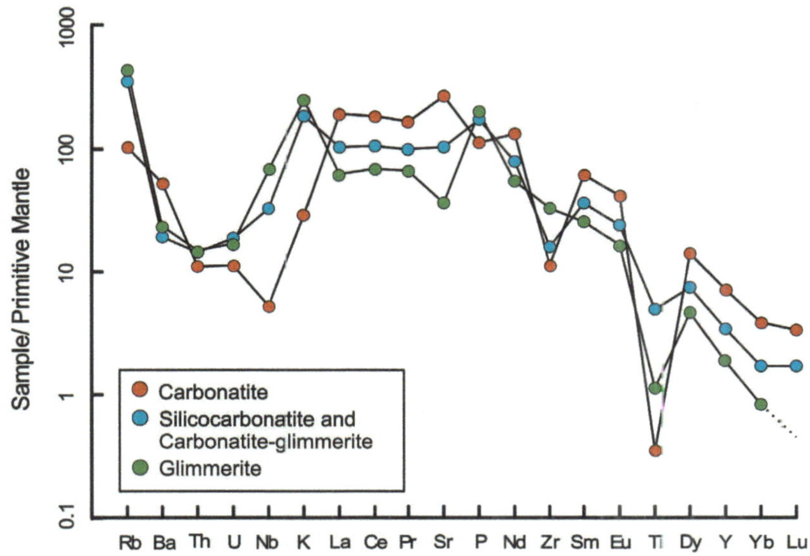

FIGURE 4.3.11 Primitive mantle-normalized diagram showing average compositions of the different ore type rocks.

Source: *Normalizing values from Sun and McDonough (1989).*

14-PTP-05 described in the preceding (Table 4.3.1). In this case, it appears that there is too little P_2O_5 to have produced sufficient apatite and too much FeO, mostly as magnetite, for which there are no known cumulate counterparts in the ore body. Nevertheless, the dike rock is a fine-grained, approximate mineralogic analog to the main intrusion and, thus, theoretically contains all the necessary components to form the main glimmerite-carbonatite cumulate body. These preliminary results warrant further studies including additional sampling of dikes and information such as mineral chemistry and whole-rock chemistry (including isotopic data) to identify any parental magmas in the area.

The mantle-normalized incompatible trace element patterns for the average carbonatite, glimmerite, and intermediate rock are shown in Fig. 4.3.11. Several features can be highlighted:

- A negative Zr anomaly in the carbonatite is not seen in the glimmerite, although this may be a nugget effect as zircon is generally rare.
- The large negative anomaly in Ti can have a number of origins, but likely candidates include Ti-magnetite or titanite fractionation. This may also explain the negative Nb anomaly in the carbonatite.
- The overall REE content of these rocks is not high considering the apatite and calcite content in the complex. Clearly the primary magma was not very REE-enriched.

The isotopic composition of Siilinjärvi has been determined by a number of researchers. Tichomirowa et al. (2006) measured the isotopic composition of amphibole, mica, carbonate, and apatite from a representative suite of Siilinjärvi rock types and reported a primary carbon and oxygen composition of $\delta^{13}C = -3.7$ ‰, $\delta^{18}O = 7.4$ ‰. This composition is very uniform throughout the complex, and it plots within the field of mantle-derived primary igneous carbonatites determined by Taylor et al. (1967) and also overlaps with the data from Sokli (Demeny et al., 2004). The lowest Sr_i = (0.70137) (equivalent to εSr = 0), εNd of −2 to +5, and εHf of −1.4 to +0.4 at 2.61 Ga were measured on Siilinjärvi drill core and hand samples by Tichomirowa et al. (2006, 2013).

Corroboration of the Sm-Nd data comes from a richterite-apatite-phlogopite-whole rock isochron of Zozulya et al. (2007) with an inferred age of T = 2615 ± 57 Ma, providing a well-constrained initial εNd of +0.4 ± 0.2. The isotopic composition of the Siilinjärvi rocks are rather homogenous, and very much in line with the bulk of carbonatites globally, plotting near the bulk earth composition, and fully within the Ocean Island Balsalt (OIB) isotopic compositional field (see Fig. 4.1.3 and the discussion in Subchapter 4.1).

GENESIS OF THE SIILINJÄRVI GLIMMERITE-CARBONATITE

As described for the general case in Subchapter 4.1, the Siilinjärvi glimmerite-carbonatite complex probably represents a plutonic complex formed as the result of passage of highly potassic magmas into and through a magma chamber, and the consequent accumulation of crystallizing minerals, a process that was active over the lifetime of the magma chamber. The importance of studying the potassic dikes was outlined earlier, and calculations of the total carbonate budget need to be made.

The observation of rather uniform apatite and tetraferriphlogopite compositions, which appears to be independent of ore rock type, suggest that the minerals crystallized in a large, well-stirred magma chamber. However, the ore rocks are clearly not uniformly distributed, and the process by which the diffusely laminated glimmerite-carbonatite texture of the central portion of the main cumulate body formed is still an open question. The similarity of orientation of the laminations and the late carbonatite veins along the same vertical to subvertical north–south trends strongly suggests their formation is linked.

SIILINJÄRVI MINE

The Siilinjärvi mine has been active since 1979. Currently the mine comprises two pits; the relatively large southern Särkijärvi pit is shown in Fig. 4.3.12. It is approximately 250 m deep, with a bench height of 28 m. The overall blast rate at the mine is 600,000 t per week (450 kt from the Särkijärvi pit and 150 kt from the Saarinen pit). Almost all of the glimmerite-carbonatite series rocks constitute ore, while the fenites and crosscutting diabases are stockpiled separately. Exceptions include late apatite-poor carbonatite veins and certain blocks of carbonatite-glimmerite with <0.5 wt% P_2O_5 that seemingly have had their apatite removed by some unknown process, perhaps related to metamorphism and fluid flow. Understanding this process and facilitating identification of these low-grade blocks is an important ongoing topic of research.

The Siilinjärvi mine is the only operating phosphorus mine in Western Europe, with the closest competition located in the Kola alkaline province of northwest Russia. The main product is apatite, which is processed by flotation in the concentrator near the Särkijärvi pit (seen at the left side of aerial photograph in Fig. 4.3.12) to phosphoric acid using sulfuric acid derived currently from pyrite of the Pyhäsalmi mine, about 130 km northwest of the Siilinjärvi mine. By-products from the

FIGURE 4.3.12 Aerial photograph of Siilinjärvi mine, southern Särkijärvi pit.

Source: *Yara Suomi Oy.*

Siilinjärvi mine include mica concentrate and calcite concentrate. Since the beginning of full operation, almost 400 Mt of rock have been mined, 260 Mt tons of this being ore. The mine optimization to the year 2035 gives ore reserves as of January 2014 of 234 Mt, and waste rock of 157 Mt. Current production is roughly 11 Mt of ore per year, while the average in situ grade is 4.0 wt% of P_2O_5. A new optimization and reserve update is planned for 2015 following completion of a major Särkijärvi infill drilling campaign.

SUMMARY

1. Siilinjärvi represents the second largest carbonatite complex in Finland, and one of the oldest carbonatites on Earth at 2610 ± 4 Ma.
2. The complex hosts a significant ore deposit of apatite, mined since 1979, and through 2014 had produced 21.4 Mt of apatite concentrate out of ~260 Mt of ore.
3. Mineral compositions, particularly apatite and phlogopite, do not show any systematic compositional variability based on rock type. This is consistent with the glimmerite-carbonatite rocks representing an equilibrium assemblage of cumulate minerals. The Siilinjärvi complex probably represents a deeper level magma chamber than at Sokli; this is depicted for plutonic complexes in Subchapter 4.1, Fig. 4.1.2.
4. Comparing the tetraferriphlogopite at Siilinjärvi and Sokli suggests that the parental magma for the Siilinjärvi glimmerite-carbonatite complex was moderately evolved, and this fact can help identify potential parental magmas.

5. Tetraferriphlogopite was converted to biotite and chlorite in shear zones, and by inference, its dominance in the main ore zone suggests that the laminated, alternating glimmerite-carbonatite texture is a primary magmatic feature. These laminated-textured rocks are cut by numerous carbonatite veins and dikes that represent the final stages of magmatic influx into the Siilinjärvi system.
6. Further work will likely provide a larger variety of ultramafic dikes that may lead to a better understanding of the primary magma(s) of this system. This includes the melasyenite and ultramafic lamprophyre from the northern portion of the intrusion, and the glimmerite dikes with primary carbonate segregations along the railroad cut and in drill core.
7. Fenite formed around the subvolcanic magma system as K and Na-rich fluids, produced through crystallization in the magma chamber, were forced into the surrounding bedrock. Acting over a significant period of time, the process converted country rock gneisses into a variety of fenites dependent on the fluid flux, composition, and distance from the fluid source.

ACKNOWLEDGMENTS

We would like to thank Yara Suomi Oy for the rights to publish Chapter 4.3 and to use the aerial photograph of the mine in this chapter and on the cover of the book. R.E. Harmer and W.D. Maier reviewed the section and made many helpful comments.

REFERENCES

Al-Ani, T., 2013. Mineralogy and Petrography of Siilinjärvi Carbonatite and Glimmerite Rocks, Eastern Finland. Geological Survey of Finland Archive Report 164/2013, 15p.

Basu, A.R., Puustinen, K., 1982. Nd-isotopic study of the Siilinjärvi carbonatite complex, eastern Finland and evidence of early Proterozoic mantle enrichment. Geological Society of America. Abstracts with Programs 14 (7), 440.

Bayanova, T.B., 2006. Baddeleyite: a promising geochronometer for alkaline and basic magmatism. Petrologiya 14, 203–216.

Belousova, E.A., Griffin, W.L., O'Reilly, S.Y., Fisher, N.I., 2002. Apatite as an indicator mineral for mineral exploration: trace-element compositions and their relationship to host rock type. Journal of Geochemical Exploration 76, 45–69.

Demeny, A., Sitnikova, M.A., Karchevsky, P.I., 2004. Stable C and O isotope compositions of carbonatite complexes of the Kola alkaline province: Phoscorite–carbonatite relationships and source compositions. In: Wall, F., Zaitsev, A. (Eds.), Phoscorites and Carbonatites from Mantle to Mine: The Key Example of the Kola Alkaline Province. The Mineralogical Society Series 10, pp. 407–431.

Härmälä, O., 1981. Siilinjärven kaivoksen mineraaleista ja malmin rikastsmineralogisista ominaisuuksista. University of Turku, unpublished Master's thesis. p. 127 (in Finnish).

Hornig-Kjarsgaard, I., 1998. Rare earth elements in sövitic carbonatites and their mineral phases. J. Petrol 39, 2105–2121.

Karhu, J.A., Mänttäri, I., Huhma, H., 2001. Radiometric ages and isotope systematic of some Finnish carbonatites. University Oulu, Res. Terrea, Ser. A, No. 19, 8.

Kontinen, A., Paavola, J., Lukkarinen, H., 1992. K-Ar ages of hornblende and biotite from Late Archaean rocks of eastern Finland—interpretation and discussion of tectonic implications. Bulletin Geological Survey of Finland 365, 31p.

Kouvo, O., 1984. GTK internal report to H. Lukkarinen 4p.

Poutiainen, M., 1995. Fluids in the Siilinjärvi carbonatite complex, eastern Finland: Fluid inclusion evidence for the formation conditions of zircon and apatite. Bulletin of the Geological Society of Finland 67, 3–18.

Puustinen, K., 1971. Geology of the Siilinjärvi carbonatite complex, Eastern Finland. Bulletin of the Geological Society of Finland 249, 1–43.

Puustinen, K., 1972. Richterite and actinolite from the Siilinjärvi carbonatite complex. Finland. Bulletin of the Geological Society of Finland 44, 83–86.

Puustinen, K., 1974. Tetraferriphlogopite from the Siilinjärvi carbonatite complex. Finland. Bulletin of the Geological Society of Finland 44, 35–42.

Sun, S.S., MsDonough, W.F., 1989. Chemical and isotopic systematic of oceanic basalt: implications for mantle composition and processes. In: Saunders, A.D., Norry, M.J. (Eds.), Geological Society of London, Special Publication 42, pp. 313–345.

Taylor, H.P., Frechen, J., Degens, E., 1967. Oxygen and carbon isotope studies of carbonatites from the Laacher See District, West Germany and the Alnö District in Sweden. Geochimica et Cosmochimica Acta 31, 407–430.

Tichomirowa, M., Grosche, G., Götze, J., et al., 2006. The mineral isotope composition of two Precambrian carbonatite complexes from the Kola alkaline province—alteration versus primary magmatic signatures. Lithos 91, 229–249.

Tichomirowa, M., Whitehouse, M.J., Gerdes, A., et al., 2013. Different zircon recrystallization types in carbonatites caused by magma mixing: Evidence from U-Pb dating, trace element and isotope composition (Hf and O) of zircons from two Precambrian carbonatites from Fennoscandia. Chemical Geology 353, 173–198.

Rukhlov, A.S., Bell, K., 2010. Geochronology of carbonatites from the Canadian and Baltic shields, and the Canadian Cordillera: clues to mantle evolution. Mineralogy and Petrology 98, 11–54.

Zozulya, D.R., Bayanova, T.B., Serov, P.N., 2007. Age and isotopic geochemical characteristics of Archaean carbonatites and alkaline rocks of the Baltic shield. Doklady Earth Sciences 415A (6), 874–879.

CHAPTER

KIMBERLITE-HOSTED DIAMONDS IN FINLAND

4.4

H. O'Brien

ABSTRACT

The levels of brilliance (brightness and contrast), fire (flashes of rainbow color), and scintillation (intense sparkles when moved) of diamonds are unmatched by any other gemstone. Also diamonds of gem size and quality are relatively rare. As a result, gem diamonds are extremely valuable, yet the supply of diamonds is ultimately limited. This reality has pushed diamond exploration and mining into extreme environments, from the far Arctic North to the deserts of southern Africa and onto the ocean bed off the coast of Namibia. About two-thirds of the annual production of diamonds by weight comes from ancient volcanoes that consist of the rock types kimberlite, orangeite, or lamproite. Tracking down the remnants of these small volcanoes requires sophisticated and efficient collection and processing of samples for kimberlite indicator minerals (i.e., peridotite constituent minerals) and evaluation of enormous amounts of mineral data to constrain the diamond prospectivity of a region, cluster of pipes, or particular diatreme. The exploration sampling stage is usually followed by aero- or ground-geophysical measurements, target evaluation, and, finally, drill testing. Diamond exploration is expensive, but the rewards can be great.

Diamond exploration in Finland started in 1985, and has been continuous, albeit with varying levels of activity, since that time. As a result, diamondiferous rocks have been found in three regions—namely, the Kuhmo-Lentiira area hosting a group of 1200 Ma orangeites, the Kuusamo-Hossa area containing several 760 Ma kimberlites, and the Kaavi-Kuopio area with a cluster of ~600 Ma kimberlites. Driven by the needs of these exploration activities, our understanding of the makeup of the Karelian craton, and our understanding of the magmas that have transported diamonds to the surface in this part of the world have benefitted enormously.

Keywords: kimberlite; orangeite; lamproite; diamond; diatreme.

INTRODUCTION

Aside from microdiamonds found in ultrahigh-pressure terranes (e.g., Massonne, 2003) or those associated with astroblemes (e.g., Masaitis, 2013), diamonds do not form in the Earth's crust. Gem-sized diamonds only crystallize at depths of around 140–150 km or greater (depending on the geotherm), where the mantle has been stable over billions of years and there is the right balance of heat, pressure, and low oxygen conditions for carbon to exist as diamond. Complex zoning in diamonds attests to the fluid processes involved as diamond grows in the mantle (Kaminsky and Khachatryan, 2004) and in rare cases encapsulates tiny grains of Mg-rich pyrope, chromite, and sulfide that record the particular types of peridotite and eclogite that host diamond in the mantle (Stachel and Harris, 2008). However, no peridotite or

Mineral Deposits of Finland. http://dx.doi.org/10.1016/B978-0-12-410438-9.00014-5

eclogite massif that has been moved tectonically to the Earth's surface has been found to contain diamond, although rarely, diamond pseudomorphs are known from peridotites (e.g., Beni Bousera; Pearson et al., 1989). Instead, diamonds have been transported to the Earth's surface by rapidly moving magma.

There are only a few magma types that are derived from sufficient depth to pass through mantle where diamond is stable, and that ascend rapidly enough to the surface so that the diamond xenocrysts they entrained are not fully oxidized to graphite on the way up. These magmas are all alkaline in composition, and comprise kimberlites, orangeites, lamproites, and ultramafic lamprophyres (definitions follow). Nearly all hard-rock diamonds mined are found in kimberlites and orangeites, whereas there is currently only one operating mine in lamproite (Argyle in northwestern Australia) and none in ultramafic lamprophyre. Thus, upward of two-thirds of the annual world production of 130 million carats is sourced from kimberlite and orangeites.

This subchapter provides a brief discussion of diamond-bearing rocks in general, and then focuses on diamond exploration and deposits in Finland. Appendix A provides further information about exploration results in the Kaavi-Kuopio area, as well as further data on the Lahtojoki kimberlite, the highest grade kimberlite in Finland.

AGE AND OCCURRENCE

Diamond-bearing igneous bodies have been emplaced from the Holocene (e.g., Igwisi Hills kimberlite volcanoes in Tanzania; Brown et al., 2012) to as far back as the Proterozoic (the Kuruman kimberlites in South Africa erupted ~1.8 Ga; Donnelly et al., 2011). Although examples of diamond mines based on Proterozoic kimberlite (e.g., the Cullinan mine on the 1.2 Ga Premier pipe, South Africa) and olivine lamproite (e.g., the Argyle mine on the 1.2 Ga Argyle pipe, Western Australia) exist, the bulk of the larger diamond mines are built on pipes that are Paleozoic or younger, probably due to the poor preservation of older diatremes (Brown and Valentine, 2013).

Current age data for kimberlite magmatism show that they have been concentrated in "pulses" within the Earth's history, with particularly important intrusion events at around 1100–1200 Ma, 550–650 Ma, 360–380 Ma, 120 Ma, 90 Ma, and 60 Ma. These pulses have been related to periods of supercontinent formation or breakup, while periods with relatively few kimberlites correlate with stable supercontinents (Jelsma et al., 2009). As an example, the Devonian kimberlite event coincides with continent breakup in the Kola and Siberian craton, but a stable Gondwana at that time was almost unaffected by kimberlite magmatism, as described in Jelsma et al.

Kimberlite-related rocks have not been found in oceanic settings, and are not as common in continental areas where the crustal basement rocks are younger than Precambrian. However, this is likely due to the fact that most kimberlites are discovered by exploration companies that focus on cratonic areas understood to be most prospective for diamonds. There are probably more kimberlites in younger terranes, but they are less likely to be found.

Sm-Nd isotope studies of inclusions of Mg-pyrope and diopside (e.g., Richardson, 1986) and Re-Os on sulfide inclusions (e.g., Pearson et al., 1998) have proven that diamonds are generally very old. Some date back to 3.8 Ga, and the vast majority, although not all, are much older than the rocks that contain them. Gurney et al. (2010) reviewed the age information on diamond inclusions (diamonds themselves cannot be dated as they are composed mainly of carbon, and some nitrogen and other impurities) and concluded that there are two main age groups for peridotite-hosted diamonds: 3520–3200 Ma and 2030–1900 Ma. The range of ages for eclogite diamonds is 2900–580 Ma. These ages are older than nearly all of the diamondiferous kimberlites and orangeites in existence, and thus reinforce the fact that diamonds are xenocrysts in these magmas.

DEFINITIONS

Kimberlites are formed from ultramafic, volatile-charged, incompatible element-rich magmas that represent a mixture of small degree partial melts from the mantle, mantle peridotite, and eclogite detritus carried from depth, and typically contain minerals of the megacryst suite such as Ti-pyrope, Mg-ilmenite, and subcalcic clinopyroxene. There are two end-member rocks that have been called kimberlite in the past: Group I has abundant large, rounded grains (macrocrysts) of olivine, in a matrix of subhedral to euhedral olivine, monticellite, perovskite, spinel, mica, calcite, and serpentine, whereas Group II typically has abundant phlogopite ± olivine in a matrix of phlogopite, K-richterite, and other diagnostic minerals.

Due to the lack of evidence for a genetic relationship between these two kimberlite groups, Mitchell (1995) reinstated the name *orangeites,* originally coined by Wagner (1928), for the Group II kimberlites. Mineralogically, the orangeites are more akin to olivine lamproites (except for the lack of carbonate in the latter), and it is now generally agreed that orangeites are the southern African equivalents of lamproites, representing metasomatized lithospheric mantle melts (Mitchell, 2006). The Karelian craton also contains rocks that can be called orangeites, and these are discussed in a later section. The fourth rock type that can contain sufficient diamond to constitute an ore deposit is ultramafic lamprophyre (UML), namely the varieties aillikite and alnöite. However, as mentioned earlier, no lamprophyre is currently mined. The mineralogical differences, noted in the following, between these rock types are summarized in Table 4.4.1:

- Kimberlites and orangeites carry more mantle-derived material than lamproyes and lamprophyres, including xenoliths, xenocrysts, and olivine crystals.
- Phlogopite occurs in all of these rocks, consistent with their alkaline, fluid-rich compositions, but the composition of phlogopite varies significantly between the magma types.
- The CO_2-to-water ratio of the magmas decreases from kimberlite to orangeite to lamproite, with ultramafic lamprophyres showing a wide range of CO_2/H_2O ratios.
- Evolved orangeites and lamproites are mineralogically very similar.

FORM

Kimberlites form dikes, sills, diatremes, and, exclusively in young occurrences, lavas and tuffs (e.g., Igwisi Hills; Brown et al., 2012). The most important kimberlites in terms of volume are those that form diatremes as part of explosive maar-diatreme volcanoes. A diatreme is the carrot-to-cylindrical shaped body that extends downward from the crater (Fig. 4.4.1). It is important to point out that diatremes form from a wide variety of magma types, ranging in composition from alkali basalts to nephelinites to carbonatites to kimberlites. When referring to kimberlites, diatremes are often called "pipes." The exact mechanism by which kimberlite diatremes form is still open to much debate.

Table 4.4.1

		Kimberlites	Orangeites	Lamproites	Ultramafic Lamprophyres
Mantle	*Xenoliths*	common	common	rare	rare
	Xenocrysts	common	common	rare	rare
Olivine	*Macrocrysts*	common	common	rare	rare
	Xenocrysts	common	common	common	common
Mica	*Macrocrysts*	common	common, phlogopite	common, phlogopite to	common
	Phenocrysts	phlogopite		Ti-phlogopite	phlogopite
	Groundmass	common, phlogopite kinoshitalite	common, tetraferriphlogopite	common, Ti-tetraferriphlogopite	common, Al-biotite
Spinels		abundant, Mg-chromite to Mg-ulvöspinel	rare, Mg-chromite to Ti-magnetite	rare, Mg-chromite to Ti-magnetite	common, Mg-chromite to Ti-magnetite
Monticellite		common	--------------	--------------	common
Diopside		--------------	common, Al- + Ti-poor	common, Al- + Ti-poor	common, Al- + Ti-rich
Perovskite		common, Sr- + REE-poor	rare, Sr- + REE-rich	rare, Sr- + REE-rich	common, Sr- + REE-poor
Apatite		common, Sr- + REE-poor	abundant, Sr- + REE-rich	common, Sr- + REE-rich	common, Sr- + REE-poor
Primary Serpentine		abundant	common	--------------	--------------
Calcite		abundant	common	--------------	common
Sanidine		--------------	rare groundmass	common, phenocrysts + groundmass	--------------
K-richterite		--------------	rare groundmass	common, phenocrysts + groundmass	--------------
K-Ba-titanates		very rare	common	common	--------------
Zr-silicates		very rare	common	common	--------------
Mn-ilmenite		rare	common	very rare	rare
Leucite		--------------	rare pseudomorphs	common, phenocrysts	--------------

-------------- = absent | critical characteristic | important characteristic | characteristic of evolved member

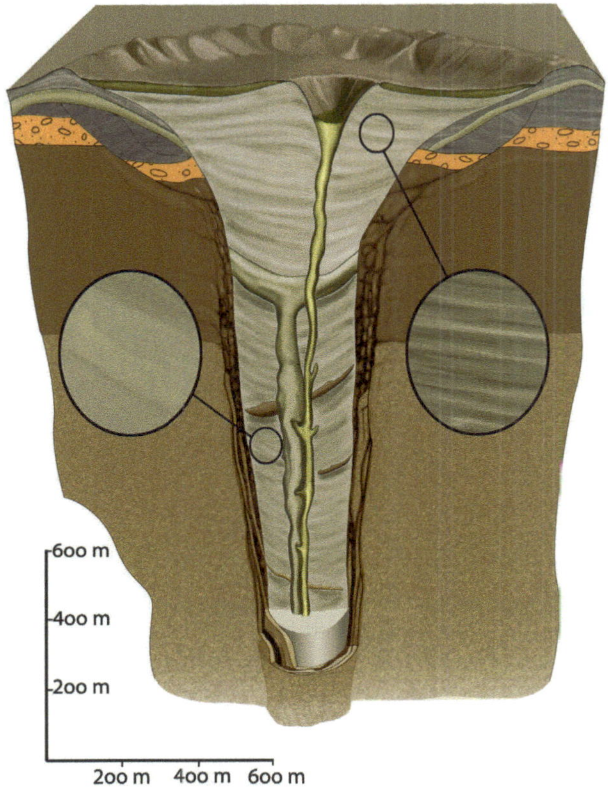

FIGURE 4.4.1 Cross section of a kimberlite diatreme based on the Missouri Breaks diatremes, Montana.

Source: *After Hearn (1968).*

In question is the role of groundwater in kimberlite diatreme formation, with some authors favoring a phreatomagmatic model (Lorenz, 1975; Lorenz and Kurszlaukis, 2007) also used to explain the formation of maar-diatremes of more common, roughly basaltic compositions. Others prefer a purely magmatic gas model wherein intense gas exsolution from uniquely volatile-rich kimberlite magma provides the explosive power to disrupt surrounding country rocks and form the diatreme (e.g., Gernon et al., 2009, 2012). The fragmental rock that fills most kimberlite diatremes in most of the pre-2000 literature was termed *tuffisitic kimberlite breccia* (TKB), but the current term is *volcaniclastic kimberlite* (Scott-Smith et al., 2013).

This fragmental rock represents a mixture of kimberlite and pulverized country rock produced as the diatreme grows downward (Fig. 4.4.1). In well-preserved diatremes, including kimberlite and other magma types, it is common to find the upper diatreme fill showing relatively fine-scale layering, whereas deeper into the diatreme the infill takes on a more massive texture, with only vague large-scale layering developed (see Fig. 4.4.1 and White and Ross, 2013). This, and many other similarities among diatremes of diverse magma types, argues for similarity of processes in their formation.

Most of the diamond mines globally exploit diatremes, as they represent large volumes of diamondiferous rock. However, because the volcaniclastic kimberlite diatreme fill is considerably diluted in mantle-derived material compared to the kimberlite feeder dike magmas, the latter can produce the richer diamond ore. For example, Petra Diamonds' Helam dike has a diamond grade of 2.7 ct/t but only 1.5 Mt of reserves. Nevertheless, such small deposits are economically challenging relative to large diamondiferous pipes. For example, the Orapa mine has a grade of 0.587 ct/t but 146.1 Mt of ore reserves as of December 2012 (DeBeers Annual Report, 2013).

MANTLE ASSIMILATION

As previously described, diamond-bearing alkaline rocks form from volatile-rich potassic magmas that in most cases have incorporated variable amounts of lithospheric mantle-derived peridotite (lherzolite, harzburgite, and dunite) and, in many localities, eclogite. The key to these magmas containing diamond is that as they ascend through the mantle, they fracture and break off meter-sized pieces of the mantle conduit (Fig. 4.4.2). This material disaggregates as the magma ascends, subsequently becoming incorporated and forming a main component of the kimberlite magma.

In some kimberlites, the xenolith material may represent 70% of the magma (Patterson et al., 2009). As a consequence, most of the olivine in kimberlite is xenocrystic and the degree of mantle assimilation is critical in rendering a kimberlite economic because diamond occurs preferentially in harzburgitic peridotite and Mg-poor eclogites. Thus, both mineralogy and lithogeochemistry can provide a measure of the amount of mantle assimilation, which is generally positively correlated with a higher potential for diamond content.

Kimberlite indicator minerals (KIMs) are peridotite and eclogite constituent minerals that are contained in surficial sediments. In exploration samples, the presence of KIMs indicates that mantle-containing

FIGURE 4.4.2 Mantle-derived lherzolite xenolith about 0.75 m in its longest dimension from the Liqhobong mine in Lesotho, southern Africa.

material has been exposed at the surface, and that by erosion of this source material, these grains have been liberated, transported, and deposited by surficial processes. The source is located in the opposite direction of the flow of the transporting medium (e.g., glacial ice or flowing streams). The KIMs comprise mostly olivine, pyrope garnet, chromian diopside, Mg-rich ilmenite (picroilmenite), and chromite; in other words, the constituent minerals of peridotites and eclogites that have been disaggregated and partially assimilated during ascent of the kimberlite magma to the surface.

These minerals have compositions specific to mantle and hence form pathfinders to hidden kimberlite targets. Moreover, the compositions of the KIMs provide substantial information about the type of mantle that was sampled, and allow a rough estimation of the diamond prospectivity of a given region. Several compositional plots for indicator minerals for the Lahtojoki kimberlite (discussed in a later section) are provided as examples in Appendix A.

Chemical compositions of well studied hypabyssal diamondiferous kimberlites, orangeites, and lamproites are plotted on a SiO_2 versus FeO diagram in Fig. 4.4.3. The plots show the geochemical consequences of significant mantle assimilation on kimberlite magma compositions, and they can be used to screen new kimberlite discoveries with regard to their diamond prospectivity. (Note that volcaniclastic diatreme infill as described earlier cannot be plotted due to its large crustal component; see O'Brien and Tyni, 1999, for Finnish examples). Specifically, as the kimberlite whole-rock compositions converge on those of harzburgite—as a direct consequence of having assimilated greater amounts of this material—the probability of containing economic quantities of diamonds increases.

Although there have been many estimates of the so-called protokimberlite melt composition, the composition proposed by Le Roex et al. (2003) has been used here to demonstrate the degree of shift

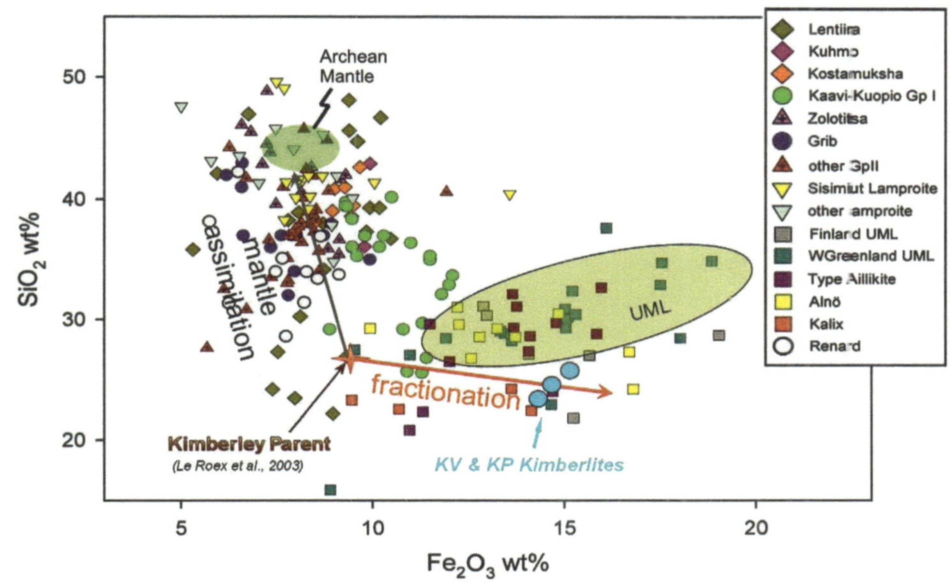

FIGURE 4.4.3 Plot of SiO_2 versus FeO for selected Finnish kimberlites and orangeites.

Shown are a number of well-studied kimberlites, orangeites, and lamproites from around the world and Finland. UML = ultramafic lamprophyre; KV and KP refer to Kuusamo kimberlites.

toward SiO_2-rich compositions that results from significant peridotite assimilation. The general paucity of orthopyroxene xenocrysts in any of the diamond-carrier rocks, and its abundance in peridotites, strongly suggests preferential resorption of this mineral during magma ascent. This model is consistent with the SiO_2-enrichment trend shown in Fig. 4.4.3. It should also be noted that the field that contains the bulk of ultramafic lamprophyre compositions lies at distinctly more Fe-rich compositions than the main cluster of diamondiferous rocks.

DIAMOND EXPLORATION IN FINLAND

As a direct result of the high level of diamond exploration activity in Finland, diamond-bearing rocks have been identified in three different areas of the country: Kaavi-Kuopio, Kuhmo-Lentiira, and Kuusamo-Hossa (see Subchapter 4.1, Fig. 4.1.4; see also locations of individual kimberlites on the GTK map server at http://gtkdata.gtk.fi/mdae/index.html). Each of these regions is described in subsequent sections following a brief review of the diamond exploration activities in Finland that led to the kimberlite discoveries and, consequently, our studies of them.

The first kimberlite in Finland was discovered in 1964 by Malmikaivos Oy, a private prospecting company based in Luikonlahti in eastern Finland. The 1-ha pipe was discovered during a regional ground magnetic survey while prospecting for base metals in the vicinity of Kaavi. Overburden was only 2 m thick and the kimberlite was exposed by trenching and drilling. Since there was no potential for copper ore, the discovery of this pipe was not considered further. However, in the late 1970s, during further base metal prospecting in the area, glacial boulders of well-preserved "almond rocks" were discovered. Samples sent to several diamond companies were identified as kimberlite and proven to contain microdiamonds. By tracing boulders in the up-ice direction, Malmikaivos Oy located a second kimberlite in 1984 under a small swamp and a third ("no. 3") in 1985 under a small lake, only 500 m away from no. 2.

In 1985, Malmikaivos Oy decided to find a partner with diamond exploration expertise, and signed an agreement with Ashton Mining Ltd. Malmikaivos exploration teams subsequently managed to find all but one of the kimberlites now known in the Kaavi-Kuopio area. Till sampling and heavy mineral separation were also used extensively by Dia Met Minerals Ltd. during its four years of exploration for diamonds in Finland from 1996–1999.

Rio Tinto Ltd. (which acquired Ashton Mining in 2000) and Dia Met Minerals relied more heavily on geophysical methods for locating kimberlites including magnetic, electromagnetic, gravimetric, and seismic methods. The success of these various methods was inconsistent and depended on the composition of the pipes (i.e., the pipes did not give equivalent responses to a given geophysical method). For example, in a 1996 helicopter aerogeophysical survey flown for Dia Met, only half of the known bodies were able to be located (Matti Tyni, personal communication, 2012). On the other hand, numerous nonkimberlite anomalies with geophysical responses similar to certain kimberlites (e.g., thick lake-bottom sediments) were found and drill tested. Information on the extent of activities by DeBeers (as Finnsearch Oy) is limited, but it appears the company worked mostly in areas of Finland other than Kaavi-Kuopio.

Near the end of operations by Dia Met and Malmikaivos, the Geological Survey of Finland (GTK) developed new tools for processing till samples for kimberlite indicator minerals (Chernet et al., 1999; Lehtonen et al, 2005). To test this new capability, GTK processed till samples from several sampling lines in the Kuhmo-Lentiira area of eastern Finland, with sufficient positive results that an invitation for tenders was published, and was won by American Mineral Fields. During this same period, Conroy Diamonds and Gold PLC (transferred to its subsidiary Karelian Diamonds Ltd.) was acquiring ground and conducted a successful sampling program in eastern Finland. North Star Diamonds AS, on the other hand, took up activities in the Kaavi-Kuopio area vacated by Ashton/Malmikaivos and Dia Met. The company spent considerable time and effort drilling earlier discoveries but ultimately turned its attention to areas near Ekati Mine in Canada, and left Finland in 2006.

In 2000, European Diamonds was incorporated to explore for diamonds in Finland, and the first target was the area canvassed earlier by American Mineral Fields. Work proceeded for several years in the area, resulting in the discovery of several orangeite dikes and diamonds in till, but the ultimate sources of the main indicator fans were not located. The company also made discoveries in the Kuusamo area, and spent several years (2003–2006) removing overburden, drilling, and sampling for microdiamonds from the Lahtojoki kimberlite in the Kaavi-Kuopio area. It extracted a 2000-ton bulk sample from this pipe and had it processed at GTK Outokumpu for diamonds. Work on this project and others in Finland were, however, rapidly ramped down in 2006 when company resources were directed instead toward developing the Liqhobong mine in Lesotho, southern Africa. Also during the period 2004–2006, Gondwana Investments drilled targets in the Kaavi-Kuopio area and did reconnaissance sampling in other Archean parts of Finland, with some very interesting indicator mineral results that have not yet been followed up.

Mantle Diamonds Ltd. joint-ventured with European Diamonds beginning in 2008 to further the Lahtojoki project. The company moved several million m^3 of overburden and built pads for auger drill sampling of the pipe, but by then a downturn in exploration investment had started and by 2010, further work on this project was halted.

In early 2005, Sunrise Diamonds PLC discovered seven kimberlite bodies in four different localities within a 20-km radius south of Kuusamo, thereby establishing the third kimberlite province in Finland (see later section). The company subsequently broadened its Finnish diamond exploration holdings by being part of the third wave to work on the kimberlites in the Kaavi-Kuopio area (except Lahtojoki) after it agreed on a joint venture with Nordic Diamonds Ltd. (formerly North Star Diamonds) and gained full control of the claims in early 2009. Work included further drilling on several kimberlites, especially no. 17, mostly for microdiamond sampling and additional till sampling in the area. Sunrise Diamonds had several 200–300 kg samples of Kaavi-Kuopio kimberlites processed by Mineral Services in South Africa for microdiamonds and found several macrodiamonds in the process (Table 4.4.2).

DIAMOND REGIONS OF FINLAND

KAAVI-KUOPIO GROUP I KIMBERLITE PROVINCE

The Kaavi-Kuopio kimberlites are found in two fields, located only ~50 and ~30 km inboard from the southwestern margin of the craton, respectively (refer to Fig. 4.1.4 in Subchapter 4.1). The 20 occurrences known are typical Group I kimberlites, with the archetypal suite of mantle-derived peridotite and eclogite xenoliths; disaggregated xenoliths (xenocrysts); and the megacryst suite minerals: Ti-pyrope, Mg-ilmenite, and Cr-diopside. They also commonly contain xenoliths of lower, mid, and upper crustal rocks through which the magmas have traversed (Peltonen et al., 2006).

These kimberlites have abundant large, rounded grains (macrocrysts) of olivine, in a matrix of euhedral olivine, monticellite, perovskite, magnesian ulvöspinel-magnetite, Ba-rich phlogopite-kinoshitalite mica, apatite, calcite, and serpentine. In addition to having typical kimberlite mineralogies, the

Table 4.4.2 Diamond results for Kaavi-Kuopio kimberlites

Claim report	No.	Claim	Latitude			Longitude			Size	micro diamond smp	micro diamond	Macro diamond smp	Result
			deg	min	sec	deg	min	sec	ha	kg	count	tons	ct/100t
No report	1	Koskeniemi	62	54	23.6	28	37	28.7	1.1	107	0[1]	---	
5287/1	2	Niilosuo	63	5	29.8	28	42	13.0	1	172	72	162.8	0.23
5287/1	3	Niilonlampi	63	5	40.9	28	42	3.2	1	---		2.2	0.91
4251/1	4	Karetinsaaret	62	47	49.9	28	44	0.4	1.5	27	18	8.2	0.21
4285/1	5	Kärenpää	63	2	12.5	28	40	1.6	2.2	122	20	185.9	0.6-7.0
4528/1	6	Teeripuro 1	63	0	40.7	28	38	32.7	1	48	18	13.4	6.6
4557/1	7	Lahtojoki	63	1	50.3	28	34	39.2	1.8	59	20	23.3[2]	30.6*
No report	8	Aviolampi	63	3	38.4	28	32	55.5	sill	---		---	
No report	9	Kalajärvi	62	58	10.0	28	31	29.6	0.5	25	0	---	
4937/1	10	Ryönä	62	57	17.8	28	0	51.7	2.2	26	5	13.3	1.8
SD	10									202.76	10		3 macros[3]
5873/1	11	Munakka	62	57	49.1	28	1	41.8	0.3	25	1	---	
4886/1	12	Kotkatniemi	62	56	41.6	27	50	1.6	1.6	161.5	70	9.4	17.11[4]
4988/1	13	Säyneenjärvi	62	58	30.9	27	47	54.1	0.9	48	3	3	14.33
5584/1	14	Kaatronlampi [5]	63	4	2.9	28	25	9.4	0.3	73	1	---	
5175/1	17	Kylmälahti	62	56	44.8	27	54	41.8	2.0	98.7	8	---	
SD	17									134.96	20		5 macros[6]
5584/1	20	Ala-Vehkalahti	63	4	9.4	28	24	39.8	0.3	21	0	---	
5495/1	21	Lapinluhta	62	57	41.8	27	48	59.5	1.6	109	37	16.6	26.65[7]
5495/2	22	Uuhilahti	62	57	28.4	27	47	28.3	0.2	60	22[8]	---	
6058/1	23	Jokiharju	62	43	2.2	27	36	30.4	0.25	154.5	positive	---	
6343/1	25	Viitasalo	63	0	8.3	28	5	56.1	0.2	63.1	5	---	

Notes
SD = Sunrise Diamonds press release
[1]Sunrise Diamonds also reported no microdiamonds
[2]also 1000 ton bulk sample giving 5.68 ct/100 t
[3]largest macrodiamond 1.7x0.93x1.57 mm
[4]5 Diamonds over 0.4 mm. Macrodiamond content range 12.50-26.06 ct/100t
[5]Contains abundant fresh olivine
[6]largest macrodiamond 1.02x0.75x0.87 mm
[7]4 Diamonds over 0.4 mm. One 1.126 ct diamond Excluding these, diamond content is 19 ct/100t.
[8]3 Diamonds over 0.4 mm.
**see Appendix A for details*

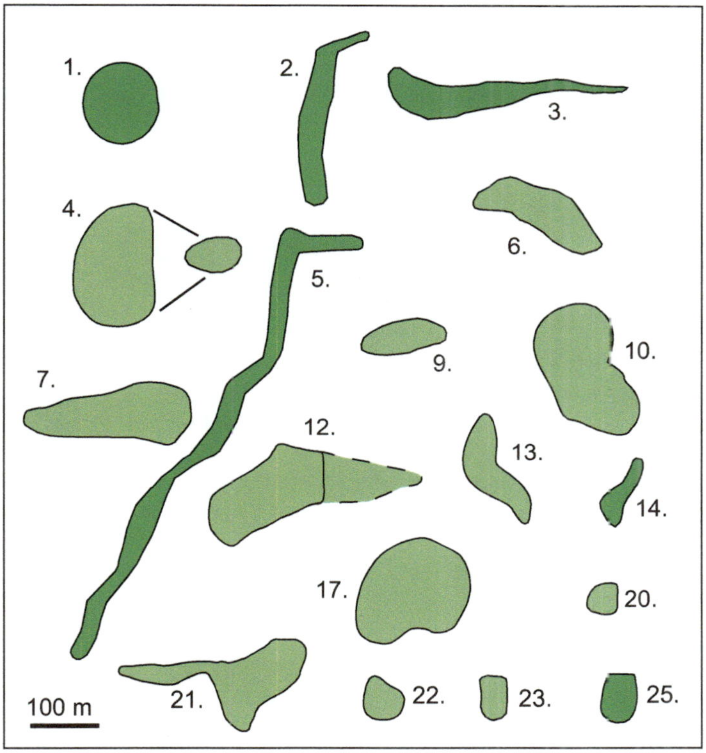

FIGURE 4.4.4 Comparison of kimberlites from the Kaavi-Kuopio area.

Dark green and light green indicate hypabyssal and volcaniclastic kimberlite, respectively. No. 11 is missing because it was not drill-delineated.

Kaavi-Kuopio pipes also have major and trace element and Sr, Nd, and Pb isotopic compositions similar to other Group I kimberlites globally (O'Brien and Tyni, 1999). Age determinations by ion microprobe analyses of U-Pb in perovskites have been made on four of these kimberlites with measured ages ranging from 589–626 Ma (O'Brien et al., 2005), but additional radiometric age dating is ongoing.

The Kaavi-Kuopio kimberlites range from hypabyssal, dike-like bodies (e.g., no. 5), to sheets (e.g., no. 25) to volcaniclastic kimberlite and volcaniclastic kimberlite breccias formed in steep-sided funnel- or carrot-shaped diatremes, as described earlier. Outlines of 19 of the 20 kimberlites from this area are shown in Fig. 4.4.4 (Pipe no. 11 is not included as it was never drill delineated, but its size is estimated at 0.3 ha). As in many kimberlite provinces hosted by crystalline basement, the kimberlites are rather small, ranging in size from narrow dikes <1 m wide to elongate bodies 500 m × 30 m in size to nearly circular diatremes up to 4 ha in size. None of the Kaavi-Kuopio pipes appear to have existing crater-facies materials due to removal of the upper portions of the diatremes by uplift and erosion.

The hypabyssal facies rocks (dark green in Fig. 4.4.4) are hard and compact, with dark gray to black matrices enclosing coarser minerals, particularly olivine, and crustal and mantle xenoliths. Pipe 1 (Fig. 4.4.5(A)) may represent the deepest exposure of root zone material because of its well-developed

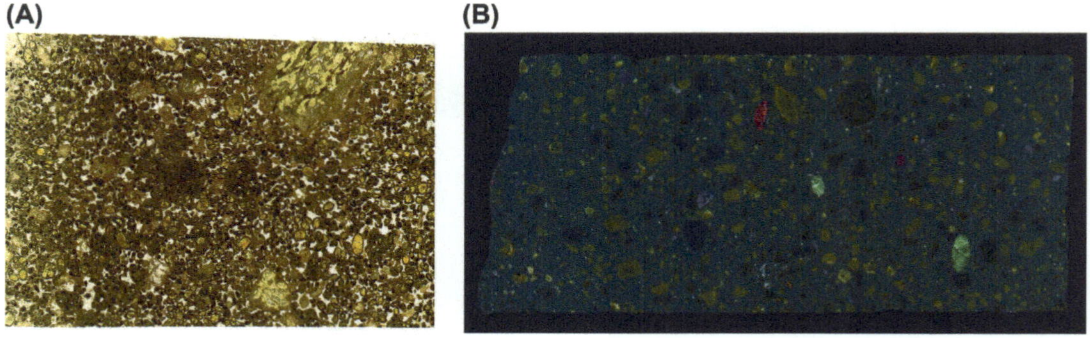

FIGURE 4.4.5 Hypabyssal kimberlites from Kaavi-Kuopio.

(A) Scan of a thin section of the margin of Koskeniemi kimberlite (Pipe no. 1) showing altered olivine xenocrysts and phenocrysts and tiny magmaclasts containing vesicles filled with carbonate, cemented together by carbonate and serpentine matrix. Vertical dimension in image is 3 cm. (B) Hypabyssal kimberlite containing abundant indicator minerals (Malmikaivos Oy, prospect no. 10; Ryönä). The indicator minerals are chrome-rich red pyrope garnet, green diopside, and steel gray magnesian ilmenite. In addition, the sample contains two generations of altered olivine grains. Mantle xenocryst compositions and the existence of diamonds demonstrate that some sampling of the mantle occurred from within the diamond stability field. Core is 3 cm wide.

segregationary texture, which in some samples, particularly near the edge of the intrusion, develops into globular segregations in which late crystallizing serpentine and calcite form irregular pools in a more uniform silicate matrix. A more typical hypabyssal kimberlite from Kaavi-Kuopio is shown in Fig. 4.4.5(B). It displays the classic suite of lithospheric mantle-derived xenocrysts:

1. Mg-ilmenite, which shows complex zoning and resorption features suggesting extensive magma mixing (see O'Brien and Tyni, 1999).
2. Pyrope garnet derived from a range of sources including high Cr, Ca-depleted harzburgite, Ca-saturated lherzolite, Ca-rich wehrlite, Ti-rich megacryst-compositions, and, at lower MgO and higher CaO contents, orange garnets derived from mantle eclogite (Lehtonen et al., 2004).
3. Clinopyroxene comprising lherzolitic, low-Cr megacrystic, and eclogitic subgroups.
4. Spinels from upper mantle spinel lherzolites and rare chromites plotting within the diamond inclusion field.

Quantities and volume ratios of mantle-derived xenocrysts vary considerably among the bodies but, in general, the compositions of these minerals do not show large intrapipe variations. For further details of the Kaavi-Kuopio hypabyssal kimberlite mineralogy, the reader is referred to O'Brien and Tyni (1999) and O'Brien et al. (2005).

The diatreme facies rocks (light green in Fig. 4.4.4) are softer and span the color spectrum from green to gray to brown to dark red. Diagnostic textures of the diatremes facies rock types include rounded pelletal lapilli, in which kernels of small crustal xenoliths or mineral grains, particularly olivine (pseudomorphs), are surrounded by a rim of magma (crystals and melt), later altered to serpentine- and carbonate-rich material. Altered juvenile pyroclasts are also common. The content of crustal material incorporated into the diatreme during formation is large, raising silica contents from original levels of ~30 wt% to ~44 wt% $SiO2$ or more (O'Brien and Tyni, 1999).

Diamond grade results for the 20 known Kaavi-Kuopio kimberlites are summarized earlier in Table 4.4.2. A short discussion for each kimberlite from the area has been extracted from the GTK Research report (Korkeakoski et al., 2004) and compiled in Appendix A. All of the macrodiamond results in the table are based on processing of the kimberlite in the Malmikaivos Oy dense media plant set up in the nearby town of Luikonlahti, except for a few additional data reported by Sunrise Diamonds. Most of the grades are relatively modest, but the value of the diamonds is critical in calculating the value per ton, and none of the kimberlites were sampled adequately to get a representative diamond sample for valuation.

The Lahtojoki Kimberlite (Pipe no. 7) is the best studied of all the kimberlites in the Kaavi-Kuopio area, mostly because it has the most promising diamond content (Table 4.4.2). It is located in the Kaavi district, eastern Finland (63° 1' 50.3" north latitude, 28° 34' 39.2" east longitude). The pipe was discovered by Malmikaivos Oy by sampling glaciogenic deposits, which in the Kaavi area consist of an ablation till overlying predominantly a single basal till layer, together forming a till cover rarely more than 3 m in thickness. Basal till samples were processed for kimberlite indicator minerals using the Luikonlahti dense media separation plant and positive samples were followed up the main ice flow direction of 335° until KIM counts dropped drastically (Matti Tyni, personal communication, 2012). At the head of the train, in what turned out to be a deep pocket in the till, a 500 m × 500 m magnetic survey identified a target that was drill tested in October 1989 and proved positive for kimberlite.

A more recent mineralogical study of the glaciogenic deposits in the vicinity of the Lahtojoki pipe designed to investigate in detail the indicator mineral fan in the down-ice direction from Lahtojoki (Lehtonen et al., 2005) concluded the following about the kimberlite indicator minerals in the till:

- Picroilmenite, garnet, and Cr-diopside are the main indicator minerals; chromite is virtually absent.
- Picroilmenite grains are usually covered by leucoxene alteration products and polygranular recrystallization is also a common morphological feature.
- Pyrope garnets are divided into two groups based on color and composition: (a) purple, red, and lilac lherzolitic and harzburgitic Cr-rich pyropes and (b) orange megacryst-composition Ti-pyropes and eclogitic garnets.
- Approximately 20% of the 0.25–2.0 mm pyrope grains are covered by kelyphitic rims, which get more common as the grain size increases and are more common on lherzolitic garnet varieties. Intense alteration and orange-peel resorption surfaces are also abundant in garnets separated from the Lahtojoki kimberlite.
- Cr-diopside grains are mostly angular and corroded.

The Lahtojoki diatreme is suboval in plan, measuring approximately 200 m (east–west) × 100 m (north–south) and is approximately 2 ha in area at the surface. Drill information suggests that at the −100 m contour, the pipe dips 80° toward the south, and is still nearly 2 ha in cross section. A representative sample of the volcaniclastic kimberlite breccia excavated as part of the Malmikaivos/Ashton bulk sample is shown in Fig. 4.4.6. From this image it is clear that at least parts of the bulk sample were quite rich in country rock clasts. Logging of the available Lahtojoki drill core suggests that the bulk of the volcaniclastic diatreme material has much lower country rock dilution

This difference in wall rock contamination may be the explanation for the large discrepancy between the mini-bulk sampling, which totaled 23.3 t and returned diamond grades ranging from 21 to 45 ct/100 t (cpht), and the bulk sample of 1000 t that returned only 5.7 cpht (see Table 4.4.2 and Appendix A).

FIGURE 4.4.6 Photograph of Lahtojoki bulk sample material tested for diamonds by Malmikaivos Oy.

This volcaniclastic kimberlite material is particularly rich in country rock granite clasts relative to the bulk of the pipe infill based on more recent drilling and core logging.

The exact same processing methods and equipment were used at Malmikaivos Oy for both sampling programs (Matti Tyni, personal communication), so the difference between the two must be related to heterogeneities in the volcaniclastic pipe infill.

The large range in mini-bulk sample results (Appendix A) may be related to the fact that diamonds occur not only as xenocrysts in the kimberlite matrix, but also as a rock-forming mineral in some of the eclogite xenoliths that have been recovered from this pipe (Peltonen et al., 2002, and Fig. 4.4.7(A)). A photograph of the larger diamonds recovered from the Malmikaivos bulk sample of the Lahtojoki kimberlite is shown in Fig. 4.4.7(B).

KUHMO AND LENTIIRA ORANGEITES AND OLIVINE LAMPROITES

Confined to a north–south zone of faults in the of Kuhmo-Lentiira area in eastern Finland (see Subchapter 4.1, Fig. 4.1.4) occurs a series of dikes (less commonly veins or stockworks) 0.5–4 m in thickness and extending up to 450 m in length. The existence of pervasive primary carbonate suggests that these rocks are more akin to orangeites than lamproites (see Table 4.4.1). In a hand specimen, the most distinctive feature of the Kuhmo potassic, ultramafic rocks is their phlogopite-rich nature. Phlogopite occurs rarely as macrocrysts, but is abundant as phenocrysts and microphenocrysts with relatively Ti-rich grain cores that are similar in composition to those of lamproite microphenocrysts, but that are zoned to rims of tetraferriphlogopite, a feature common in orangeites (Mitchell, 1995). The more primitive Kuhmo potassic rocks may also contain large amounts of olivine macrocrysts (Fig. 4.4.8(A)) and in some cases abundant xenocrysts and xenoliths of mantle peridotite.

Additional groundmass minerals include K-richterite, Mn-rich ilmenite, Cr-rich spinel zoned to Ti-magnetite, apatite, and perovskite in a calcite and serpentine matrix. More evolved versions of this rock type contain abundant olivine phenocrysts (rimmed by perovskite; Fig. 4.4.8(B)) rather than olivine

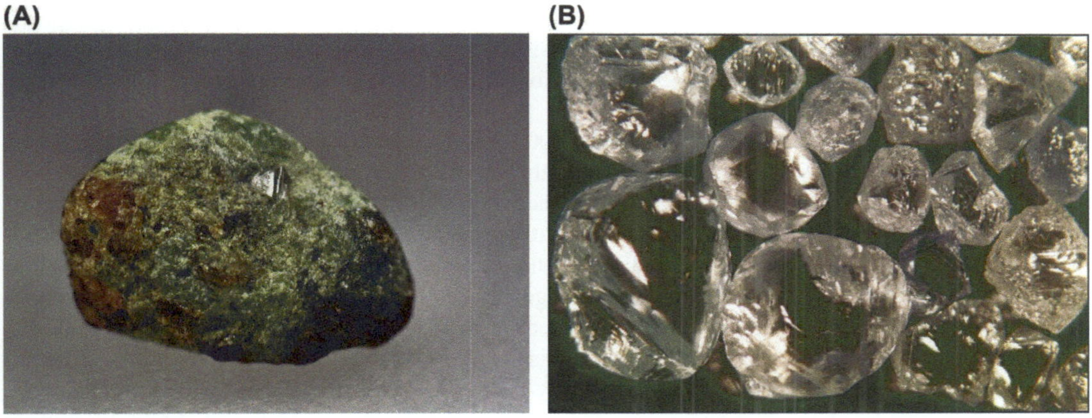

FIGURE 4.4.7 Diamonds occur not only as xenocrysts in the kimberlite matrix but also as rock-forming minerals.

(A) Diamondiferous eclogite xenolith from Lahtojoki kimberlite. Xenolith is 10 mm in the longest dimension. (B) Diamonds from Pipe no. 7, Lahtojoki kimberlite. The largest diamond is about 1 ct and 5 mm in its longest dimension.

Source: *Photograph shown in (A) was taken by Kari Kinnunen, GTK and (B) by Matti Tyni, Malmikaivos Oy.*

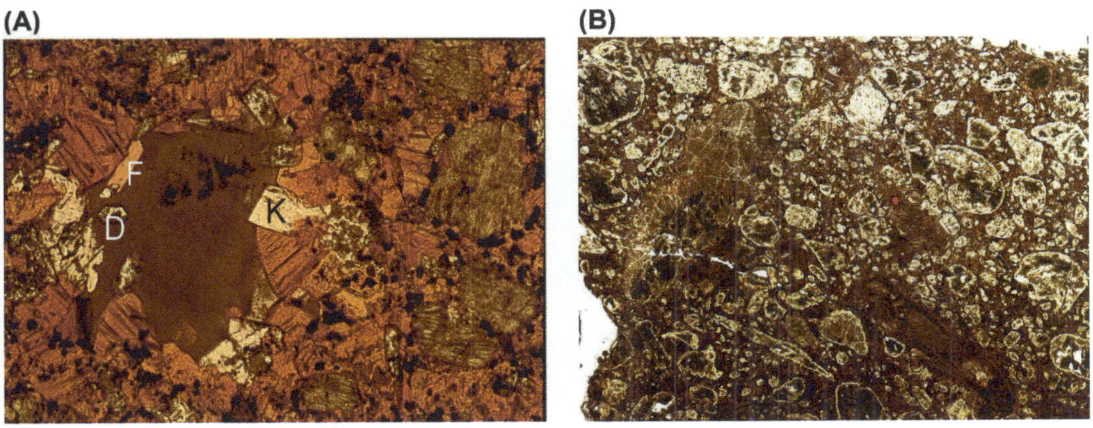

FIGURE 4.4.8 Photomicrographs of Kuhmo-Lentiira orangeites.

(A) Slightly more evolved orangeite from Kuhmo (Seitaperä). Grains surrounding former pool of late-stage liquid include tetraferriphlogopite (F), diopside (D), and K-richterite (K). Olivine pseudomorphs ringed by perovskite in a matrix of phlogopite, apatite, and calcite are also apparent. (B) Primitive olivine macrocryst-rich phlogopite orangite dike rock that also contains abundant euhedral olivine phenocrysts (serpentinized) in a matrix of Ti-rich phlogopite, K-richterite, Mn-rich ilmenite, Cr-rich spinel zoned to Ti-magnetite, apatite, perovskite, calcite, and serpentine (European Diamonds PLC, Lentiira prospect). Both images in plane polarized light; width of field is 2.55 mm (A) and 7.7 mm (B).

macrocrysts, and low-Al clinopyroxene. Mantle-derived xenocrysts are dominantly chrome-spinels zoned to Ti-magnetite, Ti-rich (megacryst-composition) pyropes, less commonly Mg-rich ilmenites and microdiamonds. Notably, mantle-derived Cr-diopside is almost completely absent from the xenocryst suite.

Ar-Ar age dates reported by O'Brien et al. (2007) on a Seitaperä drill core, a pit sample, and a sample 30 km to the north at Lentiira produced ages that are all within error of each other, at 1202 ± 3 Ma (2s), 1199 ± 3 Ma (2s), and 1204 ± 4 Ma (2s), respectively. The consistency of the step-heating results suggests that these ages represent precise determinations of the dike emplacement ages at these localities. The fact that they overlap in time, combined with the mineralogical and geochemical similarities of the dikes, suggest that these dikes represent a cogenetic suite of magmas.

Seitaperä (Pipe no. 16) is the largest of a number of dikes, stockworks, and rare pipes located in the Kuhmo-Lentiira area (see Fig. 4.1.4). It was discovered by Malmikaivos/Ashton using indicator mineral anomaly mapping of the till, followed by noting a small up-ice magnetic anomaly in the regional GTK aeromagnetic survey data. At the time, ground geophysics suggested a body of more than 4 ha in size. As of this writing, Karelian Diamond Resources PLC is the holder of the claims over Seitaperä, and the company has conducted a significant amount of drilling and sampling of the body, producing the geologic map redrafted in Fig. 4.4.9. At this stage of exploration drilling, the area containing massive orangeite and country rocks brecciated by orangeite covers as much as 7 ha. In July 2008, Karelian Diamond Resources reported 61 microdiamonds and 7 macrodiamonds from a 100-kg sample from its Seitaperä drill core, with the largest diamond measured to be 0.63 mm × 0.48 mm × 0.38 mm. More recently reported microdiamond results from other parts of Seitaperä have not been as successful, but the initial results suggest there are zones with potentially higher diamond grades.

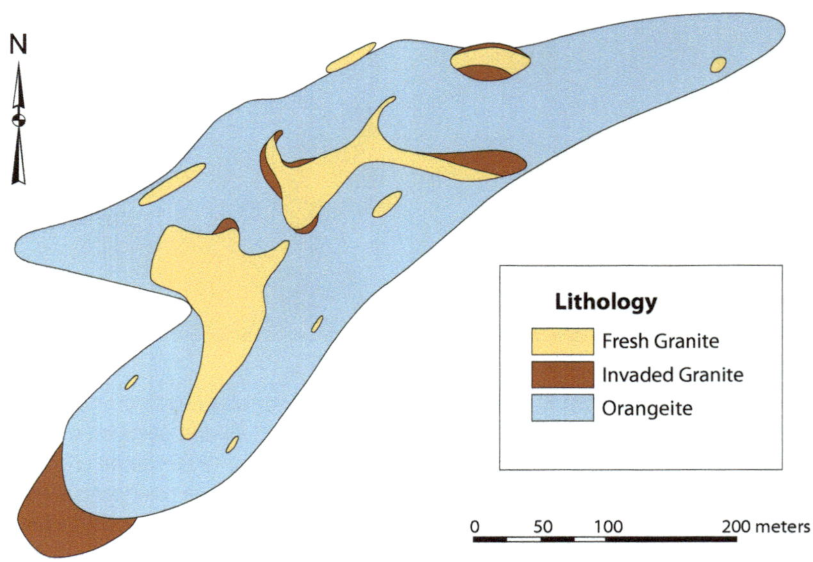

FIGURE 4.4.9 Geological map of Seitaperä based on drilling results from Karelian Diamond Resources.

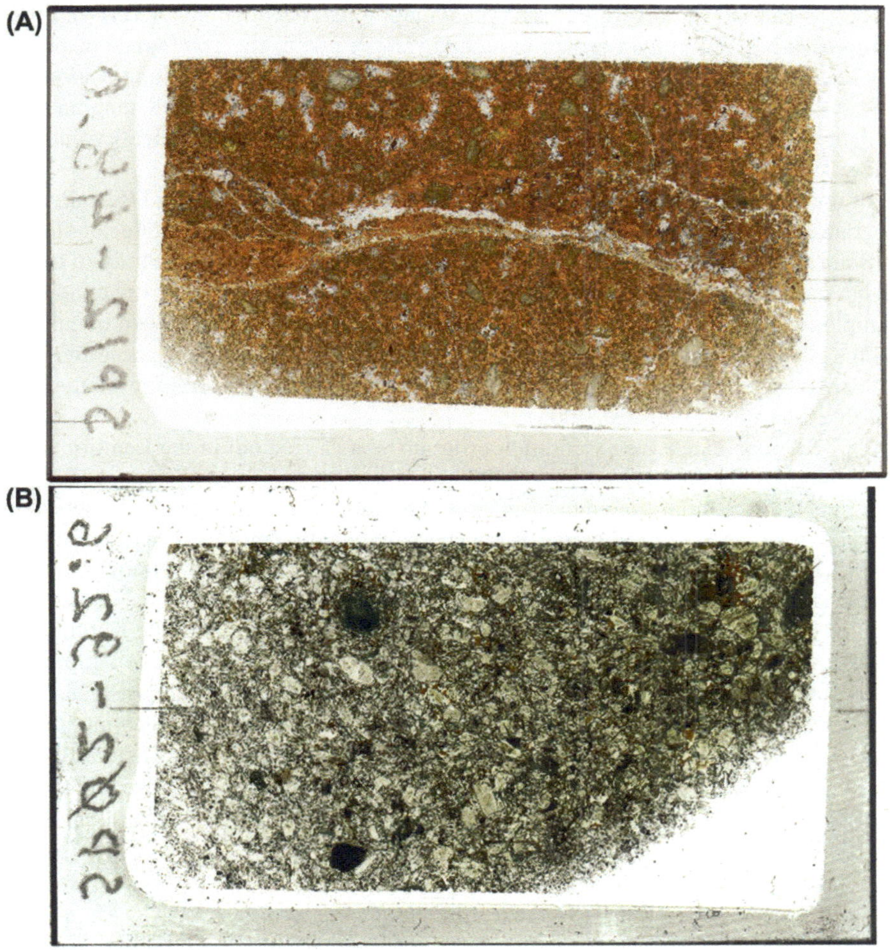

FIGURE 4.4.10 Seitaperä rock types in thin section.

(A) Mica-rich orangeite with carbonate segregations. (B) Olivine-rich orangeite with elongate phlogopite laths in a carbonate-serpentine matrix. Rock slab is 2 cm wide in minimum direction.

Petrographically, the Seitaperä rock types range from hypabyssal orangeite breccias near contacts with the granitic country rock, to porphyritic olivine-rich orangeite without significant crustal contamination. Unusual globular segregation-rich sections mapped in drill core are very similar to those found in some orangeites from southern Africa (e.g., Finsch). Whether these globular structures, many of which are olivine macrocryst or peridotite xenolith-cored, represent coarsened pelletal lapilli (indicating diatreme facies) or some other magmatic separation process (indicating hypabyssal intrusion) remains to be determined.

Although mica-rich orangeites are the most common variety at Seitaperä (Fig. 4.4.10(A)), those dominated by olivine with abundant perovskite in a rhombohedral primary carbonate, apatite, and serpentine (both clear and red-brown Fe-rich types) matrix also occur (Fig. 4.4.10(B)). Fresh samples of the mica-rich variety are golden brown in a hand sample due to the dominance of phlogopite, but more highly serpentinized varieties are thoroughly black in color. Phlogopite compositions from Seitaperä span nearly the entire compositional range of micas known from the KLK magma suite, starting within the lamproite microphenocryst field yet following the orangeite mica evolution trends to extremely low TiO_2 and Al_2O_3 tetraferriphlogopite (Fig. 4.4.11).

Studies of the numerous mantle xenoliths recovered in the Karelian Diamond Resources 2008-2011 drilling program at Seitaperä have provided some sections extremely rich in mantle xenoliths; in places, the xenoliths are tightly packed with little to no intervening magmatic matrix. Study of these xenoliths is ongoing, but preliminary results show that most are garnet-bearing, few have any clinopyroxene, and several examples of highly strained porphyroclastic textured peridotites have been obtained. Garnet in these xenoliths has been preferentially targeted by alteration, with little fresh garnet remaining in the peridotites, explaining the chromite-rich, relatively garnet-depleted mineralogy of the indicator fan down-ice of this intrusion (Lehtonen, 2005).

From 2000–2006, an extensive exploration program was carried out in the Lentiira area, 30 km to the north of Kuhmo, by European Diamonds. Although orangeites were discovered (see Fig. 4.4.8(A))—in some cases with spectacular amounts of mantle material and all with highly prospective indicator mineral compositions—none of the bodies produced significant quantities of microdiamonds. This is despite the fact that macrodiamonds were discovered by till sampling in a number of locations. The exploration possibilities in this area are still intriguing.

In the Kostomuksha region of Russian Karelia, about 40 km northeast of the Kuhmo occurrences, more than 80 dikes and small breccia pipes of similar rocks ranging from leucite lamproite to olivine lamproite to orangeite have been identified and studied (Nikitina, 1999; Kargin et al., 2014). Their age appears to be slightly older than the Kuhmo-Lentiira rocks, with an Rb-Sr mineral isochron age of 1231 ± 8.9 (Belyatskii et al., 1997). The combined Kuhmo-Lentiira-Kostomuksha orangeite region is more than 500 km^2 in size, and undoubtedly contains additional undiscovered intrusions of this rock type.

KUUSAMO KIMBERLITE PROVINCE

The Kuusamo kimberlites mark the third, and most recently discovered, kimberlite region in Finland, located in the northern part of the western terrane of the Karelian craton (refer to Fig. 4.1.4). Six kimberlites occur in two discrete groups. The Kattaisenvaara (KV), Kalettomanpuro (KP), and Lampi hypabyssal kimberlites consist of serpentinized olivine macrocrysts and phenocrysts in a fine-grained matrix composed of microphenocrysts of phlogopite, perovskite, apatite, and spinel with late-stage carbonate and serpentine filling grain interstices (Fig. 4.4.12). The other three kimberlites, named 47, 45, and 45 South, occur close to each other. Pipe no. 45 contains both hypabyssal and diatreme-facies kimberlite, whereas only diatreme material was recovered from Pipe no. 47.

The volcaniclastic kimberlite in these small pipes contains abundant pelletal lapilli. A seventh discovery is a mantle xenocryst-rich orangeite dike intersected at Kalettomanpuro. It appears to be another member of the 1.2 Ga phlogopite-rich dike rocks described from the Lentiira-Kuhmo-Kostomuksha area. Mantle xenocrysts in the kimberlites consist of picroilmenites, pyropes, and chrome diopsides, while chromites appear to be absent. U-Pb analyses of two perovskite fractions from one sample each of the Kattaisenvaara and Kalettomanpuro kimberlites gave weighted mean $^{206}Pb/^{238}U$ ages of 759 ± 15 Ma and 756.8 ± 2.1 Ma, respectively (O'Brien and Bradley, 2008).

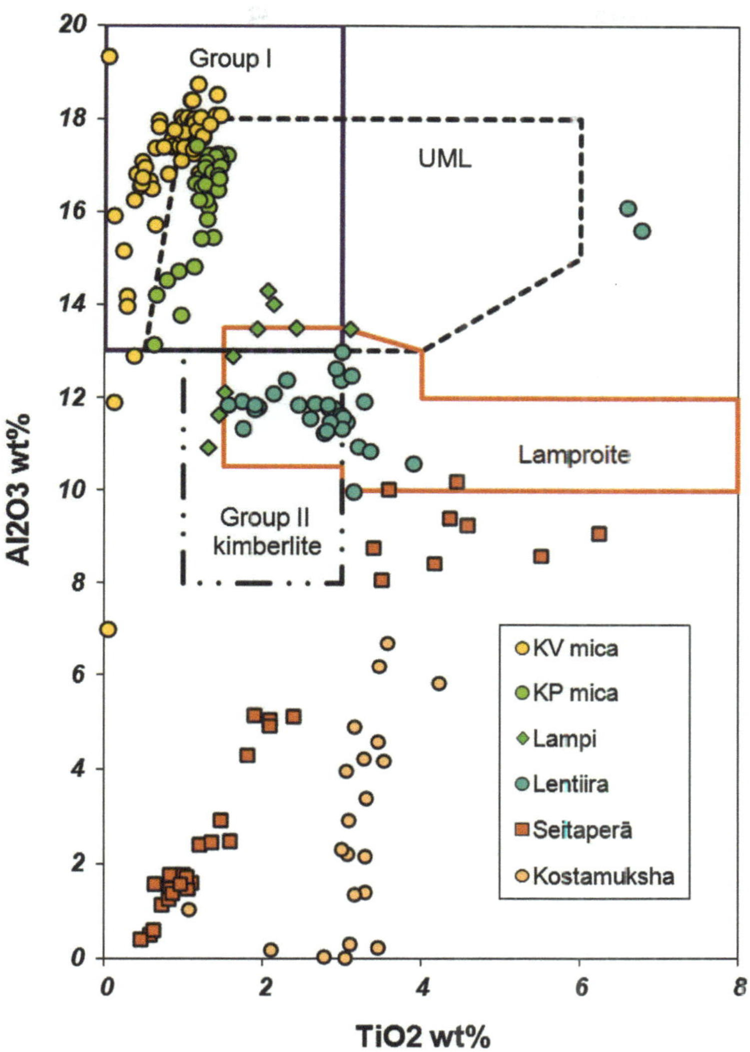

FIGURE 4.4.11 Mica compositions from orangeites.

Mica compositions from Kuhmo, Lentiira Dikes, Seitaperä, and Kostamuksha compared with Kuusamo kimberlites Kalettomanpuro (KP), Kattaisenvaara (KV), and Lampi kimberlite micas plotted in TiO_2 versus Al_2O_3. All of the KP and KV points plot directly in the archetypal kimberlite compositional field, although there is some overlap with the UML field.

Source: *Compositional fields after Mitchell, 1995.*

Mica compositions of the Kattaisenvaara and Kalettomanpuro kimberlites are highly Al- and Ba-enriched kinoshitalites, typical of kimberlite microphenocrysts, with rare tetraferriphlogopite rims detected on some Kattaisenvaara grains. The Lampi kimberlite, on the other hand, has mica that is Ba-rich, yet with only moderate Al_2O_3 and TiO_2 (10–14 wt% and 2–4 wt%, respectively); mica compositions

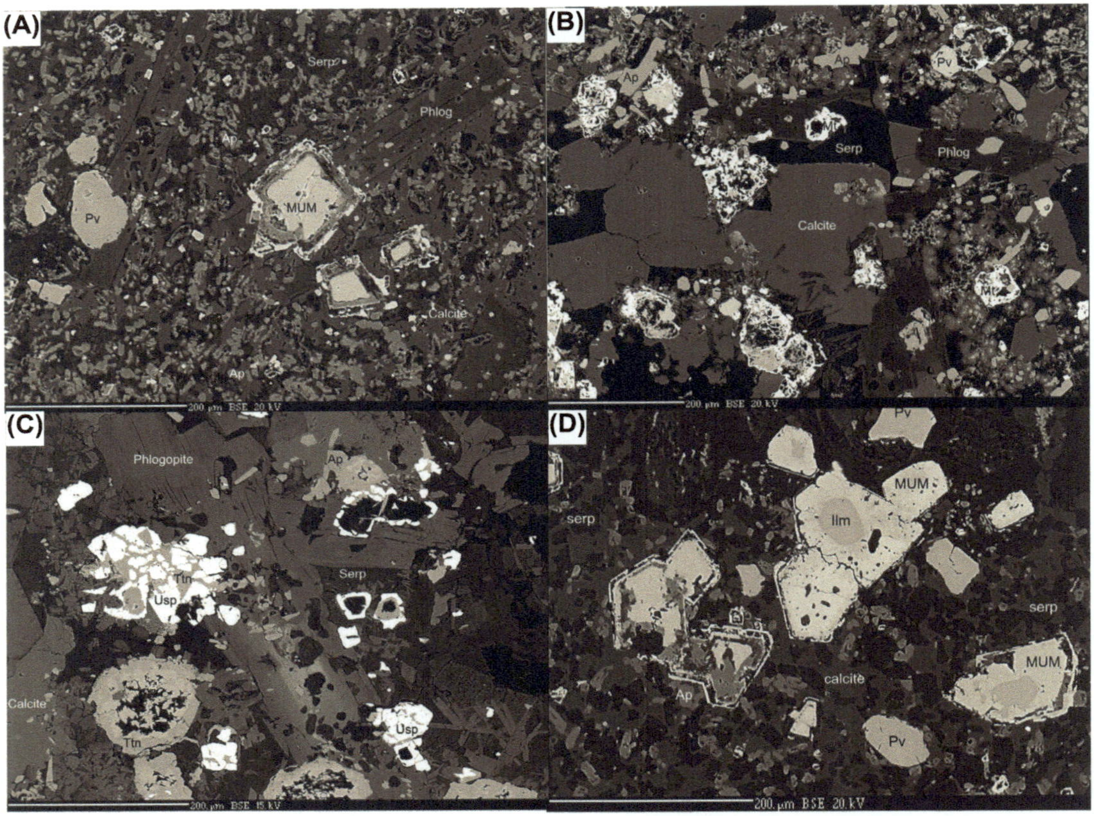

FIGURE 4.4.12 Backscattered electron images of representative samples of Kuusamo kimberlites.

(A) KP: Poikilitic Ba-Al-rich kinoshitalite phlogopite encloses atoll-textured magnesian ulvöspinel-magnetite series (MUM) spinel with late-stage Ti-magnetite rims, apatite, and patches of what are now serpentine. (B) KV: Relatively coarse zone showing highly zoned phlogopite (bottom of image), complex magnetite crystals (Mt), abundant apatite, and late crystals of phlogopite growing within and across calcite-serpentine grain boundaries. (C) Lampi: Strongly zoned Ba-rich poikilitic phlogopites enclose ulvöspine (Usp) and apatite and have abundant embayments filled with serpentine. The development of titanite (Ttn) is thought to be late-stage magmatic and appears to have replaced all original perovskite in this sample. (D) Kasma 45: Atoll structured MUM spinels with partially resorbed, high Mg-Cr picroilmenite cores are common in the Kasma kimberlites. Abundant subhedral to euhedral perovskite is contained in a matrix of subhedral serpentine patches within a carbonate and apatite matrix.

extend from the kimberlite field into the orangeite compositional field. Finally, the zoning patterns from high Al_2O_3 toward lower Al_2O_3 are coupled with decreasing TiO_2 (see Fig. 4.4.11). The KV-KP-Lampi kimberlites are moderately to highly magnetic, and this is directly related to the relatively high modal abundance of magnetite. Mineralogically, the Kasma kimberlites are distinct from KV-KP-Lampi in that they are very perovskite-rich and phlogopite is rare to absent (see Fig. 4.4.12). The Kasma kimberlites, which contain less magnetite, are consequently less magnetic, but nevertheless produce distinct anomalies in a ground magnetic survey against a relatively nonmagnetic baseline.

Major element data confirm the relatively evolved nature of the KV-KP-Lampi kimberlites. Plots such as Fe_2O_3-SiO_2 (refer to Fig. 4.4.3) and CaO-MgO (not shown) indicate that the Kuusamo kimberlites plot slightly outside the field of typical kimberlites, with a vector toward the magnetite and perovskite-enriched partial cumulates from the Benfontein sill where extreme Fe_2O_3 contents reflect accumulation of significant magnetite. Despite their evolved nature, microdiamonds have been recovered from the KV and Lampi kimberlites. Microdiamonds have also been recovered from the Kasma locality (kimberlite 45) but comparative chemical analyses have not been made due to the lack of magmatic kimberlite at this locality.

KARELIAN CRATON MANTLE DIAMOND POTENTIAL

The search for diamonds requires that a considerable amount of information be known about the underlying lithospheric mantle to determine if it has diamond potential. A number of studies (e.g., O'Brien et al., 2003; Lehtonen et al., 2004; Lehtonen, 2005; Lehtonen and O'Brien, 2009) have been made in cooperation with the exploration companies working in the area, to gain a better understanding of the structure and composition of the Karelian cratonic mantle.

Data from xenoliths and xenocrysts from Finnish kimberlites and orangeites and the indicator minerals in exploration till samples allow us to make the following interpretations:

1. P-T data indicate that the mantle root of the Karelian craton varies considerably from margin to core, thickening from about 220 km at Kuopio to nearly 250 km in the Lentiira-Kuhmo-South. Kuusamo section and then thinning again to about 220 km at the North Kuusamo locality.
2. At the craton margin, in the Kuopio and Kaavi area, the mantle is comprised of at least three distinct layers:
   - Layer A—a shallow, 60–110 km, garnet-spinel peridotite layer with high Mg and low Al.
   - Layer B—a variably depleted peridotitic horizon from 110–180 km containing harzburgitic pyropes that represent the diamond window.
   - Layer C—a deep layer, >180 km depth, composed largely of refertilized peridotites. Eclogite and clinopyroxene are common throughout this mantle section.
3. The mantle stratigraphy of the craton core, in the Kuhmo, Lentiira, and Kostomuksha areas, shows less mineralogical variation, with no evidence of Layer A. Layer B begins with the lowest temperature pyropes at an inferred depth of 70 km and continues to a depth of about 250 km, showing a relatively homogenous distribution of harzburgite and lherzolite pyropes throughout, and a high Mg/(Mg + Fe) even compared to global mantle samples. Eclogite and clinopyroxene are rare.
4. The Kuusamo area is located well within the Karelian craton and the composition of mantle xenocrysts derived from this area confirms the existence of depleted mantle underlying the area. Xenocryst pyrope chemistry shows a roughly uniform distribution of lherzolite and harzburgite down to depths of roughly 180 km, and then only lherzolite down to 220 km. The existence of relatively abundant harzburgitic pyrope, some with ultradepleted compositions, throughout most of this mantle section indicates it is similar to that in the Kuhmo region, albeit slightly thinner. Bolstering this interpretation is the paucity of eclogite material in exploration and hard rock samples from the area.

SUMMARY

1. Kimberlites and orangeites are the main source of diamonds, either directly from the igneous rocks themselves, or from the placers derived from these rocks.
2. Diamonds are derived from a specific type of subcontinental lithospheric mantle; it has to be stable and old (Archean), depleted (Mg-rich and harzburgite-bearing), and thick (the diamond window starts at about 140 km depth).
3. This type of mantle has been identified in the Karelian craton, and although varying in detail, appears to be continuous at least over the distance from Kuopio to Kuusamo.
4. Diamond exploration has led to the discovery of more than 30 kimberlites and orangeites in Finland including the 1200 Ma orangeites for the Kuhmo-Lentiira area, the 760 Ma kimberlites from the Kuusamo-Hossa area, and the ~600 Ma kimberlites from Kaavi-Kuopio area. Most of these kimberlites contain at least some diamonds and significant amounts of mantle material.
5. Kimberlite occur as diatremes and dikes in the Kaavi-Kuopio and Kuusamo-Hossa area, whereas the older orangeites from Kuhmo-Lentiira all appear to be more deeply eroded root-zone intrusions.
6. In Finland and globally, knowledge of the geology of the deep lithospheric mantle roots has been advanced immeasurably by collaboration between exploration companies and research institutes.

ACKNOWLEDGMENTS

I would like to thank all of the exploration companies that have worked in Finland and collaborated with our kimberlite and mantle studies at GTK, particularly Malmikaivos Oy/Ashton Mining, whose exploration teams put Finland on the diamond map, and whose director, Matti Tyni, was, from the very early days, very helpful, optimistic, and understanding of the need for scientific cooperation. I would also like to thank Harri Kutvonen for drafting several figures for this section, and W.D. Maier for a thorough review.

REFERENCES

Belyatskii, B.V., Nikitina, L.P., Savva, E.V., Levskii, L.K., 1997. Isotopic signatures of lamproite dikes on the Eastern Baltic Shield. Geochemistry International 35, 575–579.

Brown, R.J., Valentine, G.A., 2013. Physical characteristics of kimberlite and basaltic intraplate volcanism and implications of a biased kimberlite record. Geological Society of America. Bulletin 125, 1224–1238.

Brown, R.J., Manya, S., Buisman, I., et al., 2012. Eruption of kimberlite magmas: physical volcanology, geomorphology and age of the youngest kimberlitic volcanoes known on earth (the Upper Pleistocene/Holocene Igwisi Hills volcanoes, Tanzania). Bulletin of Volcanology 74, 1621–1643.

Chernet, T., Marmo, J., Nissinen, A., 1999. Technical note—Significantly improved recovery of slightly heavy minerals from Quaternary samples using GTK modified 3" Knelson preconcentrator. Minerals Engineering 12, 1521–1526.

Donnelly, C.L., Griffin, W.L., O'Reilly, S.Y., et al., 2011. The Kimberlites and related rocks of the Kuruman Kimberlite Province, Kaapvaal Craton, South Africa,. Contributions to Mineralogy and Petrology 161, 351–371.

Gernon, T.M., Gilbertson, M.A., Sparks, R.S.J., Field, M., 2009. The role of gas-fluidisation in the formation of massive volcaniclastic kimberlite. Lithos 112S, 439–451.

Gernon, T.M., Brown, R.J., Tait, M.A., Hincks, T.K., 2012. The origin of pelletal lapilli in explosive kimberlite eruptions. Nat. Commun 3, 832.

Gurney, J.J., Helmstaedt, H.H., Richardson, S.H., Shirey, S.B., 2010. Diamonds through time. Economic Geology 105, 689–712.

Hearn, B.C., 1968. Diatremes with kimberlitic affinities in North-Central Montana. Science 159, 622–625.

Kaminsky, F.V., Khachatryan, G.K., 2004. The relationship between the distribution of nitrogen impurity centres in diamond crystals and their internal structure and mechanism of growth. Lithos 77, 255–271.

Jelsma, H.J., Barnett, W., Richards, S., Lister, G., 2009. Tectonic setting of kimberlites. Lithos 112S, 155–165.

Kargin, A.V., Nosova, A.A., Larionova, Y.O., et al., 2014. Mesoproterozoic Orangeites (Kimberlites II) of West Karelia: Mineralogy, Geochemistry, and Sr-Nd Isotope Composition. Petrology 22, 151–183.

Korkeakoski, P., O'Brien, H., Lehtonen, M.L., Marmo, J., 2004. Kimberlite/Diamond Claim Reports Kaavi-Kuopio Area—Eastern Finland, Compilation and Review. GTK report 19/3242, 3244, 3331 3333, 4222 4311/2004/1/10.4.

Lehtonen, M.L., 2005. Rare-earth element characteristics of pyrope garnets from the Kaavi-Kuopio kimberlites—implications for mantle metasomatism. Bulletin of the Geological Society of Finland 77, 31–47.

Lehtonen, M., O'Brien, H., Peltonen, P., et al., 2004. Layered mantle at the Karelian Craton margin: P-T of mantle xenocrysts and xenoliths from the Kaavi-Kuopio kimberlites. Finland. Lithos 77, 593–608.

Lehtonen, M.L., Marmo, J.S., Nissinen, A.J., et al., 2005. Glacial dispersal studies using indicator minerals and till geochemistry around two eastern Finland kimberlites. Journal of Geochemical Exploration 87, 19–43.

Lehtonen, M.L., O'Brien, H.E., 2009. Mantle transect of the Karelian Craton from margin to core based on P-T data from garnet and clinopyroxene xenocrysts in kimberlites. Bulletin of the Geological Society of Finland 81, 79–102.

Le Roex, A.P., Bell, D.R., Davis, P., 2003. Petrogenesis of group I kimberlites from Kimberley, South Africa: Evidence from bulk-rock geochemistry. Journal of Petrology 44, 2261–2286.

Lorenz, V., 1975. Formation of phreatomagmatic maar-diatreme volcanoes and its relevance to kimberlite diatremes. Physics and Chemistry of the Earth 9, 17–27.

Lorenz, V., Kurszlaukis, S., 2007. Root zone processes in the phreatomagmatic pipe emplacement model and consequences for the evolution of maar-diatreme volcanoes. Journal of Volcanology and Geothermal Research 159, 4–32.

Massonne, H.J., 2003. A comparison of the evolution of diamondiferous quartz-rich rocks from the Saxonian Erzgebirge and the Kokchetav Massif: are so-called diamondiferous gneisses magmatic rocks? Earth and Planetary Science Letters 216, 347–364.

Masaitis, V.L., 2013. Impact diamonds of the Popigai astrobleme: Main properties and practical use. Geology of Ore Deposits 55, 607–612.

Mitchell, R.H., 1995. Kimberlites, Orangeites, and Related Rocks. Plenum, New York. 410p.

Mitchell, R.H., 2006. Potassic magmas derived from metasomatized lithospheric mantle: Nomenclature and relevance to exploration for diamond-bearing rocks. Journal Geological Society of India.

Nikitina, L.P., Levsky, L.K., Lohkov, K.I., et al., 1999. Proterozoic alkaline-ultramafic magmatism in the eastern part of the Baltic Shield. Petrology 7, 246–266.

O'Brien, H.E., Tyni, M., 1999. Mineralogy and geochemistry of kimberlites and related rocks from Finland. In: Gurney, J.J., Gurney, J.L., Pascoe, M.D., Richardson, S.H. (Eds.), Proceedings of the 7th International Kimberlite Conference, pp. 625–636.

O'Brien, H., Lehtonen, M., Spencer, R., Birnie, A., 2003. Lithospheric mantle in eastern Finland: a 250 km 3D transect. In: Extended Abstracts, Proceedings of the Eighth International Kimberlite Conference. Victoria, BC, p. 5. FLA-0261.

O'Brien, H.E., Peltonen, P., Vartianen, H., 2005. Kimberlites, carbonatites, and alkaline rocks In: Lehtinen, M., et al. (Ed.), Precambrian Geology of Finland—Key to the Evolution of the Fennoscandian Shield. Developments in Precambrian Geology 14. Elsevier, Amsterdam, pp. 605–644.

O'Brien, H., Phillips, D., Spencer, R., 2007. Isotopic ages of Lentiira-Kuhmo-Kostomuksha olivine lamproite—group II, kimberlites. Bulletin of the Geological Society of Finland 79, 203–215.

O'Brien, H., Bradley, J., 2008. New kimberlite discoveries in Kuusamo, northern Finland. In: Extended Abstracts, Proceedings of the 9th International Kimberlite Conference. Frankfurt. 9IKC-A-00346, p. 3.

Patterson, M., Francis, D., McCandless, T., 2009. Kimberlites: Magmas or mixtures? Proceedings of the 9th International Kimberlite Conference 9IKC, 191–200.

Pearson, D.G., Davies, G.R., Nixon, P.H., Milledge, H.J., 1989. Graphitized diamonds from a peridotite massif in Morocco and implications for anomalous diamond occurrences. Nature 335, 60–66.

Pearson, D.G., Shirey, S.B., Harris, J.W., Carlson, R.W., 1998. Sulfide inclusions in diamonds from the Koffiefontein kimberlite, S. Africa: Constraints on diamond ages and mantle Re-Os systematic. Earth and Planetary Science Letters 160, 311–326.

Peltonen, P., Kinnunen, K.A., Huhma, H., 2002. Petrology of two diamondiferous eclogite xenoliths from the Lahtojoki kimberlite pipe, eastern Finland. Lithos 63 (3–4), 151–164.

Peltonen, P., Männtäri, I., Huhma, H., Whitehouse, M.J., 2006. Multi-stage origin of the lower crust of the Karelian craton from 3.5 to 1.7 Ga based on isotopic ages of kimberlite-derived mafic granulite xenoliths. Precambian Research 147, 107–123.

Richardson, S.H., 1986. Latter-day origin of diamonds of eclogitic paragenesis: Nature 322, 623–626.

Scott-Smith, B.H., Nowicki, T.E., Russell, J.K., et al., 2013. Kimberlite terminology and classification. Proceedings of the 10th International Kimberlite Conference 2, 1–18.

Stachel, T., Harris, J.W., 2008. The origin of cratonic diamonds—Constraints from mineral inclusions. Ore Geology Reviews 34, 5–32.

Wagner, P.A., 1928. The evidence of the kimberlite pipes on the constitution of the outer part of the Earth. South African Journal of Sciences 25, 127–148.

White, J.D.L., Ross, P.-S., 2013. Maar-diatreme volcanoes: A review. Journal of Volcanology and Geothermal Research 201, 1–29.

APPENDIX A. BRIEF SUMMARY OF DIAMOND EXPLORATION RESULTS FROM THE KAAVI-KUOPIO KIMBERLITES

Data extracted from Korkeakoski et al., 2004. Abbreviations used: GT = garnet, PIL = picroilmenite, CP = clinopyroxene, CHR = chromite, DI ## and JÄÄ ## are the registered names given to Dia Met Minerals Ltd. diamond claims.

PIPE 1. KOSKENNIEMI

10 diamond drill holes intersected in total 175m of kimberlite. This extremely fresh and petrologically interesting kimberlite is apparently nondiamondiferous. A 107 kg sample, returned no microdiamonds, and hence no bulk sample was taken for macrodiamond testing. Sunrise Diamonds also reported a negative microdiamond result for this pipe.

PIPE 2. NIILONSUO

44 percussion drill holes and 24 diamond drill holes were used to sample this pipe. Drilling and magnetic measurements revealed a 20 – 30 m wide and 300 m long sill like body. A 172 kg sample from the drill cores yielded 72 microdiamonds. Test quarrying of 162.8 tons for macrodiamond analyses produced a very low grade of 0.23 carats per hundred ton (cpht).

PIPE 3. NIILONLAMPI

A few percussion drill holes and 54 diamond drill holes showed that the body is 10 – 50 m wide and 350 m long. A 2.2 tons sample for macro diamond analyses was obtained by mini bulk sampling but contained only two diamonds, and a calculated grade of 0.91 cpht.

PIPE 4. KARETINSAARET (AKA RIKAVESI)

7 diamond drill holes intersected 163m of kimberlite from this diatreme. Drilling was conducted through winter ice, since the kimberlite is in a lake, at a depth of about 20 m. A 27 kg sample gave 18 microdiamonds. 8.2 tons were processed for macrodiamonds, but provided a very low grade of only 0.21 cpht.

PIPE 5. KÄRENPÄÄ

23 diamond drill holes were drilled, 21 of which reached the kimberlite (combined length 1157.7 m). A 122 kg caustic dissolution sample for microdiamonds was positive. Five mini bulk samples with a total of 15.9 tons and test quarrying of 170 tons of kimberlitic material produced only very low quantities of macro diamonds. Drilling and magnetic surveying revealed an almost vertical sill like body with a width of less than 30 m and a length of approximately 800 m.

PIPE 6. TEERIPURO

30 diamond drill holes were obtained over the magnetic anomaly (combined length 633.2 m) and five mini bulk holes were made into the body. A 48 kg sample from drill core for micro diamonds and a 13.4

ton mini bulk sample for macro diamonds produced only very small quantities of either. The body is elongate, 50 m wide and 200 m long, and oriented almost vertically.

PIPE 7. LAHTOJOKI

17 till samples were collected from various depths for a heavy mineral survey. Geophysical surveys included magnetic, seismic and electric measurements. 39 diamond drill holes (combined length 1747.7 m) were used to delineate the body. These data and geophysical measurements showed the pipe to be pear-shaped, 100 m x 250 m in size. 59 kg of drill core for micro diamonds revealed 20 grains. Eight 200 mm percussion drill holes recovered a 23.3 tons mini bulk sample for macro diamonds giving an average diamond content of 30.6 ct / 100t. A 1600 ton bulk sample, of which 1000 tons was processed yielded an average diamond content of only 5.68 ct/100 t (see table A1).

Table A1 Diamond grade results for lahtojoki kimberlite, Malmikaivos Oy claim reports.

Hole number	collar label	Sample wt (t)	Diamonds >0.8 mm		Diamond total		Largest diamond (mm)	Grade >0.8 mm ct/100t
			count	ct	count	ct		
D4311-116	LU 2-6 LD-11	1.8	69	0.72	179	0.83	2.4	40.00
D4311-115	LU 2-6 LD-12	1.7	34	0.77	91	0.83	4.3	45.29
D4311-114	LU 2-6 LD-13	1.9	75	0.72	148	0.81	2.3	37.89
D4311-155	LU 2-6 LD-36	3.6	101	0.77	324	1.03	2.0	21.39
D4311-154	LU 2-6 LD-37	3.9	119	1.17	380	1.46	2.4	30.00
D4311-163	LU 2-6 LD-39	3.4	52	0.73	183	0.97	3.0	21.47
D4311-162	LU 2-6 LD-40	3.5	51	0.90	238	1.26	2.5	25.71
D4311-161	LU 2-6 LD-41	3.5	86	1.35	313	1.77	3.6	38.57
Bulk Sample		1000	2481	56.85			0.97 ct	5.68

Note that results from more recent work by European Diamonds and Mantle Diamonds on this kimberlite are not available in the public record.

PIPE 8. AVIOLAMPI

On claim DI 82 a heavy mineral survey with seven surface and six basal till samples was done but no significant indicator minerals were found. 20 grains were analysed comprising 6 pyroxenes and 14 chromites. The body is a thin, ~1 m thick sill, but with a large surface exposed due to its sheet-like shape, explaining the large amount of indicator minerals it has shed (M. Tyni, pers. comm.).

There have been five other claims in the ice flow direction SE from the sill (DI 25, DI 50, DI 83, DI 85 and DI 86). 41 surface till samples were collected from these five claims and 43 grains were found and analysed (7 CP, 10 CHR, 20 GT, 5 PIL and 1 OP).

PIPE 9. KALAJÄRVI

5 diamond drill holes hit kimberlite at Kalajärvi, although 1 intersection was thought to be just a large kimberlite boulder (Matti Tyni, pers. comm.). The total intersection of kimberlite was about 91 m. A 25 kg microdiamond sample was negative, and hence no further work was carried out on the pipe.

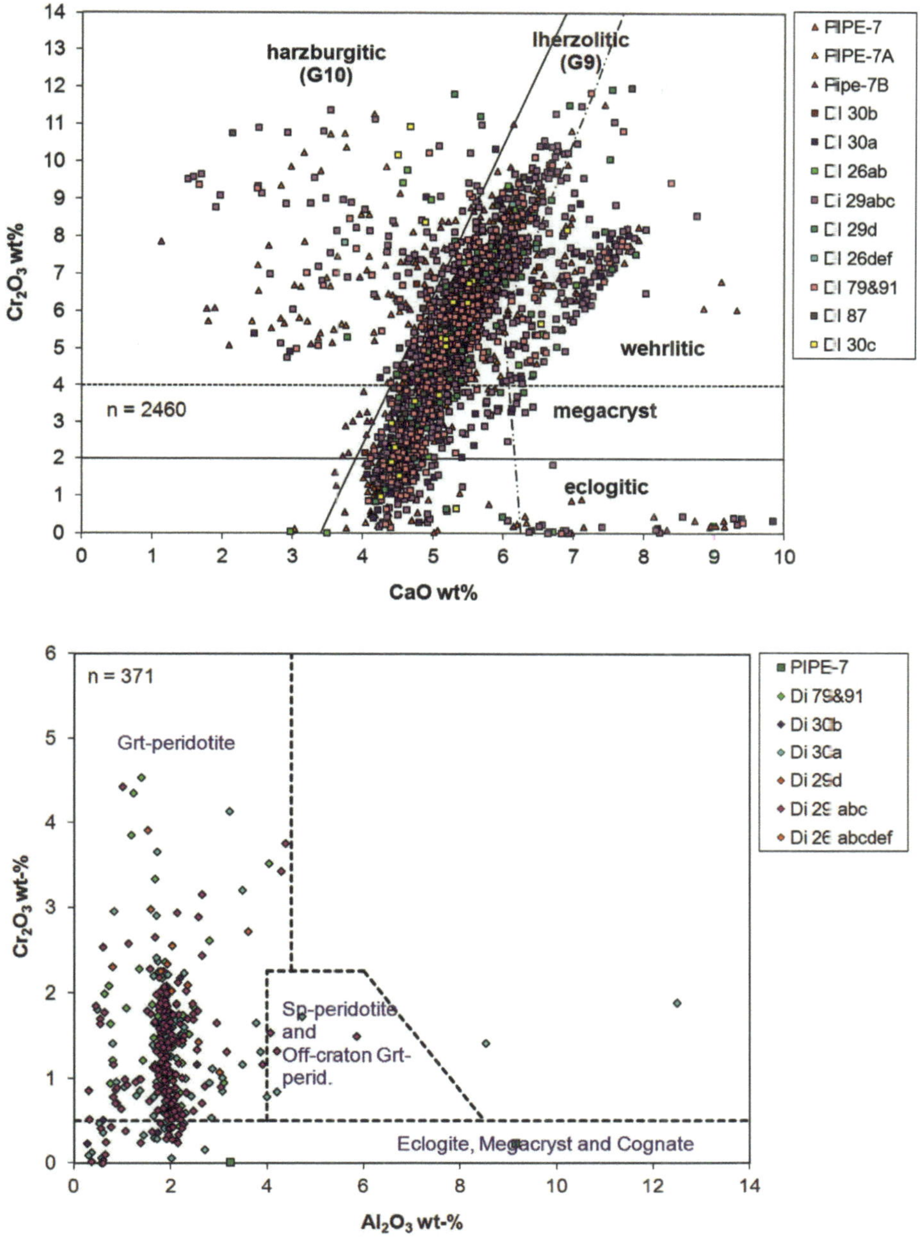

FIGURE 4.4.A1-A4

Xenocryst and kimberlite indicator mineral compositional plots for the Lahtojoki kimberlite and claims in the surrounding areas. A1. Garnet Cr-Ca diagram. A2. Clinopyroxene Cr-Al diagram. A3. Spinel Cr-Mg diagram. A4. Ilmenite Ti-Mg diagram.

Continued

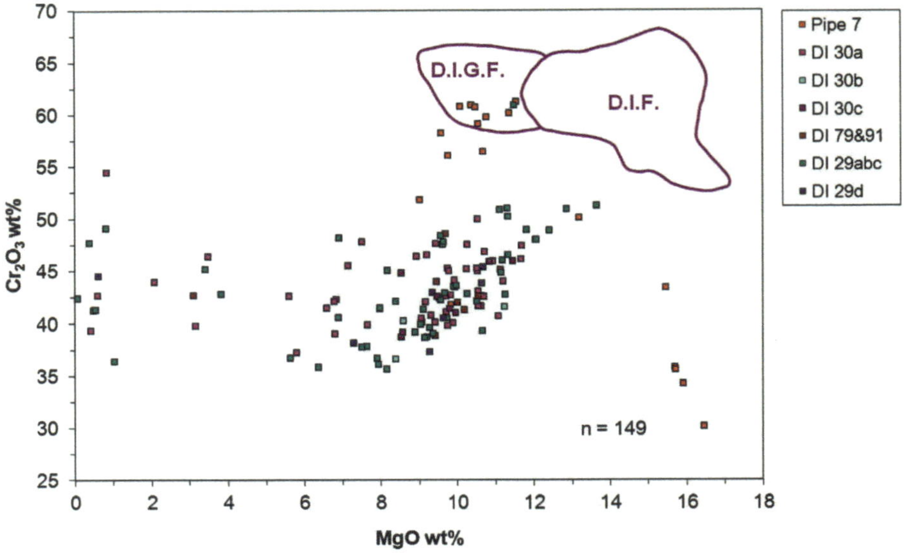

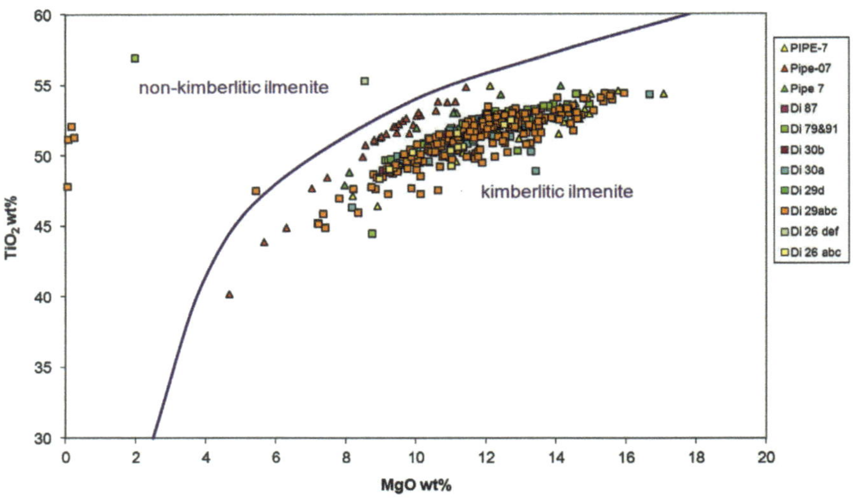

FIGURE 4.4.A1-A4 cont'd

Down-ice, SE from the pipe, there have been two claims. On DI 115, a magnetic survey was conducted, 24 surface till samples were collected and 33 grains were found and analysed (15 GT and 18 PIL). On DI 92, some surface till sampling was done and indicator minerals were found but not analysed or reported. One diamond drill hole to 23.60 m found no kimberlitic material.

PIPE 10. RYÖNÄ 1

To delineate the body, 24 diamond bore holes were drilled (combined length 901.6 m). 26 kg was sampled from the drill core for micro diamond analyses. The body was drilled with four mini bulk holes and 13.3 tons of material were recovered and analysed for macro diamonds. Micro and macro diamond contents of the analysed samples were very low. Drilling and a magnetic survey revealed an oval shaped body with an estimated area of 2.2 hectares.

Seven claims were situated down-ice from the pipe (Jää 61-66 and DI 129). On claim DI 129 a hole was drilled (total length 29.60 m) based solely on an aeromagnetic anomaly. A ground magnetic survey was made on all the JÄÄ 61-66 claims and based on the results four holes were drilled (combined length 180.7 m), one of which did not reach the bedrock.

PIPE 11. MUNAKKA 1

Two diamond drill holes tested the body, a 25 kg drill core sample was analysed for micro diamonds, but only one micro diamond was found. Additionally, the kimberlite is small, only ~0.3 hectares in size.

The single claim near the pipe (DI 103) was only tested by a ground magnetic survey.

PIPE 12. KOTKATNIEMI

19 till samples were collected from the claim area. To delimit the body, magnetic, electric and gravimetric measurements were done giving an estimate of 1.6 ha. 40 diamond drill holes (combined length 1423.4 m) and 161.5 kg of drill core were analysed for microdiamonds. Three large percussion drill holes yielding 9.4 tons of sample material were recovered and processed. 70 microdiamonds were recovered and bulk sample diamond contents ranged from 12.50 – 26.06 ct/100t. 570 microprobe analyses and 20 SEM EDS analyses were done from the pipe material.

One claim (DI 44) was taken down-ice, SE from the pipe and only magnetic measurements were done. Up-ice from the pipe, on two claims (DI 130 and Kotkatniemi 2), magnetic and electric measurements were performed and seven holes were drilled (total length 193.7 m bedrock) but no kimberlitic material was found. Basal till sampling was attempted but no till was found.

PIPE 13. SÄYNEENJÄRVI

32 till samples were collected to better locate the body which has an estimated size of 0.9 ha and is gently dipping to the northeast. Magnetic and electric surveys were conducted and 15 diamond drill holes were completed (combined length 1117 m), 13 of which hit the kimberlite. 48 kg of drill core was analysed for micro diamonds. One mini bulk sample weighting 3 tons was analysed for diamonds. Microdiamond analyses revealed three diamonds and a mini bulk sample contained 14.33 ct/100 t. 441 probe analyses and 20 SEM EDS analyses were done from the pipe minerals.

Five claims were taken near the pipe (DI 31abcd and DI 15). One surface till sample from claim DI 15 was taken, which had indicator minerals that were not analysed. From claims DI 31 a, b, c and d, 48 surface till samples were collected. 25 of these had indicator minerals of which 1001 were analysed

(155 CP, 16 CHR, 739 GT, 82 PIL and 9 Other). Six research pits were dug to investigate the ice flow direction giving reliable results of 310 - 320º. Magnetic and electric surveys were carried out and the resulting anomalies were drill tested with four holes (combined length 479.4 m), none of which hit kimberlitic material.

PIPE 14. KAATRONLAMPI (VEHKALAHTI 1)

To determine the size and shape of the body, six vertical and inclined diamond drill holes were made (combined length 298.7 m). Drillings and a magnetic survey revealed a body with an area of 0.3 ha, 20-35 m wide and 100 m long. A 73 kg sample of drill core produced only 1 micro diamond.

PIPE 17. KYLMÄLAHTI

Geophysical surveys included magnetic and electromagnetic profiles. 21 till samples were taken to investigate the heavy mineral content of the area but only picking results are available. 20 diamond drill holes were drilled to estimate the size and shape of the body. Kimberlitic material occurred in 12 drill holes. The estimated size of the body on the surface is 2.0 ha, but kimberlitic material occurs mainly between large wall rock pieces. A narrow kimberlitic dyke was found 120m east of the main body. Three separate samples were collected for micro diamond analyses (combined weight of 98.7 kg), which produced a total of 8 microdiamonds.

PIPE 20. ALA-VEHKALAHTI (VEHKALAHTI 2)

12 electromagnetic (Slingram) profiles were done in the area, 1 over the body and the rest in an area where abundant indicator minerals occur. An attempt was made to diamond drill the body, but the rock was too soft to recover any drill core. 20 percussion drill holes were done and five of them encountered kimberlite. According to percussion drillings and a gravimetric survey, the body is almost round and 0.3 ha in size. A 21 kg sample taken from the surface of the body produced no micro diamonds.

PIPE 21. LAPINLUHTA

34 till samples were collected to investigate the heavy mineral content in the area. To determine the location and size of the body, magnetic and electric surveys were concluded and showed that the body is 1.6 ha in size and irregular in shape. 62 diamond drill holes were drilled (combined length 1478.5 m). Six percussion drill holes were completed and three separate samples were sent for microdiamond analyses, with a total weight of 109 kg. Six mini-bulk samples were taken and 16.6 tons of material were processed. Microdiamond analyses revealed 37 diamonds, four were +0.4 mm. Mini bulk samples contained 26.65 ct/100t. One stone was 1.126 ct in size. Excluding this stone, the diamond content would be 19 ct/100t. 324 grains were analysed from heavy mineral sampling (259 garnets, 50 ilmenites, 7 chromites and 8 PSBK-F). An age determination gave a K-Ar age of 571 ± 5 Ma.

PIPE 22. UUHILAHTI

14 till samples were collected from the area and 41 garnets were analysed from these. To determine the location and shape of the body 16 drill holes were done. In ten of these, kimberlite was intersected.

Geophysical (magnetic and electromagnetic) measurements were done to better define the extent of the body, which is very small, ca. 0.2 ha. 60 kg of drill core was selected for microdiamond analyses, 22 diamonds were recovered, three of which were +0.4 mm.

PIPE 23. JOKIHARJU

Geophysical measurements in the claim area included magnetic and electromagnetic profiles. The resulting anomaly was drilled with five drill holes (combined length 257.0 m) to define the body, which is oval-shaped with an area of 0.25 ha and has steep, almost vertical contacts. 154.5 kg of drill core selected contained very low contents of micro diamonds.

PIPE 25. VIITASALO (CLAIMS DI 59 & DI 131)

15 till samples (11 from surface till and 4 from basal till) were collected from the area. The samples contained abundant indicator minerals, 343 were analysed (172 GT, 108 PIL, 39 PC, 4 CHR and 22 Other). To locate the source of these minerals, magnetic and electric surveys were concluded. Anomalies were drilled with 17 diamond drill holes (combined length 1208 m). From these drill cores 647 heavy mineral grains were analyzed (413 GT, 200 PIL, 19 CP and 25 Other). A 63.1 kg sample was analysed for micro diamonds, five were found. The kimberlite is estimated to be small and is under a lake.

PART

GOLD DEPOSITS

5

OVERVIEW ON GOLD DEPOSITS IN FINLAND

5.1

P. Eilu

ABSTRACT

Genetic types of gold deposits identified in Finland include Au-rich volcanogenic massive sulfide (VMS), metamorphosed high-sulfidation epithermal, porphyry gold-copper, orogenic gold, placer, and paleoplacer deposits. There also are gold deposits and occurrences whose genetic type is ambiguous, such as the Au-Co ± Cu occurrences in the Kuusamo schist belt, the Au-U occurrences in the Peräpohja schist belt, and the Pahtavaara gold deposit. In contrast to many other Precambrian shield areas, most of the known gold occurrences and resources in Finland are hosted by Paleoproterozoic sequences. All gold-mineralized environments in Finland, except the possible Archean epithermal deposits and the recent placers, can be related to supercontinent evolution of the region between ~2.75–1.77 Ga. Most of the gold deposits were formed during the main stages of crustal growth globally, at ~2.72–2.64 Ga and 1.91–1.77 Ga, during the Neoarchean and Svecofennian orogenies, respectively.

Keywords: Archean; Paleoproterozoic; Fennoscandian Shield; Finland; orogenic gold; epithermal gold; porphyry gold; supercontinents.

INTRODUCTION

Exploration for gold deposits as major targets began in Finland during the 1980s. Exploration activities soon spread across most of the country, as indicated by the discovery dates of deposits given in Table 5.1.1. Prior to the 1980s, the common perception was that northern Europe lacked major gold deposits.

Mineral Deposits of Finland. http://dx.doi.org/10.1016/B978-0-12-410438-9.00015-7

Table 5.1.1 Gold mines and deposits with a reported resource in Finland

Genetic type, deposit	District[a]	Discovery year	Period of mining	Total size (Mt)	Mined (Mt)	Grade[b]					Reference to remaining resource[d]
						Au	Ag	Co	Cu	Other metals	
Orogenic gold											
Pampalo	Ilomantsi	1990	1996, 2010–	1.85	0.55	4.1					Endomines (2013)
Valkeasuo	Ilomantsi	1992		0.68		3.3					Endomines (2013)
Kuittila	Ilomantsi	1984		0.28		2.58					Damsten (1990)
Rämepuro	Ilomantsi	1984		0.31		2.1					Endomines (2013)
Korvilansuo	Ilomantsi	1986		0.26		2.0					Endomines (2013)
Muurinsuo	Ilomantsi	1987		0.85		1.6					Endomines (2013)
Kuivisto	Ilomantsi	1993		0.18		1.4					Endomines (2013)
Kuikkapuro	Suomus-salmi	1997		0.05		14.6					Heino (2000)
Moukkori	Suomus-salmi	1990		0.02		10.6					Parkkinen (2003)
Pahkalampi	Suomus-salmi	1996		0.25		3.5					Parkkinen (2003)
Pahkosuo	Suomus-salmi	1995		0.10		1.55					Heino (2001)
Syrjälä	Suomus-salmi	1995		0.07		1.55					Heino (2000)
Kutuvuoma	CLGB	1993	1998, 2000	0.07	0.01	6.7					Korkalo (2006)
Suurikuusikko	CLGB	1986	2008–	64.22	4.27	4.14					Agnico-Eagle (2013)
Soretialehto	CLGB	1989		0.01		3.5					Keinänen (1994)
Hirvilavan-maa	CLGB	1986		0.11		2.9					Hulkki and Keinänen (2007)
Kuotko	CLGB	1986		1.82		2.89					Agnico-Eagle (2013)
Vesiperä	R-H	1984		0.3		2.5					Sipilä (1988)
Laivakangas	R-H	1982	2011–2014	26.65	1.53	1.68					GSA Global (2012)
Hirsikangas	R-H	2004		5.68		1.25					GSA Global (2012)

Ängesneva	R-H	1987		3.85		1.19					Belvedere Resources (2010)
Sikakangas	SO	1989		0.17		1.32					Isomaa et al. (2010)
Osikonmäki	SS	1986		4.84		2.0					Belvedere Resources (2011)
Kaape-linkulma	Pirkanmaa	1986		0.16		6.2					Dragon Mining (2009)
Jokisivu	Pirkanmaa	1985	2009–	2.13	0.35	5.2					Dragon Mining (2012a)
Satulinmäki	Häme	1990		0.36		2.34					Kärkkäinen et al. (2006)
Orogenic gold with anomalous metal association											
Kaaresselkä	CLGB	1987		0.3		5					Hulkki and Pulkkinen (2007)
Saattopora	CLGB	1985	1988–1995	2.16	2.16	2.9			0.25		
Kettukuusikko	CLGB	1977		0.44		1.8					Taranis Resources (2011)
Riikonkoski[c]	CLGB	1966		9.56		?			0.45		Yletyinen and Nenonen (1972)
Ängeslampi	R-H	1986		0.27		3.1			0.14		Sipilä (1990)
Pirilä	SS	1983		0.3		6.5	32		0.18	0.76 Pb, 0.11 Zn	Parkkinen (2003)
Kalliosalo	SO	1977		0.3		1	0.7			0.41 Sb	Saltikoff (1980), Tyni (1983)
Sivakkaharju[c]	Kuusamo	1986		0.05		7.5		0.03	0.12		Dragon Mining (2011)
Hangaslampi[c]	Kuusamo	1988		0.40		5.1		0.06	0.1		Dragon Mining (2012b)
Apajalahti[c]	Kuusamo	1970s		0.1		5		0.02	0.05	0.04 W	Lahtinen (1980)
Juomasuo[c]	Kuusamo	1985	1992	1.97	0.018	4.9		0.15	0.03		Dragon Mining (2011)
Pohjasvaara[c]	Kuusamo	1985		0.13		4.0		0.09	0.3		Dragon Mining (2011)
Iso-Rehvi[c]	Kuusamo	1988		0.04		4		0.05	0.1		Vanhanen (1991)
Meurastuk-senaho[c]	Kuusamo	1984		0.89		2.3		0.2	0.1		Dragon Mining (2011)
Säynäjävaara[c]	Kuusamo	1983		0.4		1		0.06	0.02		Pankka and Vanhanen (1992)
Kivimaa[c]	Peräpohja	1965	1969	0.02	0.018	5.3			1.87		

Continued

Table 5.1.1 Gold mines and deposits with a reported resource in Finland—cont'd

Genetic type, deposit	District[a]	Discovery year	Period of mining	Total size (Mt)	Mined (Mt)	Grade[b]					Reference to remaining resource[d]
						Au	Ag	Co	Cu	Other metals	
VMS											
Pahtavaara[c]	CLGB	1985	1996–	7.74	4.98	2.62					Lappland Goldminers (2013a)
Iilijärvi	Uusimaa	1757	1788, 1884	0.05	0.003	4	30		0.6	0.6 Pb, 1.3 Zn	Mäkelä (1989)
Metsämonttu	Uusimaa	1946	1951–1974	1.51	1.51	1.43	25		0.28	0.74 Pb, 3.34 Zn	
Haveri	Tampere	1737	1942–1962	26.3	1.56	1			0.5		Lappland Goldminers (2008)
Outokumpu	Outokumpu	1910	1910–1989	28.5	28.5	0.8	8.9	0.24	3.8	0.12 Ni, 1.07 Zn	
Vihanti	V-P	1946	1954–1992	37.10	27.94	0.44	25		0.48	0.36 Pb, 4 Zn	Outokumpu Oy unpubl. data (1992)
Pyhäsalmi	V-P	1958	1962–	67.38	50.86	0.3	14		0.8	2.2 Zn	Inmet Mining (2013)
Epithermal											
Kutemajärvi	Tampere	1982	1994–	3.19	2.03	5.2					Dragon Mining (2012a)
Kylmäkangas[c]	Oijärvi	1999		1.9		4.11	31				Agnico-Eagle (2013)
Porphyry											
Kopsa	R-H	1939		16.8		0.81	2.2		0.16		Belvedere Resources (2013)
Kedonojan-kulma	Häme	2010		1.8		0.12	11		0.4	0.0018 Mo	Tiainen et al. (2013)
IRG?											
Mäkärärova[c]	N. Finland	1949		0.08		2.1					Nurmi et al. (1991)

Included are those polymetallic deposits where gold potentially is among the main commodities. Also included are the three largest VMS mines (Outokumpu, Pyhäsalmi, Vihanti), as these jointly produced most of the pre-2000 gold in Finland. All deposits (except Pyhäsalmi and Outokumpu) are open at depth, most deposits open also along the strike. Mined tonnages are given as of the end of 2012. Production figures are from the Finnish Mining Registry.

[a]*District names as in Fig. 5.1.2; CLGB = Central Lapland Greenstone Belt, R-H = Raahe-Haapajärvi, SO = Southern Ostrobothnia, SS = Southern Savo, V-P = Vihanti-Pyhäsalmi.*

[b]*Grade in g/t for Au and Ag, other metal grades in percent; ? = poor coverage of data, estimated to range 0.1–5 g/t.*

[c]*Genetic type unclear.*

[d]*No reference: no reported remaining resource.*

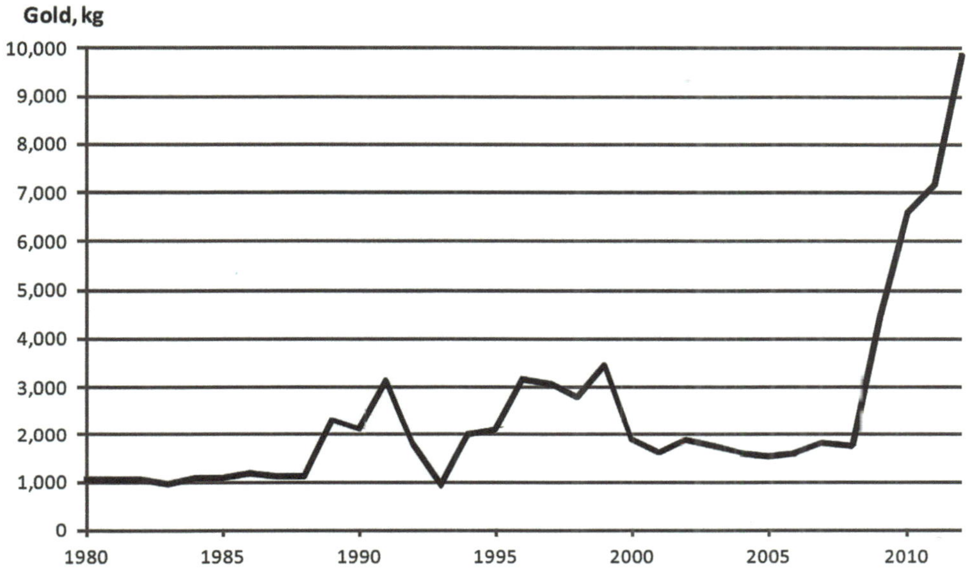

FIGURE 5.1.1

Annual gold production from Finnish mines from 1980 to 2012. Data mainly from Finland's Mining Registry statistics (summarised until 2001 by Puustinen, 2003). The 2012 data is from mining company annual reports. By-product gold from the base metal mines Pyhäsalmi, Vihanti, and Vuonos (Chapter 8) is estimated using the mined ore tonnage, average gold content in the ore and an estimated 80 % recovery grade, except for 1986–2001 where detailed gold production data exists from these mines. No gold grade nor production data exists for the Hammaslahti Cu mine (Urpo Kuronen, pers. comm. 23 Jan 2013) which produced, perhaps, 100–500 kg gold annually during 1973–1986.

The increase in the gold price, major breakthroughs in analytical methods for low-level trace element concentrations, and an improved understanding of the geology and genesis of gold deposits during the 1980s and 1990s helped to generate a much more optimistic exploration environment in Finland. Before commercial production at the Kittilä mine (Suurikuusikko deposit) started in 2009, production of gold had never played a major role in the Finnish mining industry (Puustinen, 1991, 2003). At present (2014), there are five producing gold mines (Table 5.1.1), and Finland has become one of the main gold producers in Europe. However, on a global scale, Finland remains a small player in gold mining; the annual production is close to 10 t (Fig. 5.1.1), representing <0.5 % of global production (USGS, 2012).

There are currently more than 200 drill-indicated gold deposits and occurrences in Finland (Fig. 5.1.2) and new occurrences are discovered every year. About 15–20 companies are currently actively exploring for gold, and more information is flowing in from the earlier discoveries (Eilu and Pankka, 2009; Saalmann and Niiranen, 2010; Grönholm and Kärkkäinen, 2012; Eilu and Niiranen, 2013). Research on various gold deposit types throughout the world continuously improves our ability to explore for these targets and our understanding of the petrogenesis of the deposits. Therefore, many of the Finnish deposits are now assigned to a different genetic type than they were just 10 years ago, the time of publication of the previous in-depth review (Eilu et al., 2003).

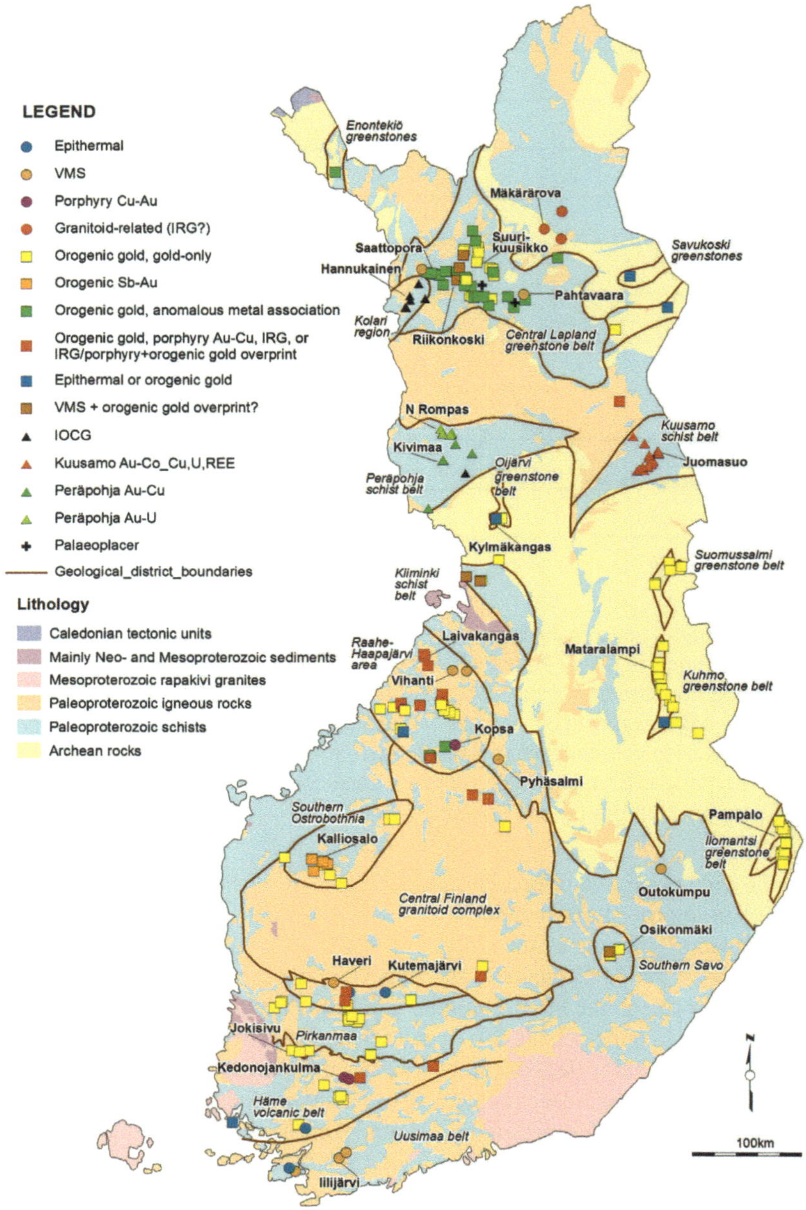

FIGURE 5.1.2

Gold deposits and occurrences in Finland, and the three major VMS-type base metal mines which have produced significant gold (Outokumpu, Pyhäsalmi, Vihanti). The deposit data are based on Eilu and Pankka (2009), Grönholm and Kärkkäinen (2012), Eilu and Niiranen (2013), and on reports referred to in these compilations. Geology is based on the GTK digital bedrock map database as of October 2013. The map is drafted by Kirsti Keskisaari, GTK.

The main aim of this subchapter is to describe the various genetic types of gold deposits in Finland. Several ambiguous cases in which the genetic type is not obvious are also included. All deposit types are described with an emphasis on general and diagnostic features that are characteristic of the Finnish examples. Details of deposits are only given where this serves to illustrate typical features of a certain deposit type. The Suurikuusikko (Kittilä mine) and Pampalo deposits, as well as the gold occurrences in the Rompas area are described in detail in other chapters (i.e., 5.2, 5.3, and 5.4) of this book.

GENETIC TYPES OF GOLD DEPOSITS RECOGNIZED IN FINLAND

Genetic types of gold deposits identified in Finland include Au-rich volcanogenic massive sulfide (VMS), metamorphosed high-sulfidation epithermal, gold-copper porphyry, orogenic gold, placer, and paleoplacer deposits (Fig. 5.1.2). In addition, certain gold occurrences such as Kopsa (Belvedere Resources, 2012) and Kuusamo (Slack et al., 2010) have recently been classified as intrusion-related (IRG) and Blackbird-type deposits. However, conclusive evidence for these models are still lacking. The iron-oxide–copper–gold (IOCG) deposits in the Kolari region and, possibly, in the Peräpohja belt (Fig. 5.1.2) are described in detail in Chapter 6 (Moilanen and Peltonen, 2015) and will not be discussed here. The large variety of deposit types, with the dominance of orogenic gold deposits (Eilu and Pankka, 2009), is typical for Precambrian shield areas and other younger metamorphic terranes (Goldfarb et al., 2005; Groves et al., 2005; Huston et al., 2012).

In general, gold deposits of various types are genetically linked to major tectonic processes such as submarine rifting and volcanism, intracratonic rifting, arc magmatism, arc and microcontinent accretion, continent–continent collision, and multistage regional deformation and metamorphism. It is a function of both geodynamic processes and preservation potential that define which deposit type(s) dominate in a region. Of all the gold deposit types, orogenic gold deposits have the best preservation potential because they form in relatively deep midcrustal regimes, particularly in terranes dominated by greenschist to midamphibolite facies rocks (Goldfarb et al., 2005). This is the case in Finland too, where they are the most common of the gold deposit types (Eilu and Pankka, 2009; Grönholm and Kärkkäinen, 2012): Orogenic gold deposits have been detected in nearly all Finnish greenschist- to amphibolite-facies supracrustal belts, whereas other genetic types are found in only some of the belts and regions in Finland, and in much smaller numbers. Also, gold occurrences are much less abundant both in the high-grade terranes and in regions containing only intrusive rocks.

OROGENIC GOLD DEPOSITS

Epigenetic gold deposits hosted by orogenic belts and formed by syn- to late-orogenic fluids are called *orogenic*. Their characteristics have been summarized by Böhlke (1982), Sibson et al. (1988), Groves et al. (1998), McCuaig and Kerrich (1998), Eilu et al. (1999), Eilu and Groves (2001), Goldfarb et al. (2001, 2005), and Pitcairn et al. (2006). Mineralization typically takes place in accretionary and collisional plate-tectonic settings under compressional to transpressional deformation regimes. The mineralizing agents are low-salinity H_2O-CO_2 fluids containing minor amounts of CH_4, N_2, and H_2S. The fluids carry gold dominantly as bisulfide complexes. The gold is mainly deposited in quartz veins by pressure fluctuations during seismic events and adjacent to veins by reaction between the fluid and the wall rocks. Most of the stable and radiogenic isotope data from ores, as well as trends in gold and associated trace element concentrations in metamorphosed belts, point toward progressive metamorphic dehydration, mostly at midcrustal levels, as being the main process releasing the fluids and metals, with no detectable contributions from local rocks. On the other hand, the data exclude meteoric fluids (modified or not) and fluids derived from local granitoids or the mantle as significant metal sources. The auriferous veins are enveloped by alteration halos characterized by K-mica and CO_2 alteration.

The alteration mineral assemblages reflect the metamorphic grade of the host rocks and ore mineral assemblages, and local structures and textures indicate that mineralization and alteration commonly took place under postpeak metamorphic conditions along the retrograde PT curve. The deposits have a distinct structural control; despite occurring near major crustal structures (first-order faults), they are dominantly hosted by second- to fourth-order faults and shear zones. Deposits may be hosted by any supracrustal rock type within a metamorphic belt, or by intrusions within or adjacent to such a belt. The locally most competent and/or reactive lithological units tend to host most of the ore. Typically, the Au/Ag ratio of the ore is 1–5; elements enriched in the ore and in the alteration halo include As, Au, Bi, K, Rb, Sb, Te, and W; and gold is the sole economic commodity. Although quartz veining is one of the diagnostic features, Si enrichment of the host rocks is rare. Dating of the deposits and their host rocks indicates that mineralization took place significantly after the formation of the host rocks; typically the time gap is 10–100 Ma.

There are about a dozen Archean and 30 Paleoproterozoic orogenic gold deposits in Finland for which both grade and tonnage data are available (Table 5.1.1); their known size varies from 0.04 to 64 Mt and gold grade is 1–15 g/t (FODD, 2013, and references therein). The two largest deposits are Suurikuusikko, with a premining resource of 266 t Au, and Laivakangas with 44 t Au. Half of the orogenic deposits in Finland contain less than 1 t Au. Sufficient resource data to make statistical comparisons with orogenic gold belts elsewhere only exist for the Central Lapland Greenstone Belt (CLGB), the Ilomantsi greenstone belt, and the Raahe-Haapajärvi area (Table 5.1.1, Fig. 5.1.2). Only in the CLGB does the deposit size distribution, with one anomalously large and many smaller deposits, resemble productive parts of greenstone belts elsewhere, such as in Zimbabwe or Western Australia (Hokka, 2011, and references therein). This could mean that (1) large and medium-sized deposits in many regions of Finland remain to be discovered; (2) some of the known deposits are much larger than their presently defined resource; or (3) many of the Finnish gold belts are anomalous in that they lack large gold deposits. Particularly intriguing is the fact that no >1 t gold deposit is known from the Kuhmo and Suomussalmi greenstone belts (KSGB), whereas the geological setting of these belts (i.e., their combined surface area, age, rock types, and metamorphic grade) is very similar to that of the Southern Cross greenstone belt in Western Australia that hosts several >30 t Au deposits and has yielded >300 t Au (Geological Survey of Western Australia, unpublished data, 2013). Another interesting detail is that the grade and tonnage distribution of Finnish orogenic gold deposits is statistically similar to the Paleoproterozoic gold deposits of northern Australia (Eilu et al., 2015). The metal grades and Au/Ag ratios in the Finnish deposits are also similar to those of deposits in other Precambrian and Phanerozoic regions. A number of Finnish orogenic gold deposits contain additional potential economic metals (Co, Cu, Ni, Sb; Table 5.1.1); these are discussed separately in the section 'Orogenic gold deposits with anomalous Ag, Cu, Co, Ni or Sb,' below.

Isotopic age data are available from only a few Finnish orogenic gold occurrences, but the present data and abundant information on structural relationships and mineral textures from the ores and their host rocks clearly suggest that mineralization took place soon after the main regional deformation. Mineralization within the Archean greenstone belts took place close to the D3 or D4 stages of regional deformation, between 2.72 and 2.64 Ga (Sorjonen-Ward, 1993; Rasilainen, 1996), or more probably between 2.68 and 2.64 Ga (Hölttä et al., 2012). This is the time of the global peak of Archean orogenic gold mineralization, of global Neoarchean orogenic activity, and of peak crustal growth (Groves et al., 2005). The second age group of orogenic gold deposits in Finland is related to the Svecofennian composite orogenies, at 1.91–1.77 Ga, coinciding with the second Precambrian global peak of orogenic gold mineralization (Groves et al., 2005). Most of the Paleoproterozoic mineralization probably took place during a continent–continent collision event at 1.84–1.78 Ga, during the regional D3–D4 (or later) stages of deformation (Patison, 2007; Saalmann et al., 2009, 2010; Saalmann and Niiranen, 2010; Eilu et al., 2013). However, some mineralization may be related to the earlier compressional stage representing accretion of igneous arcs and microcontinents at 1.91–1.87 Ga (Mänttäri, 1995; Lahtinen et al., 2012). Unpublished Re-Os isotope data of auriferous arsenopyrite from Suurikuusikko provide support for the earlier, syn-accretionary timing of mineralization (Geospec Consultants Limited, 2008). Indirect structural, textural, and mineral paragenetic evidence from the Häme, southern Savo, and Pirkanmaa belts of southern Finland suggest an ~1.88–1.87 Ga timing for some of the orogenic gold occurrences (e.g., Kontoniemi, 1998).

Some parts of the Archean greenstone belts show an overprint of Svecofennian deformation. There is, however, no recognized Paleoproterozoic gold mineralization event in the Archean rocks in Finland, despite the Proterozoic metamorphic recrystallization and disturbances of isotopic systems (Kontinen et al., 1992; O'Brien et al., 1993; Hölttä et al., 2012). Several generations of Paleoproterozoic dolerites, subsequently metamorphosed at greenschist facies conditions, crosscut the Archean gold ores in the Ilomantsi and KSGB belts, but show no indication of post-Archean gold mineralization (Nurmi et al., 1993; Eilu, 2003).

Two of the Finnish Paleoproterozoic gold deposits (Table 5.1.2), and a small number of occurrences of the same age, but without a publicized resource estimate, are hosted by upper-amphibolite facies rocks (Eilu and Pankka 2009). Orogenic gold mineralization under upper-amphibolite to granulite facies conditions is considered unlikely (e.g., Tomkins and Grundy, 2009). Thus, it seems possible that a few of the Finnish Paleoproterozoic orogenic gold deposits have been metamorphosed, as also suggested by textures present in these occurrences. In both the southern Savo area and the westernmost parts of the Häme belt (Fig. 5.1.2), there are indications of postmineralization metamorphism at upper-amphibolite to lower-granulite facies conditions (Kontoniemi, 1998; Kärkkäinen et al., 2012). Elsewhere within the Svecofennian domain (e.g., in the region to the southwest of the Archean parts of Finland; Fig. 5.1.2), one may also expect some effects of postmineralization metamorphism as the region has experienced not just one, but several spatially overlapping orogenies (Lahtinen et al., 2005, 2012). Such an overlap may result in local (mm to cm scale) remobilization of Au (± As, Bi, Sb, Te) into late fractures, leading to possible misinterpretation of the timing of mineralization. One should also keep in mind that the rather common, late to postorogenic shearing and brittle fracturing may also result in local remobilization of earlier ductile gold grains into new fractures in any orogenic gold occurrence.

All orogenic gold deposits in Finland are structurally controlled. They typically occur within 0.5–3 km from a major, first-order fault, which, in most cases, follows the main strike of the supracrustal belts. The hosting structure is generally a second- or third-order fault or shear zone branching from the major structure. Examples include the faults branching to the northwest and northeast from the main north-trending fault along the KSGB (Luukkonen et al., 2002) and the west-northwest- and west-southwest-trending shear zones associated with the regional northwest-trending shear zone at Jokisivu (Saalmann et al., 2010). All the controlling structures show more than one stage of deformation. The change from earlier high-angle thrusting to later strike-slip movement may be critical for the migration of the mineralizing fluid from the deeper prograde metamorphic source areas, via the main fault, to shallower, retrograde metamorphic crustal levels and to lower-order structures where the fluid would precipitate gold (Sibson et al., 1988; Patison, 2007). The favorable locations of mineralization are those with low or minimum local stress; these include intersections between faults, between a fault and a fold hinge, a slight bend in a fault along a lithological contact, a deviation of the general trend of a fault, or a combination of any of these. The relatively late formation of these structurally favorable sites is one

Table 5.1.2 Ages and regional metamorphic grades of gold deposits and occurrences with a reported resource in Finland

Genetic type, deposit	District[a]	Age (Ga)	Main host rock(s)[b]	Metamorphic grade[e]	Reference
Orogenic gold					
Pampalo	Ilomantsi	2.72–2.65	Intermediate tuffite	Greenschist-amphibolite tr.	Sorjonen-Ward (1993)
Valkeasuo	Ilomantsi	2.72–2.65	Intermediate tuffite	Greenschist-amphibolite tr.	Sorjonen-Ward (1993)
Kuittila	Ilomantsi	2.72–2.65	Tonalite pluton	Greenschist-amphibolite tr.	Sorjonen-Ward (1993)
Rämepuro	Ilomantsi	2.72–2.65	Tonalitic porphyry	Greenschist-amphibolite tr.	Sorjonen-Ward (1993)
Korvilansuo	Ilomantsi	2.72–2.65	Mica schist	Greenschist-amphibolite tr.	Sorjonen-Ward (1993)
Muurinsuo	Ilomantsi	2.72–2.65	Mica schist	Greenschist-amphibolite tr.	Sorjonen-Ward (1993)
Kuivisto	Ilomantsi	2.72–2.65	Intermediate tuffite	Greenschist-amphibolite tr.	Sorjonen-Ward (1993)
Kuikkapuro	Suomussalmi	2.72–2.65	Basalt	Mid amphibolite	Luukkonen et al. (2002)
Moukkori	Suomussalmi	2.72–2.65	Intermediate volcanic rock	Lower amphibolite	Luukkonen et al. (2002)
Pahkalampi	Suomussalmi	2.72–2.65	Amphibolite	Mid amphibolite?	Luukkonen et al. (2002)
Pahkosuo	Suomussalmi	2.72–2.65	Tholeiitic basalt	Mid amphibolite?	Luukkonen et al. (2002)
Syrjälä	Suomussalmi	2.72–2.65	Intermediate volcanic rock	Mid amphibolite	Luukkonen et al. (2002)
Kutuvuoma	CLGB	1.91–1.79	Komatiite, albitized phyllite[c]	Low to mid greenschist	Eilu et al. (2007)
Suurikuusikko	CLGB	1.91–1.79	Tholeiitic basalts, albitized phyllite[c]	Low to mid greenschist	Patison et al. (2013a)
Soretialehto	CLGB	1.91–1.79	Komatiite	Low to mid greenschist	Keinänen (1994)
Hirvilavanmaa	CLGB	1.91–1.79	Komatiite, mafic lava or dolerite	Low to mid greenschist	Hulkki and Keinänen (2007)
Kuotko	CLGB	1.91–1.79	Mafic tholeiites	Low to mid greenschist	Eilu et al. (2007)
Vesiperä	R-H	1.89–1.79	Plagioclase porphyry	Mid amphibolite?	Sipilä (1988)
Laivakangas	R-H	1.89–1.79	Quartz diorite	Mid amphibolite?	Mäkelä (1984)
Hirsikangas	R-H	1.89–1.79	Felsic schist	Mid amphibolite?	Kontoniemi and Mursu (2006)
Ängesneva	R-H	1.89–1.79	Plagioclase porphyry	Upper amphibolite	Nurmi et al. (1991)
Sikakangas	SO	1.89–1.79	Plagioclase porphyry	Mid amphibolite?	Isomaa et al. (2010)
Osikonmäki	SS	1.89–1.79	Tonalite pluton	Upper amphibolite	Kontoniemi (1998)
Kaape-linkulma	Pirkanmaa	1.89–1.79	Quartz diorite	Lower amphibolite	Saalmann et al. (2009)
Jokisivu	Pirkanmaa	1.89–1.79	Mafic volcanic rocks	Mid amphibolite	Saalmann et al. (2010)
Satulinmäki	Häme	1.89–1.79	Felsic to intermedite volcanic rocks	Lower amphibolite	Kärkkäinen et al. (2006)

Orogenic gold with anomalous metal association					
Kaaresselkä	CLGB	1.91–1.79	Albitized phyllite[c]	Upper greenschist	Hulkki and Pulkkinen (2007)
Saattopora	CLGB	1.91–1.79	Albitized phyllite[c]	Low to mid greenschist	Korvuo (1997)
Kettukuusikko	CLGB	1.91–1.79	Komatiite	Low to mid greenschist	Eilu et al. (2007)
Riikonkoski[d]	CLGB	2.1–1.79?	Albitized phyllite[c]	Upper greenschist	Yletyinen and Nenonen (1972)
Ängeslampi	R-H	1.91–1.79	Plagioclase porphyry	Lower amphibolite	Sipilä (1990)
Pirilä	SS	1.91–1.79	Intermediate volcanosedimentary rock	Upper amphibolite	Makkonen and Ekdahl (1988)
Kalliosalo	SO	1.91–1.79	Plagioclase porphyry	Mid amphibolite?	Saltikoff (1980), Appelqvist (1993)
Sivakkaharju[d]	Kuusamo	1.9–1.8?	Albitized sedimentary rocks	Upper greenschist	Vanhanen (2001)
Hangaslampi[d]	Kuusamo	1.9–1.8?	Albitized sedimentary rocks	Upper greenschist	Vanhanen (2001)
Apajalahti[d]	Kuusamo	1.9–1.8?	Sericite quartzite (metasilt?)	Upper greenschist	Vanhanen (2001)
Juomasuo[d]	Kuusamo	1.9–1.8?	Sericite quartzite, basalt	Upper greenschist	Vanhanen (2001)
Pohjasvaara[d]	Kuusamo	1.9–1.8?	Albitized sedimentary rocks	Upper greenschist	Vanhanen (2001)
Iso-Rehvi[d]	Kuusamo	1.9–1.8?	Albitized sedimentary(?) rocks	Upper greenschist	Vanhanen (1991, 2001)
Meurastuk-senaho[d]	Kuusamo	1.9–1.8?	Albitized metasiltstone	Upper greenschist	Vanhanen (2001)
Säynäjävaara[d]	Kuusamo	1.9–1.8?	Sericite schist, mafic volcanic rock	Upper greenschist	Vanhanen (2001)
Kivimaa[d]	Peräpohja	1.9–1.8?	Dolerite	Upper greenschist	Rouhunkoski and Isokangas (1974)
VMS					
Pahtavaara[d]	CLGB	2.10–2.05?	Komatiite	Upper greenschist	Korkiakoski (1992)
Iilijärvi	Uusimaa	1.90–1.89	Felsic volcanic rocks	Lower amphibolite	Mäkelä (1989)
Metsämonttu	Uusimaa	1.90–1.89	Felsic and mafic volcanic rocks	Lower amphibolite	Mäkelä (1989)
Haveri	Tampere	~1.905	Tholeiitic basalt	Lower amphibolite	Eilu (2012)
Epithermal					
Kutemajärvi	Tampere	1.90–1.89	Intermediate volcanic rocks	Lower amphibolite	Poutiainen and Grönholm (1996)
Kylmäkangas[d]	Oijärvi	~2.8?	Quartz-feldspar porphyry	Upper greenschist	Juopperi et al. (2001)

Continued

Table 5.1.2 Ages and regional metamorphic grades of gold deposits and occurrences with a reported resource in Finland—cont'd

Genetic type, deposit	District[a]	Age (Ga)	Main host rock(s)[b]	Metamorphic grade[e]	Reference
Porphyry					
Kopsa	R-H	1.91–1.85	Tonalitic intrusion	Lower amphibolite	Gaal and Isohanni (1979)
Kedonojan-kulma	Häme	1.89–1.88	Tonalitic intrusion	Mid amphibolite	Tiainen et al. (2013)
IRG?					
Mäkärärova[d]	N. Finland	1.9–1.75?	Granitoid gneiss	Upper amphibolite	Härkönen (1987)

Question marks indicate uncertain information.
[a]*District names as in Fig. 5.1.2; CLGB = Central Lapland Greenstone Belt, R-H = Raahe-Haapajärvi, SO = Southern Ostrobothnia, SS = Southern Savo.*
[b]*All hosts are metamorphosed.*
[c]*The "phyllite" includes originally fine-grained rocks from dominantly clastic sedimentary to dominantly volcaniclastic origin, all types commonly with a minor graphite content.*
[d]*Genetic type unclear.*
[e]*"Greenschist-amphibolite tr." = metamorphic grade is close to the transition from greenschist to amphibolite facies.*

of the main indications for the late timing of orogenic gold mineralization, such as along and near the Sirkka thrust zone in the CLGB (Patison, 2007; Saalmann and Niiranen, 2010).

A further important control on gold mineralization is defined by the competency and reactivity of the rock types within an area. Commonly, it is the locally most competent lithological unit that hosts the ore, because such rocks behave in the most brittle manner during deformation, thus creating more open space when the fluid pressure exceeds the lithostatic pressure. If a sudden pressure drop follows, then much of the metal in the fluid will rapidly precipitate; this may be repeated hundreds to thousands of times over a few million years (Sibson et al., 1988; Weatherley and Henley, 2013), forming the common banded ribbon textures of ore-hosting veins at many deposits. Competent hosts may include the most felsic rocks of an area, such as quartz porphyries in pelitic or mafic–ultramafic sequences (Eilu and Ojala, 2007), or even originally ductile rocks that have been hardened by pervasive alteration before mineralization. Ground preparation by premineralization alteration in the CLGB is common within a few kilometers of the Sirkka thrust zone, with intensely albitized and carbonatized volcaniclastic sedimentary (Kaaresselkä, Kutuvuoma, Levijärvi-Loukinen, Saattopora, Sirkka) and carbonatized ± albitized komatiitic units (Hirvilavanmaa, Kutuvuoma, Levijärvi-Loukinen, Soretialehto) (Keinänen, 1994; Korvuo, 1997; Holma and Keinänen, 2007; Hulkki and Keinänen, 2007; Hulkki and Pulkkinen, 2007). Clearly, the Sirkka thrust had been a major fluid conduit before gold mineralization, such that earliest fluid events made the surrounding rocks much more competent. Similar ground preparation has also taken place along some of the north-, northwest-, and northeast-trending faults of the CLGB, such as the Hanhimaa, Kuotko, and Suurikuusikko (Kiistala) faults. One would also expect extensive ground preparation by alteration in other supracrustal sequences with an extensive preorogenic evolution in intracontinental and marginal basin settings, especially in the Kuusamo and Peräpohja belts. This is because such settings potentially provided not just the typical low-salinity metamorphic fluids, but also brines (Yardley and Graham, 2002; Yardley and Cleverley, 2013) that were potentially able to alter the supracrustal sequence before the orogeny.

The rocks that are most reactive relative to the typical orogenic gold fluids are those with the highest Fe/(Fe + Mg) ratio (Phillips, 1986). Together with their enhanced competence, this makes iron-rich tholeiitic units the most prospective ore hosts in many areas of Finland. In places, felsic rocks of favorable Fe/(Fe + Mg) can also be reactive, as can pelitic or volcanogenic sedimentary units that contain reduced carbon (typically graphite). Ores hosted in reactive rocks, such as those at Suurikuusikko, are often called *rock hosted*; that is, most of the gold occurs with disseminated pyrite, pyrrhotite, and/or arsenopyrite in the altered host rock, with much less gold and sulfides occurring in the associated quartz veins.

The ore bodies in orogenic gold deposits typically have a flat lensoid shape. The plunge of the ellipsoid follows the lineation produced by the latest regional deformation in the region. The hosting fault and the ore bodies mostly have a steep to subvertical dip. Shallow-dipping ores also occur, such as at Osikonmäki (Kontoniemi, 1998), but typically form a minor part of a deposit. The ore bodies have lengths from tens of meters to more than 1 km along strike and plunge, and widths of 0.5–10 m. The deposits typically comprise several ore bodies (lodes) along the structure. At the Kittilä mine, the Suurikuusikko deposit and the adjacent ore bodies form a continuum >4 km long, which remains open to the north along the strike. Subparallel lodes may also be present, particularly where the hosting structure is wider than 20 m. Even where an outcropping ore body apparently terminates, additional, blind ore bodies may be present at depth as, for example, at Suurikuusikko, where extensions of the deposit extend beyond 1.5 km depth (Patison et al., 2013a). Most of the orogenic gold deposits in Finland have a reported extent along the strike that is significantly less than 1 km and extend to a maximum depth of 50–200 m. In most cases, however, this limited extent is likely due to limited drilling and exploration, and the deposit is open both along the strike and at depth. For example, this appears to be the case for all deposits in the Neoarchean Ilomantsi greenstone belt (see the Pampalo section, Subchapter 5.3) and at the Paleoproterozoic Jokisivu deposit (Dragon Mining, 2013).

Orogenic gold deposits typically contain 1–5 vol% sulfides and sulfarsenides. Pyrite ± arsenopyrite are dominant in greenschist-facies rocks, whereas pyrrhotite ± löllingite and arsenopyrite characterize deposits in amphibolite facies rocks (Eilu and Pankka, 2009). In the rocks of higher metamorphic grade, any löllingite is typically rimmed by arsenopyrite. Chalcopyrite, sphalerite, and galena are often observed in those parts of the ores that are metamorphosed to the highest grade, but these minerals are of relatively minor abundance, and mass balance evaluations consistently show no significant base metal mobility, although a few notable exceptions are described in the section 'Orogenic gold deposits with anomalous Ag, Cu, Co, Ni or Sb.' Scheelite and Bi-, Sb-, Se-, and Te-bearing minerals have been detected in many occurrences studied by microprobe (e.g., Kontoniemi et al., 1991; Kojonen et al., 1993; Luukkonen, 1994; Kärkkäinen et al., 2006, 2012; Etelämäki, 2007). However, these minerals only occur in minor to trace amounts, quite commonly in volumes equal or less than that of the gold. The main exceptions are the Sb-rich deposits in the Southern Ostrobothnia area, such as the Kalliosalo deposit where native antimony, stibnite, and aurostibite are common (Appelqvist, 1993).

Most of the gold is in its native form, present in fractures within and between the sulfides and the gangue minerals, and in some cases closely associated with Bi-, Sb-, and/or Te-bearing minerals. Gold fineness typically varies between 820 and 950. In a few deposits, including Suurikuusikko, gold is dominantly refractory, situated in the lattice of, or as submicroscopic inclusions in, arsenopyrite and pyrite (Kojonen and Johansson, 1999). In midamphibolite facies and higher-grade rocks, gold may occur as microscopic inclusions in arsenopyrite and as "invisible gold" in löllingite (Etelämäki, 2007). This suggests exsolution of gold from the löllingite lattice during retrograde replacement of löllingite by the arsenopyrite (Neumayr et al., 1993). Another common style of gold occurrence in high-grade rocks comprises composite grains of native gold, and auriferous Te- and Bi-bearing minerals (Kärkkäinen et al., 2012).

The styles of alteration, chemical changes, element enrichments, and the relationship between alteration mineral assemblages and metamorphic grade in the Finnish orogenic gold deposits are similar to those recorded in other parts of the world. Alteration includes sericitization and carbonatization at lower-greenschist to mid-greenschist facies conditions, biotitization and carbonatization at upper-greenschist to lower-amphibolite facies conditions, and biotitization and formation of K feldspar and calc-silicates in rocks of higher metamorphic grades. In addition, host rock sulfidation has taken place under all PT conditions of orogenic gold mineralization. Altered wall rocks may be associated with enrichments of Ag, As, Au, Bi, CO_2, K, Rb, S, Sb, Se, Te, and/or W (e.g., Nurmi et al., 1991; Rasilainen, 1996).

An example of orogenic gold-related alteration and associated chemical changes is provided by the Mataralampi occurrence in the northern part of the Kuhmo greenstone belt, hosted by a competent calc-alkaline quartz-feldspar porphyry (Eilu, 2003; Eilu and Ojala, 2007). Distal alteration is characterized by partial sericitization of biotite, partial to total replacement of titanite by rutile and epidote by calcite + quartz, and hydrothermal pyrite. With increasing degree of alteration and deformation closer to the deposit, muscovite gradually becomes more abundant than biotite and forms continuous, 0.1- to 1-mm-wide shear bands. In addition, calcite, quartz, and rutile gradually replace epidote and titanite, and feldspar phenocrysts are partially replaced by quartz + calcite + muscovite. Proximal alteration, directly adjacent to the auriferous quartz veins, is characterized by pervasive foliation and by the mineral assemblage quartz-albite-muscovite-calcite-rutile-pyrite. All feldspar phenocrysts are replaced by a fine-grained mass of albite + quartz + muscovite ± K-feldspar. The locations of the original quartz phenocrysts are still distinguishable by their sharp boundaries.

Pyrite occurs in disseminated form throughout the alteration halo, but is most abundant in and near quartz veins. Mass balance calculations show that Al, Cr, Ni, P, Ti, and Zr were immobile; Ba, Bi, CO_2, Cu, K, Rb, S, Sb, Te, and W were enriched; and Li, Na, and Sr were depleted throughout the alteration halo. Silver and gold are enriched only within the proximal alteration zone. Mataralampi is anomalous relative to most Finnish deposits because there is no enrichment of As in the former; in veins and in altered wall rock, As concentrations remain at typical background levels of less than 10 ppm. Only 1 km to the east, at the komatiite-hosted Timola occurrence, the As concentrations are as high as 2000 ppm in gold-mineralized zones (Hartikainen, 2001). This difference between two localities probably reflects the effect of host rock reactivity relative to arsenic solubility in the ore fluids: in the ultramafic rocks, As precipitated with gold, whereas in the more oxidized felsic wall rocks, As remained soluble in the gold depositing fluid.

The mineralizing fluid documented in most of the studied Finnish orogenic gold occurrences (see exceptions in the section 'Orogenic gold deposits with anomalous Ag, Cu, Co, Ni or Sb') is typical of this genetic type; it is a low-salinity H_2O-CO_2 fluid with significant H_2S (Kontoniemi, 1998; Poutiainen and Partamies, 2003). In addition to Au, the fluid probably also carried Ag, As, Bi, K, Rb, Se, and Te, as suggested by the ore and gangue mineral assemblages and geochemistry. Stable and radiogenic isotope work carried out on some of the Finnish occurrences (Bornhorst and Wilkin, 1993; Mänttäri, 1995; Kontoniemi, 1998; Hölttä and Karhu, 2001) suggests that the fluids were of similar composition to those in orogenic gold deposits elsewhere. The fluids contained no meteoric or seawater component and were interpreted to be dominantly of crustal origin derived from metamorphic dehydration. They possibly contained minor distal magmatic and/or mantle-derived components, but there are no obvious indications of local sources.

OROGENIC GOLD DEPOSITS WITH ANOMALOUS Ag, Cu, Co, Ni, OR Sb

There are several gold deposits in Finland that resemble typical orogenic gold deposits, but are anomalous because they contain Ag, Cu, Co, Ni, or Sb as potential commodities in addition to gold. These deposits are listed in Tables 5.1.1 and 5.1.2 under the title "Orogenic gold with anomalous metal association". In addition to pyrite, pyrrhotite, and/or arsenopyrite, they may also include significant amounts of chalcopyrite, cobaltite, pentlandite, gersdorffite, and/or stibnite (Vanhanen, 2001; Eilu et al., 2007). Goldfarb et al. (2001) suggest that orogenic gold deposits with an anomalous metal association form where Paleoproterozoic tectonism included deformation of older, intracratonic basins. According to Goldfarb et al., the resulting ore fluids were anomalously saline and in some cases the orogenic lodes are notably base metalrich. They cite examples that include the ore-hosting strata of the Transvaal basin in the Kaapvaal craton of South Africa and the Arunta, Tennant Creek, and Pine Creek inliers of northern Australia.

It is possible that such deposits reflect mobilization of basinal fluids under moderate to high-grade metamorphic conditions (Yardley and Graham, 2002; Yardley and Cleverley, 2013) with the metals possibly (but not necessarily) enriched prior to an orogeny in the fluid source areas by, for example, seafloor hydrothermal and/or diagenetic processes. A similar crustal evolution characterizes the Karelian domain of Finland where supracrustal sequences, possibly including evaporates, formed in intracratonic basins between 2.45 and 1.95 Ga (Perttunen and Vaasjoki, 2001; Rastas et al., 2001; Lahtinen et al., 2005, 2012). They were episodically intruded by magmas, locally resulting in alteration of the supracrustal rocks prior to regional metamorphism. It even seems possible that some (mostly uneconomic), local, base metal ± gold enrichment occurred before the orogeny. The extensive albitization and carbonatization in the Kuusamo, Peräpohja, and Central Lapland belts, and the regional scapolitization in northern Finland and northern Sweden, are the most obvious results of such preorogenic processes (Eilu, 1994; Vanhanen, 2001; Kyläkoski, 2009a, 2009b; Billström et al., 2010).

The issue is further complicated in the CLGB by the presence of multiply deformed, VMS-like, base metal occurrences and the unusual features of the Pahtavaara gold deposit. It remains unsolved how much these early processes affected the availability of gold and other metals now present in the orogenic Au-Cu ± Co, Ni, Sb deposits and occurrences. The geology and petrogenesis of the Pahtavaara gold deposit is discussed in more detail in the section 'Gold-rich VMS deposits'. Whatever the sources of metals and saline fluids, all the mentioned supracrustal sequences experienced multiple deformation, orogenic mineralization, and alteration during 1.91–1.77 Ga (Vanhanen, 2001; Lahtinen et al., 2005; Billström et al., 2010; Lahtinen 2012).

There are more than 20 drilled base metal-rich gold occurrences within the CLGB that can be best classified as base metal-rich orogenic gold deposits (Nurmi et al., 1991; Eilu et al., 2007; Eilu and Pankka, 2009), including the Saattopora deposit where copper was mined along with the gold (Table 5.1.1). In addition to gold, all these occurrences have copper as a potential commodity, and some of those hosted by or near komatiites have significant Ni and Co grades. Other common features include premineralization carbonatization and albitization, and moderate ore fluid salinities of 10–25 % NaCl equivalents. The abundance of Ni- and Co-bearing sulfide minerals in ores in or near the komatiites (e.g., Holma and Keinänen, 2007) suggests a local origin for the base metals. Structural relationships suggest that mineralization mostly took place late during the regional deformation. The Bidjovagge deposit in Norway and the Pahtohavare deposit in Sweden resemble Saattopora and the other Au-Cu occurrences in the CLGB (Eilu et al., 2007; Sandstad, 2012).

The Raahe-Haapajärvi area hosts several Au-Cu occurrences that may best be classified as Cu-rich orogenic gold deposits (Tables 5.1.1 and 5.1.2, and Fig. 5.1.2). Notably, the tectonic setting of this area would not appear to be favorable for the presence of saline metamorphic fluids because there are no indications of intracratonic basins nor evaporates, and the region seems to have undergone relatively rapid accretionary orogenic evolution (Lahtinen et al., 2005). A few Paleoproterozoic porphyry Cu ± Au occurrences, including the Kopsa deposit, have been detected in this area (Gaál and Isohanni, 1979). They are metamorphosed and appear to be overprinted by later orogenic gold mineralization. Such an overprint may explain the Cu-Au association in the region, where early porphyry deposits were tectonically buried and subjected to later mesozonal auriferous hydrothermal activity. In Fig. 5.1.2, such occurrences are included in the category "Orogenic gold, porphyry Au-Cu, IRG, or IRG/porphyry + orogenic gold overprint." Note, however, that not all of the occurrences of the mentioned category show an enrichment in Cu; for example, Laivakangas is a gold-only deposit (Table 5.1.1).

There also exist a few base metal-rich (Cu ± Zn, Ag, Pb) syngenetic VMS and, possibly, SEDEX occurrences that appear to be overprinted by orogenic gold, both within the CLGB (e.g., at Riikonkoski in CLGB) and elsewhere in Finland. These are indicated in Fig. 5.1.2 as "VMS + orogenic gold overprint?" Such overprinting and associated genetic issues are difficult to unravel in multiply deformed and metamorphosed terranes without radiometric dating of the mineralization processes, and thus classification remains far from certain. There are not enough data from any of the deposits marked in Fig. 5.1.2 as possible overprinted syngenetic ores to be sure of their genetic classification. In the past, many Finnish deposits, including those with very little precious metals, have been considered to be products of two different styles of mineralization. For example, the Haveri deposit has been suggested to fall into this "overprint class," but in the section 'Gold-rich VMS deposits', a different interpretation is advanced that solely favors a deformed syngenetic deposit.

KUUSAMO Au-Co ± Cu ± U ± LIGHT RARE EARTH ELEMENTS

The Kuusamo schist belt (KSB) is characterized by Au-Co ± Cu ± U ± LREE mineralization (Pankka, 1992; Pankka and Vanhanen, 1992; Vanhanen 2001). The reported tonnage and Au, Co and Cu grades of the deposits in the belt show considerable variation (Table 5.1.1). In a few cases, the resource also includes 0.01–0.03 % U and 0.1 % total rare earth elements (REE; Vanhanen, 2001; Dragon Mining, 2013). The main economic interest in the Kuusamo deposits is in Au and Co, whereas Cu and the REE have been regarded as potential by-products and U as a problematic waste product (Dragon Mining, 2013). The available data, mainly from Vanhanen (2001), suggest that most of the reported total REE grades in these occurrences comprise light rare earth elements (LREE). None of the KSB occurrences have so far been proven as economic ores, and only one (Juomasuo) has been test mined, in 1992 when 104 kg gold and 24.7 t cobalt were extracted from 17,635 t of ore (Puustinen, 2003).

The Kuusamo occurrences are hosted by a clastic sedimentary sequence deposited between 2.44 and 2.21 Ga, which also contains basaltic lavas and indications of evaporates (Vanhanen, 2001). The sequence was intruded by basaltic dikes and sills prior to regional deformation. All occurrences have a distinct structural control; most of them are hosted at the intersection of a regional northeast-trending anticline with northwest-trending faults. The largest deposit, Juomasuo, is located in a doubly-plunging part of the northeast-trending anticline.

Alteration in the KSB is extensive and occurred at several stages. The rocks in the belt were affected by three regional alteration events (Pankka, 1992; Pankka and Vanhanen, 1992; Vanhanen, 2001): (1) diagenetic, partial albitization of feldspars and sericitization of clay minerals in all sedimentary units; (2) local, partial to total albitization of clastic sedimentary units and spilitization of volcanic units, either related to the ~2.21 Ga intrusion of the mafic sills and dikes, or during the early stages of subsequent orogenic evolution; and (3) overprinting of the early spilitization and albitization by much more extensive albite and scapolite alteration and local carbonatization. The intensity of the last alteration varied from weak (<10 vol% albite) to strong, locally resulting in almost pure albite rocks (99 vol% albite, with traces of carbonate, rutile, and quartz).

Syndiagenetic, alkaline, oxidizing fluid activity has also been proposed for a clastic sedimentary sequence, 2.3–2.1 Ga in age, located in a rifted craton margin setting a few hundred km to the south of Kuusamo (Lahtinen et al., 2013). Such fluids can aggressively alter the rocks with which they interact, as indicated by the disturbed REE patterns in the area (Lahtinen et al., 2013). In addition to the regional alteration, local alteration surrounding the metallic mineralization in the KSB includes carbonatization, sericitization, biotitization, and sulfidation. The deposits are enriched in As, Au, Bi, Co, CO_2, Cu, K, LREE, Mo, Rb, S, Se, Te, U, and W. Somewhat less enriched are Fe, heavy REE (HREE), Mn, Mo, Ni, Pb, and Th, and the Au/Ag ratio is consistently higher than 1 (Pankka, 1992; Vanhanen, 2001). The main ore minerals in the KSB deposits are pyrite, pyrrhotite, cobaltite, cobaltian pentlandite, and chalcopyrite. Native gold occurs in free form within gangue and also associated with Bi- and Te-bearing minerals that are present as inclusions and in fractures in pyrite, pyrrhotite, cobaltite, and uraninite (Pankka, 1992; Vanhanen, 2001).

The metallic occurrences in the KSB have been historically classified into various deposit classes, including orogenic gold deposits with atypical metal associations, iron-oxide–copper–gold deposits, Blackbird-type deposits, and syngenetic deposits (e.g., Pankka, 1992; Pankka and Vanhanen, 1992; Vanhanen, 2001; D.I. Groves, personal communication, 2006; Slack et al., 2010; Slack, 2012). No similar deposits have yet been discovered elsewhere in Finland; therefore, they are classified as "Kuusamo-type" deposits in Fig. 5.1.2. However, as most of the available evidence suggests a late-orogenic timing of mineralization, they are included into the class of "Orogenic gold with anomalous metal association" in Tables 5.1.1 and 5.1.2. The timing of mineralization in the KSB seems to be consistent with both the orogenic gold and Blackbird-type deposit models. Alteration style, metal association, and the nature of the mineralizing fluid(s) fit best with the Blackbird-type model. The nature of the mineralizing fluid(s), the metal association, and the rift–shelf setting are consistent with features of a Blackbird-type model, or an orogenic gold deposit model with an atypical metal association. Structural control and gold fineness are not uncharacteristic of any of the proposed models. Most of the data suggest that the KSB deposits are epigenetic, and that premetamorphic alteration prepared the ground by hardening soft rocks and rendering them competent, and providing brines to transport the metals. Overall, the deposit characteristics are most consistent with the orogenic gold model, although with an atypical metal combination.

PERÄPOHJA Au-Cu AND Au-U

Two types of gold occurrences have been discovered in the Peräpohja schist belt (PSB). One occurs as Au- and Cu-rich quartz vein deposits hosted by 2.3–2.1 Ga metadolerites and their immediate clastic metasedimentary and mafic metavolcanic wall rocks (Rouhunkoski and Isokangas, 1974; Perttunen and Vaasjoki, 2001; Eilu et al., 2007). The second type includes nuggety gold ± uraninite in multiply veined and altered supracrustal rocks of probably both clastic sedimentary and volcanic origin (Nebocat, 2012). Both types are listed under "Orogenic gold with anomalous metal association" in Tables 5.1.1 and 5.1.2, although not enough is known to be sure of the genetic type(s) of these clearly epigenetic occurrences.

The Au-Cu deposit type could be regarded as orogenic gold, but again with an atypical metal association. The PSB occurrences, such as the Kivimaa deposit (Rouhunkoski and Isokangas, 1974), share many features with orogenic gold deposits. The occurrences are epigenetic, hosted by competent lithologies, and brittle to brittle-ductile in structural style. The gold is hosted in quartz veins and their immediate wall rocks associated with sulfides and silicate gangue; the alteration halo, typical of orogenic gold systems in upper-greenschist facies rocks, includes proximal biotite-calcite ± sulfides and distal biotite zones (Eilu et al., 2007). The Au-Cu metal association fits well with an intracratonic rift basin environment that contains evaporates, is filled by clastic sedimentary and mafic volcanic rocks, is intruded by mafic dikes and sills, and has a 570 Ma evolutionary history (~2.45–1.88 Ga) before regional deformation and metamorphism (Perttunen and Vaasjoki, 2001; Kyläkoski, 2009a, 2009b). This environment could produce fluids with uncommonly high salinities that are able to leach, transport, and deposit gold and copper during orogeny (Goldfarb et al., 2001). However, other models have also been suggested for these PSB deposits, such as distal mineralization in red bed copper systems (Kyläkoski, 2009b).

The Au ± U type of mineralization is even more difficult to classify. The discovery of these occurrences was made as recently as 2010, in the Rompas area, in the northwestern part of the PSB (Nebocat, 2012). The host rocks, possibly 2.3–1.95 Ga in age, are so intensely and extensively altered that their primary character is often impossible to define in outcrop or in core logging, and so not much is yet known of them. Thus, the deposits are termed Peräpohja-type Au-U (Fig. 5.1.2). More information about the Rompas occurrences is presented in Subchapter 6.1.

GOLD-RICH VOLCANOGENIC MASSIVE SULFIDE DEPOSITS

In some VMS deposits and prospects in Finland, Au grades are sufficiently high (up to 4 g/t; Table 5.1.1) to be considered a major commodity. Some of the deposits could be originally gold-poor, base-metal occurrences that were overprinted by orogenic gold mineralization, as mentioned previously (the section 'Orogenic gold deposits with anomalous Ag, Cu, Co, Ni or Sb'), but others are clearly Au-rich VMS deposits. Examples of the latter occur in many VMS belts in Finland, and include the Iilijärvi Au-Ag-Cu-Zn-Pb, Haveri Au-Cu, and Pahtavaara Au deposits.

The Iilijärvi deposit has a style of alteration that indicates strongly acidic conditions of mineralization, which characterize high-sulfidation epithermal ores (Hedenquist et al., 1996), but other features, such as a submarine setting and the presence of adjacent VMS deposits, clearly indicate that they are seafloor VMS deposits (Mäkelä, 1989). None of the Iilijärvi-style occurrences have produced any ore to date; Iilijärvi itself was test mined in the eighteenth and nineteenth centuries (Table 5.1.1). The 1.90–1.89 Ga host rocks have been altered to Si-Al rich rocks, which were metamorphosed to a mineral assemblage of quartz + Al silicates, the latter being typically andalusite ± sericite. The host rocks can comprise any lithological unit of the local supracrustal sequence, but most commonly are felsic to intermediate volcanic rocks (Mäkelä, 1989). The Au/Ag ratio is generally <1, and typically about 0.1. The metal grades are 1–10 g/t Au, 10–100 g/t Ag, and 0.1–5 % for Cu, Zn, and Pb.

The Haveri Au-Cu deposit is located in the western part of the Paleoproterozoic Tampere schist belt (TSB) in southwest Finland. During 1942–1962, 1.559 Mt of ore was mined at grades of 2.85 g/t Au and 0.39 % Cu (Puustinen, 2003). Recent exploration has produced an inferred resource of 24.7 Mt at 0.89 g/t Au (Lappland Goldminers, 2008). Kähkönen and Nironen (1994), Eilu (2012), and Ruotoistenmäki (2012) have described the deposit as chiefly hosted by massive and pillowed mafic tholeiitic lavas that are ~1.905 Ga in age and formed in a back-arc setting. Lithological, structural, geochemical, and mineralogical evidence from ore and alteration assemblages support a VMS deposit model. There appears to be no support for an IOCG model (c.f., Strauss, 2004), and only minor support for an orogenic gold overprint on a Cu-only VMS deposit.

The submarine hydrothermal system is thought to have initially produced chlorite- and albite-rich domains that were metamorphosed to amphibole- and plagioclase-dominated assemblages, respectively. Most of the sulfides and gold, which were partially remobilized during deformation, and the locally abundant magnetite, can also be shown to have been originally deposited in a submarine hydrothermal system. The Haveri deposit is dominated by a subseafloor stringer-style mineralization common to the lower parts of VMS deposits (e.g., Poulsen and Hannington, 1996; Schardt et al., 2001), although significantly affected by multistage deformation and regional metamorphism.

The Pahtavaara gold-only deposit near Sodankylä, CLGB, is hosted by pillowed and massive komatiitic volcanic rocks of possibly 2.10–2.05 Ga in age. The deposit is genetically problematic and has commonly been grouped into the orogenic gold category (e.g., Korkiakoski, 1992). It comprises a swarm of subparallel lodes with nearly all gold disseminated in free native form. The deposit has many of the alteration features of lower-amphibolite facies orogenic gold deposits and an obvious structural control. Lodes are folded and plunge steeply to the west or west-southwest. However, there also are a number of unusual features, as summarized by Patison et al. (2013b). There is a tremolite-quartz-barite-dolomite-magnetite-biotite-pyrite-gold mineral association and a very high fineness (>995) of gold. The geometry of high-gold-grade (10–20 g/t) quartz-barite lenses and amphibole-dominated ore bodies relative to biotite-rich alteration zones is also atypical of orogenic gold deposits, as are the $\delta^{13}C$ values of carbonate alteration minerals.

Pahtavaara could be interpreted as a metamorphosed seafloor hydrothermal system with ore lenses being either carbonate- and barite-bearing cherts or quartz-carbonate-barite veins. The gold may have been introduced later, but its small grain size, textural position (nearly all occurs as free native grains with silicates instead of sulfides), and high fineness point to a prepeak metamorphic genesis. In addition, the interpreted ore zone geometry is compatible with folding of a syngenetic alteration domain, remobilization of gold, and the formation of ore-hosting later veins in fractures in folds or in shear zones. In 2012, ore bodies were discovered that had a style of mineralization different from that mined previously. These recent discoveries have abundant sulfides and significant copper in addition to gold (Lappland Goldminers, 2013b). How these discoveries fit into the picture of syn- or epigenetic mineralization is yet to be explained.

EPITHERMAL GOLD DEPOSITS

Epithermal gold deposits (e.g., Hedenquist et al., 1996; Simmons et al., 2005) form in subaerial (above 2 km depth) magmatic hydrothermal systems. They occur most commonly 50–700 m below the water table, in zones of fluid upflow, and typically form at temperatures of 160–270 °C, although in some examples as high as 350 °C. The tectonic setting is most consistently one of convergent plate margins, commonly in association with magmatism in a back-arc rift. The deposits are controlled by early faults in areas of local extension. Epithermal deposits have probably formed since the onset of plate-tectonic processes on Earth. For example, epithermal gold has been identified in two localities of Archean rocks in Western Australia (Huston et al., 2001; McFarlane et al., 2007). On the other hand, the near-surface subaerial ore formation in a rapidly uplifting arc setting suggest a poor preservation potential. Thus the vast majority of mined epithermal deposits occur in late Mesozoic to Cainozoic rocks. Significant pre-Mesozoic epithermal gold deposits have probably only been preserved where the ores were quickly buried under younger supracrustal sequences and, thus, were protected from being eroded (Dube et al., 1998). Another problem with correctly classifying epithermal deposits in Precambrian terranes is that, with any superimposed regional metamorphism and deformation, many diagnostic, delicate features are destroyed beyond recognition. Thus, the emphasis that follows is on those indicators of epithermal gold deposits that have the best preservation potential.

Epithermal deposits can be subdivided into low-, medium- and high-sulfidation subclasses (LS, IS and HS, respectively) according to their sulfidation state reflecting the ambient pH and f_S of the system (Hedenquist et al., 1996; Simmons et al., 2005). The most intense and distinct alteration is produced in an HS setting, where fluid pH may range from <2 in the core to near neutral in the most distal parts of the hydrothermal system. This results in the following alteration zoning from the distal parts to the core, independent of primary host lithology: propylitic → argillic → advanced argillic → vuggy silica. The mineral assemblages produced are: chlorite ± illite, epidote + remains of primary silicates → illite or smectite + remains of primary silicates → alunite + chalcedony ± kaolinite ± pyrophyllite → chalcedony ± quartz, respectively. The ore is mainly hosted by the vuggy silica rock and, in a number of cases, also by the advanced argillic alteration zone. The main ore minerals are gold, electrum, pyrite, and/or other high-sulfur sulfides, such as enargite, luzonite, covelline, and fahlore. The Au/Ag ratio can vary from <0.01–10; the large Cainozoic HS gold deposits typically have more gold than silver. The chemical composition of the rock reflects the style of alteration. The vuggy silica rock typically contains 95–99.9 wt% SiO_2, with the remainder of the rock characterized by minerals of the least-mobile elements, such as Ti-bearing phases, and ore minerals. In the advanced argillic zone, SiO_2 and Al_2O_3 form as much as 99 wt% of the rock. During alteration, all Na, K, Ca, Mn, and Mg, and nearly all Fe, are removed; some Fe may be fixed into the ore minerals. No mafic silicates, muscovite, feldspars, or carbonates remain in the advanced argillic or the vuggy silica zones.

The most intense alteration and related chemical changes of the HS systems can be recognized even with a significant regional metamorphism and deformation, possibly even in granulite-facies rocks (Hallberg, 1994; Dube et al., 1998). In such settings, one may see an epigenetic quartz rock enveloped by quartz-Al silicate rock, both potentially hosting the ore. For example, at Enåsen, Sweden, the host to the ore is a topaz-bearing quartz-sillimanite gneiss having (SiO_2 + Al_2O_3) > 85 wt% and (Na_2O + K_2O + CaO + MgO) < 1.5 wt% (Hallberg, 1994).

In Finland, the most probable HS epithermal gold deposit is Kutemajärvi in the TSB (Fig. 5.1.2). The features described in the following for Kutemajärvi are summarized form Luukkonen (1994), Poutiainen and Grönholm (1996), and Talikka and Mänttäri (2005). The TSB formed in a volcanic-arc setting with extensive calc-alkaline to alkaline magmatism and rapid orogenic evolution of magmatism related to accretion of the Tampere arc to a microcontinent in the north. The Kutemajärvi deposit is hosted by altered intermediate volcanic rocks and comprises several pipe-like, vertical ore bodies 2- to 20-m across and hundreds of meters long; the two main ore bodies are open at a depth of 700 and 1000 m, respectively (Dragon Mining, 2013). The ore bodies are hosted by quartz-andalusite-pyrophyllite rock (SiO_2 80–90 wt%), with minor to trace amounts of topaz, fluorite, lazulite, kaolinite, rutile, apatite, tellurides, and gold. Sulfides, mainly pyrite, are more common in the alteration halo than in the ore. Proximal alteration enveloping the ore is characterized by the assemblage andalusite-pyrophyllite-quartz, whereas the distally altered rock contains the propylitic assemblage quartz-muscovite-chlorite-pyrite-rutile ± chalcopyrite.

The unaltered host rock shows an assemblage of quartz-biotite-plagioclase ± K feldspar, hornblende, epidote, and magnetite. This probably represents the primary sequence from the most proximal, originally vuggy quartz rock with some kaolinite or pyrophyllite, through the advanced argillic kaolinite-alunite-quartz, and to the distal propylitic zone, all metamorphosed under lower-amphibolite facies conditions. It is distinctly similar to that described at the Hope Brook gold deposit in Newfoundland, Canada (Dube et al., 1998). No carbonatization, potassic or sodic alteration, nor auriferous quartz veining have been identified at Kutemajärvi. It is essentially a gold-only deposit with average grades of 5–12 g/t Au, a Au/Ag ratio of 2–10, and Au/Te < 1 (Luukkonen, 1994). Other elements enriched in the ore include As, Bi, F, Sb, Se, and Si, whereas Ba, Ca, Co, Cu, Fe, K, Mn, Mg, Ni, Rb, and Sr are relatively depleted. The presence of F-rich minerals suggests a significant magmatic component in the fluid, typical of a high-sulfidation epithermal environment. The source for the heat and the acidic fluids could be the hypabyssal Pukala porphyry intrusion located <500 m to the north of the deposit.

There are several other Au ± Ag–base-metal occurrences in Finland that may also belong to the epithermal class, and there is a group of less certain deposits for which not enough conclusive data are available (Eilu and Pankka, 2009). These are indicated as "Epithermal" and "Epithermal or orogenic," respectively, in Fig. 5.1.2. For example, the Järvenpää Au-Ag-base metal occurrence and the Stenmo Au occurrence in the Tampere and Uusimaa belts, respectively, are classified here as epithermal. The best example of the less-certain cases is Kylmäkangas, in the Archean Oijärvi greenstone belt (Table 5.1.1), where the host to the ore is, possibly, a quartz rock whereas the local quartz veins are barren (Juopperi et al., 2001).

PORPHYRY COPPER-GOLD AND INTRUSION-RELATED GOLD

Porphyry-type deposits form in geological settings similar to those of epithermal deposits. The main difference with regard to preservation potential is that the former form at deeper levels of the crust, between 1–5 km below the surface (e.g., John et al., 2010; Sillitoe, 2010), hence their slightly better preservation. Nevertheless, compared to many other tectonic environments, arcs offer relatively poor preservation potential, as indicated by the paucity of Precambrian porphyry deposits. In Finland and Sweden, several porphyry-style Cu-Au occurrences are known from the Paleoproterozoic Svecofennian domain, including the very large Aitik deposit in northern Sweden, which has been in production since 1968 and has a current mineral reserve of 702 Mt at 0.14 g/t Au, 1.6 g/t Ag, 0.25 % Cu, and 0.003 % Mo (Bejgarn et al., 2011; Wanhainen et al., 2012; FODD, 2013).

In Finland, the best known probable porphyry Cu-Au deposits are Kopsa and Kedonojankulma in west and southwest Finland, respectively (Table 5.1.1; Fig. 5.1.2). Both are hosted by I-type, calc-alkaline, composite, porphyritic intrusions of dominantly tonalitic composition, formed within an active igneous arc probably during periods of terrane accretion, and overprinted by regional deformation and amphibolite-facies metamorphism. They show multistage veining predating regional deformation; typical porphyry-style potassic, sericitic, and propylitic alteration, and silicification; and ore minerals occurring both in veins and as dissemination. Kedonojankulma also shows metal zoning typical of porphyry deposits, with Cu-Au-Ag-As-Mo in the core, Mo and Cu in quartz veins outside of the core, and Zn-Cu-Ag in the outermost zone. The host intrusion at Kopsa is located at the intersection of major early(?) faults, also a typical feature for porphyry Cu-Au systems.

Because Precambrian porphyry Cu-Au deposits commonly occur in metamorphic terranes, it is conceivable that they are porphyry Cu deposits overprinted by orogenic gold or other mineralizing processes (Wanhainen et al., 2012). However, no clear evidence has been found in the Finnish deposits for such overprinting. Common deformation-related late structures host some mineralization, but textures seem to suggest localized remobilization of porphyry gold, rather than much later overprinting mineralization.

It has recently been suggested that the Kopsa deposit may be an intrusion-related gold deposit (IRG; Belvedere Resources, 2012), probably because the gold grade is low (0.8 g/t), the mineralization is synchronous with emplacement of the host granitoid, there is some shearing in the ore, and the deposit is hosted by a granitoid in an orogenic belt. Economic intrusion-related gold deposits are globally relatively rare (Lang and Baker, 2001; Goldfarb et al., 2005; Groves et al., 2005, 2009). It is unclear if such occurrences exist at all in Finland. For example, the diagnostic criteria of (1) tectonic setting well inboard of convergent plate boundaries in deformed shelf sequences; (2) regional association with W or Sn ores and reduced (Cu-poor) magmas; (3) postorogenic timing; and (4) metal zoning from Au-Bi-Te ± W, Mo, As, in sheeted veins in the source intrusion cupola, to hornfels-hosted Au-As ± W, Sn, Sb (within 2–3 km outside of the intrusion), and distal, low-temperature Au-As-Sb-Hg ± Ag, Pb, Zn, and Ag-Pb-Zn (Lang and Baker, 2001; Goldfarb et al., 2005; Groves et al., 2009) are missing for the gold occurrences in the Svecofennian domain of Finland. In addition, the following features, suggested to be typical for IRGs (e.g., Lang and Baker, 2001), are not diagnostic, as all these also characterize many orogenic gold systems: (1) anomalous concentrations of Bi, Te, and W; (2) low-salinity H_2O-CO_2 fluids; (3) postpeak metamorphic, but still synorogenic timing; (4) shallow ore formation; and (5) spatial association with reduced granitoids (Goldfarb et al., 2005).

Some of the criteria that are diagnostic for the IRGs possibly apply to some deposits in northern Finland. For example, this area contains deformed shelf sequences that are characteristic of IRG deposits. Along the southwest margin of the Lapland granulite belt, there are several unusual gold occurrences whose age is late relative to the ductile deformation stages of the Paleoproterozoic orogeny (Mäkärärova, and others). It is only these occurrences that are marked as "Granitoid-related (IRG?)" in Fig. 5.1.2.

Although these do not resemble the Fort Knox deposit (Alaska, U.S.A.), the only economic example of the IRG class, they may well be postorogenic because they are located close to postorogenic granitoids, and they have metal associations (Au-Fe-Ba) and ore and gangue minerals (baryte–hematite ± pyrite ± chalcopyrite) not typical of orogenic gold (Härkönen, 1987). In any case, exploration geologists should be aware that orogenic gold and porphyry Cu-Au deposits are major global gold producers, whereas the IRG occurrences are not. Many deposits around the world that have been suggested by some workers to fall into the IRG category (e.g., Pogo in Alaska, Donlin Creek in Yukon, Linglong in China, Muruntau in Uzbekistan, Vasilikovskoye in Russia, Jilau in Tajikistan, Kidston in Australia) are not IRGs, but orogenic gold or porphyry-related deposits (Groves et al., 2009; R. Goldfarb, personal communication 2013).

PALEOPLACER AND PLACER GOLD

Paleoplacer gold occurrences have been identified in the uppermost Kumpu group of the CLGB. No resource has been published about them, hence they do not appear in Tables 5.1.1 and 5.1.2. These occurrences resemble paleoplacer gold deposits elsewhere, most notably those at Tarkwa, Ghana, which also are Paleoproterozoic in age (Minter, 1991). The Finnish occurrences are hosted by both monomictic and polymictic conglomerates deposited in deltaic and fluvial fan environments after ~1.873 Ma and probably before 1.77 Ga (Härkönen, 1984; Rastas et al., 2001). Most of the gold is hosted by conglomerate lenses that are as much as 30–60 m in thickness. Within these lenses, the grade typically varies from 0.1–5 g/t Au, but locally reaches 22 g/t (Härkönen, 1984). However, only two cases are known (Fig. 5.1.2) where the Au content exceeds 1 g/t for an interval of more than 1 m in thickness. Gold in the Kumpu paleoplacers occurs mainly as free, native, detrital grains with diameters of between 0.03 and 0.4 mm and, in minor amounts, as smaller inclusions in quartz clasts (Härkönen, 1984). Other heavy minerals associated with the gold are magnetite, hematite, uraninite, pyrite, ilmenite, rutile, silver, tourmaline, monazite, titanite, and zircon (Härkönen, 1984).

A large number of small placer gold occurrences are known from northern Lapland, in Pleistocene to Holocene regolith, till, sand, and gravel. Despite the numerous occurrences, more than 140 years of artisanal exploitation has officially produced less than 2000 kg of gold (Puustinen, 1991; Finnish Mining Registry data). The metal endowment of these occurrences is difficult to estimate; grades commonly are below 1 g/m^3, but in the best pay streaks may reach several tens g/m^3 (A. Peronius, personal communication 2013). In any case, all presently available information suggests meager possibilities for profitable industrial-scale gold production from these occurrences. On the other hand, these occurrences remain a potentially useful tool in exploration for hard-rock gold deposits in parts of northern Finland, as they indicate areas of primary gold mineralization in the region.

SUMMARY: GOLD MINERALIZATION IN FINLAND AND ITS RELATIONSHIP TO SUPERCONTINENT EVOLUTION

Finland contains a large number of Au deposits and occurrences, due in large part to the diversity of tectonic processes that affected the region (Table 5.1.3, Figs. 5.1.3 and 5.1.4). The oldest gold deposits in Finland may be epithermal, as suggested by indications from the Oijärvi greenstone belt. Subsequently, orogenic gold mineralization took place from 2.72–2.64 Ga in probably all Archean greenstone belts and was related to the collisional processes during the assembly of the Neoarchean

Table 5.1.3 Gold mineralization events in Finland and their possible relationship to supercontinent evolution stages (Lahtinen, 2012; Lahtinen et al., 2012)

Metallogenic event	Metal association[a]	Deposit type[b]	Geological domains[b]
Kenorland assembly			
2.8–2.72 Ga Igneous arcs?	Au-Ag	Epithermal?	Oijärvi?
2.72–2.64 Ga Continent–continent collision	Au	Orogenic gold	All Archean greenstone belts
Paleoproterozoic rifting stages of the Archean continents: Kenorland breakup			
2.45 Ga Failed rifting	–	No Au mineralization	
2.2–2.0 Ga Intracontinental rifting	Au + base metals?	Metal enrichment in red bed and VMS systems?; ground preparation by alteration?	CLGB?, Kuusamo?, Peräpohja?
2.1–1.95 Ga Ocean opening	Au	Au-rich VMS?	CLGB (Pahtavaara)?
Paleoproterozoic igneous arcs and accretion: pre-Columbia arcs to Columbia initial assembly			
1.91–1.90 Ga Igneous arcs	Au-Cu	Au-rich VMS	Tampere
1.91–1.90 Ga Igneous arc and microcontinent accretion	Au	Orogenic gold?	CLGB (Suurikuusikko)?
1.89–1.88 Ga Igneous arcs	Au, Au-Cu	Au-rich VMS, epithermal, porphyry	Uusimaa, Häme, Tampere, Raahe-Haapajärvi
1.885–1.875 Ga Igneous arc and microcontinent accretion	Au, Au-Cu ± Ni, Co?	Orogenic gold, orogenic gold with anomalous metal association?	Häme?, Southern Savo?, Raahe-Haapajärvi? Kuusamo?, CLGB?, Peräpohja?
Paleoproterozoic collisional: Columbia assembly			
1.84–1.79 Ga Continent–continent collision	Au, Au-Cu ± Ni, Co?, Fe-Cu-Au	Orogenic gold, orogenic gold with anomalous metal association?	Häme, Southern Savo?, Raahe-Haapajärvi?, Kuusamo?, CLGB?, Peräpohja?
1.79–1.77 Ga Late collision	Au?	IRG?	North of the CLGB?

Supercontinent stages are according to Mertanen and Pesonen (2012). Geological domains are named as in Fig. 5.1.2. Timing overlap of metallogenic events is due to different processes simultaneously taking place in difference parts of the region. For the Paleoproterozoic evolution, see also Figs. 5.1.3 and 5.1.4.
[a]*Main auriferous metal association.*
[b]*Uncertain timings of mineralization and uncertain genetic types are indicated by the question mark. The problematic genetic types include the Archean epithermal, 2.1–1.95 Ga Au-rich VMS, and the IRG type. The rest are timing issues: For example, the anomalous metal association of orogenic gold mineralization took place at 1.89–1.88 or 1.84–1.79 Ga or during both time intervals within the CLGB, Kuusamo, and Peräpohja. Orogenic gold in the Häme and Pirkanmaa belts may have taken place both during the 1.885–1.875 Ga and 1.84–1.79 Ga orogenic peaks.*

supercontinent Kenorland. The episodic disintegration of Kenorland from 2.45 or 2.2 to 1.95 Ga (Mertanen and Pesonen, 2012), first with formation of intracontinental basins hosting red beds and evaporates and later with ocean opening (pink in Fig. 5.1.3), provided a setting for multiple hydrothermal systems that may have enriched gold and other metals in the supracrustal sequences and, during the oceanic stages, possibly formed Au-rich deposits such as Pahtavaara in submarine VMS-like settings.

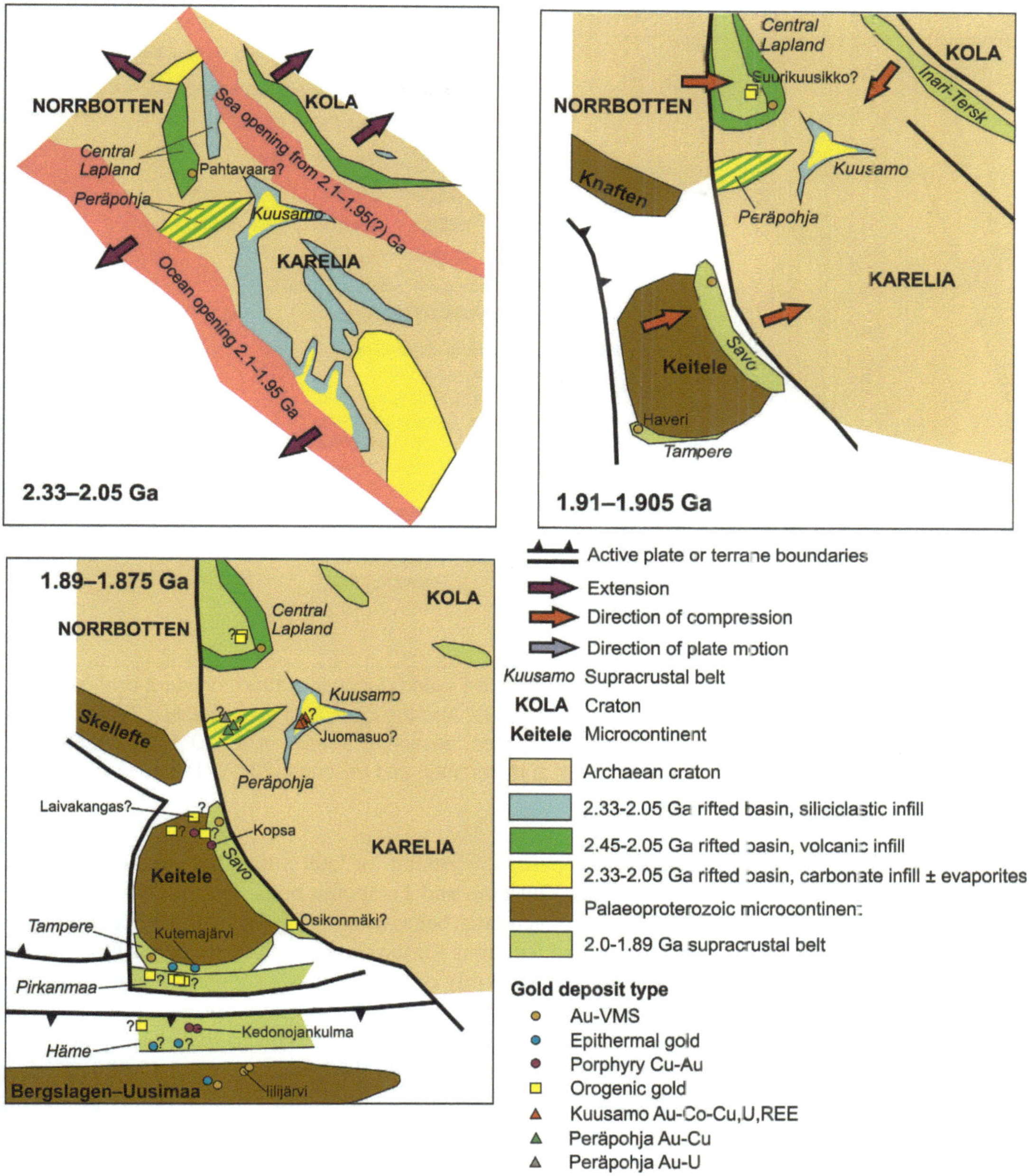

FIGURE 5.1.3

General tectonic setting of Fennoscandia and gold mineralisation in Finland from ca. 2.33 to 1.875 Ga. Only the episodes directly related to gold mineralisation are shown. Based on Lahtinen et al. (2005, 2013), Eilu and Pankka (2009), Grönholm and Kärkkäinen (2012), and Mertanen and Pesonen (2012). The question mark in the maps indicates cases whose timing is uncertain; this uncertainty is most prominent for the deposits in northern Finland.

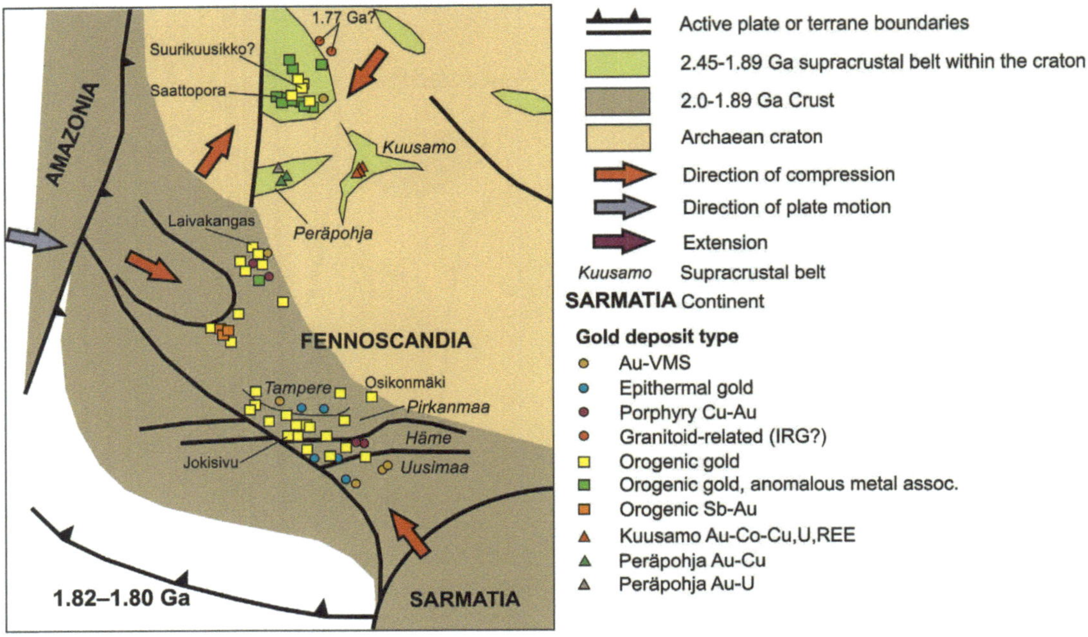

FIGURE 5.1.4

Tectonic setting of Fennoscandia and gold mineralisation in Finland during the latter, continent-continent collisional stage of the Svecofennian orogeny, 1.82–1.80 Ga. Also the location of the possibly 1.77 Ga, IRG-related(?), gold occurrences in northern Finland is indicated. Based on Lahtinen et al. (2005, 2013), Eilu and Pankka (2009), Grönholm and Kärkkäinen (2012), and Mertanen and Pesonen (2012). The question mark indicates cases whose timing is unclear.

Extensive albitization and carbonatization prepared the ground for later mineralizing processes in the intracontinental basins of the CLGB and the Kuusamo and Peräpohja belts. Essential in this ground preparation was the hardening of soft rocks, which later behaved as competent lithological units providing structurally and lithologically controlled locations for the orogenic fluids to deposit the metals.

The onset of the Svecofennian orogeny and assembly of the Columbia supercontinent resulted in multiple gold-mineralization processes (Table 5.1.3). The VMS deposits were related to back-arc basins that were common from 1.93–1.89 Ga. At least one of these, at Haveri, produced a gold-rich seafloor deposit at ~1.905 Ga in the Tampere belt, and several others in the western Uusimaa belt at 1.90–1.89 Ga. At the same time, terrane accretion was taking place in the north, possibly producing the first orogenic gold deposits in the CLGB (Suurikuusikko). Both the Paleoproterozoic deposits and the older Archean deposits may be the sources of gold in the CLGB paleoplacers. Evolution of igneous arcs at 1.89–1.88 Ga introduced epithermal gold (Tampere and Uusimaa belts) and porphyry copper-gold (Raahe-Haapajärvi region and Häme belt) within central and southern Finland. The arcs and juvenile microcontinents amalgamated with each other and with the Archean cratons from 1.89 Ga onward. This may have resulted in extensive orogenic gold mineralization across Finland in 1.885–1.875 Ga and possibly constituted a further gold source for the paleoplacers of the CLGB. The second major stage of Columbia assembly, with continent–continent collision between Fennoscandia, Sarmatia (in the southeast), and Amazonia (in the west), was at 1.84–1.77 Ga. Collision resulted in the formation of orogenic gold deposits in southern Finland and, possibly, IRG mineralization in northernmost Finland in the final stages of the orogeny.

The orogenic gold deposits of anomalous metal association of the CLGB, and the Kuusamo Au-Co ± Cu deposits, can be seen as products of a combination of both preorogenic and synorogenic to late-orogenic processes. They required a sequence of tectonic events related to supercontinent breakup followed by supercontinent assembly, including the following elements: (1) formation of intracontinental rifts, with clastic ± evaporitic sedimentation (serving as fluid ± metal sources), and intracratonic igneous activity (providing heat ± metals) during breakup; (2) ascent of diagenetic and basinal brines, which caused extensive albitization and possibly preenrichment of metals; and (3) formation of orogenic gold deposits involving both early brines and metamorphic fluids during the Svecofennian orogeny and Columbia assembly. The Peräpohja Au-Cu and Au ± U occurrences probably also relate to the evolution of the Paleoproterozoic intracratonic basin and subsequent orogenesis, but the details of this relationship are yet to be unraveled.

ACKNOWLEDGMENTS

Sincere thanks to Rich Goldfarb for a critical and comprehensive review of this contribution. Wolfgang Maier and Eero Hanski are thanked for reviewing the text. Kirsti Keskisaari, at GTK, drew Fig. 5.1.2.

REFERENCES

Agnico-Eagle, 2013. Deposit Reserves and Resources. Accessed online at www.agnicoeagle.com/en/Operations/Reserves-and-Resources/Pages/default.aspx.

Appelqvist, H., 1993. Ore mineralogy in the Proterozoic Kalliosalo gold occurrence, western Finland, with preliminary data on an unspecified mineral, Ag3Au3Sb10S10. Geological Survey of Finland. Special Paper 18, 37–44.

Bejgarn, T., Årebäck, H., Weihed, P., Nylander, J., 2011. Geology, petrology and alteration geochemistry of the Palaeoproterozoic intrusive hosted Älgträsk Au deposit, northern Sweden. Geological Society of London Special Publications 350, 105–132.

Belvedere Resources, 2010. Media release 2 June 2010. Accessed online at http://www.belvedere-resources.com/belvedere-announces-indicated-resource-estimate-for-the-k1angesneva-occurrence-on-the-kiimala-gold-property/.

Belvedere Resources, 2011. Media release 11 October 2011. Accessed online at http://www.belvedere-resources.com/belvedere-significantly-increases-gold-resources-on-their-osikonmaki-prospect-in-finland/.

Belvedere Resources, 2012. Media Release 18 January 2012. Accessed online at http://www.belvedere-resources.com/further-high-grade-gold-drill-results-and-commencement-of-bulk-metallurgical-testing-at-kopsa-finland/.

Belvedere Resources, 2013. Media Release 2 October 2013. Accessed online at http://www.belvedere-resources.com/assets/news02oct2013final.pdf.

Billström, K., Eilu, P., Martinsson, O., et al., 2010. IOCG and related mineral deposits of the northern Fennoscandian Shield. In: Advances in the Understanding of IOCG deposits. In: Porter, T. (Ed.), Hydrothermal Iron Oxide-Copper-Gold & Related Deposits: A Global Perspective, vol. 4. PGC Publishing, Adelaide, pp. 367–400.

Böhlke, J.K., 1982. Orogenic (metamorphic-hosted) gold-quartz veins. U.S. Geological Survey Open-File Report 82–795, 70–76.

Bornhorst, T.J., Wilkin, R.T., 1993. Fluid inclusion data for selected quartz from gold prospects in the late Archaean Hattu schist belt, Ilomantsi, eastern Finland. Geological Survey of Finland. Special Paper 17, 317–322.

Damsten, M., 1990. Tutkimustyöselostus Ilomantsin kunnassa valtausalueella Kuittila 1 3, kaiv. rek. n:ot 3808 ja 3956, suoritetuista malmitutkimuksista. Geological Survey of Finland, Report M06/4244/ 90/1/10. p. 8. (in Finnish).

Dragon Mining, 2009. Annual Report 2008. Dragon Mining Limited, Perth. p. 104. Accessed online at http://www.dragon-mining.com.au/sites/default/files/2008_-_full.pdf.

Dragon Mining, 2011. Media release 3 November 2011. Accessed online at http://www.dragon-mining.com.au/sites/default/files/2011-11-03_juomasuo_resource_update_v2.pdf.

Dragon Mining, 2012a. Media release 31 January 2012. Accessed online at http://www.dragon-mining.com.au/sites/default/files/2012-01-31_december_2012_quarterly_report.pdf.

Dragon Mining, 2012b. Media release 19 June 2012. Accessed online at http://www.dragon-mining.com.au/sites/default/files/2012-06-19__hangaslampi_resource_update_v2.pdf.

Dragon Mining, 2013. Annual Report 2012. Dragon Mining Limited, Perth. p. 100. Accessed online at http://www.dragon-mining.com.au/sites/default/files/2013-03-27_dra_annualreport_2012.pdf.

Dube, B., Dunning, G., Lauziere, K., 1998. Geology of the Hope Brook mine, Newfoundland, Canada: a preserved Late Proterozoic high-sulphidation epithermal gold deposit and its implications for exploration. Economic Geology 93, 405–436.

Eilu, P., 1994. Hydrothermal alteration in volcano sedimentary rocks in the Central Lapland greenstone belt, Finland. Geological Survey of Finland, Bulletin 374, 145.

Eilu, P., 2003. Primary and alteration geochemistry at Härmänkylä (Mataralampi), Kuhmo greenstone belt, eastern Finland. Geological Survey of Finland, Report CM19/4412/2003/4/10, p. 34.

Eilu, P., 2012. Haveri copper-gold deposit: genetic considerations. Geological Survey of Finland. Special Paper 52, 255–266.

Eilu, P., Groves, D.I., 2001. Geology, geochemistry and primary geochemical dispersion haloes of Archaean orogenic gold deposits in the Yilgarn Craton: the preweathering scenario. Geochemistry: Exploration, Environment. Analysis 1, 183–200.

Eilu, P., Niiranen, T. (Eds.), 2013. Gold deposits in northern Finland. Excursion guidebook FIN1. 12th SGA Biennial Meeting, August 12–15. Geological Survey of Sweden, Uppsala, p. 56.

Eilu, P., Ojala, V.J., 2007. Mass balance and geochemical anomalies related to the Mataralampi orogenic gold occurrence, Archaean Kuhmo greenstone belt, eastern Finland. Proceedings of the 23rd International Applied Geochemistry Symposium, Oviedo, Spain, June 14–19. Extended abstracts. University of Oviedo. 64–76.

Eilu, P., Pankka, H., 2009. FINGOLD—A public database on gold deposits in Finland. Version 1.0. Geological Survey of Finland. Digital data product 4. Optical disc (CDROM).

Eilu, P., Mathison, C.I., Groves, D.I., Allardyce, W., 1999. Atlas of Alteration Assemblages, Styles and Zoning in Orogenic Lode-Gold Deposits in a Variety of Host Rock and Metamorphic Settings. Geology and Geophysics Department (Centre for Strategic Mineral Deposits) & UWA Extension. The University of Western Australia. Publication 30, p. 64.

Eilu, P., Niiranen, T., Lauri, L., 2013. Geological and tectonic evolution of the northern part of the Fennoscandian shield. Excursion guidebook FIN1. Proceedings of the 12th SGA Biennial Meeting, August 12–15. Geological Survey of Sweden, Uppsala, pp. 8–21.

Eilu, P., Pankka, H., Keinänen, V., et al., 2007. Characteristics of gold mineralisation in the greenstone belts of northern Finland. Geological Survey of Finland, Special Paper 44, 57–106.

Eilu, P., Rasilainen, K., Halkoaho, T., Huovinen, I., Karkkainen, N., Kontoniemi, O., Lepisto, K., Niiranen, T. & Sorjonen-Ward, P. 2015. Quantitative assessment of undiscovered resources in orogenic gold deposits in Finland. Geological Survey of Finland, Report of Investigation 216. p. 318.

Eilu, P., Sorjonen-Ward, P., Nurmi, P., Niiranen, T., 2003. Review of gold mineralisation in Finland. Economic Geology 98, 1329–1353.

Endomines, 2013. Media release 18 March 2013. Accessed online at http://www.endomines.com/index.php/release/?releaseID=1274530.

Etelämäki, T., 2007. Tammelan Riukan Au-mineralisaatio. M.Sc. thesis, Department of Geology, University of Helsinki, p. 97 (in Finnish).

FODD, 2013. Fennoscandian Ore Deposit Database. Accessed online at http://en.gtk.fi/informationservices/databases/fodd/index.html.

Gaal, G., Isohanni, M., 1979. Characteristics of igneous intrusions and various wall rocks in some Precambrian porphyry copper-molybdenum deposits in Pohjanmaa, Finland. Economic Geology 74, 1198–1210.

Geospec Consultants Limited, 2008. Re-Os Analyses, Arsenopyrite Geochronology for Geological Survey of Finland. Analytical Report No 2.

Goldfarb, R.J., Baker, T., Dube, B., et al., 2005. Distribution, character, and genesis of gold deposits in metamorphic terranes. Economic Geology 100, 407–450.

Goldfarb, R.J., Groves, D.I., Gardoll, S., 2001. Orogenic gold and geologic time: a global synthesis. Ore Geology Reviews 18, 1–75.

Grönholm, S., Kärkkäinen, N. (Eds.), 2012. Gold in Southern Finland: Results of GTK studies 1998–2011. Geological Survey of Finland. Special Paper 52, p. 276.

Groves, D.I., Bierlein, F.P., Goldfarb, R.J., 2009. Some irks with IRGS and snags with TAGs. In: Williams, P.J., et al. (Eds.), Smart Science for Exploration and Mining. Proceedings of the 10th Biennial SGA Meeting, Townsville, Australia, 17–20.8.2009. EGRU. James Cook University, Townsville, pp. 356–358.

Groves, D.I., Goldfarb, R.J., Gebre-Mariam, M., et al., 1998. Orogenic gold deposits: A proposed classification in the context of their crustal distribution and relationship to other gold deposit types. Ore Geology Reviews 13, 7–27.

Groves, D.I., Condie, K.C., Goldfarb, R.J., et al., 2005. Secular changes in global tectonic processes and their influence on the temporal distribution of gold-bearing mineral deposits. Economic Geology 100, 203–224.

GSA Global, 2012. Summary report—mineral resource and ore reserve update. Laiva gold deposit. 7 December. GSA Global, Perth, p. 23.

Hallberg, A., 1994. The Enåsen gold deposit, central Sweden. 1. A Palaeoproterozoic high-sulphidation epithermal gold mineralization. Mineralium Deposita 29, 150–162.

Härkönen, I., 1984. The gold bearing conglomerates of Kaarestunturi, Central Finnish Lapland. In: Foster, R.P. (Ed.), Gold '82; the Geology, Geochemistry and Genesis of Gold Deposits. Geological Society of Zimbabwe. Special Publication 1, 239–247.

Härkönen, I., 1987. Kultatutkimukset Mäkärärovan alueella Sodankylässä vuosina 1980–1985. Geological Survey of Finland, Report M19/3724/-87/1/10, p. 18 (in Finnish).

Hartikainen, A., 2001. A research report of studies concerning the claim of Timola 1 (register number of claim 5451/1), located in the village of Härmänkylä, Kuhmo, eastern Finland. Geological Survey of Finland, Report M06/4412/2001/1/10, p. 10.

Hedenquist, J.W., Izawa, E., Arribas, A., White, N.C., 1996. Epithermal gold deposits: Styles, characteristics, and exploration. Resource Geology Special Publication 1, 16.

Heino, T., 2000. Mineral Resource Assessment of the Kuikkapuro 1 (mine reg. N:o 6748/1), Kuikkapuro 2 (N:o 6841/1), Kuikkapuro 3 (N:o 6880/1), Kuikkapuro 4 (N:o 6880/2) and Kuikkapuro 5 (N:o 6912/1) gold mineralizations. Geological Survey of Finland Report CM06/4511/2000/1, p. 22.

Heino, T., 2001. Mineral Resource Assessment of the Pahkosuo gold mineralization, claims Pahkosuo (mine reg. N:o 5621/1), Pahkosuo 2 (N:o 6272/1), Pahkosuo 3 (N:o 6272/2) and Pahkosuo 4 (N:o 6410/1). Geological Survey of Finland, Report M19/4513/2001/1, p. 3.

Hokka, J., 2011. Malmiesiintymämalli Suomen orogeenisille kultaesiintymille. M.Sc. thesis. Department of Geology, University of Helsinki. p. 62 (in Finnish).

Holma, M., Keinänen, V., 2007. The Levijärvi-Loukinen gold occurrence: An example of orogenic gold mineralisation with atypical metal association. Geological Survey of Finland, Special Paper 44, 163–184.

Hölttä, P., Karhu, J., 2001. Oxygen and carbon isotope compositions of carbonates in the alteration zones of orogenic gold deposits in central Finnish Lapland. Geological Survey of Finland, Special Paper 31, 25–29.

Hölttä, P., Heilimo, E., Huhma, H., et al., 2012. The Archaean of the Karelia Province in Finland. Geological Survey of Finland, Special Paper 54, 21–73.

Hulkki, H., Keinänen, V., 2007. The alteration and fluid inclusion characteristics of the Hirvilavanmaa gold deposit, Central Lapland Greenstone Belt, Finland. Geological Survey of Finland, Special Paper 44, 135–151.

Hulkki, H., Pulkkinen, E., 2007. Exploration history of the Kaaresselkä gold-copper occurrence, Central Lapland. Geological Survey of Finland, Special Paper 44, 155–164.

Huston, D., Blewett, R.S., Champion, D.C., 2012. Australia through time: a summary of its tectonic and metallogenic evolution. Episodes 35, 23–43.

Huston, D.L., Blewett, R.S., Sweetapple, M., et al., 2001. Metallogenesis of the north Pilbara granite–greenstones, Western Australia—a field guide. Western Australia Geological Survey, Record 2001/11, p. 87.

Inmet Mining, 2013. Media release February 13, 2013. Accessed online at http://ir.inmetmining.com/Files/ac/acc06f0d-b2db-474b-97b4-89243c3d46d2.pdf.

Isomaa, J., Koistinen, E., Kärkkäinen, N., 2010. Sikakangas gold prospect at Seinäjoki, Western Finland. Geological Survey of Finland, Report M19/2222/2010/55, p. 37.

John, D.A., Ayuso, R.A., Barton, M.D., et al., 2010. Porphyry copper deposit model, chap. B of Mineral deposit models for resource assessment. U.S. Geological Survey Scientific Investigations Report 2010-5070-B, p. 169.

Juopperi, H., Karvinen, A., Rossi, S., 2001. The Oijärvi gold prospects in Northern Finland. Summary. Geological Survey of Finland, Report CM06/3521/2000/1, p. 12.

Kähkönen, Y., Nironen, M., 1994. Supracrustal rocks around the Paleoproterozoic Haveri Au-Cu deposit, southern Finland: evolution from a spreading center to a volcanic arc environment. Geological Survey of Finland, Special Paper 19, 141–159.

Kärkkäinen, N., Jokinen, T., Koistinen, E., 2006. Satulinmäki gold prospect at Somero, SW Finland. Geological Survey of Finland, Report M19/2024/2006/1/10, p. 55.

Kärkkäinen, N., Lahtinen, R., Pakkanen, L., 2012. Discovery and mineralogy of gold occurrence at Velkua, Southwestern Finland. Geological Survey of Finland, Special Paper 52, 115–130.

Keinänen, V., 1994. Shear zone-related Soretialehto gold occurrence in green carbonate rocks in the Central Lapland Greenstone Belt. Kittilä, Finnish Lapland. 21:a Nordiska Geologiska Vintermötet, Luleå. Abstracts, p. 99.

Kojonen, K., Johansson, B., 1999. Determination of refractory gold distribution by microanalysis, diagnostic leaching, and image analysis. Mineralogy and Petrology 67, 1–19.

Kojonen, K., Johansson, B., O'Brien, H.E., Pakkanen, L., 1993. Mineralogy of gold occurrences in the late Archaean Hattu schist belt, Ilomantsi, eastern Finland. Geological Survey of Finland, Special Paper 17, 233–271.

Kontinen, A., Paavola, J., Lukkarinen, H., 1992. K-Ar ages of hornblende and biotite from Late Archaean rocks of eastern Finland – interpretation and discussion of tectonic implications. Geological Survey of Finland, Bulletin 365, 31.

Kontoniemi, O., 1998. Geological setting and characteristics of the Palaeoproterozoic tonalite-hosted Osikonmäki gold deposit, southeastern Finland. Geological Survey of Finland, Special Paper 25, 39–80.

Kontoniemi, O., Mursu, J., 2006. Hirsikangas gold prospect in Himanka, western Finland. Claim areas Hirsi 1 and 2 (7847/1 and 8036/1). Geological Survey of Finland, Report M19/2413/2006/1/10, p. 32.

Kontoniemi, O., Johanson, B., Kojonen, K., Pakkanen, L., 1991. Ore mineralogy of the Osikonmäki gold deposit, Rantasalmi, southeastern Finland. Geological Survey of Finland, Special Paper 12, 81–89.

Korkalo, T., 2006. Gold and copper deposits in Central Lapland, northern Finland, with special reference to their exploration and exploitation. Acta Univ. Oulensis. A Scientiae Rerum Naturalium 461, p. 122.

Korkiakoski, E., 1992. Geology and geochemistry of the metakomatiite hosted Pahtavaara gold deposit in Sodankylä, northern Finland, with emphasis on hydrothermal alteration. Geological Survey of Finland, Bulletin 360, 96.

Korvuo, E., 1997. The Saattopora gold ore and the Pahtavuoma Cu-Zn-U occurrences in the Kittilä region, northern Finland. Geological Survey of Finland, Guide 43, 21–25.

Kyläkoski, M., 2009a. Basin scale alteration features and their implications for ore formation in the Palaeoproterozoic Peräpohja schist belt, Northwestern Finland. In: Smart Science for Exploration and Mining: Proceedings of the 10th Biennial SGA Meeting, Townsville, Australia, August 17–20. James Cook University, Townsville, pp. 457–459.

Kyläkoski, M., 2009b. Varhaisproterotsooista laakiobasalttikerrostumaa edustavan Jouttiaavan muodostuman nikkeli-, kupari- ja platinaryhmäpotentiaali. Phil Lic thesis. Department of Geology, University of Oulu, p. 209 (in Finnish).

Lahtinen, J., 1980. Väliraportti Au-mineralisaatiosta. Outokumpu Oy Malminetsintä, Report 001/4522/JJL/80, p. 13 (in Finnish).
Lahtinen, R., 2012. Main geological features of Fennoscandia. Geological Survey of Finland, Special Paper 53, 13–18.
Lahtinen, R., Hallberg, A., Korsakova, M., et al., 2012. Main metallogenic events in Fennoscandia. Geological Survey of Finland, Special Paper 53, 397–401.
Lahtinen, R., Huhma, H., Lahaye, Y., et al., 2013. Long-lived LREE mobility in the cratonic, rift and foredeep to foreland sedimentary cover margin of the Karelia Province. Lithos 175–176, 86–103.
Lahtinen, R., Korja, A., Nironen, M., 2005. Palaeoproterozoic tectonic evolution of the Fennoscandian Shield. In: Lehtinen, M., Nurmi, P.A., Rämö, O.T. (Eds.), Precambrian Geology of Finland—Key to the Evolution of the Fennoscandian Shield. Developments in Precambrian Geology 14. Elsevier, Amsterdam, pp 418–532.
Lang, J.R., Baker, T., 2001. Intrusion-related gold systems: the present level of understanding. Mineralium Deposita 36, 477–489.
Lappland Goldminers, 2008. Haveri project, Finland. Accessed online at http://www.lapplandgoldminers.com/system/visa.asp?HID=1345&FID=1154&HSID=25481&ActMenu=25557&ActSubMenu=25578&ActSubSubMenu=25632.
Lappland Goldminers, 2013a. Mineral Resources at 1st January 2013. Accessed online at http://www.lapplandgoldminers.com/system/visa.asp?HID=1345&FID=1154&HSID=25479&ActMenu=25538&ActSubMenu=25542.
Lappland Goldminers, 2013b. Media release in 15 November 2013 Available online at http://mb.cision.com/Main/8073/9499979/184669.pdf.
Luukkonen, A., 1994. Main geochemical features, metallogeny and hydrothermal alteration phenomena of certain gold and gold-tin-tungsten prospects in southern Finland. Geological Survey of Finland, Bulletin 377, 153.
Luukkonen, E., Halkoaho, T., Hartikainen, A., et al., 2002. Itä-Suomen arkeeiset alueet -hankkeen (12201 ja 210 5000) toiminta vuosina 1992–2001 Suomussalmen, Hyrynsalmen, Kuhmon, Nurmeksen, Rautavaaran, Valtimon, Lieksan, Ilomantsin, Kiihtelysvaaran, Enon, Kontiolahden, Tohmajärven ja Tuupovaaran alueella. Geological Survey Finland, Report M19/4513/2002/1, p. 265 (in Finnish).
Mäkelä, M. 1984. Raahen Laivakankaan kvartsi-arseenikiisukultaesiintymä. M.Sc. thesis. University of Turku, p. 105 (in Finnish).
Mäkelä, U., 1989. Geological and geochemical environments of Precambrian sulphide deposits in southwestern Finland. Annales Academiae Scientiarum Fennicae. Series A. III. Geologica-Geographica 151, 102.
Makkonen, H., Ekdahl, E., 1988. Petrology and structure of the early Proterozoic Pirilä gold deposit in southeastern Finland. Bulletin of the Geological Society of Finland 60, 55–66.
Mänttäri, I., 1995. Lead isotope characteristics of epigenetic gold mineralization in the Palaeoproterozoic Lapland greenstone belt, northern Finland. Geological Survey of Finland, Bulletin 381, 70.
McCuaig, T.C., Kerrich, R., 1998. P-T-t-deformation-fluid characteristics of lode gold deposits: evidence from alteration systematics. Ore Geology Reviews 12, 381–453.
McFarlane, C.R.M., Mavrogenes, J.A., Tomkins, A.G., 2007. Recognizing hydrothermal alteration through a granulite facies metamorphic overprint at the Challenger Au deposit, South Australia. Chemical Geology 243, 64–89.
Mertanen, S., Pesonen, L.J., 2012. Paleo-Mesoproterozoic assemblages of continents: paleomagnetic evidence for near Equatorial supercontinents. Lecture Notes in Earth System Sciences 137. Springer, Berlin. pp. 11–35.
Minter, W.E.L., 1991. Ancient placer deposits. In: Foster, R.P. (Ed.), Gold Metallurgy and Exploration. Blackie and Sons Ltd., Glasgow, pp. 283–308.
Moilanen, M., Peltonen, P., 2015. Hannukainen Fe-Cu-Au deposit, western Finnish Lapland: Deposit model updated. In: Maier, W.D., O'Brien, H.,. Lahtinen, R. (Ed.), Mineral Deposits of Finland. Elsevier, Amsterdam. pp. 485–504.
Nebocat, J., 2012. Progress report on the geology, mineralization and exploration activities on the Rompas gold-uranium property, Southern Lapland, Finland. Resources Ltd. PGS Pacific Geological Services, p. 58.
Neumayr, P., Cabri, L.J., Groves, D.I., et al., 1993. The mineralogical distribution of gold and relative timing of gold mineralization in two Archean settings of high metamorphic grade in Australia. Canadian Mineralogist 31, 711–725.

Nurmi, P., Lestinen, P., Niskavaara, H., 1991. Geochemical characteristics of mesothermal gold deposits in the Fennoscandian Shield, and a comparison with selected Canadian and Australian deposits. Geological Survey of Finland, Bulletin 351, 101.

Nurmi, P.A., Sorjonen-Ward, P., Damsten, M., 1993. Geological setting, characteristics, and exploration history of mesothermal gold occurrences in the late Archaean Hattu schist belt, Ilomantsi, eastern Finland. Geological Survey of Finland, Special Paper 17, 193–231.

O'Brien, H.E., Nurmi, P.A., Karhu, J.A., 1993. Oxygen, hydrogen, and strontium isotopic compositions of gold mineralization in the late Archean Hattu schist belt, eastern Finland. Geological Survey of Finland, Special Paper 17, 291–306.

Pankka, H., 1992. Geology and mineralogy of Au-Co-U deposits in the Proterozoic Kuusamo volcanosedimentary belt, northeastern Finland. Ph.D. dissertation. Geology, Michigan Technological University, p. 233.

Pankka, H.S., Vanhanen, E.J., 1992. Early Proterozoic Au-Co-U mineralization in the Kuusamo district, northeastern Finland. Precambrian Research 58, 387–400.

Parkkinen, J., 2003. Update of Quality Control of Mineral Resource Estimate, 2003. Geological Survey of Finland, Report CM19/2332/2003/1/10.

Patison, N.J., 2007. Structural controls on gold mineralisation in the Central Lapland Greenstone Belt. Geological Survey of Finland, Special Paper 44, 105–122.

Patison, N., Ojala, J., Eilu, P., Niiranen, T., 2013a. Kittilä gold mine (Suurikuusikko deposit). Excursion guidebook FIN1. Proceedings of the 12th SGA Biennial Meeting, August 12–15. Geological Survey of Sweden, Uppsala, pp. 31–42.

Patison, N., Eilu, P., Niiranen, T., Ojala, J., 2013b. Pahtavaara gold mine. Excursion guidebook FIN1. Proceedings of the 12th SGA Biennial Meeting, August 12–15. Geological Survey of Sweden, Uppsala. pp. 24–28.

Perttunen, V., Vaasjoki, M., 2001. U-Pb geochronology of the Peräpohja schist belt, northwestern Finland. Geological Survey of Finland, Special Paper 33, 45–84.

Phillips, G.N., 1986. Geology and alteration in the Golden Mile. Kalgoorlie Economic Geology 81, 779–808.

Pitcairn, I.K., Teagle, D.A.H., Craw, D., et al., 2006. Sources of metals and fluids in orogenic gold deposits: insights from the Otago and Alpine Schists, New Zealand. Economic Geology 101, 1525–1546.

Poulsen, K.H., Hannington, M.D., 1996. Volcanic-associated massive sulphide gold. In: Eckstrand, O.R., Sinclair, W.D., Thorpe, R.I. (Eds.). Geology of Canadian Mineral Deposit Types. Geological Survey of Canada, 8. Geology of Canada, pp. 183–196.

Poutiainen, M., Grönholm, P., 1996. Hydrothermal fluid evolution of the Palaeoproterozoic Kutemajärvi gold telluride deposit, Southwest Finland. Economic Geology 91, 1335–1353.

Poutiainen, M., Partamies, S., 2003. Fluid inclusion characteristics of auriferous quartz veins in Archean and Paleoproterozoic greenstone belts of eastern and southern Finland. Economic Geology 98, 1355–1369.

Puustinen, K., 1991. Gold deposits of Finland: Journal of Geochemical Exploration 39, 255–272.

Puustinen, K., 2003. Suomen kaivosteollisuus ja mineraalisten raaka-aineiden tuotanto vuosina 1530–2001, historiallinen katsaus erityisesti tuotantolukujen valossa. Geological Survey of Finland, Report M10.1/2003/3, p. 578 (in Finnish).

Rasilainen, K., 1996. Geochemical alteration of gold occurrences in the late Archean Hattu schist belt, Ilomantsi, eastern Finland. Geological Survey of Finland, Bulletin 388, 80.

Rasilainen K., Eilu P., Halkoaho T., et al. in press. Quantitative assessment of undiscovered resources in orogenic gold deposits in Finland. Geological Survey of Finland, Report of Investigation.

Rastas, P., Huhma, H., Hanski, E., et al., 2001. U–Pb isotopic studies on the Kittilä greenstone area, Central Lapland, Finland. Geological Survey of Finland, Special Paper 33, 95–141.

Ruotoistenmäki, T., 2012. Local- and regional-scale geophysical characteristics of the Haveri mine. Geological Survey of Finland, Special Paper 52, 245–254.

Rouhunkoski, P., Isokangas, P., 1974. The copper-gold vein deposit of Kivimaa at Tervola, N-Finland. Bulletin of the Geological Society of Finland 46, 29–35.

Saalmann, K., Niiranen, T., 2010. Hydrothermal alteration and structural control on gold deposition in the Hanhimaa shear zone and western part of the Sirkka Line. Geological Survey of Finland, Report M19/2741/2010/58, p. 30.

Saalmann, K., Mänttäri, I., Ruffet, G., Whitehouse, M.J., 2009. Age and tectonic framework of structurally controlled Palaeoproterozoic gold mineralization in the Häme belt of southern Finland. Precambrian Research 174, 53–77.

Saalmann, K., Mänttäri, I., Peltonen, P., et al., 2010. Geochronology and structural relationships of mesothermal gold mineralization in the Palaeoproterozoic Jokisivu prospect, southern Finland. Geological Magazine 147, 551–569.

Saltikoff, B., 1980. Nurmon Kalliosalon antimoniesiintymän malmimäärä. Geological Survey of Finland, Report M19/2222/-80/2/10, p. 4 (in Finnish).

Sandstad, J.S., 2012. Kautokeino Au-Cu. Geological Survey of Finland. Special Paper 53, 116–118.

Schardt, C., Cooke, D.R., Gemmell, J.B., Large, R.R., 2001. Geochemical modeling of the zoned footwall alteration pipe, Hellyer volcanic-hosted massive sulfide deposit, Western Tasmania, Australia. Economic Geology 96, 1037–1054.

Sibson, R.H., Robert, F., Poulsen, H., 1988. High-angle reverse faults, fluid-pressure cycling, and mesothermal gold-quartz deposits. Geology 16, 551–555.

Sillitoe, R.S., 2010. Porphyry copper systems. Economic Geology 105, 3–41.

Simmons, S.F., White, N.C., John, D.A., 2005. Geological characteristics of epithermal precious and base metal deposits. Society of Economic Geologists. Economic Geology 100th Anniversary Volume 485–522.

Sipilä, E., 1988. Kultatutkimukset Haapaveden Vesiperällä ja sen ympäristössä 1985–1988. Geological Survey of Finland, Report M19/2433/-88/1/10, p. 4 (in Finnish).

Sipilä, E., 1990. Kultamineralisaatioiden tutkimuksista valtausalueilla Kiimala 1 ja 2 Haapavedellä 1988–1989. Geological Survey of Finland, Report M19/2433/-90/3/10, p. 22 (in Finnish).

Slack, J.F., 2012. Strata-bound Fe-Co-Cu-Au-Bi-Y-REE deposits of the Idaho cobalt belt: multistage hydrothermal mineralization in a magmatic-related iron oxide-copper-gold system. Economic Geology 107, 1089–1113.

Slack, J.F., Causey, J.D., Eppinger, R.G., et al., 2010. Co-Cu-Au deposits in metasedimentary rocks. A preliminary report. U.S. Geological Survey Open-File Report 2010–1212, 13. Available online at http://pubs.usgs.gov/of/2010/1212.

Sorjonen-Ward, P., 1993. An overview of structural evolution and lithic units within and intruding the late Archaean Hattu schist belt, Ilomantsi, eastern Finland. Geological Survey of Finland, Special Paper 17, 9–102.

Strauss, T.A.L., 2004. The geology of the Proterozoic Haveri Au-Cu deposit, southern Finland. Ph.D. thesis, Rhodes University.

Talikka, M., Mänttäri, I., 2005. Pukala intrusion, its age and connection to hydrothermal alteration in Orivesi, southwestern Finland. Bulletin of the Geological Society of Finland 77, 165–180.

Taranis Resources, 2011. Media release 12 October. Accessed online at http://www.taranis.us/News_Releases/Taranis_10_12_2011nr.pdf.

Tiainen, M., Molnar, F., Koistinen, E., 2013. The Cu-Mo-Au mineralization of the Paleoproterozoic Kedonojankulma intrusion, Häme belt, southern Finland. In: Johnsson, et al. (Eds.), Mineral Deposit Research for a High-Tech World. Proceedings of the 12th Biennial SGA Meeting, August 12–15, Uppsala. pp. 892– 895, Sweden.

Tomkins, A.G., Grundy, C., 2009. Upper temperature limits of orogenic gold deposit formation: Constraints from the granulite-hosted Griffin's Find deposit, Yilgarn Craton. Economic Geology 104, 669–685.

Tyni, M., 1983. Nurmon Kalliosalon antimoni-kultaesiintymän malmiarvio. Myllykoski Oy. report (in Finnish).

USGS, 2012. Mineral commodity summaries 2012. U.S. Geological Survey. p. 198. Accessed online at http://minerals.usgs.gov/minerals/pubs/mcs/2012/mcs2012.pdf.

Vanhanen, E., 1991. Tutkimustyöselostus Kuusamossa valtausalueella Iso-Rehvi 1, kaiv. rek. n:o 4442, malmitutkimuksista vuosina 1988–1989. Geological Survey of Finland, Report M06/4611/-91/1/10, p. 4 (in Finnish).

Vanhanen, E., 2001. Geology, mineralogy and geochemistry of the Fe-Co-Au-(U) deposits in the Paleoproterozoic Kuusamo schist belt, northeastern Finland. Geological Survey of Finland, Bulletin 399, 229.

Wanhainen, C., Broman, C., Martinsson, O., Magnor, B., 2012. Modification of a Palaeoproterozoic porphyry-like system: integration of structural, geochemical, petrographic, and fluid inclusion data from the Aitik Cu-Au-Ag deposit, northern Sweden. Ore Geology Reviews 48, 306–331.

Weatherley, D.K., Henley, R.W., 2013. Flash vaporization during earthquakes evidenced by gold deposits. Nature Geoscience 6, 294–298.

Yardley, B.W.D., Cleverley, J.S., 2013. The role of metamorphic fluids in the formation of ore deposits. Geological Society, London. Special Publications 393, doi:10.1144/SP393.5.

Yardley, B.W.D., Graham, J.T., 2002. The origins of salinity in metamorphic fluids. Geofluids 2, 249–256.

Yletyinen, V., Nenonen, E., 1972. Selostus malmitutkimuksista Kittilän Riikonkoskella vuosina 1969–1972. Geological Survey of Finland, Report M19/2734/72/1/10, p. 32 (in Finnish).

CHAPTER

THE SUURIKUUSIKKO GOLD DEPOSIT (KITTILÄ MINE), NORTHERN FINLAND

5.2

N.L. Wyche, P. Eilu, K. Koppström, V.J. Kortelainen, T. Niiranen, J. Välimaa

ABSTRACT

Suurikuusikko is a gold deposit in central Finnish Lapland which had a premining gold endowment of 7.9 million ounces. The deposit is hosted by tholeiitic mafic volcanic rocks of the ~2.02 Ga Kittilä group of the Central Lapland Greenstone Belt (CLGB). Gold is refractory occurring in arsenopyrite and pyrite, and mineralization is associated with intense pre-gold albite and syngold carbonate(-sericite) alteration. The host rock sequence is deformed by the subvertical to steeply east-dipping Kiistala shear zone. Numerous ore lenses are distributed along and within this north- to north-northeast-trending structure. Individual ore lenses have experienced multiple phases of deformation and generally have a moderate northerly plunge. Mineralized intervals have been found for more than a 5 km strike length of the host structure, and from surface to a depth of >1.5 kms. Four stages of sulfide formation have been detected at Suurikuusikko. Gold is associated with the second stage of arsenopyrite and pyrite growth. An Re–Os age of 1916 ± 19 Ma has been obtained from gold-bearing arsenopyrite. This suggests that mineralization took place 60–100 Ma after Kittilä group deposition and before the end of collision-related sedimentation in the CLGB. This age for Suurikuusikko is similar to the southwest-directed thrusting event related to the CLGB. Suurikuusikko has nearly all features typical for an orogenic gold deposit. The relatively early apparent timing of this deposit in the local orogenic evolution of the CLGB, albitized host rocks, fine-grained carbon in volcanic host rocks, and dominance of refractory gold are less commonly documented features of orogenic gold deposits, but do not suggest an alternative genetic type for this deposit and are not inconsistent with an orogenic gold system.

Keywords: Suurikuusikko; Kittilä Mine; Central Lapland Greenstone Belt; orogenic gold; refractory gold; arsenopyrite.

INTRODUCTION

Suurikuusikko is by far the largest of the many epigenetic gold (±copper) deposits of the Central Lapland Greenstone Belt (CLGB). It is located about 40 km northeast from the town of Kittilä, in the middle of the CLGB. Nearly all of the CLGB gold occurrences can be classified into the orogenic gold category—the deposit class defined by Böhlke (1982), Groves et al. (1998), and Goldfarb et al. (2001) in which epigenetic gold deposits hosted by orogenic belts and formed by syn- to late-orogenic fluids are called *orogenic*. The gold deposits within the CLGB have a few unusual characteristics. These features include pre-gold mineralization, albitization, and significant enrichment of Cu ± Co ± Ni. Similar features in deposits elsewhere caused Goldfarb et al. (2001) to call such deposits *orogenic gold*

Mineral Deposits of Finland. http://dx.doi.org/10.1016/B978-0-12-410438-9.00016-9

with *anomalous metal association*. At Suurikuusikko, no other commodities (except the gold), are present, whereas the host rocks do show distinct, pre-gold albitization.

The issues of regional to localized albitization and the polymetallic nature of some of the orogenic gold deposits in Finland is addressed in Subchapter 5.1 (Eilu, 2015). It is sufficient to note here that the Au-Cu ± Ni ± Co occurrences in the CLGB probably do not belong to the iron-oxide–copper–gold category of mineral deposits. Exceptions to this are the few cases in the westernmost part of the region, where iron oxides indeed are part of the mineral assemblage in the ores (Niiranen et al. (2007). These are described in detail in Chapter 6 (Moilanen and Peltonen, 2015).

Exploration in the Suurikuusikko area was initiated by the Geological Survey of Finland (GTK) after visible gold was discovered in a quartz vein in a road cut 4 km south-southwest of Suurikuusikko in 1986. The deposit was discovered in 1987 during diamond drilling by GTK on the deposit's host structure, the Kiistala shear zone (KSZ). Previous reports on the deposit have focused on reporting exploration activities, beneficiation, and giving brief overviews of deposit geology, structure, and ore mineralogy (Aho, 2009; Kojonen and Johanson, 1999; Chernet et al. 2000; Patison et al., 2006, 2007; Saloranta, 2011). In this chapter, we update the previous work with recent, yet unpublished, data from the deposit.

GOLD RESOURCE AND MINE DEVELOPMENT HISTORY

In April 1998, the Suurikuusikko deposit was acquired by Riddarhyttan Resources AB. This company's exploration activities identified gold mineralization over a 5-km strike length of the KSZ, and increased the known gold resource to more than 2 million ounces (Moz). In 2004, Agnico Eagle Mines Limited (AEM) acquired a 14 % ownership interest in Riddarhyttan, and commenced acquisition of the project in 2005.

Commercial gold production began in May 2009 via an open pit operation (Kittilä Mine), with underground production beginning in November 2012. The current (late 2014) predicted life for the Kittilä mining operation is 20 years. Proven and probable gold reserves at the end of 2014 were 4.5 Moz or 28.5 Mt of ore with an average grade of 4.9 g/t Au; with additional resources of 2.6 Moz of gold with an average grade of 3.5 g/t Au (Agnico Eagle Mines Ltd., 2015). In addition, the mine had produced 0.9 Moz of gold (recovery grade at about 90%) by the end of 2014. This makes The Kittilä Mine the largest currently active gold mine in Europe. Table 5.2.1 provides examples of the grade and width of mineralized intervals at Suurikuusikko.

REGIONAL GEOLOGICAL SETTING

Suurikuusikko is hosted by the Paleoproterozoic CLGB (Figs. 5.2.1 and 5.2.2) in northern Finland. The CLGB is part of the Karelian craton, a component of the Fennoscandian Shield. Supracrustal rocks of the CLGB are divided into seven lithostratigraphic groups (Lehtonen et al., 1998; Hanski and Huhma, 2005; Bedrock of Finland–DigiKp, 2014). Felsic volcanic rocks (2.44 Ga) overlying the Archean basement comprise the oldest exposed unit in the eastern and southeastern part of the belt, the Salla group.

The Vuojärvi group quartz-sericite schists and gneisses of uncertain origin represent the oldest lithostratigraphic unit in the southern part. The Salla and Vuojärvi groups are overlain by extensive mafic volcanic rocks of the Kuusamo group. Initial rifting of the Archean basement, and related volcanic activity represented by the Salla and Kuusamo groups, was followed by a more tranquil phase and deposition of a thick sequence of epiclastic sedimentary and minor mafic volcanic rocks, which

Table 5.2.1 Suurikuusikko deposit—examples of gold intercepts from drill core

Zone	Drill hole number	Mineralized section length (m)	Averaged grade of section (g/t Au)
Ketola	02114	6.40	4.20
Ketola	02107	7.00	11.10
Ketola	02107	3.20	7.10
Ketola	02104	10.70	4.00
Etelä	R407	7.00	7.50
Etelä	01802	5.60	8.60
Etelä	02039	8.10	9.50
Main	R473	14.00	10.40
Main	R504	10.80	9.10
Main	00717	14.30	10.60
Main	R478	18.20	5.10
Main	99002	18.20	16.50
Main	R479	26.80	17.30
Main	00730	18.90	9.10
Main	98004	29.60	11.90
Main	00903	46.20	8.90

comprise the Sodankylä group. Outcrop observations and crosscutting relationships indicate that both the Kuusamo and Sodankylä groups were deposited from 2.2–2.44 Ga, and that the latter is younger than the former group. These are overlain by the Savukoski group, a 2.05–2.2 Ga cratonic to cratonic-margin volcanosedimentary sequence of phyllites, graphite- and sulfide-bearing schists, and tholeiitic volcanic rocks, suggesting deepening of the depositional basin.

The uppermost rocks of the Savukoski group consists of komatiitic to picritic volcanic rocks indicating further rifting of the basin. The 2.02 Ga Kittilä group (discussed in more detail later) represents the youngest rifting phase and is bound by tectonic contacts with the older units of the CLGB. The <1.89 Ga Kumpu group, consisting of immature clastic sedimentary rocks and minor felsic to intermediate volcanic rocks, is the youngest unit and was deposited during a collisional tectonic stage. All rocks in the region are metamorphosed; hence, the prefix "meta" is implied but omitted from the names of the rock types in this chapter.

Intrusive rocks of the CLGB consist of 2.44 Ga Koitelainen and 2.05 Ga Kevitsa layered intrusions, 2.2 Ga and 2.1 Ga mafic sills and dikes and 2.05 Ga mafic to felsic dikes (Fig. 5.2.2). These all indicate repeated rifting of the depositional basement (Lehtonen et al. 1998; Rastas et al. 2001; Hanski and Huhma, 2005). The 2.13 Ga Nilipää suite granites in the southern margin of the CLGB indicate that felsic intrusive activity took place during or between the extensional phases. The 1.91–1.86 Ga Haparanda suite calc-alkaline felsic to mafic intrusives and ~1.80 Ga granites represent the syn- and late-collisional intrusive activity in the area, respectively (Lehtonen et al. 1998; Rastas et al. 2001). The post-collisional 1.79–1.77 Ga Nattanen suite granites are the youngest intrusives dated from the region (Heilimo et al., 2009).

The Suurikuusikko deposit is hosted by the Kittilä group rocks, a volcanosedimentary rock package dominated by tholeiitic mafic volcanic rocks (Figs. 5.2.1 and 5.2.3A). The Kittilä group includes four

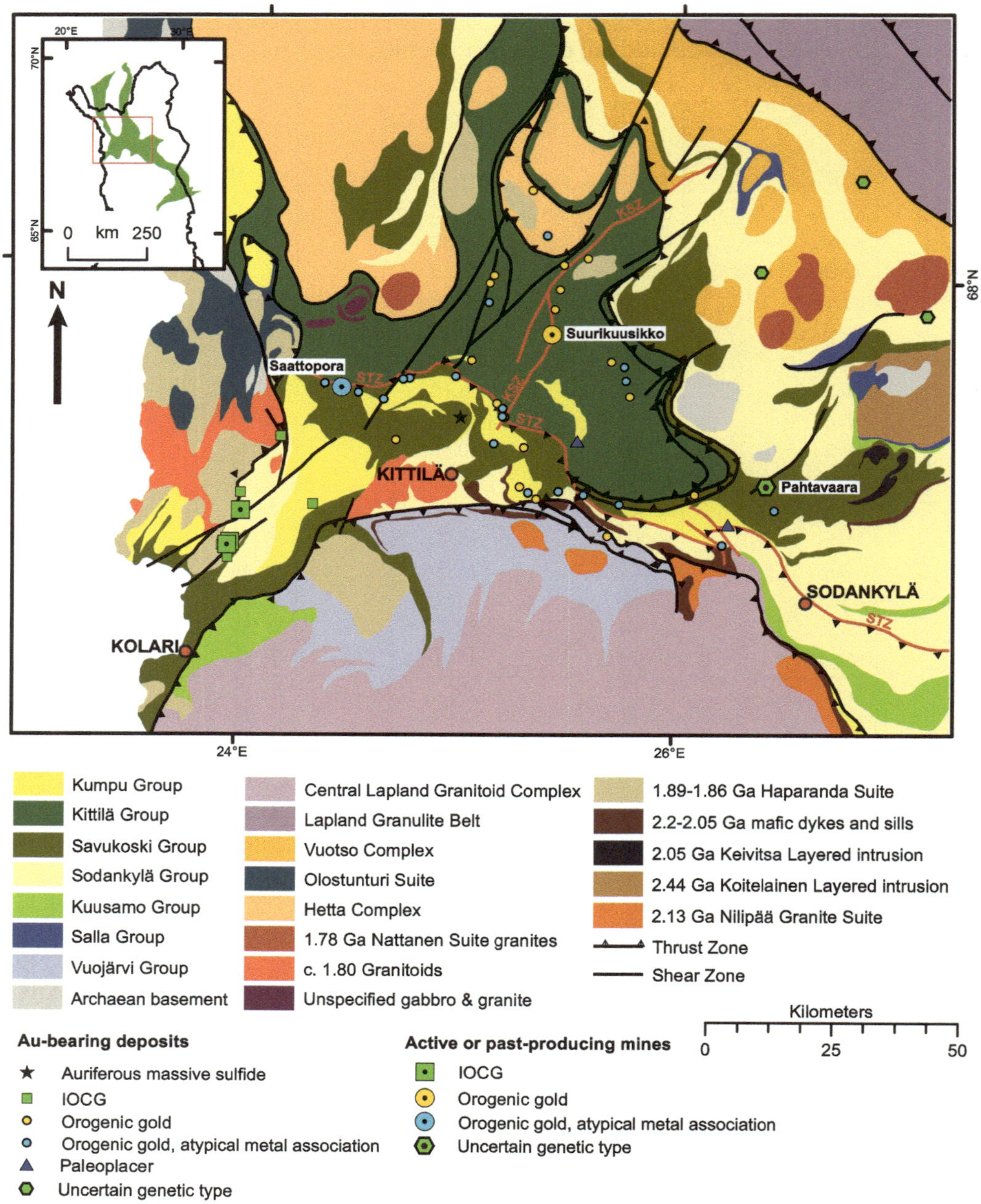

FIGURE 5.2.1 Regional geology map of central Lapland.

Known gold deposits and occurrences, and the location of Kittilä town. KSZ = Kiistala shear zone, STZ = Sirkka thrust zone. Inset: Location of the study area and location and extent of the CLGB (green). Coordinate system in both WGS84, north up.

Source: *Modified after Bedrock of Finland–DigiKp (2014).*

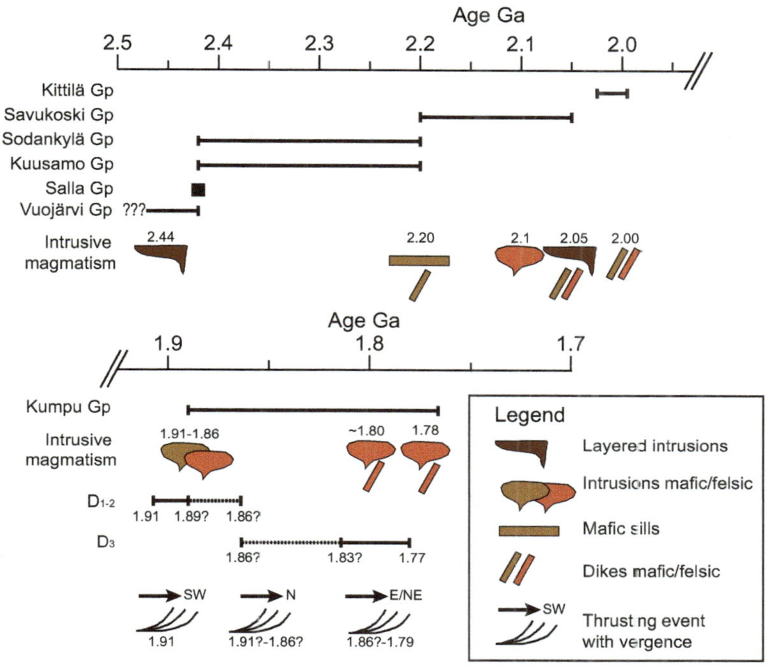

FIGURE 5.2.2 A schematic sequence of the lithostratigraphic groups, intrusive stages, and deformation for the CLGB.

Source: *Based on Lehtonen et al. (1998), Rastas et al. (2001), Hanski and Huhma (2005), Tuisku and Huhma (2006), Hölttä et al. (2007), Niiranen et al. (2007), and Patison (2007).*

formations. Tentatively from oldest to youngest, these are: (1) Kautoselkä formation containing iron-tholeiitic volcanic rocks; (2) Porkonen formation containing oxide and carbonate facies banded iron formation, and iron sulfide-bearing phyllites and schists; (3) Vesmajärvi formation containing magnesium-tholeiitic volcanic rocks; and (4) Pyhäjärvi formation dominated by sedimentary schists (Lehtonen et al., 1998). The regional distribution of each formation shown on published geology maps has been extrapolated from only a few geochemical samples by using the typical magnetic properties of each formation in aeromagnetic data. The Kautoselkä formation is interpreted where mafic rocks have a stronger magnetic response, and the Vesmajärvi formation where the total magnetic field is weaker (Fig. 5.2.4A). Mixed packages of conductive and magnetic rocks are generally assigned to the Porkonen formation (gray areas in Fig. 5.2.4A).

Geochemical heterogeneity among the Kittilä group rocks (for example, mafic rocks with enriched mid-ocean ridge basalt (E-MORB), normal mid-ocean ridge basalt (N-MORB), ocean island basalt (OIB) and island arc tholeiite (IAT) affinities) has been interpreted to indicate that the group is a composite of arc terranes and oceanic plateaus amalgamated during oceanic convergence (Hanski and Huhma, 2005). The Sm–Nd data of the Vesmajärvi formation tholeiites indicate a depleted mantle source and lack of crustal contamination, whereas the slightly negative εNd values of the Kautoselkä formation indicate either a subcontinental lithospheric mantle (SCLM) source or crustal contamination of a depleted mantle source (Hanski and Huhma, 2005). The geochemical signature and isotope data of the Kittilä group tholeiites suggest that it

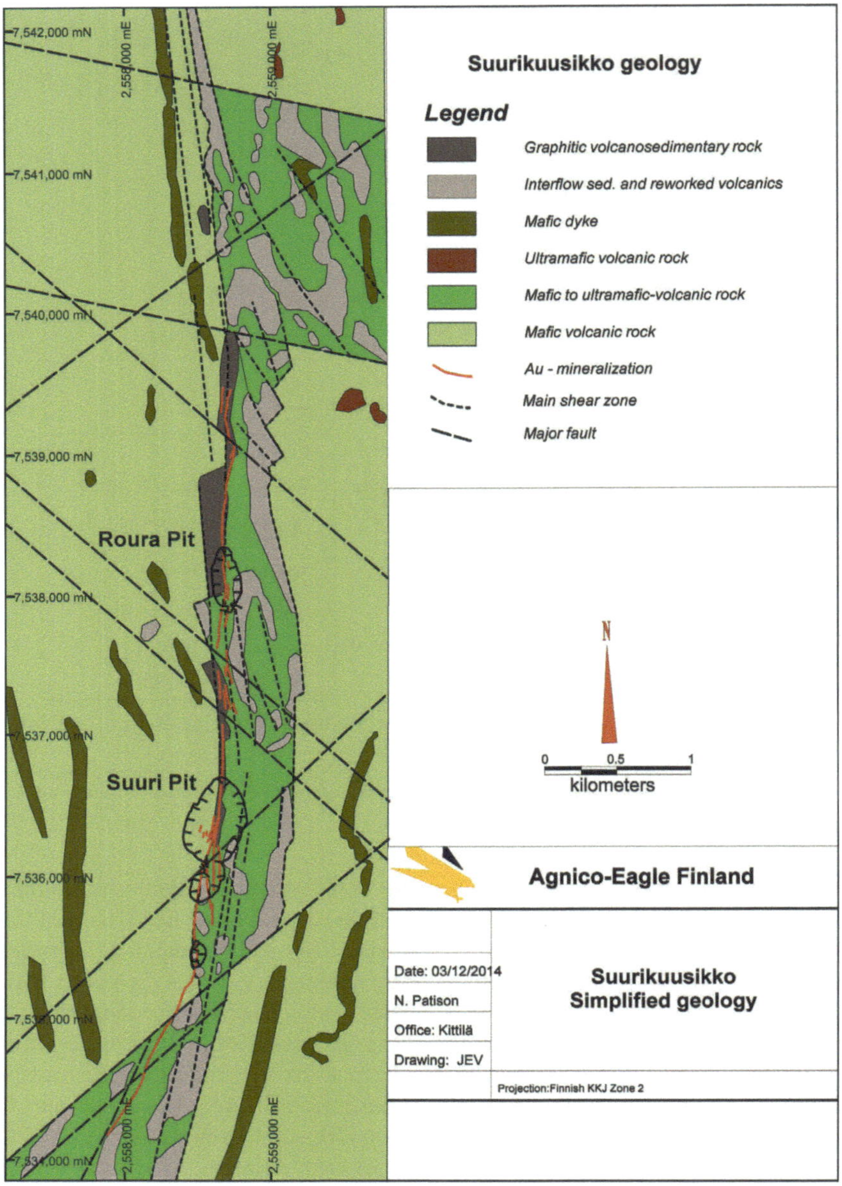

FIGURE 5.2.3 Suurikuusikko trend geology.

Interpreted geology of the Suurikuusikko area with open pit outlines. Coordinate system is Finland KKJ2, values in metres.

may partly represent juvenile oceanic crust. At its current position, the underlying rocks are most likely either Archean, of the surrounding Proterozoic sequences, or both. If the Kittilä group is partly or wholly allochthonous, the proposed tectonic emplacement timing is constrained by the 1.92 Ga Nyssäkoski felsic dike and the 1.91 Ga Ruoppapalo intrusion, both of which intrude the Kittilä group rocks (Hanski et al., 1998). The Kittilä group is also an anomalous area of low-grade (greenschist facies) metamorphism surrounded by higher grade (amphibolite to granulite facies) metamorphic rocks (Hölttä et al., 2007).

Interpretations of reflection seismic data across the CLGB also suggest that geology assigned to the Kittilä group potentially contains a number of distinct crustal blocks (Patison et al., 2006) and, thus, the published extent of this group requires revision. The maximum interpreted thickness of the Kittilä group is about 9 km (Niiranen et al., 2014). Felsic intrusive rocks in the region near Suurikuusikko include the 1.91 Ga Ruoppapalo granodiorite and ~2.02 Ga felsic porphyry dikes throughout the area (Rastas et al., 2001; Hanski and Huhma, 2005).

Multiple extensional events of various timing relative to compressive deformation are inferred due to the presence of extensive mafic dike swarms and layered intrusions (2.44–2.05 Ga intrusions in Figs. 5.2.1 and 5.2.2), but no extensional structures have been mapped in detail. The earliest mapped deformation phases (D1, D2) involve roughly synchronous north- to north-northeast-directed and south- to southwest-directed thrusting at the southern (e.g., Sirkka thrust zone (STZ) in Fig. 5.2.1) and northeastern margins of the CLGB, respectively (Ward et al., 1989). In the northeast, the CLGB is overthrust by the Paleoproterozoic Lapland granulite belt. At its southern margin, the CLGB is overthrusted by volcanosedimentary rocks associated with the Central Lapland Granitoid Complex. The western edge of the CLGB in Finland is interpreted as a cratonic boundary (Berthelsen and Marker, 1986a, 1986b; Lahtinen et al., 2015). All boundaries of the Kittilä group are interpreted as tectonic (Fig. 5.2.1).

Northwest-, north-, and northeast-trending D3 strike-slip shear zones, including the KSZ, cut early folding and thrusting, but may also reflect reactivation of older structures (e.g., transfer faults between the boundary thrusts of the Kittilä group). Post-D3 events are limited to brittle, low-displacement faults. More detailed discussions of regional deformation affecting the CLGB are provided by Gaál et al. (1989), Ward et al. (1989), Sorjonen-Ward (1993), Väisänen et al. (2002), Bergman (2003), Nironen and Mänttäri (2003), Hölttä et al. (2007), and Patison et al. (2007).

Direct age determinations on deformation events are scant. Based on correlations with other areas, potential D1–D2 ages include 1.93–1.91 Ga for southwest-directed thrusting in the northeast (Fig. 5.2.2; Daly et al., 2001; Tuisku and Huhma, 2006); and 1.89–1.87 Ga for north to northeast-directed thrusting (particularly in the southern CLGB; Sorjonen-Ward 1993; Hölttä et al. 2007). Possible D3 deformation ages include 1.89–1.88 Ga, and a minimum age of 1.77 Ga (for an undeformed felsic dike intruding D1 or D2 thrust-related folds within the CLGB; Väisänen 2002). Post-collisional Nattanen-type granites (Fig. 5.2.1) also have an age of 1.79–1.77 Ga (Lehtonen et al. 1998). Peak metamorphic conditions were reached during D1–D2 compressional deformation (Holttä et al. 2007), and a general thermal resetting event is reported for the CLGB in the period 1.79–1.77 Ga (Lahtinen et al. 2005).

GEOLOGY OF THE SUURIKUUSIKKO AREA

Suurikuusikko is located in the central part of the Kittilä group area (Figs. 5.2.1, 5.2.3, 5.2.4), about 15 km east-southeast from the thickest part of the group (Niiranen et al. 2014). Patison (submitted) has produced a bedrock geology model for a 15 × 35 km region including the Suurikuusikko deposit, summaries

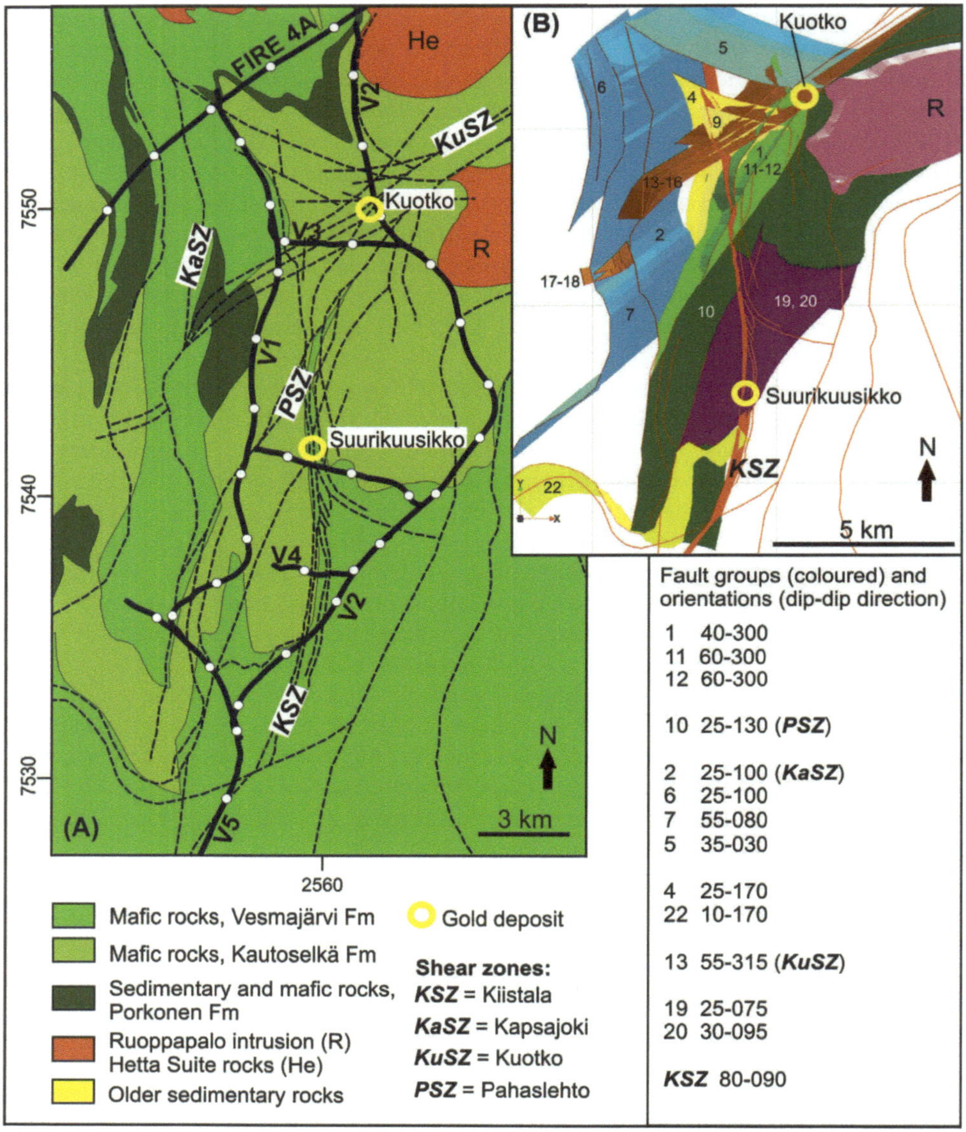

FIGURE 5.2.4 Surface geology of the Suurikuusikko area.

A. Geology map for the Suurikuusikko region. Selected faults are shown as black dashed lines. Thick black lines (V1 to V5) show the high-resolution reflection seismic (HIRE) transects. Transparent labels are seismic line numbers. Labels on white backgrounds are structure names.

B. Modeled faults in the Suurikuusikko region, roughly covering the area of Fig. 5.2.4A. Pink solid (R) is the Ruoppapalo intrusion. Red lines are surface traces of unmodeled faults. Red solid is the Kiistala shear zone (KSZ). Numbers for dip-dip direction are degrees; for example, "40–300" means "40°/300°." Coordinate system is Finland KKJ2, values in km.

of which are presented here. This model was interpreted from surface geology maps, drilling data, 3D inversion models of magnetic and electromagnetic data, and 2D inversions of magnetotelluric data, and is heavily based on interpretations of high-resolution reflection seismic data collected from an 86.1-km grid over the area extending to a depth of approximately 5 km (line locations shown in Fig. 5.2.4A).

Modeled stratigraphy (rock packages of contrasting density) is shown in Fig. 5.2.5A (blue and purple solids). The interpreted orientation of contacts between rock packages has a low to moderate dip to the northeast. In the east–west sections shown in Fig. 5.2.5A (lines E1 and V4), open folding is interpreted. Similar structures have been mapped immediately adjacent to mineralized zones outside of intensely sheared rock (e.g., Fig. 5.2.5F). Directly beneath the deposit, the modeled bedrock has an antiformal appearance. This may be a thrust-related structure or the result of movements on the Kiistala shear zone (KSZ). The stratigraphic contacts in Fig. 5.2.5A also correlate with modeled faults, interpreted as thrusts with dips of 25° in directions ranging from 075° to 130° (e.g., green Pahaslehto shear zone (PSZ) in Fig. 5.2.4 Patison submitted), complicating the crustal structure in this area prior to the development of the KSZ.

Cross-sections based on drill core data are shown in Figs. 5.2.5C and 5.2.5E. These show similarly oriented bedrock to the model in Fig. 5.2.5A, but also a steepening of dips and a tightening of folding as the KSZ is approached. Cross-section interpretations of drill core data are difficult as the local stratigraphy contains numerous lava flows with ambiguous boundaries, and distinct units that can be used as marker horizons are few. Alteration overprinting provides further complications.

In the cross-sections and in the seismic profile over Suurikuusikko (line V4 in Figs. 5.2.4 and 5.2.5A), the KSZ occurs at an apparent faulted fold hinge. However, this interpretation is not always readily apparent when viewing drill core, and instead different geology is usually mapped on either side of the KSZ. This inconsistency in interpretation might be due to the scales of each interpretation (i.e., single drill holes and mappable exposures vs. 5 km-deep seismic profiles) and/or because of vertical displacement across the KSZ juxtaposing different lithological units.

Local rock types are predominantly thick mafic lava packages (Fig 5.2.4), including pillow lavas. Coarser-grained shallow sills may also occur. Mafic rocks are typically iron tholeiitic in composition (Aho, 2009). In some areas, there is a significant component of mafic volcaniclastic breccias, with little to no reworking during deposition. These breccias are only occasionally polymictic. Primary breccias are in places difficult to recognize due to substantial overprinting by hydrothermal and structural brecciation.

A heterogeneous rock package between lava-dominated domains also occurs. This rock package has been a focus for both deformation and associated alteration. It includes volcaniclastic rock, intermediate to felsic volcanic rocks (andesite flows of Powell, 2001), and carbonaceous ("graphitic") sedimentary intercalations containing chert, argillitic units, and sulfide-rich sedimentary units occurring with mafic volcanic rock. Ultramafic rocks are also present in the more distal parts of the host sequence. The deposit's host structure (KSZ; Figs. 5.2.4, 5.2.5A, and 5.2.5B) mostly follows this heterogeneous rock package. Minor, as yet uneconomic, ore shoots occur on stratigraphic contacts that are adjacent to but at an angle to the KSZ, where elevated gold values are associated with sedimentary and volcaniclastic layers within unmineralized mafic lavas.

The estimated orientation of the KSZ in the deposit area is 80°/090° (dip/dip direction; red plane in Figs. 5.2.5A and 5.2.5C). This is consistent with the orientation mapped for this structure in mine exposures (Figs. 5.2.5B and 5.2.5D) and that interpreted from drill core. The strike of the KSZ changes slightly over its length, but the dip is consistently steep to subvertically eastward. The KSZ has a strike length of at least 25 km (Figs. 5.2.1 and 5.2.4).

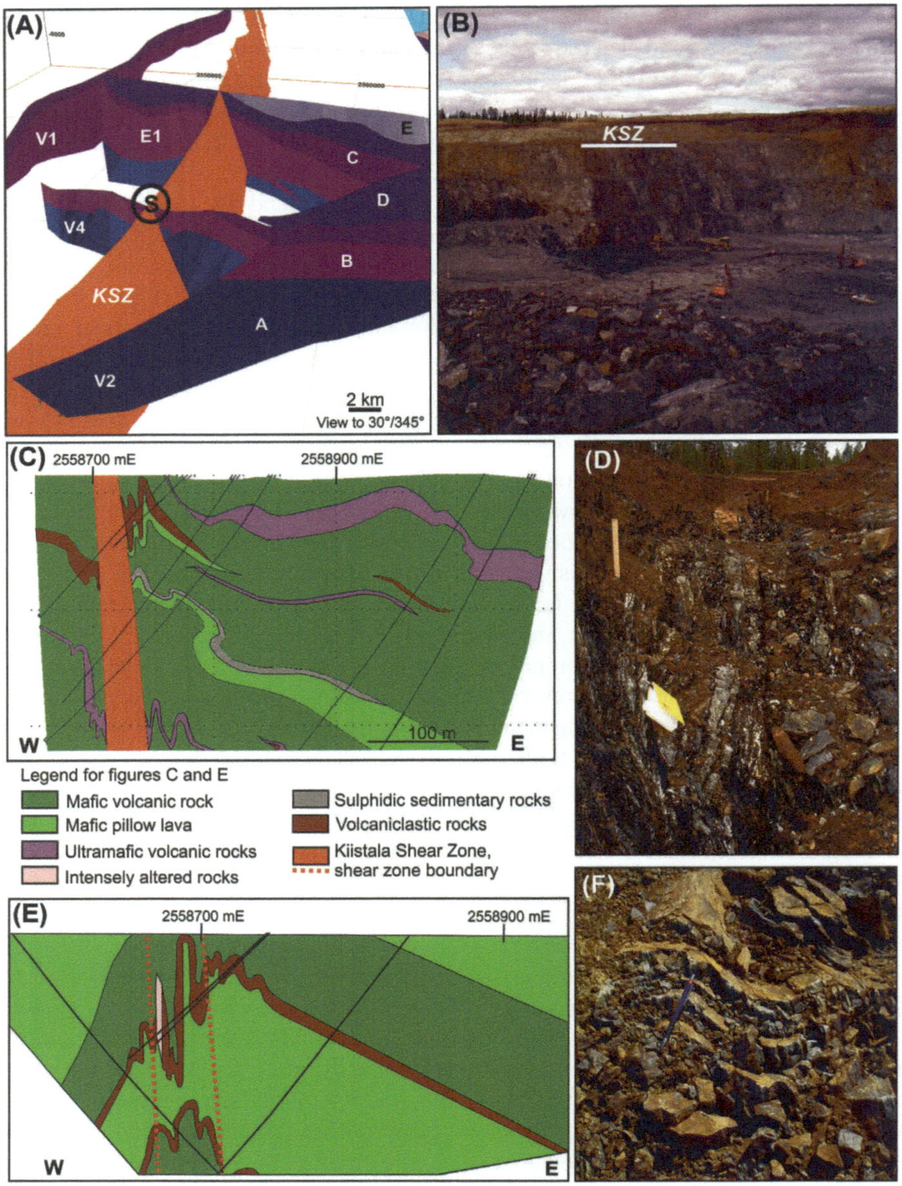

FIGURE 5.2.5 Modelled (A, C, E) and actual (B, D, F) Suurikuusikko geology.

(A) Modeled bedrock layers (blue and purple, labeled A to E) in the Suurikuusikko area, and the Kiistala shear zone (red solid). Suurikuusikko area circled. Interpretation based on HIRE seismic lines E1, V2, and V4. Datum for vertical sections is +250 m and the depth of the interpretation is 2.5 km. The viewing direction for this 3D fence diagram is toward 30°–345° (the plunge and plunge direction of viewing angle).

Known mineralization occurs within north-trending and less frequently northeast-trending (e.g., the Ketola occurrence) shear zone segments. It is a complex structure, recording several phases of movement. A minor degree of west-up movement has occurred, but most deformation has occurred by flattening accompanied by some strike-slip movement. Aeromagnetic images of the KSZ indicate apparent early sinistral strike-slip movement along the zone. Immediately above the widest mineralized zones, late dextral strike-slip movements are recorded on shear planes bounding mineralized zones. An apparent positive correlation exists between points of more intense shearing within the KSZ and the amount of gold present in host rocks.

SUURIKUUSIKKO GOLD DEPOSIT

MINERALIZED ROCK

Mineralized rock at Suurikuusikko has undergone varying degrees of brecciation and veining. Figure 5.2.6 illustrates the typical appearance of the ore. These are altered mafic rocks, probably lava, which show pervasive albite alteration and carbonate-quartz veining. In some areas, brittle deformation has produced breccias Fig. 5.2.6C in which completely fragmented rock has an extensive carbonate-quartz hydrothermal matrix.

Other mineralized rock intervals have a combination of lithologies intermingled by shearing, an example of which is shown in Fig. 5.2.5D; the black material in this picture has a combination of sedimentary and hydrothermally produced or remobilized carbon in a shear-interleaved mix of black schist and volcanosedimentary rock. The pale mineralized rock at this locality may be altered mafic rock, but occasionally it is an altered and mineralized intermediate to felsic rock. Sedimentary units rarely host more than low-grade mineralization, with pyrite. Intensely sheared rock related to mineralization is typically healed (the breccia veins filled mainly by quartz), whereas later shear movements have generated the fractures evident in Fig. 5.2.5D.

ALTERATION

The alteration described next is based on published reports (Härkönen, 1997; Kojonen and Johanson, 1999; Chernet et al., 2000; Patison et al., 2007; Saloranta, 2011; Koppström, 2012) and on findings during ongoing exploration at the mine and its immediate surroundings. Mineral assemblages in the region are typical for greenschist facies rocks. In unaltered mafic rocks, the mineral assemblage is actinolite-chlorite-albite-epidote ± pyrrhotite. Most of the rocks in the hosting shear zone are albitized

◀ (B) Suurikuusikko open pit in August 2009. Darker area is mineralized zone bound by the Kiistala shear zone (KSZ in the photo). Viewing direction approximately north-northeast.
(C) Vertical section geology interpretation based on drill core for section 7539100 mN (= seismic line V4 in Figs. 5.2.4A and 5.2.5A).
(D) Sheared gold-mineralized rock with albitized fragments (pale color) and dark "graphitic" shear planes (location KKJ2 7536,180 mN, 2558,595 mE).
(E) Vertical section geology interpretation based on drill core for section 7539,400 mN (= seismic line E1 in Figs. 5.2.4A and 5.2.5A). The pink polygon, "Intensely altered rocks" indicates area where the original rock types has been so much altered (mostly by albitization) that it is not possible to identify the primary rock type.
(F) Gently north-plunging folds in a chert or an intensely silicified volcanic rock immediately east of sheared ore zones (location KKJ2 7535,935 mN, 2558,635 mE).

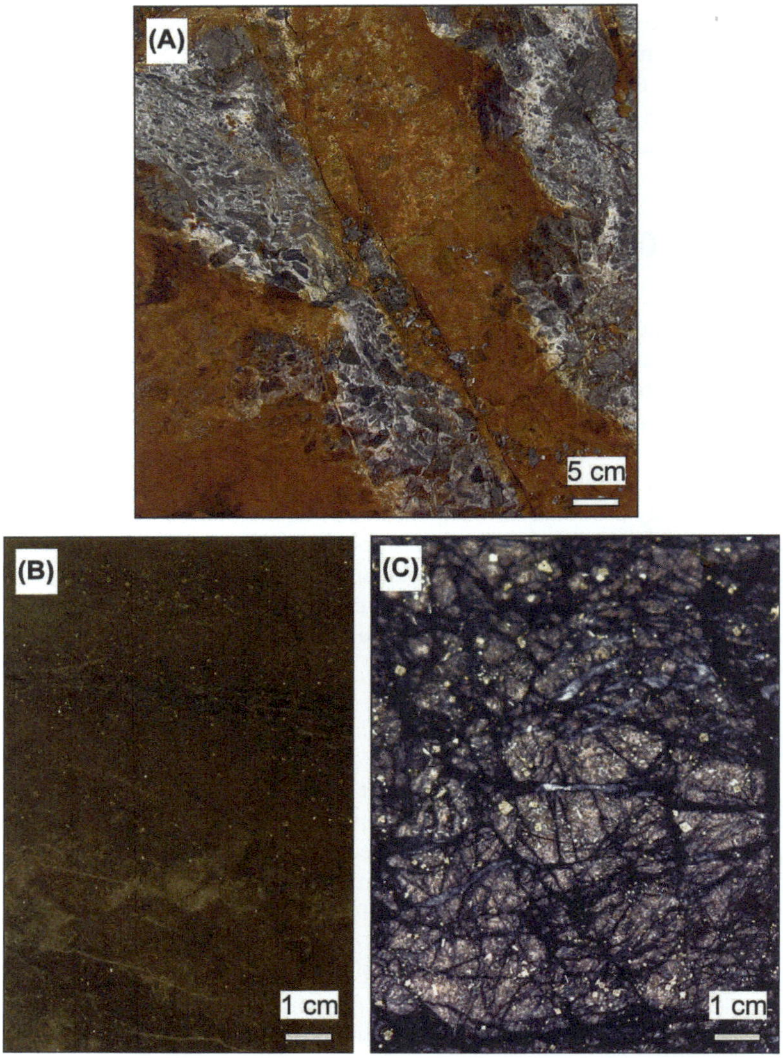

FIGURE 5.2.6 Examples of mineralized rocks.

(A) Pit exposure of hydrothermal breccia with carbonate-quartz veining. Fragments are albitized mafic rock (Rouravaara open pit, 2010).
(B) Bleached, albitized, and veined massive mafic volcanic rock with disseminated arsenopyrite and pyrite.
(C) Sheared, brecciated, and albitized mafic volcanic rock with disseminated arsenopyrite and pyrite, amorphous carbon-carbonate-quartz alteration, and veining.

to a variable degree. Largely, albitization is pervasive in the KSZ and so obvious that mapping albitization is part of core logging routine at the mine.

Albitization is almost always present in the ore, but also in the barren sedimentary and volcanic rocks within the KSZ. This fact and textural relationships in the ore and its wallrocks, such as carbonates replacing all the detected stages of albite porphyroblasts, and ore minerals apparently overprinting albite (Saloranta, 2011), suggest that albitization predated gold mineralization. Intense albitization occurs in both the altered rock itself, replacing other feldspars and mafic silicates, and as albite brecciating micro-veinlets. The amount of albite (up to 82.6 wt%, as calculated from vol.% data achieved by Mineral Liberation Analyzer (MLA) measurements from thin sections) has a strong positive correlation with elevated gold grades (Koppström, 2012). The relationship between albitization and mineralization is further assessed in the "Discussion and summary" section of this chapter.

The alteration halo directly related to mineralization has been divided into three concentric zones: distal, intermediate, and proximal (Koppström, 2012). Of these, the extent of the intermediate zone seems to be most difficult to define, possibly because it is the hardest to recognize in routine core logging. The distal alteration zone is from 100 to more than 300 m wide, extends beyond the KSZ, at least locally, and is characterized by chlorite, rutile ("leucoxene"), minor quartz(?) and calcite replacing amphibole, epidote, and titanite. Calcite also occurs in distal veining (with quartz).

The proximal zone includes all ore and varies in width from the typical 10–50 m to up to 100 m in places. There are also parts of the KSZ with a set of subparallel ore bodies (Fig. 5.2.7) and proximal alteration, with distally altered barren rock between the ore bodies. Proximal alteration and ore gangue (Fig. 5.2.6) include albite, chlorite, muscovite, dolomite, quartz, and lesser amounts of rutile, sericite, amorphous carbon associated with shearing, and relict chlorite and calcite. Dolomite or ankerite veins and hydrothermal breccia matrix characterize the proximal alteration domain, including the ore.

The intermediate alteration zone is characterized by the assemblage albite-calcite-quartz(?)-rutile-pyrite. This assemblage is not visually obvious when compared to assemblages diagnostic for the distal and proximal zones, and during core logging, intermediate alteration might be included either into the proximal or distal alteration zone.

A peculiar feature within the domain of the ore and proximal alteration is the presence of fine-grained reduced carbon species with a graphitic appearance, but having a low degree of crystallization. This mineral is often called"graphite" or "amorphous carbon" at the mine. The relative amount of "graphite" is nowhere large, it occurs everywhere as an accessory mineral, in both the altered rock and in quartz-carbonate veins, but is easy to recognize both in drill core and outcrop. It is clearly of secondary origin, as it occurs not only in lithological units of sedimentary origin, but also in volcanic rocks hosting and enveloping the ore, and in veins.

ORE MINERALOGY

The majority of gold at Suurikuusikko occurs as solid-solution lattice substitutions in arsenopyrite (73.2%) and arsenian pyrite (22.7%) grains; inclusions of native Au and alloys (4.1%) are observed in pyrite and arsenopyrite grains, at grain boundaries and in silicates, and occasionally in association with chalcopyrite (relative amounts defined in mineralogical assessments of the deposit based on drill core samples; Kojonen and Johanson 1999). The composition of gold inclusions includes various alloys with silver and mercury (Chernet et al., 2000). Textures indicate that the Au-rich inclusions and free gold grains are late. They were formed during post-mineralization deformational events, as a product

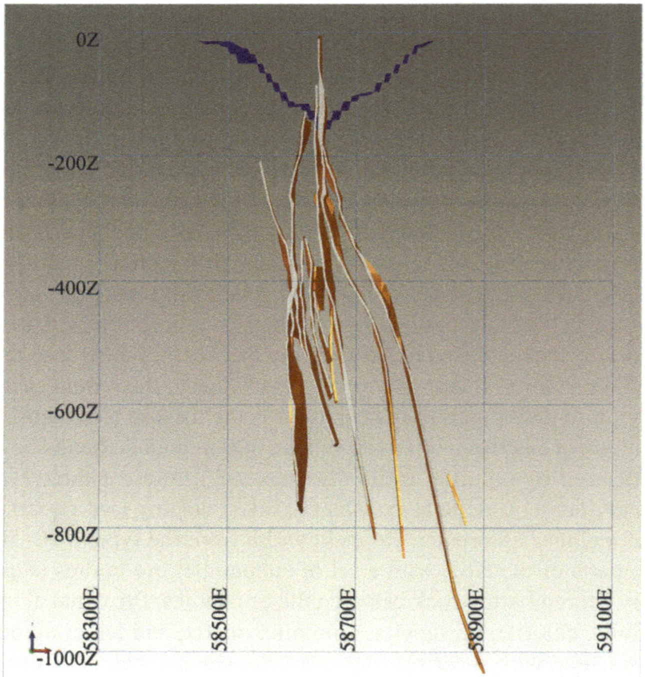

FIGURE 5.2.7 Assay-based modeled east–west section across a set of subparallel ore bodies (brown) at Suurikuusikko.

Outline of the Suuri open pit in this section is marked by blue. View to the north; grid 200 m.

of later-stage alteration or by introduction of late fluids (Koppström, 2012). For the ore-related elements at Suurikuusikko, gold and arsenic exhibit a strong positive correlation, whereas gold and antimony correlate negatively. However, a rare later stage of stibnite veining, which overprints the ore-related sulfides, contains a high gold content.

Pyrite and arsenopyrite commonly form aggregates and intergrowths. Koppström (2012) used electron probe micro-analyzer (EPMA) analysis and laser-ablation inductively-coupled plasma mass spectrometry (LA-ICP-MS) to divide arsenopyrite and pyrite grains into four generations distinguished by textural, compositional, and temporal differences (Figs. 5.2.8A–C). For both minerals, the second generation contains the highest gold content (Fig. 5.2.8B).

The first-generation arsenopyrite (Fig. 5.2.8A) contains inclusions of rutile and other gangue materials, and these arsenopyrites are typically richer in antimony and poorer in arsenic than the average values for all Suurikuusikko arsenopyrites (Härkönen, 1997; Parkkinen, 1997; Kojonen and Johanson, 1999; Chernet et al., 2000; Koppström, 2012). The second arsenopyrite generation mantles the first generation and differs from it by a higher arsenic and a lower antimony content (Fig. 5.2.8A). The third arsenopyrite generation is interpreted as a recrystallization product of the previous generations (Fig. 5.2.8A). The fourth generation is observed only in contact with relatively late quartz-carbonate veins and may be a product of late recrystallization or late fluid introduction events similar to the fourth pyrite generation (Härkönen, 1997; Chernet et al., 2000; Koppström, 2012).

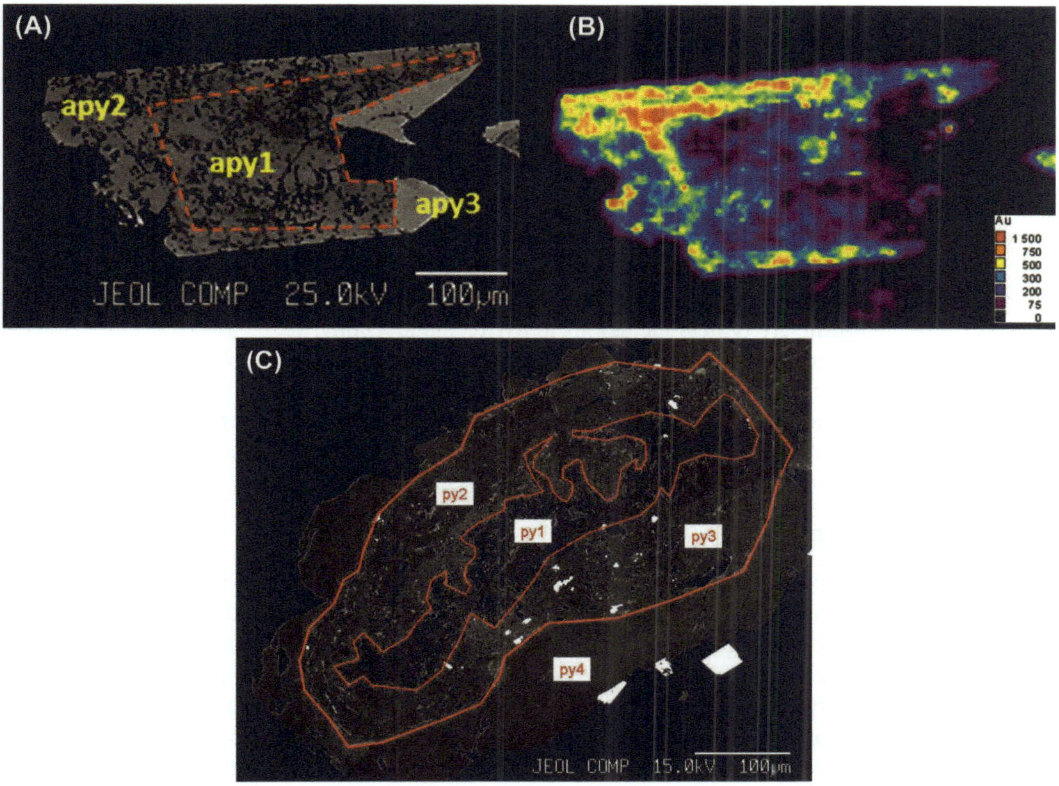

FIGURE 5.2.8 Element distribution maps of an arsenopyrite and a pyrite grain. Growth phases of each grain are labeled.

(A) Back-scatter image of an arsenopyrite grain. The red line shows the border between the first (apy1), the second (apy2), and the third (apy3) generation arsenopyrites.

(B) LA-ICP-MS map of Au distribution in an arsenopyrite grain. The brighter colors represent higher Au concentrations. Note the relation of gold to growth zoning of arsenopyrite. Same scale as in Fig. 5.2.8A. The legend indicates count-detector values representing the relative amount of gold.

(C) Back-scatter image of a zoned pyrite grain that includes four different stages of pyrite formation. The first stage, generation one (py1), is the innermost, porous, and inclusion-rich part of the grain. The remnants of the second generation (py2) are seen as light gray patches within the third generation (py3). The second stage pyrite is defined as the arsenian-generation pyrite. The third generation is the darker gray material inside the py2-py3 area. The fourth generation (py4) forms the outermost gray and only slightly fractured area with almost no inclusions. White inclusions inside the pyrite are mostly antimony minerals and in lesser amounts arsenopyrite.

Source: *After Koppström, 2012.*

Early pyrite grains (py1 in Fig. 5.2.8C) contain inclusions of carbonates, quartz, and rutile. The second pyrite generation (py2 in Fig. 5.2.8C) is arsenian, and has the highest arsenic content of all pyrite generations (Koppström, 2012). The third pyrite generation (py3 in Fig. 5.2.8C) is most likely formed via recrystallization of the second pyrite generation. Inclusions are arsenopyrite (potentially mobilized from py2), antimonides, and gangue minerals. The fourth pyrite generation (py4 in Fig. 5.2.8C) is the least altered. Coarse-grained pyrite of this stage is commonly rimmed by quartz, carbonates, chlorite, and sericite. (Härkönen, 1997; Chernet et al., 2000; Koppström, 2012).

Other ore minerals detected at Suurikuusikko include gersdorffite, chalcopyrite, chalcocite, sphalerite, pyrrhotite, monazite, ullmannite, berthierite, gudmundite, bournonite, galena, jamesonite, native antimony, stibnite, bismuth, bornite, chromite, talnahkite, native gold, gold-silver alloys, and tetrahedrite (Kojonen and Johanson, 1999; Chernet et al., 2000; Aho, 2009). These are present only in trace amounts and are not auriferous except for the native gold and gold-silver alloys. Chalcopyrite, chalcocite, tetrahedrite, and rutile are typical inclusions in pyrite and arsenopyrite.

Abundant, fine-grained carbon with a poorly organized (amorphous) graphitic lattice structure characterizes most of the ore: Gold-bearing sulfides are common on shear planes, stylolithic cleavage, and fractures containing amorphous carbon. Unpublished carbon isotope data (trends in the $\delta^{13}C$ value) suggest that this material is sourced from C-rich sedimentary units within the host sequence.

ORE ZONE GEOMETRY

Modeled ore zones strike north, in some cases northeast, parallel to the local trend of the KSZ (Fig. 5.2.4). The dip of ore bodies is variable. It is generally subvertical and parallel to that of the KSZ (Fig. 5.2.7), but may contain steps with a lower dip angle. The ore bodies have a moderate plunge to the north. The control on the northerly plunge is not completely resolved. Potential causes include the role of intersections between multiple shear planes, and of the intersections of lithological unit contacts and shear planes. The orientation of regional fold axes (similar to axes in Fig. 5.2.5F) may also have a role in determining favorable sites for mineralization during shearing. Sulfides in the ore show evidence for deformation relating to post-mineralization movements on host shear planes. Post-mineralization brittle faults crosscut mineralized zones but are not known to cause significant displacement of ore lenses (Fig. 5.2.4).

AGE OF MINERALIZATION

Rhenium–osmium arsenopyrite geochronology has been used to obtain a direct age of 1916 ± 19 Ma for gold-bearing arsenopyrites at Suurikuusikko (Geospec Consultants Limited, 2008). No age for rocks in the immediately adjacent area has been obtained, but comparing the Re–Os age to the stratigraphic ages quoted, mineralization took place 60–100 Ma after Kittilä group deposition and before the end of collision-related sedimentation in the CLGB (<1.89 Ga). The age for Suurikuusikko is most similar to minor felsic to intermediate intrusive rocks 15–20 km from the deposit: 1919 ± 8 Ma Nyssäkoski porphyry dike (Rastas et al., 2001) and 1914 ± 3 to 1905 ± 5 Ma Ruoppapalo intrusion (Rastas et al., 2001; Ahtonen et al., 2007) that post-date the Kittilä group rocks or (in the case of Ruoppapalo) are early synorogenic in relative age. This age for Suurikuusikko is similar to the southwest-directed thrusting event in the region, based on the suggested age of the overthrusting of the Lapland Granulite Belt, 1.92–1.90 Ga (Tuisku and Huhma, 2006). This thrusting event may have more relevance to mineralization at Suurikuusikko than the not-so-voluminous intrusive activity in the region at that time, as discussed below.

DISCUSSION AND SUMMARY

Most of the characteristics of the Suurikuusikko gold deposit are those typical for an orogenic gold deposit, the deposit class originally suggested by Böhlke (1982), first fully defined by Gebre-Mariam et al. (1995), and further elaborated by Groves et al. (1998) and Goldfarb et al. (2001, 2005). These characteristics include the following:

1. The deposit is hosted by an orogenic belt (a greenstone belt) in rocks metamorphosed under greenschist-facies conditions.
2. Local structures and host rock textures, as well as the arsenopyrite Re–Os age, indicate that the timing of mineralization is during an orogeny.
3. The deposit has a distinct structural control by a long-lived shear zone (KSZ) where the siting, extent, and plunge of individual ore bodies seem largely to be controlled by the structural evolution of the shear zone. Admittedly, there are several structures that have only very recently become obvious, such as most of those shown in Figs. 5.2.3 through 5.2.5, which are yet to be explained and put into the structural evolution sequence of the region. This is an issue of future research at Suurikuusikko.
4. The lithological control appears to be the pre-gold albitized rocks, that is, the locally most competent lithological units.
5. Alteration directly related to gold mineralization has produced a proximal mineral assemblage of albite-muscovite (sericite)-dolomite or ankerite-quartz-rutile-arsenopyrite-pyrite. This mineral assemblage is enveloped by the calcite-albite-chlorite-rutile assemblage with no titanite nor amphibole present even in the distally altered rock. This is diagnostic for an orogenic gold system and reflects a reduced, near-neutral, low-salinity H_2O–CO_2 fluid with H_2S and arsenic as minor constituents, a fluid able to carry and deposit gold under greenschist-facies conditions.
6. The mineralizing fluid probably also contained CH_4, as suggested by the extensively present "graphite" (the amorphous carbon). This reduced carbon may have been sourced from the local graphitic sedimentary units, as is suggested by unpublished carbon isotope data. Another open issue is how did the carbon precipitation affect the gold mineralization process. Did it result in a significant change in redox conditions of the fluid causing S, As, and Au to precipitate? Only careful geochemical modeling may answer these questions. In any case, the close spatial relationship between the "graphite" and auriferous sulfides suggest a close genetic relationship.
7. From the preceding items 4 and 5, we can derive the conclusion that the mineralizing fluid was a typical orogenic fluid, which can be derived from progressive metamorphic devolatilization at depth, typically at the transition from greenschist to amphibolite facies conditions (Goldfarb et al., 2005; Phillips and Powell, 2010). The relatively abundant "graphite" may suggest that the fluid contained more CH_4, and/or was slightly more reducing than the average fluid of an orogenic gold system.
8. The fact that nearly all gold is refractory in arsenopyrite and pyrite suggests that (1) the gold was transported in the mineralizing fluid as both bisulfide and arsenate (or thioarsenate) complexes (Mikucki, 1998; Phillips and Powell, 2010; Williams-Jones and Migdisov, 2014), and (2) reaction between host rock iron and the S and As ligands were among the significant agents (if not the most significant) in gold precipitation (Goldfarb et al., 2005; Phillips and Powell, 2010). Ferrous iron for the sulfidation reactions was readily available, especially from the local tholeiitic basalts.
9. Mass balance has not been evaluated for Suurikuusikko rocks. The existing geochemical raw data indicate that the components enriched during mineralization include, at least, As, Au, CO_2, and

S, whereas there are no indications of base-metal mobility. This fits well with the supposed type of the mineralizing fluid. Only robust mass balance calculations can show how much real Na enrichment is related to the pre-gold albitization, and if any Na, K, or any minor or trace-elements except As, Au, and S were mobile and related to gold mineralization.

A number of features at Suurikuusikko could be seen as unusual for an orogenic gold system, or even suggesting some other genetic type of mineralization. These include refractory gold, albitized host rocks, similar ages for mineralization and local granitoid intrusions, and the possibly early-orogenic timing of mineralization relative to main orogenic events of northern Finland.

The dominance of refractory gold over free gold is uncommon in orogenic gold occurrences in Finland, but is not totally exceptional. A smaller part of gold occurs in pyrite and/or arsenopyrite or löllingite in a large number of gold occurrences and deposits in Finland (Eilu and Pankka, 2009). Globally, refractory gold is not uncommon. A number of small to very large orogenic gold deposits are known to contain significant, or be dominated by, refractory gold in various host rocks of varied mineralization ages. These include the metasedimentary-hosted Obuasi in Ghana and Sukhoi Log in Siberia (Oberthür et al., 1996; Zhang et al., 2008); mafic volcanic-hosted Bulletin in Western Australia and Giant in Yellowknife, Canada (Eilu and Mikucki, 1998; Shelton et al., 2004); the ultramafic-hosted Haimur in Egypt and Jian Cha Ling in China (Emam and Zoheir, 2013; Vielreicher et al., 2003).

Host rocks are intensely albitized at Suurikuusikko. Similar albitization is also present in many barren parts of the KSZ. Textural relationships in the ore and its wallrocks suggest that albitization predated gold mineralization (Saloranta, 2011). Similar relationships are found in the Hanhimaa shear zone 10 km to the west of the KSZ (Saalmann and Niiranen, 2010). These suggests that albitization prepared ground for mineralization by producing the locally most competent rocks and was not part of the gold-mineralizing event. The locally hardest rocks behave in the most brittle manner during deformation, thus creating more open space when the fluid pressure exceeds the lithostatic pressure, hence serving as physical traps for later fluids to deposit silica and ore minerals (Sibson et al., 1988; Weatherley and Henley, 2013).

The CLGB has a long pre-orogenic sedimentary and volcanic depositional history in a rifted intracontinental basin (Hanski and Huhma, 2005). Such an environment is favorable for saline connate fluids to form, with or without evaporates (Yardley and Graham, 2002). These pre-orogenic fluids may have leached and altered the fluid flow channels extensively with albitization, carbonatization, or, at higher temperatures, even scapolitization of any rock type. A similar environment has been suggested for other intracontinental rifted basins of the same age in Finland, Kuusamo, and Peräpohja (Vanhanen, 2001; Kyläkoski, 2012). Saline connate fluids may survive during an orogeny (Yardley and Graham, 2002; Yardley and Cleverley, 2013), causing structurally controlled alteration including albitization during the orogeny.

Albitization has been seen as one of the typical, even diagnostic, features of the iron-oxide–copper–gold (IOCG) class of mineralizing systems (e.g., Oliver et al., 2004; Lopez et al., 2014). Other features regionally typical for the CLGB, and elsewhere connected to the IOCG concept (e.g., Hitzman, 2000; Duncan et al., 2014), include epigenetic copper-gold mineralization, medium- to high-salinity fluids, multistage alteration, and extensive occurrence of scapolite in higher metamorphic grade domains around most of the CLGB.

All of this has been used to argue that many, if not all, of the CLGB polymetallic gold and gold-only deposits could go into the IOCG category (Frietsch et al., 1997). However, with the exception of the extreme western part, no epigenetic hydrothermal iron oxide mineralization, high-salinity oxidizing fluids related to alteration and mineralization, nor metal zoning within a deposit has been detected within the CLGB. All features deviating from the characteristics of the gold-only type of orogenic mineralization within the CLGB can be explained by the Goldfarb et al. (2001) model, which suggests that "where Paleoproterozoic tectonism included deformation of older, intracratonic basins, the resulting ore fluids were anomalously saline and in some cases the orogenic lodes are notably base metal-rich". This resulted in the model of "orogenic gold with anomalous metal association" (Goldfarb et al., 2001). At Suurikuusikko, it is only the albitized rocks that need to be explained as anomalous compared to most deposits of the orogenic gold model.

In the regional geology summary just provided, we mention granitoids with ages close to the arsenopyrite Re–Os age from Suurikuusikko and occurring relatively close to the deposit (Fig. 5.2.9): the 1.92 Ga Nyssäkoski felsic dike and the 1.91 Ga Ruoppapalo granodiorite intrusion (Hanski et al., 1998; Rastas et al., 2001; Ahtonen et al., 2007). This could be taken as an indication of a genetic relationship between granitoid magmatism and gold mineralization. Intrusion-related gold (IRG) or thermal aureole gold (TAG) have been suggested for many regions globally, mainly based on fluid and stable isotope indications, by a close spatial relationship to a granitoid, and in some places by an obvious case of a metamorphosed, low metal grade, porphyry Au(-Cu) occurrence (e.g., Hall et al., 2001; Hart et al., 2002; Duuring et al., 2007; Helt et al., 2014).

The TAGs and IRGs have many features similar to orogenic gold, such as fluid inclusion types, stable isotope ratios, element mobility, and local structural control. However, in most cases the evidence is not conclusive for a genetic relationship to an intrusion, and all features can be explained by the normal orogenic gold model (Goldfarb et al., 2005). Within the CLGB we have not detected any direct diagnostic evidence of intrusion-related mineralization processes such as metal zoning within a deposit or between deposits located concentrically around intrusions, gold mineralization hosted by upper carapaces of granitoids or by a granitoid of any kind, or vertical extensional vein sets (e.g., sheeted veins) within intrusions; post-orogenic gold mineralization, or contemporaneous magmatic Sn or W deposits (cf., Hart et al., 2002; Goldfarb et al., 2005). The IRG concept, as defined by Hart et al. (2002), appears not applicable as, in addition to features listed in above, the CLGB gold deposits (Suurikuusikko included) are clearly not post-orogenic.

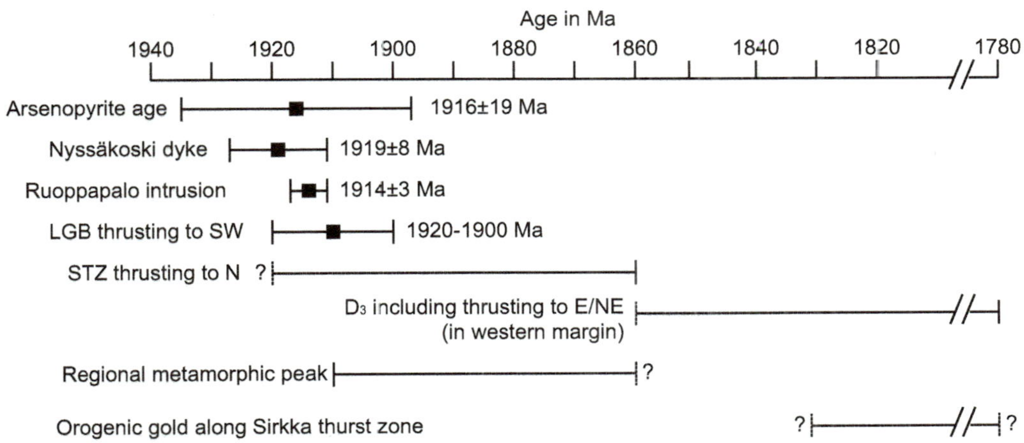

FIGURE 5.2.9 A suggested sequence of events for the CLGB, covering the Paleoproterozoic orogenic events in northern Finland from 1940–1780 Ma, and with a focus on Suurikuusikko and its surroundings.

Thrusting of the Lapland granulite belt (LGB) from the northwest and along the Sirkka thrust zone (STZ) from the south, seem to overlap with mineralization at Suurikuusikko.

Source: *The diagram is chiefly based on data in Hanski et al. (1998), Rastas et al. (2001), Hanski and Huhma (2005), Tuisku and Huhma (2006), Hölttä et al. (2007), Patison (2007), and Niiranen et al. (2014).*

Four generations of pyrite and arsenopyrite have been detected at Suurikuusikko. Microtexture details suggest that the first two generations may reflect separate stages of sulfidation; that is, two stages when S and As were introduced into the KSZ. Mineral composition data suggest that the second stage sulfidation was the main gold-mineralizing event. On the other hand, the third and fourth generations of pyrite and arsenopyrite appear to indicate stages of local remobilization of S and As, if the textures are correctly interpreted. These later stages might also be those when free gold was formed at Suurikuusikko by remobilization from the arsenopyrite and pyrite lattices. Some gold may also have been introduced by a late external fluid, related to the formation of the rare stibnite veins, which clearly are late in the local paragenetic sequence.

Two to five stages of veining and ore mineral formation have been described from most orogenic and many other types of gold deposits investigated in detail (e.g., Zhang et al., 2008; Helt et al., 2014; Shelton et al., 2004). In practically all cases we know of (except the rare occasions of clear overprint of a syngenetic occurrence by orogenic gold) one hydrothermal stage has introduced most, if not all, of the gold. This seems to be the case also at Suurikuusikko. An open question remains regarding the absolute timing of all the alteration and sulfidation stages.

The arsenopyrite Re–Os age from Suurikuusikko may fit well into the regional orogenic evolution of the CLGB during the Paleoproterozoic, as depicted in Fig. 5.2.9. From the mineral texture details to the larger tectonic scale, it seems gold mineralization took place during the regional metamorphic peak, roughly contemporaneously with minor granitoid intrusions. During that time, the Lapland granulite belt (LGB) was overthrust from the northeast, and the KSZ could then have acted as a transform fault. This early timing would mean that there were two major stages of gold mineralization within the CLGB, an early one shown by Suurikuusikko and a late one indicated by many of the gold occurrences along the Sirkka thrust zone, such as Saattopora (Patison, 2007; Niiranen et al., 2014).

ACKNOWLEDGMENTS

The comprehensive review by Nick Oliver helped us best present the available data for our contribution. Many thanks go to Raimo Lahtinen for gently pushing the authors toward eventually rewriting the manuscript, leading to the final product.

REFERENCES

Agnico Eagle Mines Ltd, 2015. Deposit Reserves and Resources Accessed online at www.agnicoeagle.com/en/Operations/Reserves-and-Resources/Pages/default.aspx.

Aho, P., 2009. Malmimineraloginen ja geokemiallinen vaihtelu kahdella Suurikuusikon kultakaivosalueen tutkimusprofiililla. M.Sc. thesis. University of Oulu. p. 91 (in Finnish).

Ahtonen, N., Hölttä, P., Huhma, H., 2007. Intracratonic Palaeoproterozoic granitoids in northern Finland: prolonged and episodic crustal melting events revealed by Nd isotopes and U–Pb ages on zircon. Bulletin of the Geological Society of Finland 79, 143–174.

Bedrock of Finland–DigiKp. Digital map database [electronic resource]. Geological Survey of Finland [accessed 5 November, 2014].

Bergman, S., 2003. Regional Precambrian geology of northern Norrbotten county. In: Eklund, O. (Ed.), Excursion Guide to Finnish and Swedish Lapland 1–7.9.2003. Turku University, Åbo Akademi University Geocenter, Report 20.

Berthelsen, A., Marker, M., 1986a. Tectonics of the Kola collision suture and adjacent Archaean and Early Proterozoic terrains in the northeastern region of the Baltic Shield. Tectonophysics 126, 31–55.

Berthelsen, A., Marker, M., 1986b. 1.9–1.8 Ga old strike-slip megashears in the Baltic Shield, their plate tectonic implications. Tectonophysics 128, 163–181.

Böhlke, J.K., 1982. Orogenic (metamorphic-hosted) gold-quartz veins. U.S. Geological Survey Open-File, Report 82-795, 70–76.

Chernet, T., Kojonen, K., Pakkanen, L., 2000. Applied mineralogical study on the near-surface Suurikuusikko refractory gold ore, Kittilä, Western Finnish Lapland (Phase 1). Geological Survey of Finland Report C/MA9/2743/2000/10. p. 22.

Daly, J.S., Balagansky, V.V., Timmerman, M.J., et al., 2001. Ion microprobe U–Pb zircon geochronology and isotopic evidence for a trans-crustal suture in the Lapland–Kola orogen, northern Fennoscandian Shield. Precambrian Research 105, 289–314.

Duncan, R., Hitzman, M., Nelson, E.P., Togtokhbayar, O., 2014. Structural and lithological controls on iron oxide-copper-gold deposits of the southern Selwyn–Mount Dore corridor, Eastern Fold Belt, Queensland, Australia. Economic Geology 109, 419–456.

Duuring, P., Cassidy, K.F., Hagemann, S.G., 2007. Granitoid-associated orogenic, intrusion-related, and porphyry-style metal deposits in the Archean Yilgarn Craton, Western Australia. Ore Geology Reviews 32, 157–186.

Eilu, P., Mikucki, E.J., 1998. Alteration and primary geochemical dispersion associated with the Bulletin lode-gold deposit, Wiluna, Western Australia. Journal of Geochemical Exploration 63, 73–103.

Eilu, P., Pankka, H., 2009. FINGOLD – A public database on gold deposits in Finland. Version 1.0. Geological Survey of Finland. Digital data product 4 Optical disc (CDROM).

Eilu, P., 2015. Overview on gold deposits in Finland. Mineral Deposits of Finland. Elsevier, Amsterdam. pp. 377–403.

Emam, A., Zoheir, B., 2013. Au and Cr mobilization through metasomatism: microchemical evidence from ore-bearing listvenite, South Eastern Desert of Egypt. Journal of Geochemical Exploration 125, 34–45.

Frietsch, R., Tuisku, P., Martinsson, O., Perdahl, J.-A., 1997. Early Proterozoic Cu-(Au) and Fe ore deposits associated with regional Na-Cl metasomatism in northern Fennoscandia. Ore Geology Reviews 12, 1–34.

Gaál, G., Berthelsen, A., Gorbatschev, R., et al., 1989. Structure and composition of the Precambrian crust along the POLAR profile in the northern Baltic Shield. Tectonophysics 162, 1–25.

Gebre-Mariam, M., Hagemann, S.G., Groves, D.I., 1995. A classification scheme for epigenetic Archaean lode-gold deposits. Mineralium Deposita 30, 408–410.

Geospec Consultants Limited, 2008. Re-Os Analyses, Arsenopyrite Geochronology for Geological Survey of Finland. Analytical Report No 2 Unpublished.

Goldfarb, R.J., Groves, D.I., Gardoll, S., 2001. Orogenic gold and geologic time: a global synthesis. Ore Geology Reviews 18, 1–75.

Goldfarb, R.J., Baker, T., Dube, B., et al., 2005. Distribution, character, and genesis of gold deposits in metamorphic terranes. Economic Geology 100th Anniversary Volume, 407–450.

Groves, D.I., Goldfarb, R.J., Gebre-Mariam, M., et al., 1998. Orogenic gold deposits: A proposed classification in the context of their crustal distribution and relationship to other gold deposits. Ore Geology Reviews 13, 7–27.

Hall, G.A., Wall, V.J., Massey, S., 2001. Archaean pluton-related (thermal aureole) gold: the Kalgoorlie exploration model. In: Williams, P.J. (Ed.), A Hydrothermal Odyssey May 17–19, Townsville. Extended abstracts. EGRU and JCU. 66–67.

Hanski, E., Huhma, H., 2005. Central Lapland Greenstone Belt. In: Lehtinen, M., Nurmi, P.A., Rämö, O.T. (Eds.), Precambrian Geology of Finland: Key to the evolution of the Fennoscandian Shield Developments in Precambrian Geology 14, 139–193.

Hanski, E.J., Huhma, H., Lehtonen, M.I., Rastas, P., 1998. 2.0 Ga old oceanic crust in northern Finland. In: Hanski, E., Vuollo, J. (Eds.), International Ophiolite Symposium and Field Excursion: Generation and Emplacement of Ophiolites through Time, August 10–15, Oulu, Finland Abstracts and Excursion.

Härkönen, I., 1997. Tutkimustyöselostus Kittilän kunnassa valtausalueilla Suurikuusikko 2 ja Rouravaara 1–10 (kaivosrekisterinumerot 5965/1, 6160/1, 6288/1–6288/9) suoritetuista kultatutkimuksista vuosina 1987–1997. Geological Survey of Finland, Report M 06/2743/97/1. p. 47 (in Finnish).

Hart, G.J.R., McCoy, D.T., Goldfarb, R.J., et al., 2002. Geology, exploration, and discovery in the Tintina gold province, Alaska and Yukon. Economic Geology Special Publication 9, 241–274.

Heilimo, E., Halla, J., Lauri, L., Rämö, T., 2009. The Paleoproterozoic Nattanen-type granites in northern Finland and vicinity—a postcollisional oxidized A-type suite. Bulletin of the Geological Society of Finland 81, 7–38.

Helt, K.M., Williams-Jones, A.E., Clark, J.R., et al., 2014. Constraints of the genesis of the Archean oxidized, intrusion-related Canadian Malartic gold deposit, Quebec, Canada. Econ. Geol. 109, 713–735.

Hitzman, M.W., 2000. Iron oxide-Cu-Au deposits: what, where, when, and why. In: Porter, T.M. (Ed.), Hydrothermal Iron Oxide Copper-Gold and Related Deposits: A Global Perspective. Australian Mineral Foundation. Adelaide. pp. 9–25.

Hölttä, P., Väisänen, M., Väänänen, J., Manninen, T., 2007. Paleoproterozoic metamorphism and deformation in Central Finnish Lapland. Geological Survey of Finland Special Paper 44, 109–120.

Kojonen, K., Johanson, B., 1999. Determination of refractory gold distribution by microanalysis, diagnostic leaching and image analysis. Mineralogy and Petrology 67, 1–19.

Koppström, K., 2012. Electron microprobe and LA-ICP-MS study on the distribution of gold and other elements in pyrite and arsenopyrite from the Suurikuusikko gold deposit, northern Finland. M.Sc. thesis, University of Oulu. p. 131.

Kyläkoski, M., Hanski, E., Huhma, H., 2012. The Petäjäskoski Formation, a new lithostratigraphic unit in the Paleoproterozoic Peräpohja Belt, northern Finland. Bulletin of the Geological Society of Finland 84, 85–120.

Lahtinen, R., Korja, A., Nironen, M., 2005. Proterozoic tectonic evolution. In: Lehtinen, M., Nurmi, P.A., Rämö, O.T. (Eds.), Precambrian Geology of Finland: Key to the Evolution of the Fennoscandian Shield Developments in Precambrian Geology 14, 481–531.

Lahtinen, R., Huhma, H., Lahaye, Y., et al., 2015. New geochronological and Sm–Nd constraints across the Pajala shear zone of northern Fennoscandia: reactivation of a Paleoproterozoic suture. Precambrian Research 256, 102–119.

Lehtonen, M., Airo, M.-L., Eilu, P., et al., 1998. The stratigraphy, petrology and geochemistry of the Kittilä greenstone area, northern Finland: a report of the Lapland Volcanite Project. Geological Survey of Finland. Report of Investigations 140, p. 144.

Lopez, G.P., Hitzman, M.W., Nelson, E.P., 2014. Alteration patterns and structural controls of the El Espino IOCG mining district, Chile. Mineralium Deposita 49, 235–259.

Mikucki, E.J., 1998. Hydrothermal transport and depositional processes in Archean lode-gold systems: A review. Ore Geology Reviews 13, 307–321.

Moilanen, M., Peltonen, P., 2015. Hannukainen Fe-(Cu-Au) deposit, Western Finnish Lapland: Deposit model updated. Mineral Deposits of Finland. Elsevier, Amsterdam. pp. 485–504.

Niiranen, T., Poutiainen, M., Mänttäri, I., 2007. Geology, geochemistry, fluid inclusion characteristics, and U–Pb age studies on iron oxide–Cu–Au deposits in the Kolari region, northern Finland. Ore Geology Reviews 30, 75–105.

Niiranen, T., Lahti, I., Nykänen, V., Karinen, T., 2014. Central Lapland Greenstone Belt 3D modeling project final report. Geological Survey of Finland Report of Investigation 209, p. 78.

Nironen, M., Mänttäri, I., 2003. Structural evolution of the Vuotso area, Finnish Lapland. Geological Society of Finland. Bulletin 75, 93–101.

Oberthür, T., Mumm, A.S., Vetter, U., et al., 1996. Gold mineralization in the Ashanti belt of Ghana: genetic constraints of stable isotope geochemistry. Economic Geology 91, 289–301.

Oliver, N.S., Cleverly, J.S., Mark, G., et al., 2004. Modeling the role of sodic alteration in the genesis of iron oxide-copper-gold deposits, eastern Mount Isa block, Australia. Economic Geology 99, 1145–1176.

Parkkinen, J., 1997. The Suurikuusikko gold deposit. Mineral resource estimate. Geological Survey of Finland Report M19/2743/97/1, p. 20.

Patison, N.L., 2007. Structural controls on gold mineralisation in the Central Lapland Greenstone Belt. Geological Survey of Finland. Special Paper 44, 107–124.

Patison N.L. (submitted). A 3D regional geology model for the Central Lapland gold district (northern Finland). Geological Survey of Finland, Special Paper.

Patison, N.L., Korja, A., Lahtinen, R., Ojala, V.J., the FIRE Working Group, 2006. FIRE seismic reflection profiles 4, 4A, and 4B: Insights into the crustal structure of northern Finland from Ranua to Näätämö. Geological Survey of Finland. Special Paper 43, 161–222.

Patison, N.L., Salamis, G., Kortelainen, V.J., 2007. The Suurikuusikko gold deposit: Project development summary of northern Europe's largest gold resource. Geological Survey of Finland. Special Paper 44, 109–120.

Powell, W., 2001. Petrographic Report on Suurikuusikko rock types. Report for Riddarhyttan Resources AB, November.

Phillips, G.N., Powell, R., 2010. Formation of gold deposits: a metamorphic devolatilization model. Journal of Metamorphic Geology 28, 689–718.

Rastas, P., Huhma, H., Hanski, E., et al., 2001. U–Pb isotopic studies on the Kittilä greenstone area, central Lapland, Finland. Geological Survey of Finland 33, 95–141 Special Paper.

Saalmann, K., Niiranen, T., 2010. Hydrothermal alteration and structural control on gold deposition in the Hanhimaa shear zone and western part of the Sirkka Line. Geological Survey of Finland. Report M19/2741/2010/58, p. 30.

Saloranta, J., 2011. Albite alteration at Suurikuusikko, northern Finland, and its relation to gold deposition. M.Sc. thesis, University of Helsinki, p. 95.

Shelton, K.L., McMenamy, T.A., van Hees, E.H.P., Falck, H., 2004. Deciphering the complex fluid history of a greenstone-hosted gold deposit: fluid inclusion and stable isotope studies of the Giant mine. Yellowknife, Northwest Territories, Canada. Economic Geology 99, 1643–1663.

Sibson, R.H., Robert, F., Poulsen, H., 1988. High-angle reverse faults, fluid-pressure cycling and mesothermal gold-quartz deposits. Geology 16, 551–555.

Sorjonen-Ward, P., 1993. Structural history, alteration and gold mineralization in the Lapland Greenstone Belt, Finland. Mid- to lower-crustal metamorphism and fluids conference. Geological Society of Australia, Abstracts 35, 88–90.

Tuisku, P., Huhma, H., 2006. Evolution of migmatitic granulite complexes; implication from Lapland granulite belt; Part II, isotopic dating. Bulletin of the Geological Society of Finland 78, 143–175.

Väisänen, M., 2002. Structural features in the Central Lapland Greenstone Belt, northern Finland. Geological Survey of Finland Report, K21.42/2002/3, p. 20.

Vanhanen, E., 2001. Geology, mineralogy and geochemistry of the Fe-Co-Au-(U) deposits in the Paleoproterozoic Kuusamo Schist Belt, northeastern Finland. Geological Survey of Finland Bulletin 399, p. 229.

Vielreicher, R.M., Vielreicher, N.M., Hagemann, S.G., Jones, G., 2003. Fault zone evolution and its controls on ore-grade distribution at the Jian Cha Ling gold deposit, western Qinling region, central China. Mineralium Deposita 38, 538–554.

Ward, P., Härkönen, I., Nurmi, P.A., Pankka, H.S., 1989. Structural studies in the Lapland greenstone belt, northern Finland and their application to gold mineralization. Geological Survey of Finland 10, 71–77 Special Paper.

Weatherley, D.K., Henley, R.W., 2013. Flash vaporization during earthquakes evidenced by gold deposits. Nature Geoscience 6, 294–298.

Williams-Jones, A.E., Migdisov, A.A., 2014. Experimental constraints on the transport and deposition of metals in ore-forming hydrothermal systems. Society of Economic Geologists Special Publication 18, 77–95.

Yardley, B.W.D., Graham, J.T., 2002. The origins of salinity in metamorphic fluids. Geofluids 2, 249–256.

Yardley, B.W.D., Cleverley, J.S., 2013. The role of metamorphic fluids in the formation of ore deposits. Geological Society, London, Special Publications 393.

Zhang, C., Large, R.R., Maslennikov, V., 2008. Sulfur isotopes in sediment-hosted orogenic gold deposits: Evidence for an early timing and a seawater sulfur source. Geology 36, 971–974.

CHAPTER

EXPLORATION TARGETING AND GEOLOGICAL CONTEXT OF GOLD MINERALIZATION IN THE NEOARCHEAN ILOMANTSI GREENSTONE BELT IN EASTERN FINLAND

5.3

P.B. Sorjonen-Ward, A. Hartikainen, P.A. Nurmi, K. Rasilainen, P. Schaubs, Y. Zhang, J. Liikanen

ABSTRACT

The Hattu schist belt in easternmost Finland has been systematically explored for gold over several decades and the Pampalo mine commenced operations during the latter half of 2010. Here we document the exploration history of this region, which represents the first Archean gold province to be exploited in Fennoscandia. This review focuses on exploration strategies as well as geological characterization of gold occurrences and deposits, with particular emphasis on till geochemical exploration, which proved particularly effective, and the structural controls on mineralization.

The Hattu schist belt supracrustal sequence is notable for the relative abundance of felsic volcanic and epiclastic deposits. Isotopic data indicate that deposition, deformation, and granitoid intrusion were very closely related in time, the ages of the earliest supracrustal units, at 2.75 Ga, effectively overlapping with those of syntectonic granitoids. All exposed contacts between the Hattu schist belt and these granitoids are intrusive, or else tectonically modified, and hence the granitoids cannot represent depositional basement to the greenstone belt. In spite of locally intense and complex deformation, the Hattu schist belt has retained a high degree of stratigraphical coherence, which has enabled the overall topology of lithic units and structures to be further clarified.

Mineralization does not appear to occur preferentially in any particular lithology, although lithological transitions may be favored, due to associated chemical and rheological gradients. Pervasive fluid flow is, however, also indicated by broad, highly deformed alteration zones and disseminated rather than vein-style mineralization at some prospects. Alteration characteristics of gold mineralization throughout the Hattu schist belt gold occurrences are broadly similar, if less intense than the hydration, carbonation, and potassic alteration generally reported for Archean lode gold deposits. The albite-carbonate alteration mineralogy so characteristic of gold deposits in Archean mafic and ultramafic provinces is far from ubiquitous in the Hattu schist belt, although it is well developed at the largest known deposit, at Pampalo. Instead, chlorite, muscovite, and tourmaline dominate alteration parageneses in mineralized metasediments. Native gold is fine grained (mostly <15 μm) and occurs largely in free-milling form between silicate grains, associated with pyrite, pyrrhotite, and minor arsenopyrite and rutile; it is typically intergrown with Bi, Pb, Ag, Fe, and Au tellurides and native bismuth.

Gold mineralization at most prospects is geochemically distinctive, with notable enrichments of Te, B, Bi (and locally Mo and W), compared to many other Archean gold provinces, reflected in the abundance of tourmaline, tellurides and locally, scheelite. The syntectonic Kuittila tonalite contains an early molybdenite-scheelite vein system overprinted by sheared biotite-muscovite-calcite-scheelite-pyrite-gold veins and demonstrates the syntectonic

Mineral Deposits of Finland. http://dx.doi.org/10.1016/B978-0-12-410438-9.00017-0

timing of mineralization. Extensive alteration of granites and country rocks in the northern part of the schist belt is typified by microcline-muscovite-pyrite alteration and is perhaps more reminiscent of epithermal systems. Therefore, it is possible that gold mineralization in the Hattu schist belt does record some kind of interaction between magmatic and metamorphic influences during a continuum of arc-related and orogenic tectonic processes. These issues are the subject of ongoing investigations, to be reported elsewhere.

The structural architecture of the Hattu schist belt is characterized by upward-facing, generally steeply dipping structures, and it is possible to establish a close, sequential relationship between tightening of folds, attenuation of fold limbs, development of shear zones with strike-slip displacements, and the propagation of new folds due to strain incompatibilities between shear zones. Structural controls on alteration and mineralization are a consequence of strain partitioning due to rheological contrasts between rock units and interaction with large scale structures. This is most apparent in the distribution of disseminated mineralization in the hanging wall above the western margin of the Kuittila tonalite and in the location of the Pampalo gold deposit within the backrotated toe of a strike slip duplex, recording a progressive transition from contractional to oblique extensional behavior. Numerical simulations of regional structural patterns with FLAC3D™ have been performed to assess the importance of variations in far-field stress configuration in controlling rock failure and localizing mineralization, in particular addressing the issue of whether the inferred kinematics represent a local response to orthogonal shortening and compression, rather than deformation within a regional strike-slip regime. Results indicate that northwest–southeast directed deformation favors simultaneous failure and dilation in both the northeast-trending Pampalo zone and northwest-trending Kuittila zone, which can be attributed to their somewhat anomalous orientations and attitudes. Ongoing research involves assessment of results of modeling against field-based geometric and kinematic constraints, to better understand and predict structural controls on mineralization.

Keywords: gold; Archean; greenstone belt; Pampalo; till geochemistry; pathfinder elements; deformation; numerical modeling.

INTRODUCTION

Neoarchean crustal growth and reworking were closely associated with the formation of a significant proportion of known and mined gold deposits, most notably in the Witwatersrand Basin of the Kaapvaal craton in southern Africa, the Abitibi subprovince of the Superior craton in Canada, and the Eastern Goldfields terrane of the Yilgarn craton in Western Australia. However, although Archean rocks in the Fennoscandian Shield are distributed over an area comparable to that of the Yilgarn craton, there is so far little indication that gold resources might be comparable. Whether this is a fundamental issue of gold endowment, relating to essential differences in subcontinental lithospheric characteristics and Neoarchean orogenic processes, remains to be seen. Here we review the exploration history and geological context of mineralization for the only Archean greenstone belt in Fennoscandia where gold is currently being mined, namely the Hattu schist belt in the Ilomantsi district in eastern Finland. Our purpose is to both document the nature of mineralization, comparing and contrasting with Archean gold deposits elsewhere, and to demonstrate the exploration targeting approaches that have been most useful; this approach also reflects the interaction between data collection and mapping, and our evolving understanding of the key features relating to mineralization. For this reason, particular emphasis is given to till geochemical exploration, which was critical in appreciating the potential of the region and in identifying prospective targets, and to structural geometry, as during the course of investigations, the importance of structural controls on mineralization became more apparent.

GEOLOGIC OUTLINE OF THE HATTU SCHIST BELT

In recent reviews of the Archean of Fennoscandia, the terms *Ilomantsi terrane* (Sorjonen-Ward and Luukkonen, 2005) and *Ilomantsi* (Hölttä et al., 2012) have been used to refer to the greenstone belts and surrounding granitoids at the western edge of the Neoarchean Karelian craton in southeastern Finland (Fig. 5.3.1). The Ilomantsi complex includes the Ilomantsi greenstone belt, which consists of a western branch, known as the Kovero belt, and an eastern segment, which has been informally referred to as the Hattu schist belt (Nurmi et al., 1993; Sorjonen-Ward, 1993a).

The Kovero belt contains a greater proportion of mafic and ultramafic rocks than the Hattu belt, which is dominated by feldspathic epiclastic sediments and felsic volcanics, with sporadic magnetite banded iron formation (BIF) and mafic and ultramafic volcanic units (Nurmi et al., 1993; Sorjonen-Ward, 1993a). The Kovero belt also contains older rock units than the Hattu belt (Huhma et al., 2012) and consists of two bifurcating northwest- and northeast-trending branches (Fig. 5.3.2), which appear

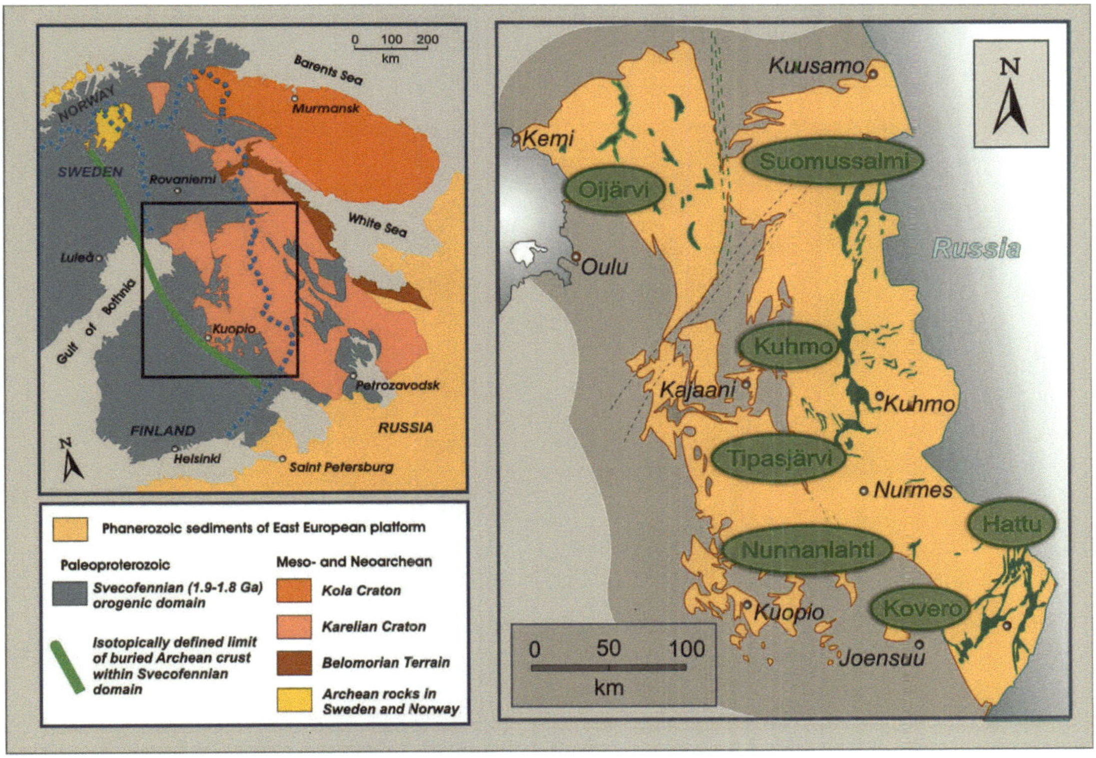

FIGURE 5.3.1 Context of Hattu schist belt within Fennoscandian Shield.

Left: Principal Archean crustal domains and significant boundaries, in part defined isotopically. The Kola and Karelian cratons are separated by the Belomorian terrane, which records both Neoarchean and Paleoproterozoic orogenic events. Black rectangle indicates location of diagram on right panel. *Right:* The extent of the Karelian craton in Finland and the principal greenstone belts.

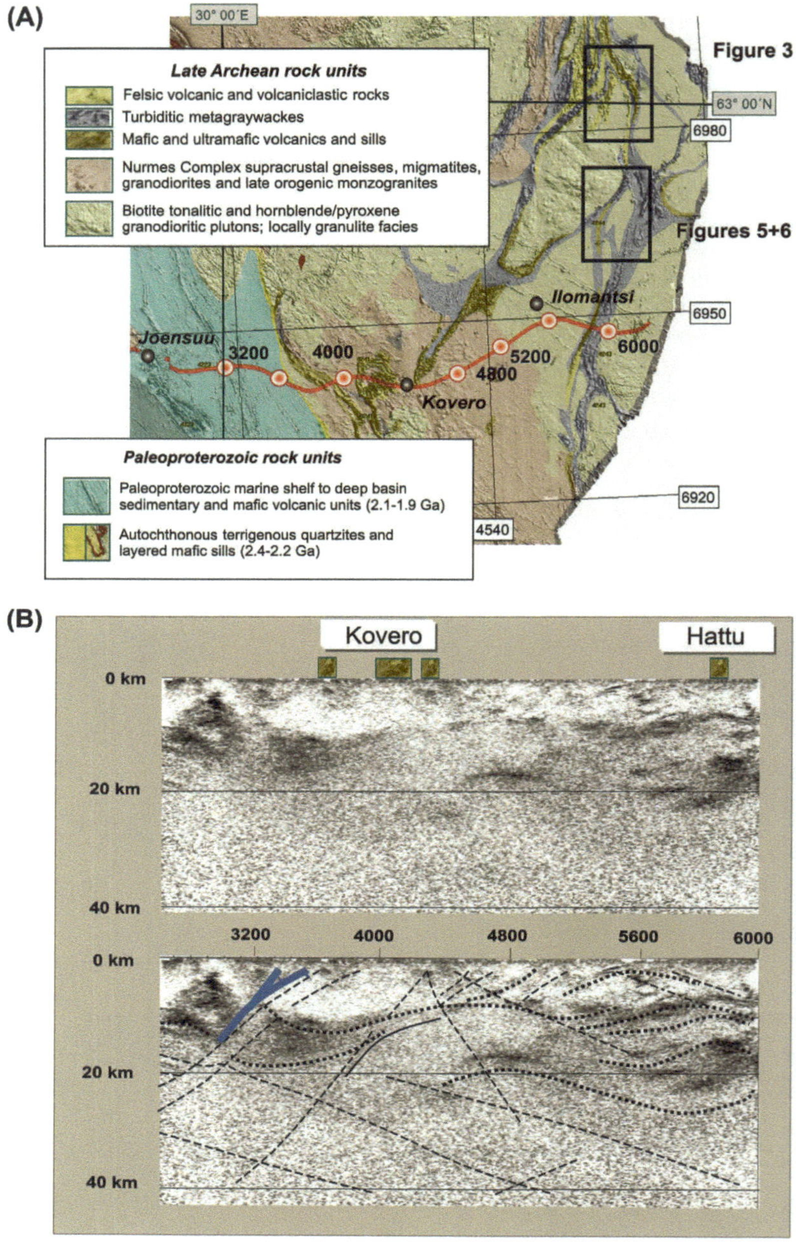

FIGURE 5.3.2 Geological extent and depth expression of the Hattu schist belt.

(A) Simplified regional geology superimposed on total magnetic intensity across the Archean–Proterozoic boundary zone, showing the Hattu schist belt and location of the FIRE 3A seismic profile. Black rectangles indicate location of figures. (B) Crustal seismic reflection profile, to a depth of 40 km, across the area shown in (A), from Sorjonen-Ward (2006). Note prominent subhorizontal reflectivity down to 20 km depth, but surface trace of greenstone belts do not seem to be discernible in the image. West-dipping heavy blue line marks position of Proterozoic boundary.

favorable for orogenic Au mineralization in terms of structural architecture and metamorphic grade (cf. Goldfarb et al., 2001). However, while reconnaissance till geochemical surveys by the Geological Survey of Finland (GTK) indicated anomalous Au, particularly in the southwestern part of the area (Fig. 5.3.4), subsequent studies and outcrop investigations have shown very little evidence for structurally constrained hydrothermal alteration or anomalous concentrations of gold or associated pathfinder elements.

In contrast, the Hattu belt is known to be prospective for gold and has been mapped systematically, such that the overall structural architecture and lithostratigraphic units are well defined, providing a framework for exploration and interpretation of controls on mineralization (Sorjonen-Ward, 1993a,b). The Hattu schist belt supracrustal sequence consists predominantly of felsic pyroclastic and epiclastic deposits with laterally persistent but minor tholeiitic intercalations, and some komatiites in the upper part of the succession. This lithostratigraphical interpretation is considered robust, being well constrained in the area west of the Pampalo Gold Mine (Fig. 5.3.3). Isotopic constraints indicate that the supracrustal rock units were deposited and erupted during a brief time around 2.76–2.75 Ga and was intruded by felsic to intermediate granitoids and porphyries with ages from 2.75–2.72 Ga (Vaasjoki et al. 1993, Sorjonen-Ward and Luukkonen 2005; Huhma et al., 2012).

The greenstone sequence thus represents one of the youngest Archean supracrustal units in the Fennoscandian Shield, and was deformed and metamorphosed simultaneously with and shortly after sediment deposition. Microstructural studies documenting dynamic recrystallization and peak metamorphic assemblages overprinting alteration, in combination with titanite and monazite ages from some granitoids, intersecting concordia at 2.70 Ga, have been used to infer a minimum age constraint on the timing of gold mineralization (Sorjonen-Ward, 1993a; Vaasjoki et al., 1993).

In spite of locally intense and complex deformation, the Hattu schist belt has retained a high degree of stratigraphical coherence, which has enabled formal lithostratigraphic units to be defined. It has also been possible to interpret lateral facies variations as delineating two distinct and partially overlapping felsic volcanic complexes, with associated subvolcanic intrusions, developed within sporadically emergent but generally turbidite-dominated basins (Sorjonen-Ward, 1993a). Tholeiitic basalt flows and a single clastic-textured komatiite unit, exhibiting autoclastic brecciation as well as massive textures, punctuate the stratigraphic sequence and form useful marker horizons, enabling the structural geometry to be further clarified.

An abundance of depositional younging criteria in the metasediments indicate that the major folds in the greenstone belt are upward facing, militating against interpretations invoking large-scale early recumbent folding. No evidence of unconformities or any kind of depositional substrate to the Hattu schist belt has been found, although progradation of the turbidites and volcaniclastic deposits of the Hattu supersequence, over a mafic volcanic substrate, has been deduced, based on lithofacies transitions observed in the easternmost part of the schist belt (Sorjonen-Ward, 1993a).

A full description of lithostratigraphic units and the basis for subdivision is beyond the scope or purpose of this review, but an important key to understanding the geological context and structural controls on gold mineralization derives from mapping and geophysical interpretation of the area west of the Pampalo Gold Mine, summarized earlier in Fig. 5.3.3. Recognition of the regional scale, northerly plunging Pihlajavaara anticline allows an eastward younging stratigraphical series to be defined, from the Sivakkojoki formation in the core of the anticline through the Tiittalanvaara formation to the Pampalo formation, which is the lithological host to the Pampalo ore. The Poikopää formation is lithologically diverse, with abundant epiclastic feldspar-phyric sediments with local evidence for

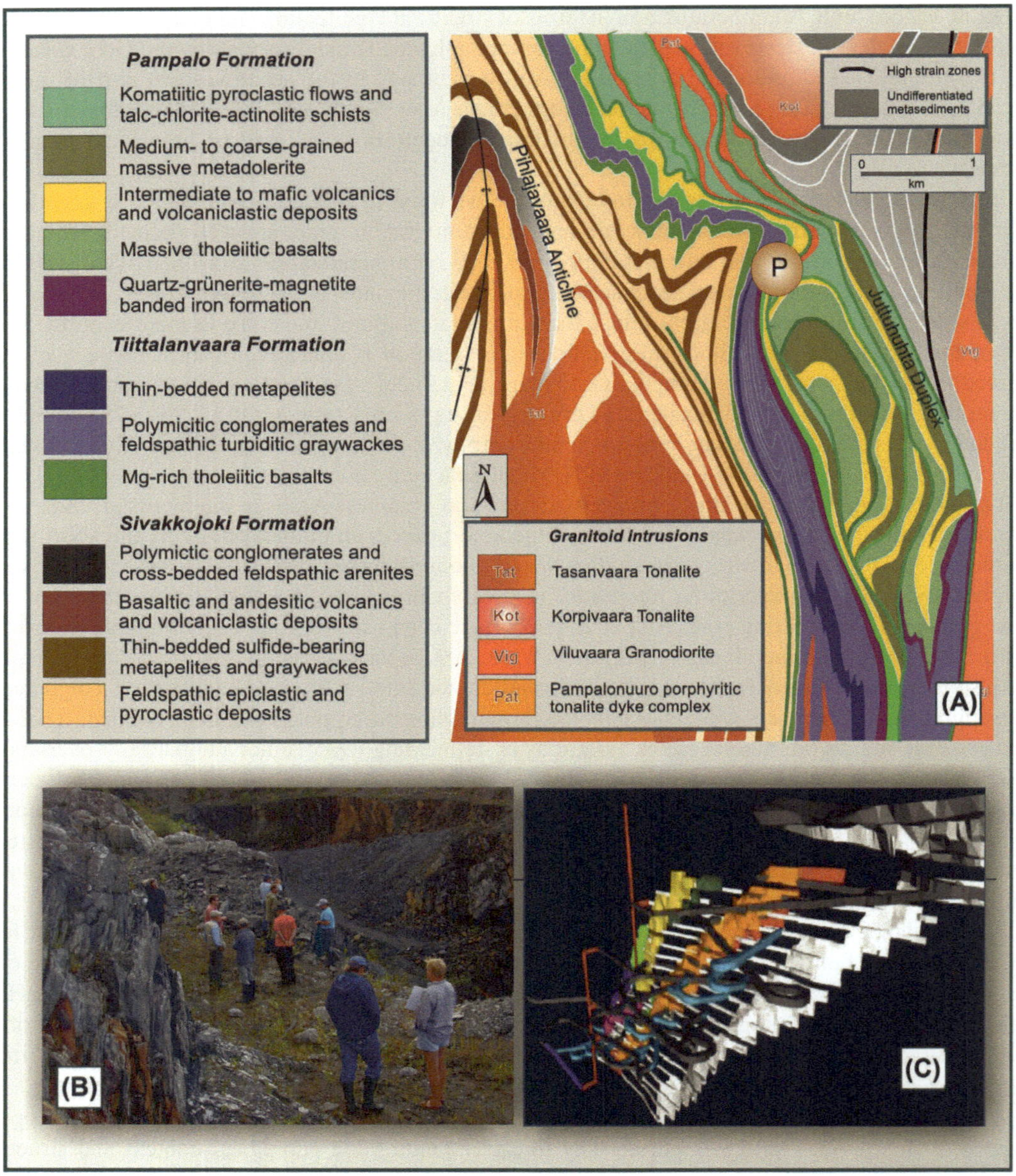

FIGURE 5.3.3 Lithostratigraphic setting of the Pampalo gold deposit.

(A) Simplified geology of the Pampalo zone, indicating the position of the Pampalo ore body at the toe of the Juttuhuhta strike-slip duplex. (B) View southward over open pit prior to construction of processing plant by Endomines Oy. Foreground comprises talc-chlorite schists intruded by porphyritic tonalite dikes; far side of pit consists of metabasalt defining footwall to the main lodes. (C) Northward view of model of the three main north-northeast–plunging ore lodes at Pampalo, showing decline and stopes to a depth of nearly 400 m.

shallow-water reworking, as well as evidence for eruptive deposits, including some basaltic flows. The overall thickness of this sequence, accounting for strain, could exceed several kilometers.

The Tiittalanvaara formation records a transition through coarse-clastic deposits, including well-sorted polymictic conglomerates to finer-grained turbidites, punctuated by coarser debric flows and culminating in banded iron formation, transitional to the overlying basalts, autoclastic komatiites, and felsic to intermediate volcanics of the Pampalo formation. The Pampalo formation appears to record a brief if diverse phase of volcanism, apparently, but not conclusively, overlain by a rather monotonous sequence of metasediments with only sporadic basaltic intercalations (Sorjonen-Ward, 1993a,b).

Published U-Pb zircon ages of the supracrustal rock units range from 2754 ± 6 Ga, for an intermediate hyaloclastic to peperitic unit, at one of the lowest stratigraphic levels, within the Poikopää formation, to 2726 ± 15 Ma for porphyry clasts in a conglomerate higher up in the sequence, within the Tiittalanvaara formation (Vaasjoki et al., 1993). These ages overlap statistically with those for surrounding syntectonic granitoids (2757 ± 4 to 2725 ± 6 Ma) which intrude the belt (Vaasjoki et al., 1993). Although this indicates rapid crustal evolution, there is increasing evidence of heterogeneous and xenocrystic zircon populations, suggesting the presence of an older crustal component in detrital sediments, even if most of the Hattu schist belt sediments represent reworking of penecontemporaneous volcanogenic deposits.

Isotopic data from plutonic zircons also indicate the presence of older crustal material in the source regions of some granitoids, with SIMS zircon studies on the Silvevaara granodiorite and associated porphyritic dike swarms having revealed inheritance from a protolith up to 3.1–3.3 Ga in age (Sorjonen-Ward and Claoué-Long, 1993; Heilimo et al, 2011; Huhma et al., 2012). Anatexis of older continental crust is also evident from some highly evolved, peraluminous granitoids, including tourmaline-muscovite leucogranites, which appear to be analogous to S-type granitoids in Phanerozoic collisional belts (Sorjonen-Ward, 1993a; O'Brien et al., 1993). Titanite and monazite ages with concordia intercepts in the range 2710–2696 are considered to provide a minimum age estimate of crustal metamorphism and reworking (Vaasjoki et al., 1993), which correlates broadly with the widespread deposition of graywackes that were rapidly converted to migmatites and paragneisses throughout the Archean of eastern Finland (Käpyaho et al., 2007; Kontinen et al., 2007).

EXPLORATION TECHNIQUES AND HISTORY

By the late 1970s, airborne and ground geophysical surveys (Airo, 2005) and till geochemical mapping had become well-established and were being routinely used in Finland, in exploration programs managed by both the Geological Survey of Finland and Outokumpu Oy, which at that time was the dominant exploration and mining company operating in the country. Exploration was focused primarily on base metals, particularly nickel, copper, and zinc, which had led to some investigations in the Kovero domain of the Ilomantsi greenstone belt (Kurki, 1980; Hartikainen and Salminen, 1982; Salminen and Hartikainen, 1985; Männikkö et al., 1987). In addition, a comprehensive review of iron ore resources in Finland included a provisional assessment of magnetite in banded iron formations in the Hattu schist belt (Lehto and Niiniskorpi, 1977; Hugg and Heiskanen, 1983).

However, the global resurgence in gold exploration and mining in the early 1980s also awakened a general awareness of the potential for gold discoveries in the Paleoproterozoic and Archean terrane in Finland (Nurmi et al., 1993; Eilu, 2007). Thus, the initial stimulus for gold exploration in the Hattu schist belt was in 1982, when anomalous arsenic values were recognized in till geochemical data, acquired during the GTK regional geochemical mapping program (Hartikainen and Salminen, 1982). Based on the known association between arsenic and gold from studies in Canada, follow-up heavy mineral studies were undertaken in 1983, but concentrates were found to contain scheelite rather than gold. Subsequent comprehensive analysis of till samples revealed coincident anomalies in W and Mo over an area of some 20 km^2, instigating a more detailed sampling survey (16 samples/km^2). Encouraged by the results, several exploration trenches were excavated in the Kuittila area, in the southern part of the Hattu schist belt, during the northern summer of 1984, confirming the presence of scheelite and molybdenite within quartz vein networks in tonalite. Additional sampling and analysis of altered and pyritic quartz veins from the trenches also revealed the presence of anomalous gold, while elevated Au concentrations were simultaneously reported from further regional till surveys (Hartikainen and Damstn, 1991; Salminen and Hartikainen, 1985, 1986).

Therefore, a comprehensive exploration program was established by GTK in 1985, with the twin aims of assessing the gold potential of the Hattu schist belt while systematically evaluating optimal sampling and analytical techniques that might also be applied to exploration elsewhere (Hartikainen and Damstén, 1991). Geochemical mapping was supplemented by bedrock lithogeochemical analysis, from outcrop and drill core, ground geophysical surveys, and detailed outcrop mapping to define structural architecture and the nature of lithostratigraphic units (Fig. 5.3.3). Results of these investigations were subsequently published in the Geological Survey of Finland Special Paper series (Nurmi et al., 1993; Sorjonen-Ward, 1993a,b).

Meanwhile, Outokumpu Oy had also independently become interested in assessing gold potential in Finland and began a comprehensive program of assaying archived sulfide-bearing samples. One such sample, from a solitary, small riverbank exposure in the midst of an extensive swamp, just a couple of kilometers south of the village of Hattuvaara, had originally been submitted by an amateur prospector, Timo Poikonen, in 1970. Upon analysis, this sample was found to contain a remarkable 93 ppm Au and 75 ppm Ag (Pekkarinen, 1988; Ojala et al., 1990). Outokumpu Oy, operating at that time as Outokumpu Finnmines Ltd., therefore commenced exploration in 1985, and excavations around the discovery outcrop and the first analyses across the contact zone between metagraywackes and tourmaline-sericitic altered porphyritic tonalite also yielded an average Au grade of 6.7 ppm. Ground magnetic and electromagnetic surveys assisted in tracing this critical contact zone over a strike length of nearly a kilometer, with subsequent induced polarization surveys being the most useful in delineating areas characterized by disseminated sulfides (Pekkarinen, 1988; Ojala et al., 1990). This occurrence was initially known as Hattuvaara, but was later named Rämepuro, after the stream containing the discovery outcrop.

By 1994, on completion of an additional sampling and drilling survey by GTK in the northern parts of the Hattu schist belt (Heino et al., 1995; Hartikainen and Niskanen, 1995), bedrock mineralization had been confirmed along nearly 40 km of strike length. After submission of GTK technical reports to the then Ministry of Trade and Industry, the subsequent tendering process resulted in the southern part of the belt and the Pampalo concession being granted to Outokumpu Mining Oy, which also still held the Rämepuro lease. The area from Pampalo northward was awarded to the junior company Endomines Oy in 1996. Outokumpu Mining carried out regional exploration further south, as far as Mutalahti near the Russian border, but focused on Pampalo, conducting beneficiation tests and opening a small production pit in 1996, and commencing selective underground mining in 1999. A total of 114,372 tons of ore was transported to Outokumpu Oy's processing facilities at Pyhäsalmi and Vammala, yielding 1755 kg gold at an average grade of 15.3 ppm.

With a major shift in corporate policy, Outokumpu relinquished its exploration and mining activities in the region in 2003, selling its assets to Polar Mining Oy, the Finnish subsidiary of Australia's Dragon

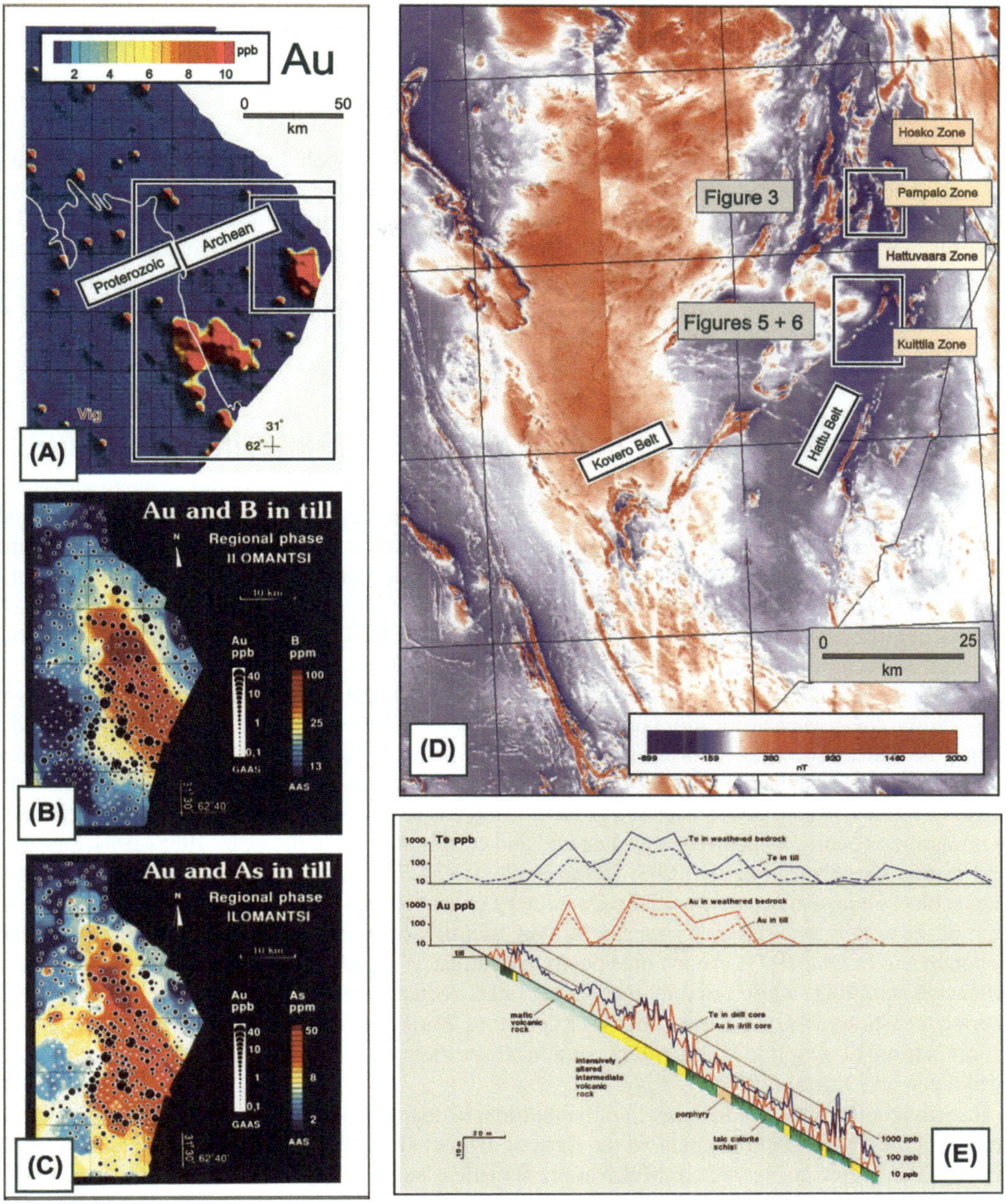

FIGURE 5.3.4 Regional geophysical signatures of rock units in the Hattu schist belt and till geochemical expressions of gold distribution.

(A) Regional survey data (1 sample/4 km^2) shows a distinct anomaly coincident with Hattu schist belt, but also with the Kovero belt, and straddling the Archean–Proterozoic boundary; the latter anomalies are not yet satisfactorily explained. Larger black rectangle indicates area of geophysical image in (D); smaller rectangle shows area illustrated in (B) and (C). (B) Distribution of Au and B in till; compare with (C) which shows Au and As; both As and B defined large anomaly field coincident with that of Au in (A). (D) Total magnetic intensity image of area shown in (A); broad geometry of greenstone belt is evident but correlation between magnetic and regional Au geochemical anomalies is weak, or misleading. (E) Example of close correlation between till anomalies for Te and Au, also demonstrating coincidence with bedrock mineralization and, thus, applicability as a targeting tool in exploration.

Source: *From Hartikainen and Nurmi (1993).*

Mining NL. In October 2006, all of the mining licenses and claims held by Polar Mining, as well as all exploration results from the southern part of Hattu schist belt, together with the Pampalo deposit, were transferred to Endomines Oy. Underground mining commenced at the Pampalo deposit in early 2011, with crushing and flotation processing onsite; at the time of writing, additional ore was being supplied by expansion of the open pit at Pampalo, and the commencement of mining at Rämepuro. Exploration in the near-mine environment and throughout the Hattu belt is continuing, and a list of resource and reserve estimates is summarized in Table 5.3.1.

TILL GEOCHEMISTRY AS AN EXPLORATION AND TARGETING TECHNIQUE

Geochemical surveying by the Geological Survey of Finland was initially undertaken at reconnaissance scale (on average, 1 composite sample/16 km^2) for the Karelian region as a whole, and then regionally over the Hattu schist belt (during 1976, with composite samples comprising 10 successive samples taken at 100-m intervals), and finally in selected areas on a regular grid with a sampling density of 16 samples/km^2. Anomalous areas were then sampled along profiles with a sampling density of 100–400 analyses per km^2. Particular efforts were made to sample the till directly overlying the bedrock interface, on the assumption that dispersion of material over distances greater than several hundred meters would generally lead to dilution below detection limits for gold and critical pathfinder elements. Accordingly, the presence of anomalous metal values in the analyzed <0.06 mm fraction could be inferred to indicate proximity to mineralized bedrock. In most cases, rock chips were also obtained from the bedrock surface and a strong correlation with till analytical data was observed (Fig. 5.3.4(E)).

This approach was used to select and guide drilling targets and proved successful: more than half of the mineralized bedrock occurrences delineated and confirmed during the years 1985–1994 were identified in this way (Nurmi et al., 1993; Heino et al., 1995). The effectiveness of this technique in the Hattu schist belt is also in part due to the Quaternary glacial and postglacial sedimentary history, in that much of the area is covered by a single, relatively thin till sequence, commonly only 5 m in thickness (Nenonen and Huhta, 1993). An average depth to bedrock of about 5 m also facilitated rapid and comprehensive sampling, while reliable analysis and data reduction was ensured by the ability to obtain accurate measurements of elements, down to detection limits of 0.1 ppb for Au and 1 ppb for Te, using a combination of graphite furnace and atomic absorption spectrometry (GAAS), as described by Hartikainen and Nurmi (1993).

It seems remarkable that insights into the nature and distribution of mineralizing processes are discernible in the till geochemical data, in the form of distinct elemental associations of anomaly patterns (Fig. 5.3.3 and later in Fig. 5.3.6). Multivariate statistical analysis of these associations of anomalous elements enabled several factors to be identified that proved useful in target selection and in interpretation and also correlated with pathfinder element associations identified in lithogeochemical data from drill core (Hartikainen and Nurmi, 1993; Rasilainen et al., 1993; Rasilainen, 1996). Together with detailed mineralogical and microprobe analyses (Kojonen et al., 1993), these studies also confirmed a close association between gold and a variable range of trace elements, notably Te, Bi, Ag, As, W, B, and Mo, while altered rocks enclosing mineralization are characterized by additions of volatiles and K, and removal of Na (see Fig. 5.3.7 later).

Fig. 5.3.4(A) shows regional scale data for Au, while Figs. 5.3.4(B) and (C) illustrate the most significant features of regional-scale (1 sample/4 km^2) till geochemical surveys over the Hattu schist

Table 5.3.1 Gold deposits with reserve or resource estimates from the Hattu schist belt

Name	*X*-coordinate	*Y*-coordinate	Class	Tonnage (Mt)	Grade (Au, g/t)	Contained metal (Au, t)	Reference
Hosko	710900	7001000	Proven reserves	0.048	7.9	0.380	Heino et al., (1995); Parkkinen (2003); Geoconsulting Parkkinen cited in Endomines Oy Ab (2014)
Hosko	710900	7001000	Probable reserves	0.049	5.5	0.271	Heino et al., (1995); Parkkinen (2003); Geoconsulting Parkkinen cited in Endomines Oy Ab (2014)
Hosko	710900	7001000	Indicated resources	0.604	1.5	1.160	Heino et al., (1995); Parkkinen (2001, 2003); Geoconsulting Parkkinen cited in Endomines Oy Ab (2014)
Hosko	710900	7001000	Inferred resources	0.153	1.4	0.220	Heino et al., (1995); Parkkinen (2003); Geoconsulting Parkkinen cited in Endomines Oy Ab (2014)
Pampalo	715900	6991250	Stockpile	0.006	2.1	0.013	Geoconsulting Parkkinen cited in Endomines Oy Ab (2014)
Pampalo	715900	6991250	Proven reserves	0.230	3.0	0.690	Geoconsulting Parkkinen cited in Endomines Oy Ab AB (2014)
Pampalo	715900	6991250	Probable reserves	0.108	2.4	0.260	Geoconsulting Parkkinen cited in Endomines Oy Ab AB (2014)
Pampalo	715900	6991250	Pillars	0.021	2.0	0.043	Geoconsulting Parkkinen cited in Endomines Oy Ab AB (2014)
Pampalo	715900	6991250	Inferred resources	0.634	2.0	1.249	Geoconsulting Parkkinen in Endomines Oy Ab AB (2014)
Pampalo East	716150	6991200	Probable reserves	0.140	1.3	0.181	JV Kaivossuunnittelu Oy (2012) cited in Endomines Oy Ab (2014)
Pampalo East	716150	6991200	Inferred resources	0.195	1.2	0.234	Outotec (2012) cited in Endomines Oy Ab (2014)
Rämepuro	716700	6981200	Probable reserves	0.169	2.1	0.358	Pekkarinen (1988); Geoconsulting Parkkinen cited in Endomines Oy Ab (2014)
Rämepuro	716700	6981200	Inferred resources	0.136	2.3	0.309	Pekkarinen (1988); Geoconsulting Parkkinen cited in Endomines Oy Ab (2014)

Source: *Compiled from published reports and press releases by Endomines Oy.*

belt; visual comparison reveals that at this scale, arsenic and boron anomalies are more or less congruent, and a positive correlation between them is evident, irrespective of whether they are genetically related or of independent origin. Similar patterns are evident in Fig. 5.3.7 from which it may be deduced that enrichments in some elements, for example Bi and As, do not spatially correlate with anomalies in Mo and W. Tungsten anomalies are somewhat independent from gold, apart from the discrete anomaly in the Kuittila area in the southeastern part of the images. Moreover, although scheelite is abundant and closely associated with gold at the Pampalo mine, this association is not seen in the regional sampling data. This high degree of correlation was also observed in detailed studies and is also reflected in the abundance of tourmaline veins and disseminations in altered and mineralized bedrock outcrops. Tellurium was found to be a very reliable proxy for gold (Figs. 5.3.3(E) and 5.3.6(A)) because background concentrations are somewhat higher, while telluride phases typically occur as fine-grained minerals and alloys, intimately associated with gold. In contrast, gold may be present as isolated large grains that, through the so-called nugget effect, can skew and obscure interpretation of data. It was also noted that Te was sometimes highly abundant in sulfide-bearing zones that lacked evidence of anomalous Au; such zones were also easily discriminated on the basis of elevated chalcophile element concentrations.

The principal element associations identified in multivariate analysis of till and bedrock data (Hartikainen and Nurmi, 1993; Rasilainen et al., 1993) proved relevant both as pathfinders in exploration and in considering mineralizing processes. Factor analyses of chemical data from till and gold occurrences were found to be characterized by several distinct element enrichment assemblages, notably Au-Te-Bi-Ag, Mo-W-CO_2, and Cu-S. The Mo-W-CO_2 assemblage is typically allied to vein networks in Kuittila suite plutons, predating the main stage of mineralization, but other enrichment associations are more independent of host rock lithology (cf. Fig. 5.3.7). Prominent regional As and B anomalies are evident, but this correlation often breaks down at deposit scale, and while tourmaline is usually present, gold is mostly as tellurides or free grains between silicates or within pyrite; a notable exception to this is the turbidite-hosted Hosko prospect where arsenopyrite is closely associated with gold.

DESCRIPTIVE CHARACTERISTICS OF MINERAL DEPOSITS

Bedrock drilling, mostly on the basis of till geochemical anomalies (Hartikainen and Nurmi, 1993; Nurmi et al., 1993), has confirmed the presence of gold in almost all rock types, indicating that mechanical and compositional contrasts within and between rock units have been influential in localizing fluid-rock interaction and mineralization. Hydrothermal alteration occurs within well-defined domains but differs in some respects from typical Archean orogenic Au deposits in that there is evidence for superimposition on early, perhaps volcanic-related alteration (Rasilainen et al., 1993; Sorjonen-Ward, 1993a; Rasilainen, 1996). This is particularly evident in the northernmost part of the belt, where extensive potassic alteration has affected turbiditic sediments and volcaniclastic deposits, as well as some granitic rocks interpreted as subvolcanic intrusions. At present there is no conclusive evidence that gold would have been introduced by synvolcanic epithermal or porphyry-type mineralizing events, since early sericite-microcline alteration assemblages appear to have been deformed and foliated prior to vein-hosted gold mineralization and associated alteration.

Another distinctive feature of gold mineralization in the Hattu schist belt is the abundance of scheelite and molybdenite, tourmaline, and numerous minerals containing bismuth and tellurium

(Kojonen et al., 1993), particularly within and proximal to mineralization in granitoids. This is suggestive of the possible involvement of synorogenic magmatic fluids, consistent with, but not proven by, the results of isotopic studies (Vaasjoki et al., 1993; Stein et al., 1998; Molnar et al., 2013), and is reminiscent of the class of intrusive-hosted gold deposits. On the other hand, combined paragenetic and structural studies of mineralized veins within the Kuittila tonalite indicate that auriferous shear zones and veins, with characteristic sericite-biotite-carbonate alteration, were superimposed on an earlier set of veins containing molybdenite and scheelite (Nurmi et al., 1993; Sorjonen-Ward, 1993a).

A further connection with a specific association of granitoids is also intriguing and might indicate a larger-scale geodynamic link between orogenic processes and gold mineralization. Granitoids intruding and surrounding the Hattu schist belt fall into distinct categories on the basis of field relationships and composition, with ages from 2757 ± 4–2725 ± 6 Ma (Vaasjoki et al., 1993). The Kuittila suite forms a distinct group of small, elongate plutons aligned within the structural trend of the schist belt and which appear to have been constructed by coalescing steeply dipping *en echelon* sheets intruded during highly partitioned transpressive deformation. Compositions range from biotite tonalite to trondhjemite; hornblende is absent and inclusions are very rare, while associated plagioclase-phyric porphyry dikes indicate that differences in chemistry result in progressive fractionation with retention of plagioclase earlier and deeper.

U-Pb zircon studies, Sm–Nd model ages, and Re–Os studies on molybdenite indicate a short crustal residence time for the protoliths to these rocks. Recent and more detailed studies of the petrogenesis of these plutons reveal a sanukitoid affinity and show that they are characteristic of the earliest stages of convergent deformation in the Karelian craton, between 2.74 and 2.71 Ga in age, potentially representing hybridization of magmas derived from mantle affected by subduction-derived fluids (Halla, 2005; Lobach-Zhuchenko et al., 2005; Heilimo et al., 2010). While a spatial association between these types of intrusions and the prospective terrane in the Ilomantsi complex is discernible, it is not yet clear whether they have general potential as regional exploration indicators. The fact that the Ilomantsi sanukitoids appear to have intrusive ages some 10–20 Ma older than similar rocks in the eastern part of the Karelian craton has been considered puzzling and anomalous, and therefore of interest with regard to the coupling of geodynamic processes and gold potential. However, until the precise timing of mineralization is established, either closer to 2740 Ma and the emplacement of the Kuittila suite of sanukitoids, or later, more related to the 2700 crustal reworking and migmatization event, the potential significance of this remains purely speculative.

The issue of timing is also inherently more problematic in the Hattu schist belt for two reasons. First, the evidence of porphyroblast growth and dynamic recrystallization of altered and mineralized assemblages indicated that the thermal metamorphic peak postdated gold mineralization, which was inferred as late Archean, on the basis of cooling ages from monazite and titanite in several granitoids intruding the Hattu schist belt (Vaasjoki et al., 1993). This inference was at variance with the postpeak timing generally advocated for gold mineralization in the Yilgarn craton at that time, although recognition of synmetamorphic mineralization in amphibolites facies domains was emerging, ultimately leading to the metamorphic continuum model (Groves, 1993). Moreover, a Paleoproterozoic thermal event, albeit with low strain, was clearly superimposed on the Hattu schist belt; resetting of Archean biotite, or growth of new biotite at around 1800 Ma is recorded in K–Ar data (Kontinen et al., 1992), while Rb–Sr and O data from tourmaline, quartz, and mica in mineralized zones also show the effects of Proterozoic disturbance (O'Brien et al., 1993).

Halla and Heilimo (2009) further demonstrated regional-scale Paleoproterozoic Pb isotopic disturbance in K-feldspar phenocrysts in granitoids, including those adjacent to the Hattu schist belt. Moreover, the proximity to metamorphosed and deformed Proterozoic quartzites that unconformably overlie Archean basement, some 50 km to the west and 30 km to the east of the Hattu schist belt, as well as the recrystallization of mineral assemblages in Paleoproterozoic mafic dikes, demonstrates that temperatures had attained greenschist facies conditions between 1.9 and 1.8 Ga, as a consequence of Svecofennian orogenic deformation. However, the lack of significant deformatione of Paleoproterozoic mafic dikes, being restricted to brittle-ductile strain at the dike margins, combined with microstructural studies of recrystallization, provides strong evidence against significant remobilization or introduction of gold during Paleoproterozoic time (Sorjonen-Ward, 1993a).

DESCRIPTIONS OF INDIVIDUAL GOLD OCCURRENCES

Based on distribution of mineralization and anomalies, and geological and structural features, it has proven useful to consider the Hattu schist belt in terms of four distinct mineralized segments or anomalous zones, each of which contains one or more gold deposits (where estimates of resources or reserves have been attempted) or occurrences (where drilling or outcrop sampling has identified the presence of anomalous gold). From south to north these are, respectively, the Kuittila, Hattuvaara, Pampalo, and Hosko anomaly zones (Figs. 5.3.2 and 5.3.8 later). Detailed descriptions of structural architecture of these zones and the structural setting and alteration, mineralogy, and of gold occurrences are reported in Sorjonen-Ward (1993a) and Nurmi et al. (1993). Table 5.3.1 summarizes the gold occurrences and deposits at the time of writing. The following summaries are intended to provide an overview and background for considering the overall structural controls on mineralization and the application of numerical modeling to exploration targeting.

KUITTILA ZONE

The Kuittila zone is centered on and surrounds the small, elongate pluton known as the Kuittila tonalite (Figs. 5.3.4 and 5.3.5). It was named after the location of the first exploration trenches where exposed quartz-vein networks and sericitic and carbonated shear zones containing pyrite and gold (in addition to scheelite and molybdenite) were found. Till geochemical studies subsequently indicated the presence of anomalous gold with metasedimentary rocks along the western margin of the Kuittila tonalite, but not along the eastern side, which might have been expected if gold were exclusively sourced from the tonalite itself (Figs. 5.3.4(A) and 5.3.6). However, the eastern margin of the tonalite intrusion is poorly exposed, and while no evidence for alteration has been found in outcrop, it has not yet been rigorously assessed by till geochemical or geophysical surveying. Despite this lack of information, there are still several geological features that suggest that mineralization may be asymmetrically distributed, preferentially along the western margin. The Kuittila zone is readily delineated on the basis of till geochemical anomalies (Figs. 5.3.3(A) and 5.3.6), in particular those due to Au, Te, Na, and loss of ignition, the last two reflecting the intensity of hydrothermal processes in the metasedimentary country rocks (Hartikainen and Nurmi, 1993).

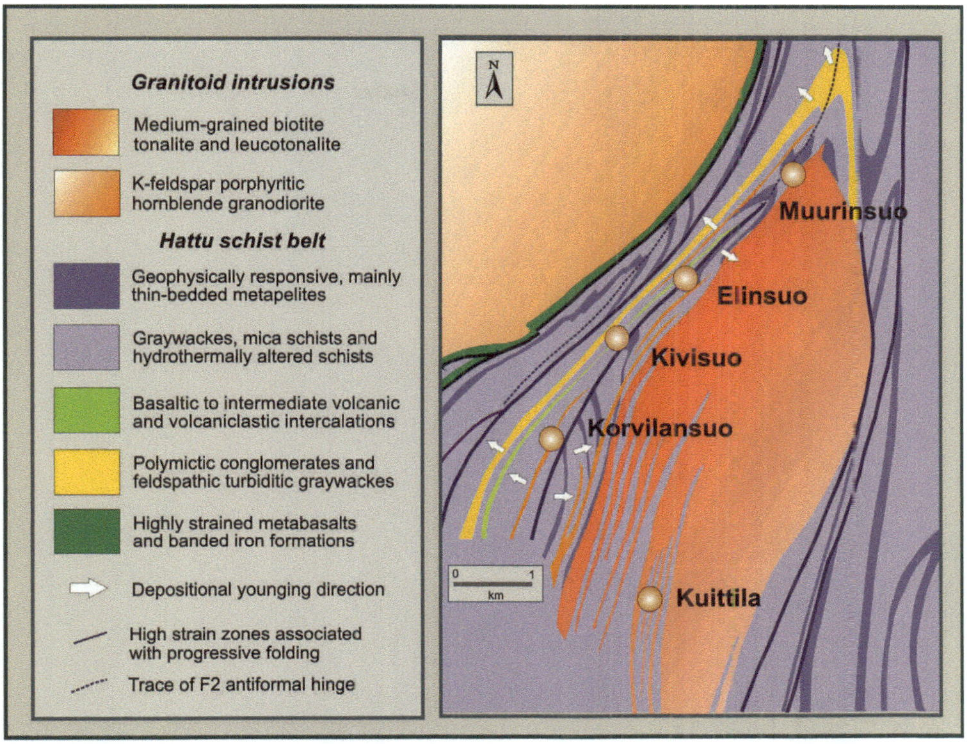

FIGURE 5.3.5 Essential geological features of the Kuittila zone, showing location of gold occurrences several hundred meters from the contact with the Kuittila tonalite.

Note alternation of screens of metasediment and tonalitic dikes between Kuittila and Korvilansuo, constrained by outcrop mapping, drill core, and till–bedrock interface sampling. Note also oblique transaction of fold hinges along the western margin of Kuittila tonalite between Korvilansuo and Muurinsuo, from which close interaction between emplacement and deformation is deduced.

Detailed ground geophysical surveys cover the entire Kuittila zone but gold mineralization is not directly evident, neither can the metasediments and tonalite be distinguished from one another geophysically (Figs. 5.3.5, 5.3.6), except to a limited extent in the induced polarization (IP) survey data. This is because the Kuittila tonalite is a weakly magnetic ilmenite-series granitoid, and the low abundances of sulfides and oxide minerals result in generally weak IP responses, whereas fine-grained pyrrhotite disseminations in mica schists commonly generate distinct IP anomalies.

Mineralization in metasedimentary rocks in the Kuittila zone

The Kuittila zone was further subdivided by Nurmi et al. (1993) into a southerly subzone, where the contact with the Kuittila tonalite and foliation and lithic layering in metasediments trends almost north–south, and a northern subzone, where the contact is northeast-trending (Figs. 5.3.4(B) and 5.3.5). The southern subzone contains the Kuittila deposit, which is the only example of gold

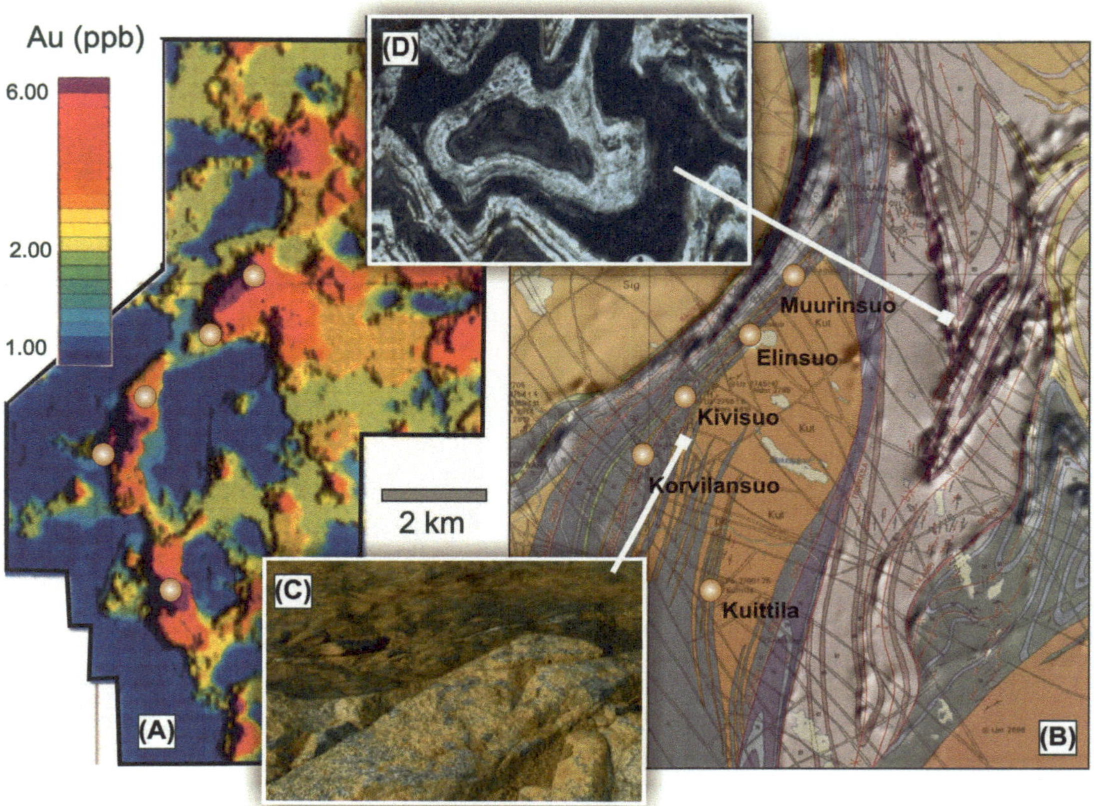

FIGURE 5.3.6 Geochemical anomalies and Kuittila zone geology.

Gold occurrences and deposits are indicated by golden dots. (A) Gold concentrations in till at detailed sampling scale, from Hartiakinen and Nurmi (1993). Compare the distribution of the most intense Au anomalies, which define an arcuate zone surrounding the Kuittila tonalite, with the geological map in (B). (C) Ductile strain at contact of Kuittila tonalite and metasediments; no indication of alteration or mineralization at the contact. Note also the lack of magnetic anomalies along the western margin of the Kuittila tonalite. (D) Refold interference pattern in banded iron formation resembles map of Australia. Note correlation with prominent magnetic anomalies in (B).

mineralization known within the main body of the pluton, and the Kelokorpi occurrence, which comprises diverse rock types dominated by mica schists and metagraywackes variably altered to sericite- or biotite-dominated schists and intercalated with sporadic greenish chloritic schists and strongly recrystallized porphyritic tonalite dykes. Tourmaline is abundant in mineralized zones, which are characterized by weak, yet pervasive, sulfide disseminations containing pyrrhotite, sphalerite, and chalcopyrite. Additional phases include arsenopyrite, molybdenite, tellurides and argentopentlandite, and rare native gold (Kojonen et al., 1993).

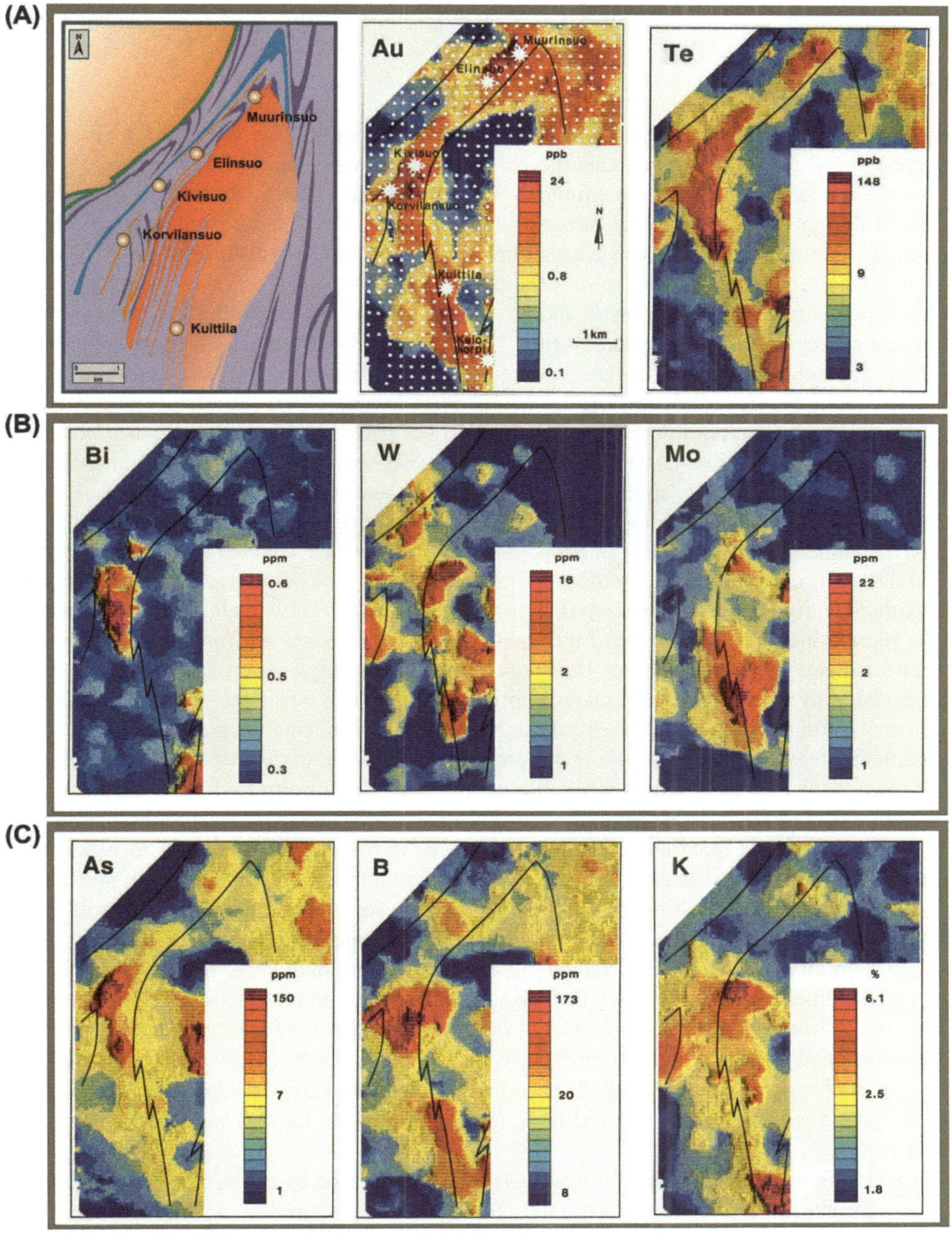

FIGURE 5.3.7 Till geochemical maps for selected elements in the Kuittila zone.

(A) Simplified geology surrounding the Kuittila tonalite, Au distribution and sampling grid, and Te anomalies, which correspond closely with Au. (B) The congruence of Mo and W anomalies and their location within the Kuittila tonalite, contrasting with the Bi anomaly centered on the Korvilansuo area outside the pluton; this not only indicates the precision of till anomalies with respect to bedrock mineralization but is also consistent with the existence of independent enrichment processes for Bi and Mo. (C) The similarity between As and Bi in (B), the intensity of tourmaline in the Korvilansuo area, and the possibility that K alteration in the margin of the Kuittila tonalite was associated with Mo and W enrichment.

Source: *From Hartikainen and Nurmi (1993).*

The Kelokorpi occurrence is, thus, similar in character to the mineralization found within the northern subzone, which specifically includes the Korvilansuo, Kivisuo, and Muurinsuo deposits, and the Elinsuo occurrence, which may be contiguous with Muurinsuo. These were discovered, drilled, and documented during the GTK research program (Nurmi et al., 1993) but Korvilansuo and Muurinsuo have been drilled more recently since acquisition by Endomines Oy, with further clarification of mineral resources and some preliminary reserve estimates (see Table 5.3.1).

The prospectivity of the Korvilansuo and Kivisuo areas was apparent after the earliest detailed till geochemical surveys (Hartikainen and Nurmi, 1993) and confirmed when arsenopyrite-bearing bedrock samples assaying 4–8 ppm Au were recovered from an exploration pit excavated over a conspicuous anomaly at Kivisuo. Shallow drilling was also undertaken along several east–west traverses across the schist belt. Mineralized lithologies consisted mostly of sheared, silicified, and partly tourmalinized porphyry dikes and tourmaline-bearing mica schists showing chlorite-sericite alteration. Despite alteration, poor exposure, and locally high strain (Figs. 5.3.5(B) and (C)), primary depositional grading has been recognized in some graywacke units, in outcrop and in drill core. In combination with foliation-lithology, asymmetry and truncations of geophysical anomalies have enabled the structural architecture to be defined with some degree of confidence. The mineralized zone coincides with an area where foliation and lithic layering dips are less steep than in most of the Hattu schist belt, between 55 and 70°, and fold axes plunge moderately northward throughout the zone. Several antiformal and synformal pairs have been delineated, truncated by, or transposed into oblique high-strain zones that are discernible through excision of lithostratigraphic marker horizons (Fig. 5.3.6). Sorjonen-Ward (1993a) interpreted this geometry in relation to emplacement of the Kuittila tonalite during progressive deformation.

A distinctive feature at Korvilansuo is the presence of tonalitic dikes that show intensive silicification and replacement by tourmaline, while highly strained, locally ultramylonitic metasediments can contain gold either as sporadic disseminations or within tourmaline-quartz vein arrays assaying up to 10–15 ppm Au. It is likely that tourmaline was introduced in two stages, the first being related to alteration of the tonalite, while the later stage involved vein-type alteration associated with gold deposition. Korvilansuo exemplifies the distinctive ore mineral parageneses of the Hattu schist belt; sulfides include pyrrhotite, chalcopyrite, and pyrite with minor arsenopyrite, galena, and sphalerite, while fine-grained native gold is commonly associated with Au-Ag-Bi-Pb tellurides and native bismuth.

The metasediments at Kivisuo and Korvilansuo commonly contain sericitic pseudomorphs after andalusite porphyroblasts and microstructural studies indicate that they overprint early foliation and, in some places, sulfide minerals, suggesting that dynamic recrystallization under peak metamorphic conditions postdated hydrothermal alteration. Kyanite has also been observed and may indicate that mineral assemblage has in fact recrystallized later, in response to the Paleoproterozoic thermal event, as discussed earlier.

Mineralization at Muurinsuo and Elinsuo was initially revealed by prominent anomalies in Au and Te and less distinct As, loss on ignition (LOI), and negative Na anomalies, which led to delineation of a continuous northeast-trending anomaly zone within the schist belt, located about 200 m from the northwestern contact of the Kuittila tonalite (Figs. 5.3.5 and 5.3.6). As at Kivisuo and Korvilansuo, the host rocks are intensely foliated mica schists with graywacke and mafic tuffitic interbeds, invariably altered to sericite and sericite-chlorite schists. Pyritic dissemination is typical in quartz-and plagioclase-phyric tonalitic porphyry dikes, which are abundant, particularly in the southwestern part of the deposit. The mineralization is situated near the hinge zone of the Muurinsuo F2 antiform (Fig. 5.3.6), but the discontinuous nature of mineralization, at least with the presently available drilling density, favours interpretation as a number of separate lenses that could well be elongated parallel to the dominant shape fabric lineation, which in this region plunges at 60–70 degrees towards the north-northeast. Mineralization is disseminated and invariably hosted by altered mica schists, mafic volcanics, and porphyry dikes. Gold again occurs in association with Bi, Au, and Ag tellurides, with minor arsenopyrite, chalcopyrite, pentlandite, and galena (Kojonen et al., 1993).

Mineralization in the Kuittila tonalite

The Kuittila tonalite is typically medium-grained and equigranular, with randomly oriented subhedral plagioclase grains and oriented aggregates of fine-grained biotite; hornblende is absent and K-feldspar is uncommon. Mafic enclaves are sporadically present, with irregular shapes attesting to low strain during crystallization, consistent with interpretations from quartz vein networks. Alteration of the tonalite does not invariably correlate with increasing strain intensity, although the most highly mineralized zones are intensely sericitized, with complete destruction of plagioclase and biotite, and intensely deformed. In drill core, it is evident that the tonalite commonly has a sheared and bleached appearance over distances of several meters around the quartz lodes, reflected in a mineral assemblage containing quartz, albite, sericite, calcite, K-feldspar, biotite, and epidote; where alteration is most intense, quartz, sericite, and carbonate dominate. This visible alteration correlates with a lithogeochemical anomaly surrounding the deposit, with distinct enrichments in Mo, W, CO_2, Cu, S, Rb, K, and Si and loss on ignition, indicative hydration, carbonation, and potassic alteration; conversely, Na and Sr are depleted in comparison with unmineralized parts of the pluton (Nurmi et al., 1993). Native gold occurs principally as inclusions or intergrowths with pyrite in association with sporadic tellurides (Kojonen et al., 1993).

The overall form of the Kuittila tonalite is an asymmetric lenticular, elongate pluton with dips of about 60°E to the west, below the schist belt, on its western margin, whereas the eastern contact is evidently subvertical. The pluton is compositionally zoned, with a distinctly more leucocratic trondhjemitic central phase, compared to the generally medium-grained biotite tonalite comprising the bulk of the pluton. The western margin is also characterized by porphyritic dikes with euhedral plagioclase and in places, quartz phenocrysts. Although mutual intrusive relationships between the porphyry dikes and the main tonalite phase have not been demonstrated in outcrop, nor in drill core. They are however, very similar chemically, with a notable exception being positive Eu-anomalies in dikes contrasting with the negative anomaly in the main tonalite intrusion (O'Brien et al., 1993), which may be attributed to the abundance and retention, or accumulation of, plagioclase phenocrysts in the dikes.

Drilling within and adjacent to the southwestern margin of the pluton, between the Korvilansuo and Kuittila deposits, revealed alternating screens of metasediments between tonalitic dikes (Figs. 5.3.4(B) and 5.3.5). Together with structural interpretations of folding and shearing in the country rocks, this permits an interpretation for construction of the Kuittila tonalite by successive emplacement and coalescence of dikes during progressive deformation. A plausible scenario for explaining the present geometry is that the intrusion represents a releasing bend within an extensive north–south trending deformation zone. Observations of quartz vein networks at several locations within the pluton are of relatively uniform orientation and, though recrystallized, are not highly strained, implying that the present pluton shape is original, and not a product of post-emplacement shearing (Sorjonen-Ward, 1993a).

Following the discovery of mineralization in the metasediments that structurally overlie the steep, westward-dipping tonalite contact, it was also natural to consider whether the mineralization could have been associated with magmatic to hydrothermal fluid infiltration in the hanging wall of the cooling pluton. Alternatively, the mechanical contrast between the pluton and the country rock could have induced strain partitioning, with gold mineralization occurring in brittle-ductile fault meshes in the more homogenous tonalite, contrasting with pervasive alteration along shear zones in the more anisotropic sediments. The role of plutons as mechanically rigid bodies that locally influence rock permeability and fluid flow, rather than primary sources of heat and metals, has been controversial and continues to be debated for many orogenic gold systems (Goldfarb et al., 2001).

Three groups of observations favor the former interpretation in the case of the Kuittila tonalite. First, structural relationships with respect to folding in the country rocks favor emplacement in the ductile regime under greenschist to amphibolites facies conditions, rather than at high crustal levels in which a porphyry or epithermal magmatic to hydrothermal regime could have been established. Second, at the Kuittila deposit, the alteration and deformation zones associated with gold mineralization characterized by intense veining, silicification, and sericitization have been superimposed upon northwest to west-northwest trending subvertical quartz vein networks containing molybdenum and scheelite; this indicates that gold was introduced in zones of structurally enhanced permeability after solidification of the pluton. This suggests that after cooling sufficiently to deform by brittle failure, the pluton was subjected to only small amounts of regionally homogenous strain.

Microstructural observations of recrystallization within and at the margins of molybdenite-scheelite-bearing quartz veins also indicate that penetrative foliation developed in the tonalite after, if not during, vein development (Sorjonen-Ward, 1993a). Third, while the zones of pervasive alteration and mineralization associated with the Korvilansuo, Kivisuo, and Muurinsuo deposits trend subparallel to the margin with the Kuittila tonalite, they are nevertheless located several hundred meters from the contact; if the regional permeability structure had been conducive to the pluton generating convective cells with discharge and infiltration zones, more evidence of alteration immediately adjacent to, and within the margin of, the pluton would surely be expected.

While it is plausible that the same structural zones that facilitated emplacement of the Kuittila tonalite were active in transferring and circulating hydrothermal fluids, the role of the pluton as a thermal driver can further be disputed by considering that if it were emplaced into country rocks at greenschist facies conditions, then the thermal gradient generated across the contact would have been rather modest, and even more so if the emplacement occurred in the form of successive sheet-like bodies.

HATTUVAARA ZONE

The Hattuvaara zone is simply defined as the predominantly north-trending segment of the schist belt, passing through the village of Hattuvaara, from the northern end of the Kuittila zone to the Pampalo zone (refer to Figs. 5.3.2 and 5.3.8 later). The schist belt in this area is in places less than a kilometer in width, and consists of highly strained and tightly folded epiclastic metagraywackes, although north of Hattuvaara, there is a transition to mafic rock units of the Pampalo formation, exposed in the core of a tight northerly plunging F2 synform (Sorjonen-Ward, 1993a,b). Southward of Hattuvaara, in the vicinity of the Rämepuro deposit, it appears that the limbs of the synform are attenuated and transposed, perhaps accompanied by major shear displacement. Apart from several till anomalies and isolated mineralized glacial boulders found along the zone (Nurmi et al., 1993), Rämepuro is the only known gold deposit; following extensive evaluation and feasibility tests and environmental permitting, mining commenced at Rämepuro in 2014.

As described earlier, the discovery of Rämepuro was due to analysis—after a delay of more than a decade—of a chalcopyrite-bearing sample collected by a local prospector in 1970. When Outokumpu Finnmines Oy commenced geophysical surveys and a drilling program was initiated in 1985, a small exploration pit excavated around the discovery outcrop revealed a shear zone some 10-m wide at the boundary between a porphyritic tonalite dike and metasediments, which assayed about 7 ppm Au. In contrast to mineralization within the Kuittila zone, it was found that the mineralized shear zone at Rämepuro generated a distinct IP anomaly, 20–30 m in width and several hundreds of meters in length. Conversely, Rämepuro is one of the few gold occurrences in the Hattu schist belt that is not associated with significant till anomaly patterns on geochemical maps, even at a sampling density of 16 samples/km^2 (Hartikainen and Nurmi, 1993; Hartikainen and Niskanen, 2001). This is perhaps surprising as drilling by Outokumpu during the 1985–1987 campaign (Pekkarinen, 1988) and more recently by Endomines Oy Ab (2014) confirms that the bedrock surface is mineralized along strike over a distance of at least 500 m.

The Rämepuro deposit is located at the boundary between two different supracrustal rock units, where a porphyritic tonalite dike up to 30 m in thickness has intruded along the contact. The sequence to the west of the dike consists predominantly of well-preserved graywackes with some polymictic conglomeratic intercalations (Endomines Oy Ab, 2014; Ojala et al., 1990; Pekkarinen, 1988; Sorjonen-Ward, 1993a). On the eastern side of the dike, more homogenous, fine-grained intermediate biotite-plagioclase-quartz schists prevail, with local banded iron formation horizons; protoliths may have been of intermediate pyroclastic origin (Ojala et al., 1990) but their affinities with rocks elsewhere in the schist belt are unclear.

Gold at Rämepuro is typically found with dynamically recrystallized and strained quartz-tourmaline-sulfide veins within the intensely deformed tonalite dike and adjacent metagraywackes, indicating that mineralization accompanied, or more probably post-dated, dike emplacement. Silicate alteration assemblages comprise quartz, sericite, biotite, albite, chlorite, and tourmaline. Native gold is fine-grained, occurs interstitially between quartz and tourmaline grains and intergrown with pyrite and pyrrhotite, and shows a broad spatial correlation with chalcopyrite and sphalerite; metallic bismuth and hedleyite are also characteristic (Kojonen et al., 1993).

PAMPALO ZONE

The Pampalo zone has more of a northwesterly trend than the Hattuvaara zone to the south and Hosko zone to the north and northeast (Figs. 5.3.2 and 5.3.7); this change in structural geometry between the Pampalo and Hosko zones is attributed to strain partitioning and accommodation of deformation within a shear system, which has had a complex history. Geometrically, however, the system is relatively simple, if viewed essentially as a progressively evolving regional-scale antiformal and synformal fold pair subjected to oblique sinistral transpression (Sorjonen-Ward, 1993a). As can be seen from Fig. 5.3.3, the Pihlajavaara anticline, with well-constrained younging determinations, lies to the west of the Pampalo zone and provides a key to both stratigraphic and structural interpretation. The plunge of this anticline and subsidiary minor folds varies in the range of 40–70° toward the north in the southern part of the area, which is typical for much of the Hattu schist belt, but becomes progressively gentler in the northern part of the zone, with southerly plunges and lineations being characteristic of the boundary between the Pampalo and Hosko zones. This geometrical pattern can be explained by transpressive shear, resulting in asymmetric doubly plunging dome and basin folds in the adjacent Hosko zone, while within the Pampalo zone, discrete shear zones appear to have excised some of the stratigraphic sequence along the eastern limb of the Pihlajavaara anticline.

In the southern part of the Pampalo zone, the boundary with the Hattuvaara zone is somewhat arbitrary, but the structural architecture can be interpreted in terms of tightening of the regional synform complementary to the Pihlajavaara anticline, which is again supported by abundant depositional younging criteria (Sorjonen-Ward, 1993a,b). In this area, plunges are steep toward the north and the resultant geometry is explained as a consequence of variations in response to different rock types to deformation, in particular the presence of the largest occurrence of mafic and ultramafic rocks observed in the schist belt. These rocks are assigned to the Pampalo formation which, as indicated in Fig. 5.3.3, is the uppermost stratigraphical unit recognized in the northern part of the Hattu schist belt, overlying conglomerates and graywackes of the Tiittalanvaara formation and including, at the base, a banded iron formation unit (Sorjonen-Ward, 1993a,b).

The anomalous thickness of the Pampalo formation in this area is interpreted as a strike-slip duplex formed by oblique shearing along the steeply dipping eastern limb of the Pihlajavaara anticline, resulting in imbrication of mafic units of the Pampalo formation by detachment along the anisotropic banded iron formation, which separates the Pampalo formation from the underlying sediments of the Tiittalanvaara formation (Fig. 5.3.3(A)). Although further resolution of details is still desirable, this interpretation satisfies the first-order geometrical and structural constraints based on field mapping and interpretation of geophysical imagery. For a detailed description and analysis of the Juttuhuhta duplex and its relationship to gold mineralization, the reader is referred to Nurmi et al. (1993) and Sorjonen-Ward (1993a).

The prospective part of the Pampalo zone appears to be 1–2 km wide and is distinctive in that ultramafic rocks are exposed over much of its length, so that regionally, the zone coincides with prominent Ni and As geochemical anomalies in till; although the latter element is also commonly associated with Au, arsenopyrite has also been found in hydrothermally altered mafic sediments that are devoid of gold, and which are more likely to represent subseafloor alteration processes. Several additional gold occurrences have been delineated in proximity to the Pampalo deposit, namely Pampalo Northwest and Pampalo East, and bedrock mineralization was also demonstrated in an outcrop several kilometers along the strike from Pampalo, at Korpilampi, adjacent to the Korpivaara tonalite (Nurmi et al., 1993).

In the northern part of the Pampalo zone, till anomalies have also been drilled and bedrock mineralization has subsequently been confirmed in extensively sericitized feldspathic wackes or pyroclastic deposits, with delineation of the Kuivisto and Kuivisto East occurrences as steeply dipping mineralized zones concordant with respect to enclosing lithological units (Heino et al., 1995). This sericitic alteration demonstrably predated intensive foliation development and may also be related to subvolcanic alteration processes, demanding careful paragenetic studies in distinguishing this from later structurally controlled alteration related to gold mineralization.

The Korpilampi occurrence, between Kuivisto and Pampalo, was perhaps the most accessible and obvious target in the entire region, being one of the few road cuttings in the area. The exposure comprised talc-chlorite schists and ultramafic rocks intruded by pegmatites with extensive tourmaline alteration and with abundant sulfides. Potassic alteration is also indicated by biotite replacement of talc-chlorite assemblages (Nurmi et al., 1993). Mineralized pegmatite veins contained erratic anomalous gold, with arsenopyrite and pyrite, as well as bismuth minerals, pyrrhotite, marcasite, sphalerite, and chalcopyrite (Kojonen et al., 1993), but drilling failed to intersect significant mineralization. Nevertheless, the area is of potential interest given the presence of till geochemical anomalies on both sides of the outcrop and the general lithological similarities with those at Pampalo, which remains, at the time of writing, the most significant deposit in the Pampalo zone.

Pampalo Gold Mine

Attention was first drawn to the region surrounding the Pampalo deposit on the basis of prospective analogy; the Pampalo zone is the only part of the Hattu schist belt where mafic lithologies are relatively abundant, as in many gold districts, but more importantly, the distinctive and deviant structural geometry appeared, by analogy with structurally controlled gold deposits elsewhere, to be favorable for mineralization (Sorjonen-Ward, 1993a; Nurmi et al., 1993). It was therefore fortuitous indeed that a structurally anomalous outcrop in this area, trending almost orthogonally with respect to the main shear system, was found to contain visible gold. In addition, it was realized that distinct and prominent till geochemical anomalies had been recorded from several profiles that terminated some distance to the west of the outcrop (Hartikainen and Nurmi, 1993). The first exploratory drilling across the area in 1990 consisted of three holes, the second of which yielded ore grade intersections (17 ppm over 7 m and 3.3 ppm over 12.6 m), indicating that the discovery outcrop was situated at the eastern edge of the main ore zone, in an area that is now known and mined as the Pampalo East deposit. Three *en echelon* elongate or lenticular ore zones, plunging moderately at 40–50°E to the north-northeast, were defined after the earliest phase of drilling and this interpretation has been subsequently confirmed and formed the basis for mine planning (Fig. 5.3.3(C)).

A critical and significant feature of the Pampalo deposit is its structural setting, having a northerly to northeasterly trend, which is highly discordant to the overall northwesterly trend of the Pampalo zone. This can be explained by sinistral transpressive folding within the Pampalo zone, even though the larger scale map geometry gives the impression that the Pampalo zone represents a sinistral dilational jog or left-stepping releasing bend architecture between the northerly trending Hattuvaara and Hosko zones. In contrast, the transpressive, contractional interpretation places the Pampalo deposit within the toe of the oblique sinistral, Juttuhuhta duplex and would seem to imply, if not require, progressive changes in relative strain rate along the bounding shear zones (Sorjonen-Ward, 1993a). A detailed assessment and understanding of the evolution of these structures is important in developing structurally oriented exploration strategies, as it is critical to establish whether *en echelon* mineralization zones are present elsewhere, oriented within the axial planes of these transpressive folds, rather than parallel to the main northwest-trending Pampalo zone shear system.

Porphyritic dikes, including those at the discovery outcrop, are tightly folded with the country rocks, as well as being mineralized so that the relationships between deformation, magmatic events, and mineralization share much in common with the Kuittila zone. Indeed, there is an evident mirror symmetry when comparing the relationship between the Tasanvaara tonalite and Pampalo zone with the Kuittila tonalite and the Kuittila zone, the former having a northeasterly trend and latter a northwesterly trend. An important difference, however, which may be of significance in assessing the possible role of granitoids in mineralization, is that the Tasanvaara tonalite is situated further away from the mineralized zone. On the other hand, the enveloping surface to the Pampalo shear system and associated porphyry dikes are effectively dipping moderately eastward beneath the contact with the Korpivaara tonalite.

The Pampalo deposit is hosted by a diverse range of rock types, with a reasonably well-defined stratigraphy based on correlation with surrounding areas (Fig. 5.3.3(A)). The eastward younging transition from turbiditic graywackes and conglomerates of the Tiittalanvaara formation to banded iron formations marking the base of the Pampalo formation is well-documented from drilling, and this has also been taken as the likely floor thrust to the Juttuhuhta duplex (Sorjonen-Ward, 1993a). The internal lithological subdivisions recognized within this latter formation include massive basalts and dolerites, separated by a distinctive intermediate to mafic clastic unit which appears to be the most favorable host rock for mineralization. Although younging criteria are lacking, the uppermost part of the Pampalo formation comprises bimodal felsic and ultramafic units, the latter having been altered to talc-chlorite-carbonate schist in the eastern part of the deposit; the ultramafic rocks are not mineralized, but folded and boudinaged porphyry dikes contain gold. Distinct vitreous felsic units, which may have been rhyolitic flows or subvolcanic, are extensively albitized with abundant tourmaline and are pervasively mineralized, forming one of the main rock types mined in the Pampalo East deposit.

The nature of structural control and the timing of introduction of gold, the possible role of magmatic as well as metamorphic fluids, and the extent of recrystallization and remobilization at the Pampalo deposit are not yet adequately understood. Visible gold occurs within narrow anastomozing mylonitic seams that show evidence for both brittle and ductile behavior, in equilibrium with biotite and in some cases actinolite, but it is not clear if this represents the remobilization of earlier mineralization, or actually represents the main stage of mineralization. Resolution of these issues is an important matter for exploration if, for example, it were shown that gold was locally remobilized into the northeasterly trending shear zones during folding, having originally been introduced within the main northwest-trending shear system, or was actually introduced within an axial planar orientation during the later folding event.

Gold tends to be disseminated, with sulfides, within the altered clastic unit, or in dynamically recrystallized veins and fractures in porphyritic dikes, indicative of contrasting mechanical behavior for the two rock types. In general, discrete mineralized quartz veins arrays are rare, suggesting that wall-rock sulfidation has been a more significant process than fluid mixing or phase separation.

The main sulfide mineral phases accompanying gold at Pampalo are pyrite, with pyrrhotite, chalcopyrite, pyrrhotite, galena, and sphalerite. As elsewhere, bismuth and Pb-, Ag-, and Au-tellurides are characteristic (Kojonen et al., 1993). Scheelite is remarkably common within the ore and a useful qualitative indicator of proximity to mineralization underground and in drill core. Gold tends to occur as free grains, fracture fillings in hydrothermally altered mineral grains, such as microcline, or as inclusions in pyrite; this has proven very favorable for beneficiation with efficient recovery by gravitational and flotation processes.

HOSKO ZONE

This zone represents the northeastern part of the Hattu schist belt, and is distinctive in terms of the asymmetric doubly plunging folds and the extensive seritic alteration of epiclastic or pyroclastic deposits, which evidently predates regional foliation development, and is therefore likely a product of submarine volcanic-related hydrothermal processes. Despite the extensive alteration, primary depositional units and clastic features are commonly preserved, facilitating mapping of structural features and outlining the dome and basin topology. The Kartitsa granite also displays dynamically recrystallized to quartz-muscovite-microcline assemblages, and, locally, fluorite, implying that it may be a subvolcanic intrusion, with later sulfide-bearing vein systems superimposed (Sorjonen-Ward, 1993a). Accordingly, in assessing gold potential, careful microstructural studies are necessary to discriminate between early synvolcanic and later structurally controlled synmetamorphic alteration assemblages.

Detailed till geochemical data are available for the southern part of the zone (Hartikainen and Nurmi, 1993; Hartikainen and Niskanen, 2001) and on the basis of till anomalies in the Valkeasuo area, the Hosko deposit was identified and bulk samples (760 kg) were collected from a test pit by Endomines Oy in 1999 for beneficiation tests. Recovery up to 90% was obtained; however, the ore mineral assemblage at Hosko differs somewhat from other mineralization in the Hattu schist belt in that it is more refractory, with abundant arsenopyrite. Metamorphic grade, however, is not higher than at other deposits, and the host rocks are thin- to thick-bedded turbidites with dark pelitic intercalations, and show varying degrees of sericitic alteration. Tourmaline is ubiquitous in mineralized veins, and is also texturally replacive in the pelitic layers, which also typically contain abundant laminated mineralized quartz-tourmaline-pyrite-arsenopyrite carbonate veins. In addition to arsenopyrite, pyrrhotite, marcasite, and scheelite are present (Heino et al., 1995; Käpyaho et al., 2013).

Mineralization has been delineated over about 600 m of strike length, and varies from 2–15 m in width, with sporadic drill hole intersections yielding exceptionally high Au abundances. Variations in grain size and the erratic distribution of mineralization have also made resource estimation more difficult. The mineralized zone corresponds to a distinct IP anomaly, which can be traced southward from Hosko for several kilometers toward the Kuivisto deposit in the Pampalo zone, but the presence of mineralization along the entire strike length has yet to be confirmed.

STRUCTURAL ANALYSIS AND EXPLORATION TARGETING—NUMERICAL SIMULATION OF FAVORABLE STRUCTURAL ARCHITECTURE AND CONTROLS ON MINERALIZATION

Orogenic gold deposits characteristically show strong structural control, or are at least associated with networks of crustal scale and subsidiary shear zones (Goldfarb et al, 2001). As described earlier, gold mineralization in the Hattu schist belt also displays distinct structural controls, although the orientation and intensity of hosting structures varies from place to place. It is therefore natural to consider whether some structural features can be identified that would provide insights into the distribution of mineralization and perhaps prove to be of predictive value.

Structurally enhanced permeability during orogenic deformation is widely invoked and accepted as a critical factor in the formation of lode gold deposits, driven by feedback between rock material properties and structural architecture, fluid pressure, and orientation of far-field stresses (Cox et al., 2001; Ord and Oliver, 1997; Sibson, 2001). In recent decades, numerical simulations of mechanical processes and coupled fluid–rock interaction have been used to analyze and predict structural controls on gold mineralization at various scales (Schaubs and Zhao, 2002; Sorjonen-Ward et al., 2002; Zhang et al., 2013). Therefore, we decided to perform some numerical simulations of deformation relevant to the known structural geometry and rock types present in the Hattu schist belt.

Because mineralization is known to occur within tonalities, as at Kuittila, as well as in enclosing metasediments, or in high-strain zones near the contact between plutonic and supracrustal rocks, a useful model should be capable of discriminating between at least these three rock categories. We therefore developed some 3D models based on a highly simplified emulation of the geometry of rock units in the Hattu schist belt (Fig. 5.3.8(A)). Although simple, the mesh does incorporate several fundamental geological features, namely the less steep northwesterly and northeasterly dips of respective enveloping surfaces in the Kuittila and Pampalo zones. Moreover, because mineralization occurs in various structural orientations, it was considered necessary to test the influence of systematic variations in stress fields as well as mechanical contrasts in rock units. This allows the testing of various hypotheses, for example, whether results would suggest that structures having different orientations could be equally and effectively permeable under the same stress regime, or whether mineralization in different orientations represents superimposed or unrelated events.

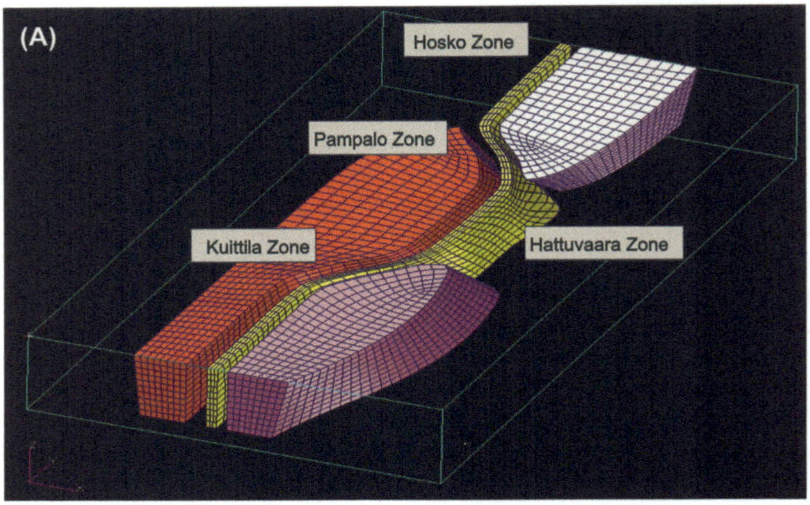

Rock Unit	Density	Young's modulus	Poisson's ratio	Cohesion	Tensile strength	Friction angle	Dilation angle	Porosity	Permeability
Shear Zone	2500	4e10	0.2	5e6	0.5e6	15	3	0.1	1e-15 m2
Metasediments	2650	5e10	0.25	1e7	1e6	30	3	0.1	5e-17 m2
Granitoids	2650	6e10	0.25	3e7	3e6	35	3	0.1	1e-17 m2

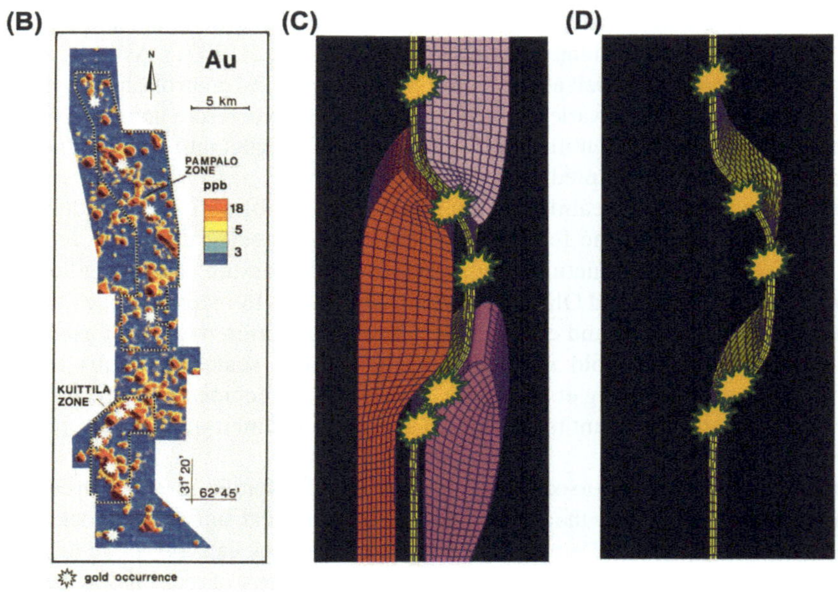

FIGURE 5.3.8 Mesh geometry, material properties, and boundary conditions of FLAC3D™ model.

(A) 3D perspective of model mesh, representing three domains simulating stiff plutonic intrusions in a uniform anisotropic matrix, apart from a simulated shear zone transecting the model from north to south. A significant element of the model is the northeast dip of the northwest-trending Pampalo zone and conversely, the northwest dip of the northeast-trending Kuittila zone. Mesh simulates a volume of rock 40 km (north–south) × 20 km (east–west) × 5 km. Simulations are based on Mohr–Coulomb rheology with and fluid coupling according to Darcy's law, with pore fluid pressure sublithostatic. (B) Gold anomaly distribution along the Hattu schist belt, from Hartikainen and Nurmi (1993), to illustrate the complexity of patterning represented by the simplified meshes in (C) and (D).

The simulations were performed with the finite element code FLAC3D, which was originally developed for engineering applications by the Itasca Consulting Group (Itasca, 2005). FLAC3D allows simulation of fluid flow patterns in deforming rock masses and has been widely applied to the study of structurally controlled hydrothermal processes. The material properties assigned to the rock units in the numerical simulations are based on the interpreted relative strength of the rocks and, in the absence of properties derived via geotechnical testing of the actual rocks, average representative values were obtained from literature sources (Lama and Vutukuri, 1978). The granitoid rocks were interpreted to be the most competent, and were therefore assigned mechanical strength properties higher than those of the metasediments.

Shear zones, because of their pronounced anisotropy, were interpreted to be much weaker than the granitoids, and metasediments and properties were assigned accordingly. Shear zones were interpreted to be fluid pathways and therefore the permeability assigned to them is much higher than in the granitoids and metasediments. It should also be noted that these are bulk properties averaged over time, which is considered a reasonable approximation for an active shear zone subject to episodic overpressuring and failure (cf. Cox et al., 2001; Sibson, 2001). Testing the sensitivities of these parameters was beyond the scope of this study. Nevertheless, comparison of variations and contrasts between model results, for example, where stress vectors vary but material properties and duration of run are held constant, can provide valid insights, irrespective of the absolute precision of model parameters.

In FLAC3D, fluid flow is predominantly governed by permeability and, as such, this parameter is used to differentiate the flow conduit (shear zone) and less permeable host rocks. This qualitatively reflects field observations, for example, the tendency for dispersed hydrothermal alteration throughout the Korvilansuo-Kivisuo-Elinsuo-Muurinsuo deposits in metasediments parallel to the contact with the less mineralized Kuittila tonalite pluton. Porosity only plays a minor role in flow in crystalline bedrock and accordingly is not linked to permeability here, due to the complexity of the relationship between these two parameters, in particular, the degree of pore space connectivity. Uniform porosity was, therefore, used in the models to highlight the importance of permeability.

When a FLAC3D model mesh is subjected to simulated stresses, elements that fail according to Mohr–Coulomb rheology undergo positive or negative dilation, which is linked to changes in permeability, which in turn affects fluid pressure and flow vectors. In this way, the simulation reflects episodic failure and fluid transport in natural deformation zones. As the model simulation progresses, we observe where volumetric or shear strain dilation is greatest, as a proxy for rock masses that are favorable for failure and fluid-rock interaction, and by inference, for hydrothermal mineralization.

The aim of these simulations was to test whether we could predict, or at least understand, the distribution of gold mineralization (Fig. 5.3.8(B) through (D)), by comparing results of modeling with known gold occurrences (Schaubs et al., 2012). The ability to test model predictions by validation against available field data, or by seeking new information in the field, is one of the main reasons for modeling, and also requires that the model and its boundary conditions are carefully designed.

Accordingly, we may define three types of model scenarios, in which the model geometry and material properties remain uniform, but principal stress orientations are varied, to establish:

- Whether the known geometry and orientation could be explained by a single far-field stress configuration.
- Whether rotation of the structures with respect to an external reference frame during progressive deformation led to preferential activation of some orientations.
- Whether the different orientations represented discrete superimposed processes.

The first two cases assume that mineralization was essentially coeval and, by implication, derived from a favorable set of circumstances with respect to metal supply, transport efficiency, and precipitation mechanisms in permeable structures. The third case, if shown to be feasible, or even necessary, would then provide further stimulus for seeking evidence for overprinting of structural fabrics and modification or remobilization of mineral assemblages; recall that isotopic evidence for recrystallization and local remobilization at the Pampalo deposit was indicated by Vaasjoki et al. (1993) and substantiated by Molnar et al. (2013).

In these models, we were particularly interested in the orientation of maximum compressive stresses that would allow the formation of veins and shear zones in the observed orientations (Fig. 5.3.9(A)). This was done by systematically and incrementally varying stress directions across the models, using 15° increments. Results were then compared with known distribution of high-strain zones and gold-bearing structures. In all simulations, the same initial state was chosen for model mesh and material properties.

Results for some of the simulations are presented in Fig. 5.3.9(B), which illustrates cumulative volumetric strain increment as the model mesh is subject to shortening, with varying principal stress directions. The most interesting conclusion, or speculation, that can be made from these results is that the northeast-dipping orientation of the Pampalo zone is shown to be most favorable for tectonic activation and dilation in many different stress configurations. By implication, such areas may be the loci of repeated mineralizing events and therefore have cumulative potential for greater endowment. However, it may seem counterintuitive that west-northwest to east-southeast to northwest-southeast directed maximum compressive stress orientations that are subparallel to the Pampalo zone show less of a tendency for dilation, while the Kuittila zone volume strain is conversely greater.

Although sensitivity to changes in mechanical properties should ideally be tested as well, it is nevertheless noteworthy that when the maximum principal stress orientation is centered around north-northwest to south-southeast, dilation is significant in both the Pampalo and Kuittila zones. There is, however, no known reason why we should expect mineralization to have occurred simultaneously along the entire schist belt, rather than forming from discrete events in isolated short-lived hydrothermal cells. We may further speculate that the structural significance of both of these zones lies in their deviation from vertical orientation (Fig. 5.3.8(A) and (D)), which should favor the formation of oblique gently dipping or plunging structures, with opportunities for both reverse-sense and strike-slip components oblique to the main shear zone trends. Under this stress configuration, dilation within north to south shear zones, with generally steep plunges would be anticipated, as indeed is the case for mineralization in the Hosko zone and within the Hattuvaara zone at Rämepuro.

Some concluding points illustrating the interaction between model results and field verification include the evidence for a zone of dilation propagating from the Pampalo zone toward the northeast, around the southern margin of the Korpivaara tonalite, for maximum stress orientations between north-northwest to south-southeast and north-northeast to south-southwest. Another rather unexpected zone of dilation is present to the east of the Kuittila tonalite, where the maximum compressive stress is oriented east to west. Assessment of such potential targets can be made with reference to till geochemical surveys and structural analysis in outcrop, to confirm or refute predictions. For example, although there are no bedrock exposures in the area to the east of the Kuittila tonalite, making structural analysis difficult, Figs. 5.3.4(A) and 5.3.8(B) both indicate at least some anomalous Au in till. Conversely, regional scale till geochemical anomalies are somewhat more erratic in the area northeast of Pampalo, but there are reasonable constraints on structure and rock type due to better bedrock exposure (Sorjonen-Ward, 1993b). Finally, it should be noted that favorable structural architecture alone need not imply high proespectivity, if there were no associated processes generating fluids of appropriate composition during a deformation event.

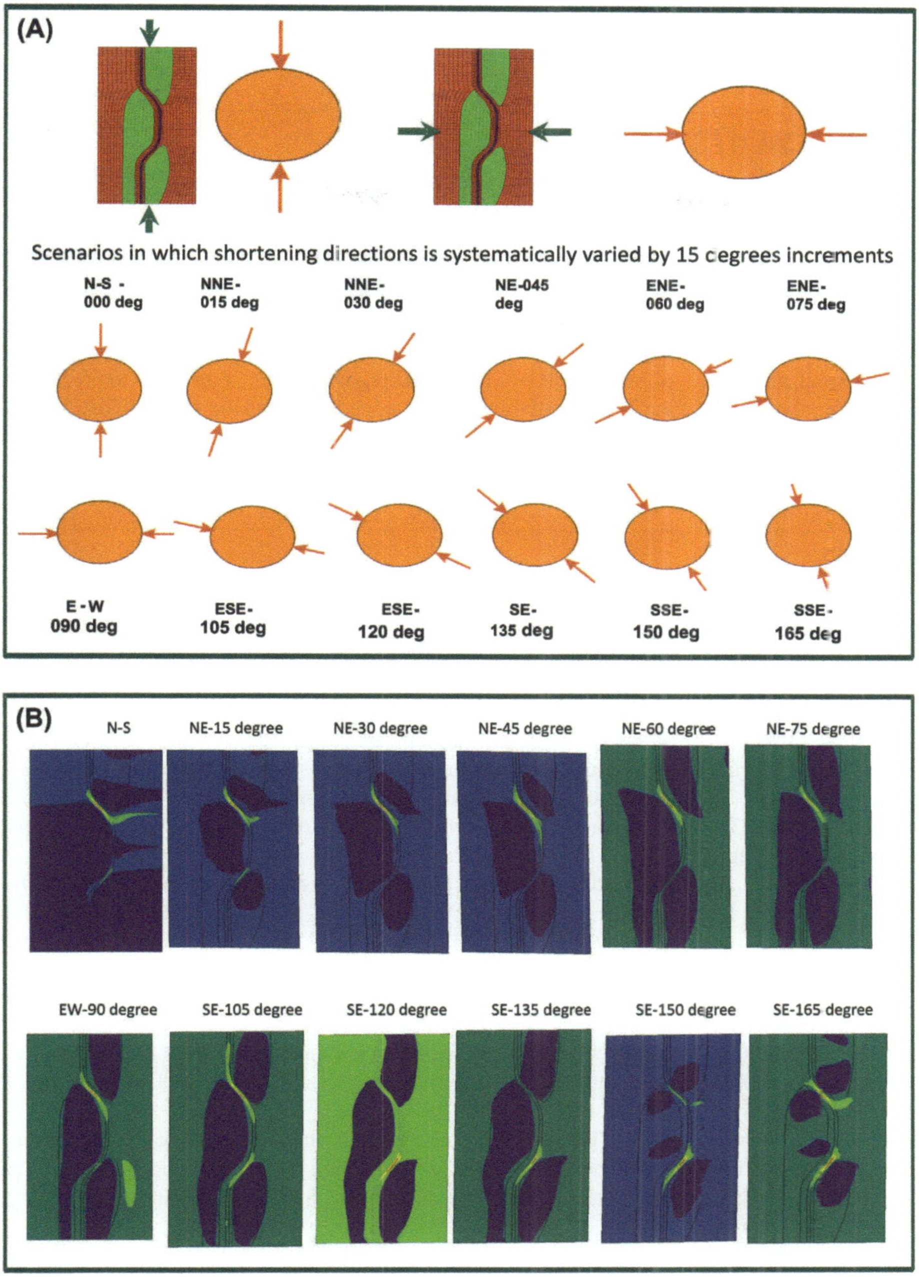

FIGURE 5.3.9 FLAC3D simulations of varying stress field orientation.

(A) Illustration of systematic variation of principal stress orientations, or velocity conditions imposed on model, by 15° increments. (B) Results of shortening simulations showing contours of volumetric strain in plan-view, on a section plane 500 m below top surface of model. Blue shades correspond to negative volume strain, with volume reduction in the range of 2–3%, while yellow to red colors indicate positive volume strain or dilation, with values ranging from 3–8% and indicate sites of preferential rock failure and fluid infiltration. The Pampalo zone is dilatational under most stress configurations, whereas dilation occurs simultaneously in the Pampalo and Kuittila zones when principal compressive stress is between northwest to-southeast and north to south.

ACKNOWLEDGMENTS

The senior author was very much in the junior category when the research described here was carried out and would very much like to express indebtedness to many colleagues, in particular, the late Professors Kalevi Kauranne and Jouko Talvitie, who encouraged what was, at the time, a rather novel and eventually fruitful model for research. Acknowledged with gratitude are the insights into the diversity and metallurgical significance of ore minerals from Kari Kojonen, and isotopic studies with Hannu Huhma, Hugh O'Brien, and the late Matti Vaasjoki, as well as practical support from retired colleagues Martti Damstén and Martti Saastamoinen, and the staff of Endomines Oy at the Pampalo Mine. Recent discussions with Asko Käpyaho, Ferenc Molnar, and Thomas Wagner, who have been embarking on the next stage of ore deposit research in the region, are also very much appreciated. Financial support from the Renlund Foundation is also gratefully acknowledged, enabling the visit to CSIRO Earth Resource and Engineering in Perth in 2011 for the purposes of FLAC3D model development and numerical simulations.

REFERENCES

Airo, M.L. (Ed.), 2005. Aerogeophysics in Finland 1972–2004: Methods, system characteristics and applications. Geological Survey of Finland, Special Paper 39, 197.

Cox, S.F., Knackstedt, M.A., Braun, J., 2001. Principles of strucutral control on permeability and fluid flow in hydrothermal systems. In: Richards, J.R., Tosdal, R.-M. (Eds.), Structural Controls on Ore Genesis, Reviews in Economic Geology, 14, pp. 1–24.

Eilu, P., 2007. FINGOLD: Brief descriptions of all drilling-indicated gold occurrences in Finland—the 2007 data. Geological Survey of Finland, Report of Investigation 166, 135 pp.

Endomines Oy Ab, 2014. Karelian Gold Line Mineral Resources and Ore Reserves – Estimate Summary. Extracted on 2014.03.31 from Endomines.com website at: http://www.endomines.com/index.php/reports-gold/reports-and-drilling-gold.

Goldfarb, R.J., Groves, D.I., Gardoll, S., 2001. Orogenic gold and geologic time: a global synthesis. Ore Geology Reviews 18, 1–75.

Groves, D.I., 1993. The crustal continuum model for late-Archaean lode-gold deposits of the Yilgarn Block, Western Australia. Mineralium Deposita 28, 366–374.

Halla, J., 2005. Late Archean high-Mg granitoids (sanukitoids) in the southern Karelian domain, Eastern Finland. Lithos 79, 161–178.

Halla, J., Heilimo, E., 2009. Deformation-induced Pb isotope exchange between K-feldspar and whole rock in Neo-archean granitoids: Implications for assessing Proterozoic imprints. Chemical Geology 265, 303–312.

Hartikainen, A., Salminen, R., 1982. Tohmajärven karttalehtialueen geokemiallisen kartoituksen tulokset. Summary: The results of the geochemical survey in the Tohmajärvi map-sheet area. Geologian Tutkimuslaitos, Geokemiallisten karttojen selitykset, Lehti 4232. Geological Survey of Finland, Explanatory Notes to Geochemical Maps, Sheet 4232, p. 57.

Hartikainen, A., Damstén, M., 1991. Applications of till geochemistry to gold exploration in Ilomantsi, Finland. Journal of Geochemical Exploration 39, 323–342.

Hartikainen, A., Niskanen, M., 1995. Maaperägeokemialliset kultatutkimukset Hatun liuskejaksolla Ilomantsissa vv. 1983–1995. Geological Survey of Finland. Unpublished Report S/41/4244/1/2001 Karttalehdet 4244 ja 4333, p. 22. With 26 appendices (in Finnish).

Hartikainen, A., Nurmi, P.A., 1993. Till geochemistry in gold exploration in the late Archean Hattu schist belt, Ilomantsi, eastern Finland. In: Nurmi, P.A., Sorjonen-Ward, P. (Eds.), Geological Development, Gold Mineralization and Exploration Methods in the Late Archean Hattu Schist Belt, Ilomantsi, Eastern Finland, Geological Survey of Finland, Special Paper 17, 323–352.

Hartikainen, A., Niskanen, M., 2001. Maaperägeokemialliset kultatutkimukset Hatun liuskejaksolla Ilomantsissa vv. 1983–1995. Geological Survey of Finland, Unpublished Report S/41/4244/1/2001 Karttalehdet 4244 ja 4333 (in Finnish), 22 p. with 26 appendices.

Heilimo, E., Halla, J., Hölttä, P., 2010. Discrimination and origin of the sanukitoid series: Geochemical constraints from the Neoarchean western Karelian Province (Finland). Lithos 115, 27–39.

Heilimo, E., Halla, J., Huhma, H., 2011. Single-grain zircon U-Pb age constraints of the western and eastern sanukitoid zones in the Finnish part of the Karelian province. Lithos 121, 87–99.

Heino, T., Hartikainen, A., Koistinen, E., Niskanen, M. 1995. Report on exploration in Ilomantsi in 1992–1995. Geological Survey of Finland. Unpublished report M06/4333/-95/1/10, p. 17.

Hölttä, P., Heilimo, E., Huhma, H., et al., 2012. Archaean complexes of the Karelia province in Finland. Geological Survey of Finland 54, Special Paper, 9–20.

Hugg, R., Heiskanen, V., 1983. Suomen rautamalmiesiintymät, malmiutumat ja malmiviitteet. Rautaruukki Oy OU 2/79, RAETSU Report 8.2.1983, p. 412 (in Finnish).

Huhma, H., Mänttäri, I., Peltonen, P., Kontinen, A., Halkoaho, T., Hanski, E., et al., 2012. The age of the Archaean greenstone belts in Finland. Geological Survey of Finland, Special Paper 54, 74–175.

Itasca, 2006. FLAC3D: Fast Lagrangian Analysis of Continua in 3 Dimensions, User Manual, version 3.1. Itasca Consulting Group, Inc., Minneapolis.

Käpyaho, A., Hölttä, P., Whitehouse, M.J., 2007. U-Pb zircon geochronology of selected Archaean migmatites in eastern Finland. Bulletin of the Geoogical Society of Finland 78, 121–141.

Käpyaho, A., Molnar, F., Sakellaris, G., 2013. Mineralogical characteristics of a metasedimentary hosted Hosko gold deposit from the Archean Hattu schist belt, eastern Finland. In: Mineral Deposit Research for a High-Tech World. Proceedings of the 12th Biennial SGA Meeting, August 12–15, Uppsala, pp. 1120–1123, Sweden.

Kontinen, A., Paavola, J., Lukkarinen, H., 1992. K-Ar ages of hornblende and biotite from Late Archaean rocks of eastern Finland - interpretation and discussion of tectonic implications. Geological Survey of Finland, Bulletin 365, 31 pp.

Kojonen, K., Johanson, B., O'-Brien, H.E., Pakkanen, L., 1993. Mineralogy of gold occurrences in the late Archaean Hattu schist belt, Ilomantsi, eastern Finland. Geological Survey of Finland, Special Paper 17, 233–271.

Kontinen, A., Käyaho, A., Huhma, H., et al., 2007. Nurmes paragneisses in eastern Finland, Karelian craton: Provenance, tectonic setting and implications for Neoarchaean craton correlation. Precambrian Research 152 (2007), 119–148.

Kurki, J., 1980. Kaivoslain 19§:n mukainen tutkimustyöselostus, Ilomantsi, Huhus, kaiv.rek n:o 2713/1. Outokumpu Oy Exploration, Report 080/4242 12C/JAK/80, p. 3 (in Finnish).

Lama, R.D., Vutukuri, V.S., 1978. Handbook on Mechanical Properties of Rocks—Testing Techniques and Results, Volume II, Trans Tech Publications, Clausthal-zellerfeld, Germany. 515 pp.

Lehto, T., Niiniskorpi, V., 1977. The iron formations of northern and eastern Finland. Geological Survey of Finland Report of Investigations 23, p. 49.

Lobach-Zhuchenko, S.B., Rollinson, H.R., Chekulaev, V.P., et al., 2005. The Archaean sanukitoid series of the Baltic Shield: Geological setting, geochemical characteristics and implication for their origin. Lithos 79, 107–128.

Männikkö, K.H., Määttä, T.T., Ojala, V.J., Tuukki, P.A., 1987. Otravaaran rikkikiisumalmin ja sen ympäristön geologia. University of Oulu, Department of Geology. Pohjois-Karjalan malmiprojekti. Report 4, p. 78 (in Finnish).

Molnar, F., O'Brien, H.E., Lahaye, Y., et al., 2013. Signatures of overprinting mineralisation processes in the orogenic gold deposit of the Pampalo mine, Hattu schist belt, eastern Finland. In: Mineral Deposit Research for a High-Tech World. Proceedings of the 12th Biennial SGA Meeting. August 12–15, Uppsala, Sweden, pp. 1160–1163.

Nenonen, J., Huhta, P., 1993. Quaternary glacial history of the Hattu schist belt and adjacent parts of the Ilomantsi, district, eastern Finland. Geological Survey of Finland, Special Paper 17, 185–191.

Nurmi, P.A., Sorjonen-Ward, P., Damstén, M., 1993. Geological setting, characteristics and exploration history of mesothermal gold occurrences in the late Archaean Hattu schist belt, Ilomantsi, eastern Finland. Geological Survey of Finland, Special Paper 17, 193–231.

O'Brien, H.E., Nurmi, P.A., Karhu, J.A., 1993. Oxygen, hydrogen and strontium isotope compositions of gold mineralization in the late Archean Hattu schist belt, Ilomantsi, eastern Finland. Geological Survey of Finland, Special Paper, 17, 291–306.

Ojala, V.J., Pekkarinen, L.J., Piirainen, T., Tuukki, P., 1990. The Archaean gold mineralization in Rämepuro, Ilomantsi greenstone belt, eastern Finland. Terra Nova 2, 240–244.

Ord, A., Oliver, N.H.S., 1997. Mechanical controls on fluid flow during regional metamorphism. Some numerical models: Journal of Metamorphic Geology 15, 345–359.

Outotec, 2012. Mineral Resource and Ore Reserve Update, Pampalo Gold Mine (Prepared by Lovén, P. and Meriläinen, M., for Outotec Ltd.), 14 p. Extracted on 2014.03.31 from Endomines.com website at http://www.endomines.com/index.php/reports-gold/reports-and-drilling-gold.

Parkkinen, J., 2003. Update of Quality Control of Mineral Resource Estimate, 2003. Geological. Survey of Finland, Report CM 19/2332/2003/1/10.

Pekkarinen, L.J., 1988. The Hattuvaara gold occurrence, Ilomantsi; a case history. Annales Universitatis, Turkuensis, Series 24C–67, pp. 79–87.

Rasilainen, K., Nurmi, P.A., Bornhorst, T.J., 1993. Rock geochemical implications for gold exploration in the late Archaean Hattu schist belt, Ilomantsi, eastern Finland. Geological Survey of Finland, Special Paper, 17, 353–362.

Rasilainen, K., 1996. Alteration geochemistry of gold occurrences in the late Archean Hattu schist belt, Ilomantsi, Eastern Finland. Academic dissertation: synopsis and four research papers. Geological Survey of Finland, p.140.

Salminen, R., Hartikainen, A., 1985. Glacial transport of till and its influence on the interpretation of geochemical results in North Karelia, Finland. Geological Survey of Finland 48 Bulletin 335.

Salminen, R., Hartikainen, A., 1986. Tracing of gold, molybdenum and tungsten mineralization by use of a step by step geochemical till study in Ilomantsi, eastern Finland. In: Phillips, W.J. (Ed.), Prospecting in Areas of Glaciated Terrain. Institute of Mining and Metallurgy, London, pp. 201–211.

Schaubs, P.M., Zhao, C., 2002. Numerical models of gold-deposit formation in the Bendigo-Ballarat Zone, Victoria, Australia. Australian Journal of Earth Sciences 49, 1077–1095.

Schaubs, P.M., Sorjonen Ward, P., Zhang, Y., 2012. Kinematic framework of Neoarchean gold mineralization in Finland—constraints from bedrock mapping and numerical simulations. Proceedings of the 34th International Geological Congress: Unearthing our Past and Future—Resourcing Tomorrow, Abstracts, 822.

Sibson, R., 2001. Seismogenic Framework for Hydrothermal Transport and Ore Deposition. In: Richards, J.R., Tosdal, R.M. (Eds.), Structural Controls on Ore Genesis, Reviews in Economic Geology, Volume 14, pp. 25–50.

Sorjonen-Ward, P., Claoué-Long, J., 1993. A preliminary note on ion probe results for zircons from the Silvevaara Granodiorite, Ilomantsi, eastern Finland, Geological Survey of Finland, Special Paper, 18, 25–29.

Sorjonen-Ward, P., 1993a. An overview of structural evolution and lithic units within and intruding the late Archaean Hattu schist belt, Ilomantsi, eastern Finland. Geological Survey of Finland, Special Paper 17, 9–102.

Sorjonen-Ward, P., 1993b. Geological map of the Hattu schist belt, Ilomantsi district, eastern Finland. Geological Survey of Finland, Special Paper 17, Appendix 1.

Sorjonen-Ward, P., Luukkonen, E.J., 2005. Archean rocks. In: Lehtinen, M., Nurmi, P.A., Rämö, O.T. (Eds.), Precambrian Geology of Finland—Key to the Evolution of the Fennoscandian Shield. Elsevier B.V., Amsterdam, pp. 19–99.

Sorjonen-Ward, P., Zhang, Y., Zhao, C., 2002. Numerical modeling of orogenic processes and gold mineralization in the southeastern part of the Yilgarn craton, Western Australia. Australian Journal of Earth Sciences 49, 935–964.

Stein, H.J., Sundblad, K., Markey, R.J., et al., 1998. Re-Os ages for Archean molybdenite and pyrite, Kuittila-Kivisuo, Finland and Proterozoic molybdenite, Kabeliai, Lithuania: Testing the chronometer in a metamorphic and metasomatic setting. Mineralium Deposita 33, 329–345.

Vaasjoki, M., Sorjonen-Ward, P., Lavikainen, S., 1993. U-Pb age determinations and sulfide Pb-Pb characteristics from the late Archaean Hattu schist belt, Ilomantsi, eastern Finland. Geological Survey of Finland, Special Paper 17, 103–131.

Zhang, Y., Schaubs, P.M., Sheldon, H.A., et al., 2013. Modelling fault reactivation and fluid flow around a fault restraining step-over structure in the Laverton gold region, Yilgarn craton, Western Australia. Geofluids 13, 127–139.

CHAPTER

THE ROMPAS PROSPECT, PERÄPOHJA SCHIST BELT, NORTHERN FINLAND

5.4

E. Vanhanen, N.D.J. Cook, M.R. Hudson, L. Dahlenborg, J.-P. Ranta, T. Havela, J. Kinnunen, F. Molnár, A.R. Prave, N.H.S. Oliver

ABSTRACT

The Rompas prospect in the Ylitornio municipality is one of the most exciting recent discoveries in Finland. Spectacular bonanza grades, with abundant visible gold in association with uraninite, are hosted within deformed dolomite–calc-silicate–quartz veins in metabasalts along a 6-km-long ridgeline. The highest gold grades are associated with texturally late, sulfur-bearing pyrobitumen inferred to have formed at least 150° cooler than peak metamorphic conditions, approximately 150 million years after the crystallization of the vein-hosted uraninite.

Keywords: Finland; Peräpohja; Rompas; gold; uranium; uraninite; pyrobitumen; dolomite; calc-silicate; Svecofennian orogeny.

INTRODUCTION

The Rompas gold-uranium prospect is located in Ylitornio municipality, northern Finland, at 66.45 °N and 24.75 °E. It lies some 50 kilometers (km) to the west of Rovaniemi, the administrative capital of Lapland. Rompas was discovered by AREVA Resources Finland Oy in September 2008 as part of regional uranium exploration (Fig. 5.4.1). Limited exploration was undertaken until Mawson Resources purchased the exploration assets of AREVA in 2010 and subsequently defined more than 300 surficial bonanza-grade Au grab and channel samples from vein style mineralization over a north-trending strike of 6 km within the Peräpohja schist belt. Diamond drilling in 2012 and 2013 produced numerous Au-U rich intersections, of which 6 m at 617 g/t Au remains the highlight (in drill hole ROM0011). High fineness native gold and uraninite are the key economic minerals discovered to date.

The Peräpohja schist belt is not yet recognized as a mineral belt of high potential. The only active mine is the Kemi chromium mine, and that is within basal-layered mafic and ultramafic rocks (Halkoaho and Iljina, 2012). Kyläkoski et al. (2012a) describe quartz-carbonate vein-hosted and dolerite-hosted Cu-Au occurrences in addition to the Vähäjoki deposit ascribed to the iron oxide-copper-gold (IOCG) grouping (see also Eilu et al., 2007). In the northeastern Peräpohja schist belt, small tremolite and serpentine skarn magnetite occurrences are potentially of the IOCG class (Niiranen et al., 2003; Niiranen, 2012, and references therein). Details of the Rompas prospect to date have been largely restricted to internal Mawson Resources reporting and press releases available on the Mawson Resources website. Since discovery of the Rompas prospect, the Palokas prospect has been discovered approximately 8 km to the east. Gold

Mineral Deposits of Finland. http://dx.doi.org/10.1016/B978-0-12-410438-9.00018-2

mineralization at Palokas is a different style from Rompas, with gold and bismuth tellurides associated with pyrrhotite, ilmenite, magnetite, and scheelite within a variety of magnesian silicate minerals.

Mawson holds 833 claims and claim applications for 75,340 hectares at the Rompas Project. A total of 110 exploration claims that cover a surface area of 10,580 ha and form the core claims at Rompas came into legal force on October 15, 2012. The Natura 2000 network (European-defined biodiversity areas) is extensive in north Finland and has limited drill access to targets below the strongest Au and U anomalies. More than 90% of the strike extent of the Rompas prospect lies within the Natura 2000 designated areas. Applications have been made to the relevant Finnish granting authorities to allow diamond-drill testing of targets. Thus the drill results used here are from 84 diamond drill holes of less than 200 m down-hole depth, with the majority less than 110 m. The drill areas form only a very small portion of the prospect area (estimated at less than 5%). We expect that the understanding of the prospect will change significantly following better drill access in the coming years.

The descriptions and interpretations in this chapter are largely confined to field and drill core observations combined with geochemical and petrographic descriptions to confirm the nature of the major rock types. Detailed petrologic and isotopic work in progress by Ferenc Molnár and coworkers at GTK (Geological Survey of Finland) and Harry Oduro of the University of St Andrews (Scotland) is fast advancing the knowledge of the relationship of uraninite and gold at Rompas.

REGIONAL GEOLOGY

The Rompas prospect lies within the Peräpohja schist belt (PSB), a Paleoproterozoic supracrustal sequence of quartzites, mafic volcanics and volcaniclastics, carbonate rocks, black shales, mica schists, and graywackes that unconformably overlies Archean rocks of the Pudasjärvi complex (Fig. 5.4.1). Deposition of volcanics and sediments of the Peräpohja commenced after 2.44 Ga (Huhma et al., 1990) with the youngest rocks deposited after approximately 1.91 Ga based on inherited zircon populations in the Martimo formation (Ranta, 2012; Ranta et al., in prep.). We believe it is likely that the bulk of the mapped lowermost PSB was deposited on a relatively low gradient surface, either emergent (regionally extensive quartz sandstones) or shallow water (stromatolitic carbonate rocks). Also within the sequence are inferred evaporites of the Petäjäskoski formation (Kyläkoski et al., 2012b). Toward the end of the history of the PSB, however, traction current related sediments, reflecting basinal development, dominated the sequence (Korkiavaara, Pöyliövaara, and Martimo formations). The nature of the depositional environment and the interpreted long history of the PSB makes the likelihood of regional unconformities very high.

Correlation of rocks of the PSB sequence with the apparently similar Russian stratigraphy described in the Far Deep Project volumes (Melezhik et al., 2013) has been made by Prave (2013, internal company reporting) who proposed three major basin stages (Fig. 5.4.2). First, a continental margin dominated by nonmarine siliciclastic rocks and mafic volcanism with potentially two glacial episodes included: the Sompujärvi and the younger Kaisavaara glacials. The second basinal stage involves a change to shallow to nonmarine deposition of carbonate rocks and quartzites and is recognized as including the Lomagundi-Jatuli episode of enhanced $\delta^{13}C$ deposition (Karhu, 1993; Huhma et al., 2011). The third depositional stage involves development of considerable topographic relief. Basinal deepening with arkosic and quartzose turbidite fans are synchronous with mafic and felsic volcanism (e.g., Hanski et al., 2005). We propose that strike-slip tectonics are driving both transpressional and transtensional stress regimes resulting in both the uplift and the basin formation during this third depositional stage. This is likely synchronous with metamorphism and the emplacement of igneous intrusives.

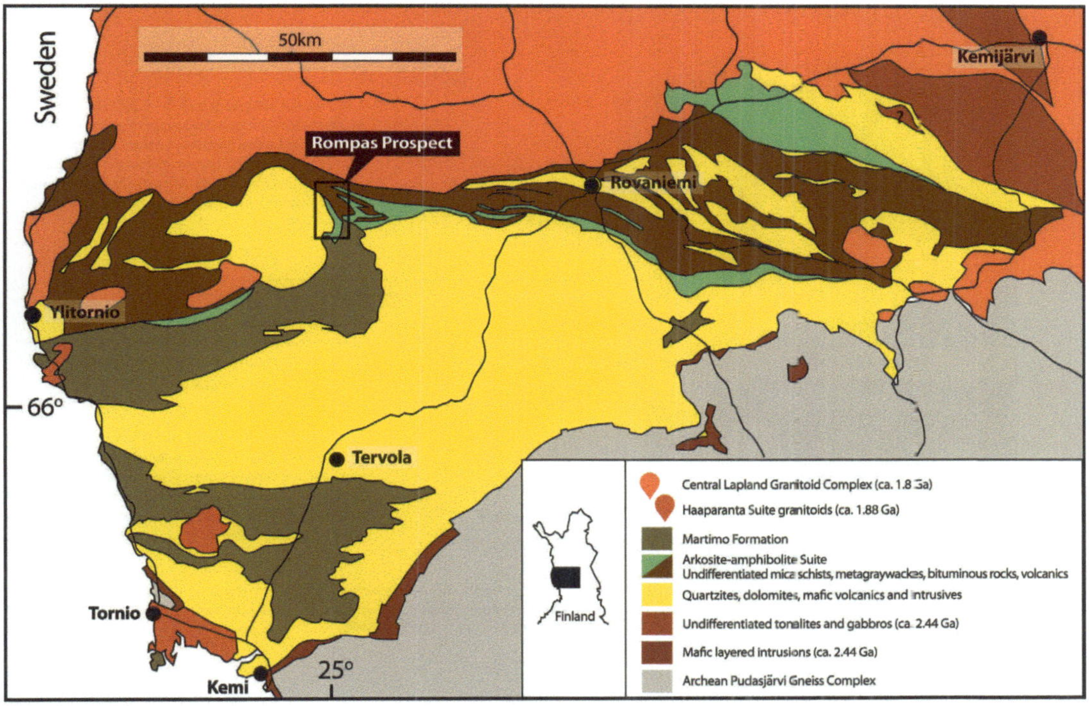

FIGURE 5.4.1 Regional geology and location details of Rompas prospect, Ylitornio, northern Finland.

This map is modified largely from Hanski et al. (2005) by grouping the arkosic-amphibolite suite with the formations inferred to be below the Martimo formation, but within the "late basin stages" as detailed in the text and Fig. 5.4.2. Lauri (personal communication, 2013) has also indicated the presence of approximately 2.44 Ga tonalities and gabbros in the Kemijärvi area (northeast corner of the map).

Metamorphism ranges from mid or upper greenschist facies in the south of the PSB to amphibolite facies conditions in the northern part of the belt, where Rompas is also located. To the north of Rompas, the pelitic compositions contain abundant migmatites and are intruded by granitoids of the Central Lapland Granite Complex (e.g., Perttunen et al., 1996; Lauri, 2012). The Sm-Nd data on tourmaline granites intruding the migmatites give ages of approximately 1.78 Ga (Ranta, 2012).

Deformation in the Peräpohja schist belt varies in style, largely related to the metamorphic grade. In the southern, greenschist facies portions, open to tight folds dominate, with a strong indication of thrust-related folding where axial surfaces are commonly broken by small thrust faults. As metamorphic grade increases to the north toward the Central Lapland Granite Complex, the fold intensity and strain increase dramatically and is typified by isoclinal to tight folds, with the highest strain occurring in the upper parts of the sequence that is dominated by thinly bedded sediments, mafic, and carbonate rocks. This may reflect the reactivation of basin-controlling structures. We believe that in the Rompas area, the rocks of the uppermost second basinal and third basinal stages (Fig. 5.4.2) have taken much of the shortening strain during the amphibolite facies metamorphism. Repeated thrust faulting and folding is evident west of Rompas where repetition of thin conductive graphitic rocks marks repeats of the stratigraphic sequence.

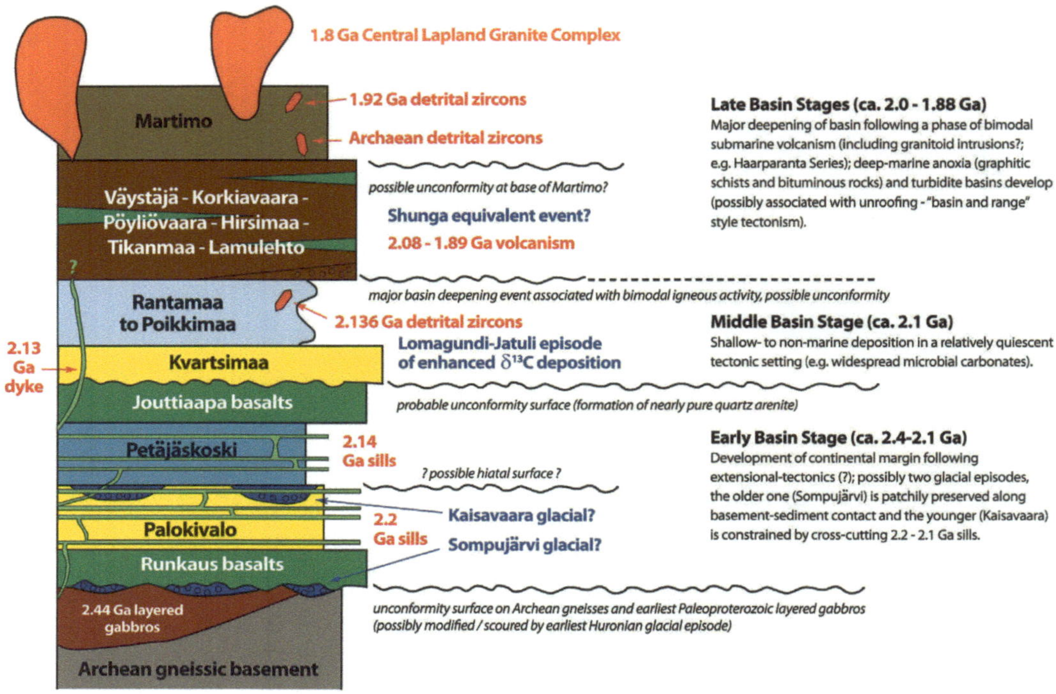

FIGURE 5.4.2 Proposed stratigraphic subdivision of the Peräpohja schist belt.

We interpret the Rompas prospect to lie in rocks deposited near the top of the middle basin stage, within the platform carbonate and quartzite rocks above the Jouttiaapa basalts, and in close contact with the overlying late basin stage rocks (metamorphosed turbiditic clastic sediments, carbonates, carbon-rich sediments, and mafic rocks). The stratigraphic column shows formation names for the PSB, largely based on Perttunen and Hanski (2003) and Kyläkoski et al. (2012b).

ROMPAS AREA GEOLOGY

The Rompas prospect area (Fig. 5.4.1) is dominated by a north-trending resistant ridge formed by metabasalt, carbonate-rich, and carbonate-albite rocks (Fig. 5.4.3). On the ridge crest, these rocks either crop out or are covered by a thin glacial till (generally less than 1 m). On either side of the ridge, small lakes, peat bogs, and thicker till are abundant with isolated bedrock outcrops. However, Au-U mineralized rock in the metabasalts is known to continue under deeper till cover on both sides of the ridge (Fig. 5.4.4). The metabasalts are generally fine to medium grained, dominated by the assemblage of plagioclase, quartz, amphibole, and biotite. Original igneous textures are absent from South Rompas (e.g., Fig. 5.4.5), and variably preserved at North Rompas (Fig. 5.4.6). At North Rompas, we interpret some textures to represent interflow sediment remnants and deformed amygdales, now rich in biotite.

The albitic calc-silicate-bearing rocks that lie between the mafic units are generally paler, owing to more abundant albite and quartz and lesser biotite and amphibole. Where highly foliated and veined with calc-silicate-carbonate-quartz, the metabasalts can be hard to distinguish from the metasediments.

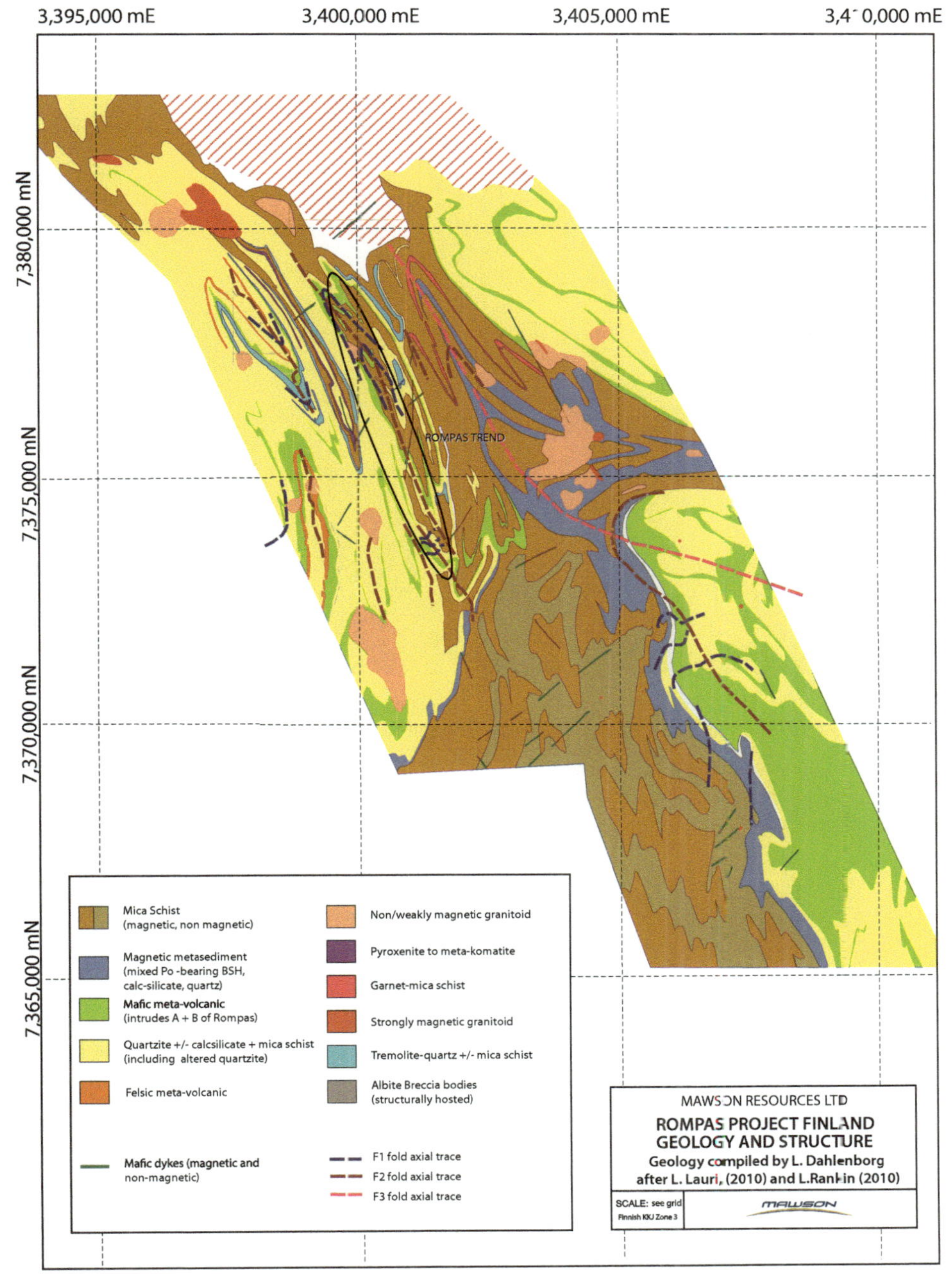

FIGURE 5.4.3 Geological map centered on Rompas prospect based on outcrop observations and interpretation of aeromagnetic images.

The Rompas trend is shown within the black ellipse.

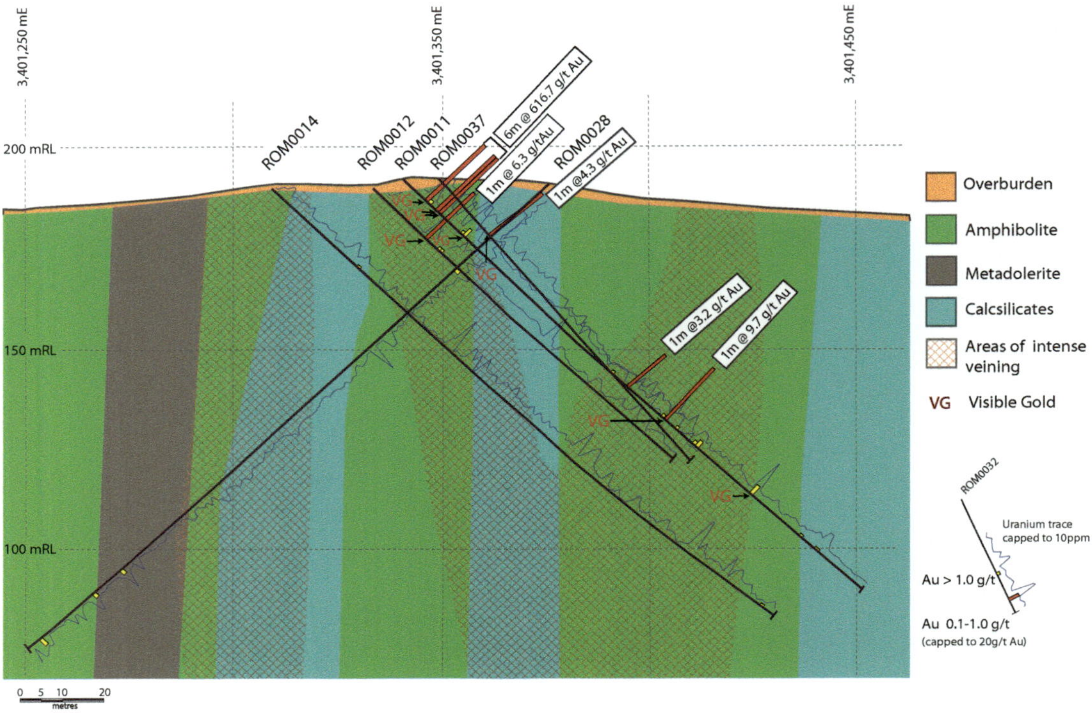

FIGURE 5.4.4 Cross section from South Rompas passing through ROMO011.

Note the metabasalt units within the calc-silicate (+ albite) rocks and metadolerite in the western part of the section. Gold intersections and the uranium traces are placed on the drill holes. The red hatched area indicates the most intense mineralization, but note that the high gold values are only within the mafic rocks.

FIGURE 5.4.5 Metabasalt with strong foliation defined by biotite-amphibole-rich zones (dark) and lighter albitic alteration bands at South Rompas.

Some of the light bands appear to be marginal to highly attenuated dolomite–calc-silicate veins. The drill core is 48 mm wide.

Using the TiO_2 content (Table 5.4.1) as an indicator of mafic precursor has proved successful to gain confidence in visual logging of the drill core.

Within the metabasalts, variably oriented larger amphibole grains up to 8 mm long grow across the S2 foliation throughout the North and South Rompas prospect areas. The amphibole compositions may vary from cummingtonite to anthophyllite, in addition to pargasitic hornblende. Indeed, individual samples may

FIGURE 5.4.6 Mafic rock preserving a doleritic igneous texture at North Rompas.

The former euhedral feldspars are recrystallized and lie within a matrix of amphibole and biotite. The drill core is 48 mm wide.

contain three amphiboles, namely hornblende, cummingtonite, and anthophyllite. Fine-grained pyrrhotite is the dominant sulfide away from mineralized veins, with rarer chalcopyrite and scarce pentlandite. Magnetite is largely absent from South Rompas, but is widespread in the mafic rocks at North Rompas. Rare fine graphite lies within the foliation planes adjacent to the mineralization at ROM0011, although the distribution of graphite in the mafic rocks has not been carefully checked in other locations. Assemblages in the main foliation in the metabasalts are compatible with amphibolite facies metamorphism.

Within both the mafic units and the calc-silicate-albite rocks are common metamorphosed dolomite ± quartz veins, now comprising largely dolomite, calcite, diopside, quartz, and actinolite. These veins are the hosts to the uraninite-gold mineralization, but only where they occur within metabasalt. Of particular importance, however, is the observation that these calc-silicate-carbonate-quartz veins are as abundant in the enclosing metasediments (Fig. 5.4.7) as they are within the mafic rocks. Alteration on the margins of both mineralized and unmineralized veins, such as in ROM0034, is typically biotite and albite rich, totally destroying the primary igneous textures in the metabasalt.

The western boundary of the Au-U occurrence in mafic rocks is marked by a thin, highly conductive graphitic unit of siltstone or mudstone protolith. This lies within a sequence of interbedded aluminous-biotite schists and arkosic to moderately quartzose metamorphosed siltstones and sandstones. Interspersed with these clastic metasediments are mafic units and carbonate rocks. Mineral assemblages in the aluminous schists include muscovite-biotite-feldspar-quartz ± garnet (South Rompas; Fig. 5.4.8) to sillimanite, cordierite, and garnet-bearing metapelites (Fig. 5.4.9) adjacent to the North Rompas occurrence. Metamorphic grade is therefore proposed to increase over the 6 km from South Rompas to conditions above the muscovite stability field at North Rompas.

Three significant deformational events are recognized at Rompas, although these are yet to be correlated with fold generations in the less metamorphosed, southern portions of the PSB. North, Central, and South Rompas lie on a largely linear trend (see Fig. 5.4.3), generated by the interaction of folds inferred to be related to D2 and D3 deformational events. The D1 event in the Rompas area is difficult to identify, but regional work (e.g., Terry Lees and Leigh Rankin, unpublished reports to Mawson Resources) suggests it is now present as transposed isoclinal folds with the foliation defined by amphibolite facies assemblages parallel or subparallel to the D2 foliation. D2 formed upright N-striking tight to isoclinal folds. Of significance at Rompas is that the D2 foliation is quite variable in intensity, with an increase toward South

Table 5.4.1 A representative selection of whole-rock analyses from the Rompas prospect, including mineralized and calc-silicate veined mafic rocks (mb) and the enclosing calc-silicate (Cs) rocks

Hole ID	ROM0011	ROM0011	ROM0011	ROM0011	ROM0014	ROM0014	ROM0014	ROM0014	ROM0028	ROM0028
Sample ID	226218	226219	226241	226297	226484	226547	226552	226642	229315	229446
Depth from (1 m samples)	11	12	33	87	7	67	71	159	16	141
Rock type	mb	mb	cs	mb	cs	mb	cs	mb	cs	mb
SiO_2[a]	54	51	62	54	59	52	58	56	61	50
TiO_2	1.62	1.10	0.67	1.51	0.46	2.12	0.44	1.21	0.63	1.50
Al_2O_3	12.96	9.09	14.21	11.62	11.07	14.15	11.49	11.70	14.28	12.13
FeO	10.97	7.89	5.18	9.73	5.80	14.60	2.42	2.83	3.49	11.03
MnO	0.10	0.14	0.02	0.07	0.06	0.09	0.05	0.04	0.03	0.16
MgO	5.41	9.45	2.67	7.31	7.45	6.92	6.88	7.46	3.33	8.64
CaO	5.96	15.18	4.70	8.00	8.54	3.79	12.06	12.30	7.39	11.18
Na_2O	2.76	2.16	6.98	3.24	4.19	3.59	6.17	5.93	7.98	2.39
K_2O	3.43	1.66	0.89	1.45	1.24	0.31	0.39	0.17	0.19	0.98
P_2O_5	0.22	0.14	0.13	0.18	0.10	0.26	0.11	0.20	0.13	0.18
S	0.30	0.01	0.02	0.05	0.01	0.19	0.00	0.00	0.00	0.07
Au	3540	137	<0.001	9.58	0.013	0.007	0.003	0.001	0.01	0.35
Pb	780	252	4.5	117.5	4.3	3	4.9	3.7	3.9	7.6
U	1770	650	2.3	338	2.6	0.7	3.2	2.1	2.5	6
Ag	12.55	0.84	0.03	0.35	0.04	0.08	0.02	0.05	0.05	0.11
As	79.4	35	<0.2	1.2	<0.2	<0.2	<0.2	0.3	0.5	1.5
Be	1.04	0.7	0.96	0.78	0.78	0.55	0.8	0.48	0.66	0.7
Bi	22.4	0.66	0.01	0.06	0.08	0.02	0.02	0.03	0.01	0.22
Ce	47	50	55.4	34.5	49.3	52.2	50.1	30.7	60.8	52.9
Co	54.3	39	3.7	43.5	20.2	62.5	3.1	6.4	2.1	62.4
Cs	6.26	2.22	1.17	2.28	1.31	0.42	0.15	0.07	0.18	1.29
Cu	120	4.6	1.6	64.3	1.8	138.5	3.3	4.1	1.2	59.6
Ga	22.6	15.05	19.1	19.75	16.55	22.7	18.6	16.5	22.9	19.85
Ge	0.24	0.18	0.13	0.26	0.14	0.25	0.14	0.13	0.19	0.15
Hf	4.9	3.4	4.8	3.9	3.4	6	3.5	3.6	3.8	4.5

In	0.082	0.065	0.006	0.055	0.056	0.106	0.007	0.012	<0.005	0.102
La	21.4	23.8	29.3	16.6	24.8	24.2	26	15.5	31	27.1
Mo	15.4	25.2	0.13	3.15	0.51	1.51	0.15	0.57	0.14	3.44
Nb	8.4	5.3	7.9	6.7	6.1	9.6	5.1	5	7.8	7.3
Ni	195.5	117.5	43.7	68.1	63.8	57.3	38.4	23.9	27.2	82.1
Rb	104	44.2	26.2	45.4	31.8	7.6	7.1	2.8	4.5	27.2
Re	0.042	0.036	<0.002	0.004	<0.002	0.003	<0.002	<0.002	<0.002	0.004
Sb	4.38	0.83	0.27	0.21	0.12	0.1	0.26	0.35	0.42	0.19
Sn	1.4	1.2	1.9	0.7	1.4	1.9	1.1	2.1	1.9	1.1
Ta	0.46	0.31	0.67	0.4	0.53	0.6	0.43	0.31	0.62	0.44
Te	103	11.45	<0.05	0.17	0.06	0.05	<0.05	<0.05	c0.05	0.25
Th	14.9	4.2	11.3	5.9	7.9	4.9	8.6	5.3	10.5	3.6
Tl	0.48	0.17	0.06	0.15	0.1	0.03	0.03	<0.02	<0.02	0.07
V	300	300	113	284	107	286	94	179	124	254
W	4.4	2.1	1.1	2.3	1	0.7	3.9	16.1	5	0.8
Y	34.8	37.6	16.4	24.6	18.9	28.2	18.4	20.2	15.4	39.8
Zn	89	79	8	28	21	24	17	8	7	30
Zr	175.5	123.5	162	146.5	122	213	123.5	129.5	151.5	154
Sr (%)	0.005	0.010	0.004	0.005	0.004	0.003	0.004	0.005	0.005	0.013
Cr (%)	0.011	0.008	0.016	0.008	0.012	0.009	0.010	0.006	0.011	0.009
Ba (%)	0.020	0.012	0.008	0.004	0.010	0.003	0.003	0.001	0.002	0.008
Albite (%)	23.38	18.25	59.08	27.37	35.47	30.34	52.24	50.19	67.52	20.19

Note: Refer to Fig. 5.4.3 for the location of the samples. The analyses were conducted by Australian Laboratory Services (ALS) using ICP-MS, excluding gold, which was analyzed using fire assay techniques. Negative numbers refer to the detection limits of the individual elements. All trace elements with the exception of Sr, Cr, and Ba are reported in parts per million.

[a]*SiO_2 is estimated by difference as the acid digest and ICP method prevented analysis. Albite percentage has been estimated on the assumption of all Na_2O is contained within albite.*

FIGURE 5.4.7 Albite-altered metasediments with dolomite-diopside-actinolite-quartz-calcite veins.

Note the biotite-rich selvages on the veins. Note that these metasediments enclosing the mineralized Rompas mafic rocks lack any peraluminous minerals other than biotite and typically contain some fine-grained calc-silicate minerals (ROM0034 drill hole, core diameter is 48 mm).

FIGURE 5.4.8 Possible graded bedding within biotite-muscovite metapelites from South Rompas.

These rocks are characteristic of the metasediments from basin stage 3. Drill core is 48 mm across.

FIGURE 5.4.9 Garnet porphyroblasts in sillimanite-biotite gneiss from North Rompas.

Diamond saw cut across outcrop was for sampling purposes. Scribe is 150 mm long.

Rompas to such an extent that any protolith textures are destroyed. Relatively low-strain rocks are described at North Rompas, preserving vesicles and pillow structures in mafic rocks, whereas at South Rompas, in particular, total strain is very high and relic early structures are difficult to identify.

Transposition of lithologic contacts in the amphibolite facies rocks is observed at both outcrop and regional scales (from interpretation of aeromagnetic data). Superimposed on the D2 foliation at South Rompas is significant extension-producing boudinage of veins (Fig. 5.4.10) and subvertically plunging, cigar-shaped dolomite pods. These dolomite pods are important as they are a common host to large porphyroblastic uraninite and extreme gold values, with some samples at South Rompas exceeding 10,000 g/t Au! Unfortunately, the linear nature and the lack of predictability in location of these pods within the mafic rocks makes for a difficult drill target. Weathering of the carbonate in these pods produces deep brown soils with abundant residual gold that is recoverable with panning (Fig. 5.4.11).

FIGURE 5.4.10 Examples of boudinaged veins from the Rompas trend.

Top: South Rompas drilling area (field of view 50 cm). *Bottom:* Western end of "The Wall" at North Rompas (field of view 1 m). Cavities within the vein are dolomite and calcite with uraninite.

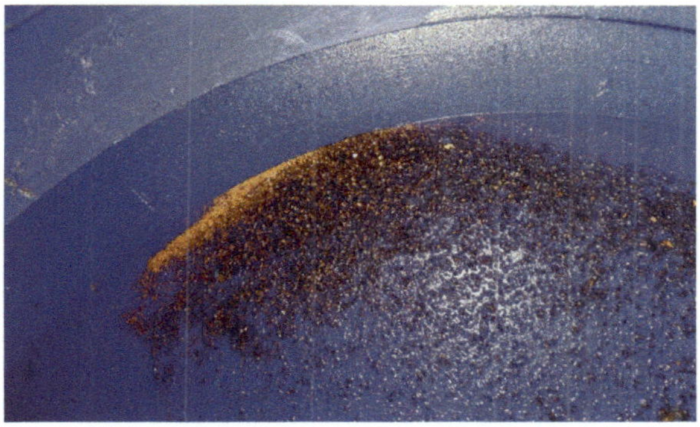

FIGURE 5.4.11 The results of gold panning above a subvertical weathered dolomitic pod.

These are clearly evident in this photo. Approximately 1 kg of sample was panned. Field of view is 10 cm.

At least one brittle, post-metamorphic deformation episode is proposed. Late fractures filled with calcite and quartz are common, and some may be synchronous with the last stages of gold mineralization and brittle fracturing of uraninite. In one location, a concretion-like mass of pyrobitumen is enclosed in a calcite-filled fracture, compatible with the observations on mineralization detailed in the following.

Understanding the geochemistry of the host mafic sequence to the mineralization is complicated by variable alkali, Mg, and S mobility and dilution of less mobile components. A selection of representative analyses of the mafic host rocks and albite–calc-silicate metasediments is presented in Table 5.4.1 (note that SiO_2 in these analyses is unreliable as it is not determined by the inductively coupled plasma mass spectrometry (ICP-MS) method but has been calculated here by difference and should be regarded as an approximation only). Many 1-m intervals contain significant dolomite–calc-silicate–quartz veining and some data trends are a function of dilution by dolomite, rather than an inherent original chemical variation in the mafic precursor. This is evident, for example, in the data from ROM0011—the 1-m interval commencing at 12 m (down-hole depth) is diluted approximately 40% by veining. The corresponding decrease in TiO_2 is a function of this dilution, rather than protolith variation. Rare earth element (REE) studies of Rompas mafic rocks are confined to Mustonen (2012), who regarded the REE profiles to most closely match that of the Runkaus basalts, the lowermost mafic volcanics of the Peräpohja Belt. This interpretation is, however, questioned by us, as the Rompas host rocks appear to be at a higher stratigraphic position than the Runkaus formation rocks (Fig. 5.4.2).

MINERALIZATION

In terms of exploration for gold and uranium, distribution within the initial discovery area at Rompas is best described as nuggety. The discovery of gold at the surface is directly associated with the ability to find uraninite using a scintillometer or spectrometer, and a till cover of more than 1 m will generally reduce the radioactivity to background levels. Visible gold largely occurs within uraninite, with the coarsest grains occurring within and associated with the dolomite pods described previously (Fig. 5.4.12; South Rompas). Gold occurs as fracture-controlled veinlets, in places extending beyond the uraninite and onto cleavage surfaces on carbonate, and more rarely on quartz. At North Rompas, visible gold is

FIGURE 5.4.12 A small vein of gold within dolomite and fine dark amphibole.

Note the finely disseminated gold throughout the sample as the likely "hard rock" source of the gold recovered by panning in the in the soils above occurrences of weathered carbonate rocks (South Rompas). The sample is 1.5 cm long.

also present as equigranular aggregates within actinolite, spatially associated with porphyroblastic uraninite grains (Fig. 5.4.13). Visible gold surrounding pyrobitumen with associated molybdenite has been observed in ROM0084, drilled immediately below the spectacular gold intersection in ROM0011 (Fig. 5.4.14). Porphyroblastic uraninite typically has a distinctive pink to reddish albite alteration halo extending from the carbonate–calc-silicate veins into albitic margins (Fig. 5.4.15).

No geophysical responses have been detected to the mineralization. Fine pyrrhotite is a ubiquitous trace component of the metabasalt, resulting in a stronger IP chargeability response than the adjacent rocks, but the intensity of response is not directly associated with gold. It is useful, however, for mapping the occurrence of metabasalt under the till.

Thin and very high-grade drill intercepts are common, such as for the sample 11–12 m down-hole depth shown in Table 5.4.1 and Fig. 5.4.14, matching the many hundred surface trench

FIGURE 5.4.13 Spectacular gold within actinolite.

Actinolite is dominant across all except the lower right corner, where a dark, large uraninite grain occurs. The uraninite is difficult to see in the photo, but is responsible for the secondary yellow uranium products on the extreme lower right fractured surface. Thin-section examination shows a polygonal equigranular intergrowth of gold and actinolite indicating coeval formation or recrystallization. The sample is from the North Rompas diamond sawn trench at surface and is 10 cm in its longest dimension.

FIGURE 5.4.14 Dolomite–calc-silicate vein with bonanza gold grades from ROM0011.

The grade for this 1 m at 11–12 m down-hole depth is 3540 g/t, part of the 6 m at 617 g/t intersection. The drill core is 63 mm across.

FIGURE 5.4.15 Uraninite grain 15 mm across within a dolomite–calc-silicate vein with thin biotite selvage enveloped by albite-rich alteration of metabasalt.

Note the reddish albite, a common feature adjacent to uraninite.

exposures of bonanza grades found over the 6 km trend. Full details of intersections are reported on the Mawson Resources website (www.mawsonresources.com). What is clear from the drilling and trenching at North and South Rompas is the strong host rock control on mineralization—the metabasalt. Uraninite and gold are found within or marginal to carbonate–calc-silicate–quartz veins in the metabasalt. Vein densities in the adjacent metasedimentary albitic calc-silicate rocks are similar to the metabasalt, but to date lack any significant Au or U. Thus, interaction with some component in the basalt is inferred to be the key control on uraninite precipitation, with a secondary mechanism favored to precipitate gold at a later stage.

On the basis of bulk rock analyses, the gold itself is of high fineness, recording more than 95% Au, with minor Ag and Cu. Drilling up to 100 m vertical depth has not revealed any variation in the fineness. Petrographic examination by prior workers reveals that the gold is paragenetically later than the uraninite, commonly occupying cracks within uraninite (Figs. 5.4.16 and 5.4.17; plus descriptions by Gaillard, 2012; and Ashley, internal company reporting). Some secondary U-Th minerals such as coffinite and thorite are also present. Paragenetically late gummite is locally present in surface samples occurring between uraninite grains (Gaillard, 2012). In most outcrops where uraninite has been recorded, secondary yellow uranium minerals have been observed on fractures and surface coatings.

As noted before, gold occurs in two modes, the first within fractures in uraninite within the calc-silicate veins, and the second generally as crusts on the spherical to subrounded undeformed pyrobitumen. Pyrobitumen also occurs as rims and nodular accretions around uraninite in textures reminiscent of radiolytic polymerization.

The uraninite within the pyrobitumen nodules contains more than five times the ThO_2 and Y_2O_3 contents of the coarse-grained uraninites in the dolomite–calc-silicate veins and the Pb contents are a function of the alteration content. Preliminary U-Th-Pb age determinations show a clustering of ages of the least altered uraninite grains reflecting the peak of metamorphism of the host sequence, but we are uncertain of the significance of this in terms of the regional tectonic history of the Fennoscandian Shield.

FIGURE 5.4.16 A large (15 mm) uraninite grain within dolomite–calc-silicate vein.

Note gold within diagonally trending fracture across the uraninite and disseminated gold toward top of photo.

FIGURE 5.4.17 Reflected light photomicrograph of intensely fractured uraninite (light gray) with gold filling fracture networks.

Some larger "blobs" of gold on the outside of uraninite grains are with carbonate and silicate matrix (dark gray). Note the variable reflectivity of the uraninite—there are both strongly reflective, "clear" uraninite grains and many with a cloudy appearance. These have different trace element and Pb concentrations. Field of view is 5 mm.

The gold within the coarse porphyroblastic uraninite is associated with galena, altaite, nickeline, cobaltite, and rare Pb-bearing maldonite, hunchunite, and auricupride. In contrast, the gold associated with pyrobitumen is more closely related to a pyrite, pyrrhotite, chalcopyrite, galena, altaite, molybdenite, Bi-telluride, and titanite assemblage (Figs. 5.4.18, 5.4.19, and 5.4.20). Reflectance measurements conducted on the undeformed pyrobitumen nodules containing gold indicates a likely maximum maturation temperature of between 270° and 340 °C, which is probably reflecting the formation temperature of the gold mineralization. The Pb-bearing minerals (e.g. galena, altaite), associated with

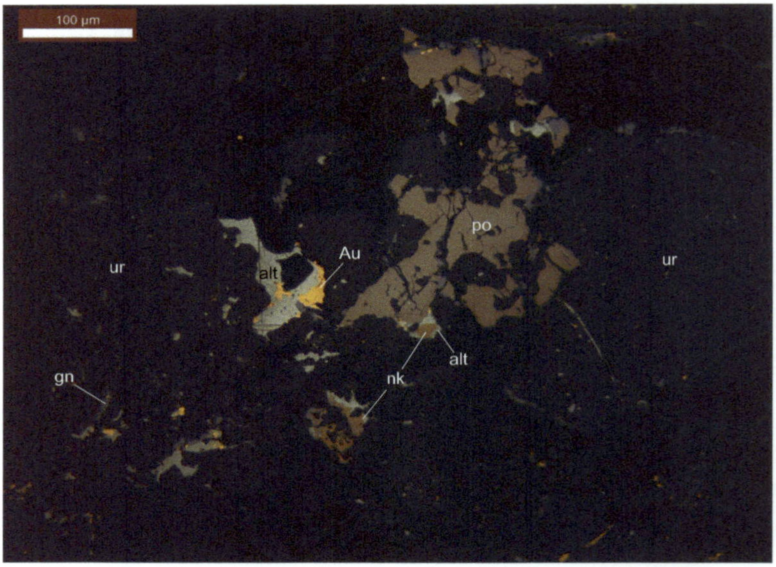

FIGURE 5.4.18 Large mid-gray uraninite (Ur) grains with very fine galena (Gn) throughout.

The more reflective light-bluish-white grains are altaite (alt), which is host to gold in the center part of the picture. Pyrrhotite (po) is the light brown mass in the center and upper part of the image. A small grain of nickeline hosted by altaite is attached to the pyrrhotite in the center of the picture. The groundmass hosting the crushed uraninite grains is carbonate. Field of view is 0.65 mm.

Source: *Photo by F. Molnár.*

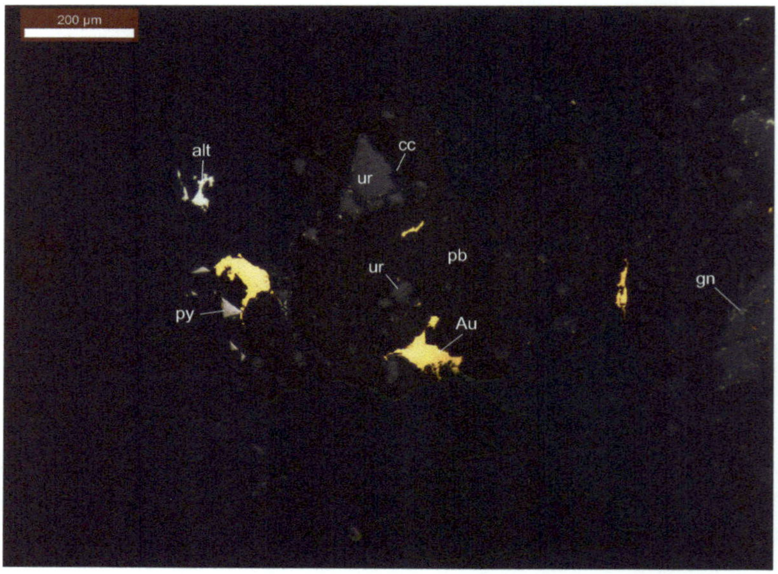

FIGURE 5.4.19 Nodular "blobs" of pyrobitumen within a matrix of carbonate (dark gray).

The medium gray fragments throughout the pyrobitumen are uraninite, including the larger, triangular fragment in the upper center of photo. The dark gray crust on some of the uraninite grains (e.g. the triangular one in the upper center) is carbonate. The very bright white grain with triangular shape next to the gold at left-center is pyrite. The bright fields above this gold grain are altaite (alt) grains (in calcite). Field of view is 1.25 mm.

Source: *Photo by F. Molnár.*

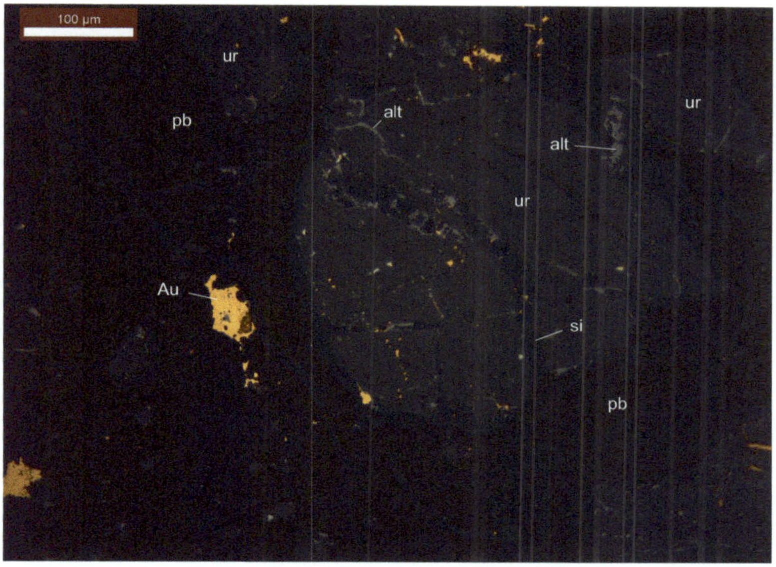

FIGURE 5.4.20 Amorphous silica (Si) filling cracks in uraninite grains enclosed within slightly desiccated pyrobitumen.

The bluish white phase in fractures is altaite (alt). Note the gold within microfractures in uraninite and associated with the pyrobitumen and amorphous silica. Field of view is 0.65 mm.

Source: *Photo by F. Molnár.*

the first stage of gold formation, are highly radiogenic. We infer that a sulfur-bearing fluid carrying gold has precipitated the Pb minerals, scavenging radiogenic lead from the uraninite. In addition, early indications of the age of molybdenite associated with pyrobitumen (H. Stein, personal communication) indicate an age close to 1.78 Ga, representing some 150-million-year difference between the formation of the uraninite and the gold-pyrobitumen (+ molybdenite). These later dates are compatible with the ages of the Central Lapland Granite Complex, concentrated around the 1.81 to 1.78 Ga range (e.g., Lauri, 2012).

SUMMARY

We consider it most likely that the uraninite with the dolomite–calc-silicate veins at Rompas was metamorphosed to amphibolite facies conditions and deformed approximately 1.95 Ga ago. This matches unpublished data of Cathelineau et al. (CREGU internal report to Mawson Resources, and in prep.). The "pre-metamorphic" age of the uraninite remains uncertain, but is evidently older than 1.95 Ga. Critically, a much younger age of approximately 1.78 Ga is proposed by us as the age of the gold mineralization, based on the textural, geochemical, and preliminary isotopic data. At least two gold-bearing sulfurous fluids, the first reacting with radiogenic lead from uraninite, and a second, lower temperature fluid, reacted with pyrobitumen, resulted in gold formation with close spatial relationships to uraninite. Further exploratory drilling and laboratory studies will no doubt provide more detailed and more robust insights into the formation of the Rompas uranium-gold occurrences.

ACKNOWLEDGMENTS

Very few mineral occurrences are found by the efforts of one person alone, and Rompas is no exception. The combination of regional geologic evaluation, through district-scale research to detailed ground checking, is acknowledged here. The Mawson exploration team presently comprises six geologists directly working on the project (Erkki Vanhanen, Nick Cook, Mike Hudson, Tuomas Havela, Janne Kinnunen, and Lars Dahlenborg). Former staff on the project, consultants, and summer students have added significantly to the understanding of the Rompas-Rajapalot project and include, but are not limited to, Jukka-Pekka Ranta, Terry Lees, Jan-Anders Perdahl, Claude Caillat, Leigh Rankin, Gerald Purvis, Nicolas Gaillard, Pertti Sarala, Eelis Pulkkinen, David McInnes, Hans Thunehed, and Marcus Tomkinson. In addition, research groups based at the GTK (Ferenc Molnár, Esa Pohjolainen, Antonin Richard, Laura Lauri) and at Nancy/CEGRU/AREVA (Michel Cathelineau, Marc Brouand) have been instrumental in allowing Mawson to further develop an understanding of the mineralization at Rompas.

REFERENCES

Eilu, P., Pankka, H., Keinänen, V., Kortelainen, V., et al., 2007. Characteristics of gold mineralisation in the greenstone belts of northern Finland. Geological Survey of Finland, Special Paper 44, 57–106.

Gaillard, N., 2012. Les concentrations en or et uranium dans les niveaux de calc-silicates du nord de la Finlande (Rovaniemi). M.Sc. thesis. Universite Henri Poincare (Nancy), p. 42.

Halkoaho, T., Iljina, M., 2012. Portimo Cr, PGE-Ni. Geological Survey of Finland Special Paper 53, 294–298.

Hanski, E., Huhma, H., Perttunen, V., 2005. SIMS U-Pb, Sm-Nd isotope and geochemical study of an arkosite-amphibolite suite, Peräpohja schist belt: evidence for ca. 1.98 Ga A-type volcanism in northern Finland. Bulletin of the Geological Society of Finland 77, 5–29.

Huhma, H., Cliff, R., Perttunen, V., Sakko, M., 1990. Sm-Nd and Pb isotopic study of mafic rocks associated with early Proterozoic continental rifting: the Peräpojha schist belt in northern Finland. Contributions to Mineralogy and Petrology 104, 367–379.

Huhma, H., O'Brien, H., Lahaye, Y., Mänttäri, 2011. Isotope geology and Fennoscandian lithosphere evolution. In: Nanonen, K., Nurmi, P.A. (Eds.), Geological Survey of Finland, Special Paper, 49, pp. 35–48.

Karhu, J., 1993. Paleoproterozoic evolution of the carbon isotope ratios of sedimentary carbonates in the Fen noscandian Shield. Geological Survey of Finland, Bulletin 371, 87.

Kyläkoski, M., Eilu, P., Perdahl, J.-A., 2012a. Peräpohja Cu-Co. Geological Survey of Finland Special Paper 53, 301–302.

Kyläkoski, M., Hanski, E., Huhma, H., 2012b. The Petäjäskoski formation, a new lithostratigraphic unit in the Paleoproterozoic Peräpohja belt, northern Finland. Bulletin of the Geological Society of Finland 84, 85–120.

Lauri, L., 2012. Temporal and Hf isotopic geochemical evolution of southern Finnish Lapland from 2.77 Ga to 1.76 Ga. Bulletin of the Geological Society of Finland 84, 121–140.

Mustonen, M., 2012. Geochemistry and petrography of the volcanic rocks in the Rompas area, Ylitornio, northern Finland. M.Sc. thesis, University of Oulu, p. 63.

Niiranen, T., 2012. Misi Fe. Geological Survey of Finland Special Paper 53, 304–306.

Niiranen, T., Hanski, E., Eilu, P., 2003. General geology, alteration, and iron deposits in the Palaeoproterozoic Misi region, northern Finland. Bulletin of the Geological Society of Finland 75, 69–92.

Perttunen, V., Hanski, E., 2003. Koivu and Törmäsjärvi. Explanation to the Geological Map of Finland 1 100 000, pre-Quaternary rocks, sheets 3631 and 2633. Geological Survey of Finland, p. 88. (in Finnish, with English summary).

Perttunen, V., Hanski, E., Väänänen, J., et al., 1996. Rovaniemin kartta-alueen Kallioperä. Geological Map of Finland 1 100 000. Explanation to the maps of pre-Quaternary rocks, sheet 3612.

Ranta, J.-P., 2012. Peräpohjan liuskealueen pohjoisosan yksiköiden zirkoniajoitus U-Pb-menetelmällä. M.Sc. thesis. Department of Geosciences. University of Oulu, p. 89 (in Finnish).

PART 6

THE HANNUKAINEN FE-(CU-AU) DEPOSIT, WESTERN FINNISH LAPLAND: AN UPDATED DEPOSIT MODEL

6

M. Moilanen, P. Peltonen

ABSTRACT

The Hannukainen deposit in the Kolari area, western Finnish Lapland, is the largest magnetite deposit in Finland, which also contains significant Cu and Au. Together with its Kuervitikko satellite, the deposit occurs along a major structural zone and appears to be located at flexures along the major structure. The Fe-Cu-Au ore bodies are flat lenses 10–50 m in thickness hosted by calc-silicate skarn rocks, which were formed through metasomatic replacement along the contact zone between a monzonite intrusion and metabasalts. Geophysical studies indicate that the mineralized body at Hannukainen has an extensive down-plunge continuation, probably as deep as 2 km. Together with its Kuervitikko satellite deposit, the total measured and indicated open pit and underground resources are 187 Mt at an average grade of 30.04 wt% Fe, 0.18 wt% Cu, 0.11 ppm Au, and the inferred resource is 63 Mt at an average grade of 32.05 wt% Fe, 0.15 wt% Cu, and 0.05 ppm Au. A recently completed definitive feasibility study indicates that the deposit is economically viable for production of high-purity (70 wt% Fe) magnetite fines supplemented by Cu-Au concentrate.

Hannukainen and more than 10 broadly similar magnetite deposits in the Kolari area formed ~1800 Ma, when local branches of the Bothnian megashear were reactivated and the inflow of hydrothermal saline fluids occurred due to breakdown of evaporites that were earlier deposited within an intracontinental rift zone between the Norrbotten and Karelian cratons. It is hypothesized that these hydrothermal fluids progressively leached Fe, Cu, and Au from the supracrustal sequence, composed mainly of metabasalts, and then deposited metals through

Mineral Deposits of Finland. http://dx.doi.org/10.1016/B978-0-12-410438-9.00019-4

mixing with more reduced fluids. Textural analysis of mineralized rocks indicates that magnetite was deposited first, followed by precipitation of copper sulfides and gold.

Keywords: IOCG deposits; iron; copper; gold; skarn; Kolari; Finland.

INTRODUCTION

The Hannukainen project comprises two Fe-(Cu-Au) deposits: Hannukainen and Kuervitikko. They are part of an elongated cluster of magnetite deposits occurring in the Kolari-Pajala district, located in Finland and Sweden. To date, as many as 30 magnetite deposits have been identified in this district, which covers approximately 1600 km^2 (40 × 40 km). Hannukainen is the largest Fe-(Cu-Au) deposit on the Finnish side of the border and is located 25 km northeast of the municipal center of Kolari. In Sweden, Fe-(Cu-Au) deposits include the Stora Sahavaara and Tapuli deposits. The styles of mineralization in these deposits display considerable variation ranging from semimassive, skarn-hosted (calc-silicate rocks) magnetite bodies to albitite breccia-hosted magnetite-pyrite-chalcopyrite bodies and disseminated magnetite-chalcopyrite-pyrite deposits in altered mafic volcanic rocks. Copper and gold grades also vary widely among these deposits.

The Kolari-Pajala district is part of a much larger iron ore province in northern Fennoscandia. The Kiruna iron deposit, located 150 km northwest of the Kolari-Pajala area, was discovered in 1696 and has been mined on a regular basis since 1900. The Kiruna deposit consists of a 5-km-long, up to 100-m-thick, steeply dipping layer of iron oxide body with the current sole mineral of economic interest being magnetite. The proven mineral reserves are 536 Mt at 48.6 wt% Fe and probable reserves 146 Mt at 46.4 wt% Fe (LKAB, 2012). The iron oxide bodies are hosted by felsic volcanic rocks within a Svecofennian supracrustal sequence known as the Kiruna porphyry group (Offerberg, 1967). Another currently operating iron mine is the Malmberget magnetite-apatite deposit located at Gällivare in Sweden, about 100 km northwest of Kolari-Pajala. The Malmberget deposit consists of about 20 ore bodies in an area of about 5 × 2.5 km; 7 of the bodies are currently being mined. The mineralization is similar to that at Kiruna. Proven mineral reserves in Malmberget are 168 Mt at 42.3% Fe and probable reserves are 103 Mt at 41.2% Fe (LKAB, 2012). The Malmberget deposit is hosted by highly metamorphosed felsic volcanic rocks, which are considered to be metamorphosed equivalents of the Kiruna porphyry group rocks.

REGIONAL GEOLOGICAL SETTING

Several crustal-scale structures are known in the Fennoscandian Shield. One of these is the north-northeast–south-southwest-striking Kolari-Pajala shear zone, which straddles the border between Finland and Sweden, and hosts the Hannukainen and Kuervitikko deposits. The Kolari-Pajala shear zone (Fig. 6.1) is up to 50 km wide and at least a 150 km long and forms the boundary between the Norrbotten craton in the west and Karelian craton in the east. It has been suggested that the Kolari-Pajala shear zone was formed during the continent–continent collision of the Norrbotten and Karelian cratons at 1.89–1.86 Ga and was subsequently reactivated during later orogenic events between 1.83 and 1.79 Ga (Lahtinen et al., 2005).

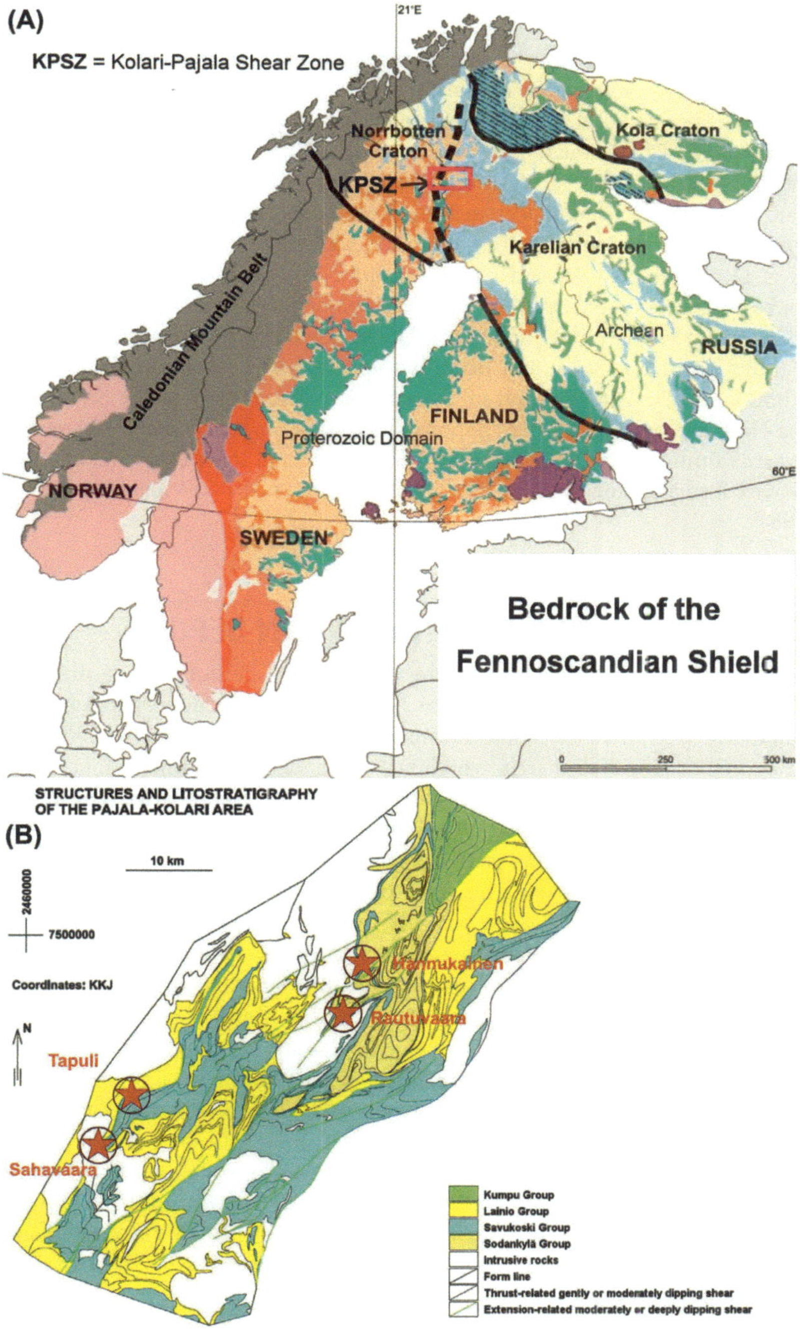

FIGURE 6.1

(A) Bedrock map of the Fennoscandian Shield; (B) Structural and lithostratigraphical map of the Kolari-Pajala region.

Sources: *(A) Modified after Koistinen et al. (2001). (B) Northland Exploration (unpublished report, 2010).*

The Kolari-Pajala area is located in the western part of the Central Lapland Greenstone Belt. The belt was formed during prolonged stages of rifting of the Archean craton, with deposition of Karelian sedimentary and volcanic rocks in intracratonic and cratonic margin rift settings between 2.5 and 2.0 Ga, and was subjected to multiphase deformation and metamorphism during orogenic events. The Haparanda suite intrusive rocks of ~1.86 Ga also occur in the Kolari-Pajala area (Lahtinen et al., 2005).

In the Kolari area, the Karelian supracrustal sequence consists mainly of gently westward-dipping quartzites, phyllites, and mica schists belonging to the Sodankylä group with an age of more than 2.22 Ga (Lehtonen et al., 1998; Hanski and Huhma, 2005). They are stratigraphically and structurally overlain by rocks of the Savukoski group, which are older than 2.06 Ga, the age of the crosscutting Kevitsa layered intrusion (Mutanen and Huhma, 2001). The Savukoski group includes pelitic metasediments, black schists, tholeiitic metabasalts, and primitive volcanic rocks of komatiitic and picritic affinity (Hanski and Huhma, 2005). The Rautuvaara formation, which is part of the Savukoski group, is the host unit for the Hannukainen Fe-(Cu-Au) deposit and other iron-base metal occurrences in the Hannukainen-Rautuvaara mining camp. The Rautuvaara formation is in structural contact with and lies below younger monzonite intrusions of the Haparanda suite (Hiltunen, 1982; Väänänen and Lehtonen, 2001; Väänänen, 1998).

EXPLORATION HISTORY OF THE HANNUKAINEN DEPOSIT

The Kolari area has been known for its iron ore resources for centuries. The first historical records indicate that in the late seventeenth century, the Juvakaisenmaa Fe deposit provided ore feed to the Köngäs ironworks at Pajala, Sweden, some 20 km west of Kolari. During World War II, Vuoksenniska Oy carried out Fe exploration in the area, which was continued by Suomen Malmi Oy in 1956–1960. The Otanmäki Oy Company continued the feasibility studies of the nearby Rautuvaara deposit until 1967 when it was merged with Rautaruukki Oy. From late 1969 to early 1970, Rautaruukki Oy reevaluated the early plans to exploit the Rautuvaara Fe deposit and a formal decision to open the mine was made in May 1970, with mining commencing in 1975. Meanwhile, Rautaruukki Oy continued exploration in the surrounding area and in 1974 exploration was focused on the Hannukainen area.

At Hannukainen, ground magnetic surveys led to the discovery of several ore bodies (Kuervaara, Laurinoja, Lauku, and Vuopio). Because the overburden thickness and stripping ratio were most favorable at the Kuervaara ore body, open-pit mining was started in May 1978. Ore was hauled 10 km to the Rautuvaara mine site to provide additional feed for the Rautuvaara plant. The most valuable ore body by in situ value was Laurinoja, which contained a significant amount of copper ore. The decision to exploit Laurinoja was made by Rautaruukki Oy in June 1981, with a flotation plant built at Rautuvaara. Production of Cu concentrate from Hannukainen started in June (Mining Magazine, 1982).

Altogether, between 1978 and 1990, Rautaruukki Oy mined approximately 4.5 Mt of iron ore from the two Hannukainen open pits. Between 1990 and 1995, the Rautuvaara mill and Hannukainen deposit were leased to Outokumpu Mining Oy, which mined an additional 0.45 Mt of magnetite ore from the Laurinoja open pit, and also processed ore from the company's other deposits (e.g., Saattopora and Juomasuo) in the Rautuvaara plant.

Since 2005, Hannukainen has been studied in great detail by Northland Resources S.A. In 2014, Northland completed the Hannukainen definitive feasibility study, which shows that the Hannukainen Fe-Cu-Au project is economically feasible.

GEOLOGY OF THE HANNUKAINEN DEPOSIT

The Hannukainen deposit consists of five ore bodies: Laurinoja, Kuervaara, Lauku, Kivivuopio, and Vuopio, as well as a satellite deposit at Kuervitikko (Figs. 6.2 and 6.3). The ore bodies form flat lenses that are oriented approximately parallel with the regional structural foliation. The location of the mineralization is strongly structurally controlled; the Fe-Cu-Au ore lenses are located at bends of the semiductile/ductile shear and thrust zone (Fig. 6.3). To the south of the deposit, the Äkäsjoki strike-slip shear zone crosscuts the thrust zone. The reverse thrust at Hannukainen dips 20–30° to the west. At Hannukainen, the lineation plunges about 30° toward the southwest.

A typical local lithological sequence in the Hannukainen deposit is from top to bottom: monzonite, diorite, different kinds of skarn rocks, amphibolite/mafic metavolcanic rocks, gneiss, and quartzite (Figs. 6.4 and 6.5). The hanging wall rocks consist of Svecofennian Haparanda suite monzonite and diorite, with the latter reinterpreted as strongly altered metabasalts/mafic metavolcanic rocks in this study, while the footwall rocks comprise supracrustal rocks, mainly mafic metavolcanic rocks (amphibolites). In association with the amphibolite units, there are thin beds of quartz-feldspar schists, which are commonly pyrrhotite-bearing and contain variable amounts of graphite. A sequence of mica schists or gneisses occurs in the eastern parts of the deposit beneath the amphibolite unit, both of which in turn overlie Sodankylä group quartzites. Quartzites form the lowermost drilled unit in the local stratigraphy. Pegmatite, aplite, and granite dikes with varying thicknesses commonly crosscut the whole sequence.

ALTERATION

In general, all rocks in the Hannukainen deposit, excluding the youngest granites, are intensively altered. Alteration has formed a deposit-scale zoned pattern where three different alteration zones can be distinguished around the ore bodies in both the hanging wall and footwall rocks. The alteration assemblages vary somewhat depending on the primary rock type, but the general pattern is that albite ± scapolite is the dominant alteration assemblage in distal zones; biotite and K-feldspar dominate the intermediate zones; and the proximal alteration zone, which encompasses the magnetite bodies, is characterized by clinopyroxene, with varying amounts of magnetite, amphibole, and calcite. Most of the sulfides occur in the proximal alteration zones, but both iron and copper sulfides can be locally present in all altered lithologies regardless of the alteration assemblage.

The main minerals in skarn rocks are clinopyroxene and hornblende. Also biotite, albite, scapolite, garnet, calcite, actinolite-tremolite, and serpentine occur in minor amounts. The main oxide mineral is magnetite, which usually occurs as disseminated grains and bands, but also massive and semimassive zones are present (see Fig. 6.9 later).

The zoned alteration pattern may repeat itself at different scales throughout the sequence. Some zones, particularly the intermediate one, may be poorly developed or even missing. Similarly, the proper magnetite skarn core is locally missing or it is very thin and pinches out. Niiranen et al. (2007) reported that the typical proximal alteration type in the Hannukainen deposit is calc-silicate dominated (Ca-clinopyroxene, actinolite, or hornblende, magnetite ± scapolite, calcite, biotite, and albite). The intermediate alteration involves biotite, K-feldspar ± albite, and scapolite, whereas the distal alteration is formed by albite ± scapolite. Figure 6.6 shows schematically typical mineral parageneses in the Laurinoja ore body.

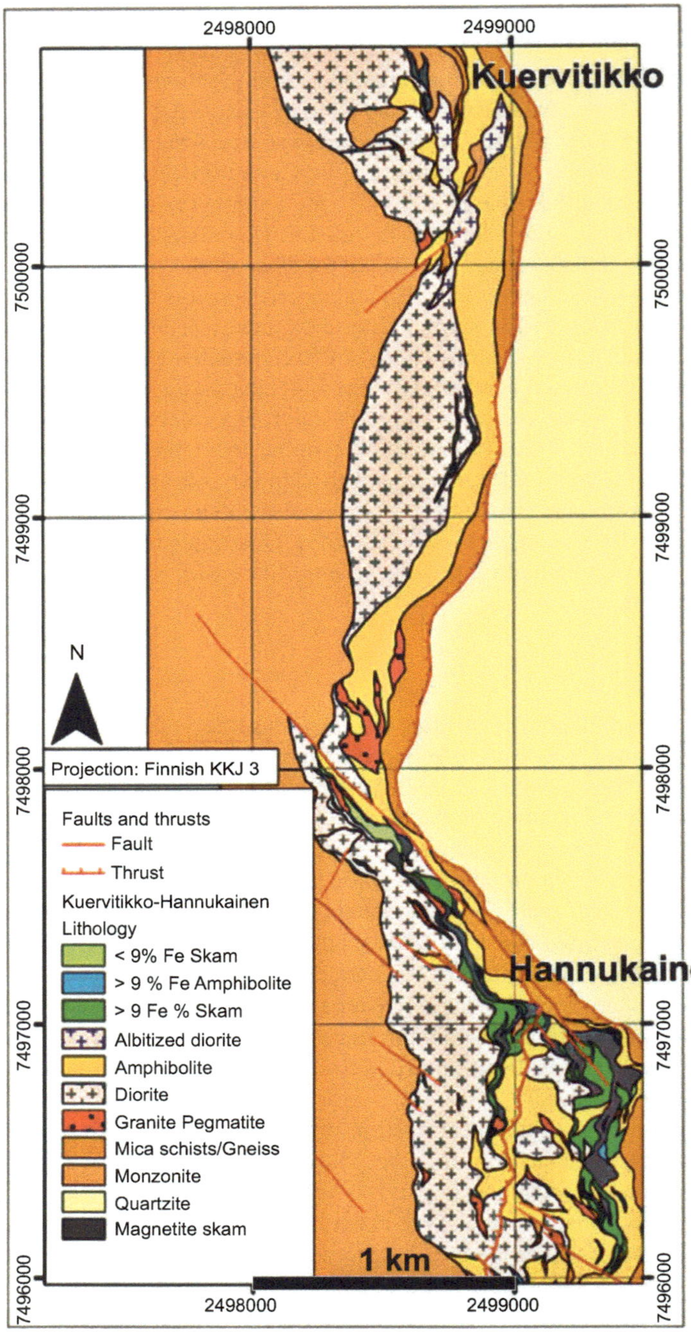

FIGURE 6.2 Surface geological map of the Hannukainen and Kuervitikko deposits.

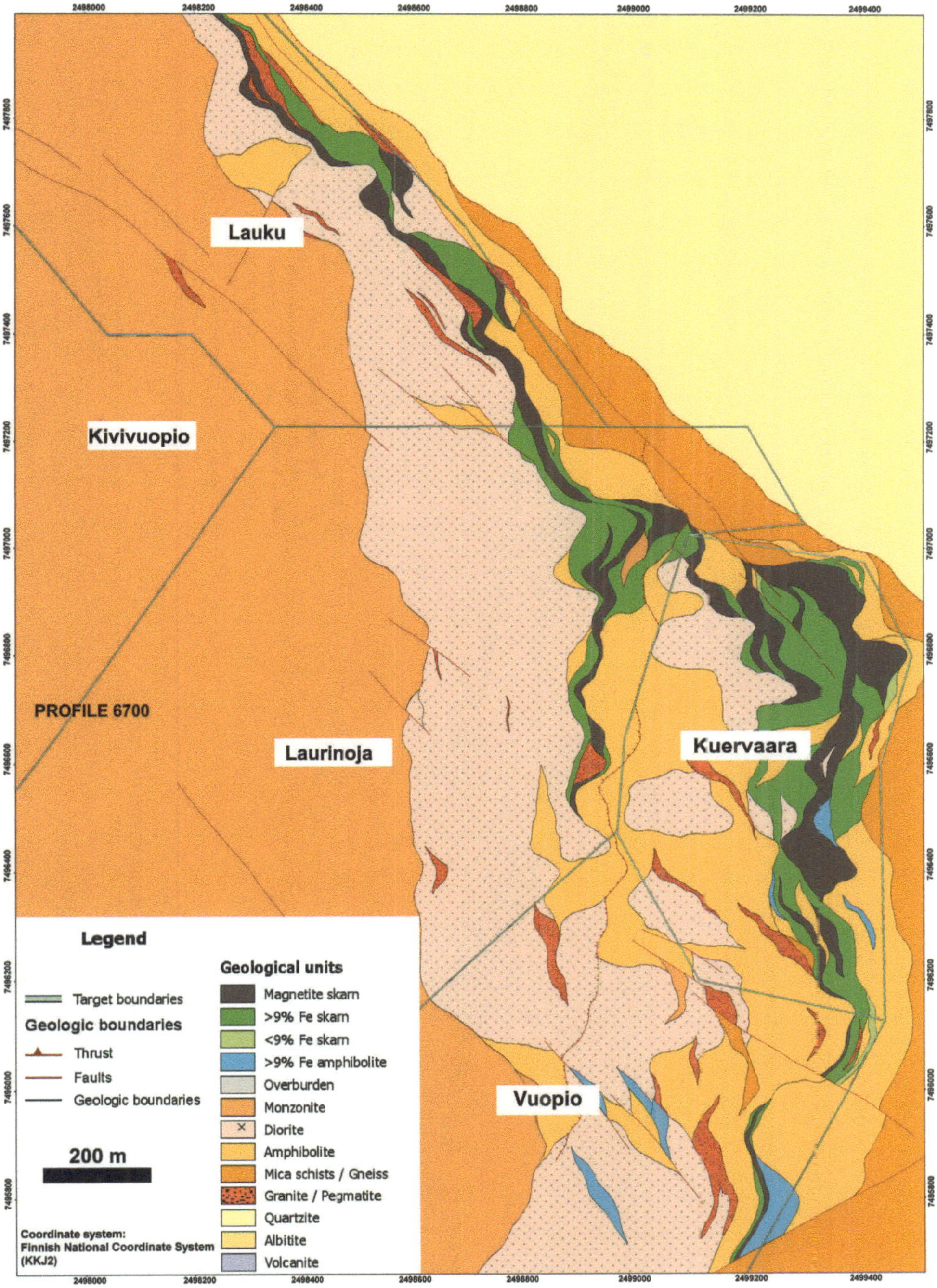

FIGURE 6.3 Surface geological map of the Hannukainen deposit.

Note that the "diorites" are not diorites sensu stricto, but have in this contribution been reinterpreted as strongly altered metabasalts/mafic metavolcanic rocks (see the following).

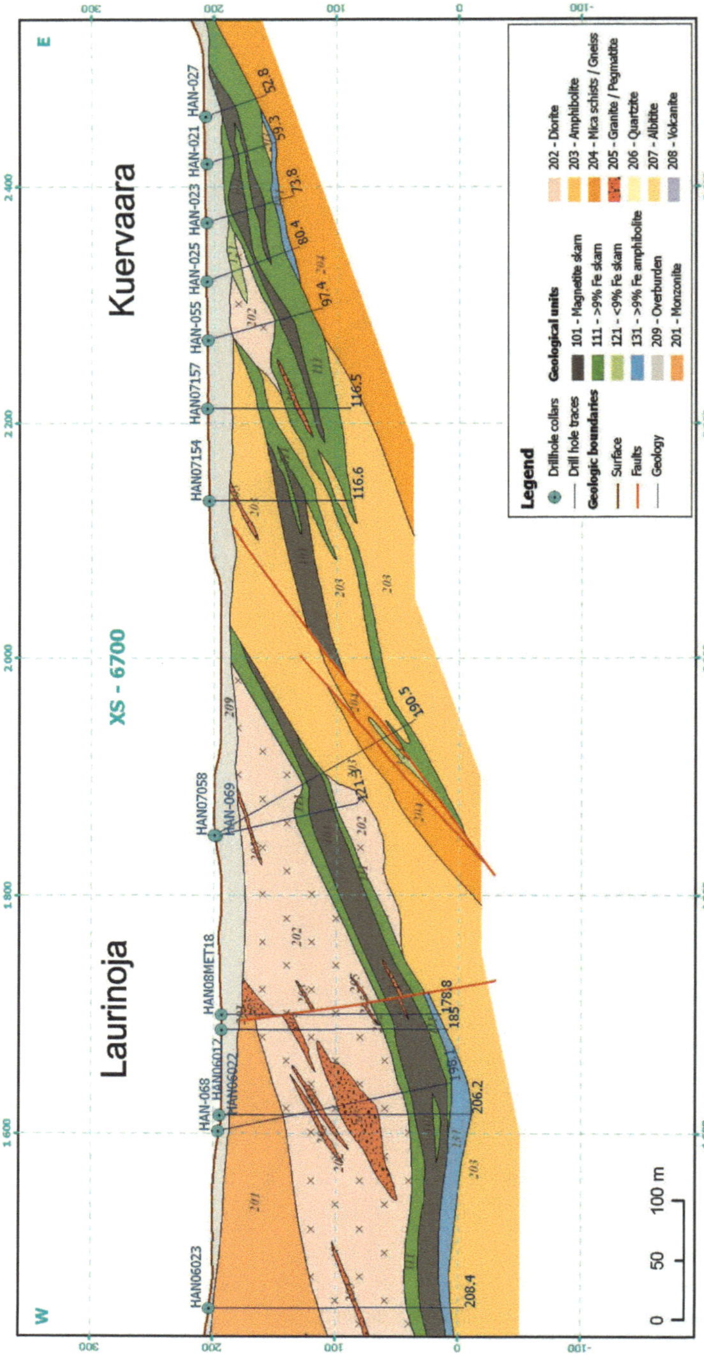

FIGURE 6.4 East–west cross section across the Hannukainen deposit

Source: Profile 6700, northing: 7496700.

FIGURE 6.5 Photographs of major rock types from the Hannukainen deposit

(A) hanging wall monzonite, (B) hanging wall "diorite" (strongly altered metabasalts/mafic metavolcanic rock), (C) clinopyroxene skarn, (D) magnetite skarn with disseminated sulfides (chalcopyrite, pyrite, pyrrhotite), (E) footwall amphibolite, and (F) footwall quartzite.

The major rock types in the Hannukainen and Kuervitikko deposits are similar and share a common stratigraphy. The alteration assemblage in the Kuervitikko deposit defines a similar zoning as described for Hannukainen earlier. The most significant differences are that at Kuervitikko, K-feldspar and biotite are less common in the hanging wall distal alteration zone, and there is an albitite (± quartz) unit between the diorite and clinopyroxene-amphibole skarn zone. In addition, amphibole is slightly more abundant as a proximal alteration mineral than at Hannukainen. Sulfides also appear to be more abundant in the distal alteration zones, especially in the albitite.

MINERALIZATION

The Fe-(Cu-Au)-mineralized lenses are hosted by skarn rocks (calc-silicate rocks), which were formed through metasomatic replacement reactions at the lithological contact and structural shear zone between the hanging wall monzonite in the west and footwall mafic metavolcanic rocks/basalts (amphibolites) in the

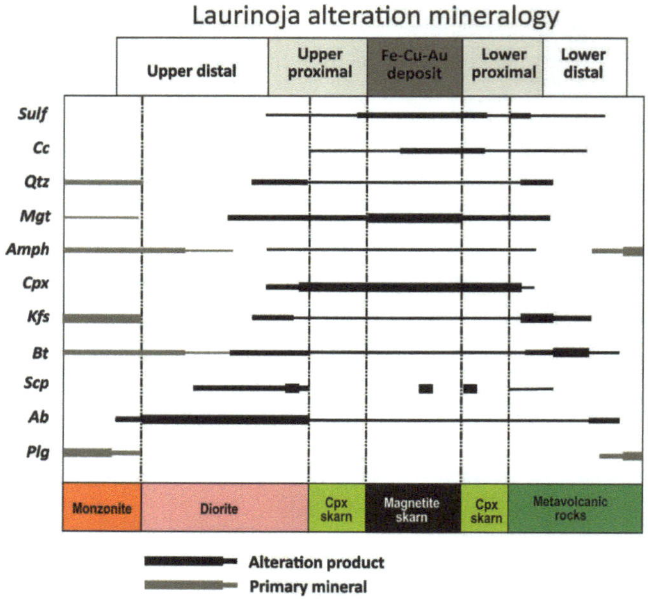

FIGURE 6.6 Schematic paragenetic diagram for minerals and alteration zones in the Laurinoja ore body, Hannukainen deposit.

Abbreviations: Sulf = sulfides, Cc = calcite, Qtz = quartz, Mgt = magnetite, Amph = amphibole, Cpx = clinopyroxene, Kfs = K-feldspar, Bt = biotite, Scp = scapolite, Ab = albite, Plg = plagioclase.

Source: *Modified after Niiranen (2007).*

Table 6.1 Physical dimensions of the mineralized ore bodies/lenses of the Hannukainen deposit and Kuervitikko satellite deposit

Mineralized lens/body	Length (m)	Down-dip width (m)	Maximum thickness (m)	Average thickness (m)
Kuervaara/Vuopio	1900	800	50	10–30
Laurinoja/Lauku/Kivi-vuopio	2100	2000	50	10–30
Kuervitikko	1200	500	50	10–20

east. Typical host rocks for the mineralization are magnetite and clinopyroxene skarns. Skarn units are well defined, with a high-grade magnetite skarn zone in the middle. The high-grade core is surrounded by a lower-grade iron mineralization, usually related to clinopyroxene skarn rocks. The average thickness of the mineralized bodies varies between 10 and 30 m, with the maximum thickness reaching approximately 50 m.

Table 6.1 summarizes the physical dimensions of the mineralized ore bodies at Hannukainen. Figure 6.7 is a 3D illustration of the Hannukainen and Kuervitikko ore bodies together with high resolution reflection seismics for ore exploration (HIRE)-line E1. Figure 6.7 implies that seismic reflectors coincide very well with the known Hannukainen Fe-Cu-Au ore lenses. This suggests that the strong

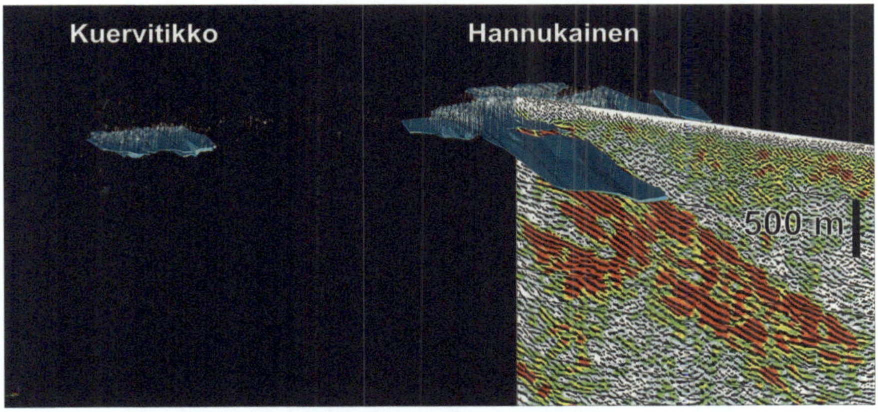

FIGURE 6.7 3D models of the Hannukainen and Kuervitikko ore bodies, diamond drill holes, and reflection seismic HIRE-line E1.

A deep ore body coincides with the top of the seismic reflector, implying that the mineralization may extend much deeper than indicated by the present drilling results. Image looking toward north-northeast.

reflectivity is due to the strong velocity contrast between monzonite and the skarn- and amphibole-dominated lithological assemblages. These seismic reflectors also correlate well with magnetic and electrically conductive layers (Kukkonen et al., 2009). Distribution of iron, copper, and gold as ore grade bodies at Hannukainen is shown in Fig. 6.8. The most recently published mineral reserves of the Hannukainen deposit are shown in Table 6.2 (SRK, 2014).

The ore mineralogy in the Hannukainen deposit is relatively simple. Magnetite is the only observed oxide mineral. Hematite has not been observed. Magnetite usually occurs as a dissemination, veins, and semimassive to massive. It is coarse grained with the grain size exceeding 0.3 mm. In sulfide-bearing zones, the dominant sulfide minerals are pyrite, pyrrhotite, and chalcopyrite. Chalcopyrite is the only significant copper mineral. The grain size of chalcopyrite is usually small but also some coarse grains and grain aggregates occur (diameter 0.02–0.03 mm). A minor amount of molybdenite occurs in association with chalcopyrite. Sulfides usually occur as a dissemination, veins, and fracture fillings (Fig. 6.9). They appear to be texturally later than magnetite, because sulfides occur as veins and veinlets around and crosscutting magnetite, and sulfide inclusions are not common in magnetite. Native gold has been detected in the silicate gangue, chalcopyrite, and magnetite. However, 95% of gold occurs as solid solution in sulfides. The copper-gold mineralization is usually hosted by magnetite skarn rocks, or, to a lesser extent, by surrounding clinopyroxene skarn rocks.

GEOCHEMISTRY AND AGE DETERMINATIONS

Average chemical compositions of major rock types in the Hannukainen deposit are listed in Table 6.3. Box-and-whisker percentile diagrams and their chondrite-normalized REE patterns are shown in Figs. 6.10 and 6.11, respectively. The monzonite intrusions that occur west of Hannukainen are metaluminous, low in Nb and Y, and have a calc-alkaline affinity (Niiranen et al., 2007). Their chondrite-normalized REE

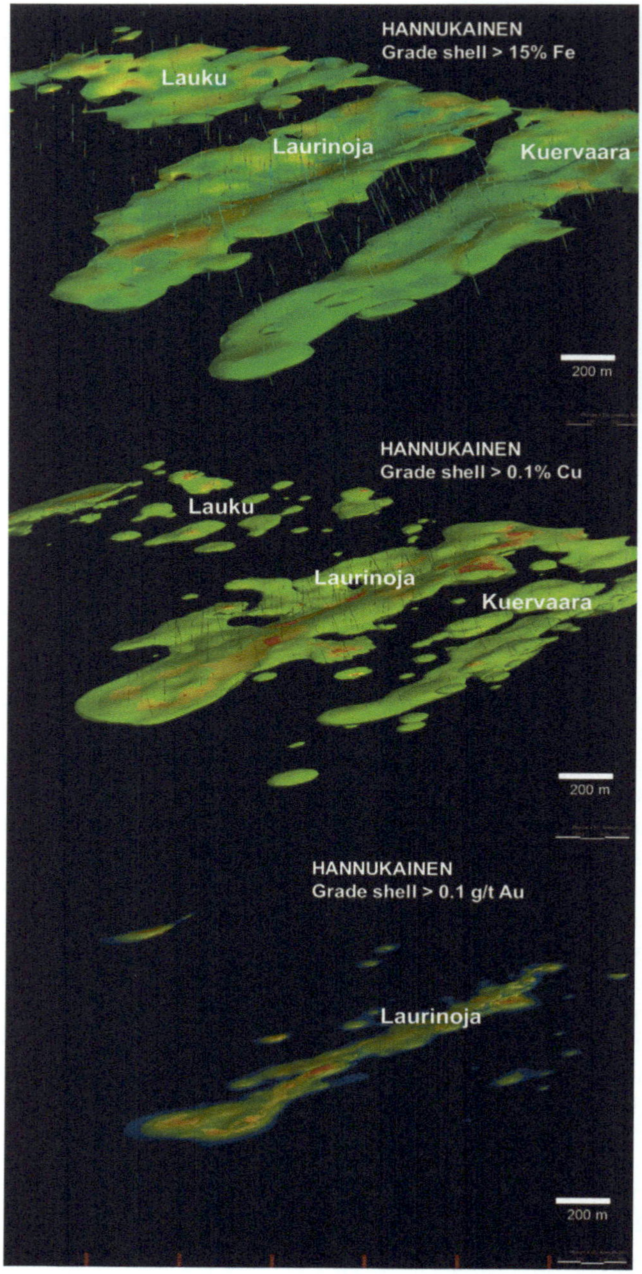

FIGURE 6.8 3D image showing the Hannukainen ore-grade lenses for iron (Fe >15 wt%), copper (>0.1 wt%), and gold (Au >0.1 g/t).

Pictures viewing toward the north. Note the prominent concentration of Cu and Au within the Laurinoja ore body. Au, in particular, is distributed along a linear feature parallel to the regional plunge direction.

Table 6.2 Open pit mineral reserves of the Hannukainen deposit

Category	Tonnes (Mt)	Fe (%)	Cu (%)	Au (ppm)
Hannukainen				
Proven	91.8	32.2	0.186	0.088
Probable	0.8	32.6	0.148	0.060
Kuervitikko				
Proven	21.9	23.6	0.183	0.216
Probable	0.3	23.8	0.177	0.194
Total				
Proven	113.7	30.5	0.185	0.112
Probable	1.1	30.0	0.157	0.100
Total	114.8	30.5	0.185	0.112

Source: *SRK, 2014.*

pattern (Fig. 6.11) shows enrichment in LREE and a weak negative Eu anomaly. The REE pattern of diorite is similar to that of monzonite with the exception that diorite has a slightly positive Eu anomaly and lower general REE abundances. Different kinds of skarn rocks in the Hannukainen deposit are the most altered rocks. Niiranen et al. (2007) showed that Ti/Al, Zr/Al, and Zr/Ti in the clinopyroxene skarn and magnetite skarn are similar, which suggests a common protolith for these rock types. Some variation in the immobile element ratios could indicate that the protoliths of these skarn rocks were locally heterogeneous and could have consisted of several different rock types. The only main difference between the clinopyroxene and magnetite skarns is that the latter contain more iron than the former.

Niiranen et al. (2007) classified mafic volcanic rocks of the Hannukainen area into two chemical groups. Based on the different Zr/Ti ratios, they distinguished type 1 (lower Zr/Ti) and type 2 (higher Zr/Ti). Type 1 mafic volcanic rocks have Zr/Ti ratios typical for basalts, whereas type 2 mafic rocks have Zr-Ti ratios that are between basalts and andesites. Mafic volcanic rocks in the distal alteration zones are rich in Na ± K, Cl, and Ba, whereas proximally altered mafic volcanic rocks are enriched in Ca ± Fe, S, Ag, Au, Bi, C, Ce, Co, Cu, La, and Te (Niiranen et al., 2007).

Magmatic zircon grains from the hanging wall monzonite and mafic volcanic rocks of the Hannukainen deposit have given U-Pb ages of 1862 ± 3 and 1864 ± 6 Ma, respectively (Hiltunen 1982; Niiranen et al., 2007). A granite dike, which brecciates and crosscuts the ore zone, has yielded a U-Pb zircon age of 1760 ± 3 Ma (Niiranen et al., 2007). These dates thus bracket the age of the alteration/mineralization event. Zircon within a skarn rock (calc-silicate rock) has been dated at 1797 ± 5 Ma, while titanites in altered wall rocks and skarns range in age between 1810 and 1870 Ma. These minerals could be original magmatic minerals or they can be hydrothermal minerals (Hiltunen, 1982; Niiranen, 2005; Niiranen et al., 2007). The dating results imply that the ore formation at Hannukainen is not directly related to the emplacement of the ~1860 Ma Haparanda suite monzonites, but probably took place close to ~1800 Ma. Zircon grains in skarn rocks may give a true age for the associated thermal event, whereas titanite ages provide a series of cooling ages ranging from the protolith age of ~1870 Ma to that of the mineralization event at ~1800 Ma (Väänänen, 1998; Väänänen and Lehtonen, 2001; Hannu Huhma, personal communication 2014).

FIGURE 6.9 Different styles of oxide and sulfide occurrences in the Hannukainen deposit.

(A) Sulfide (pyrrhotite and chalcopyrite) veins and dissemination in magnetite skarn; (B) chalcopyrite dissemination in magnetite skarn; (C) chalcopyrite veins, fracture filling; (D) close-up of disseminated magnetite and sulfides; (E) banded-textured magnetite skarn; (F) semimassive and massive (inside yellow arrows) magnetite skarn.

FLUID INCLUSIONS, AND O AND C ISOTOPES

Niiranen et al. (2007) identified three main types of fluid inclusions by using a combination of textural, phase proportional, and microthermic observations. Type 1 are formed by complex aqueous or aqueous-carbonic liquid-vapor-multisolid inclusions and are subdivided into type 1A and type 1B based on their occurrence and solid contents. Type 1A occur as solitary inclusions or clusters in quartz that are

Table 6.3 Average chemical composition of major rock types in the Hannukainen deposit

	Monzonite	Diorite	Amphibolite	Skarn	Magnetite skarn
Major elements (wt.%)					
SiO_2	61.87	55.33	50.68	31.00	24.00
TiO_2	0.62	0.42	0.66	0.24	0.18
Al_2O_3	15.24	16.59	13.43	4.54	4.28
Fe_2O_3	6.31	11.11	14.01	45.53	58.64
MnO	0.06	0.07	0.12	0.26	0.33
MgO	2.06	2.15	4.67	4.98	4.62
CaO	3.48	4.96	7.16	9.88	4.67
Na_2O	5.60	5.85	4.30	1.51	1.32
K_2O	3.14	1.88	1.71	0.47	0.78
P_2O_5	0.22	0.21	0.21	0.13	0.15
Trace elements (ppm)					
Cr	74	68	141	85	31
Ni	24	36	69	82	81
Co	19	48	63	149	160
Cu	97	607	572	1913	1601
V	100	70	166	100	88
Pb	32	55	51	49	34
Zn	46	40	56	85	90
Nb	13	3.9	4.3	3.8	3.8
Zr	321	86	97	64	51
Y	20	12	17	9.0	6.1
Sr	486	348	228	104	60
Rb	111	68	71	37	61
Sc	13	11	17	9.8	10
La	39	21	27	35	33
Ce	88	53	54	64	60
Ba	706	778	348	84	107
Ga	23	18	18	17	22

associated with sulfides. Type 1B inclusions occur as networks of discontinuous inclusion trails in quartz and calcite that are associated with sulfides and also together with type 2 inclusions. Type 2 inclusions are aqueous ± carbonic liquid-vapor inclusions occurring as trails in quartz and calcite, and are associated with sulfides. Type 3 represent single phase, liquid CO_2-rich inclusions that occur in quartz as trails and around sulfides. They also occur together with type 1B inclusions. The fluid inclusion data from Niiranen et al. (2007) suggest that ore deposition took place in two distinct stages in the Laurinoja ore body.

The compositions of the first-stage fluids corresponded to moderately saline Na-Ca ± K, Fe-bearing H_2O fluids (12–22 wt% NaCl equivalent). The second stage fluids were highly saline (32–56 wt% NaCl equivalent), Na-Ca ± K-bearing H_2O-CO_2-fluids (32–56 wt% NaCl equivalent). Estimated

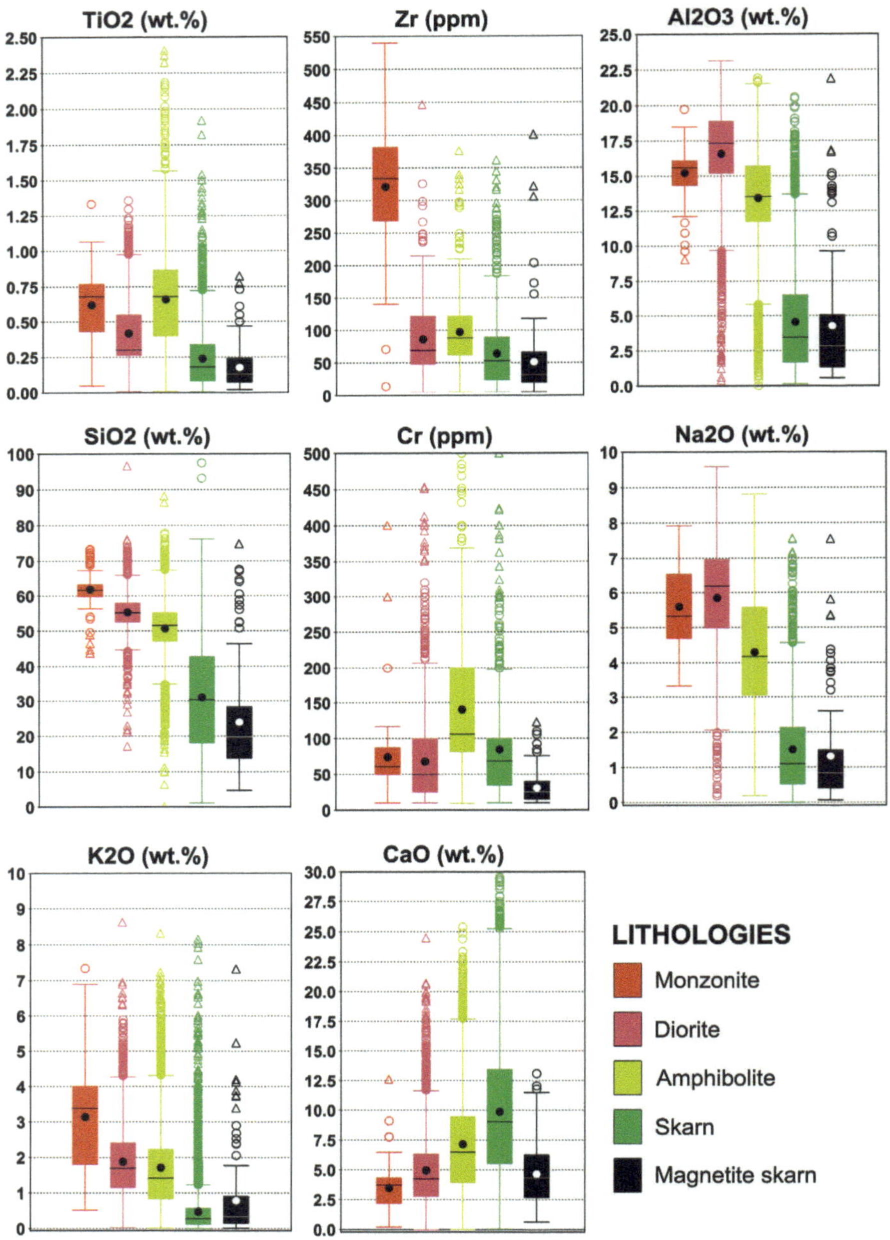

FIGURE 6.10 Box-and-whisker percentile diagrams of major elements and major rock types from the Hannukainen deposit.

The median is defined by a horizontal line within a box. The mean is represented as a black/white circle within a box. Sample points off the whiskers area are assumed to be outliers.

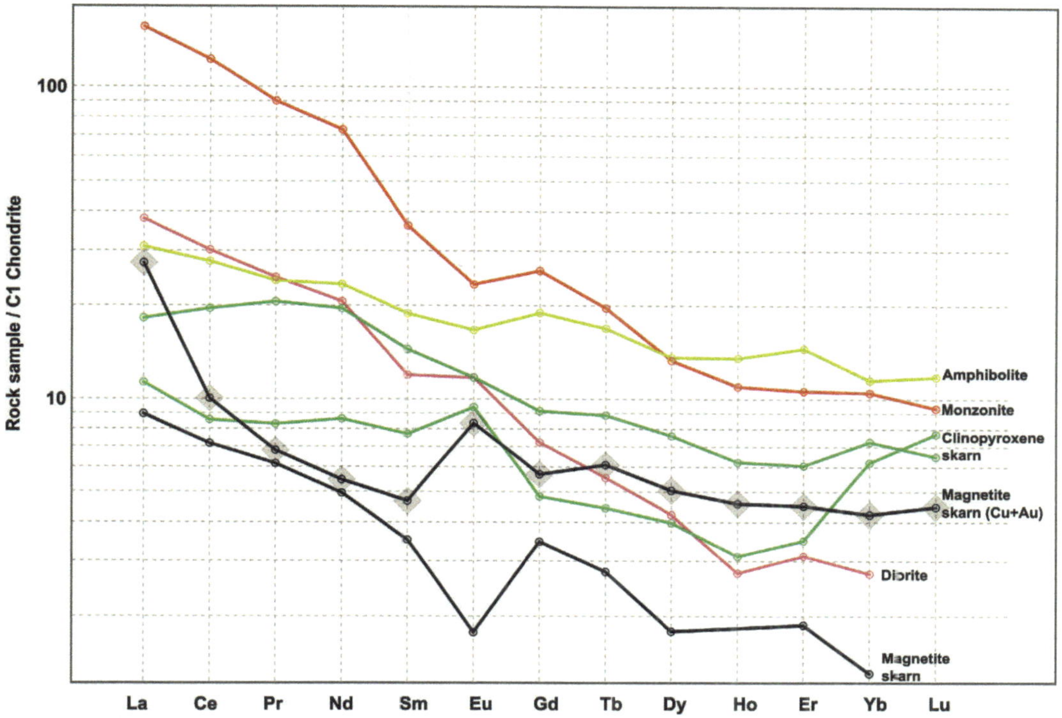

FIGURE 6.11 Examples of chondrite-normalized REE element patterns of major rock types at Hannukainen.

Red = monzonite, purple = diorite (albitized), light green = amphibolite/mafic volcanic rock (weakly albitized), green = clinopyroxene skarn (upper proximal alteration zone), black = magnetite skarn, and black with light gray squares = magnetite skarn (Cu-Au-bearing).

Source: *Normalizing values from McDonough and Sun (1995). Data from Niiranen et al. (2007).*

temperatures were in the range of 450–500 °C in the first stage and 290–370 °C in the second stage, and the pressure during both stages was between 1.5 and 3.5 kbar. Niiranen et al. (2007) and Niiranen (2005) also reported oxygen isotope compositions for oxides, silicates, and carbonates from the Laurinoja ore body. These suggest that the $\delta_{18}O$ values of the fluid phase at 500 °C are between +7.7 and +12.7 ‰ and the $\delta^{13}C$ values of the carbonates range between −3.4 and −6.9 ‰. This implies that fluids in the mineralization events were moderately to highly saline, aqueous-carbonic in composition, and that temperatures were between 290 and 500 °C, with pressure in the range of 1.5–3.5 kbar.

PROTOLITHS OF THE ORE-BEARING ROCKS

The wall rocks and host rocks of the Hannukainen Fe-Cu-Au deposit are highly altered, and hence we need to rely on immobile elements in identifying the protolith types. According to Niiranen et al. (2007), mass balance calculations indicate that Zr, Al_2O_3, and TiO_2 were immobile during the alteration event in the Hannukainen deposit. Figure 6.12 presents abundances of TiO_2 and Zr for major lithologies

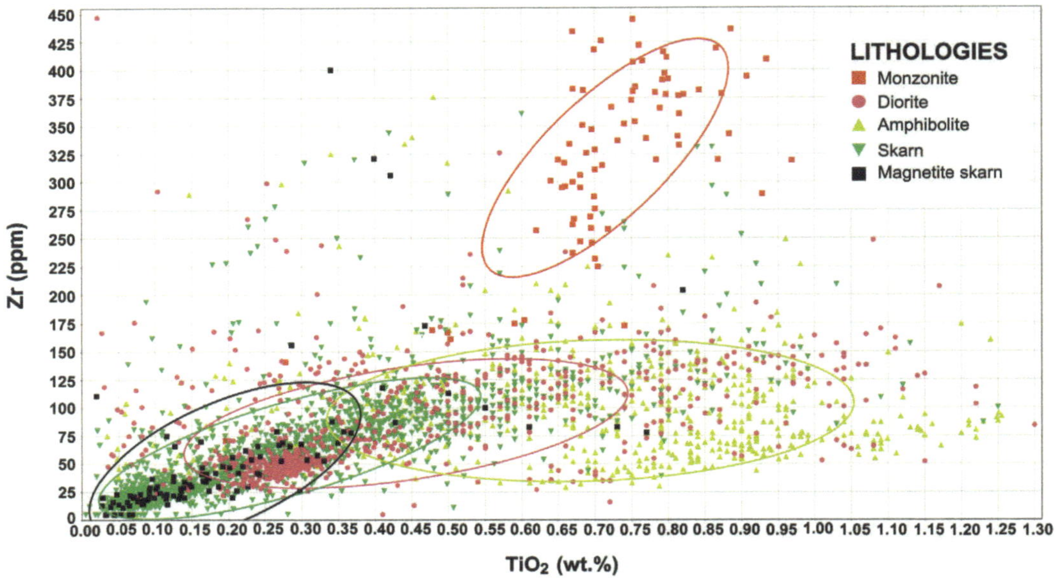

FIGURE 6.12 TiO_2 (wt.%) versus Zr (ppm) diagram for the main lithologies at Hannukainen.

of the Hannukainen deposit. It shows that magnetite skarn, skarn, diorite (strongly altered mafic metavolcanic rock), and amphibolite samples all plot along the same trend and are overlapping each other. This may indicate that these lithologies had similar protoliths (mafic volcanic rocks). On the other hand, monzonite samples plot above all other rock types and form their own distinct group. Niiranen et al. (2007) showed that there are several different protoliths for skarn rocks in the Rautuvaara-Hannukainen area. Immobile trace element ratios suggest that the dominant protolith was mafic volcanic rock. Most likely, some intercalated carbonate rocks were also present in the supracrustal sequence. They no longer exist at Hannukainen due to completion of skarn-forming decarbonation reactions, but their former presence is indicated by dolomites in other similar deposits in the region, such as Kuervitikko, Cu-Rautuvaara, Mannakorpi, and Tapuli.

Based on mass balance calculations, Kokko (2013) showed that at the Cu-Rautuvaara deposit, a similar Fe-Cu-Au deposit occurring 8 km south-southwest of the Hannukainen deposit, the protoliths for the ore-bearing skarns were also metavolcanic rocks. Ullgren (2013) came to the same conclusion for another deposit, the nearby Rautuoja Fe-Cu-Au deposit. According to Ullgren (2013), immobile element ratios in the Rautuoja deposit indicate that diorites have similar immobile element characteristics to those of the Hannukainen metavolcanic rocks. This implies a common origin for diorites and metavolcanic rocks in the Rautuoja deposit. Field evidence and chemical composition of the rocks suggest that the main protoliths for the ore-bearing skarn rocks in these deposits are metavolcanic rocks/amphibolites. Diorites in the hanging wall of the Hannukainen deposit (map shown in Fig. 6.3) are accordingly interpreted to be volcanic in origin and represent metabasalts/amphibolites altered into "diorites." Logically, along the strike of the shear zone, the diorite unit attains its maximum thickness at Hannukainen, where the strongest hydrothermal cells were located.

SUMMARY

Historically, the Fe ± Cu ± Au deposits in the Kolari-Pajala region have been classified as stratabound Ca-Mg-silicate and Mg-silicate skarn-hosted deposits that were formed by contact metasomatic events related to the intrusion of monzonites and diorites (Hiltunen, 1982), or as metamorphosed banded iron formations (e.g., Frietsch et al., 1995; Väänänen, 1998). However, as pointed out by Niiranen et al. (2007), the Hannukainen mineralization is younger than the Haparanda suite monzonite intrusions that form the hanging wall to the mineralized shear zones, and, as such, is indisputably epigenetic relative to the supracrustal sequence and plutonic rocks. The deposits comprise several bands or lenses of magnetite skarn, concordant with the surrounding skarn and sedimentary host rocks with associated Cu and Au mineralization.

The polymetallic Hannukainen Fe-(Cu-Au) deposit displays several features typical of iron oxide-copper-gold (IOCG) deposits:

- It is a magmatic-hydrothermal deposit that contains economic Cu + Au grades. The copper- and gold-rich part of the Laurinoja ore body at Hannukainen contains an average grade of 0.50 wt% Cu and 0.39 ppm Au (based on the estimation of the Laurinoja high-grade Cu ore body, above a 0.25 wt% Cu cutoff; SRK, 2014).
- The brecciated textures commonly observed in the mineralized zones indicate that magnetite is epigenetic in origin.
- Metal association Fe-Cu-Au ± Ag, Bi, Ba, Co, Mo, Sb, Se, Te, Th, U, and LREE is typical of IOCG deposits.
- Alteration styles in structurally controlled deposits define a zoning where the distal zones are characterized by sodic-potassic (Na ± Cl and K-rich) assemblages, proximal zones by calcic-iron (Ca-Fe ± CO_2-rich) assemblages, and the core, which is probably related to a structural base, is rich in magnetite (e.g., Hannukainen). In the deposits where the zoning is well developed, the Cu-Au mineralization is located within the magnetite-rich core (e.g., the Cu-Au-rich Laurinoja ore body at Hannukainen), whereas in the deposits lacking clear zoning, the Cu-Au mineralization is located in the magnetite-disseminated albitite (such as the Kuervitikko satellite deposit).
- The spatial correlation to the shear zone implies that the mineralization is structurally controlled and contains brecciated rocks.
- The deposit shows a clear temporal, but often not close spatial relationship to major magmatic intrusions.
- Although syngenetic iron formations are known in the Karelian sequence further west of the Pajala area, none are known to be present in the Kolari-Pajala district.
- Available isotopic data from the mineralization at Hannukainen suggest that the mineralization formed relatively "late," being thus consistent with an epigenetic rather than syngenetic origin.
- In the surrounding area, the Tapuli, Sahavaara, and Pellivuoma deposits are devoid of any significant Cu or Au mineralization. Nevertheless, they display some other features of an IOCG-type system. One typical feature in the known IOCG belts (for example, Cloncurry in Australia) is that numerous epigenetic magnetite and/or hematite deposits occur together with Fe-Cu-Au deposits.
- The deposit was formed from highly oxidized, saline aqueous-carbonic fluids that circulated in the system during the main alteration and mineralization event and during the subsequent brittle stage.

We interpret that Hannukainen belongs to the broad class of IOCG deposits and that most of the metals are wallrock-derived. Most likely, the Fe-Cu and Fe-Au ratios in the Hannukainen ore bodies are controlled by the wallrock background Cu and Au contents.

At Hannukainen, magnetite is stable in the skarn rocks and associated with a feldspar-actinolite-scapolite-epidote-sulfide assemblage. This assemblage is unrelated to the hematite-stable albitization. This texture-destructive albitization affects a variety of foliated and nonfoliated magmatic rocks of monzonite to diorite composition, and therefore postdates most of the magmatic activity and ductile shearing. Albite alteration is cut by the magnetite-actinolite-scapolite suite, yielding albite-actinolite rocks in the transitional zones.

The IOCG model calls for inflow of regional fluids with albitization seen as a prograde leaching alteration. During mixing with a more reduced fluid or when reacting with reduced rock, these fluids become magnetite-saturated and precipitated skarn assemblage minerals; residual dissolved metals (Cu, Au, and Na) were deposited on retrograde cooling.

For the purpose of a simple mass-balance calculation, we assume that the albitization cell around Hannukainen is approximately 3 km × 200 m × 500 m in size, corresponding to 750 Mt of rock (SG = 2.5). Because an average prealbitization Savukoski group volcanic rock contains 15.91 wt% Fe_2O_3 (Lehtonen et al., 1998), the alteration process could yield 80 Mt of dissolved Fe in hydrothermal fluids, or ~257 Mt Hannukainen-type magnetite ore (containing 31.07 wt% Fe). As the Hannukainen deposit is open at depth and has global resources of ~250 Mt at shallow levels, this mass balance seems plausible. If an average Savukoski metavolcanic rock contains 206 ppm Cu (Lehtonen et al., 1998), the same calculation would result in 155,000 t of copper metal as a total resource.

Again, this is approximately in line with the order of magnitude of Cu contained at Hannukainen (370,000 t Cu). Because there is more copper in the Hannukainen deposit than what the mass balance calculation gave, it is possible that additional copper and gold have been derived from more distal sources. The interpretation outlined here suggests that the Haparanda suite is not involved in mineralization, as also argued by Niiranen et al. (2007). Instead, the IOCG interpretation demands that the altered metasedimentary suite (Savukoski group) was once evaporite-bearing (Barton and Johnson, 1996), and perhaps was originally deposited in a rift zone located along the present Kolari-Pajala shear zone (Bothnian megashear).

ACKNOWLEDGMENTS

The authors acknowledge the contribution and effort of staff of the Northland Exploration. Eero Hanski, Murray Hitzman, and Marko Holma are thanked for useful and constructive reviews and comments, which greatly improved this chapter. We are grateful to Hugh O'Brien for his editorial work and comments. Thanks to Tim Ireland for his thoughtful comments during a site visit. We also thank Hannu Huhma for providing important geochronological information.

REFERENCES

Barton, M.D., Johnson, D.A., 1996. Evaporatic-source model for igneous-related Fe oxide-(REE)-Cu-Au-U) mineralization. Geology 24, 259–262.

Frietsch, R., Billström, K., Perdahl, J.A., 1995. Sulphur isotopes in Lower Proterozoic iron and sulphide ores in northern Sweden. Mineralium Deposita 30, 275–284.

Hanski, E., Huhma, H., 2005. Central Lapland Greenstone Belt. In: Lehtinen, M., Nurmi, P.A., Rämö, O.T. (Eds.), Precambrian Geology of Finland—Key to the Evolution of the Fennoscandian Shield. Developments in Precambrian Geology 14. Elsevier, Amsterdam, pp. 139–194.

Hiltunen, A., 1982. The Precambrian geology and skarn iron ores of the Rautuvaara area, northern Finland. Geological Survey of Finland, Bulletin 318, 133.

Koistinen, T., Stephens, M.B., Bogatchev, V., et al., 2001. Geological map of the Fennoscandian Shield, scale 1:2,000,000 (comp.) Geological Survey of Finland Special Maps 48.

Kokko, S., 2013. Ore mineralogy and host-rock geochemistry of the Cu-Rautuvaara iron oxide copper-gold deposit, Kolari, Finland. M.Sc. thesis, Department of Geosciences and Geography. University of Helsinki. 127.

Kukkonen, I., Heikkinen, P., Heinonen, S., 2009. HIRE Working Group of the Geological Survey of Finland. HIRE Seismic Reflection Survey in the Hannukainen-Rautuvaara Fe-Cu-Au exploration area, Northern Finland. Geological Survey of Finland. Report Q23/2009/50, p. 55.

Lahtinen, R., Korja, A., Nironen, M., 2005. Paleoproterozoic tectonic evolution. In: Lehtinen, M., Nurmi, P.A., Rämö, O.T. (Eds.), Precambrian Geology of Finland—Key to the Evolution of the Fennoscandian Shield. Developments in Precambrian Geology 14. Elsevier, Amsterdam, pp. 481–532.

Lehtonen, M., Airo, M.-L., Eilu, P., et al., 1998. Kittilän vihreäkivialueen geologia. Lapin vulkaniittiprojektin raportti. English summary: The stratigraphy, petrology and geochemistry of the Kittilä greenstone area, northern Finland: a report of the Lapland Volcanite Project. Geological Survey of Finland, Report of Investigation 140, 144.

LKAB, 2012. Annual Report. Available at http://www.lkab.com/Global/Documents/Finansiella%20rapporter/Annual%20Report%20eng/LKAB_2012_Annual_and%20Sustainability_Report.pdf.

McDonough, W.F., Sun, S.-S., 1995. The composition of the Earth. Chemical Geology 120, 223–253.

Mining Magazine1982. Rautaruukki and Hannukainen Mines. pp.101–110.

Mutanen, T., Huhma, H., 2001. U-Pb geochronology of the Koitelainen, Akanvaara and Keivitsa layered intrusions and related rocks. Geological Survey of Finland. Special Paper 33, 229–246.

Niiranen, T., 2005. Iron oxide-copper-gold deposits in Finland: Case studies from the Peräpohja schist belt and the Central Lapland Greenstone Belt. University of Helsinki, Publications of the Department of Geology D6. p. 27.

Niiranen, T., Poutiainen, M., Mänttäri, I., 2007. Geology, geochemistry, fluid inclusion characteristics, and U-Pb age studies on iron oxide-Cu-Au deposits in the Kolari region, northern Finland. Ore Geology Reviews 30, 75–105.

Offerberg, J., 1967. Beskrivning till berggrundskartbladen Kiruna NV, NO, SV, SO (Explanation to the geological map sheets Kiruna NW, NE, SW, SE) Sveriges geologiska undersökning. Serie Af 1–4, 146 (in Swedish).

SRK Consulting (U.K.) Limited (SRK), 2014. Technical Report on the Hannukainen iron-copper-gold project, Kolari district, Finland, January, prepared for Northland Resources S.A. Available at http://sedar.com

Ullgren, A., 2013. Hydrothermal alteration and host rock characteristics of the Rautuoja iron oxide-copper-gold deposit. M.Sc. thesis, Department of Geosciences and Geography, University of Helsinki, p. 52.

Väänänen, J., Lehtonen, M., 2001. U-Pb isotopic age determinations from the Kolari-Muonio area, western Finnish Lapland. In: Vaasjoki, M. (Ed.), Radiometric age determinations from Finnish Lapland and their bearing on the timing of Precambrian volcano-sedimentary sequences. Geological Survey of Finland. Special Paper 33, 85–93.

Väänänen, J., 1998. Explanation to the maps of pre-Quaternary rocks of the Kolari and Kurtakko map-sheet areas. Geological Survey of Finland Espoo.

PART

7

THE VIHANTI-PYHÄSALMI VMS BELT

7

T. Mäki, M. Imaña, J. Kousa, J. Luukas

ABSTRACT

The Vihanti-Pyhäsalmi belt contains the most important volcanogenic massive sulfide (VMS) deposits in Finland. The volcano-sedimentary host rocks belong to a 1.93–1.92 Ga island arc that occurs along the northwestern parts of the Raahe-Ladoga shear complex. The stratigraphy in the region is divided into two units, a lower bimodal volcanic unit conforming to the Pyhäsalmi group and an upper volcano-sedimentary association defined as the Vihanti group. The U-Pb zircon ages of the two groups are similar.

The Vihanti-Pyhäsalmi belt hosts two large VMS deposits, Pyhäsalmi in the Pyhäsalmi group and Vihanti in the Vihanti group, as well as a number of small sulfide deposits located around them. The Pyhäsalmi deposit is hosted by rhyolitic and basaltic volcanic rocks, contains 75.7 Mt of pyritic mineralization grading 0.9% Cu, 1.9% Zn, 0.4 g/t Au, and 14.1 g/t Ag, and hosts the only currently producing mine (2014) in the belt. The Vihanti mine operated from 1954–1992, producing 28 Mt at 5.12% Zn, 0.48% Cu, 0.36% Pb, 25 ppm Ag, and 0.49 ppm Au. The Vihanti deposit and three minor satellite mineralizations are hosted in a sequence dominated by intermediate and felsic metavolcanic rocks.

Keywords: stratigraphy; VMS-deposit; Paleoproterozoic; Svecofennian; Pyhäsalmi; Vihanti; Finland.

Mineral Deposits of Finland. http://dx.doi.org/10.1016/B978-0-12-410438-9.00020-0

REGIONAL GEOLOGY OF THE VIHANTI-PYHÄSALMI BELT

The bedrock of the Vihanti-Pyhäsalmi belt forms the easternmost part of the Paleoproterozoic Svecofennian domain, bordered by Archean basement complexes (3.1–2.6 Ga) in the east and the Central Finland Granitoid Complex and related supracrustal formations (1.93–1.87 Ga) of the Svecofennian domain in the west (Fig. 7.1). The area belongs to the northwest–southeast trending Raahe-Ladoga zone (RLZ) that represents the collisional border of Paleoproterozoic island arcs against Archean terranes (Korsman et al., 1997). The RLZ is considered the main sulfide ore belt in Finland (Kahma, 1973). Table 7.1 shows the production data of the most important deposits.

The Vihanti and Pyhäsalmi volcanogenic massive sulfide (VMS) deposits are genetically related to the Paleoproterozoic island arc that formed in an ocean located on the western side of the Archean basement complex 1.93–1.92 Ga ago (Mäki, 1986; Gaál and Gorbatschev, 1987; Ekdahl, 1993; Lahtinen, 1994; Kousa et al., 1994; Roberts et al., 2004 Eilu et al., 2013). Supracrustal rocks along the RLZ are mainly turbiditic metasedimentary rocks that locally contain abundant graphite-bearing schists and minor mafic metavolcanic interlayers. This sequence is defined as the Näläntöjärvi suite by Laajoki and Luukas (1988) and it has been proposed that it is a depositional basement for the volcanic rocks of the island arc (Kousa, 1997). The volcanic arc-related rocks of the Vihanti-Pyhäsalmi belt contain several separate volcanic centers, which now appear in narrow and discontinuous belts in the western part of the Näläntöjärvi suite (Fig. 7.1). The narrow volcanic configuration is the result of intrusions of voluminous postvolcanic Svecofennian plutons. The bimodal metavolcanic rocks of the Pyhäsalmi group have distinct lithogeochemical signatures on which basis they can be separated from the 1.89–1.88 Ga Ylivieska-type metavolcanic rocks in the west (Mäki, 1986; Rasilainen and Västi, 1989; Kousa et al., 1994). The regional metamorphic grade in the Vihanti-Pyhäsalmi belt is mainly from lower amphibolite facies to upper amphibolite facies. Locally, for example, in the Kiuruvesi area, the metamorphic grade is higher and reaches granulite facies (Korsman et al., 1997).

The volcanic belt appears separated into two blocks by major southwest–northeast and southeast–northwest trending faults, namely the Oulujärvi shear zone (Kärki et al., 1993) and the Revonneva and Ruhaperä shear zones (Luukas, 1997) (Fig. 7.1). Both blocks show distinct lithological features that have been used to define the Vihanti and Pyhäsalmi lithological associations. In the Vihanti block, intermediate and felsic metavolcanic rocks are the dominant protoliths but mafic metavolcanic rocks are minor and sparsely distributed. In the Pyhäsalmi-Kiuruvesi block, a wider range of volcanic compositions exists with a bimodal felsic–mafic metavolcanic association being the most characteristic. We propose a schematic informal lithostratigraphical classification for the Vihanti and Pyhäsalmi groups (Fig. 7.1). This interpretation should be regarded as tentative because of the sparse outcrop relations of the belt and overlapping dating results of the metavolcanic rocks of both groups. Nevertheless, these two different lithological associations indicate at least two different volcanic environments despite their similar age of 1.93–1.92 Ga (Kousa et al., 2013). The Pyhäsalmi group represents a lower bimodal felsic-mafic metavolcanic sequence over which the Vihanti group, a series of felsic and intermediate metavolcanic rocks with calc-silicate and graphite tuffite interlayers, was deposited during continuous evolution of the arc. This stratigraphic succession is based on observations in the Pyhäsalmi-Kiuruvesi-Pielavesi area where the oldest Pyhäsalmi group rocks are exposed in antiformal dome centers and the Vihanti group rocks are exposed in synformal basins between the antiforms.

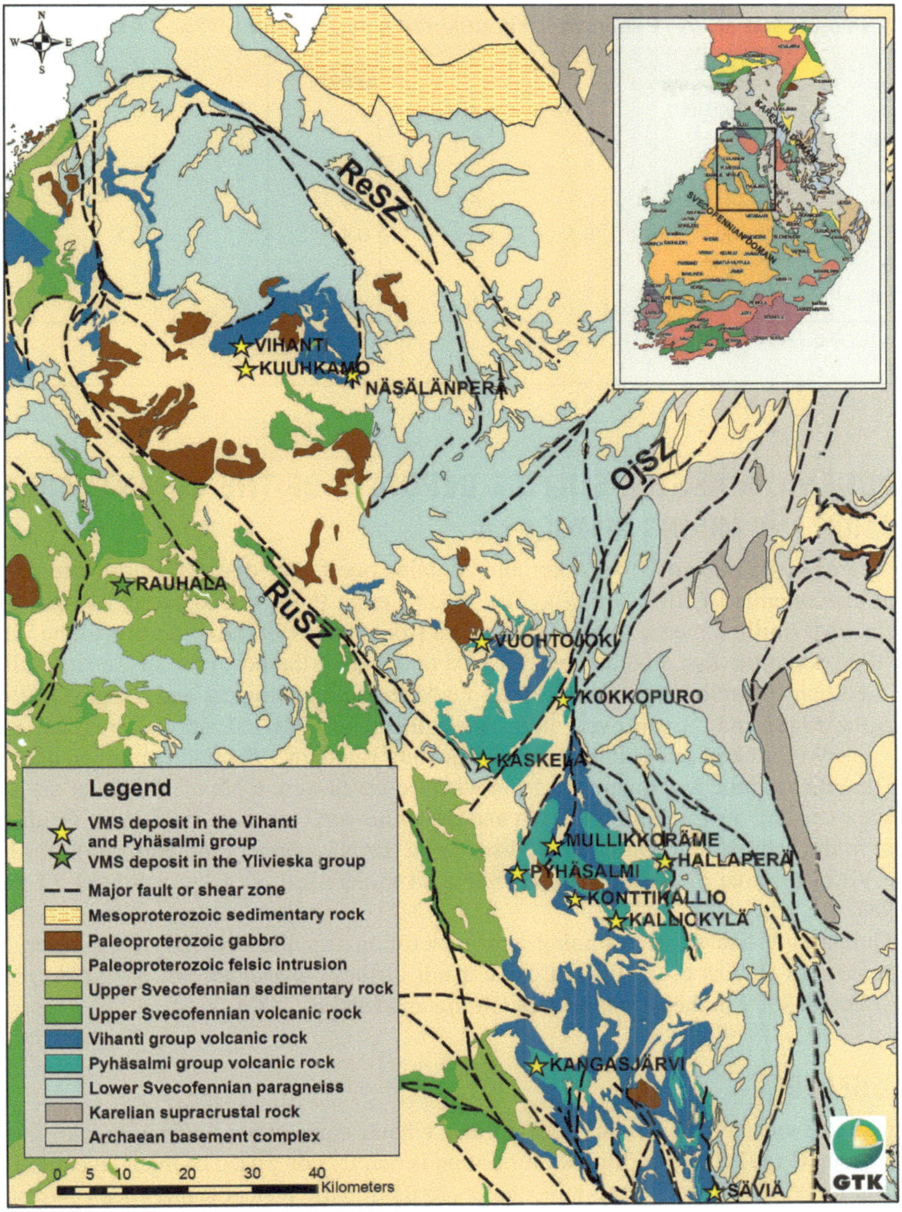

FIGURE 7.1 General geology of the northwestern part of the Raahe-Ladoga zone.

The most important Zn-Cu deposits of the Vihanti and Pyhäsalmi area are shown with stars. The main structural features are represented by black dashed lines (OjSZ = Oulujärvi shear zone, ReSZ = Revonneva shear zone, RuSZ = Ruhaperä shear zone).

Source: *The base map is modified after Bedrock Map of Finland–DigiKP (2013).*

Table 7.1 Production of ore, Pyhäsalmi, Vihanti, and satellite mines, 1962–2012

Deposit	Production (t)	Cu%	Zn%	Pb%	S%	Au g/t	Ag g/t
Pyhäsalmi[a]	58,500,000	0.94	2.38		38	0.4	14
Kangasjärvi[b]	80,000	0.1	5.4		38		
Ruostesuo[b]	240,000	0.4	2.7		30	0.3	10
Mullikkoräme[b]	1,148,000	0.3	6.1	0.9		1	45
Vihanti[b]	28,100,000	0.5	5.1	0.4		0.5	25
Kuuhkamo	150,000		4				
Näsälänperä	100,000		2				15

[a]*Production and reserves*
[b]*Production*

VOLCANOGENIC MASSIVE SULFIDE DEPOSITS OF THE VIHANTI AREA

EXPLORATION AND MINING HISTORY

The Vihanti area covers about 10 × 20 km and contains one major massive sulfide ore deposit with three small uneconomic satellite deposits. The Vihanti mine (known also as the Lampinsaari mine) produced 28.1 Mt (million metric tons) of sulfide ore from 1954–1992. Exploration in the Vihanti area started in 1936 when the first studies were focused on locating the host rock of Zn-rich sulfide boulders. The Geological Survey of Finland (GTK) made the first geological and geophysical maps and drillings in the 1940s and the discovery holes were drilled in 1946 at Lampinsaari. The Outokumpu mining company (Outokumpu) obtained the prospecting rights in 1951 and continued exploration in the area. The Vihanti mine was opened in 1954 when reserves were 6 Mt of 12.5% Zn ore. An underground mine extended down to 800 m depth and the mine was closed in 1992. As a result of intense exploration in the nearby area, Outokumpu discovered the Näsälänperä mineralization further southeast in the Vilminko area in 1978 and the Kuuhkamo mineralization 4 km southeast of the Vihanti mine in 1987.

Once Outokumpu had stopped mining and exploration activities in the Vihanti mine, the GTK restarted geological studies in the district. As a result of these investigations, the Kokkoneva sulfide body was found in 1999 in the Vilminko area (Kousa and Luukas, 2004).

GENERAL GEOLOGY

The 1.92 Ga metavolcano-sedimentary rocks of the Vihanti area belong to the Vilminko formation of the Vihanti group (Fig. 7.2). The Vilminko formation is mainly composed of thick layers of intermediate and felsic metavolcanic rocks and minor interstratified mafic metavolcanic rocks. A distinct calc-silicate horizon is intercalated with dolomites and graphite-bearing felsic metatuffs forming a continuous layer, known as the U-P horizon in reference to the uranium and phosphorous trace components (Rehtijärvi et al., 1979). This intercalation, known also as the Lampinsaari association (Rouhunkoski, 1968), is the key horizon for the massive sulfide deposits in the Vihanti area. The electromagnetic properties of the graphite and sulfide components of this horizon make it easy to delineate on geophysical maps. A massive graywacke sequence forms the uppermost lithological unit

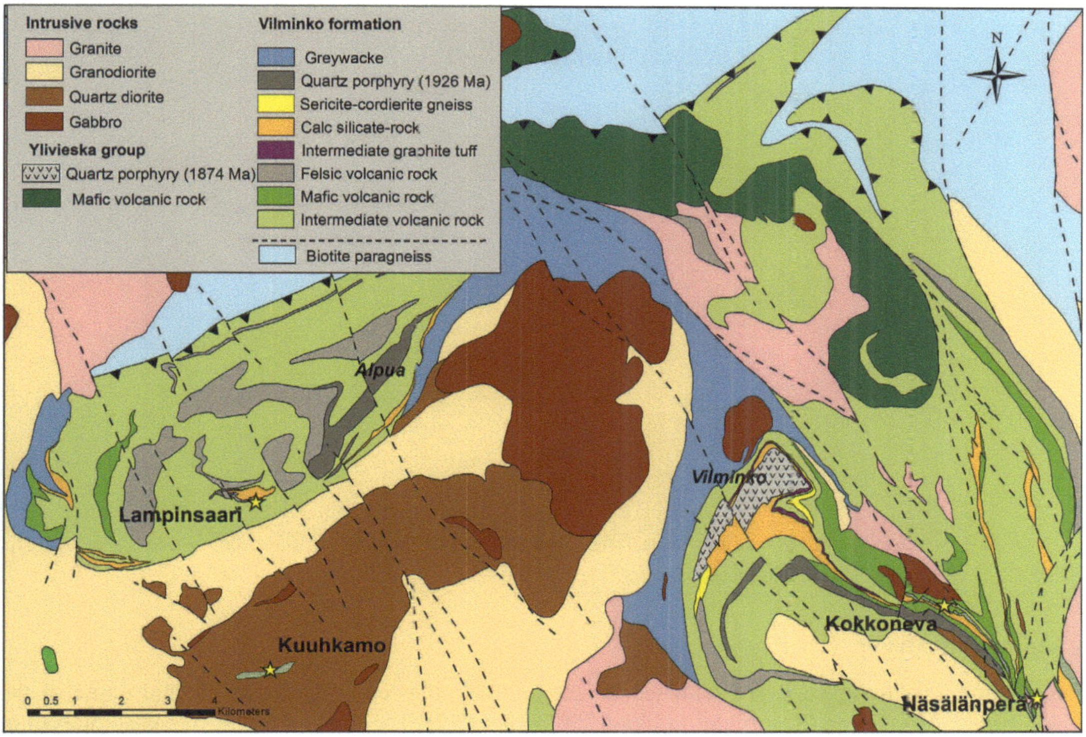

FIGURE 7.2 Geology of the Vihanti area.

Source: *The map is modified after Bedrock of Finland–DigiKP (2013).*

of the Vilminko formation. A younger sequence of volcanic rocks and volcaniclastic sediments (Ylivieska group) were deposited on the Vihanti group rocks at 1.89–1.87 Ga ago (Rasilainen and Västi, 1989; Kousa and Luukas, 2007).

The magmatic history of the Vihanti area covers a timespan from approximately 1.93–1.87 Ga (Vaasjoki and Sakko, 1988; Kousa et al., 2013). In the volcanic sequence, two different porphyry sills have been dated. A quartz-plagioclase porphyry sill (the Kokkoneva porphyry) gives an age of 1926 ± 6 Ma, which is similar to ages of the metavolcanic rocks in the Pyhäsalmi area (Kousa et al., 2013). This sill is identified in the vicinity of the Vihanti, Kuuhkamo, and Vilminko mineralizations. At all locations, the Kokkoneva porphyry is completely or partly altered and contains a remarkable amount of disseminated pyrite. These metamorphosed altered rocks are expressed as quartz rocks or cordierite-sillimanite-sericite rocks. It is suggested that a genetic connection exists between this porphyry type and VMS ore forming processes in the Vihanti area (Nikander et al., 2002; Luukas and Kousa, 2013). A younger quartz porphyry sill (1874 ± 3 Ma) intruded the Vilminko formation in the Vilminko area (Kousa and Luukas, 2007).

Structural interpretations indicate that metavolcanic rocks of the Vilminko formation form an open anticline in the Vihanti mine area and an open syncline in the Näsälänperä area. The stratigraphic model (Fig. 7.3) of the Vihanti group is based on the lithological sequences from the Vilminko area.

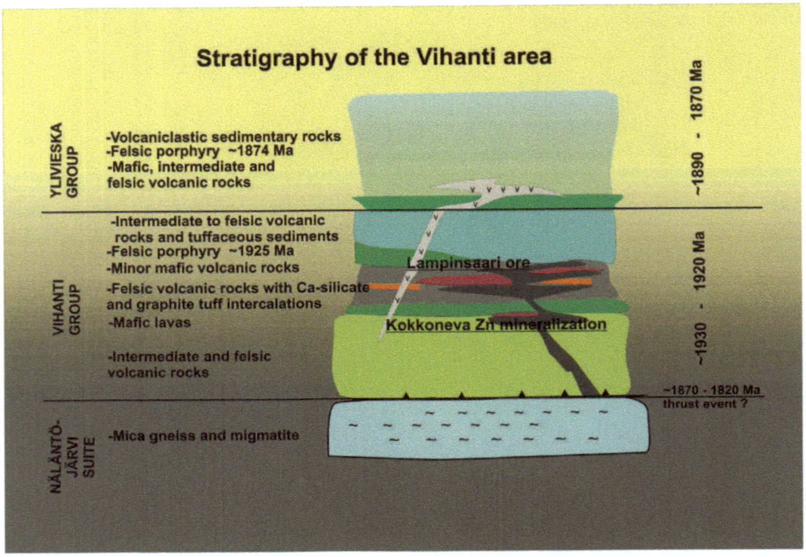

FIGURE 7.3 Stratigraphy of the Vihanti area.

VIHANTI DEPOSIT

The Vihanti Zn-Pb-Ag deposit is situated on the southwestern edge of the antiformal dome of the Lampinsaari-Alpua area (Kousa and Luukas, 2004) (Fig. 7.2). The volcanic sequence is composed of felsic to intermediate metavolcanic rocks with minor mafic metavolcanic rocks. The sequence is intruded by porphyritic granitoids and gabbros to the south. Sulfide mineralization is focused along a more than 100-m-thick and 1500-m-long sequence of altered felsic metavolcanic rocks, calc-silicate rocks, dolomites, and graphite-bearing metatuffs (U-P horizon). The volcanic stratum is cut at a low angle by a 50–100-m-thick altered quartz porphyry sill, causing alteration in the vicinity of Lampinsaari, identified now as cordierite-altered volcanic beds. Mineralized incompetent calcareous rocks and massive sulfide ore lenses have been intensely folded on the southern edge of the Lampinsaari anticline (Fig. 7.4). The lower part of the ore (between +350 and +800 mine levels) forms a fold limb that dips about 45° southward. The general fold axis dips south–southwest.

There are 5 different ore types in the Vihanti deposit and about 20 separate ore bodies that have been mined out. The metal content and size varies greatly between different bodies (Rouhunkoski, 1968). The most important ore types are the Zn ores of which the three largest (Ristonaho, Välisaari, and Lampinsaari) contained about 75% of the total mined ore (Autere et al., 1991). Minor Zn ore bodies were mined out along the southern fold limb down to the +800 mine level. The Zn-ores typically occur in tectonically thickened calc-silicate-dolomite-serpentinite beds in the upper levels of the mine. Although Zn-ores contain some Ag and Au, a separate disseminated Pb-Ag ore type occurs in close connection to the Zn-ores. The best Pb-Ag ores were found in the western part of the mine. Pyrite- or pyrrhotite-rich ores, which are situated in the stratigraphically upper parts of the mineralized horizon, are termed *pyrite ores*. The Hautaräme and Hautakangas ore bodies on the southeasterly dipping limb area are massive pyrite ores hosted by felsic metavolcanic and calc-silicate rocks. Cu ores occur in the felsic metavolcanic rocks. They are disseminated ore types and are closely related to the pyrite ores.

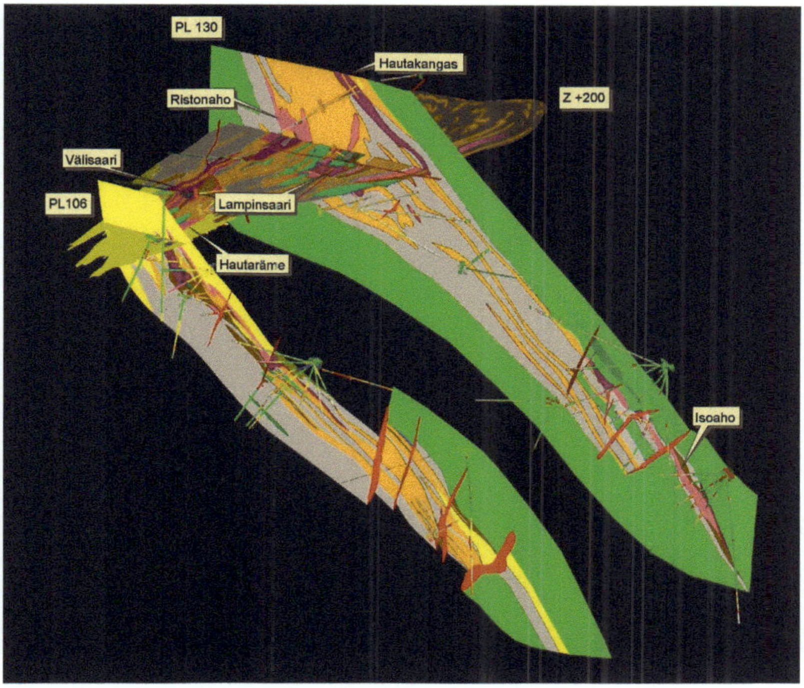

FIGURE 7.4 Geology of the Vihanti mine and its main ore bodies.

Viewing direction is from the south. Rock types: green = intermediate metavolcanic rocks; gray = calc-silicate banded felsic metavolcanic rock; orange = calc-silicate rock, dolomite; lilac = ore; yellow = cordierite-sillimanite gneiss; red = younger dikes. Distance between cross sections 106 and 130 is 600 m.

Source: *Mine level +200 is modified after Rouhunkoski (1968).*

The fifth separate ore type is uranium-phosphorous ore, which occurs as a discontinuous layer in the upper part of the U-P horizon along the southeast-dipping limb area. This unexploited ore type is hosted by banded calc-silicate bearing felsic metavolcanic rocks (Rehtijärvi et al., 1979). The underground production of the Vihanti mine was 28 Mt ore, which contained 5.1% Zn, 0.5% Cu, 0.4% Pb, 25 ppm Ag, and 0.5 ppm Au.

KUUHKAMO MINERALIZATION

In the 1950s, Outokumpu located a Lampinsaari-type lithological sequence at a distinct EM-anomaly located 4 km south of the Vihanti mine. However, the Kuuhkamo zinc mineralization was not discovered until the 1980s. The Kuuhkamo mineralization resembles the Vihanti deposit lithologically. It occurs as a small window of graphite-bearing intermediate metavolcanic rock inside a quartz diorite intrusion (Fig. 7.2). Structurally, the Kuuhkamo mineralization forms a tight east–west trending anticline with a hinge zone that is exposed at the surface (Fig. 7.5). The fold axis plunges westward at an angle of 40°. The total resources of the Kuuhkamo deposit is estimated to be less than 0.25 Mt, of which 0.15 Mt at 4% Zn is in the main lode (http://tupa.gtk.fi/karttasovellus/mdae/raportti/525_Kuuhkamo.pdf, 5.5.2015; Nikander et al., 2005a).

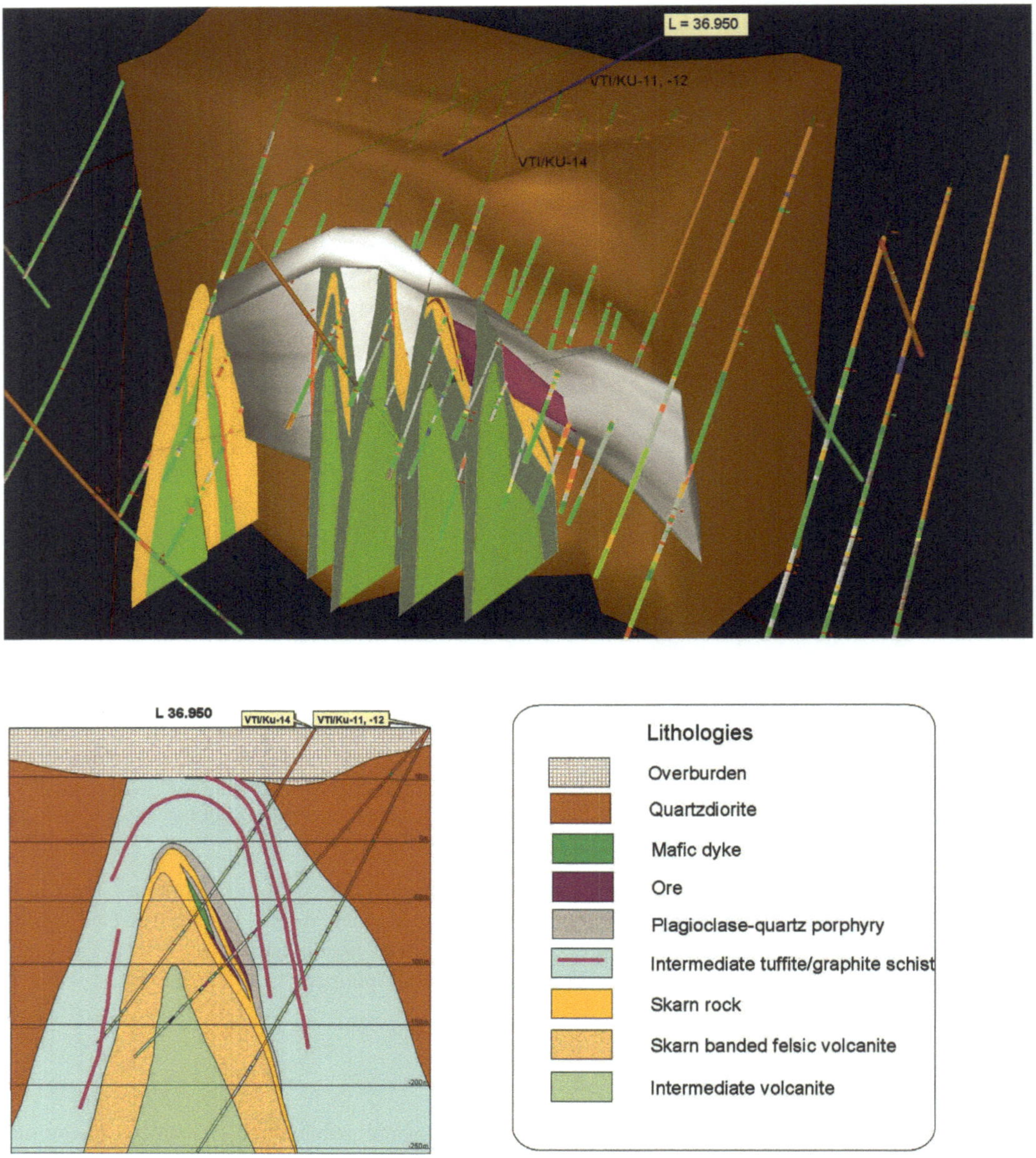

FIGURE 7.5 Geology of the Kuuhkamo deposit and profile.

L = 36.950.

The lowermost part of the volcanic pile consists of intermediate metavolcanic rocks, which are followed by a 10–50-m-thick horizon of felsic metavolcanic rocks, calc-silicate rocks, serpentine dolomites, and minor graphite-bearing metatuffs (U-P horizon). This horizon is covered by a new layer of intermediate metavolcanic rocks and graphite-bearing metatuffites, which are the source of strong electromagnetic anomalies. This volcanic sequence is cut by several 5–20-m-thick quartz porphyry sills that are similar to the Lampinsaari porphyry. The volcanic strata were folded and then cut by voluminous quartz diorite intrusions, the extent of which is also outlined from interpretation of seismic data (Heinonen, 2013).

Mineralization is mainly found on the southern limb of the antiform. The sulfide ore is related to a synvolcanic porphyry sill, a medium-grained quartz plagioclase porphyry that intruded between the U-P horizon and the overlying intermediate metatuffites. In the proximity of the mineralization, the porphyry has been altered and metamorphosed to a cordierite gneiss analogous to that of the Lampinsaari deposit.

MINERALIZATIONS AT VILMINKO-NÄSÄLÄNPERÄ AREA

Two minor sulfide mineralizations are situated in the Vihanti formation in the Vilminko-Näsälänperä area (Fig. 7.2). Lithologically the area resembles the Lampinsaari-Alpua area in the west. Outokumpu located a small zinc mineralization at Näsälänperä in 1978. The highly sheared mineralization is composed of two narrow pyrrhotite-sphalerite bearing ore lodes hosted by calc-silicate bearing felsic metavolcanic rocks. The ore lodes are 1–2 m thick and about 200 m long. The estimated resource is 100,000 t of ore, which contains about 2% Zn and 15 ppm Ag (http://tupa.gtk.fi/karttasovellus/mdae/raportti/528_Näsälänperä.pdf, 5.5.2015).

Another minor sulfide mineralization is situated 2.5 km northwest of Näsälänperä at the southwest limb of the Peuraneva syncline. This Kokkoneva mineralization was found during the GTK's regional studies in the Vihanti area in the late 1990s. Sulfide mineralization is located between the Kokkoneva quartz porphyry sill and mafic metavolcanic rock (Fig. 7.3). According to the stratigraphic model, the subvolcanic porphyry sill has intruded along a contact zone between intermediate and mafic volcanic rock below the calc-silicate bearing U-P horizon. In this respect, the Kokkoneva mineralization represents an example of mineralization under the typical U-P horizon. The total length of the sill is approximately 8 km and the thickness less than 100 m. The upper part of the porphyry sill is altered and metamorphosed to quartz K-feldspar or cordierite-sillimanite rock. The mineralization is mainly disseminated, but breccia ore textures also exist. Typical ore minerals are pyrrhotite, pyrite, sphalerite, chalcopyrite, and magnetite. The mineralization is about 250 m long and at ground surface the thickness is 1–2 m. The wedge-like ore body is cut by faults at 30–40 m depth. The ore is interpreted to have formed by hydrothermal replacement of volcanic strata adjacent to the felsic porphyry sill (Nikander et al., 2002; Nikander et al., 2005b).

ORE MODEL

Rouhunkoski (1968) explained cordierite gneiss and some calc-silicate rocks in Lampinsaari mine as products of extensive Mg-metasomatism by the ore-forming solutions. Recent studies show that an altered quartz porphyry sill, now metamorphosed to cordierite-sillimanite rock, has an important role in the ore-forming processes (Nikander et al., 2002). A two-stage ore-forming process could explain the diversity of ore types in the Vihanti area. The first stage represents primary sulfide ore formation by hydrothermal processes at the seafloor. In the Vihanti area, such primary sulfide deposits could be located in the upper part of the felsic metavolcanic to calc-silicate rock sequence (e.g., Hautaräme and Hautakangas lodes). During the second stage, a felsic subvolcanic sill intruded into mineralized horizons and in some way facilitated the formation of replacement ores in the nearby thick calc-silicate rocks.

VOLCANOGENIC MASSIVE SULFIDE DEPOSITS OF THE PYHÄSALMI AREA

EXPLORATION AND MINING

The Pyhäsalmi area and the surrounding regions have been under intense exploration since the 1950s, mainly by Outokumpu and GTK. The Pyhäsalmi deposit was found in August of 1958 by a local farmer who was digging a well in his yard in the Ruotanen village of the Pyhäjärvi municipality. Soon after discovery, Outokumpu acquired the mineral rights from the family and commenced exploration. The first drilling intersection was 38.5 m of massive sulfide grading 5.7% Zn, 0.9% Cu, and 48% S. The first estimation of reserves in February 1959 indicated 12.2 Mt of ore with 0.81% Cu, 2.93% Zn, and 36.8% S. Mine construction started in August 1959 and production from an open pit commenced on March 1, 1962. Since 1976 all production has been from the underground mine.

The mine has been deepened gradually and reached a depth of 1000 m in 1996 (Fig. 7.6). A new ore body was found in December 1996 and the decision to build a new mine was made in 1998. Production via a new shaft commenced on July 1, 2001. The Canadian mining company Inmet Mining Corporation

FIGURE 7.6 Aerial and axonometric view of the Pyhäsalmi mine and location map.

acquired the Pyhäsalmi mine in 2002, and in 2013 another Canadian company, First Quantum Minerals, took over Inmet Mining Corporation and also the Pyhäsalmi mine. The Pyhäsalmi mine is one of the most efficient underground mines in its size class. The deepest point of the mine is 1444 m below surface, making it the deepest base metal underground mine in Europe as of 2014.

During the exploration period after the initial opening of the Pyhäsalmi mine, several small massive sulfide deposits (Kalliokylä or Ruostesuo, Kangasjärvi, Hallaperä, Vuohtojoki, and Kokkopuro) were found in the Pyhäsalmi area (Fig. 7.1). The first satellite mine was the Kangasjärvi mine (Fig. 7.1) in Keitele, which GTK discovered in 1964. The second satellite mine was Ruostesuo (Kalliokylä) about 16 km southeast of Pyhäsalmi. The Ruostesuo deposit was discovered by Outokumpu in 1959 and was in production from 1988–1989. The Mullikkoräme deposit (Figs. 7.1 and 7.8), located 10 km to the northeast of Pyhäsalmi, was discovered by Outokumpu in 1987. This satellite mine was in production from 1989–1990 and again from 1996–2000. Three other satellite deposits in the Pyhäsalmi area are the Hallaperä deposit, the Vuohtojoki deposit, and the Kaskela deposit (Fig. 7.1). These massive sulfide deposits have been uneconomic and have not been mined (Ekdahl et al, 1997).

Exploration has continued around the Pyhäsalmi deposit and two new mineralizations have been found: Konttikallio between Pyhäsalmi and the Ruostesuo deposit, and Kokkopuro located 25 km north of Pyhäsalmi (Fig. 7.1). Both are small and uneconomic.

PRODUCTION AND RESOURCES

Total production at Pyhäsalmi since the beginning of mining has been 51 Mt (end of 2013) grading 0.92% Cu, 2.45% Zn, and 37.4% S. The ore also contains, on average, 0.4 g/t Au and 14 g/t Ag. Along with production from the Pyhäsalmi mine, three satellite deposits have also been processed in the

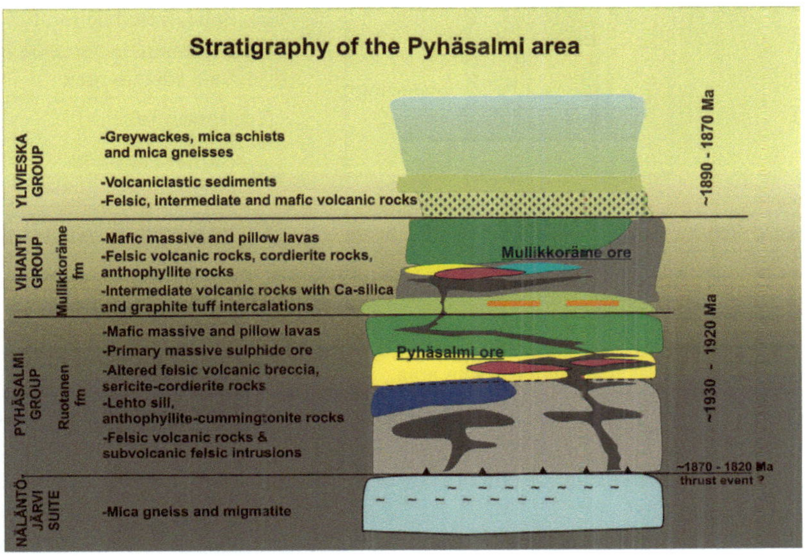

FIGURE 7.7 Stratigraphy of the Pyhäsalmi area.

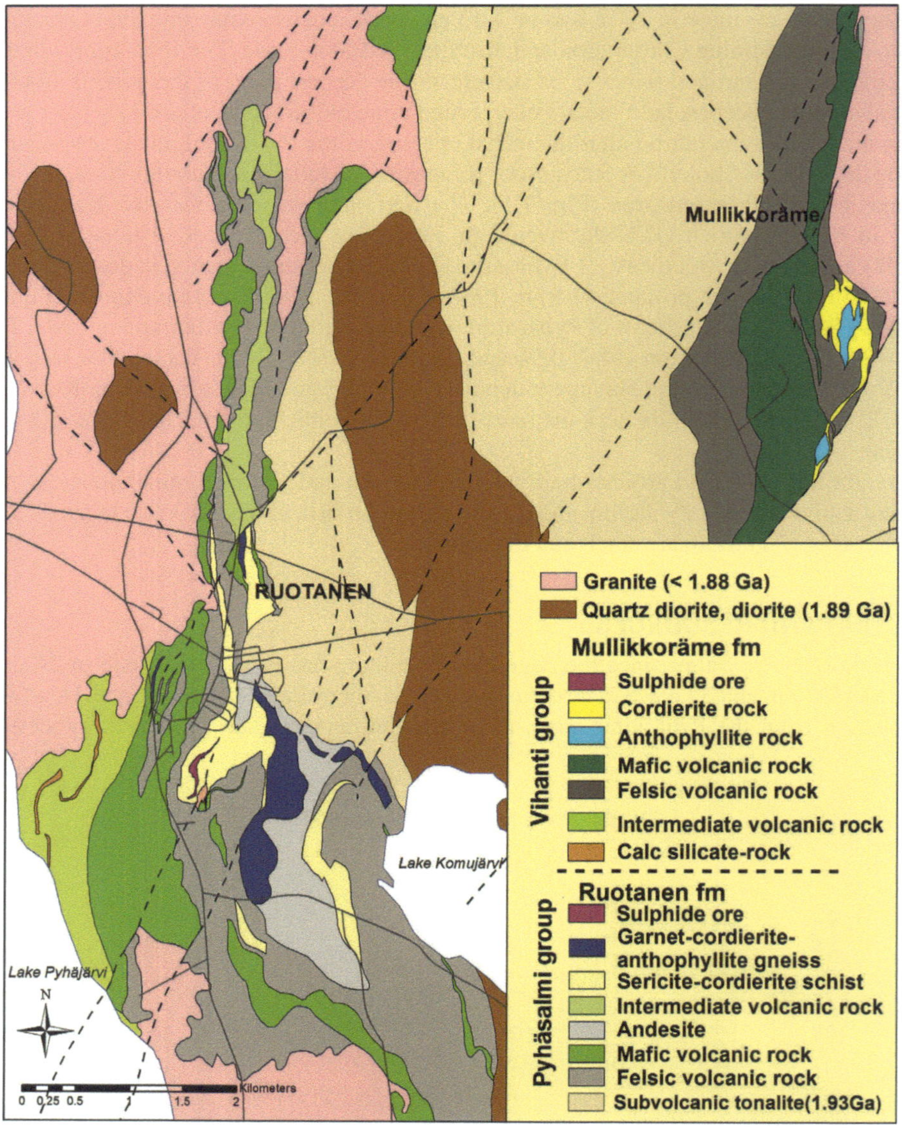

FIGURE 7.8 Geology of the Pyhäsalmi-Mullikkoräme area.

Source: *Map is modified after Bedrock of Finland–DigiKP (2013).*

Pyhäsalmi mill. Ore and concentrate production from Pyhäsalmi and its satellites are shown in Tables 7.1 and 7.2.

Mining of the Pyhäsalmi ore has been challenging because of high horizontal rock stress, which is two times greater than the vertical stress. Amphibolite facies regional metamorphism has led to sulfide recrystallization and formation of very coarse grained, strained, and deleterious-free ore assemblages.

Table 7.2 Production of concentrates and metals, 1962–2012

Product	
Copper concentrate	1,681,627 t
Copper metal	429,629 t
Zinc concentrate	2,307,459 t
Zinc metal	1,205,109 t
Pyrite concentrate	27,138,657 t
Au in copper concentrate	6987 kg
Ag in copper concentrate	349,431 kg

The Pyhäsalmi mine is also a very important pyrite producer and delivers 800,000 t of pyrite concentrate annually (Mäki, 2013). Annual ore hoisting is about 1.4 Mt. According to the latest mineral reserve estimation, the mine will operate until 2019 (Mäki, 2013). New resource exploration inside the mine lease is in progress.

REGIONAL GEOLOGY

The 1.93–1.92 Ga bimodal metavolcanic rocks of the Pyhäsalmi region belong to the Pyhäsalmi group, which is inferred to be the lowermost exposed volcanic unit in the Svecofennian domain. The depositional basement of the Pyhäsalmi group is unknown. The slightly younger or coeval Vihanti group with intermediate to felsic metavolcanics, minor calc-silicate–dolomite interlayers and minor mafic intercalations is deposited on the top of Pyhäsalmi type rocks. A schematic stratigraphic model of the Pyhäsalmi area is presented in Fig. 7.7.

In the present configuration, the metavolcanic rocks of the Pyhäsalmi group are situated in antiformal domes, whereas the overlying metavolcanic and metasedimentary rocks of the Vihanti group occupy synformal basins. A specific intrusion type of subvolcanic tonalite that is situated in the antiformal structure is closely related to the 1.93–1.92 Ga volcanic event. Later in the geological evolution, voluminous Svecofennian intrusive rocks (1.89–1.87 Ga) penetrated the volcanic sequences. The metavolcanic rocks of the Pyhäsalmi group in the Pyhäsalmi mine area belong to the Ruotanen formation (Puustjärvi, 1999). According to the inferred stratigraphy, a voluminous pile of felsic metavolcanic rocks form the lowermost part of the formation. The pinkish gray Na-rich quartz plagioclase-phyric rhyolitic rocks contain abundant mafic dikes.

The Kettuperä gneiss unit (sample A0751 in Helovuori, 1979) in the Pyhäsalmi group is interpreted here as the oldest felsic volcanic member of the Ruotanen formation. A new zircon U-Pb dating gives an age of 1924 ± 3 Ma to this sample, which is considered as a reliable age for the volcanism of the Pyhäsalmi group (Kousa et al., 2013). This unit is overlain by voluminous felsic breccias that are strongly altered and metamorphosed into cordierite-sericite schists. This pyroclastic unit is the host for the main sulfide ore body and presents abundant pyrite dissemination. The felsic metavolcanic rocks are overlain by massive mafic metalavas, which form the uppermost part of the Ruotanen formation. Intermediate metavolcanic rocks with minor calc-silicate interlayers (Vihanti group) overlie the Ruotanen formation.

In the Mullikkoräme area, about 8 km northeast of the Pyhäsalmi mine, the felsic and mafic metavolcanic rocks are named the Mullikkoräme formation, and are proposed to be representatives of the

Vihanti group in the Pyhäsalmi region (Bedrock of Finland–DigiKP, 2013). The zircon U-Pb age of 1921 ± 2 Ma from the Riitavuori metarhyolite and the age of 1925 ± 4 Ma from a metarhyolite at the Mullikkoräme mine area are both from the Mullikkoräme formation. These results indicate that both the Pyhäsalmi group and the Vihanti group were both formed within a short period of only a few million years (Kousa et al., 2013).

The Pyhäsalmi Deposit

The Pyhäsalmi deposit is the largest VMS deposit in the Vihanti-Pyhäsalmi belt. It is hosted in a metamorphosed volcanic sequence composed of lapilli tuffs, coherent lava flows, and sill-shaped intrusions of varied composition (Figs. 7.7 and 7.8). Rhyolitic volcaniclastic rocks are the most common host rock near the sulfide deposit. Voluminous pegmatites are emplaced along the main subvertical foliation near the eastern tectonized contact of the Pyhäsalmi deposit. Further to the east, sericite-altered felsic metavolcanic rocks are gently dipping to the east and at depth they extend under the granitoids (Heinonen, 2013). The upper part of the Pyhäsalmi deposit consists of subvertical intercalations of Zn- and Cu-bearing pyritic ore (Helovuori, 1979). In general, the upper part of the deposit shows a strong sheared character that denotes local tectonic transposition with intercalation of banded massive sulfides, enclaves of altered metavolcanic rocks, and tectonic rafts of country rock separated by sulfide mylonites and shear zones.

Pyrite is the most voluminous sulfide component of the system, whereas variable amounts of sphalerite, chalcopyrite, dolomite, calcite, and barite occur in different ore types (Imaña, 2003). Pyrrhotite is a common sulfide observed in areas of pegmatite contact metamorphism and ductile shear zones. Despite high grade regional metamorphism, secondary pyrrhotite formation has only minimum development along pyrite grain boundaries in the central parts of the deep ore body (Imaña, 2003).

CHEMOSTRATIGRAPHY

Lithogeochemical classification of whole-rock data obtained from systematic drill core sampling has allowed the identification of the chemical protoliths of the altered rocks. The severe obliteration of primary textures due to alteration and metamorphism has made correlations of original lithologies difficult; therefore chemical protolith identification becomes relevant to improve the stratigraphic understanding of the mine area. The Pyhäsalmi massive sulfide deposit and hydrothermally altered rocks are formed within a thick volcanic package of transitional magmatic affinity. The main components are related to felsic protoliths named Rhyolite A and Rhyolite B (Barrett, 2010), an intrusion named Mafic A-2, and andesites (Lehto sill) (Fig. 7.9). In the mine area, the original proximal part of the hanging wall sequence has been removed by faulting and does not appear in direct contact with the mineralized stratigraphy. However, rocks assigned to the hanging wall part of the deposit are interpreted to be formed from a tholeiitic sequence composed of coherent and brecciated mafic volcanic rocks (Mafic A-1). A mineralized unit of tholeiitic affinity (Rhyolite X) occurs within this mafic tholeiitic package (Barrett, 2010). There appears to be no major break in volcanic activity during the VMS formation as there is no evidence of pelagic sedimentation, laterally continuous chemical precipitates, or reworked components at the change from the transitional to tholeiitic volcanic episodes.

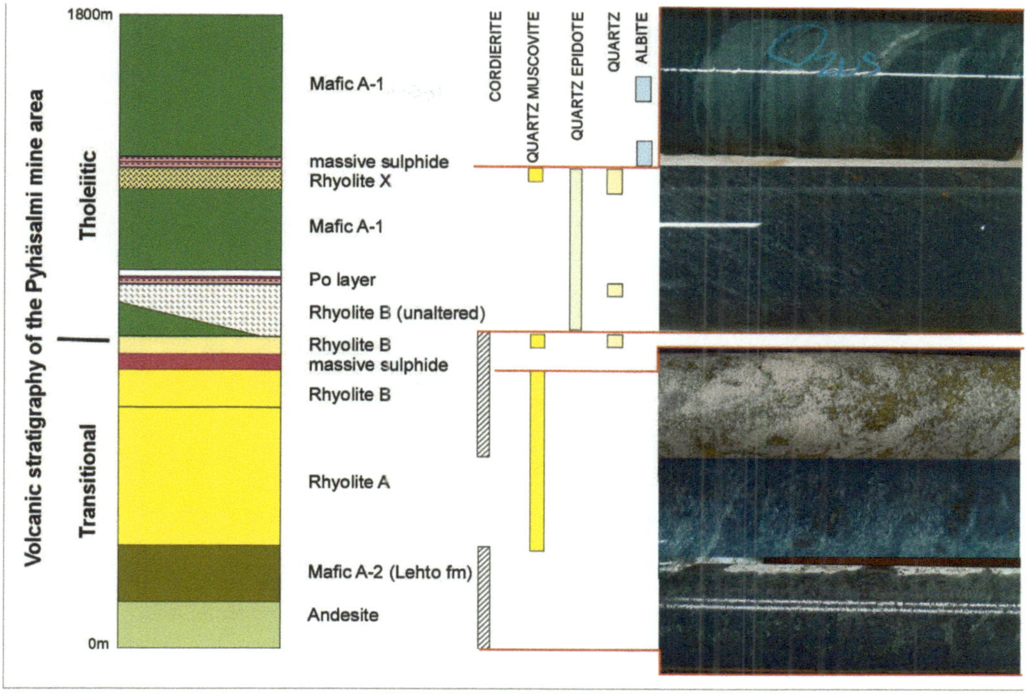

FIGURE 7.9 Chemostratigraphy and alteration of the Pyhäsalmi mine area.

PYHÄSALMI DEPOSIT AND MINERAL ZONATION

Typically, VMS deposits are characterized by well-developed mineral zonation, which normally grade from an inner core consisting of a high temperature pyritic assemblage to Cu-rich sulfide zones and then Zn sulfides in the periphery (Galley et al., 2007). In the upper portion of the Pyhäsalmi deposit, mineral zonation was not observed due to intense deformation and post-VMS tectonic stacking of different parts of the original massive sulfide lens. However, the deep mine ore body, which is located at the end of the upper ore body slab, forms a potato-shaped blob of massive sulfide ore from 1000 m down to 1400 m depth, within which mineral zonation is well preserved (Mäki and Puustjärvi, 2003; Imaña, 2003).

Much of the deep ore body is massive pyrite, which occupies the core of the potato-shaped deposit and contains uneconomic Cu tenors. A Cu-rich pyrite zone occurs around the inner barren pyrite core and is in turn surrounded by a distinct Zn-barite–rich zone toward the periphery of the deposit. The Zn ore surrounds the lower ore body on all sides except the top. This suggests the form of a downward-facing sheath fold with the upper ore body forming the upper limb of the fold. The lower limb of the fold is short and curtailed, giving the overall deposit a hook-shaped geometry in cross section (Lickorish, 2012) (Fig. 7.10). Ore types and mineral assemblages in the deep ore body were described in Imaña, 2003 and are represented in Fig. 7.11.

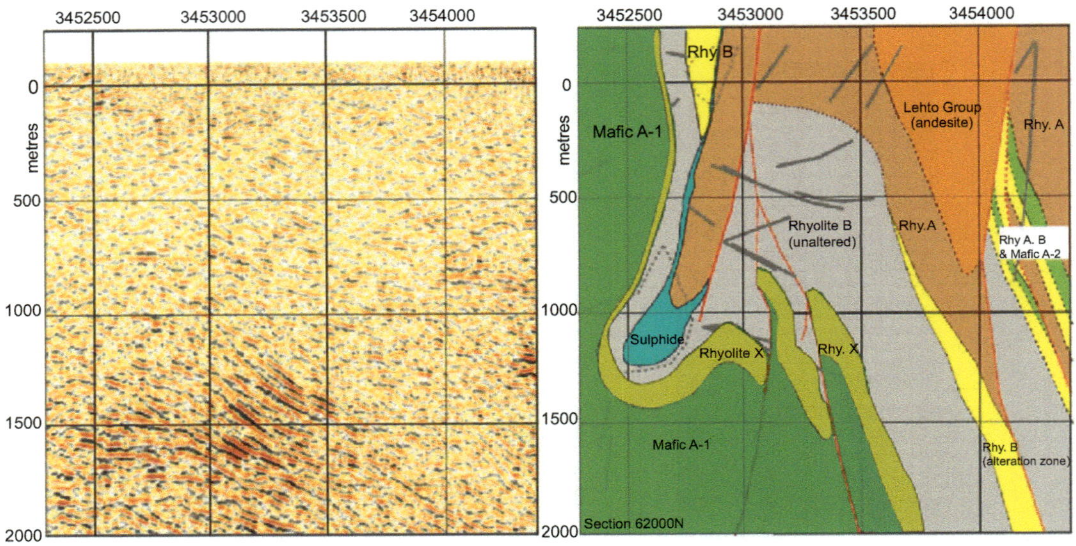

FIGURE 7.10 Seismic profile and simplified geological cross section 62000 N across the middle part of the deep ore body.

The gently dipping reflectors on the seismic profile to the left represent the location of felsic and mafic intercalations of Rhyolite X (and other felsic horizons not shown in the cross section) and Mafic A-1.

Source: *Modified after Lickorish, 2012.*

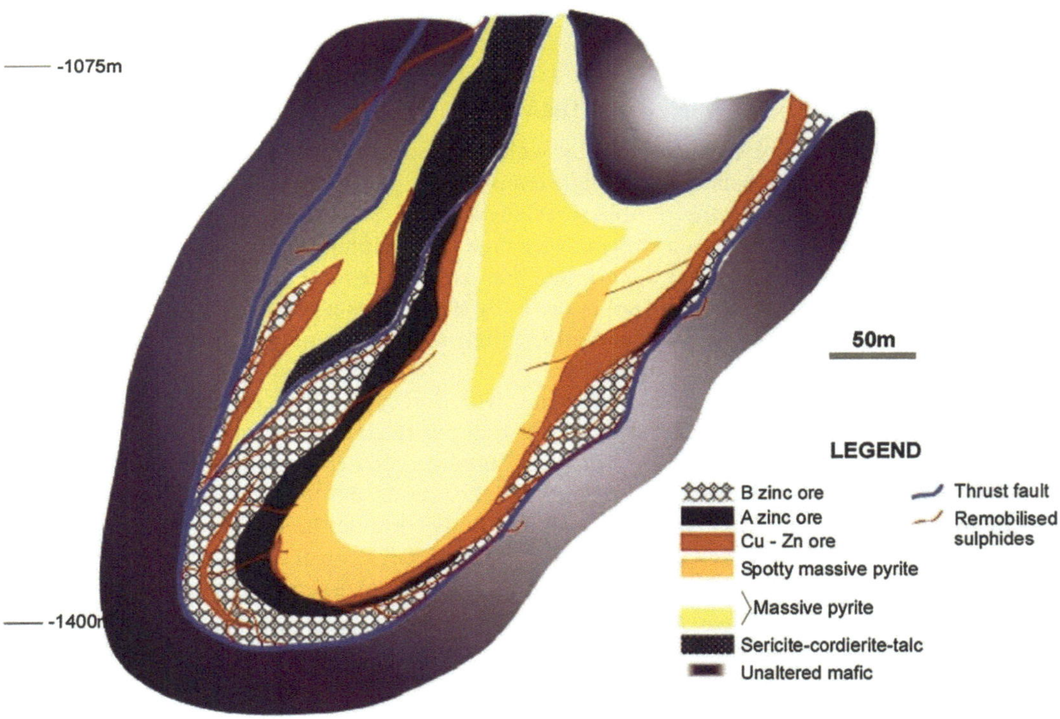

FIGURE 7.11 Ore zonation in the deep ore body (Imaña, 2003).

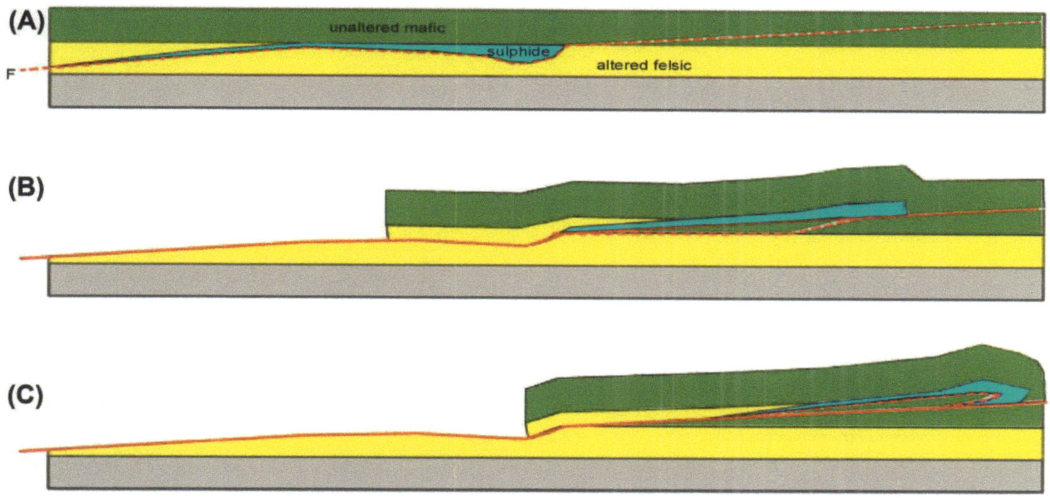

FIGURE 7.12 Separation of the massive sulfide deposit from its host alteration zone by thrusting.

(A) Original stratigraphic template, with incipient thrust fault. (B) Movement on thrust carries sulfide deposit over unaltered material. (C) A small "shortcut" fault in the footwall is all that is needed to introduce some unaltered material into the core of the VMS fold closure.

STRUCTURE AND CROSS SECTIONS

The overall structure in the Pyhäsalmi deposit is controlled by a subvertical, strongly sheared, inverted anticline structure with a sheath fold at its lower end. The sheath fold in the deep ore body is attributed to continuous high strain deformation of folded material of contrasting competency along a shear zone. In Pyhäsalmi, the existence of the massive sulfide lens would have formed a singularity in the stratigraphy that would have concentrated strain, producing a higher amplitude structure that by continual deformation would have stretched out to form a sheath fold localized on the massive sulfide (deep ore body).

Such deformation resulted in the deep ore body being tectonically transposed onto less-altered volcanic rocks of the hanging wall mafic sequence, which by subsequent folding and tilting formed the unusual aspect of a massive sulfide deposit enclosed in unaltered rocks (Fig. 7.12). Shear structures (ore mylonites) are especially common along the periphery of the deep ore body, especially within the Zn ores.

HYDROTHERMAL ALTERATION

Much of the strong sulfide mineralization in Pyhäsalmi occurs within altered parts of Rhyolite B. Alteration within this unit is pervasive and widespread for about 4 km in a north–south direction and to more than 1 km at depth. This unit presents geochemical patterns related to strong alkali depletion and significant enrichment in Fe, K, Mg, Si, and Ba accompanying base metal enrichment. Metamorphism of the original alteration assemblages has resulted in formation of mineral assemblages of muscovite, cordierite, quartz, biotite, and barite. Except for areas where tectonic dislocation has happened (the deep ore body), massive sulfides are spatially associated with this mineral assemblage.

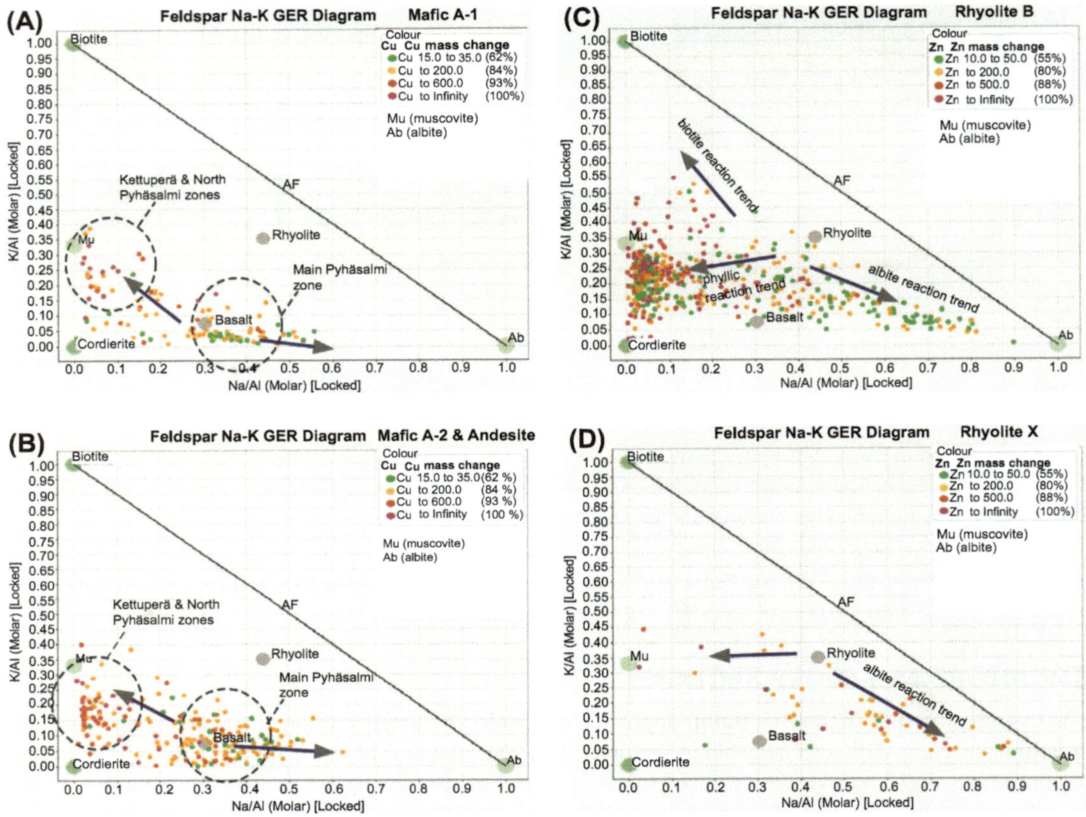

FIGURE 7.13 General element ratio (GER) scatterplots of K/Al vs. Na/Al molar ratios are representing the silicate mineralogy for different volcanic protoliths in the Pyhäsalmi mine area.

The molar plot is colored by net metal mass gains. (A) Cu enrichment in sericitized hanging wall Mafic A-1 unit is significant in the Kettuperä area, but shows very weak Cu gain near the Pyhäsalmi deposit. (B) Basalts and andesites of the Lehto unit are only weakly enriched near the Pyhäsalmi deposit but more enriched and altered in the northern part and Kettuperä zones. (C) Rhyolite B shows distinct alteration trends associated with variable Zn enrichment; stronger metal gains are associated with biotite-, muscovite-, and cordierite-bearing assemblages. (D) Rhyolite X, found near the deep ore body, is mostly albitized with only weak to moderate gains in Zn.

Exhalative style of mineralization is not common in the Pyhäsalmi deposit, however, the exclusively hydrothermal composition of the Zn-rich parts of the deposit (Imaña, 2003) and strong enrichment of barium and silica along certain pyritic horizons could represent hydrothermal precipitates (Barrett, 2010).

The silicate alteration mineralogy has been confirmed by plotting whole-rock assays from different lithologies in a series of K/Al versus Na/Al molar ratio plots. Calculated mass changes of Cu and Zn are also presented in these plots to see the relationship between metal enrichment and hydrothermal alteration (Fig. 7.13). It was shown that the alteration and metal enrichment of altered units are stronger in the northernmost parts of the system, which are located more than 2 km away from the actual ore body position. This separation also confirms the interpretation that the massive sulfides have been detached from their proximal alteration zones in the north during severe thrusting and deformation.

Several late, unaltered mafic dikes cut through the deep massive ore body, with only minor thermal reactions occurring along dike margins. These mafic dikes are generally enriched in Au and As as a consequence of metal remobilization during metamorphism and deformation. A second ore horizon, Rhyolite X contains a widespread and relatively lower-temperature alteration assemblage than altered Rhyolite B. Rhyolite X is replaced by albite and silica and is devoid of Cu, with only minor Zn enrichment. Significant Cu and Zn contents occur within locally sericitized parts.

SUBSEAFLOOR REPLACEMENT

Subseafloor VMS formation is an effective mechanism of retaining large portions of the metalliferous budget within permeable strata before it is expelled onto the seafloor. Favorable conditions and diagnostic evidence of this process were explained by Doyle and Allen (2003).

The Pyhäsalmi ore body is entirely hosted by felsic rocks interpreted to represent rapidly emplaced lapilli tuff and coherent lava flow units. Numerous relicts of altered rock are observed within the massive sulfide, mainly associated with the inner Cu-bearing, barite-poor portions of the ore body. Studies aimed at differentiating and explaining the origin of diverse rock enclaves in the ore body (Miettinen, 2011) found that many enclaves in the deposit were not just tectonic rafts of country rock assimilated during deformation. On the contrary, many enclaves show evidence of being very early relicts of an altered host rock. Lithogeochemical protolith characterization of these enclaves found striking similarities to altered rocks (Rhyolite B) in the main alteration zone outside the ore body. In distal parts of the deposit, drill holes intersect strong hydrothermal alteration that extends 25–30 m both up-hole and down-hole from the pyritic massive sulfide lens.

Altered enclaves of Rhyolite B have not been observed in the Zn ores, so it is conceivable that the Zn-rich sulfides formed by direct venting onto the seafloor. Textural and lithogeochemical evidence demonstrate that the pyrite, Cu ores and Zn ores were subjected to a zone refinement process (Mäki and Puustjärvi, 2003) consisting of successive high temperature infiltration and replacement of permeable volcaniclastic strata.

SATELLITE DEPOSITS

KANGASJÄRVI

The Kangasjärvi deposit occurs 25 km south of the Pyhäsalmi mine. It was mined in 1986 by Outokumpu and was processed in the Pyhäsalmi mill (see Fig. 7.1). Total production from the open pit was 86,000 t of ore grading 5.4% Zn, 0.1% Cu, and 38% S. The ore deposit was found by the GTK in 1964. After production, the Kangasjärvi area was explored by several companies but no new resources have been found (Mäki and Puustjärvi, 2003).

The Kangasjärvi ore and alteration have been described by Rehtijärvi (1984), Rasilainen (1991), and Roberts et al. (2004). The ore deposit is hosted by a succession of altered mafic, intermediate, and felsic metavolcanic rocks. Host rocks include cordierite-sillimanite-garnet-anthophyllite gneisses and cordierite-bearing gneisses (Rasilainen, 1991) (Fig. 7.14). The main ore minerals are pyrite and

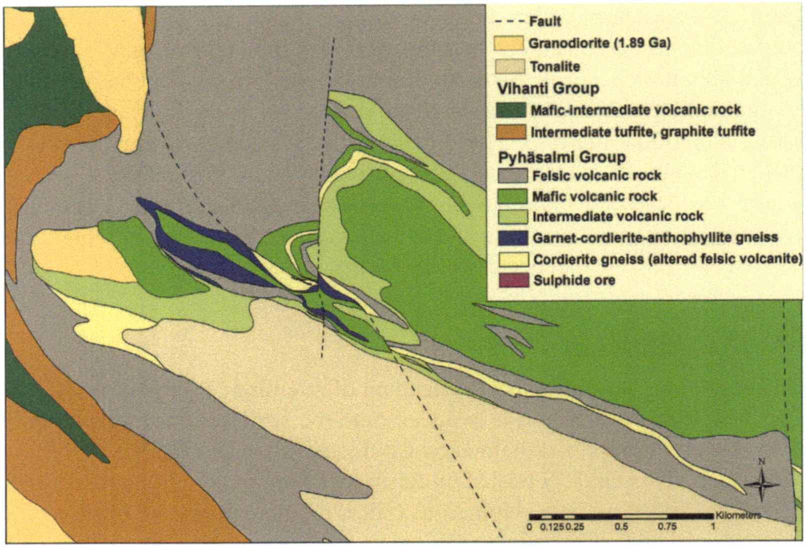

FIGURE 7.14 Geology of the Kangasjärvi area.

Source: *Map is modified after Bedrock of Finland–DigiKP (2013).*

sphalerite and the main gangue minerals are quartz and barite. Magnesium enrichment and the absence of carbonate minerals are consistent with a proximal Kuroko-type environment of deposition (Rehtijärvi, 1984). The Kangasjärvi deposit has similarities to the Pyhäsalmi ore deposit in terms of lithology, alteration, and metal abundance.

RUOSTESUO (KALLIOKYLÄ)

The Ruostesuo Zn-Cu deposit was in production from 1988–1989, and the ore was processed in the Pyhäsalmi mill. The deposit is part of a large mineralized system in the Kalliokylä volcanic formation about 15 km southeast of Pyhäsalmi (Fig.15). The total resource is estimated to be 2.7 Mt grading 1.7% Zn, 0.4% Cu, 10 g/t Ag, and 0.4 g/t Au (Ekdahl et al., 1997). The mineralizations were found by Outokumpu in 1959 using airborne geophysical measurements (Fig. 7.1).

The ores occur in the altered portion of a rock suite that contains felsic, intermediate, and mafic metavolcanic rocks (Fig. 7.15). The sequence is strongly folded and metamorphosed to granulite facies (Huhtala, 1979). Metarhyolites and both Mg-rich and Fe-rich metabasalts are the primary volcanic protolith in the area. Hydrothermal alteration proximal to the sulfide lens is represented by strong depletions of Ca, Na, and K with a characteristic metamorphic orthoamphibole-cordierite-garnet assemblage that grades outwards into Mg- and Fe-rich tholeiitic metabasalts (Roberts, 2002; Roberts et al., 2004). The ores are brecciated or massive pyrite-pyrrhotite occurrences with some chalcopyrite and sphalerite. The ore lodes in the Kalliokylä formation differ from each other geochemically, particularly in their zinc content. In the southeast, ores hosted within cordierite-anthophyllite rocks are poor in Zn, assaying 0.2–0.4% Zn and 0.4–0.6% Cu. In the northeast, the Ruostesuo mine contains ores hosted by felsic metatuffites, amphibolites, and cummingtonite gneisses. These ores assay over 2% Zn and 0.4% Cu (Huhtala, 1979).

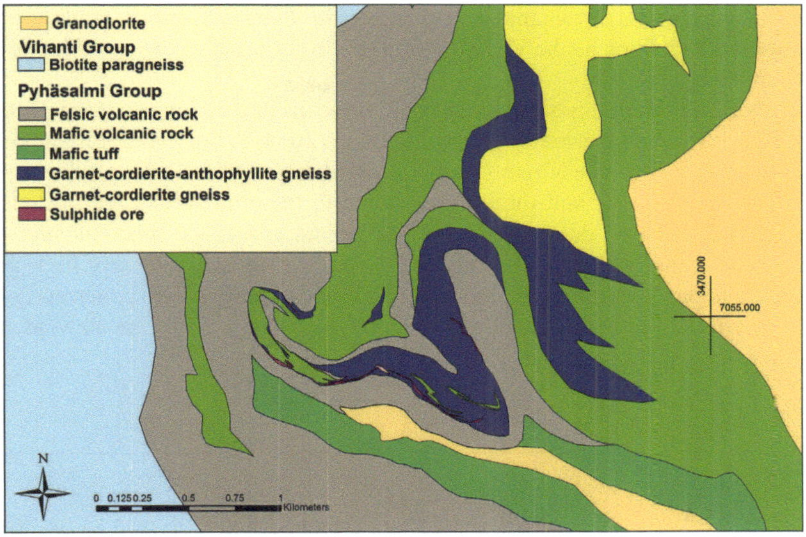

FIGURE 7.15 Geology of the Kalliokylä mine site and Ruostesuo deposit.

Source: *Map is modified after Bedrock of Finland–DigiKP (2013).*

MULLIKKORÄME

In 1983, a research program was commenced by Outokumpu to study lithogeochemical halos around the Pyhäsalmi deposit (Mäki, 1986). According to this study, Outokumpu reevaluated several old exploration targets. The Mullikkoräme deposit was discovered using lithogeochemistry and geophysics in 1987.

The Mullikkoräme formation is located about 10 km northeast of Pyhäsalmi (Fig. 7.1). It is a north–south trending, bimodal volcanic formation composed of basaltic pillow metalavas (west) and rhyolitic felsic metavolcanic rocks (east). The metamorphic grade is lower than in Pyhäsalmi, mostly in greenschist facies. Metamorphic alteration minerals include: chlorite, sericite, phlogopite, talc, quartz, and pyrite (Ekdahl et al., 1997).

The Mullikkoräme deposit is a typical small size polymetallic sulfide deposit. It was in production in two periods, first between 1990 and 1993 and the deeper parts from 1996 to 2000. The total production was 1.15 Mt of 6.1% Zn, 0.3% Cu, 0.9% Pb, 45 ppm Ag, and 1 g/t Au (Pelkonen, 2000). The mineral assemblage of carbonate sphalerite and galena suggests that mineralization in Mullikkoräme represent lower-temperature processes than those in the Pyhäsalmi system. Sulfide lenses are small, Zn and Pb rich, Cu poor and are hosted by hydrothermal carbonates and talc schist. Some ore lenses were massive and pyrite rich but these commonly have very low base metal tenors (Pelkonen, 2000).

SUMMARY

The Vihanti-Pyhäsalmi belt contains the most important volcanogenic massive sulfide (VMS) deposits in Finland. The volcano-sedimentary host rocks belong to a 1.93–1.92 Ga island arc that occurs along the northwestern parts of the Raahe-Ladoga shear complex. The stratigraphy in the region is divided into two units, a lower bimodal volcanic unit forming the Pyhäsalmi group and an upper volcano-sedimentary association defined as the Vihanti group. U-Pb radiometric dating suggests that these two groups have a similar age.

The Vihanti-Pyhäsalmi belt hosts two large VMS deposits and a number of small sulfide deposits located around them. The only currently producing mine (2014) is the Pyhäsalmi deposit, containing 75.7 Mt of pyritic Zn-Cu ore. The Vihanti mine operated from 1954–1992, producing 28 Mt of Zn-Pb-Ag ore. The Vihanti deposit and three minor satellite mineralizations are hosted in a volcanic sequence of the Vihanti group dominated by intermediate and felsic metavolcanic rocks. Recent exploration projects at the old mine sites show that both areas are still very high potential VMS exploration targets. In particular, new exploration techniques, such as deep penetrating EM-measurements, seismic methods, and deep drilling with 3D modeling, will give new ideas for further studies.

REFERENCES

Autere, I., Pelkonen, K., Pulkkinen, K., 1991. Outokumpu Finnmines Oy:n Vihannin kaivos. Vuoriteollisuus 2, 81–82 (in Finnish).

Barrett, T.J., 2010. Lithogeochemistry of 2500 samples from the Pyhäsalmi mine: unpublished internal report for Pyhäsalmi mine Oy.

Doyle, M.G., Allen, R.L., 2003. Subsea-floor replacement in volcanic-hosted massive sulfide deposits. Ore Geology Reviews 23, 183–222.

Eilu, P., Bergman, T., Bjerkgård, T., Feoktistov, V., et al., 2013. Metallic Mineral Deposit Map of the Fennoscandian Shield 1:2,000,000. Revised edition (comp.). Geological Survey of Finland, Geological Survey of Norway, Geological Survey of Sweden, the Federal Agency of Use of Mineral Resources of the Ministry of Natural Resources of the Russian Federation.

Ekdahl, E., 1993. Early Proterozoic Karelian and Svecofennian formations and the evolution of the Raahe-Ladoga Ore Zone, based on the Pielavesi area, central Finland. Geological Survey of Finland 137 Bulletin 373. Espoo: Geologian tutkimuskeskus.

Ekdahl, E., Mäki, T., Pelkonen, K., 1997. Geology and mineral deposits of the central Ostrobothnia. VHMS-deposits. In: Weihed, P., Mäki, T. (Eds.), Research and exploration—where do they meet? 4th Biennial SGA Meeting, August 11–13, pp. 49–61 Turku, Finland. Excursion guidebook A2: volcanic hosted massive sulphide and gold deposits in the Skellefte district, Sweden and western Finland. Geologian tutkimuskeskus. Opas 41. Espoo: Geologian tutkimuskeskus.

Bedrock of Finland–DigiKP, 2013. Digital map database [electronic resource]. Espoo: Geological Survey of Finland [referred to 01.01.2013] Version 1.0.

Gaál, G., Gorbatschev, R., 1987. An outline of the Precambrian evolution of the Baltic Shield. Precambrian Research 35, 15–52.

Galley, A.G., Hannington, M.D., Jonasson, I.R., 2007. Volcanogenic massive sulphside deposits. In: Goodfellow, W.D. (Ed.), Mineral deposits of Canada, 5. Geological Association of Canada, Mineral Deposits Division, pp. 141–161. Special Publication.

Heinonen, S., 2013. Seismic reflection profiling for massive sulfide exploration in Finland. Doctoral thesis, Helsinki University, p. 109.

Helovuori, O., 1979. Geology of the Pyhäsalmi ore deposit, Finland. Economic Geology 74, 1084–1101.

Huhtala, T., 1979. The geology and zinc-copper deposits of the Pyhäsalmi-Pielavesi district, Finland. Economic Geology 74 (5), 1069–1083.

Imaña, M., 2003. Petrography, mineralogy, geochemistry and 3D modelling of the A zinc ore in the Pyhäsalmi Zn-Cu VMS deposit, central Finland. Masters thesis, University of Turku, Department of Geology and Mineralogy, p. 72.

Kahma, A., 1973. The main metallogenic features of Finland. Geological Survey of Finland 28 Bulletin 265. Otaniemi: Geologinen tutkimuslaitos.

Kärki, A., Laajoki, K., Luukas, J., 1993. Major Palaeoproterozoic shear zones of the central Fennoscandian Shield. In: The Baltic Shield. Precambrian Research 64 (1–4), 207–223.

Korsman, K., Koistinen, T., Kohonen, J., et al. (Eds.), 1997. Suomen kallioperäkartta—Berggrundskarta över Finland—Bedrock map of Finland 1:1,000,000. Geological Survey of Finland, Espoo, Finland.

Kousa, J., Marttila, E., Vaasjoki, M., 1994. Petrology, geochemistry and dating of Paleoproterozoic metavolcanic rocks in the Pyhäjärvi area, central Finland. In: Geochemistry of Proterozoic supracrustal rocks in Finland. IGCP Project 179 Stratigraphic methods as applied to the Proterozoic record and IGCP Project 217 Proterozoic geochemistry 7–27 Geological Survey of Finland. Special Paper 19. Espoo: Geologian tutkimuskeskus.

Kousa, J., 1997. Geology and mineral deposits of the central Ostrobothnia. Regional geology. In: Weihed, P., Mäki, T. (Eds.), Research and Exploration—Where Do They Meet? Proceedings of the 4th Biennial SGA Meeting, August 11–13, Turku, Finland, pp. 43–46 Excursion guidebook A2: volcanic hosted massive sulphide and gold deposits in the Skellefte district, Sweden and western Finland. Geologian tutkimuskeskus. Opas 41. Espoo: Geologian tutkimuskeskus.

Kousa, J., Luukas, J., 2007. Piippolan ja Rantsilan kartta-alueiden kallioperä [electronic resource]. Summary: Pre-Quaternary rocks of the Piippola and Rantsila map-sheet areas Suomen geologinen kartta 1:100,000: kallioperäkarttojen selitykset lehdet 3411 ja 3412. Espoo: Geologian tutkimuskeskus p. 65.

Kousa, J., Luukas, J., Huhma, H., Mänttäri, I., 2013. Paleoproterozoic 1.93–1.92 Ga Svecofennian rock units of the Raahe-Ladoga Zone, Central Finland. Geological Survey of Finland Report of Investigation 198, 91–96.

Kousa, J., Luukas, J., 2004. Vihannin ympäristön kallioperä-ja malmitutkimukset vuosina 1992–2003. 142 s., 1 liite (toim.) Geologian tutkimuskeskus, arkistoraportti, M10.4/2004/2 (in Finnish).

Laajoki, K., Luukas, J., 1988. Early Proterozoic stratigraphy of the Salahmi-Pyhäntä area, central Finland, with an emphasis on applying the principles of lithodemic stratigraphy to a complexly deformed and metamorphosed bedrock. Bulletin of the Geological Society of Finland 60 (2), 79–106.

Lahtinen, R., 1994. Crustal evolution of the Svecofennian and Karelian domains during 2.1–1.79 Ga, with special emphasis on the geochemistry and origin of 1.93–1.91 Ga gneissic tonalites and associated supracrustal rocks in the Rautalampi area, central Finland. Geological Survey of Finland Bulletin 378. Espoo: Geologian tutkimuskeskus p.128.

Lickorish, W.H., 2012. Structure of the Pyhäsalmi VMS deposit, Finland. Unpublished internal report for Pyhäsalmi mine Oy p. 41.

Luukas, J., 1997. Geology and mineral deposits of the central Ostrobothnia. Deformation history. In: Weihed, Mäki (Ed.), Research and Exploration—Where Do They Meet? Proceedings of the 4th Biennial SGA Meeting, August 11–13, Turku, Finland, pp. 46–47 Excursion guidebook A2: volcanic hosted massive sulphide and gold deposits in the Skellefte district, Sweden and western Finland. Geologian tutkimuskeskus. Opas 41. Espoo: Geologian tutkimuskeskus.

Luukas, J., Kousa, J., 2013. The major Palaeoproterozoic (1.93–1.92 Ga) VMS-deposits in the northwestern Raahe-Ladoga zone, Central Finland. Geological Survey of Finland 91–96 Report of Investigation 198.

Mäki, T., 1986. The Lithogeochemistry of the Pyhäsalmi Zn-Cu-Pyrite Deposit, Finland. In: Prospecting in areas of glaciated terrain symposium, Sept. 1–2, Kuopio. Finland. Institute of Mining and Metallurgy, London. 69–82.

Mäki, T., 2013. Report on Estimated Mineral Reserves and Resources, 2012, Pyhäsalmi Mine. Finland 43 (company report).

Mäki, T., Puustjärvi, H., 2003. The Pyhäsalmi massive Zn-Cu-pyrite deposit, Middle Finland—a Paleoproterozoic VMS-class "giant." In: Ashton, J., et al. (Ed.), Europe's Major Base Metal Deposits. Irish Association for Economic Geology, Dublin. pp. 87–91.

Miettinen, E., 2011. Detailed geology of the level—1275, Pyhäsalmi Mine, central Finland and genetic implications of rock inclusions within the ore body. Unpublished master's thesis, Helsinki University.

Nikander, J., Luukas, J., Ruotsalainen, A., Kousa, J., 2002. Kallioperä-ja malmitutkimukset Vihannin Vilmingon ja Rantsilan Pelkoperän välisellä alueella vuosina 1993–2002. 71 s. Geologian tutkimuskeskus, arkistoraportti, M19/2434, 3412/2002/1/10 (in Finnish).

Nikander, J., Luukas, J., Ruotsalainen, A., 2005a. Vihannin Lampinsaaren ympäristön ja Kuuhkamon kairaukset karttalehdellä 2434 05 vuosina 2004–2005. 18 s. Geologian tutkimuskeskus, arkistoraportti, M19/2434/2005/2/10 (in Finnish).

Nikander, J., Luukas, J., Ruotsalainen, A., 2005b. Rantsilan Peuranevan ja Vihannin Vilmingon alueiden kairaustutkimukset vuonna 2004 karttalehdellä 2434 08. 28 s. Geologian tutkimuskeskus, arkistoraportti, M19/2434, 3412/2005/1/10 (in Finnish).

Pelkonen, K., 2000. Mullikkoräme Mine (company report; in Finnish).

Puustjärvi, H. (Ed.), 1999. Pyhäsalmi modeling project. Technical Report. 13.5. 1997-12. 5. 1999. p. 251. Geologian tutkimuskeskus, arkistoraportti, M19/3321/99/1/10.

Rasilainen, K., 1991. Geochemistry and wall-lrock alteration at the Kangasjärvi massive sulphide deposit, central Finland. Current Research 1990, Geological Survey of Finland, Special Paper 12, 107–110.

Rasilainen, K., Västi, K., 1989. Geochemistry, wall rock alteration and metal zonality at the Rauhala Zn-Cu-Pb sulphide deposit. In: The early Proterozoic Zn-Cu-Pb sulphide deposit of Rauhala in Ylivieska, western Finland. Geological Survey of Finland 43–58 Special Paper 11. Espoo: Geologian tutkimuskeskus.

Rehtijärvi, P., 1984. Distributions of phosphorus, sulphur and sulphur isotopes in a strata-bound base metal deposit, Kangasjärvi, Finland. Geologian tutkimuskeskus Tutkimusraportti 65. Espoo: Geologian tutkimuskeskus p. 16.

Rehtijärvi, P., Äikäs, O., Mäkelä, M., 1979. A middle Precambrian Uranium- and Apatite-Bearing Horizon Associated with the Vihanti Zinc Ore Deposit, Western Finland. Economic Geology 14, 1102–1117.

Roberts, M.D., 2002. Architecture, geochemistry and dynamics of Paleoproterozoic seafloor hydrothermal systems preserved in a high-grade metamorphic terrane, Vihanti-Pyhäsalmi district, central Finland. James Cook University of North Queensland, Townsville. 411.

Roberts, M.D., Oliver, N.H.S., Lahtinen, R., 2004. Geology, lithogeochemistry and paleotectonic setting of the host sequence to the Kangasjärvi Zn-Cu deposit, central Finland: Implications for volcanogenic massive sulphide exploration in the Vihanti-Pyhäsalmi district. Bulletin of the Geological Society of Finland 76 (1–2), 31–62.

Rouhunkoski, P., 1968. On the geology and geochemistry of the Vihanti zinc ore deposit, Finland. Bull. Comm. Géol. Finlande 236, 1–121.

Vaasjoki, M., Sakko, M., 1988. The evolution of the Raahe-Ladoga zone in Finland: isotopic constraints. In Korsman, K., (Ed.), Tectono-meamorphic evolution of the Raahe-Ladoga zone, Geological Survey of Finland Bulletin 343, 7–32.

FURTHER READING

Mineral Deposits and Exploration database, http://tupa.gtk.fi/karttasovellus/mdae/raportti/526_Kuuhkamo.pdf, 5.5.2015; http://tupa.gtk.fi/karttasovellus/mdae/raportti/528_Näsälänperä.pdf, 5.5.2015..

PART

8

MINERAL DEPOSITS RELATED TO GRANITIC ROCKS

8

I. Haapala, O.T. Rämö

ABSTRACT

In the Archean basement of eastern Finland, some Mo (Cu ± Au) occurrences, including the exhausted Mätäsvaara Mo deposit, are associated with Neoarchean granitoids of the tonalite-trondhjemite-granodiorite (TTG) association and microcline granites. In the extensive Paleoproterozoic Svecofennian domain, a number of Mo, Mo-Cu, Cu-W, Cu-Au, and Au occurrences are found in close spatial association with 1.89 to 1.86 Ga synorogenic calc-alkaline granitoids (tonalities, granodiorites, granites) and show, in several cases, the characteristics of porphyry-Cu (Mo, Au) deposits. The exhausted Ylöjärvi Cu-Mo deposit, a mineralized tourmaline breccia, is described in detail. The Kutemajärvi Au deposit is associated with a synorogenic hypabyssal granodiorite pluton. Late-orogenic pegmatite provinces host interesting Li, Ta, Nb, and Be prospects. The 1.65–1.54 Ga anorogenic rapakivi granite batholiths and stocks of southern Finland and Russian Karelia contain topaz-bearing granite as late-stage intrusion phases. Greisen-type Sn-Be-W-In-Zn mineralization is in many cases associated with these highly evolved late-stage granites, but only the skarn-type deposits of the Pitkäranta area in Russian Karelia have been exploited.

Keywords: granitoid; rapakivi granite; greisen; mineralization; porphyry copper; tourmaline breccia; Archean; Proterozoic; Finland.

Mineral Deposits of Finland. http://dx.doi.org/10.1016/B978-0-12-410438-9.00021-2

OVERVIEW OF THE PRECAMBRIAN GRANITOIDS IN FINLAND

Granitic rocks, ranging in age from 3.5 to 1.5 Ga, are the prevailing constituents of the Finnish bedrock. On the basis of their geological context and isotopic ages, these granitoids can be divided into the following main groups (Nironen, 2005; Lahtinen et al., 2009; Huhma et al., 2011; Lahtinen, 2012) (Fig. 1.1):

1. *Granitoids of the Archean basement in eastern Finland:* These variably foliated rocks mainly belong to the tonalite-trondhjemite-granodiorite (TTG) association, but also sanukitoids and, less commonly, true granites are encountered. The oldest granitoids are up to 3.5 Ga in age (the Siurua trondhjemite gneiss), but most granitoids of the TTG association have ages of 2.83–2.78 Ga and 2.76–2.72 Ga, the sanukitoids 2.74–2.72 Ga (Heilimo et al., 2012), and the granites typically ~2.7 Ga (Huhma et al., 2011). Some minor cratonic A-type granitic–syenitic intrusions, ~2.44 Ga in age, are found associated with coeval mafic layered intrusions in northern Finland (Lauri et al., 2012).
2. *Granitoids related to the Paleoproterozoic composite Svecofennian orogeny (~1.92–1.79 Ma):* These granitoids can be divided into four groups:
   - Preorogenic Paleoproterozoic granitoids (gneissic tonalities and granodiorites) of central Finland (1.93–1.91 Ma): These are calc-alkaline and peraluminous to metaluminous and their chemical and isotopic compositions (positive initial ε_{Nd}) are consistent with juvenile primitive island arc magmatism preceding accretion or collision (Vaasjoki et al., 2003; Lahtinen et al., 2009; Huhma et al., 2011).
   - Synorogenic granitoids (1.89–1.86 Ma) that are found in central Finland and in the surrounding Svecofennian schist belts, as well as in northern Finland: These are variably foliated calc-alkaline tonalites, granodiorites, and granites. In the voluminous granite complex of central Finland there are also slightly younger, postkinematic, unfoliated or weakly foliated quartz monzonitic, granodioritic, and granitic plutons that intersect the foliated granites and are more evolved and alkaline in composition (Rämö et al., 2001).
   - Late-orogenic granites (1.85–1.80 Ga) of southern and northern Finland: These potassium-rich peraluminous granites are present in southern Finland and are confined to a 100-km-wide, east–northeast-trending zone as sheet-like migmatizing plutons, which commonly contain garnet and cordierite (Kurhila, 2011). In northern Finland, the Central Lapland Granitoid Complex consists mainly of late-orogenic granites (Nironen, 2005).
   - Postorogenic granitoids (~1.81–1.77 Ga) in southern and northern Finland: These are discordant, rounded (diameter 2–15 km), single or multiphase intrusions, and, in some cases, dike-like bodies. In southern and central Finland, they have a large compositional variation from monzonite to granite and are metaluminous. In northern Finland, the Nattanen-type granites show a moderate compositional variation only. Lamprophyre dikes and magma mingling are associated with postorogenic granitoid plutons of Ahvenanmaa in southern Finland (Eklund et al., 1998). In the Nattanen granites, the only hint for possible bimodal magmatism are some quartz dioritic dikes in the Riestonvaara pluton (see Front et al., 1989). The majority of the Nattanen-type granites are oxidized A-type granites, and their isotopic compositions imply a major Archean source component (Huhma, 1986; Heilimo et al., 2009).
3. *Rapakivi granites (1.65–1.54 Ga) in southern Finland*: These granites form large, epizonal, multiphase batholiths and stocks. At least the large batholiths contain granite varieties that display the rapakivi texture (Vorma, 1976). Chemically, they are ferroan, alkali-calcic, metaluminous to

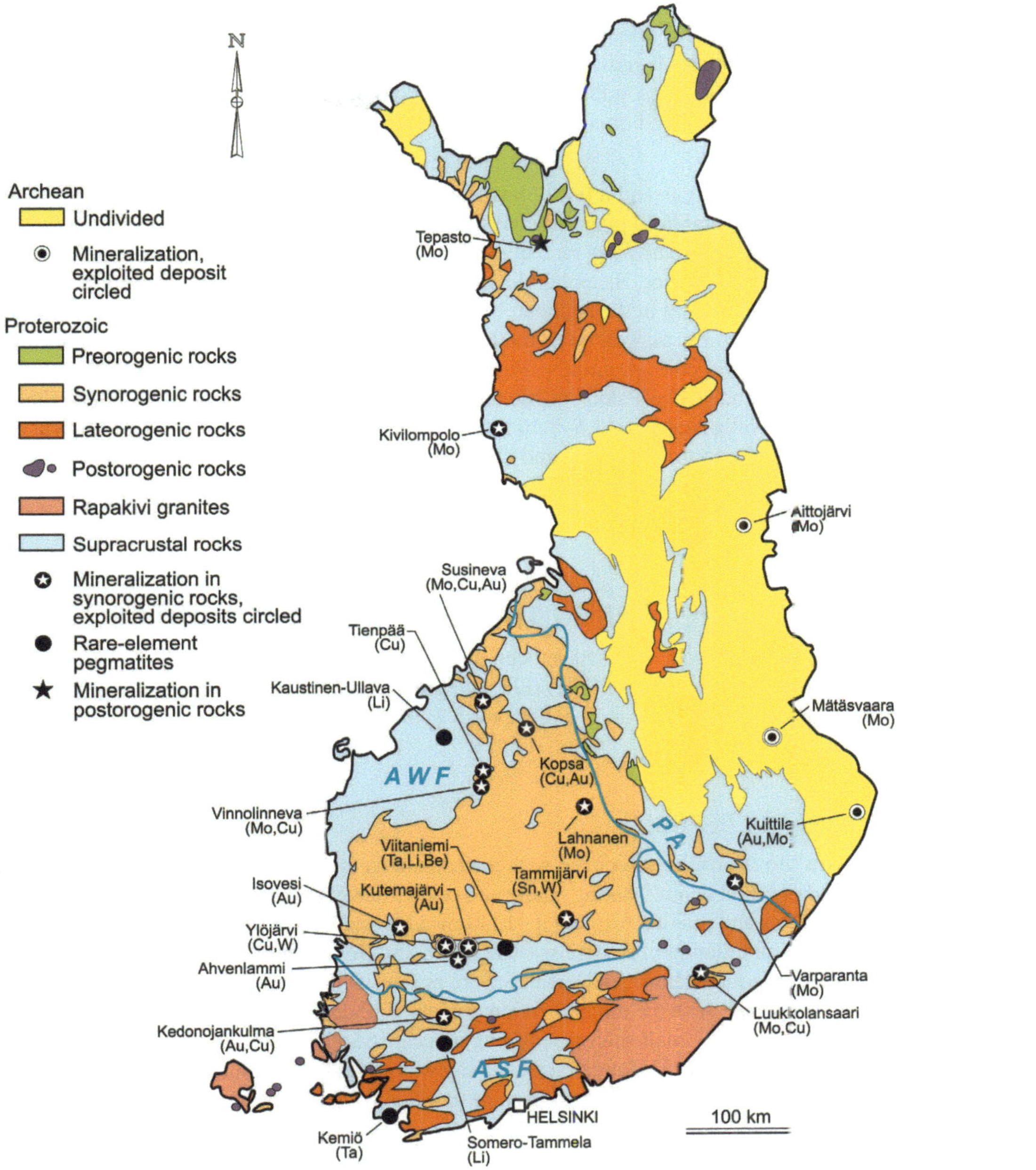

FIGURE 8.1 Simplified geological map of Finland showing the location of selected mineral deposits and showings associated with granitic rocks.

The mineral occurrences are mainly from Haapala (1983a) and Nurmi and Haapala (1986). Terrace boundaries between the Paleoproterozoic arch complexes are from Korsman et al. (1997): PA = primitive arch complex; AWF = arch complex of western Finland, ASF = arch complex of southern Finland.

marginally peraluminous rocks with A-type and within-plate characteristics (Haapala and Rämö, 1992). Bimodal character is indicated by the close spatial and temporal association with mafic igneous rocks (diabase dikes, gabbros, anorthosites, basalts). Chemical and isotopic data indicate that both mantle-derived and crustal components have been involved in the genesis of the rapakivi granites (Rämö, 1991).

A number of ore showings and a few deposits are related, directly (magmatic-hydrothermal mineralization) or indirectly (granitoid magma acted as a heat source for rocks and ore-forming fluids, or the brittle granitoids providing pathways and space for the migrating ore-forming fluids), to the Precambrian granitoids of Finland (see Haapala, 1983a; Nurmi and Haapala, 1986). Porphyry-type Mo, Cu, and Au occurrences are found in the Archean basement of eastern Finland in association with the Neoarchean gneissic granitoids, as well as in the Paleoproterozoic Svecofennian domain associated with synorogenic and postorogenic granitoids (Fig. 8.1). Greisen and vein-type Sn, Be, W, and In occurrences have been found in the Paleo- to Mesoproterozoic rapakivi granites of southern Finland and adjacent Russian Karelia (see Figs. 8.6–8.10 later). In the following pages, a short geological review is given on the more important prospects and showings. A literature-based review is given also on the skarn-type Sn-Cu deposits of Pitkäranta, Russian Karelia, because they, in an important way, complement the metallogenic picture of the Fennoscandian rapakivi granites.

MINERALIZATION RELATED TO THE NEOARCHEAN GRANITOIDS

Several Mo, Cu, and Au occurrences have been discovered in association with the Neoarchean gneissic tonalite-trondhjemite-granodiorite (TTG) suite and greenstone belts of eastern Finland. The associated gold deposits are described separately in Chapter 5.3; in this chapter only the Mo-Cu ± W occurrences are included (Fig. 8.1).

MÄTÄSVAARA Mo DEPOSIT

The Mätäsvaara Mo deposit is associated with a Neoarchean tonalite gneiss and microcline granite. The ore is composed of molybdenite-bearing quartz and pegmatite veins and impregnations in the contact zone of the tonalite gneiss and granite, forming a ~1–30-m-wide and 800-m-long northwest-trending ore zone parallel to the general foliation of the gneiss (Kranck, 1945). Besides molybdenite, small amounts of chalcopyrite, pyrite, pyrrhotite, galena, and sphalerite are present. Silicification and, to some extent, K-feldspatization were associated with the mineralization process. The deposit was discovered in 1902, and it was exploited in several stages: Small-scale test mining was carried out before and during World War I, as well as in 1920–1922 and 1932–1933. Larger-scale mining was carried out by Vuoksenniska Oy in 1940–1947, producing 1.15 million tons of ore with an average grade of 0.14 wt% MoS_2 (Zeidler, 1949).

AITTOJÄRVI Mo PROSPECT

The Aittojärvi Mo prospect in Suomussalmi, northeastern Finland, is associated with a north–south-trending complex of altered gneissic granitoids (tonalite, granodiorite, granite) and amphibolite in the northern part of the Kuhmo-Suomussalmi greenstone belt (Kurki, 1989; Fig. 8.2). Isotopic Re-Os ages of molybdenite, determined in several laboratories, range from 2791 to 2809 Ma (Selby and Creaser,

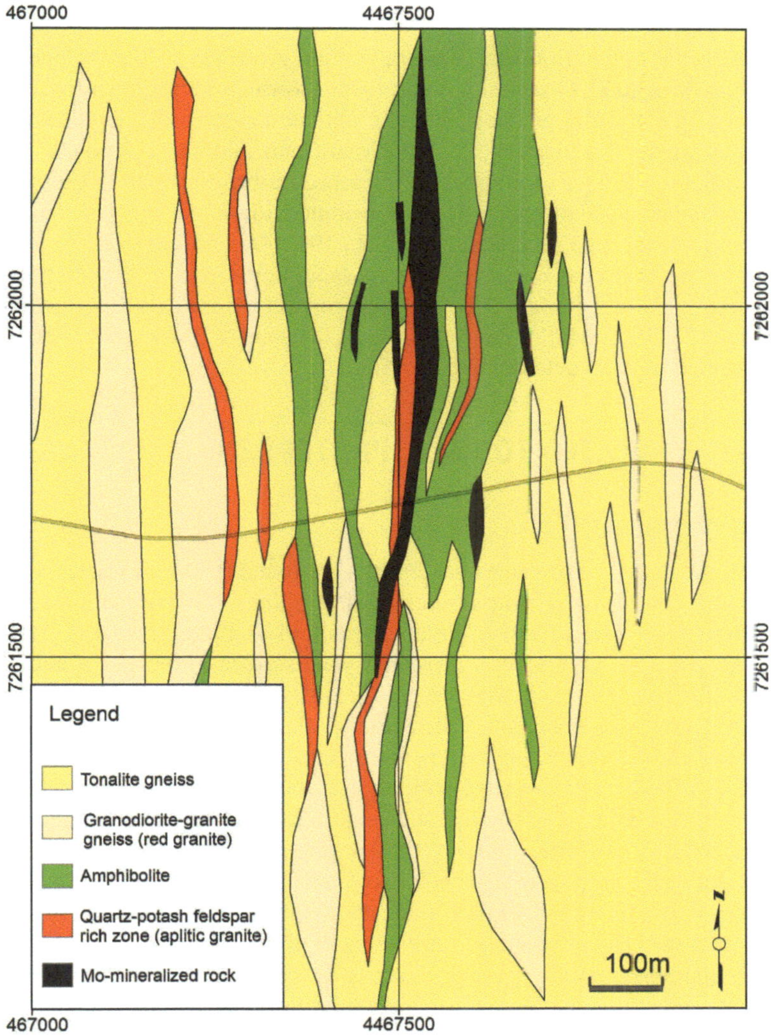

FIGURE 8.2 The Aittojärvi Mo prospect at the northern end of the Kuhmo-Suomussalmi greenstone belt (Kurki, 1989).

2004). The main ore body trends in a north–south direction and is 4–25-m wide and ~800-m long. Molybdenite and other ore minerals (chalcopyrite, pyrite, sphalerite, pyrrhotite, magnetite) are found as a dissemination in host rocks as well as in quartz veins and as fracture fillings.

Metasomatic K-feldspathization, silicification, albitization, and muscovitization of the host rocks are related to the mineralization. In the richest parts of the ore body, the host rock is mainly leucocratic, K-feldspar-rich aplitic granite, whereas in the poorer, disseminated parts of the ore body the main host rock is reddish albite-rich granite. Based on deep drillings and cutoff grade of 0.05% Mo, the assessed resources are 8.5 mt of rock averaging 0.10 wt% Mo.

KUITTILA Mo-W-Cu-Au PROSPECT

In the 2.75 Ga Hattu schist belt in Ilomantsi, eastern Finland, Au ± Mo ± Cu ± W deposits and showings are found in association with TTG-series plutonic rocks (Nurmi and Sorjonen-Ward, 1993). Some of the occurrences (e.g., the Kuittila prospect) contain notable amounts of molybdenite, scheelite, and chalcopyrite resembling in this respect the Mätäsvaara and Aittojärvi deposits. The hypothermal tonalite-hosted Kuittila occurrence (molybdenite and scheelite-bearing quartz vein systems) appears to have a more direct (magmatic-hydrothermal) genetic connection to the granitoid magmatism than the ordinary gold deposits of the area (e.g., Kojonen et al., 1993). According to Heilimo et al. (2012), the Kuittila tonalite has the chemical characteristics of sanukitoids. The ^{187}Re-^{187}Os isochron ages of molybdenite and pyrite (2778 ± 8 Ma) of the Kuittila prospect are in reasonably good agreement with the U-Pb zircon age (2753 ± 5 Ma) of the tonalite host (Stein et al., 1998). The deposits of the Hattu schist belt are described in more detail in Chapter 5.3.

MINERALIZATION RELATED TO PALEOPROTEROZOIC OROGENIC GRANITOIDS

METALLOGENIC OVERVIEW

Several Mo, Cu, W, Au, and As showings and deposits are found in close connection with the synorogenic, 1.89–1.85 Ga Svecofennian granitoid stocks and batholiths. Most of them are found in the schist belts around the large synorogenic Central Finland Granitoid Complex (Fig. 8.1). The mode of occurrence and magmatic association of several prospects were studied in the 1970s and 1990s (Gaál and Isohanni, 1979; Luukkonen, 1994), as was the geochemistry of the plutonic rocks (Gaál et al., 1981; Nurmi, 1983, 1985; Nurmi and Isohanni, 1984; Nurmi et al., 1984).

Good representatives of these porphyry-type occurrences in central western Finland are the Susineva and Taipale Mo prospects within the Rautio diorite-monzodiorite-granodiorite-granite batholith in the municipality of Kalajoki. Molybdenite and small amounts of other ore minerals (chalcopyrite, sphalerite, pyrite, pyrrhotite, vallerite, and covellite) are found in narrow quartz veins, as fracture fillings and as disseminations in porphyritic granodiorite (Gaál and Isohanni, 1979; Marmo and Hyvärinen, 1953). Potassic and propylitic alterations are associated with the mineralization. The stockwork-type Kopsa Au-Cu prospect in Haapajärvi comprises a network of gold-bearing quartz veins and dissemination in a tonalite-granodiorite porphyry stock. The major ore minerals are chalcopyrite, arsenopyrite, and pyrrhotite. Löllingite, cubanite, and pyrite are found in minor amounts, and sphalerite, metallic gold, scheelite, and molybdenite are found as accessory minerals. Potassic alteration comprising growth of secondary microcline and propylitic alteration also characterize this prospect (Gaál and Isohanni, 1979).

The Lahnanen Mo occurrence in the municipality of Viitasaari (Fig. 8.1) is associated with a hydrothermally altered hypersthene granodiorite stock of the Central Finland Granitoid Complex, and the Luukkolansaari (Puumala) Mo-Cu prospect in southeastern Finland occurs in shear zones and granite dikes within the Käkövesi tonalite-granodiorite batholith (Nurmi et al, 1984; Nurmi and Haapala, 1986). A network of molybdenite-bearing quartz veins has been found in silicified and sericitized gneiss at the margin of the granitoid complex of central Lapland against the Peräpohja schist belt at Kivilompolo in the municipality of Ylitornio (Fig. 8.1; Yletyinen, 1967; Saltikoff et al., 2006), but the genetic relationship to the granitoids are not clear.

In south-central Finland, several hydrothermal prospects and deposits are found in the Tampere schist belt, which is intruded by stock or batholith-sized synorogenic granitoid plutons (Fig. 8.3). This belt contains mainly volcanic-arc type rocks and is bordered in the north by the large Central Finland Granitoid Complex and in the south by the turbidite-dominated Pirkanmaa migmatite belt. A comprehensive geological and mineralogical survey of these occurrences and their geological environment is presented in Luukkonen (1994), and more recent reviews can be found in Eilu (2012a, b). The exhausted Ylöjärvi or Paronen deposit in Ylöjärvi is a porphyry-type, high-temperature hydrothermal W-Cu-Sn-Au occurrence in a tourmaline breccia (Figs. 8.1 and 8.3); the Tammijärvi prospect in the municipality of Luhanka (Fig. 8.1) has a similar ore mineral association in metagraywacke-hosted quartz veins. The Kutemajärvi deposit (Orivesi) as well as the Järvenpää (Ylöjärvi) and Isovesi (Suodenniemi, Fig. 8.1) prospects are essentially (metamorphosed) epithermal Au occurrences.

In the Kutemajärvi deposit (1) the zonal hydrothermal alteration, (2) presence of a comb-layered quartz zone at the top of the synorogenic Pukala porphyry (Fig. 8.3) against the hydrothermally altered domain hosting the ore deposit, and (3) isotopic ages suggest close genetic connection between the hypabyssal Pukala granodiorite-trondhjemite-granite porphyry pluton and the deposit (Talikka and Mänttäri, 2005; Kinnunen, 2008). The Jokisivu deposit in the municipality of Huittinen, in the Pirkanmaa migmatite belt (Fig. 8.1), consists of several auriferous quartz vein arrays hosted by ~1880 Ma synorogenic diorite. The mineralization occurred at ~1800 Ma during regional granitic magmatism and

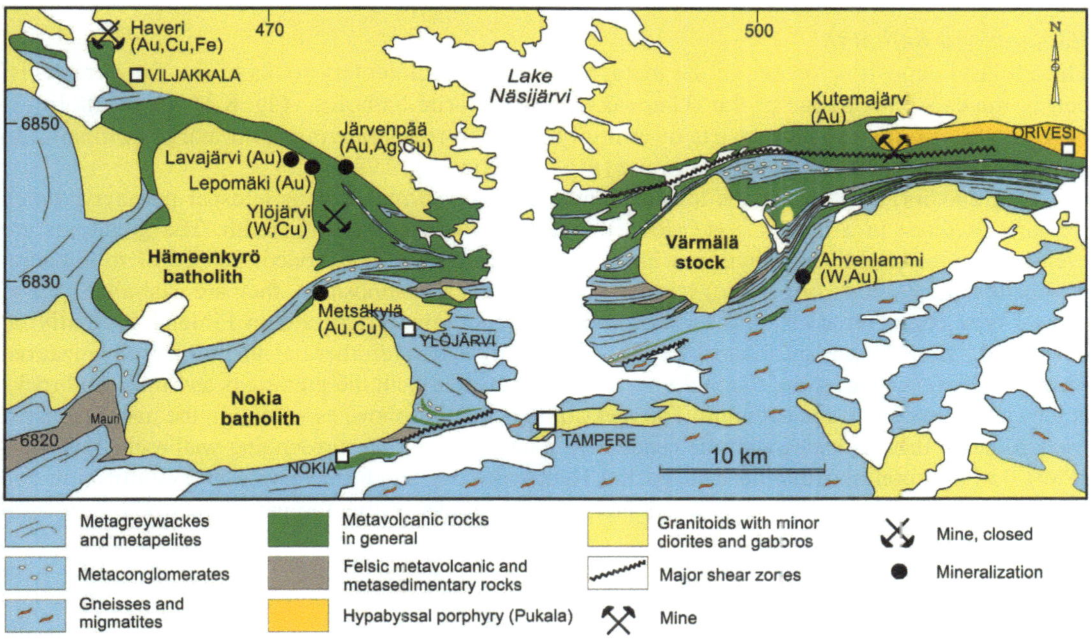

FIGURE 8.3 Geological map of the Tampere schist belt showing locations of hydrothermal ore deposits and prospects associated with granitoids.

Source: *The lithology of the schist belt is simplified from Kähkönen (2005). The mineral occurrences are from Luukkonen (1994) and Eilu (2012a,b).*

shearing. The deposit is classified as orogenic (Luukkonen, 1994; Eilu, 2012b; Eilu and Kärkkäinen (2012). In the Häme belt there are some granitoid-related occurrences, for example, the Kedonojankulma Cu-Au prospect (Fig. 8.1), which is associated with a synorogenic tonalite-granite complex and resembles the Kopsa prospect (Tiainen et al., 2012). The gold deposits are described in Chapter 5.1, and in this chapter a more detailed account is presented only of the Ylöjärvi Cu-W deposit.

No important metallic ore deposits are known to be related to the late-orogenic Svecofennian granites of Finland, but the significant rare-element pegmatites in Kemiö (Eilu, 2012c), Somero-Tammela (Aurola, 1963; Ahtola, 2012a), Eräjärvi (Volborth, 1954; Lahti, 1981; Eilu, 2012d), and Kaustinen-Ullava (Ahtola, 2012b) are, according to current knowledge, late-orogenic (Nurmi and Haapala, 1986) (Fig. 8.1). Some Mo showings and prospects are associated with the evolved members (e.g., the Tepasto stock in Kittilä; Fig. 8.1) of the postorogenic, oxidized A-type Nattanen granites of Lapland (Haapala et al., 1987; Front et al., 1989; Heilimo et al., 2009).

YLÖJÄRVI Cu-W DEPOSIT

In 1937, the Geological Survey of Finland received a sample of sericite-rich sulfide ore taken from an outcrop in Ylöjärvi, 18 km northwest of Tampere, in central Finland. Field studies led in 1938 to the discovery of a copper ore in a tourmaline breccia in the Tampere schist belt. The deposit was exploited in 1940–1964 by the mining company Outokumpu Oy.

Hämeenkyrö Batholith

The Ylöjärvi deposit is located in the metavolcanic-metasedimentary rocks of the Tampere schist belt in the exocontact zone of the Hämeenkyrö granodiorite batholith (Fig. 8.3). The batholith is 14 km × 15 km in size and consists of multiply-intruded rocks ranging from marginal tonalite to prevailing granodiorite and central granite (Fig. 8.4; Nironen, 1989).

The granodiorite brecciates the tonalite. The most evolved rock type is feldspar porphyry, which is found as dikes in a narrow zone along the eastern contact of the batholith. The granitoids are even-grained or slightly porphyritic with subhedral microcline and zoned, subhedral to euhedral plagioclase (andesine to oligoclase) megacrysts. In chemical composition, they are calc-alkaline and show I-type characteristics, as the synorogenic Svecofennian granitoids in Finland generally do (Gaál et al., 1981; Nurmi and Haapala, 1986). Several types of hydrothermal alteration (carbonization, scapolitization, chloritization, tourmalinization) have affected both the granitoids and their wall rocks at the eastern margin of the batholith. The rocks of the batholith show, especially in the marginal parts, weak foliation that coincides with the penetrative foliation S1 of the surrounding wall rocks (Nironen, 1989). U-Pb zircon and titanite data on the Hämeenkyrö granodiorite imply a crystallization age of 1885 ± 2 Ma. The zircon ages of the metavolcanic rocks of the Tampere schist belt vary from 1880 to 1904 (Kähkönen, 2005).

Ore Deposit

Tourmaline breccia bodies and tourmaline veins as well as sericitized rocks are found in several localities in the metavolcanic rocks in the immediate vicinity of the eastern contact of the Hämeenkyrö batholith. The largest breccia body is the northeast-striking Parostenjärvi breccia pipe (Fig. 8.4), which is ≤80 m wide and 1 km long on the surface, shows a subvertical dip, and thins to the southwest, extending to the batholith contact. This breccia body is the only one that has been mined.

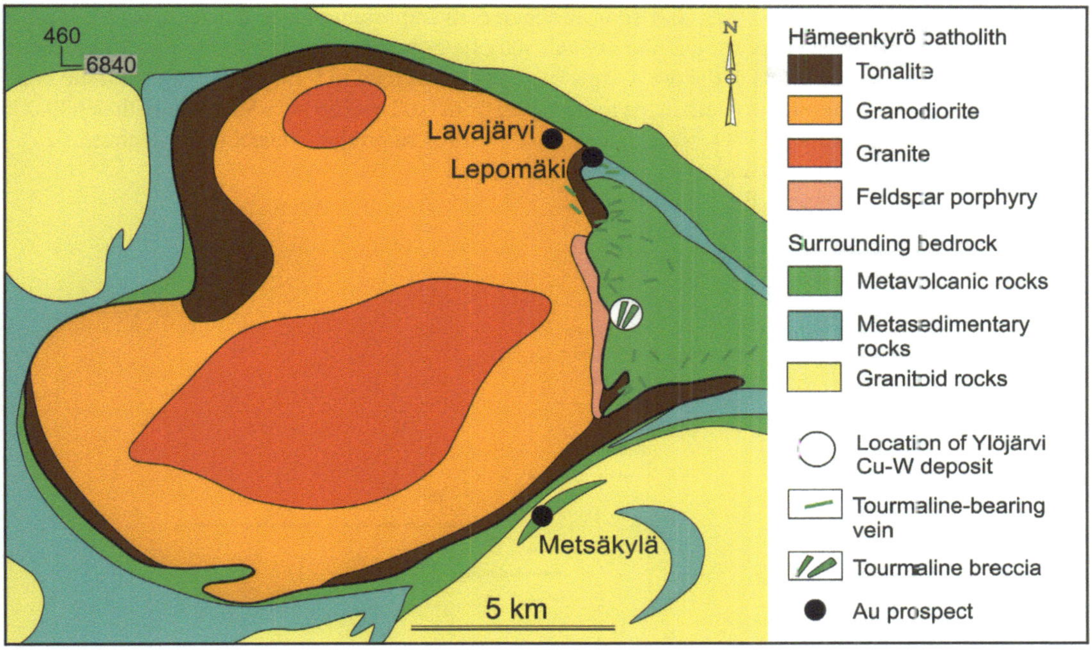

FIGURE 8.4 Geological map of the Hämeenkyrö granitoid batholith, showing the associated Ylöjärvi Cu-W deposit and tourmaline-bearing breccias and veins.

Source: *Based on Himmi et al. (1979), Gaál et al. (1981) and Nironen (1989).*

The host rocks of the Parostenjärvi breccia are mafic to intermediate metavolcanic rocks (tuffite, porphyrite, agglomerate). The breccia fragments vary in size from small pebbles to boulders several meters in diameter, averaging 20–50 cm. Most of the fragments are tuffitic, but porphyrite fragments are present as well. The fragments are enveloped by a pale, silicified alteration rim, and small fragments are completely silicified. The matrix is composed of iron-rich tourmaline and variable amounts of ore and other minerals (quartz, apatite, chlorite, rutile, etc.). The breccia grades through weakly brecciated to unbrecciated rock, and only the southeastern margin is marked by shearing.

The Ylöjärvi ore deposit is composed of four minable, subparallel and vertical ore bodies, all located in the widest (80 m) northeastern part of the breccia zone (Clark, 1965; Himmi et al., 1979). The most important one is ore body 1, which follows the southeastern boundary of the breccia zone. Its horizontal length is 190 m and average width 14 m, and it extends from the surface down to a depth of 330 m. The ore bodies do not exhibit any clear geological boundaries; they are distinguished from the rest of the tourmaline breccia only by higher sulfide grades. The main ore minerals are chalcopyrite, arsenopyrite, scheelite, pyrrhotite, magnetite, cassiterite, and sphalerite. The presence of metallic silver, gold, and bismuth is also noteworthy. Two ore mineral associations can be distinguished: (1) the arsenopyrite mineralization that is found in the chloritized wall rocks and in the chloritized and silicified breccia fragments, and (2) the chalcopyrite-arsenopyrite-pyrrhotite assemblage that prevails in the breccia matrix. The paragenesis of the main stage of the mineralization is shown in Fig. 8.5.

Scheelite was identified in 1945, and soon it became an important side-product of the mine. It is distributed randomly in the deposit, and the crystals may be grouped into bands, larger accumulates or nests, or occur as single crystals in the ore. Larger concentrations tend to follow the margins of the breccia zone. Although the tungsten content in the mine varied from 0.02 to 0.25 wt% WO_3, more than 670,000 tons of ore averaging 0.12 wt% WO_3 were treated in 1948–1961 to produce scheelite concentrate.

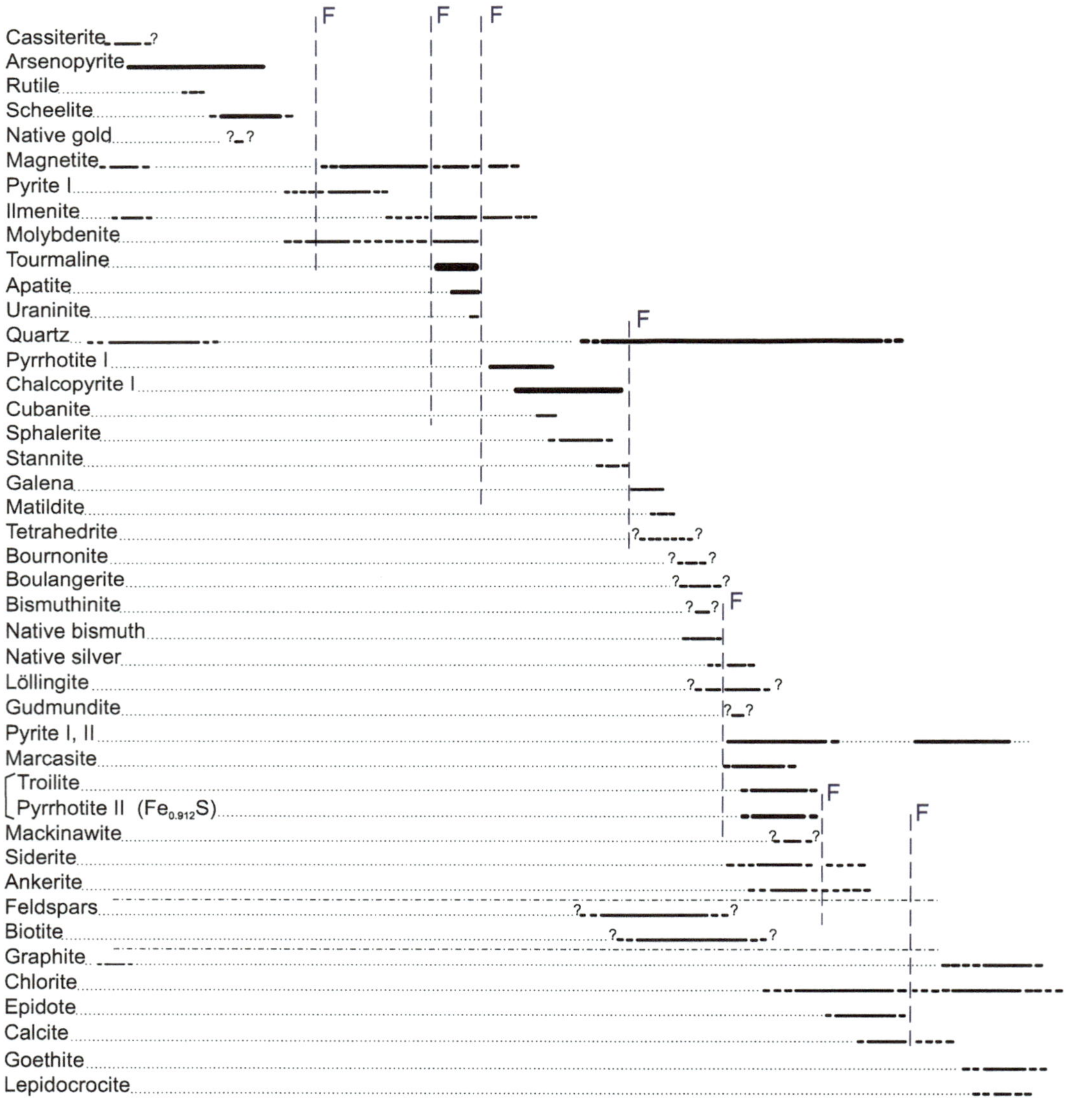

FIGURE 8.5 Paragenesis of the ore minerals in the Ylöjärvi Cu-W deposit.

Source: *After Clark (1965).*

The contents of ore elements vary in the deposit from one ore body to another. The mill feed of the years 1958–1965 serves as an average composition of the ore: S 1.34 wt%, Cu 0.68 wt%, As 0.46 wt%, WO_3 0.04 wt%, Co 0.02 wt%, Zn 0.017 wt%, Sn 0.013 wt%, and Mo 0.12 wt%; Ag 17 g/tn, Au 0.13 g/tn, and Se 12 g/tn (Himmi et al., 1979). Altogether, from 1840 to 1966 ~4.01 Mt of ore with 0.76 wt% Cu were hoisted, producing 28,000 t of Cu, 427 t of W, 49 t of Ag, and 270 kg of Au (Himmi et al., 1979; Kukkonen, 1967).

Ore Genesis

The genesis of the Ylöjärvi Cu-W deposit has been unanimously linked to the Hämeenkyrö granodiorite batholith and comparison has been made to the Tertiary tourmaline breccia pipes of the Central Andes (Clark, 1965; Himmi et al., 1979; Gaál et al., 1981; Nironen, 1989). The similar ages of the Ylöjärvi granitoids and the tourmaline breccias, the structure and texture of the tourmaline breccia, and the mineral composition of the ore are all compatible with the model that the breccia was generated by explosion of mineralizing hydrothermal fluids from the crystallizing Hämeenkyrö granitic magma, in the same way that the Chilean tourmaline breccia pipes and dikes were formed (see, e.g., Frikken et al., 2005). The repeated weaker fracturing and mineralization in different stages in the Ylöjärvi deposit can be explained by stages of later, weaker fluid invasions. The shear zone at the southeastern border of the breccia may have controlled the explosion and migration of the mineralizing fluids that generated the Ylöjärvi deposit, as suggested by Nironen (1989).

Sn-Be-W-Zn-In MINERALIZATION ASSOCIATED WITH RAPAKIVI GRANITES OF FINLAND AND RUSSIAN KARELIA

INTRODUCTION

In southern Finland and adjacent Russian Karelia, large rapakivi granite batholiths (Salmi, Wiborg, Laitila, Vehmaa, Åland) and smaller plutons intrude the Paleoproterozoic and Archean crust of the Fennoscandian Shield (Fig. 8.6).

Although skarn-type tin-copper-zinc and iron deposits were mined over a period of 60 years (1842–1904) in the Pitkäranta area near the western contact of the Salmi rapakivi batholith (Trüstedt, 1907), the rapakivi granites have been, and often still are, considered as barren rocks in regard to ore potential. In the late 1960s and the 1970s, the Geological Survey of Finland carried out reconnaissance exploration for tin in the rapakivi granite areas, based on apparent similarities between the rapakivi granite complexes and tin-bearing granite complexes. The general procedure was to first find the potential late-stage tin granites and then, by detailed exploration, the associated mineral occurrences. Even-grained and porphyritic granites marked on the existing geological maps were sampled and analyzed for selected elements, including Sn, F, and Rb, to test if they represent late-stage phases similar to the known tin granites. A number of greisen and lode-type tin-beryllium-zinc occurrences were found in association with topaz-bearing late-stage granites in the Laitila, Vehmaa, and Wiborg rapakivi granite areas (Haapala and Ojanperä, 1972; Haapala, 1977a,b, 1988). These topaz granites showed the mineralogical and geochemical characteristics of the tin granites (cf., Tischendorf, 1977).

The steel and mining company Rautaruukki Oy discovered greisen-type tin mineralization in the Ahvenisto anorthosite-rapakivi complex (Edén, 1991), and tin-bearing greisen veins were also found in the Åland batholith (Bergman, 1981). In the 2000s, indium mineralization was found in greisen veins in the Wiborg batholith (Cook et al., 2011) and in skarn rocks of the Salmi rapakivi granite area (Ivashchenko et al., 2011). These examples, together with the large tin deposits associated with late-stage phases of the Proterozoic rapakivi granite complexes in Brazil (Bettencourt et al., 2005; Costi et al.,

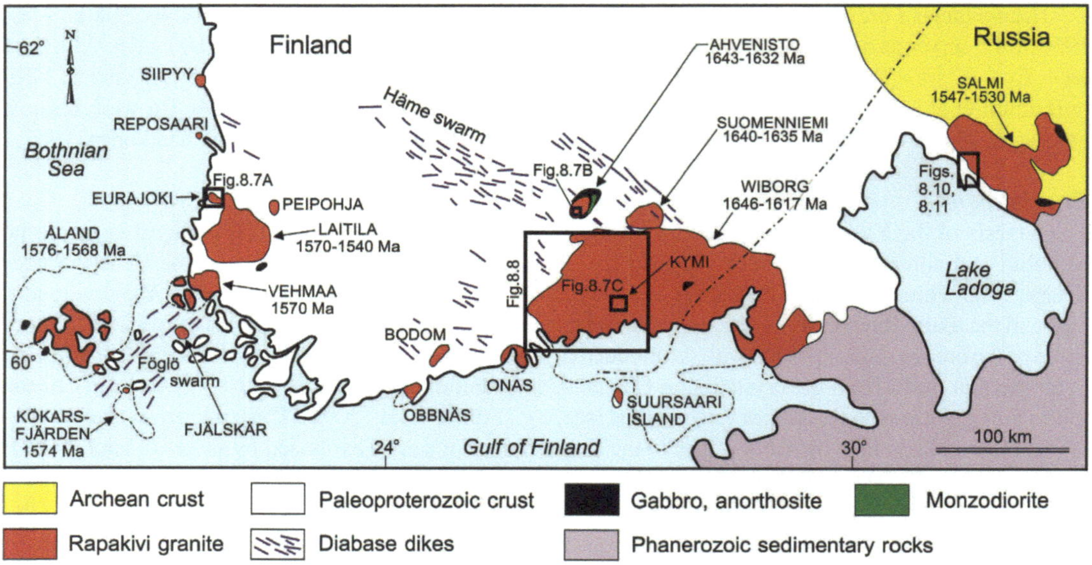

FIGURE 8.6 Location and ages of rapakivi granite batholiths and stocks in southern Finland and Russian Karelia.
The boxes indicate the areas for which more detailed maps are presented in Figs. 8.7, 8.9, and 8.10.

2009) and elsewhere (Haapala, 1995), indicate that the rapakivi complexes of Finland have considerable potential for tin-beryllium-tungsten-indium-zinc deposits. Because the Pitkäranta ore field in Russian Karelia complements the metallogenic picture of Finnish rapakivi granites with important skarn-type deposits, a short description of these deposits is included.

Long known are some small epigenetic, galena-rich quartz veins in the Wiborg rapakivi granite area, and some of them (Säkkijärvi, Hopiala in Luumäki) have been excavated in the past under irregular times (Vaasjoki, 1977). They are not discussed in this chapter.

EURAJOKI STOCK

The Eurajoki stock (8–9 km in diameter) in southwestern Finland is a satellite of the Laitila rapakivi granite batholith. It is composed of a marginal, even-grained fayalite-bearing biotite-hornblende granite known as the Tarkki granite, and a younger, more evolved central granite mass called the Väkkärä granite (Fig. 8.7A).

The Väkkärä granite consists of an even-grained biotite granite and a porphyritic topaz granite, in which the biotite is lithium siderophyllite. Various petrographic features (Haapala, 1974, 1977a, 1997) and melt inclusion studies (Haapala and Thomas, 2000) show that topaz in this granite is mainly magmatic, and to a small degree, secondary. Typical accessory minerals in the topaz granite include monazite, xenotime, bastnäsite, Nb- and Ta-rich cassiterite, columbite, zircon, and thorite. Miarolitic cavities and pegmatite-lined druses suggest that the topaz granite crystallized from a volatile-saturated magma. Rhyolitic and hybrid porphyry dikes cut the Tarkki granite and the biotite granite. The topaz granite is geochemically highly evolved showing high Rb, F, Sn, Nb, and Ga and low Ti, Zr, Ba, and Sr (Table 8.1; Haapala, 1977a).

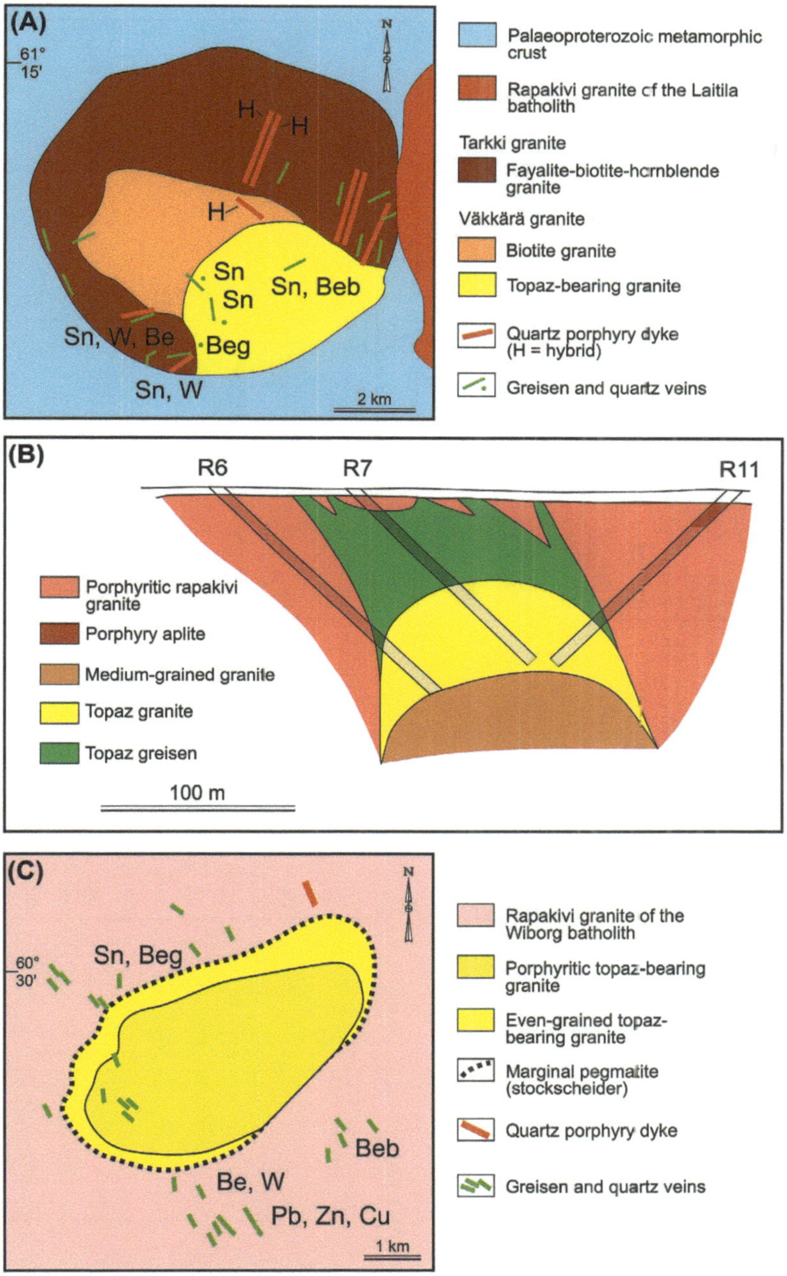

FIGURE 8.7 Lithologic variation in late-stage intrusion phases of the Finnish rapakivi granite complexes and associated greisen mineralization.

(A) The Eurajoki stock (Haapala, 1977a) (B) the Kuusisuo late-stage granite and greisen (simplified from Edén, 1991), and (C) the Kymi stock (Haapala and Ojanperä, 1972). In (A) and (B), important ore elements are indicated. Beryllium is bound partly in beryl (Be), partly in genthelvite (Beg) and bertrandite Beb.

Table 8.1 Chemical analyses of wiborgite and pyterlite from the Wiborg batholith, of the normal Laitila rapakivi granite, and of the topaz-bearing granites from the Eurajoki, Ahvenisto, Kymi, and Salmi complexes

	1	2	3	4	5	6	7	8
SiO_2	69.93	75.69	71.16	75.78	73.98	74.12	73.83	75.60
TiO_2	0.49	0.22	0.42	0.02	0.02	0.08	0.03	0.07
Al_2O_3	13.41	11.70	14.24	13.62	14.03	12.76	14.83	13.60
Fe_2O_3	0.45	0.31	0.68	0.19	0.83	0.88	0.27	0.38
FeO	3.4	1.5	2.54	0.65	1.04	1.22	0.47	1.79
MnO	0.06	0.03	0.06	0.05	0.04	0.02	0.02	0.03
MgO	0.27	0.10	0.20	0.00	0.01	0.13	0.07	0.22
CaO	2.07	1.08	1.62	0.64	0.70	1.17	0.73	0.97
Na_2O	2.97	2.49	2.63	3.70	2.82	2.84	4.24	2.23
K_2O	5.55	5.76	5.31	4.35	4.88	5.22	4.46	4.45
P_2O_5	0.13	0.03	0.08	0.02	0.01	0.01	0.01	<0.03
F	0.24	0.43	0.22	1.19	1.43	0.85	1.22	1.10
Trace elements in ppm								
Li	na	na	34	na	210	102	242	686
Rb	271	349	260	1050	740	634	1103	893
Be	5	5	na	16	na	37	14	na
Sr	155	76	120	8	20	43	20	21
Ba	1144	541	1180	28	na	304	105	50
Ga	27	24	na	60	na	48	68	na
Zr	460	300	318	51	70	231	22	52
Sn	8	9	na	110	40	37	9	na
Nb	26	26	na	70	80	166	60	79.3
Ta	2	2	na	23	na	24	43	30.5
W	3	3	na	9	na	22	18	na

[1] *Wiborgite from the Wiborg batholith, sample 1A/2001 (Haapala and Lukkari, 2005).*
[2] *Pyterlite from the Wiborg batholith, sample 2A/IH/2001 (Haapala and Lukkari, 2005).*
[3] *Normal Laitila rapakivi granite, mean of five analyses (analyses 3–7, Appendix 2, Vorma, 1976).*
[4] *Topaz-bearing Väkkärä granite, Eurajoki, sample 5/IH/2001 (Haapala et al., 2005).*
[5] *Topaz-bearing granite, Ahvenisto complex, mean of 26 analyses (Edén, 1991).*
[6] *Porphyritic topaz-bearing granite, Kymi stock, mean of five analyses (Haapala and Lukkari, 2005).*
[7] *Equigranular topaz-bearing granite, Kymi stock, mean of three analyses (Haapala and Lukkari, 2005).*
[8] *Coarse-grained albite-lithian siderophyllite granite, sample 403-13, western margin of the Salmi batholith, Russian Karelia (Larin, 2009).*
[na] *Not analyzed*

Greisen-type mineralization is found in different parts of the Eurajoki stock. The mode of occurrence and mineral composition of the greisen bodies vary markedly depending on the locality and host rock. In the Tarkki granite, the greisens are present as veins that often have a central quartz veinlet. Subparallel, usually ≤20-cm-wide greisen veins form swarms or wider lodes, which have been followed in several places hundreds of meters in the outcrops or drill holes. The ubiquitous gangue minerals are quartz, dark mica (brown or green), white mica, chlorite, topaz, and fluorite; and locally, garnet and gahnite are abundant. The metallic ore minerals include cassiterite, wolframite, sphalerite, molybdenite, chalcopyrite, galena, pyrite, arsenopyrite, pyrrhotite, stannite, roquesite $CuInS_2$, sakuraiite $(Cu,Zn,Fe)_3(In,Sn)S_4$, and argentite. Roquesite is found as inclusions in chalcopyrite, sakuraiite in sphalerite (Kari Kojonen, written communication in 2015). Sphalerite is nearly always present in the greisens, whereas beryl and wolframite are found in the central veinlets. Examples of typical zonal structures of the veins are shown in Fig. 8.8. Although there are ore shoots with tin contents up to 20 wt%, no minable deposits have been found.

In the topaz-bearing Väkkärä granite, the greisen is found as irregular lenses and zones that may be a few meters wide, as small rounded or irregular patches, and as irregularly oriented veins. Cassiterite is frequently present, but the greisens have ≤0.5 wt% Sn. Of special interest are the Be-minerals

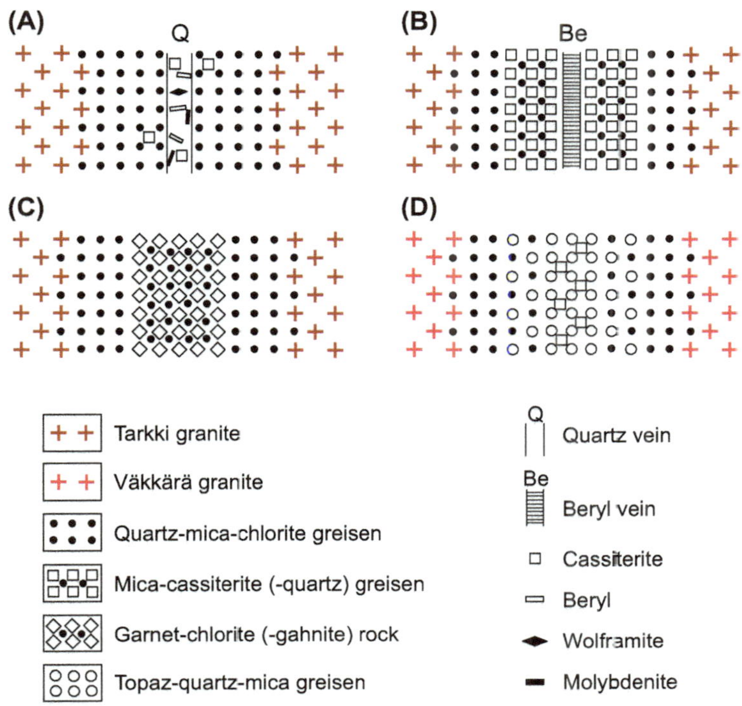

FIGURE 8.8 Examples, without scale, of the zonal structure of the greisen veins in the Eurajoki stock. The greisen veins of Figs. A-C are hosted by the Tarkki granite, of Fig. D by the Väkkärä granite.

Source: *Modified from Haapala (1977a).*

genthelvite (Zn, Fe, Mn)$Be_3Si_3O_{12}S$, and bertrandite $Be_4(OH)_2Si_2O_7$, which locally are among the major minerals in the greisens (Haapala and Ojanperä, 1972; Haapala, 1977a). Cassiterite is also found in small pegmatite pockets and pegmatite veins in the topaz granite.

The few available observations of the contact relations suggest that the topaz-bearing Väkkärä granite forms a cupola-shaped intrusion into the Tarkki granite. Obviously, the more massive greisen zones of the cupola have been eroded away. Yet it is possible that mineralized subsurface topaz granite cupolas are present below the greisen vein swarms found in the Tarkki granite.

AHVENISTO ANORTHOSITE-RAPAKIVI GRANITE COMPLEX

The satellitic Ahvenisto (Mäntyharju) complex on the northwestern side of the Wiborg batholith consists mainly of granitic rocks (70%), gabbroic rocks (25%), and monzodiorites (≤5%) (Alviola et al., 1999; Heinonen et al., 2010). The granites include hornblende granite, hornblende-biotite granite, biotite granite, topaz granite, porphyritic aplite, and aplite. In the late 1970s, the Rautaruukki Oy company discovered, on the basis of tin anomaly in till, tin-bearing greisen veins and a hidden greisenized topaz-granite–biotite-granite cupola at Kuusisuo, inside the Ahvenisto complex (Edén, 1991; Fig. 8.7B). The Kuusisuo granites, especially the topaz granite, are characterized by high F, Rb, Li, Sn, Nb, U, and Th, and by low TiO_2, Fe, MgO, CaO, MnO, P_2O_5, La, Ba, and Sr (see Table 8.1). Cassiterite is an accessory mineral in both granites. The greisen at the Kuusisuo granite cupola is tens of meters thick and contains approximately 50% quartz, 30% mica minerals and chlorite, and 15% topaz. Ore minerals include cassiterite, arsenopyrite, chalcopyrite, sphalerite, and pyrite, but metal contents are too low for mining. Some greisen veins also contain elevated Ag and Bi contents.

WESTERN MARGIN OF THE WIBORG BATHOLITH

In the western part of the Wiborg batholith, there are several stocks of porphyritic or even-grained biotite granite. In the exploration work carried out by the Geological Survey of Finland in the late 1960s and 1970s, geochemically evolved and, in many cases, topaz-bearing late-stage granites were found in some of these stocks. Greisen veins (Fig. 8.9) were discovered in or near such granites in Liljendal, Artjärvi, and Sääskjärvi (see Lukkari, 2002).

Lukkari and Holtz (2007) and Lukkari et al. (2009) have studied the magmatic evolution of the topaz granites within the Wiborg batholith utilizing experimental petrological and melt inclusion studies. The greisen veins contain sulfides (mainly sphalerite) and cassiterite as ore minerals, but not in ore grade. Significant skarn-type deposits have not been found. The old Perheniemi limestone quarry at the endocontact of the batholith contains Sn-enriched vesuvianite in skarn, but no cassiterite was detected.

New greisen veins, magnetite-sphalerite veins and lenses, and sulfide-bearing quartz veinlets enriched in indium were discovered in the 2000s along the western margin of the Wiborg batholith (Cook et al., 2011). Indium is partly contained in sphalerite (up to 0.5 wt% In), and in part it forms an own sulfide mineral, roquesite ($CuInS_2$). Roquesite is found in the Sarflaxviken area in Pernaja, in 0.5–2 cm-wide sulfide-quartz veinlets together with sphalerite, chalcopyrite, pyrite, stannoidite, arsenopyrite, cassiterite, and wolframite. Also of interest is the 3-m-wide magnetite-sphalerite lode, which is located between rapakivi granite and ignimbrite in Getmossmalmen, Pernaja. The mean of four analyses showed 10.3 wt% Zn, 26.7 wt% Fe, and 38 ppm In (Cook et al., 2011).

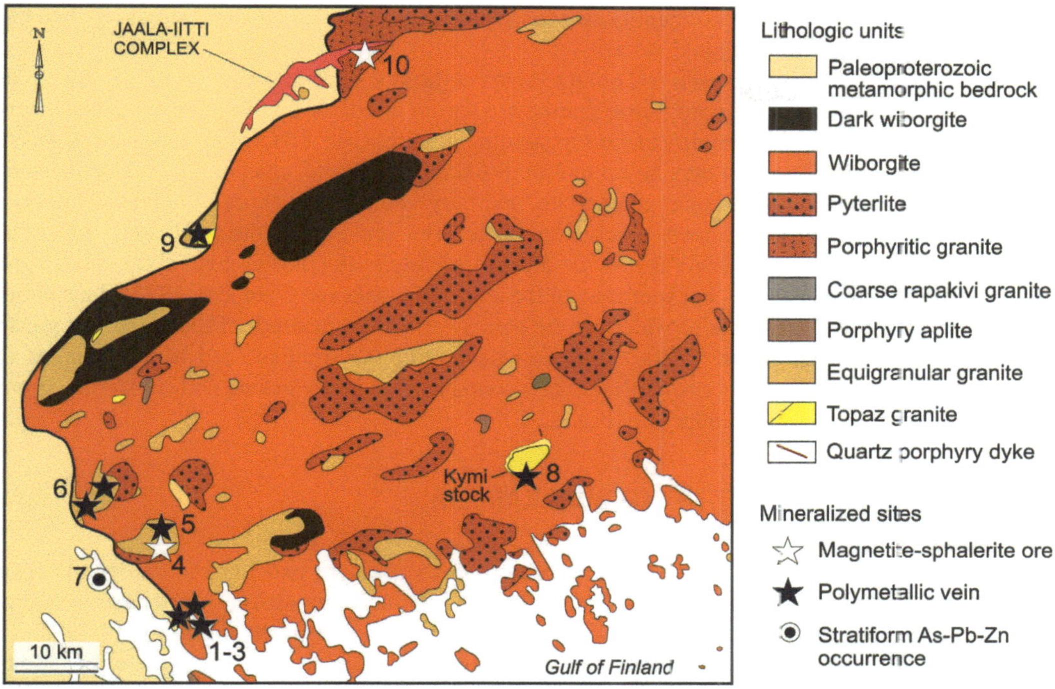

FIGURE 8.9 Geological map of the western margin of the Wiborg rapakivi granite batholith showing sites of hydrothermal mineral occurrences.

The numbered stars refer to greisen-type and vein-type mineralization: 1–3, Sarvlaxviken area, quartz veins with Cu-In-As-Sn mineralization; 4, Getmossmalmen, In-bearing magnetite-sphalerite ore lenses; 5, Jungfrubergen, greisen vein with Zn-Cu-Pb minerals; 6, Liljendal area, greisen vein swarm; 8, Kymi, Sn-Be-W- mineralized greisen veins; 9, greisen veins in Artjärvi and Sääskjärvi; 10, Pahasaari, sphalerite and magnetite-bearing greisen vein.

Source: *Slightly modified from Cook et al. (2011).*

KYMI TOPAZ GRANITE STOCK

The Kymi stock in the central western part of the Wiborg batholith is a 5-km-long and 3-km-wide topaz granite cupola that intrudes older rapakivi granites including wiborgite, pyterlite, and a sparsely porphyritic granite. The stock consists of a porphyritic central granite and an even-grained marginal granite, and the cupola is rimmed by a marginal topaz pegmatite (stockscheider) (Fig. 8.7C). Both granites are leucocratic, topaz-bearing late-stage rocks, but the even-grained granite is younger and more evolved. The biotite is siderophyllite in the porphyritic granite, and lithian siderophyllite in the even-grained granite (Haapala and Lukkari, 2005; Lukkari et al., 2009). The accessory minerals in the porphyritic granite are topaz, fluorite, zircon, ilmenite, anatase columbite, monazite, thorite, and molybdenite, and in the even-grained granite topaz, fluorite, monazite, bastnäsite, columbite, microlite, and apatite (Haapala, 1974). As other topaz-bearing rapakivi granites, these granites, especially the even-grained one, are geochemically highly evolved (Table 8.1). However, the tin content is not as high as in the topaz granite of the Eurajoki stock.

Northwest–north-trending greisen and quartz veins cut the stock granites and the surrounding older granites. The veins found in the porphyritic granite are barren quartz or quartz-fluorite veins, whereas mineralized greisen veins are common in the surrounding granites (Fig. 8.7C). The width of the greisen veins is commonly 5–30 cm, but may be up to 3 m. Sulfides (sphalerite, chalcopyrite, galena, arsenopyrite, pyrite), cassiterite, wolframite, genthelvite, and beryl are present as ore minerals. Genthelvite occurs in small quantities in two greisen veins on the northwestern side of the stock, and more abundantly in an irregular, approximately 3-m-wide phengitic muscovite-chlorite-genthelvite-quartz greisen vein located approximately 1 km southeast of the stock (Haapala and Ojanperä, 1972). An example of normal hydrothermal zoning is provided by the vein swarm 1–2 km south of the stock: the greisen veins closer to the topaz granite contact contain arsenopyrite, wolframite, and beryl, whereas the 1-m-wide quartz vein farther away from the contact contains abundantly galena and sphalerite, as well as some chalcopyrite and rare genthelvite (Fig. 8.7C).

PITKÄRANTA ORE DISTRICT IN RUSSIAN KARELIA

Introduction

Greisen and vein-type tin-polymetallic ores are found in the western exocontact zone of the Salmi batholith (Fig. 8.10). Copper was discovered in the area in 1810, and later also cassiterite was identified. Production of tin and copper started in 1842 and continued, with short interruptions, until 1904. The total production from that time was ~6600 t of Cu, 490 t of Sn, 11 t of Ag, and 100,000 t of Fe, and the average metal contents 1.35 wt% Cu, 0.1 wt% Sn, and 40 wt% Fe (Trüstedt, 1907; Korsakova, 2012). The exploitation of the deposits took place during the time when Finland, separated from Sweden in 1809, was united to Russia as a grand duchy. The ore district was studied thoroughly by Trüstedt (1907). After Finland attained independence in 1917, the ore district was studied by Palmunen (1939), Saksela (1951), and Eskola (1951). After Ladogan Karelia was again united to the Soviet Union (Russia) in 1940, the area was studied intensely by Russian geoscientists utilizing geophysical, geological, mineralogical, and geochemical methods and deep drillings (Khasov, 1973; Amelin et al., 1991 and 1997).

Salmi Batholith

The Salmi rapakivi granite–anorthosite batholith is situated at the contact between the Archean granite gneiss–greenstone belt basement and Paleoproterozoic Svecofennian orogenic belt. On the eastern side of the batholith is the satellitic Uljalegi rapakivi granite pluton. The batholith is approximately 85 km long, 35 km wide, and 10 km thick. It is composed of six main rock types: (1) anorthosite-gabbronorite-gabbro association, (2) monzonite-quartz syenite-syenite association, (3) biotite-amphibole syenogranite (Uljalegi pluton), (4) amphibole-biotite rapakivi granites (wiborgite and pyterlite), (5) biotite granite, and (6) topaz-bearing albite-lithian sideophyllite ("protholithionite") granite (Amelin et al., 1991, 1997). The gabbroic rocks, monzonite, and syenite dominate the buried southern end of the batholith. According to isotopic U-Pb datings, the rocks of the batholith crystallized in several episodes within a timespan of 17 million years. The gabbroic rocks and monzonites are the oldest, ~1547 Ma, and the granitic rocks crystallized between 1543 and 1530 Ma. The highly evolved albite-lithian siderophyllite granites (Table 8.1) are present as dikes and small stocks in the amphibole-biotite and biotite granites as well as in the metamorphic country rocks.

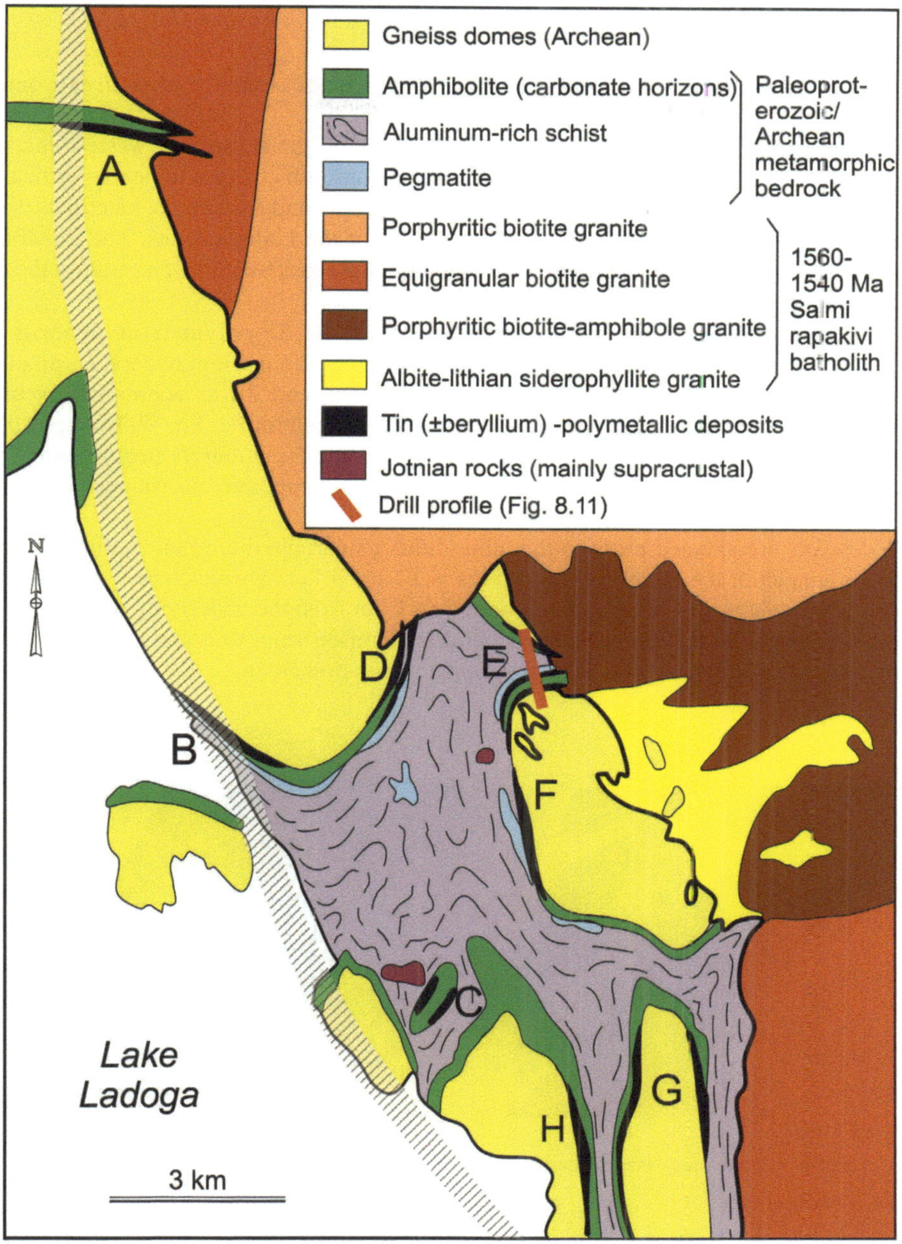

FIGURE 8.10 Geology of the western margin of the Salmi rapakivi granite batholith in Russian Karelia.

The letters refer to the mineral deposits: A–C, tin-polymetallic deposits (A, Kitelä; B, Old Ore Field; C, Heposelkä); D–H, beryllium-tin-polymetallic deposits (D, New Ore Field; E, Hopunvaara; F, Lupikko; G, Uuksu; H, Ristiniemi). The shadowed zone is a projection of the western margin of the west-dipping Salmi batholith to the present erosion level.

Source: *Simplified from Amelin et al. (1991).*

Ore Deposits

The ore deposits are found in a 50-km-long and 4–5 km wide belt along the western exocontact of the batholith (Fig. 8.10).

The western margin of the mineralized belt coincides with the subsurface western margin of the batholith, where the gently westward dipping roof of the batholith changes to nearly vertical (Amelin et al., 1991). The mineralization is mainly restricted to the upper and lower carbonate rock/skarn layers of the Pitkäranta suite below a thick pile of mica schists of the Ladoga series. The Pitkäranta suite, which also includes an amphibolite bed between the carbonate rock/skarn layers, frames the deformed Archean granite-gneiss domes.

Two types of ore mineralization are distinguished: (1) Be-Sn-polymetallic deposits with Be, Sn, Cu, Zn, and fluorite as important components (New Ore Field, Hopunvaara, Lupikko, Uuksu, Ristiniemi), and (2) Sn-polymetallic deposits with Sn, Cu, and Zn as economically interesting metals (Kitelä, Old Ore Field, Heposelkä). Both types also contain Fe, Pb, W, Bi, Ag, Au, Cd, and In. Beryl, chrysoberyl, bavenite, and helvite (group) are the Be minerals present, and In is contained as a minor component in sphalerite or as a major component in roquesite (Ivashchenko et al., 2011).

Type 1 deposits are situated close to the albite-lithian siderophyllite granites that are found at or near the upper contact of the Salmi batholith (Figs. 8.10 and 8.11), whereas type 2 deposits are generally found at larger distances, in some cases more than 1 km from the underlying roof of the batholith, near the outer edge of the batholith. Some mineralized greisen veins and greisen zones are met also within the rapakivi batholith, as well as in the granite gneiss domes.

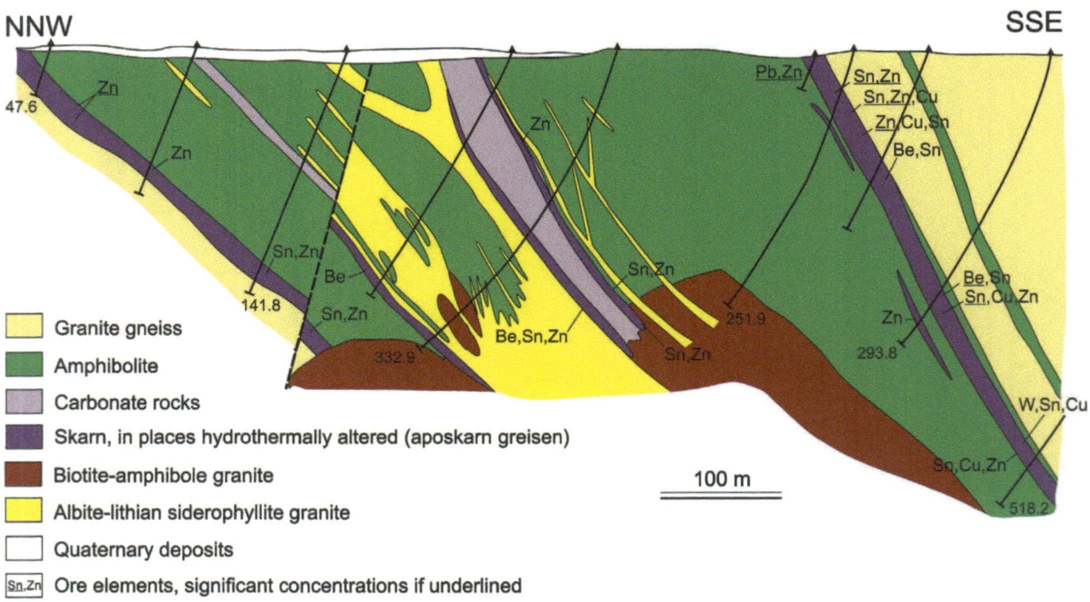

FIGURE 8.11 Geological profile of the Hopunvaara deposit.

Source: *Slightly modified from Amelin et al. (1991).*

The mineralization has taken place in several stages, and different types of metasomatites can be distinguished: magnesian skarns and lime skarns, fluorite-vesuvianite-magnetite metasomatites, greisenized skarns (aposkarn greisens), propylite, and so on. Outside of the contact halo of the batholith barren synmetamorphic (1950 Ma), magnesian skarns are common. There is no current mining of the Pitkäranta ores. However, the list of ore deposits with in situ metal contents and total ore tonnages published by Korsakova (2012) shows that there are still remarkable ore reserves in the Pitkäranta ore district.

REMARKS

The working principle in the reconnaissance exploration studies carried out by the Geological Survey of Finland in the late 1960s and 1970s was to find first the typical tin granites and then the related mineral occurrences. This method led to the discovery of a number of tin-polymetallic occurrences, but it had an inherent weakness. The best tin deposits are in many areas situated at cusps or carapaces of the tin granite intrusions rather than in the deeper, inner parts of the plutons. Such small apical sections of the late-stage intrusions would only rarely be visible on outcrops.

Our recommendation for further prospecting in the rapakivi granite areas is to proceed with systematic till geochemical surveying in promising areas such as the western part of the Wiborg batholith and its satellites. When an interesting anomaly is found, the nature of its source (hydrothermal/magmatic) can be estimated from the Nb and Ta contents of the cassiterite grains of the till samples (Haapala, 1983b). Till geochemistry and heavy mineral studies have been applied successfully to exploration for tin in the Eurajoki and Ahvenisto areas (Lehmuspelto, 1973; Peuraniemi and Heinänen, 1985). If there are carbonate rocks near evolved late-stage rapakivi granites, there also are possibilities for skarn-type deposits. The variable mineral associations found in the occurrences indicate that, besides tin, beryllium, indium, tungsten, copper, and zinc may also be important ore elements in rapakivi granite terrains.

ACKNOWLEDGMENTS

Professors Eero Hanski and Roberto Dall'Agnol provided useful comments on the manuscript, which is gratefully acknowledged. Markku Tiainen and Kirsti Keskisaari are thanked for their kind help in preparing some of the figures, and Kari Kojonen for identification of the indium minerals from the Eurajoki stock.

REFERENCES

Ahtola, T., 2012a. Somero Li. In: Eilu, P. (Ed.), Mineral Deposits and Metallogeny of Fennoscandia. Geological Survey of Finland, pp. 217–218. Special Paper 53.

Ahtola, T., 2012b. Emmes Li. In: Eilu, P. (Ed.), Mineral Deposits and Metallogeny of Fennoscandia. Geological Survey of Finland, pp. 217–218. Special Paper 53.

Alviola, R., Johanson, B.S., Rämö, O.T., Vaasjoki, M., 1999. The Proterozoic Ahvenisto rapakivi granite massif- type anorthosite complex, southwestern Finland: petrography and geochronology. Precambrian Research 95, 89–107.

Amelin, Yu., Beljaev, A., Larin, A., et al., 1991. Salmi batholith and Pitkäranta ore field in Soviet Karelia. In: Haapala, I., Rämö, O.T., Salonsaari, P.T. (Eds.), IGCP Project 315 Symposium Rapakivi Ganites and Related Rocks. Geological Survey of Finland, p. 57. Guide 33.

Amelin, Y.V., Larin, A.M., Tucker, R.D., 1997. Chronology of multiphase emplacement of the Salmi rapakivi granite-anorthosite complex, Baltic Shield: implications for magmatic evolution. Contributions to Mineralogy and Petrology 127, 353–368.

Aurola, E., 1963. On the pegmatites in the Torro area, southwestern Finland. Bulletin Commission géologique Finlande 206, 32.

Bergman, L., 1981. Berggrunden inom Signilskär, Mariehamn och Geta kartblad. Signilskärin, Maaranhaminan ja Getan kartta-alueiden kallioperä. Summary: Pre-Quaternary rocks of the Signilskär, Mariehamn and Geta map-sheet areas. Kallioperäkarttojen selitykset – Explanation to the maps of Pre-Quaternary rccks, lehdet – sheets 0034 + 0043, 1012, and 1021. Geological map of Finland 1:100 000, Geological Survey of Finland, p. 72.

Bettencourt, J., Leite Jr., W.B., Goraieb, C.L., et al., 2005. Sn-polymetallic greisen-type deposits associated with late-stage rapakivi granites, Brazil fluid inclusion and stable isotope characteristics. Lithos 80, 363–386.

Clark, A.H., 1965. The composition and conditons of formation of arsenopyrite and löllingite in the Ylöjärvi copper-tungsten deposit, southwest Finland. Geological Survey of Finland, p. 56 Bulletin 217.

Cook, N.J., Sundblad, K., Valkama, M., et al., 2011. Indium mineralization in A-type granites of southeastern Finland: Insight into mineralogy and partition between coexisting minerals. Chemical Geology 284, 62–73.

Costi, H.T., Dall'Agnol, R., Pichawant, M., Rämö, O.T., 2009. The peralkaline tin-mineralized Madeira cryolite albite-rich granite of Pitinga, Amazonian craton, Brazil: Petrography, mineralogy and crystallization processes. Canadian Mineralogist 47, 1301–1327.

Edén, P., 1991. A specialized topaz-bearing rapakivi granite and associated mineralized greisen in the Ahvenisto complex, SE Finland. Bulletin of the Geological Society of Finland 63, 25–40.

Eilu, P., 2012a. Tampere Au, Cu. In: Eilu, P. (Ed.), Mineral Deposits and Metallogeny of Fennoscandia. Geological Survey of Finland, pp. 224–227. Special Paper 53.

Eilu, P., 2012b. Gold mineralization in southwestern Finland. In: Grönholm, S., Kärkkäinen, N. (Eds.), Gold in Southern Finland. Geological Survey of Finland, pp. 11–22. Special Paper 52.

Eilu, P., 2012c. Kemiö Ta. In: Eilu, P. (Ed.), Mineral Deposits and Metallogeny of Fennoscandia. Geological Survey of Finland, p. 212. Special Paper 53.

Eilu, P., 2012d. Eräjärvi Ta-Li-Be. In: Eilu, P. (Ed.), Mineral Deposits and Metallogeny of Fennoscandia. Geological Survey of Finland, p. 224. Special Paper 53.

Eilu, P., Kärkkäinen, N., 2012. Pirkkala Au. In: Eilu, P. (Ed.), Mineral Deposits and Metallogeny of Fennoscandia. Geological Survey of Finland, pp. 222–223. Special Paper 53.

Eklund, O., Konopelko, D., Rutanen, H., et al., 1998. 1.8 Ga Svecofennin post-collisional shoshonitic magmatism in the Fennoscandian Shield. Lithos 45, 87–108.

Eskola, P., 1951. Around Pitkäranta. Annales Academiae Scientiarum Fennicae A III 27 90.

Frikken, P.H., Walshe, J.L., Archibald, D., et al., 2005. Mineralogical and isotopic zonation in the Sur-Sur tourmaline breccia, Rio Blanco-Los Bronces Cu-Mo deposit, Chile: Implications for ore genesis. Economic Geology 100 (5), 935–965.

Front, K., Vaarma, M., Rantala, E., Luukkonen, A., 1989. Keski-Lapin varhaisproterotsooiset Nattas-tyypin granitittikompleksit, niiden kivilajit, geokemia ja mineralisaatiot. Summary: Early Proterozoic Nattanen-type granite complexes in central Lapland: rock types, geochemistry and mineralization. Geological Survey of Finland, p. 75 Report of Investigation 85.

Gaál, G., Isohanni, M., 1979. Characteristics of igneous intrusions and various wall rocks in some Precambrian porphyry copper-molybdenum deposits in Pohjanmaa, Finland. Economic Geology 74, 1198–1210.

Gaál, G., Front, K., Aro, K., 1981. Geochemical exploration of a Precambrian batholith, source of Cu-Mo mineralization of the tourmaline breccia type in southern Finland. Journal of Geochemical Exploration 15, 683–698.

Haapala, I., 1974. Some petrological and geochemical characteristics of rapakivi granite varieties associated with greisen-type Sn, Be, and W mineralization in the Eurajoki and Kymi areas, southern Finland. In: Stemprok, M. (Ed.), Metallization Associated with Acid Magmatism I. Ustrední ustav geologicky, pp. 159–169 Praha.

Haapala, I., 1977a. Petrography and geochemistry of the Eurajoki stock, a rapakivi-granite complex with greisen-type mineralization in southwestern Finland. Geological Survey of Finland 128 Bulletin 286.

Haapala, I., 1977b. The controls of tin and related mineralizations in the rapakivi-granite areas of southeastern Fennoscandia. Geologiska Föreningen i Stockholm Förhandlingar 99, 130–142

Haapala, I., 1983a. Metallogeny of the Precambrian granitoids of Finland. In: Beus, A.A. (Ed.) Dokempriyskikh Granitoidov, Nauka, Moscov, pp. 25–71 (in Russian, with an English Summary).

Haapala, I., 1983b. Composition of cassiterite and its application in geochemical prospecting. In: Björklund, A., Koljonen, T. (Eds.), Proceedings of the 10th International Geochemical Exploration Symposium—Third Symposium on Methods of Geochemical Prospecting, Espoo/Helsinki, Finland, pp. 35–36 Abstracts.

Haapala, I., 1988. Metallogeny of the Proterozoic rapakivi granites of Finland. In: Taylor, R.P., Strong, D.F. (Eds.), Recent Advances in the Geology of Granite-related Mineral Deposits. Canadian Institute of Mining and Metallurgy, pp. 124–132 Special Volume 39.

Haapala, I., 1995. Metallogeny of the rapakivi granites. Mineralogy and Petrology 54, 149–160.

Haapala, I., 1997. Magmatic and postmagmatic processes in tin-mineralized granites: topaz-bearing granite in the Eurajoki rapakivi granite stock, Finland. Journal of Petrology 38, 1645–1659.

Haapala, I., Front, K., Rantala, E., Vaarma, M., 1987. Petrology of Nattanen-type granite complexes, northern Finland. Precambrian Research 35, 225–240.

Haapala, I., Lukkari, S., 2005. Petrological and geochemical evolution of the Kymi stock, a topaz granite cupola within the Wiborg rapakivi batholith. Finland. Lithos 80, 347–362.

Haapala, I., Ojanperä, P., 1972. Genthelvite-bearing greisens in southern Finland. Geological Survey of Finland 22 Bulletin 259.

Haapala, I., Rämö, O.T., 1992. Tectonic setting and origin of the rapakivi granites of southeastern Fennoscandia. Transactions of the Royal Society of Edinburgh 165–171 Earth Sciences 83.

Haapala, I., Thomas, R., 2000. Melt inclusions in quartz and topaz of the topaz granite from Eurajoki, Finland. Journal of the Czechoslovakian Geological Society 45, 149–154.

Heilimo, E., Halla, J., Lauri, L.S., et al., 2009. The Paleoproterozoic Nattanen-type granites in northern Finland and vicinity—a postcollisional oxidized A-suite. Bulletin of the Geological Society of Finland 81 7–38 Part I.

Heilimo, E., Halla, J., Mikkola, P., 2012. Overview of Neoarchaean sanuktoid series in the Karelia Province, eastern Finland. In: Hölttä, P. (Ed.), The Archaean of the Karelia Province in Finland. Geological Survey of Finland, pp. 214–225. Special Paper 54.

Heinonen, A.P., Rämö, O.T., Mänttäri, I., et al., 2010. Formation and fractionation of of high-Al tholeiitic magmas in the Ahvenisto rapakivi granite—massif-type anorthosite complex, southeastern Finland. Canadian Mineralogist 48, 969–990.

Himmi, R., Huhma, M., Häkli, A., 1979. Mineralogy and metal distribution in the copper-tungsten deposit of Ylöjärvi, Southwest Finland. Economic Geology 74, 1183–1197.

Huhma, H., 1986. Sm-Nd, U-Pb and Pb-Pb isotopic evidence for the origin of the Early Proterozoic Svecokarelian crust in Finland. Geological Survey of Finland, p. 48 Bulletin 337.

Huhma, H., O'Brien, H., Lahaye, Y., Mänttäri, I., 2011. Isotope geology and Fennoscandian lithosphere evolution. Geological Survey of Finland 35–48 Special Paper 49.

Ivashchenko, V.I., Valkama, K., Sundblad, K., et al., 2011. New data on mineralogy and metallogeny of scarns in the Pitkyaranta ore region. Doklady Earth Sciences 440, 1307–1311.

Kähkönen, Y., 2005. Svecofennian supracrustal rocks. In: Lehtinen, M., Nurmi, P.A., Rämö, O.T. (Eds.), Precambrian Geology of Finland – Key to the Evolution of the Fennoscandian Shield. Elsevier B.V., Amsterdam, pp. 343–406.

Kinnunen, A., 2008. A Palaeoproterozoic high sulphidation epithermal gold deposit at Orivesi, southern Finlad. Acta Universitatis Ouluensis A 507, p. 178.

Khasov, R.A., 1973. Geologicheskie osbennosti olovyannogo orudeneniya Severnogo Priladozh'ya. [Peculiarities of tin ore formation in northern part of Ladoga region]. Akad. Nauk SSSR, Karel. Filial Inst. Geol. Tr 15, p. 87.

Kojonen, K., Johanson, B., O'Brien, H.E., Pakkanen, L., 1993. Mineralogy of gold occurrences in the late Archean Hattu schist belt, Ilomantsi, eastern Finland. In: Nurmi, P.A., Sorjonen-Ward, P. (Eds.), Geological Development, Gold Mineralization and Exploration Methods in the Late Archean Hattu Schist Belt, Ilomantsi, Eastern Finland. Geological Survey of Finland, pp. 233–271. Special Paper 17.

Korsakova, M., 2012. Northern Ladoga Sn-Zn-Pb, U, Au, W. In: Eilu, P. (Ed.), Mineral Deposits and Metallogeny of Fennoscandia. Geological Survey of Finland, pp. 389–390. Special Paper 53.

Korsman, K., Koistinen, T., Kohonen, J., et al. (Eds.), 1997. Bedrock Map of Finland 1:1,000,000. Geological Survey of Finland. Espoo.

Kranck, E.H., 1945. The molybdenum deposit at Mätäsvaara in Carelia (E. Finland). Geologiska Föreningen i Stockholm Förhandlingar 67, 325–350.

Kukkonen, M., 1967. Ylöjärven kaivos 1943–1966. Summary: The Ylöjärvi mine 1943–1966. Vuoriteollisuus 28–38.

Kurhila, M., 2011. Late Svecofennian leucogranites of southern Finland—Chronicles of an orogenic collapse. Ph.D. thesis. Department of Geosciences and Geography A8, University of Helsinki.

Kurki, J., 1989. Granitoids and ore mineralization at Aittojärvi. Symposium Precambrian Granitoids, IGCP Project 217, Proterozoic Geochemistry. In: Gaál, G. (Ed.), Archean Granitoids and Associated Mo-W and Au Mineralization in Eastern Finland. Geological Survey of Finland, pp. 16–18. Guide 25.

Lahti, S., 1981. On the granitic pegmatites of the Eräjärvi area in Orivesi, southern Finland. Geological Survey of Finland 82 Bulletin 314.

Lahtinen, R., 2012. Evolution of the bedrock of Finland. In: Haapala, I. (Ed.), From the Earth's Core to Outer Space. Springer -Verlag, Berlin/Heidelberg, pp. 47–60.

Lahtinen, R., Korja, A., Nironen, M., et al., 2009. In: Earth Accretionary Systems in Space and Time. Geological-Society, London, pp. 237–256. Special Publications 318.

Larin, A.M., 2009. Rapakivi granites in the geological history of the Earth, Part 1, Magmatic associations with rapakivi granites: age, geochemistry, and tectonic setting. Stratigraphy and Geological Correlation 17 (3), 235–258.

Lauri, L., Mikkola, P., Karinen, T., 2012. Early Paleoproterozoic felsic and mafic magmatism in the Karelian province of the Fennoscandian Shield. In: Dall'Agnol, R., Frost, C.D., Rämö, O.T. (Eds.), A-type Granites and Related Rocks through Time, pp. 74–82 Lithos 151.

Lehmuspelto, P., 1973. Tinan esiintymisestä moreenissa Eurajoen rapakivigraniittialueella. Summary: Distribution of tin in glacial till in the Eurajoki rapakivi granite area, SW Finland. Geologi 25, 1–5.

Lukkari, S., 2002. Petrography and geochemistry of the topaz-bearing granite stocks in Artjärvi and Sääskjärvi, western margin of the Wiborg rapakivi granite batholith. Bulletin of the Geological Society of Finland 74, 115–132.

Lukkari, S., Holtz, F., 2007. Phase relations of a F-enriched peraluminous granite: an experimental study of the Kymi topaz granite stock, southern Finland. Contributions to Mineralogy and Petrology 153, 273–288.

Lukkari, S., Thomas, R., Haapala, I., 2009. Crystallization of the Kymi topaz granite stock within the Wiborg rapakivi granite batholith, Finland: evidence from melt inclusions. Canadian Mineralogist 47, 1359–1374.

Luukkonen, A., 1994. Main geological features, metallogeny and hydrothermal alteration phenomena of certain gold and gold-tin-tungsten prospects in southern Finland. Geological Survey of Finland, Bulletin 377, p. 153, plus 8 appendices of the Kymi topaz granite stock, southern Finland Contributions to Mineralogy and Petrology 153, 273–288.

Marmo, V., Hyvärinen, L., 1953. Molybdenum-bearing granite and granodiorite, Rautio, Finland. Economic Geology 48, 704–714.

Nironen, M., 1989. Emplacement and structural setting of granitoids in the early Proterozoic Tampere and Savo schist belts, Finland – implications for contrasting evolution. Geological Survey of Finland, p. 83 Bulletin 346.

Nironen, M., 2005. Proterozoic orogenic granitoid rocks. In: Lehtinen, M., Nurmi, P.A., Rämö, O.T. (Eds.), Precambrian Geology of Finland—Key to the Evolution of the Fennoscandian Shield. Elsevier, Amsterdam, pp. 443–479.

Nurmi, P.A., 1983. Trace element variations in the mid-Proterozoic Rautio batholith, Finland: petrogenetic implications. In: Augusttithis, S.S. (Ed.), The Significance of Trace Elements in Solving Petrogenetic Problems and Controversies. Theophrastus Publications S.A, Athens, pp. 353–376.

Nurmi, P.A., 1985. Lithogeochemistry in exploration for Proterozoic porphyry-type molybdenum and copper deposits, southern Finland. Journal of Geochemical Exploration 23, 209–228.

Nurmi, P., Front, K., Lampio, E., Nironen, M., 1984. Etelä-Suomen svekokarjalaiset, porfyyrityyppiset molybdeeni- ja kupariesiintymät; niiden granitoidi-isäntäkivet ja litogeokemiallinen etsintä. Summary: Svecokarelian porphyry-type molybdenum and copper occurrences in southen Finland; their granitoid host rocks and lithogeochemical exploration Geological Survey of Finland, p. 88 Report of Investigation 67.

Nurmi, P.A., Haapala, I., 1986. The Proterozoic granites of Finand: granite types, metallogeny and relation to custal evolution. Bulletin of the Geological Society of Finland 58 203–233 Part 1.

Nurmi, P.A., Isohanni, M., 1984. Rock, till, and stream-sediment geochemistry in search for porphyry-type Mo-Cu-Au deposits in the Proterozoic Rautio batholith, western Finland. Journal of Geochemical Exploration 21, 209–228.

Nurmi, P.A., Sorjonen-Ward, P. (Eds.), 1993. Geological Development, Gold Mineralization and Exploration Methods in the Late Archean Hattu Schist Belt, Ilomantsi, Eastern Finland. Geological Survey of Finland, p. 386. Special Paper 17.

Palmunen, M.K., 1939. Pitkäranta vv. 1934–1938 suoritettujen vuoriteknillisten tutkimusten valossa. Suomen Geologinen Toimikunta, Geoteknllisiä Julkaisuja 44, p. 154.

Peuraniemi, V., Heinänen, K., 1985. Mineralogical investigations in the interpretations of heavy-mineral geochemical results from till. Journal of Geochemical Exploration 23, 315–328.

Rämö, O.T., 1991. Petrogenesis of the Proterozoic rapakivi granites and related basic rocks of southeastern Fennoscandia: Nd and Pb isotopic and general geochemical constraints. Geological Survey of Finland, p. 161 Bulletin 355.

Rämö, O.T., Vaasjoki, M., Mänttäri, I., et al., 2001. Petrogenesis of postkinematic magmatism of the Central Finland Granitoid Complex I; radiogenic isotope constraints and implications of crustal evolution. Journal of Petrology 42, 1971–1993.

Saksela, M., 1951. Zur Endstehung der Pitkäranta-Erze. Bulletin of the Geological Survey of Finland 154, 181–231.

Saltikoff, B., Puustinen, K., Tontti, M., 2006. Metallogenic zones and metallogenic mineral deposits in Finland. Geological Survey of Finland, p. 66 Special Paper 35.

Selby, D., Creaser, R.A., 2004. Macroscale NTIMS and microscale La-Mc-ICP-MS Re-Os isotopic analysis of molybdenite: Testing spatial restrictions for reliable Re-Os age determinations, and implications for the decoupling of Re and Os within molybdenite. Geocimica et Cosmochimica Acta 68, 3897–3908.

Stein, H.J., Sundblad, K., Markey, R.J., et al., 1998. Re-Os ages for Archean molybdenite and pyrite, Kuittila-Kivisuo, Finland and Proterozoic molybdenite, Kabeliai, Lithuania: testing the chronometer in metamorphic and metasomatic setting. Mineralium Deposita 33, 329–345.

Talikka, M., Mänttäri, I., 2005. Pukala intrusion, its age and connection to hyrothermal alteration in Orivesi, southwestern Finland. Bulletin of the Geological Society of Finland 77, 165–180.

Tiainen, M., Kärkkäinen, N., Koistinen, E., et al., 2012. Discovery of the Kedonojankulma Cu-Au occurrence, hosted by a Svecofennian porphyritic granitoid in southern Finland. In: Grönholm, S., Kärkkäinen, N. (Eds.), Gold in Southern Finland. Geological Survey of Finland, pp. 73–90. Special Paper 52.

Tischendorf, G., 1977. Geochemical and petrographic characteristics of silicic magmatic rocks associated with rare-element mineralization. In: Štemprok, Burnol, L., Tischendorf, G. (Eds.), Metallization Associated with Acid Magmatism, Vol. 2. Geological Survey, Prague, pp. 41–96.

Trüstedt, O., 1907. Die Erzlagerstätten von Pitkäranta am Ladoga-See. Bulletin de la Commission geologiqué de Finlande 19, 333.

Vaasjoki, M., 1977. Rapakivi granites and other postorogenic rocks in Finland: their age and the lead isotopic composition of certain associated galena mineralizations. Geological Survey of Finland 64 Bulletin 294.

Vaasjoki, M., Huhma, H., Lahtinen, R., Vestin, J., 2003. Sources of Svecofennian granitoid rocks in light of ion probe measurements on the zircons. Precambrian Research 121, 251–262.

Volborth, A., 1954. Phosphatminerale aus dem Lithiumpegmatit von Viitaniemi, Eräjärvi, Central-Finnland. Annales Academiae Scientiarum Fennicae A 39, 90 III.

Vorma, A., 1976. On the petrochemistry of rapakivi granites with special reference to the Laitila massif, southwestern Finland. Geological Survey of Finland 98 Bulletin 285.

Yletyinen, V., 1967. Ylitornion Kivilompolon molybdeenihohde-esiintymästä. Abstract: the molybdenite occurrence at Kivilompolo, Ylitornio 21 Geoteknillisiä julkaisuja 73.

Zeidler, W., 1949. Om gruvdriften I Mätäsvaara åren 1940–47. Vuoriteollisuus – Berghanteringen 1947 (1), 20–33.

PART

OTHER TYPES OF MINERAL DEPOSITS 9

THE TALVIVAARA BLACK SHALE-HOSTED Ni-Zn-Cu-Co DEPOSIT IN EASTERN FINLAND 9.1

A. Kontinen, E. Hanski

ABSTRACT

The Talvivaara Ni-Zn-Cu-Co ore deposit, currently the largest sulfidic Ni resource under exploitation in Western Europe, is hosted by amphibolite-facies black shales of the Talvivaara formation in the Kainuu schist belt, eastern Finland. It is a low-grade deposit averaging 0.22 wt% Ni, 0.50 wt% Zn, 0.13 wt% Cu, and 0.02 wt% Co, but with a huge Ni content exceeding 1 million tons as calculated with a cutoff of 0.07 wt% Ni. The deposition of the Talvivaara formation took place in the Lower Kaleva stage in a continental margin rift basin devoid of syn-rift magmatism. The clastic material is derived mainly from older Paleoproterozoic (Jatulian) mature sediments and intercalated mafic sills and volcanic rocks. Available detrital zircon data and other geological constraints demonstrate that the depositional age of the nickeliferous black shales falls between 2.0 and 1.9 Ga. Deformation and metamorphic processes upgraded the proto-ore by fold-thickening, grain-size coarsening, and development of Ni-rich peak-metamorphic monosulfide solid solution (mss), which exsolved into the easily oxic-acid soluble pyrrhotite ± pentlandite assemblage holding >80% of the ore's metal value.

Sedimentary aspects of Talvivaara metal muds, including thick (up to 1 m), compositionally very homogenous beds, support deposition dominantly in turbidity flows, suggesting recycling of metal-enriched muds from shallower parts of the depositional basin. High S and C abundances, layers of syngenetic pyrite and sulfur isotope data imply deposition in an environment with high organic production and rate of microbial sulfate reduction and thus anoxic-sulfidic conditions at least in the bottom parts of the water column. On the other hand, abundant sulfate supply, high

Mineral Deposits of Finland. http://dx.doi.org/10.1016/B978-0-12-410438-9.00022-4

concentrations of authigenic Mo (typically 50 ppm) and other oxyanionic trace elements in the Talvivaara formation/ore and other Kalevian black shales suggest deposition under well-oxygenated (>10% PAL) atmosphere, which is consistent with the generation of the black shales after the Great Oxidation Event. Given the huge amounts of S and "excess" metals in Talvivaara and coeval black shales in eastern Finland, the origin of the deposit seems to be related to a basin connected at least for its upper parts to oxygenated global ocean and thus closely linked to global atmo-hydrospheric evolution. The high Ni-Zn-Cu-Co content of the ore is likely due to elevated contents of these metals in contemporary ocean water, high influx of MnO-FeOOH particles (+absorbed metals), and their recycling across well-developed, steady redoxcline in the depositional basin. Metals liberated from the oxide particles by dissolution in an anoxic water column became absorbed on decaying organic matter and uptaken by syngenetic sulfides.

Keywords: Black shale; Ni-Zn-Cu ore; Paleoproterozoic; Talvivaara; Finland.

INTRODUCTION

Black shales are dark-colored, usually thinly laminated mudstones containing appreciable organic matter (>0.5 wt% C), authigenic iron sulfides and, silt- and clay-sized detrital particles that in most cases have been accumulated under anoxic bottom water and/or bottom sediment conditions in marine or continental sedimentary basins (Huyck, 1990; Wignall, 1994). Compared to other shales or the average crust, black shales are commonly highly enriched in many redox-sensitive and/or sulfide-forming metals and metalloids, including Cr, V, Mo, Re, U, As, Cd, Sb, Se, Ag, Cu, Ni, Zn, Co, Pb, and P, and in some cases also Hg, Li, Au, PGE, Y, or rare earth elements (REE; Vine and Tourtelot, 1970; Quinby-Hunt et al., 1989; Ketris and Yudovich, 2009). Various synsedimentary to early diagenetic mechanisms of metal concentration have been proposed, but generation of anoxic conditions and H_2S by bacterial reduction of sulfate in an environment rich in decaying organic matter and subsequent reaction between H_2S and metals are generally regarded as essential factors in metal enrichment. This can take place by direct metal uptake in sulfides as well as adsorption on particles such as dead organic matter, clay, sulfides, and oxides, in the water column and bottom sediment (e.g., Wignall, 1994; Robb, 2005). Among the best-known modern examples of sedimentary basins where fine-grained, sulfide-bearing metal-enriched sediments accumulate are the Black Sea (Holland, 1984), the Cariaco Basin on the Venezuela shelf (Piper and Dean, 2002), and Framvaren Fjord in southern Norway (Skei, 1983).

Because of their often large extent, ancient black shale strata represent one of the most significant crustal reservoirs of organic carbon and sulfur, and potential resources of many metals, including so-called critical metals. In some cases, the metal concentrations of black shales can reach sufficiently high levels so as to generate an ore deposit. Characteristically, these deposits are polymetallic, such as the Paleoproterozoic Talvivaara Ni-Zn-Cu-Co deposit in Finland, the subject of this chapter, and the Early Cambrian Mo-Ni-Zn-Au-PGE mineralization in China (Steiner et al., 2001; Lehmann et al., 2007). Black shales may also contain economically significant amounts of uranium (Anderson et al., 1985; Schovsbo, 2002) or vanadium (Li et al., 2010). On the other hand, a clear distinction between black shale-hosted deposits and volcanogenic massive sulfide (VMS) Cu-Zn-Pb-Ag-Au, sedimentary-exhalative (SEDEX) Pb-Zn-Ag, or sediment-hosted epigenetic (Kupferschiefer, Katanga) Cu-Zn-Pb-Ag deposits cannot always be drawn, as the host rocks of these deposits may contain black shales influenced by migrating hydrothermal fluids (Coveney, 2003).

In the Fennoscandian Shield, one main time period with abundant Paleoproterozoic black shale sedimentation was broadly related to the so-called Shunga event (Lepland et al., 2013a). Stratigraphically, these black shale-bearing sedimentary sequences in Finland have commonly been assigned to the Upper Jatuli (marine Jatulian) and Kaleva (Laajoki, 2005; Hanski and Melezhik, 2012). The included black shales (as metamorphosed and schistose, usually called black schists) are widespread in a zone extending from southeasternmost Finland to Lapland in the north. Current geochronological data indicate that their deposition took place between ~2.06 and 1.92 Ga (Mutanen and Huhma, 2001; Lahtinen et al., 2010). In terms of metallogenesis, Fennoscandian black shale successions are important in several ways: they host the huge Talvivaara Ni-Zn-Cu-Co sulfide deposit (Kontinen, 2012; Loukola-Ruskeeniemi and Lahtinen, 2013), are closely associated with the ophiolitic rocks hosting the Outokumpu-type Cu-Co-Zn-Ni sulfide deposits (Peltonen et al., 2008), are potential host rocks of flake graphite deposits (Lehtinen, 2015), provide favorable S-bearing country rocks for formation of magmatic Ni-Cu deposits (e.g., Santaguida et al., 2015), and in the Onega region, Russian Karelia, comprise V-U mineralization (Korsakova, 2012) and accumulations of extremely C_{org}-rich rocks, interpreted to mark an ancient petrified oil field (Melezhik et al., 2004).

The Talvivaara black shale-hosted Ni-Zn-Cu-Co deposit occurs in eastern Finland (Fig. 9.1.1) and is currently the largest sulfidic Ni resource under exploitation in Western Europe. It is a low-grade (0.22 wt% Ni, 0.50 wt% Zn, 0.13 wt% Cu, 0.02 wt% Co, 0.0017 wt% U at Ni cutoff of 0.07 wt%) but very large deposit with total ore resources (measured + indicated + inferred) reaching 2.053 Gt (Talvivaara Mining Company Plc, 2014). Since the early 1990s, several articles have been published on the deposit and its host rocks by Kirsti Loukola-Ruskeeniemi and her coworkers (e.g., Loukola-Ruskeeniemi, 1991; Loukola-Ruskeeniemi et al., 1991; Loukola-Ruskeeniemi and Heino, 1996; Loukola-Ruskeeniemi, 1999; Loukola-Ruskeeniemi and Lahtinen, 2013), with their studies centered mainly on chemical characterization of the nickeliferous black shales and their comparison with black shales in the

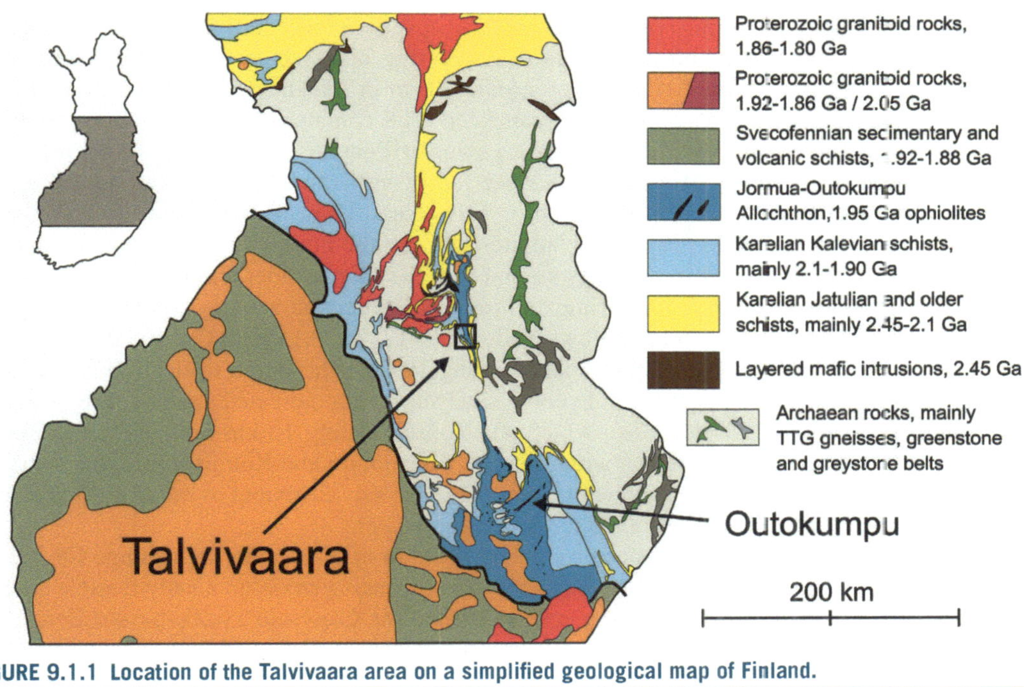

FIGURE 9.1.1 Location of the Talvivaara area on a simplified geological map of Finland.

Outokumpu Cu-Co-Zn mining area (Fig. 9.1.1). In previous works, deposition of the Talvivaara black shales and concurrent precipitation of the metals at submarine hot springs within oceanic spreading ridge or related hydrothermal environments has been considered the most likely genetical scenario (e.g., Loukola-Ruskeeniemi, 1995; Loukola-Ruskeeniemi and Heino, 1996).

Despite the long study history, many aspects of the geology and genesis of the Talvivaara ore, including those related to its stratigraphical and structural setting, geochronology, mineralogy, and metamorphic history, are not yet sufficiently established. Even the favored hydrothermal interpretation of the metal enrichment in the deposit can be considered still an improperly tested hypothesis (cf. Kontinen, 2012). Recent fieldwork and laboratory studies carried out by the Geological Survey of Finland (GTK) have resulted in a lot of new information on sulfide mineral compositions, zircon grain ages, Sm-Nd isotopes, and the geochemistry of previously insufficiently studied elements, such as REE. Furthermore, analysis of exploration drill core data in cooperation with Talvivaara Mining Company geologists has improved our understanding of the spatial distribution and stratigraphy of the deposit (Kontinen, 2012; Kontinen et al., 2013b). This chapter summarizes core findings and interpretations from these studies.

Description of methods used in electron microprobe and geochemical whole-rock analyses for this work are provided in the supplementary online material.

SHORT EXPLORATION AND MINING HISTORY

The first indications of the potential existence of Ni-bearing black shales in the southern part of the Kainuu schist belt appeared in the 1930s when glacially transported erratic boulders of Ni-enriched black shale were discovered approximately 40 km southeast of the belt. In 1962, a junior company, Suomen Malmi Oy, made the first drill holes in Ni-bearing black shales in the Talvivaara area, penetrating a low-grade sulfide dissemination around 10 m in thickness 2.5 km southeast of the proper Talvivaara deposit (Kolmisoppi). A few years later, GTK commenced an exploration campaign in the schist belt, under the leadership of state geologist Pentti Ervamaa, which, among other things, resulted in a discovery of a 20-m-thick section of mineralized black shale in 1972, at a location approximately 40 km north-northeast of Talvivaara. These findings encouraged further exploration in areas occupied by Kalevian graphite-rich, sulfidic metasediments. Investigations in the Talvivaara area were started in the early summer of 1977 by boulder tracking, and already in late 1977, drillings revealed an ore-bearing rock unit with the best section containing 55 m of partly highly sulfidic black shale with 0.30 wt% Ni, 0.47 wt% Zn, 0.14 wt% Cu, and 0.02 wt% Co. By the end of 1982, GTK had drilled 95 holes totaling 19.6 km in length in the Talvivaara area. These drillings together with mapping and geophysical surveying indicated the presence of a zone of Ni-enriched (>0.1 wt%) black shales, which was approximately 10 km long and up to 500 m wide, with the two main mineralized subareas termed Kuusilampi and Kolmisoppi in the southern and northern parts of the zone, respectively (Fig. 9.1.2). The surface extent of the deposit was well delineated but the depth dimensions remained still largely open.

In 1986, the mineral rights to the deposit were transferred from GTK to the Outokumpu Co., which carried out further studies of the deposit, with an emphasis on finding an economically feasible method for metal extraction from the low-grade and graphite-rich deposit. Closest to a workable solution to this challenge came from the bioheap leaching technology (Puhakka et al., 2007), which was further

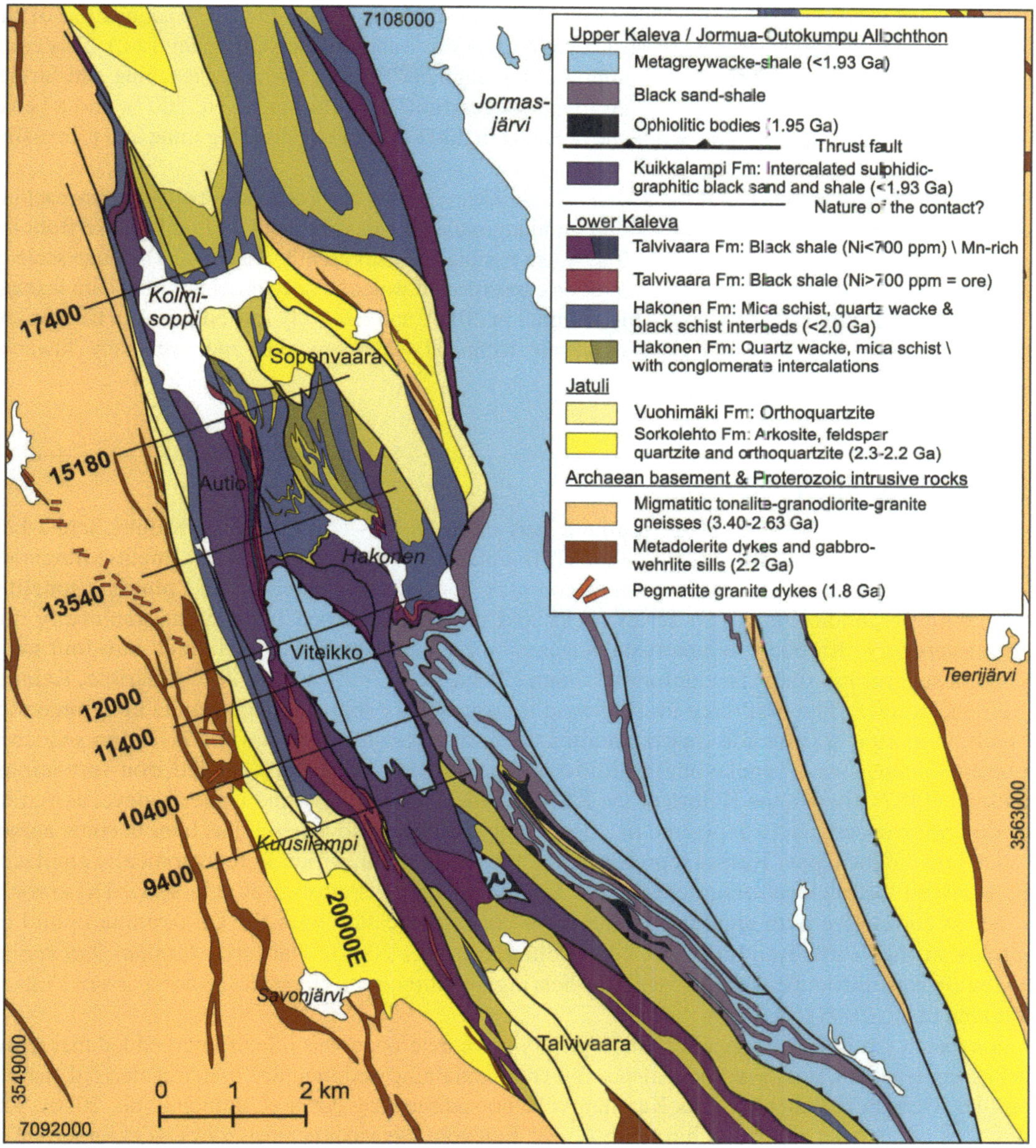

FIGURE 9.1.2 Generalized geological map of the Talvivaara area.

Black lines with numbers referring to the local mine grid indicate the locations of the vertical cross sections shown in Fig. 9.1.4. The stratigraphic units are shown with the same colors as in the lithostratigraphical column in Fig. 9.1.3.

developed and experimented in the Talvivaara Project, established since the 2004 acquisition of the mineral rights to the Talvivaaran Kaivososakeyhtiö Oyj (Talvivaara Mining Company Ltd.). This company was founded by the former Outokumpu mining engineer Pekka Perä. After leaching and further metal processing experiments, which were found successful (Riekkola-Vanhanen, 2007), and a bankable feasibility study, the company made the decision in 2007 to start constructing mine and processing facilities.

The production of merchantable Ni, Zn, and Co sulfides started in late 2008, and at present, far-reaching preparations have also been made for starting uranium extraction and production of yellowcake from the leach solution. However, the future of the Talvivaara mining seems presently uncertain. The huge scale of both the mining and mineral processing operations, unexpected water-balance problems, and dam leakage (the last two related to the exceptionally high rainfall in 2012), and relatively low Ni prices in the last few years, have resulted in a combination of severe technical, environmental, and eventually financial problems.

GEOLOGICAL SETTING

Talvivaara deposit is located in the southern part of the north–south-trending Kainuu schist belt (KSB) (Fig. 9.1.1). The Paleoproterozoic KSB comprises mainly siliciclastic metasedimentary rocks deposited on the Archean basement complex consisting of gneissic and plutonic tonalite-trondhjemite-granodiorite (TTG) series rocks and minor greenstone belts. Stratigraphically, the Paleoproterozoic "Karelian" supracrustal succession has traditionally been divided into four main units of a supergroup status (unconformity-bound "tectofacies" of Laajoki, 2005): Sariola, Kainuu, Jatuli, and Kaleva. The first three units show a transgressive evolution from fluvial conglomerates to shallow-marine arkosic and quartz arenites to shelf dolomites, whereas the Kalevian sequence represents deeper-water wacke and argillitic metasediments and locally banded iron formations. Kontinen (1987) divided the Kaleva of the KSB into the Lower Kaleva and Upper Kaleva, assuming that they were successively deposited in a deepening rift basin. However, it has later become apparent that most of the Upper Kaleva represents allochthonous strata, which were tectonically emplaced together with the enclosed fragments of the Jormua ophiolite (Peltonen et al., 1998). These rocks are hence correlative with similar allochthonous rocks associated with the Outokumpu ophiolitic complex in the North Karelia schist belt (Peltonen et al., 2008). The term *Jormua-Outokumpu allochthon* was introduced for the ophiolite-bearing allochthonous Upper Kaleva in eastern Finland (Peltonen et al., 2008).

In general, the Lower Kaleva is a lithologically more heterogeneous pile of interbedded metagraywackes, metapelites, and quartzitic metasediments containing sedimentary material derived mainly from the Archean basement and its Karelian volcano-sedimentary cover (Lahtinen et al., 2010). The Talvivaara ore deposit occurs within these Lower Kalevian metasedimentary rocks. It is important to note that there is little evidence for synsedimentary magmatism in the Lower Kaleva sequences, neither in the Talvivaara area nor even in the whole KSB.

The Upper Kaleva constitutes a relatively homogenous sequence of graywackes, which contains abundant interlayers of graphitic schists in its apparent lower part and is typically devoid of psephitic materials or any coeval volcanic or hypabyssal igneous rocks. In contrast to the Lower Kaleva, the Upper Kaleva contains a large Paleoproterozoic detrital component as revealed by zircon grain populations having ages in the range of 2.00–1.92 Ga (Lahtinen et al., 2010).

LITHOSTRATIGRAPHY OF THE TALVIVAARA AREA

GENERAL STRATIGRAPHIC DIVISION

In the area around the Talvivaara ore deposit (see Fig. 9.1.2), three main Paleoproterozoic stratigraphic units can be distinguished (Kontinen, 1987): (1) Jatulian platformal cover arenites, (2) Lower Kalevian autochthonous riftogenic sediments, and (3) Upper Kalevian deeper-water turbiditic wackes-shales. These units are further divided into formations as shown by the schematic lithostratigraphical column of Fig. 9.1.3 and the simplified geological map of Fig. 9.1.2.

Division of the Kaleva-type rocks in the Talvivaara area into meaningful lithostratigraphic units is not straightforward due to the multiply folded and faulted nature of the strata, but the abundant drilling

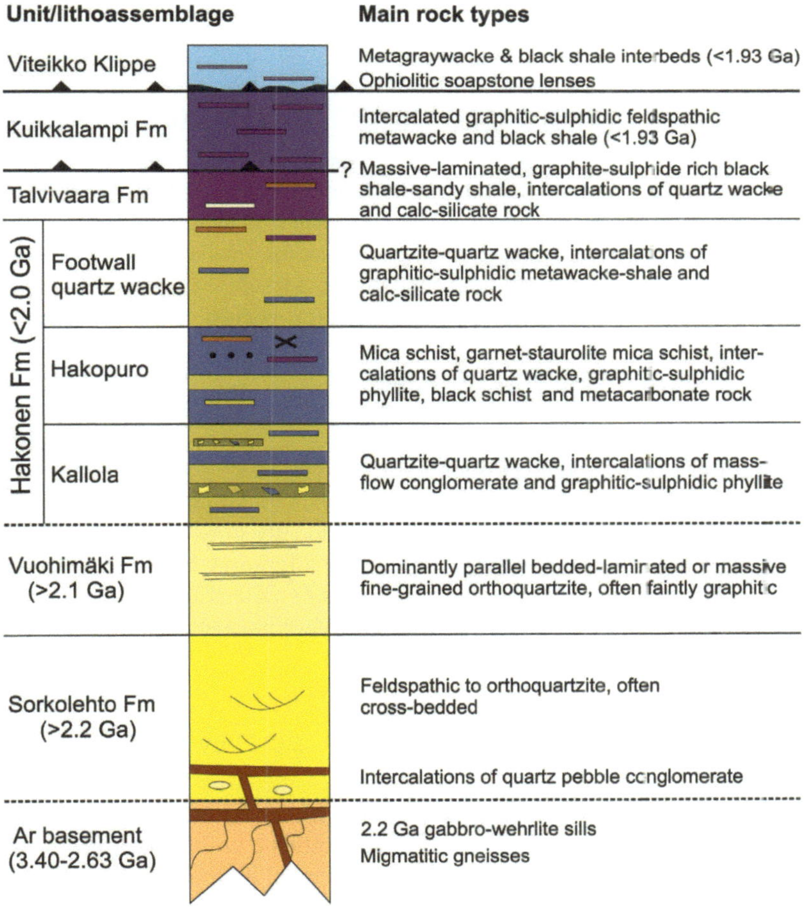

FIGURE 9.1.3 Lithostratigraphical column of the Talvivaara area.

The stratigraphic units are shown with the same colors as in the map in Fig. 9.1.2.

in the area and our recent field, petrographic, geochemical, and isotopic studies have considerably improved our understanding of the stratigraphy of the rock units, though it is still somewhat tentative. In unison with Kontinen et al. (2013b) we divide the Kaleva-type metasediments into the following four units: (1) the autochthonous Hakonen, (2) Talvivaara, and (3) Kuikkalampi Formations, and (4) the allochthonous Viteikko schists in the Viteikko klippe. The Talvivaara Formation is dominated by the black shales, which are currently mined, and the Hakonen Formation combines all the Lower Kaleva strata that are thought to have deposited before the Talvivaara Formation.

The criteria to regard the Hakonen and Talvivaara Formations as Lower Kalevian include their quartz-rich, yet wacke-type, dominantly massive sands, exclusively Archean detrital zircon populations and ε_{Nd} (1900 Ma) of −4.6 ± 1.6, which characterize Lower Kalevian sediments elsewhere in the KSB (Kontinen, 1986, 1987; Kontinen et al., in prep., this work p. 569). The Kuikkalampi and Viteikko metasediments are considered Upper Kalevian because of their immature, graywacke-type sandstones, dominantly Paleoproterozoic zircon grains, and more positive ε_{Nd} (1900 Ma) values of about −2 (Kontinen et al., in prep., this work p. 569).

JATULIAN METASEDIMENTS: SORKOLEHTO AND VUOHIMÄKI FORMATIONS

The dominantly quartz arenitic Jatulian metasediments fringing the basement block west of the Talvivaara deposit are divided into the Sorkolehto and Vuohimäki Formations (Kontinen et al., 2013b). Even though the contacts of both formations are presently mostly fault-displaced, it seems likely that the Sorkolehto Formation was initially deposited autochthonously on Archean rocks and the Vuohimäki Formation is a younger conformable unit deposited on the Sorkolehto Formation. The latter has been intruded by ~2.22 Ga differentiated gabbro-wehrlite sills (cf. Hanski et al., 2010) giving a minimum age for the formation.

The Sorkolehto Formation consists of cross- and planar-bedded arkosic, felspathic, and quartz arenites, now largely recrystallized to quartzites, which probably deposited in stable aeolian-fluvial to shallow marine environments. The Vuohimäki Formation comprises more mature and fine-grained and dominantly ultrapure quartz arenites. Laminar to thin-bedded, planar, massive, and graded bedsets are the dominant sediment type, indicating deposition in shallow-marine, high-energy costal to shelf settings. The Vuohimäki Formation quartzites are locally gray or dark bluish gray due to the presence of minor graphite, possibly after hydrocarbon migration from the adjacent black shales during the burial–early diagenetic metamorphism. Tectonic slices of basement rocks and similar mature quartzites to those found in the Sorkolehto and Vuohimäki Formations occur in the middle part of the study area.

LOWER KALEVIAN METASEDIMENTS: HAKONEN AND TALVIVAARA FORMATIONS

Hakonen Formation

In the area north of the Hakonen Lake (Fig. 9.1.2), the formation comprises two distinct lithofacies, of which the eastern one is called the Kallola lithoassemblage and the western one the Hakopuro lithoassemblage. The former is probably older and dominated by graded to mostly massive quartzite-quartz wacke with intercalations of mass-flow conglomerate ranging from meters to tens of meters in thickness and carbonaceous-sulfidic mica schist-phyllite centimeters to tens of meters in thickness. Most clasts in the conglomerate layers or lenses are feldspathic to quartz arenite, being probably derived from Jatulian sources. In addition, there are sulfidic-graphitic schist clasts, which probably are intraformational. So far, no unequivocal granitoid/gneiss phenoclasts have been observed in the conglomerates in the Talvivaara area although such are present in correlative conglomerates elsewhere in the KSB (Kontinen, 1986). The Hakopuro lithoassemblage consists mainly of thick- to thin-bedded to laminar, relatively lowly carbonaceous-sulfidic mica schist-phyllite with interbeds-bedsets of quartzite-quartz wacke varying from 5 cm to 5 m in thickness There are also occasional 10-cm- to 2-m-thick calc-silicate intercalations. The more graphitic-sulfidic schists in the Hakopuro lithoassemblage also show somewhat elevated metal contents (especially Zn) compared to the background levels in similar rocks in the KSB.

In the Kolmisoppi area, the eastern, footwall contact of the Talvivaara Formation is typically against Hakopuro-type rocks but the contact is clearly faulted. At Kuusilampi, the Talvivaara Formation is underlain by a >200-m-thick sequence of dominantly quartzite-quartz wacke We regard this "footwall quartzite" as the third and likely youngest lithofacies of the Hakonen Formation. The structure in the central part of the Kuusilampi area seems to be defined by an antiform, with the footwall quartzites in its core being enveloped by Talvivaara Formation rocks (cf. Fig. 9.1.4). The quartz wackes are intercalated by beds–bedsets of sulfidic-graphitic mica schist-phyllite from <5 cm to several meters in thickness. There are also calc-silicate-dominated interlayers.

Talvivaara Formation

Based on our drill-core observations from tectonically least disturbed sections of the Talvivaara Formation, it has the following features: The lower contact with the underlying footwall quartzite is gradational over a short interval, typically within <10–20 m, from sparsely phyllite-intercalated quartz wacke to dominantly Ni-Zn-Cu-Co-enriched (Ni > 1000 ppm), massive mudstones, which have occasional interbeds of quartz sands showing grading and load casts. The lower parts of the Talvivaara Formation are dominated by massive, fine-grained, C-S-rich, originally probably silty mudstones intercalated with occasional more coarse-grained, 5–50-cm-thick, often graded sandy beds (Fig. 9.1.5), calc-silicate rocks, and variable amounts of distinctly laminated, partly heavily pyritic mudstones. The tremolite ± diopside ± carbonate-dominated calc-silicate rocks comprise about 5% of the ore volume. These mostly sharply bound layers, which range from 5–200 cm in thickness, seem to be more common and thicker in the lower parts of the formation. The pyrite-dominated, typically thinly bedded laminated mudstone layers, which comprise about 10 vol% in the more complete and better preserved sections, range in thickness from <1 cm up to 5 m and are clearly more common and thicker toward the assumed top of the formation.

Other characteristics of the upper part of the formation include frequent sharply bound, deep black laminae of fine-grained, apatite-bearing and very graphite-rich (C_{org} up to 30 wt%) mudstones, most apparent when present in pyrite-dominated intervals. It is intriguing that the fine-grained, thinly bedded laminar pyritic mudstones also contain occasional thin (often <5 cm) interbeds of poorly sorted sandstones containing sometimes clasts up to granule size. The Talvivaara Formation also contains Mn-enriched beds, characterizing especially the upper part of ore-containing rock sequence, which were previously separated as a distinct chemotype (Loukola-Ruskeeniemi and Heino, 1996). Our observations indicate that both the lower and upper parts of the formation comprise layers/parts that are manganoan, with Mn being incorporated dominantly in spessartine garnet in the former and alabandite (MnS) in the latter. The Mn enrichment is partly related to the presence of calc-silicate layers.

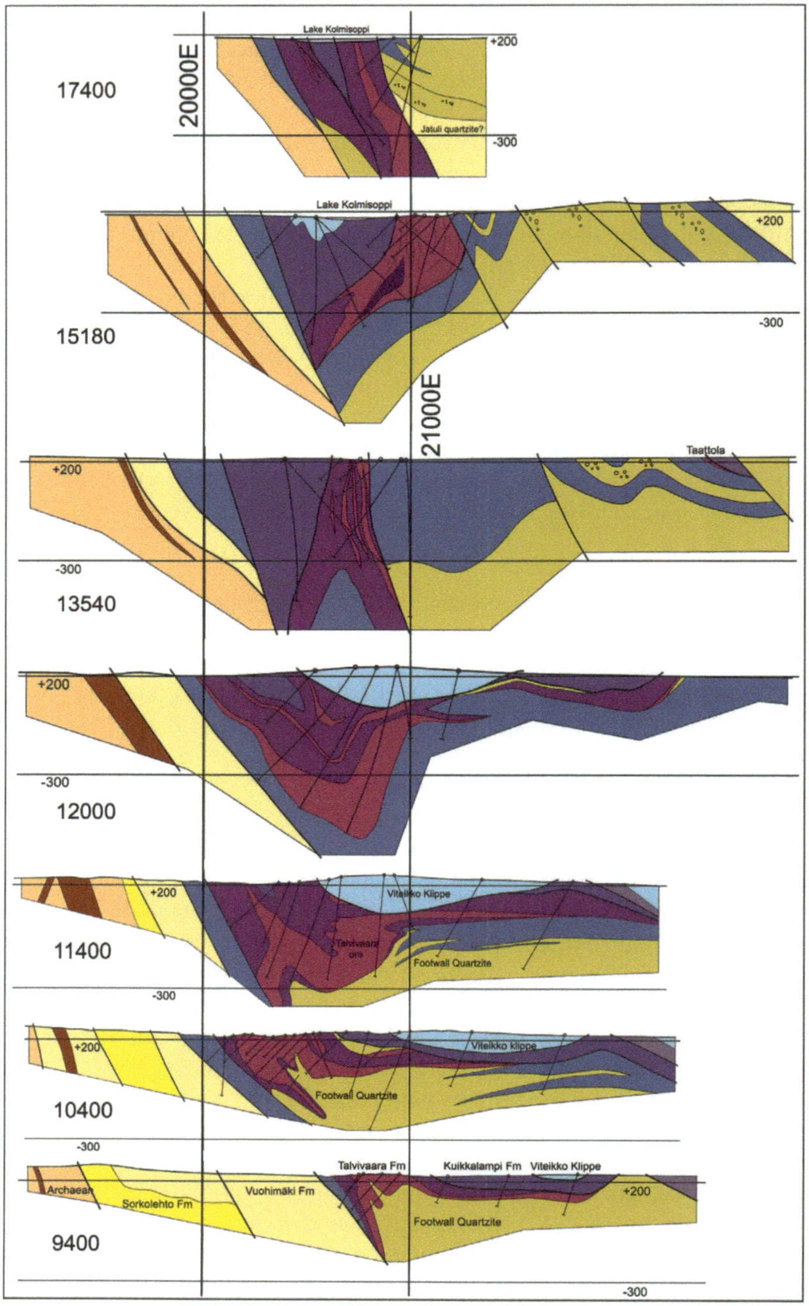

FIGURE 9.1.4 A stack of vertical cross sections over the Talvivaara deposit.

For location of the sections, see Fig. 9.1.2. The easting lines of the local mine grid are spaced 1 km apart.

FIGURE 9.1.5 Black shale with a sand intercalation from the stratigraphically bottom part of the Talvivaara Formation/ore at Heittimensuo, Kuusilampi.

Length of the pen is 14 cm.

So far, we have not been able to define the exact nature of the upper contact of the formation because, in the drill cores that we studied, the contact appeared markedly different and in some cases disturbed by late faulting. However, in all drill cores dissecting the apparent upper contact, the contact was with rocks of the Kuikkalampi Formation (described later). From the presently available evidence, there seem to be two options: (1) The contact is defined by a sharp thrust fault, which is a common case at the base of the Outokumpu-Jormua allochthon, or (2) the contact involves an abrupt shift in sedimentation from the Talvivaara type to the Kuikkalampi type in terms of the material source, but with continuing high organic productivity and hence anoxic conditions in the water column.

UPPER KALEVIAN METASEDIMENTS: KUIKKALAMPI FORMATION AND VITEIKKO KLIPPE SCHISTS

The major part of the rocks mapped as black shales in the Talvivaara area (Loukola-Ruskeeniemi and Heino, 1996; Kontinen and Eskelinen, 2005) are graphitic, black to dark gray sandstones intercalated with graphite- and sulfide-rich shales belonging to the Kuikkalampi Formation. The type area for the Kuikkalampi Formation is immediately to the southeast of the Viteikko klippe (Fig. 9.1.2), where these rocks form a <100-m-thick layer between the klippe and underlying footwall quartzite with a thin sliver of the Talvivaara Formation in-between. However, the major occurrence is in the "Autio synform" west of the ore zone (cf. Fig. 9.1.2). Differing from the quartz-rich and largely plagioclase-devoid and sodium-poor (Na_2O typically <0.5 wt%) Lower Kalevian sandstones of the Hakonen and Talvivaara Formations, the Kuikkalampi Formation sandstones are plagioclase-bearing and relatively sodium-rich (Na_2O typically 2–4 wt%). Relatively immature nature of the material is also supported by zircon grains that in the Kuikkalampi sandstones look for the most part only weakly abraded with prismatic shapes and sharp edges.

The Kuikkalampi sandstones share a similar chemistry with sandstones in the Outokumpu-Jormua allochthon and the black shales are also similar to the Outokumpu black shales, recording a strong enrichment in redox-sensitive metals, such as Mo, U, and V; moderate enrichment in Zn, Cu, Ni; and only slightly elevated Co. Furthermore, in common with some black shale-dominated parts of the Outokumpu allochthon, the Kuikkalampi Formation is locally associated with ophiolite lenses. Based on all these observations, we assign the Kuikkalampi Formation to the Upper Kaleva.

The lower contact of the <150-m-thick Viteikko klippe is defined by an upright-folded, flat-lying thrust fault controlling the location of thin tectonic lenses of peridotite-derived talc-carbonate to talc schist. The northern and southern parts of the klippe are underlain by rocks of the Kuikkalampi Formation, whereas in the middle part, the basal contact is directly to rocks of the Talvivaara Formation. The main constituents in the Viteikko klippe are typical Upper Kalevian mica schists derived from massive, medium- to thin-bedded metagraywackes-pelites with some relatively thin and weakly graphitic-sulfidic black sand-wackepelite intercalations. In terms of their petrological and sedimentological features, the rocks of the klippe are indistinguishable from the metawackes–pelites of the Nuasjärvi "basin" that form the core of the KSB east of the Talvivaara area.

DEPOSITIONAL AGE OF THE KALEVIAN ROCKS IN THE TALVIVAARA AREA

In the absence of syndepositional volcanic or intrusive rocks, which by U-Pb zircon dating would constrain the depositional age of the Talvivaara sedimentary rocks, one alternative approach is to employ U-Pb dating of detrital zircon grains, preferably using an in situ "point-and-shoot" method, such as laser ablation inductively coupled plasma mass spectrometry (LA-ICP-MS) or secondary ion mass spectrometry (SIMS). Recently, zircon grains from 16 Jatulian and Lower and Upper Kalevian sedimentary samples from the eastern part of the KSB, including those from the Hakonen, Talvivaara, and Kuikkalampi Formations and the Viteikko klippe, were analyzed by LA-ICP-MS in the geochronological laboratory of the GTK at Espoo. The analytical results will be published and discussed in detail elsewhere (Kontinen et al., in prep.) and only the implications pertinent to the timing of the Talvivaara ore deposit are summarized briefly below. A graphical presentation of the main features of the data is given in Fig. 9.1.6.

Lower Kalevian rocks in the Hakonen and Talvivaara Formations contain predominantly, or perhaps even exclusively, Archean detrital zircon, as excluding data from rare (<1.9 Ga) metamorphic rims, nearly 100% of the large number of laser-ablated zircon grains from 6 samples have only yielded Archean ages (Fig 9.1.6). Most of the grains are rounded and thus likely recycled through high-energy Jatulian environments/quartzitic sediments. A maximum age of ~2.0 Ga for Lower Kalevian formations can be deduced based on the youngest detrital zircon population in a sandy sample from ~50 km north of Talvivaara, from the Kotila Formation (Kontinen, 1986), which is regarded a broad correlative of the Hakonen Formation. The overlying Uva Formation contains similar Ni-Cu-Zn-rich black shales as those in the Talvivaara Formation (Kontinen, 1986).

Detrital zircon grains from two sandy black shale samples from the Kuikkalampi Formation contain zircon populations that are dominantly Paleoproterozoic in age (1.92–2.0 Ga), as is the case with the Upper Kalevian sedimentary material in general (cf. Lahtinen et al., 2010). Based on one representative sample from the bottom part of the Viteikko klippe, these wackes also contain dominantly Paleoproterozoic 1.93–2.0 Ga zircon grains (Fig. 9.1.6). Similar to Upper Kalevian sandstones elsewhere, the Viteikko sandstones contain zircon grains that are typically fairly coarse and mostly sharp-edged, nearly unabraded, hence representing evidently first-cycle detrital grains. From this information, and assuming that the allochthonous Upper Kaleva was emplaced at ~1.90 Ga (Peltonen et al., 2008), we can conclude that the depositional age of the Kuikkalampi Formation and the rocks of the Vitikko klippe is constrained somewhere between 1.90–1.93 Ga and the ore-bearing Talvivaara Formation is somewhat older, yet younger than ~2.0 Ga.

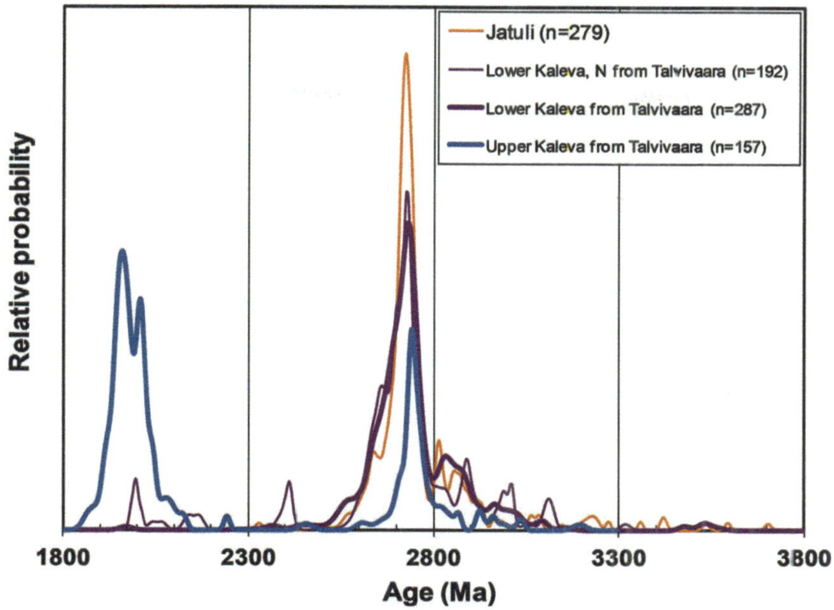

FIGURE 9.1.6 Detrital zircon ages obtained by LA-ICP-MS from Kalevian and Jatulian metasedimentary rocks in the Kainuu schist belt.

Note that the ~2.0 Ga population in the Lower Kaleva data is from a location ~70 km north of Talvivaara. These grains are from the Kotila Formation, which broadly correlates with the Hakonen Formation, being also overlain by metal-enriched black shales of the Uva Formation (Kontinen, 1986).

DEFORMATION AND METAMORPHISM IN THE TALVIVAARA AREA

The Kainuu schist belt has been deformed and metamorphosed under greenschist to middle-amphibolite facies (garnet-staurolite-kyanite) conditions (Korsman and Glebovitsky, 1999). In the Talvivaara area, almandine garnet, staurolite, chlorite, biotite, and in some samples, also late cordierite are common minerals in aluminous pelites. Ultramafic rocks are usually altered to talc–carbonate rocks that locally contain porphyroblastic olivine extensively retrogressed to lizardite ± chrysotile. Taken together, the mineral assemblages around Talvivaara indicate amphibolite facies peaking probably at about 550°C (Säntti et al., 2006). Pressures are more difficult to estimate but the presence of garnet and staurolite succeeded by cordierite in some metapelites indicates a clockwise path down from >4 kb, agreeing with minimum pressures of 4–6 kb based on the sphalerite geothermometry (Törnroos 1982a; this study).

The structure of the Talvivaara area is complicated because of at least one early major thrusting event affecting the KSB, at least two subsequent major folding events, and considerable faulting. From regional bedding plane and schistosity observations, the belt appears to be an upright-folded and faulted structural synform, with the strata at its western margin dipping generally at low angles to the east and at the eastern margin steeply toward the west. The allochthonous Kaleva rocks in the middle part of the belt have been affected by intense upright to overall sinistral north–south shearing, but still preserve evidence for an upright-folded, flat-lying sheet. However, it is possible that the allochthonous units contain large-scale recumbent folds, in a similar way as in the allochthonous Upper Kaleva in North Karelia (Koistinen, 1981).

Open, shallow, north or south-plunging mesoscale folds characterize almost every outcrop in the area. The open folds and related axial plane schistosity have been affected by sinistral shear and related "hook folds." The flat thrusting preceding the open folds is best evidenced by the Viteikko klippe in the middle part of the study area. The klippe, which has been intersected by drilling in many places, consists of allochthonous Kalevian metawackes with some thin, weakly carbonaceous and sulfidic intercalations. The basal contact of the klippe is associated with numerous thin (<1-m- to 50-m-thick) lenses of talc-carbonate-talc schist representing highly altered residual peridotites similar to those occurring in the Jormua ophiolite 40 km north of Talvivaara.

The generalized cross sections over the Talvivaara deposit presented in Fig. 9.1.4 show that the southern Kuusilampi part of the deposit is located in an antiformal structure with considerable thickening of the hinge area due to the folding. The ore plunges below the Viteikko klippe where the structural pattern is not yet well known, although a synformal pattern en echelon with respect to the Kuusilampi antiform seems obvious in some of the better drilled sections, such as 12,000 outlined in Fig. 9.1.4. In the limbs of the Kuusilampi antiform and synform below the Viteikko Klippe, the ore layer rapidly thins out, with a concurrent drop in the metal grades (Ni-Co-Cu-Zn) of the contained more strained and reworked ore.

Although no systematic structural analysis has been done yet, our observations suggest that, for example, the thickening of the Kolmisoppi ore at Hovilahti involves complex disharmonic internal folding and layer repetition. As layer repetition is probably common over the entire deposit, stratigraphical reconstructions are difficult and interpreting drill sections as simple stratigraphic sections is problematic. This concern is also true for drill intersections through the crest of the Kolmisoppi antiform, previously considered to dissect a simple homocline (e.g., Loukola-Ruskeeniemi and Heino, 1996).

TALVIVAARA ORE DEPOSIT

To be compatible with the classification used in earlier studies (Loukola-Ruskeeniemi and Heino, 1996), the "Talvivaara ore" or ore-grade in this chapter is simply equated with those parts of the Talvivaara Formation that contain Ni abundances of >0.1 wt%, although the Talvivaara Mining Company has used a somewhat lower Ni cut-off of 0.07 wt% in its resource estimates.

MAIN LITHOTYPES IN THE TALVIVAARA ORES

Ni-rich black shales

The black shales in the ore zone vary greatly, depending on the degree of deformation and recrystallization. A rough division into three main structural types can be made: (1) massive, thickly bedded, (2) thinly bedded to laminar, and (3) products of (1) and (2) that have been structurally reworked to a different degree causing mobilization and concentration of sulfides into blebs and veins and quartz ± feldspar-dominated segregation/veining (Fig. 9.1.7). All these types vary in their total sulfide content (<5–60%) and pyrrhotite-pyrite ratio with a tendency of the apparently least deformed rocks to be

FIGURE 9.1.7 Images of diamond-sawn slabs of the main types of the Talvivaara ore.

(A) Massive pyrrhotitic ore/mud with 4800 ppm Ni. Slab length 9 cm. (B) Bedded-laminar pyritic ore/mud with 1500 ppm Ni, slab height 20 cm. (C) Sulfide-laminar mud with pyrrhotite-pyrite monosulfide solid solution (mss) blebs. The mud part contains 1000–2000 ppm Ni and the blebs 5000–15,000 ppm Ni. Slab length 9 cm. (D) Quartz-feldspar-phlogopite-veined, strongly reworked mud/ore with 1500 ppm Ni. Slab length 8 cm.

dominated by pyrite. The thinly-bedded, laminar mudstones, especially in the upper part of the Talvivaara Formation, are often highly pyritic up to being banded-laminar, semimassive-massive pyrite ore (Fig. 9.1.7B). The strongly reworked types have experienced significant coarsening of the grain size and also intense segregation of quartz + mica or more "granitic" quartz + phlogopite ± microcline ± anorthite blebs and veins (Fig. 9.1.7D). Locally, quartz and quartz feldspar veins sharply brecciate the black shale, but more often form irregularly folded and veined gneisses.

Although most of the deposit is made of structurally moderately to strongly reworked rocks, having lost most of their primary textural features, domains of less strained/reworked rocks do occur locally, both in the Kuusilampi and Kolmisoppi parts of the deposit. Microscopic study of the most massive, fine-grained-looking samples reveals, however, that even these rocks frequently contain more coarse-grained (Ø 0.1–0.4 mm) porphyroblasts of rutile and/or titanite and scattered coarse mica flakes indicative of pervasive metamorphic recrystallization/reequilibration and advanced grain size coarsening, assuming that the protholiths were mostly clay- or silt-sized (<0.063 mm) shales/muds (Fig. 9.1.8A). While recrystallization in fine muds has been mostly pervasive, intercalated more coarse-grained sandy layers still preserve some traces of the original clastic texture (Fig. 9.1.8B).

Due to the difficulties in studying very fine-grained, graphite- and sulfide-rich, and thus largely opaque samples, by transmitted-light optical microscopy, we applied mineral liberation analysis (MLA) to three fine-grained and most "primary" looking but mineralogically apparently contrasting ore-grade black shale samples (1-4-ATK/2180 ppm Ni, 2TY-ATK-05/2540 ppm Ni, 7-2A-ATK-05/4180

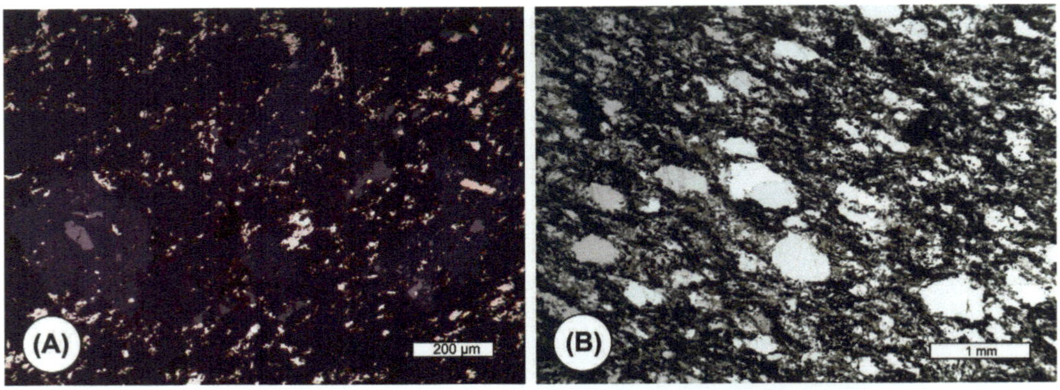

FIGURE 9.1.8 Photomicrographs of textures in metasediments of the Talvivaara Formation/ore.

(A) Coarse titanite porphyroblasts (gray) in massive, Ni-rich (4800 ppm Ni) mud. The same sample as in Fig 9.1.6A. Reflected light. (B) Coarse clastic texture in sandy mud at the base of sand- to mud-graded bed within a section of good-grade ore in the hole DDKL-201. Transmitted, plane-polarized light.

ppm Ni). Results of the MLA analyses and corresponding modal compositions calculated (wt%) on the basis of whole-rock and mineral compositional data (CBM) are listed in Table 9.1.1, together with an average modal composition of ore-bearing mudstone (TV-Ni). The MLA and CBM modes consistently show that main minerals in the three samples are, in greatly different proportions, quartz, microcline, anorthite, phlogopitic biotite, and muscovite plus pyrrhotite, pyrite, sphalerite, pentlandite, and alabandite. Carbon occurs mainly as well-crystalline, tiny graphite flakes after complete denaturation of the proto-organic material. None of the samples contains significant carbonate, tremolite, garnet, or chlorite. However, more generally, minor to major amounts of tremolite and garnet may occur in association with calc-silicate rock intercalations. Common minor to trace minerals include apatite, monazite, rutile, titanite, tourmaline, and zircon. Retrograde alteration of phlogopitic biotite to chlorite is observed locally. The Fe-Ti oxides magnetite and ilmenite are characteristically absent due to the highly reduced nature of the black shales.

Consistent with the first appearance of phosphate-rich marine sediments in the global geological record at ~2.0 Ga (Lepland et al., 2013b), phosphatic layers are found in the Talvivaara Formation (Loukola-Ruskeeniemi and Heino, 1996). They are thin (<2 cm) beds and laminae, which are typically composed of fine-grained, dense black material without any preserved hints of possible pelletoid structures. It seems that the deposition of these layers took place directly from the water column during distinct, short-lived, C_{org}-dominated sedimentation events.

Mean compositions of the main minerals in Talvivaara black shales, calculated on the basis of our and previously published data, are presented in the supplementary online material. Due to the low to very low whole-rock Na_2O (<0.07–1.0 wt%) contents, plagioclase is anorthitic with some variation toward the more sodic bytownite-labradorite. Extensive analyses of biotite from both the Kuusilampi and Kolmisoppi areas (from 11 samples) show that it has an almost uniform major and trace element composition in the whole deposit (Kontinen et al., 2014). The high MgO (22.3 ± 1.5 wt%) and low FeO (1.49 ± 0.48 wt%) contents are characteristic of phlogopitic biotite, but the relatively high Al_2O_3 (18.0 ± 1.4 wt%) and high Al_2O_3/K_2O (1.86) indicate a significant eastonite component. Biotite is relatively low

Table 9.1.1 Mineral composition of three apparently little-deformed but mineralogically contrasting massive, fine-grained Talvivaara ore/mud samples as obtained by mineral liberation analysis (MLA) and calculation from whole rock and mineral compositions (CBM)

	1-4-ATK-05		2TY-ATK-05		7-2A-ATK-05		TV-Ni
	MLA (wt%)	CBM (wt%)	MLA (wt%)	CBM (wt%)	MLA (wt%)	CBM (wt%)	CBM (wt%)
Quartz	22.81	23.92	19.67	20.31	16.94	17.91	24.80
Microcline	3.47	0.22	38.51	26.91	0.00	0.00	10.24
Albite	0.09	0.00	0.00	0.00	0.12	2.91	2.86
Anorthite	2.57	4.44	3.86	7.18	48.53	32.63	9.59
Tremolite	–	0.00	–	0.00	–	1.29	0.33
Biotite	4.68	7.82	12.91	13.57	11.74	18.33	13.55
Muscovite	13.65	28.08	1.06	6.29	0.00	0.00	7.08
Chlorite	0.00	0.00	0.01	0.00	0.35	0.00	0.00
Titanite	0.55	0.21	0.47	0.13	3.20	0.98	0.22
Rutile	0.61	0.74	0.42	0.45	0.08	0.38	0.42
Apatite	0.49	0.43	0.49	0.65	0.40	0.45	0.56
Pyrite	36.94	19.44	9.77	7.42	1.14	0.57	10.97
Pyrrhotite	9.28	3.93	8.84	8.50	11.94	12.86	9.60
Sphalerite	0.94	0.77	0.76	0.77	0.30	0.45	0.99
Chalcopyrite	0.47	0.35	0.34	0.36	0.56	0.41	0.43
Pentlandite	0.03	0.51	0.18	0.60	0.03	1.06	0.70
Violarite	0.17	–	0.09	–	0.63	–	–
Alabandite	–	0.11	–	0.23	–	0.16	0.26
Uraninite	0.00	–	0.00	–	0.00	–	–
Unclassified	0.14	–	0.06	–	0.05	–	–
Graphite	3.11	11.80	2.57	7.30	4.00	10.40	7.78
Total	100.00	102.77	100.00	100.67	100.00	100.79	100.38
MLA points	10,402		11,205		10,869		

Note: For comparison average modal composition of samples for TV-Ni (see Table 9.1.3) is also presented.
Dash indicates a component not observed by MLA or assumed in the CBM calculation. Plagioclase in the CBM modes is given divided in its end-member components and violarite included in pentlandite. Graphite in the MLA modes is by grayscale segmentation, in CBM modes by noncarbonate carbon. It is to be noted that MLA struggles with the most fine-grained components such as pyrite, which is dominantly in 1-4-ATK-05, and graphite in all samples.

in TiO_2 (1.14 ± 0.15 wt%) and, on average, contains 0.12 wt% Cr_2O_3, 0.54 wt% V_2O_3, and 0.15 wt% ZnO. Even though containing less than 2 wt% of Fe, biotite is, besides sulfides, the only significant Fe-bearing mineral in the Talvivaara ore.

Muscovite is an important host for Al in part of the Talvivaara ore. Our few analyses show that muscovite contains significant V_2O_3 (0.41 wt%) and some Cr_2O_3 (0.06 wt%). Calculations suggest that >90% of whole-rock V and Cr reside in biotite and muscovite. Other hosts of V are rutile (average 1.68 wt% V_2O_3) and titanite (average 0.56 wt% V_2O_3). Titanite, which contains 0.60 wt% Y_2O_3, is, together with apatite, an important host for whole-rock Y and heavy rare earth elements (HREE). In many samples, Ti and Nb reside near completely in rutile, which is refractory to acid-based digestion. Consequently, some of the previously published whole-rock analyses have revealed too low concentrations for Ti and/or Nb (e.g., Loukola-Ruskeeniemi and Heino, 1996; Loukola-Ruskeeniemi and Lahtinen, 2013).

Mn-enriched black shales

The Mn-enriched rocks in the Talvivaara formation can be divided into at least three main lithotypes: (1) spessartine-bearing black shales, (2) alabandite-bearing black shales, and, (3) alabandite-bearing calc-silicate rocks. The spessartine-bearing black shales are largely restricted to the lower parts of the ore layer and along contacts of calc-silicate layers. The amount of the Mn-garnet in these rocks varies from <5 vol% up to >50 vol%, with the highest abundances occurring in black shales along calc-silicate layers. Garnet analyzed from two samples is very rich in the spessartine component, containing 31.0 ± 1.8 wt% MnO and 6.1 ± 0.6 wt% CaO, but only 2.3 ± 0.9 wt% FeO and 1.9 ± 0.9 wt% MgO.

The alabandite-bearing rocks are diverse. Some of them are very rich in graphite (up to 35 wt%), and fine-grained alabandite may easily be overlooked in such cases. More commonly, alabandite enrichment is related to veins and blebs of coarse-grained, remobilized Fe-sulfides, chalcopyrite, and sphalerite, where alabandite is also coarse-grained and contains abundant graphite and silicate inclusions, and is thus undoubtedly of a metamorphic origin. In many cases, the alabandite-sulfide veins seem to occur in the originally low-Mn sandstones-shales. We stress that Mn enrichment and the occurrence of alabandite is by no means restricted to high-Mn zones but, in a lesser scale, is a ubiquitous feature of the Talvivaara ore.

Calc-silicate rocks

The calc-silicate rock intercalations representing less than 5 vol% of the Talvivaara ore zone are coarse- to very coarse-grained, massive or schistose rocks, and besides occasional indistinct layering, usually show no other preserved sedimentary structures or textures. Most of the calc-silicate layers are dominated by graphite-pigmented, black to bluish black (sometimes iridescent) tremolite (Fig. 9.1.9A), but locally abundant green-gray diopside occurs (Fig. 9.1.9B). The rocks usually contain variable amounts of calcite, anorthite, and minor phlogopitic biotite and occasionally microcline. The graphite content is typically 7–14 wt%. In tremolite-dominant varieties, graphite occurs mainly as tiny flakes and dust or "pigment" distributed in tremolite crystals and other main rock-forming minerals, including sulfides. In diopside-dominant samples, a large part of the graphite is found concentrated in up to cm-sized domains of massive graphite occupying interstices of coarse diopside crystals, in which the graphite is frequently extensively replaced by sulfides, especially pyrrhotite. Tremolite and diopside were analyzed from a couple of representative samples, yielding 2.0 ± 0.8 and 0.6 ± 0.4 wt% Al_2O_3; 1.2 ± 0.6 and 1.7 ± 0.9 wt% MnO; and 0.84 ± 0.48 wt% and 0.39 ± 0.26 wt% FeO, respectively.

FIGURE 9.1.9 Images of diamond-sawn slabs from calc-silicate rocks in the Talvivaara Formation/ore.

(A) Pyrrhotite-rich tremolite rock with 2200 ppm Ni. Slab height 12 cm. (B) Alabandite-rich (black clots mainly alabandite) diopside rock containing 400 ppm Ni and 2.7 wt% Mn. Slab height 12 cm.

The sulfide content and mineralogy varies greatly from sample to sample, apparently without any clear dependence on the host mineralogy. Sulfides are found both as inclusions in silicates and carbonates and as a network interstitial to them. In some layers, sphalerite and/or alabandite may be very abundant up to several vol%, and on average, the calc-silicate-bearing rocks are enriched in alabandite/Mn relative to average Ni-rich black shale. Common minor to trace minerals include apatite, biotite, titanite, rutile, and zircon, and less frequently scheelite.

ORE MINERALOGY

Sulfide textures and chemical compositions

The main sulfide mineral assemblages in all the main rock types in the Talvivaara Formation and ore are rather simple, containing variable proportions of pyrite, pyrrhotite, sphalerite, chalcopyrite, pentlandite, and alabandite. In addition, trace amounts of galena, gersdorffite, hessite, ullmannite, and stannite are observed in some samples. The trace sulfides were not studied for this work but some information is available in Törnroos (1982a) and Törmälehto (2008).

Pyrite is the most abundant sulfide in the Talvivaara ore although it may be nearly absent in some of the less sulfur-rich, pyrrhotite-dominated samples. It is present in a wide variety of textural types. The three most common types, illustrated in Fig. 9.1.10, are: (1) fine-grained (<30 μm), spherical-irregular to euhedral pyrite (FPY) dominating massive, thickly bedded and banded-laminated pyritic mudstones; (2) medium to coarse (usually 30–200 μm), solitary anhedral or euhedral pyrite grains or their aggregates (SPY) spotting samples with abundant fine-grained pyrite and/or pyrrhotite; and (3) typically coarse (200 μm up to >10 mm), mostly idioblastic or resorbed idioblastic pyrite (MSSPY) in pyrrhotite-dominated blebs and veins of mss (monosulfide solid solution) origin. In addition, there are various minor generations of usually coarse pyrites, for example, in pyrite-dominated, irregularly crosscutting veins, in stripes along schistosity planes or fracture fillings.

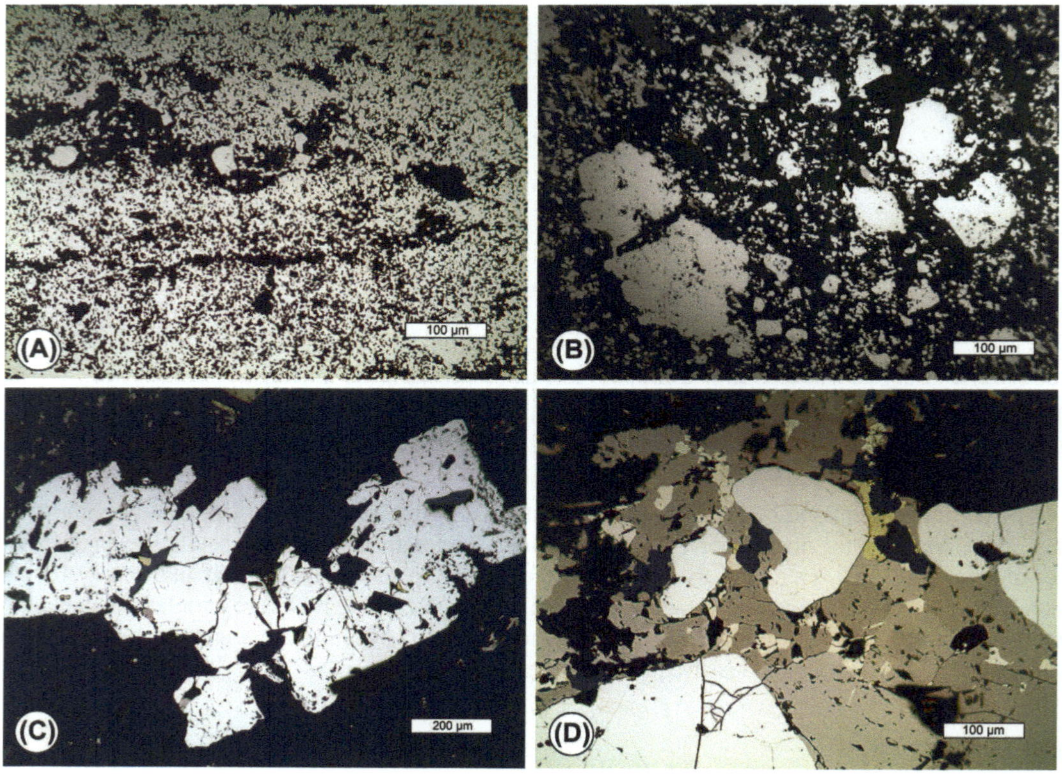

FIGURE 9.1.10 Reflected-light photomicrographs of pyrite grains representing the main pyrite types in the Talvivaara Formation/ore.

(A) FPY type fine-grained (<30 μm), anhedral to euhedral, probably synsedimentary grains in a bedded-laminar highly pyritic ore/mud. (B) FPY pyrite grains and coarser-grained, SPY-A type diagenetic or early metamorphic pyrite grains in a massive pyrrhotite- and pyrite-rich ore/mud. (C) Large SPY-B type metamorphic pyrite porphyroblast showing mica (black), sphalerite (gray), and chalcopyrite (yellow) inclusions in pyrrhotite dominated ore/mud. (D) Mss type coarse, resorbed pyrite grains (white) enclosed in pyrrhotite (brown) with inclusions of grain-type pentlandite (cream yellow) in strongly structurally reworked, pyrrhotite dominated ore.

The FPY grains, especially those in pyrite-rich layers, are usually considered a "primary" phase and reported as *spheroidal pyrite* (e.g., Loukola-Ruskeeniemi and Lahtinen, 2013; Young et al., 2013). However, spheroidal pyrite sensu stricto (see Rickard, 2012) is very rare and even spherical pyrite forms are subordinate as most fine-grained pyrite grains vary from anhedral-irregular to euhedral in their shape. The FPY grains are optically homogenous without any signs of possible spheroidal or framboidal organization in their internal structure or grouping. Roundish, lensoid to streak-form aggregates of FPY grains occur in some samples but are mostly results from shear transposition and related disintegration of laminas of disseminated FPY.

Many of the SPY grains, especially in samples with abundant disseminated or laminated FPY, could be considered diagenetic to burial metamorphic in origin. However, the usual occurrence of abundant inclusions of well-crystalline graphite, rutile, and metamorphic silicates in these grains implies that, in most cases, they represent either diagenetic grains that were recrystallized, grown, and variably annealed during metamorphism and/or grains that are purely metamorphogenic, porphyroblastic in origin.

The MSSPY comprise, to reiterate, solitary or aggregated, euhedral or resorbed, usually coarse (0.05–10 mm) porphyroblasts in pyrrhotite-dominated, small to large sulfide blebs and veins, interpreted to represent postpeak metamorphic mss + iss (iss = intermediate solid solution), from which the present bleb and vein sulfides exsolved after the temperature peaking (Kontinen et al., 2013a).

Average compositions of the three pyrite types (FPY, SPY, MSSPY) are listed in Table 9.1.2. Of the measured trace metals, only Ni, Co, and As were consistently analyzed and found mostly above the detection limits (in most probe sessions: Co and Ni 60 ppm, As 120 ppm). The whole dataset is plotted on Co versus Ni and Co/As versus Ni/As diagrams in Figs. 9.1.11 and 9.1.12, respectively. The FPY grains (<30 μm) show a large variation in their As, Ni, and Co contents, both between and within samples, to an extent that morphologically similar-looking adjacent FPY grains may have contrasting Co and Ni contents. Consequently, in the Co versus Ni plot, the FPY data cover a large part of the whole diagenetic-syngenetic pyrite field constrained by the "all-time" sedimentary pyrite data by Large et al. (2014).

Part of the FPY grains have high Co/Ni of >1, whereas some show distinctly low Co/Ni of <0.1, suggesting metamorphic Co-Ni mobility/redistribution. However, there is a discrete cluster in the FPY data presented in Fig. 9.1.11 that is defined by grains with 500–3000 ppm Ni and 100–1000 ppm Co. If the data are further cleared from analyses with <500 or >3000 ppm As and suspect outliers are screened, in Fig. 9.1.12, a dense cluster just in the middle of the all-time sedimentary pyrite field

Table 9.1.2 Average composition of the main pyrite types in black shales of the Talvivaara Fm/ore

	n s/m/t	Fe (wt%)	S (wt%)	Ni (ppm)	Co (ppm)	As (ppm)
FPY	8/55/134	60.0 ± 3.5	46.1 ± 2.9	1980 ± 2120	810 ± 1500	1940 ± 2060
SPY	15/142/211	52.3 ± 1.0	46.9 ± 0.8	980 ± 1020	4100 ± 4410	1780 ± 1450
MSSPY	8/83/83	52.8 ± 0.7	46.0 ± 0.9	160 ± 140	14310 ± 6460	1290 ± 2460

FPY = fine-grained, presumably originally syngenetic-diagenetic pyrite (<30 μm), SPY = "coarse-grained" (>30 μm) diagenetic and/or metamorphic (porphyroblastic) pyrite, MSSPY = dominantly coarse-grained (>200 μm) idiomorphic and/or resorbed idiomorphic pyrite in pyrrhotite-pyrite blebs and veins (after peak-metamorphic mss). s = number of samples, m = number of Fe and S analyses, t = number of As, Co, and Ni analyses. ± gives standard deviation by MS Excel function stdev.

Note: For analytical details, see supplementary online material.

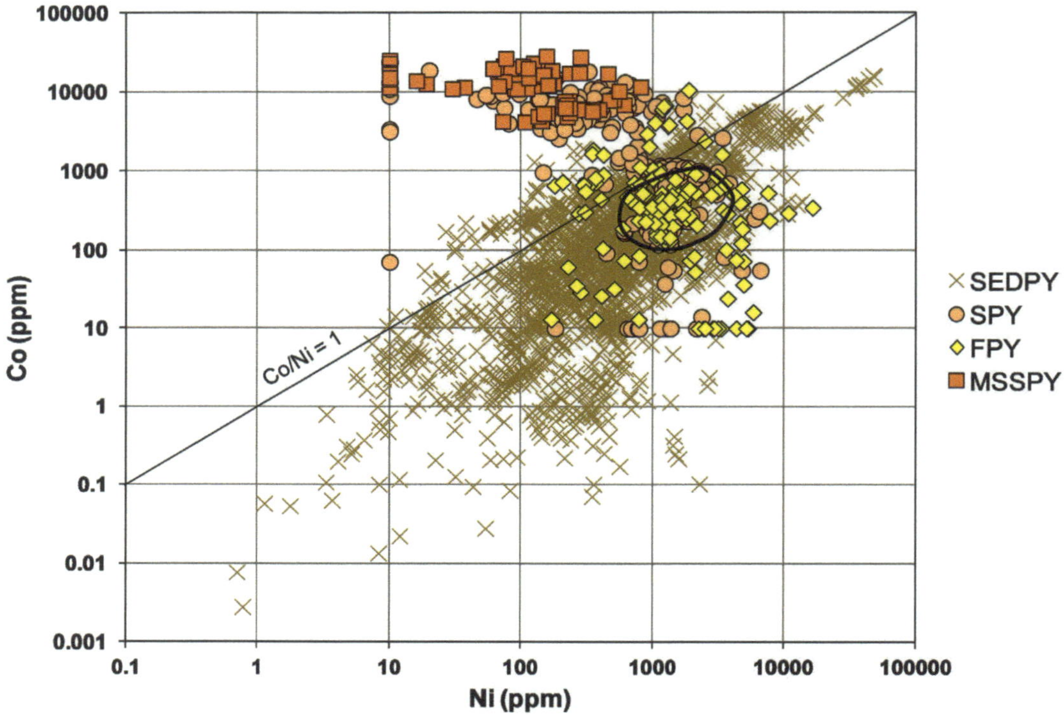

FIGURE 9.1.11 Co versus Ni diagram for the main pyrite types in black shales of the Talvivaara Formation/ore.

FPY = early synsedimentary-diagenetic, fine-grained, <30 μm pyrite; SPY = diagenetic and metamorphic, coarse (>30 μm) pyrite; MSSPY = pyrite exsolved from peak-metamorphic monosulfide solid solution. A field of FPY pyrites thought to be "most primary" is outlined by a black line using criteria explained in the text.

Source: *Data for "all-time" sedimentary pyrites (SEDPY) from Large et al. (2014) are shown for comparison.*

appears, providing the most probable estimate of the "primary" FPY composition. These pyrites contain on average 1300 ppm Ni, 390 ppm Co, and 1540 ppm As.

In the MSSPY type, Co is high (970–28520 ppm), Ni is low (<60–1170 ppm), and As varies greatly (<120–12950 ppm). Importantly, Co/Ni and Co/As are systematically >1. Preliminary probe traverses across MSSPY grains demonstrate that they display distinct metal zoning. Based on our MSSPY data, it seems well grounded to suggest that all Talvivaara pyrite grains with >1000 ppm Co and Co/Ni >1 record metamorphic compositions. The SPY type is compositionally most heterogeneous. In Figs. 9.1.11 and 9.1.12, the analyses spread over the entire spectrum of the Talvivaara pyrite analyses, though there is a clear bimodal clustering in the MSSPY-dominated (Co/Ni >1) and FPY-dominated fields (Co/Ni <1). The grains of the subtype with >1000 ppm Co and Co/Ni >1 (SPY-B) are interpreted as metamorphic porphyroblasts equilibrated with the peak-metamorphic mss, whereas the other subtype (SPY-A) with <1000 ppm Co and Co/Ni <1, especially for those grains plotting in the area defined by the screened FPY grains, probably represents diagenetic-low metamorphic pyrite. These grains seem to be largely restricted to the pyrite-richest samples.

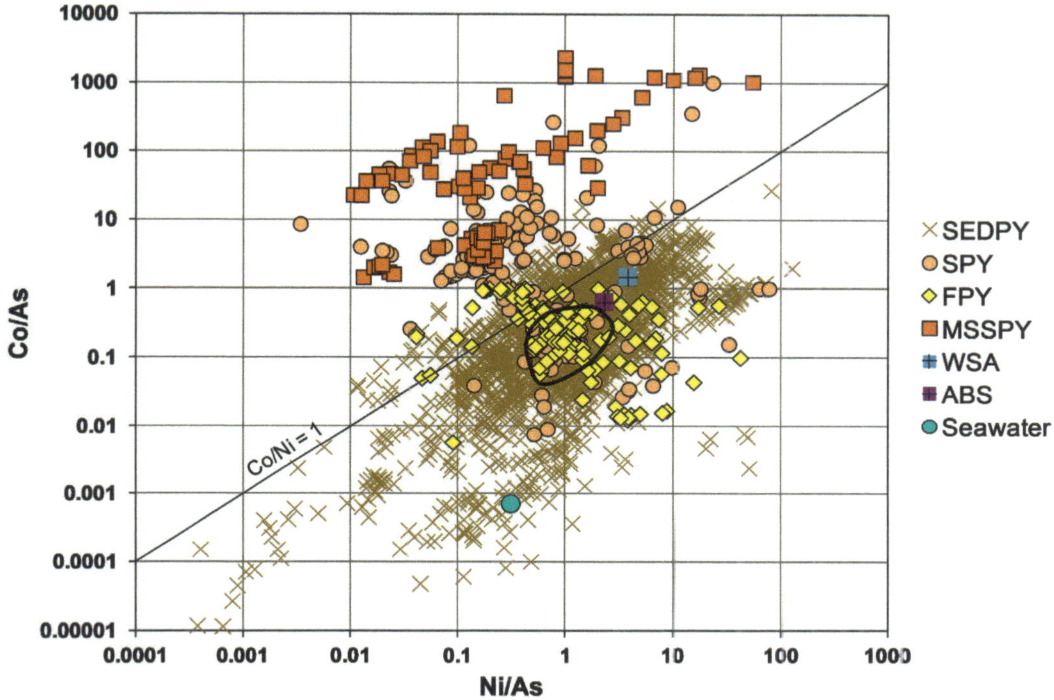

FIGURE 9.1.12 Co/As versus Ni/As diagram for the main pyrite types in black shales of the Talvivaara Formation/ore.

FPY = early synsedimentary-diagenetic, fine-grained, <30 μm pyrite; SPY = diagenetic and metamorphic, coarse (>30 μm) pyrite; MSSPY = pyrite exsolved from peak-metamorphic monosulfide solid solution. A field of FPY pyrites thought to be "most primary" is outlined with a black line.

Source: *For comparison, ratios of average black shale (ABS) from Ketris and Yudovich (2009), average shale (WSA) and seawater from Li (2000), and the "all-time" sedimentary pyrite data (SEDPY) from Large et al. (2014) are also shown.*

High Ni contents in some Talvivaara FPY grains were pointed out earlier (e.g., Loukola-Ruskeeniemi and Lahtinen, 2013). In most samples with significant FPY, our microprobe analyses also revealed FPY and SPY-A-type grains with exceptionally high Ni contents (3000–16,400 ppm, mean 4790 ± 2460 ppm). The Ni-rich varieties occupy about 12% of all analyzed FPY + SPY grains. The Co content of these grains varies between <100 and 1800 ppm (mean 480 ± 70 ppm) and the As content between <100 and 2000 ppm (mean 1270 ± 920 ppm).

The grain size and shape of pyrrhotite varies greatly in both ore-grade and subgrade rocks and between and within samples, depending on the original sediment mineral composition, type, and, first of all, degree of structural reworking and recrystallization. The textures range from minute (<30μm), disseminated grains in massive and bedded-laminar ore to aggregates of coarse grains in mss blebs and veinlets in the more reworked mudstones. A tendency to develop mobilized blebs and veins is evident, at least under the microscope, even in the most massive and fine-grained mudstones, regardless of whether they are dominated by pyrrhotite or pyrite. With the aggregate size increasing, increasingly coarser grains of pyrite, sphalerite, chalcopyrite, and, locally, also alabandite appear in the mss blebs and veins.

In all ore-grade black shales, both in the Kuusilampi and Kolmisoppi areas, pyrrhotite grains contain variable amounts of pentlandite exsolutions, irrespective of their grain size or morphology. Pyrrhotite has a very uniform composition throughout the ore, corresponding to what could be expected for pyrrhotite + pentlandite resolved by cooling from Ni-enriched peak-metamorphic mss (Kontinen et al., 2013a). Accordingly, pyrrhotite averages 4930 ± 690 ppm in Ni whereas it is consistently very low in Co (average <64 ppm, max. 240 ppm). The As content is variably low (average <330 ppm, max. 1270 ppm) and Se is variable (<200–2480 ppm). It should be noted that some older studies have reported significantly higher Co (500–1500 ppm) for pyrrhotite (e.g., Loukola-Ruskeeniemi and Heino, 1996; Törmälehto, 2008), but our reanalyses of pyrrhotite in some of the polished sections studied by them indicate only low cobalt contents (<100 ppm).

Sphalerite is a ubiquitous phase in Talvivaara ore-grade black shales. Its mode of occurrence varies from minute (<30 μm), apparently "primary" grains in samples of massive and laminar, fine-grained mudstones to very coarse (>1 mm) grains, as in association with the pyrrhotite-dominated mss blebs and veinlets in structurally reworked matrices. Coarse sphalerite in the sulfide aggregates and veins may be rimmed by chalcopyrite and/or contain chalcopyrite as small inclusions. Also the association of sphalerite with alabandite is common. Our analyses and those in Törmälehto (2008) indicate that sphalerite has a nearly constant composition with 8.1 ± 1.0 wt% Fe, 4.7 ± 1.3 wt% Mn, and 0.13 ± 0.08 wt% Cd over the whole deposit. The near-constant Fe content provides strong evidence for its metamorphic equilibration with the associated silicates and sulfides. Applying the calibration of Lusk and Ford (1978) and assuming a locking temperature of 550°C, the sphalerite geobarometer indicates a pressure of ~5.6 kbar, which seems reasonable considering the common presence of staurolite + garnet in the local nonsulfidic metapelites. However, the result should be taken with caution because the sphalerite analyses were not performed with the aim of geothermobarometry, and pyrrhotite is variably hexagonic and/or monoclinic.

Chalcopyrite is mechanically the weakest of the Talvivaara main sulfides, which is revealed by its common occurrence in thin seams filling late fractures, minor faults, and coating their slip surfaces. Calcopyrite is the dominant Cu holder at Talvivaara as the other main sulfides are nearly free of Cu (Riekkola-Vanhanen 2007; Langwaldt and Kalapudas, 2007; Törmälehto, 2008). Typically for chalcopyrite in amphibolite-grade metamorphic rocks, the Talvivaara chalcopyrite has a close-to-stoichiometric composition (Törmälehto, 2008).

Pentlandite is economically the most important sulfide in the Talvivaara deposit, with its Ni accounting for >75% of the metal value of the deposit (Riekkola-Vanhanen, 2007; Langwaldt and Kalapudas, 2007). As described previously, the pyrrhotite grains nearly always contain variable amounts of exsolved pentlandite, mostly as granular grains enclosed in, veining, and/or rimming pyrrhotite grains (Fig. 9.1.10D). Along with the granular type, minute flame-type exsolutions are also present but usually in subordinate abundances. The amount of pentlandite exsolutions in pyrrhotite depends on the whole-rock Ni content and pyrrhotite abundance. Expectedly, richest in pentlandite are the pyrrhotite grains in relatively pyrrhotite-poor but Ni-rich samples. Nonaltered pentlandite has a nearly constant Ni content of 33.9 ± 1.7 wt% and a relatively low cobalt content of 1.21 ± 0.71 wt%. In surface parts of the deposit (0–50 m below surface), pentlandite is commonly variably altered/oxidized, ultimately to violarite.

Alabandite is rather common at Talvivaara. In addition to Mn-enriched black shales, it occurs in minor quantities in the Ni-rich black shales, consistently with their typical elevated MnO content of 0.3–0.6 wt% and highly reduced nature. In least-deformed, massive, fine-grained mudstones, the alabandite abundance and grain size are usually both small. In the zones of more Mn-enriched black shales and calc-silicate rocks, alabandite occurs dominantly in blebs and veins of coarse, mobilized sulfides where it is also mainly coarse grained (>1 mm) and often contains plentiful inclusions of graphite and other sulfides, especially pyrrhotite. Alabandite is typically slightly ferroan with 4.2 ± 0.9 wt% Fe (Törnroos 1982b; Törmälehto, 2008; this study).

Uraninite

A significant part of U in the black shales occurs as uraninite, which is, together with tiny sulfide grains, enclosed in microscopic nodules of semilustrous carbonaceous material forming "thucholite balls." These are commonly considered a paragenetically early feature, generated when the host sediments were in the oil-producing window (Schidlowski, 1981; Rasmussen, 2005). However, in the Talvivaara case, such an interpretation would include an unlikely assumption that the "bitumen" in the thucholites withstood metamorphism, for some mysterious reason, better than the totally graphitized surrounding organic matter. Therefore, the poorly crystalline carbonaceous matter in the thucholites postdates rather than predates the enclosed uraninite grains. The uraninite associated with thucholite has been dated at ~1.88 Ga (Lecomte et al., 2014), which means that the thucholite "bitumen" was formed subsequently, probably via radiolysis on CO_2-CH_4-bearing metamorphic fluids generated by dehydration-decarbonation reactions in the carbonate rock intercalated with black shales in the Talvivaara formation.

WHOLE-ROCK GEOCHEMISTRY

In this chapter, we review major aspects of the geochemistry of the Talvivaara Ni-rich black shales and also briefly summarize the geochemical features of the low-Ni black shales, Mn-enriched varieties, and calc-silicate interlayers.

An average composition of the Talvivaara ore-grade (Ni > 0.1wt%) black shales (TV-Ni), as well as associated low-Ni (Ni <0.1 wt%) black shales (TV-low-Ni), calc-silicate rocks (TV-CSR), manganese-enriched black shales (TV-Mn), and quartz-rich sands (TV-QSA), are compared with average compositions of representative Lower Kaleva low-C-S pelites from the Kainuu schist belt (LKAPE12), Upper Kaleva pelites from the Outokumpu area (OPE20), and Outokumpu black shales (OBS31). Also listed is the world shale average (WSA; Li, 2000), which is often used as a geochemical reference. The Talvivaara sample set produces an average composition that is for most elements very close to that reported by Loukola-Ruskeeniemi and Lahtinen (2013) for the Talvivaara ore (Table 9.1.3) and we believe that TV-Ni in the table characterizes the ore very well. The compositions of the Talvivaara high-Ni, low-Ni, and Outokumpu black shales are graphically compared with the WSA in Fig. 9.1.13.

Major elements

The bulk of the Ni-rich black shales show a surprisingly invariable composition with 45–53 wt% SiO_2, 10–13 wt% Al_2O_3, 6–10 wt% C, and 5–9 wt% S as the four most abundant constituents. The SiO_2 versus Al_2O_3 plot presented in Fig. 9.1.14 illustrates two departures from this basic pattern of monotonous compositions, one trend being toward elevated SiO_2 related to addition of quartz (quartz sand-silt), and the other trend toward the origin, related to the addition of extra sulfide.

To allow more direct comparison of the clastic component in the TV-Ni and the reference material presented in Table 9.1.3, we have normalized the major element compositions to 100% without C_{org} and S, and adjusted Fe and Mn to yield Fe/Al of 0.54, Mn/Al of 0.01, and P/Al of 0.0 as in the WSA (Table 9.1.4). The resulting average TV-Ni composition resembles more LKA12 than OPE20 or the

Table 9.1.3 Average composition of ore-grade (Ni >0.1 wt%) black shale (TV-Ni) and associated materials in the Talvivaara Formation

	TV-Ni	TV-Ni	TV-Ni	TV ore	TV-low Ni	TV-Mn	TV-CSR	TV-QSA	LKAPE12	OPE20	OBS31	WSA
	No. of analyses	Average	stdev	Median	Average	Average	Average	Average	Average	Average	Average	Average
No. of samples		58		556	18	10	12	9	12	20	31	
SiO_2_xrf (wt%)	58	46.7	9.8	38.1	45.3	43.0	30.1	76.5	58.2	59.1	47.6	58.90
TiO_2_xrf	58	0.65	0.19	0.33	0.63	0.41	0.19	0.32	1.09	0.85	0.57	0.77
Al_2O_3_xrf	58	10.79	2.47	9.49	12.31	9.02	4.24	7.93	14.00	16.62	10.90	16.60
Fe_2O_3t_xrf	58	16.41	8.01	16.38	15.75	16.12	12.63	2.98	11.26	8.50	14.07	6.75
MnO_xrf	58	0.32	0.14	0.36	0.37	5.20	1.61	0.16	0.09	0.07	0.06	0.11
MgO_xrf	58	3.09	1.02	3.30	3.52	5.65	12.02	3.46	4.82	3.78	5.80	2.50
CaO_xrf	58	2.15	1.41	2.03	2.20	2.49	16.55	1.66	1.25	2.04	5.52	2.20
Na_2O_xrf	58	<0.41	0.23	0.46	0.52	0.42	<0.12	<0.64	1.08	2.51	1.05	0.80
K_2O_xrf	58	3.63	1.88	2.83	3.52	2.59	0.42	2.14	4.13	4.07	2.23	3.20
P_2O_5_xrf	58	0.24	0.28	0.16	0.18	0.37	0.29	0.17	0.15	0.16	0.16	0.16
S_xrf	41	9.71	6.07	–	8.92	10.76	5.92	1.37	1.01	0.36	6.41	0.24
S_leco	51	10.05	7.25	8.22	10.64	11.86	6.86	1.86	–	–	7.33	
C_tot_leco	58	8.12	2.55	–	9.08	7.6	10.54	2.86	0.77	0.46	7.73	0.55
C_graf_leco	46	8.10	2.24	–	8.63	8.05	7.32	2.63	–	–	6.93	0.20
C_carb_leco	46	<0.59	0.72	–	0.72	0.25	3.00	0.26	–	–	0.44	0.35
Cl_xrf	40	<0.01	–	–	<0.01	0.02	<0.01	<0.01	0.02	0.01	0.02	0.018
Ba_xrf (ppm)	58	255	162	–	315	174	<26	128	238	827	299	580
Rb_xrf&icpms	58	99	36	53	93	117	24	71	138	161	101	140
Sr_xrf	58	65	26	38	80	56	34	51	67	170	150	170
Y_xrf	39	<27	13	–	36	–	22	<15	22	25	35	26
Y_icpms	56	29	13	16	38	68	21	14	21	24	34	
Zr_xrf	39	105	32	–	128	72	31	94	160	165	117	160
Zr_icpms	56	105	29	39	134	74	34	87	140	145	113	
Hf_icpms	56	2.9	0.78	–	3.7	2.0	0.92	2.4	3.7	4.1	3.2	5
Nb_icpms	56	7.6	2.5	–	9.4	5.3	2.3	2.9	13	12	8.3	11
Ta, icpms	56	0.67	0.33	–	0.66	0.49	<0.27	0.40	0.91	0.82	0.62	1
U_xrf	39	<17	5	–	17	–	<12	<10	<10	<10	<17	2.7
U_icpms	56	17	6	17	20	11	6.4	2.2	2.6	2.9	17	
Th_icpms	56	5.6	1.6	5	8.7	4.2	1.5	3.6	6.2	10.2	8.3	12

Sc_icpms	56	19	5.3	7	17	16	6.2	10	28	24	17	13
Cr_xrf	58	133	40	146	151	96	46	58	145	147	127	90
V_xrf&icpms	58	589	172	660	682	426	248	279	204	188	647	130
Mo_xrf	53	<50	23	–	56	<11	<12	<10	<10	<10	65	2.6
Mo_aqr	35	54	23	58	53	–	<13	3.3	–	–	–	
Co_icpms	56	186	112	189	71	117	158	21	31	24	31	19
Ni_xrf	58	2983	1140	2415	445	1087	2416	211	113	83	406	50
Cu_xrf	58	1429	868	1340	1082	1314	1085	177	115	62	312	45
Zn_xrf	58	5171	2307	4955	3011	4357	4809	1420	371	166	1771	95
Pb_xrf	41	<82	134	–	36	50	<52	<27	<24	<30	<37	20
Pb_aqr	35	75	105	47	43	–	23	<7.9	–	–	18	
As_xrf	58	<162	232	–	173	214	<61	<49	<30	<30	<78	13
As_aqr	41	174	222	86	131	–	63	36	–	0.6	32	
Bi_aqr	35	0.7	0.3	–	0.7	–	0.8	<0.1	–	0.2	0.5	0.43
Cd_aqr	35	15	8.0	16	14	–	23	6.2	–	0.1	14	0.3
Sb_aqr	35	6.1	8.4	–	6.8	–	2.0	0.8	–	<0.1	0.9	1.5
Se_aqr	35	46	18	–	28	–	31	6.5	–	0.2	15	0.6
Ag_aqr	35	2.2	2.4	2.0	1.5	–	4.7	0.2	–	0.1	1.0	0.07
La_icpms	56	21	8.9	20	39	25	5.8	16	21	32	26	43
Ce_icpms	56	36	14	–	72	39	9.5	28	41	65	47	82
Pr_icpms	56	5.3	2.1	–	9.3	5.7	1.6	3.7	5.1	7.7	6.4	9.8
Nd_icpms	56	22	8.7	–	37	23	7.5	15	20	30	26	33
Sm_icpms	56	4.5	1.7	–	7.4	5.2	2.0	3.0	3.8	5.6	5.3	6.2
Eu_icpms	56	1.2	0.41	–	1.8	1.6	0.51	0.69	0.97	1.1	1.2	1.2
Gd_icpms	56	5.0	2.0	–	8.0	8.2	2.7	3.0	4.1	5.3	5.8	5.1
Tb_icpms	56	0.77	0.31	–	1.2	1.5	0.44	0.43	0.62	0.78	0.89	0.84
Dy_icpms	56	4.6	2.0		6.7	11	2.8	2.3	3.5	4.2	5.2	4.7
Ho_icpms	56	0.97	0.42	–	1.4	2.3	0.63	0.44	0.70	0.82	1.1	1.1
Er_icpms	56	2.9	1.3	–	4.1	6.7	1.9	1.3	2.0	2.3	3.3	3.0
Tm_icpms	56	0.41	0.18	–	0.61	0.96	0.28	0.20	0.29	0.33	0.48	0.44
Yb_icpms	56	2.8	1.2	–	4.2	6.3	1.9	1.3	1.8	2.2	3.3	2.8
Lu_icpms	56	0.44	0.17	–	0.67	0.95	0.32	0.22	0.27	0.33	0.53	0.42

Note: For comparison, median composition of Talvivaara ore (TV ore) from Loukola-Ruskeeniemi and Lahtinen (2013) and compositions of LKAPE12, OPE20, OBS31, and WSA are listed. For explanation of the abbreviations, see text.

xrf = XRF on pressed powder pellets, icpms = ICP-MS finish after multiacid + microfusion digestion, aqr = ICP-MS or ICP-OES finish after hot aqua regia dissolution (for analytical details, see supplementary online material, except for TV ore and WSA, original data sources). stdev = standard deviation by MS Excel function stdev. < indicates an average, which includes cases below detection limit (dl), and for which dl was used as input.

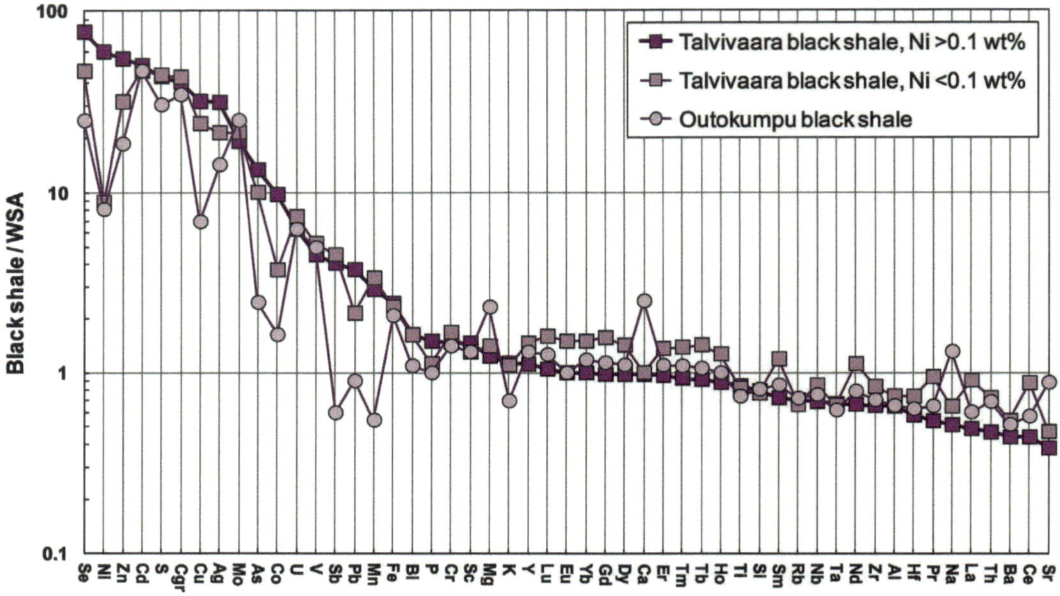

FIGURE 9.1.13 Shale-normalized spidergram for Talvivaara high-Ni and low-Ni black shales and Outokumpu area black shales.

Source: *Elemental abundances in world shale average (WSA) from Li (2000) were used as normalizing values.*

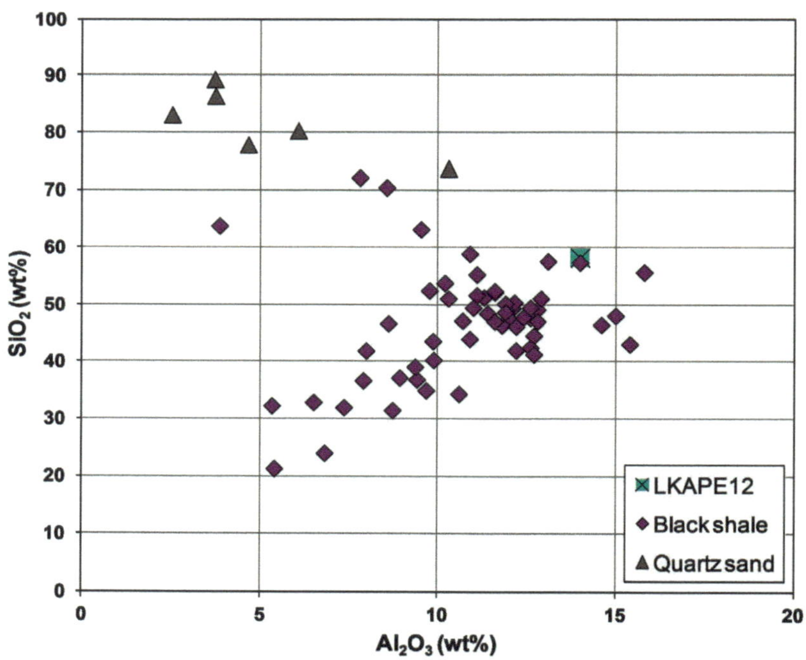

FIGURE 9.1.14 SiO_2 versus Al_2O_3 diagram for Talvivaara black shales (>0.1 wt% Ni) and their quartz sand intercalations.

Two main trends from an end-member mud around 13 wt% Al_2O_3 and 48 wt% SiO_2 are observed; one toward low SiO_2 and Al_2O_3, controlled mainly by sulfide, and the other toward high SiO_2 and low Al_2O_3, controlled by quartz. Also plotted is the average composition (LKAPE12) of 12 representative low-C-S Lower Kaleva pelites.

Table 9.1.4 Major element compositions of Talvivaara and other rocks presented in Table 9.1.3 calculated to 100% without C, S, and authigenic Fe, Mn, and P

	TV-Ni	TV-low Ni	TV-Mn	TV-CSR	TV-QSA	LKAPE12	OPE20	OBS31	WSA
SiO_2 (wt%)	64.73	61.61	63.59	45.04	79.73	64.26	61.47	60.57	63.99
TiO_2	0.90	0.86	0.60	0.30	0.33	1.21	0.89	0.72	0.84
Al_2O_3	15.07	16.91	13.29	6.59	8.22	15.47	17.33	13.93	18.03
Fe_2O_3t	6.18	6.93	5.45	2.70	3.37	6.34	7.10	5.71	7.39
MnO	0.10	0.11	0.09	0.04	0.05	0.10	0.11	0.09	0.12
MgO	4.35	4.86	8.74	18.52	3.61	5.33	3.95	7.48	2.72
CaO	2.97	3.10	3.60	25.87	1.72	1.37	2.13	7.16	2.39
Na_2O	<0.60	0.73	0.59	0.19	<0.66	1.19	2.60	1.32	0.87
K_2O	4.95	4.73	3.92	0.68	2.22	4.56	4.25	2.86	3.48
P_2O_5	0.15	0.17	0.13	0.07	0.08	0.15	0.17	0.14	0.18
Total	100.00	100.00	100.00	100.00	100.00	100.00	100.00	100.00	100.00

WSA, which contain more Al_2O_3 and, in the case of OPE20, also more Na_2O. The low Na_2O content in TV-Ni is consistent with similarly low Na_2O also in nonmineralized Lower Kaleva rocks elsewhere in the KSB. Compared to TV-Ni, OBS31 contains significantly more MgO and CaO, suggesting more carbonate (now tremolite) in the protolith muds.

A particularly useful approach to analyze the major element composition in a mineralogical framework is to plot $K_2O/(Al_2O_3 + CaO + Na_2O)$ versus $Al_2O_3/(Al_2O_3 + CaO + Na_2O + K_2O)$, of which the former could be called K-index and the latter corresponds to the chemical index of weathering (all components in wt%). This diagram (Fig. 9.1.15) comprises the compositions of kaolinite, montmorillonite, phlogopite, K-feldspar, muscovite, tremolite-dolomite, and plagioclase, while omitting those of quartz, graphite, and sulfides, which can be considered as "diluents." As expected from optical and MLA studies, the Talvivaara samples plot in a triangle cornered by microcline, anorthite, and muscovite. There is a large variation in the K-index that cannot be ascribed to any combination of source-sorting, but requires significant postsedimentation alteration and element redistribution. This is

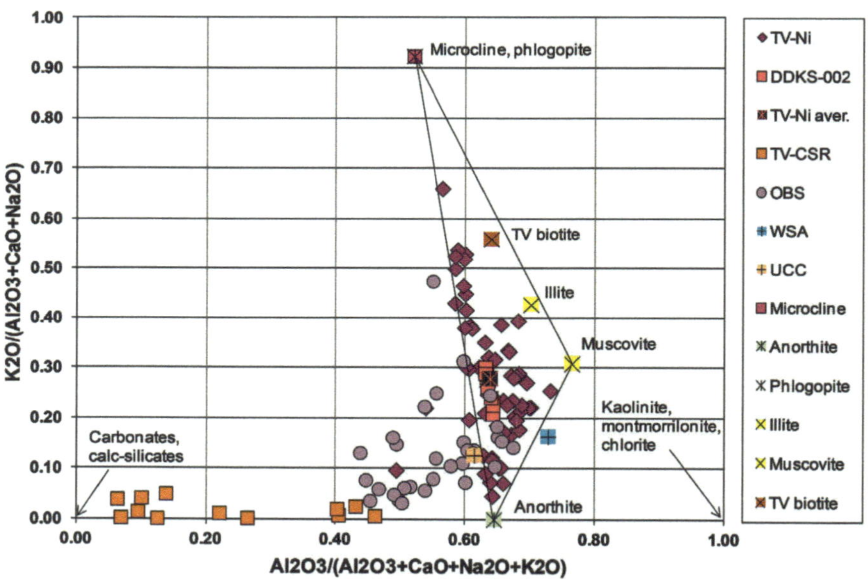

FIGURE 9.1.15 $K_2O/(Al_2O_3 + CaO + Na_2O)$ versus $Al_2O_3/(Al_2O_3 + CaO + Na_2O + K_2O)$ diagram for ore-grade black shale samples (TV-Ni) from the Talvivaara Formation/ore.

Also shown are compositions of various relevant clastic and metamorphic minerals as well as average upper crust (UCC) and shale (WSA). Remarkably, 10 samples across a structurally little reworked, fine-grained, ~1 m thick Ni-rich (4673 ± 189 ppm) massive mud bed (DDKS-002) define a coherent major element composition matching closely with the TV-Ni average composition. This, together with field and petrographic evidence, suggests that most of the Talvivaara muds deposited with a major element composition similar to that of DDKS-002, and that a large part of the observed variation in major elements reflects metamorphic segregation processes (see text). Also plotted are data for Talvivaara calc-silicate rocks and Outokumpu black shales (OBS), which define a clearly different trend, obviously because of their originally much higher carbonate contents compared to typical Talvivaara ore muds.

compatible with rock textures suggesting that variable metamorphic segregation to microcline-rich and microcline-poor lithosomes, in a scale exceeding hand sample scale, is the main cause of the high variation in the K-index.

Major elements are not well suited for source/provenance analysis, although some useful information can be gained from the immobile/poorly soluble/nonhydrothermal and, in sorting, relatively similarly behaving elements Ti and Al. Fig. 9.1.16 shows a plot of TiO_2/Al_2O_3 versus SiO_2/Al_2O_3, which is sensitive to weathering maturation and sorting. Compared to the WSA (0.046) and Upper Kaleva pelite compositions (0.051), Talvivaara samples have somewhat higher TiO_2/Al_2O_3 (0.059) suggesting a more mafic source. Also the Upper Kaleva black shales from the Outokumpu area (OBS) have, on average, lower TiO_2/Al_2O_3 (0.052) than TV-Ni. The Lower Kalevian low-C-S pelite samples have even higher, partly basalt-like TiO_2/Al_2O_3 (0.056–0.102, average 0.078) and thus were probably derived from an even-more mafic-dominated source than the Talvivaara Ni-rich black shales.

The majority of the Talvivaara samples plot on the mud side of the mud-sand divider (at SiO_2/Al_2O_3 = ~4.5) in the Upper Kaleva distribution, which suggests that most of the sedimentary material in Talvivaara ore was silt-clay–dominated mud. Mostly very low to low whole-rock Na_2O and petrographic observations indicate that coarse silt- and sand-sized clasts in the sandy intercalations are

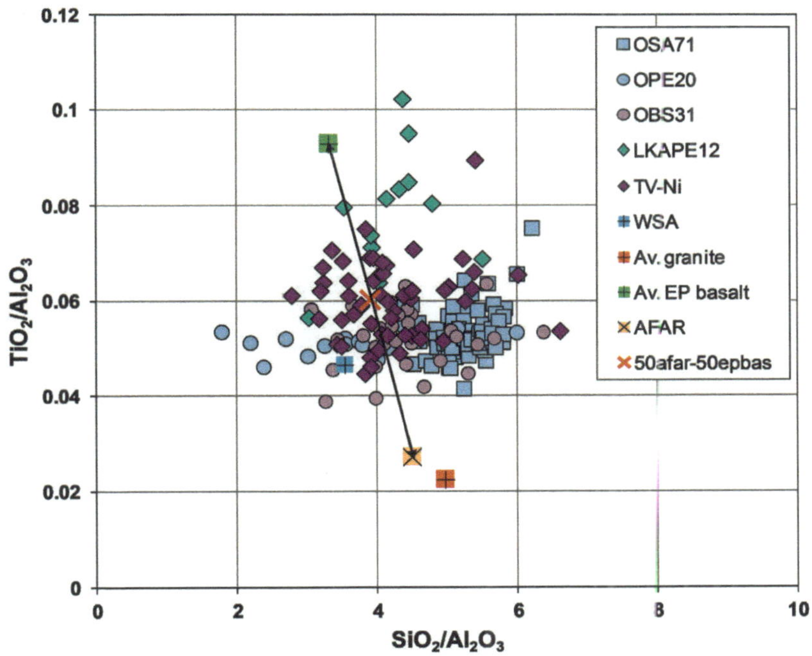

FIGURE 9.1.16 TiO_2/Al_2O_3 versus SiO_2/Al_2O_3 diagram for black shale samples from the Talvivaara Formation/ore.

Source: *For comparison, compositions of representative Outokumpu (Upper Kaleva) sand (OSA71), pelite (OPE20), and black shale (OPE31) samples as well as various reference materials including average granite (Li, 2000), Early Proterozoic basalt (Condie, 1993), and Finnish Karelian domain Archean rock (AFAR; from Rasilainen et al., 2008, Table 82) are shown. Talvivaara mud compositions are explained by a 50–50% blend of average Early Proterozoic basalt (Condie, 1993) and AFAR.*

usually almost exclusively quartz. The mud end of the Talvivaara Ni-rich black shale distribution in the TiO_2/Al_2O_3 and SiO_2/Al_2O_3 diagram can be modeled by 50–50% mixture of the average early Proterozoic basalt of Condie (1993) and average Finnish (Karelian Domain) Archean rock (Rasilainen et al., 2008, Table 82). Considering the high maturity of the Talvivaara black shales and their regional setting, the actual main end-members were Jatulian cratonic sandstones-shales recycled from the Archean basement and interleaved/dissecting mafic lavas, dikes, and sills. The high amount of basalt in the source, apparently not present in the preserved Jatuli, not at least in the KSB, implies that a dominant part of it probably was in the upper part of the overall Jatulian sequence and was largely removed by erosion during or since the Lower Kaleva deposition.

Iron in the Talvivaara black shales is strongly elevated and manganese notably elevated over the typical level in shales, both in terms of average concentrations (Fe 11.47 ± 5.60, Mn 0.25 ± 0.11) and ratios with respect to Al or Ti. The Fe/Al versus Fe/Al plot in Fig. 9.1.17 shows that, compared to the WSA, LKAPE12, or OPE20 pelites, nearly all Talvivaara samples are significantly enriched in Fe. This implies a significant nondetrital "excess" Fe, which seems to have been added for the most part in syngenetic-diagenetic pyrite. The variation in the Fe content can be explained by mixing of a mud end-member with variable Fe-sulfide, quartz, organic carbon, and carbonate. The mud end-member had

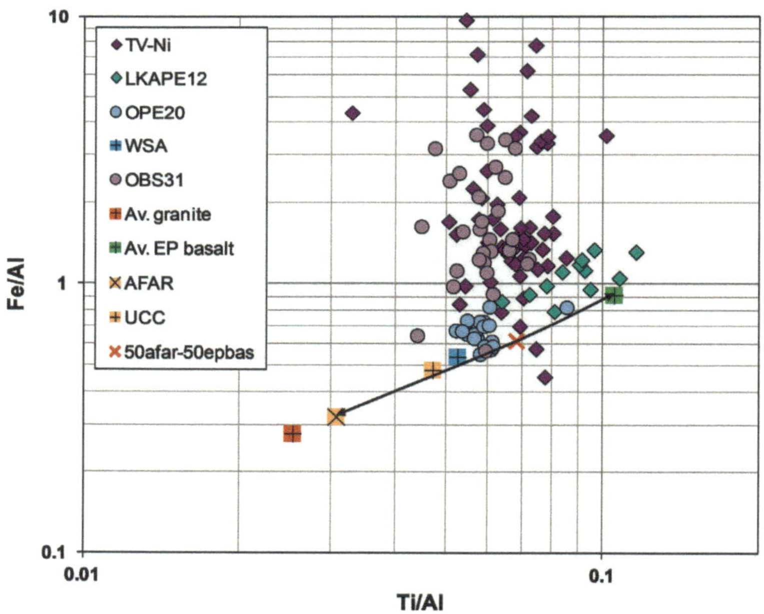

FIGURE 9.1.17 Fe/Al versus Ti/Al diagram for Talvivaara black shales (>0.1 wt% Ni).

The diagram shows that Talvivaara black shales are strongly enriched in Fe compared to average shale (WSA) and also Outokumpu (OPE20) and Lower Kaleva (LKAPE12) low-C-S pelites. They also show mostly higher Ti/Al than WSA and Outokumpu area pelites and black shales, for example. Most Talvivaara rocks cluster around Ti/Al = 0.07, corresponding to about 50–50% blend of AFAR and average Early Proterozoic basalt.

rather invariable Al (7–8 wt%) and other clastophile elements but variable Ni (1500–5000 ppm) and Cu, Zn, and Co. The Fe content of the mud end-member, which was graphically estimated based on the assumed mixing scenario, seems to have been restricted at 7–8 wt%. As we already noted before, presently nearly all Fe and, in many cases, also all Mn are located in the sulfide phase, leaving the silicate fraction nearly devoid of Fe and Mn. Overall, only <1–4% of the Fe is presently hosted by nonsulfide minerals. This is a strong indication of extensive diagenetic and/or later metamorphic sulfidation of iron in detrital minerals.

The Mn/Fe ratio is an important parameter as these two elements are not notably fractionated in magmatic differentiation or sand-shale producing processes (Lepp, 1963). Consequently, Mn/Fe in igneous rocks and sandstones-shales is near to that (~0.02) in the upper crust and mantle (Rudnick and Gao, 2003; McDonough and Sun, 1995). Against this background, the Talvivaara rocks show a remarkably large spread in Mn/Fe from 0.001 to 1.0. Plotting Ni and Ni/Fe versus Mn/Fe reveals three end-members, which are (1) often massive and thick bedded Ni-rich muds with high Ni/Fe (0.05) and Mn/Fe (0.034), (2) distinctly laminated-bedded Fe–rich pyrite muds with low Ni (<1500 ppm), Ni/Fe (<0.01), and Mn/Fe (<0.006), and (3) Mn-rich muds (and calc-silicate rocks) with low Ni (<1000 ppm), low Ni/Fe (<0.01), and high Mn/Fe (0.3–1.0). The trend from Ni mud to pyrite mud can be explained by the addition of syngenetic pyrite that was low in Ni and Mn, while the control(s) of the trend toward high Mn/Fe and Mn enrichment is difficult to deduce, because of the metamorphic overprint and insufficient studies of the Mn-enriched rocks.

The mean P_2O_5 content of the Talvivaara Ni black shales is 0.24 wt%, resulting in a mean P/Al ratio of 0.017, which demonstrates that the rocks are enriched in P by a factor of 2.5 compared to the WSA. However, they are not different from black shales in general, which average 0.32 wt% P_2O_5 (Ketris and Yuodowich, 2009), and are even low in P compared to Phanerozoic metalliferous black shales (0.3–3 wt%) and their phosphorite (P_2O_5 >18 wt%) interlayers (e.g., Coveney and Glascock, 1989). Also in the Talvivaara Formation, part of the P enrichment is confined to thin, P-C–rich beds and laminas. One sample from a 2-cm-thick layer was analyzed for this study, giving 5.67 wt% P_2O_5 and 14.80 wt% C_{org}.

Carbon and sulfur

Talvivaara black shale samples are characterized by a fairly uniform graphite-bound (organic derived) carbon content of 8.1 ± 2.2 wt%. As most Talvivaara black shales contain only negligible carbonate, the average total carbon content is equally 8.1 ± 2.6 wt%. $\delta^{13}C$ values of graphite are between −28 and −24 (Loukola-Ruskeeniemi, 1999) and thus well within the range from about −40 to −15 observed for 2.6–1.6 Ga organic carbon by Karhu and Holland (1996). On average, sulfur occurs at about the same level as carbon (Loukola-Ruskeeniemi, 1991) but in the most pyrite-rich layers may reach values up to 26 wt%. In contrast to many Phanerozoic black shales, which show a high bed-lamina-related variation in C/Al (<1 to >12), the majority of the Talvivaara samples have C/Al rather consistently near 1.3. Even stranger is the restricted variation in C/Al within individual, often thick to very thick, massive beds. The major factors contributing to the variation in the C content are clearly the dilution effects of quartz and especially sulfide (pyrite). From trace metals only those that enter early in the clay phase, such as V (0.56) and Cr (0.43), show significant correlation with C. Especially poor correlations are observed with the more redox-sensitive, thiophile elements, such as Cd, Se and Sb, probably due to postdepositional sulfide phase related mobility effects.

Redox-sensitive trace elements Cr, V, Mo, U, Se, As, Sb, and Cd

In common with black shales in general, the Talvivaara samples are high in redox-sensitive trace metals and metalloids compared to common shales, with the average WSA-based enrichment factors being: Se117, Cd 77, Mo 32, As 19, U 9.4, V 7.0, Sb 6.2, and Cr 2.3. The mean Cr content of the Talvivaara Ni-rich black shales is 133 ± 40 ppm, which is slightly elevated compared to 90 ppm in the WSA. Also Cr/Th ratio in TV-Ni (24) is higher than in the average upper continental crust (UCC) (7.0) and WSA (7.5). However, as Cr/Sc in TV-Ni (7.3) is similar to that in the WSA (6.9), there is little support for enrichment in authigenic or ultramafic Cr but the elevated Cr and Cr/Th can be solely explained by the relatively large basaltic component in TV-Ni.

Vanadium has a high mean concentration of 589 ± 172 ppm as compared to 130 ppm in the WSA. None of the examined sample sets from Talvivaara ore show any simple correlation of Ni or Cu with V, likely due to their contrasting mobilities since burial, with V being located dominantly in phyllosilicates and Cu and Ni in sulfides. Molybdenum occurs as tiny molybdenite (MoS_2) flakes associated with mica and graphite, with its abundance varying from <10–110 ppm (average 50 ± 23 ppm). Its mode of occurrence is reflected in the modest correlations with respect to C_{org} (0.42) and mica-hosted elements Cr (0.44) and V (0.31). The average Mo content and Mo/C is clearly higher than the values measured for Archean black shales (Mo <2–20 ppm), which is consistent with the model that after the Great Oxygenation Event (GOE), oxidative weathering of sulfides raised the molybdenum/molybdate anion (MoO_4^{2-}) content of ocean water (e.g., Lyons et al., 2014).

Uranium contents in the Talvivaara black shales range from 5 to 37 ppm and average 17 ± 6.1 ppm. The average U/Th 3.1 ± 0.6 is more than 10 times higher than that in the WSA (0.23), indicating the presence of significant authigenic U, up approximately 95% of the total U as calculated from the U-Th relations. Uranium correlates well with V (r^2 = 0.75), Cr (0.60), and Al (0.55), but surprisingly, the U-Mo and U-C_{org} correlations are significantly poorer (r^2 = 0.34 and 0.29, respectively), probably reflecting postdepositional mobility of the organic derived carbon.

The main control of the arsenic distribution is pyrite, which, in its all three main morphotypes, contains significant As (average 1200–1900 ppm). In low-pyrite samples, As is very low (<5–30 ppm), even in cases where there is abundant S and pyrrhotite ± pentlandite. The mean As content of the dominantly pyrrhotitic Ni-mud end-member (<10 wt% S) is only <60 ppm whereas pyritic end-member (>20 wt% S) samples contain 570 ppm As on average. In comparison, mean As concentrations in Phanerozoic metal-enriched black shales vary between 28 and 115 ppm (Quinby-Hunt et al., 1989; Gavshin and Zakharov, 1996; Perkins et al., 2008).

The Talvivaara black shales have a rather high selenium concentration of 46 ± 18 ppm resulting in an elevated mean Se/S ratio of 540 ± 232 × 10^{-6}, which are typical for many metal-enriched black shales (Hatch and Leventhal, 1992; Schultz and Coveney, 1992; Gavshin and Zakharov, 1996; Perkins et al., 2008). In comparison, average MOR sulfides show 97 ppm Se and Se/S of 285 × 10^{-6} (Herzig et al., 1998) and the Outokumpu ore contains 24 ± 25 ppm Se and has Se/S of 97 × 10^{-6} (Peltonen et al., 2008). At Talvivaara, the usefulness of Se as a geochemical tracer is complicated by the fact that Se clearly prefers pyrrhotite + pentlandite in the mss blebs, in which the Se concentration may exceed 100 ppm. The pyrite laminated muds richest in "primary" FPY (>50 wt% pyrite) have "only" 38–40 ppm Se and Se/S <250 × 10^{-6}.

The mean antimony and cadmium contents in the Talvivaara black shales are 6.1 ± 8.4 ppm and 15.0 ± 7.8 ppm, respectively, both well above the WSA values, though remaining lower than in some Phanerozoic metal-enriched black shales (Gavshin and Zakharov, 1996; Coveney et al., 1989; Schultz and

Coveney, 1992) or recent sapropels (Niejenhuis et al., 1999). Strangely enough, even though the Cd level in OBS31 is similar to that in TV-Ni, the level of Sb is 7x lower (0.9 ppm). Cadmium in the Talvivaara black shales correlates well with Zn ($r^2 = 0.67$), reflecting the fact that nearly all Cd resides in sphalerite.

Chalcophile metals Ni, Zn, Cu, Co, and Pb

Here we consider Ni, Zn, Cu, Co, and Pb, of which the first four are the principal pay metals of the mine. Nickel, zinc, and copper are nearly 100% hosted by metamorphogenic/metamorphosed sulfide grains. Nickel is incorporated in mss exsolved into pyrrhotite + pentlandite, whereas Zn occurs in sphalerite and Cu in chalcopyrite, both often in blebs and ragged veins, in which they seem retrogressively exsolved from iss. In pyrrhotite-dominated samples, Co is dominantly in pentlandite but strongly prefers pyrite when it is present. Optical studies suggest that most of Pb is located in tiny galena inclusions in the main sulfides.

Excluding Pb, base metal enrichment in the Talvivaara black shales is extraordinary high even on a global all-time perspective; the average contents are: 2980 ± 1140 ppm Ni, 1430 ± 870 ppm Cu, 5170 ± 2310 ppm Zn, 186 ± 112 ppm Co, and 75 ± 105 ppm Pb. These abundances are 3–7 times higher than in the Outokumpu metalliferous black shales (OBS) and 10–60 times higher (except Pb 4x) than in the WSA (Fig. 9.1.13). The highest concentrations analyzed in this study reach 5740 ppm for Ni, 5560 ppm for Cu, and 11,360 ppm for Zn. Note that despite the high concentrations, the average values of base metal ratios in the Talvivaara Ni-rich black shales, such as Ni/Zn (0.67), Ni/Cu (2.53), and Cu/Zn (0.33), are very similar to those in WSA: Ni/Zn (0.53), Ni/Cu (2.50), and Cu/Zn (0.47). Average Co/Ni (0.062) and Pb/Ni (0.028) are, however, much lower in the Talvivaara rocks compared to WSA: Co/Ni (0.38) and Pb/Ni (0.40). In terms of these base element ratios, Talvivaara black shales are similar to black shales in general. Base metals in the Talvivaara black shales correlate poorly with each other and also with most other elements, including S and C_{org}. To a large extent, this is because of the syntectonic sample-scale differential mobility of the principal host sulfides, and consequently only metals with strong partitioning into specific base metal sulfides, such as Cd and Zn into sphalerite, would show strong correlation.

Precious metals

The concentrations of Au, Pt, and Pd measured in this study were below the detection limits of 10 or 20 ppb for most samples. Higher Au (up to 115 ppb) and Pd contents (up to 54 ppb) were observed only in samples from drill core R305 (south of Kolmisoppi), which has previously been noted to intersect relatively Au-Pd–rich black shales, but without marked Pt enrichment (Loukola-Ruskeeniemi and Heino 1996). Previous more extensive fire-assay-ICP-MS analyses have indicated median concentrations of 2.3 ppb Au, 12 ppb Pd, and 8 ppb Pt for Talvivaara rocks (Ni >0.1 wt%) and similar contents (Au 3.0 ppb, Pd 18 ppb, Pt 8 ppb) for low-Ni (<0.1 wt%) black shales (Loukola-Ruskeeniemi and Heino, 1996), which are in line with our data. The Ag contents in 34 ore-grade black shale samples analyzed in this study vary from 0.3 to 13 ppm and give an average of 2.2 ppm. Overall, precious metals seem to be not particularly enriched in the Talvivaara Ni-rich black shales compared to average black shale estimates (Crocket, 1991; Ketris and Yudovich, 2009). Nevertheless, it should be mentioned that a thin (<3 cm) graphite-pyrrhotite–rich band, which was discovered in a drill core by handheld X-ray fluorescence (XRF) during the present study owing to its exceptionally high Sb (260 ppm) and Pb (1700 ppm), shows 643 ppb Au and 75 ppb Pd.

Rare earth elements

The average TV-Ni contains less ΣREE (108 ± 41 ppm) than the WSA (194 ppm) and ΣREE/Al (18.6) is also slightly lower than that of WSA (22.1). The ΣREE level ranges considerably, depending mainly on the total sulfide and P_2O_5 contents: samples with a high sulfide content show lower and samples with elevated P_2O_5 higher ΣREE concentrations. Shale-normalized REE patterns of the Talvivaara samples are remarkably invariable, indicating somewhat lower light REE/heavy REE compared to the WSA or LKAPE12 (Fig. 9.1.18). The elevated Yb/Al ratios in the figure are indicative of a minor, probably marine authigenic component in the HREE budget of most of the Talvivaara samples. No or only minor negative Eu anomalies (Eu/Eu*) are observed in chondrite-normalized patterns and mostly minor positive anomalies in shale-normalized patterns. In some cases, negative Ce anomalies (Ce/Ce*) are observed in shale-normalized plots (Fig. 9.1.18). Neither Eu/Eu* nor Ce/Ce* show any correlation with the base metal concentrations or redox proxies such as V/Al, Mo/Al, or U/Th.

Other trace elements (Zr, Sc, Th, Nb, Rb, Ba, Sr)

Elements Zr, Sc, Th, and Nb show good correlations with Al (r^2 0.65–0.74) in Talvivaara black shales and thus their abundance is mainly controlled by the detrital component. The Sc/Th ratio, which is sensitive to the ratio of detritus derived from mafic and felsic sources, averages 3.4 ± 0.7 in TV-Ni, being somewhat lower than in LKAPE12 (4.6 ± 0.8) but higher than in OPE20 (2.3 ± 0.4) or especially WSA (1.1). Relatively high Sc/Th ratios in Talvivaara rocks are consistent with the

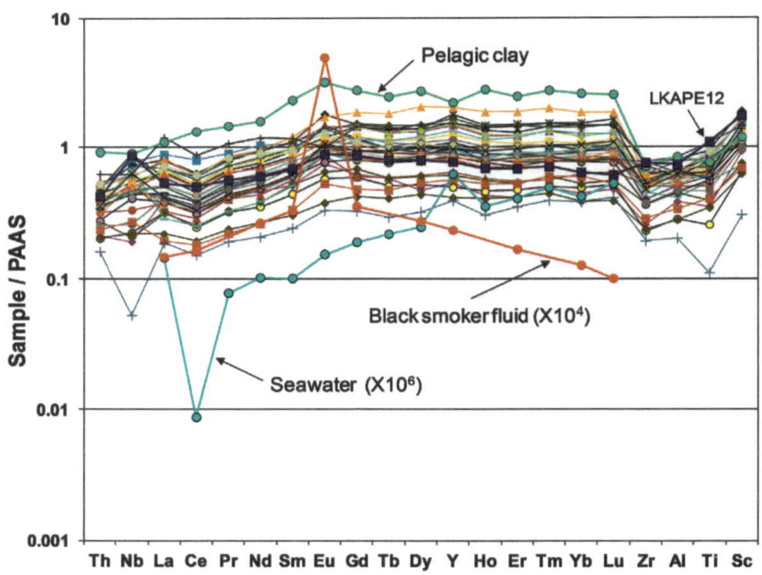

FIGURE 9.1.18 Post-Archean average Australian shale (PAAS) normalized extended REE diagram for black shales from the Talvivaara Formation/ore (all >0.1 wt% Ni).

For comparison, patterns of LKAPE12, pelagic clay (Li, 2000), black smoker fluid from the Trans-Atlantic Geotraverse hydrothermal field (TAG) field (Bau and Dulski, 1999), and seawater (Li 2000) are shown.

Ti/Al ratios (Fig. 9.1.19), with both indicating a markedly more mafic source for the Talvivaara than for the WSA and UCC. A small amount of well-rounded, small zircon grains contributes up to 170 ppm Zr (on average 105 ppm) to the whole-rock compositions. The advanced sediment maturity and source enriched in mafic components are reflected in that muds are richer in Zr than intercalated sands.

In the Talvivaara Ni-rich black shales, niobium contents and more importantly Nb/Al (ppm/%) are very similar (7.6 ± 2.5 ppm/1.3) to those in the WSA (11 ppm/1.3), OPE20 (12/1.3), or the median black shale (11 ppm) of Ketris and Yudovich (2009). This implies that the abundant mafic component in the Talvivaara rocks was from Nb-"normal" rather than Nb-rich sources such as alkali basaltic rocks. The Ba content is rather low for shales (avg. 255 ± 160 ppm vs. 580 ppm of the WSA), which probably reflects a relatively mafic provenance. The Rb content averages 99 ± 36 ppm giving a relatively high average Rb/Ba ratio of 0.65 compared to 0.24 in the WSA, implying a source-related phenomenon or postdepositional K-metasomatism. The average Sr content of TV-Ni is low, 65 ± 26 ppm (170 ppm in the WSA), and Sr/Al is also systematically low 11 (19 in the WSA) suggesting a plagioclase-poor, matured nature of the protolith muds.

Sulfur isotopes

An extensive program for determining the S isotopic composition of Talvivaara sulfides was carried out by GTK in the early 1980s, covering ore-grade to barren black shales, skarn rocks, mica schists, and

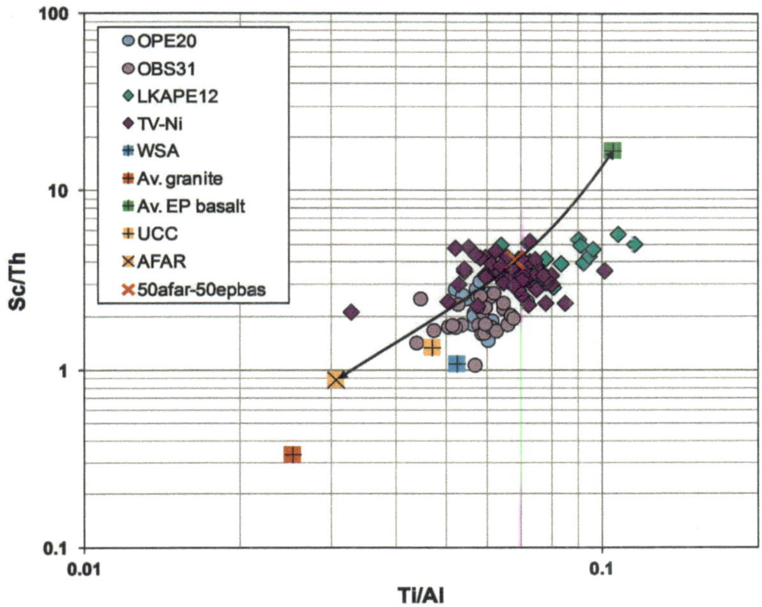

FIGURE 9.1.19 Sc/Th versus Ti/Al diagram for black shale samples from the Talvivaara Formation/ore.

Talvivaara samples plot at clearly higher Sc/Th and Ti/Al than Outokumpu (Upper Kaleva) pelites (OPE20) and black shales (OBS31), indicating a more mafic source explained by a ~50–50% blend of AFAR and average Early Proterozoic basalt.

quartz wackes from both the Kuusilampi and Kolmisoppi parts of the deposit. Analytical methods are described in Mäkelä and Tammenmaa (1977) and part of the results has been published (presented in graphs) in Loukola-Ruskeeniemi (1996). Most of the analyses were done on coarse-grained pyrrhotite and pyrite and yielded a rather large variation in $\delta^{34}S$ from −13.4 to +38.5 ‰, although the great majority of the values, for both the ore-grade and subore-grade samples, fell between −8 and +4 ‰. No clear correlation between $\delta^{34}S$ and the Ni or other base metal contents was found. In the obtained dataset, the mean $\delta^{34}S$ of all analyzed pyrite grains is −1.9 ‰ ($n = 233$) and that of pyrrhotite grains −1.2 ‰ ($n = 218$). Per-sample pyrite and pyrrhotite compositions show a good correlation, and the approximately 0.5‰ higher average $\delta^{34}S$ of pyrite broadly corresponds to the expected equilibrium pyrite-pyrrhotite fractionation at ~550°C (Ohmoto and Rye, 1979; see also Oliver et al., 1992). On average, sulfides from ore-grade samples have approximately 2‰ heavier S isotope composition compared to sulfides from barren samples.

Analyses of extracts obtained in the GTK isotope project by the Thode solution from fine-grained, pyrite-dominated laminas suggest that they contain more ^{34}S-depleted sulfur ($\delta^{34}S$ down to −20‰) than the analyzed coarser sulfide grains. Recently, Virtasalo et al. (2014) applied in situ SIMS analysis to fine-grained, semimassive-massive pyrite laminas from a drill-core section rich in such pyritic beds, and found that the $\delta^{34}S$ values of their pyrite are really rather low, ranging from −12 to −7‰.

It has been proposed that the approximately 2‰ heavier sulfur in the ore-grade than barren black shales would indicate hydrothermal origin of the mineralization (e.g., Loukola-Ruskeeniemi and Heino, 1996). In that work, isotope data from drill core 3344R329 were presented as an example of how the heavier sulfur characterizes the ore zone. However, our reexamination of the data from the same drill core (a full set of isotope and metal assay data provided by T. Heino) indicates no statistically significant correlation between the sulfide-sulfur isotope and whole-rock Ni (or other base metal) values. In fact, a great deal of the samples with positive $\delta^{34}S$ (+2 to +21 ‰) come from a section (221.25–256.87 m) of relatively weak ore (Ni <1500 ppm) with tremolite skarn and quartz wacke interlayers, the latter being low in Ni (80–350 ppm) and sulfur (<1–4 wt%) but showing some of the highest $\delta^{34}S$ values (+6.2 to +12.8 ‰). In contrast, samples from the two widest and best mineralized sections, at intervals 139.15–156.80 m (average Ni 3010 ppm) and 258.40–275.60 m (average Ni 3330 ppm), yield significantly lower $\delta^{34}S$ values of −6.3 to −2.2‰ and −4.8 to −2.9 ‰, respectively. Our review of the GTK isotope data revealed that similar $\delta^{34}S$ fluctuation without a relationship to the Ni content is also common elsewhere in the deposit.

Recently, Young et al. (2013) published new multiple sulfur isotope data from Talvivaara. Their $\delta^{34}S$ data show slightly lower mean values of −4.9‰ (range −11.4 to +9.8‰) and −3.9‰ (range −10.5 to + 9.1‰) for AVS (acid volatile sulfides, including alabandite, pentlandite, pyrrhotite, and sphalerite) and CRS (chromium reducible sulfides, including pyrite and chalcopyrite), respectively, than the mean pyrite (−1.9‰) and pyrrhotite values (−1.2‰) in the GTK data. This difference may be partly related to their preference to sample fine-grained, pyrite-dominated rocks with >10 wt% total sulfur and <2500 ppm Ni, whereas samples from pyrrhotite-dominated, good ore (S 8–10 wt%, Ni >2500 ppm) seem to be rare in their sample set. Interestingly, Young et al. (2013) reported the presence of minor mass-independent sulfur isotope fractionation (MIF-S) in Talvivaara sulfides, which is abnormal for post-2.4 Ga sedimentary rocks (e.g., Johnston, 2011). Anomalous $\Delta^{33}S$ values were obtained for about 20% of the samples and range from −0.55 to +0.57‰ with one outlier at +1.25‰. As in the case of $\delta^{34}S$, there is no correlation between the $\Delta^{33}S$ values and Ni contents.

Low-Ni black shales, Mn-enriched black shales, and calc-silicate rocks

This work was concentrated on the main facies of the Talvivaara Formation/ore (i.e., Ni-rich (>0.1 wt%) black shales), and hence only a subordinate number of samples from low-Ni, Mn-enriched, and calc-silicate rocks were analyzed. Average compositions of the ore-grade black shale are listed in Table 9.1.3.

In most respects, the low-Ni black shales show a similar average composition to that of the Ni-rich black shales (refer to Fig. 9.1.13), with the major difference being much higher Ni (2980 vs. 445 ppm) and elevated Zn (5170 vs. 3010 ppm), Cu (1430 vs. 1080 ppm), Co (186 vs. 71 ppm), and Pb (75 vs. 43 ppm) in the latter. The low-Ni samples yield higher average ΣREE (193 ppm) than the high-Ni samples (108 ppm), but the normalized patterns are very similar. The relatively high average Cu content of the low-Ni samples reflects probably syntectonic mobility of Cu/chalcopyrite, as our samples come from inside or margins of the ore. In fact, there is a possibility that at least some of the low-Ni samples gained their base metal characteristics by syntectonic sulfide mobility, in the first-place loss of Ni.

The number of samples studied in this work from the diverse (rock type, spatial distribution) group of Mn-enriched (<0.8 wt% MnO) black shales is too low to enable relevant overview of their geochemistry, which seems to be highly varied. Besides the elevated Mn, there are, however, some other common aspects in their geochemistry. For example, Fe and S are generally high; MgO and MgO/CaO are elevated; Ni is mostly but not always low (<0.1 wt%); Cu and Cu/Ni are mostly high; and Zn is variable.

Samples from calc-silicate layers are high in MgO (7–17 wt%) and CaO (7–25 wt%) and variably low in Al_2O_3 (<1.5–7.5), as one would expect for originally carbonate-, probably dolomite-rich epiclastic sediments. The usually rather high SiO_2 (20–45 wt%) requires detrital quartz and/or diagenetic silica. Very low K and K/Al in many of the more aluminous samples is problematic, apparently requiring either kaolinite-montmorrillonite, authigenic alumina, or diagenetic-metamorphic loss of K (if protolithic clay phase was illite dominated). The "excess Al" is presently in anorthite-rich plagioclase. TiO_2 (<0.10–0.50 wt%) and Zr contents (<35 ppm) are mostly low, allowing a 20–30% Ni-mud component in the typical case, which would only result in up to 1500 ppm of Ni if all Ni was in this mud phase. Nickel and other base metal contents are yet distinctly variable from sample to sample (200–7900 ppm), suggesting, together with other aspects such as regularly pyrrhotite-rich/dominated nature of the Ni-richest samples, that mss mobility at least in the sample scale was common in the calc-silicate rocks as well as in quartz sand layers with significant clastic or matrix carbonate. Decarbonation loss of volume in these rocks obviously created room/pathways for synmetamorphic sulfide migration.

Calc-silicate rocks show low ΣREE contents (average 38 ± 25 ppm) but frequently relatively high Yb/Al and Yb/La ratios, suggesting a considerable nonclastic, probably seawater-derived component in the REE budget. The overall shale-normalized patterns resemble those of present deep seawaters showing also slight Ce minimas and positive Y spikes.

DISCUSSION

SEDIMENT SOURCES AND DEPOSITIONAL TIME AND ENVIRONMENT

Lithostratigraphic and sedimentological features of the Lower Kalevian strata in the Talvivaara area and elsewhere in the KSB (Kontinen 1986, 1987) suggest deposition in a continental rift environment after the apparently shield-wide deposition of platformal carbonates in the preceding Jatuli stage, dated at ~2.11–2.06 Ga (Karhu et al., 2007; Melezhik et al., 2007). Our new isotope data bracket the depositional time of the Lower Kalevian strata, including the ore-bearing Talvivaara Formation, within a still nonprecise range of ~2.00–1.90 Ga. Previously it has been proposed (Kontinen 1986, 1987) that the Lower Kalevian rocks accumulated in a marginal intracontinental rift environment on variably tilted and differentially eroded and sediment-buried half-graben blocks, explaining why the sedimentary sequences show variation from one location to another, albeit with a similar general upward development in the sediment types, with the uppermost units being metal-enriched black shales as exemplified by the Talvivaara Formation.

The Hakonen Formation underlying the Talvivaara Formation had its sources mainly in the platformal Jatulian quartzites and carbonate sediments and mafic volcanic or dike rocks, as at least phenoclasts directly from Archean basement rocks seem to be absent or very rare. This situation seems to have also persisted during most of the deposition of the Talvivaara Formation. The observed Th-Sc-Cr relations in the mean ore composition (TV-Ni) suggest that the detrital fraction can be explained by mixing of 15% quartzite, 45% shale, and 40% mafic rocks and additional weathering effects. The source rocks seem to have radically changed in the overlying, petrographically and geochemically Upper Kaleva-type Kuikkalampi Formation, which had a major Paleoproterozoical provenance, as testified by detrital zircon populations falling in the age range of 2.00–1.93 Ga. However, the exact nature of the transition from the Talvivaara Formation to the Kuikkalampi Formation remains an unresolved major issue. Either the contact involves a conformable transition with an abrupt provenance shift or a cryptic thrust plane placing the Kuikkalampi Formation above the Talvivaara Formation. In any case, the Viteikko structure, containing rocks of a similar provenance, is a minor klippe with lenses of ophiolitic mantle-derived ultramafic rocks at its basal contact.

The depositional environment was clearly anoxic-euxinic as evidenced by high concentrations of C_{org} and S coupled with high levels of redox-sensitive elements, such as Mo, V, Cd, Se, and U, and their metal/Al ratios (compared to the WSA composition). Highly reducing conditions are also supported by the presence of alabandite. The frequent intervals of fine-grained pyrite-laminated mudstones suggest pyrite precipitation directly from the water column. The abundance of pyrite in the laminated mudstones reflects a high Fe influx and accumulation rate of organic material in the basin, and the P-C–rich bands probably represent in situ deposition, possibly related to sudden high mortality episodes in the water column. The dominance of massively bedded, resedimented muds and presence of platform-derived calc-turbidite interbeds implies that the environment was dynamic, with much of the organic-rich and mineralized mud being redeposited from shallower parts of the basin and these shallower water intermittent depositories were likely below the basinal redoxcline. Therefore, it seems that only the very uppermost part of the basinal water column was oxygenated.

ATMO-HYDROSPHERIC CONDITIONS

The Talvivaara ore deposit is a very large, apparently synsedimentary metal resource hosted entirely in C_{org}-Fe_{sulf}–rich sediments with high contents of redox-sensitive metals and metalloids. The deposit is related to regionally extensive and equally C-S–rich black shale units, which also display relatively high metal contents. This suggests that the metal enrichment did not taken place just in an isolated local basin but was more likely closely linked to the global atmospheric-hydrospheric evolution. The current view of Earth's atmospheric evolution implies that before ~2.5 Ga, the atmosphere remained extremely low in free oxygen, only with some temporary "whiffs" of oxygen in the late Archean atmosphere (Anbar et al., 2007).

A significant rise in atmospheric oxygen at around 2.3 Ga, called the Great Oxygenation Event (GOE; Holland, 2002, 2006), resulted from a complicated interplay between the oxygen production by cyanobacterial photosynthesis and diminished countereffects of different oxygen sinks. By ~2.0 Ga, the GOE produced a permanently higher oxygen level, which, depending on the modeler, is estimated to have stayed between ~10^{-4} × PAL and 10^{-1} × PAL (PAL = present atmospheric level) until the next substantial increase at the end of the Precambrian (e.g., Lyons et al., 2014).

The ~2.0–1.9 Ga time of formation of the Talvivaara black shales means that their deposition took place after the GOE. The elevated Mo, U, and other redox-sensitive metal concentrations at Talvivaara provide strong evidence for the presence of free oxygen in the atmosphere and probably also in the upper parts of hydrosphere (cf. Holland, 1984; Crusius et al., 1996; Algeo and Rowe, 2012; Tissot et al., 2013). In fact, the Mo and U contents and Mo/C_{org} in Talvivaara rocks reach levels observed in many Phanerozoic black shales and recent sapropels (e.g., Brumsack, 1991; Piper and Isaacs, 1995; Lipinski et al., 2003; Rimmer, 2004; Piper and Dean, 2002). The oxygenated atmosphere also created the prerequisite for redox-controlled metal enrichment in terms of considerable sulfate concentrations in seawater (cf. Strauss et al., 2013).

TECTONOTHERMAL REEQUILIBRATION

The Talvivaara rocks have taken part in at least three major folding events involving significant slip-dip displacements within the folded units. Our observations of mineral assemblages in the local Al-pelites and ultramafic rocks imply metamorphic temperature peaking at ~550°C (see also Säntti et al., 2006). Others have inferred that temperatures during the Svecofennian orogeny remained above 500–300°C between 1.9 and 1.8 Ga (Kontinen et al., 1992). Pervasive moderately high-T equilibration of the Talvivaara rocks over a long period of time concerns both silicate-sulfide and sulfide-sulfide assemblages and is reflected in the relatively uniform compositions of minerals such as biotite, pyrrhotite, pentlandite, sphalerite, and alabandite. With the exception of zircon, all silicate minerals are metamorphogenic. Pyrrhotite and pentlandite represent exsolved phases from metamorphic mss and sphalerite and chalcopyrite from iss, with further modification by tectonic reworking and differentiation at elevated temperatures and strain. Many parts of the deposit demonstrate the generation of leucosome-style quartz-mica-plagioclase-microcline segregations and mobilization of this material into breccias and veins. Large zones of the ore and adjacent black shales have been converted to quartz ± phlogopite veined schists-gneisses very poorly preserving any finer primary features.

In terms of preservation, the pyrite-richest beds represent an important exception because of the high thermal stability of pyrite (up to 710°C) and its high abundance, preventing complete consumption and equilibration in reactions with other sulfides, silicates, and fluid during metamorphism. Nearly all whole-rock Fe was sulfidized in diagenetic processes or, at the latest, under peak-metamorphic conditions by reactions such as this:

$$2\text{Annite} + 6\text{Pyrite} + 3\text{C}_{\text{org}} = 2\text{K - feldspar} + 12\text{Pyrrhotite} + 2\text{H}_2\text{O} + 3\text{CO}_2$$

These reactions consumed pyrite and produced pyrrhotite with significant syngenetic-diagenetic pyrite being only preserved in rocks which contained pyrite in excess of that required to balance the relocation of the silicate-hosted ferrous iron into pyrrhotite.

Given the total destruction of the "primary" mineral assemblages, they can only be indirectly deduced based on the present mineral assemblages and whole-rock chemical data. Our graphical analysis and calculations indicate that, most probably, the dominant minerals in the detrital fraction of the protolith silts and muds were quartz and illite with minor chlorite and carbonate. The average composition TV-Ni as calculated C- and S-free can be produced by a mixture containing 60 wt% illite, 22 wt% quartz, 7 wt% clinochlore, 3 wt% calcite, 2 wt% kaolinite, and 4 wt% Fe oxide. Samples with more than the >0.5 wt% Na_2O may have originally contained some sodic feldspar.

PYRITE

Pyrite shows several genetic types with clear compositional identities. Of the three main types distinguished in this work, part of the FPY and SPY-A grains seem to represent potential relic diagenetic/syngenetic grains. Although morphological criteria cannot be used due to pervasive recrystallization, the fine grain size and apparently bedding-controlled high abundance of FPY in the banded-laminated pyrite mudstones (up to >50 wt%) attest to its dominantly syngenetic, most probably framboidal origin. One could thus hypothetize that the screened FPY pyrite composition we presented above, including the contents of Ni (1300 ppm), Co (390 ppm), and As (1540 ppm), was close to "primary." However, considering the amphibolite facies conditions and the presence of 40–60 wt% of mudstone component, even in the cleanest studied pyrite mudstones, this seems unlikely. More probably, there was variable diffusive exchange of these metals between the mudstone matrix and pyrite grains during the burial and further heating. Nevertheless, the median Co/As and Ni/As ratios of the FPY and SPY-A pyrites are very similar, with the median of all-time sedimentary pyrites as calculated from the dataset of Large et al. (2104). The close compositional similarity of the FPY with the clearly diagenetic-early metamorphic SPY-A pyrite suggests that the present FPY compositions record diagenetic modification and the mean As, Co, and Ni values of the present FPY grains can be considered maximum values for the concentrations of these metals in the primary syngenetic pyrite.

Pyrite in the Talvivaara ore has been reported to contain 1.4–8% of the whole-rock Ni (Riekkola-Vanhanen, 2007; Langwaldt and Kalpudas, 2007). Our calculations indicate that this portion is on average less than 8%. Assuming that all S in the ore was initially incorporated mostly into pyrite, as has previously been proposed (e.g., Loukola-Ruskeeniemi and Heino, 1996; Loukola-Ruskeeniemi, 1999), the average S content of the ore converts to ~15.5 wt% of pyrite. If this pyrite was primarily the only Ni host, it had to contain ~1.56 wt% Ni to explain the present Ni content in the ore. On the other hand, if the primary pyrite corresponded to the screened FPY with 1300 ppm Ni, the average ore would only contain ~200 ppm Ni, which happens to be ~8% of the Ni in the present ore. From these numbers, it is evident that either just a small fraction of the Ni was introduced in the ore in pyrite, or alternatively, extensive relocation of Ni from the high-Ni primary pyrite must have taken place during the metamorphism, the latter case involving near total conversion of pyrite to pyrrhotite±pentlandite in large part of the deposit.

Handheld XRF and electron probe microanalyzer (EMPA) studies demonstrate that in pyrite-laminated muds, much more Ni resides in the dark, mud-dominated laminas (Ni typically 1500–3500 ppm) than in the pyrite-dominated ones (Ni typically <1500 ppm), with the mud laminas being compositionally similar to the thick (up to 1 m), massive mud beds in the massive, mud-dominated Ni-richest facies of the ore. If pyrite was just about an environmental feature, and not a primary or not even an important carrier of Ni, in which form or material was Ni deposited during the mud intervals? Although there is, due to small-scale mobility, only poor correlation between Ni and C, the high organic carbon (not so much S) in the mineralized mud layers indicates that organic matter likely was essential in the metal enrichment process. Metals could have been introduced by absorption on organic matter and/or as organometallic complexes. However, the total denaturalization of the primary organic matter and metal association to crystalline graphite and secondary metamorphic pyrrhotite ± pentlandite effectively hampers the study of the original metal carrier.

Assuming that the high Ni enrichment in the source area muds was related to absorption on organic matter (by dead bacteria falling in the water column, or by algal mats on the seafloor) and possible other surface reactive particles (e.g., Mn-oxides or clay particles) falling through the water column and concurrent accumulation in the bottom sediment would obviously require much higher Ni (and Cu, Co, Zn) in ambient seawater than in the present ocean waters. Systematically high Ni, Cu, and Zn compared to most Phanerozoic black shales characterizes many Paleoproterozoic black shales over southern Finland (Loukola-Ruskeeniemi et al., 1997; Loukola-Ruskeeniemi, 1999; Västi, 2008), suggesting that base metal concentrations in Proterozoic seawater may indeed have been significantly higher than in the present seawater. But even these generally higher metal concentrations were likely not sufficient for metal concentration to such an extent as has happened in Talvivaara mudstones.

ORIGIN AND ROLE OF THE MONOSULFIDE SOLID SOLUTION (PYRRHOTITE ± PENTLANDITE)

Using average sulfide compositions and whole-rock chemical data, it can be calculated that pyrrhotite and pentlandite hold ~92% of the Ni in the Talvivaara deposit: 17% in pyrrhotite and 75% in pentlandite. The best parts of the ore with 3000–5000 ppm Ni are always dominated by pyrrhotite, and in these parts, nearly all Ni is located in pyrrhotite ± pentlandite. Solid-state mobility of mss has taken place in a cm–dm scale, generating blebby and veiny pyrrhotite ± pentlandite (see Fig. 9.1.7C) in structurally reworked parts of the deposit. However, the frequently observed strictly bedbound/layered nature of the occurrence of the mss-dominated (pyrrhotite ± pyrite) ore argues against the mms mobility as a major, deposit-wide process. With respect to solid-state mss mobility, samples from structurally less-reworked, thick, massive, and nonblebby beds can thus be treated as closed systems, with the reservation that the more "volatile" elements, such as As and Sb, may have been variably lost into the departing dehydration/decarbonation fluid.

While a peak-metamorphic mss origin of the present pyrrhotite ± pentlandite is obvious, no concrete evidence for the precursor phases of the mss is present in the rocks. One explanation to the presence of pyrrhotite-dominated units could be that their sulfide content reflects deposition from Fe-dominated waters under conditions where sulfide availability was restricted to allow supersaturation of Fe-monosulfide, which could have been Ni-rich (perhaps mackinawite) or precipitated with some Ni-rich monosulfide such as millerite. Taking into account that the whole-rock Fe is now almost 100% in pyrrhotite and other minor sulfides, this model would require that the associated clastic component sedimented nearly Fe-free. However, an Fe-free matrix is most unlikely as the high-Ni pyrrhotite ores are typically among those richest in Al and Ti and thus represent some of the originally silt-clay–richest parts in the ore. Generation of the mss simply must have involved transfer of the silicate Fe from the mud into sulfide phase. The simplest explanation for the mss would be that the now pyrrhotite-dominated beds originally contained fine-grained (syngenetic-diagenetic) pyrite, some Ni-rich phase(s) (millerite, polydymite, organometals), and these components reacted when the temperature rose and produced Ni-bearing mss.

IRON ENRICHMENT

Earlier in this chapter we point out that high Fe/Al in the Talvivaara rocks indicates significant excess Fe, which is best explained by syngenetic-recycled pyrite that has reacted with the silicate Fe to produce pyrrhotite in large parts of the deposit. However, where from and how the excess iron in the pyrite was introduced remains unexplained. One possibility is that the Fe was introduced by submarine hydrothermal or land/self-sourced differentiated flux of Fe-Mn-oxyhydroxide particles, which settled down and dissolved in anoxic bottom water with the metals precipitating as sulfides in deeper sulfidic bottom seawater, in a rifted shelf or slope-rise setting. We note here the presence of Fe-rich black shale and intercalated BIF just below the metal-enriched black shale in the stratigraphy of the Kainuu schist belt (KSB). A common source of Fe could thus be argued but seems unlikely as there is little evidence for strong base metal enrichment in the iron formations and their black shale interbeds. Thus, there seems to have been a change in the Fe-rich sedimentation, involving a shift to more base metal-rich deposition by the end of the Lower Kaleva deposition. We speculate that this could have been related to a deepening of the Lower Kaleva basins and enhanced access to open ocean circulation that brought hydrothermal Fe-Mn and nutrients from the distant deep, the former perhaps from ocean spreading, ridge-linked sources. Low δ^{56}Fe values between −0.5 and −1.5‰ measured for FPY pyrites (Virtasalo et al., 2014) are within the range of values observed for Fe from major seafloor hydrothermal or continental margin reductive sedimentary sources (e.g., Rouxel et al., 2008; Severmann et al., 2010).

It could be suggested that, similar to the fine-grained pyrite, pyrrhotite was also mainly a syngenetic or early diagenetic phase, indicating either pulses of dominantly high-T hydrothermal fluids and/or alternatively, as we noted earlier, periods of limited availability of sulfur during deposition and diagenesis, prohibiting conversion of iron monosulfide to the more stable pyrite. However, for the reasons elucidated previously and due to the overall high to very high sulfur content in the Talvivaara rocks, also in many pyrrhotite-dominated rocks, limited sulfur availability seems an unlikely explanation for the locally abundant pyrrhotite. Also, the S isotope data do not indicate that the pyrrhotite-dominated rocks received their sulfur (or iron) from either a distinct hydrothermal or sulfur/sulfate-limited source.

S ISOTOPES AND S SOURCE

The observation by Young et al. (2013) of small mass-independent S isotope fractionation in black shale samples from Talvivaara is enigmatic because of the post-GOE deposition of the Talvivaara Formation, as the appearance of free oxygen in the atmosphere likely created an ozone shield preventing photochemical fractionation of S isotopes by solar ultraviolet radiation (Farquhar et al., 2001) and thus eliminated the source of MIF-S signatures in the sedimentary record (Johnston, 2011). Young et al. (2013) interpreted the anomaly in terms of thermochemical fractionation related to supposedly dominantly hydrothermal deposition of the Talvivaara sulfides. However, because evidence for hydrothermal contribution to the formation of the Talvivaara ore deposit seems to be generally very weak, the MIF-S signature more probably reflects recycling of Archean MIF-sulfur or is otherwise entirely a result of heterogeneous metamorphic reactions. If recycling of Archean sulfur did happen, a future study comparing Δ^{33}S from the younger Kuikkalampi Formation and the older Hakonen and Talvivaara Formation, dominated by Paleoproterozoic and Archean detrital material, respectively, would be informative.

The present δ^{34}S data, though abundant, are mainly restricted to coarse metamorphic pyrite and pyrrhotite. These data indicate that most of the Talvivaara Formation has sulfide with δ^{34}S between −6 and

0‰ (in about 70% of cases) for both pyrrhotite and pyrite. Pyrite and pyrrhotite data correlate well, defining a linear trend with pyrite being, on average, 0.5‰ higher than pyrrhotite. This suggests metamorphic equilibration among the coarse-grained sulfides. The recent in situ $\delta^{34}S$ data acquired by Virtasalo et al. (2014) for FPY from pyrite-laminated layers suggest that there may be a difference in $\delta^{34}S$ of sulfides between sulfur- and pyrite-rich and S-poorer, pyrrhotite-dominated beds in the ore. On the basis of presently available data, on average, the sulfur in pyritic mudstones seems to be 5‰ lighter than sulfur in pyrrhotite-dominated mudstones. However, this estimate is subject to change as $\delta^{34}S$ data for fine-grained sulfides in pyrrhotite-dominated muds are still few. The $\delta^{34}S$ values reported by Virtasalo et al. (2014) for fine-grained pyrite, ranging from −12 to −7‰, are markedly lower than $\delta^{34}S$ of the contemporaneous seawater, which we assume to have had $\delta^{34}S$ values close to the range of +9 to +11‰ as observed in the 2.1 Ga evaporates in the Onega region, Russian Karelia (Reuschel et al., 2012). This is a strong indication of the role of bacterial seawater sulfate reduction in the sulfide origin.

The origin of Talvivaara sulfides by bacterial sulfate reduction in a sulfate-restricted environment has been proposed previously (e.g., Loukola-Ruskeeniemi, 1999). In this case, the $\delta^{34}S$ values would approach the $\delta^{34}S$ of the sulfate supply. However, as we noted earlier, given the systematically high Fe-sulfide content of the deposit, sulfur/sulfate-limited environment of formation for the sulfides (pyrite or pyrrhotite) seems to be an unlikely overall scenario. We are rather inclined to interpret that sulfate contents in the source basins of the resedimented mudstones and in the Talvivaara basin were >200 mmol, which is usually set as a basic limit for sulfate-unlimited conditions. This would be in accord with the significant average Mo content of 60 ppm Mo in the Talvivaara black shales, with peaks up to 100 ppm, suggesting extensive continental oxidative weathering of Mo-bearing sulfides, principally pyrite (cf. Miller et al., 2001).

Hydrothermal contribution to the sulfur (and the metal budget) has been a favorite idea for explaining the genesis of the Talvivaara deposit (Loukola-Ruskeeniemi, 1991, 1995, 1999; Loukola-Ruskeeniemi and Heino, 1996; Young et al., 2013). However, the lithological association lacks any obvious source rocks for supplying the required enormous quantities of metal- and sulfur-rich hydrothermal solutions. Furthermore, there is no evidence for hot springs or any syngenetic or epigenetic metal zonation (in Ni-Co-Cu-Zn) at Talvivaara, and no syndepositional or closely succeeding magmatic or other obvious heat/energy sources for hydrothermal celling of fluids. Significant sulfur and/or metal supply from remote deep sources such as midocean ridges is unlikely as hydrothermal sulfur, especially in deep anoxic-euxinic ocean, would have been deposited (and immobilized) as sulfides immediately at the fluid sources/hot springs (Stüeken et al., 2012). In addition, the huge total amount of sulfur and metals in the Talvivaara formation and other Kalevian units in eastern Finland suggests that the supply of Fe, S, and base metals was related to global-scale fluctuation in crustal weathering, volcanism, and related oceanic processes, not to any local hydrothermal springs. It is worth noting that the period 2.0–1.9 Ga was likely characterized by worldwide intensified plume-related mafic–ultramafic plutonism and volcanism (Condie, 2001). Related gas emissions and weathering solutions could have been a major source of the S and base metals.

ORIGIN OF THE TALVIVAARA DEPOSIT

In the previous hydrothermal models of ore genesis, metals, and especially Ni, are thought to be derived through hydrothermal leaching from ultramafic rocks (Loukola-Ruskeeniemi, 2011; Loukola-Ruskeeniemi and Lahtinen, 2013) such as the small altered peridotite lenses at the base of the Viteikko Klippe. In this context, the Talvivaara sulfides have often been correlated with and seen as a facies of the ophiolite-associated massive sulfides in the Outokumpu area (Mäkelä, 1981; Loukola-Ruskeeniemi et al., 1991). Problems in this model have recently been listed by Kontinen (2012) and Kontinen et al. (2013b). First, the Talvivaara ore-bearing black shales occur within an autochthonous sedimentary unit that was deposited on a nonmagmatically rifted continental basement, and thus far from, for example, any mid-ocean ridge exposing ultramafic rocks as oceanic crustal cumulates or mantle rocks.

Second, the Ni-dominated nature of the mineralization is a strong argument against hydrothermal origin. Experimental studies have shown that Ni sulfides have a low solubility in saline fluids (Liu et al., 2012; Tian et al., 2012) and thus probably all common geothermal fluids, explaining the much lower Ni contents compared to other base metals such as Co, Cu, and Zn in VMS deposits through the geological time. Even in the ultramafic-floored sulfide systems, both the fluids and sulfide accumulations are enriched, in descending order, in Zn, Cu, Co, Ni, Pb (e.g., Douville et al., 2002).

Third, the correlation of the Talvivaara deposit with the ophiolite-associated Outokumpu sulfides is problematic because of the autochthonous nature of the former and allochthonous nature of the latter (Peltonen et al., 2008). Fourth, there was no obvious heat source in the Talvivaara sedimentary setting to run such a huge-scale hydrothermal system that would have been required to extract the high tonnages (>1 Mt of Ni in the Talvivaara ore alone) of metals from the mostly very Ni-poor footwall rocks of the Talvivaara formation. Fifth, there is no evidence for large-scale, oxidative fluid flow and related metal extraction in the Talvivaara footwall strata as observed, for example, in the extensive "Rote Fäule" associated with the Kupferschiefer deposit in Poland (Bechtel et al., 2002).

Sixth, there is no lateral metal zoning in the Talvivaara deposit that tends to characterize both exhalative and epigenetic hydrothermal deposits, controlled by fluid temperature and redox gradients, for example. In addition, we can mention that the whole Talvivaara formation is relatively low in Ba (<700 ppm) with an average content of approximately 260 ppm. Ba/Al is also very low, approximately 45, compared to 66 of the WSA and 75 in Upper Kalevian metasediments, probably reflecting more mature character and a more mafic source of the Talvivaara formation. So there is no evidence of hydrothermal barite deposition that is a characteristic of many VMS and SEDEX and also some black shale-hosted sulfide deposits.

Kontinen (2012) proposed that the Talvivaara deposit represents a largely redox-controlled variant of black shale-hosted, syngenetic mineralization. His principal supporting arguments are the following:

(1) The mineralized black shales show very similar shale-normalized enrichment patterns of most redox-sensitive metals compared to many (less) metal-enriched black shales (Fig. 9.1.20), such as Outokumpu black shales, Bazhenov black shales (Gavshin and Zakharov, 1996), and U.S. Mid-Continent black shales (Schultz and Coveney, 1992).
(2) Despite the high contents of Ni and other base metals at Talvivaara, exceeding several times those in common black shales, the ranges in the ratios such as Zn/Ni and Cu/Ni and Co/Ni correspond to those observed in typical black shales.

One clue to the origin of metals seems to lie in the exceptional, minor to strong enrichment of Talvivaara black shales in Mn, suggesting that Mn and Fe oxides-oxyhydroxides probably were instrumental in the supply of Ni. Here we should also note that Ni, Cu, Zn, Co (and Pb), which all occur dominantly as cations in present oxic seawater (Hein et al., 2000), belong to trace elements that are most influenced by Mn-Fe recycling (Tribovilard et al., 2006). The crust-/shale-like Zn/Ni, Cu/Ni, and

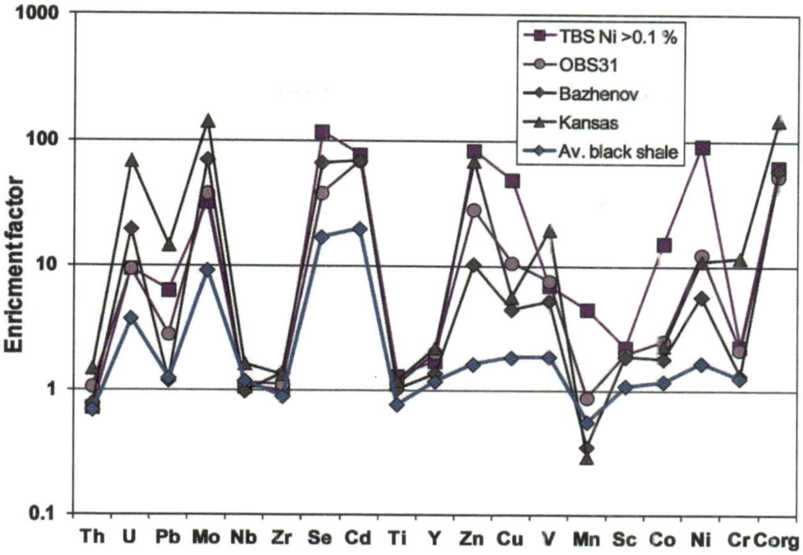

FIGURE 9.1.20 Enrichment factor diagram for selected redox sensitive elements, base metals, and clastophile trace elements in average Talvivaara ore-grade (>0.1 wt% Ni) black shale.

Also shown are some other metal-enriched black shales, including Outokumpu (OBS31), Bazhenov Formation (Gavshin and Zakharov, 1996), and Kansas cyclothem (Algeo and Maynard, 2004) black shales. Average black shale from Ketris and Yudovich (2009) is shown for comparison. The Talvivaara composition displays a typical black shale enrichment distribution but with exceptional enrichment in Zn, Cu, Mn, Co, and Ni.

Cu/Zn ratios in the Talvivaara Ni-rich rocks suggest continental runoff as the ultimate main metal source, while the low Co/Ni and Pb/Ni ratios imply that the flux was via oxic seawater (very low in Co and Pb) rather than high-T hydrothermal fluids.

We note that the Talvivaara Mn-enriched muds have Mn/Fe, Ni/Mn, and Ni/Fe ratios and Ni content close to those in the more distal facies of 0–65 Ma basal metalliferous Fe-Mn-rich sediments found in the southeast Pacific area (cf. Gurvich, 2006). Therefore, we propose that the supply of extra Mn and Fe came in MnO-FeOOH particles from a remote hydrothermal source, distant enough to enable high enrichment of particles in seawater-derived Ni-Zn-Cu-Co before their settling and dissolution in the anoxic-sulfidic bottom parts of the Talvivaara basin. Dissolution of the originally hydrothermal particles and dispersion by ocean circulation removed most of the Mn whereas the other metals mainly precipitated by adsorption and sulfide-controlled scavenging on the bottom of the basin. The dissolved Fe was incorporated in syngenetic pyrite forming in the H_2S-rich basin water. Base metal contents in seawater were likely higher than presently, possibly related to superplume events characterizing the time period (2.0–1.9 Ga) of the deposition of the Talvivaara rocks. The similarity of the Talvivaara mudstones to the OBS in terms of their redox-sensitive elements (Se, Mo, V, Cd, U) indicates that the redox conditions and oxyanionic metal supplies were the same, for this part of metal enrichment was clearly controlled by normal black shale mechanisms to levels typical in anoxic sediments at that time. The muds with high base metals/Mn probably reflect basinal conditions most favorable to solution of the Mn oxides, in deeper anoxic-sulfidic parts of the water column, with enhanced scavenge of Ni by organic matter and clay particles and, to some extent, syngenetic sulfides such as ZnS, $CuFeS_2$, and perhaps NiS. Effective recycling of Mn across well-developed, areally wide-ranging redoxclines may have contributed to the metal enrichment. More manganese and lesser amounts of base metals deposited during occasional periods of decreased anoxia, which may have been related to initially clastic-dominated turbidite flows bringing whiffs of oxygen from shallower, better oxygenated parts of the basin.

Besides extreme enrichment in base metals, the bulk of the Talvivaara mudstones differ from typical Phanerozoic metal-enriched black shales also in their compositional homogeneity concerning organic carbon and metal contents and their ratios. A significant part of the mineralization occurs as massive and compositionally rather uniform beds 1–100 cm in thickness. The overall variation in Mo/Al and V/Al (Fig. 9.1.21) or Ni/Al versus C_{org}/Al (Fig. 9.1.22), for example, is remarkably limited compared to typical Phanerozoic metal-enriched black shales. The presence of obvious layer-dependent variation in metal contents and completely unmineralized sand and mudstone interlayers in the ore militates against

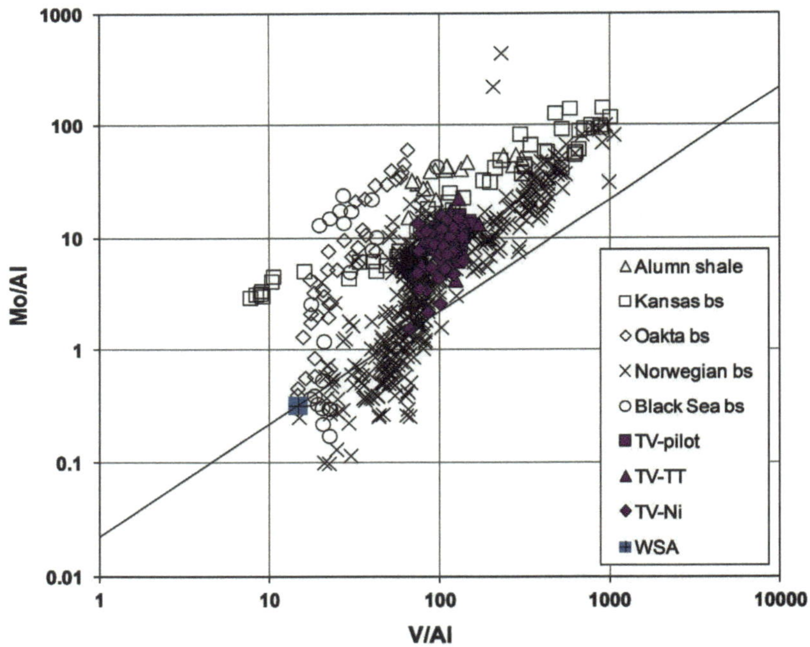

FIGURE 9.1.21 Mo/Al versus V/Al diagram for black shales from the Talvivaara Formation/ore.

TV-pilot, TV-Ni, and TV-TT refer to large (2–3 m dd-core; data from Talvivaara Mining Company), medium (5–40 cm; data from this work), and small (3–5 cm; data from Törmälehto, 2008) samples, respectively. Note the remarkably restricted composition of the Talvivaara rocks compared to those of typical metal-enriched black shales and organic-rich sediments represented by data from Cambrian alumn shale (Leventhal, 1991), middle Devonian Oakta Creek formation (Werne et al., 2002), Carboniferous Kansas-type cyclothems (Algeo and Maynard, 2004), Jurassic/Cretaceous Norwegian Shelf and Barents Sea strata (Lipinski et al., 2003), and recent Black Sea bottom sediments (Brumsack, 1989).

significant postdepositional diagenetic or metamorphic metal mobility and homogenization. To explain the often strikingly monotonous thick bedding, Kontinen et al. (2013b) suggested that most of the mineralization is in resedimented/recycled "turbiditic" carbonaceous mudstones where the organic matter, primary sulfides, and associated metals were redeposited from a source containing highly metalliferous C_{org}-S–rich sediments as one main component. The resedimentation hypothesis would explain the homogeneity of most of the individual massive mudstone beds and the subtle differences in metal enrichment and lithophile element contents observed between the homogenous layers/beds. Only the pyrite banded-laminated mudstones and included phosphatic beds-laminas can be interpreted as representing sedimentation directly from the overlying water column. Given the relatively low Ni content in these facies, it is thus apparent that most of the Ni in the massive muds is recycled from the inferred, shallower-water precursor muds rich in metals and organic matter. As there is nothing preserved in the regional rock record from the precursor rocks, their nature and depositional setting cannot be easily depicted.

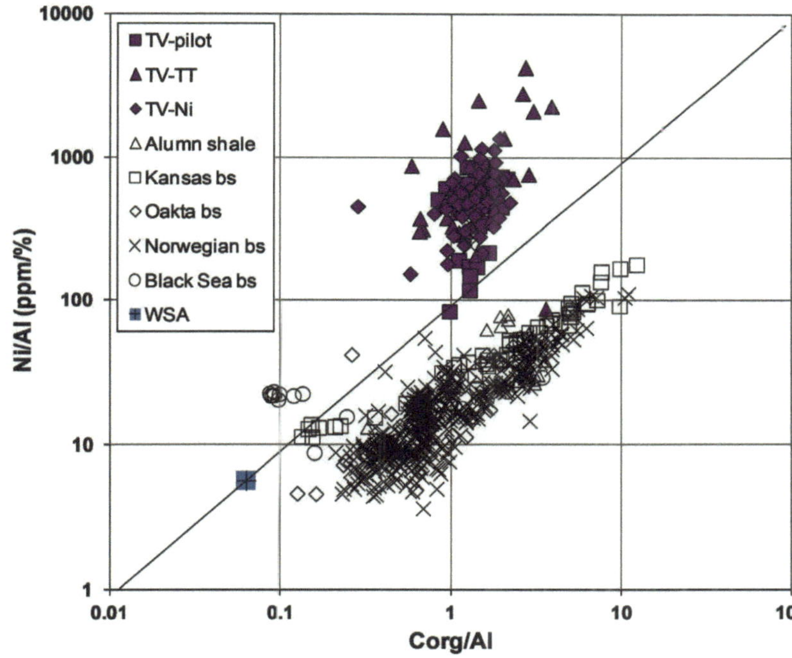

FIGURE 9.1.22 Ni/Al versus C/Al diagram for black shales from the Talvivaara Firmation/ore.

TV-pilot, TV-Ni, and TV-TT refer to large (2–3 m dd-core; data from Talvivaara Mining Company), medium (5–40 cm; data from this work), and small (3–5 cm; data from Törmälehto, 2003) samples, respectively. Note the much higher Ni/C in the Talvivaara samples but also their remarkably constant C_{org}/Al compared to typical metal-enriched black shales and organic-rich sediments represented by data from Cambrian alumn shale (Leventhal, 1991), middle Devonian Oakta Creek formation (Werne et al., 2002), Carboniferous Kansas-type cyclothems (Algeo and Maynard, 2004), Jurassic/Cretaceous Norwegian Shelf and Barents Sea strata (Lipinski et al., 2003), and recent Black Sea bottom sediments (Brumsack, 1989).

In conclusion, the generation of the Talvivaara deposit was, as that of many other exceptionally large ore deposits, by "normal" but enhanced Earth processes, in this case black shale metal enrichment processes. This involved principally synsedimentary scavenging of metal from seawater, but in an exceptionally favorable tectonic setting and under environmental conditions that may not recur often. The two most essential factors were: (1) an already oxidized atmosphere (probably >10% PAL) and hydrosphere allowing development of a redox-stratified depositional basin and sulfate supply, and (2) a high supply of reactive Fe-Mn oxyhydroxide particles by long transport through base metal enriched oxic upper ocean water. The high Fe-Mn and base metal conditions were probably reflecting worldwide intensified mafic–ultramafic volcanism and related weathering during the given time period. The free access to global ocean facilitated abundant supply of nutrients and trace elements to promote high basinal organic production. The basin of deposition was adjacent to or within continental margin in a setting where downslope recycling/resedimentation in deeper-water final turbidite sink enabled accumulation of a thick unit of the dominantly organic and metal-rich mud. It needs to be added that Talvivaara would not be a viable metal resource without syntectonic-metamorphic processes that upgraded the deposit by further thickening and piling up the metalliferous mud layers. The metamorphic processes also led to overall coarsening of the grain size and generated an easily oxic-acid soluble, mss-derived pyrrhotite + pentlandite assemblage, in which most of the metal value is located.

Black shale-hosted deposits, which are an important source of Cu-Co-Ag-Zn-Pb (Kupferschiefer and Katangan type) and Au (Sukhoi Log type), have been proposed (e.g., Robb, 2005; Salpeteur, 2007; Jowitt and Keays, 2011) to provide a potential major future source of many other metals as well (Co, Ni, Mo, Sb, Se, V, Zn, REE, PGE, etc.). Talvivaara is often used as an example of this potential being realized. The typically low metal grades in many of the potential deposits mean that large-scale mining and low-cost metal extraction methods, such as the open-sky heap leaching applied at Talvivaara, would be necessary for their economic utilization. However, the presence of typically high contents of potentially environmentally harmful substances in black shales, such as Fe-sulfides, Cd, Se, Sb, U, and Ra, together with high fluxes of aggressive chemicals and water in hydrometallurgical processing, results in a difficult-to-manage combination with high environmental risks and costs. The past six years of mining and metal extraction history at Talvivaara have been fraught with costly problems in almost every step of the production (crushing, heap leaching, metal extraction, etc.) but especially in managing the effects of fluctuations in the environmental conditions (low winter temperatures, high rainfalls) to the heap leaching and consequent emissions into the environment. As a result, the Talvivaara Mining Company is currently seeking for financial restructuring. Thus, considerable future technological improvements and probably also a permanent hike in the level of metal prices seem to be necessary preconditions before metalliferous black shales, at Talvivaara or elsewhere, could be economically utilized, forming an "unlimited" source of metals as presented in some mining company promotions.

ACKNOWLEDGMENTS

We greatly appreciate the permission by the Talvivaara Mining Company to use drill core data and other research material from Talvivaara. Risto Juhava, Hannu Lahtinen, Taija Lahtinen, Teemu Törmälehto, and Jukka Pitkäjärvi are sincerely thanked for their help in core logging, sampling, and discussions, without which this contribution would not have been possible. Equally important has been the analytical work done by Hannu Huhma, Hugh O'Brien, Yann Lahaye, and others in the GTK isotope laboratory, and Bo Johanson, Jukka Laukkanen, Lassi Pakkanen, and Mia Tiljander in the GTK electron microprobe laboratories.

REFERENCES

Algeo, T.J., Maynard, J.B., 2004. Trace-element behavior and redox facies in core shales of Upper Pennsylvanian Kansas-type cyclothems. Chemical Geology 2006, 289–318.

Algeo, T., Rowe, H., 2012. Paleoceanographic applications of trace-metal concentration data. Chemical Geology 324–325, 6–18.

Anbar, A.D., Duan, Y., Lyons, T.W., et al., 2007. A whiff of oxygen before the Great Oxidation Event? Science 317, 1903–1906.

Andersson, A., Dahlman, B., Gee, D.G., Snäll, S., 1985. The Scandinavian alum shales. Sveriges geologiska undersökning ser. Ca 56, 50 pp.

Bau, M., Dulski, P., 1999. Comparing yttrium and rare earths in hydrothermal fluids from the Mid-Atlantic ridge: Implications for Y and REE behavior during near-vent mixing and for the Y/Ho ratio of Proterozoic seawater. Chemical Geology 155, 77–90.

Bechtel, A., Gratzer, R., Püttmann, W., Oszczepalski, S., 2002. Geochemical characteristics across the oxic/anoxic interface (Rote Fäule front) within the Kupferschiefer of the Lubin-Sieroszowice mining district (SW Poland). Chemical Geology 185, 19–31.

Brumsack, H.-J., 1989. Geochemistry of recent TOC-rich sediments from the Gulf of California and Black Sea. Geologische Rundschau 78, 851–882.

Brumsack, H.-J., 1991. Inorganic geochemistry of the German 'Posidonia Shale': Palaeoenvironmantal consequences. Geological Society, London. Special Publications 58, 353–362.

Condie, K.C., 1993. Chemical composition and evolution of the upper continental crust: Contrasting results from surface samples and shales. Chemical Geology 104, 1–37.

Condie, K.C., 2001. Mantle Plumes and Their Record in Earth History. Cambridge University Press, Cambridge, 306 pp.

Coveney, R.M. Jr., 2003. Metalliferous Paleozoic black shales and associated strata. In: Lentz, D.R. (Ed.), Geochemistry of Sediments and Sedimentary Rocks: Evolutionary Considerations to Mineral Deposit-Forming Environments. Geological Association of Canada, pp. 135–144. Mineral Deposits Division, GeoText 4.

Coveney, R.M. Jr., Glascock, M.D., 1989. A review of the origins of metal-rich Pennsylvanian black shales, central U.S.A., with an inferred role for basinal brines. Applied Geochemistry 4, 347–367.

Coveney, R.M. Jr., Watney, W.L., Malpes, C.G., 1989. Rates and durations for accumulation of Pennsylvanian black shales in the Midwestern United States. Kansas Geological Survey. Subsurface Geology Series 12, 73–76.

Crocket, J.H., 1991. Distribution of gold in the Earth's crust. In: Foster, R.P. (Ed.), Gold Metallogeny and Exploration. Chapman and Hall, London, New York, pp. 1–36.

Crusius, J., Calvert, S., Pedersen, T., Sage, D., 1996. Rhenium and molybdenum enrichments in sediments as indicators of oxic, suboxic and sulfidic conditions of deposition. Earth and Planetary Science Letters 145, 65–78.

Douville, E., Charlou, J.L., Oelkers, E.H., et al., 2002. The Rainbow vent fluids (36° 14′N, MAR): the influence of ultramafic rocks and phase separation on trace metal content in Mid-Atlantic Ridge hydrothernmal fluids. Chemical Geology 184, 37–48.

Farquhar, J., Savarino, J., Airieau, S., Thiemens, M.H., 2001. Observation of wavelength-sensitive mass-independent sulfur isotope effects during SO_2 photolysis: Application to the early atmosphere. Journal of Geophysical Research 106, 32829–32840.

Gavshin, V.M., Zakharov, V.A., 1996. Geochemistry of the Upper Jurassic-Lower Cretaceous Bazhenov Formation, West Siberia. Economic Geology 91, 122–133.

Gurvich, E.G., 2006. Metalliferous Sediments of the World Ocean—Fundamental Theory of Deep-Sea Hydrothermal Sedimentation. Springer-Verlag Berlin Heidelberg, 410 pp.

Hanski, E., Huhma, H., Vuollo, J., 2010. SIMS zircon ages and Nd isotope systematics of the 2.2 Ga mafic intrusions in northern and eastern Finland. Bulletin of the Geological Society of Finland 82, 31–62.

Hanski, E.J., Melezhik, V.A., 2012. Litho- and chronostratigraphy of the Karelian formations. In: Melezhik, V.A., Prave, A.R., Hanski, E.J., et al. (Eds.), Reading the Archive of Earth's Oxygenation. vol. 1.: The Palaeoproterozoic of Fennoscandia as Context for the Fennoscandian Arctic Russia—Drilling Early Earth Project. Springer-Verlag, Berlin/Heidelberg, pp. 39–110.

Hatch, J.R., Leventhal, J.S., 1992. Relationship between inferred redox potential of the depositional environment and geochemistry of the Upper Pennsylvanian (Missourian) Stark shale Member of the Dennis Limestone, Wabaunsee county, Kansas, U.S.A. Chemical Geology 99, 65–82.

Hein, J.R., Koshinsky, A., Bau, M., et al., 2000. Cobalt-rich ferromanganese crusts in the Pacific. In: Cronan, D.S. (Ed.), Handbook of Marine Mineral Deposits. CRC Press, Boca Raton, pp. 239–279.

Herzig, P.M., Petersen, S., Hannington, M.D., 1998. Geochemistry and sulphur-isotopic composition of the TAG hydrothermal mound, Mid-Atlantic Ridge, 26°N. Proceedings of the Ocean Drilling Program. Scientific Results 158, 47–70.

Holland, H.D., 1984. The Chemical Evolution of the Atmosphere and Oceans. Princeton University Press, Princeton, 582 pp.

Holland, H.D., 2002. Volcanic gases, black smokers, and the Great Oxidation Event. Geochimica et Cosmochimica Acta 66, 3811–3826.

Holland, H.D., 2006. The oxygenation of the atmosphere and oceans. Philosophical Transactions of the Royal Society B 361, 903–915.

Huyck, H.L.O., 1990. When is a metalliferous black shale not a black shale? In: Grauch, R.I., Huyck, H.L.O. (Eds.), Metalliferous Black Shales and Related Ore Deposits—Proceedings, 1989 United States Working Group Meeting, Denver, International Geological Correlation Program Project 254. U.S. Geological Survey, Circular 1058, pp. 42–56.

Jowitt, S.M., Keays, R.R., 2011. Shale-hosted Ni-(Cu-PGE) mineralization: a global overview. Applied Earth Science (Transactions of the Institute of Mining and Metallurgy B) 120, 187–197.

Johnston, D.T., 2011. Multiple sulfur isotopes and the evolution of Earth's surface sulfur cycle. Earth-Science Reviews 106, 161–183.

Karhu, J.A., Holland, H.D., 1996. Carbon isotopes and the rise of atmospheric oxygen. Geology 24, 867–870.

Karhu, J., Kortelainen, N.M., Huhma, H., et al., 2007. New time constraints for the end of the Paleoproterozoic carbon isotope excursion. Proceedings of the 7th International Symposium on Applied Isotope Geochemistry, Abstracts: Stellenbosch, South Africa, pp. 76–77.

Ketris, M.P., Yudovich, Y.E., 2009. Estimations of clarkes for carbonaceous biolithes: World averages for trace element contents in black shales and coals. International Journal of Coal Geology 78, 135–148.

Koistinen, T.J., 1981. Structural evolution of an early Proterozoic stratabound Cu-Co-Zn deposit, Outokumpu, Finland. Transactions of the Royal Society of Edinburgh. Earth Sciences 72, 115–158.

Kontinen, A., 1986. Early Proterozoic stratigraphy and sedimentation in the Hyrynsalmi area, eastern Finland. In: Sokolov, V.A., Heiskanen, K.I. (Eds.), Early Proterozoic of the Baltic Shield, Proceedings of the Finnish—Soviet Symposium held in Petrozavodsk August 19–27, 1985. Karelian Science Centre of the USSR Academy of Sciences, Petrozavodsk, pp. 75–103.

Kontinen, A., 1987. An early Proterozoic ophiolite—the Jormua mafic-ultramafic complex, northern Finland. Precambrian Research 35, 313–341.

Kontinen, A., 2012. F029 Talvivaara Ni-Zn-Cu. In: Eilu, P. (Ed.), Mineral Deposits and Metallogeny of Fennoscandia. Geological Survey of Finland, Special Paper 53, 276–280.

Kontinen, A., Eskelinen, J., 2005. Laakajärvi. Geological Map of Finland 1:100,000. Pre-Quaternary Rocks. Geological Survey of Finland. Sheet 3344.

Kontinen, A., Paavola, J., Lukkarinen, H., 1992. K-Ar ages of hornblende and biotite from late Archaean rocks of Eastern Finland—Interpretation and discussion of tectonic implications. Geological Survey of Finland, Bulletin 365, 31.

Kontinen, A., Johanson, B., Pakkanen, L., 2013a. Ni and Co in Talvivaara pyrrhotite. In: Hölttä, P. (Ed.), Current Research: GTK Mineral Potential Workshop, Kuopio, May 2012. Geological Survey of Finland, Report of Investigation 198, p. 80.

Kontinen, A., Sorjonen-Ward, P., Huhma, H., Lahtinen, H., 2013b. Lithostratigraphy, sedimentary environment and origin of the Talvivaara black schist-hosted Ni-Zn-Cu-Co deposit. In: Hölttä, P. (Ed.), Current Research: GTK Mineral Potential Workshop, Kuopio, May 2012. Geological Survey of Finland, Report of Investigation 198, pp. 89–90.

Kontinen, A., Johanson, B., Pakkanen, L., Tiljander, M., 2014. Talvivaara biotite has stories to tell. In: Lauri, L.S., Heilimo, E., Leväniemi, H., et al. (Eds.), Current Research: Proceedings of the Second GTK Mineral Potential Workshop, Kuopio, Finland, May. Geological Survey of Finland, Report of Investigation 207, pp. 72–73.

Korsakova, M., 2012. R041 Onega V, Ti, U, PGE, Cu, Au. In: Eilu, P. (Ed.), Mineral Deposits and Metallogeny of Fennoscandia. Geological Survey of Finland, Special Paper 53, 386–389.

Korsman, K., Glebovitsky, V. (Eds.), 1999. Structure–lithology, metamorphism and metallogeny of the Raahe-Ladoga Zone, Map 2: Metamorphism 1:1,000,000. Geological Survey of Finland, Espoo.

Laajoki, K., 2005. Karelian supracrustal rocks. In: Lehtinen, M., Nurmi, P.A., Rämö, O.T. (Eds.), Precambrian Geology of Finland—Key to the Evolution of the Fennoscandian Shield. Elsevier, Amsterdam, pp. 279–342.

Lahtinen, R., Huhma, H., Kontinen, A., et al., 2010. New constraints for the source characteristics, deposition and age of the 2.1–1.9 Ga metasedimentary cover at the western margin of the Karelian Province. Precambrian Research 176, 77–93.

Langwaldt, J., Kalapudas, R., 2007. Bio-benefication of multimetal black shale ore by flotation. Physicochemical Problems of Mineral Processing 41, 291–299.

Large, R.R., Halpina, J.A., Danyushevskya, L.V., et al., 2014. Trace element content of sedimentary pyrite as a new proxy for deep-time ocean–atmosphere evolution. Earth and Planetary Science Letters 389, 209–220.

Lecomte, A., Cathelineau, M., Deloule, E., et al., 2014. Uraniferous bitumen nodules in the Talvivaara Ni-Zn-Cu-Co deposit (Finland): Influence of metamorphism on uranium mineralization in black shales. Mineralium Deposita 49, 513–533.

Lehmann, B., Nägler, T.F., Holland, H.D., et al., 2007. Highly metalliferous carbonaceous shale and Early Cambrian seawater. Geology 35, 403–406.

Lehtinen, M., 2015. Industrial minerals of Finland. In: Maier, W.D., O'Brien, H., Lahtinen, R. (Eds.), Mineral Deposits of Finland. Elsevier, Amsterdam. pp. 685–705.

Lepland, A., Hanski, E.J., Filippov, M.M., Melezhik, V.A., 2013a. World-wide record of Palaeoproterozoic carbonaceous sediments representing the Shunga Event with emphasis on the Fennoscandian Shield. In: Melezhik, V.A., Fallick, A.E., Strauss, H., et al. (Eds.), Reading the Archive of Earth's Oxygenation. Vol. 3: Global Events and the Fennoscandian Arctic Russia—Drilling Early Earth Project. Springer-Verlag, Berlin/Heidelberg, pp. 1196–1197.

Lepland, A., Melezhik, V.A., Papineau, D., et al., 2013b. The earliest phosphorites: Radical change in the phosphorus cycle during the Palaeoproterozoic. In: Melezhik, V.A., Fallick, A.E., Strauss, H., et al. (Eds.), Reading the Archive of Earth's Oxygenation. Vol. 3: Global Events and the Fennoscandian Arctic Russia—Drilling Early Earth Project. Springer-Verlag, Berlin/Heidelberg, pp. 1275–1296.

Lepp, H., 1963. The relation of iron and manganese in sedimentary iron formations. Economic Geology 58, 515–526.

Leventhal, J.S., 1991. Comparison of organic geochemistry and metal enrichment in two black shales: Cambrian Alumn Shale of Sweden and Devonian Chattanooga Shale of United States. Mineralium Deposita 26, 104–112.

Li, C.-X., Wei, C., Deng, Z.-G., et al., 2010. Recovery of vanadium from black shale. Transactions of Nonferrous Metals Society of China 20, 127–131.

Li, Y.-H., 2000. A Compendium of Geochemistry: From Solar Nebula to the Human Brain. Princeton University Press, Princeton. 440 pp. Available at www.soest.hawaii.edu/oceanography/faculty/yhli/datatables.html.

Lipinski, M., Warning, B., Brumsack, H.-J., 2003. Trace metal signatures of Jurassic/Cretaceous black shales from the Norwegian Shelf and the Barents Sea. Palaeogeography, Palaeoclimatology. Palaeoecology 190, 459–475.

Liu, W., Migdisov, A., Williams-Jones, A., 2012. The stability of aqueous nickel(II) chloride complexes in hydrothermal solutions: Results of UV-visible spectroscopic experiments. Geochimica et Cosmochimica Acta 94, 276–290.

Loukola-Ruskeeniemi, K., 1991. Geochemical evidence for the hydrothermal origin of sulphur, base metals and gold in Proterozoic metamorphosed black shales, Kainuu and Outokumpu areas, Finland. Mineralium Deposita 26, 152–164.

Loukola-Ruskeeniemi, K., 1995. Origin of the black-shale-hosted Ni–Cu–Zn deposit at Talvivaara, Finland. Geological Survey of Finland, Special Paper 20, 31–46.

Loukola-Ruskeeniemi, K., 1999. Origin of black shales and serpentinite-associated Cu-Zn-Co ores at Outokumpu, Finland. Economic Geology 94, 1007–1028.

Loukola-Ruskeeniemi, K., 2011. Graphite- and sulphide-bearing schists in the Outokumpu R2500 drill core: Comparison with the Ni-Cu-Co-Zn-Mn-rich black schists at Talvivaara, Finland. Geological Survey of Finland, Special Paper 51, 229–252.

Loukola-Ruskeeniemi, K., Heino, T., 1996. Geochemistry and genesis of the black shale-hosted Ni-Cu-Zn deposit at Talvivaara, Finland. Economic Geology 91, 80–110.

Loukola-Ruskeeniemi, K., Lahtinen, H., 2013. Multiphase evolution in the black-shale-hosted Ni-Cu-Zn-Co deposit at Talvivaara, Finland. Ore Geology Reviews 52, 85–99.

Loukola-Ruskeeniemi, K., Kuronen, U., Arkimaa, H., 1997. Geochemical comparison of metamorphosed black shales associated with the Vihanti zinc deposit and prospects in western Finland. In: Autio, S. (Ed.), Geological Survey of Finland, Current Research 1995–1996. Geological Survey of Finland, Special Paper 23, 5–13.

Loukola-Ruskeeniemi, K., Heino, T., et al., 1991. Base-metal-rich metamorphosed black shales associated with Proterozoic ophiolites in the Kainuu schist belt, Finland: a genetic link with the Outokumpu rock assemblage. Mineralium Deposita 26, 143–151.

Lusk, J., Ford, C.E., 1978. Experimental extension of the sphalerite geobarometer to 10 kbar. American Mineralogist 63, 516–519.

Lyons, T.W., Reinhard, C.T., Planavsky, N.J., 2014. The rise of oxygen in Earth's early ocean and atmosphere. Nature 506, 307–315.

Mäkelä, K., 1981. Outokumpu-tyyppisten malmien esiintymismahdollisuuksista Itä- ja Pohjois-Suomessa. Summary: on the potential of finding Outokumpu-type ore deposits in East- and North-Finland. Geologi 33, 17–20.

Mäkelä, M., Tammenmaa, J., 1977. A system for precise sulfur isotope analysis by a small mass spectrometer. Geological Survey of Finland, Report of Investigation 20, 23 pp.

McDonough, W.F., Sun, S.-S., 1995. The composition of the Earth. Chemical Geology 120, 223–253.

Melezhik, V.A., Filippov, M.M., Romashkin, A.E., 2004. A giant Palaeoproterozoic deposit of shungite in NW Russia: Genesis and practical applications. Ore Geology Reviews 24, 135–154.

Melezhik, V.A., Huhma, H., Condon, D.J., et al., 2007. Temporal constraints on the Paleoproterozoic Lomagundi-Jatuli carbon isotopic event. Geology 35, 655–658.

Miller, C.A., Peucker-Ehrenbrink, B., Walker, B.D., Marcantonio, F., 2001. Re-assessing the surface cycling of molybdenum and rhenium. Geochimica et Cosmochimica Acta 75, 7146–7179.

Mutanen, T., Huhma, H., 2001. U-Pb geochronology of the Koitelainen, Akanvaara and Keivitsa mafic layered intrusions and related rocks. In: Vaasjoki, M. (Ed.), Radiometric Age Determinations from Finnish Lapland and Their Bearing on the Timing of Precambrian Volcano-Sedimentary Sequences. Geological Survey of Finland, Special Paper 33, 229–246.

Niejenhuis, I.A., Bosch, H.J., Sinninghe Damsté, J.S., et al., 1999. Organic matter and trace element rich sapropels and black shales: a geochemical comparison. Earth and Planetary Science Letters 169, 277–290.

Ohmoto, H., Rye, R.O., 1979. Isotopes of sulfur and carbon. In: Barnes, H. (Ed.), Geochemistry of Hydrothermal Ore Deposits. John Wiley and Sons, New York, pp. 509–567.

Oliver, N.H.S., Hoering, T.C., Johnson, T.W, et al., 1992. Sulfur isotopic disequilibrium and fluid-rock interaction during metamorphism of sulfidic black shales from the Waterville-Augusta area, Maine, USA. Geochimica et Cosmochimica Acta 56, 4257–4265.

Peltonen, P., Kontinen, A., Huhma, H., 1998. Petrogenesis of the mantle sequence of the Jormua Ophiolite (Finland): Melt migration in the upper mantle during Palaeoproterozoic continental break-up. Journal of Petrology 39, 297–329.

Peltonen, P., Kontinen, A., Huhma, H., Kuronen, U., 2008. Outokumpu revisited: New mineral deposit model for the mantle peridotite-associated Cu-Co-Zn-Ni-Ag-Au sulphide deposits. Ore Geology Reviews 33, 559–617.

Perkins, R.B., Piper, D.Z., Mason, C.E., 2008. Trace-element budgets in the Ohio/Sunbury shales of Kentucky: Constraints on ocean circulation and primary productivity in the Devonian-Mississippian Appalacian Basin. Palaeogeography, Palaeoclimatology, Palaeoecology 265, 14–29.

Piper, D.Z., Dean, W.E., 2002. Trace-element deposition in the Cariaco Basin, Venezuela Shelf, under sulfate reducing conditions—a history of the local hydrography and global climate, 20 ka to the present. U.S. Geological Survey Professional Paper 1670. 41 pp.

Piper, D.Z., Isaacs, C.M., 1995. Geochemistry of minor elements in the Monterery Formation, California: Seawater chemistry of deposition. U.S. Geological Survey Professional Paper 1566. 41 pp.

Puhakka, J.A., Kaksonen, A.H., Riekkola-Vanhanen, M., 2007. Heap leaching of black schist. In: Rawlings, D.E., Johnson, D.B. (Eds.), Biomining. Springer, Berlin/Heidelberg, pp. 139–151.

Quinby-Hunt, M.S., Wilde, P., Orth, C.J., Berry, W.B.N., 1989. Elemental geochemistry of black shales—Statistical comparison of low-calcic shales with other shales. In: Grauch, R.I., Leventhal, J.S. (Eds.), Metalliferous Black Shales and Related Ore Deposits. U.S. Geological Survey Circular 1037, pp. 8–15.

Rasilainen, K., Lahtinen, R., Bornhorst, T.J. 2008. Chemical Characteristics of Finnish Bedrock—1:1,000,000 Scale Bedrock Map Units. Geological Survey of Finland, Report of Investigation 171, 94 pp.

Rasmussen, B., 2005. Evidence for pervasive petroleum generation and migration in 3.2 and 2.63 Ga shales. Geology 33, 497–500.

Reuschel, M., Melezhik, V.A., Lepland, A., et al., 2012. Isotopic evidence for a sizeable seawater sulfate reservoir at 2.1 Ga. Precambrian Research 192, 78–88.

Rickard, D., 2012. Sulfidic Sediments and Sedimentary Rocks. Developments in Sedimentology 65. Elsevier, Oxford. p. 801.

Riekkola-Vanhanen, M., 2007. Talvivaara black schist bioheapleaching demonstration plant. Advanced Materials Research 20–21, 30–33.

Rimmer, S.M., 2004. Geochemical paleoredox indicators in Devonian-Mississippian black shales, Central Appalachian Basin (U.S.A.). Chemical Geology 206, 373–391.

Robb, L., 2005. Introduction to Ore-Forming Processes. Blackwell Publishing, Malden, MA, 373 pp.

Rouxel, O., Shanks III, W.C., Bach, W., Edwards, K.J., 2008. Integrated Fe- and S-isotope study of seafloor hydrothermal vents at East Pacific Rise 9–10°N. Chemical Geology 252, 214–227.

Rudnick, R.L., Gao, S., 2003. Composition of the continental crust. In: Rudnick, R.L. (Ed.), The Crust. Treatise on Geochemistry, vol. 3. Elsevier, Amsterdam, pp. 1–64.

Salpeteur, I., 2007. Major base metal districts favorable for future bioleaching technologies. A BRGM contribution to the EEC funded Bioshale project WP6. Final report RP-55610-FR. 71 pp.

Santaguida, F., Lappalainen, M., Ylinen, J., et al., 2015. The Kevitsa Ni-Cu-PGE deposit, Central Lapland Greenstone Belt, Finland. In: Maier, W.D., O'Brien, H., Lahtinen, R. (Eds.), Mineral Deposits of Finland. Elsevier, Amsterdam. pp. 195–209.

Säntti, J., Kontinen, A., Sorjonen-Ward, P., et al., 2006. Metamorphism and chromite in serpentinized and carbonate–silica-altered peridotites of the Paleoproterozoic Outokumpu–Jormua ophiolite belt, eastern Finland. International Geology Review 48, 494–546.

Schidlowski, M., 1981. Uraniferous constituents of the Witwatersrand conglomerates: Ore microscope observations and implications for Witwatersrand metallogeny. In: Armstrong, F. (Ed.), Genesis of Uranium and Gold-Bearing Quartz Pebble Conglomerates. U.S. Geological Survey Professional Paper 1161-N, pp. 1–29.

Schultz, R.B., Coveney, R.M., 1992. Time-dependent changes for Midcontinent Pennsylvanian black shales, U.S.A. Chemical Geology 99, 83–100.

Schovsbo, N.H., 2002. Uranium enrichment shorewards in black shales: a case study from the Scandinavian Alum Shale. GFF 124, 107–115.

Severmann, S., McManus, J., Berelson, W.M., Hammond, D.E., 2010. The continental self benthic iron flux and its isotope composition. Geochimica and Cosmochimica Acta 74, 3984–4004.

Skei, J., 1983. Geochemical and sedimentological considerations of a permanently anoxic fjord–Framvaren, South Norway. Sedimentary Geology 36, 131–145.

Steiner, M., Wallis, E., Erdtmann, B.-D., et al., 2001. Submarine-hydrothermal exhalative ore layers in black shales from South China and associated fossils—Insights into a Lower Cambrian facies and bio-evolution. Palaeontography, Palaeoclimatology, Palaeoecology 169, 165–191.

Strauss, H., Melezhik, V.A., Reuschel, M., et al., 2013. Abundant marine calcium sulphates—Radical change of seawater sulphate reservoir and sulphur cycle. In: Melezhik, V.A., Fallick, A.E., Strauss, H., et al. (Eds.), Reading the Archive of Earth's Oxygenation. vol. 3. Global Events and the Fennoscandian Arctic Russia—Drilling Early Earth Project. Springer-Verlag, Berlin/Heidelberg, pp. 1169–1194.

Stüeken, E.E., Catling, D.C., Buick, R., 2012. Contributions to late Archaean sulphur cycling by life on land. Natural Geoscience 5, 722–725.

Talvivaara Mining Company Plc, 2014. Talvivaara 2013, Annual Report. www.talvivaara.com/investors/financial_information/Annual-reports..

Tian, Y., Etschmann, B., Liu, W., et al., 2012. Speciation of nickel (II) chloride complexes in hydrothermal fluids: in situ XAS study. Chemical Geology 334, 345–363.

Tissot, F., Dauphas, N., Reinhard, C.T., et al., 2013. Mo and U geochemistry and isotopes. In: Melezhik, V.A., Fallick, A.E., Strauss, H., et al. (Eds.), Reading the Archive of Earth's Oxygenation, vol. 3. Global Events and the Fennoscandian Arctic Russia—Drilling Early Earth Project. Springer-Verlag, Berlin/Heidelberg, pp. 1500–1505.

Törmälehto, T., 2008. Ore mineralogy and geochemistry of the Kuusilampi ore deposit in Sotkamo. Unpublished Master's thesis. University of Oulu. 84 pp. (in Finnish).

Törnroos, R., 1982a. Sphalerite geobarometry of some metamorphosed sulphide deposits in Finland. Geological Survey of Finland. Bulletin 323, 42 pp.

Törnroos, R., 1982b. Properties of alabandite; alabandite from Finland. Neues Jahrbuch für Mineralogie, Abhandlung 144, 107–123.

Tribovillard, N., Algeo, T.J., Lyons, T., Riboulleau, A., 2006. Trace metals as paleoredox and paleoproductivity proxies: An update. Chemical Geology 232, 12–32.

Västi, K., 2008. Chemical composition of metamorphosed black shale and carbonaceous metasedimentary rocks at selected targets in the Vihanti area, western Finland. Geological Survey of Finland, Report of Investigation 173, 22 pp.

Vine, J.D., Tourtelot, E.B., 1970. Geochemistry of black shale deposits—a summary report. Economic Geology 65, 253–272.

Virtasalo, J.J., Laitala, J., Lahtinen, R., Whitehouse, M.J., 2014. Microbial mat influence and event deposition of pyritic beds in the Palaeoproterozoic Talvivaara formation. Proceedings of the 1st Finnish National Colloquium of Geosciences, Abstract Book, Espoo, March 19–20. Geological Survey of Finland. Guide 58, 107–108.

Werne, J.P., Sageman, B.B., Lyons, T.W., Hollander, D.J., 2002. An intergrated assessment of a "type euxinic deposit": Evidence for multiple controls on black shale deposition in the Middle Devonian Oakta Creek Formation. American Journal of Science 302, 110–143.

Wignall, P.B., 1994. Black Shales. Clarendon Press, Oxford. 127 pp.

Young, S.A., Loukola-Ruskeeniemi, K., Pratt, L.M., 2013. Reactions of hydrothermal solutions with organic matter in Paleoproterozoic black shales at Talvivaara, Finland: Evidence from multiple sulfur isotopes. Earth and Planetary Science Letters 367, 1–14.

CHAPTER

HIGH-TECH METALS IN FINLAND

9.2

O. Sarapää, N. Kärkkäinen, T. Ahtola, T. Al-Ani

ABSTRACT

The demand for high-tech metals such as Li, Ti, Co, Ga, Ge, Nb, In, Sb, Ta, Pd, Pt, and rare earth elements (REE) has increased in recent years especially in green energy technology. In this chapter, we concentrate on REE, titanium, and lithium deposits, which are among the most important high-tech metal deposits in Finland.

The known REE deposits are associated with carbonatite dikes at Sokli, alkaline gneisses at Katajakangas, and monazite granites at Kovela. The Korsnäs Pb deposit contains rare earth elements in apatite and monazite, probably due to enrichment by hydrothermal and weathering processes. In the Sokli carbonatite complex, REEs are concentrated in late magmatic carbonatite veins in the surrounding fenite area. The Katajakangas Nb-Y deposit is enriched in heavy REE in zircon, bastnaesite, and columbite. At Kovela, late-orogenic S-type granite contains monazite with thorite inclusions.

The Otanmäki V-Ti-Fe, Koivusaarenneva Ti, Kauhajoki Ti-P-Fe, and Karhujupukka Fe-Ti-V deposits are associated with magmatic mafic layered intrusions. A common feature is that ilmenite and magnetite occur as individual grains and thus can be separated economically. Pure ilmenite and magnetite grains are primary magmatic minerals at Koivusaarenneva and Kauhajoki, but metamorphic at Otanmäki and Karhujupukka. In the future, these gabbro-hosted deposits could be important producers of Ti, Fe, V, and P with low environmental impact.

The Kaustinen Li province contains several economically interesting albite-spodumene pegmatite deposits such as at Länttä, Emmes, Outovesi, Syväjärvi, Leviäkangas, and Rapasaaret, and includes Li mineral resources for several decades. In the Somero-Tammela Li province with albite-petalite-spodumene pegmatites, exploration activity has been smaller, although the deposits contain high-quality petalite and spodumene in economic concentrations.

Keywords: high-tech metals; rare earth elements; spodumene; petalite; ilmenite; Finland.

INTRODUCTION

High-tech metals are essential for the production of high-tech devices such as computers, mobile phones, advanced weapons, solar panels, wind turbines, fuel cells, and batteries. Generally, high-tech metals include Li, Ti, Co, Ga, Ge, Nb, In, Sb, Ta, Pd, Pt, and rare earth elements (REEs), which are mostly produced as by-products of other metals. In this section, we focus on the rare earth elements, titanium and lithium. Cobalt and platinum-group metals are described in another part of this volume. Indium, gallium, and germanium are less common in the Finnish bedrock and are not discussed in this chapter. Magmatic REE deposits in carbonatite, alkaline intrusions and peralkaline granites, lithium deposits in LCT (Li, Cs, Ta) pegmatites, and Ti deposits in anorthosites and noritic gabbros occur in the Fennoscandian Shield (Eilu, 2012).

Mineral Deposits of Finland. http://dx.doi.org/10.1016/B978-0-12-410438-9.00023-6

A variety of techniques have been used in the exploration of REE, Ti, and Li deposits in Finland. Boulder fans (Li, REE), till- and lithogeochemistry (REE, Li), magnetic + gravity (Ti, REE), and radiation (REE) measurements have been the main exploration methods for these deposits.

RARE EARTH ELEMENT DEPOSITS

The REEs, comprising the lanthanide series from lanthanum to lutetium (atomic numbers 57 to 71) plus geochemically similar yttrium and scandium, occur in minor quantities in common rocks, with an average total abundance of 183 ppm in the upper crust (Rudnick and Gao, 2003), but economically mineable REE deposits are relatively rare. Carbonatites, ion-adsorption type deposits in weathered rocks, and heavy mineral placer deposits are the main sources for REE (Long et al., 2010). Primary magmatic or hydrothermal REE types of mineralization are mostly found with carbonatites and peralkaline igneous rock (Castor and Hedrick, 2006; Long et al., 2010).

Carbonatites are characterized by high abundances of light REE (LREE, La to Eu) while more valuable heavy REE (HREE, Gd to Lu) are concentrated in some alkaline rocks, fractionated granites and pegmatites, and ion-adsorption deposits. China currently produces at least 95% of the world's supply of rare earth elements. Examples of the tonnages and grades of exploited REE deposits are 48 Mt at 6 wt% RE_2O_3 in Bayan Obo Fe-REE-ore, China (Wu, 2008); 16.7 Mt at 7.98 wt% REO in Mountain Pass carbonatite, U.S.A. (Castor, 2008; Mariano and Mariano, 2012); and 0.05–0.3 wt% REO in ion-absorbed clays, 2.8 Mt (counted as REO) in Jiangxi, southern China (Chi and Tian, 2008). With current technology, only bastnaesite, monazite, loparite, and kaolinite with adsorbed REE have sufficient concentrations of REE, favorable lanthanide distribution, and processability for production. Other potentially exploitable minerals include xenotime, eudialyte, fergusonite, and apatite.

There are currently no known economic REE deposits in Finland. Nevertheless, REEs have been extracted in Finland: in the 1950s from apatite concentrate imported from the Kola Peninsula, and since 1963, from apatite concentrate mined as a by-product in the Korsnäs Pb mine in western Finland (Lounamaa, 1972). In addition to the Korsnäs Pb-REE deposit associated with a carbonate dike, the alkaline gneissic granite at Otanmäki, central Finland, hosts a small Nb-REE deposit (Hugg, 1985).

The Fennoscandian Shield hosts several REE deposits in the Devonian Kola alkaline province in Russia (Arzamastsev et al., 2008; Korsakova et al., 2012). The Sokli carbonatite complex (total area 20 km^2) is part of the province (Vartiainen, 1980; O'Brien et al., 2005) and recent studies have confirmed the high REE potential of the area. The Geological Survey of Finland (GTK) has focused on the fenite aureole and associated late-stage crosscutting carbonatite dikes that seem to have the highest potential for REE mineralization in the Sokli area (Sarapää et al. 2013).

Thorium and REE-rich mineralization of Kovela in southern Finland represents an REE ore type associated with late-orogenic granite.

KORSNÄS Pb-REE DEPOSIT

Galena-bearing boulders found by a layman initiated an exploration program by GTK in the Korsnäs area, and in 1955 extensive deep drilling led to the discovery of the ore in a gravity low (Isokangas, 1975). The Outokumpu Oy mining company operated the Korsnäs mine from 1961–1972 and produced 45,000 tons of lead and 36,000 tons of lanthanide concentrate (Himmi, 1975). The grade of the ore was 3.57%

Pb and 0.91% RE_2O_3 . The major ore minerals were galena, apatite, monazite, and allanite (Papunen and Lindsjö, 1972). The REEs were extracted from the apatite concentrate at Typpi Oy, Oulu.

The deposit comprises the Svartören Pb-REE-bearing carbonate dike which hosts the main ore body, measuring 5–30 m in width, up to 1.5 km in length, and extending down to a depth of around 350 m. The Svartören dike occurs in a north–south-trending fracture zone, dips to the east at an angle of 40–60°, and crosscuts ~1.9 Ga migmatitic mica gneiss (Isokangas, 1975; Lehtonen, et al., 2005) of the poorly exposed south Pohjanmaa schist belt, western Finland (Figs. 9.2.1 and 9.2.2). Thermal ionization mass spectrometry (TIMS) titanite U-Pb age of the Svartören dike is 1.825 Ga (Papunen, 1986a). It represents the largest dike in a network of narrow carbonate veins and dikes that occur in an area of

FIGURE 9.2.1 Simplified geological map of Finland showing the distribution of rare earth elements, titanium, and lithium deposits.

Source: *Bedrock according to GTK DigiKP.*

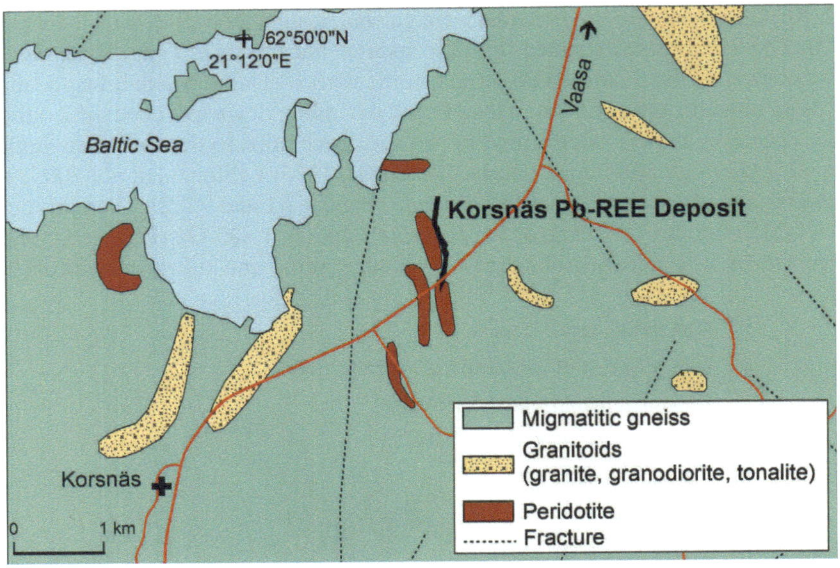

FIGURE 9.2.2 Geological map of the Korsnäs area showing the location of the Pb-REE ore.

Source: *Geology based on Bedrock of Finland–DigiKP.*

approximately 10 km^2. Some boulders appear to be derived from still-undiscovered sources, and testing gravity lows may be the best method to find new Pb-REE-bearing carbonate veins in the area. According to Torppa and Karhu (2013), the $\delta^{13}C$ isotopic values from −17 to −18‰ measured for Korsnäs rocks suggest a crustal source for carbon, and the positive δO^{18} values from +10 to +12‰ indicate probably a meteoritic origin for calcite. Pb isotopic compositions of the Korsnäs galena plot with the rest of the galena from Svecofennian supracrustal rocks studied by Vaasjoki (1981). Taken together, these data strongly suggest the carbonate is derived from crustal sources, probably hydrothermal in origin, and is not a carbonatite, sensu stricto (Mitchell, 2005).

The deposit consists of mineralized zones in pegmatite and the hydrothermal carbonate in calcareous scapolite-diopside-barite–bearing rocks. The ore zone is strongly sheared and weathered, and the wall rock of carbonate is kaolinized. Apatite and monazite are heterogeneously distributed in the ore but follow the occurrence of galena. Monazite occurs as small, discrete crystals or as fine-grained inclusions in apatite within the zones of the weathered rocks (Isokangas, 1975; Sarapää et al., 2013). The RE_2O_3 content in apatite is up to 6.16 wt% and contains more HREE than monazite and allanite (Papunen and Lindsjö, 1972). However, the chondrite-normalized REE patterns for drill-core samples from the Korsnäs Pb-REE deposit are very similar in shape to those of the Sokli carbonatite due to the concentration of LREE in monazite and allanite. The weathered ore is clearly more enriched in REEs than unweathered carbonate rock (Fig. 9.2.3).

KATAJAKANGAS Nb-REE DEPOSIT

The Katajakangas Nb-REE deposit is located within alkaline gneissic granite in the Otanmäki area, central Finland (Fig. 9.2.1 and Fig. 9.2.4). Similar small mineralizations have also been discovered

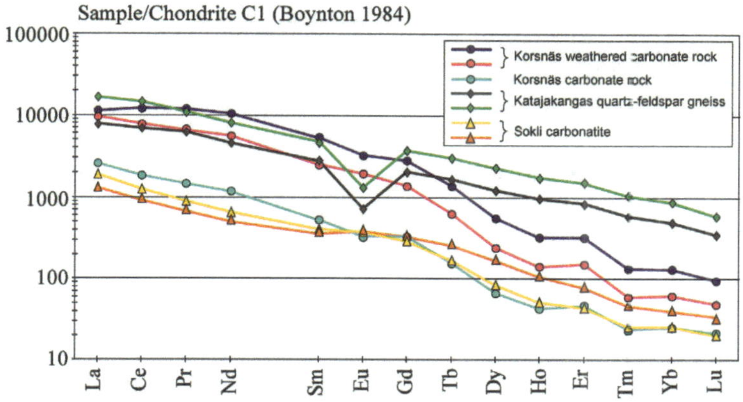

FIGURE 9.2.3 Chondrite-normalized REE abundances for selected carbonatites of Sokli, the carbonate dike at Korsnäs, and alkaline gneiss of Katajakangas.

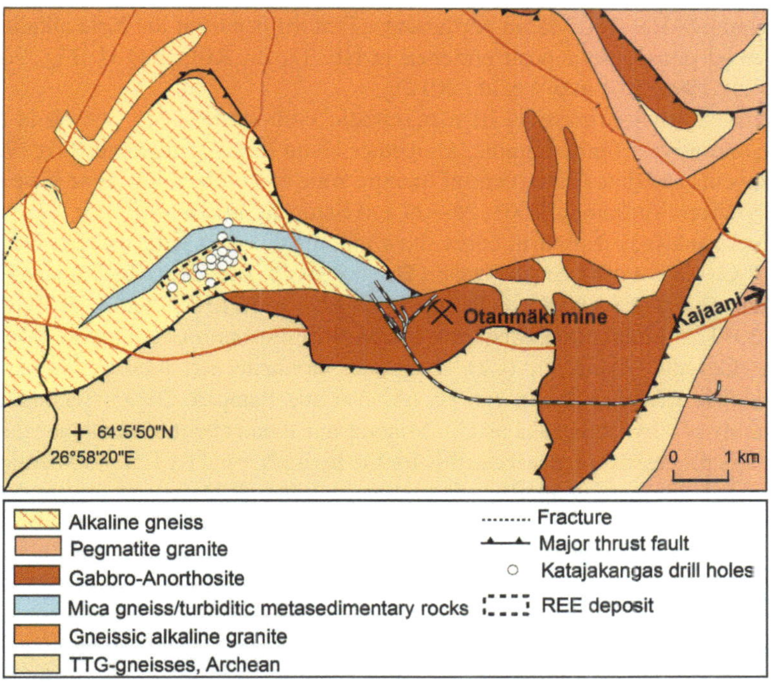

FIGURE 9.2.4 Geological map of the Otanmäki area, showing locations of the closed Fe-Ti-V mine and Katajakangas REE deposit.

Source: *Geology based on Bedrock of Finland–DigiKP.*

elsewhere in the area (Hugg, 1985), which is dominantly composed of Archean granitic gneisses and alkali-granites and ~2.05 Ga gabbro-anorthosites with Fe-Ti-V mineralizations (Lindholm and Anttonen, 1980; Talvitie and Paarma, 1980; Kontinen et al., 2013). The Nb-REE mineralization occurs in the contact zone between turbiditic metasediments and gneisses (Marmo et al., 1966; Puumalainen, 1986). The mineralization consists of a few-meters-wide lenses or layers in sheared Katajakangas quartz-feldspar gneiss with riebeckite and alkaline pyroxene. The main ore minerals are zircon, bastnaesite, columbite, and thorite (Hugg and Heiskanen, 1986; Äikäs, 1990). The narrow mineralized zone contains high concentrations of Nb, Zr, Y, Th, and REE, with an estimated Nb-YREE resource of 0.46 Mt at 2.4% RE_2O_3, 0.31% Y_2O_3, and 0.76% NbO. The mineralization also contains 0.7–1.5% Zr and 0.1–0.2% Th (Hugg and Heiskanen, 1986).

Drill-core samples have relatively high HREE contents compared with samples from carbonatites (Sarapää et al., 2013; Fig. 9.2.3). At Katajakangas, the La/Yb ratio is between 16 and 20 and the maximum Dy content is up to >700 ppm. A metasomatic origin is obvious for the Katajakangas mineralization. The alkaline gneisses from the Otanmäki area have the same kind of mineral and chemical compositions as peralkaline granites of an Ocean Island Basalt (OIB) affinity (Kontinen et al., 2013).

SOKLI CARBONATITE VEINS

The Sokli carbonatite (~360–380 Ma) in northeastern Finland is part of the Kola alkaline province and hosts an unexploited phosphate deposit enriched in Nb, Ta, Zr, REE, and U (Fig. 9.2.1; Vartiainen, 1980; Kramm et al., 1993; Korsakova et al., 2012).

The carbonatite intrusion consists of a magmatic carbonatite core, which is surrounded by metacarbonatite and a wide fenite aureole, altogether about 9 km in diameter (Fig. 9.2.5). The late-stage carbonatite veins and dikes in the central fracture zone and in the fenite zone have a high potential for REE mineralization (Vartiainen, 2001; Al-Ani and Sarapää, 2013).

Recent chemical analyses from drill cores show that the 0.5–1 m wide carbonatite dikes in fenites are enriched in P_2O_5 (19.9 wt%), Sr (1.9 wt%), Ba (6.8 wt%), and Zn (0.3wt%) and also have a high total REE content of 0.5–1.83 wt%, including 0.11–1.81 wt% LREE and 0.01–0.041 wt% HREE (Sarapää et al., 2013). Dominant REE-bearing minerals in the Jammi carbonatite dikes are REE carbonates ancylite-(Ce) and bastnaesite-(Ce), Sr-apatite, monazite, strontianite, barite, and brabantite, which are enriched in LREE, P, F, Sr, and Ba (Al-Ani and Sarapää, 2013). Secondary alteration of monazite is probably of a hydrothermal origin. Mineralogical and chemical evidence demonstrates that late-stage magmatic processes were responsible for the hydrothermal REE mineralization in the Jammi carbonatite veins, where igneous apatite and carbonate minerals were partially replaced by various assemblages of REE-Sr-Ba minerals.

KOVELA MONAZITE GRANITE

The late-orogenic Kovela monazite-bearing granite, which shows a strong positive aeroradiometric gamma-radiation anomaly, penetrates migmatitic gneisses in the Svecofennian Uusimaa belt, which is, according to Kähkönen (2005), 1.91–1.88 Ga in age (Fig. 9.2.1). Two very strong positive thorium anomalies were drilled by GTK in 2011 and 2012, and the holes penetrated two northwest–southeast-trending zones of REE mineralization in monazite- and garnet-bearing, coarse-grained microcline granite.

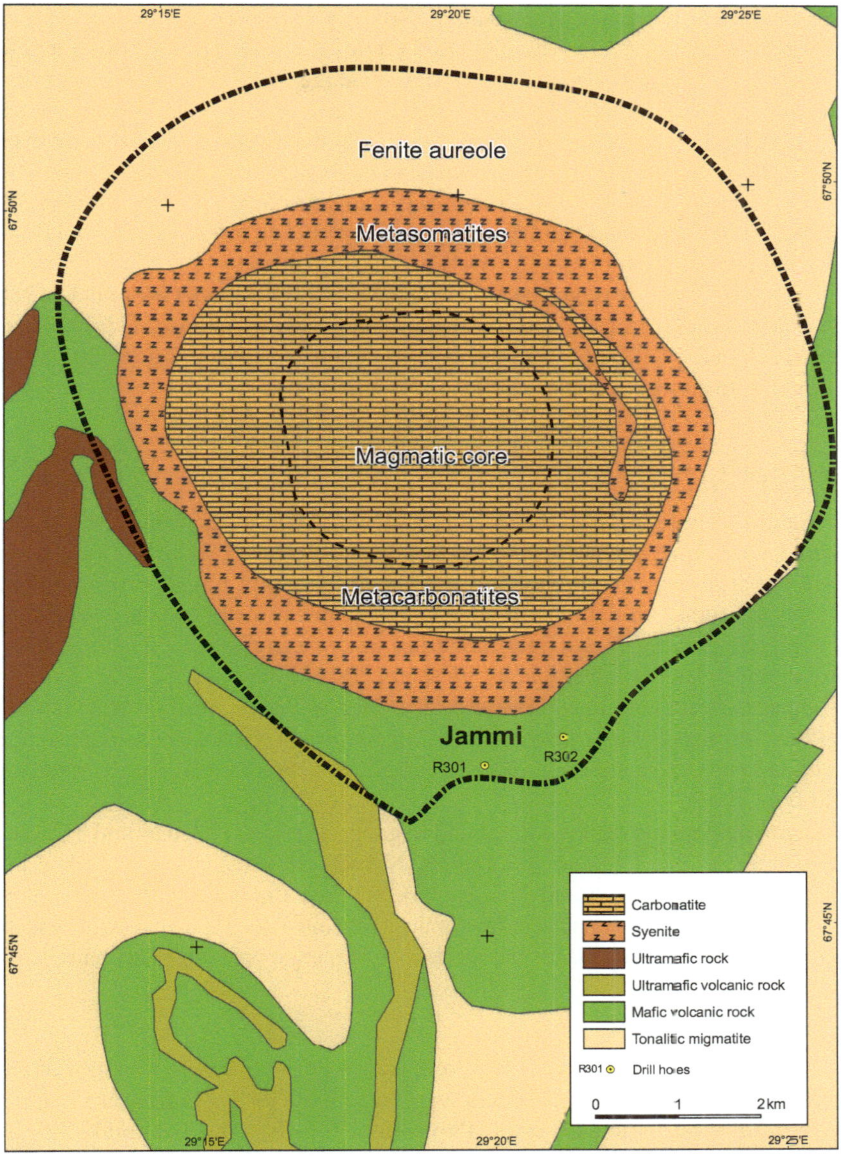

FIGURE 9.2.5 Geological map of the Sokli complex showing rocks from the oldest to the youngest: fenite aureole, metasomatic silicate rocks, metasomatic carbonatites, and magmatic carbonatite core.

The surrounding rocks are Archean tonalitic migmatites and mafic and ultramafic volcanic rocks.

Source: *Modified from Vartiainen (1980) and Bedrock of Finland–DigiKP.*

The Kovela monazite granite is a peraluminous, S-type granite according to the definition of Shand (1943), with molecular (Al_2O_3/CaO + Na_2O + K_2O) ratios (A/CNK) greater than 1.2. The wall rock, K-rich calc-alkaline granite, has a lower REE content than the REE- and Ca-rich pegmatitic monazite granite. Monazite is the dominant REE-mineral, and accessory minerals include zircon, xenotime, and thorite (Al-Ani and Grönholm, 2011). Anhedral to subhedral crystals of monazite are evenly disseminated throughout the rock. The size of the monazite crystals is usually 100–1000 μm, with the largest being up to 0.3 mm long. The crystals are zoned and inclusions of thorite are common. The monazite crystals contain Ce, La, Nd, Pr, and Sm averaging around 26.6, 9.6, 9.5, 2.5, and 1.5 wt%, respectively, and significant amounts of ThO_2 (14.9–22.1 wt%).

The chemical analysis of four monazite granite samples show 1800–17,700 ppm Ce, 786–8100 ppm La, 739–7000 Nd, 211–2000 ppm Pr, 120–1180 ppm Sm, and 87.6–620 ppm Gd. The thorium content varies between 1110 and 10,100 ppm and the uranium content between 3 and 320 ppm. Whole-rock REE contents in 2-m drill-core sections vary from 0.6–4.3 %.

TITANIUM DEPOSITS

The major titanium minerals are anatase, brookite, ilmenite, leucoxene, perovskite, rutile, and sphene, but only ilmenite, leucoxene, and rutile have economic importance (USGS, 2013). Titanium is a light metal compared to iron, has a high strength-to-weight ratio, and high corrosion resistance. Most titanium (95%) is consumed as titanium dioxide (TiO_2) pigment for a white material in paints, paper, and plastics. High refractive index and light-scattering ability impart excellent hiding power and brightness. Titanium dioxide also has a possible use in solar cells.

The most important titanium mineral is ilmenite, as well as Rutile. Most of the commercial ilmenite and rutile deposits are sedimentary heavy mineral concentrates in coastal environments, ocean beaches, or shoreline eolian dunes; only a few are located in an alluvial environment such as riverbeds and deltas (Murphy and Frick, 2006).

Primary, magmatic ilmenite deposits are found in gabbroic rocks related to mid-Proterozoic massif-type anorthosites in Lac Tio, Canada, and Tellnes, Norway. Ilmenite is also enriched in large mafic layered intrusions (e.g., Bushveld and Koillismaa), but true ilmenite deposits are rare because Ti usually occurs as an ulvöspinel component or fine-grained ilmenite lamellas in magnetite, and the separation of ilmenite from magnetite grains makes the recovery of ilmenite presently uneconomic.

OTANMÄKI V-Ti-FE DEPOSIT

The first Fe-Ti-V ore boulders from the area were found in 1937 at Sukeva, 60 km south of Otanmäki. The following year, the source of the boulders was traced by GTK geologists to a magnetic anomaly found while mapping the Otanmäki area (Papunen, 1986a,b). The Otanmäki Fe-Ti-V deposit, composed mainly of magnetite and ilmenite, was mined from 1953 to 1985, and produced 31 Mt ore with 32–34 wt% Fe, 5.5–7.6 wt% Ti, 0.26 wt% V, and some pyrite as a by-product. The deposit contains two ore bodies, Otanmäki and Vuorokas, with the remaining reserves being estimated at 16 Mt with additional resources of 3 Mt (Puustinen, 2003; Vuorokas Oy, 2013).

The Otanmäki magnetite-ilmenite deposit is hosted by Paleoproterozoic (2.06 Ga) layered gabbro-anorthosite (Fig. 9.2.3) that intruded between the Archean Pudasjärvi and Iisalmi blocks (Talvitie and Paarma, 1980; Kuivasaari et al., 2012). The ore deposit consists of hundreds of lenses up to 200 m in length and 3–50 m in width (Illi, 1985). Massive ore lenses are situated in a heterogeneous contact zone between amphibolite and anorthosite. Fractional crystallization and accumulation of Ti-magnetite formed the proto-ore while regional metamorphism at amphibolite facies caused recrystallization of igneous ulvöspinel or Ti magnetite with fine ilmenite exsolution lamellae forming independent grains of ilmenite and magnetite (Pääkkönen, 1956). Consequently the main ore minerals are ilmenite and vanadium-rich magnetite (0.62 wt% V). High-grade ore contains 30–40% magnetite and 28–30% ilmenite. Chlorite, hornblende, some plagioclase, and pyrite (up to 1–2%) make up the gangue. Two main components for TiO_2-pigment production, ilmenite ore from Otanmäki and sulfuric acid, were the domestic raw materials for the then active titanium pigment factory founded in Pori in 1957.

KOIVUSAARENNEVA TI DEPOSITS

The small ilmenite-magnetite-gabbros (1881 Ma) at Koivusaarenneva, Kairineva, Peräneva, Lylyneva, and Riutta make up a chain of igneous bodies in tonalite in the northwestern part of the Central Finland Granitoid Complex (Fig. 9.2.6) (Kärkkäinen and Bornhorst, 2003; Sarapää et al., 2001). Gravity and magnetic highs, and anomalies in Electromagnetic (EM) measurements indicate the locations of these buried deposits, composed mainly of ilmenite and lesser magnetite in gabbronorite and pyroxenite of

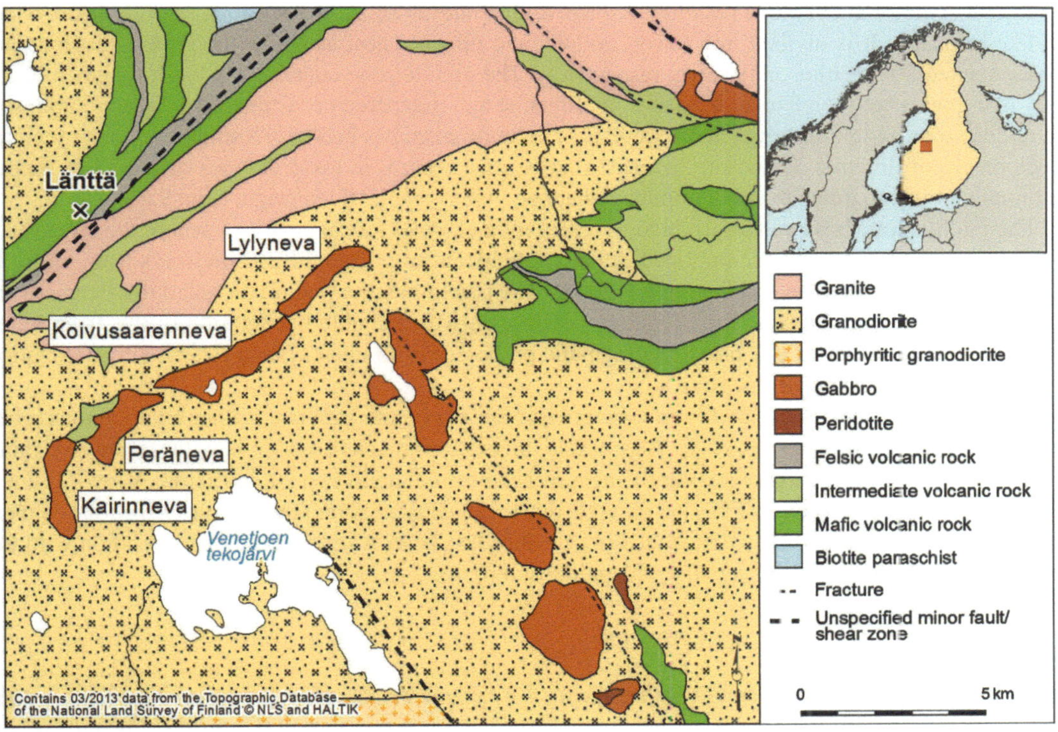

FIGURE 9.2.6 Location of Koivusaarenneva Ti deposits and Länttä Li deposit.

Source: *Geology based on Bedrock of Finland–DigiKP.*

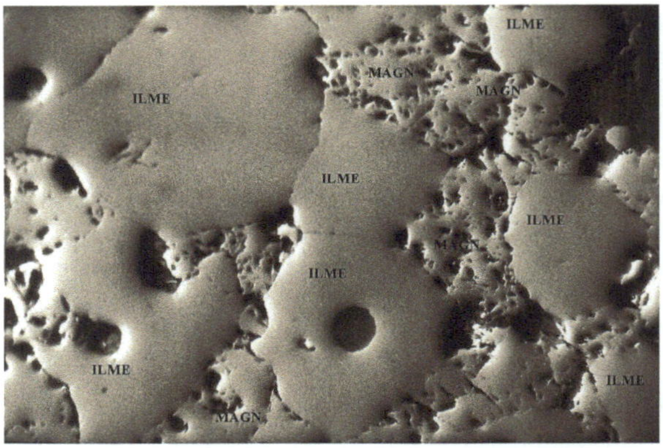

FIGURE 9.2.7 Photomicrograph of Koivusaarenneva ilmenite ore (grains 0.5–3 mm).

Source: *Photo by Niilo Kärkkäinen.*

the layered mafic intrusions. The resource estimate is 74 Mt ore with 8 wt% TiO_2 (or 16% ilmenite) and 5.2 wt% magnetite (Kärkkäinen, 2012a; Endomines, 2015).

The most important of the intrusions is the Koivusaarenneva gabbro, a 3-km-long and 0.5–1-km-thick, sill-like intrusion. It is divided into lower, middle, and upper zones, in which the typical minerals are ilmenomagnetite, ilmenite, and apatite, respectively. The middle zone contains a 2-km-long ilmenite ore body that grades between 8 and 48% ilmenite. Ilmenite and magnetite occur mainly as discrete grains in the middle zone and in the upper zone. Regional metamorphism has had only a small role in the textural development of ilmenite. Based on the clear stratigraphic control in textural types of the Ti-Fe oxides within the intrusion, ilmenite in the middle and upper zone is magmatic in origin (Fig. 9.2.7).

The middle zone has been interpreted to represent a channel for magma flow where the large mass of oxides were trapped and accumulated in favorable localities within the channelways from multiple magma pulses (Kärkkäinen and Bornhorst, 2003). The parent magma was fractionated in a deep crustal reservoir under dry and low-oxygen fugacity conditions, which enabled generation of the Ti-enriched mafic magma. Subsequent low pressure magmatic processes in an open system initiated crystallization of ilmenite from Ti-enriched magmas.

The type of bedrock is critical for the existence of these ilmenite gabbros. Rather homogenous felsic and intermediate intrusive rocks or high-grade metamorphic environments with relatively small amounts of OH-bearing minerals favor the occurrence of ilmenite in a mafic intrusion. A low-f_{O2} environment enables enrichment of Ti in the magma, due to delayed titanomagnetite crystallization (Fenner-type crystallization series). The abundance of iron sulfides shows a strong positive correlation with the abundance of ilmenite in this type of Fe-Ti oxide-rich rocks.

KAUHAJOKI Ti-P-FE

The Ti-P-Fe deposits in the Perämaa (or Peräkorpi), Kauhajärvi, and Lumikangas layered mafic intrusions were discovered by drilling gravity and magnetic highs in the western part of the Central Finland Granitoid Complex (CFGC) (Kärkkäinen and Appelqvist, 1999; Sarapää et al., 2005). The CFGC represents a primitive Paleoproterozoic arc complex and is composed mainly of collision-related felsic intrusive rocks (Korsman et. al., 1997). Mafic intrusions of the Kauhajoki area intrude granitoids of the CFGC and are closely associated with the K-rich Lauhanvuori granite that implies a mature postorogenic magmatism (Rämö et al., 2001; Peltonen, 2005). Ti-P-Fe gabbros form an enormous low-grade mineral resource of more than 500 Mt mineralized rock with a total amount of apatite, ilmenite, and ilmenomagnetite ~20 wt% (Kärkkäinen, 2012b). The Perämaa deposit contains maximum values of 8–9 wt% TiO_2, 2.5–4.0 wt% P_2O_5, and 20–27 wt% Fe; the Kauhajärvi deposit contains 4–8 w% TiO_2, 1–3.6 wt% P_2O_5, and 13–28 wt% Fe; and the Lumikangas deposit contains 8.7 wt% ilmenite (maximum 21 wt%), 5.4 wt% apatite (maximum 17 wt%), and 4.8 wt% magnetite (maximum 17 wt%). Perämaa and Kauhajärvi show a complete differentiation series from peridotite to anorthosite. The main zone of the Kauhajärvi gabbro crystallized under rather high f_{O2} conditions, from a fractioned P-Ti-Fe–rich mafic magma, which enabled apatite to crystallize at the same time as ilmenite and Ti magnetite in the earliest olivine- and pyroxene-rich cumulates (Kärkkäinen and Appelqvist, 1999). Lumikangas is composed of homogenous layered gabbro and monzogabbro with a high normative alkali feldspar content and shows coeval crystallization of apatite, ilmenite, magnetite, and mafic minerals (Sarapää et al. 2005).

KARHUJUPUKKA FE-TI-V DEPOSIT

In 1988, GTK discovered the Karhujupukka deposit by drilling into a positive magnetic anomaly, which was completely covered by till and weathered bedrock (Karvinen et al., 1989). The Karhujupukka area belongs to the western part of the Central Lapland Greenstone Belt composed of tholeiitic and komatiitic volcanic rocks, mica gneisses, gabbros, and granitoids (Fig. 9.2.1). The host rock of the deposit is amphibolite occurring between hanging wall anorthosite-gabbro and footwall metasediments. The Ti-V-Fe ore body has a thickness of 10–50 m and a total length of 550 m.

The ore contains 40 wt% Fe, 5.5 wt% Ti, 0.3 wt% V. Ilmenite occurs as anhedral grains (0.5–2.0 mm) with minute inclusions of hematite. Granoblastic magnetite contains minor ilmenite exsolution lamellae. Sulfides between ilmenite and magnetite grains include pyrrhotite, chalcopyrite, pentlandite, violarite-polydymite, pyrite, and marcasite. The gabbroic rocks were enriched in Fe, Ti, and V during magmatic differentiation and a granoblastic texture was developed during recrystallization under amphibolite facies metamorphic conditions. Due to the granoblastic texture of the ore and the large grain size of ore minerals, concentration tests on the ore gave high-quality magnetite and ilmenite concentrates. The magnetite concentrate contained 92–94 wt% Fe_3O_4 tot, 0.84 wt% V_2O_3, and the ilmenite concentrate 47.6 wt% TiO_2, 46.5 wt% FeO tot.

LITHIUM DEPOSITS

Global lithium consumption has increased throughout the 2000s. The main uses of lithium are ceramics and glass 30%, batteries 22%, lubricating greases 11%, metallurgical 5%, air treatment 4%, polymers 3%, pharmaceuticals 2%, primary aluminum production 1%, and other uses 23% (USGS, 2013). The growth rate is highest for batteries, because the properties of lithium make it the most suitable of all the elements for battery materials. Lithium is produced from brine deposits in Chile, Argentina, and China; granites in China; and pegmatites in Australia, China, Brazil, Portugal, and Zimbabwe (Linnen et al., 2012, USGS, 2013). It occurs in 145 minerals, but only spodumene, petalite, amblygonite, lepidolite, and eucrytite have been commercial Li sources.

According to the classification of Černý and Ercit (2005), lithium-bearing pegmatites belong to the LCT (Li, Cs, Ta) family of pegmatites, and generally occur in fault zones in areas in which the metamorphic grade corresponds to greenschist or amphibolites facies. According to Selway et al. (2005),

LCT pegmatites are associated with late-tectonic S-type, peraluminous (A/CNK <1.0) quartz- and feldspar-rich granites. Fertile granites have Mg/Li <10 and Nb/Ta <8. S-type granites and are derived from a magma produced by partial melting of sedimentary rocks; extreme fractional crystallization of these magmas concentrates rare elements in residual melts that form the pegmatites (Linnen et al., 2012). In Finland, LCT pegmatites are common in many places in southern Ostrobothnia, including Kaustinen, Somero-Tammela, Kitee-Tohmajärvi, Haapaluoma-Kaatiala, Eräjärvi, Seinäjoki, Heinola, Kisko, Kemiö, and Kalajoki (Alviola, 2012). The most prospective Li provinces are Kaustinen and Somero-Tammela (see Fig. 9.2.1), which have been the main focus of exploration in recent years.

KAUSTINEN Li PROVINCE

In the Kaustinen Li province, spodumene boulders were first discovered in 1959. Since then, several companies have carried out Li exploration in the area. Suomen Mineraali Oy started the first studies in the early 1960s, continued by Paraisten Kalkki Oy in the 1980s, while Keliber Oy started in 1999 and GTK in 2003. These studies have led to the discovery of dozens of spodumene pegmatite dikes and hundreds of ore boulders. Keliber Oy is planning to begin lithium carbonate production at Länttä within the next few years (Fig. 9.2.8).

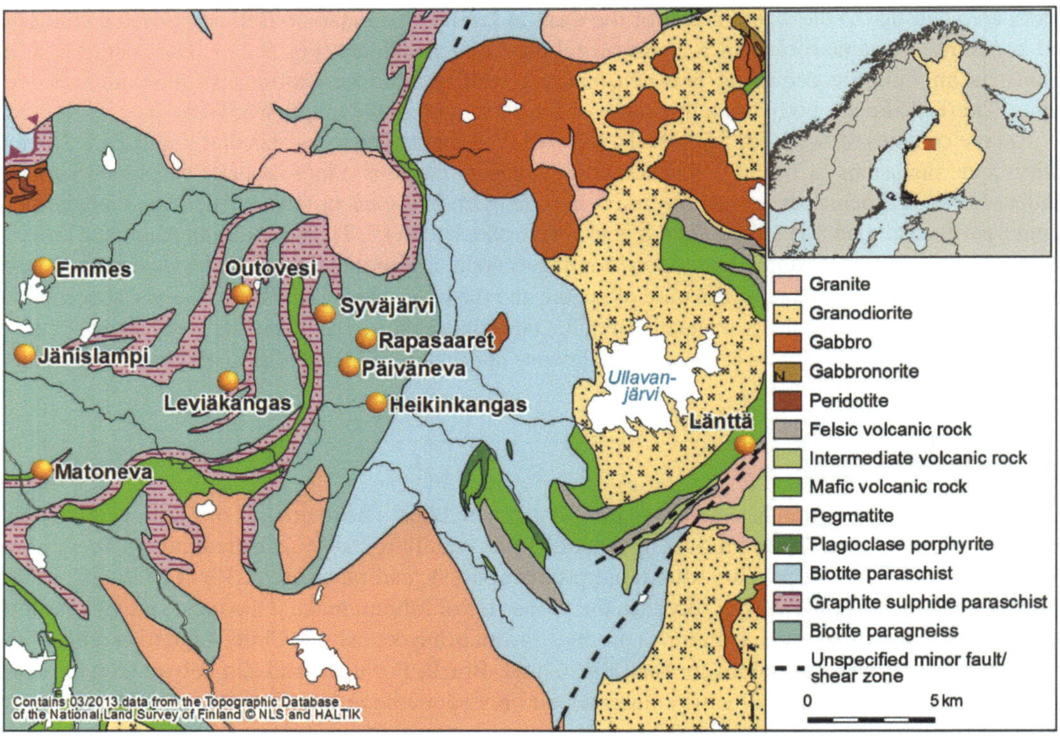

FIGURE 9.2.8 Geological map of Kaustinen Li province showing the most important Li deposits of Kaustinen.

Source: *Geology based on Bedrock of Finland–DigiKP*

The Kaustinen Li province in the Pohjanmaa schist belt is part of the Paleoproterozoic Svecofennian accretionary arc complex in central and western Finland (Kähkönen, 2005). The Pohjanmaa belt is bordered by the Central Finland Granitoid Complex in the east and the Vaasa migmatite complex in the west. The Kaustinen Li province covers 500 km^2, or even more according to new Li analyses on regional till samples (Kontoniemi, 2013).

The most common rock types within the Kaustinen province are mica schists and mica gneisses, intercalated with metavolcanic rocks. Sedimentation, which took place after 1.92 Ga, was followed by amphibolite facies metamorphism ~1.9 Ga (Williams et al., 2008) with peak metamorphism 1.89–1.88 Ga. U–Pb dating of columbite from the Ullava albite-spodumene pegmatite gave an age of 1.79 Ga, significantly post peak metamorphism (Alviola et al., 2001).

The Li pegmatites of the Kaustinen province belong to the albite-spodumene type and intrude both pyroclastic metavolcanic and metasedimentary rocks, mostly conformably. At least 16 separate Li pegmatite occurrences are known. They are not exposed, and the contact relationship can be seen only in erratic boulders and drill cores. According to Martikainen (2012), the Kaustinen pegmatite granite is the most probable source of the spodumene pegmatites, but to confirm this, age determinations of the granitic pegmatites are still required.

The Länttä Li deposit is an example of a homogenous albite-spodumene pegmatite. It consists of two main boudinized dikes at the contact between metavolcanic rock and graywacke schist. The vertical dikes, with a thickness of up to 10 m, trend northeast and dip 70° to the southeast. The total mineral resource is 1.3 Mt, 1.08 wt% Li_2O with a cutoff value at 0.5 wt% Li_2O (Keliber, 2013).

GTK investigations from 2003–2012 led to the discovery of the Rapasaaret deposit and new resources for the previously known Leviäkangas and Syväjärvi deposits, which are all located in a geologically similar area (Ahtola et al., 2010a,b; Käpyaho et al., 2007; Kuusela et al., 2011). The only visible evidence of the Li pegmatites are glacially transported boulder fans down-ice (to the southeast) of all of the discovered occurrences.

The pegmatites have intruded into mica and graywacke schists and plagioclase-phyric rocks or intermediate volcanic rocks. Locally scheelite-garnet-tremolite skarns and quartz veins are present in the wall rocks. The presence of pyrite and pyrrhotite in fracture zones and disseminations in mica schist cause magnetic and electromagnetic anomalies. With the influx of graphite and sulfides, the schists were transformed into black schists.

Spodumene, albite, quartz, K-feldspar, and muscovite are the main minerals in the Li pegmatites (Ahtola et al., 2010b; Al-Ani and Ahtola, 2008). Spodumene contains 6.5–7.4 wt% Li_2O and 0.1–0.6 wt% FeO and is often altered to muscovite. The accessory minerals are apatite, beryl tourmaline, Li- and Mn-Fe phosphates, graphite, garnet, Nb-Ta oxides, arsenopyrite, cassiterite, sphalerite, zinnwaldite, zeolite, and cookeite. The spodumene content increases toward the center of the dike and the c-axes of spodumene crystals are generally oriented perpendicular to the wall rock contact.

The Leviäkangas spodumene pegmatite dike has a length of 500 m and a thickness of 1–20 m, with a northwest–southeast strike, dipping 40–60° to the west (Ahtola et al., 2010a). The deposit contains 2.1 Mt at 0.70 wt% Li_2O. The Syväjärvi northwest–southeast striking dikes dip 30-40° west, have a thickness of 1–22 m, and resources of 2.6 Mt at 0.98 wt% Li_2O (Ahtola et al., 2010b). The Rapasaaret deposit consists of at least two dike swarms with dike thicknesses ranging from 1 to 24 m. After an extensive till-sampling program, drilling resulted in a resource estimate of 3.7 Mt at 1.02 wt% Li_2O (Kuusela et al., 2011). Also, tantalium, beryllium, and niobium contents are relatively high (Table 9.2.1).

Table 9.2.1 Lithium, tantalium, niobium, and beryllium contents in the Kaustinen spodumene deposits studied by GTK; analyses by ICP-AES

		Li_2O %			Ta_2O_5 ppm			Nb_2O_5 ppm			BeO ppm		
	n	avg.	max	min	avg.	max	min	avg.	max	min	avg.	max	min
Leviäkangas	101	0.74	2.13	0.02	72	337	8	87	312	12	185	494	77
Syväjärvi	200	1.00	2.09	0.03	26	119	4	36	149	11	148	497	67
Rapasaaret	159	1.18	3.36	0.05	53	547	3	58	209	13	502	1912	141

Spodumene pegmatites have also been drilled at the Matoneva, Heikinkangas, and Päiväneva targets (Ahtola et al., 2012). Uninvestigated spodumene pegmatite boulder fans and recent till geochemistry from the region further indicate that there is a great potential for new discoveries (Kontoniemi, 2013; Wik et al., 2013).
Source: *Ahtola et al. (2010a,b) and Kuusela et al. (2011).*

SOMERO-TAMMELA RE PEGMATITES

The Somero-Tammela RE pegmatite area in the Häme belt, southern Finland, is 400 km^2 in total size and comprises at least 56 complex pegmatites enriched in Li, Nb, Ta, Be, Sn, Cs, P, and B. The Häme belt is composed of mafic volcanic rocks and mica schists into which syntectonic gabbro, diorite, granodiorite, late-tectonic microline granite, and finally LCT pegmatites dikes have intruded (Alviola, 2003; Ahtola, 2012). RE pegmatites include lithium silicates and phosphates such as cookeite, elbaite, heterosite-siclerite, lepidolite, lithiophilite, petalite, spodumene, triphylite, and Li-Fe-micas (Vesasalo, 1959; Alviola, 1993).

The largest and best-known Li pegmatites are the Hirvikallio petalite pegmatite and the Kietyömäki spodumene pegmatite. Hirvikallio was drilled by GTK in 1958 and later studied by Lohja Oy/Partek Oy in 1974–1996. The dike is 170 m long, 5–25 m wide, and contains 0.2 Mt with 1.8 wt% LiO_2 to the depth of 50 m. The Hirvikallio petalite pegmatite occurs in a weakness zone along the contact between mica schists and amphibolites (Vesasalo, 1959). Petalite contains 4.74 wt% Li_2O and a very low iron content of 0.01 wt% Fe_2O_3. The Kietyönmäki dike swarms were drilled by GTK in 1987–1988. Petalite occurs only in one dike; elsewhere it has completely been converted to spodumene and quartz by hydrothermal alteration (Alviola, 1993). The main dike contains 0.4 Mt with 1.5% Li_2O, 0.016% Sn, 0.003% Ta. It is highly likely that several undiscovered RE pegmatites still occur in the Somero-Tammela area.

DISCUSSION

The world-class REE deposits, such as Bayan Obo, Mountain Pass, and Mount Weld, are associated with carbonatite complexes. However, the most promising HREE- deposits are closely associated with alkaline and peralkaline rocks such as Norra Kärr in Sweden and Kvanefjeld in Greenland. In Finland, the Korsnäs Pb-REE–bearing dike has similar REE patterns to those in carbonatites, but isotopic evidence proves that it is derived from crustal sources. Late-magmatic, hydrothermal, and weathering processes may have caused the enrichment of REEs in apatite. The Katajakangas Nb-YREE deposit occurs in alkaline gneisses and shows clear enrichment of HREE contained in the minerals zircon, bastnaesite, and columbite.

The mineralogical and chemical studies of the Sokli carbonatite veins indicated that the REE minerals rarely form during the primary crystallization of carbonatite. Postcrystallization hydrothermal solutions, metasomatism, metamorphism, and weathering can all result in remobilization of REEs and changes in the original mineralogy. REE-bearing minerals concentrated especially in late carbonatite veins are ancylite, bastnaesite, monazite, allanite, and REE-apatite. During late-stage processes, apatite and carbonate minerals were partially replaced by various assemblages of REE-Sr-Ba minerals. In the Kovela late-orogenic, S-type granite, monazite is the dominant REE mineral and the accessory minerals are zircon, xenotime, and thorite. Monazite granites of southern Finland, such as the Kovela pluton, may be potential Th and REE sources in the future.

The texture of Fe-Ti oxides is an important factor in evaluating the utilization potential of ilmenite deposits as ilmenite should occur as discrete grains instead of lamellae in magnetite; the latter cannot be separated economically. In the Otanmäki massive Fe-Ti oxide ore, there are almost equal amounts of magnetite and ilmenite, whereas in the Koivusaarenneva deposit, ilmenite is clearly more abundant than magnetite.

Ilmenite is a primary igneous mineral at Koivusaarenneva and Kauhajoki, whereas at Otanmäki and Karhujupukka, discrete ilmenite, and magnetite grains developed by breakdown of the ulvöspinel-magnetite solid solution during regional metamorphism. A typical feature of Otanmäki, Karhujupukka, and Kälviä is that magnetite has a high vanadium concentration (averaging 0.4–0.6 wt% V), whereas at Kauhajoki there is a large amount of apatite (up to 8 wt%) along with Ti-Fe oxides and Fe-Mg-silicates. The low-grade Ti-Fe-P deposits at Kauhajoki may also be potential phosphorus sources for fertilizer industries in the future.

All the Finnish magmatic ilmenite deposits are hosted by Paleoproterozoic mafic intrusive rocks, but the internal structure of deposits is as diverse as their geotectonic setting and age. Karhujupukka (~1.9 Ga) is closely related to a greenstone belt, the Otanmäki (2.05 Ga) gabbro was intruded into an Archean granite-gneiss complex, the Koivusaarenneva (1.88 Ga) igneous body occurs in an intermediate igneous environment below arc-type volcanic rocks, and the host rocks of the Kauhajärvi (1.87 Ga) deposits were emplaced in an environment containing diverse granitoids. However, a common feature of these intrusions is a relatively dry crustal emplacement environment that seems to favor enrichment of Ti and crystallization of ilmenite at the expense of Ti magnetite. This contrasts with Ni-enriched Svecofennian mafic–ultramafic intrusions that were intruded into supracrustal belts and are often closely associated with graphite- and sulfide-rich schists.

The Kaustinen spodumene pegmatites and Somero-Tammela petalite-spodumene pegmatites contain potential Li resources for the battery industry of the European Union countries. At Kaustinen, current known mineral resources are sufficient for several decades. The quality of spodumene and petalite is good owing to high lithium and low iron contents.

The Li pegmatites belong to the LCT family. They are characterized by a significant enrichment in Li, Cs, and Ta, as well as Rb, Be, Sn, B, and F and are fractionation products of S-type, peraluminous granites (Černŷ and Ercit, 2005). The P and T conditions during mineral formation controlled the Li mineralogy in the deposits. At Kaustinen, spodumene crystallized as the first Li mineral during a late magmatic stage. However, the Somero-Tammela Li pegmatites occur in a weakness zone where the pressure was probably lower, and primary magmatic petalite was later converted to spodumene and quartz by hydrothermal processes, except at Hirvikallio.

SUMMARY

Finland has several different potential sources of rare earth elements, namely in the carbonate veins at Korsnäs (0.9 wt% REE) and the alkaline gneiss at Katajakangas, central Finland (2.4 wt% REE); in the fenite zone of the Sokli carbonatite complex (1–2 wt% REE), northern Finland; and in the Kovela monazite-granite in southern Finland (0.5–4.3 wt% REE). The LREE minerals dominate, including ancylite, bastnaesite, monazite, allanite, and apatite.

The Ti potential of small mafic intrusions of Finland is prominent in various geotectonic settings, based on large ilmenite deposits (50 to more than 100 Mt) at Otanmäki and Koivusaarenneva. One economic benefit of the gabbro-hosted deposits is that they also have potential for Fe and V, or P in addition to Ti. The Otanmäki mine was a globally important producer of vanadium and nationally important source of iron for the steel industry. In the future, the Kauhajärvi-type large, low-grade Ti-P-Fe deposits may be important producers of phosphorous in addition to iron and titanium. Gabbro-hosted ilmenite deposits have a low environmental impact because they show low abundances of harmful elements (U, Th, and S) and the wall rocks usually serve as good construction materials.

Recently, the lithium resources of several Li-bearing spodumene pegmatite occurrences at Kaustinen have been estimated by GTK and Keliber Oy. The high potential of the area is supported by areas with boulder fans without known sources and recent Li analyses of regional till samples.

ACKNOWLEDGMENTS

The authors wish to acknowledge professor Eero Hanski and Dr. Hugh O'Brien for constructive reviews that markedly improved the manuscript.

REFERENCES

Ahtola, T., 2012. Somero Li. Geological Survey of Finland, Special Paper 53, 217–218.

Ahtola, T., Kuusela, J., Koistinen, E., et al., 2010a. Report of investigations on the Leviäkangas lithium pegmatite deposit in Kaustinen, Western Finland. Geological Survey of Finland, Report M19/2323/2010/32, p. 59.

Ahtola, T., Kuusela, J., Koistinen, E., et al., 2010b. Report of investigations on the Syväjärvi lithium pegmatite deposit in Kaustinen, Western Finland. Geological Survey of Finland, Report M19/2323/2010/44, p. 49.

Äikäs, O., 1990. Torium-ja niobi-lantanidimalmiaiheet Vuolijoen Otanmäessä. Geological Survey of Finland, Report M19/3431/-90/1/60, p. 28 (in Finnish).

Al-Ani, T., Ahtola, T., 2008. Mineralogy of spodumene pegmatites, Kaustinen, Western Finland. Geologian tutkimuskeskus, arkistoraportti, Geological Survey of Finland, Report M19/2323/2008/6, p. 26.

Al-Ani, T., Grönholm, S., 2011. Ore mineralogy of REE potential Kovela Th granite in Nummi-Pusula, Southern Finland. Geological Survey of Finland, Report 49/2011.

Al-Ani, T., Sarapää, O., 2013. Geochemistry and mineral phases of REE in Jammi carbonatite veins and fenites, southern end of the Sokli complex, NE Finland. Geochemistry: Exploration, Environment 217–224 Analysis 13.

Alviola, R., 1993. Tammelan Kietyömäen litium-esiintymää koskevat tutkimukset vuosina 1985–1993. Geological Survey of Finland, Report M19/2024/-93/1/85, p. 38.

Alviola, R., 2003. Pegmatiittien malmipotentiaalista Suomessa. Geologian tutkimuskeskus, arkistoraportti, Geological Survey of Finland unpublished report M10/-03/1/85, p. 7.

Alviola, R. 2012. Distribution of rare element pegmatites in Finland. In: Lithosphere 2012: Seventh Symposium on the Structure, Composition and Evolution of the Lithosphere in Finland, Espoo, November 6–8: Programme and Extended Abstracts. Institute of Seismology, University of Helsinki, Report S-56. 1–4.

Alviola, R., Mänttäri, I., Mäkitie, H., Vaasjoki, M., 2001. Svecofennian rare-element granitic pegmatites of the Ostrobothnia region, western Finland: their metamorphic environment and time of intrusion. In: Mäkitie, H. (Ed.), Svecofennian granitic pegmatites (1.86–1.79 Ga) and quartz monzonite (1.87 Ga), and their metamorphic environment in the Seinäjoki region, western Finland. Geological Survey of Finland, Special Paper 30, pp. 9–29.

Arzamastsev, A., Yakovenchuk, V., Pakhomovsky, Y., Ivanyuk, G., 2008. The Khibina and Lovozero alkaline massifs: Geology and unique mineralization. 33 IGC excursion No 47 Geological Institute of the Russian Academy of Science, Apatity, Russia. p. 58.

Bedrock of Finland - DigiKP. Digital map database [Electronic resource]. Espoo: Geological Survey of Finland [12.10.2014]. Version 1.0.

Boynton, W.V., 1984. Geochemistry of the rare earth elements: meteorite studies. In: Henderson, P. (Ed.), Rare Earth Element Geochemistry. Elsevier, Amsterdam, pp. 63–114.

Castor, S.B., Hedrick, L.B., 2006. Rare earth elements. In: Kogel, J.E., Trivedi, N.C., Barker, J.M., Krukowski, S.T. (Eds.), Industrial Minerals volume, 7th ed.: Society for Mining, Metallurgy, and Exploration. Colorado, Littleton, pp. 769–792.

Castor, S.B., 2008. Rare earth deposits of North America. Resource Geology 58, 337–347.

Černý, P., Ercit, T.S., 2005. The classification of granitic pegmatites revisited. Canadian Mineralogist 43, 2005–2026.

Chi, R., Tian, J., 2008. Weathered Crust Elution-Deposited Rare Earth Ores. Nova Science Publishers, New York, p. 288.

Endomines, 2015. Exploration and Development of Ilmenite Deposits. Available at http://www.endomines.com/index.php/kaelviae-ilmenite.

Himmi, R., 1975. Outokumpu Oy:n Korsnäsin ja Petolahden kaivosten vaiheita. Vuoriteollisuus–Bergshanteringen 33, 35–38 (in Finnish with English abstract).

Hugg, R., 1985. Katajakangas. Geologinen malmiarvio. Report OU 12/85, Rautaruukki Oy Malminetsintä, p. 13.

Hugg, R., Heiskanen, V., 1986. Otanmäen alueen niobi-lantaniditutkimukset, tilanne 31.12.1985 (Lisäys 15.5.1986 L. Pekkarinen). Report OU 28/85, Rautaruukki Oy Malminetsintä, p. 14.

Illi, J., Lindholm, O., Levanto, U.-M., et al., 1985. Otanmäen kaivos. Summary: Otanmäki mine. Vuoriteollisuus 43, 98–107.

Isokangas, P., 1975. The mineral deposits in Finland. Licentiate thesis. Helsinki University, Department of Geology, p. 128.

Kähkönen, Y., 2005. Svecofennian supracrustal rocks. In: Lehtinen, M., Nurmi, P.A., ja Rämö, O.T. (Eds.), Precambrian Geology of Finland—Key to the Evolution of the Fennoscandian Schield. Elsevier, Amsterdam, pp. 343–406.

Käpyaho, A., Ahtola, T., Alviola, R., 2007. Whole-rock geochemical characteristics of a lithium pegmatite dike: a case example from Vintturi (Kaustinen), W-Finland. August 20–23. In: Digging Deeper: Proceedings of the 9th Biennial Meeting of the Society for Geology Applied to Mineral Deposits. Irish Association for Economic Geology, Dublin, pp. 1493–1496.

Kärkkäinen, N., Appelqvist, H., 1999. Genesis of a low-grade apatite-ilmenite-magnetite deposit in the Kauhajärvi gabbro, western Finland. Mineralium Deposita 34 (8), 754–769.

Kärkkäinen, N.K., Bornhorst, T.J., 2003. The Svecofennian gabbro-hosted Koivusaarenneva magmatic ilmenite deposit, Kälviä, Finland. Mineralium Deposita 38 (2), 169–184.

Kärkkäinen, N., 2012a. Koivusaarenneva Ti. Geological Survey of Finland, Special Paper 53, 261–266.

Kärkkäinen, N., 2012b. Peräkorpi Ti. Geological Survey of Finland, Special Paper 53, 230–231.

Karvinen, A., Kojonen, K., Johanson, B., 1989. Geology and mineralogy of the Karhujupukka Ti-V-Fe deposit in Kolari, northern Finland. In: Geological Survey of Finland. Current Research 1988. Geological Survey of Finland, Special Paper 10, pp. 95–99.

Keliber, 2013. Geology and deposits: Länttä. http://Keliber.fi/lantta.

Kontinen, A., Huhma, H., Lahaye, Y., O'Brien, H., 2013. New U-Pb zircon age, Sm-Nd isotope and geochemical data for Otanmäki suite granites in the Kainuu area, Central Finland. Geological Survey of Finland, Report of Investigation 198, 65–69.

Kontoniemi, O., 2013. Kaustisen alueen Li-potentiaali – vanhojen moreeninäytteiden uudelleenanalysointi, vaihe 2. 9 s. +4 liitekarttaa Geologian tutkimuskeskus. Geological Survey of Finland, unpublished report 52/2013, p. 9 (in Finnish).

Korsakova, M., Krasotkin, S., Stromov, V., et al., 2012. Metallogenic areas in Russian part of the Fennoscandian Shield. Geological Survey of Finland, Special Paper 53, 343–395.

Korsman, K., Koistinen, T., Kohonen, J., et al. (Eds.), 1997. Bedrock Map of Finland 1:1,000,000. Geological Survey of Finland.

Kramm, U., Kogarko, L.N., Kononova, V.A., Vartiainen, H., 1993. The Kola alkaline province of the CIS and Finland: Precise Rb-Sr ages define 380–360 Ma age range for all magmatism. Lithos 30, 33–44.

Kuivasaari, T., Torppa, A., Äikäs, O., Eilu, P., 2012. Otanmäki V-Ti-Fe. Geological Survey of Finland, Special Paper 53, 284–286.

Kuusela, J., Ahtola, T., Koistinen, E., et al. 2011. Report of investigations on the Rapasaaret lithium pegmatite deposit in Kaustinen-Kokkola, Western Finland. Geological Survey of Finland, report 42/2011.

Lehtonen, M.I., Kujala, H., Kärkkäinen, N., et al., 2005. Etelä-Pohjanmaan liuskealueen kallioperä. Summary: Pre-Quaternary rocks of the South Ostrobothnian schist belt. Geologian tutkimuskeskus Tutkimusraportti 158, p. 155, plus 1 app. map.

Lindholm, O., Anttonen, R., 1980. Geology of the Otanmäki mine. In: Häkli, T.A. (Ed.), Precambrian ores of Finland, Proceedings on the 26th International Geological Congress, Guide to Excursions 078A+C, Part 2 (Finland). Geological Survey of Finland, pp. 25–33.

Linnen, R.L., Van Lichtervelde, M., Cérny, P., 2012. Granitic pegmatites as sources of strategic metals. Elements 8, 275–280.

Lounamaa, N., 1972. Utvinning av lantanider. Kemisk Tidskrift 5, 40–42.

Long, K.R., Van Gosen, B.S., Foley, N.K., Cordier, D., 2010. The principal rare earth elements deposits of the United States—A summary of domestic deposits and a global perspective. U.S. Geological Survey Scientific Investigations Report 2010, p. 96.

Mariano, A.N., Mariano Jr., A., 2012. Rare earth mining and exploration in North America. Elements 8, 369–376.

Marmo, V., Hoffren, V., Hytönen, K., et al., 1966. On the granites of Honkamäki and Otanmäki, Finland, with special reference to the mineralogy of accessories. Bulletin de la Commission Géologique de Finlande 221, p. 34.

Martikainen, A., 2012. Kaustisen–Ullavan litiumpegmatiittien alueelliset geokemialliset piirteet ja lähdegranitoidit. Unpublished Master's thesis, University of Helsinki, p. 77.

Mitchell, R.H., 2005. Carbonatites and carbonatites and carbonatites. The Canadian Mineralogist 43, 2049–2068.

Murphy, P., Frick, L., 2006. Titanium. In: Kogel, J.E., Trivedi, N.C., Barker, J.M., Krukowski, S.T. (Eds.), Industrial Minerals volume, 7th ed.: Society for Mining, Metallurgy, and Exploration. Colorado, Littleton, pp. 987–1003.

O'Brien, H.E., Peltonen, P., Vartiainen, H., 2005. Kimberlites, carbonatites, and alkaline rocks. In: Lehtinen, M., Nurmi, P., Rämö, O.T. (Eds.), Precambrian Geology of Finland—Key to the Evolution of the Fennoscandian Shield, pp. 605–644.

Pääkkönen, V., 1956. Otanmäki, the ilmenite-magnetite ore field in Finland. Bulletin de la Comission Géologique de Finlande 171, p. 71.

Papunen, H., 1986a. Suomen metalliset malmiesiintymät. In: Papunen, H., Haapala ja, I., Rouhunkoski, P. (Eds.), Suomen malmigeologia: metalliset malmiesiintymät. Geological Society of Finland. 133–214 (in Finnish).

Papunen, H., 1986b. One hundred years of ore exploration in Finland. In: The Development of Geological Sciences in Finland, Bulletin of the Geological Society of Finland, 336, pp. 165–203.

Papunen, H., Lindsjö, O., 1972. Apatite, monazite and allanite: three rare earth minerals from Korsnäs, Finland. Bulletin of the Geological Society of Finland 44 (2), 123–129.

Peltonen, P., 2005. Svecofennian mafic-ultramafic intrusions. In: Lehtinen, M., Nurmi, P., Rämö, O.T. (Eds.), Geology of Finland—Key to Evolution of the Fennoscandian Shield. Elsevier, Amsterdam, pp. 407–442.

Puumalainen, V.-M., 1986. Otanmäen liuskealueen ja sen ympäristön kivilajit ja rakenne. Pro gradu-tutkielma. University of Oulu, Department of Geology, p. 53 (in Finnish).

Puustinen, K., 2003. Suomen kaivostoiminnan varhaiset vaiheet. Geologi 55 (9–10), 254–257.

Rämö, O.T., Vaasjoki, M., Mänttäri, I., et al., 2001. Petrogenesis of the post-kinematic magmatism of the Central Finland Granitoid Complex I, Radiogneic isotope constraints and implications for crustal evolution. Journal of Petrology 42, 1972–1993.

Rudnick, R.L., Gao S., 2003 Composition of the continental crust. In: R.L. Rudnick (Ed.), The Crust, vol. 3 Treatise on Geochemistry (eds. H.D. Holland, K.K. Turekian). Elsevier-Pergamon, Oxford, pp. 1–64.

Sarapää, O., Al-Ani, T., Lahti, S., et al., 2013. Rare earth exploration potential in Finland. Journal of Geochemical Exploration 133, 25–41.

Sarapää, O., Reinikainen, J., Seppänen, H., et al., 2001. Industrial minerals exploration in southwestern and western Finland. Geological Survey of Finland, Special Paper 31, 31–40.

Sarapää, O., Kärkkäinen, N., Chernet, T., et al., 2005. Exploration results and mineralogical studies on the Lumikangas apatite-ilmenite gabbro, Kauhajoki, western Finland. Geological Survey of Finland, Special Paper 38 31–40.

Selway, J.B., Breaks, F.W., Tindle, A.G., 2005. A review of rare-element (Li-Cs-Ta) pegmatite exploration techniques for the Superior Province, Canada, and large worldwide tantalum deposits. Exploration and Mining Geology 14, 1–30.

Shand, S.J., 1943. The Eruptive Rocks, 2nd ed. John Wiley, New York, p. 444.

Talvitie, J., Paarma, H., 1980. Precambrian basic magmatism and the Ti-Fe ore formation in central and northern Finland. Bulletin of the Geological Society of Finland 307, 98–107.

Torppa, A., Karhu, J.A., 2013. Stable isotope and trace element constraints to the origin of carbonate rocks in the Korsnäs Pb-REE deposit, western Finland. Proceedings of the 12th SGA Biennial Meeting 4, 1746–1749.

USGS, 2013. Mineral commodity summaries 2013. Available at http://minerals.usgs.gov/minerals/pubs/mcs/2013/mcs2013.pdf.

Vaasjoki, M., 1981. The lead isotopic composition of some Finnish galenas. Bulletin of the Geological Society of Finland 316, 30.

Vartiainen, H., 1980. The petrography, mineralogy and petrochemistry of the Sokli carbonatite massif, northern Finland. Bulletin of the Geological Society of Finland 313, 126.

Vartiainen, H., 2001. Sokli carbonatite complex, northern Finland. Res Terrae. Series A 20, University of Oulu. 8–24.

Vesasalo, A., 1959. On the petalite occurrences of Tammela, SW-Finland. Bulletin de la Comission Géologique de Finlande 31, 59–74.

Oy, Vuorokas, 2013. Otanmäki mine. available at www.otanmaki.fi/.

Williams, I.S., Rutland, R.W.R., Kousa, J., 2008. A regional 1.92 Ga tectonothermal episode in Ostrobothnia, Finland: Implications for models of Svecofennian accretion. Precambrian Research 165, 15–36.

Wik, H., Laxström, H., Ahtola, T., Kuusela, J., 2013. Emerging Li-potential in Kaustinen. Mineral deposits research for a high-tech world. Proceedings of the 12th SGA Biennial Meeting Vol. 4, 1750–1753.

Wu, C., 2008. Bayan Obo controversy: Carbonatites versus iron oxide-Cu-Au-(REE-U). Resource Geology 58 (4), 348–354.

CHAPTER

THE TAIVALJÄRVI Ag-Au-Zn DEPOSIT IN THE ARCHEAN TIPASJÄRVI GREENSTONE BELT, EASTERN FINLAND

9.3

T. Lindborg, H. Papunen, J. Parkkinen, I. Tuokko

ABSTRACT

The Taivaljärvi Ag-Au-Zn deposit is associated with felsic metavolcanic rocks in the Archean Tipasjärvi greenstone belt of eastern Finland. The current owner, Sotkamo Silver Oy, has named it the Silver Mine deposit. The deposit is stratiform and consists of four mineralized layers: the A, B, C, and D bodies. The deposit contains 3.34 Mt of ore reserves (according to the JORC 2012 code) with 0.71% Zn, 0.34% Pb, 102 g/t Ag, and 0.29 g/t Au, as part of 3.5 Mt measured and indicated resource. The content of sulfide ore minerals is low, 5% on average. The main ore minerals are sphalerite, galena, chalcopyrite, pyrite and pyrrhotite, dyscrasite, freibergite, electrum, and native Ag. The deposit displays disseminated and vein-type ore textures. The volcanic host rocks are deformed and metamorphosed at upper amphibolite facies conditions, and the metamorphic mineral assemblages reflect the variation in chemical compositions of the rock sequence. The intense hydrothermal alteration associated with mineralization predated the regional metamorphism, but it can still be recognized in the chemical compositions of the rock sequence. The $\delta^{34}S$ values fluctuate from 2.0–3.8 per mil, indicating a volcanic origin of sulfur. The deposit is a product of hydrothermal activity associated with volcanoclastic felsic eruption, and the type of deposition is assumed to be epithermal in character.

Keywords: Taivaljärvi; Sotkamo; silver; epithermal; Archean; greenstone.

INTRODUCTION

The Taivaljärvi Silver Mine Ag-Au-Zn deposit in eastern Finland is located in the Archean Tipasjärvi greenstone belt (Fig. 9.3.1), in the southern part of a north-trending, about 200-km long and from 3–5 km wide, Tipasjärvi-Kuhmo-Suomussalmi greenstone complex that is a part of the Archean Kianta terrane (Sorjonen-Ward and Luukkonen, 2005). Several studies discuss the geology and geochemistry of the Tipasjärvi belt (e.g., Vartiainen, 1970; Blais et al., 1978; Martin et al., 1983; Taipale, 1983; Martin and Querre, 1984; Luukkonen, 1985, 1988; Piirainen, 1985; Luukkonen and Lukkarinen, 1986; Sorjonen-Ward and Luukkonen, 2005; Papunen et al., 2009; Pietikäinen et al., 2008). Isotope geology and radiometric ages of the area are included in the works of Vaasjoki et al. (1999) and Huhma et al. (2012a).

Mineral Deposits of Finland. http://dx.doi.org/10.1016/B978-0-12-410438-9.00024-8

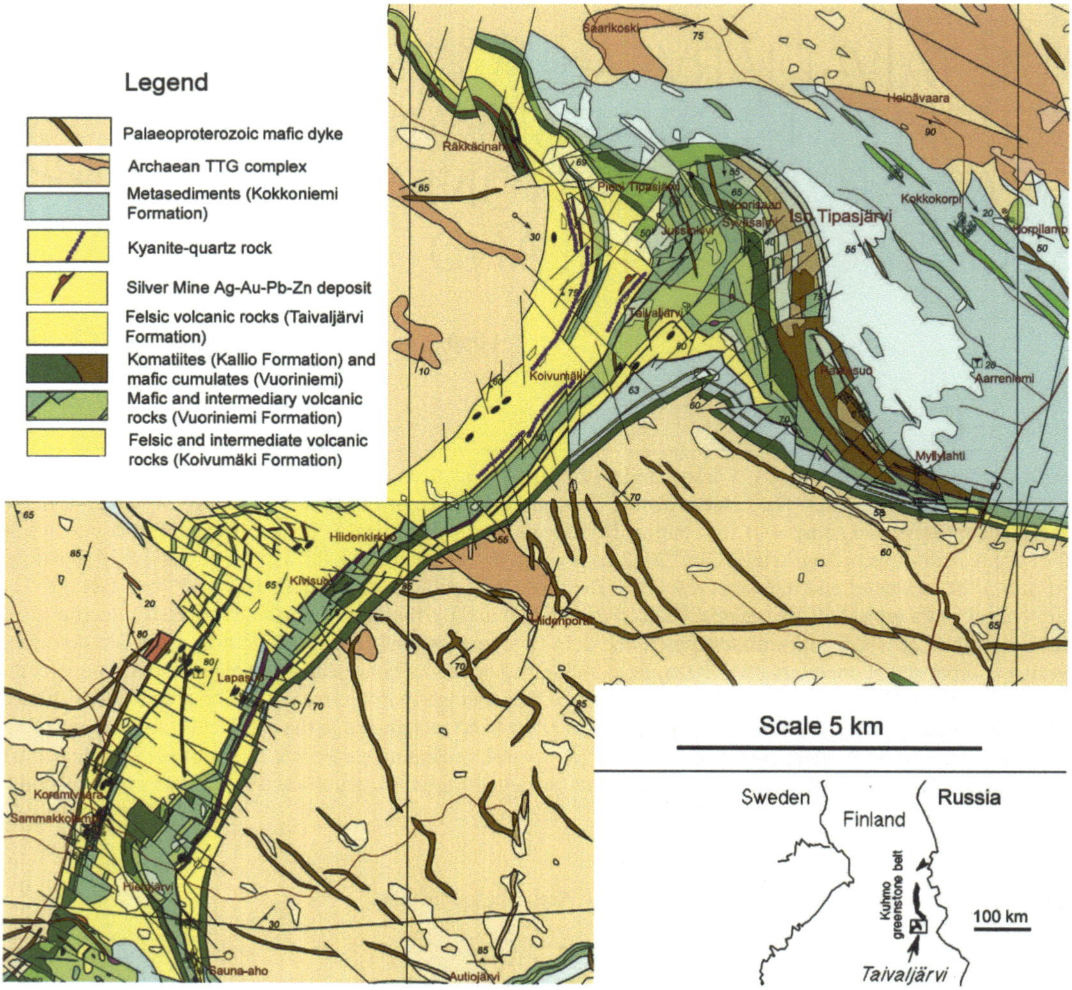

FIGURE 9.3.1 Geology of the Tipasjärvi area and location of the Taivaljärvi Silver Mine deposit.

Source: *The map is from Pietikäinen et al. (2008), with the legend and location of quartz-kyanite rocks added by the authors.*

As late as in the 1970s, the geology and metallogeny of the Archean areas in Finland were poorly known. To initiate further mineral exploration, the Ministry of Trade and Industry financed a research program at Oulu University targeted to basic geological and geochemical studies on the Finnish Archean greenstone belts. During the 1980 summer field season, several sphalerite- and galena-bearing glacial boulders were discovered in the Tipasjärvi area, and shortly thereafter a mineralized subcrop was found in the up-ice direction from the glacial float. At the time of this discovery, the exploration department of Kajaani Oy conducted a geochemical exploration program in the Tipasjärvi area and noted local Zn anomalies in the till. A combination of the boulder and mineralized outcrop information with the geochemical anomaly data guided Kajaani Oy to start diamond drilling, which soon led to the discovery of the deposit. The 400-m-long and up to 110-m-wide subcrop of the Taivaljärvi Zn-Au-Ag ore body was nearly completely covered with till and bog.

In 1988, the prospect was transferred to a joint venture between UPM-Kymmene and Outokumpu Mining Oy, called *Taivalhopea*. From 1988–1991, the Joint Venture (JV) performed additional inventory drilling, concentration tests, mine planning, and completed a feasibility study. To test the beneficiation processes, the JV also constructed a 2600-m-long production decline down to a depth level of 350 m and opened a ventilation shaft to the 340-m level. Due to low metal prices in 1991, the project was put on hold.

In 2005, Silver Resources Oy (at present, Sotkamo Silver Oy) was established. The company acquired the mineral rights and all previous investigation material. It performed additional surface and underground drilling, which now totals approximately 51 km. The company also performed pilot processing tests with excellent results and completed the mineral resource estimate. The mining license was granted in 2011, the environmental permit in 2013, and the detailed technical mine planning is now underway. Production is anticipated to start in 2016. With updated mineral resources, the present owner named the occurrence the Silver Mine deposit.

The area around the Silver Mine deposit is potential for further exploration. From 2005–2007, the Geological Survey of Finland (GTK) performed an extensive exploration program in the Tipasjärvi greenstone belt (Fig. 9.3.1; Pietikäinen et al., 2008). It pinpointed a number of prospects in the voluminous felsic rocks along the western margin of the Tipasjärvi greenstone belt. The Kokkokorpi gold occurrence, the Sauna-Aho zinc occurrence, and the Hiidenkirkko Zn-rich ore boulders and geochemical gold anomalies were considered as the most promising.

REGIONAL GEOLOGY

The Archean granitoid area around the greenstone belt (Fig. 9.3.1) is heterogeneous, varying from stromatic and nebulitic migmatitic gneisses to discrete felsic plutons from tonalite to trondhjemite and monzogranite in composition (Sorjonen-Ward and Luukkonen, 2005). Traditionally, the tonalite-trondhjemite-gneiss (TTG) and granodiorite-granite associations, which surround the Tipasjärvi greenstone belt, have been considered as the basement of greenstone-belt rocks, although they have been reworked and remobilized during metamorphism and deformation (e.g., Taipale, 1983; Papunen et al. 2009; Fig. 9.3.2). However, single-crystal zircon and monazite U-Pb analyses performed with secondary ion mass spectrometry (SIMS) and laser ablation multi-collector inductively coupled plasma mass spectrometry (LA-MC-ICPMS) together with multigrain thermal ionization mass spectrometry (TIMS), have recently become available from the greenstone belts (Huhma et al., 2012a). It is now evident that the bulk of granitoids are younger than the greenstone belt. Signs of older crust have been obtained from migmatitic gneisses at Lylyvaara, east of the Kuhmo greenstone belt (Luukkonen, 1985; Käpyaho et al., 2007). The Sm-Nd data available on the greenstone belt and most granitoid rocks suggest a relatively short crustal residence time for the granitoids (Huhma et al., 2012b).

The greenstone belt was metamorphosed and deformed in several phases, the youngest phase being Proterozoic. Although the rock-forming mineral assemblages are metamorphic, the primary textures are preserved to such an extent that the primary character of the rocks can still be determined. About 80% of the greenstone belt is composed of mafic, ultramafic, and felsic volcanic rocks. The rest is metasedimentary rocks, mica gneisses, banded iron formation, black schists, and mafic–ultramafic intrusive rocks (Piirainen, 1985).

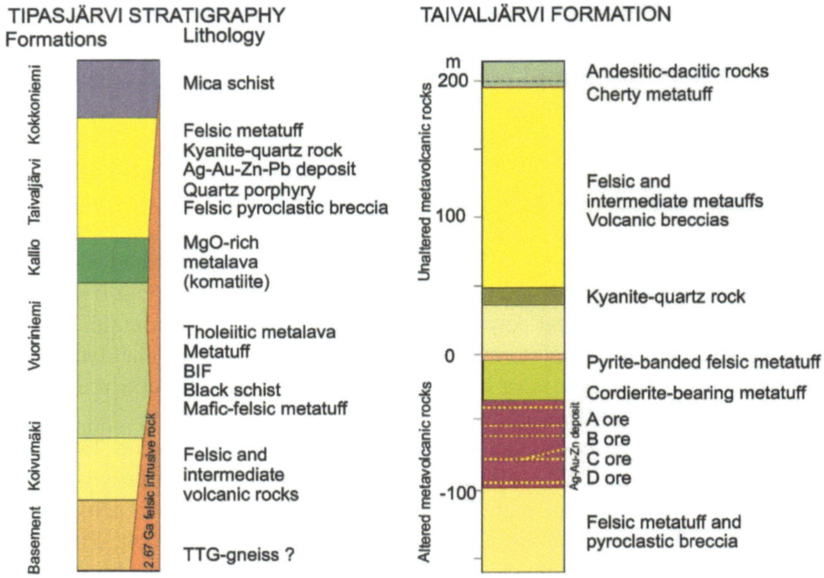

FIGURE 9.3.2 Stratigraphic scheme of the Tipasjärvi greenstone belt and lithologic succession of the Taivaljärvi formation.

The Tipasjärvi section of the greenstone belt was traditionally divided into four lithostratigraphic formations (Taipale, 1983; Papunen et al., 1989, 2009). The lowermost formation, mainly composed of felsic metatuffs, is called the Koivumäki formation, where also the Taivaljärvi felsic volcanics were included (Fig. 9.3.2). The Vuoriniemi formation overlies it, and its lithologic composition varies within a broad range from tholeiitic metatuffs and lavas to banded iron-formation, graphitic black schists, rare dacitic metatuffs, and minor amounts of mafic metalavas. The Kallio formation, characterized by ultramafic komatiitic lavas, overlies the Vuoriniemi formation. The uppermost Kokkoniemi formation is composed of mica schists interpreted as weathering products of the underlying volcanic rocks. This stratigraphic scheme was based on field observations on structures of stratified lithologic units and interpretation of the successive deformation phases (Luukkonen, 1985).

The present (TIMS) U-Pb ages from zircons (Huhma et al., 2012a) set the limits for the stratigraphic succession. The oldest unit dated at 2828 ± 3 Ma is the felsic Koivumäki formation dated from samples from the southern part of the Tipasjärvi belt (Huhma et al., 2012a), and the maximum age of the Kokkoniemi sedimentary unit is set at 2.75 Ga. However, a quite extensive felsic volcanic unit of the belt is dated at 2798 ± 2 Ma, and this unit is considered to be younger than the mafic–ultramafic volcanic rock suite. It terminated the volcanic succession of the greenstone belt and this age group also includes the felsic volcanics of the Taivaljärvi area that host the Ag-Au-Zn deposit. In the stratigraphic scheme, this radiometric age places this felsic formation between the mafic and ultramafic Kallio formation and the sedimentary Kokkoniemi formations and in the present context it is tentatively called the Taivaljärvi formation.

The rocks of the Taivaljärvi formation are either massive quartz-porphyries or, more commonly, volcanic breccias, as well as layered felsic tuffs and tuffites indicating shallow water or subaerial eruption (Taipale, 1983). The layering of the folded sequence is now subvertical to steeply dipping eastwards, and the few available observations of graded bedding in drill cores indicate an eastward younging direction. The Silver Mine Ag-Au-Zn deposit is located in the middle part of the volcanic succession where a number of quartz veins characterize the ore zone. Papunen et al. (1989) divided the deposit into four mineralized layers with different base and precious metal ratios. These layers are roughly parallel to primary stratigraphic layering. However, Parkkinen (2012) indicated in the mining area that the sequence is intensely isoclinally folded, and that the fold structures may control the metal ratios of the deposit.

Geochemistry and mineralogy of the host rocks prove potassic and magnesian hydrothermal alteration. An extensive quartz-kyanite rock layer is present some 100 m stratigraphically above the mineralized zone. The felsic rocks between the silver deposit and the layer of quartz-kyanite rock are K-Mg altered, whereas such alteration is not present in the volcanic rocks stratigraphically above (to the east of) the quartz-kyanite rock. The quartz-kyanite rock is interpreted as the metamorphic equivalent of an intensely leached cap-rock developed in an epithermal hydrothermal system (Papunen et al., 1989). The upper contact of the Taivaljärvi formation is tectonic, and the intense isoclinal folding and overthrusting has brought the older greenstone rocks into the upper contact of the Taivaljärvi succession.

The basal contact of the Taivaljärvi formation with the underlying TTG gneisses is not exposed in outcrop, but drilling reveals that the felsic fragmentary succession is underlain by homogenous felsic porphyry, which grades transitionally into TTG granitoid.

The Juurikkaniemi felsic volcanics in the southern part of the Kuhmo greenstone belt 23 km to the north of Taivaljärvi are lithologically very similar to the Taivaljärvi formation. The multigrain U-Pb zircon age of the felsic volcanic rock of Juurikkaniemi (Katerma) is 2798 ± 15 Ma (Hyppönen, 1983).

THE SILVER DEPOSIT AS A PART OF THE TAIVALJÄRVI FORMATION

LITHOLOGY OF THE FOOTWALL SEQUENCE

Graded bedding with an eastward stratigraphic top direction was locally observed in the drill cores of fine-grained volcanoclastic rocks. This indicates that the basement of the sequence is to the west of the deposit. The footwall of the ore consists of a homogenous, weakly mineralized and, at least, 150-m-thick unit of felsic volcanic rock (metarhyolite) containing quartz, sericite, and biotite as main minerals with thin bands containing garnet and tremolite and locally ankerite together with sulfides. Accessory minerals are chlorite, epidote, cordierite, and rutile. The texture is porphyritic with quartz as phenocrysts (2–4 mm). To the west of the deposit, the footwall rock is intensely foliated but northward it is more massive.

THE ORE DEPOSIT

The till-covered subcrop of the deposit is 400 m long and from 5–110 m wide, averaging 40 m. The deposit dips at 65° to the southeast, plunges 60° to the south-southwest, and drilling-indicated ore extends to a depth of at least 600 m (Fig. 9.3.3). Geophysical deep-penetrating "Sampo" surveys, conducted in spring 2013 by GTK (Niskanen, 2013), indicate that the mineralized zone continues down to 1.5 or even a 2 km depth. This interpretation and the synthesis of geophysics and structural observations of the depth extension were done by Parkkinen (2013).

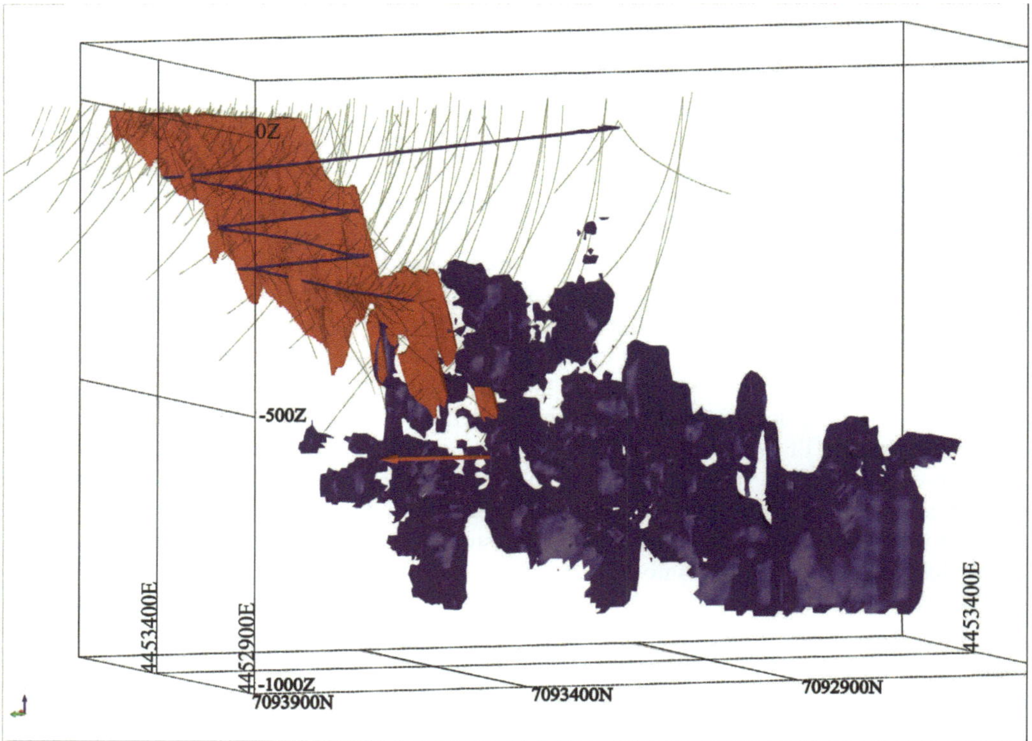

FIGURE 9.3.3 3D model of the Taivaljärvi Silver Mine deposit looking southeast.

Measured, indicated, and inferred resource is red, while cyan is the extension of mineralization inferred with Sampo soundings. Mine decline is blue. Green thin lines indicate drill hole traces. A 200-m-long red arrow points to the north. Finnish National Coordinate System (KKJ) in meters.

Source: *Modified from Niskanen (2012).*

Wardell Armstrong International (2012) performed for Sotkamo Silver Oy a feasibility study of the deposit, and Parkkinen (2013) updated the estimate of mineral resources according to the Australian Code for reporting of exploration results, mineral resources and ore reserves (JORC-code 2012); on this basis Engtec (2014) updated the feasibility study. With a cutoff grade of 50 g/t Ag (eq), the deposit contains 3.52 Mt of measured and indicated resource with 0.71% Zn, 0.34% Pb, 102 g/t Ag, and 0.29 g/t Au. An additional inferred resource contains 1.5 Mt with equal composition. The total inferred mineral resource calculated according to the national instrument for the standards of disclosure for mineral projects within Canada (NI-43101) with a cutoff grade of 30 g/t Ag is 13 Mt of mineralized rock with 0.5% Zn, 0.2% Pb, 65 g/t Ag, and 0.2 g/t Au. Ag accounts for 75–80% of the total commercial value of the deposit, the rest being due to Au, Zn, and Pb.

According to Kopperoinen and Tuokko (1988) and Papunen et al. (1989), the heterogeneous mineralized section is composed of several parallel mineralized layers named from the footwall upward the D, C, B, and A layers, each having specific compositions and metal ratios (Fig. 9.3.4). However, they vary

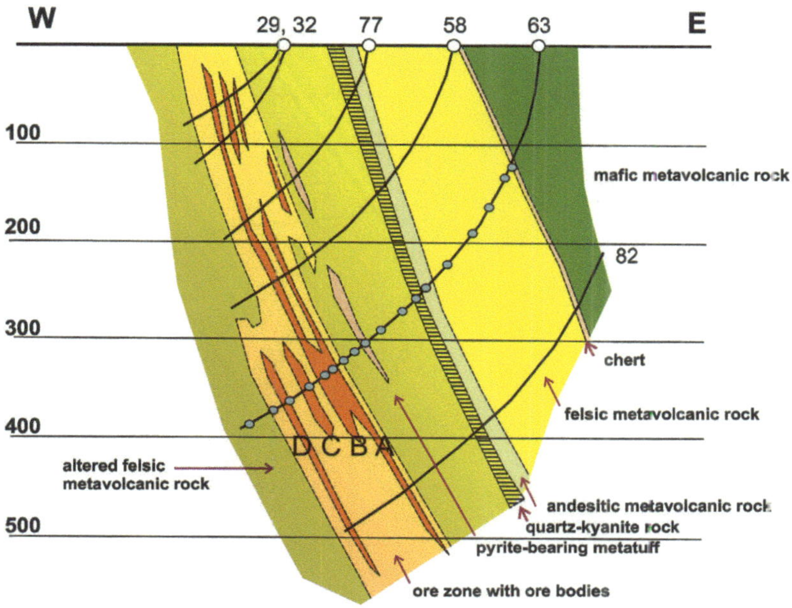

FIGURE 9.3.4 A section across the Taivaljärvi Silver Mine deposit looking northeast.

Locations of the analyzed samples in Table 9.3.1 indicated in the drill core 63 trace.

laterally in thickness and locally join up with the neighboring zones. Parkkinen (2012, 2013) interpreted the deposit as a tightly folded single layer (Fig. 9.3.5), but recently (CTS Engtec, 2014) he suggested a modification of the model with six layers (Fig. 9.3.6). The model needs further justification. The rocks between the mineralised layers are not are not totally barren, but contain a weak dissemination of ore minerals, and the mineable ore contains parts of the entire mineralized section (Fig. 9.3.6). Thick zones are more heterogeneous than thin ones. In addition to quartz and sericite, the felsic metavolcanic host rock close to the footwall contains abundant garnet and biotite, but the abundance of these minerals diminishes toward the hanging wall, and the rock passes to light-colored quartz-sericite rock. Abundant carbonate minerals, forming both disseminations and veins, characterize the deposit (Fig. 9.3.7(A)). Accessory minerals are chlorite, tourmaline, rutile, barite, epidote, and tremolite, which occur locally as bands.

The main layers forming the mineable ore are the A and B zones. The A layer is stratigraphically the uppermost mineralized zone. The thickness varies from 4 to 8 m but locally attains 15 m. It is discontinuous and has a break at a depth of 150–200 m. At the northern edge, close to the surface, the body is banded and Zn-Pb rich with a high amount of iron sulfides. As a whole, however, the ore body is Ag rich, has lower tenors of Zn and Pb compared to other mineralized layers, and contains abundant quartz and ankerite veins.

The B layer is 4–8 m thick, locally attaining a thickness of 16 m. On the surface it is relatively uniform, has a minor break at a depth of 50–100 m, and then continues to a depth of 500 m. The host rock is banded quartz-sericite schist with abundant quartz-carbonate veins (20–30%). Sphalerite and galena occur as banded dissemination together with pyrite and pyrrhotite in quartz-ankerite veins. Fig. 9.3.7(B) represents these structures in the test tunnel mined through B ore layer at level +150.

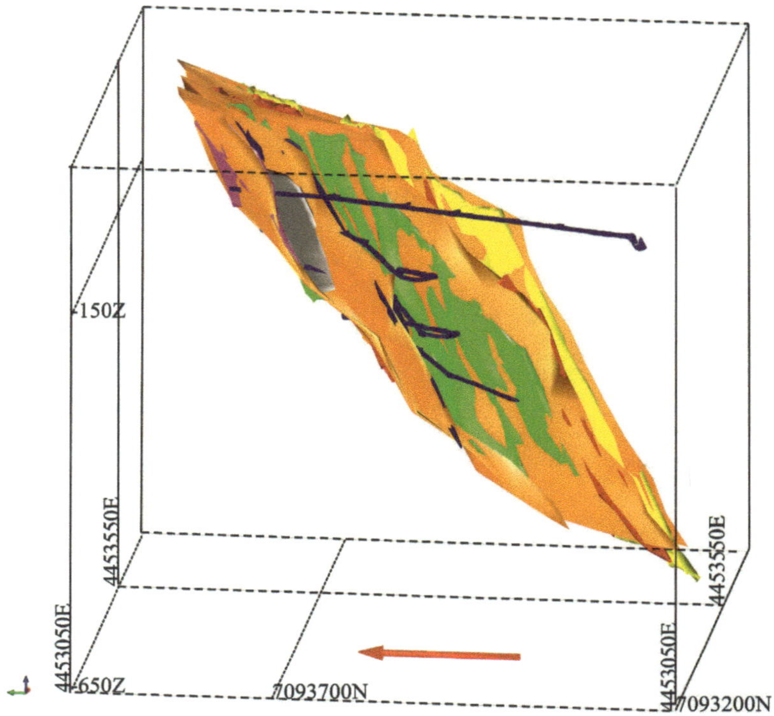

FIGURE 9.3.5 3D model of the isoclinally folded pattern of the Taivaljärvi Silver Mine ore deposit looking east-northeast.

The anticipated six ore layers are indicated with purple, gray, blue, green, red, and yellow colors and the proposed fold with brown color; mine decline is blue. The red arrow represents 200-m length and points to the north. Finnish National Coordinate System (KKJ) in meters.

Source: *From Parkkinen (2012).*

The C layer is on average 5 m thick, but is not as widespread as the other zones. Here the sulfides occur mostly as thin bands. Stratigraphically, the lowermost mineralized D layer is up to 5 m thick and characterized by calc-silicate bands and relatively abundant sulfide minerals, sphalerite, pyrrhotite, chalcopyrite, galena, and pyrite. The main gangue minerals are quartz, carbonates, tremolite, garnet, and biotite.

Deformation has been limited to intense small-scale folding and pinch and swell structures of the layers. Probably simultaneous to the folding, a generation of quartz veins was formed, in places mineralized. This phase was followed by brittle deformation with another generation of barren quartz veins. The geometry of quartz-bearing veins, faults, and shear fractures, which deviate from the directions of bedding and foliation, may indicate a minimum principal stress in northeast-southwest direction (Fig. 9.3.8), or they may also reflect the predeformation orientation of hydrothermal quartz veins across the primary layering.

A structural analysis was focused on the geometry of bedding, foliation, and fold axis observations measured in the 2.5-km-long mine decline in 1990–1991. According to Parkkinen (2012, 2013), the

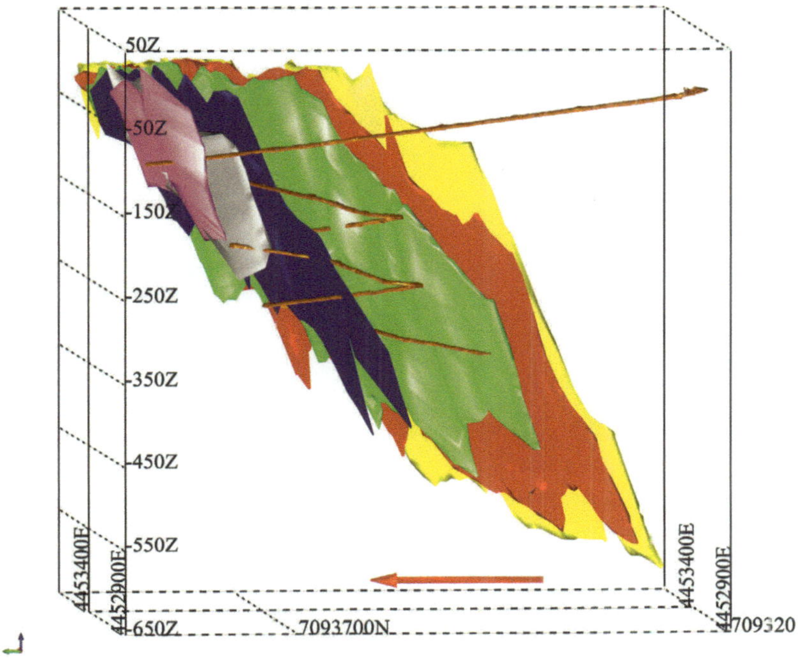

FIGURE 9.3.6 3D model of the Taivaljärvi Silver Mine looking northeast.

Six ore layers, indicated with different colors, and the mine decline (brown). The red arrow represents 200-m length and points to the north. Finnish National Coordinate System (KKJ) in meters.

structural elements give a uniform and simple geometry illustrated in stereographic projections (Fig. 9.3.8). They indicate that the latest foliation phase (local S_2) both parallels the bedding and also forms on average a 10° angle clockwise to bedding. This might be interpreted as sinistral folding within the deposit.

The structural analysis proves that the deposit follows the direction of lineation, and that that direction has a major potential to find extensions of the deposit (Parkkinen, 2010, 2012). Geophysical deep-penetrating Sampo surveys, which GTK conducted in 2011 (Niskanen, 2012) and 2013 (Niskanen, 2013), confirmed the idea, and the potential for the deposit to extend to the depth of 1.5 km (Parkkinen, 2013).

LITHOLOGY OF THE HANGING WALL

The hanging-wall contact of the deposit is almost a plane. It is overlain by a quartz-sericite rock with quartz phenocrysts. Carbonates, which are the main minerals in the ore zone, are very rare above the ore zone. Some layers contain local, ghostlike, fragments suggestive of lapilli tuff or agglomerate. Iron sulfide bands and veins are common, having local tenors of Zn, Pb, and Ag slightly above the background level. Cordierite is present as pinitized porphyroblasts in some layers.

A layer above the ore deposit proper is characterized by abundant pyrite and pinitized cordierite porphyroblasts, and is marked with a zero in the stratigraphic column (refer to Figs 9.3.2 and to 9.3.9

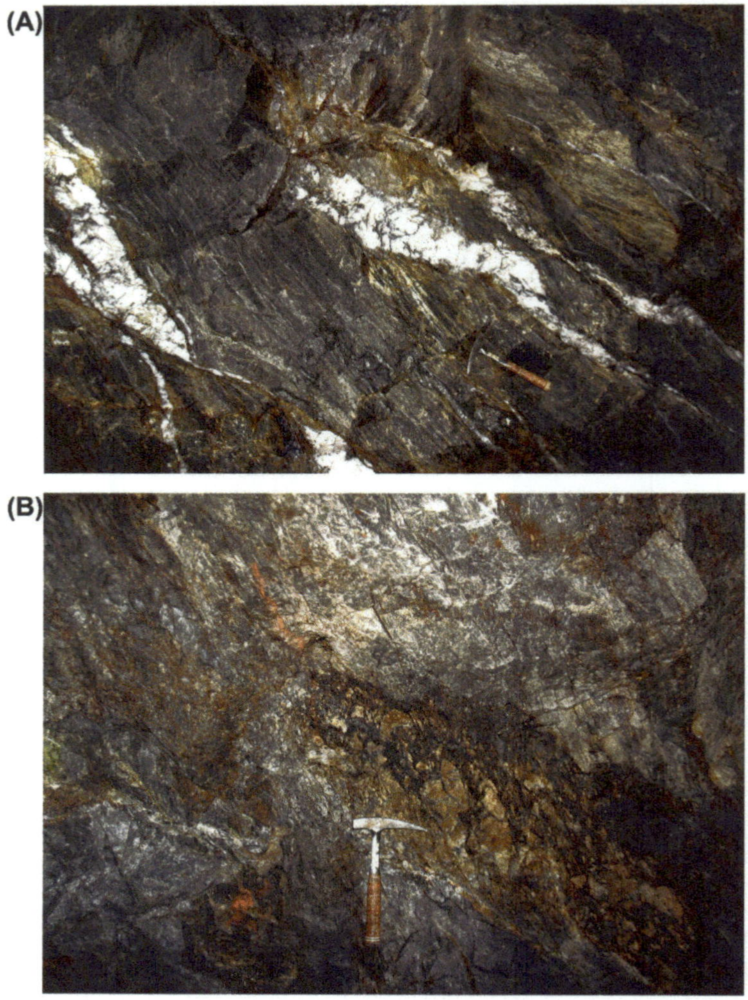

FIGURE 9.3.7 Structure of the ore deposit.

(A) Ore in test tunnel level +150 m; quartz-carbonate veins with sulfide bands; sphalerite and galena accumulation in the upper part of the photo. (B) Ore in test tunnel level +265 m; massive carbonate-sulfide breccias.

later in this section). In addition to pyrite and cordierite, the layer contains sericite, phlogopite, and tourmaline. To the east of this layer, the abundances of K-feldspar and plagioclase increase gradually; to the west, plagioclase is totally absent. The first indications of plagioclase are remnants of plagioclase phenocrysts, which are almost totally altered to K feldspar and sericite. Such remnants are relatively common above the pyrite-banded cordierite-rich layer.

A layer composed of Al silicates, mainly kyanite and quartz, exists about 80–90 m above the Ag-Au-Zn ore. Layers of this rock type are very extensive in the Taivaljärvi formation, south of the deposit, in

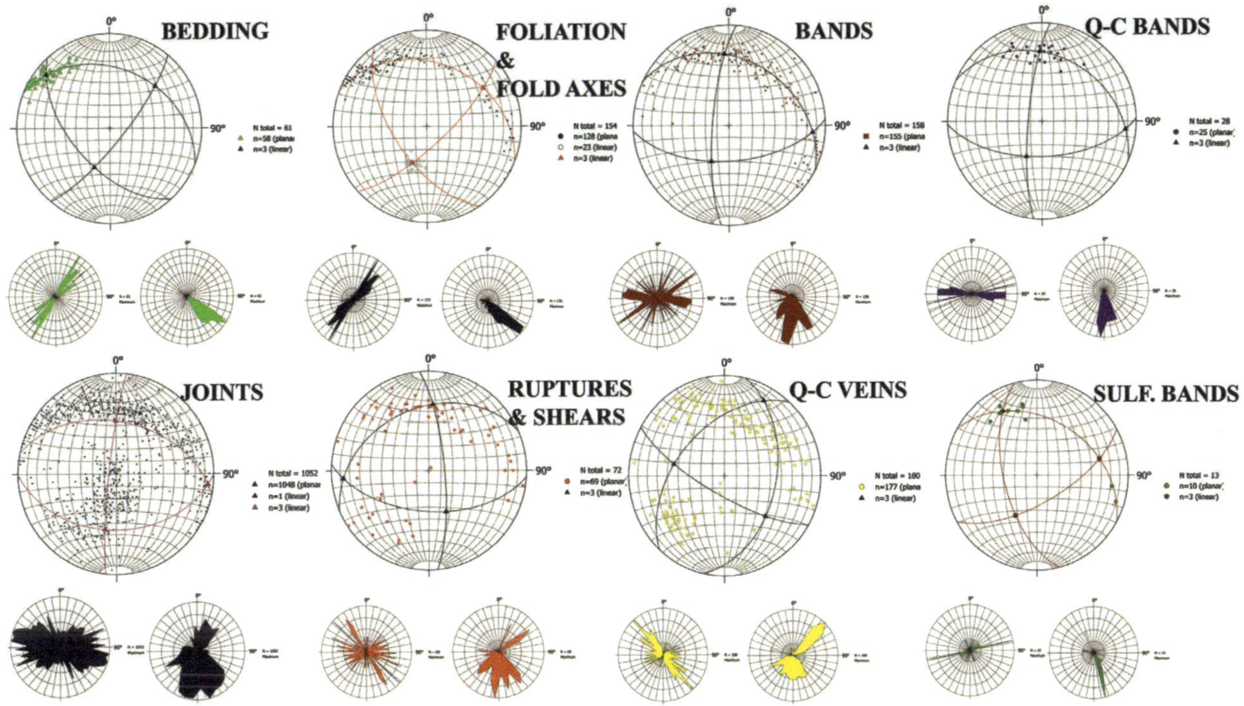

FIGURE 9.3.8 Stereographic projections (Stereo 32, lower hemisphere, equal area) of the Taivaljärvi Silver Mine bedding, foliation and fold axes, banding, quartz-carbonate bands (Q-C BANDS), joints, ruptures and shears, quartz-carbonate veins (Q-C VEINS), and sulfide bands.

Small projections depict the distribution of strike and dip of the elements. The elements were measured from the decline tunnel.

the Kivisuo area where it locally attains a thickness of 80 m; in the Taivaljärvi section, the quartz-kyanite rock is only 5–25 m thick. The quartz-kyanite rocks are characterized not only by quartz and kyanite but also by sillimanite and, locally, by staurolite, cordierite, muscovite, tourmaline, rutile, chlorite and chloritoid, zircon, and apatite. The structure of the original fragmental felsic volcanic rock is still visible on the weathered surface in some of the outcrops.

The rock sequence to the east of the quartz-kyanite rock of the Taivaljärvi formation comprises intermediate and felsic metavolcanic rocks with quartz and plagioclase phenocrysts. In contrast to the felsic metatuffs to the west of the quartz-kyanite rock, the plagioclase phenocrysts are not corroded or replaced by potassic feldspar. A 15–25-m-thick interlayer of intermediate (dacitic) metatuff above the quartz-kyanite rock is characterized by garnet porphyroblasts and abundant magnetite, which is rare elsewhere in the sequence. The total thickness of the intermediate-felsic sequence east of the quartz-kyanite layer is about 90 m and, before the mafic rocks of the Vuoriniemi formation, it terminates in a 10–15-m-thick layer of a cherty metatuff. This quartz-rich rock is very fine-grained and contains bands of disseminated pyrite.

MINERALOGY OF THE DEPOSIT

The following sections on mineralogy and geochemistry are based on the works of Kopperoinen and Tuokko (1988), Papunen et al. (1989), and Laukkanen (2011).

CARBONATE MINERALS

Carbonates occur in intersecting quartz-carbonate-sulfide veins and conformable bands of the A and B zones, in calc-silicate interlayers of the D zone, and in the footwall. In the A and B zones, the content of carbonates varies from 6–8%, and in the C and D zones from 1–2%.

According to microprobe analyses (Papunen et al., 1989), most of the carbonates are mangano-ankerites with MnO contents of 3–6 wt%. The FeO content of manganoankerite increases from the A zone toward the footwall: in the A zone FeO is 5.27 wt%, in the D zone, 6.42 wt%, and in the footwall, 7.55 wt%; the MgO content decreases correspondingly. Manganocalcite with 3 wt% MnO is associated with ore minerals in the quartz-carbonate veins in the B zone.

ORE MINERALS

The most important ore minerals are freibergite, dyscrasite, pyrargyrite, native silver, electrum, sphalerite, galena, and chalcopyrite. The following minerals have also been identified: pyrite, pyrrhotite, arsenopyrite, cubanite, covellite, gudmundite, acanthite, miargyrite, freieslebenite, bournonite, scheelite, native Sb, and native Bi. The grain size of the common sulfides varies from 0.1–0.5 mm and that of the Ag-bearing minerals from 0.01–0.1 mm. Galena and the associated Ag minerals are more abundant in the stratigraphically uppermost ore zones, where they occur in quartz-carbonate veins and bands. The sulfide content in the ore varies between 5 and 8%, and more than 50% of them are iron sulfides pyrite and pyrrhotite. Silver mineralogy stays basically constant in the whole deposit: more than 95% of silver is in sulfides and antimonides, but in silver-rich parts of the ore, galena contains about 0.1 wt% silver and also native silver and silver-gold alloys occur. Since native silver is quite a rare mineral, a grade of >700 g/t was used as the upper cutoff grade for silver to control the nugget effect in the resource estimates.

Iron sulfides and arsenopyrite

Pyrite is the dominant iron sulfide mineral. It occurs in all ore types and in the wall rock as euhedral cubes. In the quartz-ankerite veins and in the C and D ores, pyrrhotite is an important sulfide mineral. Arsenopyrite occasionally occurs as banded disseminations. It does not show any correlation with the gold content.

Sphalerite

The grain size of sphalerite varies from 0.05–0.3 mm, in places up to 0.5 mm. The mineral exists throughout the deposit, mainly as bands together with iron sulfides and minor galena, and in quartz-ankerite veins together with galena and other ore minerals. Three grains of sphalerite analyzed by microprobe contained 5–7 wt% Fe and 0.26–0.35 wt% Cd (Papunen et al., 1989).

Galena

This occurs mainly as dissemination in quartz-ankerite veins and in bands together with sphalerite, chalcopyrite, and iron sulfides. Galena commonly fills the cracks between the other minerals or occurs as inclusions in ankerite. In the Ag-rich part of the ore body, galena contains inclusions and intergrowths of dyscrasite, freibergite, native Ag, antimony, freieslebenite, and bournonite. The Ag content of galena varies from 0.01–1.7 wt%, the galena in the D ore being poorest in Ag (Papunen et al., 1989). High Ag values are due to submicroscopic inclusions of Ag-bearing minerals.

Chalcopyrite

This is a common accessory ore mineral throughout the deposit. The grain size varies from 0.02 to 0.1 mm. It occurs in Ag-rich parts of the ore intergrown with Ag minerals and in the calc-silicate layers together with pyrrhotite, sphalerite, and galena. Some of the chalcopyrite grains in the Ag-rich part of the ore tarnish green, blue, or gray soon after polishing. The tarnished mineral assays up to 6.8 wt% Ag (Papunen et al., 1989). Chen et al. (1980) described a similar phenomenon and attributed it to Ag diffusion onto the surface of polished Ag-rich chalcopyrite, where it forms a thin layer of acanthite, (Ag2S).

Freibergite [$(Ag,Cu)_{12}(Sb,As)_4S_{13}$] and Dyscrasite [$Ag_3Sb$]

Freibergite is the most common Ag mineral in the deposit. It occurs together with galena in disseminated ore or as intergrowths with galena and dyscrasite in the quartz-ankerite veins. It has also been found in veins together with galena and native Ag as a coating on the grains of galena. The grain size is between 0.03 and 0.06 mm. The Ag content varies from 20–50 wt% (Papunen et al., 1989), the lowest concentrations being in freibergites of the C zone. Silver replaces copper in freibergite-tetrahedrite series, that is, when Ag increases, the Cu content decreases. The Ag-rich freibergites of the B zone plot in the Riley (1974) diagram close to the pure Ag end-member of the tetrahedrite-freibergite series (Papunen et al., 1989). Ag-rich freibergites (up to 50 wt% Ag) occur in the B zone intergrown with normal freibergite (22–25 wt% Ag). Together with freibergite, dyscrasite is the most important carrier of Ag in all ore types. It commonly occurs in grain clusters with galena, freibergite, chalcopyrite, and native Sb in the quartz-ankerite veins and as inclusions in ankerite. Dyscrasite exists throughout the deposit together with galena and the other Ag minerals. The grain size varies from 0.01–0.1 mm. Analyses of dyscrasites are presented in Papunen et al. (1989). The dyscrasite of sample D H55/136.50 contains 0.66 wt% As, which possibly replaces Sb in the mineral.

Freieslebenite (PbAgSb3)

This is a rare Ag mineral and has been recognized only in the A and B ores as inclusions in galena. The chemical compositions of freieslebenite indicate obvious Ag-Cu replacement (Papunen et al., 1989).

Pyrargyrite (Ag3SbS3)

Pyrargyrite exists as thin coatings on the fracture surfaces of the galena-ankerite-quartz veins. The red color typical of the mineral makes it easy to detect, even with the naked eye. The grain size varies between 0.01 and 0.03 mm. The mineral also occurs as inclusions in and along the grain boundaries of galena.

Native Ag and Ag-dominant intermetallic alloys with Au and Sb

As with the other constituents these are common accessories in the ore (Papunen et al., 1989). The alloys commonly occur together with galena and the other Ag minerals, but they also fill the fractures between the other ore minerals or exist as fine-grained dust in carbonate. The grain size is between 0.01 and 0.05 mm. A couple of 1- to 2-mm-thick native Ag veins filling tension cracks have been observed in drill cores; native Ag in the footwall D layer contains 94.1 wt% Ag, 4.8 wt% Sb, and 0.19 wt% Au. Electrum is a rare mineral and has been found as inclusions in pyrite and sphalerite. In the analyzed grains, the concentration of silver exceeds that of gold (51.7 wt% Ag, 43.4 wt% Au). The minerals called "dyscrasite" and "Au-bearing dyscrasite" in Papunen et al. (1989) are complicated intergrowths of different phases of Ag-Sb-Au minerals.

GEOCHEMISTRY OF THE SEQUENCE

The major and trace element compositions of the felsic lithologic sequence of the Taivaljärvi formation are presented according to Papunen et al. (1989) in Table 9.3.1 and Fig. 9.3.9. Geochemically, the rocks can be divided into two major units: the rocks to the east of the kyanite-quartz rock and those in and to the west of the layer. On the basis of the SiO_2 content, the felsic metavolcanic rocks in the eastern part of the sequence range in composition from felsic to intermediate with numerous interlayers of dacites and chert, although their TiO_2 contents are too high for average Archean dacites and rhyolites (Condie, 1981). The rocks of the eastern part of the sequence differ in composition from the corresponding rocks of the western part, having slightly higher MgO, Ba, U, and Th contents but lower TiO_2, $Al2O_3$, Na_2O, P_2O_5, Co, Sr, CaO, Cr, V, Sc, Li, Cs, and Y contents (Fig. 9.3.9). However, the western part of the Taivaljärvi formation contains several rock units, for example, the mineralized rocks, the quartz-kyanite rocks, and the pyrite-banded metatuff, which have specific geochemical characteristics differing from those of the pyroclastic metavolcanic rocks.

The tenors of Cu, Sb, and Au are highest in the D ore body, near the western footwall, in the stratigraphically lower mineralized portion of the Ag-Zn deposit. The central part of the deposit, the B and C ore bodies, are characterized by high tenors of Ag, Zn, As, Se, Pb, and Cu, and the A ore body close to the hanging wall contains abundant carbonates, Ag, and Zn. The content of Ni is also highest in the hanging wall part, although the values are only twice the average background. The contents of Mn, Mg, and Ca show clear positive correlations with that of CO_3.

Table 9.3.1 Stratigraphic heights, rock types and chemical compositions of the samples collected from the Taivaljärvi Formation

Nr	1	2	3	4	5	6	7	8	9	10	11	12	13	14	15	16	17	18	19	20	21	22
Height	208	203	197	167	142	104	71	56	43	16	0	−16	−28	−33	−38	−50	−56	−64	−74	−96	−110	−141
Rock	andes	chert	dacite	dacite	dacite	andes	andes	dacite	k-qz-r	rhyol	py-tuff	crd-tuff	rhyol	crd-tuff	A ore	felsic	B ore	felsic	C ore	D ore	rhyol	rhyol
(wt %)																						
SiO_2	66.10	84.90	73.00	65.40	71.00	62.50	65.40	70.00	76.60	72.10	49.80	72.00	73.30	64.70	50.00	72.70	57.10	65.20	56.50	75.90	76.20	72.90
TiO_2	0.43	0.65	0.65	0.73	0.75	0.71	0.74	0.43	0.31	0.36	0.56	0.17	0.25	0.56	0.13	0.27	0.10	0.25	0.19	0.20	0.25	0.24
Al_2O_3	15.80	9.69	15.30	15.60	16.20	15.70	15.80	14.70	15.00	14.20	23.20	11.10	12.10	12.40	5.79	12.80	4.37	12.00	8.64	8.81	11.40	10.80
Fe_2O_3	4.34	0.83	0.72	6.05	1.13	7.87	7.64	3.78	1.88	3.15	8.10	2.50	4.72	5.61	8.35	3.10	7.09	<0.01	11.40	2.93	3.20	4.64
MnO	0.11	<0.02	<0.02	<0.02	0.02	0.07	0.11	0.14	0.02	0.06	0.06	0.21	0.06	0.44	0.80	0.36	1.48	0.49	0.56	0.14	0.28	0.31
MgO	1.50	0.11	0.40	0.77	0.61	0.49	0.72	1.26	0.97	2.08	1.46	2.17	0.77	2.65	6.84	1.38	3.73	3.60	1.28	1.15	1.42	2.30
CaO	3.34	0.05	1.31	0.63	1.37	0.97	1.86	2.04	1.35	2.40	3.84	4.12	0.35	3.82	12.50	1.56	8.78	1.44	0.84	1.31	0.35	1.29
Na_2O	3.30	0.20	5.44	0.92	5.65	1.84	3.34	1.14	1.43	0.51	1.20	0.19	0.07	0.12	0.17	0.10	<0.01	0.15	<0.01	<0.01	0.06	0.06
K_2O	2.48	2.51	2.00	4.55	2.19	4.72	3.32	2.65	0.61	1.55	4.62	3.05	3.93	4.31	2.48	4.21	1.52	4.37	2.63	2.88	3.95	3.92
P_2O_5	0.12	0.14	0.14	0.16	0.14	0.16	0.17	0.10	0.03	0.07	0.08	0.03	0.06	0.11	0.07	0.05	0.03	0.06	0.06	0.06	0.06	0.06
L.O.I.	1.60	1.10	1.10	5.00	1.10	4.40	1.10	3.10	2.00	2.60	7.20	2.50	3.80	3.30	5.30	2.40	4.50	2.70	6.30	2.70	2.20	2.00
CO_2	0.87	0.02	0.20	0.02	0.12	0.15	0.12	0.09	0.07	0.04	0.38	2.70	0.23	3.10	9.60	1.20	7.40	7.10	0.56	1.00	0.19	1.00
Na/(K+Na)	0.571	0.074	0.731	0.168	0.721	0.280	0.502	0.301	0.701	0.248	0.206	0.059	0.018	0.027	0.064	0.023	0.007	0.033	0.004	0.003	0.015	0.015
(ppm)																						
Ag	1.5	<0.5	<0.5	<0.5	<0.5	<0.5	<0.5	<0.5	<0.5	<0.5	4	<0.5	<0.5	1.5	10.0	1.5	2700	19.0	220.0	33.0	5.0	10.0
Au (ppb)	15	<5	<5	<5	<5	<5	<5	<5	<5	<5	45	<5	5	15	33	<5	16000	220	530	500	18	57
Sb	0.6	0.3	1.8	1.4	2	0.3	0.5	0.8	1.3	1.2	1.8	0.4	0.7	1.7	8.7	0.8	390	7.4	230	15	2.9	9.6
As	<1	26	4	17	2	16	2	21	120	10	120	18	77	74	30	9	950	13	88	23	16	6
Bi	1.9	<0.1	<0.1	<0.1	0.3	<0.1	0.4	<0.1	<0.1	1.8	0.4	<0.1	<0.1	<0.1	<0.1	<0.1	1.7	<0.1	2.1	<0.1	0.2	0.4
Se	<0.7	<0.7	1.4	2.6	<1.2	<1.3	3.8	<0.7	<0.7	1.1	3.3	2.9	<1.7	3.5	2.9	<1.0	21	2.3	12	<0.6	1	<0.5
Cd	<0.2	<0.2	<0.2	<0.2	<0.2	<0.2	<0.2	<0.2	<0.2	<0.2	<0.2	<0.2	<0.2	1.2	34	<0.2	32	8.8	240	68	1.2	2
Pb	28	12	20	8	8	3	4	12	12	8	36	16	16	32	480	40	8600	180	9300	520	96	470
Zn	100	11	34	57	55	37	64	55	170	120	86	71	46	530	7400	220	5900	2000	46000	13000	400	570
Cu	36	4.5	7	8.3	11	12	14	17	19	13	220	20	6.5	38	190	110	700	180	370	46	6.5	27
Ni	9	34	6	16	6	25	6	35	15	16	13	6	9	13	85	9	14	5	14	8	4	7
Co	10	6	4.5	17	9.7	27	7.1	11	4.9	5.7	7.8	2.1	3.5	10	15	1.9	3.6	5.8	8.4	2.5	2.1	3
Cr	7.8	0.8	4.5	8.4	6.7	6.2	7	23	10	9.2	7.1	<0.5	1.2	4	<0.5	<0.5	7.5	2.8	3.1	2.3	<0.5	2
V	66	4	36	44	40	42	42	46	28	40	38	12	10	32	14	6	8	8	6	8	6	6
Sc	6.94	3.42	8.32	9.97	8.69	8.45	9.33	7.77	5.55	7.41	12.3	3.87	6.57	7.1	3.69	6.72	3.56	6.11	4.83	3.22	6.1	5.58
Y	10	20	30	40	30	30	30	30	50	40	50	20	50	50	<10	<10	<10	10	<10	<10	40	<10

Continued

Table 9.3.1 Stratigraphic heights, rock types and chemical compositions of the samples collected from the Taivaljärvi Formation—cont'd

Nr	1	2	3	4	5	6	7	8	9	10	11	12	13	14	15	16	17	18	19	20	21	22
Zr	70	210	200	240	210	230	200	190	170	170	470	140	260	170	90	270	70	270	160	200	270	240
Hf	3.2	7	5.6	7.3	6.6	6.6	6.3	6.6	7.3	5.6	14	5.5	8.4	5.9	3.8	8.6	3.4	6.5	4.9	4.8	6.5	5.9
Nb	<10	10	10	10	10	20	30	<10	20	10	30	20	10	<10	20	10	30	20	50	20	20	20
Ta	<0.5	0.8	1.4	<0.5	<0.5	<0.5	0.6	0.7	0.9	0.6	1.3	0.8	0.8	0.8	<0.5	0.7	<0.5	0.6	<0.5	0.7	0.6	0.7
Th	2.1	7.9	6.8	6.9	7.3	6.9	7.7	6.5	10	8.7	13	11	9.3	7	4.4	10	3.5	9.7	7.6	7.1	11	9.5
U	0.5	2.2	2.1	1.9	2	2.2	0.8	1.8	3.5	2	4.1	3.5	2.9	1.9	2.4	3.3	1.4	2.6	3.2	2.5	3.1	2.8
Li	18	6	14	26	17	22	18	14	20	30	12	14	10	14	16	12	16	18	13	14	12	14
Rb	100	60	90	150	70	130	120	70	20	60	200	140	140	160	120	149	10	160	120	100	140	150
Cs	8	1.8	2.6	8.4	2.9	2.3	7.3	2.8	1	1.8	6.3	3.5	2.7	5.3	6.5	2.5	3.2	6.1	2.9	1.1	1.9	2.8
Be	2	2	2	3	2	4	3	2	2	2	4	3	3	3	2	1	2	3	4	1	3	2
Sr	230	50	130	40	130	120	160	60	110	60	50	<10	<10	40	60	20	30	10	<10	10	<10	20
Ba	480	200	420	490	400	500	540	300	100	160	560	410	760	630	240	830	200	500	630	580	1100	920
B	20	20	40	30	20	20	20	30	170	20	460	30	20	40	<10	20	<10	30	90	90	20	20
Cl	<50	50	50	<50	<50	150	<50	<50	<50	<50	50	50	50	200	50	50	50	<50	<50	50	50	<50
Br	2.3	<0.5	0.5	<0.5	<0.5	0.6	1.4	<0.5	<0.5	<0.5	<0.5	0.5	0.5	1.2	0.6	<0.5	<0.5	<0.5	0.5	<0.5	0.5	0.6
La	14	25.5	33.9	33.5	31.5	35.7	37.6	28.1	37.2	30.6	24.3	37.6	42.8	27.2	25	42.2	24	39.1	29.2	24.8	33.5	29
Ce	27	46	58	60	52	62	65	48	67	53	44	66	72	48	39	76	25	78	61	47	70	59
Nd	10	20	26	26	25	27	27	19	25	19	17	26	31	22	17	28	10	30	23	20	27	25
Sm	2.2	4.52	5.07	5.09	4.63	5.18	5.02	3.67	5.92	4.5	3.87	5.55	6.1	4.28	3.98	6.41	2.32	6.22	4.96	3.95	5.64	4.93
Eu	0.77	0.58	0.68	0.52	0.68	0.68	0.132	0.74	0.48	0.33	1.01	0.71	0.88	0.57	0.78	0.66	1.29	1.31	1.22	0.73	0.85	0.83
Tb	0.3	0.7	1	0.9	0.6	0.9	0.8	0.5	1.1	0.7	1.1	1.2	1.1	0.6	0.9	1.1	0.5	0.9	1.1	0.5	0.9	0.7
Yb	0.64	2.06	1.99	2.88	2.08	2.26	2.41	1.95	3.19	2.12	6.39	2.93	2.79	2.46	2.82	3.42	1.87	3.51	6.51	1.39	4.43	2.05
Lu	0.11	0.33	0.33	0.48	0.34	0.4	0.42	0.34	0.57	0.36	1.12	0.52	0.48	0.41	0.49	0.58	0.26	0.57	1	0.22	0.74	0.36

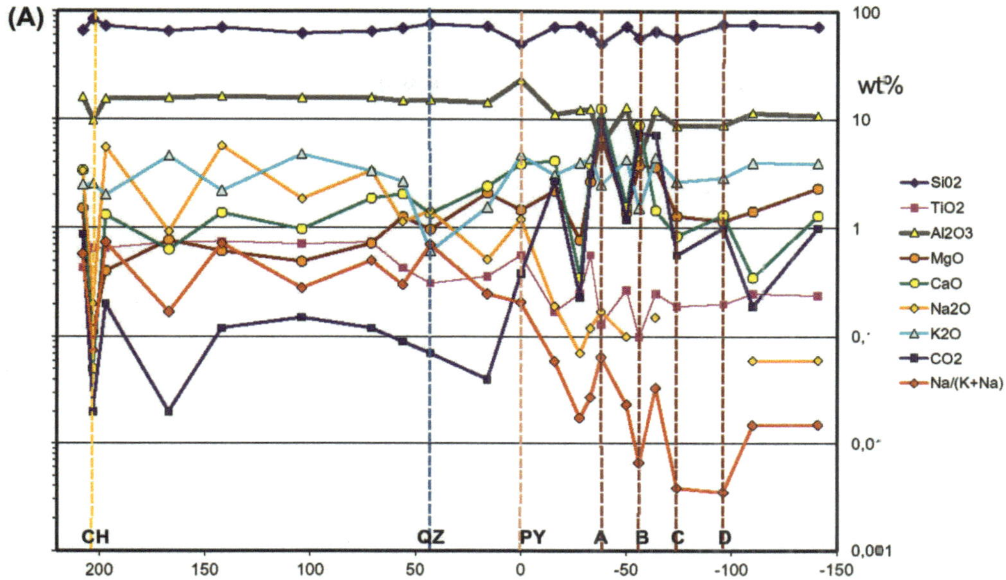
(A)
100
wt%
10
1
0,1
0,01
0,001
CH
QZ
PY
A
B
C
D
200
150
100
50
0
-50
-100
-150
Si02
TiO2
Al2O3
MgO
CaO
Na2O
K2O
CO2
Na/(K+Na)

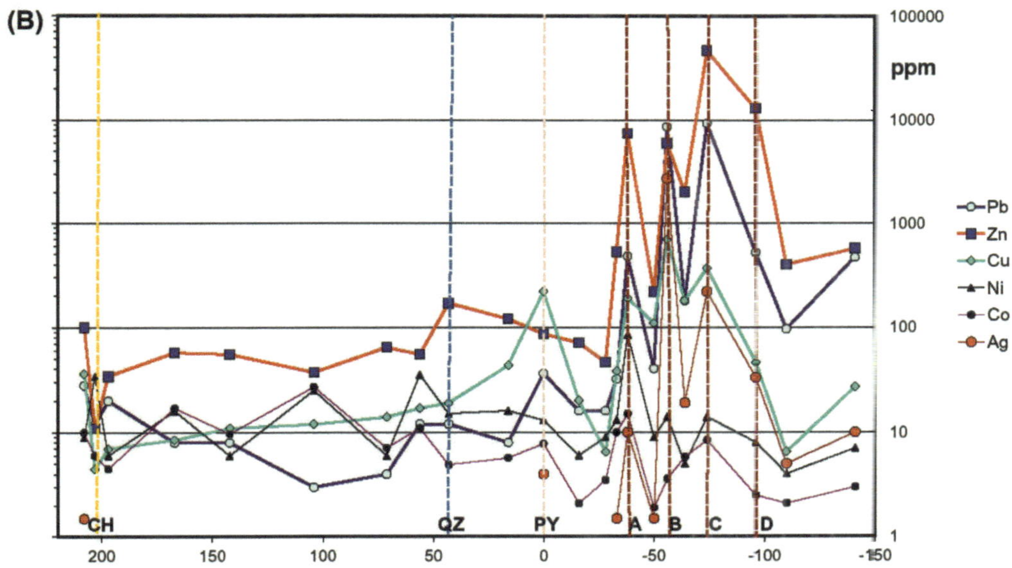
(B)
100000
ppm
10000
1000
100
10
1
CH
QZ
PY
A
B
C
D
200
150
100
50
0
-50
-100
-150
Pb
Zn
Cu
Ni
Co
Ag

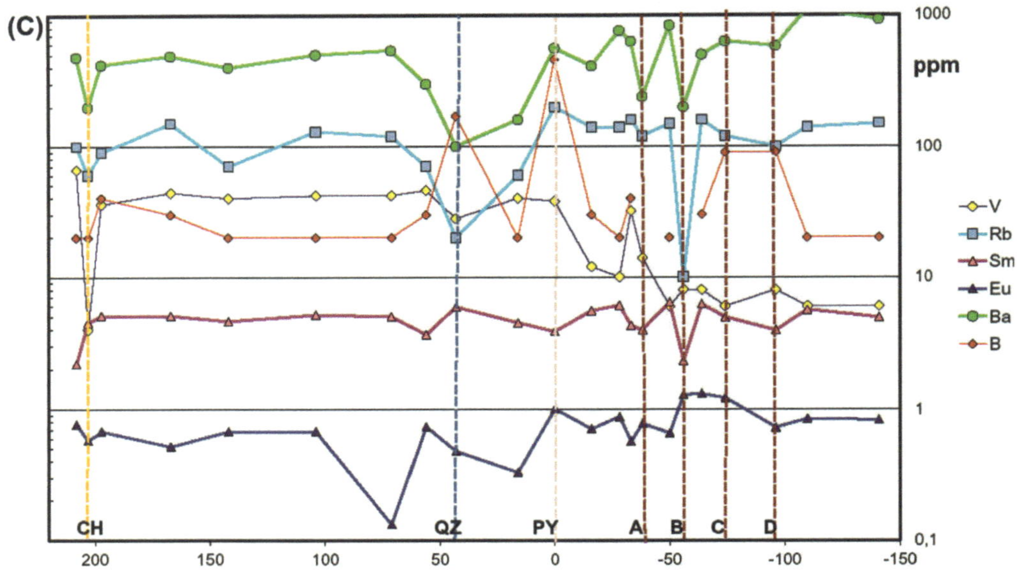

FIGURE 9.3.9 Variation of the chemical compositions versus stratigraphic height at Taivaljärvi.

Table 9.3.1 and Figs. 9.3.2 and 9.3.4 are referred to for the location of the samples in stratigraphic scheme. (A) Variation of the main elements; (B) the ore elements; (C) trace elements. CH = chert layer; QZ = quartz-kyanite layer; PY = pyrite-bearing metatuff; A, B, C, D = ore layers.

Of the mineralogically distinct layers of the sequence, the pyrite-banded felsic metatuff, about 30 m above the A ore body, is characterized by higher Al and K and lower Si contents than the felsic metatuffs on average. The concentrations of ore elements Cu, As, Pb, and Sb are enriched in this layer relative to the barren metavolcanic rocks, and the high Fe content and loss on ignition are due to pyrite dissemination. Also the abundances of the trace elements Zr, Hf, Sc, and B are elevated in this layer.

The quartz-kyanite rock (QZ in Fig. 9.3.9) is characterized by low abundances of all the main elements except SiO_2 and Al2O_3. Of the trace elements, only Y and B show slightly higher contents in this rock type than in the average felsic volcanic rock; the contents of all the other elements are reduced. Cr has the highest peak in the felsic metatuff just above the quartz-kyanite rock.

The chert layer at the eastern contact of the Taivaljärvi formation against the Vuoriniemi formation displays a high SiO_2 content, but the contents of all the other major elements are lower than in the felsic metavolcanics of the upper part of the formation. The abundances of Ti and P and of the trace elements Mo, Y, Zr, Hf, Th, and U are at about the same level as in the felsic metavolcanic rocks, but all the other trace elements display lower contents.

There are no marked differences in the total rare earth element contents between the felsic metavolcanic rocks from the lower and upper parts. Fractionation of rare earth elements is somewhat more advanced in the upper part than in the lower part of the sequence, the La/Yb ratios being 14.9 and 12.1, respectively. Rare earth element distributions correspond well to the F2 type of Archean felsic volcanic rocks described by Condie (1981). Most of the samples display a negative Eu anomaly in the chondrite-normalized distribution pattern but, in the mineralized part of the sequence, the anomaly is not as prominent as in the felsic metavolcanic rocks stratigraphically above and below it (Papunen et al., 1989). In terms of Eu/Eu* ratio, the negative anomaly is largest in the dacitic metavolcanic rock about 50 m above the quartz-kyanite rock. The intermediate metavolcanic rock in the eastern part of the Taivaljärvi formation displays a positive Eu anomaly and lower total rare earth element contents than the felsic rocks of the Taivaljärvi formation.

Papunen et al. (1989) presented sulfur isotope compositions for 11 pyrite samples extracted from the Taivaljärvi formation. The samples represent the mineralized parts of the formation and the barren iron sulfide layers above the ore deposit. The $\delta^{34}S$ values vary from 2.0–3.8 per mil, with the largest variation in the ore deposit. The average value is 2.64 per mil. Pyrites of the felsic metavolcanic rocks display upward-decreasing values, although the differences among the samples are very small.

MINERAL PROCESSING, METALLURGY, AND MINING

First metallurgical research and processing tests of the Taivaljärvi Silver Mine ore were undertaken as a part of the feasibility study completed in 1991. The processing tests were performed in the 1980s on a laboratory scale at the State Technical Research Centre, VTT; this laboratory at Outokumpu is now called the Mineral Processing Laboratory of the Geological Survey of Finland, or GTK MinTek. The results of the early tests generally demonstrated that the ore is amenable to concentrate with standard flotation techniques. Early in 2007, GTK MinTek was commissioned to do a review and upgrade of historic metallurgical tests and mineral processing flow charts for Sotkamo Silver Oy. An updated report was received at the end of August 2007. Connected with the feasibility study, GTK MinTek conducted additional metallurgical tests of the Ag-Au-Zn ore in the spring 2011 at the Mini Pilot scale (Peltoniemi, 2011).

According to the tests, the Ag-Au-Zn ore is amenable to flotation processing to achieve silver-lead and zinc concentrates at acceptable recoveries from a grind size of 80% passing 75 μm. Gravity processing does not give any significant advantages over flotation. In the locked cycle test, the total recovery of silver was 76.2% and 90% for gold. Approximately 7% of the silver was recovered in the zinc concentrate grading 544 g/t Ag. Zinc recovery of 90.8% was achieved in a concentrate grading 51 wt% Zn. With the production of a pyrite concentrate, the sulfur content of the tailings could be reduced from 0.7–1.1 wt% to 0.1–0.2 wt% S, which is beneficial for the tailings disposal. Silver grade in the sulfur (pyrite) concentrate ranged from 200 to 300 g/t Ag and the recovery of Ag in pyrite concentrate is about 6%. GTK MinTek concluded that the ore is, in general, easily amenable to concentrating using simple processes and reasonable reagent consumptions. Fair concentrate grades can be yielded with satisfactory recoveries. Based on the pilot plant tests, the silver recovery in the feasibility study was readjusted to 86–91%.

Ore processing is planned to be a traditional crushing-grinding-flotation-filtering scheme, and the underground mining will be a sublevel stoping with backfill.

DISCUSSION AND SUMMARY

ORIGIN OF THE ROCK TYPES

Despite regional metamorphism and deformation, the Taivaljärvi formation shows fragmentary and layered structures typical of pyroclastic felsic volcanic rocks. The primary porphyritic textures of the rocks are still visible locally, providing support for the volcanic origin of the sequence. The silicate mineral assemblages (kyanite, staurolite, cordierite, garnet) indicate medium to upper amphibolite facies conditions of metamorphism. In this context, metamorphism is considered isochemical and the metamorphic mineral assemblages reflect the chemical compositions caused by premetamorphic processes.

The interdependence of the chemical composition of the rock and metamorphic mineral assemblage is clearly visible. For example, the Al_2O_3 excess, which is necessary for the formation of kyanite in the quartz-kyanite rock, is a result of depletion of alkalies and all major elements except Al, which remained unchanged. Correspondingly, the cordierite in the metatuff above the ore deposit crystallized in a rock that contained more MgO than felsic volcanic rocks on average.

On the basis of SiO_2 versus Zr/Ti diagrams by Winchester and Floyd (1977), the felsic metavolcanic rocks in the upper (eastern) part of the Taivaljärvi formation have andesitic to dacitic compositions, and the alkali ratios correspond to those of primary volcanic rocks (Hughes, 1973). In contrast, the metavolcanic rocks stratigraphically below the quartz-kyanite rock are rhyolitic, and the Na/K ratios are far from magmatic (Hughes, 1973). TiO_2 contents correspond well to the andesitic to dacitic primary composition of the upper part, and the rhyolitic composition of the lower part. The felsic metavolcanic rocks have average compositions of Archean rhyolites and dacites, and the rare earth element distribution patterns indicate Fe_2-type felsic volcanic rocks in the classification of Condie (1981). However, the total variation in the Na/K ratios cannot be attributed to magmatic processes alone, and in the following section, the chemical compositions are examined in terms of hydrothermal alteration.

Certain primary textures can survive the recrystallization associated with regional metamorphism. For example, relics of plagioclase phenocrysts are common in the upper, dacitic part of the sequence above the pyrite-banded metatuff, where the abundance of plagioclase phenocrysts increases upward (to the east) parallel to the Na content of the rock. Plagioclase phenocrysts do not exist in the lower, rhyolitic part of the sequence. This is because the plagioclase phenocrysts either were lacking initially or were destroyed before metamorphism by hydrothermal decomposition. The best explanation for the anomalously low Na/K ratio in the lower part of the section is the decomposition of the primary plagioclase and dissolution of its components, particularly Na^+ and Ca^{++} under low pH conditions with muscovite as a stable mineral (Helgeson et al., 1978). On a wide scale, hydrothermal potassic alteration was more intense in the rhyolitic lowermost (western) part of the Taivaljärvi formation.

In the upper part of the section, only certain layers of the sequence are intensely altered. For example, the pyrite-banded felsic metatuff is depleted in silica but enriched in ore elements, especially Fe and S, and Al, K, Sc, Zr, Hf, and B if compared with the metavolcanic rocks above and below the metatuff layer. The variation in composition may be a result of premetamorphic hydrothermal alteration of the tuff layer to clay minerals, especially to K micas (sericite alteration), which may absorb trace elements—for example, boron—on account of their ion exchange capability (Potter et al., 1963). High concentration of pyrite and the ore elements indicate conditions similar to in the ore layers, as do also the peaks of Eu and Yb. High concentrations of Zr and Hf may indicate accumulation of zircon as a residual heavy mineral.

A relative increase in Mg is seen in the cordierite-bearing felsic metavolcanic rock layers above the deposit. High concentrations of Mg exist also above and below the pyrite-banded metatuff. Although in the upper part of the ore deposit Mg, Mn, and Ca are mainly incorporated in carbonates, the existence of abundant garnet, cordierite, and amphibole minerals indicates redistribution of Mg and Fe and suggests a premetamorphic, altered Mg-Fe mineral assemblage (chlorite alteration) in the rocks of the ore zone.

The quartz-kyanite rock is depleted in all major elements except SiO_2 and $A12O_3$. Of the trace elements, there is a slight increase in Y, Th, U, and B. The probable premetamorphic minerals, silica and kaolinite, were deposited as a result of intense leaching by an acidic fluid (low pH). The peak of Cr in the Taivaljärvi sequence close to the kyanite-quartz rock layer (sample 8) is of interest. Schreyer (1982) considers that the fuchsite-aluminum silicate rocks in the Archean greenstone belts of Zimbabwe and Western Australia represent metamorphosed alunite-bearing deposits, where Cr was primarily incorporated in the alunite and clay minerals. An alunitic origin can well be postulated also for the Taivaljärvi quartz-kyanite rocks. The quartz-kyanite rock, an extensive key unit of the sequence, has been found in a stratigraphically similar position as much as 8 km southwest of the Taivaljärvi Silver Mine deposit. It could be interpreted as the cap rock of a hydrothermally altered pyroclastic sequence in which acid vapor leached all the minerals and deposited alunite, kaolinite, and silica.

On exsolution from magma, an aqueous fluid can consist of a single phase of intermediate salinity and density (Hedenquist and Taran, 2013). During ascent to a low-pressure regime, the fluid may intersect its solvus, where a low-salinity vapor and dense, hypersaline liquid separate. At this point, the solubility of the elements carried in the fluid changes, leading to precipitation of minerals such as carbonates and sulfides, depending on the primary anion concentrations. The saline fluid also causes alteration of the host rock to hydrous minerals (kaolinite, chlorite, montmorillonite, etc.).

Permeable lithologic units guide the condensates of magmatic vapor that contain SO_2 and HCl to flow laterally and form a subhorizontal blanket of alteration zones (Hedenquist and Henley, 1985; Hedenquist and Taran, 2013). As a lithocap, the advanced hydrothermal argillic alteration can extend to wide areas and provide a guide to districts where near-surface boiling zones may host precious and base-metal deposits. The dacitic volcanic rocks overlying the quartz-kyanite rock did not experience as intense hydrothermal alteration as did the underlying sequence, as the vapors causing the alteration were channeled laterally along the permeable layer and prevented from reacting with the rocks above. The quartz-rich layer at the top of the Taivaljärvi formation shows enrichment of silica and some lithophile elements. The rock could be interpreted as chemical sediment, chert, in origin.

The incompatible elements have often been classified as immobile (e.g., LREE, Ti, Zr, Y, Hf) and mobile (e.g., U, Th, Rb, Cs) in relation to hydrothermal processes. In altered volcanic rocks, Ti and Al are considered to be largely unaffected by regional alteration (Bence and Taylor, 1985) and their trends reflect magmatic processes. In the Taivaljärvi formation, the volcanic suite was originally heterogenous, varying from rhyolitic to andesitic-dacitic in composition. The marked difference in Ti content between the lower, altered part and the upper part of the Taivaljärvi formation is ascribed to this primary compositional difference. The upper part could have been primarily intermediate in composition, but secondary hydrothermal enrichment of SiO_2 altered the rocks, making them more silicic in composition yet not changing the primary high Ti content. The marked depletion of Y and Ti in the host rocks of the deposit indicates reactions caused by hydrothermal fluids.

Lemiére et al. (1986) have observed partial leaching of Y and Nb and total leaching of Mg, Ca, Na, K, and P_2O_5 in aluminous rocks formed by acid leaching. This finding may well apply to quartz-kyanite rocks, too. In the volcanic suite of the Taivaljärvi formation the altered rocks are slightly depleted in light relative to heavy rare earth elements. Michard et al. (1983) noted enrichment of light over heavy rare earth elements by a factor of ten in 13°N East Pacific Rise hydrothermal fluids. Preferential extraction of light rare earth elements can thus cause fractionation of rare earth elements. The negative Eu anomaly observed in the chondrite-normalized distribution patterns in the volcanic suite could be ascribed to plagioclase fractionation from the melt.

Bence and Taylor (1985), however, have noted that negative Eu anomalies can be generated during prograde alteration reactions, and positive Eu anomalies during retrograde reactions when the temperature drops and hydrothermal phases are precipitated. Eu enrichment has been observed in the active hydrothermal vents at 13°N on the East Pacific Rise (Michard et al., 1983). Bence and Taylor (1985) observed Eu enrichment relative to the adjacent rare earth elements in the vent alteration zone of the Balaklala mine, Shasta district, California. This is in harmony with the Taivaljärvi rare earth element distribution in which the negative Eu anomalies in the mineralized rocks are not as prominent as in other altered metavolcanic rocks.

ORIGIN OF THE DEPOSIT

The silver deposit is stratiform, with well-developed compositional layering. The geochemistry of the host rocks indicates zoning of premetamorphic hydrothermal potassic, chloritic, sericitic, and argillic alteration. Sulfides and other ore minerals characterize the ore zone as stratiform bands, dissemination, and fracture fillings. The carbonates in the upper part of the ore zone mainly fill fractures in the rock, and abundant Ca, Fe, Mn, and Mg characterize the geochemistry of this part. The observed carbonate mineral species, manganocalcite and manganoankerite, explain well the interdependence between the elements. The ore zone is depleted in the trace elements Sc and Y but enriched in Nb, W, and Mo. The barren interlayers in the deposit exhibit high silica contents and low carbonate contents.

From the anomalous ore metal contents, the ore layer can be traced laterally along the sequence to a considerable distance from the actual ore deposit, thus referring to the stratabound character of the ore. Carbonate minerals were deposited together with sulfides, sphalerite, and chalcopyrite, in particular, in the fractures of the upper part of the ore deposit. The existence of ore minerals and carbonates in the fractures of the metamorphosed host rock indicate late remobilization of sulfides and carbonates, but some observations also demonstrate that the fractures were folded together with the host rock and hence the fracture filling could partly be syngenetic with premetamorphic hydrothermal processes.

According to Kopperoinen and Tuokko (1988) and Papunen et al. (1989), mineralization took place during periods of hydrothermal activity following cycles of explosive volcanism. The fluids separated from the magma reservoir below the felsic volcanic rocks carried metal ions together with sulfur and carbonate species, and were homogenous at the time of separation from magma. During ascent as a hydrothermal plume along rock fractures to lower pressure levels, the fluids attained the boiling level where an acidic vapor separated from the fluid, escaped along permeable layers, and along these channel ways caused intense leaching of host rock. Due to boiling, the solubility of the ions in the primary fluid was reduced, and resulted in precipitation of sulfides and carbonates in the fractures. Precipitation of hydrothermal minerals healed the rock fractures that acted as channel ways for the hot mineralizing fluids and formed a network of sulfides, carbonates, and quartz.

Sulfur isotopes are generally similar throughout the sequence with the slight decrease in $\delta^{34}S$ values upward attributed to the temperature differences in the fluids between the ore horizons and the overlying sequences. The $\delta^{34}S$ values correspond well with those observed in the volcanic environment and, thus, the sulfur may be volcanic in origin.

Structural analysis by Parkkinen (2012) indicates that the ore deposit is located in a tightly isoclinally folded synclinorium structure, and the minor fold structures control the distribution of ore minerals. The geophysically indicated deep extension of the ore deposit may follow the fold structure, which plunges southwest, but at deep levels the plunge turns almost horizontal.

The deposit as a whole displays geochemical signatures that can be observed also in the pyrite-bearing metatuff layer east of the main ore. Unfortunately, sulfur was not analyzed from exactly the same samples as the trace elements and, thus, the S/Se ratios were approximated from the ore analyses.

In the deposit, the S/Se ratio fluctuates between 1400 and 4800, but in the barren pyrite-bearing metavolcanic rocks above the ore bodies, the ratio is between 15,000 and 18,000. The relationship is about the same as that observed by Auclair et al. (1987) between the high-temperature hydrothermal base-metal sulfides and the iron-rich sulfides of the low-temperature regime in the recent seafloor hydrothermal sulfide deposits at 13°N East Pacific Rise. Hence, the pyrite-banded metatuff might be of the same stratigraphic layer as the ore deposit, but more distal and of a low-temperature regime. In the folded sequence, a minor anticlinorium can be constructed to the east of the ore-bearing syncline, raising the possibility that the pyrite-bearing metatuff may be the same primary strata as the ore layer. The compositional difference is due to lateral difference between the proximal (ore) and distal (pyritic tuff) locations in a primary hydrothermal fluid plume.

The ore type can be defined on the basis of the data presented earlier. The association with felsic pyroclastic rocks, the stratiform setting, and the compositional alteration features around the deposit are consistent with a low-sulfidation epithermal deposit associated with a geothermal system. Also, the other features such as the scarcity of iron sulfides, the disseminated and stringer-vein-type ore textures, the abundant carbonate gangue minerals, the vertical arrangement of alteration zones with potassic alteration at the lowest stratigraphic levels overlain by chloritic and sericitic zones, and, finally, the zone of advanced argillic or acid-leached type of alteration above the ore deposit are all typical of the epithermal deposit type (Henley and Hedenquist, 1986). The quartz-kyanite rock represents the primary argillic lithocap alteration zone that formed along a permeable lithology suitable for the leaching of the host rock by acidic vapors, and left residual silica, aluminous clays, and alunite that recrystallized to quartz and kyanite during metamorphism.

The wide extent of the quartz-kyanite rock in the area is typical for lithocaps related to hydrothermal alteration plumes above felsic intrusions. In this sense, it is worth noting the observation (Papunen et al., 1989) that in drill cores, the Taivaljärvi volcanic sequence grades westwards to a homogenous felsic porphyry, which further turns to a felsic plutonic rock. In the magnetic map, an oval anomaly to the west of Taivaljärvi could indicate a partly magnetized intrusion. It is an open question whether the felsic volcanic rocks are associated with this intrusion, and if the Silver Mine deposit is related to a porphyry type system.

ACKNOWLEDGMENTS

The authors thank Hannu Huhma, Hugh O'Brien, and an anonymous reviewer for thorough reviews and improvements on the text.

REFERENCES

Auclair, G., Fouquet, Y., Bohn, M., 1987. Distribution of selenium in high-temperature hydrothermal sulfide deposits at 13° north, East Pacific Rise. Canadian Mineralogist 25, 577–587.

Bence, A.E., Taylor, B.E., 1985. Rare earth element systematics of the West Shasta metavolcanic rocks: Petrogeneis and hydrothermal alteration. Economic Geology 80, 2164–2176.

Blais, S., Auvray, B., Capdevila, R., et al., 1978. The Archaean greenstone belts of Karelia (Eastern Finland) and their komatiitic and tholeiitic series. In: Windley, B.F., Naqvi, S.M. (Eds.), Archaean Geochemistry. Elsevier, Amsterdam, p. 87–107.

Chen, T.T., Dutrizac, J.E., Owens, D.R., Lafiamme, J.H.G., 1980. Accelerated tarnishing of some chalcopyrite and tetrahedrite specimens. Canadian Mineralogist 18, 173–180.
Condie, K.C., 1981. Archean greenstone belts. Developments of Precambrian Geology 3, 434.
Engtec, C.T.S., 2014. Bankable Feasibility Study Sotkamo Silver Mine, p. 137. Accessible online at www.silver.fi.
Hedenquist, J.W., Henley, R.W., 1985. Hydrothermal eruptions in the Waiotapu geothermal system, New Zealand: Their origin, associated breccias and relation to precious metal mineralization. Economic Geology 80, 1640–1668.
Hedenquist, J.W., Taran, Y.A., 2013. Modeling the formation of advanced argillic lithocaps: volcanic vapor condensation above porphyry intrusions. Economic Geology 108, 1523–1540.
Helgeson, H.C., Delany, J.M., Nesbitt, H.W., Bird, D.K., 1978. Summary and critique of the thermodynamic properties of rock-forming minerals: American Journal of Science 278A, 1–229.
Henley, R.W., Hedenquist, J.W., 1986. Introduction to the geochemistry of active and fossil geothermal systems. Berlin-Stuttgart, Borntrager. Mineral Deposits Monograph 6, 1–22.
Hughes, C.J., 1973. Spilites, keratophyres and the igneous spectrum. Geology Magazine 109, 513–527.
Huhma, H., Mänttäri, I., Peltonen, P., et al., 2012a. The age of the Archaean greenstone belts in Finland. In: Hölttä, P. (Ed.), The Archaean of the Karelia Province in Finland. Geological Survey of Finland, Special Paper 54, 74–175.
Huhma, H., Kontinen, A., Mikkola, P., et al., 2012b. Nd isotopic evidence for Archaean crustal growth in Finland. In: Hölttä, P. (Ed.), The Archaean of the Karelia Province in Finland. Geological Survey of Finland, Special Paper 54, 176–213.
Hyppönen, V., 1983. Explanation to the maps of pre-Quaternary rocks, sheets 4411 Ontojoki, 4412 Hiisjärvi, and 4413 Kuhmo. Geological Map of Finland 1:100 000. Geological Survey of Finland, p. 60.
Käpyaho, A., Hölttä, P., Whitehouse, M.J., 2007. U-Pb zircon geochronology of selected Archaean migmatites in eastern Finland. Bulletin of the Geological Society of Finland 79 (1), 95–115.
Kopperoinen, T., Tuokko, I., 1988. The Ala-Luoma and Taivaljärvi Zn-Pb-Ag-Au deposits, eastern Finland. Geological Survey of Finland, Special Paper 4, 131–144.
Laukkanen, J., 2011. Silver mineralogy of the locked cycle products of Taivaljärvi silver deposit. Geological Survey of Finland, Report C/MT/2011XX, p. 20.
Lemiére, B., Delfour, J., Maine, B., et al., 1986. Hydrothermal alteration and the formation of aluminous haloes around sulfide deposits. Mineralium Deposita 21, 147–155.
Luukkonen, E.J., 1985. Structural and U-Pb isotopic study of late Archaean migmatitic gneisses of the Presvecokarelides, Lylyvaara, eastern Finland. Royal Society Edinburgh Earth Science Transactions 76, 401–410.
Luukkonen, E.J., 1988. The structure and stratigraphy of the northern part of the late Archaean Kuhmo greenstone belt, eastern Finland. Geological Survey of Finland, Special Paper 4, 71–96.
Luukkonen, E., Lukkarinen, H., 1986. Explanation to the stratigraphic map of middle Finland. Geological Survey of Finland, Report of Investigations 74, p. 47.
Martin, H., Querrè, G., 1984. A 2.5 Ga reworked sialic crust: Rb-Sr ages and isotopic geochemistry of late Archaean volcanic and plutonic rocks from eastern Finland. Contributions to Mineralogy and Petrology 85, 292–299.
Martin, H., Chauvel, C., Jahn, B.M., 1983. Major and trace element geochemistry and crustal evolution of Archaean granodioritic rocks from eastern Finland. Precambrian Research 21, 159–180.
Michard, A., Albarede, F., Michard, G., et al., 1983. Rare earth elements and uranium in high-temperature solutions from the East Pacific Rise hydrothermal vent field (13°N). Nature 303, 795–797.
Niskanen, M., 2012. Electromagnetic SAMPO soundings at Taivaljärvi 2011. Geological Survey of Finland Report M185K2011.
Niskanen, M., 2013. Electromagnetic SAMPO soundings at Taivaljärvi 2013. Geological Survey of Finland Report M178K2012.
Papunen, H., Kopperoinen, T., Tuokko, I., 1989. The Taivaljärvi Ag-Zn deposit in the Archean greenstone belt, eastern Finland. Economic Geology 84, 1262–1276.

Papunen, H., Halkoaho, T., Luukkonen, E., 2009. Archaean evolution of the Tipasjärvi-Kuhmo-Suomussalmi greenstone complex, Finland. Geological Survey of Finland, Bulletin 403, p. 68.

Parkkinen, J., 2010. Rakenneanalyysin ja geostatistiikan synteesi (English summary: Taivaljärvi Ag-Zn-Pb-Au deposit, Synthesis of structural analysis and geostatistics). Geologi, Helsinki. 62, 238–243.

Parkkinen, J., 2012. Unfolding Taivaljärvi. Geologi, Helsinki. 64, 70–74.

Parkkinen, J., 2013. Silver Mine Deposit at Sotkamo: Geological Model of EM Sampo Soundings. Accessible online at www.silver.fi.

Peltoniemi, S., 2011. Pilot Plant on Taivaljärvi ore. GTK Mintek Research Report, p. 47.

Pietikäinen, K., Halkoaho, T., Hartikainen, A., 2008. Tipasjärven vihreäkivivyöhykkeen malmivarojen kartoitushankkeen (2901006) toiminta vuosina 2005–2007 Sotkamon, Valtimon, Kuhmon ja Nurmeksen alueilla. Geological Survey of Finland. Report M10.4/4322/2008/27, p. 65. (in Finnish).

Piirainen, T. (Ed.), 1985. Arkeeisten alueiden malmiprojektin loppuraportti (Progress report on the ore research project on the Archaean areas). University of Oulu, Department of Geology. Report 28, p. 183 (in Finnish).

Potter, P.E., Shrimp, N.E., Whitter, J., 1963. Trace elements in marine and fresh-water argillaceous sediments. Geochimica et Cosmochimica Acta 27, 669–694.

Riley, J.F., 1974. The tertahedrite-freibergite series, with reference to the Mt. Isa Pb-Zn-Ag orebody. Mineralium Deposita 9, 117–124.

Schreyer, W., 1982. Fuchsite-aluminium silicate rocks in Archaean greenstone belts: Are they metamorphosed alunite deposits? Geologische Rundschau 71, 347–360.

Sorjonen-Ward, P., Luukkonen, E.J., 2005. Archaean rocks. In: Lehtinen, M., Nurmi, P.A., Rämö, O.T. (Eds.), Precambrian Geology of Finland: Key to the Evolution of the Fennoscandian Shield. Elsevier., Amsterdam, pp. 19–99.

Taipale, K., 1983. The geology and geochemistry of the Archaean Kuhmo greenstone-granite terrain in the Tipasjärvi area, eastern Finland. Acta Universitatis Ouluensis, Ser A5, 98.

Vaasjoki, M., Taipale, K., Tuokko, I., 1999. Radiometric ages and other isotopic data bearing on the evolution of Archaean crust and ores in the Kuhmo-Suomussalmi area, eastern Finland. Bulletin of the Geological Society of Finland 71, 155–176.

Vartiainen, H., 1970. Schist belt of Tipasjärvi in the parish of Sotkamo, Finland. Bulletin of the Geological Society of Finland 42, 13–22.

Wardell Armstrong International, 2012. Taivaljärvi mine bankable feasibility study, summary. Accessible online at www.silver.fi.

Winchester, J.A., Floyd, P.A., 1977. Geochemical discrimination of different magma series and their differentiation products using immobile elements. Chemical Geology 20, 325–343.

CHAPTER

URANIUM DEPOSITS OF FINLAND

9.4

E. Pohjolainen

ABSTRACT

Most areas with elevated uranium potential in Finland are associated with Paleoproterozoic metamorphic supracrustal schist belts. The main uranium deposits are located in the Kolari-Kittilä area, the Peräpohja and the Kuusamo schist belts, and the Koli and Uusimaa areas. The most common types of uranium deposits are intrusive, vein-type, quartzite-hosted, and polymetallic Au-Co-Cu-Fe-U deposits. Known uranium deposits are small, low-grade, and uneconomic for exploitation. However, there is potential for new discoveries and uranium is still relatively underexplored in Finland compared to many other commodities.

In addition to uranium deposits, there are some uraniferous polymetallic deposits that have potential to host uranium as a by-product. Uranium is associated with some gold deposits, especially in the Kuusamo and Peräpohja schist belts. The most significant uraniferous polymetallic deposits in Finland are the Talvivaara low-grade black schist hosted Ni-Zn-Cu-Co-U deposit within the Paleoproterozoic Kainuu schist belt, the Juomasuo Au-Co-U deposit in the Kuusamo area, and the Rompas Au-U occurrence in the Peräpohja schist belt. However, by-product uranium potential is dependent on main commodity production and economics, deposit size, grade of uranium, uranium characteristics in the production process, licensing, and access to the resources (e.g., within or close to nature conservation areas).

Keywords: uranium; late orogenic granite; vein; quartzite-hosted; polymetallic.

INTRODUCTION AND HISTORICAL REVIEW OF URANIUM EXPLORATION AND MINING IN FINLAND

HISTORICAL REVIEW

Pilot-scale uranium mining and milling were carried out by Atomienergia Oy in Paukkajanvaara, eastern Finland between 1958 and 1961. The Paukkajanvaara mine operated as a pilot mine and plant, producing some 30 tU. A total of 40,000 t of ore was hoisted, grading 0.14% U. The Paukkajanvaara mine site was restored under the Radiation and Nuclear Safety Authority's (STUK) control in the early 1990s. In addition, Imatran Voima Oy (IVO) performed milling and concentration tests in Lakeakallio, Eastern Uusimaa from 1958–1959. The total historical uranium production in Finland calculated from the Mining Register statistics is no more than 41 tU between 1958 and 1961 (OECD, 2012).

Uranium exploration was carried out in Finland from 1955–1989 by several organizations, including Atomienergia Oy, Imatran Voima Oy, Outokumpu Oy, and the Geological Survey of Finland (GTK). The GTK abandoned its uranium exploration activities in the late 1980s. Exploration began again in the 2000s by AREVA and some junior companies. AREVA NC closed its subsidiary

Mineral Deposits of Finland. http://dx.doi.org/10.1016/B978-0-12-410438-9.00025-X

AREVA Resources Finland Oy and sold the Finnish uranium exploration projects to Mawson Resources Ltd. in 2010. However, Mawson Resources is currently focused on gold exploration.

There is currently no uranium production in Finland and only small-scale exploration projects are underway. Known uranium deposits are small, low-grade, and uneconomic for exploitation. Finland's reasonably assured conventional uranium resources are 1500 tU in two deposits, Palmottu and Pahtavuoma-U (OECD, 2012). Unconventional by-product uranium resources are dominated by the Talvivaara polymetallic deposit.

REGULATORY REGIME

Uranium mining is regulated by the Mining Act, the Nuclear Energy Act, and the Radiation Act. The Mining Act regulates exploration and mining activities. Licenses are granted by the mining authority Tukes. Exploration licenses are required for uranium exploration and drilling or trenching, and none of these activities are permitted until the exploration licenses are in legal force. Production of uranium or thorium needs a license from the government according to the Nuclear Energy Act. Uranium mine permit applications under the Mining Act and Nuclear Energy Act are handled jointly and decided on in a single decision by the government. In addition, municipality approval is required for a uranium mine.

In uranium exploration and production, the radiation safety of the workers is regulated according to the Radiation Act and Decree. The Radiation and Nuclear Safety Authority (STUK) is the regulatory body for uranium production and radiation exposure to the workers and public. STUK also regulates environmental monitoring, radioactive waste management, and emergency preparedness. Uranium concentrate export is controlled by the Ministry for Foreign Affairs and international safeguards. Additionally, environmental impact assessment (EIA) is applied to all uranium mining projects and an environmental permit is needed.

AREAS WITH URANIUM POTENTIAL IN FINLAND

So far, there have been no major discoveries in Finland of the most economically significant uranium deposit types, such as unconformity-related, sandstone-hosted, and breccia deposits. Based on current knowledge, the most prospective areas for the discovery of uranium deposits are the Kolari-Kittilä, the Peräpohja and the Kuusamo schist belts, and the Koli and Uusimaa areas (Fig. 9.4.1). Most of the areas with elevated uranium potential in Finland are associated with Paleoproterozoic metamorphic supracrustal schist belts. The main uranium deposit types are intrusive, vein-type, quartzite-hosted, and polymetallic Au-Co-Cu-Fe-U deposits.

The Uusimaa area is dominated by intrusive-type uranium deposits, which are mostly hosted by late orogenic (1.85–1.79 Ga) granites; however, indications of hydrothermal and younger uranium remobilization processes are also present, especially in Eastern Uusimaa. The Koli and Kolari-Kittilä areas are characterized by vein-type and quartzite-hosted deposits. The Kuusamo schist belt is dominated by metasomatic and polymetallic Au-Co-Cu-Fe-U deposits, which are associated with albitization in many places. Uraniferous phosphorites and structurally controlled hydrothermal vein-type uranium occurrences with untypical metal association are present in the Peräpohja schist belt.

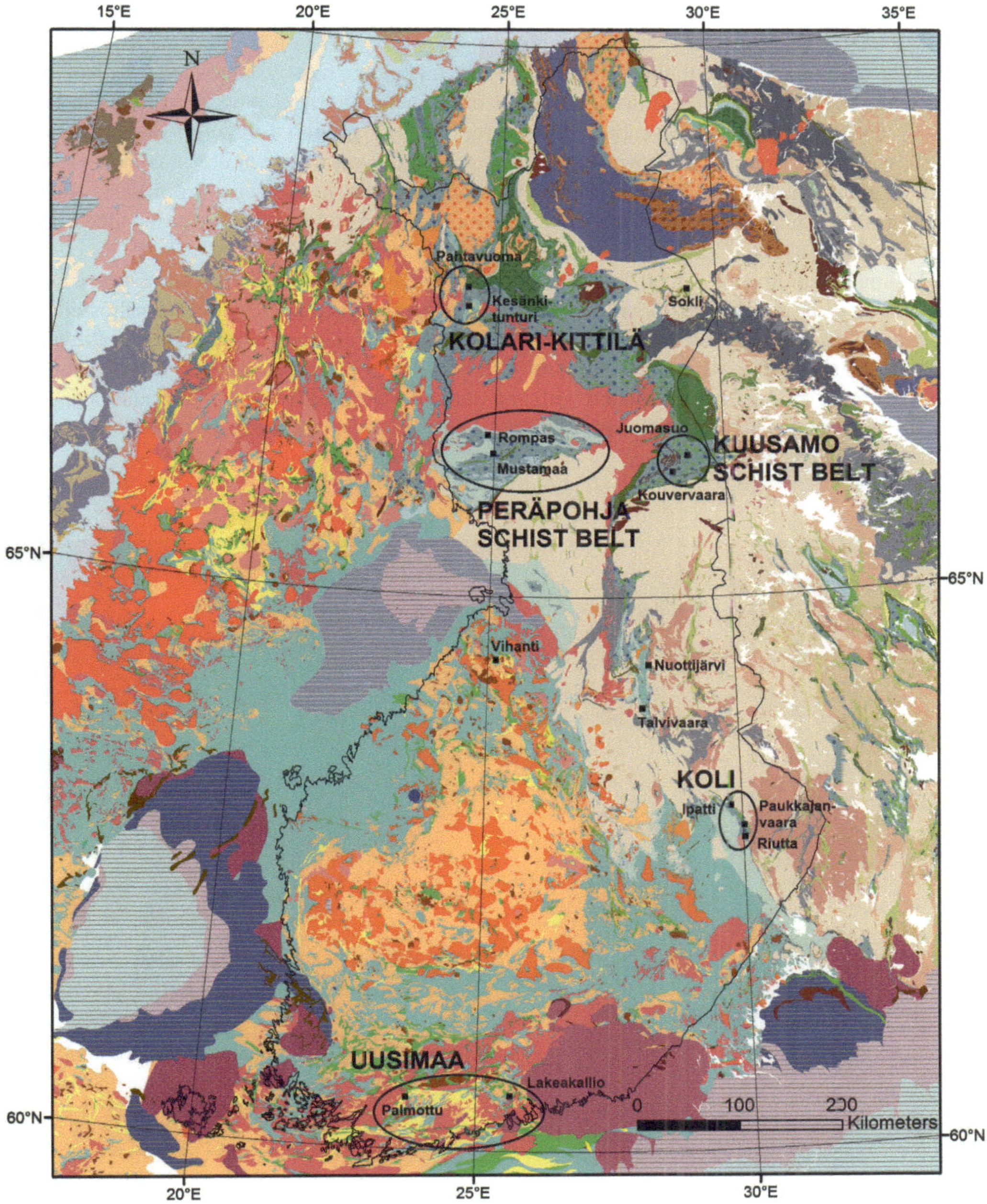

FIGURE 9.4.1 Uranium potential areas (circled) in Finland showing some of the most significant uranium deposits on the geological map of the Fennoscandian Shield.

Source: *Geology is based on Koistinen et al. (2001).*

UUSIMAA AREA

Intrusive-type uranium deposits in the Uusimaa supracrustal belt are mostly associated with late orogenic granites (1.85–1.79 Ga) and migmatites. Numerous uranium occurrences have been found in Uusimaa (Fig. 9.4.2) and most of them have been discovered through ground follow-up of aeroradiometric anomalies in granitic areas. All known uranium deposits and occurrences are small and low-grade (Pohjolainen and Äikäs, 2012). Some U-Pb analyses of uraninites and allanites from Eastern Uusimaa (Alho, Lakeakallio, and Särkijärvi) indicate uranium remobilization and enrichment during the Ordovician, ~450 Ma ago (Vaasjoki et al., 2002).

The Uusimaa uranium area can be divided into two subareas: Western Uusimaa and Eastern Uusimaa. Western Uusimaa is dominated by intrusive-type uranium occurrences, which are hosted by late orogenic granites (1.85–1.79 Ga), whereas in Eastern Uusimaa both intrusive-type and hydrothermal-style uranium with younger remobilization are found. Characteristics of this younger Eastern Uusimaa type include uranium mineralization associated with hematite alteration and shearing in granitoids, showing geologically young ages between 600 and 400 Ma inside the Paleoproterozoic Svecofennian granitoids and migmatites (Äikäs, 2003).

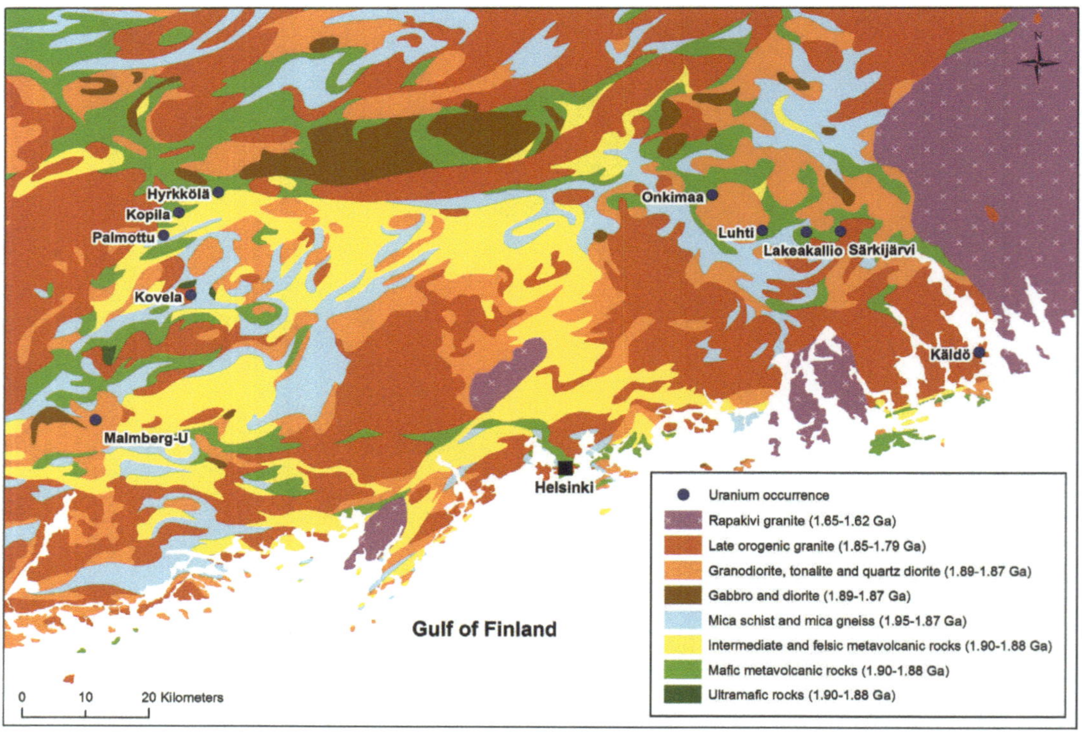

FIGURE 9.4.2 Geology of the Uusimaa area with selected uranium occurrences (Kovela is Th-REE occurrence).

Source: *Geology is based on Koistinen et al. (2001).*

The Geological Survey of Finland (GTK) carried out uranium exploration in the Uusimaa area mainly in the early 1980s. In Western Uusimaa, GTK had a drilling program in Palmottu, while in Eastern Uusimaa it tried to find the bedrock source of the high-grade Alho boulders and studied many other small occurrences. No positive exploration results were achieved.

Most of the uranium deposits and occurrences in Western Uusimaa are related to the granitic dikes and lenses in migmatitic mica gneisses. They are located along the marginal zone of the late orogenic Perniö granite and in gneisses forming country rocks to the granite. The gneisses mostly consist of garnet-bearing mica gneiss, quartz-feldspar gneiss, and pyroxene-bearing charnockite. In late orogenic granites, uraninite (UO_2) and monazite are commonly associated with biotite-rich seams and clusters, which are represented by strongly radioactive pockets in granitic outcrops.

There are two main stages of uranium mineralization in Eastern Uusimaa. Primary intrusive-type uranium deposits and occurrences mainly occur in late orogenic granites and in granitic neosomes of migmatites with age ranging from 1.85–1.79 Ga. The emplacement of the Wiborg rapakivi batholith took place at about 1640 Ma but significant primary igneous-type uranium mineralization has not been discovered in the Rapakivi area. The second stage is related to possibly supergene uranium enrichment in the Phanerozoic at 440 Ma (Äikäs, 2003).

In addition, potential unconformity-related type of uranium mineralization at the interface between Phanerozoic sedimentary cover (now eroded) and Proterozoic crystalline basement in the Eastern Uusimaa area is suggested (Äikäs, 2003). The source of uranium would be the primary uraninite and refractory U-bearing minerals in the granitoids of the crystalline basement. According to this model, the circulation of fluids and possible traps are controlled by faults and fracture zones in the basement.

PALMOTTU

The Palmottu intrusive-type uranium deposit is located in Lohja (80 km northwest of Helsinki) in the Uusimaa supracrustal belt. The deposit was discovered by GTK in 1979 during ground follow-up of an airborne radiometric uranium anomaly. Uranium exploration was carried out in Palmottu from 1979–1984 and 62 exploration drill holes (9093 m) were drilled by the GTK. The deposit has an estimated 1083 tU with an average grade of 0.11% U in the total resources 1.02 Mt (Räisänen, 1991).

Palmottu consists of several granite dikes (2–10 m) interlayered with migmatitic mica gneisses. There are two types of uraniferous dikes: (1) granite pegmatites, and (2) sheared, quartz-rich granite with biotite accumulations (Räisänen, 1989). The main ore body is discontinuous (Fig. 9.4.3) with a total length of 400 m and thickness from 1–15 m (Räisänen, 1991). The ore body continues from the surface to a depth of 400 m. Palmottu is located on the southern limb of a large fold, which has a vertical, approximately east–west trending axial plane (Kuivamäki et al., 1991). The fractures now hosting uranium-rich granitic dikes may have been opened during folding and the dikes were probably derived from the late orogenic Perniö granite (Räisänen, 1989).

The main uranium-bearing minerals at Palmottu are uraninite and monazite. Uranium is mainly incorporated in uraninite, which is typically associated with biotite accumulations in granite dikes within migmatitic mica gneiss. Alteration of uraninite has resulted in coffinite ($USiO_4$) rims around uraninite grains. In addition to uranium, uraninite contains thorium and radiogenic lead (average 17.9% PbO). High thorium content (average 7.9% ThO_2) and euhedral crystals indicate igneous origin and high-temperature conditions during uraninite crystallization. The only identified secondary U^{6+} mineral is fracture-fill uranophane (Kaija et al., 2000). Molybdenite is found in small amounts. Conventional

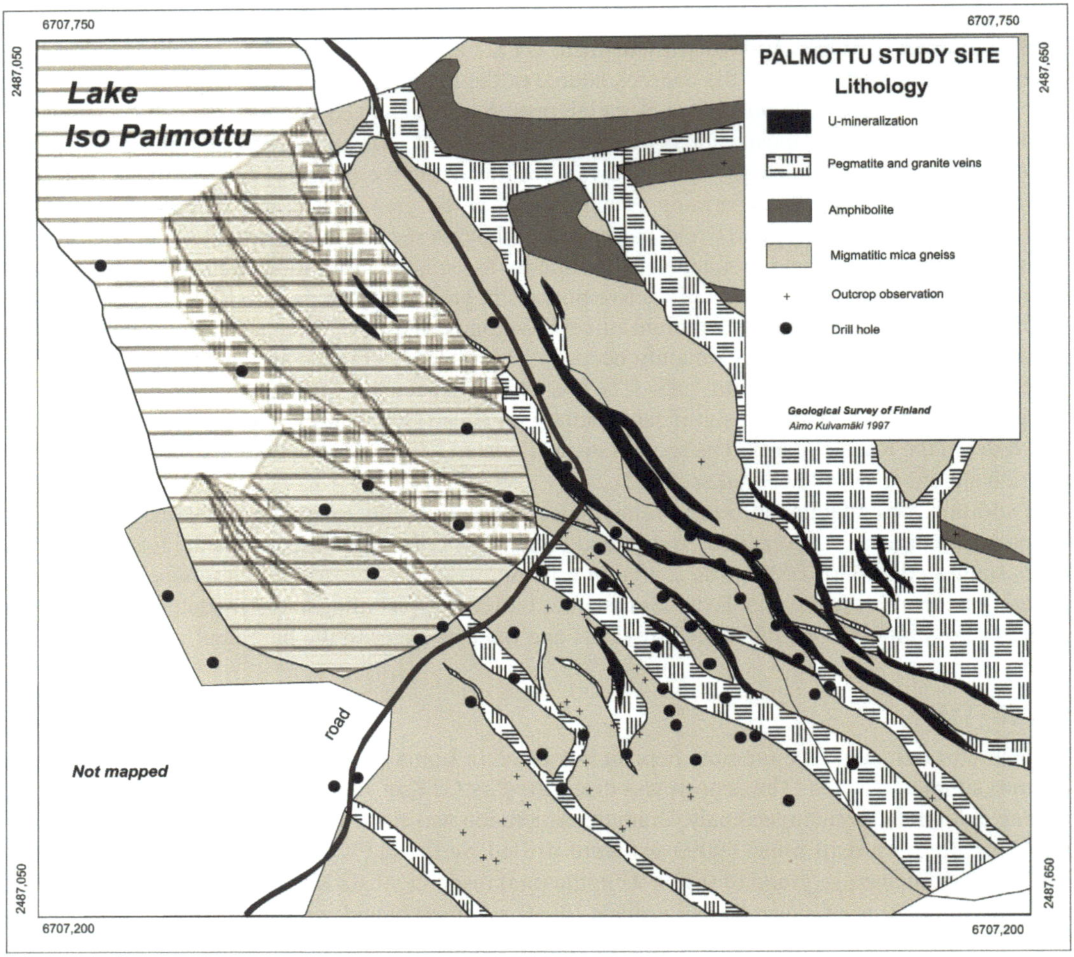

FIGURE 9.4.3 Geological map of the Palmottu uranium deposit.

Source: *Kaija et al. (2003).*

U-Pb dating of uraninite gave discordant ages (indicating the loss of lead) between 1678 and 1741 Ma (Ruskeeniemi et al., 1994). Monazite from Palmottu gives a concordant age of 1793 Ma, which may be considered as the age of uranium mineralization (Vaasjoki, 1996).

OTHER URANIUM OCCURRENCES IN WESTERN UUSIMAA

In addition to Palmottu, some minor uranium and thorium occurrences have been discovered in Western Uusimaa, such as Hyrkkölä, Malmberg-U, Kopila, and Kovela. The Hyrkkölä uranium occurrence is located 10 km northeast of Palmottu. Uraninite of Hyrkkölä has been dated at 1774 Ma, which is younger than the late orogenic granites (1830 Ma) in the area (Räisänen, 1989). Native copper has been found in Hyrkkölä as very fine films between feldspar grains.

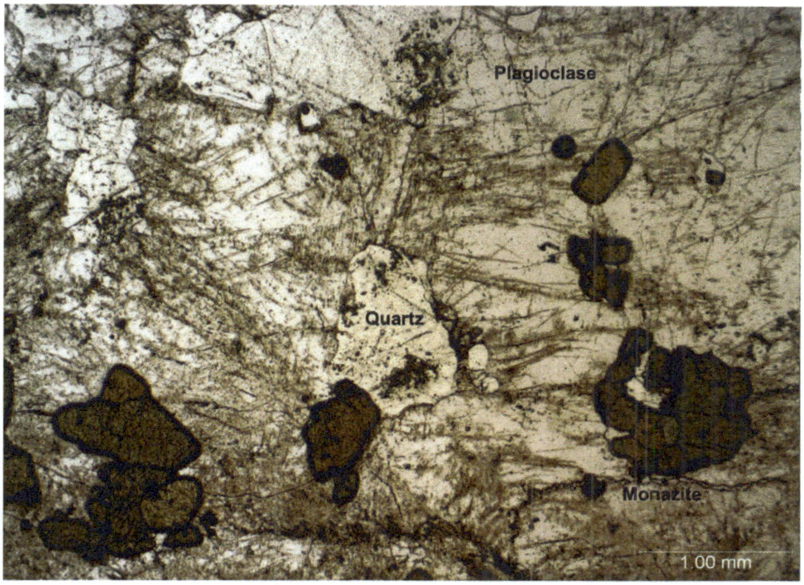

FIGURE 9.4.4 Thin section image (plane polarised light) of monazite clusters in tonalite of the Kovela occurrence.

Photo by E. Pohjolainen.

Atomienergia Oy held a claim for uranium over the ancient iron mine Malmberg in Kisko, Western Uusimaa from 1956–1959. Uranium contents of the Malmberg uranium occurrence are given for four samples, 0.29–26.86% U_3O_8 (Atomienergia Oy, 1957). The Kopila uranium showing in Somero was discovered by GTK in 1988 through ground follow-up of aeroradiometric uranium anomalies. Kopila is located 5 km northeast of the Palmottu deposit and consists of radioactive granite in association with mica gneiss and amphibolite. The showing is located at the marginal zone between amphibolite and quartz-feldspar schist. Uranium-bearing minerals at Kopila are related to a narrow biotite-rich band (Seppänen, 1991).

The Kovela monazite occurrence in Lohja is hosted by late orogenic granitoid, with a composition ranging from granite to tonalite. Monazite is rich in thorium with an average grade of 17.2% ThO_2 (Pohjolainen, 2012). Monazite occurs as clusters especially in leucocratic parts of the granitoid (Fig. 9.4.4). The Kovela monazite gives U-Pb laser ablation multicollector inductively coupled plasma mass spectrometry (LA-MC-ICPMS, GTK) concordia age of 1814 ± 6 Ma (Pohjolainen, 2012, unpublished data).

ALHO BOULDERS

Two uranium-rich glacial erratics were found by Imatran Voima Oy (IVO) in Alho, northwest of Porvoo, in 1956–1957. During carborne radiometric survey, IVO engineers found a radioactive boulder known as IVO-9 (Fig. 9.4.5) close to the road between the village of Myrskylä and the town of Porvoo. Uranium content of boulder IVO-9 is 36% U_3O_8 and the main minerals are uraninite, calcite, quartz, hematite, chlorite, and accessory pyrite (Appelqvist and Kinnunen, 1977). Uraninite is botryoidal and usually surrounded by carbonate. Fractures in the vein quartz are partly filled by uraninite, whereas cracks in

FIGURE 9.4.5 The remaining sample of Alho boulder IVO-9, photographed at the GTK Loppi storage.

The mineral with metallic reflection is uraninite. The diameter of the European 5 cent coin is 2.1 cm.

Photo by O. Äikäs.

uraninite are filled with carbonate or hematite (Äikäs, 2003). Both uraninite and hematite occur as grains and as fracture fillings.

The terrane around the boulder IVO-9 was studied carefully by IVO within a radius of a couple of hundreds of meters with radiometric ground measurements at 1-m spacing (Äikäs, 2003). In 1957, this work resulted in the discovery of a second smaller boulder, called boulder IVO-99. The uranium content of boulder IVO-99 is 12% U_3O_8 and its mineralogy is similar to boulder IVO-9.

The presence of uraninite in the Alho boulders indicates hydrothermal uranium mineralization, related to fault and shear zones in the bedrock. Uraninite dating gives a concordant $^{207}Pb/^{206}Pb$ age 456 Ma (A664A) from the boulder IVO-9 (Vaasjoki et al., 2004). Despite the field studies by IVO in the 1950s–60s and by the GTK in the 1970s–80s, the bedrock source of the Alho boulders has not been discovered.

LAKEAKALLIO

Imatran Voima Oy (IVO) carried out uranium exploration activities in the Eastern Uusimaa area between 1956 and 1959. One of the discoveries was the Lakeakallio uranium occurrence in Askola. IVO had claims and a mining concession at Lakeakallio with a field laboratory for milling and concentration tests. At Lakeakallio, IVO excavated two separate pits and other pits in the surrounding areas (e.g., at Luhti). From these ore bodies, the company mined 557 t of uranium ore with a grade 0.11% U_3O_8 and treated the ore at the Lakeakallio laboratory from 1958–1959. Milling and concentration tests at the IVO Lakeakallio field laboratory were also carried out on material from other occurrences in Eastern Uusimaa. The workings at Lakeakallio were restored in the 1980s (Äikäs, 2003).

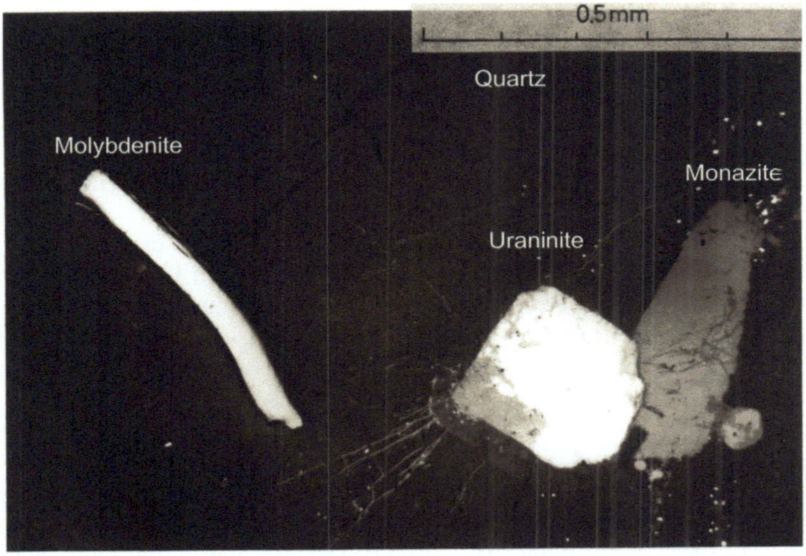

FIGURE 9.4.6 Euhedral uraninite crystal of the Lakeakallio uranium occurrence, Eastern Uusimaa.

Source: *Modified after Appelqvist (1982).*

The rocks at Lakeakallio are Svecofennian migmatitic mica gneiss and hornblende gneiss that are intruded by granitic dikes. The uraniferous host rock is granite that contains coarse-grained garnet and biotite. The mineralized rocks are sheared and hematized. The ore minerals are uraninite (Fig. 9.4.6), molybdenite, and pyrite, and accessory minerals include zircon and monazite. Remobilized uranium has been found in microcracks around the primary uraninite. The Lakeakallio zircon dating gives an age of 1835 Ma, which is considered as the age of primary Svecofennian crystallization of the late orogenic granites. Uraninite is syngenetic with the host granitoids but it was partially remobilized at about 600 Ma and experienced episodic lead loss at about 440 Ma (Vaasjoki et al., 2002).

KÄLDÖ

The Käldö uranium occurrence was discovered by A. Viento in 1956. Radioactive areas occur along a 3-km trend on the island of Käldö and in areas north and south of Käldö, at Pernaja. The bedrock of Käldö consists of granite with narrow enclaves of mica gneiss. Granite is cut by a mylonite zone 10–50 m wide, striking 30°. The uranium mineralization follows sheared fracture zones and mylonite zones. Radioactivity occurs in the mylonite zone for 300 m beginning from the southwest shore, and again for 50 m on the northeast shore of Käldö. The mylonite is fine to coarse-grained, containing enclaves of granite and supracrustal rocks. Many parts contain uraninite as thin layers on cleavage planes. The mylonite samples contain 0.1–0.3% U (Äikäs, 2003). An isotopic date has been determined from uraninite by Vaasjoki (1977), suggesting an age of 587 ± 39 Ma. Exploration activities in Käldö also included ore dressing tests at the IVO Lakeakallio field laboratory.

ONKIMAA

The Onkimaa uranium deposit in Mäntsälä was discovered by the GTK in 1972 (Appelqvist, 1980). The Svecofennian Onkimaa granodiorite is circular in plan with concordant contacts against the surrounding metasedimentary and metavolcanic rocks. The deposit is located at the northwest margin of the Onkimaa granodiorite stock. The country rock on the northwest side of the granodiorite is migmatitic mica gneiss, intruded by coarse-grained granite. Close to the contact, the granodiorite is silicified (Äikäs, 2003).

The most uniform part of the mineralized zone was delineated as a 20 m × 500 m body down to the depth of 50 m, containing 320 t U_3O_8 (in situ) at 0.01% U (Appelqvist, 1974). The mineralized rock is coarse-grained granite forming the neosome of the migmatite. The main uranium-bearing mineral is uraninite, which occurs as euhedral grains (0.15–0.25 mm in size). In addition, molybdenite is found in small amounts. Uranium mineralization of the Onkimaa deposit is syngenetic with the neosome of the migmatite and contemporaneous with the late orogenic granites.

OTHER URANIUM OCCURRENCES IN EASTERN UUSIMAA

Naarkoski is an exposed uranium showing that lies on the bottom of the Porvoonjoki River, in the rapids below the dam of a hydro power station (Äikäs, 2003). The host rock is migmatitic mica gneiss, with abundant crushed granitic material. Numerous cracks in granite are filled with hematite and these hematized portions contain several hundreds of ppm of uranium.

The Särkijärvi uranium occurrence is located 12 km west of the margin of the Wiborg rapakivi batholith. The rocks at Särkijärvi are Svecofennian granite and hornblende gneiss. The occurrence comprises several radioactive boulders and a small uraniferous patch in bedrock in a narrow shear zone at the bottom of a trench. Radioactive boulders mostly consist of hornblende gneiss, more or less with reddish stain. Uranium occurs as uraninite, partly bound into goethite, possibly as submicroscopic grains of uraninite. Native copper occurs in small amounts. The altered granitic boulders and their uranium mineralization have features common with the Alho boulders. Shearing, fracturing, and hydrothermal alteration along brittle fractures have possibly resulted in the enrichment of uranium in these structures. In Särkijärvi, the suggested sources of uranium are the primary uranium minerals in the Svecofennian granites and granitic neosomes of the migmatites (Äikäs, 2003).

KOLI AREA

The Koli uranium area is characterized by vein-type and quartzite-hosted uranium deposits. The uranium mineralization ranges from stratiform conglomerate and quartzite-hosted occurrences to epigenetic veins and breccia infills in the Proterozoic host rocks (Äikäs, 2012). The main uranium ore mineral is uraninite. The Koli uranium area follows the boundary between the Paleoproterozoic Höytiäinen belt and the Archean Eastern Finland Complex. The area comprises basal quartzites and metaconglomerates in the lowermost part of the Höytiäinen belt, above the unconformity against the Archean basement rocks.

Archean granitoids near the Proterozoic unconformity are composed of monzogranites and stromatic migmatites with potassic granite leucosomes, the youngest granitoids being 2.63 Ga in age (Äikäs, 2012). Above the unconformity, the metasedimentary sequence of the Koli belt is composed of quartzites of the Jatulian Herajärvi group. The mafic Koli sill complex, which intrudes both the Archean basement and part of the overlying Herajärvi group, was emplaced at 2.22 Ga (Äikäs, 2012). Tholeiitic dike swarms of 2.1 Ga and 1.97 Ga age groups crosscut the Archean basement and the overlying basal units of the Höytiäinen belt.

Several small quartzite-hosted uranium deposits in the metasedimentary rocks of the Herajärvi group have been discovered in the Koli area. These deposits include Ipatti, Martinmonttu, and Ruunaniemi with total 250 tU, grading 0.08–0.14% U (OECD, 2008). In some places, a close spatial association between uranium and mafic dikes is found, for instance at Paukkajanvaara. Structural and thermochronological constraints indicate that the Koli uranium area has experienced lower-amphibolite facies pressure-temperature conditions, and remained at elevated temperatures until at least 1.82 Ga (Äikäs, 2012). Late orogenic retrograde hydrothermal processes are a plausible mechanism for mobilization of uranium (Sorjonen-Ward and Äikäs, 2008).

RIUTTA

The Riutta hydrothermal vein-type uranium occurrence is located in Eno, North Karelia, at the unconformity between Archean granitoids and Paleoproterozoic metasedimentary rocks. The occurrence was discovered by Atomienergia Oy in 1958. Exploration has been conducted by Atomienergia Oy (1958–1959), Outokumpu Oy (1960–1963), the GTK (1983–1988), and AREVA (2008–2009). In total, 65 drill holes have been drilled (6275 m) and more than 500 mineralized boulders have been located at Riutta.

The bedrock at Riutta is composed of Archean granitoids and gneisses (Fig. 9.4.7), which are overlain by Paleoproterozoic metasedimentary rocks of the Jatulian Herajärvi group. At Riutta, the basal units of the Herajärvi group are mainly composed of quartzites, which are mostly separated from the Archean basement gneisses by sericite-quartz schist of the Hokkalampi formation. Sericite-quartz schist at Riutta represents either Hokkalampi paleoregolith formed in the upper parts of the Archean granitoids or highly altered and sheared Archean basement rocks. Mafic to ultramafic dikes cut across these rocks in the Koli area, representing the three main sets of mafic intrusive rocks: the 2.2 Ga layered intrusions or sills, the 2.1 Ga Fe-tholeiitic diabases, and the 1.97 Ga tholeiitic diabases (Äikäs et al., 2009).

Riutta is characterized by structurally controlled uranium ores, which are high-grade, but fragmentary and small in size. The high-grade uranium zones are mainly hosted by sericite-quartz schist (Fig. 9.4.8). The zones have structural (fracture zones and faults), mineralogical (sulfides), and lithological (sericite-quartz schist and lithological boundaries) controls. The most radioactive zones and pockets are associated with faults and fracture zones.

The main uranium mineral uraninite is related to brittle deformation in many places. Disseminated to massive uraninite occurs as veins, clusters, fracture fillings, and in the cement of hydraulic breccias. The largest and highest grade zones are found at Ristimonttu, related to fracture zones in sericite-quartz schist. Other significant ores include Unimonttu and Kolmosviuhka, both located southeast of Ristimonttu. The highest grade zone at Ristimonttu is located at the faulted and transitional contact between sericite-quartz schist and quartz diorite.

The highly radioactive zones and pockets at Riutta are commonly related to sulfides, which are in close association with fracture zones. Uraninite generally occurs with sulfides such as pyrrhotite, chalcopyrite, pyrite, molybdenite, and galena (Äikäs et al., 2009). Barren sericite-quartz schist is usually sulfide poor outside the fracture zones. Thorium is absent from unaltered hydrothermal uraninite.

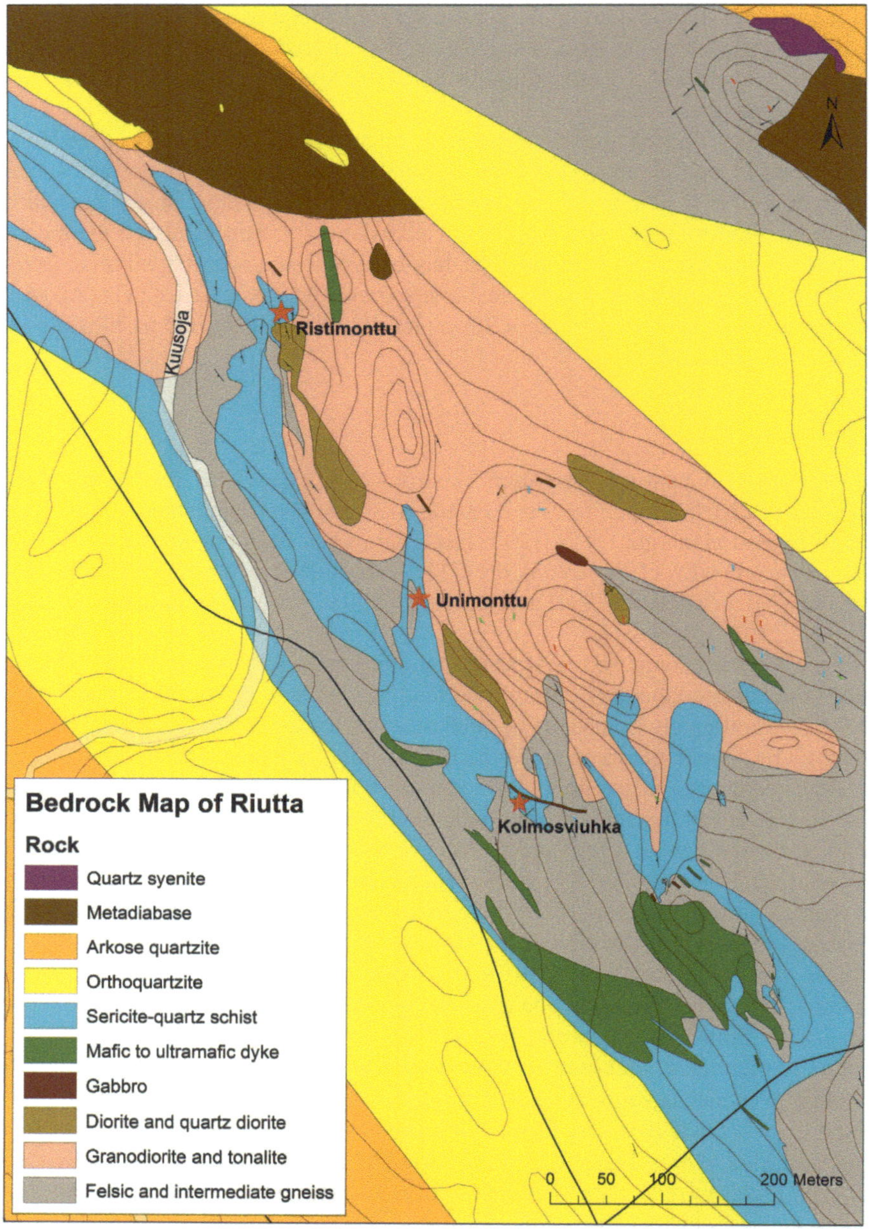

FIGURE 9.4.7 Geological map of the Riutta area with selected mineralized zones at Ristimonttu, Unimonttu, and Kolmosviuhka.

Source: *Geology is based on Äikäs et al. (2009).*

FIGURE 9.4.8 Uraninite (black) in pyrrhotite-bearing sericite-quartz schist.

Yellow secondary uranium minerals occur along the fracture surface.

Photo by E. Pohjolainen.

Strongly radioactive sections, fracture zones, and sulfides are spatially associated with each other, especially at Ristimonttu. Field observations suggest that oxidizing U-bearing solutions moved along faults and fracture zones and precipitated uraninite mainly in sericite-quartz schist. Fluid interaction with the country rocks or mixing with reducing fluids caused physicochemical changes to the hydrothermal fluid. A decrease in the oxidation state of the fluid probably led to uranium deposition.

After precipitation, uraninite was strongly altered to coffinite ($USiO_4$) by silica-rich hydrothermal solutions during the late stage of the Svecofennian orogeny. According to electron microprobe analyses, uraninite alteration was associated with strong depletion of radiogenic lead. Released lead has precipitated as grains of galena in many places. In addition, uraninite alteration resulted in a decrease in uranium and strong increase in SiO_2. Uraninite veins and infills are also fractured with later infillings of quartz and sulfides, mainly chalcopyrite (Fig. 9.4.9).

AREVA had a small drilling program at Riutta in 2008. The best drill result (DH 1) intersected 11.3 m at 0.68% U_3O_8 including 3.7 m at 1.53% U_3O_8 at Ristimonttu (Mawson Resources, 2011). This is probably the best uranium intersection in Finland and indicates the potential for high-grade and near-surface uranium ore at Riutta.

PAUKKAJANVAARA

The Paukkajanvaara uranium occurrence in North Karelia was discovered by Atomienergia Oy in 1958. Pilot-scale mining and milling were carried out at Paukkajanvaara from 1958–1961, with 0.04 Mt ore at 0.14% U treated from 1960–1961 (Äikäs, 2012). The occurrence is located at the unconformity between the Archean basement rocks and Paleoproterozoic metasedimentary rocks, intruded by mafic dikes. The uranium ore bodies are located at the contact between a mafic dike and the country rocks,

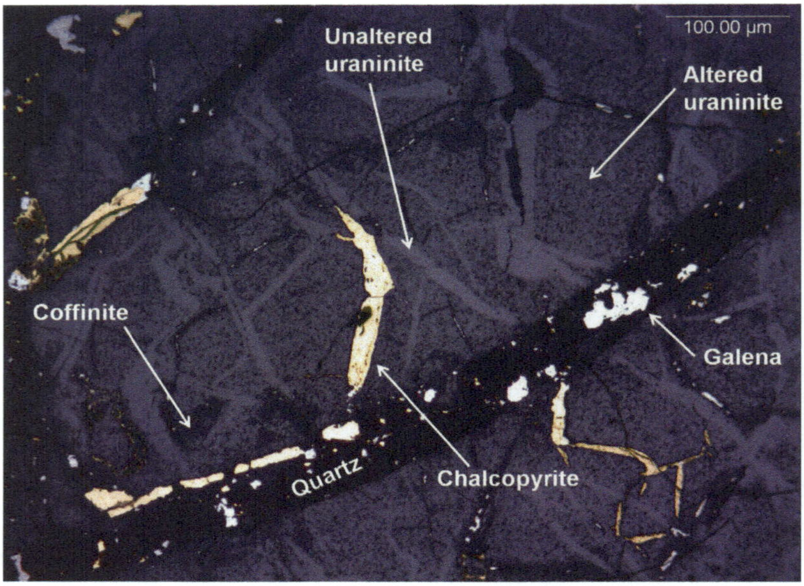

FIGURE 9.4.9 Uraninite from Kolmosviuhka, Riutta.

Uraninite (partly altered to coffinite) contains a network of uraninite veinlets, indicating remobilization of uranium. Uraninite is also fractured with later fillings of quartz, chalcopyrite, and galena. Reflected light.

Photo by E. Pohjolainen.

which are composed of quartzite, metaconglomerate, sericite-quartz schist, and Archean basement gneisses. Quartzite and metaconglomerate basal units of the Herajärvi group are separated from the Archean basement gneisses by sericite-quartz schist of the Hokkalampi paleoregolith, probably representing the same paleoregolithic or altered and sheared schist that is found at Riutta.

Paukkajanvaara consists of two main ore bodies: Mårtensson and Kunnansuo. The main uranium ore minerals are uraninite (pitchblende) and secondary uranophane. Additionally, brannerite and coffinite are accessory uranium minerals (Makkonen et al., 1988). Uranium minerals occur mainly as fracture fillings and uranium deposition postdates the emplacement of the mafic dikes.

Paukkajanvaara might be included in the subtype of sandstone uranium deposits, known as mafic dikes/sills in Proterozoic sandstones, in the International Atomic Energy Agency (IAEA) classification system of uranium deposit types. In addition to the Mårtensson and Kunnansuo ore bodies, a uranium showing known as the Kunnansuo vein occurs in a mafic dike within the Archean basement gneiss, about 100 m southeast of the Paukkajanvaara Kunnansuo pit.

KOLARI-KITTILÄ AREA

The Kolari-Kittilä uranium area is located in the southwestern part of the Central Lapland Greenstone Belt (CLGB). Uranium deposits are mostly represented by quartzite-hosted and vein types. However, styles and features of uranium occurrences are rather variable in northwestern Finland and include uranium mineralization hosted by pegmatites, albite diabase dikes, stratabound quartzites, metaconglomerates, pebbles of older uraniferous schist in metaconglomerates, stratabound biotite schists, and uraniferous veins in graphite- and sulfide-bearing schists. A mineral resource has been defined for Kesänkitunturi and Pahtavuoma-U deposits.

The Kesänkitunturi uranium deposit is hosted by Paleoproterozoic sericite quartzite of the Ylläs formation, which is discordantly overlain by younger metasedimentary rocks of the Kumpu group (Lehtonen et al., 1998). At Kesänkitunturi, uranium occurs as uraninite in the cement between quartz blastoclasts. The rock contains torbernite as a secondary uranium mineral. In addition, chalcopyrite and pyrite are present. In situ resources of Kesänkitunturi are 950 tU, with an average grade of 0.065% U for 1.45 Mt of ore (Sarikkola, 1972). The deposit is inside the Pallas-Yllästunturi National Park.

The Pahtavuoma vein-type uranium deposit consists of three ore bodies in fine-grained greenstone-associated metasedimentary graphitic schists in the Matarakoski formation of the 2.15–2.05 Ga Savukoski group of the CLGB. The lodes are thin, subvertical sets of veins and lenses, with thicknesses varying from a few centimeters to several meters. The lengths and depths of the lodes are approximately 100 m. Each lode is composed of more than one parallel vein, where uranium occurs in rich pods. In situ resources in the two best bodies are 500 tU, with an average grade of 0.39% U. The main uranium-bearing mineral is uraninite, which forms intergrowth structures with pyrrhotite and molybdenite (Korkalo, 2006). In addition to uranium, the 0.14 Mt deposit is enriched in Ag (avg. 24 g/t), Co (0.01%), Cu (0.24%), and Pb (0.09%). The uraniferous veins at Pahtavuoma are associated with shear zones and fractures younger than the main deformational event of the CLGB. Pahtavuoma is also within the Pallas-Yllästunturi National Park.

Uraninite-bearing veins are also present at Laavivuoma, 1.5 km west of Pahtavuoma. Uraninite occurs as roundish or angular grains in veins with brown amphibole, quartz, and molybdenite. The uraninite grains are partly fractured due to subsequent deformation, and grain fractures are filled with quartz. Uraninite also occurs as bands, in the cement of the microbreccia, and as inclusions with magnetite and ilmenite inside brown amphibole. Colloform uraninite (pitchblende) occurs in veins and clusters with silicates and fine-grained molybdenite (Pääkkönen, 1988).

The Aakenusvaara uranium occurrence is located 10 km east of the Pahtavuoma deposit. Before discovery, pebbles of uraniferous biotite schist were identified in outcrops of radioactive conglomerate about 5 km to the southeast from the Aakenusvaara occurrence. A subsequent search for the provenance of this type of mineralized rock led to the discovery of Aakenusvaara. The occurrence is located in the metasedimentary and metavolcanic rocks of the Pittarova formation of the Savukoski group, 2 km southeast of the Saattopora gold mine. Uraninite occurs as roundish grains with a variable degree of alteration, as dissemination, and as grain chains in biotite schist layers. Uraninite fills microcracks that cut all preceding mineral phases in the biotite schist (Pääkkönen, 1989).

KUUSAMO SCHIST BELT

Uranium is associated with many gold deposits in the Kuusamo area. The Kuusamo schist belt is characterized by polymetallic Au-Co-Cu-Fe-U deposits in the Paleoproterozoic quartzites and mafic metavolcanic rocks that are intensely albitized. Au-Co-Cu-Fe-U deposits characteristically show a complex pattern of alteration and mineralization. The host units for the polymetallic deposits include sericite quartzites, sericite schists, and mafic-felsic metavolcanic rocks. Numerous occurrences and deposits have been located during the past 50 years of exploration. In the 1980s, uranium exploration by the GTK led to the discovery of a cluster of small Au-Co-Cu-Fe-U deposits, including, for example, Juomasuo, Konttiaho, Hangaslampi, and Sivakkaharju.

Dragon Mining Ltd. is investigating the possibility of developing a gold mining operation with several mines and a central processing facility. These deposits are Juomasuo, Hangaslampi, Pohjasvaara, Meurastuksenaho, and Sivakkaharju. Uranium contents are highly variable between and within deposits. Uraninite is commonly found together with gold in fracture fillings and shear seams. Based on chemical analyses given in the Environmental Impact Assessment Report (Dragon Mining, 2013), elevated uranium contents are found in Juomasuo (158 ppm U), Hangaslampi (194–347 ppm U), and Sivakkaharju (184 ppm U), whereas in Pohjasvaara and Meurastuksenaho, uranium contents are low. Average U grade is currently available only for Juomasuo. Unconventional uranium resources (in situ) of the Juomasuo deposit, calculated from the EIA report, are not more than 210 tU in the measured and indicated mineral resources of 1.5 Mt. Currently, Dragon Mining is not planning to recover uranium as a by-product.

The suggested mineralization models for the Au-Co-Cu-Fe-U deposits in the Kuusamo schist belt include orogenic gold with untypical metal association and iron oxide-copper-gold (IOCG). Uraninite is the main uranium mineral; brannerite is also found at Hangaslampi and Konttiaho. Migration of uranium possibly happened at multiple stages with deposition and remobilization differing according to the structural control and metamorphic conditions in the different parts of the schist belt. Age datings indicate deposition of uranium at approximately 1.85–1.80 Ga, after the peak metamorphism of the Svecofennian orogeny (Vanhanen, 2001).

JUOMASUO

Juomasuo is the largest known gold deposit in the Kuusamo schist belt, located 37 km north of the town of Kuusamo. The deposit is hosted by mafic-felsic metavolcanic rocks with intercalations of ultramafic rocks and metasedimentary units. Uranium is incorporated in uraninite at many places within the gold lodes and sometimes within the cobalt mineralized zone. Gold is mainly as inclusions in pyrite, pyrrhotite, cobaltite, molybdenite, and silicates. Gold, which represents later stage precipitation or mobilization, also occurs in the cracks of uraninite, commonly with tellurides or sometimes bordering the uraninite grains (Vanhanen, 2001). The most common telluride mineral at Juomasuo is altaite (PbTe), which occurs mostly as anhedral grains in microfractures of uraninite.

Intense albitization of the host rocks took place before the main stage of mineralization at Juomasuo. After albitization, hydrothermal activity comprised successive Mg-bearing, silicic, Fe-bearing, and increasingly potassic fluid flows (Vanhanen, 2001). Subsequent fluids evolved to a sulfur-saturated stage with high Fe, Au, Co, Se, and Te concentrations. According to Vanhanen (2001), uraninite precipitated mostly during Mg-Fe metasomatism and K-metasomatism.

KOUVERVAARA

Kouvervaara is probably the largest known uranium occurrence in the Kuusamo schist belt, although there is no resource estimate. The stratabound occurrence, located 500 m south of the Kouvervaara Co-Cu-Au-deposit, is hosted by sericite quartzite. The east–west oriented ore body is more than 3 km long, whereas the thickness varies from few centimeters up to some meters. The main alteration types are sericitization and albitization. Uranium was possibly partly remobilized due to alteration and albitization (Vanhanen, 2001). The main uranium mineral is uraninite; brannerite is also found. Dust-like uraninite occurs in the matrix of the host rock along grain boundaries. Minor ore minerals include pyrrhotite, chalcopyrite, and molybdenite. Uranium content of the Kouvervaara U occurrence is mostly less than 0.1% U.

PERÄPOHJA SCHIST BELT

Uranium exploration was carried out in the Paleoproterozoic (2.4–1.9 Ga) Peräpohja schist belt in northern Finland in the 1970s–80s and again in the 2000s. Uranium occurrences discovered in the 1970s and 80s include Mustamaa, Ranta-Tulkkivaara, and Kuohunki. The most important recent uranium discoveries are Rompas-Rajapalot, Rumavuoma, and Asentolamminoja. These prospects were located by ground follow-up of airborne radiometrics and confirmed with surface sampling.

Mustamaa and Ranta-Tulkkivaara are uraniferous phosphorites. Kuohunki is located in the transition zone between the Peräpohja schist belt and the Archean Pudasjärvi complex. The Asentolamminoja uranium occurrence in Ranua is located within the Pudasjärvi complex. The mineralized zone extends from the previously known Kuohunki uranium occurrence to Asentolamminoja and beyond, with a total length of 25–30 km. In Asentolamminoja, hundreds of mineralized boulders have been identified and the area is considered prospective for vein-type hydrothermal uranium ores within granitoids. Although the Kuohunki and Asentolamminoja occurrences are hosted by Archean rocks, the deposition of uranium occurred in the Paleoproterozoic time. Uraninite ages point to a ~2.0 Ga mineralization event and possible remobilization of uranium at 1.76 Ga (Lauri and Mänttäri, 2013).

ROMPAS

The Rompas gold-uranium occurrence is located in the Ylitornio municipality, approximately 50 km west of the town of Rovaniemi. The initial discovery area, Rompas, was discovered by AREVA Resources Finland Oy in 2008 as part of regional uranium exploration. Mawson Resources Ltd. purchased the exploration assets of AREVA in 2010. Rompas is defined over a 6-km strike and 200–250-m width. Diamond drilling results comprise narrow but high-grade Au-U intersections, including 6 m at 617 g/t Au (Eilu and Niiranen, 2013). The Rajapalot target, 8 km east of Rompas, is a new discovery made by Mawson Resources Ltd.

The Rompas structurally controlled hydrothermal vein-type gold and uranium ores are mostly hosted by calc-silicate veins in mafic metavolcanic rocks. Gold is in many places intimately associated with uraninite and sometimes with carbonaceous pyrobitumen (Molnár et al., 2014). Gold typically occurs in microfractures of uraninite (Fig. 9.4.10) and postdates uraninite.

Thorium is in many places present in uraninite and its content is mostly low, but higher concentrations are also found. Radiogenic lead content is about 20 wt% PbO in unaltered uraninite. Uraninite typically occurs as coarse-grained and fragmented grains surrounded by carbonate, with compositions varying from ankerite and ferroan dolomite to calcite. Locally, fragmented uraninite grains or small euhedral crystals of uraninite also occur in pyrobitumen nodules. Pyrobitumen formation was probably controlled by radiolytic polymerization of hydrocarbons around uraninite grains. Pyrobitumen-free uraninite might represent grains that never received a hydrocarbon flux. Gold also occurs in cracks of pyrobitumen and along contacts between pyrobitumen and carbonate (Molnár et al., 2014).

The precipitation of gold appears to have been preceded by fracturing of uraninite, which was accompanied or followed by the introduction of gold-bearing fluids (Pohjolainen et al., 2014). Native gold occurs along late brittle fractures and in microfractures of uraninite, typically accompanied by galena (PbS), altaite (PbTe), hunchunite (Au_2Pb), nickeline (NiAs), and cobaltite CoAsS (Molnar et al., 2014; Fig. 9.4.11). The total amount of sulfides is rather small at Rompas.

FIGURE 9.4.10 Gold in microfractures of uraninite, Rompas, under reflected light.

Photo by E. Pohjolainen.

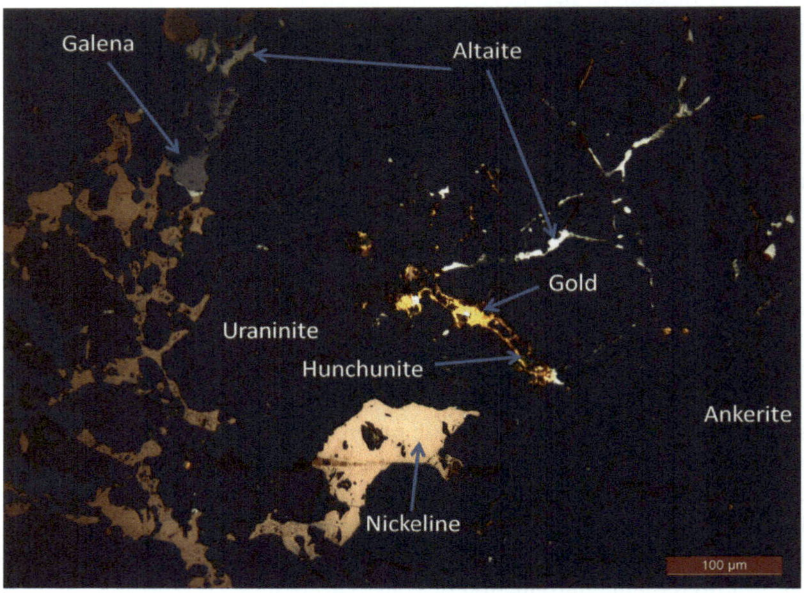

FIGURE 9.4.11 Au-, Pb-, Te-, and As-bearing minerals in microfractures of uraninite under reflected light.

Photo by E. Pohjolainen.

Mineral paragenesis and textural setting suggest that gold was transported by hydrothermal fluids that were enriched in gold, tellurium, arsenic, nickel, cobalt, bismuth, silver, copper, and sulfur. Gold precipitation from reducing hydrothermal fluids was probably associated with desulfurization and fluid interaction with radiogenic lead of uraninite, which led to galena, altaite, and hunchunite crystallization and gold precipitation from fluid (Pohjolainen et al., 2014). This model is supported by highly radiogenic Pb contents in galena, altaite, and hunchunite in close association with uraninite and gold (Molnar et al., 2014). However, the model requires significant time between uraninite crystallization and Au-bearing fluid flow due to the long half-life of U, eventually decaying into stable Pb.

URANIUM DEPOSITS NOT INCLUDED IN ANY SPECIFIC URANIUM POTENTIAL AREAS

TALVIVAARA

The Talvivaara black schist-hosted Ni-Zn-Cu-Co-U deposit is located in the Paleoproterozoic Kainuu schist belt in Sotkamo, eastern Finland. The ore is hosted by metamorphosed and intensively folded black schists of the Lower Kaleva sequence. Deposition of sulfide- and metal-rich carbonaceous wacke-pelite sediments took place in a riftogeneous setting approximately 2.0–1.9 Ga. The basin was characterized by a high sedimentation rate of organic matter, leading to anoxic conditions when organic material was preserved in the bottom sediments. The sediments were deformed and metamorphosed to lower amphibolite facies during the 1.91–1.79 Ga Svecofennian orogeny. Graphitic-sulfidic metasedimentary units are strongly enriched in redox-sensitive metals (As, Fe, Mo, Sb, Se, V, U) and base metals (Cu, Zn, Ni) relative to average upper crust (Kontinen, 2012).

Black schists of Talvivaara are characterized by the homogenous distribution and simple mineralogy of uranium. The ore contains 17 ppm U on average and most of the uranium is incorporated in uraninite, which is mainly surrounded by carbonaceous nodules (Fig. 9.4.12). Grain shapes of uraninite are typically angular but can range from rounded to euhedral. Carbonaceous nodules are typically rounded and globular, sometimes irregular. Sulfur content of nodules is highly variable, mostly between 5 and 21% SO_3 based on microprobe analyses (Pohjolainen, 2013). Small amounts of uranium are also hosted by monazite and xenotime, which are metamorphic phosphate minerals in these rocks.

In addition to uranium, uraninite contains thorium and radiogenic lead. Uranium content varies between 57.5 and 69.3 wt% UO_2 (average 64.1% UO_2). Lead content is between 11.8 and 22.4% PbO (average 18.2% PbO) and thorium between 2.8 and 10.9% ThO_2 (average 6.7% ThO_2). Uraninite alteration has resulted in the strong depletion of lead, an increase in Si, and partial replacement by coffinite (Pohjolainen, 2013). In addition, alteration is associated with a decrease in U and Th contents and a decrease in the analyses totals. The carbonaceous nodules have to some extent protected uraninite from alteration (Lecomte et al., 2014).

The primary enrichment of uranium in shales was synsedimentary. Uranium precipitated from the water column into the sediments as microscale uranium oxide by adsorption and reduction processes (Lecomte et al., 2014) in oxic-anoxic stratified restricted-marine basin (Kontinen, 2012). Primary dispersed uranium crystallized as coarser grained uraninite during metamorphism related to the Svecofennian orogeny at 1.88–1.87 Ga (Lecomte et al., 2014). In addition, high Th content and euhedral crystals indicate high T conditions during uraninite crystallization under amphibolite facies conditions.

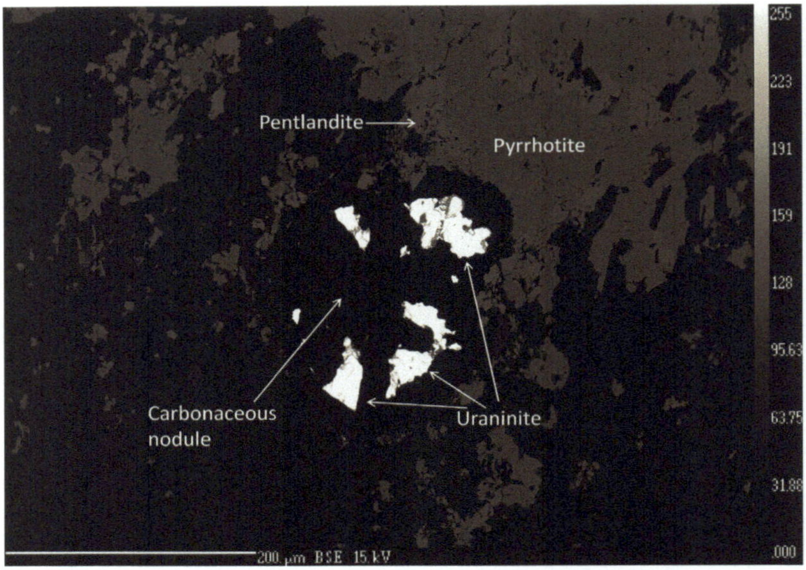

FIGURE 9.4.12 BSE image of fragmented uraninite grains surrounded by carbonaceous nodule in the Talvivaara black schist.

Carbonaceous matter of the nodule contains 20.2% SO_3.

Source: *BSE image by L. Pakkanen, GTK.*

The origin of uraninite and the host carbonaceous nodules is not fully understood. Two main hypotheses have been forwarded: (1) coprecipitation and (2) postmetamorphic formation of nodules. According to the coprecipitation hypothesis, uranium partly existed in urano-organic complexes during diagenesis. During subsequent metamorphism, uranium crystallized as uraninite leaving the organic matter to form the carbonaceous nodules. According to the second hypothesis, postmetamorphic formation of nodules is related to radiolytic polymerization of light hydrocarbons around existing uraninite grains. Hydrocarbons migrated through microfractures and were fixed by uraninite due to alpha and beta radiation-induced polymerization of hydrocarbons. Hence, uraninite grains without carbonaceous rims at Talvivaara may record sites that never received a hydrocarbon flux (Pohjolainen, 2013). Radiolytic polymerization and postmetamorphic formation of carbonaceous nodules is supported by undeformed and rounded form of nodules.

The Talvivaara open pit mine, operated by Talvivaara Sotkamo Ltd., began production in 2008. The company has applied heap leaching to extract metals from black schist ore. In heap leaching, uranium dissolves in the pregnant leach solution (PLS) along with the main base metals (Ni, Zn). Dissolved uranium has mostly ended up in the gypsum pond tailings and partly in the Ni-Co sulfide concentrate product. In 2010, Talvivaara released its plans to recover uranium as a by-product using a solvent extraction process. The licensing process for U production was unfinished as of December 2014. In November 2014, Talvivaara Sotkamo Ltd. filed for bankruptcy due to financial problems. Unconventional by-product uranium resources of Talvivaara are approximately 22,000 tU (OECD, 2014) in the measured and indicated resources of 1305 Mt, and about 35,000 tU in the total mineral resources of 2053 Mt, calculated from the resource update by Talvivaara Mining Company Plc (Talvivaara, 2012).

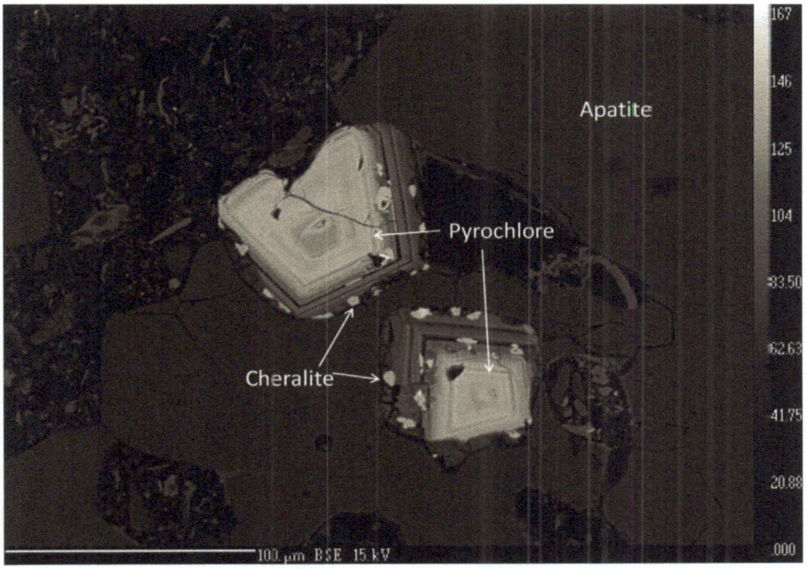

FIGURE 9.4.13 Back-scattered electron image of euhedral, thorium-bearing pyrochlore crystals as inclusions in gray-colored apatite of the Sokli regolithic phosphate ore.

BSE images, coupled with results of wavelength-dispersion analyses, reveal compositional zoning in pyrochlore. Small, light-colored grains are cheralite crystals. Radial cracks around pyrochlore crystals suggest volume increase of metamict pyrochlore.

Source: *BSE image by L. Pakkanen, GTK.*

SOKLI

Yara International ASA is planning to begin phosphate mining in Sokli, northeastern Finland. Sokli is located in Savukoski, eastern Lapland, 12 km from the Finnish-Russian border. The Sokli phosphorus deposit is hosted by the Devonian Sokli carbonatite complex (365 Ma), which is the westernmost intrusion of the Kola alkaline province. The Sokli phosphorus ore is hosted by the surface weathered zone (regolith) of the carbonatite, with the ore mainly consisting of apatite. The ore is enriched in uranium and thorium relative to average upper crust and uranium is a potential by-product at Sokli (OECD, 2014).

In the hardrock carbonatite, uranium-bearing pyrochlore occurs in specific zones, which, at a grade of 0.01% U, have been evaluated to contain 2500 tU (OECD, 2012). Uranium and thorium mostly occur in pyrochlore-group minerals in the Sokli hardrock carbonatite. Uranium partly dissolved from pyrochlore during weathering processes and was rebound in secondary apatite of the regolithic phosphate ore (Lindqvist and Rehtijärvi, 1979). Despite the complex weathering, leaching, recrystallization, and lithification processes, most uranium is still bound by pyrochlore-group minerals (Fig. 9.4.13) in regolith. As a weakly soluble element, thorium occurs in primary minerals (pyrochlore-group and monazite-group minerals) of the carbonatite. In addition to pyrochlore-group minerals, the minor uranium-bearing minerals in regolith are secondary apatite, fine-grained goethite, and altered residual magnetite (Rehtijärvi and Lindqvist, 1978).

URANIFEROUS PHOSPHORITES

Sedimentary phosphorites contain low concentrations of uranium in fine-grained apatite. Uranium of this type is considered as an unconventional resource. Uraniferous phosphorites are associated with sedimentary carbonates of the Paleoproterozoic sequences. More than 20 occurrences of uraniferous phosphorite and phosphatic rocks are known in Finland. Uranium mineralized zones are stratabound, mostly occurring in Paleoproterozoic calcareous metasediments in the Svecofennian and Karelian domains. The most significant occurrences of uraniferous phosphorites are Vihanti-U (0.03% U, 700 tU) in the Northern Ostrobothnia region and Nuottijärvi (0.04% U, 1 000 tU) in Paltamo. Other uraniferous phosphorite occurrences in Finland include Temo in Nilsiä, Savijärvi in Pielavesi, Toso in Siilinjärvi, Harjakangas in Noormarkku, Ruotsalo in Kälviä, Isokylä in Veteli, Ylipää-S in Pattijoki, Manderbacka in Kruunupyy, Losonvaara in Sotkamo, and Näärinki in Juva.

SUMMARY

Uraninite (ideally UO_2) is the main ore mineral in the uranium deposits of Finland. Uranium mineralization types can be divided into two main categories according to uraninite characteristics: magmatic and hydrothermal. Primary intrusive-type uranium deposits are characterized by magmatic, commonly euhedral uraninite, which has crystallized from granitic melts. In addition to uranium and radiogenic lead, magmatic uraninite usually contains thorium. This results from quite similar crystal chemistry of uranium and thorium as relatively insoluble 4+ valence states in magmatic systems. Palmottu is the best-known intrusive-type uranium deposit in Finland. In some cases, magmatic uranium-bearing minerals are dominated by other minerals (e.g., uranpyrochlore in the Sokli carbonatite).

Hydrothermal uraninite typically occurs as veins, sometimes in massive or colloform and botryoidal (pitchblende) form. Ore-forming processes are controlled by timing, source, fluids, and the trap for uranium. A potential source of uranium is primary uraninite and refractory U-bearing minerals (e.g., monazite) in granitoids and metasedimentary rocks. In hydrothermal systems, transport of hexavalent uranium (U^{6+}) typically occurs as soluble uranyl $(U^{6+}O_2)^{2+}$ complexes in oxidizing hydrothermal solutions along faults. Uraninite precipitation is often controlled by fluid interaction with country rocks or mixing with reducing fluids, causing physicochemical changes to a hydrothermal fluid, especially a decrease in the oxidation state of a fluid. Thorium is usually low or absent from purely hydrothermal uraninite, because thorium does not adopt the distinctive highly soluble 6+ valence state of uranium. The hydrothermal uranium ores are commonly structurally controlled and related to fracture and fault zones. Hydrothermal uraninite occurs especially in vein-type uranium ores in Finland, represented e.g., by the Riutta occurrence in North Karelia.

Other mineralization mechanisms include synsedimentary uranium enrichment with later reworking and crystallization of uranium minerals during regional metamorphism, for instance in the cases of the uraniferous phosphorites and Talvivaara black schists. Ore-forming processes relating to Paleoproterozoic quartzite hosted uranium deposits include uranium deposition and reduction from oxidizing uranium-bearing solutions controlled by reductants in sedimentary sequence.

Generally, uranium ores in Finland have not been studied in detail due to lack of economic deposits. Therefore, more detailed research is needed for understanding ore-forming processes. An indication of rather young uranium remobilization in Eastern Uusimaa is not well understood. In the Peräpohja and Kuusamo schist belts, supracrustal sequences with evaporite signatures are potential fluid sources for saline and oxidizing hydrothermal uranium-bearing solutions. Many base and precious metal deposits in Finland contain uraninite, which is surrounded by carbonaceous nodules, but the formation mechanism of these nodules or their relation to the ore-forming processes are not fully understood. Additionally, gold occurs with uraninite in some gold deposits in Finland, and therefore precipitation mechanisms of gold in these ore bodies should be studied more closely.

ACKNOWLEDGMENTS

The author is grateful to Olli Äikäs, Laura Lauri, and Hugh O'Brien for constructive comments on the manuscript.

REFERENCES

Äikäs, O., 2003. Uranium in the Itä-Uusimaa area, Finland. Geological Survey of Finland, unpublished report CM60/2003/2, p. 34.

Äikäs, O., 2012. Koli U. Geological Survey of Finland 53, 254–255 Special Paper.

Äikäs, O., Pohjolainen, E., Sorjonen-Ward, P., Damstén, M., 2009. Uranium exploration for AREVA Resources Finland Oy in the Eno-Kontiolahti study area. Geological Survey of Finland. archive report, M60/2009/72.

Appelqvist, H., 1974. Selostus geologisen tutkimuslaitoksen uraanitutkimuksista Mäntsälän Onkimaalla vuosina 1972–1974. Geological Survey of Finland, unpublished report M19/2044/74/4/10, p. 12 (in Finnish).

Appelqvist, H., 1980. Kaivoslain mukainen tutkimustyöselostus kaiv. rek. n:o 2440/1–2. KTM Report n:o 107/460/74 (in Finnish).

Appelqvist, H., 1982. Uraanitutkimukset Askolan Lakeakallion alueella vuosina 1978–1980. Geological Survey of Finland, unpublished report M19/3022/-82/1/60 (in Finnish).

Appelqvist, H., Kinnunen, K., 1977. Raportti geologisen tutkimuslaitoksen uraanitutkimuksista Askolan-Porvoon alueella 1976–1977. Geological Survey of Finland, unpublished report M19/3022/77/1/10, p. 12 (in Finnish).

Atomienergia Oy, 1957. Malmberg 1 (Malmbergin kaivos). Unpublished report. Mining register number 1160/1 (in Finnish).

Dragon Mining, Oy, 2013. Kuusamon kultakaivoshankkeen ympäristövaikutusten arviointiselostus. (in Finnish).

Eilu, P., Niiranen, T., 2013. Mineral deposit research for a high-tech world. In: Proceedings of the 12th Biennial SGA Meeting, August 16–18. Gold deposits in northern Finland: excursion guidebook FIN1, Uppsala, Sweden.

Kaija, J., Blomqvist, R., Suksi, J., Rasilainen, K., 2000. The Palmottu Natural Analogue Project, Summary Report 1996–1999. The behaviour of natural radionuclides in and around uranium deposits, Nr. 13. Geological Survey of Finland. Report YST-102.

Kaija, J., Rasilainen, K., Blomqvist, R., 2003. IAEA Coordinated Research Project (CRP): The use of selected safety indicators (concentrations, fluxes) in the assessment of radioactive waste disposal. Report 6: Site-Specific Natural Geochemical Concentrations and Fluxes at the Palmottu U-Th Mineralisation (Finland) for Use as indicators of Nuclear Waste Repository Safety. Geological Survey of Finland. Report YST-114.

Koistinen, T., Stephens, M.B., Bogatchev, V., et al. (comps.), 2001. Geological map of the Fennoscandian Shield 1:2,000,000. Geological Survey of Finland, Espoo.

Kontinen, A., 2012. Talvivaara Ni-Zn-Cu. Geological Survey of Finland 53, 276–280 Special Paper.

Korkalo, T., 2006. Gold and copper deposits in Central Lapland, northern Finland, with special reference to their exploration and exploitation. Acta Universitatis Oulensis, Series A, Scientiae Rerum Naturalium 461.

Kuivamäki, A., Paananen, M., Kurimo, M., 1991. Structural modelling of bedrock around the Palmottu U-deposit: progress report 1990. Geological Survey of Finland.

Lauri, L.S., Mänttäri, I., 2013. Paleoproterozoic U-mineralization in the Archean Pudasjärvi complex, northern Fennoscandian Shield. Mineral deposit research for a high-tech world. Proceedings of the 12th Biennial SGA Meeting, August 12–15, Uppsala, Sweden, pp. 1598–1600.

Lecomte, A., Cathelineau, M., Deloule, E., et al., 2014. Uraniferous bitumen nodules in the Talvivaara Ni-Zn-Cu-Co deposit (Finland): influence of metamorphism on uranium mineralization in black shales. Mineralium Deposita 49, 513–533.

Lehtonen, M., Airo, M.-L., Eilu, P., et al., 1998. The stratigraphy, petrology and geochemistry of the Kittilä greenstone area, northern Finland: a report of the Lapland Volcanite Project. Geological Survey of Finland. Report of Investigation, p. 140.

Lindqvist, K., Rehtijärvi, P., 1979. Pyrochlore from the Sokli carbonatite complex, northern Finland. Bull. Geol. Soc. Finland 51, 81–93.

Makkonen, H.T., Mikkonen, J.A., Ruotanen, K.E., et al., 1988. Paukkajanvaaran uraanimineralisaatiot. Pohjois-Karjalan malmiprojekti. Raportti 17. Oulu: Oulun yliopisto. p. 39 (in Finnish).

Mawson Resources Ltd, 2011. Mawson Intersects 11.3 Metres at 0.68% U3O8 at Riutta in Finland. News Release, September 20.

Molnár, F., Pohjolainen, E., Johanson, B., et al., 2014. Association of bonanza grade gold with pyrobitumen in the Rompas Au-U prospect, Peräpohja schist belt, northern Finland. Abstract volume: Proceedings of the 21st meeting of the International Mineralogical Association. September 1–5, Johannesburg, South Africa. p. 87.

OECD, 2008. Uranium 2007: Resources, production and demand. A Joint Report Prepared by the OECD Nuclear Energy Agency and the International Atomic Energy Agency.

OECD, 2012. Uranium 2011: Resources, production and demand. A Joint Report Prepared by the OECD Nuclear Energy Agency and the International Atomic Energy Agency.

OECD, 2014. Uranium 2014: Resources, production and demand. A Joint Report Prepared by the OECD Nuclear Energy Agency and the International Atomic Energy Agency.

Pääkkönen, K., 1988. Tutkimustyöselostus Kittilän ja Muonion kunnissa valtausalueilla Kolvakero 1–3, kaiv.rek. n:o 3315 sekä Laavivuoma 1–2 ja Kolvakero 4, kaiv.rek. n:o 3572 suor. malmitutkimuksista vv. 1982–1985. Geological Survey of Finland, unpublished report M06/2741/88/1/60 (in Finnish).

Pääkkönen, K., 1989. Uraanimalmitutkimukset Aakenusvaaran länsiosassa Kittilässä 1981–82. Geological Survey of Finland, unpublished report M19/2741/-89/1/60 (in Finnish).

Pohjolainen, E., 2012. Preliminary REE potential mapping in the Puumala-Ruokolahti area, South-Eastern Finland. Geological Survey of Finland. archive report, 54/2012.

Pohjolainen, E., 2013. Uranium as by-product from heap leaching of Talvivaara black schist nickel ore, Finland. Unpublished MS PowerPoint file of the presentation given in the IAEA Technical Meeting on Classification of World Uranium Deposits, Vienna, February 19–22. Geological Survey of Finland.

Pohjolainen, E., Äikäs, O., 2012. Palmottu U. Geological Survey of Finland 53, 213–215 Special Paper.

Pohjolainen, E., Molnár, F., Sorjonen-Ward, P., 2014. Uraniferous polymetallic mineral deposits in Finland. Unpublished MS PowerPoint file of the presentation given in the IAEA Technical Meeting on Uranium as a By-Product with an Emphasis on Base and Precious Metal Deposits, Vienna, November 24–26. Geological Survey of Finland.

Räisänen, E., 1989. Uraniferous granitic veins in the Svecofennian schist belt in Nummi-Pusula, southern Finland. In: Uranium deposits in magmatic and metamorphic rocks: Proceedings of a technical committee meeting, Salamanca, September 29–October 3, 1986. International Atomic Energy Agency, Vienna. pp. 37–44.

Räisänen, E., 1991. Tutkimustyöselostus ja malmiarvio Nummi-Pusulan kunnassa valtausalueilla Palmottu 1–2, kaiv. rek. N:o 3243 suoritetuista uraanimalmitutkimuksista. Geological Survey of Finland, unpublished report M06/2023/-91/10 (in Finnish).

Rehtijärvi, P., Lindqvist, K., 1978. Uraani ja torium eräissä uraaniesiintymien näytteissä: tiivistelmä menetelmistä ja tutkimustuloksista. Helsingin yliopisto. Geologian laitos. Tiedonanto 7. Helsinki: Helsingin yliopisto. Geologian laitos (in Finnish).

Ruskeeniemi, T., Blomqvist, R., Ahonen, L., 1994. Uraninite and its alteration at Palmottu—a possible natural analogue for spent fuel under reducing conditions. The behaviour of natural radionuclides in and around uranium deposits, Nr. 7. In: The Palmottu Analogue Project, Progress Report 1993. Geological Survey of Finland. YST-85, 53–64.
Sarikkola, R., 1972. Väliraportti, Kolari, Kesänkitunturi. Outokumpu Oy, unpublished report 050/2732/RSA-72, p. 34 (in Finnish).
Seppänen, H., 1991. Raportti uraanitutkimuksista Someron Kopilassa vuosina 1988–1990. Geological Survey of Finland. archive report, M19/2024/-91/1/60 (in Finnish).
Sorjonen-Ward, P., Äikäs, O., 2008. Structural framework and event history relevant to uranium mineralization near the Palaeoproterozoic-Archaean unconformity in southeastern Finland. Proceedings of the Thirty-Third International Geological Congress (Abstract, MRD 19 Uranium Deposits), Oslo, August.
Talvivaara Mining Company Plc, 2012. Talvivaara Operational and Resource Update. Stock Exchange Release, November 28.
Vaasjoki, M., 1977. Phanerozoic resetting of U-Pb ages in some South-Finnish uraninites. In: ECOG V. Fifth European Colloquium of Geochronology. Cosmochronology and Isotope Geology. Pisa, September 5–10.
Vaasjoki, M., 1996. U-Pb dating of monazite from Palmottu. Geological Survey of Finland. Report 15.8.
Vaasjoki, M., Appelqvist, H., Kinnunen, K.A., 2002. Paleozoic remobilisation and enrichment of Proterozoic uranium mineralisation in the East-Uusimaa area, Finland. In: Lithosphere 2002: 2nd Symposium on the Structure, Composition and Evolution of the Lithosphere in Finland: Programme and Extended Abstracts, Espoo, Otaniemi, November 12–13. Institute of Seismology, University of Helsinki. Report S-42, p. 139.
Vaasjoki, M., Sundblad, K., Alm, E., Huhma, H., 2004. Paleozoic supergene processes in the Precambrian Fennoscandian Shield: radiogenic isotope evidence. In: Mertanen, S. (Ed.), Proceedings of the 5th Nordic Paleomagnetic Workshop: Supercontinents, Remagnetizations and Geomagnetic Modeling Extended Abstract Volume. Geological Survey of Finland Report Q29.1/2004/1, 129–132.
Vanhanen, E., 2001. Geology, mineralogy and geochemistry of the Fe-Co-Au-(U) deposits in the Paleoproterozoic Kuusamo schist belt, northeastern Finland. Geological Survey of Finland. Bulletin 399.

CHAPTER

INDUSTRIAL MINERALS AND ROCKS 9.5

M.J. Lehtinen

ABSTRACT

Mining of industrial minerals and rocks in Finland dates back at least to the year 1329 when the first known mining for limestone in southwestern Finland took place. It has developed into an important industry in Finland, and from 1949–1966 and 1983–2009, mining of industrial minerals and industrial rocks exceeded tonnages of the metallic ores. However, this trend has reversed recently, and by 2013 it totaled 15.4 Mt, or about 42% of the total ore mining in Finland. Most of the currently exploited deposits were discovered before 1960. Since the 1950s, important technological advances—often in collaboration with domestic clients such as the paper industry—have been made in mineral processing and product development for a number of commodities such as talc, apatite, TiO_2, feldspars, and wollastonite. The domestic restructuring of the nonmetallic sector began in the late 1980s, and while the entry of Finland into the European Union (EU) in 1995 opened new export markets, it also brought about increased competition, resulting in the takeover of most of the main Finnish producers by leading foreign companies. Some earlier domestic production has been replaced by substitutes, or even raw material imports, but the huge influx of foreign capital and investments seen in the metallic mining sector has not materialised to the same degree in the nonmetallic sector.

Keywords: industrial minerals; industrial rocks; geology; production; Finland.

INTRODUCTION

The term *industrial minerals* is still rather purely defined, varying at extremes between the academic and end-user terminologies. According to Harben and Bates (1990): "The industrial minerals, broadly defined, include all those materials that man takes out of the earth's crust except for fuels, metallic ores, water, and gemstones. A commonly used synonym is *nonmetallics*; a somewhat more precise one is *industrial minerals and rocks*," which will be used in this text. A more recent review of the classifications was completed by Jeffrey (2006). Pondering the limitations and difficulties of the evolving classifications, he concludes: "A robust classification system must address the geological, compositional, economic, and end-use properties of each commodity." What makes industrial minerals and rocks different from their metallic counterparts is the critical need for development work with the end-users, by which mineral products are tailored based on the variable geological possibilities of each deposit. Good examples of this are found in Finland.

Mineral Deposits of Finland. http://dx.doi.org/10.1016/B978-0-12-410438-9.00026-1

The product range of industrial minerals is large, varying from low-value, locally used bulk products, to high-value niche products (e.g., those used in nanotechnology). Companies are also increasing their efforts to convert as much of all mined rock into marketable products as possible. This chapter will not cover all industrial minerals and rocks, because carbonatites and kimberlites, as well as the high-tech metals, are described in separate chapters. Also, even though the main Ti-oxide minerals are metallic minerals, they are mostly classified as industrial minerals, as 95% of their use is as high-brightness TiO_2 pigments. Even chromite has important industrial mineral uses, but is clearly thought of as metallic. In this book, all of these minerals are described in detail in Subchapters 3.4 and 9.2.

Industrial rocks are rocks utilized as such, without specific mineral separation processes, as raw materials for making products. As examples, the rocks used in making stone wool or soapstone for fireplace manufacturing can be classified as industrial rocks. They can also be classified as natural stone, whereas talc and magnesite floated from soapstone, are industrial minerals. These increasingly important types of materials are also classified as industrial rocks.

The publications on *Industrial Minerals and of Rocks in Finland* include some older summaries (e.g. Aurola, 1954a, 1964a, 1964b, Boström, 1986), Industrial Minerals and Industrial Rocks in Finland by Haapala (1988), and a number of more recent comprehensive review papers covering the broader topic of the Nordic countries. These include Lehtinen (2006), Johansson et al. (2008), and the latest by Kananoja et al. (2013). The industrial mineral deposit map of the Fennoscandian Shield (1:2,000,000) was published in 2013 (Gautneb et al., 2013). In terms of available data, the mining statistics of industrial minerals and rocks in Finland before 1946 are incomplete, but very good thereafter. The availability of product data for this sector is good from the 1960s to 2011. Since 1974, annual production summaries have also been published in *Kemia* (formerly *Kemia-Kemi*) magazine. However, after 2011, the situation noticeably worsened, with far fewer data available to the public. In this chapter, the mining data are, with few exceptions, based on those published annually by the Ministry of Employment and the Economy in Materia (formerly Vuoriteollisuus-Bergshanteringen). Only the most important, potentially important, or previously exploited, commodities are presented in this chapter.

Localities of all the known industrial mineral and rock mining operations in Finland from the 1300s to 2013 (inclusive), with their respective commodity symbols and total ore extraction classes (<1000, 1000–99,999, 100,000–999,999, and >1,000,000 tons) are shown in Fig. 9.5.1. While specific reference may not be made to this map in the commodity descriptions that follow, it is suggested that the reader refer to it on a regular basis. The main trends of ore mining from 1965–2013, are shown in Fig. 9.5.2. The largest commodities, "apatite ore" (including the carbonates from that mining), "carbonate rocks" (of sedimentary origin), and "talc ore," are also shown as separate trend lines. All of the other industrial minerals (IMs) shown in the figure exclude the carbonates, apatite ore, and talc ore.

Trend lines of ore mining of soapstones quarried for fireplace manufacturing (Fp soapstones), and industrial rocks (IRs)—that is, the stone wool rocks and diabase quarried for the cement industry—show that these materials are mined only in relatively small quantities. The trend line of metallic ore mining is shown for comparison to the trend line for all industrial minerals mining. The economic recession years, 1991 and 2009, are clearly reflected in the trend lines of the nonmetallic commodities.

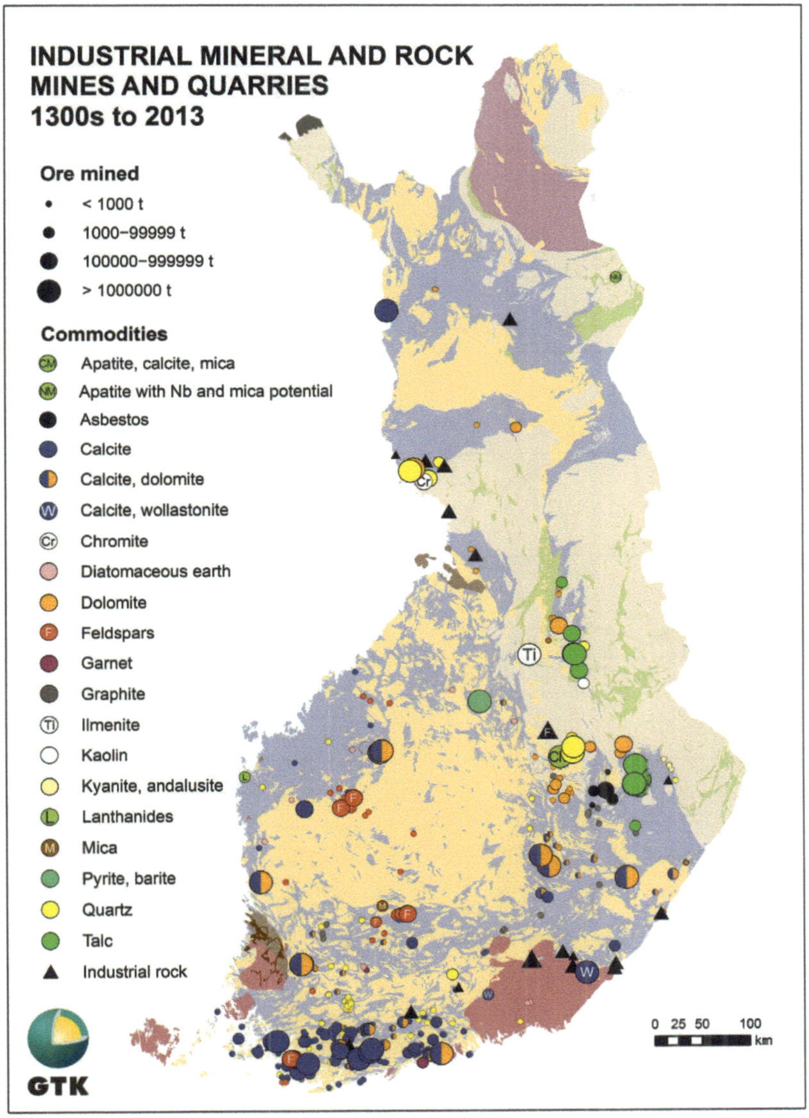

FIGURE 9.5.1 Locations of all the known industrial mineral and rock mining operations in Finland.

Shown are those from the 1300s to 2013 (inclusive), with their respective commodity symbols and total ore extraction classes: <1000, 1000–99,999, 100,000–999,999, and >1,000,000 tons.

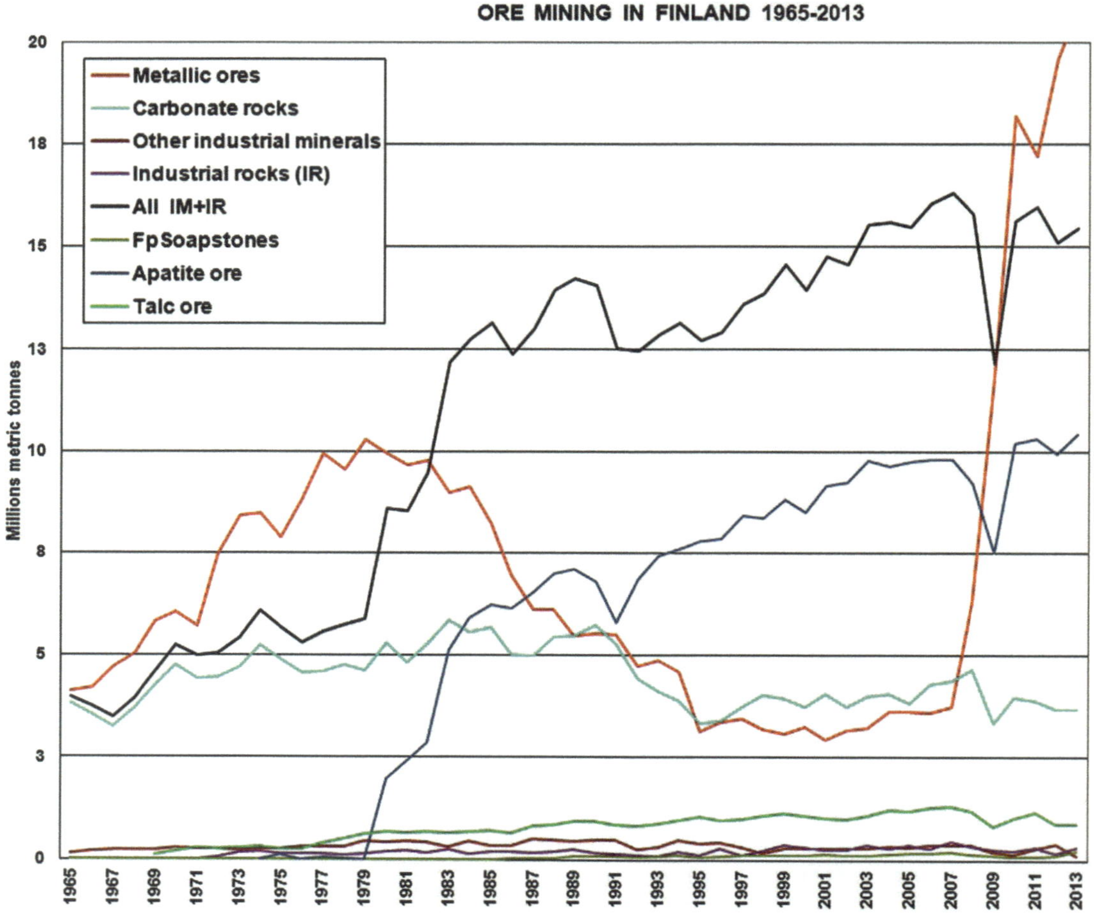

FIGURE 9.5.2 The main trends of ore mining from 1965 –2013.

CALCITE AND DOLOMITE

GEOLOGY AND MINING

Nearly all of the sedimentary carbonate rocks of mainland Finland are Paleoproterozoic marbles. Of these, most of the calcite-rich varieties occur in the Svecofennian domain, whereas the dolomite-rich ones dominate in the Karelian sedimentary rocks deposited on the Archean basement (see Fig 9.5.1). Eskola et al. (1919) and Pekkala (1988) recognized this rough geographical division and provided additional defining geochemical characteristics of these two groups. Eskola et al. (1919) also subdivided the types of the carbonate rocks by mineralogy within their respective geographic division into the following lithological types: (1) quartz-carbonate rocks, (2) tremolite-carbonate rocks, (3) diopside-carbonate rocks, and (4) wollastonite-carbonate rocks. These mineralogical differences reflect the dominant metamorphic grade with the marbles showing more mineralogical complexity in the higher metamorphic grade areas.

Mined deposits, both calcite and dolomite varieties, have been described by several authors in the compendium *The Mines and Quarries of Finland* edited by Aurola (1954a). A number of papers have been written concerning the Jatulian carbonate rocks of Finland and nearby Russian Karelia, including Metzger (1945), Perttunen (1985), Sokolov (1987), Melezhik et al. (2001), and Laajoki (2005). Later carbon isotope studies by Karhu (1993) confirmed earlier models regarding depositional environments and provided new ideas, especially in his division of the Karelian sequences into cratonic and marginal types. Reinikainen (2001) focused on the petrogenesis of four marble areas in the Svecofennian domain: two in the Uusimaa Belt, and two in the Virtasalmi District.

The carbonate rocks of the Svecofennian domain, mainly in southwestern and southern Finland, are ~1900 Ma old. In contrast, the Karelian occurrences found in the North Karelia, west Kainuu, Peräpohja, and the Kuusamo belts represent the upper part of the Jatulian sedimentary succession, which is ~2100–2060 Ma in age (Hanski and Melezhik, 2012). Upper Jatulian dolomite marbles, especially in the Rantamaa formation in the Peräpohja belt, contain well-preserved stromatolite structures similar to those described from Russian Karelia. Their presence suggests an intertidal to supratidal marine environment. The lower age limit of the Rantamaa formation can be constrained by a U-Pb zircon age of ~2106 ± 8 Ma obtained for the underlying mafic tuffs (Karhu et al., 2007).

Deposition of the Svecofennian carbonate rock suites are interpreted to have occurred in deeper basin environments compared to the Jatulian ones, which were deposited in shallow, mostly closed basins (Pekkala, 1988). This depth of deposition is seen in the predominance of calcite (deep) or dolomite (shallow), although later metamorphism and alteration have, at least in the higher grade areas, overprinted the mineralogy of the original sedimentary precursor. In practice, it is difficult to ascertain how much the present marbles retain their primary sedimentary mineral compositions. For example, at Ihalainen in Lappeenranta (Fig 9.5.3), the calcite marble with highest modal calcite has anomalously high Sr content compared with the other types, which might indicate that $CaCO_3$ precipitated originally in a shallow sea as aragonite, as aragonite tends to incorporate Sr more easily than calcite (Lehtinen, 1995). Recrystallization during metamorphism has produced a calcite marble with a distinctive trace element signature.

Younger carbonate rocks occur only in very limited or difficultly accessed areas in Finland. These include the seabed of the Selkämeri portion of the Gulf of Bothnia; Ordovician limestones under the sea in the Lumparn fjärd, southeastern Åland; the very impure dolomitic carbonate rocks in the parautochthonous Jerta or Valddejohkka plate in the Caledonian part of northwestern Finnish Lapland; and some from ice-age erosion sheltered carbonate rock remnants. Today these are known only from erratics on continental Finland (Lehtovaara, 1982).

Economically, the most important calcite marbles in the Svecofennian domain are located at Ihalainen, Lappeenranta (Lundén, 1979; Lehtinen, 1995, 1999), Limberg-Skräbböle, Pargas (Metzger, 1945; Fjäder, 1991), Tytyri (with Törmä and Solhem ore bodies), Lohja (Parras and Tavela, 1954), and Louhi, Savonlinna (Heiskanen, 1954; Lehtinen, 1999) on the northeastern border area of Svecofennia. Other, smaller, active, and/or recently mined or investigated sites are located at Förby (Saarmaa, 1983) and Hyypiämäki in Salo; Mustio and Kuovila in Raasepori (Sarapää et al., 2001, 2003); and Kalkkiranta in Sipoo (Tavela, 1954; Pakarinen, 1992). Of these, Tytyri, Louhi, Kalkkiranta, and Förby are underground mines (Förby closed in 2009).

A large number of abandoned, smaller quarries and some underground mines also exist in this Western Uusimaa–Kimito–Pargas carbonate rock province ("the leptite belt" by Eskola et al., 1919; Sarapää et al., 2001, 2003), which also hosts many deposits classified as mixed calcite-dolomite deposits that have also been exploited for agricultural products. Nearly all of these calcite marbles occur southwest

FIGURE 9.5.3 Ihalainen in Lappeenranta.

A view into the northern part of the Ihalainen open pit mine. White-light grey rocks are calcite marbles. Dark rocks with bedding conformable marbles, are leptites (rhyodacitic composition). Cross-cutting these are variably dark, granitic and composite dykes; the darkest are diabases. Average bench height is ca. 15 m.

Photo M.J. Lehtinen, 2014.

of the Savonlinna–Kuopio–Middle Ostrobothnia "line," but one large unexploited occurrence is geographically distinct, that being the rather dark calcite marble at Äkäsjoensuu in Kolari, west Lapland (Nurmi, 1989).

The largest dolomite marble resources occur in the Jatulian host rocks of Peräpohja (Tervola–Tornio–Keminmaa area), Kuusamo, and Kainuu belts. The Kalkkimaa quarry in Tornio has been operated since the 1500s (Härme, 1954). Several deposits in Vimpeli, Vampula, and Siikainen in western Finland and Ankele (earlier also Montola at Pieksämäki), Louhi, Juuka, Paltamo, and Kesälahti in eastern and southeastern Finland have been active sites for dolomite-based agrilime production, which continues to the present day. Predominantly dolomitic or dolomite-rich deposits in Siikainen, Kesälahti, Huutokoski, Muhos, and Utajärvi were not mentioned in the monograph of Eskola et al. (1919) because they had not yet been discovered.

Of the 652 carbonate rock deposits and occurrences documented by Eskola et al. (1919), Puustinen (1999, with a personal communication, 2013) lists 328 carbonate rock deposits that have been known to have been exploited in Finland up to 2013. The estimate of the total amount of carbonate rock mined in Finland up to that time is approximately 288 Mt. Only 15 of these have had a total of >1 Mt of carbonate rock mined, and 15 others were in the area ceded to the Soviet Union in 1944. The four sites with the largest volume of carbonate ore mining (excluding Siilinjärvi carbonatite) have been from Pargas (~98 Mt), Ihalainen, Lappeenranta (~63 Mt), the Lohja area mines (present Tytyri and the former Ojamo and Pitkäniemi mines; ~46.7 Mt), and from Louhi (previously named Ruokojärvi) (~13.3 Mt).

These, together with the Äkäsjoensuu deposit area (4.5 Mt ore mined), have the largest future calcite marble reserves. Due to production restructuring, increased competition with imports abroad, and markedly lower sales to the domestic agriculture, the annual mining of carbonate rocks—calcitic and dolomitic combined—has declined from 5.1 Mt/a in the period of 1970–1991 to an average of 3.7 Mt/a between 1992–2013. The import of foreign limestone and dolomite, 1.5–1.8 Mt/a and 0.045–0.092 Mt/a, respectively, represents a marked increase for the period 2007–2013.

CARBONATE PRODUCTION AND USAGE

Indirect information of the likely oldest quarrying and utilization of crystalline limestone in Finland comes from the "Black Book" of the Turku Cathedral in 1329, at that time under construction, and relates to the Krakanes limestone hill in the Förby area, southwestern Finland (Boström, 1986). The first industrial-scale lime production at Förby was started by Karl Forsström Ab in 1882 and ended in 2009, serving its last decades practically only as a source of calcite marble for the nearby paper pigment production. The first commercial cylinder lime kiln was built in 1862 in Helsinki (Boström, 1986). This kiln was fed with calcite marble brought from Pargas.

Many other similar kilns were erected thereafter in the 1800s. Nowadays, domestic calcite marble is being burned only by Nordkalk Oy Ab using rotary kilns in Lohja, Louhi, and Lappeenranta (ended in 2014). The calcite content of such calcite marble must usually be >88%, and HCl-soluble MgO <2 wt%. For floated, wet-ground $CaCO_3$ (GCC), calcite contents in the feed above 80% are feasible, if the brightness of calcite is high. Based on the demand of the steel industry and precipitated $CaCO_3$ (PCC) manufacturing, the amount of imported quicklime has grown remarkably since the end of the 1980s. Nordkalk and SMA Mineral Oy also burn imported limestone in shaft kilns in Finland.

The first cement plant was built in 1869, and operated until 1894 at Savio in Kerava. Larger scale Portland cement production began at Pargas in 1914 (Eriksson, 2014). The number of cement plants continued to increase, located near larger calcite marble deposits including Virkkala in Lohja (1919–1994), Lappeenranta (1938–present), and Kolari (1968–1989; Kitunen, 1970). The present cement producer, with an annual production of ~1.4 Mt, is Finnsementti Oy, which is owned by the Irish company CRH. Different grades of calcite marble are delivered to the cement plants where they will be optimally mixed with other needed mineral-based ingredients to achieve the desired chemical composition for the raw mix for Portland cement production.

After World War II, rebuilding of the country required a rapid expansion in cement and lime production, as well as increased lime demand for the growing pulp and paper industry. Begun in 1940, the initially sporadic production of PCC at the Tervakoski paper mill has been continuous since 1967. Increasing demand for PCC from 1990 to the early 2000s resulted in quicklime burned from Ihalainen and later from Louhi calcite marbles also being delivered to other PCC plants in Finland, until the early 2000s, when the use of domestic raw material for PCC gradually ended. Of much more importance has been the production of GCC for paper and paperboard coating and filling by Omya at Förby (1982–present) and Suomen Karbonaatti at Lappeenranta (1983–present). Nowadays the Förby plant uses white calcite marble from Tytyri and Hyypiämäki as its raw material. In the comparison made by Wilson (2008), the brightness of GCC produced from Ihalainen calcite marble (see Fig. 9.5.3) was the highest (95.0, ISO 457) among the major European GCC production sites.

Since the 1990s, GCC has gradually replaced imported kaolin as the main paper pigment. Domestic, annual GCC consumption has recently been 1–1.3 Mt (dry tons), and roughly half of that is imported GCC. In contrast to the United States, PCC has by use stayed well behind GCC in Finland, 0.3–0.5 Mt (dry), and in other European countries. The highest consumption of paper pigments (GCC, PCC, chalk, kaolin, talc, gypsum, TiO_2) was between 1997 and 2008, and peaked in 2004 at ~3.5 Mt (dry). Of the total paper pigment consumption between 1990–2012, ~41% was kaolin, ~31% GCC, ~11% PCC, ~10% talc, ~5% chalk, ~2% gypsum, and «1% TiO_2. Of these, all kaolin and chalk were imported, most quicklime for PCC was burned from foreign limestone in Finland, or was imported as quicklime, and all TiO_2 was produced in Finland during that period from foreign raw materials. Digitalization of media is an increasing threat to paper pigment consumption, both in Europe and North America. White dry-ground calcite fillers are also produced from Finnish calcite marbles for different end-uses at the main production sites.

MAGNESITE

In addition to calcite and dolomite, magnesite is a carbonate mineral with economic interest in Finland, especially when obtained as a potential by-product of talc production from soapstones. This carbonate mineral forms the $MgCO_3$–$FeCO_3$ solid solution and in Finland is mostly ferroan magnesite (breunnerite) in composition.

In 2013, global magnesite production was about 25 Mt (USGS estimate), with the macrocrystalline (sparry), "Veitsch-type" magnesite from ancient marine platform carbonate deposits (e.g., in the Chinese-North Korean province, Austria, and Slovakia) fully dominating. This type has been found mainly from dolomitic carbonate rock occurrences of Archean to Carboniferous ages (Ebner and Wilson, 2006). According to Wilson (2013), of the world's 13.1 bn tons of magnesite resources, as estimated in 2012, 93% of exploited magnesite was obtained from macrocrystalline and 7% from the microcrystalline deposits. Deposits of the latter type (also known as cryptocrystalline) include: (1) ultramafic-hosted, tectonically controlled magnesite (Kraubath-type, mined, for example, in western Turkey, the Balkans, Greece, Iran, and Pakistan); and (2) deposits in clastic, lacustrine sediments near ophiolite complexes (Bela Stena-type).

Melezhik et al. (2001) described a Paleoproterozoic sedimentary magnesite mineralization from the ~2000-Ma-old stromatolite-dolomite-"red bed" sequence in the Tulomozerskaya formation, Russian Karelia, which they interpreted to have formed in a combination of shallow-marine and nonmarine, evaporitic environments. Later Morozov et al. (2010) documented a discovery of several hundred meters of Jatulian magnesite-bearing anhydrite-halite strata in the Onega region of Russian Karelia. Potential for this type also exists in the upper Jatulian sequences in Finland. From 1986–1988, plant-size tests ending with a short production period were carried out at Luikonlahti by Myllykoski Oy to float a pure magnesite product (this resulted in 83–96% magnesite with 16.1% dolomite) for the fertilizer industry. Other markets were also studied, but the product (total <10,000 t) proved to be uneconomical.

TALC

Commercial Finnish talc deposits occur in ~2000 Ma ophiolitic ultramafic rocks that are significantly altered to talc-magnesite rocks, called soapstones (Fig. 9.5.4). The main minerals of these ultramafic rocks are talc (40–65%) and ferroan magnesite (35–55%). The most common other minerals include dolomite, chlorite, and in less pure soapstones, some serpentine (antigorite). Pyrrhotite and pentlandite are the most common sulfides, the latter also having economic importance in some locations. Ferritchromite and magnetite represent the main oxides. Talc schists also occur, especially along the border zones between the ultramafic bodies and their wall rocks (typically gradational black schists to mica

FIGURE 9.5.4 Commercial Finnish talc deposits.

Talc (flaky), with sparry magnesite (larger) and some lighter dolomite grains. Lipasvaara open pit mine, Polvijärvi. Scale bar 1 cm.

Sample: M.J. Lehtinen, photo: J. Väätäinen.

schists), also along which thin skarns are commonly developed. Vesasalo (1965) described more than 100 Finnish talc occurrences, some of which Wiik (1953) already had discussed in his description of the origin of soapstone. A more recent review was given by Niemelä (2001).

The first production of talc in Finland used dry processing by Suomen Mineraali Oy (later Paraisten Kalkki Oy) at Kinttumäki, Outokumpu in the 1940s, at Jormua, Kajaani from 1951 to 1971, and by Liperin Talkki Oy at Leppälahti, Liperi (1950–1953, 1955) (Aurola and Nieminen, 1954).

Based on quality factors, talc production has until now been restricted to the soapstones in the Karelian schist belts in Polvijärvi, North Karelia, and Sotkamo and Paltamo, western Kainuu. To the west of Polvijärvi, the higher metamorphic grade, along with the increased occurrence of amphiboles (cf. asbestos), has limited the usability of those soapstones for talc extraction. To the east of Polvijärvi in the Archean soapstones, the mineralogy (e.g., more oxides, and chlorite ± serpentine) and the texture of the main minerals, based on tests, do not favor the production of high brightness talc products.

Between 1969 and 2013, using modern froth flotation-based processing, 15.6 Mt of talc concentrate was produced out of ~35.8 million tons of ore from 11 different open pit mines. This production was started by Suomen Talkki Oy, with its main owners United Paper Mills Ltd. and Lohjan Kalkkitehdas Oy. The Lahnaslampi plant was the first talc flotation plant in Europe. Later names for this producer have been Finnminerals Oy and Mondo Minerals. Other producers were Myllykoski Oy (1979–1988) and Partek Corp. (1988–1990) with their mining at Repovaara and Lipasvaara in Polvijärvi, and processing at Luikonlahti. During its production time, from 1969–2010, the largest mine, Lahnaslampi in Sotkamo, alone produced ~17.6 Mt of talc ore. In addition to Lahnaslampi, >1 Mt talc ore has been mined from Horsmanaho, Pehmytkivi, Lipasvaara, and Vasarakangas in the Polvijärvi area, and from Punasuo in Sotkamo. At Mieslahti, Paltamo, remarkable talc reserves exist, but environmental constraints have thus far prevented any mining. An additional important, yet unmined, talc reserve is at Alanen in southernmost Sotkamo.

For the first time in Finland, a talc deposit other than soapstone type was test mined in 2010 by Mondo Minerals B.V., which extracted 5600 tons of ore from altered dolomites at Pihlajavaara, Puolanka in the Kainuu schist belt. Globally, the biggest commercial talc deposits are of this type; examples are in China (e.g., Liaoning, Shandong), the United States (Montana), Western Australia (Three Springs), and the French Pyrenees (Trimouns near Luzenac).

Finnish talc has been mainly used in paper, paperboard, plastics, and pitch control (adsorption of organic impurities in pulp and paper processes). The brightness range has been 84–90 (ISO 457), depending on the mineralogy and the grain size of the product. The recent production cuts of the European paper industry have hit talc production hard. Finland and France have been the largest European talc producers, and Mondo Minerals is globally the second largest single producer. Nickel concentrate has been a by-product of Finnish talc production since 1969. Until recent years, ~10,000 tons of nickel concentrate have been produced annually. Additionally, the potential of a magnesite by-product still holds economic potential for the future.

Of the Archean soapstone bodies in east Finland, the most important ones are those in the Nunnanlahti area, Juuka, and in the Suomussalmi–Kuhmo greenstone belt in eastern Kainuu. During the "new era" (1985–2013) of soapstone fireplace manufacturing in Finland, ~3.2 Mt of soapstone ore were mined from those areas. These quarries are not shown earlier in Fig. 9.5.1; however, the trend line of mining of fireplace soapstones is shown in Fig. 9.5.2.

QUARTZ

The earliest quartz production for small local glass factories was from the Somero-Tammela area, southwestern Finland, for the Åvik glass factory (1748–1833). Later production was from many other pegmatites in southwestern (Kimito island), southern (Orivesi), and western (Kuortane, Seinäjoki) Finland. The same ~1800 Ma, late-orogenic granite-related pegmatites have more recently turned out to be the main source of feldspars exploited in Finland.

The ~2300–2060 Ma Jatulian sedimentary rocks in eastern and northern Finland include thick quartz-dominated sequences (Paavola, 1984; Saikkonen, 1986; Laajoki, 2005). Some of these units consist of very little else besides quartz. This is the case with the Kinahmi quartzite in the Nilsiä, Kuopio area, the dominant quartz source in Finland. Kinahmi quartzite has been exploited since 1914 and since the production modernization and enhancement by Lohja Corp. in 1976 and 1980; the processing is based on flotation. Quartzite at Kinahmi is white, light gray, or a bit reddish in color (Fig. 9.5.5). The quartz content of the exploitable quartzite is 96% on average and the most abundant impurity is 2–8% sericite. Sheared areas contain kaolinite, and in places pyrophyllite. The total accessory mineral content is usually less than 1%. The quartzite ores are classified into three types: soft ("super quality"), semihard, and hard. The Fe_2O_3 and Al_2O_3 contents of the soft ore can be even lower than 0.015 wt% and 0.15 wt%, respectively.

Low titanium is also a benefit for special applications. The hard ore is too high in iron to be suitable for glass manufacturing, but finds use in refractory masses, plasters, and other special products. The main part of the Nilsiä quartz was delivered for glass-making (flat and package glass), but closures of

FIGURE 9.5.5 Quartzite at Kinahmi.

A view to "Super" quartz ore horizon in Kinahmi quartz quarry, Nilsiä (Kuopio). Mobile hydraulic hammer serves as a scale.

Photo: M. Häggman, Sibelco Nordic, 2010.

the Lahti flat glass and Karhula package glass factories in 2009 have strongly hit this end-use segment. Other end uses of Nilsiä quartz have been in induction ovens, sandblasting, refractory and white plasters, tiles, water glass, and crushed materials. Companies with production at Nilsiä have shifted over time: 1992–1997 by Partek Corp., 1998–2009 by SP Minerals Oy Ab, and since 2010 by Sibelco Nordic Oy Ab.

Very pure Jatulian quartzites also occur in the middle part of the Peräpohja belt of southwest Finnish Lapland. Here the <200-m-thick Kvartsimaa formation occurs between the overlying tuffitic Tikanmaa formation and underlying mafic volcanic rocks of the Jouttiaapa formation (Perttunen, 1991). Quartz in this orthoquartzite occurs as granoblastic mass, in which clastic grains can hardly be recognized from the matrix, where the sericite content is very low. This formation also contains minor dolomite and conglomerate intercalations. These quartzites were mined in 1966, 1980–1982, and 1984 from Tikanmaa (Tikka); in 1993–1995 from Kvartsimaa; and continuously between 1993 and 2012 from Ristimaa in Tornio. This mining supplies lump ore for the nearby stainless steel industry. More recently, production in Tornio has been by SMA Mineral Oy Ab. Since 2009, new lump quartz production for the same end use has been made available by Morenia Oy and comes from Juokuanvaara in Keminmaa, and also part of the Jatulian quartzite formations.

Since 1914, ~10 Mt of quartz ore from quartzites has been mined in Finland, of which quarries at Nilsiä alone account for ~86%. Of the total amount, lump ore as a product accounts for ~1.3 Mt (~14%). Altogether, 8.7 Mt of other, floated quartz products have been produced. This number also includes quartz produced as a by-product of feldspar production from Kimito Island. Tailings (i.e., fine-grained quartz and in Nilsiä also sericite) serve as additional potential credits if markets for these materials can be found.

FELDSPARS

The complex pegmatites which were the earliest sources for feldspar production in Finland occur at Somero-Tammela (Mäkinen, 1912), Kimito island (Pehrman, 1945; Lindroos et al., 1996), South Ostrobothnia (Haapala, 1966; Alviola et al., 2001; Mäkitie et al., 2001), and Orivesi (Volborth, 1954; Lahti, 1981). Similar pegmatites also exist elsewhere in Finland, including the Valkeakoski, Kisko, and Kitee areas and have been the focus of a great number of mineralogical investigations. The Li-pegmatites (albite-spodumene type) of the Ullava–Kaustinen–Kruunupyy area, western Finland, planned for mining in the near future, are described in Chapter 9.2.

The 1800 Ga pegmatites of Kimito Island represent beryl pegmatites (beryl-columbite subgroup) and albite pegmatites, except the Lövböle RE-pegmatite, which belongs to the gadolinite subgroup. These pegmatites are mostly hosted by gabbros (Lindroos et al., 1996) and their distribution is shown in Figure 5 of Eilu et al. (2012). Feldspar quarrying in Kimito followed previous quartz extraction for glass and iron works, presumably since about 1686. In 1737, the occurrence of tantalite was described from the Skogsböle pegmatite. Better organized production was between 1917–1954, when Suomen Mineraali Oy dry-processed Kimito feldspar, mainly quarried by local small enterprises.

FIGURE 9.5.6 Ala-Aulis.

Ala-Aulis pegmatite quarry, Island of Kimito. Reddish rock is pegmatite, dark one gabbro. Quarry length 490 m, width 70-140 m.

Photo: M. Häggman, Sibelco Nordic, 2014.

This was followed by Lohja Corp., which in the early 1960s conducted more extensive drilling in the Fröjdböle, Skogsböle, and Brokärr areas, resulting in 1966 in the first flotation-based feldspar plant in Europe with by-product quartz production. Present-day K-feldspar production in Finland is limited to Kimito, depending on the situation, and comes from pegmatite or both pegmatite (Ala-Aulis; Fig. 9.5.6) and late-orogenic (1840–1830 Ma) microcline granite (Kyrkoberget). Depending on the product, the K_2O content varies between 4 wt% and 7 wt%, K_2O/Na_2O ~1, and Fe_2O_3 ~0.1% (SP Minerals Brochure, 2001). The main end-uses of feldspar have been in the ceramic and glass industries, and for the Kimito quartz, in the domestic glass fiber and ceramic industries.

Of the Ostrobothnian ~1800 Ma RE-pegmatites, the complex pegmatites of Haapaluoma and Kaatiala have been classified by Alviola et al. (2001) as members of the spodumene subgroup. The Kaatiala dike (30 m wide, >200 m long) in Kuortane is a spodumene pegmatite, and the Haapaluoma dike (10–30 m × 500 m) in Seinajoki is a spodumene-lepidolite pegmatite. The country rocks for these pegmatites are foliated granodiorite and sillimanite-muscovite-biotite gneiss. Approximately 152,000 tons of K-feldspar, 44,000 tons of quartz, 700 tons of mica, and several tons of beryl, columbite, loellingite, and spodumene were extracted from ~516,000 tons of ore from the zoned Kaatiala pegmatite mined in Kuortane by Suomen Mineraali Oy between 1942 and 1968 (Puustinen, 2003). Production by Lohja Minerals and later by Partek Industrial Minerals Oy from the Haapaluoma pegmatite in Seinäjoki, from 1962–1997, was ~650,000 tons of feldspar + quartz ore (Puustinen, 2003). The Haapaluoma feldspar ore was famous for its high K_2O (~8 wt%) and low Fe_2O_3 (~0.04 wt%) content. Quartz was also obtained as a by-product.

The historical production of quartz and feldspar from the ~400 km^2 Somero-Tammela area, with more than 100 pegmatitic granites and pegmatites (1850–1750 Ma), indicates good potential for many rare element minerals as sources of Ta, Li, Be, Cs, and Rb (Teertstra et al., 1993; Alviola, 2001; Eilu et al., 2012). Exploration and testing have been carried out by the GTK and several companies.

Combining official production data for 1965–2013 with best estimates of older production (Puustinen, 2003), the main feldspar (+ quartz + in some cases small amounts of beryl, columbite, tantalite, etc.) ore-mining regions in Finland are ranked by Puustinen as follows:

- Kimito Island, southwest Finland: 5.82 Mt pegmatite (diverse quarries) and (pegmatitic) granite (Kyrkoberget) ore
- Seinäjoki-Kuortane (nearly all from Haapaluoma and Kaatiala) in western Finland: 1.17 Mt pegmatite ore
- Orivesi, southern Finland: 0.15 Mt pegmatite ore, mainly from the Viitaniemi pegmatite, and from at least 23 pegmatite quarries since the 1920s

Finnish anorthosites and anorthositic rocks have been studied (Sotka, 1988), tested, and used as raw materials for different purposes, including dimension stones (Ylämaa in Lappeenranta, Jaala in Kouvola, and Angeli in Inari), lapidary material ("spectrolite" from Ylämaa), and in larger tonnages as industrial rocks and minerals (Lapinlahti). In 1984, Kemira Corp. tested production of aluminum sulfate from Teerisuo in Lapinlahti by mining 27,166 tons of rock from the anorthositic part of the ring-structured orogenic (~1895 ± 15 Ma, from gabbro pegmatoid) Lapinlahti gabbro intrusion (Paavola, 1988). The bytownitic plagioclase (An_{75-85}) yielded to the whole rock the Al_2O_3 content of 27.5–30.5 wt%, and although Al solubility was favorable, 85–90%, production was found to be uneconomic for the planned water purification end use.

Since 1998, Paroc Oy Ab has used anorthositic rock from the same area for stone wool production in its factories around the Baltic Sea. The use of Joutsenenlampi anorthosite has grown faster than any other stone wool rock type in Finland, and between 1998–2013 ~2.07 Mt of ore was produced. The reason for that growth is the demand of better biosolubility of mineral wool fibers. The EC directive

97/69/EC of December 5, 1997, excluded certain mineral wools as a potential carcinogen: "Man-made vitreous (silicate) fibers with random orientation with alkaline oxide and alkali earth oxide (Na_2O + K_2O + CaO + MgO + BaO) content greater than 18% by weight."

Additionally, the International Agency for Research on Cancer concluded in 2001 that mineral wool insulation material was "unclassifiable as to its carcinogenicity to humans." Toward the end of the 1990s, SP Minerals Oy Ab (now Sibelco Nordic Oy Ab) also tested the suitability of Finnish anorthosites as another raw material for its Kimito feldspar plant. It started to buy the <40 mm fraction from Paroc's Lapinlahti crushed material and now sells the wet-processed product mainly to the fiberglass industry (refer to "F" inside the industrial rock symbol in Fig. 9.5.1). Besides Lapinlahti anorthosite, only anorthosite from Gudvangen in Aurland, west Norway, has been exploited for this type of use, although a potential White Mountain project in southwest Greenland is in the testing phase.

WOLLASTONITE

Wollastonite is a minor industrial mineral with a USGS-estimated worldwide crude ore production of ~600,000–630,000 tons in 2012, and product sales in the range of 500,000–540,000 tpa. Finnish Nordkalk Oy Ab was globally among the first two commercial producers of wollastonite and the pioneer in its wet processing. In 2012, China was the dominant wollastonite producer (~51%), followed by India (~24%), the United States (~12%), and Mexico (~9%) (BGS, World Mineral Statistics), with Finland still among the seven countries that produced natural wollastonite. Two deposits have been exploited in Finland, namely, Ihalainen in Lappeenranta and Perheniemi in Iitti.

At Ihalainen, wollastonite-rich carbonate rocks have been traced at least 1.5 km, mainly conforming to the roughly north–south strike and 65° eastward dip of this large calcite marble deposit, ~3 km long. Until now, mining for wollastonite has only taken place in the main wollastonite zone on the western and south-southwestern part of the present open pit mine, where the zone is widest (up to 65 m) and most massive (see App. 2 in Lehtinen, 1995, and Figure 1 in Lehtinen, 1999). Two types of wollastonite ore can be distinguished. The main ore type consists of sets of calc-silicate bands (wollastonite and diopside ± quartz), each band up to several centimeters wide, in a matrix of white to bluish calcite (Fig. 9.5.7). The wollastonite content of this ore type seldom exceeds 30%. Wollastonite is predominantly short-needled and, to the naked eye, the bands it forms look dense and sugary.

Calcite is, however, by quantity the main economic mineral (55–75%) in this ore type. This formation of wollastonite seems to be older than the rapakivi magmatism (~1600 Ma), because rapakivi dikes clearly crosscut this silicate banding, and an age determination from the titanite fraction from this type yielded 1858 ± 20 Ma (unpublished data, determination by H. Huhma; Lehtinen, 1995). The variable fluorescence and phosphorescence is less definitive, but indicative of differences between the two ore types. This older wollastonite ore type is thought to have its origin in upper medium- to high-grade metamorphism, when tectonic activity and temperature at a greater depth fractured this originally more impure siliceous limestone zone, allowing fluids to flow and CO_2 to escape.

The second, younger skarn-type wollastonite ore is at its best massive with a wollastonite content normally exceeding 30%. Its occurrence is related to the later thermal effects of larger granite dikes emanating from the nearby rapakivi intrusion, especially in the south-southwestern part of the open pit mine. In smaller scale and volume, younger wollastonite can also be seen along the main zone as wollastonite-dominated reaction skarns on the contacts with the mafic and intermediate volcanic rocks,

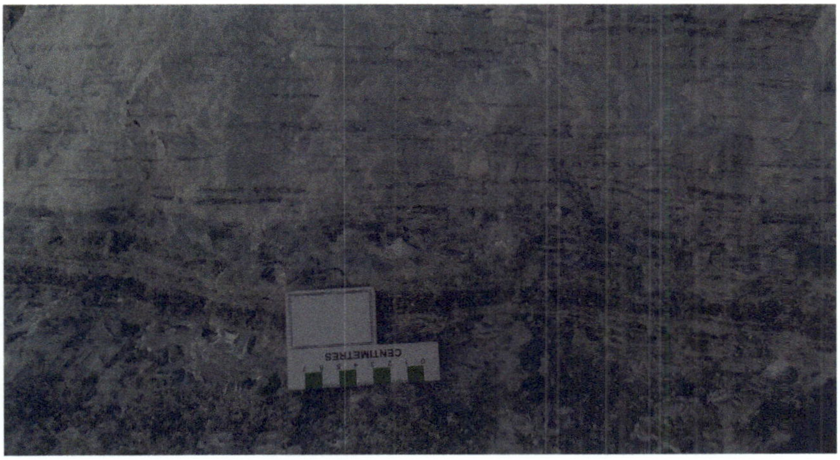

FIGURE 9.5.7 The main ore type of wollastonite consisting of sets of calc-silicate bands (wollastonite, diopside ± quartz).

Typical banded wollastonite ore (main type) in calcite marble from Ihalainen open pit mine, Lappeenranta. Dense white silicate bands consist of wollastonite+diopside ± quartz. Calcite is bluish.

Photo: M.J. Lehtinen, 2014.

very likely also as a result of the thermal effect of the rapakivi intrusion. In this younger ore type, besides wollastonite, calcite, diopside, and quartz, in places also grossular, vesuvianite, titanite, and fluorite occur. Wollastonite in this ore type has the highest aspect ratio of any in the deposit. Late overprint by hydrothermal alteration was related to the final stages of the rapakivi magmatism. In addition to some Mg hydrosilicates, fluorapophyllite, fracture-filling graphite, and pectolite characterize this late evolutionary stage. It is important to mention that in most cases there is no wollastonite skarn at the contact with the rapakivi pluton itself.

Until 1947, wollastonite at Ihalainen was seen as an impurity mineral in the calcite marble. In the 1950s, lump wollastonite was variably used as a small component in local stone wool manufacturing; since 1956 it was used for specific wall tiles of Turun Kaakeli Oy; and regular exports started in 1965. Production from the first wollastonite-specific beneficiation flotation plant began in 1966/1967, followed by flotation capacity enhancements in 1972 and 1986. Lump ore deliveries ended in 1977. Production from the present wollastonite-calcite plant initially went mainly for ceramics and metallurgy, but now is mainly for plastic filler end uses.

Until 2009, Lappeenranta was the sole continuously operated natural wollastonite producer in Europe. Besides Nordkalk, the only flotation-based processor of this mineral worldwide is NYCO Minería. The period of highest wollastonite output in Lappeenranta (1988–1997) produced in excess of 20,000 tons, peaking at 31,436 tons in 1989.

Another deposit that has been exploited for wollastonite in Finland is located at Perheniemi in Iitti (Arhe, 1980). This calcite marble with wollastonite skarns was first utilized as raw material for lime, supplied for the building of the Medieval Häme Castle, believed to have been built at the end of the thirteenth century. Wollastonite and diopside dominate in thin (<0.5 m) reaction skarns between the calcite marble and the viborgitic rapakivi granite intrusion. The wollastonite itself is of excellent quality, in part acicular, in part fine, platy "Tafelspat." Original mineralogical zoning and later alteration have been described by Arhe (1980). In the 1950s, Suomen Mineraali Oy quarried the easily exploitable reserves and sold ~400–500 tons as lump ore, mainly to Denmark.

The largest known unexploited wollastonite deposit with economic potential is at Kaihtula, Savitaipale, 35 km northeast from Lappeenranta. It likely represents the same carbonate rock system as the one at Ihalainen that has been distorted and disrupted by intrusion of the younger rapakivi granite. It was found in 1984 by Partek, when tracing the source of numerous glacier-deposited carbonate rock boulders, and finally with geophysics and drilling. There are no outcrops and the deposit is under a rather thick, mostly wet overburden. In 1995, the deposit was again drilled and evaluated, and the wollastonite ore reserves were re-estimated to be ~14 million tons. Additionally, the higher Mg content of the wollastonite-carbonate rock (more dolomite and serpentine than in Lappeenranta) is a negative finding (Internal reports of Partek companies, 1982–1995, Lundén, 1988).

In the Lake Saimaa district, there are numerous small occurrences of wollastonite, but their development is unlikely due to environmental reasons.

APATITE AND MICAS

Since 1979, production of apatite and mica (tetraferriphlogopite) in Finland has been from the Archean Siilinjärvi carbonatite (U-Pb 2609 ± 6 Ma; O. Kouvo, cited in O'Brien et al., 2005), which is described in Subchapter 4.3. Including the test production, 21.85 Mt of apatite concentrate out of ~260.4 Mt of ore were recovered between 1975 and 2013 (Kemira Corp., 1979–2007, Yara International ASA, 2007–present). Production since 2010 has been ~0.85 Mt/a. Of the by-products of the apatite production, the carbonate products are most voluminous. The mica production, initiated in 1985 by Kemira Corp. and since 2005 owned by LKAB Minerals (former Minelco Corp.), was ~195,000 t for 1990–2013, and ~12,500 tpa since 2010. The end uses of this mica continue to expand, for example, in the construction, polymer, and electronic sectors. Its growth potential is large, depending on the development of global markets.

Rather than gold miners, as is the case today, the first mineral claim holders of the Kutemajärvi mining area were instead industrial mineral companies. Already in 1947, Renlundin Tiili Oy evaluated local Svecofennian sericite schist as a raw material for bricks, finding sericite reserves at depths to 50 m of 18–20 Mt. Kemira Corp. explored the area in 1966–1974 for Al and K and in their claim area discovered reserves of 2.5 Mt sericite, 1.5 Mt topaz, 1.4 Mt kaolin and andalusite, and 8.5 Mt quartz (Ollila et al., 1990). The sericite content in that schist is 40%, with several percent topaz, and an Al_2O_3 content of ~20 wt% (Sotka, 1988). From 1975–1990, Lohja Corp. explored for sericite and quartz, and in 1982 discovered the epithermal Kutemajärvi gold ore.

Sericite (muscovite) has good potential as a by-product both from the Kinahmi (Kuopio) and Kutemajärvi (Orivesi) mines, but such production still remains unrealized, in part because it requires additional investments to further purify the raw mica-rich material due to variable amounts of pyrophyllite, kaolinite, and some nonphyllosilicate minerals, and to find markets for such by-products. According to Talikka (2007), the pervasive alteration at Kutemajärvi resulted from acidic hydrothermal fluids of magmatic origin at ~1895–1890 Ma, after which the rocks were deformed and metamorphosed during the Svecofennian orogeny.

Aurola (1957) described the Maaninkavaara vermiculite occurrence in Posio. The regolithic phosphate ore above the Devonian Sokli carbonatite in Savukoski, northeast Lapland (O'Brien et al., 2005), holds the best-known potential for vermiculite in Finland, besides its potential for Fe, Mn, Nb, Ta, and rare earth elements (REEs).

CLAY MINERALS

The geological record in Finland suggests there have been many suitable periods for weathering of the crust that may have formed laterites with clay minerals, but the multiple pulses of continental ice sheets have scoured away most such deposits.

The kaolin deposit at Pihlajavaara, Puolanka, in the Kainuu belt, was known already in the 1920s (Väyrynen, 1929). In the 1930s, small amounts of kaolin were extracted for ceramics from Prolanvaara, Soanlahti (ceded to the Soviet Union in 1944), and Pihlajavaara. When buying Turun Kaakelitehdas Oy in 1943, Paraisten Kalkkivuori Oy received the rights to the Pihlaja deposit. After some periods of small-scale kaolin test extraction, it proved to be uneconomic due to transportation costs and easier access to imported kaolin. Weathered kaolinite-bearing zones related to Jatulian quartzites were also test mined at Ruma in southern Sotkamo. Morenia Oy recently tested the usability of Pihlajanvaara kaolin and had a test delivery in 2012.

In 1978, GTK started exploration for kaolin in central Finnish Lapland, continuing from 1980 on in the Puolanka area, in part jointly with Lohja Corp., resulting in seven kaolin deposits and six kaolin weathering showings discovered. The total amount of kaolinized material of variable quality was estimated to be ~25 Mt, of which 8 Mt were of white grade and with ~6 Mt alone in Kerkkä. A kaolinite content of 5–20% is typical for these deposits, reflecting the low primary alumina contents of their protoliths (i.e., lower Paleoproterozoic arkoses and sericite quartzites). The deposits were grouped into in situ weatherings and shear zone kaolins, the first group being volumetrically more important.

Continuation of the GTK kaolin exploration and evaluation projects in Lapland, through the 1980s and 1990s in the Sodankylä area and into the early 2000s in the Salla area, resulted in discoveries of only ceramic-grade material at best. The only known sedimentary kaolin deposit in Finland is located at Saarijärvi, Taivalkoski. It is large, but because it consists of dark-colored, Fe-rich kaolinitic claystone forming intercalating layers with sandstone (Venäläinen, 1988; Pekkala and Sarapää, 1989), it is uneconomical.

Between 1986 and 1992, GTK carried out an extensive survey on kaolin deposits in the Virtasalmi area in Pieksämäki, and described 10 kaolin deposits from an area of 20 $km^2 \times 5$ km^2, of which 6 were evaluated suitable as raw material for paper filling and coating (Sarapää, 1996). Samples were evaluated by several companies, including Kemira Corp., which conducted additional testing on kaolin from two deposits. The dimensions of individual deposits range from 500–2000 m (length), 50–400 m (width), with an average thickness of 30–40 m, up to at most 100 m. Covering these is a 20–30-m-thick Quaternary overburden. The lighter kaolins are altered in situ from quartz-feldspar gneiss, mica gneiss, or tonalite, whereas the mainly iron- and graphite-colored kaolins are weathered from amphibolite or mica schist, respectively.

The kaolinite content varies between 40 and 75%, the remainder consisting mainly of quartz, muscovite, and feldspars. The ISO brightness for samples classified as white kaolin measured in the <20 μm fraction to range between 60 and 86% and <60% for colored kaolins. The amount of <20 μm fraction is reported to be ~60%, whereas that of <2 μm is ~30%. In the processing tests of Kemira, brightnesses of >85% were achieved for paper coating and 79–85% for the filler grade.

Colored kaolin might be suitable for tile production. The probable reserves of light kaolin were estimated to be ~18 Mt, of colored kaolins 16 Mt, and possible additional kaolin reserves ~15 Mt. The age of this kaolinization has been determined to be Mesoproterozoic (K-Ar 1180 Ma from illite; Sarapää, 1996). Subsequently, halloysite, $Al_2Si_2O_5(OH)_4$, (5–50%) has been found in 10 samples taken from the Litmanen and Kahdeksaisiensuo deposits, which fits with earlier observations of in part high viscosities of kaolin suspensions, and also supports a hydrothermal origin of these deposits (Al-Ani et al., 2006). With increasing applications in nanotechnology, halloysite, previously only rarely exploited for ceramics (New Zealand and Turkey), has now met a growing interest.

GRAPHITE

Finnish graphite deposits have been described by Frauenfelder (1924), Laitakari (1925), and Sarapää (1988). Laitakari (1925) lists ~150 graphite occurrences and concludes that graphite is regularly associated with paragneisses, mica gneisses, and metamorphosed limestones (i.e., those with sedimentary protoliths). Only some small, mainly flake graphite-type lenticular deposits from Svecofennian mica gneisses have been mined. According to Laitakari (1925), the largest of those was the Kärpälä deposit in Mäntyharju, from which a maximum of a few thousand tons of graphite ore were mined (Laitakari, 1925; Puustinen, 2003). The sorted ore was sold to England and Sweden.

Aurola (1964b) lists 15 small deposits with carbon contents ≥10%, of which only 4 contain ≥30% carbon: Rääpysjärvi (Kuopio) 60.8%, Soukko (Vammala) 49.6%, Kärpälä (Mäntyharju) 39.0%, and Laivonsaari (Kuopio) 32.6%. Later investigations (Sarapää, 1988) focused in part on the potential energy source from graphite-rich schists, of which the best evaluated deposits are at Hyypiä, Kiihtelysvaara in Joensuu (1 Mt rock with 28.6 wt% C, 2.3 wt% S) and Polvela, Juuka (2.5 Mt with 18 wt% C, 1.5–2 wt% S), both in North Karelia. Volumetrically, most graphite in Finland occurs in the marine uppermost Jatulian sedimentary rocks and in the Kalevian black schists (Sarapää, 1988; Arkimaa et al., 2000).

With improved processing and tailored products, this microcrystalline graphite might become an economically viable by-product (e.g., from certain black schist-related metallic ores) though the carbon content, morphology of the graphite, and sulfur content present critical constraints. However, the best Finnish graphite deposits only represent future potential, if the production from Asia stays at its present level. In 2012, the world production of graphite, 2.12 Mt, was dominated by China (~85%), followed by India (~6%), and Brazil (~4%). Of the European countries, Norway produced ~7000 t and Ukraine 4600 t (BGS, World Mineral Statistics).

SILLIMANITE GROUP MINERALS

These minerals include the Al_2SiO_5 polymorphs sillimanite, kyanite, and andalusite, all theoretically with 62.92 wt% Al_2O_3. Although sillimanite is known to occur in many high-grade gneisses and in some quartzites, no potential deposits have yet been found in Finland. Aurola (1959) summarized the most important kyanite and pyrophyllite deposits. Small amounts of andalusite and kyanite have been used as industrial rocks practically without any mineral processing. Between 1973 and 1981, Partek Corp. mined 83,715 t of andalusite-bearing mica schist from Mantovaara, Sodankylä and used it as an aluminum source in its Kolari cement plant (1968–1989). Andalusite raw materials also have been investigated at Rantakylä, Tohmajärvi; Leteensuo, Hattula; and Kutemajärvi, Orivesi.

Between 1979 and 1984, the Geological Survey of Finland carried out detailed investigations at Kontiolahti, Joensuu where andalusite and kyanite (Fig. 9.5.8) occur together in the lower Jatulian

FIGURE 9.5.8 Kontiolahti, Joensuu.

Coarse kyanite (bluish), quartz (grey), and mica (mainly muscovite) from Kapteeninautio quarry, Kontiolahti (Joensuu). Scale bar in nature is 23 mm.

M.J. Lehtinen, 2001.

Hokkalampi paleosol (Marmo, 1988). The total mineral reserves of the six mineralized zones that were evaluated contain at least 25 Mt of rock averaging 13% kyanite + andalusite and at least 5 Mt of rock with ≥20% kyanite + andalusite (Marmo, 1988). Processing and application tests have been made from raw materials from Hokkalampi and Hirvivaara, Joensuu . In the 1990s, more tests and shallow drilling were conducted at Kapteeninautio near Hokkalampi. Keramia Oy used a part of that 20,000 t mined rock in 2000 as one of the base components of its refractory tiles.

Other studied quartzite and quartzite schist-related Paleoproterozoic kyanite occurrences are at Höllärinvaara, Joensuu and Hallavaara in Sotkamo. Kyanite occurrences in Archean mica schists and gneisses are located at Tetrilampi, Nurmes; Kivisuo, Sotkamo; Peurasuo, Salla; and Mutsoiva in Savukoski, but none of these has proved to be economic.

OTHER NONMETALLIC INDUSTRIAL MINERALS

The use of asbestos in pottery on the shores of Lake Juojärvi, eastern Finland, dates back at least 4500 years. Between 1904 and 1975, ~586,000 t of anthophyllite asbestos were produced from Paakkila, Tuusniemi, and 28,861 t from 1943–1953 from Maljasalmi, Outokumpu (Aurola, 1954b; Palomäki and Halonen, 1968). In this area, asbestos was formed in the local ultramafic rocks due to the heating by dikes from the Paleoproterozoic Maarianvaara granite. Globally, the end uses for asbestos were manyfold, particularly for insulation, flooring, and roofing due to its resistance against heat, fire, and most acids. However, growing data of occupational risks related to asbestos led to a total ban of all types of utilization of asbestos in the European Union (EU) after January 1, 2005. Since 2012, the only practical production of (chrysotile) asbestos has been in Russia, China, Brazil, and Kazakhstan.

Between 1983 and 1989, 51,370 t of barite concentrate were produced by Outokumpu Oy from the Pyhäsalmi mine, central Finland, as were 47,000 t of REE (lanthanides)-bearing apatite concentrate by Outokumpu Oy from 1966–1971 from the Korsnäs lead mine, western Finland. Small-scale production of diatomaceous earth occurred between 1962 and 1970 from several places in central and southeastern Finland, totaling 13,794 t.

Almandine-pyrope–rich garnets are very common in large areas of the high-grade gneissic regions of the Svecofennian domain, as well as in the granulite belt of northern Lapland. Almandine crystals from amphibolite at Isopää, Kalvola area, Hämeenlinna were already famous in the 1800s (Holmberg, 1858). From 1946–1947, Rudus Oy test processed between 1000 and 3000 t (Laitakari, 1947; Puustinen, 2003) of andradite garnet rock from Laajasalo, Helsinki, for floor plates and abrasives. Processing of almandine from Ruostesuo, Kiuruvesi has been tested for an end use as abrasives (Kirjavainen, 1980). Garnets from sulfide deposits tend to have harmful sulfide inclusions, but in suitable mineral parageneses, at least by-product garnet might be economic. The major uses of garnets are for abrasive purposes, including waterjet cutting, but as long as India and China continue to dominate the world garnet production, there seems to be little opportunity for a competitive European production.

The average nepheline content of the Iivaara alkaline complex in Kuusamo is ~40%, with an Al_2O_3 content of 14.2 wt%. According to Sotka (1988), at least 7–8 tons of nepheline-bearing rock needs to be processed to yield 1 t of Al_2O_3 by a calc-sintering process utilized in the Kola Peninsula. More cost-effective, alternative aluminum-rich raw materials are available globally, leaving the Iivaara nepheline intact.

Minor occurrences of different zeolite species are known from Finland, but so far no commercial-scale deposits have been encountered. Of the notable occurrences, the laumontite occurrence in the altered migmatite of the northwest–southeast-trending major shear zone from Sarvasjärvi, Pälkäne, was described by Eskola (1935).

Of the proper industrial rocks, excluding the soapstones, most have been mined for stone wool manufacture. Their composition varies between anorthosite and peridotite, with most being gabbros. Between 1970 and 2013, 8.5 Mt of usable rocks were mined by Paroc Oy Ab, from 21 quarries. Of this amount, 6.9 Mt were stone wool plant feed. Diabase and andalusite rock have been mined for cement manufacture, and phyllite for roofing material.

METALLIC INDUSTRIAL MINERALS

The following commodities are mainly handled in other chapters in this book, but also need to be mentioned briefly here.

Pyrite concentrate, on average ~610,000 tpa and in total ~25 Mt produced from 1973–2013, continues to be sold from the Pyhäsalmi mine for sulfuric acid production, mainly (~350,000 tpa) to Siilinjärvi, but is also exported. Chromite sand from Outokumpu's Kemi mine was periodically sold for industrial minerals end uses (molding sand, etc.), but sales ended in 1997.

The base for the Finnish TiO_2 pigment production was the 3.8 Mt of ilmenite concentrate from Otanmäki, Kajaani. This ilmenite was extracted, along with 7.6 Mt of iron concentrate, 0.2 Mt of sulfur concentrate, and 44,545 t of V_2O_5 in the years 1953–1985 by Otanmäki Oy (1968–Rautaruukki Oy) from 33 Mt of Fe-Ti-V ore mined from the Otanmäki and Vuorokas ore bodies (Paarma, 1954; Pääkkönen, 1956; Illi et al., 1985). In 1961, Vuorikemia Oy began the TiO_2 production by sulfate route method at Pori, which still continues, but since 1986 has been based on imported raw materials, mainly ilmenite concentrate from Tellnes, Norway. Since 2013, this production has been in foreign ownership (actually Huntsman Corp.).

Halsua and Kälviä, in western Finland, have been targets of exploration for potential magmatic ilmenite deposits by GTK and Kalvinit Oy-Endomines (Sarapää et al., 2003). These 1881 ± 10 Ma deposits are based on ilmenite-magnetite ores in layered gabbros at Kairineva, Koivusaarenneva, and Peräneva. From 2012–2013, Cove Resources Ltd. examined the feasibility of exploiting these deposits as well as the Peräkorpi apatite and ilmenite-bearing gabbro at Honkajoki, near Pori, discovered in 1959.

SUMMARY

As shown in the preceding, and verified by past and present mining, Finland's bedrock offers further good potential for the mining of many industrial minerals and rocks. Of the critical raw materials classified by the EU (European Commission, 2010a,b) and the Minerals Strategy of Finland (GTK, 2010) as "economically important," at least talc, "limestone" (= the white calcite marbles), quartz (silica), and feldspar have significant measured reserves and future resources in Finland. Of those 20 critical raw materials included in the classification update (EC, 2014), phosphate rock and chromium (besides the non-industrial minerals PGMs and cobalt) have mineable reserves and proven potential in Finland. Titanium and lithium have shown discovery potential.

Suitable geological environments and even exploited or evaluated deposits of magnesite (by-product, iron-bearing), graphite, and niobium, and in much lesser amounts, the REEs and beryllium, are known in Finland, but to establish their real economic value requires more research, exploration, evaluation, and knowledge of their end uses and markets. The same also applies to many other commodities, such as tantalum, sillimanite group minerals, and garnet, which are actually not produced in Finland. There is much work left to be done to improve the recovery of many economic minerals in present production chains, to minimize the amount of unused mined rock, and, in general, to convert tailings and by-products recovered from the chemical processes into saleable products.

ACKNOWLEDGMENT

The comments of two anonymous reviewers are appreciated as well as the assistance provided by Mr. J. Pokki (GTK) in producing Fig. 9.5.1.

REFERENCES

Al-Ani, T., Sarapää, O., Lehtinen, M.J., 2006. Occurrence of halloysite in Virtasalmi kaolin deposits and its effect on rheological properties. In: The 27th Nordic Geological Winter Meeting, January 9–12, Oulu, Finland. Abstract Volume. Bulletin of the Geological Society of Finland Special Issue 1, 6.

Alviola, R., 1989. The granitic pegmatites of the Somero-Tammela area. In: Symposium Precambrian granitoids. Petrogenesis, geochemistry and metallogeny, August 14–17, 1989, Helsinki, Finland. Excursion C 1: Lateorogenic and synorogenic Svecofennian granitoids and associated pegmatites of southern Finland. Geological Survey of Finland. Guide 26, 16–25.

Alviola, R., Mänttäri, I., Mäkitie, H., Vaasjoki, M., 2001. Svecofennian rare-element granitic pegmatites of the Ostrobothnia region, western Finland: their metamorphic environment and time of intrusion. Svecofennian Granitic Pegmatites (1.86–1.79 Ga) and Quartz Monzonite (1.87 Ga), and Their Metamorphic Environment in the Seinäjoki Region, Western Finland. Geological Survey of Finland, Special Paper 30, 9–29.

Arhe, M., 1980. The Perheniemi carbonate rock-skarn occurrence and its geology in Iitti. Unpublished M.Sc. thesis. University of Helsinki, 116 pp. (in Finnish).

Arkimaa, H., Hyvönen, E., Lerssi, J., Loukola-Ruskeeniemi, K., Vanne, J., 2000. Proterozoic Black Shales and Aeromagnetic Anomalies in Finland. Scale 1 : 1 000 000. Geological Survey of Finland, Espoo.

Aurola, E. (Ed.), 1954a. The Mines and Quarries of Finland. Geoteknisiä Julkaisuja 55, 123 pp.

Aurola, E., 1954b. The asbestos occurrences of Paakkila and Maljasalmi. In: Aurola, E. (Ed.), The Mines and Quarries of Finland. Geoteknisiä Julkaisuja 55, 95–99.

Aurola, E., 1957. On the vermiculite of Maaninkavaara. Geoteknisiä Julkaisuja 60, 1–24 (in Finnish with English summary).

Aurola, E., 1959. The kyanite and pyrophyllite occurrences in North Karelia. Geoteknisiä Julkaisuja 63, 1–36 (in Finnish with English summary).

Aurola, E., 1964a. Industrial minerals and industrial rocks. In: Rankama, K. (Ed.), Geology of Finland. Kirjayhtymä, Helsinki, pp. 189–237 (in Finnish).

Aurola, E., 1964b. Industrial mineral and industrial rock occurrences in Finland. Vuoriteollisuus–Bergshanteringen 2, 21–26 (in Finnish).

Aurola, E., Nieminen, K., 1954. The talc industry. In: Aurola, E. (Ed.), The Mines and Quarries of Finland. Geoteknisiä Julkaisuja 55, 102–104.

Boström, R., 1986. The history of the stone and mineral industry in Finland. Geological Survey of Finland Bulletin 336, 273–298.

British Geological Survey, 2012. World mineral production 2008-12. (British Geological Survey, Keyworth, Nottingham.) 115 pp.

Ebner, F., Wilson, I., 2006. Magnesit–globales Potenzial und geologische Lagerstättencharakteristik. BHM Bergu. Hüttenmänn. Monatshefte 151 (4), 164–174.

Eilu, P., Ahtola, T., Äikäs, O., Halkoaho, T., Heikura, P., Hulkki, H., Iljina, M., Juopperi, H., Karinen, T., Kärkkäinen, N., Konnunaho, A., Kontinen, A., Kontoniemi, O., Korkiakoski, E., Korsakova, M., Kuivasaari, T., Kyläkoski, M., Makkonen, H., Niiranen, T., Nikander, J., Nykänen, V., Perdahl, J.-A., Pohjolainen, E., Räsänen, J., Sorjonen-Ward, P., Tiainen, M., Tontti, M., Torppa, A., Västi, K., 2012. Metallogenic areas in Finland. In: Mineral deposits and metallogeny of Fennoscandia. Geological Survey of Finland, Special Paper 53, 207–342.

Eriksson, H.C., 2014. From Portland to plus—a history of cement in Finland. In: Tompuri, V. (Ed.), From Portland to Plus—100 Years of Finnish Cement Finnsementti Oy, 226–233 (in Finnish with English summary).

Eskola, P., Hackman, V., Laitakari, A., Wilkman, W.W., 1919. Limestones in Finland. Geoteknisiä tiedonantoja 21. Valtioneuvoston kirjapaino, Helsinki. 265 pp. (in Finnish with English summary).

Eskola, P., 1935. Kuhmoisten zeoliittiesiintymä ja sen mahdollinen suhde Päijänteen murroslaaksoihin. Summary: An occurrence of zeolite in the parish of Kuhmoinen, central Finland and its possible relation to the fracture lines in the basin of Lake Päijänne. Terra 47, 171–178.

European Commission (EC), 2010a. Critical raw materials for the EU. Report of the ad-hoc working group on defining critical raw materials. Version of July 30, 85pp. Available at http://ec.europa.eu/enterprise/policies/raw-materials/files/docs/report-b_en.pdf.

European Commission, 2010b. Annex V to the report of the ad-hoc working group on defining critical raw materials. 220 pp. Available at http://ec.europa.eu/enterprise/policies/raw-materials/files/docs/annex-v_en.pdf.

European Commission, 2014. The European Critical Raw Materials review. MEMO-14-377-EN. Brussels. May 26, 2014.

Fjäder, K., 1991. The stratigraphy and tectonics of Proterozoic supracrustal rocks at Pargas, SW Finland. Unpublished M.Sc. thesis, Åbo Akademi, 63 pp. (in Swedish).

Frauenfelder, K.O.H., 1924. Der Graphit in Finnland, seiner Entstehung und Verwertung. Geoteknisiä Julkaisuja 38, 51 pp.

Gautneb, H., Ahtola, T., Bergman, T., Gonzalez, J., Hallberg, A., Litvinenko, V., Shchiptsov, V., Voytekhovsky, Y., 2013. Industrial mineral map of the Fennoscandian Shield. In: Proceedings of the 12th Biennial SGA Meeting, Uppsala 2013, vol. IV, pp. 1767–1769.

GTK, 2010. Finland's mineral strategy, 19 pp. Available at www.mineraalistrategia.fi.

Haapala, I., 1966. On the granitic pegmatites in the Peräseinäjoki-Alavus area, South Pohjanmaa, Finland. Bulletin de la Commission Géologique de Finlande 224, 98 pp.

Haapala, I. (Ed.), 1988. Industrial Minerals and Industral Rocks in Finland. Yliopistopaino Helsinki, 168 pp. (in Finnish).

Hanski, E.J., Melezhik, V.A., 2012. Litho- and chronostratigraphy of the Karelian formations. In: Melezhik, V., Prave, A., Hanski, E., Fallick, A., Lepland, A., Kump, L., Strauss, H. (Eds.), Reading the Archive of Earth's Oxygenation. Volume 1: The Palaeoproterozoic of Fennoscandia as Context for the Fennoscandian Arctic Russia - Drilling Early Earth Project. Springer-Verlag, Berlin, Heidelberg, pp. 39–110.

Harben, P.W., Bates, R.L., 1990. Industrial minerals. Geology and World Deposits. Industrial Minerals Division. Metal Bulletin Plc, London, 312 pp.

Härme, M., 1954. The Kalkkimaa dolomite quarry. In: Aurola, E. (Ed.), The Mines and Quarries of Finland. Geoteknisiä Julkaisuja 55, 87–88.

Heiskanen, E.V., 1954. The limestone quarries of Ruskealan Marmori Oy. In: Aurola, E. (Ed.), The Mines and Quarries of Finland. Geoteknisiä Julkaisuja 55, pp. 84–86.

Holmberg, H.J., 1858. Materialer till Finlands geognosi. Bidrag till Finlands naturkännedom, etnografi och statistik 4. Heft, 1–254.

Illi, J., Lindholm, O., Levanto, U-M., Nikula, J., Pöyliö, E., Vuoristo, E , 1985. Otanmäki mine. Vuoriteollisuus 43 (2), 98–107 (in Finnish with English summary).

Jeffrey, K., 2006. Classification of industrial minerals and rocks. In: Kogel, J.E., Trivedi, N.C., Barker, J.M., Krukowski, S.T. (Eds.), Industrial Minerals and Rocks: Commodities, Markets, and Uses, 7th ed. SME, Littleton, CO, pp. 7–11.

Johansson, K., Larsen, R.B., Lehtinen, M.J., Persson, L., Räisänen, M., Pedersen, Schack Pedersen, S.A., Weihed, P., Wik, N-G., 2008. Industrial minerals and rocks, aggregates and natural stones in the Nordic countries. Episodes 31, 133–139.

Kananoja, T., Pokki, J., Ahtola, T., Hyvärinen, J., Kallio, J., Kinnunen, K., Luodes, H., Sarapaa, O., Tuusjärvi, M., Törmänen, T., Virtanen, K., 2013. Geologisten luonnonvarojen hyödyntäminen Suomessa vuonna 2011. Summary: Geological resources in Finland, production data and annual report 2011. Geological Survey of Finland, Report of Investigation 203, 38 pp.

Karhu, J.A., 1993. Paleoproterozoic evolution of the carbon isotope ratios of sedimentary carbonates in the Fennoscandian Shield. Geological Survey of Finland Bulletin 371, 87 pp.

Karhu, J., Kortelainen, N.M., Huhma, H., Perttunen, V., and Sergeev, S., 2007. New time constraints for the end of the Paleoproterozoic carbon isotope excursion. Proceedings of the 7th International Symposium on Applied Isotope Geochemistry, 10th–14th September, 2007, Stellenbosch, South Africa, pp. 76–77.

Kirjavainen, V., 1980. Tutkimus Kiuruveden Ruostesuon granaatin rikastamisesta hiomateollisuuden raaka-aineeksi. Diplomityö, TKK, Vuoriteollisuusosasto, 60 pp. (in Finnish)

Kitunen, K., 1970. Paraisten Kalkki Oy:n Kolarin sementtitehdas. Vuoriteollisuus–Bergshanteringen 28 (1), 25–27 (in Finnish with English summary).

Laajoki, K., 2005. Karelian supracrustal rocks. In: Lehtinen, M., Nurmi, P.A., Rämö, O.T. (Eds.), Precambrian Geology of Finland—Key to the Evolution of the Fennoscandian Shield. Developments in Precambrian Geology 14. Elsevier, Amsterdam, pp. 279–342.

Lahti, S.I., 1981. On the granitic pegmatites of the Eräjärvi area in Orivesi, southern Finland. Geological Survey of Finland Bulletin 314, 82 pp.

Laitakari, A., 1925. Die Graphitvorkommen in Finnland und ihre Entstehung. Geoteknisiä Julkaisuja 40, 100 pp.

Laitakari, A., 1947. Eräitä uutuuksia hyödyllisten kaivannaisten alalta. Vuoriteollisuus–Berghanteringen 1–2, 16–18 (in Finnish).

Lehtinen, M.J., 1995. On the mineralogy and geology of Ihalainen formation at Lappeenranta. Unpublished Ph.Lic. thesis, University of Turku, p. 127 (in Finnish).

Lehtinen, M.J., 1999. The crystalline limestone deposits of Ihalainen (Lappeenranta) and Louhi (Kerimäki) in SE Finland and their production history. In: Industrial Minerals: Deposits and New Developments in Fennoscandia. Proceedings of the International Conference. Institute of Geology, Karelian Research Center, Petrozavodsk, pp. 41–44.

Lehtinen, M.J., 2006. The status of industrial minerals and rocks in Finland. In: Industrial Minerals Forum and Dinner: Northern lights—a mineral focus on Nordic countries and north-west Russia, November 16, Helsinki. Metal Bulletin, London. 14 pp.

Lehtovaara, J.J., 1982. Palaeozoic sedimentary rocks in Finland. Annales Academiae Scientiarum Fennicae, Series A. III. Geologica–Geographica 133. Suomalainen Tiedeakatemia, Helsinki, 35 pp.

Lindroos, A., Romer, R.L., Ehlers, C., Alviola, R., 1996. Late-orogenic Svecofennian deformation in SW Finland constrained by pegmatite emplacement ages. Terra Nova 8, 567–574.

Lundén, E., 1979. On the geology of the Lappeenranta limestone quarry. Vuoriteollisuus–Bergshanteringen 37 (1), 20–23 (in Finnish with English summary).

Lundén, E., 1988. Wollastonite. In: Haapala, I. (Ed.), 1988. Industrial Minerals and Industrial Rocks in Finland. Yliopistopaino, Helsinki, pp. 61–65 (in Finnish).

Mäkitie, H., Kärkkäinen, N., Lahti, S.I., Alviola, R., 2001. Compositional variation of granitic pegmatites in relation to regional metamorphism in the Seinäjoki region, western Finland. In: Svecofennian granitic pegmatites (1.86–1.79 Ga) and quartz monzonite (1.87 Ga), and their metamorphic environment in the Seinäjoki region, western Finland. Geological Survey of Finland, Special Paper 30, 31–59.

Marmo, J., 1988. Aluminium silicates. In: Haapala, I. (Ed.), Industrial Minerals and Industrial Rocks in Finland. Yliopistopaino, Helsinki, pp. 85–92 (in Finnish).

Melezhik, V.A., Fallick, A.E., Medvedev, P.V., Makarikhin, V.V., 2001. Palaeoproterozoic magnesite: Lithological and isotopic evidence for playa/sabkha environments. Sedimentology 48, 379–397.

Metzger Th, A.A., 1945. Beiträge zur Geologie der Insel Ålö und Kyrklandet in Pargas-Parainen, SW Finland. Acta Acad. Aboensis. Ser. Math.-Phys 15 (3), 1–103 also Medd. Åbo Akad Geol.-Mineral. Inst., 27.

Morozov, A.F., Hakhaev, B.N., Petrov, O.V., Gorbachev, V.I., Tarkhanov, G.B., Tsvetkov, L.D., Erinchek, Yu.M., Akhmedov, A.M., Krupenik, V.A., Sveshnikova, K.Yu, 2010. Rock-salts in Palaeoproterozoic strata of the Onega depression of Karelia (based on data from the Onega parametric drillhole). Transactions of the Academy of Sciences 435, 230–233 (in Russian).

Niemelä, M., 2001. Talc-magnesite deposits in Finland. In: Radvanec, M., Gondim, A.C., Nemeth, Z. (Eds.), IGCP 443: Magnesite and Talc: Geological and Environmental Correlations. Mineralia Slovaca 33, 561–566.

Nurmi, H., 1989. The Äkäsjoensuu carbonate rock deposits in Kolari. Unpublished M.Sc. Thesis. University of Turku, 81 pp. (in Finnish).

O'Brien, H.E., Peltonen, P., Vartiainen, H., 2005. Kimberlites, carbonatites, and alkaline rocks. In: Lehtinen, M., Nurmi, P.A., Rämö, O.T. (Eds.), Precambrian Geology of Finland—Key to the Evolution of the Fennoscandian Shield. Elsevier, Amsterdam. Developments in Precambrian Geology 14, 605–644.

Ollila, H., Saikkonen, R., Moisio, J., Kojonen, K., 1990. Oriveden Kutemajärven kultaesiintymä. Summary: The gold deposit in Kutemajärvi, Orivesi. Vuoriteollisuus 48 (1), 26–30 (in Finnish with England summary).

Pääkkönen, V., 1956. Otanmäki, the ilmenite-magnetite ore field in Finland. Bulletin de la Commission Géologique de Finlande 171, 1–171.

Paavola, J., 1984. Pre-Quaternary rocks of the Nilsiä map-sheet area. Geological map of Finland 1: 100,000 Explanation to the maps of pre-Quaternary rocks, Sheet 3334 Nilsiä. Geological Survey of Finland, 57 pp. (in Finnish with English summary).

Paavola, J., 1988. Pre-Quaternary rocks of the Lapinlahti map-sheet area. Suomen geologinen kartta 1:100 000. Geological map of Finland 1, 100,000 Explanation to the maps of pre-Quaternary rocks, Sheet 3332 Lapinlahti. Geological Survey of Finland, 60 pp. (in Finnish with English summary).

Paarma, H., 1954. The ilmenite-magnetite ore deposit of Otanmäki. In: Aurola, E. (Ed.), The Mines and Quarries of Finland. Geoteknisiä Julkaisuja 55, 36–42.

Pakarinen, J., 1992. The past and present of the Sipoo lime factory and mine. Vuorityö ja-tekniikka 1–2, 33–36 (in Finnish).

Palomäki, A., Halonen, O., 1968. The Paakkila antophyllite asbestos quarry. Vuoriteollisuus 26, 92–98 (in Finnish with English summary).

Parras, K., Tavela, M., 1954. The limestone deposits in Lohja. In: Aurola, E. (Ed.), The Mines and Quarries of Finland. Geoteknisiä Julkaisuja 55, 69–74.

Pehrman, G., 1945. Die Granitpegmatite von Kimito (SW-Finland) und ihre Minerale. Acta Academiae Aboensis, Series B. Mathematica et Physica 15 (2), 84 pp.

Pekkala, Y., 1988. Carbonate rocks of Finland. In: Haapala, I. (Ed.), 1988. Industrial Minerals and Industrial Rocks in Finland. Yliopistopaino, Helsinki, pp. 19–35 (in Finnish).

Pekkala, Y., Sarapää, O., 1989. Kaolin exploration in Finland. In: Autio, S. (Ed.), Geological Survey of Finland. Current Research 1988. Geological Survey of Finland, Special Paper 10, 113–118.

Perttunen, V., 1985. On the Proterozoic stratigraphy and exogenic evolution of the Peräpohja area, Finland. Geological Survey of Finland. Bulletin 331, 131–141.

Perttunen, V., 1991. Pre-Quaternary rocks of the Kemi, Karunki, Simo and Runkaus map sheet areas. Geological map of Finland 1:100,000 Explanation to the Maps of Pre-Quaternary Rocks, sheets 2541 Kemi, 2542 and 2524 Karunki, 2543 Simo, and 2544 Runkaus. Geological Survey of Finland, 80 pp. (in Finnish with English summary).

Puustinen, K., 1999. Distribution and production of Finnish limestone mines. In: Geological Survey of Finland, Current Research 1997–1998. Geological Survey of Finland. Special Paper 27, 43–47.

Puustinen, K., 2003. Suomen kaivosteollisuus ja mineraalisten raaka-aineiden tuotanto vuosina 1530–2001, historiallinen katsaus erityisesti tuotantolukujen valossa. Geological Survey of Finland, Report M10.1/2003/3, 578 pp. (in Finnish)

Reinikainen, J., 2001. Petrogenesis of Paleoproterozoic marbles in the Svecofennian Domain, Finland. Geological Survey of Finland. Report of Investigation 154, 84 pp.

Saarmaa, B., 1983. Karl Forsström Ab—100 vuotta. Vuoriteollisuus–Bergshanteringen 41 (2), 120–123 (in Finnish).

Saikkonen, R., 1986. The quartzite deposit of Nilsiä. In: Kojonen, K. (Ed.), Prospecting in Areas of Glaciated Terrain: Guide to the Field Excursion for the Symposium in Eastern Finland, September 3–5, 1986. Geological Survey of Finland. Guide 17, 14–18.

Sarapää, O., 1988. Graphite. In: Haapala, I. (Ed.), Industrial Minerals and Industrial Rocks in Finland. Yliopistopaino, Helsinki, pp. 66–73 (in Finnish).

Sarapää, O., 1996. Genesis and age of the Virtasalmi kaolin deposits, southeastern Finland. In: Proterozoic Primary Kaolin Deposits at Virtasalmi, Southeastern Finland. Geological Survey of Finland, Espoo, pp. 87–152.

Sarapää, O., Ahtola, T., Reinikainen, J., Seppänen, H., 2003. Industrial mineral potential in Finland. Geological Survey of Finland, Current Research 2001–2002, Special Paper 36, 5–12.

Sarapää, O., Reinikainen, J., Seppänen, H., Kärkkäinen, N., Ahtola, T., 2001. Industrial minerals exploration in southwestern and western Finland. In: Autio, S. (Ed.), Geological Survey of Finland, Current Research 1999–2000. Geological Survey of Finland. Special Paper 31, 31–40.

Sokolov, V.A. (Ed.), 1987. The Geology of Karelia. Nauka, Leningrad. 232 pp. (in Russian).

Sotka, P., 1988. Potential raw material for aluminium in Finland. In: Haapala, I. (Ed.), Industrial Minerals and Industrial Rocks in Finland. Yliopistopaino, Helsinki, pp. 93–109.

Talikka, M., 2007. Tectonic evolution of the Paleoproterozoic Tampere Belt during the Svecofennian orogeny, with reference to hydrothermal alteration at Kutemajärvi. In: Kojonen, K.K., Cook, N.J., Ojala, V.J. (Eds.), Au-Ag Telluride Deposits: Field Workshop, Espoo, Finland, August 26–31, 2007. Geological Survey of Finland. Guide 53, pp. 71–76.

Tavela, M., 1954. The limestone quarry of Sibbo. In: Aurola, E. (Ed.), The Mines and Quarries of Finland. Geoteknisiä Julkaisuja 55, pp. 75–77.

Teertstra, D.K., Lahti, S.I., Alviola, R., Cerny, P., 1993. Pollucite and its alteration in Finnish pegmatites. Geological Survey of Finland Bulletin 368, 39 pp.

Väyrynen, H.A., 1929. Über den Chemismus der finnischen Kaolinvorkommen verglichen mit Verwitterungssedimenten. Bulletin Commission Géologique de Finlande 87, 128–158.

Vesasalo, A., 1965. Talc schists and soapstone occurrences of Finland. Bulletin de la Commission Géologique de Finlande 216, 75 pp.

Volborth, A., 1954. Phosphatminerale aus dem Lithiumpegmatit von Viitaniemi, Eräjärvi, Zentral-Finnland. Annales Academiae Scientarium Fennicae. Series A III 39, 90 pp.

Wiik, H.B., 1953. Composition and origin of soapstone. Bulletin de la Commission Géologique de Finlande 165, 57 pp.

Wilson, I., 2008. Current and future growth of global GCC markets. In: Taylor, L. (Ed.), Proceedings of the 18th Industrial Minerals International Congress, San Francisco, 2006. Industrial Minerals/Metal Bulletin plc, United Kingdom, pp. 118–137.

Wilson, I., 2013. Global update on magnesite resources and production. Industrial Minerals 9, 56–64.

PART

EXPLORATION METHODS 10

SURFICIAL GEOCHEMICAL EXPLORATION METHODS 10.1

P. Sarala

ABSTRACT

Surficial geology, till geochemistry, and heavy mineral studies are practical exploration tools in glaciated terrains. Geochemical methods have been widely applied for mineral potential mapping and exploration in Finland and in the Northern Hemisphere for more than 50 years. Till is an effective sampling media in mineral exploration because its composition reflects the nature and composition of fresh bedrock, preglacial weathered bedrock, and older sediments of the up-ice region. From clay-sized particles to boulders, debris in till derived from a mineralized zone have been dispersed some distance from the source(s) in the direction of ice flow, resulting in dispersal trains that are generally larger than the source area exposed to glacial erosion. Several studies have demonstrated that detrital dispersal differs among glacial geomorphological areas. Subglacial processes of active, moving ice differ from passive, ablating ice, resulting in different distance of glacial transport. Furthermore, a number of case studies indicate that the distance of glacial transport varies among glaciomorphological terrains such as transverse ribbed moraines and streamlined moraines, associated to active ice conditions. Till deposits in the ice-divide zone have their own characteristics. Mineral potential studies in Finland have proven that till geochemistry supported by heavy mineral research and new, low-impact sampling methods provide effective ways to do mineral exploration.

Keywords: geochemistry; till; glacial deposits; moraine; weathered bedrock; exploration.

Mineral Deposits of Finland. http://dx.doi.org/10.1016/B978-0-12-410438-9.00027-3

INTRODUCTION

Geochemistry can be broadly defined as the science concerned with all geological studies involving chemical change (Clarke, 1924). It includes the study of the distribution of elements in minerals, rocks, and soils along with the interaction between these earth materials. In this chapter, geochemical exploration methods refer to the use of chemical properties of naturally occurring substances (including rocks, glacial debris, soils, stream sediments, waters, vegetation, and air) to find economic deposits of metals, minerals, and hydrocarbons.[1] Geochemical exploration methods are mainly based on observations of anomalous concentrations of major or trace elements that are derived from a core part of a mineral deposit itself or a wider halo surrounding the ore body (Rankama and Sahama, 1950; Horsnail, 2001). These halos or zones are useful in detecting or tracing mineral deposits, as they are typically many times larger than the ore deposit itself, and therefore represent a larger mineral exploration target. Halos can be classified as primary or secondary. A primary halo is a zone surrounding the mineralization core and represents the distribution patterns of elements in the bedrock formed during the original ore-forming processes. Primary dispersion halos vary greatly in size and shape as a result of the numerous physical and chemical variables that affect fluid movements in rocks. Secondary dispersion halos form, instead, more widespread anomaly patterns of elements surrounding the mineralization core. Elevated elemental concentrations are the products of erosion or weathering of the mineralization core and primary halo, and are transported by agents such as gravity, groundwater, streams, and glaciers. A common secondary halo in glaciogenic sediments is known as a *dispersal train*, of which shape and extent depend on the main transportation mechanisms involved (Kauranne et al., 1992).

A geochemical survey is based on the sampling of available earth materials in a given survey area. Depending on the size of the area and the sampling density, survey types can be roughly divided into continental, regional, targeting, and local scale. The design of a geochemical survey depends on its purpose: (1) characterizing regional elemental and mineralogical concentrations over a large area, (2) delineating geochemical or mineralogical anomalies related to a potential mineralized zone, or (3) focusing mineral exploration within a known mineralized zone (McMartin and McClenaghan, 2001; McClenaghan et al., 2013). The stage of mineral exploration and costs are usually the dominating factors in decision making for the methods and scales of a geochemical survey. What also must be included in the cost factor is that geochemical techniques cannot be used to pinpoint a mineralization on their own, and generally need to be integrated with geological and geophysical surveys (e.g., Plouffe et al., 2011).

Finland is located in the area of the central part of the last Scandinavian ice sheet of the Weichselian ice age (Fig. 10.1.1). This area has been glaciated several times during cold stages of the Quaternary period (e.g., Svendsen et al., 2004; Johansson et al., 2011). The glaciogenic morphology of Finland varies from subglacial to ice-marginal depositional environments and from an active, warm-based ice-lobe network in the south to cold-based, more passive ice in the ice-divide zone in the north (Johansson et al., 2011). Glacial deposits are related to several glacial phases with separate till sheets, each associated with glacial striations and till fabrics with varying orientations (Hirvas, 1991; Sarala, 2005a). Due to the glaciogenic nature of surficial sediments and their extensive cover (97% of Finland's land area) (Johansson and Kujansuu, 2005), surficial geological, and geochemical methods are essential mineral exploration tools. Till is the most common sediment type and because

[1] http://staging.ecomii.com/science/encyclopedia/geochemical-prospecting

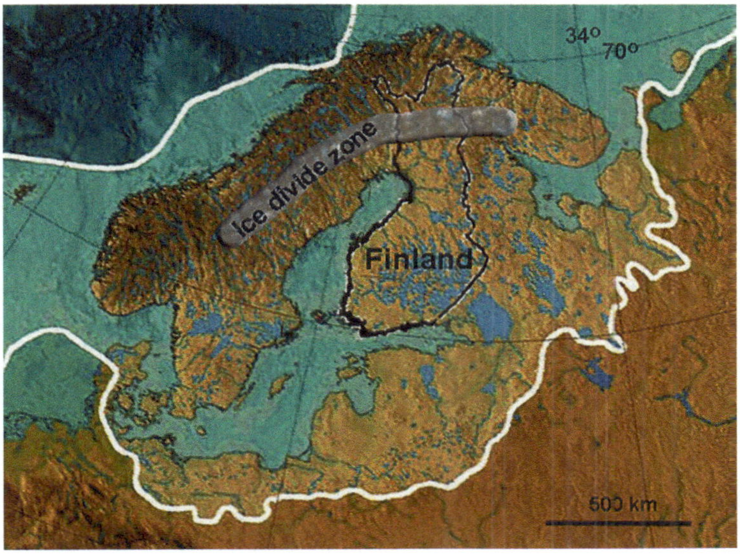

FIGURE 10.1.1 Location of Finland in northern Europe.

The white outline shows the extent of ice at the Last Glacial Maximum (LGM).

its composition reflects the nature and composition of bedrock in the up-ice region, it forms a near-perfect sampling medium for mineral exploration in glaciated terrains.

GEOCHEMICAL EXPLORATION

Geochemical surveys follow the same strategy as other mineral exploration methods: sampling density correlates with the stage of exploration. The primary objective of sparse sampling densities is to identify districts of enhanced mineral potential within which targeted exploration can be conducted. With increasing sampling density, potential mineralized zones or structures in bedrock can be identified. In this context, geochemical methods can rely on the sampling of two mediums: bedrock (lithogeochemistry) and surficial sediments (surficial geochemistry).

LITHOGEOCHEMISTRY

Lithogeochemistry is a geochemical method that involves the sampling of bedrock. It can be considered for the recognition and definition of mineralized and anomalous areas, and to distinguish dispersion halos at different mapping scales (Hawkes, 1957). In a broad sense, lithogeochemistry covers the research of igneous, sedimentary, and metamorphic petrology; hydrothermal alteration; weathering; and diagenesis. Also, it has a close connection to the mining-related fields of applied geochemistry (including exploration and environmental research and monitoring), genesis of mineral deposits, metallurgy, and deep hydrogeochemistry.

Lithogeochemical surveys can be carried out on a grid or on traverses of an area, with samples taken of all available rock outcrops or at some specific interval. After lithological mapping, one or several rock types may be selected for sampling and analyzed for various elements. Using symbols and/or contours, results are compiled in geochemical maps to allow interpretation. Under favorable conditions, mineralized zones or belts may be outlined in which more detailed work can be concentrated. In the case of large territories, geochemical provinces may be outlined.

Isotopic analyses can help build stratigraphical and geochronological models. Applicable isotopic systems include stable isotopes such as $^{16}O/^{18}O$, $^{12}C/^{13}C$, $^{32}S/^{34}S$, and common Pb and radiogenic isotopes such as samarium-neodymium and rhenium-osmium (Faure and Mensing, 2004). For example, the isotopic composition of sulfur can help identify the origin of hydrothermal fluids and the conditions within the depositional environment. Isotopic ratios are also used for age determination of rock types (geochronology) relying on radiogenic isotopes (e.g., U series, U-Pb, K-Ar, Sm-Nd) (Allègre, 2008). Although isotopic geochemistry is an important part of the geochemical research, it is beyond the scope of this chapter and not handled in detail here.

WEATHERED BEDROCK GEOCHEMISTRY

In Finland, and particularly in northern Finland, weathered bedrock has been preserved beneath glacial deposits and needs to be considered in mineral exploration because it can contain elemental enrichment unrelated to mineralization (Hirvas, 1991; Nenonen, 1995). In the ice divide zone of central Lapland, the remnants of weathered regolith up to tens of meters thick are frequently found. The thickest weathering profiles are in topographic depressions under till cover. Typically, only the saprock has been preserved, but in places, also the lower saprolite and parts of the upper saprolite are present. The saprock horizons are strongly fractured and, therefore, are zones of preferential groundwater movement with enrichment of secondary iron minerals such as goethite and some clay minerals. Elements such as Fe, Cu, Ni, Co, Zn, and Mo may have been enriched in the fine fraction of the goethitic weathering crust and their concentrations can be many times higher than in the underlying fresh bedrock (Peuraniemi, 1990a). Several other elements such as Ca, Na, Mg, Au, Br, S, and U are easily mobilized during weathering and can be enriched at the top or bottom of the weathering profile. Consequently, large amounts of secondary enriched weathered material might have been eroded by glaciers causing false geochemical anomalies in till with no relation to mineralization (Peuraniemi, 1982; Sarala and Ojala, 2008).

SURFICIAL GEOCHEMISTRY

Surficial methods are traditionally used in exploration wherever different kinds of soils and unconsolidated sediments cover the bedrock. Surficial geology, and, more specifically, drift prospecting, has been used as a practical exploration tool in glaciated areas (e.g., in North America, Europe, and Russia) since the beginning of the twentieth century (Sauramo, 1924); surficial geochemical methods have been in use since the 1950s (Kauranne, 1958; Wennervirta, 1968; Shilts, 1972; Kujansuu, 1976). Soil, stream sediments, and glacial till surveys have been used effectively in geochemical prospecting and have resulted in the discovery of a number of ore bodies. A benefit of surficial methods is the relative ease and speed of sampling and the possibility of using different sampling strategies, from regular grids to variable sampling lines and scattered models. The principal objective of a surficial geochemical survey is to detect the secondary halo associated with a mineralized zone. In addition to the traditional sampling of fresh mineral soil (C-horizon of soil profile), surficial methods include the sampling of:

- Humus, plants, and peat (i.e., biogeochemistry)
- Weathering profile (i.e., pedogeochemistry)
- Top soil together with selective and weak leach methods
- Groundwater
- Snow

Surficial geochemistry has environmental applications including the field of urban geology (e.g., study of airborne fallout).

Boulder tracing (e.g., Shilts, 1984; Salminen and Hartikainen, 1985; Salonen, 1986) and heavy mineral (i.e., indicator mineral) research (Peuraniemi, 1990c; McClenaghan, 2005) are included in the broad theme of surficial mineral exploration methods. Furthermore, geochemistry applied to hydrocarbon exploration (for coal, gas, and oil deposits) is part of surficial exploration methods (Jahn et al., 2008), but all of these topics are beyond the scope of this section.

In surficial geochemistry, the knowledge of the entire process of erosion, transportation, and sedimentation (or deposition) is a key factor in understanding the geochemical signature in different sedimentary deposits. Secondary dispersion is directly dependent on surficial processes, and, therefore, the methods of sampling, analysis, and interpretation have to take into account the sedimentary environment. For example, fluvial or lacustrine sediments usually represent much larger source areas (i.e., river basins) than glacial till and this fact should be taken into account in geochemical data interpretation. Furthermore, stratigraphical control of samples needs to be considered during sampling and data interpretation. This is particularly important in complex stratigraphic settings with multiple units being sampled. For example, in the case of complex till stratigraphy, sampling at different depths provides necessary information to recognize source areas of different till units (Hirvas and Nenonen, 1990). Till thickness and depth of sampling within a single till unit represents another variable to consider. In the case of till thickness, the general phenomenon is that on high land areas, till cover is thinner than in the valleys or depressions. In thick sequences, a sample from the bottom of a section or from the lowermost till unit will reflect a more proximal bedrock source than a sample collected from the top (Kauranne et al., 1992).

Today, there is an ever-increasing demand to use low-impact exploration methods to decrease the environmental footprint of mineral exploration activities. With this emphasis in mind, the use of humus, peat, and living plants have become more popular methods in conjunction with top soil sampling; that is, the uppermost layers (10–50 cm) of the mineral soil (e.g., Kokkola, 1977; Nuutilainen and Peuraniemi, 1977; Hamilton, 2007). Biogeochemical surveys can be divided into two types (Dunn, 2007). One type uses the trace-element content of plants to follow up secondary dispersion halos of mineralization, similar to the soil geochemical method. The second type is to use specific plants that are sensitive to certain elements in soils and can indicate potential ore mineralization. The latter type is referred to as a "geobotanical survey" and can also be considered one type of surficial geochemical methods

GLACIOLOGICAL CONTEXT

Prior to implementing a surficial geochemical survey (e.g., soil or till), it is essential to study and understand the Quaternary geological landforms in the target area and the processes by which they formed; that is, the glaciological context of the study area. This can be done by interpreting glacial morphology visible on aerial photographs, satellite imagery, or a digital elevation model (DEM), which in turn allows determination of direction of glacial transport and estimated distance of transport (Salonen, 1988; Aario and Peuraniemi, 1992; Sarala, 2005a).

Morphological interpretation is supported by field work that includes research on the characteristics of different moraines and other glacial landforms and includes geographical distribution of landforms, relation to other landforms, sediment type, internal structure, and till stratigraphy (Hirvas and Nenonen 1990; Aario and Peuraniemi, 1992). Test pit excavations, observations on natural or human-made sections, or drilling-based sampling are part of the field methods for landform research. Till geochemical and heavy mineral sampling are essential in mineral exploration and should be complemented by grain size determinations and pebble count data. Till-fabric analyses and measurements of glacial striations are necessary for the determination of ice-flow direction. Furthermore, till-fabric analyses together with stratigraphic observations can provide useful information as to the subglacial depositional environments and processes (Sarala, 2005b).

In a new study area, a relatively simple field method of excavating test pits can provide valuable information on till stratigraphy and the sediment thickness (Sarala, 2005a). This knowledge is useful later when planning geochemical sampling strategies or conducting other mineral exploration activities such as diamond drilling or mineralized boulder tracing.

TILL GEOCHEMISTRY

Various geochemical analytical methods are available for the analyses of till. The choice of a method will be dictated by the scale and objective of the geochemical survey (e.g., Shilts, 1984; Hirvas and Nenonen, 1990; Kauranne et al., 1992; McMartin and Campbell, 2009; Paulen, 2009). Regular sampling grids are used in regional and local-scale surveys, when the main goal is to map the variation of the elements in the study area. Percussion drilling with a flow-through bit or other low-impact methods (e.g., handmade test pits or hand drilling) are commonly used for till sampling (Fig. 10.1.2A). Linear "fences" are used in regional till sampling when variations have to be minimized, or when the bedrock structure and ice-flow direction are well-known. Test pits and test trenches are effective in detailed studies, when the source (or sources) of anomalies needs to be identified. Vertical variation of the elements can be studied by taking samples at regular intervals from different depth levels. By using test trenches, both horizontal and vertical elemental composition variability in till can be determined using sampling profiles at regular intervals. In the case of thick glacial overburden, soil sampling with different drilling techniques or percussion drilling are the most effective, and many times the only way to take till samples.

The normal procedures for the processing of till samples for geochemical analysis involves drying (room temperature, 40 °C and 90 °C), dry sieving, and leaching. For chemical analysis, the size fraction of <0.063 mm is commonly utilized worldwide because it represents the most homogenous part of the till matrix and includes both local and distal portions of glacially eroded, transported, and deposited materials. Coarser size or high density fractions can be analyzed if the objective is to identify locally derived components of the till.

SELECTIVE AND WEAK-LEACH GEOCHEMICAL METHODS

Selective and weak-leach analytical methods were important developments that occurred in 2000. The best known of these is probably the patented Mobile Metal Ion (MMI™) method, where upper soil

FIGURE 10.1.2 Different till sampling and analysis techniques used in mineral exploration.

(A) Percussion drilling with a flow-through bit in till geochemical sampling; note the vehicle and trailer mounted on wide tracks to reduce impact on vegetation; (B) low environmental impact soil sampling for weak leach geochemistry; note that the moss matt is replaced after sampling, leaving the ground surface undisturbed, (C) portable XRF (pXRF) analyzer for onsite till geochemical analyzes in a dug exposure, (D) heavy mineral separation using a spiral concentrator.

Source: *Photos (A), (B), and (C) by Pertti Sarala; photo (D) by Reijo Lampela.*

samples are leached with very weak acidic solutions to strip only the loosely adsorbed ions from the surface of mineral grains or organic material before analysis by ICP-MS. The method has been studied extensively, especially in North America and Australia, where it has achieved good results in areas of very thick glacial overburden or deeply weathered soils. Some testing has also been done in Finland by exploration companies and GTK (e.g., Sarala et al., 2008). The mobilization, movement, and enrichment of the ions are the sum of many factors including capillary action, diffusion, difference of electrochemical charge, and effect of gases (Smee, 1983; Cameron et al., 2004; Hamilton et al., 2004), as well as biogeochemical processes (Tack, 2010).

The MMI method has advantages including: (1) easy sampling, because the samples are taken from the near-surface top soil (Fig. 10.1.2B), (2) the sample material may be any mixture of mineral-derived soil to purely organic material, (3) it gives a direct indication of the bedrock type through the soil, (4) interpretation of results can be viewed as relative concentrations (to avoid disturbance caused by, for example, seasonal variations or local weather changes), and (5) sample processing and analysis are inexpensive compared to traditional till geochemical methods. Usability of the method has substantially increased due to analytical method development that allows ppt or even ppq detection limits for many elements. Nowadays there are a number of alternative analysis methods besides MMI, using the same theoretical background of ion mobilization from the bedrock through the soil, which broaden the range of elements and ore types that can be explored. Such methods include, for example, various weak acid extraction methods, bio- and enzyme leach, and the soil-gas-hydrogen method. These methods are being tested in GTK's ongoing project, "Ultra low-impact exploration methods for the subarctic," which is a part of the Tekes Green Mining Programme (Sarala and Nykänen, 2014a,b).

PORTABLE XRF METHODS

X-ray fluorescence (XRF) has been a standard method in laboratories for decades and has been used to determine major and minor element concentrations, as well as several precious elements in a variety of sample materials. However, conventional laboratory analyses with sample processing and powder pellet preparation are still time-consuming and increase the costs of the analysis. Significant developments have been made with portable XRF (pXRF) analyzers in the last decade including the reduction of detection limits for many elements to the ppm level. Portable equipment is lightweight and easy to take into the field or use at field camps, and that is why they are also called online XRF scanners (Fig. 10.1.2C). Applications such as ScanMobile®, where the XRF scanner is mobilized using a car-supported system, has facilitated usage of this instrument in field conditions (Sarala, 2009; Sarala and Ojala, 2008; *Mining Journal*, 2012; Sarala and Mäkikyrö, 2012). This application allows till, weathered bedrock, and drill core samples to be analyzed directly in the field or field camp. After collection, the samples are simply placed into drill core boxes for measurement, requiring little or no preprocessing. Although the portable XRF analyzers are widely used in exploration and in mining for bedrock and grab samples, their usage in mineral exploration for the analysis of till or weathered bedrock has so far been relatively limited, but this situation is changing rapidly.

HEAVY MINERAL STUDIES

In addition to geochemical methods, the study of the mineralogy of the heavy mineral fraction of till has been used extensively in mineral exploration (Peuraniemi, 1982, 1990b; Lehtonen, 2005; McClenaghan, 2005). It is based on the separation of the heavy mineral fraction from till or weathered bedrock samples. The size of bulk sediment samples collected for heavy mineral processing varies depending on the purpose of the study, the type of material sampled, and the texture of the sediment. Commonly, sample size varies from 10–12 liters (i.e., 20–25 kg), which is enough to get a representative sample of the heavy mineral composition in the sand size fraction of the material. Separation of the heaviest mineral fraction is based on gravity concentration. Preconcentration can be done with a spiral separator (gold screw panner) (Fig. 10.1.2D), by hand panning, with a shaking table, or a Knelson concentrator (Chernet et al., 1999; McClenaghan and Cabri, 2011).

After removing the magnetic fraction, heavy liquids (density 2.96 g/cm^3 or 3.31 g/cm^3) can be used to separate the heaviest minerals from the preconcentrates. Basic heavy mineral examination is done using a binocular microscope. The elemental composition and, thus, mineral identification of a selected number of grains can be done with a scanning electron microscope with energy-dispersive X-ray spectroscopy (SEM + EDS). Elemental composition of specific minerals is done using an electron microprobe (EMP). FEI's Mineral Liberation Analyzer (MLA) can also be used in mapping the heavy mineral population distribution, by doing automated mineral search and identification. Strict protocols need to be implemented to ensure the quality assurance and quality controls of indicator mineral studies (Plouffe et al., 2013).

Specific indicator minerals within the heavy mineral concentrates are useful for the detection of a variety of mineral deposits, including diamond, precious, and base metal deposits (e.g., Peuraniemi, 1982; Kauranne et al., 1992; Lehtonen, M., 2005; McClenaghan and Kjarsgaard, 2001; McClenaghan and Cabri, 2011; Paulen et al., 2011, 2013; Plouffe et al., 2014). Some lists of indicator minerals usable in mineral exploration have been published by Peuraniemi (1982) and McClenaghan (2005). Indicator minerals may come from the lighter end of the density spectrum normally used in heavy mineral research. For example, some phosphates can be used as indicators for rare earth element (REE) mineralization. A benefit of using heavy or indicator minerals is that transport distances are usually much longer than those of conventional till geochemical anomalies (Peuraniemi, 1990c). This can facilitate the reconnaissance stages of mineral exploration by allowing the use of sparse till sampling grids over a large area, leading to more rapid exploration at lower cost when compared to till geochemistry methods.

TILL GEOCHEMICAL EXPLORATION AT DIFFERENT SCALES

Continental and national geochemical sampling programs are good starting points for mineral exploration and evaluating the potential of different areas for specific elements, metals, and minerals. Sampling grid densities are critical in this view and depend on the extent of the areas under exploration (Theobald et al., 1991). DiLabio (1990) used the classifications continental, regional, local, and property-scale, where the sampling densities vary from over 100 km to under 100 m. Amount of detail usually increases according to survey scale where in the continental or national scale, for instance, the sampling density can be 1 sample/10–100 km^2, giving an indication of the major lithological units or the broad geochemical signature of regional bedrock units.

Regional scale commonly represents the sampling density of 1 sample per ±5 km^2. Regional geochemical mappings with sparse grids can serve to identify targets or anomalies for mineral exploration. A good example of this is the mapping project that was carried out in Finland in the 1980s (Salminen, 1995; Salminen and Tarvainen, 1995). The geochemical atlas for the fine fraction of till (<0.06 mm) was produced based on the same database (Koljonen, 1992). In this national compilation, till geochemistry clearly reflects broad lithological provenances (e.g., green stone and schist belts, granitoid provenances, and granulite belt). Furthermore, this national till geochemical database, with one sample per quarter km^2 is sufficient for directing mineral exploration in some places.

Mineral exploration at a local scale requires a dense sampling grid (<250 m) because geochemical anomalies can be subtle compared to background levels; it seeks mineralized bedrock sources that are small in area, leading to short dispersal trains. Sampling at a local scale (at intervals of tens to hundreds of meters) allows the characterization of small bedrock units and soil composition over a restricted area.

The shape of dispersal trains can be complex in the case of multiple glacial advances with multiple till units. Hirvas and Nenonen (1990) and Klassen (2001) have discussed the shape of dispersal trains along with their relation to glacial morphology. Glacial dynamics can also be inferred from the shape of dispersal trains (Sarala, 2005a; Trommelen et al., 2013). Furthermore, dispersal trains can be defined in three dimensions. DiLabio (1990) showed that dispersal trains rise gently within the till package in the down-ice direction, typically reaching the soil surface after 300–1000 m transportation in shallow till areas. This is the case when the basal glacial conditions are quite stable, warm-based, and the till is deposited by the lodgement process. In the central part of glaciers, under variable subglacial conditions, transport distances can vary greatly. For example, in the drumlin areas in eastern and northern Finland, till anomalies are narrow but long, usually from several kilometers to tens of kilometers, and mineralized boulders can be found far from their source(s) (Hirvas, 1989; Aario and Peuraniemi, 1992). In contrast, till anomalies are distinct and short in the ribbed moraine areas in southern Finnish Lapland, where numerous reported examples reveal very short glacial transport distances, in the order of tens to hundreds of meters down-ice from the distal side of mineralization (Peuraniemi, 1982; Aario, 1990; Sarala et al., 1998; Sarala and Rossi, 2000; Sarala, 2005b; Sarala and Peuraniemi, 2007).

TILLS AND GLACIAL LANDFORMS AS INDICATORS OF TRANSPORTATION

Till genesis can be evaluated and interpreted from (1) till facies analysis or (2) glacial morphological analysis. Identification of different till units is based on the analysis of composition and structure of till deposits. For mineral exploration purposes, a key point is to recognize basal tills such as basal lodgement and melt-out tills from the till types deposited on marginal parts of the glacier, such as flow till and waterlain till. The critical difference is that the latter have been affected by secondary, usually nonglacial processes (e.g., gravity flow) during deposition. Dreimanis (1989, 1990) has listed characteristic features of different till types and produced a classification scheme of till deposition based on the influence of primary and secondary processes. Dreimanis also described a number of till types with specific characteristics that can be related to variable glacial and depositional environments. Subglacial processes are effective at eroding, transporting, and depositing rock fragments and debris in the direction of ice flow. Consequently, lodgement till, deposited at the base of the glacier, is the best sampling media for till geochemical and mineralogical surveys. Other till types, such as deformation till, can be intermixed with a variety of other sediment types and bear evidence of deformation—for example, presence of folds and shears (e.g., Hart, 1997; Piotrowski et al., 2001; Roberts and Hart, 2005). Deformation tills are transitional with melt-out tills and it is difficult to draw a strict division between them.

A close relationship exists between till genesis and the glacial morphology. The glacial morphology provides an indication of depositional conditions and glacial transport distances both at regional and target scales as pointed out in several papers from northern Finland (e.g., Aario, 1977; Aario and Peuraniemi, 1992; Sarala et al., 1998; Sarala, 2005b,c). For mineral exploration purposes, it is important to distinguish active-ice deposits from passive-ice deposits. Active-ice deposits are those that are formed subglacially under a moving ice sheet. Typical landforms and deposits associated with active ice include drumlins, flutings, ribbed moraines, and thin basal till sheets (Fig. 10.1.3). Under passive-ice conditions, hummocky moraines and thick ablation till formations are deposited, usually in the marginal zone during the retreating phase of a glacier. This is also the environment where the end and other marginal moraines occur. Active- and passive-ice deposits have sustained different transport histories. Typically, deformation, lodgement, or basal melt-out tills associated with active-ice deposits are deposited subglacially, grain by grain or fragment by fragment, together with reworking of older deposits under high pressure. In spite of great variation between the transport distances in different formation types, the transport direction is in line with the ice-flow direction. On the other hand, till in passive-ice formations are usually transported englacially or supraglacially, so the transport distance and direction can vary considerably, and, consequently, source areas are not easily traced.

STREAMLINED, ACTIVE-ICE LANDFORMS

Drumlins, flutings, and some other radial moraine types (e.g., the Sevetti moraine in northern Lapland) are clear indicators of ice-flow direction (Aario, 1990). These landforms form typically large streamlined moraine fields consisting of several thousands of unique formations. The dimensions of drumlins are usually 5–50 m in height, 10–200 m in width, and from 100 m to several kilometers in length, and they are mainly formed under warm-based glacial conditions on the marginal zone of an ice sheet. These forms can be both depositional and erosional (Aario, 1990). Deposition happened subglacially when plastic debris was mobilized due to changing stress fields or the influence of active basal meltwater, which carves cavities beneath an ice mass that are later infilled with stratified sediments and till (Menzies and Shilts, 1996). Molding of previously deposited material in subglacial conditions, where the meltwater content is low, is an explanation for erosional forms. For that reason, in this case, the sediment bedding and structures usually represent an earlier glacial or interglacial depositional history of an area.

Flutings are mainly formed in settings similar to drumlins, but their dimensions are much smaller, with heights of 1–2 m, widths of 2–10 m, and lengths of 100–1000 m (Aario, 1990). These forms are usually obvious on aerial photographs as a group of narrow stripes in close connection with drumlins. Identification of these landforms in the field is difficult if there is vegetation cover. Furthermore, in Finland, they are usually surrounded by peat bogs, where surfaces are almost on the same level as the fluting tops.

Transport mechanisms of till in regions of streamlined landforms have been investigated widely (DiLabio and Coker, 1989; Aario and Peuraniemi, 1992; Piotrowski, 1997). In those areas, the transport direction of till is usually parallel to the orientation of the streamlined landforms (e.g., drumlins and flutings), with long transport distances: usually several kilometers, but even tens of kilometers in some places. Aario and Peuraniemi (1992) estimated that sampling till in the regions of streamlined landforms is an effective mineral exploration method. This is certainly the case where only one advance phase occurred during the drumlin formation, (i.e., ice movement was straightforward and older glacial overburden was relatively thin). In contrast, in regions of multiple glacial advances with variable flow directions, glacial dispersion can be very complex and transport distances long. For example, in the Kuusamo drumlin field in northeastern Finland, at least two different flow phases have been observed (Aario and Forsström, 1979). In eastern and southeastern Finland, in several parts of the Pieksämäki drumlin field,

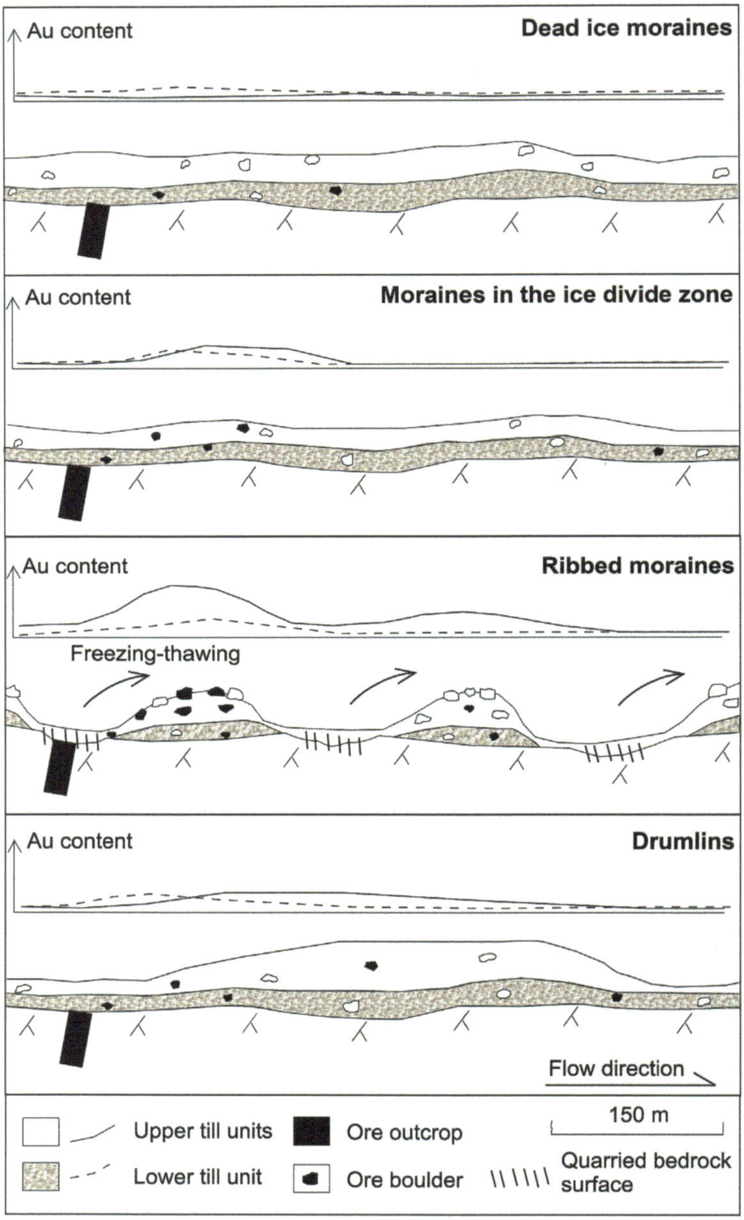

FIGURE 10.1.3 Schematic presentations of gold distribution into two till units representing glacial advance and retreat phases of glaciers in various morphological terrains.

Dead-ice moraines, moraines in the ice divide zone, ribbed moraines, and drumlins.

transport distances vary from 500 m to several km (e.g., Hirvas and Nenonen, 1990; Peuraniemi, 1990b; Hartikainen and Damsten, 1991). Although the secondary halos of the geochemical anomalies are long, they are commonly also very narrow. The same phenomenon is seen in heavy minerals, which form distinct elongated anomaly patterns in the down-ice direction (Peuraniemi, 1990b,c).

TRANSVERSE, ACTIVE-ICE MORAINE FORMATIONS

The transverse, active-ice moraine type in northern Finland is classified as *ribbed moraine* (Sarala, 2003, 2005a) following Hättestrand's (1997) classification. The ribbed moraine morphology consists of till ridges transverse to the ice-flow direction, with dimensions of 100–1000 m in length, 50–200 m in width, and 2–10 m in height. The interval between individual ridges is 100–300 m.

The term *ribbed moraine* is sometimes used as a synonym for *Rogen moraine*, but here it is used as the name for all ridge-type moraines formed by a similar process and with a similar morphology (cf. Sarala, 2003). Hättestrand (1997) has divided ribbed moraines into four subtypes: hummocky ribbed moraine, Rogen moraine, Blattnick moraine, and minor-ribbed moraine. All of these types have been formed as a result of similar conditions and formation processes during deglaciation, under a fast-flowing glacier, at the contact between cold-based and warm-based subglacial conditions. Hättestrand (1997) and Kleman and Hättestrand (1999) suggest that due to tension, subglacial deposits are broken up and transported with the flowing ice resulting in an undulating, puzzle-like ridge morphology.

Sarala (2005a) suggests that the ribbed moraine formation was a two-step process: (1) internal ice movement with tensional pressure caused fracturing of the cold-bed glacier and the subglacial sediments into blocks, which moved under the ice sheet and formed a ridge morphology; and (2) flowing ice transported the material loosened between the blocks by the freeze–thaw process and redeposited it on the surfaces of the new ridges. The deposition occurred in the transitional zone between the cold-bed and the thawed-bed glacier during deglaciation. This two-step mechanism explains the presence of locally derived boulders at the surface and in the uppermost part of the till in ribbed moraines (Fig. 10.1.4). Indeed, several case studies in southern Finnish Lapland show the extremely strong glacial quarrying activity and short transportation of till material and surficial boulders from the underlying bedrock to the top of moraine ridges. Consequently, a short distance of transport should be expected in regions of ribbed moraines common in Fennoscandia and northern America (Aario, 1977; Lundqvist, 1989; Kleman, 1994; Hättestrand, 1997; Sarala, 2005a,b).

Several mineral exploration studies have been completed in regions of ribbed moraines in southern Finnish Lapland (e.g., Sarala et al., 1998; Sarala and Rossi, 2000; Sarala, 2005a; Sarala and Peuraniemi, 2007; Sarala et al., 2007a). Mineral exploration in regions of ribbed moraines is facilitated by the composition of debris and rock fragments in the upper part of the ridges and the surficial boulders on the ridge top, which are derived from the local underlying bedrock. Furthermore, till geochemistry in those regions is characterized by narrow and sharp anomalies also related to local bedrock sources. Due to short and effective glacial erosion and subsequent transportation, fresh heavy minerals, such as sulfides and gold grains, are found to be well preserved in till of ribbed moraine. For example, in southwestern Rovaniemi, within the Paleoproterozoic Peräpohja schist belt, Cu-Au mineralization at Petäjävaara is found in a hydrothermal alteration zone in-between quartzite and mafic volcanic rocks, and is related to 1-m-wide quartz vein with associated pyrite and chalcopyrite dissemination (Sarala and Rossi, 2000). The transport distance of mineralized boulders, pebbles, and other debris is very short (only 5–50 m). Similar mineral exploration case studies with very short glacial transport have been reported in the

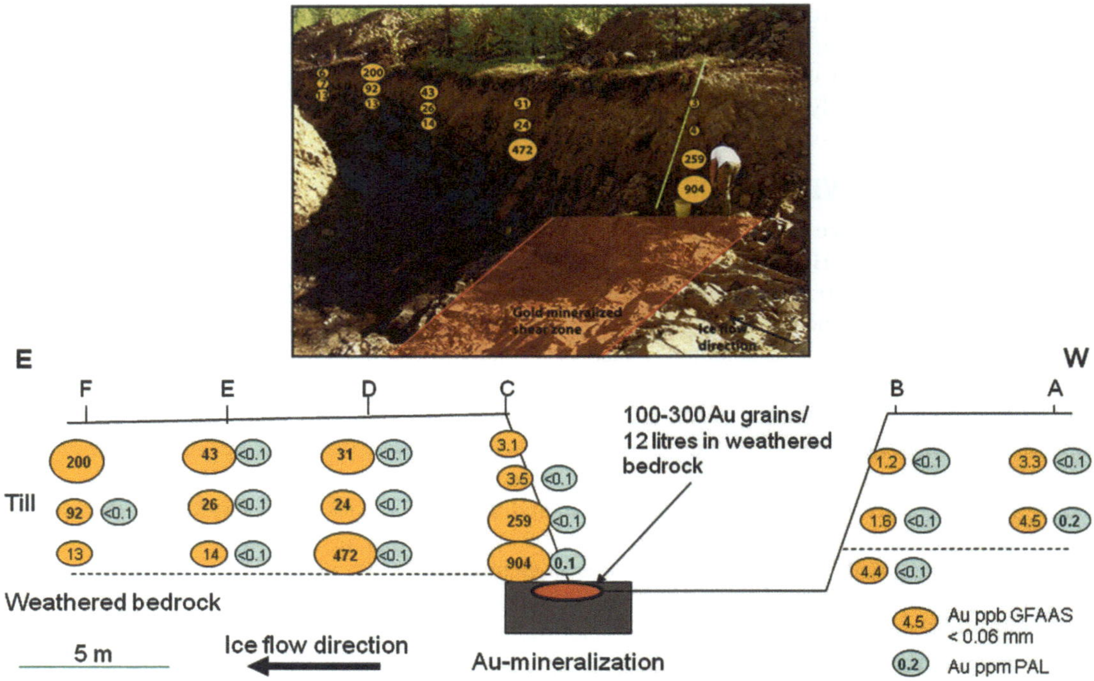

FIGURE 10.1.4 Au contents of the till in the test trench at the Petäjäselkä target, showing very short glacial transport in the central part of the last glaciation.

Yellow ovals show Au contents in the <0.06 mm size fraction after an aqua regia leach followed by graphite furnace atomic absorption spectrometry (GFAAS) analysis. Blue ovals represent Au contents in the <2 mm size fraction after a cyanide leach followed by GFAAS analysis using PAL1000 equipment (simultaneous pulverizing and cyanide leach using 0.5 kg subsample).

Portimojärvi area, north of Ranua village (Aario, 1990; Sarala and Peuraniemi, 2007), and in the Misi area in northeastern Rovaniemi (Sarala and Nenonen, 2005).

CORE AREAS OF GLACIERS

Generally, in regions covered by former ice divides, glacial erosion and transportation is weak. Minimal movement at the base of the glaciers, combined with cold-based conditions, generally resulted in limited glacial erosion, and in places only redeposition of preexisting glaciogenic deposits happens.

One of the studied areas in the ice-divide zone is the Petäjäselkä target in central Lapland, where GTK undertook gold exploration in the 2010s (Sarala et al., 2007b; Nykänen et al., 2007). In the Petäjäselkä target area, gold mineralization follows a heterogeneous and deformed north-northwest trending zone of graphic tuffs, cherts, and intermediate volcanic rocks within a mafic volcanic rock dominated domain (Hulkki et al., 2011). Till stratigraphy is simple: fresh bedrock is overlain by 1–2 m of weathered material composed of fractured bedrock or deeply weathered saprock, in turn overlain by a 1–2 m till that can be divided into two units: a bottom unit deposited during the advance phase and an upper unit deposited as ablation till during the melting phase. Till geochemistry shows very rapid upward dispersion of mineralized material from the bedrock surface into the upper till, 15–20 m down-ice (to the east) from the bedrock mineralization (see Fig. 10.1.4). The anomaly pattern is very sharp and strong, indicating gold mineralization in the underlying bedrock. The observation of a short transport distance is supported by the high number of gold grains in the heavy mineral concentrates close to mineralized bedrock.

Another example of short glacial dispersal comes from the Mäkärä area in the northern part of Sodankylä municipality. Gold and REE exploration have been undertaken here since the 1950s, most recently in 2009. The Mäkärä Au-REE target is located in the Tana belt, south of the 1.9 Ga Lapland granulite belt in northern Finland, where there are prominent lanthanum and yttrium anomalies in regional till and bedrock geochemical data (Sarapää and Sarala, 2013; Sarapää et al., 2013). Test pit and trench excavations, together with deep drilling samples, prove that the elevated gold and REE contents are related to the nearby narrow gold-hematite-quartz veins and the surrounding strongly weathered kaolinitic saprolite derived from arkosic gneiss, implying a short distance of glacial transport of around 100–200 m.

However, even in the ice-divide zone there can be small and local differential movement or reactivation between ice lobes at the end of the deglaciation phase, causing local erosion and even producing active-ice landforms. In places, this is seen particularly in the uppermost till layer, which can include distally transported debris and rounded rock fragments. In those cases, till material has been transported a relatively long distance inside the ice or englacially. After melting of ice, the material is released and deposited, forming a thin till veneer on the underlying sediments and giving no indication of underlying bedrock.In till geochemistry, it is seen as having a low content of elements and dilution of contents to background levels.

TRANSPORT DISTANCES AND DILUTION OF MINERALIZED MATERIAL IN TILL

As noticed in the cases of Petäjäselkä and Mäkärä, till composition can be used to estimate distance of glacial transport. Short and long distances of glacial transport measured from till composition needs to be considered in the interpretation of glacial transport mechanisms (e.g., Larson and Mooers, 2005). The loss of a geochemical or mineralogical signal in till at some distance from a bedrock source is related to a combination of glacial comminution (i.e., crushing of the mineralized debris to small size fraction) and dilution (i.e., unmineralized debris being eroded and incorporated in the glacial load). Therefore, general rules apply to glacial transport. For example, well-indurated intrusive bedrock can be englacially transported over longer distances compared to weathered, sheared, or poorly indurated lithologies. Calculations of glacial transport based on till matrix composition follow the same principles of boulder transport (e.g., Salonen, 1986). This can also be used in a reverse fashion to estimate the maturity of material: The higher the clay content a till has, the more mature it is, meaning the larger the source area of the till material and the longer the till material has been under the influence of glaciogenic processes. In some cases this calculation correlates with the distance from the source areas. Note that when interpreting the results, the high clay content of till can also indicate certain rock types in the bedrock (sensitive to clay-forming processes during erosion and secondary surficial processes) or existence of highly weathered bedrock surface (i.e., saprolite type weathering) as the source for the till.

For example, in the central Lapland area (the area of the last ice-divide zone), transport distances are short, as seen in the case of Petäjäselkä and Mäkärä (described earlier). In the Petäjäselkä area, gold contents dilute to background levels 15–20 m down-ice (east) from the source (see Fig. 10.1.4).

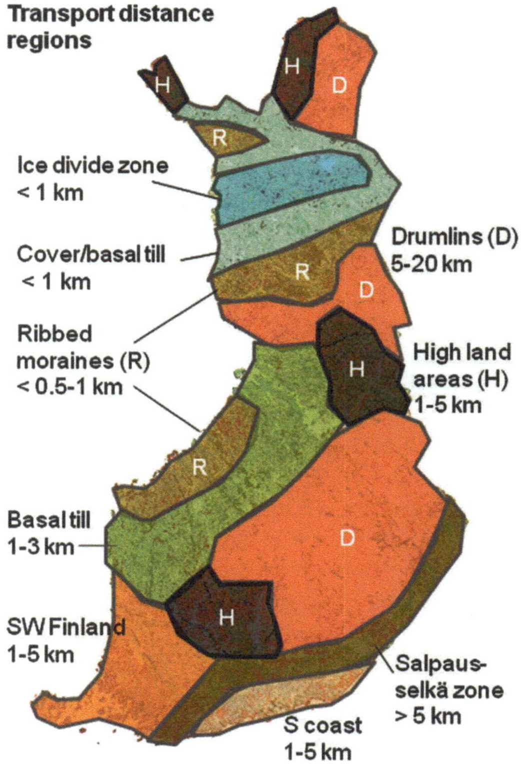

FIGURE 10.1.5 Generalized classification of Finland into eight glaciomorphological terrains and associated distance of glacial transport.

This is based on till geochemistry data and surficial boulder fans in Finland.

In the Mäkärä area, elevated concentrations of gold and REE return to background levels 100 m down-ice (toward the east) from the source (Sarapää and Sarala, 2013). In the ribbed moraine terrains, distance of glacial transport is only some tens of meters to 100 m for gold and a few hundreds of meters for copper and other indicator elements (Sarala et al., 1998; Sarala and Rossi, 2000). Pyrite and chalcopyrite grains preserved in till in those terrains attest to limited glacial comminution at short distances from the bedrock source and to weakly oxidizing conditions in the surface till, which permitted the preservation of sulfide minerals.

Linking glacial morphology and till composition in a variety of settings, it is possible to generate a rough estimation of the distance of glacial transport based on glacial terrain type. Fig. 10.1.5 presents an estimate of glacial transport distances based on glacial geomorphology. The model presented in Fig. 10.1.5 is largely based on till geochemistry and transport distances of mineralized material in till. It can be used as a starting point to interpret till geochemical data and to plan sampling programs in different terrain types. It is worth remembering that when moving into smaller research or exploration scales, the stratigraphical control should be increasingly taken into account.

SUMMARY

Geochemical methods have a long tradition of use in mineral exploration. Till geochemistry and mineralogy are practical methods for mineral exploration. Use of till as a sample material has been the cornerstone in the geochemical exploration of glaciated terrains and has led to the discovery of numerous ore deposits in Finland and abroad. Especially in the 1970s and 1980s, development of chemical analysis and sampling methods was rapid, and till sampling and mapping was very active. Extensive sampling campaigns covered all of Finland and, together with the development of fast and reliable analytical methods, national geochemical mapping was completed. Interpretation of till geochemistry and mineralogy needs to rely on a good knowledge of till stratigraphy and glacial morphology to be effective in mineral exploration. This has been proven in numerous case studies in Finland, as presented in several mineral exploration examples in this chapter from ribbed moraine, drumlin, and ice-divide terrains. Till geochemical methods can now rely on sampling and analytical methods with low environmental impact, which is especially important in exploring sensitive natural areas.

ACKNOWLEDGMENTS

The author would like to thank Dr. Alain Plouffe and Dr. Hugh O'Brien for their valuable comments and improvement of the manuscript.

REFERENCES

Aario, R., 1977. Classification and terminology of morainic landforms in Finland. Boreas 6, 87–100.

Aario, R., 1990. Morainic landforms in northern Finland. In: Aario, R. (Ed.), Glacial Heritage of Northern Finland; an excursion guide, 1. Nordia tiedonantoja, Sarja A, pp. 13–27.

Aario, R., Forsström, L., 1979. Glacial stratigraphy of Koillismaa and North Kainuu, Finland. Fennia 157, 1–49.

Aario, R., Peuraniemi, V., 1992. Glacial dispersal of till constituents in morainic landforms of different types. Geomorphology 6, 9–25.

Allègre, C.J., 2008. Isotope Geology. Cambridge University Press, Oxford.

Cameron, E.O., Hamilton, S.M., Leybourne, et al., 2004. Finding deeply buried deposits using geochemistry. Geochemistry: Exploration, Environment, Analysis 4, 7–32.

Chernet, T., Marmo, J., Nissinen, A., 1999. Technical note: Significantly improved recovery of slightly heavy minerals from Quaternary samples using GTK modified 3" Kenlson preconcentrator. Minerals Engineering 12 (12), 1521–1526.

Clarke, F.W., 1924. The Data of Geochemistry. United States Geological Survey, Bulletin 770. U.S. Government Printing Office, Washington, p. 841.

DiLabio, R.N.W., 1990. Glacial dispersal trains. In: Kujansuu, R., Saarnisto, M. (Eds.), Glacial Indicator Tracing. A.A. Balkema, Rotterdam, pp. 109–122.

DiLabio, R.N.W., Coker, W.B. (Eds.), 1989. Drift prospecting. Geological Survey of Canada, Paper 89–20.

Dreimanis, A., 1989. Tills, their genetic terminology and classification. In: Goldthwait, R.P., Matsch, C.L. (Eds.), Genetic Classification of Glacigenic Deposits. A.A. Balkema, Rotterdam, pp. 15–81.

Dreimanis, A., 1990. Formation, deposition, and identification of subglacial and supraglacial tills. In: Kujansuu, R., Saarnisto, M. (Eds.), Glacial Indicator Tracing. A.A. Balkema, Rotterdam, pp. 35–55.

Dunn, C., 2007. Biogeochemistry in Mineral Exploration. Handbook of Exploration and Environmental Geochemistry 9. Elsevier, Amsterdam.

Faure, G., Mensing, T.M., 2004. Isotopes: Principles and Applications. John Wiley, New York.

Hamilton, S.M., 2007. A Prospector's Guide to the Use of Selective Leach and Other Deep Penetrating Geochemical Techniques in Mineral Exploration. Ontario Geological Survey. Open File Report 6209.

Hamilton, S.M., Cameron, E.M., McClenaghan, M.B., Hall, G.E.M., 2004. Redox, pH and SP variation over mineralization in thick glacial overburden. Part II: field investigation at Cross Lake VMS property. Geochemistry: Exploration, Environment, Analysis 4, 45–58.

Hart, J.K., 1997. The relationship between drumlins and other forms of subglacial glaciotectonic deformation. Quaternary Science Reviews 16, 93–107.

Hartikainen, A., Damsten, M., 1991. Application of till geochemistry to exploration, Ilomantsi, Finland. Journal of Geochemical Exploration 39, 323–342.

Hättestrand, C., 1997. Ribbed moraines in Sweden—distribution pattern and paleoglaciological implications. Sedimentary Geology 111, 41–56.

Hawkes, H.E., 1957. Principles of geochemical prospecting. Geological Survey Bulletin 1000-F. U.S. Government Printing Office, Washington, pp. 225–355.

Hirvas, H., 1989. Application of glacial geological studies in prospecting in Finland. In: DiLabio, R.N.W., Coker, W.B. (Eds.), Drift Prospecting. Geological Survey of Canada, Paper 89–20, pp. 1–6.

Hirvas, H., 1991. Pleistocene stratigraphy of Finnish Lapland. Geological Survey of Finland Bulletin 354, p.123.

Hirvas, H., Nenonen, K., 1990. Field methods for glacial indicator tracing. In: Kujansuu, R., Saarnisto, M. (Eds.), Glacial Indicator Tracing. A.A. Balkema, Rotterdam, pp. 217–248.

Horsnail, R.F., 2001. Geochemical prospecting. Electronic publication, available at www.accessscience.com, http://dx.doi.org/10.1036/1097-8542.285700 last modified March 29, 2001.

Hulkki, H., Karinen, T., Sarala, P., 2011. Till geochemistry in gold exploration at Petäjäselkä, northern Finland. Programme and Abstracts. The 25th International Applied Geochemistry Symposium, August 22–26. In: Sarala, P., Ojala, V.J., Porsanger, M.-L. (Eds.). Finnish Association of Mining and Metallurgical Engineers, Rovaniemi, Finland, pp. 148–149. Serie B 92–1.

Jahn, F., Cook, M., Graham, M., 2008. Hydrocarbon Exploration and Production. Developments in Petroleum Science 55. Elsevier, Amsterdam.

Johansson, P., Kujansuu, R. (Eds.), 2005. Pohjois-Suomen maaperä. Geological Survey of Finland, Espoo.

Johansson, P., Lunkka, J.P., Sarala, P., 2011. The glaciation of Finland. In: Ehlers, J., Gibbard, P.L., Hughes, P.D. (Eds.), Quaternary Glaciations—Extent and Chronology: A Closer Look, vol. 15. Elsevier, Amsterdam, pp. 105–116. Developments in Quaternary Sciences.

Kauranne, K., 1958. On prospecting for molybdenum in the basis of its dispersion in glacial till. Bulletin de la Commission Géologique de Finlande 180, 31–43.

Kauranne, K., Salminen, R., Eriksson, K. (Eds.), 1992. Handbook of Exploration Geochemistry, vol. 5. Regolith Exploration Geochemistry in Arctic and Temperate Terrains. Elsevier, Amsterdam.

Klassen, R.A., 2001. A Quaternary geological perspective on geochemical exploration in glaciated terrain. In: McClenaghan, M.B., Bobrowsky, P.T., Hall, G.E.M., Cook, S.J. (Eds.), Drift Exploration in Glaciated Terrain. Geological Society, London, Special Publication 195. pp. 1–18.

Kleman, J., 1994. Preservation of landforms under ice sheets and ice caps. Geomorphology 9, 19–32.

Kleman, J., Hättestrand, C., 1999. Frozen-bed Fennoscandian and Laurentide ice sheets during the Last Glacial Maximum. Nature 402, 63–66.

Kokkola, M., 1977. Application of humus to exploration. In: Jones, M.J. (Ed.), Prospecting in Areas of Glaciated Terrain. Institution of Mining and Metallurgy, London, pp. 104–110.

Koljonen, T. (Ed.), 1992. The Geochemical Atlas of Finland Part 2: Till. Geological Survey of Finland, Espoo, p. 218.

Kujansuu, R., 1976. Glaciogeological surveys for ore prospecting purposes in Northern Finland. In: Legget, R.F. (Ed.), Glacial Till, 12. The Royal Society of Canada, Special Publication, pp. 225–239.

Larson, P.C., Mooers, H.D., 2005. Generation of a heavy-mineral glacial indicator dispersal train from a diabase sill, Nipigon region, northwest Ontario. Canadian Journal of Earth Science 42, 1601–1613.

Lehtonen, M., 2005. Kimberlites in Finland: Information about the mantle of the Karelian craton and implications for diamond exploration. Ph.D. thesis, Geological Survey of Finland, Espoo, p. 31, plus 4 original papers.

Lundqvist, J., 1989. Rogen (ribbed) moraine—identification and possible origin. Sedimentary Geology 62 (2), 281–292.

McClenaghan, M.B., 2005. Indicator mineral methods in mineral exploration. Geochemistry: Exploration, Environment. Analysis 5, 233–245.

McClenaghan, M.B., Cabri, L.J., 2011. Review of gold and platinum group element indicator minerals methods for surficial sediment sampling. Geochemistry: Exploration, Environment. Analysis 11, 251–263.

McClenaghan, M.B., Kjarsgaard, B.A., 2001. Indicator mineral and geochemical methods for diamond exploration in glaciated terrain in Canada. In: McClenaghan, M.B., Bobrowsky, P.T., Hall, G.E.M., Cook, S.J. (Eds.), Drift Exploration in Glaciated Terrain. The Geological Society of London, London, pp. 83–123.

McClenaghan, M.B., Plouffe, A., McMartin, I., et al., 2013. Till sampling and geochemical analytical protocols used by the Geological Survey of Canada. Geochemistry: Exploration, Environment, Analysis 13, 285–301.

McMartin, I., Campbell, J.E., 2009. Near surface till sampling protocols in shield terrain, with examples from western and northern Canada. In: Paulen, R.C., McMartin, I. (Eds.), Application of Till and Stream Sediment Heavy Mineral and Geochemical Methods to Mineral Exploration in Western and Northern Canada, Geological Association of Canada, Short Course Notes 18, pp. 75–95.

McMartin, I., McClenaghan, M.B., 2001. Till geochemistry and sampling techniques in glaciated shield terrain: A review. In: McClenaghan, M.B., Bobrowsky, P.T., Hall, G.E.M., Cook, S.J. (Eds.), Drift Exploration in Glaciated Terrain. Geological Society, Special Publications 185, London, pp. 19–43.

Menzies, J., Shilts, W.W., 1996. Subglacial environments. Past Glacial Environments. In: Menzies, J. (Ed.), Sediments, Forms and Techniques, vol. 2. Butterworth-Heinemann, Oxford, pp. 15–136. Glacial Environments Series.

Mining Journal, 2012. Finland is European hub for exploration. Mining Journal, Finland Supplement Jan 2012, 6–9.

Nenonen, N., 1995. Pleistocene stratigraphy of southern Finland. In: Ehlers, J., Kozarski, S., Gibbard, P.L. (Eds.), Glacial Deposits in North-East Europe. A.A. Balkema, Rotterdam, pp. 11–28.

Nuutilainen, J., Peuraniemi, V., 1977. Application of humus analysis to geochemical prospecting: some case histories. In: Jones, M.J. (Ed.), Prospecting in Areas of Glaciated Terrain. Institution of Mining and Metallurgy, London, pp. 1–5.

Nykänen, V.M., Ojala, V.J., Sarapää, O., et al., 2007. Spatial modelling techniques and data integration using GIS for target scale gold exploration in Finland. In: Exploration in the New Millenium. Proceedings of the 5th Decennial International Conference on Mineral Exploration. Decennial International Mineral Conferences, Toronto, pp. 911–917.

Paulen, R.C., 2009. Sampling techniques in the Western Canada Sedimentary Basin and the Cordillera. In: Paulen, R.C., McMartin, I. (Eds.), Application of Till and Stream Sediment Heavy Mineral and Geochemical Methods to Mineral Exploration in Western and Northern Canada. Geological Association of Canada, Short Course Notes 18, pp. 49–74.

Paulen, R.C., McClenaghan, M.B., Hicken, A.K., 2013. Regional and local ice-flow history in the vicinity of the Izok Lake Zn-Cu-Pb-Ag deposit, Nunavut1. Canadian Journal of Earth Sciences 50, 1209–1222.

Paulen, R.C., Paradis, S., Plouffe, A., Smith, I.R., 2011. Pb and S isotopic composition of indicator minerals in glacial sediments from northwest Alberta, Canada: Implications for Zn-Pb base metal exploration. Geochemistry: Exploration, Environment, Analysis 11, 309–320.

Peuraniemi, V., 1982. Geochemistry of till and mode of occurrence of metals in some moraine types in Finland. Geological Survey of Finland. Bulletin 322, p. 75.

Peuraniemi, V., 1990a. The weathering crust in Finnish Lapland and its influence on the composition of glacial deposits. In: Aario, R. (Ed.), Glacial heritage of Northern Finland; an excursion guide. Nordia tiedonantoja, Sarja A 1, pp. 7–11.

Peuraniemi, V., 1990b. On boulder transport in drumlins, Rogen moraines and Sevetti moraines. In: Aario, R. (Ed.), Glacial Heritage of Northern Finland: an Excursion Guide. Nordia tiedonantoja, Sarja A 1, pp. 29–32.

Peuraniemi, V., 1990c. Heavy minerals in glacial material. In: Kujansuu, R., Saarnisto, M. (Eds.), Glacial Indicator Tracing. A.A. Balkema, Rotterdam, pp. 165–185.
Piotrowski, J., (Ed.), 1997. Subglacial environments. Sedimentary Geology 111 (1–4), p. 330.
Piotrowski, J.A., Mickelson, D.A., Tulaczyk, S., et al., 2001. Were deforming subglacial beds beneath past ice sheets really widespread? Quaternary International 86, 139–150.
Plouffe, A., Anderson, R.G., Gruenwald, W., et al., 2011. Integrating ice-flow history, geochronology, geology, and geophysics to trace mineralized glacial erratics to their bedrock source, an example from south central British Columbia. Canadian Journal of Earth Sciences 48, 1113–1130.
Plouffe, A., Ferbey, T., Anderson, R.G., 2014. Till composition and ice-flow history in the region of the Gibraltar Mine: developing indicators for the search of buried mineralization. Geological Survey of Canada. Open File 7592.
Plouffe, A., McClenaghan, M.B., Paulen, R.C., et al., 2013. Processing of glacial sediments for the recovery of indicator minerals: protocols used at the Geological Survey of Canada. Geochemistry: Exploration, Environment, Analysis 13, 303–316.
Rankama, K., Sahama, T.G., 1950. Geochemistry. University of Chicago Press and Cambridge University Press, Oxford. pp. i–xvi and 912.
Roberts, D.H., Hart, J.K., 2005. The deforming bed characteristics of a stratified till assemblage in north East Anglia, U.K.: investigating controls on sediment rheology and strain signatures. Quaternary Science Reviews 24, 123–140.
Salminen, R. (Ed.), 1995. Alueellinen geokemiallinen kartoitus Suomessa 1982–1994, Summary: Regional geochemical mapping in Finland in 1982–1994. Geological Survey of Finland, Report of Investigation 130, p. 47.
Salminen, R., Hartikainen, A., 1985. Glacial transport of till and its influence on interpretation of geochemical results in North Karelia, Finland. Geological Survey of Finland, Bulletin 335, p. 48.
Salminen, R., Tarvainen, T., 1995. Geochemical mapping and databases in Finland. Journal of Geochemical Exploration 55, 321–327.
Salonen, V.-P., 1986. Glacial transport distance distributions of surface boulders in Finland. Geological Survey of Finland. Bulletin 338, p. 57.
Salonen, V.-P., 1988. Application of glacial dynamics, genetic differentiation of glacigenic deposits and their landforms to indicator tracing in the search for ore deposits. In: Goldthwait, R.P., Matsch, C.L. (Eds.), Genetic Classification of Glacigenic Deposits. A.A. Balkema, Rotterdam, pp. 183–190.
Sarala, P., 2003. Ribbed-moreenit—jäätikön liikesuunnan poikittaiset indikaattorit. Summary: Ribbed moraines—transversal indicators of the ice flow direction Geologi 55 (9–10), 250–253.
Sarala, P., 2005a. Glacial morphology and dynamics with till geochemical exploration in the ribbed moraine area of Peräpohjola, Finnish Lapland. Ph.D. thesis, Geological Survey of Finland, Espoo, p. 17 plus 6 original papers.
Sarala, P., 2005b. Till geochemistry in the ribbed moraine area of Peräpohjola, Finland. Applied Geochemistry 20, 1714–1736.
Sarala, P., 2005c. Weichselian stratigraphy, geomorphology and glacial dynamics in southern Finnish Lapland. Bulletin of the Geological Society of Finland 77 (2), 71–104.
Sarala, P., 2009. New advances in geochemical exploration in glaciated terrain—examples from northern Finland. In: Lentz, D., Thorne, K.G., Beal, K.-L. (Eds.), Proceedings of the 24th International Applied Geochemistry Symposium. Fredericton, New Brunswick, June 1–4, vol. II. The Association of Applied Geochemistry, pp. 565–567.
Sarala, P., Peuraniemi, V., Aario, R., 1998. Glacial geology and till geochemistry in ore exploration studies in the Tervola area, southern Finnish Lapland. Bulletin of the Geological Society of Finland 70, 19–41.
Sarala, P., Hartikainen, A., Sarapää, O., Iljina, M., et al. (Eds.), 2008. Mobile metal ion (MMI)—menetelmän testaus malminetsintätutkimuksissa Itä- ja Pohjois-Suomessa vuonna 2007. Geological Survey of Finland, Archive Report S44/2008/37, p. 62, 3 app. (in Finnish).

Sarala, P., Mäkikyrö, M., 2012. Mobile XRF methods in mineral exploration in glaciated terrain—examples from northern Finland. In: Proceedings of the 34th International Geology Congress, Brisbane Australia, August 5–10. Optical disc (CD-ROM).

Sarala, P., Nenonen, J., 2005. Ore prospecting in the ribbed moraine area of Misi, northern Finland. Geological Survey of Finland, Special Paper 38, 25–29.

Sarala, P., Nykänen, V., 2014a. Development of low-impact exploration methods promoting the green mining concept in Finland. Proceedings of the 31st Nordic Geological Winter Meeting, Lund Sweden, January 8–10. Geological Society of Sweden, p. 60. Also available online at http://geologiskaforeningen.se/pdf/31%20NGWM%20Program%20and%20Abstract%20Volume.pdf.

Sarala, P., Nykänen, V., 2014b. UltraLIM—Ultra low-impact exploration methods in the Subarctic. In: Sarala, P. (Ed.), 11. Geokemian Päivät 2014—Proceedings of the 11th Finnish Geochemical Meeting, 5.-6.2.2014, GTK, Espoo, Finland. Tiivistelmät—Abstracts. Vuorimiesyhdistys, Sarja B97, p. 40.

Sarala, P., Nykänen, V., Sarapää, O., et al., 2007b. Quaternary geological and till geochemical studies in verifying GIS-based prospectivity mapping in the Central Lapland Greenstone Belt, northern Finland [Electronic resource]. In: Proceedings of the 23rd International Applied Geochemistry Symposium (IAGS): Exploring Our Environment, Extended Abstracts. Oviedo, Spain, June 14–19. University of Oviedo, pp. 162–168. Optical disc (CD-ROM).

Sarala, P., Ojala, V.J., 2008. Implications of complex glacial deposits for till geochemical exploration: Examples from the central Fennoscandian ice sheet. In: Stefansson, Ó. (Ed.), Geochemistry Research Advances. Nova Publishers, New York, pp. 1–29.

Sarala, P., Peuraniemi, V., 2007. Exploration using till geochemistry and heavy minerals in the ribbed moraine area of southern Finnish Lapland. Geochemistry: Exploration, Environment, Analysis 7 (3), 195–205.

Sarala, P., Rossi, S., 2000. The application of till geochemistry in exploration in the Rogen moraine area at Petäjävaara, northern Finland. Journal of Geochemical Exploration 68, 87–104.

Sarala, P., Rossi, S., Peuraniemi, V., Ojala, V.J., 2007a. Distinguishing glaciogenic deposits in southern Finnish Lapland: implications for exploration. Applied Earth Science (Trans. Inst. Min. Metall. B) 116 (1), 22–36.

Sarapää, O., Al-Ani, T., Lahti, S.I., et al., 2013. Rare earth element occurrences and exploration potential in Finland. Journal of Geochemical Exploration 133, 25–41.

Sarapää, O., Sarala, P., 2013. Rear earth element and gold exploration in glaciated terrain—example from the Mäkärä area. Northern Finland. Geochemistry, Exploration, Environment, Analysis 13, 131–143.

Sauramo, M., 1924. Tracing of glacial boulders and its application in prospecting. Bulletin de la Commission Géologique de Finlande 67, 37.

Shilts, W.W., 1972. Drift prospecting; geochemistry of eskers and till in permanently frozen terrain. District of Keewatin, N.W.T. Geological Survey of Canada, Paper 72–45, p. 34.

Shilts, W.W., 1984. Till geochemistry in Finland and Canada. Journal of Geochemical Exploration 21, 95–117.

Smee, B.W., 1983. Laboratory and field evidence in support of the electrogeochemically enhanced migration of ions through glaciolacutrine sediment. Journal of Geochemical Exploration 19, 277–304.

Svendsen, J., Alexanderson, H., Astakhov, V., et al., 2004. Late Quaternary ice sheet history of northern Eurasia. Quaternary Science Reviews 23, 1229–1271.

Tack, F.M.G., 2010. Trace elements: General soil chemistry, principles and processes. In: Hooda, P.S. (Ed.), Trace Elements in Soils. John Wiley and Sons Ltd., West Sussex, U.K, pp. 9–37.

Theobald, P.K., Eppinger, R.G., Turner, R.L., Shen, S., 1991. The effect of scale on the interpretation of geochemical anomalies. Journal of Geochemical Exploration 40, 9–23.

Trommelen, M.S., Ross, M., Campbell, J.E., 2013. Inherited clast dispersal patterns: Implications for palaeoglaciology of the SE Keewatin Sector of the Laurentide Ice Sheet. Boreas 42, 693–713.

Wennervirta, H., 1968. Application of geochemical methods to regional prospecting in Finland. Bulletin de la Commission Géologique de Finlande 234, 91.

CHAPTER

THE OROGENIC GOLD POTENTIAL OF THE CENTRAL LAPLAND GREENSTONE BELT, NORTHERN FENNOSCANDIAN SHIELD

10.2

T. Niiranen, I. Lahti, V. Nykänen

ABSTRACT

The Paleoproterozoic Central Lapland Greenstone Belt (CLGB) hosts a number of orogenic gold deposits, the total reported gold endowment being 9.1 Moz. Genetic models of the orogenic gold deposits suggest that the gold source for these deposits is the rocks undergoing metamorphism at depth. If a metamorphic source model is applied to the Kittilä terrane, comprising the core of the Central Lapland Greenstone Belt, potentially as much as 30 times the currently known gold resource was mobilized from the up to 9-km-thick metavolcanic rock sequence comprising the terrane. The size distribution of orogenic gold deposits in gold districts with long exploration histories follows a consistent nonlinear pattern. In comparison, the size distribution of the Central Lapland Greenstone Belt is highly skewed, indicating that the number of gold deposits, including deposits in the 0.5–1.5 Moz size category, remain undiscovered in the belt. The study shows that the Central Lapland Greenstone Belt is underexplored and that the remaining orogenic gold potential of the belt is significant.

Keywords: Orogenic gold deposits; greenstone belts; Paleoproterozoic; 3D modeling; Fennoscandian Shield; Finland.

INTRODUCTION

Orogenic gold deposits are typically the dominant gold source in metamorphosed greenstone and sedimentary belts. More than 100 Moz of gold have been produced from some of the most renowned gold provinces, including the Superior Province in Canada, the West African craton, the Yilgarn craton in east Australia, and the Lachlan Fold belt in western Australia (e.g., Goldfarb et al., 2001). At camp and district scale, known deposits cluster in proximity to transcrustal or other major deformation zones that are formed synchronously with the thickening of the crust during accretionary or collisional tectonic events. In most prospective districts, the deposits were formed at mid-crustal levels as suggested by the dominant greenschist facies metamorphic assemblages of the host rocks.

Genetic models for orogenic gold deposits have been discussed in several studies (e.g., Groves et al., 1998; McCuaig and Kerrich, 1998; Goldfarb et al., 2001; Groves et al., 2003; Philips and Powell, 2010). The key aspects of these models are: (1) metals, complexing agent(s), and fluids transporting the

Mineral Deposits of Finland. http://dx.doi.org/10.1016/B978-0-12-410438-9.00028-5

metals are released from the source (or sources) at depth, (2) metal carrying fluids are focused into shear zones, and (3) the auriferous fluids migrate along structures into suitable structural and/or chemical traps where the gold and associated metals are deposited via various physicochemical reactions. The ultimate sources of these fluids, metals, and complexing ligands have been under considerable discussion, with models ranging from entirely magmatic to purely metamorphic sources to somewhere in-between. The metamorphic source models suggest that the gold and gold-carrying fluids are released from rocks undergoing prograde metamorphism related to accretionary orogenic events (e.g., Goldfarb et al., 2005).

Studies by Pitcairn et al. (2006) and Pitcairn (2012) show that the background concentrations of gold drop systematically with increasing metamorphic grade. It has also been shown that metamorphic devolatization reactions in mafic rock sequences at the greenschist–amphibolite facies transition produce significant quantities of low salinity, near-neutral H_2O-CO_2-H_2S fluids characteristic of orogenic gold deposits (Philips and Powell, 2010). Thus, it is possible that all metals, complexing ligands, and fluids related to orogenic gold deposits are products of devolatization reactions of the rocks undergoing metamorphism at depth and that magmatic sources are not involved. The role of subducting crust and enriched subcontinental lithospheric mantle as the bulk source for gold and gold-carrying fluids has been emphasized by some authors (e.g., Groves et al., 1998; Goldfarb et al., 2005) and the evidence from certain gold districts (e.g., Jiandong, China) suggests that the subducting crust appears to be the most likely source in some districts (Goldfarb and Santosh, 2014).

The orogenic gold deposits of the world include some giant (>250 t Au) and numerous world class (>100 t Au) deposits. In well-explored gold districts, the known deposits appear to follow a size distribution pattern where the total gold endowment of a district is divided among one to a few large deposits, a group of mid-sized deposits, and a large number of smaller deposits. For example, the Abitibi belt hosts three giant deposits, one world-class deposit, a total of seven 1–3 Moz deposits, and 25 deposits with <1 Moz Au (Gosselin and Dubé, 2005). Similar distribution patterns can be found in other well-explored gold districts; for example, greenstone belts in Yilgarn and Zimbabwe cratons (Gosselin and Dubé, 2005).

A number of orogenic gold deposits are known in Archean and Paleoproterozoic greenstone and schist belts of the Fennoscandian Shield (e.g., Eilu et al., 2003). One of the most prospective belts for orogenic gold, indicated by currently known total gold resources in excess of 9 Moz, is the Central Lapland Greenstone Belt (CLGB), which is the most extensive Paleoproterozoic volcano-sedimentary belt in Europe. It extends approximately 450 km from northern Norway through Finland into Russian Karelia to the south (Fig. 10.2.1). Active gold exploration in the CLGB started in the early 1980s and since then several gold deposits have been discovered within the belt (Eilu et al., 2007). The known gold deposits (e.g., Eilu et al., 2012) are most numerous within and at the southern contact of the Kittilä terrane, forming the central core of the CLGB, but gold occurrences are known throughout the belt.

Currently, the most significant gold deposit in the CLGB is the world-class Suurikuusikko deposit, with total resources at the end of 2013 of 7.67 Moz (Agnico Eagle Ltd., 2013). The Suurikuusikko deposit alone represents 84% of the total reported gold endowment of the CLGB, while the second largest deposit, Pahtavaara, has only 0.37 Moz gold (Table 10.2.1). This highly skewed size distribution most likely reflects the relatively short exploration history of the CLGB and suggests that the amount of undiscovered gold in the CLGB may be considerable.

Various methods have been used to estimate undiscovered metal resources in certain districts. These include the three-part quantitative mineral-resource assessment method (Singer, 1993; Singer and Manzie, 2010). Such estimation was carried out for orogenic gold deposits in northern Finland, suggesting that

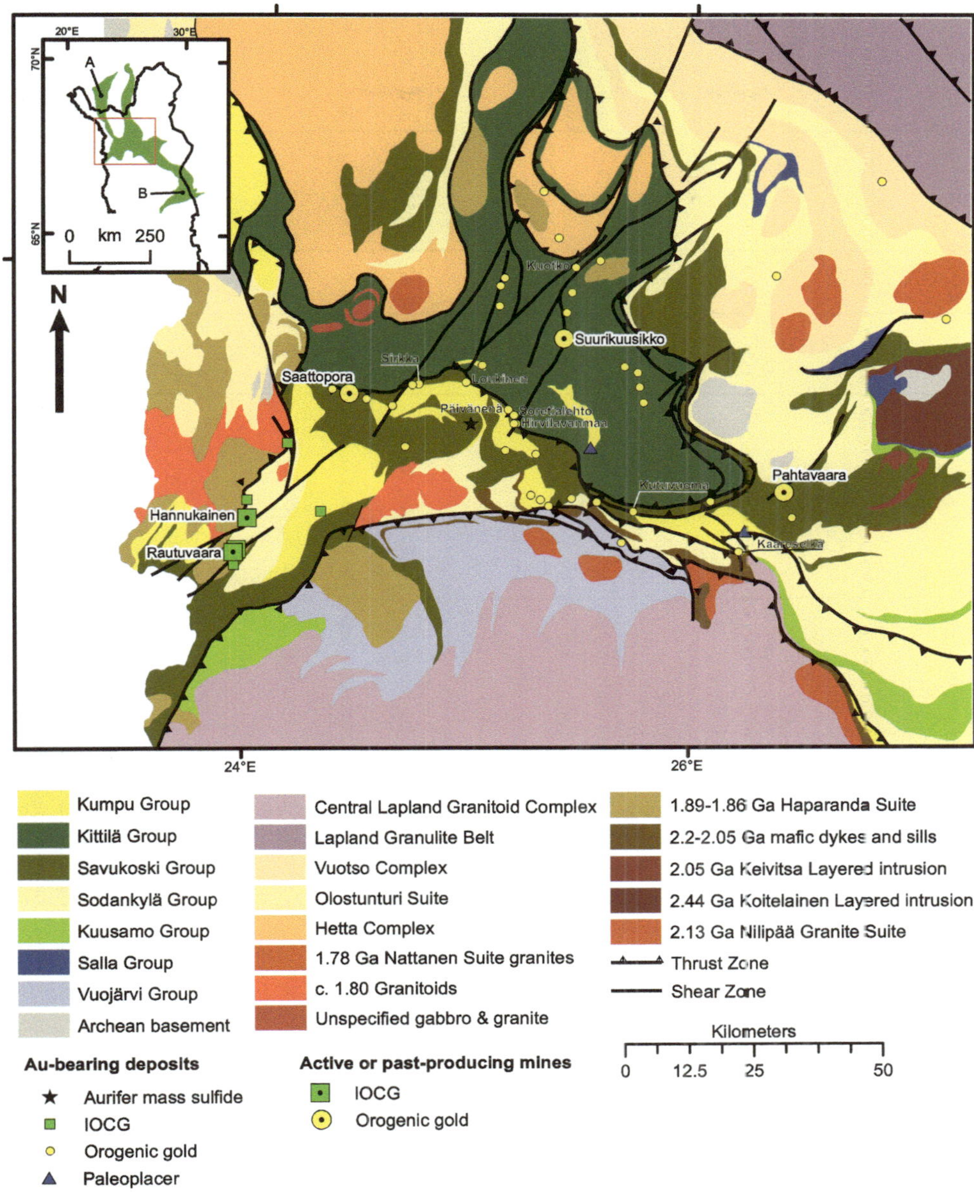

FIGURE 10.2.1 Location of the study area with extent of the Central Lapland Greenstone Belt (green) and location of (A) Bidjovagge and (B) Juomasuo deposits (inset).

Main geological units and location of the known gold deposits.

Source: *Modified from GTK Bedrock of Finland – DigiKP (October 15, 2013) and FINGOLD (October 15, 2013).*

Table 10.2.1 Orogenic gold deposits with reported resources within the CLGB

Deposit	Mineralization style	Host rock(s)	Metal association	Size (Mt)	Au (g/t)	Resources (oz)	References
Suurikuusikko	Refractory Au in disseminated arsenopyrite and pyrite	Albitized KiG metatholeiite and argillite	Au-As Ag, Bi, Co, Sb, Se, W	56.07	4.25	7,665,129	a, b
Pahtavaara[1]	Qtz-barite-Au lenses and pods	Altered metakomatiite of SaG	Au-Ba, Ag, B, Te, W	4.98	2.29	367,742	a, d
Bidjovagge	Native gold with sulfides in qtz-carb-veins and breccia	Albitized dolerite and graphitic phyllite (corresponding SaG)	Au-Cu, Te, U	3.31	2.87	306,452	d
Juomasuo[1]	Native gold associated with disseminated sulfides and uraninite	Multiply altered SoG quartzite and meta-arkose	Au-Co-Cu-U-LREE, As, Ba, Bi, Mo, Ni, Se, Te, Th, U, W, Y	1.96	4.90	305,600	a, c
Saattopora	Native gold with sulfides in qtz-carb-veins and breccia	Albitized SaG phyllite	Au-Cu, Ag, As, B, Bi, Se, Te, U, W	2.16	2.91	203,226	a, d
Kuotko	Native and refractory gold with sulfides in qtz-carb-veins and breccia	Albitized KiG metatholeiite and argillite	Au-As-Bi, Te, W	1.82	2.89	170,044	a, b, d
Kaareselkä	Native gold with sulfides in qtz-carb-veins and disseminated sulfides in altered wall rocks	Albitized SaG metatuffites and phyllites	Au-Cu, As	0.30	5.00	48,387	a, d
Kettukuusikko	Native gold with sulfides in qtz-carb-veins	Altered SaG metakomatiite	Au-Cu, Ag, Ni, Sb, Se	0.44	1.80	25,548	a, d
Kutuvuoma	Native gold with sulfides in qtz-carb-veins	Altered SaG metakomatiite and graphitic phyllite	Au-Cu, As, Co, Ni	0.07	6.70	14,697	a, d
Hirvilavanmaa	Native gold with sulfides in qtz-carb-veins and breccia	Altered SaG metakomatiite	Au, Ni, Sb, Te	0.11	2.90	10,290	a, d
Sirkka	Native gold with sulfides in qtz-carb-veins and breccia	Albitized SaG phyllite, mafic metalava, and metakomatiite	Au-Co-Cu-Ni, As	0.25	0.80	6452	a, d
Loukinen	Native gold with sulfides in qtz-carb-veins and breccia	Albitized SaG graphitic-phyllite	Au-Ni, Ag, As, Co, Cu	0.11	0.50	1839	a, d
Soretialehto	Native gold with sulfides in qtz-carb-veins and breccia	Altered SaG metakomatiite	Au, As, W	0.01	3.50	1468	a, d

Resources reported with total in situ including reserves, resources, and mined resources (where applicable). KiG = Kittilä group, SaG = Savukoski group, SoG = Sodankylä group
[1]*Other genetic type(s) proposed.*
[a]*Eilu et al. (2007)*
[b]*www.agnico-eagle.com*
[c]*www.dragon-mining.com.au*
[d]*Eilu et al. (2012)*

with 50% certainty, the total gold endowment in northern Finland as a whole is at least 290 tonnes or 9.4 Moz (Rasilainen et al., 2012). In the same study, the estimated undiscovered gold resources of the Kittilä terrane are 45 tonnes (1.4 Moz). A spatial mineral prospectivity assessment using methodologies described by Bonham-Carter (1988) was performed by Nykänen et al. (2008). This study delineated several high prospectivity areas for orogenic gold deposits within the current study area; however, it does not contain a quantitative estimate for the undiscovered gold resources.

In the work reported here, the gold potential of the Kittilä terrane and CLGB is estimated using two different approaches. The first approach applies a metamorphic source model for the orogenic gold deposits that assumes that all the metals and fluids related to the deposits in CLGB are released from the rocks undergoing metamorphism at depth. This model is applied to the Kittilä terrane for which a geologic 3D model is constructed based on the available seismic and gravity data. The total amount of gold mobilized from the Kittilä terrane rocks is estimated based on the data by Pitcairn et al. (2006) and Pitcairn (2012) for gold depletion in sedimentary and volcanic rocks during prograde metamorphism, and the volumes of Kittilä terrane rocks based on the geological 3D model created in this work.

In a second approach, the gold potential of the CLGB is estimated by a statistical comparison with some of the best-known orogenic gold-hosting greenstone belts with prolonged exploration history around the world. The size distributions of the orogenic gold deposits in the Abitibi (Canada), Norseman-Wiluna (Australia), and Zimbabwe greenstone belts are studied for any consistencies between these belts, and a comparison is made with the CLGB. The data for deposit size by Gosselin and Dubé (2005) is used for these belts. Although it is approximately 10 years old and, thus, somewhat outdated, it is probably the most coherent dataset available and still suitable for the purposes used in this work.

GENERAL GEOLOGICAL FEATURES OF THE CENTRAL LAPLAND GREENSTONE BELT

The geology of the central part of the CLGB and its immediate surroundings is shown in Fig. 10.2.1. The current geological features of the CLGB result from a multistage rifting history between 2.44 and 2.0 Ga followed by multiphase compressional and metamorphic events related to the Svecofennian orogeny at 1.91–1.79 Ga. The supracrustal units of the CLGB can be divided into two main units; the Karelian 2.44–2.0 Ga rift-related metamorphosed volcano-sedimentary sequence deposited on the Archean basement and the 1.89–1.77 Ga Svecofennian sequence dominated by clastic metasedimentary units (Lehtonen et al., 1998; Hanski et al., 2001; Hanski and Huhma, 2005). The Karelian and Svecofennian units are bounded by the migmatites and granites of Central Lapland Granitoid Complex in the south; gneisses of the Olostunturi suite, intrusions of the Haparanda suite, and ~1.80 Ga granitoids in the west; reworked Archean gneisses of the Hetta complex in the north; and high-metamorphic-grade rocks of the Lapland granulite belt and Vuotso complex in the northeast (Lehtonen et al., 1998; Hanski and Huhma, 2005; Fig. 10.2.1).

STRATIGRAPHY

The supracrustal rocks of the CLGB are divided into seven lithostratigraphical units or groups: the Vuojärvi, Salla, Kuusamo, Sodankylä, Savukoski, and Kittilä groups comprising the Karelian units and the Kumpu group in the Svecofennian unit (Fig. 10.2.2). The Salla group is the oldest unit exposed in the eastern and

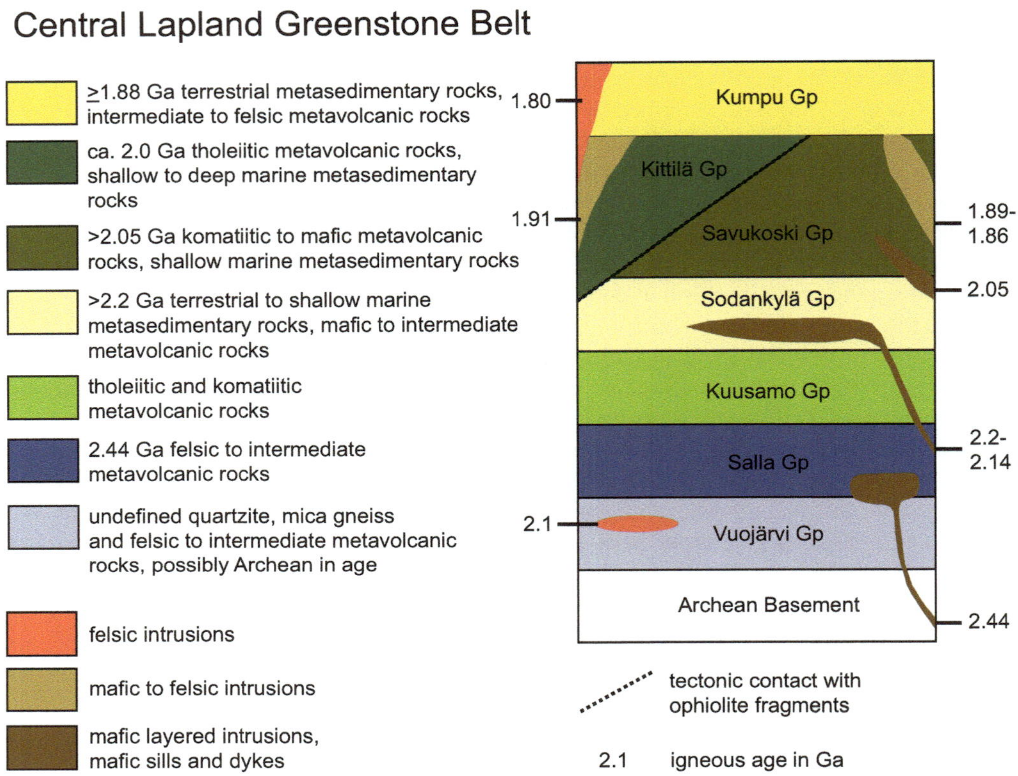

FIGURE 10.2.2 Stratigraphy and main igneous events of the Central Lapland Greenstone Belt.

Archean basement consists of tonalite-trondhjemite-granodiorite (TTG) series rocks.

Source: *Compiled after Hanski et al. (2001) and Bedrock of Finland – DigiKP (October 15, 2013).*

southeastern part of the CLGB and consists of 2.44 Ga felsic metavolcanic rocks deposited on the metamorphosed Archean basement (Lehtonen et al., 1998). The Vuojärvi group rocks, located at the southern margin of the CLGB, consist of quartz-feldspar-sericite schists and gneisses of uncertain origin. The felsic metavolcanic rocks of the Salla group, together with the 2.44 Ga layered intrusions, resulted from the initial rifting phase of the Archean in the area (Hanski and Huhma, 2005; Lahtinen et al., 2005).

The Salla and Vuojärvi groups are overlain by extensive mafic metavolcanic rocks of the 2.2–2.44 Ga Kuusamo group. This volcanic activity was followed by a more tranquil phase resulting in deposition of thick and extensive epiclastic sequences of the Sodankylä group, dominated by quartzites and meta-arkoses with only minor mafic metavolcanic rock intercalations. The 2.05–2.2 Ga Savukoski group overlying the earlier units consists of fine-grained metasedimentary rocks including phyllites, mica schists, graphite- and sulfide-bearing schists, and metatuffites, and tholeiitic metavolcanic rock intercalations suggest deepening of the depositional basin. The komatiitic to picritic metavolcanic rocks, which represent the uppermost part of the Savukoski group, and abundant 2.2 and 2.05 Ga mafic dikes and sills suggest further rifting of the basin. The Kittilä group comprises the youngest Karelian unit of the CLGB. It consists of a thick sequence of tholeiitic metavolcanic rocks with passive margin and oceanic affinity and minor metasedimentary intercalations of argillites, graphite- and sulfide-bearing schists, and banded iron formations (Lehtonen et al., 1998; Hanski and Huhma, 2005).

The age data of the Kittilä group metavolcanic rocks and felsic porphyry dikes crosscutting them indicate that the Kittilä group rocks were deposited at 2.015 ± 0.002 Ga (Hanski and Huhma, 2005). The Kittilä group is bound against the surrounding sequences by tectonic contacts (see Fig. 10.2.1). Hanski (1997) interpreted the Nuttio serpentinite pods located near the eastern margin of the Kittilä group as ophiolite fragments and suggested that the Kittilä group rocks represent an allochthonous unit that is at least partly oceanic in origin. Based on this and the distinct geochemical and isotopic features of the Kittilä group (e.g., Hanski and Huhma, 2005), it is referred to here as the Kittilä terrane. The youngest stratigraphical unit of the CLGB, the Kumpu group, consists of quartzites, conglomerates, and minor felsic metavolcanic rocks that overlie the Karelian rocks (Fig. 10.2.1; Lehtonen et al., 1998). U-Pb zircon data of the felsic metavolcanic rocks give an age of 1880 ± 8 Ma, being consistent with the ~1880 Ma detrital zircon populations obtained from the metasedimentary rocks and the 1888 ± 22 Ma age of a granitoid pebble in conglomerate from the Kumpu group (Hanski et al., 2001; Rastas et al., 2001).

INTRUSIVE MAGMATISM

The early intrusive magmatism in the area consists of 2.44 Ga and 2.05 Ga mafic layered intrusions and mafic dikes at 2.2 Ga, 2.05 Ga, and 2.0 Ga (Lehtonen et al., 1998; Hanski et al., 2001). The 2.2 Ga mafic dikes and sills occur as an east–west trending zone chiefly in the Sodankylä group rocks in the southern part of the CLGB (Fig. 10.2.1). The 2.05 Ga and 2.0 Ga magmatism consist of mafic dikes and small gabbroic intrusions, although in places coeval felsic quartz-feldspar porphyry dikes have been detected (e.g., Lehtonen et al., 1998; Hanski and Huhma, 2005). The most voluminous intrusive rocks in the area relate to the 1.91–1.86 Ga Haparanda suite felsic and mafic magmatism and 1.82–1.79 Ga granitoids. The Nilipää suite granites, which occur along the southern margin of the CLGB, have an enigmatic age of 2.1 Ga, suggesting that they were intruded during or between the extensional stages (Hanski et al., 2001).

DEFORMATION AND METAMORPHISM

The CLGB records a complex multistage deformation history which, despite numerous studies, is still not completely understood. The following sequence of deformation events is modified after the interpretations of Ward et al. (1989), Lehtonen et al. (1998), Väisänen (2002), Tuisku et al. (2006), Hölttä et al. (2007), and Patison (2007). The deformation history of the CLGB area is divided into three ductile stages followed by a completely brittle stage. The main deformation stages D1 and D2 lack clear overprinting relationships and absolute age constraints and therefore are grouped here as D1–2 stage. The D1-2 stage relates to thrust tectonics that formed the main ductile deformation features of the CLGB. The D1–2-stage features indicate south–southwest vergent transport and north–northeast vergent transport in the northern and southern part of the CLGB, respectively. The south–southwest vergent thrusting in the north relates to the collision of the Karelian and Kola cratons and thrusting of the Lapland granulite belt to the southwest. The northward-directed thrusting involved initial generation of south-dipping thrust zones at the southern margin of the area (Fig. 10.2.1). It is currently unknown if the two thrusting events were contemporaneous or successive events.

The D3-stage deformation features have highly variable vergence depending on the location, and possibly D3 consisted of several stages that were not necessarily related to each other. The D3-stage deformation features include north- to northeast-striking shear zones and local refolding of the earlier deformation features. Robust timing of the D3 stage is lacking, but a minimum age can be set as the maximum age for the subsequent brittle D4 stage at 1.77 Ga (Väisänen, 2002). The best estimates for D3 ages are 1.89–1.86 Ga or 1.82–1.77 Ga, but this may vary in different parts of the CLGB (e.g., Väisänen, 2002; Patison, 2007).

The peak metamorphic conditions in the CLGB were reached during the D1–2 stage, and represent mid-greenschist facies excluding the northernmost parts and western margin (Hölttä et al., 2007). Metamorphic grade increases south from the Kittilä terrane reaching upper amphibolite facies close to the Central Lapland Granitoid Complex rocks, and westward reaching mid-amphibolite facies at the western margin. The peak metamorphic temperature estimates for the Kittilä terrane are approximately 350 °C, although no reliable pressure estimates exist for the lowest grade metamorphic areas (Hölttä et al., 2007). The western margin of the Kittilä terrane was metamorphosed to higher temperature due to heat flow from adjacent granitoids. Thermobarometry data from this area yield a pressure-temperature (P-T) estimate of 3.2 kbar and 550 °C (Hölttä et al., 2007).

GOLD DEPOSITS

A number of gold deposits and occurrences are known within the study area (refer to Fig. 10.2.1). The deposits fall into four different genetic types, of which the orogenic type is by far the most numerous. A total of 43 orogenic gold deposits are known within the study area, although only 4 of them—Saattopora, Pahtavaara, Suurikuusikko, and Kuotko—have been mined or are planned to be mined in the near future. The remaining deposits are currently uneconomic, although some of these are under active exploration by various companies operating in the area and may prove to be economic in the future.

Most of the deposits classified as orogenic gold deposits display clear features of the deposit type as outlined by Groves et al. (1998). However, Pahtavaara is somewhat enigmatic (e.g., Eilu et al., 2003; Eilu et al., 2007; Patison et al., 2007). It was initially classified as an orogenic gold deposit by Korkiakoski (1992), but recently, alternative genetic models have been proposed (Eilu et al., 2007). The general features of the CLGB orogenic gold deposits with resource estimates are shown in Table 10.2.1. Significant gold deposits in the CLGB, but located outside of the study area, include the Bidjovagge orogenic gold deposit in the northwesternmost extension of the belt in Norway, and a group of enigmatic Au-Co-Cu-U-REE deposits, including the Juomasuo deposit in the Kuusamo district, part of the southeastern CLGB. For the Kuusamo deposits, several genetic models, including orogenic gold, have been proposed (e.g., Vanhanen, 2001; Eilu et al., 2003; Eilu et al., 2007).

DATA AND METHODS

The regional gravity survey data of the Geological Survey of Finland (GTK) was used in this work. The average observation density of the dataset is about 1–6 sites/km^2. Data acquisition has been made primarily using the Scintrex CG-3 and CG-5 Worden gravity meters. The Kittilä area observations started in 1972 and the latest measurements used in this work were taken in the summer of 2009. The total number of gravity observations used is 19,273. The density of gravity observations is primarily 1–4 sites per km^2 with the highest density being in the northern and northwestern parts of the study area.

Two sets of seismic data were used in this work (Fig. 10.2.3). The Finnish Reflection Seismic Experiment 2001–2005 (FIRE) data consist of two profiles covering 220 line-km in the study area and the High Resolution Reflection Seismics for Ore Exploration 2007–2010 (HIRE) data consist of seven profiles totaling 116 line-km. The data acquisition and processing of each dataset is described in Kukkonen et al. (2006) and Kukkonen et al. (2011). The initial 2D interpretation of the FIRE profiles was presented by Patison et al. (2006).

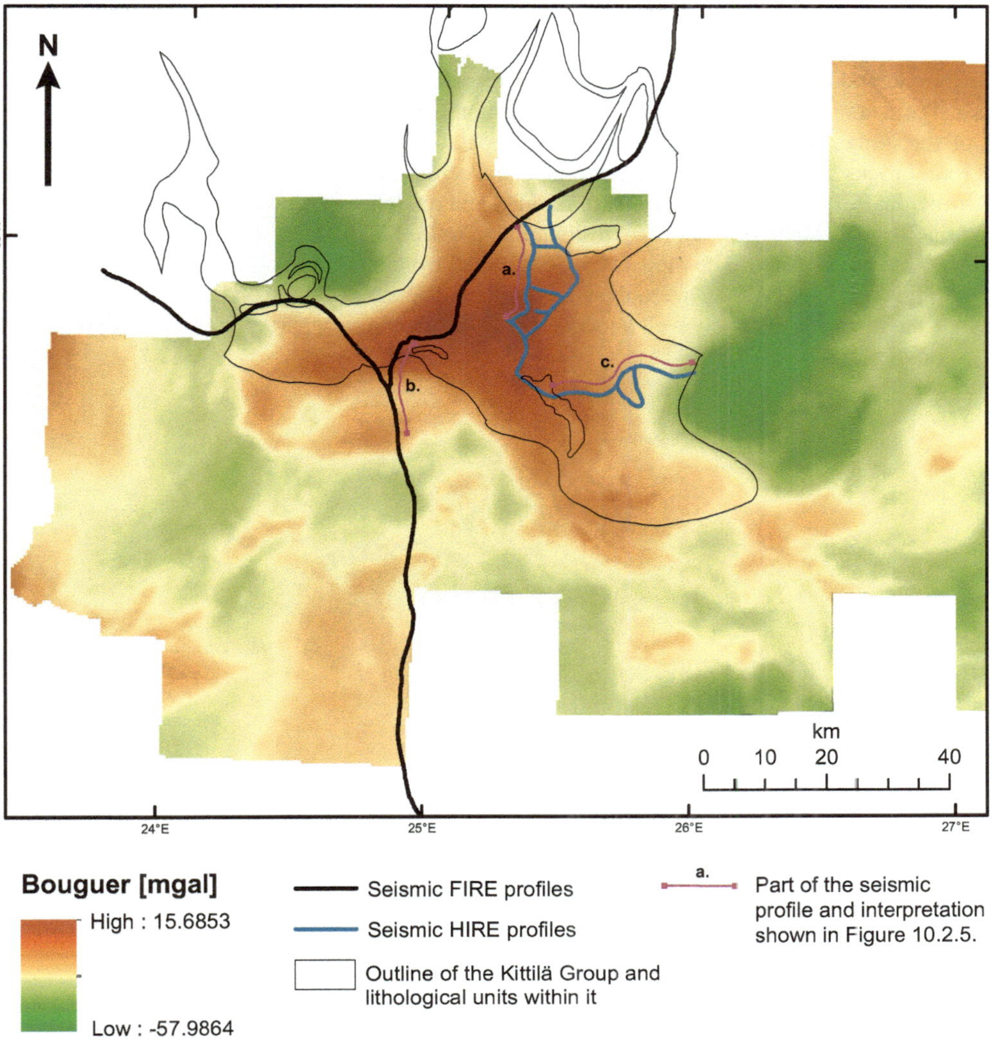

FIGURE 10.2.3 Bouguer map of the area with location of seismic profiles used in this work and outline of the Kittilä terrane.

Location of the seismic sections presented in Fig. 10.2.5.

The surface contacts of the Kittilä terrane were outlined from geological and aerogeophysical maps, and the base of the terrane from seismic profiles and using the 3D-forward model constructed using the gravity data. The final 3D model of the base of the Kittilä terrane was interpolated using the discrete smooth interpolation (DSI) method built into GOCAD® software.

3D MODEL OF THE KITTILÄ TERRANE

Three-dimensional gravity modeling was carried out to obtain information on the deep geometry of the Kittilä terrane and adjacent areas (Fig. 10.2.4). The terrane is favorable for gravity modeling as it produces a significant positive Bouguer anomaly of ~45 mGal. The anomaly is caused by thickness variations of the higher-density mafic metavolcanic rocks of the terrane in contrast to the lower-density background rocks such as granitoids, quartzites, and schists.

The regional Bouguer gravity data were gridded and modeling profiles were created by sampling the grid along 20 north–south oriented, 80-km-long lines, spaced 5 km apart. The model area is 95 km east–west by 80 km north–south. The model was constructed from a number of layer-type, three-dimensional bodies as shown in Fig. 10.2.4. Due to thousands of bedrock and drill hole observations, the surface geology of the target area is relatively well known. Therefore, the topmost part of the model is primarily based

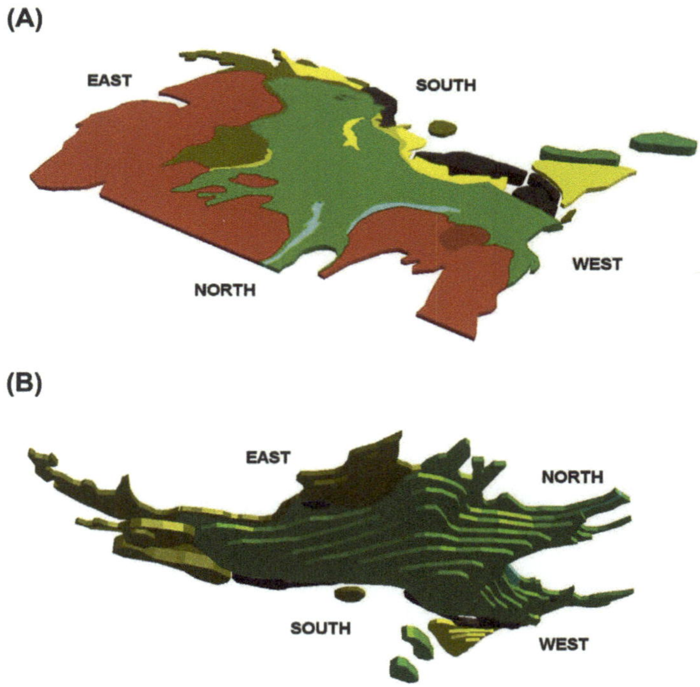

FIGURE 10.2.4 Three-dimensional gravity model of the terrane area.

Perspective view from (A) northwest and (B) northeast. Model colors denote various densities used in the modeling.

on lithological maps. Model densities, geometries, and regional trends were modified manually to improve the fit between the measured and calculated data along the modeled profiles. Although such a "trial and error" forward modeling procedure is time consuming, it allows control so that the results can be modified to best fit geological and geophysical constraints during the modeling process.

The model was revised several times in 3D together with reflection seismics to get good agreement between the dip angles and directions of dipping reflectors. In seismic profiles, the boundary between the Kittilä terrane and adjacent units shows up as an unconformable zone between the subhorizontal reflectors, and in some places as transparent nonreflective zones representing shear or thrust zones truncating the reflectors (Fig. 10.2.5). In places, the contact is located in zones with conformable reflectors or it is difficult to outline in the seismic data without the use of the gravity model (Fig. 10.2.5).

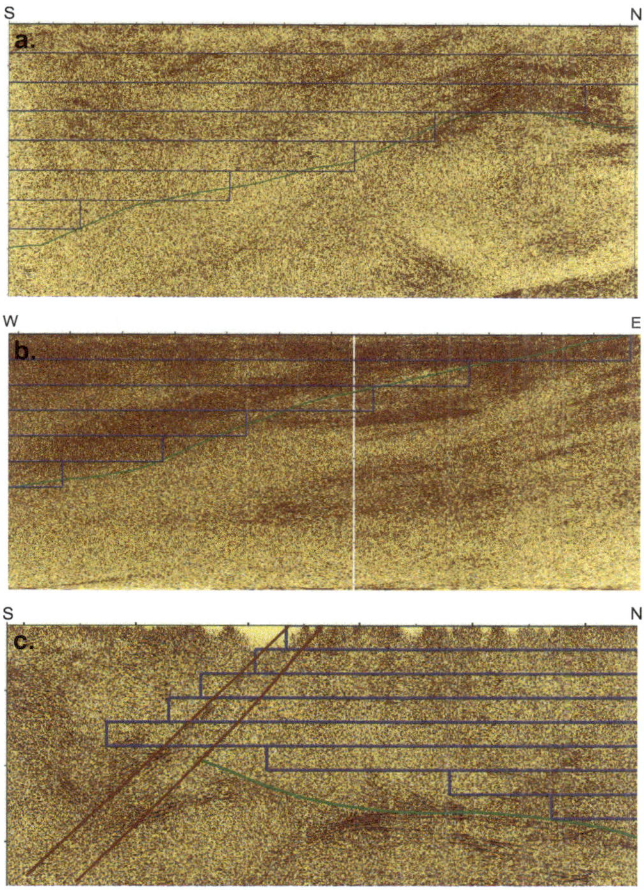

FIGURE 10.2.5 Examples of the seismic interpretation of the base of the Kittilä terrane.

The base is drawn in green, blue lines represent the 3D forward model layers, and brown lines in panel c. represent the Sirkka thrust zone structures outlining the southern margin of the unit. Locations of the sections a., b., c. shown in Fig. 10.2.3. Vertical extent of the seismic profiles is 10 km in all panels.

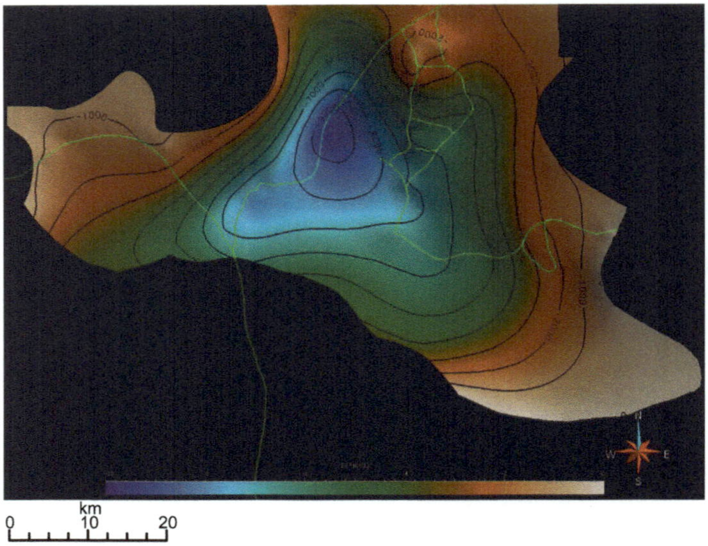

FIGURE 10.2.6 Modeled base of the Kittilä terrane with depth contours in meters below current erosional level, which is set to zero.

To get a better final modeling result, the most significant geological units around the terrane were incorporated into the 3D gravity model. The area is characterized by granitoids to the north and east, which cause negative Bouguer anomalies. Densities of 2600–2650 kg/m^3 and thicknesses of 1–6 km were used for granitic intrusions to fit these negative Bouguer anomalies. To the south of the Kittilä terrane, denser rock types such as schist, mica schist, and mafic–ultramafic intrusions create short wavelength-positive anomalies. Consequently, higher density blocks (2650–3050 kg/m^3) were added in the southern part of the model. The Kittilä terrane was modeled using a density of 2950 kg/m^3, representative of mafic metavolcanic rocks in the study area. A background density of 2780 kg/m^3 was assigned for the rest of the 3D model as this value represents an average between the mafic and felsic rock types encountered in the area. Several smaller-scale high- or low-density bodies were necessary to add to the terrane model. These anomalies are caused by Kittilä terrane banded iron formations, overlying Kumpu group quartzites and granitoid intrusions.

The final 3D model of the base of the Kittilä terrane is shown in Fig. 10.2.6. Based on the model, the thickness of the unit varies considerably: up to 9-km thick in the central parts but thinning dramatically to the west, north, and southeast. The terrane is up to 5-km thick at its southern margin where it is cut by the south-dipping Sirkka thrust zone.

ESTIMATION OF THE QUANTITY OF GOLD MOBILIZED FROM THE KITTILÄ TERRANE ROCKS

The 3D model of the Kittilä terrane, combined with the metamorphic data, allows rough estimates to be made on the potential quantity of gold mobilized from the rocks during regional metamorphism.

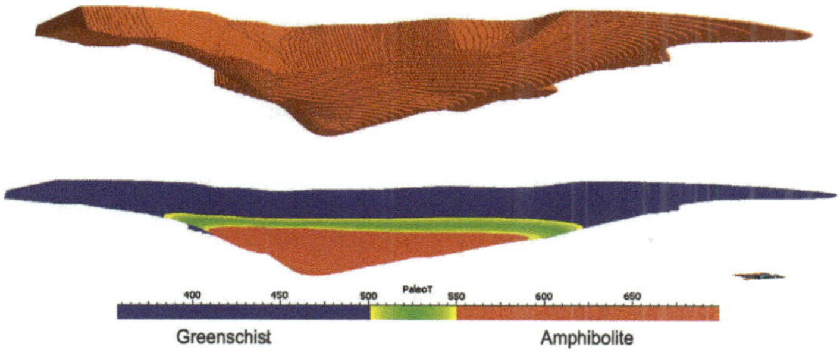

FIGURE 10.2.7 Voxet model of the Kittilä allochthon and calculated peak metamorphic conditions.

The transition zone (500–550 °C) between greenschist and amphibolite facies is shown as a green-yellow zone. Total volume of the Kittilä terrane rocks in amphibolite facies (>525 °C) is approximately 1500 km^3. View to the north.

The method for this estimate follows the studies by Pitcairn et al. (2006) and Pitcairn (2012), which show that the gold concentrations of nonmineralized rocks decrease systematically with increasing metamorphic grade and that the drop is most dramatic at the boundary between greenschist and amphibolite facies. Their data indicate that rocks metamorphosed at amphibolite facies contain between 50% and 80% less gold than their nonmineralized, unmetamorphosed protoliths. Based on the peak metamorphic estimates by Hölttä et al. (2007), the temperature gradient during metamorphism in the Kittilä area was 37 °C/km, close to a typical Barrovian-type metamorphism. The data on the metamorphic conditions for the Kittilä terrane presented by Hölttä et al. (2007) indicate that the peak metamorphic conditions were approximately 350 °C in the central and southern parts of the area. Using these values, the greenschist-amphibolite facies transition zone at 500–550 °C occurs between 4.05–5.35 km depth below the current erosional level. This transition zone is modeled in the voxet model of the Kittilä terrane shown in Fig. 10.2.7. The total volume of the Kittilä terrane below the greenschist-amphibolite facies zone is approximately 1500 km^3, which equals 4.425×10^{12} tonnes using a density of 2950 kg/m^3.

No measured data exist on the background Au concentration of the Kittilä terrane rocks. Data from the literature suggest that the mean gold concentration of the unmetamorphosed basalts (excluding continental flood basalts) varies between 0.75 and 4.7 ppb depending on the chemical affinity of the basalt (Pitcairn, 2012, and references therein). As the bulk of the Kittilä terrane consists of tholeiitic metabasalts, a background Au concentration of 2 ppb was used in calculations. Using this value and the calculated tonnage of the Kittilä terrane rocks below the greenschist-amphibolite facies boundary, some 4425–7080 tonnes or 143–228 Moz of Au may have been mobilized just from the Kittilä terrane rocks using gold mobilization efficiencies of 50–80%.

COMPARISON OF THE GOLD ENDOWMENT OF THE CLGB TO WELL-EXPLORED GREENSTONE BELTS WORLDWIDE

Besides geology and size, the known gold endowment of a certain region is highly dependent on the exploration maturity of the area. Gold exploration in the CLGB has only been active for the relatively short period of about 30 years. The currently reported gold endowment related to orogenic gold deposits is 9.1 Moz hosted by 13 deposits (Table 10.2.1). For comparison, data on orogenic gold deposits from greenstone belts in the Zimbabwe craton, Abitibi greenstone belt, and Norseman-Wiluna belt are presented in Fig. 10.2.8. All of these are well-known gold districts with 100+ years of exploration history. The extent of the greenstone belts is shown in Fig. 10.2.9. There are differences in the sizes of the belts; the Abitibi belt is the least extensive, yet the total gold endowment of it is the largest of the three. The CLGB has roughly the same extent as the Abitibi belt. The data of each district show size distributions of deposits, which appear to follow near logarithmic patterns. The districts have variably harmonic patterns: the Zimbabwe craton and Abitibi belt data are most harmonic and the Norseman-Wiluna belt is somewhat less harmonic.

The data for each belt were plotted on bivariant diagrams where the *x*-axis represents the deposit size in Moz and the *y*-axis is the ordinal number of the deposit from the largest to the smallest, normalized by the number of deposits in each dataset (Fig. 10.2.10). A regression line following the function

$$y = \mathrm{a}x^{-\mathrm{b}} \tag{10.2.1}$$

was fitted to each of the datasets, with a and b being constants of which b reflects the skew of the data and a the total gold endowment of the district.

The fit of the regression lines for the Zimbabwe craton and Abitibi and Norseman-Wiluna belt data is good, with correlation coefficients of 0.94 or higher. Based on the data of the selected districts, it appears that the size distribution of orogenic gold deposits in gold districts with prolonged gold exploration history follow a pattern similar to the Gutenberg–Richter law, describing the relationship between magnitude and total number of earthquakes in any given seismic region (Gutenberg and Richeter, 1949). In seismology, the constant b of Equation (10.2.1) varies between 0.5 and 1.5, being typically close to 1 (i.e., close to log-normal) and the Gutenberg–Richter law based regression has been used to predict numbers and magnitudes of earthquakes in given districts.

In contrast to that just described, a similar regression line fitted to the CLGB data shows a poor fit, both constants of Eq. (10.2.1) being unreasonable (Fig. 10.2.10). Thus, it is obvious that the dataset is too small. For comparison, the shaded area covering the calculated regression lines of the other datasets is shown. Poor fit at the smaller deposit end of the CLGB data to the shaded area is due to differences in the datasets, as the CLGB data includes data from subeconomic occurrences (<0.1 Moz) that are not included in other datasets. If the subeconomic deposits are omitted from the CLGB data, the smaller deposits of the curve would fit relatively well to the regression lines defined by other areas. However, the Suurikuusikko deposit plots well outside this area. For the CLGB data to better fit the distribution of the other belts, the dataset should contain a significantly higher number of deposits including ones in the 0.5–1.5 Moz size range.

DISCUSSION

The 3D model of the Kittilä terrane presented in Fig. 10.2.7 shows that thickness of the unit varies from 0–9 km. The variation is likely a result of considerable deformation of the unit during the southwest- and north-vergent thrusting events related to the main deformation phases of the Svecofennian orogeny and subsequent erosion. The Kittilä terrane has been considered to represent, at least partially, an allochthonous unit based on geochemical characteristics. The presence of serpentinite pods near the eastern margin is interpreted to represent ophiolite fragments and the tectonic or, at least, tectonized character of unit

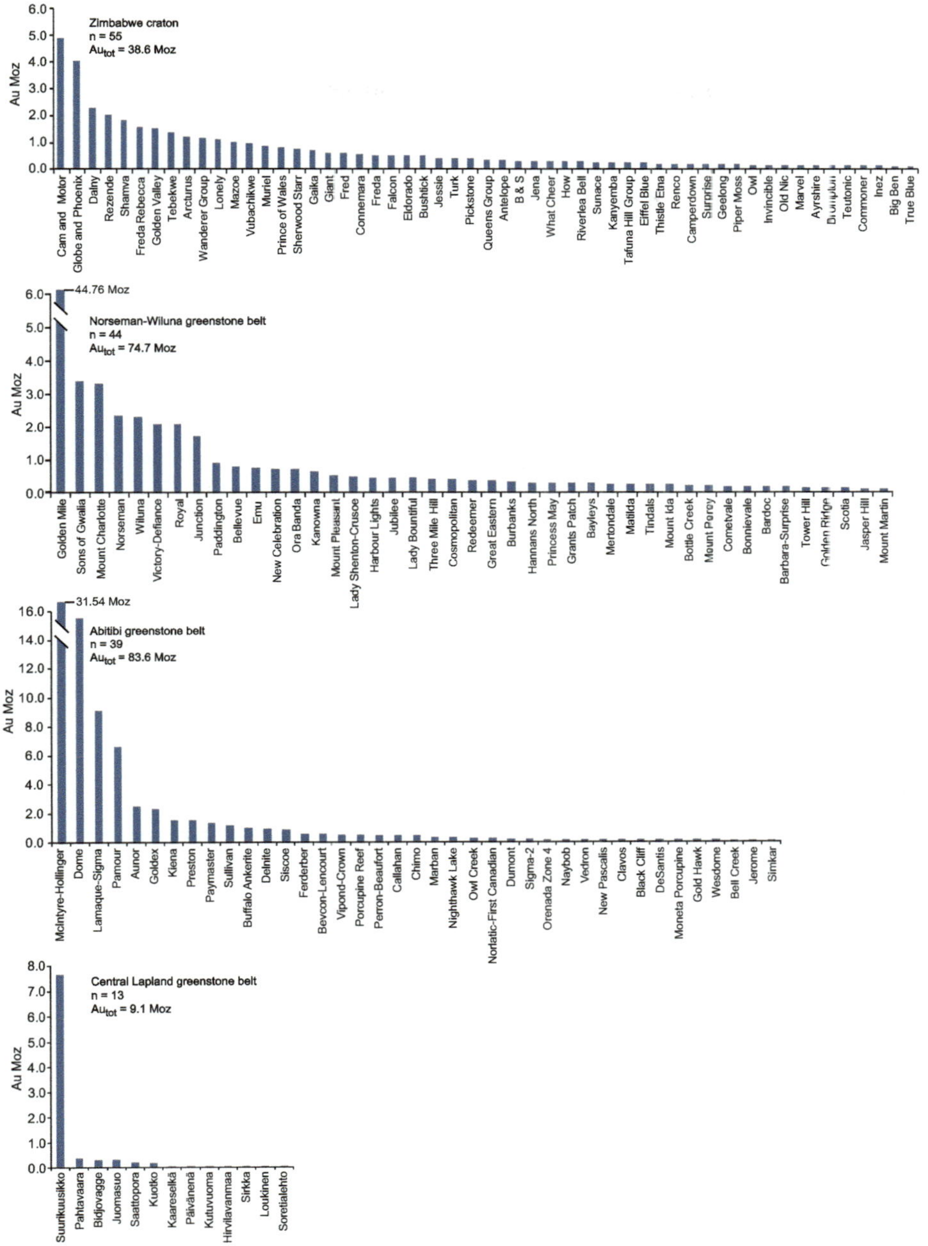

FIGURE 10.2.8 Resources of the known orogenic gold deposits in the Zimbabwe craton, Abitibi, Norseman-Wiluna, and Central Lapland greenstone belts.

Source: *Data after Gosselin and Dube (2005) except for CLGB, which is from Table 10.2.1.*

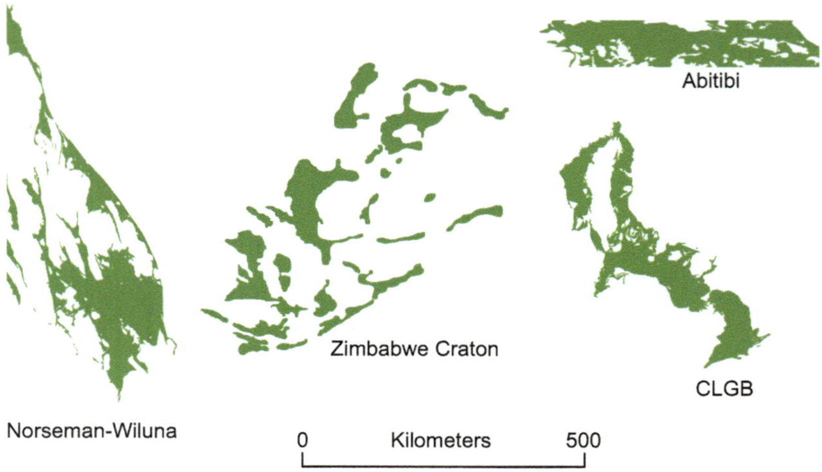

FIGURE 10.2.9 Size comparison of the Norseman-Wiluna, Abitibi, Central Lapland, and Zimbabwe Craton greenstone belts.

All belts presented are shown at the same scale.

Source: *Modified and compiled after GTK Digital Bedrock Database, Geological Survey of Western Australia (2010), Robert et al. (2005), and Pendergast (2003).*

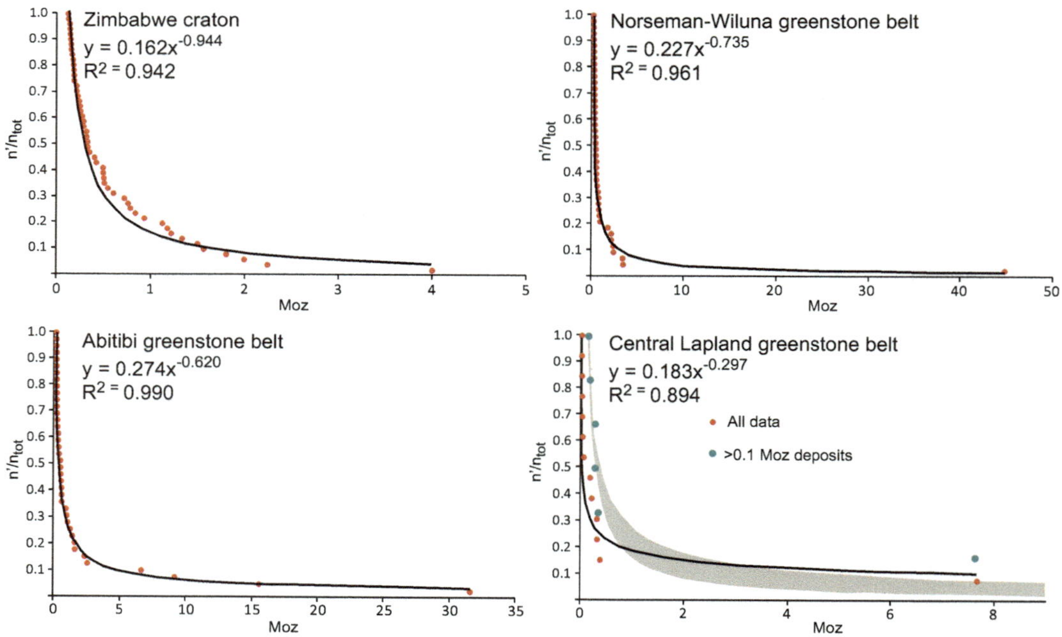

FIGURE 10.2.10 Regression plots of the gold districts and deposits shown in Fig. 10.2.8.

n' = ordinal number of a deposit in the dataset from largest to smallest; n_{tot} = total number of deposits in the dataset; n'/n_{tot} represents the probability of a deposit discovered being of same size or larger on the *x*-axis; Moz = size of the deposit in millions of troy ounces. Black lines are the regression line approximations fitted with least-squares method, formula, and correlation coefficient (R^2) shown in each case. The regression line fitted in the last panel is calculated using the whole dataset of the CLGB. The gray-shaded area in the CLGB panel shows the area covered by the fitted curves of other datasets. The CLGB data without subeconomic deposits is also plotted in the last panel, showing similarity with the data from other districts (see text).

contacts where they are exposed or drilled (e.g., Hanski, 1997; Lehtonen et al., 1998; Hanski and Huhma, 2005). If the Kittilä terrane represents a sequence of oceanic crust as suggested by Hanski (1997), the 9-km thickness in the central parts suggests a doubling of this part of the unit during deformation.

An increasing amount of evidence on the geological features of orogenic gold deposits indicates that they develop as a consequence of a regional metamorphic fluid flow event rather than being linked to local igneous activity (e.g., Groves et al., 1998; Goldfarb et al., 2005). It has been clearly demonstrated that hydrothermal fluids identical to those that form orogenic gold deposits, as well as the gold and other metals enriched in orogenic gold deposits, are released via prograde metamorphism (e.g., Pitcairn, 2006; Philips and Powel, 2010). The quantitative estimation presented in this work suggests that 20–30 times the currently known gold resources of the Kittilä terrane may have been mobilized during metamorphism from the Kittilä terrane rocks alone. Obviously, the mobilized gold needs to be focused into suitable structures and ultimately precipitated in appropriate traps. The efficiency of these processes is unknown, and some of the gold undoubtedly precipitated above the current erosion level.

In spite of the number of uncertainties and the hypothetical character of the calculations presented, it is believed that this modeling provides a good rough estimation of the gold potential of the Kittilä terrane. The calculations suggest that the orogenic gold deposits in the Kittilä terrane could have formed entirely via metamorphic processes and the sole source of gold can be the rocks that underwent metamorphism at depth. The data do not exclude the potential input of gold from intrusions or input from subducted crustal slab undergoing metamorphism at depth. All magmatic effects should have a positive effect on gold potential and gold mobility, either by adding heat, fluid, or gold to the system. Similarly, input from a potential subduction zone would have a positive effect on the gold budget. Calculations presented, however, suggest more than enough gold may have been mobilized from metamorphosed Kittilä terrane rocks alone to cover the currently known gold resources.

A statistical comparison of Abitibi, Norseman-Wiluna, and Zimbabwe greenstone belts indicate surprisingly coherent size distribution patterns between the belts. All of the districts are under active exploration and mining, and evidently there have been increases in the resources of actively explored deposits and new deposits since the database compilation of Gosselin and Dubé (2005). Although the new resources discovered in these areas will undoubtedly have minor effects on the distribution patterns, we find it very unlikely that the near-logarithmic fractal pattern detected will dramatically change. In our comparison, the differences in geological features (e.g., age, tectonic, and structural setting of the belts) was not taken in account. These features most certainly have an impact on total gold endowment of the respective district; however, based on the comparison presented, the size distribution of the deposits appears to be less affected. The comparison between the well-explored orogenic gold districts and CLGB presented in this work suggests that the size distribution of the gold deposits in the CLGB is significantly different from the other belts presented. Most likely this reflects the short exploration history and therefore a significant number of orogenic gold deposits still to be discovered in the belt. The number and size of those is unknown, but there is a clear gap between the current largest and second largest deposits of the CLGB.

SUMMARY

This study was aimed to establish the 3D shape of the Kittilä terrane, use the model generated to estimate its orogenic gold potential, and carry out a statistical comparison of the CLGB gold deposits with similar but more mature orogenic gold districts around the world.

Applying an in situ metamorphic source model for the orogenic gold deposits, the total amount of gold mobilized from the Kittilä terrane rocks during prograde metamorphism may have reached 143–228 Moz, roughly 20–30 times the amount of gold currently reported from the known deposits within the terrane. This indicates that at least in the Kittilä area, metamorphic processes alone could have potentially released enough gold to generate the known gold resources. However, magmatic input in the metal, fluid, or thermal budget of the system cannot be excluded, nor can the potential of subcrustal sources for metals and fluids be excluded. Should magmatic processes and/or additional deeper-sourced processes have been involved in the Kittilä area, they would have had a positive effect on the gold budget.

The statistics of the gold deposit sizes in the Norseman-Wiluna belt, Abitibi belt, and greenstone belts of the Zimbabwe craton indicate that the size distribution of gold deposits in certain districts appears to follow nonlinear distributions that bear similarities to the one outlined by the Gutenberg–Richter law for earthquakes. In comparison, the size distribution of the known gold deposits in CLGB is abnormal. Several gold deposits, including deposits in the size category of 0.5–1.5 Moz, need to be added to the CLGB dataset for it to follow the distribution pattern of the other datasets. This is assuming Suurikuusikko is the largest deposit. Should there be undiscovered deposits larger than Suurikuusikko, the gold potential will be considerably higher.

ACKNOWLEDGMENTS

The authors wish to thank Dr. David Groves for an insightful review, although he did not agree on all aspects of the modeling, and Dr. Hugh O'Brien for his editorial comments; both considerably improved the quality of the manuscript.

REFERENCES

Agnico Eagle Ltd, 2013. Kittilä Mine resources and reserves as of December 31, 2012. http://www.agnicoeagle.com/en/Operations/Northern-Operations/Kittila/Pages/default.aspx.

Bedrock of Finland – DigiKP. Digital map database [Electronic resource]. Geological Survey of Finland (referred 15th October 2013).

Bonham-Carter, G.F., Agterberg, F.P., Wright, D.F., 1988. Integration of geological datasets for gold exploration in Nova Scotia. Photogrammetric Engineering and Remote Sensing 54 (77), 1585–1592.

Eilu, P., Sorjonen-Ward, P., Nurmi, P., Niiranen, T., 2003. A review of gold mineralization styles in Finland. In: A group of papers devoted to the metallogeny of gold in the Fennoscandian Shield. Economic Geology 98, 1329–1353.

Eilu, P., Pankka, H., Keinänen, V., et al., 2007. Characteristics of gold mineralisation in the greenstone belts of northern Finland. In: Ojala, J. (Ed.), Gold in the Central Lapland Greenstone Belt. Geological Survey of Finland. Special Paper 44, 57–106.

Eilu, P., Ahtola, T., Äikäs, O., et al., 2012. Metallogenic areas in Finland. In: Eilu, P. (Ed.), Mineral Deposits and Metallogeny of Fennoscandia. Geological Survey of Finland, Special Paper 53, 207–342.

FINGOLD – a public database on gold deposits in Finland [Electronic resource]. Geological Survey of Finland (referred 15th October 2013)

Geological Survey of Western Australia, 2010. East Yilgarn, 2010 update: Geological Survey of Western Australia. Geological Information Series. 1:100,000.

Goldfarb, R.J., Groves, D.I., Gardoll, S., 2001. Orogenic gold and geologic time: a global synthesis. Ore Geology Reviews 18, 1–75.

Goldfarb, R.J., Baker, T., Dube, B., et al., 2005. Distribution, character, and genesis of gold deposits in metamorphic terranes. In: 100th Anniversary Volume of Economic Geology, pp. 407–450.
Goldfarb, R.J., Santosh, M., 2014. The dilemma of the Jiaodong gold deposits: Are they unique? Geoscience Frontiers 5, 139–153.
Gosselin, P., Dubé, B., 2005. Gold deposits of the world: distribution, geological parameters and gold content. Geological Survey of Canada. Open file 4895, 1 CD-ROM.
Groves, D.I., Goldfarb, R.J., Gebre-Mariam, M., et al., 1998. Orogenic gold deposits: A proposed classification in the context of their crustal distribution and relationship to other gold deposit types. Ore Geology Reviews 13, 1–28.
Groves, D.I., Goldfarb, R.J., Robert, F., Hart, G.J.R., 2003. Gold deposits in metamorphic belts: overview of current understanding, outstanding problems, future research, and exploration significance. Economic Geology 98, 1–30.
Gutenberg, B., Richter, C.F., 1949. Seismicity of the Earth and Associated Phenomena. Princeton University Press, Princeton, NJ. 273.
Hanski, E., 1997. The Nuttio serpentinite belt, Central Lapland: An example of Paleoproterozoic ophiolite mantle rocks in Finland. Ofioliti 22, 35–46.
Hanski, E., Huhma, H., Vaasjoki, M., 2001. Geochronology of northern Finland: a summary and discussion. In: Vaasjoki, M. (Ed.), Radiometric Age Determinations from Finnish Lapland and Their Bearing on the Timing of Precambrian Volcano-Sedimentary Sequences. Geological Survey of Finland, Special Paper 33, 255–279.
Hanski, E., Huhma, H., 2005. Central Lapland Greenstone Belt. In: Lehtinen, M., Nurmi, P.A., Rämö, O.T. (Eds.), Precambrian Geology of Finland—Key to Evolution of the Fennoscandian Shield. Elsevier, Amsterdam, pp. 139–194.
Hölttä, P., Väisänen, M., Väänänen, J., Manninen, T., 2007. Paleoproterozoic metamorphism and deformation in central Lapland, Finland. In: Ojala, V.J. (Ed.), Gold in the Central Lapland Greenstone Belt. Geological Survey of Finland, Special Paper 44, 7–56.
Korkiakoski, E., 1992. Geology and geochemistry of the metakomatiite-hosted Pahtavaara gold deposit in Sodankylä, northern Finland, with emphasis on hydrothermal alteration. Geological Survey of Finland, Bulletin 360, 96, 18 app.
Kukkonen, I.T., Lahtinen, R. (Eds.), 2006. Finnish Reflection Experiment—FIRE 2001–2005. Special Paper 43. Geological Survey of Finland, Espoo, p. 247. plus 15 app. (CD-ROM).
Kukkonen, I.T., Heikkinen, P., Heinonen, S., Laitinen, J., HIRE Working Group, 2011. In: Nenonen, K., Nurmi, P.A. (Eds.), Geoscience for Society 125th Anniversary Volume. Geological Survey of Finland, Special Paper 49, 49–58.
Lahtinen, R., Korja, A., Nironen, M., 2005. Paleoproterozoic tectonic evolution. In: Precambrian Geology of Finland—Key to the Evolution of the Fennoscandian Shield. Developments in Precambrian Geology 14. Elsevier, Amsterdam, pp. 481–531.
Lehtonen, M., Airo, M.-L., Eilu, P., et al. 1998. The stratigraphy, petrology and geochemistry of the Kittilä greenstone area, northern Finland. A report of the Lapland Volcanite Project. Geological Survey of Finland, Report of Investigation 140, p. 144 (in Finnish with summary in English).
McCuaig, T.C., Kerrich, R., 1998. P-T-t-deformation-fluid characteristics of lode gold deposits: evidence from alteration systematics. Ore Geology Reviews 12, 381–453.
Nykänen, V., Groves, D.I., Ojala, V.J., Gardoll, S.J., 2008. Combined conceptual/empirical prospectivity mapping for orogenic gold in the northern Fennoscandian Shield, Finland. Australian Journal of Earth Sciences 55 (1), 39–59.
Patison, N.L., Korja, A., Lahtinen, R., Ojala, V.J., 2006. FIRE Seismic Reflection Profiles 4, 4A and 4B: Insights into the crustal structure of northern Finland from Ranua to Näätämö. In: Finnish Reflection Experiment (FIRE), 2001–2005. Geological Survey of Finland, Special Paper 43, 161–222.

Patison, N.L., 2007. Structural controls on gold mineralisation in the Central Lapland Greenstone Belt. In: Ojala, V.J. (Ed.), Gold in the Central Lapland Greenstone Belt. Geological Survey of Finland, Special Paper 44, 107–124.

Pendergast, M.D., 2003. The nickeliferous Late Archean reliance komatiitic event in the Zimbabwe craton—magmatic architecture, physical volcanology, and ore genesis. Economic Geology 98, 865–891.

Philips, G.N., Powell, R., 2010. Formation of gold deposits: A metamorphic devolatilization model. Journal of Metamorphic Geology 28, 689–718.

Pitcairn, I.K., Teagle, D.A.H., Craw, D., et al., 2006. Sources of metals and fluids in orogenic gold deposits: Insights from the Otago and Alpine schists, New Zealand. Economic Geology 101, 1525–1546.

Pitcairn, I.K., 2012. Background concentrations of gold in different rock types. Applied Earth Science 120, 31–38.

Rasilainen, K., Eilu, P., Karvinen, A., et al., 2012. Undiscovered orogenic gold resources in northern Finland. In: Hölttä, P. (Ed.), Current Research: GTK Mineral Potential Workshop, Kuopio, May. Geological Survey of Finland, Report of Investigation 198, pp. 141–145.

Rastas, P., Huhma, H., Hanski, E., et al., 2001. U-Pb isotopic studies on the Kittilä greenstone area, central Lapland, Finland. In: Vaasjoki, M. (Ed.), Radiometric age determinations from Finnish Lapland and their bearing on the timing of Precambrian volcano-sedimentary sequences. Geological Survey of Finland, Special Paper 33, 95–141.

Robert, F., Poulsen, H., Cassidy, K.F., Hodgson, C.J., 2005. Gold metallogeny of the Yilgarn and Superior cratons. Economic Geology One Hundred Anniversary volume. Society of Economic Geology 1001–1034.

Singer, D.A., 1993. Basic concepts in three-part quantitative assessments of undiscovered mineral resources. Nonrenewable Resources 2, 69–81.

Singer, D.A., Menzie, W.D., 2010. Quantitative Mineral Resource Assessments: An Integrated Approach. Oxford University Press, New York. 219.

Tuisku, P., Mikkola, P., Huhma, H., 2006. Evolution of migmatitic granulite complexes: implications from Lapland granulite belt, Part 1: Metamorphic geology. Bulletin of the Geological Society of Finland 78, 71–105.

Väisänen, M., 2002. Structural features in the central Lapland greenstone belt, northern Finland. Geological Survey of Finland, Report K 21.42/2002/3; p. 20, 16 app.

Vanhanen, E., 2001. Geology, mineralogy and geochemistry of the Fe-Co-Au-(U) deposits in the Paleoproterozoic Kuusamo schist belt, northeastern Finland. Geological Survey of Finland, Bulletin 399, p. 229, 13 app., 1 app., and map.

Ward, P., Härkönen, I., Nurmi, P.A., Pankka, H.S., 1989. Structural studies in the Lapland greenstone belt, northern Finland and their application to gold mineralization. In: Geological Survey of Finland. Current Research, 1988. Geological Survey of Finland, Special Paper 10, 71–77.

PART

11

FINLAND'S MINERAL RESOURCES

OPPORTUNITIES AND CHALLENGES FOR FUTURE MINING

11

P.A. Nurmi, K. Rasilainen

ABSTRACT

This chapter provides an overview of Finland's metallic mineral resources, including past production and presently identified and assumed resources. We also discuss the exploration potential and challenges for future mining in Finland.

Forty-seven metallogenic areas have been defined in Finland, and there are more than 30 different genetic types of mineral deposits. Mining in Finland focused on Fe and Cu in past centuries, and gradually also included Zn, Ni, Cr, and Au from the 1930s to the 1980s. Currently, 12 metal mines are in operation. Ore production has increased to an all-time high for Cr, Au, and Ni, and there are interesting occurrences for numerous other commodities. The ultimate resources of various mineral commodities are impossible to accurately define. Most deposits are insufficiently studied or have not been discovered, therefore forming untapped reserves for future mining. Changing needs of raw materials and improving mining and processing technologies allow new types of deposits to be excavated.

Mine development is becoming increasingly difficult because of growing competition with other land use purposes and tightening regulations. The green mining concept was developed in Finland as a tool to promote future sustainable and acceptable mining.

Keywords: ores; metals; mineral resources; mining; exploration; sustainable development; Finland.

Mineral Deposits of Finland. http://dx.doi.org/10.1016/B978-0-12-410438-9.00029-7

INTRODUCTION

A thriving modern society is based on using mineral commodities, and their consumption correlates strongly with economic growth and urbanization. Infrastructures, logistics, food production, energy technology, information and communications technology, and consumer electronics all increasingly rely on an ever-widening array of metals and minerals. More than 40 elements are needed to produce a personal computer or a smartphone.

According to the United Nations, the world population stood at 7 billion in 2012 and will reach 9 billion in 2050, most of the growth occurring in developing countries. In the same time frame, the average urbanization will rise from the current 50% to 67% (United Nations Department of Economic and Social Affairs/Population Division, 2012). This means that we will have 3 billion new urban dwellers in 2050, and see almost 70 million new people moving to cities annually. The growing world population, accelerating urbanization, and more evenly distributed wealth have created a huge and steadily increasing demand for natural resources.

An average gross domestic product (GDP) of around US$5000 per capita is regarded as the point at which people start to rapidly increase their consumption of mineral commodities (e.g., Halada et al., 2008). Copper is a good indicator of economic development, and it plays a crucial role in modern society across the world, contributing to infrastructure, technology, and lifestyles. In developed urban societies with a high GDP, such as the U.S.A., Germany, and South Korea, copper usage has expanded, along with increasing GDP to a very high level (Fig. 11.1). In less developed societies, copper consumption is low but seems to follow a uniform path of development. Therefore, there will be an enormous growth potential for copper usage with the development of populous countries such as China and India.

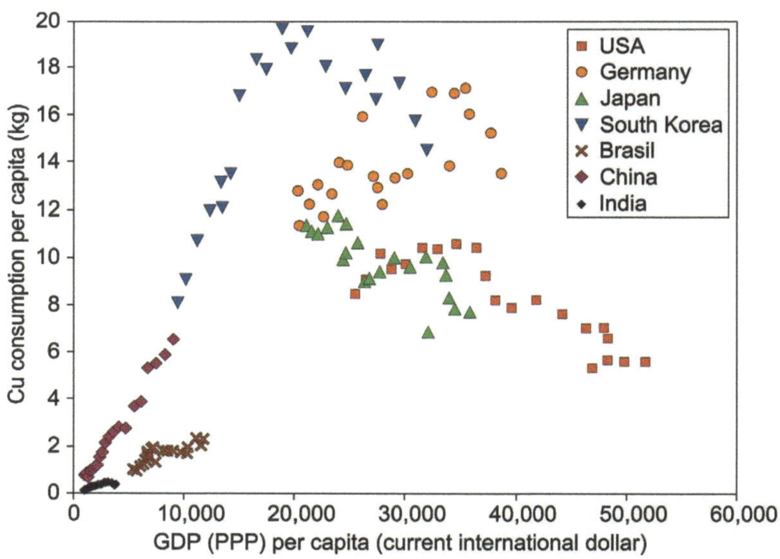

FIGURE 11.1 Copper consumption versus gross domestic product for selected countries.

Data sources: *Chilean Copper Commission (2002, 2012, 2013), International Monetary Fund (2013), and World Bank (2013).*

In developed societies, the consumption of minerals seems to reach a so-called saturation level (Fig. 11.2). Consumption no longer increases, but remains at a high level or shows a slight decrease with decreasing demand, effective recycling, and substitution. Metals used for infrastructural development, such as iron and copper, are the first ones needed in the early stages of societal development, followed by a variety of other metals and minerals. The demand for so-called high-tech metals starts to increase in the most developed societies. The present world average GDP is close to US$10,000, and needs to increase several times before the saturation level of consumption is reached. This will take at least several decades, and create a long-term growth cycle for mineral demand.

China presently consumes 40% of the global metal supply, and will continue to be the largest customer for metals in the global market in the coming years (PricewaterhouseCoopers International, 2013). Although China's high economic growth may gradually weaken, rising long-term demand in India and other populous countries will ensure an economic future for mining industry products. Indeed, the combination of a growing world population and rising living standards for an expanding share of humanity will increase the demand for metal and mineral commodities for many decades to come, although mineral demand and prices are volatile and highly dependent on economic cycles.

The Fennoscandian Shield is evidently the richest area of mineral resources in Europe, and will be an important source of mineral raw materials for the future. Finland, located in the center of the shield, is composed of a variety of geological formations, from Archean gneisses and greenstone belts to Paleoproterozoic volcano-sedimentary belts, and ultramafic to granitic intrusive rocks. These geological formations also contain a wide variety of mineral deposits, as described in other chapters of this book. Many of the deposits are still poorly known, and there is an obvious potential for new mineral discoveries.

This chapter aims to provide an overview of Finland's metallic mineral resources, including past production, presently known resources, and assumed geological resources. We will also discuss the exploration potential and challenges for future mining in Finland.

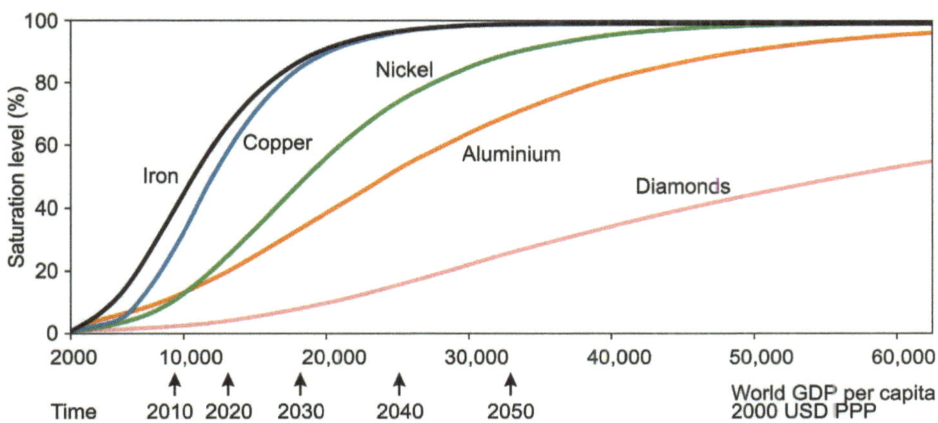

FIGURE 11.2 Saturation level of consumption for various commodities versus average world gross domestic product.

Source: *Modified from Rio Tinto (2013).*

WHAT ARE MINERAL RESOURCES?

In reporting or discussing mineral resources, confusion can arise if the terms used are undefined or have different meaning for different parties. The most important terms required for the proper understanding of this chapter are described here. A *mineral occurrence* is a concentration of any useful mineral found in bedrock in a sufficient quantity to suggest further exploration, and a *mineral deposit* is a mineral occurrence of sufficient size and grade that it might, under the most favorable circumstances, be considered to have potential for economic extraction. *Ore* is a naturally occurring material from which a mineral or minerals of economic value can be extracted at a reasonable profit (American Geosciences Institute, 2013). An *ore deposit* is a mass of material of sufficient ore content to make extraction economically feasible. Ore is therefore an economic concept tied to particular circumstances; the economic viability of a mineral deposit varies according to current needs, commodity prices, the grade and quality of useful commodities, the location of the deposit, local infrastructure, and the kinds of technology available for mining. In addition to geological, technical, and economic considerations, the mining industry is increasingly subject to the reconciliation of environmental, social, and political interests in the use of land.

Mining is a capital-intensive international commercial activity that utilizes mineral resources. Hence, it is obvious that the terms used in reporting mineral resources are regulated by national and international standards such as the JORC, CRIRSCO, NI 43-101, PERC, and UNFC codes (Joint Ore Reserves Committee of the Australasian Institute of Mining and Metallurgy, Australian Institute of Geosciences and Mineral Council of Australia, 2012; Committee for Mineral Reserves International Reporting Standards, 2013; National Instrument 43-101, 2011; Pan–European Reserves and Resources Reporting Committee, 2013; United Nations Economic Commission for Europe, 2009). All companies that are listed on the stock exchange should follow one of these standards when they report information on their mineral resources. The reports must be signed by experts who have a Qualified Person status particularly related to the deposit type reported.

A *mineral resource* is a concentration or occurrence of solid material of economic interest in or on the Earth's crust in such form, grade (or quality), and quantity that there are reasonable prospects for eventual economic extraction. The location, quantity, grade (or quality), continuity, and other geological characteristics of a mineral resource are known, estimated, or interpreted from specific geological evidence and knowledge, including sampling (Joint Ore Reserves Committee of the Australasian Institute of Mining and Metallurgy, Australian Institute of Geosciences and Mineral Council of Australia, 2012). Mineral resources are subdivided, in order of increasing geological confidence, into *inferred*, *indicated* and *measured* categories (Fig. 11.3). Mineral resources must always satisfy the requirement that there are reasonable prospects for eventual economic extraction. Geological evidence and knowledge required for the estimation of mineral resources must include appropriate sampling data; a mineral resource cannot be estimated in the absence of sampling information.

An *ore reserve* is the economically mineable part of a measured and/or indicated mineral resource. It includes diluting materials and allowances for losses that may occur when the material is mined or extracted, and is defined by studies at the prefeasibility or feasibility level, as appropriate, that include the application of modifying factors. Such studies demonstrate that, at the time of reporting, extraction could reasonably be justified (Joint Ore Reserves Committee of the Australasian Institute of Mining and Metallurgy, Australian Institute of Geosciences and Mineral Council of Australia, 2012). Ore reserves are subdivided in order of increasing confidence into *probable ore reserves* and *proved ore reserves* (Fig. 11.3). A *probable ore reserve* is the economically mineable part of an indicated, and in some circumstances a measured, mineral resource. A *proved ore reserve* is the economically mineable part of a

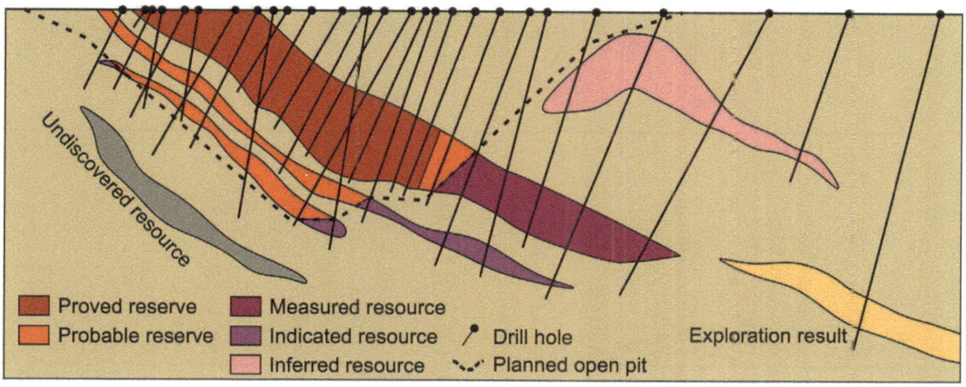

FIGURE 11.3 Schematic representation of the subdivision of mineral resources and reserves.

Representation is based on geological confidence (density of drilling data) and various modifying factors (see text).

measured mineral resource. The modifying factors mentioned are considerations used in converting mineral resources into ore reserves. They include, but are not restricted to, mining, processing, metallurgical, infrastructure, economic, marketing, legal, environmental, social, and governmental factors.

The mineral resources as previously defined refer to identified or known resources. Such resources can be observed and measured, and the category (inferred, indicated, measured) in which they fall depends on the amount and reliability of the observation and measurement data. In this chapter we also use the term *assumed mineral resources*. By this we mean the estimated geological resources that are not supported by large quantities of data or are not reported according to current standards, but are based on drill intersections of mineralized zones and interpretation of other geoscientific data that allow evaluation of the potential extent of mineralization. Assumed resources are less well established than inferred resources and fall outside the category of identified resources.

Considering the tonnage values presented in this chapter, it is very important to keep in mind that the reported resources are mostly geological in situ values, and in most instances the resource estimates are not based on existing industrial standards. Furthermore, the monetary values given are based on estimates of in situ mineral resources (not ore reserves), which do not take into account any modifying factors. Therefore, our estimated tonnage and monetary values must not be confused with the results of studies on the economic profitability of mining projects, carried out according to strict standards such as the JORC or NI 43-101.

MINING HISTORY IN FINLAND

The information in this chapter is largely based on the data in Puustinen (2003) and FODD (2013). It is noteworthy that the mined tonnages of metals discussed here are not the actual amounts of metals produced in smelters, but calculated values of metal contents of the excavated ore, based on reported average metal concentrations of the mill feed. In other words, these are theoretical total metal contents of the excavated ore, not taking into consideration the losses during enrichment and smelting.

The history of mining in Finland goes back at least to the 1500s. The Ojamo iron mine, discovered in 1530, is considered to be the oldest metal mine in Finland (Table 11.1). In the early stages, mining operations mostly produced modest amounts of iron and copper. At least 327 mines had been in

Table 11.1 Selected metal mines in Finland. Production is reported until the end of 2012.

Mine	Main metals	Discovered by	Discovery year	Operating years	Ore excavated (Mt)	Produced metals (Ag, Au, Pd, Pt: t; others: kt)	Value of production (M$)
Ojamo	Fe	Unknown	1530	1533–1564, 1609–1673, 1684–1694, 1826–1863	0.012	Fe 5.8	0.353
Orijärvi	Cu, Zn	Layman	1757	1758–1882, 1929–1955	0.92	Cu 10, Zn 24, Pb 7.6, Ag 8.4, Au 0.084	93
Jussarö	Fe	E.J. Westing, Bergstyrelsen	1834	1834–1861, 1898–1900, 1957–1967	1.6	Fe 450	28
Outokumpu	Cu, Co	GTK	1910	1910–1989	32	Cu 970, Co 56, Au 25, Zn 240, Ag 290	6982
Petsamo	Ni, Cu	GTK	1921	1936–1944	0.46	Ni 17, Cu 8.9	296
Mätäsvaara	Mo	Layman	1903	1910–1911, 1920–1922, 1932–1933, 1940–1947	1.2	Mo 0.98	34
Makola	Ni, Cu	GTK	1937	1941–1954	0.41	Ni 3.1, Cu 1.8	55
Haveri	Au, Cu	Oy Vuoksenniska Ab	1935	1942–1960	1.6	Au 4.4, Cu 6.1	99
Ylöjärvi	Cu, W	GTK	1938	1943–1966	4.0	Cu 30, W 1.6, Ag 50, Au 0.27	160
Aijala	Cu, Au, Zn	Suomen Malmi Oy	1945	1948–1961	0.84	Cu 13, Au 0.59, Zn 5.4, Ag 12	74
Otanmäki	V, Fe, Ti	GTK	1938	1949–1985	25	V 64, Fe 8600, Ti 1900	2399
Metsämonttu	Zn, Au	Suomen Malmi Oy	1946	1951–1974	1.5	Zn 45, Au 1.1, Pb 7.1, Cu 1.6, Ag 20	111
Vihanti	Zn, Cu	GTK	1947	1952–1992	28	Zn 1400, Cu 130, Pb 98, Ag 280, Au 3.3	3077
Kotalahti	Ni	Outokumpu Oy	1954	1957–1987	12	Ni 82, Cu 32	1410
Luikonlahti	Co, Cu	Ruskealan Marmori Oy	1944	1958–1983	6.9	Co 7.0, Cu 65, Zn 25	580
Korsnäs	Pb	GTK	1955	1958–1972	0.87	Pb 31	35
Kärväsvaara	Fe	Layman	1921	1921, 1937, 1958–1967	0.93	Fe 430	26
Pyhäsalmi	Zn, Cu	Layman	1958	1959–	51	Zn 1200, Cu 450, Au 7.3, Ag 240	3918

Rautuvaara	Fe, Cu	Suomen Malmi Oy	1956	1962–1988	12	Fe 5300, Cu 26	428
Raajärvi	V, Fe	Otanmäki Oy	1958	1962–1975	5.1	V 5.6, Fe 2100, Ti 5.7	266
Hitura	Ni	GTK	1963	1965–1985, 1988–2008, 2010–	17	Ni 96, Cu 36, Co 0.51, Pt 0.22	1674
Kemi	Cr	Layman	1959	1966–	40	Cr 7100	10,413
Hällinmäki	Cu	GTK	1964	1966–1984	4.2	Cu 32	124
Vuonos	Cu, Co, Ni, Zn	Outokumpu Oy	1965	1967–1986	11	Cu 120, Co 7.3, Ni 11, Zn 69, Ag 0.11	1149
Telkkälä	Ni	Outokumpu Oy	1961	1969–1970, 1988–1992	0.61	Ni 7.8, Cu 2.0	130
Hammaslahti	Cu	GTK	1968	1971–1986	5.6	Cu 62, Zn 3.8	250
Kylmäkoski	Ni, Cu	Outokumpu Oy	1962	1971–1974	0.69	Ni 2.5, Cu 1.9	47
Stormi	Ni, Cu	Outokumpu Oy	1961	1974–1995	4.6	Ni 51, Cu 32	931
Mustavaara	V	Otanmäki Oy	1967	1974–1985	13	V 27	679
Pahtavuoma	Cu, Ag	Outokumpu Oy	1970	1974–1976, 1989–1993	0.29	Cu 3.2, Ag 5.3	14
Hannukainen	Fe, Cu	Rautatuukki Oy	1974	1978–1990	4.6	Fe 1900, Cu 15, Au 0.21	175
Laukunkangas	Ni	Outokumpu Oy	1979	1984–1994	6.7	Ni 50, Cu 14	846
Saattopora	Au, Cu	Outokumpu Oy	1972	1988–1995	2.1	Au 7, Cu 5.7	139
Hälvälä	Ni	Outokumpu Oy	1970	1988–1992	0.25	Ni 3.5, Cu 0.88	59
Ruostesuo	Zn, Cu	Outokumpu Oy	1959	1988–1990	0.24	Zn 6.3, Cu 0.71, Au 0.081, Ag 2.1	15
Mullikkoräme	Zn, Au	Outokumpu Oy	1987	1989–2000	1.2	Zn 73, Au 0.0012, Ag 0.053, Cu 3.6, Pb 10	179
Kutemajärvi	Au	Lohja Oy	1981	1993–2004, 2006–	2.1	Au 14	240
Pahtavaara	Au	GTK	1987	1995–2000, 2003–	5.5	Au 11	178
Särkiniemi	Ni, Co	GTK	1994	2007–2008	0.12	Ni 0.71, Co 0.032, Cu 0.075	13
Talvivaara	Ni, Zn	GTK	1977	2008–	47	Ni 40, Zn 86	766
Kittilä	Au	GTK	1986	2006–	4.3	Au 16	272
Jokisivu	Au	Outokumpu Oy	1985	2009–	0.32	Au 0.86	14
Pampalo	Au	GTK	1990	1996–1999, 2007–2008, 2010–	0.59	Au 3.4	56

Continued

Table 11.1 Selected metal mines in Finland. Production is reported until the end of 2012.—cont'd

Mine	Main metals	Discovered by	Discovery year	Operating years	Ore excavated (Mt)	Produced metals (Ag, Au, Pd, Pt: t; others: kt)	Value of production (M$)
Laiva	Au	Outokumpu Oy	1980	2010–	1.5	Au 0.93	16
Kylylahti	Cu, Au, Zn	Outokumpu Oy	1984	2011–	0.37	Cu 4.7, Au 0.21, Zn 2.2, Ag 1.3	26
Kevitsa	Ni, Cu	GTK	1987	2011–	3.1	3.9, Cu 8.1, Pt 0.099, Au 0.17, Pd 0.088 Ni	99

Main metals are those responsible for at least 10% of the total value of production.
Ore excavated equals mill feed.
Value of production is in US dollars, based on average metal prices for 2000–2009.
Ore excavated and produced metals are rounded to two significant digits.
Short periods of test mining are not included in operating years.
Source data up to 2001: *Puustinen (2003).*
Source for newer data: *Reports and news releases of mining companies.*

operation before 1900 and had produced about 140,000 t iron and 10,000 t copper (Puustinen, 2003). In the early twentieth century, copper production began to increase due to the discovery of the Outokumpu deposit and the subsequent beginning of mining there in 1910 (Fig. 11.4). Exploitation of the Outokumpu deposits can be regarded as the beginning of the modern mining industry in Finland.

Altogether, 17 metal mines were opened in Finland from 1900–1944. After the war, the pace began to increase and was at its highest between 1955 and 1975, during which period 33 new mines started operation (Fig. 11.5). The opening rate diminished after 1975, with eight mines starting their operation in the 1980s and only four from 1990–1999, possibly reflecting the economic recession in the early 1990s and low commodity prices. There was a 12-year gap between the starting of the Pahtavaara mine in 1995 and the next significant operation at the Kittilä gold mine in 2006. However, the pace seems to be increasing again, with four metal mines opened during the 2000s, another four between 2010 and 2012, and several

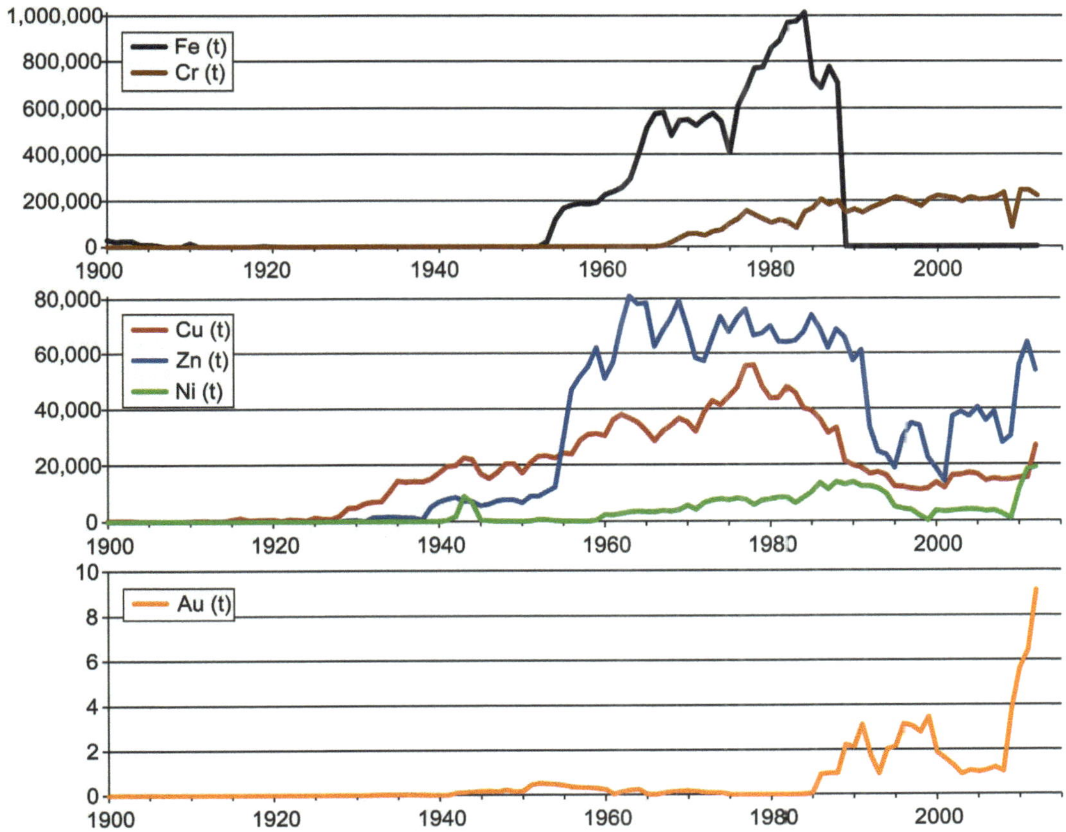

FIGURE 11.4 Annual mine production of selected metals in Finland from 1900–2012.

The figures do not represent real metal production, but are calculated total metal contents of excavated ores based on reported tonnages and average metal grades.

Source for data up to 2001: *Puustinen (2003); for newer data, see reports and news releases of mining companies.*

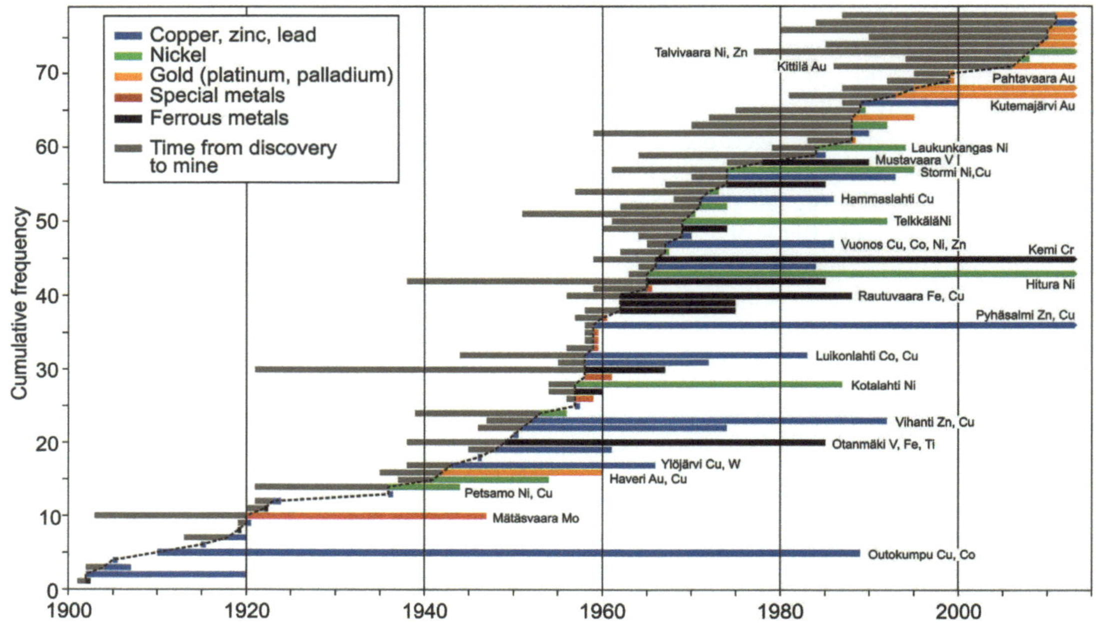

FIGURE 11.5 Mining history for metal mines in Finland from 1900–2012.

The gray bars indicate the time from discovery to the beginning of mining, while the colored bars show the duration of mining.

Data sources: *Puustinen (2003) and FODD (2013).*

advanced mine projects on the way (Fig. 11.6, Table 11.2). The average time from the discovery of a deposit to the opening of a mine remained rather constant at roughly 6 years from the 1920s to the 1960s, but has considerably increased to about 24 years for mines opened in the twenty-first century.

Until the 1940s, Outokumpu was the only important Finnish mine in operation. The Mätäsvaara molybdenite deposit was operated in several phases from 1910–1947, but the total production was only about 970 t of molybdenum. However, Mätäsvaara has been the only molybdenum mine in Finland. The Petsamo nickel deposit was discovered in 1921, but the mine only produced 17,000 t nickel and 8900 t copper from 1936–1944, before the Petsamo area was ceded to the Soviet Union after the Second World War. The nickel production of Petsamo is seen as a small peak in the early 1940s in Fig. 11.4.

The Otanmäki iron mine opened after the war in the late 1940s. Production started gradually from 1949–1954 and the mine became a significant vanadium producer responsible for about 10% of the world production in the 1960s and 1970s (Illi et al., 1985). The total production of the Otanmäki mine was about 8.6 Mt iron and 64,000 t vanadium. Other important iron mines were Kärväsvaara, Rautuvaara, Raajärvi, and Hannukainen (see Table 11.1). Mine production of iron in Finland ended when the Hannukainen mine was closed in 1990 (Fig. 11.4).

The Aijala copper-zinc mine opened after the war in the late 1940s and was soon followed by the Metsämonttu zinc and Vihanti zinc-copper mines in 1951 and 1952, respectively. Several base metal mines producing both copper and zinc were opened from 1955–1975, causing a generally increasing trend in annual copper production until 1978 (see Fig. 11.4). The most important of these mines were

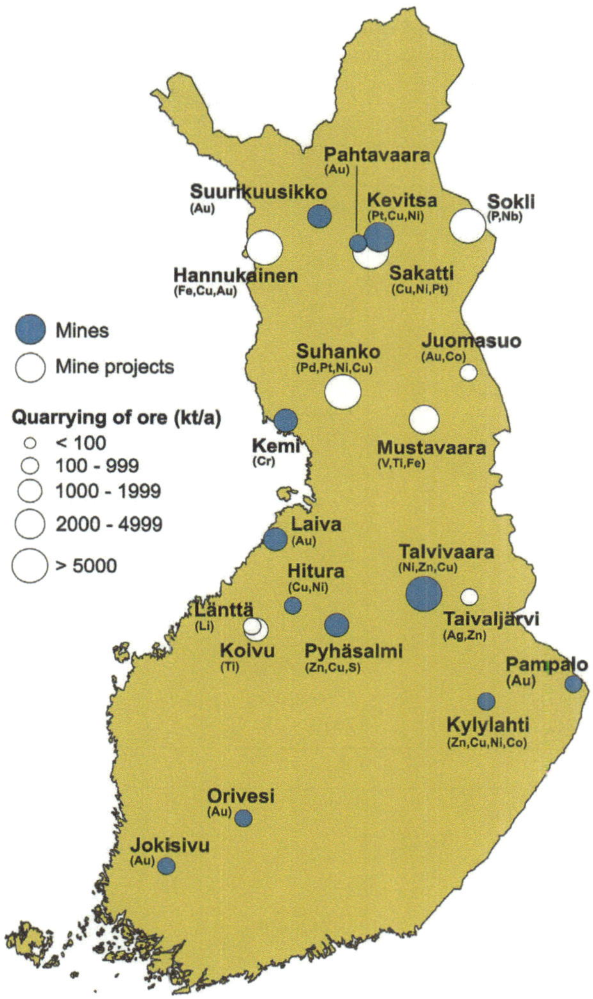

FIGURE 11.6 Mines and mine projects in Finland in 2012.

The symbol size indicates the annual amount of ore mined for operating mines and the planned annual output for mine projects.

Data source: *Pokki et al. (2014) and Geological Survey of Finland internal data.*

Pyhäsalmi and Vuonos, which, in addition to Outokumpu and Vihanti, have produced more than 100,000 t of copper (Table 11.1). The exhaustion of reserves and closure of most of the copper-producing mines led to a decrease in copper ore production from the 1980s onward, and by the beginning of the 2000s, the only mines producing copper were Pyhäsalmi and Hitura. Until the 1950s, the Orijärvi and Aijala mines were responsible for the modest zinc production in Finland. The opening of the Vihanti and Pyhäsalmi mines raised the annual mine production of zinc to the level of 60,000–80,000 t, where it remained until the closure of Vihanti in 1992. After that, Pyhäsalmi and Mullikkoräme were the only producers of zinc

Table 11.2 Operating metal mines and major deposits in Finland in 2012

Name	Main metals	Other metals	Operating years	Premining value (M$)	Premining size (Mt)	Ore excavated (Mt)
Operating mines						
Talvivaara	Ni, Co, Zn	Cu, Mo, U, Mn, Ag	2008–	124,752	2100	47
Kemi	Cr		1966–	53,490	188	40
Kevitsa	Ni, Cu, PGE	Co, Au	2011–	21,593	270	3.1
Pyhäsalmi	Zn, Cu	Au, Ag	1959–	5109	67	51
Kittilä	Au		2006–	4427	64	3.8
Hitura	Ni	Co, Cu, PGE	1965–1985, 1988–2008, 2010–	2279	21	17
Kylylahti	Co, Cu, Ni	Au, Zn	2011–	1656	8.4	0.37
Laiva	Au		2010–	801	23	1.5
Pahtavaara	Au		1995–2000, 2003–	341	7.7	5.5
Kutemajärvi	Au		1993–2004, 2006–	264	3.0	2.1
Jokisivu	Au		2009–	186	1.9	0.32
Pampalo	Au		1996–1999, 2007–2008, 2010–	127	1.8	0.59
Deposits						
Koitelainen Cr	Cr, V	Pd, Pt		23,539	72	–
Sokli Nb	Nb	Fe, U, Ta, Zr, REE		17,228	250	–
Akanvaara Cr	Cr, V	Pt, Pd		15,073	55	–
Mustavaara	V, Fe, Ti		1974–1985	7557	110	13
Suhanko	PGE, Ni, Cu	Au		6993	190	–
Hannukainen	Fe, Cu	Au	1978–1990	5549	200	4.6
Siika-Kämä	PGE, Ni	Cu, Au		3250	49	–
Konttijärvi	PGE, Ni, Cu	Au		2401	75	–
Vaaralampi	Ni, PGE, Cu	Au, Co		2494	33	–
Ruossakero Ni	Ni	Co, Cu		2009	36	–
Akanvaara V	V	Cu, Ag		1772	20	–
Sompujärvi	PGE	Au		1327	6.7	–
Kuervitikko	Fe, Cu, Au			1039	45	–
Koivu	V, Ti			921	62	–
Taivaljärvi	Ag, Zn, Au	Pb, Cu		428	13	–
Juomasuo Au	Au, Co	Cu		285	2.0	–
Juomasuo Co	Co	Au		180	3.1	–

Table 11.2 Operating metal mines and major deposits in Finland in 2012—cont'd

Name	Main metals	Other metals	Operating years	Premining value (M$)	Premining size (Mt)	Ore excavated (Mt)
Läntta	Li, Ta			55	3.0	–
Sakatti	Ni, Cu, PGE			n.a.	n.a.	–

Akanvaara Cr includes Akanvaara UC + ULC + LC.
Koitelainen Cr includes Koitelainen UL + LC.
Suhanko includes Ahmavaara + Suhanko.
Vaaralampi includes Vaaralampi + Niittylampi.
Ore excavated equals mill feed.
Premining size and ore mined are rounded to two significant digits.
Short periods of test mining are not included in operating years.
n.a.: Data not available; – : Not excavated.
Premining value is in US dollars, based on average metal prices for 2000–2009.
Source for premining size: *FODD (2013).*
Source for ore excavated: *Reports and news releases of mining companies.*

until the opening of the Talvivaara mine in 2008. Korsnäs was the only mine in Finland producing lead as a major constituent. The mine operated from 1958–1972 and produced 31,000 t lead.

Nickel production was very small after the war, until the opening of the Kotalahti and Hitura mines in 1957 and 1965, respectively. From 1960–1990, nickel ore production gradually increased due to the opening of several mines, and peaked at a little over 10,000 t of contained nickel annually between 1985 and 1993 (Fig. 11.4). The most important nickel mines were Kotalahti, Hitura, Stormi, and Laukunkangas, each of which produced at least 50,000 t of nickel. By 1995, all the other mines had stopped operation and Hitura was the only mine producing nickel in Finland. With the opening of Talvivaara in 2008, nickel mine production in Finland has started to increase again.

Seven small uranium mines operated in Finland between 1957 and 1965. All were small short-term operations, active for only one to three years. The most important of these was Paukkajanvaara, which was active from 1958–1961 and produced about 55 t of uranium.

The Kemi mine is the only mine producing chromium in Finland. The chromite deposit was discovered in 1959 and an open-pit operation started in 1966. Chromite production gradually increased to its present level of a little over 200,000 t of contained chromium metal annually. In 2006, the mine became an entirely underground operation.

Excluding the gold-panning areas in Lapland, Haveri was the only gold mine in Finland in the twentieth century, until the late 1980s. It produced 4.4 t gold from 1942–1960. Gold was a by-product in several base metal mines, but the amounts produced were minor, usually from tens to a few hundred kilograms annually. The Saattopora gold mine was opened in 1988, followed by the Kutemajärvi (Orivesi) and Pahtavaara mines in 1993 and 1995, respectively. This caused an increase in annual gold production to 2–3 t. In the early years of the 2000s, gold production decreased to about 1 t annually, until the opening of several new gold mines caused it to strongly increase from 2009 onward (refer to Fig. 11.4). The most important of these new mines was Kittilä, opened in 2006.

The selection of metals mined in Finland has changed over time. In the early stages, from the 1500s to the beginning of the 1900s, mining was mostly concentrated on copper and iron, and the amounts produced annually were small. In the early 1900s, the amounts of metals mined annually began to

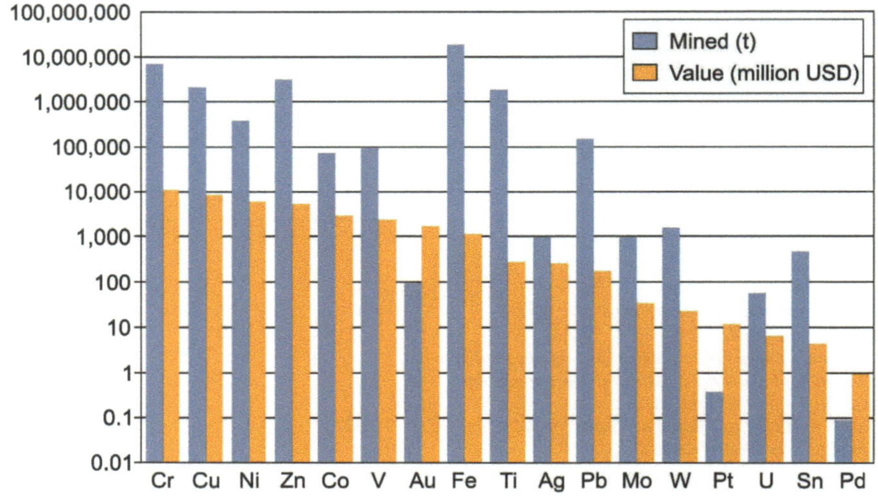

FIGURE 11.7 Total amounts of metals produced in Finland from 1501–2012.

The figures are calculated total metal contents of excavated ores based on reported tonnages and average metal grades. The nominal values (US dollars) are calculated using average metal prices for the period 2000–2009.

Source for data up to 2001: *Puustinen (2003); for newer data, see reports and news releases of mining companies.*

increase (Fig. 11.4). Copper started this trend in the late 1920s, followed by zinc in the late 1930s, nickel at the beginning of the 1960s, chromium in the late 1960s, and gold in the late 1980s. However, iron dominated the tonnages mined from the mid-1950s until the end of iron mining in Finland in 1988. By the end of the twentieth century, mining volumes of all the base metals had significantly decreased due to the closure of several major mines. Only chromium production had continued to increase steadily.

Since the beginning of the twenty-first century, the mine production of base metals in Finland has started to increase again. At the end of 2012, there were 12 operating metal mines in Finland (Tables 11.1 and 11.2). The average annual ore production from 2010–2012 had increased to an all-time high for chromium, gold, and nickel, and was at its highest in the 2000s for copper and zinc (Fig. 11.4).

Iron dominated the total tonnages of metals mined in Finland throughout history, but the calculated value of several other metals currently being mined is higher (Fig. 11.7). The total mined tonnage of ferrous metals (iron, vanadium, chromium, titanium) is 28 Mt and their calculated value is US$14 billion. Although the total mined tonnage of base metals (copper, nickel, zinc, lead, cobalt) is only 6.0 Mt, their calculated total value of US$22 billion clearly exceeds the value of the ferrous metals. The total tonnage of mined precious metals (gold, silver, platinum, palladium) is only 1.1 Mt, and the total value of this is US$2 billion.

METALLOGENY AND MINERAL RESOURCES

METALLOGENIC AREAS

Metallogeny is defined as the study of the genesis of mineral deposits, with emphasis on their relationship in space and time to regional petrographic and tectonic features of the Earth's crust. A metallogenic province, area, or belt is an area characterized by a particular assemblage of mineral deposits, or by one or more characteristic types of mineralization (American Geosciences Institute, 2013).

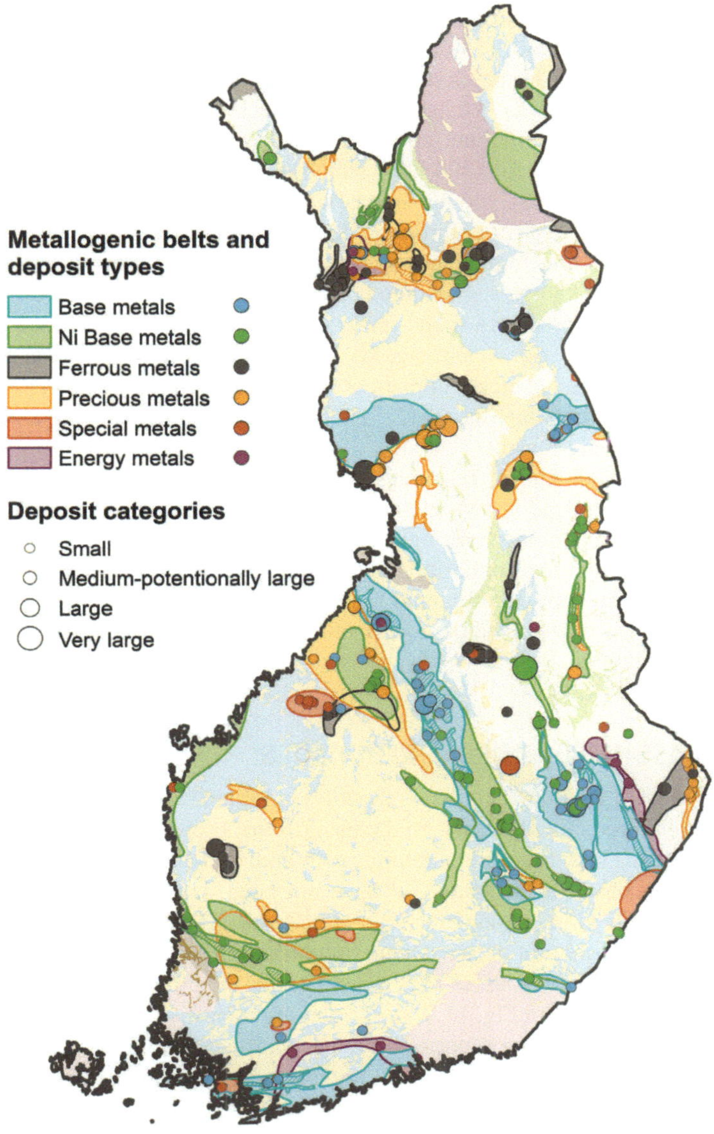

FIGURE 11.8 Metallogenic areas and known important mineral deposits and occurrences in Finland.

Source: *The metallogenic areas are modified from Eilu et al. (2009) and the deposits are based on FODD (2013).*

Altogether, 47 metallogenic areas have been delineated in Finland, and 12 additional areas extend across the Finnish border from Norway, Sweden, or Russia (Fig. 11.8). These areas are defined as domains with an indicated mining and exploration potential for one or a few genetic types of metal deposits. Metallogenic areas are delineated on the basis of the presence of metal mines, deposits, mineral occurrences, and other indications of certain types of metallic mineralization, by the local and regional bedrock geology, and by indications from geophysical and geochemical surveys (Eilu, 2012). Of the metallogenic areas in Finland, 13 are predominantly characterized by ferrous metals (iron, titanium, vanadium, chromium), 13 by precious metals (gold, platinum, palladium), 14 by nickel, 9 by other base metals (copper, zinc, lead), 7 by so-called high technology metals mostly used in advanced technologies (beryllium, lithium, niobium, rare earth metals, tantalum), and 3 by uranium. The total area of all the metallogenic belts is 103,824 km^2, which is 31% of the land area of Finland. However, the belts partially overlap (Fig. 11.8), and the area covered by at least one belt is 88,758 km^2 (26%).

More than 30 different genetic types of mineral deposits occur in the metallogenic areas, and many of the areas can contain deposits of more than one of the major groups of metals. According to the value of past production and present known and assumed resources, the most important deposit types encountered within the metallogenic areas are mafic-intrusion-hosted titanium-iron-vanadium, mafic–ultramafic-hosted chromium, iron-oxide–copper–gold, layered-intrusion-hosted nickel-copper-platinum group metals, synorogenic intrusive nickel-copper, orogenic gold, volcanogenic massive sulfide (VMS), and Outokumpu-type copper-cobalt (Eilu, 2012). The giant and unique Talvivaara nickel-zinc-copper-cobalt deposit is also highly significant.

Excluding the rather small known resources in Archean orogenic gold deposits and komatiite-hosted nickel deposits, most of the known metal resources in Finland occur in deposits formed in the Paleoproterozoic, either during multiple rifting episodes from 2.45–1.92 Ga or during the Svecofennian orogeny in 1.9–1.8 Ga. The 365 Ma Sokli niobium-rare earth element (REE) deposit is the most important post-Svecofennian metal deposit in Finland.

The metallogenic areas are described in more detail in Eilu (2012), and the reader is directed to that publication for further details.

IDENTIFIED AND ASSUMED MINERAL RESOURCES

It is no surprise that most of the known mineral resources in Finland are located within the metallogenic areas; after all, the areas are defined mainly on the basis of the known deposits and occurrences.

In addition to identified resources in operating mines and well-known and explored deposits, there are less well-defined resources in deposits that have not been thoroughly explored. These assumed resources are mostly contained in incompletely explored deposits, are not supported by large quantities of data, and have, in many cases, not been evaluated according to present standards. However, it is useful to also consider the assumed resources to acquire a better picture of the possible total endowment of metals in the Finnish bedrock. In Fig. 11.9, the remaining identified and assumed mineral resources in Finland are combined, as they are reported in the Fennoscandian Ore Deposit Database (FODD, 2013). The database contains information on all known metallic mineral deposits with a published resource estimate in the Fennoscandian Shield (Eilu et al., 2007).

The majority of the remaining zinc, copper, nickel, and cobalt resources, and a little more than half of the remaining chromium resources, reside in operating mines. Most of the remaining resources for the other metals in Fig. 11.9 are located in more or less well-known deposits that either have or have not previously been mined. Comparison with amounts mined throughout history in Finland indicates that for most of the metals, the mined amount is less than 25% of the total endowment. One must remember, however, that these figures are calculated as in situ metal contents in the deposits and calculated metal contents of the mined ore, based on reported metal grades in the ore. Although the Nb, Ta, and Be contents of the mined resources are calculated here, the metals have not actually been separated from the ore.

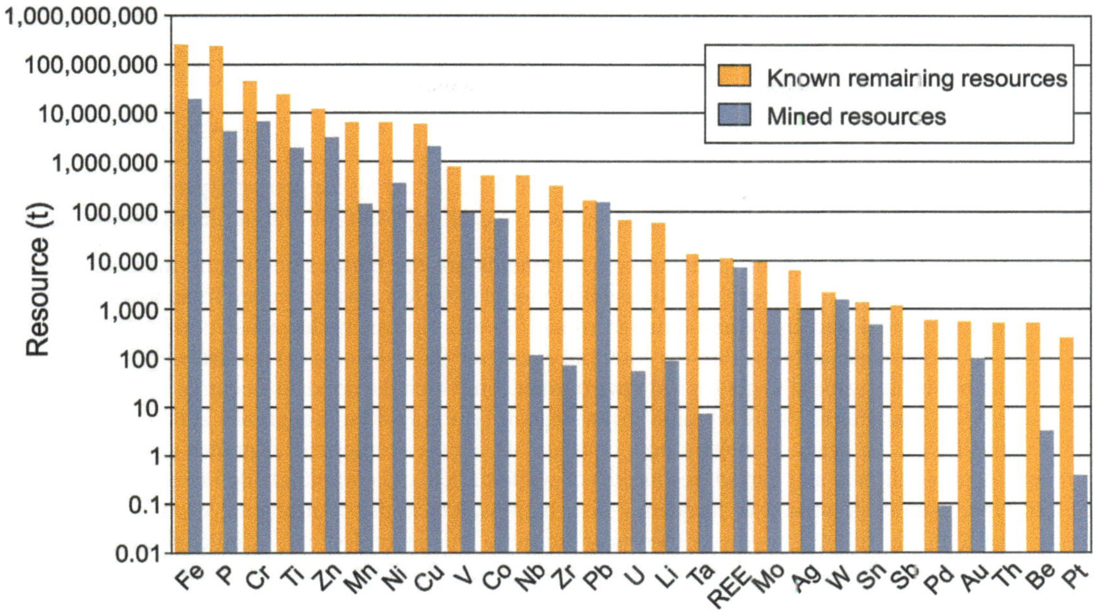

FIGURE 11.9 Remaining known and assumed in situ resources of metals in Finnish mineral deposits compared with metals contained in mined resources.

The mined resources are the calculated total metal contents of excavated ores based on reported tonnages and average metal grades.

Source: *The remaining resources, and mined resources for P, Mn, Nb, Zr, Li, Ta, REE, Sb, Th, and Be, are based on data from 351 metal deposits in Finland (FODD, 2013). The mined resources for the other metals are based on Puustinen (2003).*

UNDISCOVERED MINERAL RESOURCES

By *undiscovered resources*, we mean resources in undiscovered mineral deposits whose existence is postulated on the basis of indirect geological evidence. Unlike known resources, undiscovered resources cannot be observed or measured. They are hypothetical or speculative resources, associated with a probability of existence. The only way to verify the amount of undiscovered resources is through exhaustive exploration. However, it is possible to estimate undiscovered resources, and numerous methods have been developed and applied for this purpose over the years (e.g., Lisitsin et al., 2007, and references therein). In Finland, undiscovered resources of important metals have since 2008 been estimated by the Geological Survey of Finland (GTK) (Rasilainen et al. (2010, 2012)). The procedure selected for the GTK assessments is based on the three-part quantitative assessment method developed by the U.S. Geological Survey (USGS).

The three-part assessment method

The three-part quantitative assessment method was developed at the USGS starting in the mid-1970s (Singer and Menzie, 2010, and references therein), and has been increasingly used by the USGS and others since 1975 (e.g., Richter et al., 1975; Singer, 1975; Singer and Overshine, 1979; Drew et al., 1984; Bliss, 1989; Brew et al., 1992; Box et al., 1996; U.S. Geological Survey National Mineral Resource Assessment Team, 2000; Kilby, 2004; Lisitsin et al., 2007; Cunningham et al., 2008; Hammarstrom et al., 2010;

Sutphin et al., 2013). The method is based on statistical data analysis and integration, it treats and expresses uncertainty, it enables the use of varying amounts of objective geological data and subjective expert knowledge, and it generates reproducible assessment results. The components of the method are:

(1) Evaluation and selection or construction of descriptive models and grade-tonnage models for the deposit types under consideration.
(2) Delineation of areas according to the types of deposits permitted by the geology (permissive tracts).
(3) Estimation of the number of undiscovered deposits of each deposit type within the permissive tracts.

Deposit models designed for quantitative assessments are the cornerstone of the method. Several kinds of deposit models exist, but the two essential models that are always needed in an assessment are the descriptive model and the grade-tonnage model. A descriptive model consists of systematically arranged information describing all of the essential characteristics of a class of mineral deposits (Barton, 1993). The model describes the geological environments in which the deposits occur and lists the essential identifying characteristics by which a given deposit type might be recognized and separated from other deposit types. A grade-tonnage model consists of data on average metal grades and the total ore tonnage (past production and current resources) of well-studied and completely delineated deposits of a certain type, usually presented as frequency distributions. These distributions are used as models for grades and tonnages of undiscovered deposits of the same type in geologically similar settings.

A *permissive tract* is an area within which the geology permits the existence of mineral deposits of the type under consideration. Tract boundaries are based on geology and they should be defined so that the probability of deposits occurring outside of the tract is negligible. It is important to distinguish between areas that are favorable for the existence of deposits and permissive tracts; the former are a subset of the latter. A permissive tract does not indicate any favorability for the occurrence of deposits; neither has it anything to do with the likelihood of discovery of existing undiscovered deposits in the area.

The estimated *number of undiscovered deposits* in a permissive tract represents the probability that a certain fixed but unknown number of undiscovered deposits exist in the delineated tract. The estimates are typically made subjectively by a team of experts knowledgeable about the deposit type and the geology of the region. Uncertainty is taken into consideration and the estimates are made at several levels of confidence (typically at 90%, 50%, and 10%), from which the probability distribution of the expected number of deposits can be derived.

As the final step of the assessment, estimates of the number of deposits are combined with the grade and tonnage distributions from the grade-tonnage models, using statistical methods to achieve probability distributions of the quantities of contained metals in the undiscovered deposits. Software using Monte Carlo simulation has been developed for this purpose (Root et al., 1992; Duval, 2012).

For the assessment process to produce reliable results, it is essential that the three parts previously described are consistent with each other. The delineation of the permissive tracts should be based on criteria derived from descriptive models. The estimates of the number of undiscovered deposits must be carried out according to the deposit type and they must be consistent with the grade-tonnage models. If the consistency requirements are met, the assessment process produces reliable and repeatable probabilistic estimates of the total amount of metals in situ in undiscovered deposits. It must be emphasized that these estimates are not concordant with the present industrial standards (e.g., JORC, NI 43-101) and should never be confused with estimates of known reserves or resources. Furthermore, the modified process used in the GTK assessments does not take into account the economic, technical, social, or environmental factors that might in the future affect the potential for economic utilization of a resource. Thus, part of the estimated undiscovered resource is likely to be located in subeconomic occurrences that will never be profitable to excavate.

The undiscovered mineral resources in Finland

The undiscovered resources of the platinum group elements, nickel, copper, zinc, and gold down to the depth of 1 km in the Finnish bedrock have been estimated by the Geological Survey of Finland. The results of the platinum group element and nickel assessments have been published (Rasilainen et al., 2010, 2012) and the results for copper, zinc, and gold will be published in the GTK Report of Investigation series. As an example, the results for nickel are briefly described in the following.

The assessment was performed for: (1) Ni-Cu deposits associated with Svecofennian (~1.89–1.87 Ga) mafic–ultramafic intrusions in central and southern Finland, (2) Ni-Cu deposits associated with Archean (~2.8 Ga) komatiitic rocks in eastern and northern Finland and Paleoproterozoic (~2.05 Ga) komatiitic rocks in northern Finland, (3) Ni-Cu-PGE deposits associated with Paleoproterozoic (~2.45 Ga) mafic–ultramafic layered intrusions in northern Finland, and (4) Outokumpu-type Cu-Zn-Co deposits hosted by 1.97–1.95 Ga ophiolitic metaserpentinites, metaperidotites, and their altered derivatives in eastern Finland. The two largest known nickel deposits in Finland, Kevitsa and Talvivaara, both represent a rare and previously unknown deposit type for which not enough data are available for the construction of reliable deposit models. Hence, undiscovered nickel resources in these two deposit types could not be quantitatively estimated.

Grade-tonnage models were constructed for the deposit types using data from known deposits within the Fennoscandian Shield. Altogether, there are 40 well-known synorogenic intrusive deposits, 9 komatiitic deposits, 8 layered-intrusion-hosted deposits, and 10 Outokumpu-type deposits in the grade-tonnage models. The remaining known nickel resources in these deposits total 430,000 t of nickel (Table 11.3).

Twenty-six permissive tracts were delineated for synorogenic intrusive deposits, 30 for komatiitic deposits, 43 for layered-intrusion-hosted deposits, one for Outokumpu-type deposits, and 15 for Talvivaara-type deposits. The number of undiscovered deposits was estimated separately for each permissive tract, excluding the Talvivaara-type tracts, in a series of workshops. The mean estimate of the number of undiscovered deposits is 66 for the synorogenic intrusive type, 34 for the komatiitic type, 52 for the layered-intrusion-hosted type, and 6 for the Outokumpu type.

Metal contents of the undiscovered deposits were estimated by Monte Carlo simulation (Table 11.3). Layered intrusions are estimated to contain by far the greatest amount of undiscovered nickel, followed by orogenic intrusive and komatiitic deposit types. The estimated nickel content of

Table 11.3 Remaining known in situ resources and estimated undiscovered resources of nickel in important deposit types in Finland

Deposit type	Remaining known resources (t)	Undiscovered resources (t)
Layered intrusions	223,123	4,200,000
Orogenic intrusive Ni-Cu	128,155	480,000
Komatiitic Ni-Cu	42,003	280,000
Outokumpu	36,767	41,000
Talvivaara	4,516,160	–
Kevitsa	813,120	–
Total	5,759,328	5,001,000

– : *Not estimated*
Undiscovered resources are estimated at 50% probability.
Source: *Data from Rasilainen et al. (2010, 2012).*

undiscovered Outokumpu-type deposits is clearly small. Most of the undiscovered synorogenic intrusive nickel resources are located within the permissive tracts around the Central Finland Granitoid Complex, whereas the undiscovered komatiitic resources are evenly distributed between the greenstone belts of northern and eastern Finland (Fig. 11.10). The undiscovered layered intrusion-hosted resources and Outokumpu-type resources are restricted to northern Finland and the Outokumpu area,

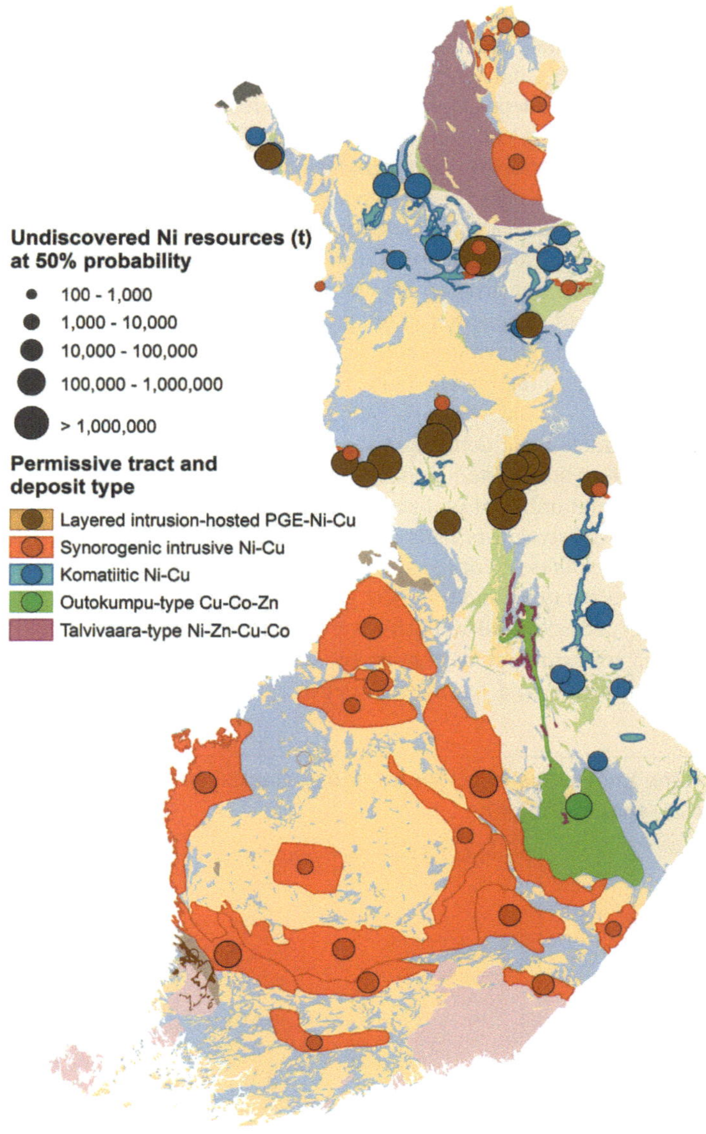

FIGURE 11.10 Permissive tracts for important nickel deposit types in Finland.

Filled circles indicate the estimated undiscovered nickel resource at the 50% probability level for each tract.

respectively. Although it is statistically not strictly correct to sum together the median estimates for the different deposit types, the results suggest that the total known remaining and estimated undiscovered nickel resources are of a roughly similar size.

CHALLENGES AND OPPORTUNITIES FOR FUTURE MINING

Mineral raw materials are unevenly distributed across the Earth and concentrated in small volumes of the crust through distinct geological processes. Mineral deposits, as such, are nonrenewable and ore reserves at existing metal mines are finite. Continuously increasing demand for mineral resources will exhaust most existing mines within the next few decades. Are we going to run out of metals, as predicted by Meadows et al. (1972) in their famous report on limits to growth for the Club of Rome? The Earth is entirely made up of minerals and there is no risk of rapid depletion of most of the raw materials. Global ore reserve data are not a good measure of remaining resources and actually give only a small fraction of the world's total mineral resources (Herrington, 2013; Mudd et al., 2013). Most mines will have a much longer lifetime than can be estimated according to present ore reserves. Drilling will move resources into reserves and new discoveries are commonly made around existing mines (see Fig. 11.3).

The Pyhäsalmi copper-zinc mine is a good example of increased mine life based on continuous and successful near-mine exploration (Fig. 11.11). Ore reserves at the start of mining in 1961 were 18 Mt. Since then, more than 50 Mt of ore have been extracted, and there are still almost 10 Mt of remaining reserves. The distinct decrease of reserves in 1982 was based on higher cutoff values related to lower commodity prices. The considerable increase of reserves in 1998 was caused by the new deep ore body.

The Kemi Cr mine is another example of extended mine life. A new estimate in 2014 increased the ore reserves by more than 50% from 33 Mt to 50.1 Mt, and the mineral resources up to 97.8 Mt (Outokumpu Oyj, 2014). The increase was based on the discovery of an extensive new ore body under one of the old open pits. Moreover, reflection seismic studies indicate the continuation of the host rocks and potential chromium deposits further downwards.

In addition to identified ore resources, a large number of subeconomic mineral deposits are known or can be assumed to contain huge resources. The subeconomic resources may become economically viable in the future, depending on their geographic location, commodity demand, and price fluctuation, and the introduction of new mining and beneficiation technologies (Tilton, 2010). Increasing demand and a more limited supply will raise commodity prices, which will make lower-grade deposits economically viable provided that environmental and social aspects are met.

New estimates of, for example, world ultimate copper resources indicate that, in spite of the growing use of copper, the assumed global resources have actually increased during the last few decades (Mudd et al., 2013). Verification of true copper ore reserves will, however, need intensive mineral exploration.

The life span of many metal products is long and metals are recyclable in most applications. Therefore, once produced, metals remain available for future generations, and sustainable societies will create effective mechanisms for recycling and decreasing the growing need for primary resources. Future manufacturing will pay special attention to product design, which will allow effective recycling of all commodities. Technological development will allow the substitution of some critical metals by other

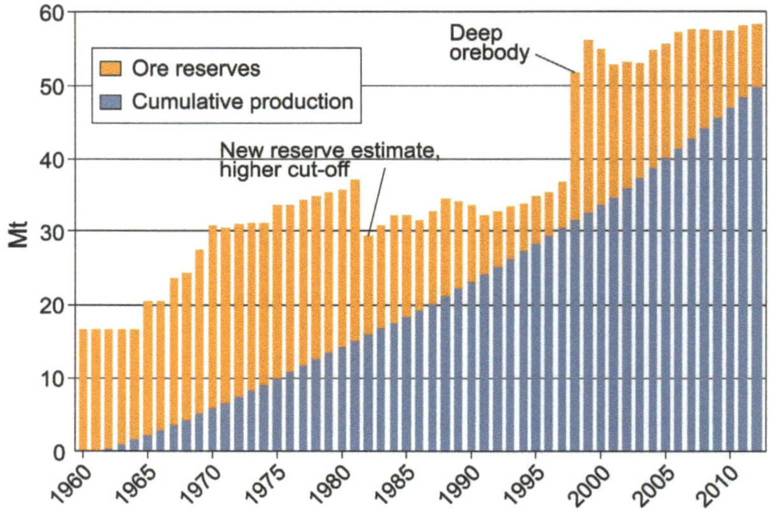

FIGURE 11.11 Development of reserves in the Pyhäsalmi mine.

Source: *Data from T. Mäki (personal communication, 2013).*

compounds. On the other hand, fast technological development and new innovations will require new raw materials for the next-generation low-carbon, high-tech society, which makes it very difficult to forecast the future critical elements (Vidal et al., 2013). Therefore, it is impossible to estimate the global need for various mineral raw materials for future generations.

Thus far, the majority of mining operations have used resources at or near the surface. A major challenge for future mineral production is the increasing difficulty of ore discovery. Deep-seated deposits remain largely unknown and it is likely that many of the remaining undiscovered deposits are located under overburden or water, or in remote or sensitive places, such as the Arctic regions. Long-term investment in geoscientific mapping and research is the basis of mineral exploration and the best way to safeguard minerals for the future. The development and effective use of deep penetrating geophysics, such as seismic, gravimetric, magnetic, electrical, and electromagnetic methods, are increasingly important. Better data and understanding of geological structures and earth processes in 3D and 4D, and the development of drilling technologies, are crucial for future discoveries. This is particularly important in countries such as Finland, where considerable resources for a wide range of commodities evidently remain undiscovered.

Mining is already technically possible today at depths of several kilometers. Underground mining can be carried out using automation, and its footprint is much more limited than that of open-pit mining. Underground mining is also environmentally and socially more acceptable, and it is evident that its role will strengthen in the future. The future smart mines will employ high-level automation, produce less waste, and be increasingly less visible.

Finland is a potential producer of a wide range of minerals. Fig. 11.12 presents the metals currently produced, those in projects under feasibility studies, and minerals that on a geological basis have an obvious potential for discoveries and future production. These commodities also include

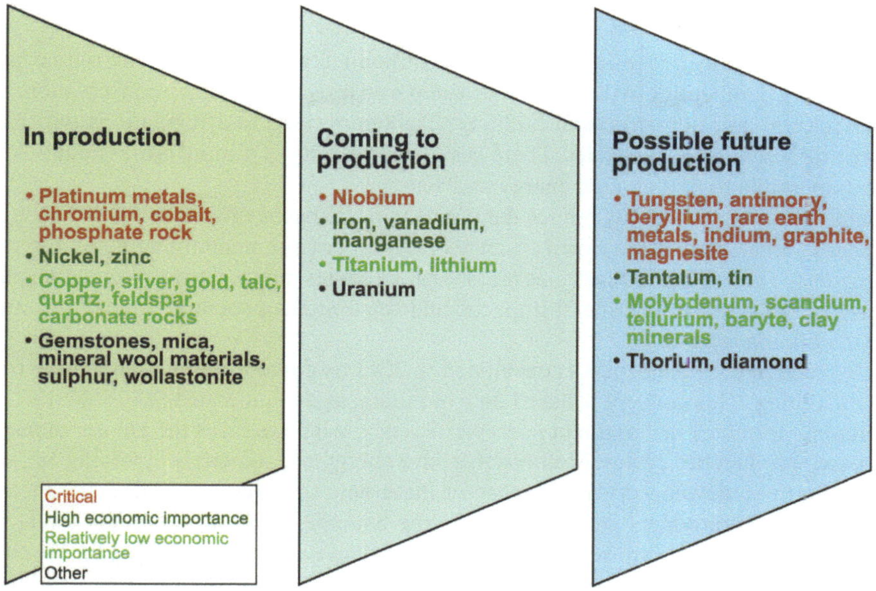

FIGURE 11.12 Commodities currently produced in Finland and those with potential for future production.

Source: *The classification of commodities is based on their criticality and economic importance at the EU level (European Commission, 2014). The group "Other" is not included in the EU classification.*

the majority of the critical and important minerals recently classified by the European Commission (2014). Technological innovations will create demand for new commodities and affect the prices of various minerals (Vidal et al., 2013). On the other hand, the development of mining and processing techniques will make new types of occurrences economically viable. Therefore, it is very difficult to forecast which commodities will be produced in Finland in the future. There are increasing efforts to recover all valuable commodities from ore mined and processed in addition to the main ore metals, and some commodities will also be produced from old mine tailings.

The ultimate resources of various mineral commodities are impossible to accurately define, because of the lack of detailed geological knowledge. In Finland, about 96% of the bedrock is covered with glacial sediments, peat bogs, and water. Although geological formations can continue to depths of several kilometers, our knowledge of geology, mineral resources, or ore potential even at depths of a few hundred meters is very limited and almost completely based on interpretations. Although the various geoscientific datasets in Finland are of high quality and we have a good understanding of regional geology, ore deposits are tiny targets and difficult to find under the cover. Therefore, even many mineral deposits reaching the bedrock surface can remain undiscovered. The recently found Sakatti Cu-Ni-PGE deposit in Sodankylä is a good example of a grass-roots discovery, and demonstrates that even huge high-grade deposits can still be discovered at the bedrock surface (Anglo American plc, 2012). Undiscovered mineral deposits, both at depth and at the bedrock surface, form untapped resources for future mining.

GREEN MINING CONCEPT

Mine development is also becoming increasingly difficult for societal and environmental reasons. There is increasing competition with other land use purposes such as nature conservation, recreation, tourism, agriculture, and infrastructural building. Tightening laws and regulations will make future mining more difficult in many countries. There is a global scarcity of mining professionals, and in some regions it is impossible to find enough energy and water.

People are not ready to radically reduce the use of mineral-based products, but increasingly oppose mining in nearby areas, within their own country or within environmentally vulnerable areas, such as the Arctic regions. The mining industry has major challenges to improve its resource efficiency, as well as its adverse environmental and social impacts, and to strengthen societal interactions and positive impacts on the mining regions.

Finland's green mining concept was developed in 2011 as a major tool to make it the front-runner in sustainable mining. This concept is based on five pillars, as demonstrated in Fig. 11.13.

Green mining promotes material and energy efficiency, which reduces the environmental footprint of mineral-based product life cycles. Methods that save energy and materials in mining and enrichment of minerals have to be developed. The purpose of these new solutions is to allow the recovery of all useful minerals and by-products and to minimize the amount of waste. Solutions for reducing raw water and energy consumption are being developed. To achieve a result that is best for the entire operation, we have to have a reliable way of measuring the material and energy efficiency and the environmental footprint during the life cycle.

Furthermore, green mining aims to ensure the availability of mineral resources for the future. Sustainable development requires that our current use of mineral resources does not endanger the ability of future generations to satisfy their needs. Mines exploit economically viable ore deposits. Although individual deposits are nonrenewable, the mineral resources in the Earth's crust are in no danger of rapidly running out. To ensure the availability of mineral resources for future needs and to fulfill the so-called mineral debt, we must continue geoscientific mapping and research, and invest in mineral exploration. The development of exploration, mining, and processing techniques is also needed to be able to discover and use new types of deposits.

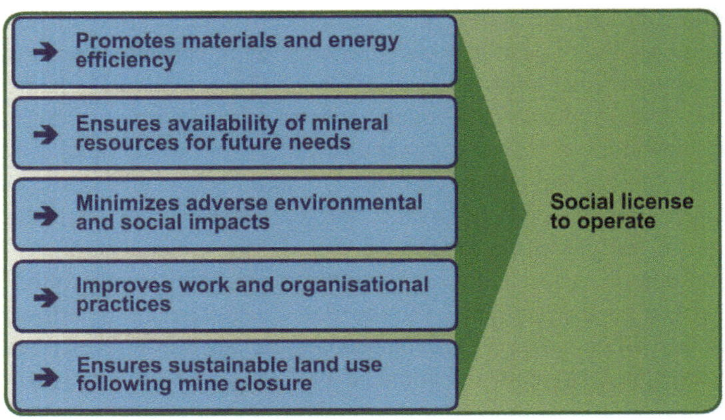

FIGURE 11.13 The green mining concept of Finland.

Mining operations always impact on the natural environment, economy, and social structure of the region. The goal of green mining is to minimize the adverse environmental and social impacts in all the stages of the operations. At the same time, the operations strive to maximize social and local benefits. Minimizing the adverse environmental impacts requires the development of better control and measurement methods that take into consideration the special characteristics of mining operations and the local natural conditions. Maximizing the beneficial societal, economic, and cultural impacts in a sustainable way requires research, communication, and methods that allow broad-based community participation. Participation is especially important on the regional level, because this allows the corporate social responsibility of the mines to be executed in the best possible way.

A wide range of technology and heavy machinery is used in mining, which increases potential safety hazards. Work must be organized in such a way that it is safe and meaningful to employees. This can be achieved by automating processes and making them more efficient, as well as by developing new practices and working methods in cooperation with the entire staff. Occupational safety aiming at zero accidents is an important starting point in all development. Operations must also be safe for local residents and the environment. Increasing automation and the development of technologies helps to reduce the need for a workforce and will improve safety. The mining organization will become lighter and most operations will be executed in mines and enrichment plants using remote control.

The operation time for individual mines can be more than 100 years, but is always limited. After this, the mining areas will be restored to make them safe and allow other kinds of land use. Planning of the controlled ending of mining operations and the proper measures for achieving this is started well before commencing mining operations, and is developed throughout the project's life cycle with the broad-based participation of local residents and other stakeholders. Closure of a mine also requires functional and tested technical and scientific methods, so that the quarries, waste areas, and other infrastructure can be restored in a way that allows further sustainable use of the area according to plans.

The actual availability of commodities is controlled not only by geological accessibility, but increasingly by availability of water and energy, and by social constraints, politics, legislation, and environmental regulations. Therefore, the mining industry needs to improve its performance in all the green mining areas to make it economically, environmentally, and socially viable and acceptable in the future.

REFERENCES

American Geosciences Institute, 2013. Glossary of Geology. Online database available at http://glossary.agiweb.org/dbtw-wpd/glossary/search.aspx, last accessed 13 August 2013.

Anglo American plc, 2012. Annual Report 2011, p. 212. Available online at http://www.angloamerican.com/~/media/Files/A/Anglo-American-PLC-V2/investors/reports/aa-annual-report-2011.pdf, last accessed 18 October 2013.

Barton, P.B., 1993. Problems and opportunities for mineral deposit models. In: Kirkham, R.V., Sinclair, W.D., Thorpe, R.I., Duke, J.M. (Eds.), Mineral deposit modelling. Geological Association of Canada. Special Paper 40, 7–13.

Bliss, J.D., 1989. Quantitative mineral resource assessment of undiscovered mineral deposits for selected mineral deposit types in the Chugach National Forest, Alaska. U.S. Geological Survey. Open-File Report 89-345, p. 25.

Box, S.E., Bookstrom, A.A., Zientek, M.L., et al. (Eds.), 1996. Assessment of undiscovered mineral resources in the Pacific Northwest: A contribution to the interior Columbia Basin ecosystem management project. U.S. Geological Survey. Open-File Report OF95-682, p. 282.

Brew, D.A., Drew, L.J., Ludington, S.D., 1992. The study of the undiscovered mineral resources of the Tongass National Forest and adjacent lands, southeastern Alaska. Nonrenewable Resources 1, 303–322.

Chilean Copper Commission, 2002. Yearbook: Statistics of copper and other minerals 1992–2001. Comisíon Chilena del Cobre, Santiago, p. 110.

Chilean Copper Commission, 2012. Yearbook: Statistics of copper and other minerals 1992–2011. Comisíon Chilena del Cobre, Santiago, p. 171.

Chilean Copper Commission, 2013. Yearbook: Statistics of copper and other minerals 1993–2012. Comisíon Chilena del Cobre, Santiago, p. 168.

Committee for Mineral Reserves International Reporting Standards, 2013. International reporting template for the public reporting of exploration results, mineral resources and mineral reserves, November 2013, p. 41. Available online at http://www.crirsco.com/templates/international_reporting_template_november_2013.pdf, last accessed 15 January 2014.

Cunningham, C.G., Zappettini, E.O., Vivallo, S.W., et al., 2008. Quantitative mineral resource assessment of copper, molybdenum, gold, and silver in undiscovered porphyry copper deposits in the Andes Mountains of South America. U.S. Geological Survey. Open-File Report 2008-1253, p. 282.

Drew, L.J., Bliss, J.D., Bowen, R.W., et al., 1984. Quantification of undiscovered mineral-resource assessment—The case study of U.S. Forest Service wilderness tracts in the Pacific Mountain System. U.S. Geological Survey. Open-File Report 84-658, p. 20.

Duval, J.S., 2012. Version 3.0 of EMINERS—Economic Mineral Resource Simulator. U.S. Geological Survey. Open-File Report 2004-1344. Available online at http://pubs.usgs.gov/of/2004/1344, last accessed 16 January 2013.

Eilu, P. (Ed.), 2012. Mineral deposits and metallogeny of Fennoscandia. Geological Survey of Finland. Special Paper 53, p. 401.

Eilu, P., Hallberg, A., Bergman, T., et al., 2007. Fennoscandian Ore Deposit Database—explanatory remarks to the database. Geological Survey of Finland. Report of Investigation 168, p. 17.

Eilu, P., Bergman, T., Bjerkgård, T., et al., 2009. Metallogenic map of the Fennoscandian Shield 1:2,000,000. Geological Survey of Finland; Geological Survey of Norway; Geological Survey of Sweden; The Federal Agency of Use of Mineral Resources of the Ministry of Natural Resources of the Russian Federation—Espoo/Trondheim/Uppsala/St. Petersburg.

European Commission, 2014. Report on critical raw materials for the EU. Report of the ad hoc working group on defining critical raw materials, May. European Commission, Enterprise and Industry, p. 41. Available online at http://ec.europa.eu/enterprise/policies/raw-materials/critical/index_en.htm, last accessed 29 May 2014.

FODD, 2013. Fennoscandian Ore Deposit Database. Geological Survey of Finland (GTK), Geological Survey of Norway (NGU), Geological Survey of Russia (VSEGEI), Geological Survey of Sweden (SGU), SC mineral. Database available online at http://en.gtk.fi/ExplorationFinland/fodd, last accessed 15 April 2013.

Halada, K., Masanori, S., Kiyoshi, I., 2008. Decoupling status of metal consumption from economic growth. Materials Transactions 49, 411–418.

Hammarstrom, J.M., Robinson Jr., G.R., Ludington, S., et al., 2010. Global mineral resource assessment—porphyry copper assessment of Mexico. U.S. Geological Survey. Scientific Investigations Report 2010-5090-A, p. 176.

Herrington, R., 2013. Road map to mineral supply. Nature Geoscience 6, 892–894.

Illi, J., Lindholm, O., Levanto, U.-M., et al., 1985. Otanmäen kaivos. Summary: Otanmäki mine. Vuoriteollisuus 43, 98–107.

International Monetary Fund, 2013. World Economic Outlook Database, October. Electronic database available online at http://www.imf.org/external/pubs/ft/weo/2013/02/weodata/index.aspx, last accessed 27 March 2014.

Joint Ore Reserves Committee of the Australian Institute of Mining and Metallurgy, 2012. Australian Institute of Geosciences and Mineral Council of Australia. Australian Code for Reporting of Exploration Results, Minerals Resources and Ore Reserves. The JORC Code, 2012 edition, p. 44. Available online at http://www.jorc.org/docs/JORC_code_2012.pdf, last accessed 14 August 2013.

Kilby, W.E., 2004. The British Columbia mineral potential project 1992–1997—methodology and results. BC Ministry of Energy and Mines. GeoFile 2004-2, p. 324.

Lisitsin, V., Olshina, A., Moore, D.H., Willman, C.E., 2007. Assessment of undiscovered mesozonal orogenic gold endowment under cover in the northern part of the Bendigo Zone. GeoScience Victoria, Department of Primary Industries, State of Victoria. Gold Undercover Report 2, p. 98.

Meadows, D.H., Meadows, D.L., Randers, J., Behrens, W.W., 1972. The Limits to Growth: A Report for the Club of Rome's Project on the Predicament of Mankind. Universe Books, New York, p. 205.

Mudd, G.M., Weng, Z., Jowitt, S.M., 2013. A detailed assessment of global Cu resource trends and endowments. Economic Geology 108, 1163–1183.

National Instrument 43-101, 2011. Standards of Disclosure for Mineral Projects, Form 43-101F1, Technical Report and Related Consequential Amendments. Canadian Institute of Mining and Metallurgy, p. 44. Available online at http://web.cim.org/standards/documents/Block484_Doc111.pdf, last accessed 14 August 2013.

Outokumpu, Oyj, 2014. Stock Exhange Release (January 13). Available online at http://www.outokumpu.com/en/news-events/press-release/Pages/Outokumpu-589375.aspx.

Pan–European Reserves and Resources Reporting Committee, 2013. PERC Reporting Standard, Pan-European Standard for Reporting of Exploration Results, Mineral Resources and Reserves, p. 61. Available online at http://www.vmine.net/PERC/documents/PERC_REPORTING_STANDARD_2013%20rev1.pdf, last accessed 14 August 2013.

Pokki, J., Aumo, R., Kananoja, T., Ahtola, T., Hyvärinen, J., Kallio, J., Kinnunen, K., Luodes, H., Sarapää, O., Selonen, O., Tuusjärvi, M., Törmänen, T. & Virtanen, K., 2014. Geologisten luonnonvarojen hyödyntäminen Suomessa vuonna 2012. Summary: Geological resources in Finland, production data and annual report 2012. Geological Survey of Finland. Report of Investigation 210, p. 67.

PricewaterhouseCoopers International, 2013. Mine—A confidence crisis: Review of global trends in the mining industry, 2013, p. 59. Available online at www.pwc.com/en_GX/gx/mining/publications/assets/pwc-mine-a-confidence-crisis.pdf, last accessed 16 October 2013.

Puustinen, K., 2003. Suomen kaivokset ja mineraalisten raaka-aineiden tuotanto vuosina 1530–2001, historiallinen katsaus erityisesti tuotantolukujen valossa. Geological Survey of Finland. Report M10.1/2003/3, p. 578, plus CD-ROM (in Finnish).

Rasilainen, K., Eilu, P., Halkoaho, et al., 2010. Quantitative mineral resource assessment of platinum, palladium, gold, nickel, and copper in undiscovered PGE deposits in mafic–ultramafic layered intrusions in Finland. Geological Survey of Finland. Report of Investigation 180, p. 338.

Rasilainen, K., Eilu, P., Äikäs, O., et al., 2012. Quantitative mineral resource assessment of nickel, copper and cobalt in undiscovered Ni-Cu deposits in Finland. Geological Survey of Finland. Report of Investigation 194, p. 521.

Richter, D.H., Singer, D.A., Cox, D.P., 1975. Mineral resource map of the Nabesna Quadrangle, Alaska. U.S. Geological Survey. Miscellaneous Field Studies Map MF-655K.

Rio Tinto, 2013. Rio Tinto Chartbook, September. Online presentation available at http://www.riotinto.com/investors/chartbook-4940.aspx, last accessed 16 Oct 2013.

Root, D.H., Menzie, W.D., Scott, W.A., 1992. Computer Monte Carlo simulation in quantitative resource estimation. Natural Resources Research 1, 125–138.

Singer, D.A., 1975. Mineral resource models and the Alaskan Mineral Resource Assessment Program. In: Vogely, W.A. (Ed.), Mineral Materials Modelling: A State-of-the-Art Review. Johns Hopkins University Press, Baltimore, pp. 370–382.

Singer, D.A., Overshine, A.T., 1979. Assessing metallic mineral resources in Alaska. American Scientist 67, 582–589.

Singer, D.A., Menzie, W.D., 2010. Quantitative mineral resource assessments: An integrated approach. Oxford University Press, New York, p. 219.

Sutphin, D.M., Hammarstrom, J.M., Drew, L.J., et al., 2013. Porphyry copper assessment of Europe, exclusive of the Fennoscandian Shield. U.S. Geological Survey. Scientific Investigations Report 2010-5090-K, p. 197.

Tilton, J.E., 2010. Is mineral depletion a threat to sustainable mining? SEG Newsletter 82, 18–20.

United Nations Economic Commission for Europe, 2009. UN Framework Classification for Fossil Energy and Mineral Reserves and Resources. ECE Energy Series 39, United Nations, New York/Geneva, p. 20.

United Nations Department of Economic and Social Affairs/Population Division, 2012. World Urbanization Prospects: The 2011 Revision. United Nations publication ST/ESA/SER.A/322, United Nations, New York, p. 302.

U.S. Geological Survey National Mineral Resource Assessment Team, 2000. 1998 assessment of undiscovered deposits of gold, silver, copper, lead, and zinc in the United States. U.S. Geological Survey. Circular 1178, p. 22.

Vidal, O., Goffé, B., Arndt, N., 2013. Metals for a low-carbon society. Nature Geoscience 6, 894–896.

World Bank, 2013. Population (total). World Bank Open Data. Available online at http://data.worldbank.org, last accessed 27 March 2014.

Index

Note: Page numbers followed by "f" and "t" indicate figures and tables respectively.

I

J

K

N

R

S

Kolibri 360 GmbH
Ferdinand-Jühlke-Straße 7
99095 Erfurt, Deutschland
produktsicherheit@kolibri360.de